UTILIZATION OF HARDWOODS GROWING ON SOUTHERN PINE SITES

Peter Koch
Southern Forest Experiment Station

Agriculture Handbook No. 605

Volume II of three volumes:

I   The Raw Material
II Processing
III Products and Prospective

U.S. DEPARTMENT OF AGRICULTURE, FOREST SERVICE

CONTENTS

VOLUME II—PROCESSING

Page

Part IV—PROCESSING

16	HARVESTING	1422
17	DEBARKING	1641
18	MACHINING	1687
19	BENDING	2283
20	DRYING	2311
21	TREATING	2463

PART IV—PROCESSING

Chapter   *Title*

- 16   HARVESTING
- 17   DEBARKING
- 18   MACHINING
- 19   BENDING
- 20   DRYING
- 21   TREATING

16
Harvesting

Major portions of data drawn from research by:

J. A. Altman
American Pulpwood
  Association
L. I. Barrett
R. C. Beltz
C. J. Biller
P. J. Bois
J. H. Buell
A. Clark III
A. Colannino
N. D. Cost
E. P. Craft
E. L. Fisher
B. W. Gibbons
S. Guttenberg
P. Hakkila
L. K. Halls

W. F. Harris
A. E. Hassan
H. C. Hitchcock III
J. R. Jorgensen
P. Koch
J. N. Kochenderfer
J. A. McClure
C. N. Mann
F. G. Manwiller
A. J. Martin
R. L. Murphy
P. O. Nilsson
R. Pennock
P. A. Peters
D. R. Phillips
C. Row
B. E. Schlaegel

R. L. Schnell
J. G. Schroeder
D. L. Sirois
W. B. Stuart
Tennessee Valley
  Authority
R. F. Thienpont
F. G. Timson
D. M. Tufts
U. S. Department of Agriculture,
  Forest Service
C. G. Vidrine
T. A. Walbridge
J. L. Wartluft
H. V. Wiant
G. E. Woodson
H. E. Young

Chapter 16

Harvesting

CONTENTS

	Page
16-1 DISTRIBUTION OF TREE BIOMASS	1428
COMPLETE TREES	1430
Trees 6 inches in dbh	1430
Complete trees 0.5 to more than 9.5 inches in dbh	1435
Complete trees 3 to 12 inches in dbh	1435
ABOVE-GROUND TREE PARTS	1439
Trees 2 inches and smaller in dbh	1439
Trees 1.0 to 4.9 inches in dbh	1440
Trees 1.0 to 10.0 inches in dbh	1440
Trees 6 inches in dbh	1440
Correlation of whole-tree weight with dbh	1446
Correlation of volume of bark-free stem with dbh and height	1447
Center of gravity for tree-length sawlogs and whole trees	1448
ASH sp.	1449
ELM Sp.	1451
HACKBERRY AND SUGARBERRY	1451
HICKORY Sp.	1451
MAPLE, RED	1452
OAK, BLACK	1453
OAKS, BLACKJACK AND CHERRYBARK	1454
OAK, CHESTNUT	1454
OAK, LAUREL	1461
OAK, NORTHERN RED	1461
OAK, POST	1464
OAK, SCARLET	1465
OAK, SHUMARD	1466
OAK, SOUTHERN RED	1466
OAK, WATER	1470
OAK, WHITE	1470
SWEETBAY	1476
SWEETGUM	1476
TUPELO, BLACK	1481
YELLOW-POPLAR	1481
CROSS REFERENCES TO BIOMASS DATA IN CHAPTER 27	1484

		Page
16-2	LOGGING RESIDUES..............................	**1487**
	RESIDUE PER TREE............................	**1488**
	Summary	**1494**
	RESIDUE PER ACRE	**1494**
16-3	FACTORS AFFECTING HARVESTING	**1498**
	SOILS AND TOPOGRAPHY	**1498**
	TRACT SIZE AND VOLUME	**1499**
	WOODS LABOR SUPPLY	**1500**
	SILVICULTURAL CONSIDERATIONS..............	**1503**
	VOLUMES PER ACRE	**1505**
16-4	ECOLOGICAL EFFECTS OF HARVESTING	**1506**
	NUTRIENT BALANCE	**1506**
	SOIL COMPACTION AND PUDDLING	**1508**
	EROSION..	**1510**
	WILDLIFE AND HABITAT	**1511**
	Habitat..	**1511**
	Wildlife..	**1513**
	RECREATION VALUES..........................	**1515**
16-5	HARVESTING SYSTEMS IN WIDE USE	**1518**
	SHORTWOOD SYSTEM	**1518**
	LONGWOOD SYSTEM	**1522**
	Longwood system using a feller-buncher and grapple skidder	**1525**
	Longwood system using a shear with carrier	**1526**
	Longwood thinning system using an inverted grapple skidder.......................	**1526**
	WHOLE-TREE CHIPPING	**1530**
	SUMMARY AND DISCUSSION	**1532**
16-6	MAXIMIZING VALUE OF TREES AND STEMS	**1537**
	LINEAR STEM MERCHANDISERS................	**1538**
	TRANSVERSE STEM MERCHANDISERS	**1544**
	Feeding infeed decks	**1544**
	Singulating stems	**1545**
	Butt indexing of stems	**1545**
	Crosscutting stems............................	**1545**
	Product separation	**1548**
	Cleaning up the trash.........................	**1548**
	Costs and productivity........................	**1548**
16-7	STUMP AND TREE PULLER-BUNCHERS.............	**1549**
	STUMP PULLER-BUNCHERS......................	**1550**
	TREE PULLER-BUNCHERS	**1554**
	DATA ON SOUTHERN HARDWOODS.............	**1560**

		Page
16-8	CABLE YARDING	1563
	HIGHLEAD YARDING	1565
	SKYLINE YARDING	1566
16-9	BALLOON AND HELICOPTER YARDING	1575
	BALLOON YARDING	1575
	HELICOPTER YARDING	1577
16-10	CLEARCUT HARVESTING ABOVE-GROUND BIOMASS OF SMALL HARDWOOD TREES	1578
	SWATHE-CUTTING FELLER BUNCHERS	1578
	SWATHE-FELLING MOBILE CHIPPERS	1581
	Felling bar harvester	1581
	Circular-saw harvester	1586
	LOWER-COST MACHINES TO HARVEST WHOLE TREES	1589
16-11	LIGHT MACHINERY TO SKID AND BUNCH LOGS FROM THINNED STANDS	1594
16-12	RESIDUE HARVESTING	1598
16-13	TRANSPORT, LOADING, AND OFF-LOADING	1600
	WEIGHT OF GREEN WOOD IN SOLID FORM	1601
	WEIGHT OF STACKED ROUNDWOOD	1601
	BULK DENSITY OF BALED CHIPS, BRANCHES, CHUNKS, AND FIREWOOD	1601
	Chips	1601
	Branches	1604
	Logging-residue chunks	1604
	Firewood	1604
	BULK DENSITY OF OTHER RESIDUES	1605
	REFERENCES PERTINENT TO TRANSPORT	1607

		Page
16-14	**ENERGY EXPENDED DURING HARVESTING**	**1608**
16-15	**SYSTEM SELECTION BY MATHEMATICAL MODEL.**	**1612**
16-16	**STORING OF PULP CHIPS**	**1613**
	CHIP DETERIORATION DURING STORAGE	1613
	CHIP HANDLING	1615
16-17	**STORING OF WHOLE-TREE CHIPS AND FUEL**	**1620**
16-18	**STORING OF ROUNDWOOD**	**1622**
	HANDLING OF PULPWOOD IN ROUNDWOOD FORM	1622
	LOG HANDLING	1623
16-19	**LITERATURE CITED**	**1626**

CHAPTER 16
Harvesting

This chapter is concerned with the central problem which deters utilization of hardwoods growing on pine sites—i.e., efficiently harvesting the trees, transporting them to a mill, storing them, and **merchandising** them at stump or mill by segmenting boles, crowns, and roots to maximize tree value. Trees are typically small in diameter, short, and crooked. Low volume per stem and per acre and highly variable species mixes from site to site and from stand to stand combine to raise harvesting costs. Five of the species, sweetgum, black tupelo, yellow-poplar, sweetbay, and sugarberry, have undivided central stems typical of **excurrent growth** (fig. 3-56); this growth pattern eases harvesting problems. Sixteen of the species (the oaks, hickories, elms, and red maple), however, have forked stems typical of **decurrent (deliquescent) growth** (fig. 3-35), which make harvesting them difficult and costly. The two ash species are intermediate in growth form, but have widely spreading crowns (figs. 3-33 and 3-34).

Moreover, much pine-site hardwood grows on terrain that is ill suited for harvesting equipment, such as the steep rocky hillsides of the Arkansas and Virginia mountains, and the soft ground of the rain-saturated, rock-free, flat to rolling coastal plains in winter.

Proposed in this chapter are only partial solutions. Significantly increased research and development work on logging systems and equipment appropriate for the resource are essential to fully adequate solutions (Boyd et al. 1977). Such work should be carried out at numerous centers because there will be many more failures than successes. The rarity of technological breakthroughs over the decades attests to the difficulty of developing efficient harvesting techniques for the pine-site hardwoods.

Harvesting methods are influenced not only by terrain features, soil characteristics, and ecological considerations, but also by stand density, diameter distribution, species mix, scale of the harvesting operation, tract size, and the purpose for which the trees are logged, i.e., for fuel, pulpwood, solid wood products, or chemical products. Basic to all considerations is knowledge of biomass distribution among tree parts. This subject will therefore be discussed first.

16-1 DISTRIBUTION OF TREE BIOMASS

Although few land managers contemplate harvesting litter on the forest floor or lateral root systems, knowledge of their quantity is useful. In section 15-2 litter in southern upland hardwoods stands is estimated at 5,000 to 11,000 pounds per acre (ovendry basis). Lateral roots, not including the central stump-root system, may total 4 or 5 tons (green-weight basis) per acre.

The remaining tree components—central stump-root mass, stemwood, stembark, branchwood, branchbark, foliage, and fruits—may be removed from the forest and utilized entirely under some management systems. Accordingly, knowledge of their weights and shapes are needed by managers concerned with harvesting.

Data on stump-root shapes are illustrated in figures 14-4 through 14-26; some species such as the elms and yellow-poplar, have shallow and widespread root systems, while others such as hickory, sweetgum, and black tupelo typically have tap roots and more compact root systems. Weights of these stump-root systems are discussed in section 14-3, tables 14-1AB and 14-2AB, and in figures 14-4 through 14-26. As noted in section 14-3, the stump-root systems in stands of upland hardwoods may total 17 to 20 tons per acre (green-weight basis).

Information on volumes and weights of stembark and branchbark is contained in section 13-3, tables 13-4 through 13-29, and figures 13-58 through 13-66. Data on foliage weight are contained in secton 15-2, in tables 15-6 through 15-12, and in figures 15-11 through 15-14. Seed yields per tree and per acre are discussed in section 15-7, in tables 15-23 through 15-26, and in figures 15-21 through 15-24.

Data relating crown diameter to tree dbh are tabulated in section 15-2.

This section will summarize weights of other tree parts, particularly stems and crowns, and their proportions of total biomass. See also table 27-97 for an index of weight tables, and page 3266 for an index of volume tables.

Young's (1978) data on mixed species stands in Maine indicate that fresh weight per acre of bark-free merchantable stemwood to a 5-inch top diameter is about equal to the fresh weight of all remaining above- and below-ground tree parts plus cull trees and saplings below merchantable size. This relationship changes with stand type and location, but it is probably a fair approximation for much commercial forest in the East and South. In general, for hardwood trees, bark-free stems contain about 50 percent of total dry weight, tops and stembark 30 to 35 percent and central stump-root systems 15 to 20 percent.

Cost's (1978) analysis of above-ground volume of wood in all hardwoods 1.0 inch in dbh and larger in the mountain region of North Carolina indicated that about 70 percent of the cubic volume of bark-free wood is in the merchantable portion of trees 5 inches dbh and larger, and the remaining volume is in the above-ground portion of stumps, tops, and limbs of trees 5 inches in dbh and larger and in saplings smaller than 5 inches in dbh (table 16-1); when below-ground tree portions, bark volume, and foliage are considered, Cost's data compare closely with those of Young.

In a study of West Virginia mountain hardwoods Frederick et al. (1979) found that whole-tree utilization (all above-ground parts) of stems larger than 2 inches in dbh increased weight yields 32 percent over conventional harvesting to a 4-inch top diameter, outside bark. Use of tops of merchantable trees alone increased utilization by more than 25 percent.

TABLE 16-1.—*Volume and proportions of above-ground wood fiber* [1] *in hardwood trees 1.0 inch dbh and larger on the 4,014,566 acres of commercial forests in the Mountain Region of North Carolina, by tree size and class of material* (Cost 1978)

Class of material and tree portion	Total	Tree size		
		Sawtimber	Pole timber	Saplings
	-----Cubic feet per acre and (percent of total)-----			
Main stems[2] of all live trees 5.0 inches in dbh and larger				
Growing stock	1,119 (59)	731 (71)	388 (64)	—
Rough and rotten	201 (11)	101 (10)	100 (16)	—
Additional wood fiber in all live trees 5.0 inches in dbh and larger				
Stump portion above ground	84 (4)	46 (4)	38 (6)	—
Top	66 (4)	16 (2)	50 (8)	—
Major limbs	20 (1)	20 (2)	—	—
Minor limbs	149 (8)	114 (11)	35 (6)	—
Stemwood and additional wood in all live trees 1.0 to 4.9 inches in dbh				
All above-ground portions	240 (13)	—	—	240 (100)
Total for all lives trees	1,879 (100)	1,028 (100)	611 (100)	240 (100)

[1]No bark or foliage included.
[2]The main stem of trees 5.0 inches and larger in dbh was defined as that portion from the top of a 1-foot stump to a 4.0-inch top diameter outside bark, or to the point where the central stem divided into limbs.

In the subsections that follow, detailed biomass data are presented for major species of hardwoods that grow on southern pine sites. When possible, these data are correlated with tree dbh.

COMPLETE TREES

Trees 6 inches in dbh.—Data on complete tree weights, including weights of below-ground parts, are scarce. Manwiller and Koch analyzed three 6-inch trees of each of 22 species (plus one tree of Shumard oak) growing on southern pine sites; their data are shown in tables 16-2 and 16-3. Average weight contributions and ranges among species of the various tree parts are as follows (the single Shumard oak was not included in these averages):

Radii at which lateral roots were severed, and tree portion	Basis of fresh green weight	Basis of ovendry weight
	----------Percent of complete tree----------	
Lateral roots severed at 3-foot radius		
Central stump-root system	21.8 (18.0-25.8)	20.6 (15.4-25.1)
Stem with bark	57.6 (38.6-69.6)	59.2 (39.6-73.7)
Branches 0.5-inch diameter and larger	11.6 (4.5-27.9)	11.9 (4.5-27.4)
Twigs	5.2 (2.5-8.8)	5.4 (2.4-10.3)
Foliage	3.8 (2.4-5.7)	2.9 (1.4-4.5)
	100.0	100.0
Lateral roots severed at 1-foot radius		
Central stump-root system	16.2 (11.9-21.2)	15.3 (11.7-19.1)
Stem with bark	61.6 (42.1-73.1)	63.0 (43.2-76.7)
Branches 0.5-inch diameter and larger	12.5 (4.8-30.4)	12.8 (4.9-29.9)
Twigs	5.6 (2.7-9.8)	5.7 (2.5-11.6)
Foliage	4.1 (2.5-6.0)	3.2 (1.5-5.0)
	100.0	100.0

TABLE 16-2—*Weights and proportions when green and ovendry, of complete small hardwood trees and tree parts of 23 species sampled on southern pine sites (Data from Manwiller and Koch 1)[2]*

Species	Tree dbh	Complete-tree weight		Stump-root[3]		Stem[4]		Branches[5]		Twigs[6]		Foliage	
		Green	Ovendry	Green	Ovendry	Green	Ovendry	Green	Ovendry	Green	Ovendry	Green	Ovendry
	Inches	----Pounds----		----------------------------------Percent of complete tree----------------------------------									
				LATERAL ROOTS SEVERED AT A RADIUS OF 3 FEET FROM TREE PITH									
Ash, green	5.9	299.1	179.7	18.3	15.4	69.6	73.7	7.1	7.1	2.5	2.4	2.4	1.4
Ash, white	6.0	407.3	262.3	19.0	17.4	61.2	64.4	11.4	10.8	4.6	4.4	3.9	2.9
Elm, American	6.1	299.1	214.7	21.4	22.5	58.7	58.6	11.4	11.9	4.7	4.7	3.8	2.4
Elm, winged	6.1	439.9	266.3	19.0	20.0	57.3	56.0	14.3	14.5	6.6	6.9	2.8	2.7
Hackberry sp.[7]	6.2	302.7	180.8	23.1	22.1	56.9	57.3	11.3	11.6	5.3	5.7	3.4	3.3
Hickory sp.	5.8	381.1	220.7	25.4	22.7	56.1	60.1	10.5	10.9	4.2	3.9	3.7	2.4
Maple, red	5.9	578.8	320.3	19.3	19.6	38.6	39.6	27.9	27.4	8.8	9.0	5.4	4.5
Oak, black	6.0	387.6	221.7	21.9	20.8	64.9	66.3	6.7	7.4	3.2	3.2	3.3	2.3
Oak, blackjack	6.4	417.3	244.7	25.8	23.4	41.9	41.6	20.1	21.8	8.5	10.3	3.6	3.0
Oak, chestnut	6.0	440.8	258.8	21.7	20.0	60.6	63.7	9.9	10.0	4.2	3.9	3.6	2.4
Oak, cherrybark	5.8	374.4	221.4	24.0	21.6	54.7	57.1	9.4	9.8	7.2	7.1	4.7	4.4
Oak, laurel	5.7	406.7	236.0	22.3	20.4	60.2	61.2	8.8	9.4	6.1	6.7	2.6	2.3
Oak, northern red	6.2	503.1	290.8	20.8	19.5	62.5	63.5	11.3	12.2	2.8	3.0	2.7	2.0
Oak, post	6.1	422.4	237.7	23.5	22.8	63.9	65.2	7.5	7.5	2.5	2.6	2.6	1.8
Oak, scarlet	5.8	433.4	239.6	22.3	20.8	61.8	63.5	8.8	9.2	3.7	3.8	3.5	2.7
Oak, Shumard	6.8	600.6	361.1	21.1	19.8	58.4	59.0	11.1	12.1	4.9	5.2	4.5	4.0
Oak, southern red	5.9	459.9	257.2	21.5	20.0	57.9	59.0	14.2	15.0	3.4	3.6	2.8	2.4
Oak, water	6.1	513.1	297.7	18.8	17.3	59.2	60.7	10.7	10.6	7.4	8.0	3.9	3.4
Oak, white	6.2	540.8	307.6	22.1	21.0	56.2	58.2	12.2	12.8	5.9	5.1	3.5	2.9
Sweetbay	6.0	426.5	210.8	21.2	20.9	51.5	52.3	15.7	16.2	6.5	6.9	5.2	3.7
Sweetgum	6.1	392.7	179.6	23.8	23.2	59.1	59.9	6.8	7.8	4.6	4.7	5.7	4.3
Tupelo, black	6.2	448.3	223.7	25.0	25.1	48.5	51.2	15.0	13.7	6.3	6.4	5.2	3.7
Yellow-poplar	5.8	342.1	163.5	18.0	17.6	66.7	68.9	4.5	4.5	5.3	5.3	5.4	3.7

See footnotes page 1433.

TABLE 16-2—*Weights and proportions when green and ovendry, of complete small hardwood trees and tree parts of 23 species sampled on southern pine sites (Data from Manwiller and Koch 1)[2]*—Continued

Species	Tree dbh	Complete-tree weight		Stump-root[3]		Stem[4]		Branches[5]		Twigs[6]		Foliage	
		Green	Ovendry	Green	Ovendry	Green	Ovendry	Green	Ovendry	Green	Ovendry	Green	Ovendry
	Inches	---*Pounds*---		--*Percent of complete tree*--									
				LATERAL ROOTS SEVERED AT A RADIUS 1 FOOT FROM TREE PITH									
Ash, green	5.9	285.2	172.7	14.2	11.9	73.1	76.7	7.5	7.4	2.7	2.5	2.5	1.5
Ash, white	6.0	383.5	248.3	13.9	12.7	65.0	68.2	12.0	11.4	4.9	4.7	4.2	3.0
Elm, American	6.1	378.9	202.5	16.8	17.7	62.3	62.2	12.0	12.6	4.9	4.9	4.0	2.5
Elm, winged	6.1	408.0	245.9	12.6	13.3	61.8	60.7	15.5	15.7	7.1	7.5	3.0	2.9
Hackberry[7]	6.2	281.9	168.9	17.4	16.6	61.0	61.3	12.1	12.4	5.7	6.1	3.7	3.5
Hickory sp.	5.8	360.5	210.0	21.2	18.7	59.3	63.1	11.2	11.5	4.4	4.1	4.0	2.5
Maple, red	5.9	530.5	293.1	11.9	12.2	42.1	43.2	30.4	29.9	9.7	9.8	5.9	5.0
Oak, black	6.0	362.6	208.1	16.7	15.7	69.2	70.5	7.2	7.9	3.4	3.5	3.5	2.4
Oak, blackjack	6.4	371.9	220.3	15.8	14.2	47.8	46.8	22.5	24.0	9.8	11.6	4.1	3.3
Oak, cherrybark	5.8	350.0	208.4	18.7	16.7	58.6	60.8	10.0	10.4	7.7	7.5	5.0	4.6
Oak, chestnut	6.0	416.1	245.7	17.2	15.8	64.1	67.1	10.0	10.5	4.4	4.1	3.8	2.5
Oak, laurel	5.7	378.0	220.6	16.5	15.0	64.7	65.4	9.4	10.1	6.6	7.1	2.8	2.4
Oak, northern red	6.2	470.6	273.2	15.5	14.5	66.7	67.5	11.9	12.8	3.0	3.2	2.8	2.0
Oak, post	6.1	400.2	225.5	19.2	18.6	67.4	68.7	7.9	8.0	2.7	2.7	2.7	1.9
Oak, scarlet	5.8	414.1	229.7	18.5	17.2	64.7	66.3	9.1	9.6	3.9	4.0	3.7	2.8
Oak, Shumard	6.8	564.4	340.7	16.0	15.0	62.2	62.5	11.8	12.8	5.2	5.5	4.8	4.2

See footnotes page 1433.

TABLE 16-2—*Weights and proportions when green and ovendry, of complete small hardwood trees and tree parts of 23 species sampled on southern pine sites (Data from Manwiller and Koch 1)[2]*—Continued

Species	Tree dbh	Complete-tree weight		Stump-root[3]		Stem[4]		Branches[5]		Twigs[6]		Foliage	
		Green	Ovendry	Green	Ovendry	Green	Ovendry	Green	Ovendry	Green	Ovendry	Green	Ovendry
	Inches	----*Pounds*----		------------------------------Percent of complete tree------------------------------									
Oak, southern red	5.9	430.3	241.8	16.1	14.8	61.9	62.7	15.3	16.1	3.7	3.8	3.0	3.8
Oak, water	6.1	478.4	279.0	13.5	12.4	62.9	64.1	11.5	11.4	8.0	8.5	4.1	3.6
Oak, white	6.2	505.9	288.6	17.0	16.1	59.9	61.8	13.0	13.6	6.4	5.4	3.8	3.1
Sweetbay	6.0	392.2	193.9	14.7	14.5	55.9	56.7	16.9	17.4	7.0	7.4	5.6	4.0
Sweetgum	6.1	367.3	168.3	18.4	18.0	63.4	64.0	7.3	8.3	5.0	5.1	6.0	4.6
Tupelo, black	6.2	416.5	207.7	18.9	19.1	52.3	55.1	16.2	14.8	6.9	6.9	5.7	4.0
Yellow-poplar	5.8	318.8	152.6	12.0	11.7	71.6	73.8	4.8	4.9	5.7	5.7	5.8	4.0

[1] Manwiller, F. G., and P. Koch. 1981. Final Report FS-SO-3201-1.59, U.S. Dep. Agric., For. Serv., South. For. Exp. Stn., Pineville, La.
[2] Each value is the average for three trees, except those for Shumard oak, which were derived from one tree. See table 16-3 for additional data on tree dimensions.
[3] Central stump-root system to 6-inch stump height, including lateral roots to the radius specified.
[4] Central bole with bark included, from 6-inch stump height to stem diameter of 0.5 inch, outside bark, or point of stem division into branches.
[5] Wood and bark of all branches 0.5 inch and larger in diameter, measured outside bark.
[6] Less than 0.5 inch in diameter.
[7] Sugarberry.

TABLE 16-3—*Dimensions of small hardwoods of 23 species sampled on southern pine sites* (Data from Manwiller and Koch[1])[2]

Species	Tree dbh	Tree[3] length	Vertical length of stump-root[4]	Stem[5] length	Stem top diameter[6]
	Inches	------------------Feet------------------			Inches
Ash, green	5.9	59.5	4.4	47.0	1.4
Ash, white	6.0	58.9	4.2	45.1	1.8
Elm, American	6.1	46.0	3.5	32.6	3.1
Elm, winged	6.1	49.1	3.0	37.8	1.9
Hackberry, sp.[7]	6.2	43.8	4.6	31.3	1.3
Hickory sp.	5.8	50.4	6.2	38.2	1.4
Maple, red	5.9	50.3	4.2	30.6	2.5
Oak, black	6.0	53.2	3.7	41.5	1.4
Oak, blackjack	6.4	29.3	2.9	21.8	1.8
Oak, cherrybark	5.8	45.9	4.8	32.4	1.6
Oak, chestnut	6.0	53.3	4.6	39.1	1.9
Oak, laurel	5.7	50.9	4.1	37.7	1.6
Oak, northern red	6.2	55.7	3.9	40.8	2.3
Oak, post	6.1	53.0	5.1	41.4	1.4
Oak, scarlet	5.8	49.8	3.5	38.5	1.5
Oak, Shumard	6.8	52.8	5.4	39.2	1.3
Oak, southern red	5.9	56.2	4.3	32.4	2.6
Oak, water	6.1	54.5	4.2	41.0	2.0
Oak, white	6.2	53.0	4.9	39.1	1.7
Sweetbay	6.0	47.9	3.9	27.7	4.1
Sweetgum	6.1	53.1	4.5	46.8	.5
Tupelo, black	6.2	45.3	6.0	34.4	1.3
Yellow-poplar	5.8	56.2	3.1	49.8	.5

[1]Manwiller, F. G., and P. Koch. 1981. Final Report FS-SO-3201-1.59, U.S. Dep. Agric., For. Serv., South. For. Exp. Stn., Pineville, La.
[2]Each value is the average for three trees except values for Shumard oak, which were based on one tree (see table 16-2 for weights of trees and tree parts).
[3]Measured from top of tree crown to the lower extremity of root system.
[4]Vertical distance from top of stump to lower extremities of roots.
[5]From stump top to stem diameter of 0.5 inch outside-bark or point of stem division into branches.
[6]Measured outside bark.
[7]Sugarberry.

Stem weights include the central bole from 6-inch-high stump to stem diameter of 0.5 inch outside bark, or to stem division into branches.

Manwiller and Koch's data (table 16-2) indicate that species with small proportions of their biomass in stump-root systems include yellow-poplar, red maple, the ashes, water oak, and winged elm; those with more massive stump-root systems include hickory, black tupelo, sweetgum, blackjack oak, and cherrybark oak.

Lowest proportions of complete-tree weight in stems were found in red maple and blackjack oak; species with highest proportion of their biomass in stems were green ash and yellow-poplar. Yellow-poplar and green ash had low proportions of branches and twigs, while red maple and blackjack oak had high proportions of these parts. Green ash had least foliage; red maple, sweetgum, and cherrybark oak had most.

Average weights (Shumard oak excepted) of the complete trees were as follows (table 16-2):

Radii at which lateral roots were severed	Green	Ovendry
	-----Pounds-----	
Lateral roots severed at 3-foot radius		
Average	419	238
Range	299-579	164-320
Lateral roots severed at 1-foot radius		
Average	396	223
Range	282-531	153-293

When harvested, the red maples were heaviest, and green ash trees weighed least; when ovendry, red maples were still the heaviest but yellow-poplar trees weighed least.

Complete trees 0.5 to more than 9.5 inches in dbh.—The mixed hardwood forests studied by Harris et al. (1973) on the Walker Branch watershed in eastern Tennessee had branch, bole, and central stump-root mass proportions that varied considerably with tree diameter class, but not greatly with forest type (table 16-4). With data on all species pooled, figure 16-1 shows the relationship of tree dbh to bole and branch weight. (See figure 15-11 for the dbh-foliage weight relationship.)

Complete trees 3 to 12 inches in dbh.—Sirois (1983) utilized a tree puller (figs. 26-25 through 26-29) to measure green and dry complete-tree weights of mockernut hickory, sweetgum, southern red oak, and white oak trees 3 to 12 inches in dbh growing in mixed natural oak-pine stands in lower Piedmont sites near Auburn, Ala. The central stump root system averaged about 18 percent of the green weight of complete trees (foliage-free) of these species. Weights, green and dry, of the various components could be expressed as simple functions of tree height and dbh^2 (table 16-5); these relationships accounted for more than two-thirds the variation observed.

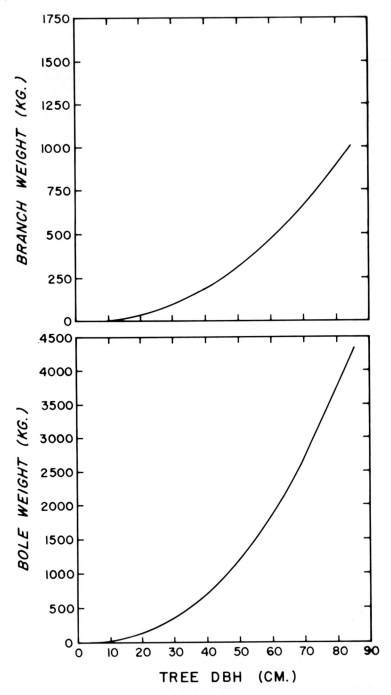

Figure 16-1.—Relationship between tree dbh and branch weight and bole weight per tree (ovendry basis, bark included). The relationship is derived from pooled data on all hardwood species sampled on the Walker Branch Watershed in eastern Tennessee. (Drawing after Harris et al. 1973.)

TABLE 16-4—*Tree part proportion of complete-tree biomass in stands of three forest types on the Walker Branch watershed in eastern Tennessee* (Data from Harris et al. 1973)

Forest type and tree part	Proportion by tree dbh class in inches[1]		
	0.5 to 3.5	3.5 to 9.5	>9.5
	------------Percent, ovendry basis------------		
Chestnut oak			
Foliage	6.9	3.4	2.1
Branches[2]	20.2	20.1	18.9
Bole[3]	50.5	62.0	69.6
Stump-root[4]	22.4	14.5	9.4
	100.0	100.0	100.0
Oak-hickory			
Foliage	6.6	3.8	2.2
Branches	22.9	20.3	18.7
Bole	52.5	62.2	69.3
Stump-root	18.0	13.7	9.8
	100.0	100.0	100.0
Yellow-poplar			
Foliage	6.8	3.7	2.2
Branches	21.2	18.9	15.3
Bole	55.1	63.7	71.8
Stump-root	16.9	13.7	10.7
	100.0	100.0	100.0

[1]Percent of above- and below-ground tree parts including large lateral roots to a radius of 60 cm. Each value is based on a sample of 24 to 44 plots.
[2]All branches with bark.
[3]Stem, including bark, above a 6-inch-high stump to 0.5-inch diameter on main stem.
[4]Central stump-root system with lateral roots to a radius of 60 cm.

TABLE 16-5—*Regression coefficients, with coefficients of determination (R^2) given in parentheses, for an equation[1] estimating green and dry weights of complete trees,[2] whole trees[3] and tree components of four species sampled from natural lower Piedmont hardwood-pine stands near Auburn, Ala,; trees measured 3 to 12 inches in diameter* (Sirois 1983)

Tree portion and moisture content	Mockernut hickory	Sweetgum	White oak	Southern red oak
Complete tree[2]				
Green	0.22822 (0.97)	0.17656 (0.99)	0.22314 (0.99)	0.23398 (0.99)
Ovendry	.14676 (.97)	.10139 (.99)	.13150 (1.00)	.13033 (.98)
Complete-tree wood				
Green	.20824 (.97)	.16823 (.99)	.21466 (.99)	.21583 (.99)
Ovendry	.13494 (.97)	.09664 (.99)	.12645 (1.00)	.11901 (.98)
Complete-tree bark				
Green	.01611 (.88)	.00758 (.96)	.00693 (.97)	.01547 (.83)
Ovendry	.00937 (.87)	.00433 (.97)	.00415 (.95)	.00987 (.81)
Whole-tree[3]				
Green	.18990 (.97)	.14360 (.99)	.17998 (.99)	.20134 (.99)
Ovendry	.12241 (.97)	.08262 (.99)	.10633 (1.00)	.11308 (.98)

See footnotes next page.

TABLE 16-4—*Tree part proportion of complete-tree biomass in stands of three forest types on the Walker Branch watershed in eastern Tennessee* (Data from Harris et al. 1973)—Continued

Forest type and tree part	Proportion by tree dbh class in inches[1]			
	0.5 to 3.5	3.5 to 9.5	>9.5	
	------------Percent, ovendry basis------------			
Whole-tree wood				
Green	.17378 (.97)	.13602 (.99)	.17305 (.99)	.18587 (.99)
Ovendry	.11304 (.97)	.07828 (.99)	.10218 (1.00)	.10321 (.98)
Whole-tree bark				
Green	.01611 (.88)	.00758 (.96)	.00693 (.97)	.01547 (.83)
Ovendry	.00937 (.87)	.00433 (.97)	.00415 (.95)	.00987 (.81)
Stem[4]				
Green	.14908 (.99)	.12536 (.99)	.14787 (1.00)	.14857 (1.00)
Ovendry	.09407 (.99)	.07281 (.99)	.08666 (1.00)	.08112 (1.00)
Stemwood				
Green	.13285 (.98)	.11909 (.99)	.14208 (1.00)	.13534 (1.00)
Ovendry	.08576 (.98)	.06903 (.99)	.08315 (1.00)	.07258 (1.00)
Stembark				
Green	.01417 (.87)	.00627 (.96)	.00579 (.98)	.01323 (.84)
Ovendry	.00831 (.86)	.00378 (.96)	.00351 (.95)	.00854 (.82)
Crown[5]				
Green	.04705 (.80)	.01824 (.84)	.03211 (.80)	.05177 (.84)
Ovendry	.03099 (.80)	.00981 (.88)	.01966 (.81)	.03197 (.83)
Crownwood				
Green	.04441 (.80)	.01692 (.83)	.03097 (.80)	.05025 (.84)
Ovendry	.02960 (.81)	.00926 (.87)	.01902 (.81)	.03060 (.83)
Crownbark				
Green	.00257 (.67)	.00132 (.93)	.00114 (.78)	.00232 (.77)
Ovendry	.00139 (.66)	.00055 (.89)	.00064 (.75)	.00137 (.72)

[1] The prediction equation is:
$Y = (a)(dbh^2)(Th)$
where:
  Y = weight, pounds
  a = coefficient in body of table
  dbh = diameter breast height, inches
  Th = total above-ground tree height, feet
Equations are based on a sample of about 16 trees of each species.
[2] All tree parts, foliage-free, including central stump-root systems to a 22-inch diameter.
[3] All tree parts, foliage-free, above a 6-inch high stump.
[4] From a 6-inch stump to a 3-inch top outside bark.
[5] Above 3-inch-diameter stem plus all live branches below the 3-inch-diameter stem.

ABOVE-GROUND TREE PARTS

Data completely descriptive of each species are not published, but the available information permits some useful extrapolation. These data include—for a few species—descriptions of trees measuring 1.0 to 10.0 inches in dbh, trees 6 inches in dbh of most of the species, some generalizations relating biomass to tree diameter, and some information identifying the balance point in hardwood stems—a statistic useful in designing log handling equipment.

Trees 2 inches and smaller in dbh.—Hitchcock (1979) found that young even-aged (2 and 6 years), well-stocked (34,000 and 20,000 stems/acre respectively), natural hardwood stands in Anderson County, Tennessee had net production of above-ground biomass—not including foliage—of 1.4 tons per acre per year, ovendry basis.

In these same Tennessee stands, Hitchcock (1978) determined the green and dry weights of individual hardwood trees measuring 0.3 to 5.0 cm in dbh. From these investigations he developed equations, based on about 15 trees of each species, that took the form:

$$\text{Log}_{10} Y = b_o + b_1 \log_{10} X \qquad (16\text{-}1)$$

where:
Y = green or dry weight, g
X = total height, m
  or, basal diameter, cm, squared x total height, m or, dbh, cm, squared x total height, m

Coefficients for equations 16-1 for species of interest are shown in table 16-6. Moisture contents of these above-ground tree portions were as follows:

Species	Moisture content
	Percent of ovendry weight
Hickory, sp.	69
Maple, red	79
Oak, chestnut	73
Oaks, red, sp.	60
Oaks, white, sp.	67
Tupelo, black	100
Yellow-poplar	102

Krinard et al. (1979) found that 5-year-old hardwoods on 10- by 10-foot spacing in minor stream bottoms of the coastal plain of Arkansas had the following average dimensions and volumes (each value based on 10 trees):

Species	Dbh	Height	Stem volume including bark	Stemwood volume proportion of total stem volume
	Inch	*Feet*	*Cubic feet*	*Percent*
Ash, green	2.1	17.4	0.30	83
Oak, cherrybark	1.6	13.9	.18	81
Oak, water	1.7	14.2	.18	82
Sweetgum	2.3	16.3	.37	83
Yellow-poplar	2.3	17.6	.41	78

Stems, including bark, accounted for about 60 percent of above-ground tree weight, dry basis.

Trees 1.0 to 4.9 inches in dbh.—Phillips and McClure (1976) studied the volume of six understory hardwood species in mature hardwood stands in the mountains of North Carolina and the Piedmont of Georgia. Twelve trees of each species were examined, i.e., three each in diameter classes of 1.0-1.9, 2.0-2.9, 3.0-3.9, and 4.0-4.9 inches. Average diameter of all trees examined was therefore between 2.9 and 3.0 inches. Species average stem cubic volume (12 trees), with bark, ranged from 0.74 to 1.13 cubic feet and made up 67 to 88 percent of whole-tree volume above ground (table 16-7). Branches contained an average of 0.13 to 0.43 cubic feet and represented 12 to 33 percent of the volume in the whole tree above ground.

Phillips (1977) analyzed weights from data in table 16-7. Total average fresh green weights of foliage-free stems and branches ranged from 63 pounds for sweetgum in the Piedmont to 87 pounds for red maple in the mountains. All other species averaged between 65 and 85 pounds. When separated into component parts, most species had 85 to 88 percent of their weight in stems and 12 to 15 percent in branches. The two notable exceptions were red maple and dogwood. Red maple had 70 to 80 percent of its weight in the stem and 20 to 30 percent in branches; dogwood had 65 to 70 percent in stem and 30 to 35 percent in branches. On an ovendry weight basis, average whole tree weight (above ground) ranged from 28 pounds for sweetgum to 50 pounds for southern red oak and hickory in the Piedmont.

Phillips (1981) used data from these North Carolina and Georgia trees, plus data from additional similar trees from South Carolina, and related total-tree green weights to dbh (table 16-8).

Trees 1.0 to 10.0 inches in dbh.—Wartluft (1977) weighed 200 whole trees, including all parts above a 6-inch-high stump, of 17 hardwood species growing near Richwood, W. Va. Average green weight per tree was 434 pounds, 80 percent of which was in material larger than 3 inches in diameter outside bark. The weighted average moisture content of the 200 trees was 69 percent of ovendry weight (table 16-9).

Trees 6 inches in dbh.—From a sample of 6-inch pine-site hardwoods collected throughout each species' southern range (see table 3-1 for tree dimensions), Manwiller obtained 10-tree averages for above-ground biomass of 22 species. Total weight of above-ground parts of the foliage-free trees averaged 244.1 pounds when green and 143.4 pounds when ovendry; variation among species was substantial even though all trees measured approximately 6 inches in dbh. Percentages of weight of the foliage-free trees, by above-ground tree part, averaged as follows (derived from table 16-10):

Tree part	Green	Ovendry
	---------Percent---------	
Stemwood	73.6	74.8
Stembark	14.5	13.1
Branchwood	8.8	9.7
Branch bark	3.1	2.4

TABLE 16-6—*Constants for regression equation 16-1 to compute green and ovendry weights of above-ground tree parts (foliage excluded) of small-diameter hardwoods in Tennessee* (Data from Hitchcock 1978)

Species	Green weight			Dry weight		
	b_o	b_1	R^2	b_o	b_1	R^2
x = above-ground height, m						
Oaks, red sp.	1.79335048	2.75322751	0.95	1.56610024	2.78687692	0.95
Oaks, white sp.	1.84414748	2.85573868	.98	1.60035602	2.88571175	.98
Oak, chestnut	1.61529642	2.86024236	.94	1.37008507	2.88199447	.94
Tupelo, black	1.66477487	2.61421901	.95	1.27005730	2.80757017	.97
Maple, red	1.62811883	2.51415373	.87	1.34289516	2.56955195	.87
Yellow-poplar	1.62687985	2.73499148	.94	1.29697358	2.77483633	.94
Hickory, sp.	1.81308953	2.72928746	.89	1.56753232	2.76332028	.90
Hard hardwoods	1.62496137	2.84100881	.92	1.34395546	2.87069714	.92
Soft hardwoods	1.66497366	2.59489835	.90	1.32730435	2.69012145	.90
x = basal diameter^2(cm) x height						
Oaks, red sp.	1.68118929	.93754582	.99	1.45264730	.94894885	.99
Oaks, white sp.	1.81296001	.89063347	.99	1.56779548	.90070457	.99
Oak, chestnut	1.82359436	.88453135	.98	1.58113440	.88998915	.98
Tupelo, black	1.78689778	.87487851	.98	1.40617602	.93435979	.98
Maple, red	1.60776256	.94667010	.99	1.31943038	.96982024	.98
Yellow-poplar	1.72857462	.86831749	.97	1.40127189	.88002638	.97
Hickory, sp.	1.67093633	.93685526	.99	1.42705981	.94587971	.99
Hard hardwoods	1.72515335	.91112679	.98	1.44367855	.92182823	.98
Soft hardwoods	1.72073248	.89225466	.97	1.38756136	.92277788	.97
x = DBH^2(cm) x height						
Oaks, red sp.	2.30904171	.81946501	.96	2.08895073	.82854453	.96
Oaks, white, sp.	2.64322802	.60522675	.93	2.41112956	.60958439	.92
Oak, chestnut	2.33797113	.76624587	.97	2.10883675	.76060803	.97
Tupelo, black	2.35614518	.68237396	.96	2.05883005	.68051994	.97
Maple, red	2.21876730	.73398700	.97	1.94000860	.75565600	.97
Yellow-poplar	2.25980029	.77085638	.95	1.91673791	.79934482	.96
Hickory, sp.	2.23890784	.81036410	.98	2.00884195	.81186082	.98
Hard hardwoods	2.38388628	.69224061	.94	2.12048949	.69200035	.94
Soft hardwoods	2.27749999	.73566095	.95	1.96660425	.75717635	.96

White oak had the greatest dry weight proportion in stemwood (80.1 percent), and blackjack oak the least (64.8 percent). Hickory and southern red oak had most stembark (16.5 and 17.4 percent, respectively); the elms and hackberry had the least (8 to 9 percent). Proportion of branchwood was greatest in hackberry (13.5 percent), and least in yellow-poplar (5.0 percent). Branchbark proportion was greatest in post oak (3.9 percent) and least in yellow-poplar (1.5 percent).

TABLE 16-7—*Volume and percentage volume of above-ground tree parts, stem density, and branch density of major understory hardwoods from the mountains of North Carolina and the Georgia Piedmont* (Data from Phillips and McClure 1976)[1]

Species	Average age	Average dbh	Tree height	Average volume[2] Stem[3]	Branches	Stem density Green	Ovendry	Branch density Green	Ovendry
	Years	Inches	Feet	Cubic feet and (percent of total)		---------Pounds/cubic foot---------			
				NORTH CAROLINA MOUNTAINS					
Dogwood[4]....	44	2.8	23	0.74 (67)	0.36 (33)	62.2	35.9	65.4	35.6
Hickory sp....	42	3.0	33	1.03 (86)	.16 (14)	59.8	38.0	56.4	34.4
Maple, red....	34	2.9	36	1.13 (81)	.27 (19)	61.5	32.6	62.5	31.9
Oak, chestnut.	25	2.9	35	1.11 (88)	.15 (12)	64.1	38.1	62.7	36.0
Oak, white ...	44	3.0	28	1.01 (88)	.14 (12)	62.2	37.0	61.4	35.3
Yellow-poplar	18	2.9	35	1.08 (88)	.14 (12)	52.5	22.5	55.2	24.7
				GEORGIA PIEDMONT					
Dogwood.....	36	3.0	27	.91 (71)	.37 (29)	66.7	36.2	69.3	36.1
Hickory sp....	36	2.9	30	.93 (76)	.30 (24)	66.0	41.3	63.5	38.2
Maple, red....	37	3.0	34	.94 (69)	.43 (31)	60.2	33.5	59.5	33.0
Oak, southern red........	33	3.1	33	1.13 (87)	.17 (13)	65.0	38.3	66.2	39.2
Oak, white ...	33	2.9	28	.90 (87)	.13 (13)	62.4	36.1	61.6	36.0
Yellow-poplar	19	3.0	32	1.10 (81)	.26 (19)	59.0	24.6	59.6	25.5

[1]Each value is the average for 12 trees with diameters in the range from 1.0 to 4.9 inches.
[2]Stem and branch volumes sum to 100 percent of above-ground tree volume.
[3]Stem with bark to top of tree.
[4]*Cornus florida* L.

TABLE 16-8—*Total green weight of foliage-free above-ground tree parts related to dbh of major understory hardwoods sampled in the mountains of North Carolina, the Piedmont of Georgia, and the Piedmont of South Carolina* (Phillips 1981)

Species	Whole-tree green weight by dbh class[1]					
	1 in	2 in	3 in	4 in	5 in	6 in
	------------------------------Pounds------------------------------					
	MOUNTAINS					
Dogwood[2].....................	6	27	66	124	202	302
Hickory.......................	4	24	64	127	219	340
Maple, red	5	29	77	153	261	404
Oak, chestnut..................	5	25	67	134	229	355
Oak, white	4	22	58	115	196	303
Sourwood[3]	5	26	67	134	227	350
Tupelo, black	5	25	64	125	210	320
Yellow-poplar..................	4	21	56	111	189	294

TABLE 16-8—*Total green weight of foliage-free above-ground tree parts related to dbh of major understory hardwoods sampled in the mountains of North Carolina, the Piedmont of Georgia, and the Piedmont of South Carolina* (Phillips 1981)—Continued

Species	Whole-tree green weight by dbh class[1]					
	1 in	2 in	3 in	4 in	5 in	6 in
	----------------------------Pounds----------------------------					
	PIEDMONT					
Dogwood	5	27	71	142	244	380
Hickory	5	25	67	132	224	345
Maple, red	5	26	69	137	233	359
Oak, post	4	20	54	108	184	285
Oak, southern red	5	25	65	128	217	335
Oak, white	4	22	58	114	193	296
Sweetgum	4	18	49	97	166	2.7
Yellow-poplar	5	24	63	126	214	332
	SPECIES AND LOCATIONS COMBINED					
Soft hardwoods	4	24	62	123	210	324
Hard hardwoods	5	25	64	127	214	328

[1]Weights predicted by regression analysis.
[2]*Cornus florida* L.
[3]*Oxydendrum arboreum* (L.) DC.

TABLE 16-9—*Whole-tree weights of green and dry small Appalachian hardwood trees and of tree parts more than 3 inches and less than 3 inches in diameter outside bark* (Data from Wartluft 1977)[1]

Tree dbh (inches)	Green[2]			Dry			Tree height
	>3 inches	<3 inches	Whole tree	>3 inches	<3 inches	Whole tree	
	----------------------Pounds----------------------						*Feet*
1	—	5	5	—	3	3	16
2	—	25	25	—	14	14	24
3	7	58	65	4	33	37	31
4	73	57	130	42	33	75	40
5	163	59	222	94	35	129	46
6	273	70	343	160	41	201	50
7	406	90	496	239	53	292	55
8	564	119	683	334	70	404	59
9	751	155	906	446	92	538	70
10	968	198	1,166	576	118	694	70

[1]Weights are for tree portions above a 6-inch-high stump; the values are computed from regression relationships based on a 200-tree sample of 17 species of hardwoods in West Virginia.
[2]Weighted average moisture content = 69% of ovendry weight.

TABLE 16-10—*Green and dry weights of foliage-free above-ground parts, and principal dimensions of 6-inch trees of 22 hardwood species collected from southern pine sites throughout each species range (Data from Manwiller*[1])[2]

Species	Tree dbh	Height of tree above stump	Height to first branch	Stem[3] top diameter	Whole-tree weight		Stemwood[3]		Stembark[3]		Branchwood[4]		Branchbark[4]	
					Green	Ovendry	Green	Ovendry	Green	Ovendry	Green	Ovendry	Green	Ovendry
	Inches	----*Feet*----		*Inches*	----*Pounds*----		----*Percent of foliage-free weight above ground*----							
Ash, green	5.9	47.4	31.3	2.2	229.3	151.0	73.6	77.1	13.8	10.6	9.2	10.0	3.5	2.4
Ash, white	5.9	49.7	29.7	.9	209.0	138.4	74.8	77.7	16.1	13.4	6.2	6.8	2.9	2.1
Elm, American	5.9	46.8	23.5	2.1	223.5	126.8	74.3	76.0	12.0	9.4	10.5	12.2	3.2	2.5
Elm, winged	6.0	45.4	20.6	1.8	238.1	143.0	75.5	76.9	9.9	8.1	11.0	12.1	3.6	2.9
Hackberry (mostly sugarberry)	6.1	44.5	22.9	2.0	201.0	118.4	74.4	74.9	9.5	9.2	13.0	13.5	3.1	2.4
Hickory sp.	6.0	44.8	24.5	1.7	234.0	150.3	69.3	72.3	19.5	16.5	8.0	8.5	3.2	2.6
Maple, red	5.8	49.6	22.4	2.0	240.5	140.2	72.3	74.3	11.5	10.3	12.4	12.6	3.8	2.7
Oak, black	6.0	48.6	26.4	1.2	280.0	169.6	69.6	70.4	17.4	16.1	9.7	10.9	3.3	2.6
Oak, blackjack	6.0	33.1	18.2	1.5	205.8	125.2	65.1	64.8	20.1	19.3	10.5	12.1	4.3	3.8
Oak, cherrybark	6.0	47.3	23.4	.8	249.1	152.2	74.8	75.8	13.8	12.8	8.9	9.6	2.5	1.9
Oak, laurel	6.0	47.8	27.1	1.7	291.1	170.3	75.5	75.1	11.8	11.5	10.0	11.4	2.7	2.0
Oak, northern red	5.9	52.4	29.5	1.1	297.9	179.4	73.7	74.9	16.4	15.4	7.3	7.6	2.6	2.0
Oak, post	6.0	39.0	17.7	1.6	222.2	137.2	68.5	70.0	16.1	13.8	10.9	12.3	4.5	3.9
Oak, scarlet	5.8	46.2	24.0	.8	258.0	155.8	72.3	73.2	16.3	14.8	8.4	10.0	3.0	2.1

TABLE 16-10—*Green and dry weights of foliage-free above-ground parts, and principal dimensions of 6-inch trees of 22 hardwood species collected from southern pine sites throughout each species range* (Data from Manwiller[1])[2]—Continued

Species	Tree dbh	Height of tree above stump	Height to first branch	Stem[3] top diameter	Whole-tree weight		Stemwood[3]		Stembark[3]		Branchwood[4]		Branchbark[4]	
					Green	Ovendry	Green	Ovendry	Green	Ovendry	Green	Ovendry	Green	Ovendry
	Inches	Feet		Inches	Pounds		Percent of foliage-free weight above ground							
Oak, Shumard	6.1	51.7	27.5	1.3	306.7	185.4	75.4	75.8	14.2	13.3	7.9	9.0	2.5	1.9
Oak, southern red	6.0	43.3	21.9	1.0	232.3	140.6	68.6	69.4	19.1	17.4	8.8	10.4	3.3	2.8
Oak, water	5.9	53.9	31.8	1.8	298.1	175.7	76.0	76.0	14.5	14.0	7.3	8.3	2.2	1.8
Oak, white	5.9	47.4	26.6	1.3	232.7	144.3	78.7	80.1	12.0	10.5	6.8	7.2	2.5	2.1
Sweetbay	5.8	44.9	30.5	1.6	246.6	122.6	75.7	77.2	13.9	12.0	7.7	8.5	2.7	2.2
Sweetgum	5.9	46.8	20.0	.9	208.4	96.8	80.2	79.0	10.9	12.2	6.4	6.8	2.5	2.0
Tupelo, black	5.8	40.3	22.3	1.8	202.6	109.0	75.1	76.3	13.7	12.6	8.2	8.8	3.0	2.2
Yellow-poplar	6.1	53.5	29.1	.8	263.3	123.4	76.3	78.9	16.7	14.1	5.0	5.5	1.9	1.5

[1] Manwiller, F. G. Unpublished data in Study file FS-SO-3201-1.59, U.S. Dep. Agric., For. Serv., South. For. Exp. Stn., Pineville, La.
[2] Each value is the average for 10 trees.
[3] The stem was defined as the bole portion above a 6-inch stump height to a stem diameter of 0.5 inch outside bark or to the point of stem division into branches. The stem top diameter was measured outside bark.
[4] Branches did not include foliage or twigs smaller than 0.5 inch diameter.

Correlation of whole-tree weight with dbh.—In tables 16-4, 16-5, 16-6, 16-8, and in later discussions of individual species, some data correlating weight of above-ground tree parts with dbh are given. To permit quick visualization of the general relationship, table 16-11 is presented, based on data from trees in Maine. The U.S. Department of Agriculture, Forest Service (1978) estimates that sawtimber and poletimber of the species described in table 16-11 have 30 percent and 40 percent, respectively of total above-stump tree weight in branches, top, and foliage. Data from Monk et al. (1970) on 11 species of southern hardwoods (table 15-7) relate dbh to the weight of whole trees (ovendry basis) above stump height.

Wiant et al. (1977) measured 19 to 22 trees 2 to 16 inches in dbh of each of eight hardwood species growing near Morgantown, West Virginia and related the weight of stemwood and branchwood to tree dbh (table 16-12); similar data on weights of stembark and branchbark of these trees are shown in table 13-11.

TABLE 16-11—*Tree dbh related to green weights of total above-ground tree portions, including foliage, of five hardwood species grown in Maine* (U.S. Department of Agriculture, Forest Service 1978)[1]

Dbh (inches)	Red maple and white ash	Elm sp.	White and northern red oaks
	---------------------Pounds--------------------		
2	14	15	18
3	44	46	55
4	97	102	122
5	177	187	223
6	287	302	363
7	431	453	544
8	608	640	768
9	819	862	1,035
10	1,067	1,122	1,347
11	1,348	1,418	1,703
12	1,695	1,783	2,140
14	2,397	2,521	3,027
16	3,253	3,422	4,108
18	4,224	4,443	5,335
20	5,295	5,570	6,687
22	6,452	6,788	8,149
24	7,683	8,083	9,704
26	8,975	9,442	11,336
28	10,314	10,851	13,027
30	11,689	12,297	14,763

[1] Weight of whole tree above a 6-inch-high stump.

TABLE 16-12—*Weight (ovendry) of stemwood and branchwood per tree of six diameters and eight species sampled in West Virginia* (Wiant et al. 1977)[1,2]

Species and wood component	Tree dbh, inches					
	6	8	10	12	14	16
	----------------------Pounds----------------------					
Hickory sp.						
Stemwood	94	248	527	974	1,639	2,572
Branchwood	44	78	151	262	412	599
Maple, red						
Stemwood	114	233	448	759	1,167	1,671
Branchwood	55	86	122	164	212	266
Oak, black						
Stemwood	129	239	421	674	1,000	1,398
Branchwood	46	78	132	209	308	430
Oak, chestnut						
Stemwood	138	289	502	777	1,115	1,516
Branchwood	42	70	117	184	270	376
Oak, northern red						
Stemwood	133	293	514	793	1,133	1,531
Branchwood	63	104	161	234	324	430
Oak, scarlet						
Stemwood	121	259	468	758	1,140	1,623
Branchwood	53	85	143	229	341	480
Oak, white						
Stemwood	126	266	474	761	1,135	1,606
Branchwood	37	51	86	144	223	325
Yellow-poplar						
Stemwood	94	216	396	634	932	1,287
Branchwood	22	19	34	67	117	184

[1]Stemwood includes wood in the stem from stump height to a 4-inch top diameter outside bark; branchwood includes that in tops and limbs smaller than 4 inches in diameter outside bark. Data are based on 19 to 22 trees of each species.

[2]See table 13-11 for weight of stembark and branchbark from these trees, and tables 27-131 and 27-132 for green and ovendry weight of branches with bark included.

Correlation of volume of bark-free stem with dbh and height—Nyland (1977) developed an equation relating dbh and height of merchantable stem to the volume of bark-free stemwood in second growth northern hardwoods in the state of New York. His equation, based on a 409-tree sample from 27 different stands of mixed species, is as follows ($R^2 = 0.97$):

$$\text{Volume in cubic feet} = -2.37745 + 0.00302(dbh^2 \times H) + 0.86434 D_t \qquad (16\text{-}2)$$

where:

dbh = diameter outside bark 4.5 feet from the ground on the uphill side of the tree, in inches

H = height of the merchantable stem, in feet

D_t = diameter outside bark at the top of the merchantable stem

This equation, while simple, should be applied with caution to mixed southern hardwoods.

Many of the references cited in following subsections describing individual southern hardwoods include regression analyses correlating weight or volume to dbh and tree height.

Meyers et al. (1980) correlated both weight and volume of above-ground tree parts to dbh and height of hickory, white ash, yellow-poplar, and black, red, and white oaks in upland southern Illinois; interested readers are referred to their publication for tabular data.

Center of gravity for tree-length sawlogs and whole trees.—In designing and using machinery for logging and trucking, it is useful to know the center of gravity of a log or tree, i.e., the point on the log or tree where it will balance. Nyland (1973) determined that cleanly limbed, single-stem logs of important northeastern species (beech-birch-maple) have balance points as follows ($R^2 = 0.91$):

$$D = 1.7419 + 0.4056 (L) \qquad (16\text{-}3)$$

where:
D = distance from the butt, in feet, to the point where the log will balance
L = total length of the tree-length log in feet

Solution of this equation yields the following values:

Log length	D
Feet	*Feet*
20	9.8
25	11.9
30	13.9
35	15.9
40	18.0
45	20.0
50	22.0
55	24.0
60	26.1

Oderwald et al. (1979) found that Appalachian hardwood whole green trees sampled near Blacksburg, Va. in summer and winter had balance points as follows (feet from butt):

Season and dbh	Total tree height, feet		
	30	50	70
Inches	------------------------------- *Feet* -------------------------------		
Summer			
4.	12.9	14.8	
6.	15.4	17.3	
8.	17.6	19.5	21.1
10.		21.5	23.1
12.		23.4	25.0
14.		25.2	26.9
16.		27.0	28.7
18.			30.4
Winter			
4.	12.2	16.9	
6.	13.0	17.7	22.2
8.		18.2	22.7
10.		18.7	23.2
12.		19.0	23.5
14.		19.3	23.8
16.		19.6	24.1
18.			24.3

ASH Sp.

See tables 16-2 and 16-10 for biomass of 6-inch trees. Schlaegel[1] has provided for ash, sugarberry, sweetgum, and elm, equations relating tree basal area and tree height to volume and to fresh and dry weights of tree portions as follows:

$$\text{Volume in cubic feet, or weight in pounds} = (A)(H)(X) \quad (16\text{-}4)$$

where:
A = basal area of tree in square feet
H = height of tree in feet
X = a form factor different for each tree part as shown in table 16-13.

---

[1] Unpublished data (1979) from B. E. Schlaegel, U.S. Dep. Agric., For. Serv., South. For. Exp. Stn., Stoneville, MS.

TABLE 16-13—*Factors for computing volume or weight of above-ground tree parts of four hardwood species that grow in Mississippi, through use of equation 16-4* (Data from Schlaegel[1])

Statistic and tree part	Sugarberry	Sweetgum	American elm	Green ash
Bole volume, cubic feet[2]				
Wood	0.342	0.329	0.323	0.291
Bark	.041	.041	.051	.055
Total	.383	.370	.374	.346
Green bole weight, pounds[2]				
Wood	20.050	21.541	20.235	15.049
Bark	2.775	2.138	2.680	3.097
Total	22.825	23.679	22.915	18.506
Ovendry bole weight, pounds[2]				
Wood	10.898	10.044	9.776	10.122
Bark	1.715	1.150	1.350	1.816
Total	12.613	11.194	11.126	11.938
Green crown weight, pounds[3]	6.956	4.867	6.338	3.710
Ovendry crown weight, pounds[3]	3.846	2.343	3.297	2.228
Green tree weight, pounds[4]	29.781	28.546	29.253	22.216
Ovendry tree weight, pounds[4]	16.459	13.537	14.423	14.166

[1]Unpublished data from B. E. Schlaegel, U.S. Dep. Agric., For. Serv., South. For. Exp. Stn., Stoneville, MS.
[2]From top 1- to 2-foot stump to tip of stem.
[3]Foliage-free branches only; does not include upper main stem.
[4]Foliage-free, above-ground tree portions above a 6-inch-high stump.

Schlaegel[1] found that the weight (green basis) of the crowns of some hardwoods (foliage-free branches only, not including upper main stem) can be predicted from knowledge of dbh (inches) and crown length (feet) according to the equation:

$$\text{Green crown weight, foliage-free, in pounds} = a + b\,(dbh^2 \times \text{crown length}) \quad (16\text{-}5)$$

where regression constants are as follows:

Species	a	b	$S_{y \cdot x}$	R^2
Green ash	−46.1	0.050	358	0.76
Sweetgum	−143.0	.067	401	.88
Sugarberry	−56.7	.078	253	.77

Weight of fresh green ash foliage is given by equation 15-1.

For equations and tabular data relating weight and volume of upland white ash in southern Illinois to dbh and tree height, see Meyers et al. (1980).

ELM Sp.

See tables 16-2, 16-10 for biomass of 6-inch trees, table 16-11 for weights, equation 16-4 and table 16-3 for weight and volume factors.

HACKBERRY AND SUGARBERRY

See tables 16-2, 16-10 for biomass of 6-inch trees, equation 16-4 and 16-13 for weight and volume factors, and equation 16-5 for crown weight factors.

HICKORY Sp.

See equation 16-1 and table 16-6 for weight constants for trees smaller than 2 inches in dbh, tables 16-7 and 16-8 for volumes, weights, and densities of understory trees, tables 16-2 and 16-10 for biomass of 6-inch trees, and table 16-5 for trees 3 to 12 inches in dbh. Table 16-12 relates dbh to weight of stemwood and branchwood in trees 6 to 16 inches in diameter.

Phillips (1977) expressed the volume and green and dry weights of the foliage-free, above-ground portions of understory hickory trees as follows:

$$\text{Log}_{10} Y = b_0 + b_1 \log_{10} (\text{dbh})^2 (H) \qquad (16\text{-}6)$$

where dbh is in inches, total tree height (H) is in feet, and the regression constants for hickory are as follows ($R^2 = 0.98$ or 0.99 for all relationships):

Statistic	North Carolina mountains	Georgia Piedmont
Total weight of green tree, pounds		
b_0	-0.45800	-0.29465
b_1	.90556	.87324
Total weight of ovendry tree, pounds		
b_0	-0.65526	-0.51369
b_1	.90460	.87835
Total volume of tree, cubic feet		
b_0	-2.20310	-0.211049
b_1	.89395	.87348

These relationships apply to trees measuring from 1.0 to 4.9 inches in dbh.

Schnell (1978) analyzed the biomass of 37 hickory trees from four eastern Tennessee locations. Trees sampled were 25 to 220 years old and measured 2 to 28 inches in dbh, and 26 to 115 feet tall. Crown diameters varied from 5 to 50 feet. Fresh leaf weight amounted to 5 percent of complete-tree weight on pole-size trees but only 1 or 2 percent on large trees. About 40 percent of fresh leaf weight was lost when foliage was ovendried.

Schnell's study indicated that complete-tree weight of a hickory 12 inches in dbh is about 1,704 pounds (ovendry basis). In hickory trees under 20 inches in dbh, weight of crown wood exceeds that of wood in the merchantable bole. For the 12-inch trees studied, merchantable bole wood comprised only 24 percent of complete-tree weight, dry-weight basis, as follows:

Component	Proportion of complete tree	Proportion of merchantable bole wood
	------------Percent------------	
Crown wood	35	151
Crown bark	10	44
Merchantable bole wood	24	100
Merchantable bole bark	5	21
Stump and root wood	20	84
Stump and root bark	4	17
Foliage	2	8
	100	

In these trees bark accounted for about 20 percent of the weight of the complete trees and about 17 percent of the merchantable boles. Bark proportion was greatest in tree components of smallest diameter, as follows (dry-weight basis):

Component diameter	Bark proportion of component
	Percent
< ½-inch	33
½ to 1 inch	33
1 to 2 inches	27
2 to 3 inches	24
3 to 4 inches	23
4 to 5 inches	21
5 to 8 inches	20
> 8 inches	17
Total tree crown	22
Merchantable bole	17

For equations and tabular data relating weight and volume of upland hickory in southern Illinois to dbh and tree height, see Meyers et al. (1980).

MAPLE, RED

See equation 16-1 and table 16-6 for weight constants for trees smaller than 2 inches in dbh, tables 16-7 and 16-8 for volumes, weights, and densities of understory trees, tables 16-2 and 16-10 for biomass of 6-inch trees, and tables 16-11 and 16-12 relating dbh to whole-tree weight of larger trees.

Phillips (1977) expressed the volume and green and dry weights of the foliage-free, above-ground portions of red maple understory trees by application of equation 16-6 (see HICKORY heading), with regression constants as follows ($R^2 = 0.99$ for all relationships):

Statistic	North Carolina mountains	Georgia Piedmont
Total weight of green tree, pounds		
b_0	−0.36125	−0.61978
b_1	.88931	.98555
Total weight of ovendry tree, pounds		
b_0	−0.70853	−0.92222
b_1	.91763	1.00528
Total volume of tree, cubic feet		
b_0	−2.15214	−2.38651
b_1	.88950	.98105

These relationships apply to trees measuring from 1.0 to 4.9 inches in dbh.

Foliage weights of red maple in Maine are given in table 15-6 (related to tree height) and table 15-10 (related to dbh).

OAK, BLACK

See tables 16-2 and 16-10 for biomass of 6-inch trees, and table 16-12 which relates dbh to dry weight of stemwood and branchwood in trees 6 to 16 inches in diameter.

The Tennessee Valley Authority (1972) analyzed complete-tree biomass of 26 black oak trees in western North Carolina, western Kentucky, and in eastern and western Tennessee. The trees measured 11.5 to 34.8 inches in dbh, and had above-ground heights of 52 to 91 feet; crowns measured 25 to 55 feet in diameter and 30 to 52 feet high above the top of the merchantable bole. Diameter inside bark at the top of the merchantable bole measured 8.0 to 21.4 inches. Stumps were 1 foot high.

Component proportions, expressed as percentages of foliage-free complete-tree weight and merchantable bole wood weight, were as follows:

Component	Proportion of complete tree	Proportion of merchantable bole wood
	Percent	
Crown wood	24	62
Crown bark	8	20
Merchantable bole wood	39	100
Bark of merchantable bole	4	10
Stumpwood and rootwood	20	52
Stump bark and root bark	5	13
	100	

Weights of stumps and roots were not measured, but were based on information from Young and Chase (1965) and Keays (1971).

Weights of complete trees, crowns, and merchantable stems of trees 12 to 36 inches in dbh are given in table 16-14. Weights of limbs (including bark) are given in table 16-15; weight percentage of bark by limb diameter class was as follows:

Limb diameter	Limb bark, as percent of limb weight including bark
Inches	*Percent*
< ½	36
1 to 2	28
2 to 4	27
4 to 8	25
> 8	22

For equations and tabular data relating weight and volume of upland black oak in southern Illinois to dbh and tree height, see Meyers et al. (1980).

OAKS, BLACKJACK AND CHERRYBARK

See tables 16-2 and 16-10 for biomass of 6-inch trees. Weight per acre of tree parts in a blackjack oak-post oak stand are given in table 15-17.

OAK, CHESTNUT

See equation 16-1 and table 16-6 for weight constants for trees smaller than 2 inches in dbh, table 16-2 for biomass of 6-inch trees, and table 16-4 for tree part proportions on the Walker-Branch watershed in eastern Tennessee. Table 16-12 relates dbh to dry weight of stemwood and branchwood in trees 6 to 16 inches in diameter.

For understory trees (1.0 to 4.9 inches in diameter) in the Southeast, volumes, weights, and densities are given in tables 16-7 and 16-8. Phillips (1977) expressed the volumes and green and dry weights of the foliage-free, above-ground portions of these chestnut oak understory trees in terms of dbh and tree height in equation 16-6 (see HICKORY heading), where the regression constants are as follows ($R^2 = 0.99$):

Statistic	b_0	b_1
Total weight of green tree, pounds	−0.25205	0.83728
Total weight of ovendry tree, pounds	− .47279	.83411
Total volume of tree, cubic feet	−2.04082	.83060

A. Clark III and J. G. Schroeder of the Southeastern Forest Experiment Station, Forest Service, U.S. Department of Agriculture, Asheville, N.C., in personal correspondence during November 1981, provided data on the wood content (table 16-16) and bark content (table 16-17) of 24 chestnut oak trees 6 to 20 inches in diameter sampled on the Nantahala National Forest in the mountains of western North Carolina. Whole-tree green and dry weights of foliage-free above-ground tree portions can be summed from the two tables mentioned.

TABLE 16-14—*Weights of components of foliage-free, 12- to 36-inch black oak trees, when green (and ovendry) (Data from Tennessee Valley Authority 1972)*[1]

Dbh (inches)	Complete tree[2]				Crown[3]				Merchantable stem[4]			
	Wood		Bark		Wood		Bark		Wood		Bark	
	------Pounds------											
12	2,579	(1,543)	515	(330)	711	(441)	215	(139)	1,231	(718)	138	(93)
14	3,576	(2,137)	702	(455)	983	(606)	297	(195)	1,710	(997)	181	(124)
16	4,747	(2,833)	919	(601)	1,300	(798)	393	(261)	2,273	(1,325)	229	(159)
18	6,094	(3,633)	1,165	(768)	1,665	(1,018)	502	(339)	2,922	(1,702)	282	(198)
20	7,619	(4,539)	1,441	(957)	2,077	(1,265)	626	(427)	3,658	(2,130)	340	(240)
22	9,326	(5,551)	1,746	(1,167)	2,536	(1,540)	764	(527)	4,481	(2,609)	403	(287)
24	11,217	(6,671)	2,080	(1,400)	3,044	(1,842)	916	(638)	5,395	(3,140)	470	(337)
26	13,292	(7,900)	2,444	(1,654)	3,601	(2,173)	1,082	(761)	6,398	(3,724)	542	(390)
28	15,554	(9,238)	2,837	(1,930)	4,207	(2,531)	1,263	(895)	7,493	(4,360)	618	(448)
30	18,005	(10,687)	3,260	(2,228)	4,862	(2,918)	1,458	(1,042)	8,680	(5,050)	698	(509)
32	20,646	(12,248)	3,712	(2,549)	5,566	(3,333)	1,668	(1,201)	9,960	(5,794)	783	(573)
34	23,478	(13,920)	4,194	(2,893)	6,321	(3,777)	1,893	(1,372)	11,334	(6,592)	872	(641)
36	26,504	(15,706)	4,705	(3,259)	7,126	(4,249)	2,133	(1,556)	12,802	(7,445)	965	(713)

[1] Values derived from data on 26 trees sampled in Tennessee, Kentucky, and North Carolina. The first value given is green weight; the next value (in parentheses) is the weight when ovendry.
[2] Includes stumpwood and rootwood estimated at 20 percent of complete-tree weight and bark from stump and roots estimated at 5 percent of complete-tree weight.
[3] The crown includes all parts above the top of the merchantable bole, foliage-free.
[4] From 1-foot stump height to top of merchantable stem, which varied in diameter from 8.0 to 21.4 inches inside bark.

TABLE 16-15—*Weights per tree of limbs, bark included, in six branch diameter classes on 12- to 36-inch black oaks when green (and ovendry) (Data from Tennessee Valley Authority 1972)*[1]

Dbh (inches)	Branch diameter class					
	> 8 inches	4 to 8 inches	2 to 4 inches	1 to 2 inches	½ to 1 inch	< ½ inch
	----------Pounds----------					
12	138 (82)	264 (162)	232 (145)	102 (64)	70 (44)	51 (31)
14	228 (136)	357 (220)	286 (180)	126 (80)	88 (55)	63 (39)
16	352 (210)	464 (288)	343 (216)	152 (96)	106 (66)	75 (47)
18	516 (307)	584 (364)	402 (254)	180 (114)	126 (79)	89 (55)
20	727 (432)	718 (448)	463 (294)	208 (132)	147 (92)	102 (64)
22	990 (588)	866 (542)	527 (335)	238 (151)	168 (106)	117 (73)
24	1,314 (779)	1,027 (645)	593 (377)	269 (170)	191 (120)	132 (82)
26	1,705 (1,010)	1,201 (756)	661 (421)	300 (191)	214 (135)	147 (92)
28	2,169 (1,284)	1,389 (876)	731 (466)	333 (212)	238 (150)	162 (102)
30	2,714 (1,605)	1,590 (1,005)	802 (513)	367 (233)	263 (166)	179 (112)
32	3,347 (1,977)	1,805 (1,142)	875 (560)	402 (255)	288 (183)	195 (122)
34	4,076 (2,406)	2,032 (1,288)	950 (609)	437 (278)	315 (200)	212 (133)
36	4,908 (2,894)	2,273 (1,444)	1,027 (659)	474 (301)	341 (217)	229 (144)

[1]Values derived from 26 trees sampled in Tennessee, Kentucky, and North Carolina. The first value gives the green weight; the value in parentheses is the weight when ovendry.

TABLE 16-16—*Average green and dry weight of wood of whole chestnut oak trees (foliage-free above-ground parts) from western North Carolina, and distribution of wood in stem and live branches* (Unpublished data from A. Clark III and J. G. Schroeder, Southeastern Forest Experiment Station, Asheville, N.C.)[1]

Dbh class (inches)	Average total height	Sample trees	Whole tree wood weight	Main stem[2]				Proportion of wood			
				Sawlog[3]	Pulpwood[4]	Topwood	Total stem	≥ 4	< 4 & ≥ 2	Branches < 2 & > 0.5	≤ 0.5	All branches
	Feet	Number	Pounds	--- Percent ---
GREEN PULPWOOD
6	52	3	234	—	72	19	91	—	—	6	2	8
8	62	3	612	—	80	5	85	—	2	12	1	15
10	75	3	1,210	—	85	3	88	—	6	6	1	13
Average	63	3	685	—	79	9	88	—	3	8	1	12
GREEN SAWTIMBER
12	91	3	1,945	64	24	1	89	2	4	4	1	11
14	98	3	2,879	70	15	.4	85	3	6	5	.5	15
16	85	3	3,475	60	18	.5	78	9	7	5	.6	22
18	96	3	4,522	66	16	.3	82	8	5	4	.8	18
20	95	3	5,937	64	10	.2	74	14	7	4	.6	26
Average	93	3	3,752	65	16	1	82	7	6	4	1	18

See footnotes on following page.

TABLE 16-16—*Average green and dry weight of wood of whole chestnut oak trees (foliage-free above-ground parts) from western North Carolina, and distribution of wood in stem and live branches* (Unpublished data from A. Clark III and J. G. Schroeder, Southeastern Forest Experiment Station, Asheville, N.C.)[1]—Continued

Dbh class (inches)	Average total height	Sample trees	Whole tree wood weight	Main stem[2]				Proportion of wood					
				Sawlog[3]	Pulpwood[4]	Topwood	Total stem	≥ 4	<4 & ≥ 2	<2 & >0.5	≤ 0.5		All branches
												Branches	
6	52	3	138	\multicolumn{4}{l}{OVENDRY PULPWOOD}						
6	52	3	138	—	72	19	91	—	—	6	2		9
8	62	3	366	—	79	5	84	—	2	13	1		16
10	75	3	718	—	85	3	88	—	6	6	1		13
Average	63	3	407	—	79	9	88	—	3	8	1		12
				\multicolumn{4}{l}{OVENDRY SAW TIMBER}						
12	91	3	1,160	64	24	1	89	2	4	4	1		11
14	98	3	1,715	70	15	.4	85	3	7	5	.5		15
16	85	3	2,066	60	18	.5	78	9	7	5	.6		22
18	96	3	2,666	65	16	.3	81	9	5	4	.8		19
20	95	3	3,461	62	10	.3	73	15	8	4	.6		27
Average	93	3	2,214	64	16	1	81	7	6	5	1		19

[1] Clark, A. III, and J. G. Schroeder. Predicted weights and volumes of selected hardwood species in North Carolina. U.S. Dep. Agric., For. Serv., South. For. Exp. Stn., Asheville, N.C. (Personal communication November 1981.)
[2] Main stem to 2-in dib top.
[3] Sawlogs to 8-in dib or merchantable top.
[4] Pulpwood in stem from butt or sawlog top to 4-in dib top.

TABLE 16-17—*Average green and dry weight of bark of whole chestnut oak trees (foliage-free above-ground parts) from western North Carolina, and distribution of bark in stem and live branches* (Unpublished data from A. Clark III and J. G. Schroeder, Southeastern Forest Experiment Station, Asheville, N.C.)[1]

Dbh class (inches)	Average total height	Sample trees	Whole tree bark weight	Main stem[2]			Proportion of bark			Branches		
				Sawlog[3]	Pulpwood[4]	Topwood	Total stem	≥ 4	< 4 & ≥ 2	< 2 & > 0.5	≤ 0.5	All branches
	Feet	*Number*	*Pounds*	----------	----------	----------	----------	*Percent*	----------	----------	----------	----------
					GREEN PULPWOOD							
6........	52	3	60	—	62	23	85	—	—	12	3	15
8........	62	3	192	—	67	6	73	—	3	20	4	27
10.......	75	3	301	—	73	5	78	—	7	13	2	22
Average..	63	3	184	—	67	11	79	—	3	15	3	21
					GREEN SAWTIMBER							
12.......	91	3	461	50	27	2	79	1	10	8	2	21
14.......	98	3	719	54	16	1	71	3	12	12	2	29
16.......	85	3	926	43	16	1	60	13	13	12	2	40
18.......	96	3	1,143	47	15	1	63	14	11	9	3	37
20.......	95	3	1,684	49	10	1	60	19	12	7	2	40
Average..	93	3	987	49	17	1	67	10	11	10	2	33

See footnotes on following page.

TABLE 16-17—*Average green and dry weight of bark of whole chestnut oak trees (foliage-free above-ground parts) from western North Carolina, and distribution of bark in stem and live branches* (Unpublished data from A. Clark III and J. G. Schroeder, Southeastern Forest Experiment Station, Asheville, N.C.)[1]—Continued

Dbh class (inches)	Average total height	Sample trees	Whole tree bark weight	Proportion of bark							
				Main stem[2]			Branches				
				Sawlog[3]	Pulpwood[4]	Topwood	Total stem	≥ 4	< 4 & ≥ 2	< 2 & > 0.5	≤ 0.5	All branches
	Feet	Number	Pounds	----------Percent----------								
				OVENDRY PULPWOOD								
6	52	3	40	—	63	22	85	—	—	10	5	15
8	62	3	119	—	70	6	76	—	3	19	2	24
10	75	3	187	—	75	5	80	—	6	12	2	20
Average	63	3	115	—	69	11	80	—	3	14	3	20
				OVENDRY SAW TIMBER								
12	91	3	280	49	28	3	80	1	10	8	1	20
14	98	3	444	55	16	2	73	2	12	11	2	27
16	85	3	571	45	16	1	62	14	12	11	1	38
18	96	3	712	48	16	1	65	14	11	8	2	35
20	95	3	1,036	48	11	1	60	20	12	7	1	40
Average	93	3	609	49	17	2	68	11	11	9	1	32

[1]Clark, A. III, and J. G. Schroeder. Predicted weights and volumes of selected hardwood species in North Carolina. U.S. Dep. Agric., For. Serv., Southeast For. Exp. Stn., Asheville, N.C. (Personal communication November 1981.)
[2]Main stem to 2-in dib top.
[3]Sawlogs to 8-in dib or merchantable top.
[4]Pulpwood in stem from butt or sawlog top to 4-in dib top.

Harvesting

Clark and Schroeder found that green wood of these 24 chestnut oaks weighed about 65 pounds per cubic foot (67 in sawlog section, 64 in pulpwood section, and 63 in topwood and branches). The green weight of wood and bark required for a cubic foot of wood was as follows:

Tree component	Green weight of wood and bark	
	Per cubic foot of wood	Per cubic foot of wood and bark
	----------Pounds----------	
Whole tree	83	62
Stem (butt to 4-in dib top)	81	63
Sawlog (butt to 8-in dib top)	79	64
Pulpwood (8- to 4-in dib top)	85	60
Topwood (4- to 2-in dib top)	101	58
Branches	93	60

Weights of foliage-free, above-ground parts of 4- to 15-inch chestnut oaks in West Virginia are given in table 16-18. Dry weight of foliage (1.3 to 102.5 pounds) and foliage proportion of above-ground biomass (1.1 to 4.6 percent) on these trees are given in table 15-11.

OAK, LAUREL

See tables 16-2 and 16-10 for biomass of 6-in trees.

OAK, NORTHERN RED

See equation 16-1 and table 16-6 for weight constants for trees smaller than 2 in in dbh, tables 16-2 and 16-10 for biomass of 6-in trees, and tables 16-11 and 16-12 relating dbh to whole-tree weight.

Clark (1978) analyzed data in the literature to derive equations to predict the dry weight of wood and bark in the main stem and crown of northern red oak; he concluded that logging this species to a 4-inch top removes 70 to 79 percent of the above-ground wood in pulpwood-size trees and 74 to 80 percent of the wood in sawtimber trees.

Phillips and Cost (1979) measured the volume of wood in crowns and stems of northern red oak and developed prediction equations based on tree diameter at breast height, and total height of tree. The data on which their equations are based are summarized in table 16-19.

TABLE 16-18—*Weight of above-ground parts of foliage-free 4- to 15-inch chestnut oak and white oak from the University Forest in north-central West Virginia (Data from Colannino 1976)*[1]

DBH (inches)	Above-ground tree parts including bark		Bark on above-ground tree parts	Foliage-free branches including bark[2]		Branch bark[2]
	Green	Ovendry	Ovendry	Green	Ovendry	Ovendry

------------------------------------Pounds------------------------------------

CHESTNUT OAK

4	65	37	6	37	21	2
5	200	117	19	64	37	6
6	365	215	34	97	56	10
7	560	331	52	136	79	15
8	785	465	73	181	106	21
9	1,040	616	97	232	136	28
10	1,325	786	124	288	169	35
11	1,641	973	153	351	206	43
12	1,986	1,178	185	420	246	52
13	2,361	1,400	220	495	290	62
14	2,766	1,641	258	575	338	73
15	3,202	1,899	299	662	389	84

WHITE OAK

4	—	—	—	22	10	—
5	141	77	8	45	24	3
6	306	170	20	74	40	7
7	501	280	34	108	59	11
8	727	408	51	148	82	16
9	982	552	69	192	107	22
10	1,267	713	90	242	135	28
11	1,583	891	113	297	167	35
12	1,928	1,086	138	357	201	42
13	2,304	1,298	165	423	238	51
14	2,709	1,527	195	493	278	60
15	3,145	1,773	226	569	322	69

[1] Values are derived from data on 21 trees of each species.
[2] Branches were defined as any material less than 4 inches dob; this includes unmerchantable tops of stems, but does not include weight of foliage.

TABLE 16-19—*Volume of wood in crowns, stems, and whole northern red oak trees[1] related to dbh and total tree height* (Phillips and Cost 1979)

Dbh (inches)	Sample trees	Total tree height	Average volume of wood		
			Crown[2]	Stem[3]	Total
	Number	Feet	----------Cubic feet----------		
6	4	60	1.4	4.5	5.9
7	6	66	1.8	6.7	8.5
8	7	69	2.4	9.4	11.8
9	5	75	3.2	12.2	15.4
10	5	74	4.0	15.4	19.4
11	4	80	4.4	20.5	24.9
12	5	83	6.9	25.6	32.5
14	5	94	12.1	39.2	51.3
16	6	96	13.4	47.3	60.7
18	4	96	16.6	59.8	76.5
20	4	101	21.0	77.8	98.8
22	5	106	36.2	104.6	140.8
24	2	96	46.2	109.9	156.1

[1]Selected from natural even-aged stands in mountains of North Carolina.
[2]All branch wood, and all crown wood above a 4-inch stem diameter outside bark.
[3]Stem wood from top of 6-inch-high stump to a 4-inch top diameter measured outside bark.

Clark et al. (1980a) sampled 36 **pulpwood-size** northern red oak 5.6 to 11.5 in dbh from a fully stocked uneven-aged oak hickory stand on the French Broad District of the Pisgah National Forest in Western North Carolina; average tree age was 49 years. Another sample of 35 dominant and codominant **sawtimber-size** trees 11.6 to 24 in dbh was taken from the Pisgah District of the same forest. Average green and dry weights of wood per tree and distribution of these weights are shown for pulpwood-size trees in table 16-20 and for sawtimber-size trees in table 16-21. For green and dry bark weights and distribution in these trees, see tables 13-13 and 13-14. Whole-tree green and dry weights of foliage-free above-ground tree portions can be summed from these two sets of tables. Green wood from these northern red oaks weighed about 65 pounds per cu ft. The green weight of wood and bark required for a cubic foot of wood was as follows:

Tree component	Green weight of wood and bark	
	Per cubic foot of wood	Per cubic foot of wood and bark
	----------Pounds----------	
Whole tree	77	65
Stem (butt to 4-in dib top)	75	65
Saw log (butt to 8-in dib top)	75	65
Pulpwood (8- to 4-in dib top)	78	65
Branches	83	63

Whole-tree chipping of above-ground tree portions of northern red oak increases wood harvested from pulpwood-size trees by 26 to 46 percent and from sawtimber-size trees by 27 to 35 percent. Whole-tree chips from above-ground,

TABLE 16-20—*Average green and dry weights of wood in whole northern red oak pulpwood trees from western North Carolina, and distribution of wood in stem and branches; see table 13-13 for weight of bark from these trees* (Clark et al. 1980a)

D.b.h. class (inches)	Average total height	Sample trees	Whole-tree wood weight	Proportion of wood in—				
				Main stem			Branches (inches d.o.b.)	
				Pulp wood[1]	Top-stem[2]	Total stem[3]	≥ 2 inches	< 2 inches	All branches
	Feet	*Number*	*Pounds*	------------------------------*Percent*------------------------------				
			GREEN					
6	61	5	381	73	11	84	4	12	16
7	66	7	547	77	9	86	3	11	14
8	69	7	766	78	6	84	6	10	16
9	75	5	999	77	5	82	8	10	18
10	76	6	1,288	79	3	82	8	10	18
11	80	6	1,566	83	2	85	6	9	15
Average	—	—	923	79	5	84	6	10	16
			DRY					
6	61	5	215	72	12	84	3	13	16
7	66	7	311	76	9	85	3	12	15
8	69	7	426	76	7	83	6	11	17
9	75	5	560	76	6	82	7	11	18
10	76	6	718	78	3	81	8	11	19
11	80	6	868	81	3	84	6	10	16
Average	—	—	517	78	5	83	6	11	17

[1] Pulpwood in stem from butt to 4-inch d.i.b. top.
[2] Stem material from 4-inch to 2-inch d.i.b. top.
[3] Main stem to 2-inch d.i.b. top.

foliage-free portions of northern red oak trees are about 84 percent wood and 16 percent bark (Clark 1978).

The proportions of above-ground wood comprised by crown wood, and above-ground bark comprised of crown bark, vary from 20 to 30 percent and 31 to 40 percent, respectively (Clark 1978).

Readers interested in equations and tabular data relating weight and volume of upland red oak species in southern Illinois to dbh and tree height will find useful the data in Meyers et al. (1980).

OAK, POST

For weights of understory post oak from the Piedmont of Georgia and South Carolina see table 16-8. For biomass of 6-inch trees see tables 6-2 and 6-10. Table 15-17 gives weight per acre of tree parts in a post oak-blackjack oak stand.

TABLE 16-21—*Average green and dry weights of wood in whole northern red oak sawtimber-size trees from western North Carolina, and distribution of wood in stem and branches; see table 13-14 for weight of bark from these trees* (Clark et al. 1980a)

D.b.h. class (inches)	Average total height	Sample trees	Whole-tree wood weight	Proportion of wood in—					
				Main stem			Branches (inches d.o.b.)		
				Saw-log[1]	Pulp-wood[2]	Total stem[3]	≥ 4 inches	< 4 inches	All branches
	Feet	*Number*	*Pounds*	------------------------- *Percent* -------------------------					
			GREEN						
12	83	5	2,139	56	23	79	4	17	21
14	94	5	3,346	65	12	77	9	14	23
16	96	6	3,957	65	13	78	10	12	22
18	98	6	5,169	69	10	79	11	10	21
20	102	5	6,389	70	9	79	11	10	21
22	106	5	9,186	67	7	74	16	10	26
24	102	5	10,631	63	10	73	16	11	27
Average	—	—	5,484	66	11	77	12	11	23
			DRY						
12	83	5	1,196	55	23	78	4	18	22
14	94	5	1,847	63	12	75	10	15	25
16	96	6	2,227	64	13	77	10	13	23
18	98	6	2,847	68	9	77	12	11	23
20	102	5	3,467	68	8	78	11	11	22
22	106	5	5,109	65	7	72	17	11	28
24	102	5	5,792	62	10	72	16	12	28
Average	—	—	3,026	65	10	75	13	12	25

[1] Saw log to 8-inch d.i.b. or saw-log merchantable top.
[2] Pulpwood in stem from saw-log top to 4-inch d.i.b. top.
[3] Main stem to 4-inch d.i.b. top.

OAK, SCARLET

See tables 16-2 and 16-10 for biomass of 6 inch trees; table 16-12 relates dbh to dry weight of stemwood and branchwood in trees 6 to 16 in in diameter.

Clark et al. (1980b) sampled scarlet oaks from the Tennessee Cumberland Plateau. They selected two to four trees from each 2-in diameter class fom 6 to 20 in dbh. The 28 trees sampled ranged from 37 to 80 years old with average age of 53 years. Green and dry weights of wood in the whole above-ground portions of the trees, and the distribution of the wood are shown in table 16-22. For green and dry bark weights and distribution in these trees see table 13-15. Whole-tree green and dry weights of foliage-free above-ground tree portions can be summed from these two tables. The main stems of these trees to a 4-in top inside bark had 15 percent of their weight in bark and 85 percent in wood, ovendry basis. In the crowns, 22 percent of the dry weight was bark and 78 percent was wood; this relationship did not vary with tree size. Branches 4 in in diameter and larger had 20 percent of their green weight in bark; those ½-in in diameter and smaller had

32 percent of their green weight in bark. Dead branches composed 11 percent of crown green weight and 13 percent of crown dry weight. The change in distribution of crown materials related to tree and branch diameter is shown in figure 16-2 (top).

Clark et al. (1980b) found that green wood of these scarlet oaks weighed about 67 pounds per cubic foot. The green weight of wood and bark required for a cubic foot of wood was as follows:

Tree component	Green weight of wood and bark	
	Per cubic foot of wood	Per cubic foot of wood and bark
	----------------------------Pounds----------------------------	
Whole tree	79	66
Stem (butt to 4-in dib top)	78	66
Saw log (butt to 8-in dib top)	77	67
Pulpwood (8- to 4-in dib top)	79	66
Topwood (4- to 2-in dib top)	82	65
Branches	84	64

OAK, SHUMARD

See tables 16-2 and 16-10 for biomass of 6-in trees.

OAK, SOUTHERN RED

See equation 16-1 and table 16-6 for weight constants for trees less than 2 in in dbh, tables 16-2 and 16-10 for biomass of 6-in trees, and table 16-5 for trees 3 to 12 inches in dbh.

For understory southern red oak (1.0 to 4.9 in in diameter) volumes, weights, and densities are given in tables 16-7 and 16-8. Phillips (1977) expressed the volumes and green and dry weights of the foliage-free, above-ground portions of these understory trees in terms of dbh and tree height in equation 16-6 (see HICKORY heading), where the regression constants are as follows ($R^2 = 0.95$):

Statistic	b_0	b_1
Total weight of green tree, pounds	−0.09254	0.77793
Total weight of ovendry tree, pounds	−.32779	.78011
Total volume of tree, cubic feet	−1.89265	.77221

Clark et al. (1980c) determined weights and volumes, above stump of 29 southern red oak trees 5 to 22 inches in dbh growing in the Highland Rim in Tennessee. Trees sampled ranged from 37 to 100 years of age and averaged 62 years; form class averaged 75. Seventy percent of the average tree's green weight was in stem material to a 4-inch top diameter inside bark, and 30 percent was in crown material. Green weight of trees sampled ranged from 279 pounds for 6-inch trees to 7,681 pounds for the 22-inch trees. Of the green weight, 13

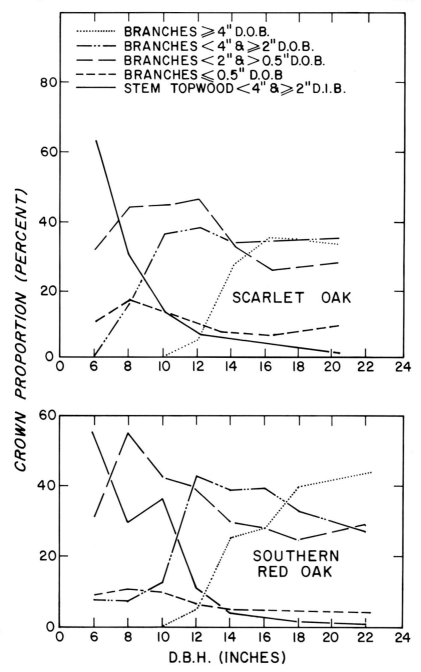

Figure 16-2.—Proportions of crown green weight in topwood and branches by d.o.b.-size classes. (Top) Scarlet oak from Tennessee Cumberland Plateau. (Bottom) Southern red oak from Highland Rim in Tennessee. (Drawings after Clark et al. 1980 bc.)

TABLE 16-22—*Average green and dry weights of wood in whole scarlet oak trees of pulpwood- and sawtimber-size from the Tennessee Cumberland Plateau, and distribution of wood in stem and branches; see table 13-15 for weight of bark from these trees* (Clark et al. 1980b)

D.b.h. class (inches)	Average total height	Sample trees	Whole tree wood weight	Proportion of wood in—							
				Main stem[1]				Live branches (inches d.o.b.)			
				Sawlog[2]	Pulpwood[3]	Topwood	Total stem	≥ 4	< 4 & ≥ 2	< 2 & > 0.5	≤ 0.5	All branches
	Feet	*Number*	*Pounds*	--- Percent ---								
GREEN PULPWOOD											
6	56	4	340	—	67	21	88	—	0	9	3	12
8	64	4	680	—	74	8	82	—	4	11	3	18
10	68	4	1,250	—	75	4	79	—	8	10	3	21
Average	—	—	757	—	72	11	83	—	4	10	3	17
GREEN SAWTIMBER											
12	69	4	1,871	44	29	2	75	2	10	11	2	25
14	82	4	3,072	52	22	1	75	7	9	8	1	25
16	86	4	3,807	56	18	1	75	9	8	6	2	25
18	87	2	4,763	54	21	1	76	9	8	5	2	24
20	84	2	5,475	51	21	0	72	10	9	7	2	28
Average	—	—	3,468	52	22	1	75	7	9	7	2	25

TABLE 16-22—*Average green and dry weights of wood in whole scarlet oak trees of pulpwood- and sawtimber-size from the Tennessee Cumberland Plateau, and distribution of wood in stem and branches; see table 13-15 for weight of bark from these trees (Clark et al. 1980b)—Continued*

D.b.h. class (inches)	Average total height	Sample trees	Whole tree wood weight	Main stem[1]			Proportion of wood in—					
				Sawlog[2]	Pulpwood[3]	Topwood	Total stem	≥ 4	Live branches (inches d.o.b.)			All branches
									< 4 & ≥ 2	< 2 & > 0.5	≤ 0.5	
	Feet	*Number*	*Pounds*						*Percent*			
DRY PULPWOOD												
6	56	4	194	—	65	22	87	—	0	10	3	13
8	64	4	395	—	73	8	81	—	4	11	4	19
10	68	4	711	—	74	4	77	—	9	11	3	23
Average	—	—	433	—	71	11	82	—	4	11	3	18
DRY SAWTIMBER												
12	69	4	1,078	43	28	2	73	2	10	13	2	27
14	82	4	1,754	50	22	1	73	8	9	9	1	27
16	86	4	2,130	54	18	1	73	9	9	7	2	27
18	87	2	2,696	53	20	1	74	9	9	6	2	26
20	84	2	3,006	49	20	0[4]	69	10	11	8	2	31
Average	—	—	1,953	50	21	1	72	8	10	8	2	28

[1] Main stem to 2-inch d.i.b. top.
[2] Saw logs to 8-inch d.i.b. or saw-log merchantable top.
[3] Pulpwood in stem from butt or saw-log top to 4-inch d.i.b. top.
[4] Less than one-half of one percent.

percent was bark, 46 percent was wood, and 41 percent was water. The proportion of bark did not vary with tree size and ranged from 19 to 21 percent, averaging 19 percent—on a green basis—slightly higher than the average for northern red oak (15 percent) and scarlet oak (16 percent). On an ovendry basis, wood made up an average of 78 percent of whole-tree weight and bark 22 percent (foliage excluded).

Clark et al. found that bark and wood are not distributed evenly throughout the tree. For example, the stem to a 4-inch top of the average sawtimber tree contained 73 percent of all the green wood in the tree (table 16-23) but only 60 percent of the bark (table 13-16). Branches, however, contained 28 percent of the green wood and 41 percent of the bark. Proportions of crown green weight in topwood and branches by branch size classes are shown in figure 16-2 (bottom).

The proportion of whole-tree green wood weight in branches increased with tree size, averaging 15 percent in pulpwood trees and 27 percent in sawtimber trees (table 16-23). For green and dry bark weights and distribution in these trees, see table 13-16. Whole-tree green and dry weights of foliage-free aboveground tree portions can be summed from these two tables.

Clark et al. (1980c) found that green wood of these southern red oaks weighed 65 to 66 pounds per cubic foot. The green weight of wood and bark required for a cubic foot of wood was as follows:

Tree component	Green weight of wood and bark	
	Per cubic foot of wood	Per cubic foot of wood and bark
	Pounds	
Whole tree	81	64
Stem (butt to 4-in dib top)........	79	64
Saw log (butt to 8-in dib top)...	79	65
Pulpwood (8- to 4-in dib top)...	82	64
Topwood (4- to 2-in dib top) ...	87	64
Branches.....................	88	64

OAK, WATER

See tables 16-2 and 16-10 for biomass of 6-in trees.

OAK, WHITE

See tables 15-7, 15-10, and 15-11 for foliage proportions of above-ground tree weight, equation 16-1 and table 16-6 for weight factors for trees smaller than 2 inches in dbh, tables 16-2 and 16-10 for biomass of 6-inch trees, and tables 16-5, 16-11, 16-12, and 16-18 relating dbh to weight.

For understory white oak 1.0 to 4.9 inches in diameter, average volume and density data are given in tables 16-7 and 16-8. Phillips (1977) expressed the volumes and green and dry weights of the foliage-free, above-ground portions of

TABLE 16-23—*Average green and dry weights of wood in whole southern red oak pulpwood- and sawtimber-size trees from the Highland Rim in Tennessee and distribution of wood in the main stem[1] and live branches (Clark et al. 1980b)*

D.b.h. class (inches)	Average total height	Sample trees	Whole tree wood weight	Proportion of wood in—								
				Main stem				Live branches (inches d.o.b.)				
				Sawlog[2]	Pulpwood[3]	Topwood	Total stem	≥ 4	< 4 & ≥ 2	< 2 & > 0.5	≤ 0.5	All branches
	Feet	Number	Pounds	---------------------------------Percent---------------------------------								
				GREEN PULPWOOD								
6	54	4	275	—	57	24	81	—	5	12	2	19
8	62	4	584	—	77	7	84	—	2	12	2	16
10	72	4	845	—	85	5	90	—	4	6	1	10
Average	—	—	568	—	73	12	85	—	3	10	2	15
				GREEN SAWTIMBER								
12	70	4	1,650	53	28	2	83	3	8	7	1	17
14	71	4	2,475	42	28	1	71	7	12	9	1	29
16	76	4	3,339	43	29	1	73	8	11	7	1	27
18	80	4	4,576	52	19	1	72	12	9	6	1	28
20	—	—	—	—	—	—	—	—	—	—	—	—
22	81	1	6,250	39	28	1	68	14	8	9	1	32
Average	—	—	3,201	46	26	1	73	10	9	7	1	27

See footnotes on following page.

TABLE 16-23—*Average green and dry weights of wood in whole southern red oak pulpwood- and sawtimber-size trees from the Highland Rim in Tennessee and distribution of wood in the main stem[1] and live branches (Clark et al. 1980b)*—Continued

D.b.h. class (inches)	Average total height	Sample trees	Whole tree wood weight	Proportion of wood in—							
				Main stem				Live branches (inches d.o.b.)			
				Sawlog[2]	Pulpwood[3]	Topwood	Total stem	≥ 4	< 4 & ≥ 2	< 2 & > 0.5	≤ 0.5	All branches
	Feet	Number	Pounds	------Percent------							
				DRY PULPWOOD							
6	54	4	160	—	56	24	80	—	8	14	3	21
8	62	4	350	—	76	7	82	—	2	13	2	17
10	72	4	488	—	84	6	89	—	6	7	1	11
Average	—	—	333	—	72	12	84	—	5	10	2	16
				DRY SAWTIMBER							
12	70	4	949	52	28	2	82	3	8	8	1	18
14	71	4	1,415	40	28	1	69	8	12	10	1	31
16	76	4	1,896	41	28	1	71	8	12	8	1	29
18	80	4	2,559	50	19	1	70	12	10	7	1	30
20	—	—	—	—	—	—	—	—	—	—	—	—
22	81	1	3,530	37	27	1	65	15	9	10	1	35
Average	—	—	1,812	45	25	1	71	11	11	8	1	31

[1] Main stem to 2-inch d.i.b. top.
[2] Saw logs to 8-inch d.i.b. or saw-log merchantable top.
[3] Pulpwood in stem from butt or saw-log top to 4-inch d.i.b. top.

these white oak understory trees in terms of dbh and height in equation 16-6 (see HICKORY heading), where the regression constants are as follows ($R^2 = 0.94$ to 0.99):

Statistic	North Carolina mountains	Georgia Piedmont
Total weight of green tree, pounds		
b_0	-0.33307	-0.23503
b_1	.86574	.81223
Total weight of ovendry tree, pounds		
b_0	$-.57181$	$-.49047$
b_1	.87060	.82009
Total volume of tree, cubic feet		
b_0	-2.09518	-1.99737
b_1	.85234	.79828

Whittaker et al. (1963) sampled ten white oaks near Oak Ridge, Tenn., averaging 96 years old, 73 feet high, and 16.1 inches in diameter; above-ground biomass distribution in these trees was as follows (ovendry weight basis):

Tree part	Proportion
	Percent
Stemwood	58.5
Stembark	12.5
Branchwood and bark	26.9
Foliage	2.1
Fruit	.02
	100.0

A. Clark III and J. G. Schroeder of the Southeastern Forest Experiment Station, Forest Service, U.S. Department of Agriculture, Asheville, N.C., in personal correspondence during May 1981, provided data on the wood content of 28 white oak trees 6 to 22 inches in diameter sampled on the Nantahala National Forest in North Carolina (table 16-24). For green and dry bark weights and distributions in these trees, see table 13-17; whole-tree green and dry weights of foliage-free above-ground portions can be summed from these two tables.

Clark and Schroeder found that green wood of these 28 white oaks weighed about 65.0 pounds per cubic foot. The green weight of wood and bark required for a cubic foot of wood was as follows:

Tree component	Green weight of wood and bark	
	Per cubic foot of wood	Per cubic foot of wood and bark
	----------*Pounds*----------	
Whole tree	77 ± 2.5	63 ± 1.7
Stem (butt to 4-in dib top)	74 ± 1.9	64 ± 1.7
Saw log (butt to 8-in dib top)	74 ± 1.9	64 ± 1.8
Pulpwood (8- to 4-in dib top)	77 ± 5.2	63 ± 2.4
Topwood (4- to 2-in dib top)	85 ± 5.1	62 ± 2.7
Branches	87 ± 4.9	61 ± 2.6

For equations and tabular data relating weight and volume of upland white oak in southern Illinois to dbh and tree height, see Meyers et al. (1980).

TABLE 16-24—*Average green and dry weight of wood of whole white oak trees (foliage-free above-ground parts) from western North Carolina, and distribution of wood in stem and live branches* (Unpublished data from A. Clark III and J. G. Schroeder, Southeastern Forest Experiment Station, Asheville, N.C.)[1]

D.b.h. class (inches)	Average total height	Sample trees	Whole tree wood weight	Proportion of wood							
				Main stem[2]			Total stem	Branches			
				Sawlog[3]	Pulpwood[4]	Topwood		≥ 4	< 4 & ≥ 2	< 2 & > 0.5	≤ 0.5	All branches
	Feet	Number	Pounds	------Percent------							
				GREEN PULPWOOD							
6	49	3	273	—	64	19	83	—	—	12	5	17
8	69	3	630	—	85	6	91	—	2	6	1	9
10	65	3	991	—	80	3	83	—	7	9	1	17
Average	—	—	631	—	76	9	85	—	4	9	2	15
				GREEN SAWTIMBER							
12	78	3	1,893	66	13	1	80	—	10	9	1	20
14	97	4	2,922	67	12	1	80	7	6	6	1	20
16	92	3	4,000	67	8	.7	75	9	9	6	1	25
18	105	3	6,779	58	13	.8	71	17	7	4	.9	29
20	92	3	6,486	51	18	.3	70	17	7	5	.8	30
22	101	3	8,687	60	8	1	69	20	7	3	.7	31
Average	—	—	5,128	61	12	1	74	12	8	5	1	26

TABLE 16-24—*Average green and dry weight of wood of whole white oak trees (foliage-free above-ground parts) from western North Carolina, and distribution of wood in stem and live branches* (Unpublished data from A. Clark III and J. G. Schroeder, Southeastern Forest Experiment Station, Asheville, N.C.)[1]—Continued

D.b.h. class (inches)	Average total height	Sample trees	Whole tree wood weight	Main stem[2]				Proportion of wood				
				Sawlog[3]	Pulpwood[4]	Topwood	Total stem	≥ 4	< 4 & ≥ 2	Branches < 2 & > 0.5	≤ 0.5	All branches
	Feet	*Number*	*Pounds*	----------Percent----------								
				OVENDRY PULPWOOD								
6	49	3	164	—	63	20	83	—	—	12	5	17
8	69	3	381	—	85	6	91	—	2	6	1	9
10	65	3	607	—	80	3	83	—	7	9	1	17
Average	—	—	384	—	76	9	86	—	3	9	2	14
				OVENDRY SAWTIMBER								
12	78	3	1,105	65	12	1	79	—	10	10	1	21
14	97	4	1,734	66	13	.9	80	7	6	6	1	20
16	92	3	2,337	65	8	.8	74	9	9	7	1	26
18	105	3	3,938	57	12	.9	69	18	7	5	1	31
20	92	3	3,802	50	18	.3	69	18	7	5	.8	31
22	101	3	4,945	58	8	1	68	21	7	4	.8	32
Average	—	—	2,977	60	12	1	73	12	8	6	1	27

[1] Clark, A. III, and J. G. Schroeder. Predicted weights and volumes of selected hardwood species in North Carolina. U.S. Dep. Agric., For. Serv., Southeast. For. Exp. Stn., Asheville, N.C. (Personal communication May 1981.)
[2] Main stem to 2-in dib top.
[3] Saw logs to 8-inch d.i.b. or saw-log merchantable top.
[4] Pulpwood in stem from butt or saw-log top to 4-inch d.i.b. top.

SWEETBAY

See tables 6-2 and 6-10 for biomass of 6-in trees.

SWEETGUM

See table 16-18 for weights of understory sweetgum in the Piedmont of Georgia and South Carolina, tables 6-2 and 6-10 for bimass of 6-in trees, equation 16-4 and table 16-13 for weight and volume factors, equation 16-5 for crown weight factors, and table 16-5 relating tree component weights to dbh and height.

A. Clark and J. G. Schroeder of the Southeastern Forest Experiment Station, Forest Service, U.S. Department of Agriculture, Asheville, N.C., in personal correspondence during November 1981, provided data on the wood content (table 16-25) and bark content (table 16-26) of 24 sweetgum trees 6 to 20 in in diameter sampled on the Oconee National Forest in the Georgia Piedmont. Whole-tree green and dry weights of foliage-free above-ground tree portions can be summed from the two tables mentioned.

Clark and Schroeder found that green wood of these 24 sweetgum trees weighed about 63 pounds per cubic foot (61 in topwood and 60 in branches). The green weight of wood and bark required for a cubic foot of wood was as follows:

Tree component	Green weight of wood and bark	
	Per cubic foot of wood	Per cubic foot of wood and bark
	------------------------------Pounds------------------------------	
Whole tree	74	60
Stem (butt to 4-in dib top).	72	60
Sawlog (butt to 8-in dib top). . . .	70	61
Pulpwood (8- to 4-in dib top) . . .	76	60
Topwood (4- to 2-in dib top) . . .	84	60
Branches	82	59

TABLE 16-25—*Average green and dry weight of wood of whole sweetgum trees (foliage-free above-ground parts) from the Georgia Piedmont, and distribution of wood in stem and live branches* (Unpublished data from A. Clark III and D. R. Phillips, Southeastern Forest Experiment Station, Asheville, N.C.)[1]

D.b.h. class (inches)	Average total height	Sample trees	Whole tree wood weight	Main stem[2]			Proportion of wood			Branches		
				Sawlog[3]	Pulpwood[4]	Topwood	Total stem	≥ 4	< 4 & ≥ 2	< 2 & > 0.5	≤ 0.5	All branches
	Feet	Number	Pounds	---Percent---								
				GREEN PULPWOOD								
6	54	3	211	—	63	26	89	—	—	10	1	11
8	67	3	534	—	80	10	90	—	—	9	1	10
10	83	3	971	—	89	4	93	—	1	5	1	7
Average	68	3	572	—	78	13	91	—	—	8	1	9
				GREEN SAWTIMBER								
12	78	3	1,438	57	23	3	83	—	2	13	2	17
14	91	3	2,775	75	10	1	86	—	3	9	2	14
16	90	3	3,214	75	10	1	86	4	3	6	1	14
18	94	3	4,644	64	11	.4	76	11	5	7	1	24
20	99	3	5,772	62	15	.3	77	8	8	6	1	23
Average	90	3	3,569	67	14	1	82	5	4	8	1	18

See footnotes on following page.

TABLE 16-25—*Average green and dry weight of wood of whole sweetgum trees (foliage-free above-ground parts) from the Georgia Piedmont, and distribution of wood in stem and live branches* (Unpublished data from A. Clark III and D. R. Phillips, Southeastern Forest Experiment Station, Asheville, N.C.)[1]—Continued

D.b.h. class (inches)	Average total height	Sample trees	Whole tree wood weight	Proportion of wood								
				Main stem[2]			Total stem	Branches				All branches
				Sawlog[3]	Pulpwood[4]	Topwood		≥ 4	< 4 & ≥ 2	< 2 & > 0.5	≤ 0.5	
	Feet	Number	Pounds	---Percent---							
				OVENDRY PULPWOOD								
6	54	3	95	—	63	25	88	—	—	11	1	12
8	67	3	238	—	79	10	89	—	—	10	1	11
10	83	3	451	—	87	5	92	—	1	6	1	8
Average	68	3	261	—	77	13	90	—	—	9	1	10
				OVENDRY SAWTIMBER								
12	78	3	678	57	22	3	82	—	4	12	2	18
14	91	3	1,326	75	10	1	86	—	2	10	2	14
16	90	3	1,540	74	11	1	86	4	3	6	1	14
18	94	3	2,200	62	12	.5	74	12	5	8	1	26
20	99	3	2,723	59	15	.4	74	9	9	7	1	26
Average	90	3	1,693	65	14	1	80	5	5	9	1	20

[1]Clark, A. III, and D. R. Phillips. Predicted weights and volumes of selected hardwood species in the Georgia Piedmont. U.S. Dep. Agric., For. Serv., Southeast. For. Exp. Stn., Asheville, N.C. (Personal communication November 1981.)
[2]Main stem to 2-in dib top.
[3]Sawlogs to 8-in dib or sawlog merchantable top.
[4]Pulpwood in stem from butt or sawlog top to 4-in dib top.

TABLE 16-26—*Average green and dry weight of bark of whole sweetgum trees (foliage-free above-ground parts) from the Georgia Piedmont and distribution of bark in stem and live branches* (Unpublished data from A. Clark III and D. R. Phillips, Southeastern Forest Experiment Station, Asheville, N.C.)[1]

D.b.h. class (inches)	Average total height	Sample trees	Whole tree bark weight	Proportion of bark							
				Main stem[2]			Total stem	≥ 4	< 4 & ≥ 2	Branches	
				Sawlog[3]	Pulpwood[4]	Topwood				< 2 & > 0.5	≤ 0.5	All branches
	Feet	Number	Pounds	----------Percent----------								
GREEN PULPWOOD												
6	54	3	43	—	46	28	74	—	—	26	2	28
8	67	3	97	—	65	10	75	—	—	21	4	25
10	83	3	154	—	70	10	80	—	4	13	3	20
Average	68	3	98	—	60	16	76	—	1	20	3	24
GREEN SAWTIMBER												
12	78	3	242	40	21	3	64	—	5	24	7	36
14	91	3	434	52	11	2	65	—	6	22	7	35
16	90	3	589	54	11	2	67	7	5	17	4	33
18	94	3	680	39	11	2	52	18	9	16	5	48
20	99	3	876	38	12	2	52	12	16	15	5	48
Average	90	3	564	45	13	2	60	7	8	19	6	40

See footnotes on following page.

TABLE 16-26—*Average green and dry weight of bark of whole sweetgum trees (foliage-free above-ground parts) from the Georgia Piedmont and distribution of bark in stem and live branches* (Unpublished data from A. Clark III and D. R. Phillips, Southeastern Forest Experiment Station, Asheville, N.C.)[1]—Continued

D.b.h. class (inches)	Average total height	Sample trees	Whole tree bark weight	Main stem[2]			Proportion of bark			Branches		
				Sawlog[3]	Pulpwood[4]	Topwood	Total stem	≥ 4	< 4 & ≥ 2	< 2 & > 0.5	≤ 0.5	All branches
	Feet	Number	Pounds				----Percent----					
					OVENDRY PULPWOOD							
6	54	3	21	—	48	28	76	—	—	22	2	24
8	67	3	50	—	68	12	80	—	—	16	4	20
10	83	3	81	—	74	10	84	—	4	10	2	16
Average	68	3	51	—	63	17	80	—	1	16	3	20
					OVENDRY SAWTIMBER							
12	78	3	119	43	23	3	69	—	4	22	5	31
14	91	3	228	58	11	2	71	—	5	18	5	28
16	90	3	354	58	12	2	72	6	4	14	4	28
18	94	3	359	43	12	2	57	16	8	14	5	43
20	99	3	455	42	13	2	57	11	15	13	4	43
Average	90	3	303	49	14	2	65	7	7	16	5	35

[1] Clark, A. III, and D. R. Phillips. Predicted weights and volumes of selected hardwood species in the Georgia Piedmont. U.S. Dep. Agric., For. Serv., Southeast. For. Exp. Stn., Asheville, N.C. (Personal communication November 1981.)
[2] Main stem to 2-in dib top.
[3] Sawlogs to 8-in dib or sawlog merchantable top.
[4] Pulpwood in stem from butt or sawlog top to 4-in dib top.

TUPELO, BLACK

See equation 16-1 and table 16-6 for weight constants for trees smaller than 2 in in dbh, table 16-8 for weights of understory black tupelo from the mountains of North Carolina, and tables 16-2 and 16-10 for biomass of 6-in trees.

YELLOW-POPLAR

See equation 16-1 and table 16-6 for weight constants for trees smaller than 2 inches in dbh, tables 16-2 and 16-10 for biomass of 6-in trees, and table 16-4 for proportions by weight of tree parts in a Tennessee stand. Table 16-12 relates dbh to dry weight of stemwood and branchwood in trees 6 to 16 inches in diameter.

For understory yellow-poplar (1.0 to 4.9 inches in dbh) volumes, weights, and densities are given in tables 16-7 and 16-8. Phillips (1977) evaluated weight and volume of above-ground portions of these yellow-poplar understory trees in terms of dbh and height in equation 16-6 (see HICKORY heading), where the regression constants are as follows ($R^2 = 0.99$):

Statistic	North Carolina mountains	Georgia Piedmont
Total weight of green tree, pounds		
b_0	−0.61138	−0.41038
b_1	.93557	.89230
Total weight of ovendry tree, pounds		
b_0	−1.03700	−.85956
b_1	.96142	.92195
Total volume of tree, cubic feet		
b_0	−2.29347	−2.23993
b_1	.91906	.91705

Whittaker et al. (1963) sampled 10 yellow-poplars near Oak Ridge, Tenn., averaging 75 years old, 75 feet high, and 13.3 inches in dbh; above-ground biomass distribution in these trees was as follows (ovendry weight basis):

Tree part	Proportion
	Percent
Stemwood	76.4
Stembark	9.2
Branchwood and bark	12.2
Foliage	1.9
Fruit	.3
	100.0

Clark (1978) derived from published data equations to predict the dry weight of wood and bark in the main stem and crown of yellow-poplar; he concluded that logging this species to a 4-inch top removes 65 to 85 percent of all the wood in pulpwood-sized trees (trees 5.0 to 10.9 inches in dbh) and 88 to 93 percent of the wood in sawtimber-size trees (trees 11 inches in dbh or larger).

Whole-tree chipping of above-ground tree portions of pulpwood-size yellow-poplar increases wood harvested by 19 to 55 percent, but for sawtimber-size trees the increase is only 10 to 15 percent. When yellow-poplar, above-ground, foliage-free tree portions are whole-tree chipped, about 84 percent of the material chipped is wood and 16 percent is bark. Additionally, Clark's (1978) analysis showed that the proportions of above-ground wood and bark from crowns varied with tree diameter as follows:

DBH	Proportion of tree wood in crown	Proportion of tree bark in crown
Inches	----------Percent----------	
6	35	39
8	23	27
10	16	20
12	11	16
14	8	13
16	8	13
18	8	13
20	9	14
22	10	14
24	11	15

Clark and Schroeder (1977) analyzed the foliage-free, above-ground biomass of 39 yellow-poplar trees 6 to 28 inches in dbh growing in natural, unevenaged mountain cove stands in western North Carolina. Merchantable top diameter to an 8.0-inch minimum inside bark averaged 9.5 inches. All material between sawlog merchantable top and a 4-inch top (inside bark) was classed as pulpwood, and the material between 4-inch and 2-inch stem diameter was classed as topwood. The tip of the stem, smaller than 2 inches diameter inside bark, was classed as branch material. Branches were cut up and separated into three categories: (1) large branches ($\geq$ 2.0 inches d.o.b.), (2) medium branches ($\geq$ 0.6 and $\leq$ 1.9 inches d.o.b.), and (3) small branches ($\leq$ 0.5 inches d.o.b.).

They found that above-ground, foliage-free whole-tree green weights ranged from 336 pounds for 6-inch trees to 14,336 pounds for a 28-inch tree. Of this weight, 51 percent was water, 41 percent was wood, and 8 percent was bark. On a dry weight basis, trees averaged 85 percent wood and 15 percent bark—proportions that varied little with diameter. The proportion of tree dry weight found in branches ranged from 7 to 12 percent, and averaged 9 percent (table 16-27). Pulpwood trees had 2 to 5 percent more of their dry weight in branches than the sawtimber trees.

Clark and Schroeder (1977) also observed that the sawlog portion of yellow-poplar trees contains 81 percent of the dry wood, but only 75 percent of the bark. Branches contain 8 percent of the dry wood compared to 14 percent of the dry bark. On average, the pulpwood section contains 10 percent of the wood and 10 percent of the bark in the whole tree.

They found that on a dry-weight basis, branches ranged from 74 to 81 percent wood, averaging 77 percent; their bark content ranged from 19 to 26 percent, and averaged 23 percent. Branches made up 7 to 12 percent of the dry weight of the whole tree, depending on tree size (table 16-27). On the average, 12 percent of the branch dry weight was in small branches, 32 percent in medium branches, and 56 percent in large branches (fig. 16-3). In this study wood specific gravity and moisture content did not vary significantly with tree size (table 16-28).

Clark and Schroeder's (1977) publication contains extensive tables of biomass weight and volume which readers should find useful; in the interest of brevity only the equations for weight (table 16-29) and volume (table 16-30) from which their tables were developed are presented here.

For equations and tabular data relating weight and volume of upland yellow-poplar in southern Illinois to dbh and tree height, see Meyers et al. (1980).

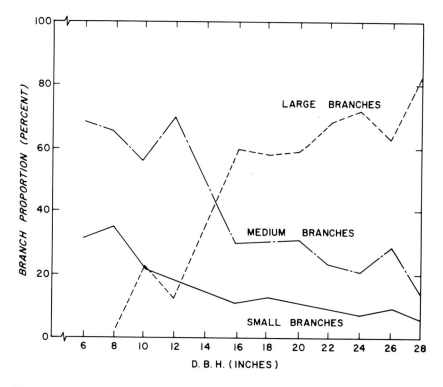

Figure 16-3.—Proportion of yellow-poplar crown in small, medium, and large branches by tree dbh. (Drawing after Clark and Schroeder 1977.)

TABLE 16-27—*Average green and dry weights of foliage-free, above-ground parts of yellow-poplar trees 6 to 28 inches in dbh from natural stands in western North Carolina, and the proportions of these trees in wood and bark, and in main stem[1] and branches[2]*
(Data from Clark and Schroeder 1977)

Dbh class (inches)	Average total height	Weight of whole tree		Tree component proportions							
				Wood		Bark		Stem		Branches	
		Green	Dry	Green	Dry	Green	Dry	Green	Dry	Green	Dry
	Feet	*Pounds*		---------Percent---------							
6	61	336	145	80	83	20	17	87	88	13	12
8	70	649	292	81	85	19	15	87	88	13	12
10	90	1,316	628	82	85	18	15	86	88	14	12
12	105	1,928	931	82	83	18	17	94	94	7	7
14	102	2,876	1,382	82	84	18	16	90	91	10	9
16	118	3,830	1,951	83	85	17	15	90	92	10	8
18	111	4,770	2,314	81	82	19	18	88	90	12	10
20	116	6,036	2,901	82	83	18	17	88	90	12	10
22	126	8,051	4,050	84	87	16	13	89	90	11	10
24	136	9,712	5,021	83	84	17	16	91	92	9	8
26	128	13,026	6,842	84	86	16	14	92	93	8	7
28	129	14,336	7,570	84	86	16	14	92	92	8	8
Average	—	3,589	1,799	83	85	17	15	90	91	10	9

[1] Stem material to 2-inch d.i.b. top.
[2] Includes all branch material and tip of main stem.

CROSS REFERENCES TO BIOMASS DATA IN CHAPTER 27

For a **list by species** of chapter 16 and chapter 27 weight tables and prediction equations, see table 27-97. Weight data on bark are in chapter 13, roots in chapter 14, and foliage in chapter 15.

Volume tables and prediction equations from chapters 16 and 27 are summarized, by species, on page 3266.

Residue weights and volumes, including tops and branches, are discussed in section 27-3.

TABLE 16-28—*Average wood and bark specific gravity, moisture content, and green weight per cubic foot for yellow-poplar trees and tree components* (Data from Clark and Schroeder 1977)[1]

Tree component	Specific gravity	Moisture content	Green weight per cubic foot
		Percent	*Pounds*
	WOOD		
Whole tree[2]	0.407 ± 0.029	104 ± 13	51.8 ± 3.5
Total stem	.407 ± .030	104 ± 14	51.8 ± 3.7
Saw log section	.406 ± .029	103 ± 13	51.5 ± 3.8
Pulpwood section	.413 ± .037	103 ± 18	52.3 ± 4.6
Topwood section[3]	.419 ± .035	111 ± 13	55.2 ± 4.4
Branches	.424 ± .026	106 ± 15	54.4 ± 3.9
	BARK		
Whole tree[2]	0.325 ± 0.028	142 ± 36	49.1 ± 8.4
Total stem	.332 ± .034	126 ± 29	46.8 ± 8.4
Saw log section	.328 ± .035	124 ± 28	46.1 ± 8.0
Pulpwood section	.354 ± .027	123 ± 35	49.3 ± 8.2
Topwood section[3]	.350 ± .035	151 ± 27	54.8 ± 6.5
Branches	.285 ± .048	247 ± 66	59.9 ± 3.1

[1] Data based on a 39-tree sample from natural stands in western North Carolina. The first value given in each column is the average; the second value is the standard deviation. Specific gravity is based on green volume and ovendry weight.
[2] Above-ground, foliage-free parts.
[3] Stem part between 2 inches and 4 inches diameter inside bark.

TABLE 16-29—*Regression equations for estimating green and dry weights of the above-ground, foliage-free biomass of yellow-poplar trees 6 to 28 inches in diameter from natural stands in North Carolina, and the weights of tree components using dbh and total height as the independent variables* (Data from Clark and Schroeder 1977)

Weight (Y)	Regression equation[1]	Coefficient of determination (R^2)	Standard error $(S_{y.x})^2$
Whole tree (excluding leaves)[3]			
Green	$Log_{10} Y = -0.69614 + 0.96067 \, Log_{10} D^2Th$	0.99	0.049
Dry	$Log_{10} Y = -1.22162 + 1.00962 \, Log_{10} D^2Th$	.99	.046
All wood in tree[3]			
Green	$Log_{10} Y = -0.83371 + 0.97283 \, Log_{10} D^2Th$	.99	.050
Dry	$Log_{10} Y = -1.31186 + 1.01330 \, Log_{10} D^2Th$	.99	.050
All bark in tree[3]			
Green	$Log_{10} Y = -1.20965 + 0.90556 \, Log_{10} D^2Th$	.98	.055
Dry	$Log_{10} Y = -1.94536 + 0.99070 \, Log_{10} D^2Th$	.98	.063
Wood and bark in stem from stump to saw-log merchantable top for trees >11.5 inches d.b.h.[4]			
Green	$Log_{10} Y = -1.40405 + 1.09136 \, Log_{10} D^2Th$	.98	.049
Dry	$Log_{10} Y = -1.92791 + 1.14055 \, Log_{10} D^2Th$	.98	.049
Wood in stem from stump to saw-log merchantable top for trees >11.5 inches d.b.h.[4]			
Green	$Log_{10} Y = -1.54143 + 1.10526 \, Log_{10} D^2Th$	.98	.051
Dry	$Log_{10} Y = -2.07644 + 1.15782 \, Log_{10} D^2Th$	.98	.053

See footnotes on following page.

TABLE 16-29—*Regression equations for estimating green and dry weights of the aboveground, foliage-free biomass of yellow-poplar trees 6 to 28 inches in diameter from natural stands in North Carolina, and the weights of tree components using dbh and total height as the independent variables* (Data from Clark and Schroeder 1977)—Continued

Weight (Y)	Regression equation[1]	Coefficient of determination (R^2)	Standard error $(S_{y \cdot x})$[2]
Bark in stem from stump to saw-log merchantable top for trees >11.5 inches d.b.h.[4]			
Green	$\text{Log}_{10} Y = -1.88576 + 1.01997 \text{ Log}_{10} D^2Th$	0.95	0.069
Dry	$\text{Log}_{10} Y = -2.32150 + 1.04433 \text{ Log}_{10} D^2Th$	.95	.071
Wood and bark in stem from stump to 2 inch d.i.b. top[3]			
Green	$\text{Log}_{10} Y = -0.81474 + 0.97694 \text{ Log}_{10} D^2Th$	.99	.040
Dry	$\text{Log}_{10} Y = -1.33025 + 1.02493 \text{ Log}_{10} D^2Th$	.99	.042
Wood in stem from stump to 2 inch d.i.b. top[3]			
Green	$\text{Log}_{10} Y = -0.92750 + 0.98533 \text{ Log}_{10} D^2Th$	.99	.043
Dry	$\text{Log}_{10} Y = -1.41983 + 1.02925 \text{ Log}_{10} D^2Th$	.99	.045
Bark in stem from stump to 2 inch d.i.b. top[3]			
Green	$\text{Log}_{10} Y = -1.42008 + 0.93339 \text{ Log}_{10} D^2Th$	.99	.051
Dry	$\text{Log}_{10} Y = -2.05248 + 1.00116 \text{ Log}_{10} D^2Th$	.98	.070
Wood and bark in branches[3]			
Green	$\text{Log}_{10} Y = -1.15689 + 0.83663 \text{ Log}_{10} D^2Th$	.80	.211
Dry	$\text{Log}_{10} Y = -1.71827 + 0.88172 \text{ Log}_{10} D^2Th$	.83	.206
Wood in branches[3]			
Green	$\text{Log}_{10} Y = -1.40010 + 0.85422 \text{ Log}_{10} D^2Th$	0.79	0.225
Dry	$\text{Log}_{10} Y = -1.76294 + 0.86612 \text{ Log}_{10} D^2Th$	.80	.219
Bark in branches[3]			
Green	$\text{Log}_{10} Y = -1.49101 + 0.79882 \text{ Log}_{10} D^2Th$	.83	.185
Dry	$\text{Log}_{10} Y = -2.60277 + 0.93682 \text{ Log}_{10} D^2Th$	.89	.172

[1]$\text{Log}_{10} Y = b_0 + b_1 \text{ Log}_{10} D^2Th$
where:
  Y = weight of tree or component in pounds,
  D = d.b.h. in inches,
  Th = total height in feet.
[2]Standard error of estimate in Log_{10} form.
[3]Regression equations based on 39 trees 6 to 28 inches d.b.h.
[4]Regression equations based on 24 trees 12 to 28 inches d.b.h.

TABLE 16-30—*Regression equations for estimating green cubic foot volume of the above-ground biomass of yellow-poplar trees from natural stands in North Carolina, and volumes of tree components using dbh and total height as the independent variables* (Data from Clark and Schroeder 1977)

Cubic foot volumes	Regression equation[1]	Coefficient of determination (R^2)	Standard error $(S_{y \cdot x})^2$
Whole tree (excluding leaves)[3]			
Wood	$\text{Log}_{10} Y = -2.60833 + 0.98740 \text{ Log}_{10} D^2Th$	0.99	0.036
Bark	$\text{Log}_{10} Y = -3.34254 + 1.01348 \text{ Log}_{10} D^2Th$	.99	.057
Wood & bark	$\text{Log}_{10} Y = -2.53544 + 0.99215 \text{ Log}_{10} D^2Th$	.99	.034
Stem from stump to a saw-log merchantable top for trees >11.5 inches d.b.h.[4]			
Wood	$\text{Log}_{10} Y = -3.36739 + 1.13164 \text{ Log}_{10} D^2Th$	.99	.036
Bark	$\text{Log}_{10} Y = -3.58003 + 1.03884 \text{ Log}_{10} D^2Th$	.95	.066
Wood & bark	$\text{Log}_{10} Y = -3.19278 + 1.11325 \text{ Log}_{10} D^2Th$	.99	.036
Stem from stump to 2 inch d.i.b. top[3]			
Wood	$\text{Log}_{10} Y = -2.70952 + 1.00212 \text{ Log}_{10} D^2Th$	.99	.031
Bark	$\text{Log}_{10} Y = -3.59784 + 1.05620 \text{ Log}_{10} D^2Th$	.98	.066
Wood & bark	$\text{Log}_{10} Y = -2.66141 + 1.01119 \text{ Log}_{10} D^2Th$	.99	.030
Branch material (excluding leaves)[3]			
Wood	$\text{Log}_{10} Y = -3.09526 + 0.84474 \text{ Log}_{10} D^2Th$	.80	.213
Bark	$\text{Log}_{10} Y = -3.34471 + 0.81757 \text{ Log}_{10} D^2Th$	.84	.180
Wood & bark	$\text{Log}_{10} Y = -2.90517 + 0.83679 \text{ Log}_{10} D^2Th$	.82	.202

[1] $\text{Log}_{10} Y = b_0 + b_1 \text{ Log}_{10} D^2Th$
where:
Y = cubic foot volume of tree component,
D = d.b.h. in inches,
Th = total height in feet.
[2] Standard error of estimate in Log_{10} form.
[3] Regression equations based on 39 trees 6 to 28 inches d.b.h.
[4] Regression equations based on 24 trees 12 to 28 inches d.b.h.

16-2 LOGGING RESIDUES

As noted previously, many of the pine-site hardwood species have widely spreading crowns (see Sec. 15-2 for a tabulation relating crown diameter to tree dbh) which contain a significant proportion of whole-tree weight. Difficult to harvest economically, tons of wood and bark in these wide-spreading crowns become unutilized residues. Additionally, cull trees, trees of less than merchantable size, and stump root systems customarily remain as residues on harvested sites.

Measurement of these residues is laborious; different methodologies have been used by various investigators. For example, Maxwell and Ward (1976ab) prepared a photographic series that quantify residues per acre in some western forest regions. Martin (1975a, 1976a) reported that the line intersect method for sampling hardwood logging residues in the Appalachians is unbiased; he provided a computer program to expedite required computations. Other investigators simply weighed all residues. Merrick and Martinelli (1952) developed a regression equation for hardwoods in Indiana that predicts cubic feet of logging waste per tree from knowledge of percentage of total tree height that is merchantable, merchantable height, and dbh.

Residue quantities are most often reported as pounds or cubic feet per tree, or as tons per acre remaining after harvest. Widely varying statistics for these data have been reported depending on logging procedures and stand characteristics. The following two sub-sections summarize the literature.

RESIDUE PER TREE

Foliage weights per tree are discussed in section 15-2. Weights of stump-root systems are given in section 14-3, tables 14-1AB and 14-2AB, figures 14-4 through 14-26, and in table 16-2. Most of the data available, however, are limited to foliage-free, above-ground residues.

Martin (1975b) developed an equation to predict the cubic feet of above-ground logging residue per tree in southwestern Virginia mixed oak stands. His equation, based on a sample of 36 trees averaging 57 years of age, 13.2 inches in dbh (range 11.1 to 17.5 inches), with bole length averaging 48.4 feet, is as follows:

(16-7)
$$V = -88.209 + 13.224D + 0.415D^2 + 0.177BL - 0.400SH$$

where:

V = above-ground residue volume per tree of all material $\geqslant 3.0$ inches in diameter outside bark, in cubic feet

D = dbh, inches

BL = Bole length in feet between the top of a 1-foot-high stump and a 4.0-inch top diameter outside bark or to a point where the central stem is terminated by branches before reaching 4.0 inches.

SH = sawlog height = distance between the top of a 1-foot-high stump and the point on the bole above which no sawlog can be produced because of excessive limbs or other defects, or to a minimum top of 9.0 inches diameter outside bark, recorded to the nearest foot.

Equation 16-7 accounted for 73 percent of the observed variation in residue volume (i.e., $R^2 = 0.73$).

Barrett and Buell (1941) determined topwood volume from 231 oaks 10 to 34 inches in diameter cut in the Bent Creek Experimental Forest, Pisgah National Forest in North Carolina. The trees were mostly scarlet and black oaks—but also included 29 northern red, 16 southern red, 13 white, and a few post and chestnut oak trees. Topwood was defined as the volume of upper stem and branches at least 5 inches in diameter inside bark and 5 feet long, but smaller than merchantable sawlog diameter (not specified). Large-diameter trees yielding few sawlogs had the largest volume of such topwood—about 100 cubic feet in a 31-inch-diameter tree containing one 16-foot and one 8-foot sawlog. As another example, a 20-inch tree yielding one 16-foot log is predicted to have a topwood volume of about 30 cubic feet; a 4-log tree of the same diameter will have only 5.8 cubic feet of topwood (table 16-31). The data in table 16-31 are appropriate for the mix of species just described; estimates for the three principal species are more accurate if multiplied by the factors below:

Species of oak	Factor
Black	0.945
Northern red	.933
Scarlet	1.119

Craft (1976) observed that total weight of logging residue in the Appalachians was about 1.8 times the weight of hardwood sawlogs removed.

Wartluft (1978) studied the weights of entire tops from 75 hardwood trees of sawtimber size from three stands in southern West Virginia and southwestern Virginia. In this study, a commercial logger felled, bucked, and topped sample trees. After the tops were on the ground (including limbs below the tops), they were weighed in their entirety. From regression analyses relating weights to tree dbh, table 16-32 was constructed. Weights of the tops when green varied from 556 pounds for 11-inch trees of soft hardwoods to 4,633 pounds for 26-inch trees of hard hardwoods. Fifty-six to 79 percent of the weight of these tops was in branches larger than 3 inches in diameter (table 16-32). The soft hardwoods included in the study were yellow-poplar, black tupelo, and cucumbertree (*Magnolia acuminata* L.). The hard hardwoods included several red oaks, chestnut oak, white oak, hickory, white ash, and sugar maple (*Acer saccharum* Marsh.). The moisture and bark contents in tops of these two groups of hardwoods were as follows:

Group	Moisture content	Bark content
	Percent of dry weight	*Percent of green weight*
Hard hardwoods	59	22
Soft hardwoods	81	27

TABLE 16-31—*Bark-free topwood[1] volume per tree in oak stands of mixed species from the Bent Creek Experimental Forest in North Carolina* (Data from Barrett and Buell 1941)[2,3]

DBH (inches)	Number of 16-foot sawlogs in stem							
	½	1	1½	2	2½	3	3½	4
	------Cubic feet inside bark------							
9	2.6	2.0	1.5	1.2				
10	3.7	2.9	2.2	1.7	1.3			
11	5.2	4.0	3.0	2.3	1.7			
12	7.0	5.3	4.0	3.1	2.3			
13	9.1	6.9	5.3	4.0	3.1			
14	11.7	8.9	6.8	5.2	4.0	3.0		
15	14.8	11.3	8.6	6.6	5.0	3.8		
16	18.4	14.0	10.7	8.1	6.2	4.7	3.6	
17	22.6	17.2	13.1	10.0	7.6	5.8	4.4	
18	27.4	20.9	15.9	12.1	9.3	7.1	5.4	4.2
19		25.1	19.1	14.6	11.1	8.5	6.5	4.9
20		29.8	22.7	17.3	13.2	10.1	7.7	5.8
21		35.2	26.8	20.4	15.6	11.9	9.0	6.9
22		41.2	31.4	23.9	18.2	13.9	10.6	8.1
23		47.9	36.5	27.8	21.2	16.1	12.3	9.4
24		55.3	42.1	32.1	24.5	18.6	14.2	10.8
25			48.4	36.9	28.1	21.4	16.3	12.4
26			55.3	42.1	32.1	24.5	18.6	14.2
27			62.8	47.8	36.5	27.8	21.2	16.1
28			71.0	54.1	41.2	31.4	24.0	18.3
29			79.9	60.9	46.4	35.4	27.0	20.5
30			89.6	68.3	52.0	39.7	30.2	23.0
31			100.2	76.3	58.2	44.3	33.8	25.8
32			111.5	84.9	64.7	49.3	37.6	28.6
33							41.7	31.8
34							46.2	35.2

[1]Topwood is that volume of upper stem and branches measuring at least 5 inches in diameter inside bark and 5 feet long, but smaller than merchantable sawlog diameter.
[2]Data based on 231 trees.
[3]To obtain volumes outside bark, multiply values in table by 1.17.

TABLE 16-32—*Treetop weights[1] for Appalachian hardwood sawtimber, by tree diameter class* (Data from Wartluft 1978)

Tree dbh class (inches)	Hard hardwoods			Soft hardwoods		
	Small[2] wood	Large[3] wood	Total	Small[2] wood	Large[3] wood	Total
	———————————————— Pounds ————————————————					
		WEIGHT WHEN GREEN				
11	379	473	852	241	314	556
12	414	540	954	262	358	620
13	451	617	1,068	283	408	691
14	491	704	1,195	306	464	771
15	534	804	1,338	331	529	859
16	580	918	1,498	356	602	958
17	629	1,048	1,677	383	686	1,069
18	681	1,197	1,878	411	781	1,192
19	736	1,366	2,102	439	890	1,329
20	793	1,560	2,353	469	1,013	1,482
21	853	1,782	2,635	499	1,154	1,653
22	915	2,034	2,949	529	1,314	1,843
23	979	2,323	3,302	558	1,497	2,055
24	1,044	2,652	3,697	587	1,705	2,292
25	1,110	3,029	4,138	614	1,941	2,556
26	1,175	3,458	4,633	639	2,211	2,850
		WEIGHT WHEN OVENDRY				
11	227	287	514	131	177	308
12	249	329	578	142	202	344
13	273	377	650	154	230	384
14	300	432	731	166	262	428
15	328	494	822	179	298	477
16	359	566	925	193	339	532
17	391	649	1,040	207	387	594
18	427	743	1,170	222	440	662
19	464	851	1,316	237	502	739
20	504	975	1,479	253	571	824
21	547	1,117	1,664	269	651	919
22	592	1,280	1,871	284	741	1,025
23	639	1,466	2,104	300	844	1,143
24	688	1,679	2,367	314	961	1,275
25	738	1,923	2,661	328	1,095	1,423
26	790	2,203	2,993	340	1,247	1,587

[1] Totals may not be exact sums of portions due to rounding of values. Values include weight of bark and include all tree parts (except foliage) above merchantable sawlog portion of the bole.
[2] Three inches or smaller in diameter outside bark.
[3] Larger than 3 inches in diameter outside bark.

Beltz (1976) compared harvested volume to inventory and logging residue volume in Midsouth logging operations. In the course of this study he found that the volume of a hardwood branch, regardless of crown type or species, can be expressed as follows ($R^2 = 0.97$):

$$\hat{Y} = 0.1672 + 0.133D^3 \tag{16-8}$$

where:
Y = Volume of the branch including bark to a minimum diameter of 1 inch outside bark, cubic feet
D = Diameter of the limb, outside bark, just beyond the limb collar, inches

Beltz observed that volume—including bark—of tops of deliquescent hardwoods (those with decurrent growth) can be estimated from the stem diameter outside bark at the base of the residual top or crown; such a tree with 6-inch diameter at crown base has a top volume of about 2.6 cubic feet, while a tree with 10-inch diameter at crown base has about 12.1 cubic feet in its crown (fig. 16-4).

Beltz (1976) also obtained data on the percentage of top volume in branches of various diameter classes for 10 deliquescent hardwoods growing in the Midsouth, as follows (the trees sampled happened to have no branches in the 6.1- to 7.0-inch class; stem diameter at the base of the residual crown varied among the 10 trees from 5.2 to 13.0 inches):

Diameter class of branch, outside bark	Branch volume in the 10 crowns including bark, by diameter class	Proportion of total branch volume
Inches	Cubic feet	Percent
1.0-2.0	1.09	1.8
2.1-3.0	8.90	15.1
3.1-4.0	9.26	15.7
4.1-5.0	12.77	21.6
5.1-6.0	20.70	35.0
6.1-7.0	—	—
7.1-8.0	6.36	10.8
	59.08	100.0

From this study, Beltz concluded that the volume of above-ground, foliage-free logging residue from hardwoods in the Midsouth is about equal in volume to half the inventoried volume established by resource survey data—i.e., that portion of the bole of growing stock trees greater than 5.0 inches in dbh from a 1-foot stump to a minimum 4-inch top outside bark or to the point where the central stem breaks into limbs. Pulpwood cutters gather slightly more than the quantity inventoried in Louisiana and Alabama hardwood stands.

In Louisiana and Alabama, Beltz (1976) observed that an average of 19 cubic feet per hardwood sawlog tree is left in the woods as logging residue. About two-thirds of this volume is in tops and the balance is in unutilized—but inventoried—upper stem sections, limbs lopped from the merchantable stem, and over-height stump sections. In these two states, hardwood trees cut for pulpwood averaged about 24 cubic feet in above-ground volume, of which 15 to 20 percent was in tree tops, 5 percent in lopped limbs, and 5 percent in stumps; the balance of 70 to 75 percent was removed as pulpwood.

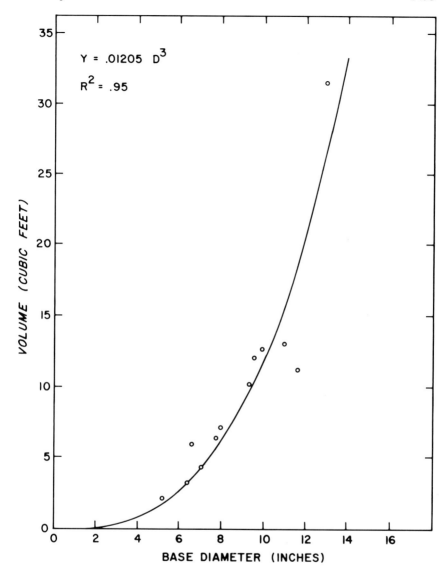

Figure 16-4.—Volume of wood and bark in tops (upper stem and branches to 1-inch diameter outside bark) of deliquescent hardwoods in the Midsouth related to diameter outside bark of the stem at the base of the crown. (Drawing after Beltz 1976.)

Porterfield and von Segen (1976) studied hardwood logging residues in Arkansas, and concluded that sawlog cutters removed 72 percent of stem volume to a 3-inch top diameter outside bark, and pulpwood cutters took 87 percent of the stem volume to a 3-inch top. Operators of tree-length logging systems left significantly less logging residue than those cutting sawlog lengths only.

Summary—To aid in obtaining data on logging residues per tree under various limits of merchantability, the following tabulation lists pertinent tables and figures, by species:

Species	Tables	Figures	Equations
Ash, white	16-11	—	—
Ash, green	16-13	—	16-4, 5
Elm, American	16-11, 13	—	16-4, 5
Hickory and oak	16-4, 5, 6, 7, 8, 12	—	16-1, 6
Maple, red	16-5, 7, 8, 11, 12	—	16-1, 6
Oak, black	16-6, 12, 14, 15	—	16-1
Oak, chestnut	16-4, 6, 7, 8, 12, 16, 17, 18	—	16-1, 6
Oak, northern red	16-6, 11, 12, 19, 20, 21	—	16-1
Oak, scarlet	16-6, 12, 22	16-2	16-1
Oak, southern red	16-5, 6, 7, 8, 23	16-2	16-1, 6
Oak, white	16-5, 6, 7, 8, 11, 12, 18, 24	—	16-1, 6
Sugarberry	16-13	—	16-4, 5
Sweetgum	16-5, 8, 13, 25, 26	—	16-4, 5
Tupelo, black	16-6	—	—
Yellow-poplar	16-4, 6, 7, 8, 12, 27, 28, 29, 30	16-3	16-1, 6

Also, table 16-2 shows weights per tree of branches, twigs, and stump-root sections for **6-in trees of 23 species**. Table 16-10 gives data permitting computation of weight per tree in branchwood and branch bark of **6-in hardwoods of 22 species**.

Chapters 13, 14, and 15 give weight data on **bark, roots,** and **foliage**.

RESIDUE PER ACRE

Logging residues per acre vary widely in the southern pinery. In general, the larger the hardwood component is in a stand, the larger the quantity of logging residue. Loggers harvesting hardwood stands of species with decurrent growth typically leave more residue than when harvesting species having the undivided stems characteristic of excurrent growth. Sawlog cutters leave more logging residue than do tree-length loggers. Summarization of a few studies illustrates the variations observed. See also section 27-3.

Harvesting

Craft's (1976) analysis of an 18-acre hardwood stand on the Monangahela National Forest near Richwood, West Virginia is instructive. The stand contained 6,700 board feet (International ¼-inch tree scale) per acre of merchantable sawtimber 12 inches dbh or larger. Average tree dbh was 16 inches. Seventy-two percent of the sawtimber volume was oak, 14 percent yellow-poplar, 8 percent beech (*Fagus grandifolia* Ehrh.), 3 percent hickory, and 2 percent maple. After the merchantable sawlogs were removed (6,667 board feet per acre International ¼-inch log scale), there was left 69.3 tons per acre of above-ground green residue or 1.8 times the weight of sawlogs. Thirty-three and one-third tons of residue were from tops of merchantable sawtimber; 36 tons were from residual trees. Intensive salvage of the residue yielded significant amounts of products (fig. 16-5). Treetop residue yielded 1,800 board feet of merchantable sawed products and 26 tons of chippable wood. The residual trees yielded 3,000 board feet of sawed products and 25.6 tons of chippable wood.

Inclusion of below-ground residues would have substantially increased residual tonnages observed by Craft (1976). Johnson and Risser's (1974) study of a post oak-blackjack oak stand in central Oklahoma showed that stemwood and bark totalled 49 tons per acre (ovendry basis) and that twigs, branches, central stump-root system, understory trees, and dead wood totalled 53 tons per acre, of which 17 tons (ovendry) was in the central stump-root systems (table 15-17).

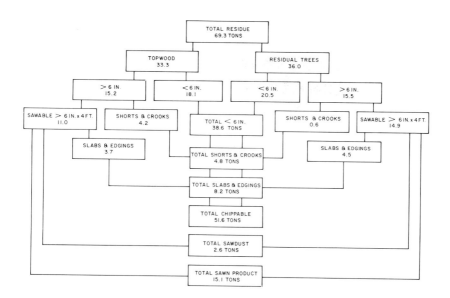

Figure 16-5.—Above-ground residue weights (green basis) after removing merchantable sawlogs from a 1-acre sample plot of mixed hardwoods in West Virginia; moisture content of residues averaged 61 percent of dry weight. (Drawing after Craft 1976.)

Under many circumstances recovery of small-diameter or below-ground residues is uneconomic. With this in mind, Martin (1975c) measured only aboveground residue material 4 inches and larger in diameter outside bark at the small end and 4 feet long or longer. On 16 1-acre plots in West Virginia, such logging residue totalled 100 to 1,300 cubic feet per acre and averaged 467 cubic feet. This amounted to approximately 15 tons per acre of green wood averaging 4.8 inches in diameter at the small end and 7.1 inches at the large end, with average piece length of 11.8 feet. The acres sampled represented a variety of harvesting conditions including clearcut and partial-cut units in both young and old hardwood stands with utilization ranging from sawtimber only to pulpwood and mine props.

To more precisely define quantities of hardwood logging residue of such size (4-inch minimum diameter and 4-foot minimum length), Martin (1976b) surveyed 20 harvest sites in West Virginia and Virginia to develop a prediction equation relating residue volume to basal area, stand age, and type of harvest. From this equation he concluded that stands 120 years of age having 140 square feet of basal area per acre, when clear cut for sawlogs only, left 2,312 cubic feet (about 74 tons green-weight basis) per acre. Other types of harvests in younger stands with less basal area resulted in much less residue (table 16-33).

Gibbons (1977) analyzed the amount of merchantable pine and hardwood pulpwood left as logging residue in natural stands in the Midsouth. Over 3,000 acres involving 20 companies were sampled; tracts studied were at least 80 acres in size. Both shortwood and tree-length operations were evaluated. Merchantable residues measured were comprised of wood down to a 4.5-inch diameter outside bark, wood in stumps above 6 inches in height, cut trees not removed from the woods, and trees within the harvesting zone that were not cut. Merchantable hardwood logging residue on all logging operations averaged 1.73 cords per acre or 46.1 percent of the original merchantable inventory, as follows:

Statistic and harvesting system	Hardwood	Pine
Total preharvest inventory, cords per acre		
Shortwood	3.56	10.09
Longwood	5.04	11.11
Longwood followed by shortwood	3.14	14.48
Residual merchantable volume after harvest, cords per acre		
Shortwood	1.54	1.61
Longwood	2.00	1.53
Longwood followed by shortwood	1.90	1.32
Merchantable residue, percent of preharvest inventory		
Shortwood	43.2	15.9
Longwood	39.6	13.7
Longwood followed by shortwood	60.5	9.1

TABLE 16-33—*Gross volume of logging residue (including bark) expected for Appalachian hardwoods for various stand densities, ages, and harvesting practices* (Data from Martin 1976b)[1]

Stand age (years)	Basal area per acre (sq. ft.)									
	50	60	70	80	90	100	110	120	130	140
	------------------------------Cubic feet per acre------------------------------									
IMPROVEMENT CUT										
60	33	63	94	124	154	184	215	245	275	305
80	101	131	162	192	222	253	283	313	343	374
100	169	200	230	260	290	321	351	381	411	442
120	237	268	298	328	358	389	419	449	480	510
SELECTION CUT										
60	296	326	357	387	417	447	478	508	538	568
80	364	394	425	455	485	515	546	576	606	637
100	432	463	493	523	553	584	614	644	674	705
120	500	531	561	591	621	652	682	712	742	773
CLEARCUT (SAWLOGS ONLY)										
60	1,835	1,865	1,895	1,926	1,956	1,986	2,016	2,047	2,077	2,107
80	1,903	1,933	1,964	1,994	2,024	2,054	2,085	2,115	2,145	2,175
100	1,971	2,001	2,032	2,062	2,092	2,122	2,153	2,183	2,213	2,243
120	2,039	2,070	2,100	2,130	2,160	2,191	2,221	2,251	2,281	2,312
CLEARCUT (OTHER)										
60	53	84	114	144	174	205	235	265	295	326
80	121	152	182	212	242	273	303	333	363	394
100	189	220	250	280	311	341	371	401	432	462
120	258	288	318	348	379	409	439	469	500	530

[1] In pieces 4.0 inches or more in diameter outside bark at the small end and 4.0 feet or more in length; at least 50 percent sound.

Gibbons (1977) concluded that longwood harvesting operations tend to leave less merchantable residue behind than do shortwood operations. High residual volumes, especially in hardwoods, can be expected after a two-stage operation in which high-value stock is removed first. Residual volumes, on a percentage basis, are much higher for hardwoods than for pines.

In addition to the residue distributed over acres that have been logged, some residues accumulate at landings where long stem sections are cut into logs and bolts prior to loading them on trucks. Goho (1976) concluded from a study in West Virginia that short lengths, poor quality, low volumes, and dirt inclusion are obstacles to utilization of this material. Some of the residue could be used for firewood, however.

Timson (1980) examined hardwood logging residues in West Virginia, Virginia, and Pennsylvania to ascertain the percentage utilizable for four quality classes of roundwood. Only 26 percent of total logging residues (i.e., residues ≥ 4 in in diameter outside bark at the small end ≥ 4 ft long, except if ≥ 2 ft long and ≥ 12 in dob) were utilizable even in the lowest quality class of roundwood, best described as local-use logs. He concluded that hardwood logging residue can best be used when extracted in the primary harvesting operation for commodities such as energy wood or pulpchips.

16-3 FACTORS AFFECTING HARVESTING

The harvesting situation is strongly affected by regional factors as well as by biomass distribution. Prominent among these factors are soils and topography, tract size and volume, woods labor suppy, and silvicultural considerations.

SOILS AND TOPOGRAPHY

Sixty percent of all pine sites in the South are on the Coastal Plains (Sternitzke 1978). The remaining sites are about evenly distributed between the Piedmont and other Highlands.[2]

The lower Coastal Plains are essentially flat, and water runoff from them is slow. Upper Coastal Plains may have slopes approaching those of the Piedmont in grade and length. The soils range from clay loams through clay to deep organic soils of extremely poor drainage. With the exception of deep swamps, these areas can be worked most of the year with rubber- and track-mounted equipment; in wet seasons, however, bearing pressure of logging equipment should not exceed 6 pounds per square inch; ground pressure of half this value would be preferable (Stuart and Walbridge 1974). There are sandy soils of considerable extent in both upper and lower Coastal Plains that are more operable in wet weather than the clays and clay loams.

The Piedmont has gently rolling terrain with some fairly steep, short slopes. Clay soils are very common; they have good bearing strength but poor tractive characteristics when wet. With the exception of large drainage areas (overflow bottoms), these soils can support tire- and track-mounted equipment year-around. Changes in location may be required during wet seasons because of rutting and failure of roads or localized terrain restrictions (Stuart and Walbridge 1974).

---

[2]Personal communication June 21, 1979 from Herbert S. Sternitzke, based on Forest Survey data.

Mountains (southern Appalachian, Ouachita, and Boston mountains) are forested at all altitudes. Slopes are long and may be steep—over 100-percent in many locations. Equipment with tracks or tires can be employed where slopes and erosion risks permit.

Lytle (1959) described the mountain soils as well-drained with loams or sandy loams on the surface, and clay subsoils. On the Ozark and Appalachian Plateaus, ridges may be rocky and slopes precipitous.

TRACT SIZE AND VOLUME

In the management and harvesting of timber, tract size is an important economic factor. Owners of small tracts pay more for services and receive less for timber harvested than do owners of large tracts. Southwide data on tract sizes are not available, but Knight (1978) has provided information on sizes of timber stands in an 18-county area of the Piedmont of South Carolina encompassing 4,528,000 acres. He defined a stand as a contiguous forest in one forest type of similar origin or age with similar ownership lines. Oak-pine stands occupied 673,000 acres, of which only 29 percent was in stands larger than 50 acres; 33 percent of the oak-pine area was in stands smaller than 10 acres. Hardwood stands, primarily oak-hickory, occupied 1,598,600 acres, of which 26 percent was in stands smaller than 10 acres (table 16-34).

TABLE 16–34—*Percentage of oak-pine and hardwood stands in the Piedmont of North Carolina by size of stand* (Data from Knight 1978)

Size of stand (acres)	Oak-pine	Hardwood[1]
	---------Percent---------	
Less than 10	32.7	26.1
10-19	19.7	12.5
20-29	7.8	8.9
30-39	6.4	4.0
40-49	4.5	2.5
50 and larger	28.9	46.0
	100.0	100.0

[1]Primarily oak-hickory.

Thienpont et al.[3] found that logging systems in the Southeast varied in mechanization with tract size. For tracts less than 50 acres in size, and yielding fewer than 500 cords, the most simple system—the bobtail truck for shortwood—predominated. For tracts 50 to 200 acres in size yielding 500 to 3,000 cords, skidder systems for longwood prevailed. Average truck haul distance to a delivery point was 15 miles for the bobtail trucks, 28 miles for simple skidder systems, and 35 miles for the most heavily capitalized and mechanized systems.

Row (1974) expressed the relationship between stumpage value and tract size as affected by volume per acre and average tree diameter. His data, based on over 2,000 sales made in National Forests in Louisiana, Mississippi, Arkansas, and Texas over a 20-year period, apply to southern pine, but these factors should affect pine-site hardwoods similarly. Row's estimated stumpage values for medium stock (500 cu. ft./acre), 6 inches dbh, ranged from $0.025 per cubic foot on 10-acre tracts to $0.070 on 2,560-acre tracts. For similar stands averaging 12 inches dbh, comparable values ranged from $0.122 to $0.167 (table 16-35). The differences due to tract size are substantial, but much less than those due to tree diameter and volume per acre. Hardwood cordwood sold at southern mills for 83 percent of the price for pine pulpwood from 1966 to 1975. Many stands of pine-site hardwoods are in small tracts, and contain low volumes of small trees; potential buyers offer little or nothing for such stumpage.

WOODS LABOR SUPPLY[4]

The southern logging industry is characterized by independent contractors faced with the problem of harvesting fragmented forest holdings. Throughout the South, these contractors stress woods labor problems and suggest that there is a significant labor shortage. In analyzing this shortage, consider that the logger must gather up his crew—sometimes at a central point, sometimes door to door—and haul them to the woods before he can begin operations. Often the journey to work is long, for loggers must range through several counties to find the large amounts of wood required to make highly capitalized systems pay off. Moreover, work is hard (fig. 16-6) and often intermittent because of weather conditions and fluctuations in the demand for wood. It is not surprising, therefore, that loggers or potential loggers are attracted by opportunities to work in air-conditioned plants on a standard work week with paid vacations, retirement programs, and other wage supplements.

---

[3]Thienpont, R. F., T. A. Walbridge, and W. B. Stuart. 1976. Tract size and timber harvesting system relationships in the Southeast. Paper presented at the Annual Meeting of the Forest Products Research Society, Toronto, Canada, July 14, 1976.

[4]Text under this heading is condensed from Guttenberg (1974), with some additions.

TABLE 16-35—*Southern pine stumpage price per cubic foot by size of tract, volume per acre, and tree dbh[1]* (Row 1974)

Tract size (acres)	Volume per acre, cubic feet					
	200	500	1,000	1,500	2,000	3,000
	Dollars per cubic foot					
	DIAMETER = 6 INCHES					
10	0.000	0.025	0.060	0.080	0.095	0.123
20	.007	.048	.071	.087	.101	.127
40	.035	.059	.077	.091	.104	.129
80	.049	.065	.080	.093	.105	.130
160	.056	.068	.081	.094	.106	.130
320	.060	.069	.082	.094	.106	.130
640	.062	.070	.082	.094	.106	.131
1,280	.063	.070	.082	.095	.107	.131
2,560	.063	.070	.083	.095	.107	.131
	DIAMETER = 12 INCHES					
10	0.047	0.122	0.156	0.176	0.192	0.219
20	.103	.144	.168	.183	.197	.223
40	.132	.156	.173	.187	.200	.225
80	.146	.161	.176	.189	.202	.226
160	.153	.164	.178	.190	.202	.227
320	.156	.166	.178	.191	.203	.227
640	.158	.166	.179	.191	.203	.227
1,280	.159	.167	.179	.191	.203	.227
2,560	.159	.167	.179	.191	.203	.227
	DIAMETER = 18 INCHES					
10	0.142	0.217	0.251	0.271	0.286	0.314
20	.198	.239	.262	.278	.292	.318
40	.226	.250	.268	.282	.295	.320
80	.240	.256	.271	.284	.296	.321
160	.248	.259	.272	.285	.297	.321
320	.251	.260	.273	.285	.297	.322
640	.253	.261	.273	.286	.298	.322
1,280	.254	.261	.274	.286	.298	.322
2,560	.254	.262	.274	.286	.298	.322

[1]Data based on over 2,000 sales made on the National Forests of Louisiana, Mississippi, Arkansas, and Texas over a 20-year period.

Figure 16-6.—A bobtail truck being loaded with 5-foot-long hardwood pulpwood through use of winch controlled at rear. The truck carries about 4.5 cords.

While operators cutting less than a thousand cords per year will persist for decades, their proportion of the annual cut will decline. At present, harvesting 2,000 cords per year or more defines a large producer; an output of at least 10,000 cords per year is more realistic for the times ahead. Such elite producers have largely been created by the pulp industry in the past, and such sponsorship is still vital. Selecting contractors with a capability for growth and encouraging independent-minded procurement foresters to enter the logging industry are good avenues.

The elite logger, be he from the ranks of foresters or of existing loggers, must be skilled in such activities as procuring stumpage, keeping expensive equipment running, scheduling operations among tracts, keeping and analyzing performance records, and recruiting, training, and keeping good men. Cottell and Barth (1974) summarized the determinants of performance by woods labor as organizational and individual, with factors as follows:

Organizational	Individual
Job design	Ability
Supervision	Motivation
Fellow workers	
Compensation/reward system	
Working conditions	
Training	
Selection	
Evaluation	

Herrick (1976) analyzed 22 characteristics of logging jobs in the East and concluded that key determinants of success were total timber harvest, hauling distance, crew size, and distance from the preceding job.

Overriding all other factors affecting logging cost is what Hamilton et al. (1961) call crew aggressiveness. One of the most successful motivational techniques to stimulate this aggressiveness is establishment of daily performance goals. The effective producer sets these goals, stays on the job with his men, and is regarded as a supervisor in addition to being a member of the crew (Plummer 1971, 1972).

As evolution of mechanized logging systems advances, managerial ability must be correlated with capital investments required by the systems. Labor can be hired and trained, but success of systems will depend on the quality of managerial talent that can be attracted to logging enterprises.

SILVICULTURAL CONSIDERATIONS

Incursions of low-grade hardwoods on southern pine sites are a deterrent to establishment of highly productive forests. Major objectives determining selection of harvesting systems include, therefore, ease in regeneration and certainty of species composition of the succeeding forest. Removal of all, or most, of the biomass facilitates establishment of the ensuing stand, and therefore much attention is given to development of systems that will accomplish virtually complete removal of residues without adversely affecting site quality, hydrological characteristics, or esthetic values.

Other silvicultural systems, however, require removal of hardwoods more selectively. Because growth rate of hardwoods on southern pine sites seldom equals that of pine, most selective harvesting systems are designed to favor pine. Systems, and circumstances in which they are appropriate, are as follows:

1. **Selection cutting** permits regeneration of forests with minimum disturbance of ecologic, hydrologic, and esthetic values. Harvest systems for such management require great flexibility, capability to handle large trees, and capability for economic operation on small cuts per acre.

2. **Timber Stand improvement** is practiced to release desirable growing stock from over-topping trees of undesirable species, quality or form. Widely accomplished in the past by girdling (fig. 16-7) or by application of selective herbicides, now banned, selective removal for use would salvage much biomass previously wasted. Such harvesting must be flexible, highly selective, and adapted to light cuts per acre.

Figure 16-7.—Southern pine reproduction released by girdling low-quality overstory hardwoods. Photo taken near Crossett, Arkansas.

3. **Weeding,** where young, overtopping hardwoods are impeding development of planted, seeded, or naturally established trees of preferred species. Harvesting requires removal of small (often 3 to 6 inches dbh) material, and great flexibility to minimize damage to growing stock.

4. **Thinning** in hardwood or mixed-pine hardwood stands, to utilize biomass that would be lost from overcrowding, and to maximize growth on superior stems. Required are flexibility and capacity to use small diameter material (fig. 16-50).

In the last three of these situations, benefits to remaining stands may justify a net loss in the harvest operation above the value of biomass harvested.

VOLUMES PER ACRE

Average volumes of hardwoods on pine sites per acre in the 12 Southern States and the region can be computed (from table 2-6 and accompanying text), as follows:

State	Pine-site acreage	Hardwood volume	Average volume/acre
	Thousand acres	Million cubic feet	Cubic feet/acre
Alabama	17,357	6,456	372
Arkansas	11,570	4,926	426
Florida	11,427	1,078	94
Georgia	19,588	6,600	337
Louisiana	9,365	3,085	329
Mississippi	12,158	3,827	315
North Carolina	15,389	6,838	444
Oklahoma	2,690	633	235
South Carolina	9,566	3,358	351
Tennessee	5,913	3,724	630
Texas (excludes post oak region)	9,266	2,593	280
Virginia	13,641	6,118	449
The South	137,930	49,236	357

Thus, **on pine sites** in the South, volumes of wood in hardwood trees 5 inches in dbh and larger (to a 4-inch top outside bark) vary by state from 94 to 630 cubic feet per acre, with Southwide average of 357 cubic feet per acre, or about 4.8 cords per acre. Such small volumes per acre are difficult to harvest profitably.

Total above-ground biomass **on all commercial forest land** in the South averages much more, and varies considerably by state. McClure et al. (1981) found that the southeastern average (Florida, Georgia, North Carolina, and Virginia) above-ground green weight of wood and bark in all live trees 1.0 inch dbh and larger (foliage-free) was 70.7 tons per acre, representing a very wide range. The average for Florida was 55.2 tons and that for Virginia 81.8 tons (table 27-98A). Three factors obviously influence the biomass on a forest acre—age, stocking, and productivity of the site. A newly established stand may contain no woody material in stems over 1 inch in diameter at breast height, while a mature hardwood sawtimber stand on a good site may contain over 250 green tons of biomass per acre. In Florida, where average biomass per acre is relatively low, more than half of the material is softwood. A large proportion of this softwood biomass is in pine plantations that are managed on short rotations. At the end of each rotation, biomass returns to near zero. In Virginia, on the other hand, more than three-fourths of the biomass is hardwood, and many of the hardwood stands are not managed for high timber production. Biomass has been allowed to accumulate in these stands for long periods, and it will continue to do so as long as hardwoods are underutilized (McClure et al. 1981).

16-4 ECOLOGICAL EFFECTS OF HARVESTING

Feller-bunchers, skidders, prehaul forwarders, and heavily loaded woods trucks may compact soils and cause ruts leading to erosion. A new generation of machines which will include swathe-cutting feller-bunchers, swathe-cutting mobile chippers, and tree pullers may accelerate loss of soil nutrients by removal of biomass now left to decay. Use of KG blades, root rakes, plows, choppers, and controlled burns during preparation for regeneration also radically alters sites. Total harvesting effects range from gross changes in soil structure and surface contour to subtle influences on nutrient balance, water relationships, and survival of residual flora and fauna. Exhaustive exploration of these effects is beyond the scope of this text, but as a minimum, forest managers must strive to maintain adequate nutrients in the soil and avoid compacting, puddling, or eroding it. At the same time, consideration must usually be given to preservation of wildlife habitat and visually pleasing landscapes on terrain modified by harvest.

NUTRIENT BALANCE

In considering the possibilities for nutrient depletion from whole- or complete-tree harvesting, it is pertinent that the annual nutrient accumulations per acre in crops such as cotton or tobacco are about equal to those in a 20-year-old natural stand of loblolly pine (Switzer et al. 1978). With annual crops, fertilization is routine; while not nearly so necessary for plants cropped at 20- to 40-year intervals, fertilization will probably also become routine for intensively managed southern tree plantations.

It is unlikely that any system now visualized for complete- and whole-tree harvesting will remove by loggers' trucks more than half of the biomass grown on an acre over a rotation. The half remaining on the site is mainly litter from foliage and twig fall; additional amounts remain from tree, brush, and grass mortality, entire root systems of grass and brush, lateral roots of crop trees, and residual standing crops of brush and grass (Koch and Boyd 1978; Koch 1980).

None of the intensive harvesting machines contemplated (e.g., whole-tree chippers, swathe-cutting feller-bunchers, swathe-cutting mobile chippers, and complete-tree pullers) are designed to collect the litter from the forest floor. Many intensive harvest systems propose that ash from branches, bark, and root portions burned in mill power plants be distributed—after addition of nitrogen and perhaps phosphorus—on the site from which these tree components were harvested.

Application of such fortified ash to a slash-free undisturbed forest floor, followed by chopping with a drum roller would likely afford better site preparation than the windrow and burn system so prevalent in the South in the 1970's. In the latter system nutrients are concentrated in the windrows, as litter and mineral soil are frequently scraped and piled with logging slash. Incomplete burning frequently makes planting there difficult. Between the windrows, nutrients may be depleted by removal of litter and some topsoil (fig. 16-8). In evaluating alternative harvesting and site preparation systems, it is doubtful that scraping wood and soil into windrows and ashing in windrows is as useful as ashing in a furnace, followed by supplementing the ash with needed nutrients and evenly distributing the resultant nutrient mix on the relatively undisturbed forest floor.

Readers interested in further reading on the impact of harvesting on forest nutrient cycling should find the following references useful:

Pierce et al. (1972)
White (1974)
Patric and Smith (1975)
Pennock et al. (1975)
Jorgensen et al. (1976)
Malkonen (1976)
Tennessee Valley Authority (1976)
Waide and Swank (1976)
Alban (1977)
Pritchett (1977)
Baker (1978)
Pritchett and Wells (1978)
Switzer et al. (1978)
Blackmon (1979)
State University of New York,
  College of Environmental Science
  and Forestry (1979)
Wells et al. (1979)

Figure 16-8.—Young southern pines on land that was site-prepared by windrowing and burning. Nutrient depletion has significantly slowed growth of trees between windrows.

SOIL COMPACTION AND PUDDLING

Adverse harvesting effects result mainly from soil disturbance and compaction, and from subsequent organic matter oxidation on the forest floor. **Puddling** of wet clay soils, i.e., destruction of soil structure and reorientation of soil particles, results from repeated churnings by passage of heavily loaded wheeled or tracked vehicles (fig. 16-9). Compaction may cause planting failures and diminishes growth of newly regenerated trees; puddling results in soil of such structure that regeneration and growth of trees is almost completely inhibited for several years.

Figure 16-9.—Repeated churnings of wet clay cause puddling of soil.

The degree of compaction is dependent on soil type and moisture content as well as frequency and bearing pressure of logging vehicles. On some sites, soil compaction is not a problem. For example, King and Haines (1979) detected no compaction at depths of 2 and 4 inches at sample points trampled seven times with wheels each exerting a pressure in excess of 17.5 psi. At the test site loamy sand topsoil about 4 in deep overlay a sandy clay loam subsoil. Moisture content at test was low—13 percent by weight. Other investigators have observed considerable compaction, however; a sampling of the literature follows.

Nyland et al. (1977) observed that soil was disturbed on 17 percent of total area during partial cuts and 28 percent in clearcuts. Severely disturbed soil occurred only in discrete segments along skid trails; deep rutting that could channelize water occurred on less than 5 percent of the areas harvested. They also reported that loamy soil on trails for wheeled skidders dragging hardwood logs was densified or compacted about 7 percent and that near-maximum bulk density of the soil was attained after only two or three trips by a skidder.

On three western Oregon soils (43 percent moisture content in first foot), Froehlich (1978) found that during the first 20 trips of a tracked skidder along a skid trail, soil density primarily increased between depths of 2 to 4 inches; density increased most during the first few trips. Two to 8-inch-deep litter and slash layers tended to remain in place for the first 20 trips under skidder ground pressure of 7 to 9 psi.

On three clearcut and site-prepared forest areas in east Texas (Nacogdoches, Cherokee, and Jasper Counties), Stransky (1981 and personal communication in 1982) measured soil bulk densities after the first, third, sixth, and seventh growing seasons since site treatments were applied. The four site treatments were: (1) control, (2) burning, (3) chopping with a Marden chopper, and (4) KG blading. Soil bulk densities were consistently higher after the mechanical treatments (chop and KG) than after the other two treatments (control and burn):

Growing season since site treatment	Site treatments			
	Control	Burn	Chop	KG
	----------------$Grams/cm^3$----------------			
First	1.29	1.28	1.33	1.40
Third	1.26	1.27	1.33	1.36
Sixth	1.29	1.31	1.34	1.34
Seventh	1.24	1.25	1.29	1.33
Mean	1.27	1.28	1.32	1.36

Gordon et al. (1981) measured site disturbance after harvesting a southern pine plantation with the swathe-felling mobile chipper (illustrated in figures 16-52 and 16-53. They found that 32 percent of the soil area was significantly compacted to a 10-cm depth. Mean bulk density at the 0 to 5-cm depth was 1.54 g/cm^3 for compacted and 1.37 g/cm^3 for undisturbed conditions, and at the 10-cm depth 1.58 and 1.42, respectively. Litter zone material showed a two-fold increase due to chips lost during harvest.

Hassan (1978) found that dry bulk density of coastal organic soils increased by as much as 50 percent during mechanical harvesting; the top zone (0-15 cm) was more compacted than lower zones (15 to 30 cm).

Dickerson (1976) found significant soil compaction in skid trails on Coastal Plains soils of northern Mississippi. In this February 1967 trial, a rubber-tired skidder made seven trips dragging three hardwood logs averaging 35 feet long and 11 inches in diameter at the small end. Bulk densities of wheel-rutted soil increased an average of 20 percent to 1.55 g/cm^3; the increase was 10 percent for the soil between the ruts which was compacted by movement of logs. Macropores were reduced 68 percent for wheel-rutted soils and 38 percent for log-disturbed soils; micropore space of both increased about 7 percent. Wheel-rutted soils required about 12 years to recover and log-disturbed soils about 8 years.

Duffy and McClurkin (1974) observed that on northern Mississippi sites where soil bulk density exceeded 1.45 g/cm^3, bar planting of loblolly pines would probably fail. Thus, initial bulk densities of both wheel-rutted and log-disturbed soils would be detrimental to the establishment and growth of loblolly pine. To assume prompt restoration of desirable soil properties, such practices as disking may be necessary following tree-length skidding (Dickerson 1976).

EROSION

By most accounts, erosion from undisturbed as well as carefully managed eastern forest land is 0.05 to 0.10 ton/acre/year (Patric 1976). Experience gained on the Fernow Watershed near Parsons, West Virginia shows that clearcutting of a hardwood forest need not destroy the characteristic functioning of its forest soil if main skid roads are properly sited and laid out parallel to, never crossing, streams (Patric 1976). From a study of hardwood harvesting operations in several eastern states, Patric (1978) concluded that proper location, drainage, traffic control, and maintenance of forest roads and skidroads is essential to soil and water stability.

In his review of the effect of harvesting systems on water quality and supply, Ursic (1978) concluded that southern pine can be clearcut and completely utilized without serious or long-lasting harm to water quality.

Few investigators disagree with the idea that intensive burning increases erosion on some sites. These increases appear to be less on most southern sites than elsewhere, however, and furthermore they usually last for only a few years (Wells et al. 1979).

Pennock et al. (1975) listed principles and practices that can be applied to logging operations to reduce soil losses from the site and prevent sedimentation problems, as follows:

- Stage the logging operation so that the minimum amount of soil is disturbed at any one time.
- Recognize soil problems such as poor drainage, rockiness, and slide-proneness; avoid them by proper road location and construction and by appropriate logging techniques.
- Promptly establish temporary or permanent erosion control practices.
- Keep soil at the source rather than trying to strain or remove sediment after it moves off-site in the water course.
- Schedule use of roads, i.e., keep off roads during spring break-up or extended periods of wet weather.
- Retire roads, trails and landings not scheduled for later use.
- If clearcutting is used, leave a buffer strip of uncut or only lightly-cut timber along streams.

Haussman (1960) and Kochenderfer (1977) have also provided guidelines for keeping erosion within acceptable levels. In general, road gradients should be kept below 10 percent and water control measures such as culverts, outsloping, and broad-based dips installed as needed. Follow-up inspections after logging should be made to insure that water control measures are functioning properly.

Where nutrient leaching is a problem, it can be moderated considerably by strip-cutting or use of other partial harvesting methods. Whenever all timber is removed, a new cover should be established rapidly. Uncut buffer strips protect streams from excessive temperature increases and from accumulations of slash and debris. They moderate siltation and nutrient leaching and provide food for many aquatic invertebrates. Strip widths vary with conditions on different watersheds. The most common widths are from 40 to 100 feet on each side of the stream. A 40-foot buffer strip may be adequate to prevent excessive temperature increases in small streams, but a strip of 66 to 100 feet is usually needed to protect the stream ecosystem. A wider streamside management zone may be needed where slope or soil conditions dictate or when windthrow or sunscald may be a problem. Although some light selection cutting can be done in the buffer strip on stable sites, tractors should be kept out. Logs should be removed in a direction away from the stream (Corbett et al. 1978).

The environmental effects of cable logging in Appalachian forests were reviewed by Patric (1980) who concluded that such logging caused less unwanted effects on forest soil, water, residual stands, wildlife, and visual appeal than did other harvest systems.

Truck-mounted cranes for yarding and loading logs in West Virginia were found by Kochenderfer and Wendel (1980) to cause about the same amount of damage to residual stands as other systems; damage was concentrated on small trees.

Forest managers charged with pollution abatement from silvicultural activities should find useful the cost effectiveness analysis procedure provided by the U.S. Department of Agriculture, Forest Service (1977).

WILDLIFE AND HABITAT[5]

Habitat—Timber harvesting practices have an immediate and longtime effect on wildlife and their habitat. Regardless of the kind and intensity of the harvest, some values are enhanced and others lessened. Seldom, if ever, is any cutting method fully compatible with all uses. Forest resource managers must therefore select management alternatives that are most likely to provide the desired goods and services.

---

[5]Text under this heading is condensed from Halls (1978).

When trees are harvested by clearcut, forage production increases dramatically; openings may produce upwards of 2,045 pounds/acre, whereas fully stocked loblolly-shortleaf stands may produce only 205 to 410 pounds/acre. Forage yields peak 5 to 8 years after a clearcut and then steadily decline as the tree canopy closes and browse grows beyond the reach of deer. The mechanical action of logging also influences forage yields. Breakage and uprooting may kill some perennial forbs and woody species, but many midstory hardwoods are broken and form new sprout growth that is available to deer and other wildlife. When hardwood tops are left in the woods they may provide a temporary source of food.

Fruit yields after a clearcut harvest consist largely of herbaceous and understory browse species, many of which are also food for songbirds. Depending on site preparation technique, fruit from browse may range from 37 pounds/acre (on plots with intensive treatment) to 110 pounds/acre (on burned plots). Cutting of dominant and codominant hardwood trees correspondingly reduces production of hard mast. (See sec. 15-7 for yield data.)

Grasses and herbaceous plants rapidly invade newly logged areas. The first year after a clearcut harvest, the **botanical composition** changes from a predominantly browse cover to a herbaceous cover of grasses, composites, and legumes. Within a few years, however, the herbaceous plant community is again dominated by woody plants.

Timber harvests affect the **food quality** and palatability of wildlife foods. Rapidly growing twigs characteristic of new sprout growth are more succulent, nutritious, and digestible than mature tissues. Fruits from many plant species, such as legumes, that invade recently cutover stands are frequently high in essential nutrients. Also, browse plants grown in the open often contain more crude protein and phosphorous than plants grown beneath a canopy.

Wildlife need **cover** for nesting, feeding, hiding, and ruminating. Because vegetation grows rapidly in the South, the forests are seldom devoid of cover for more than 2 or 3 months, and then only after a clearcut followed by intensive site preparation. The size, shape, and distribution of timber cutting units usually govern the diversity of available cover. In general, the greatest hindrance to sustained populations of wildlife occurs when cutting units are too large. Some animals are particularly vulnerable because they will not move long distances to find suitable cover, nor will they move beyond the safe limits of cover to find food. To insure a diversity of vegetation cover, timber should be regularly harvested in relatively small blocks that are well dispersed throughout the forests. For deer in loblolly-shortleaf pine forests, cutting units should be less than 100 acres, rectangular or irregular in shape, and separated by a 650- to 1300-foot strip that has been uncut for at least 10 years. In the Appalachian Mountains cutting units of 5 to 30 acres are suggested for deer. Snags and live trees with cavities are necessary as escape and nesting cover for birds and small mammals. As many as possible should be left to meet these needs.

Wildlife—The composition, distribution, density, and reproduction of wildlife species reflect changes in food and cover conditions caused by timber harvests. Certain species are inextricably linked to mature timber and when trees are harvested the habitat is no longer tenable to them. The harvest of old-growth pine trees can be devastating to the red-cockaded woodpecker, a species that nests only in live pine trees infected with red-heart fungus. Also, songbirds such as Swainson's warbler, Acadian flycather, and red-eyed vireo are likely to be found only in mature forests. In Virginia the smoky shrew, woodland jumping mouse, southern flying squirrel, gray squirrel, spotted skunk, and opposum were most prevalent in an uncut forest (Blymyer 1976).

On the other hand, many examples show that harvest cuts are inviting to some species. The cottontail rabbit frequently invades clearcuts. White-footed mice increased dramatically after clearing operations in the sandhills of Florida (Beckwith 1964). In the Coastal Plain of Virginia, the peak populations of shrews and small rodents occurred 1 year after clearcuttig, and although they decreased over the next 3 years, they were still higher than in the uncut forest 4 years after clearcutting (Trousdell 1954). In Virginia, the least shrew, eastern cottontail, and striped skunk were most prevalent in a 2-year-old clearcut and then decreased as the stand got older (Blymyer 1976).

In some cases a particular wildlife species may occupy a habitat through all phases of a timber rotation, but population density fluctuates widely with the stage of development. For example, after mature longleaf pine forests were clearcut, bobwhite quail flourished, reaching population densities never achieved before (Halls and Stransky 1971). Then as the forests were regenerated to pines the quail population declined to near extinction in local areas. Recent increases of quail populations in forests result largely from clearcuts on pine lands. Upward trends in today's white-tailed deer populations in the South stem in part from timber cutting that has increased forage in upland pine forests (fig. 16-10). In contrast, where timber management has reduced the hardwood component in pine-hardwood forests squirrel populations have declined drastically.

How far wildlife species will move in response to timber harvest practices depends largely on the inherent mobility of the species and the degree of habitat disturbance. In Florida, partially cleared areas as large as 625 acres received substantially less use by white-tailed deer than uncleared areas (Beckwith 1964). Near Nacogdoches, Tex., fox squirrels did not move from their home range even when the hardwoods, their main source of food, were eliminated from a pine-hardwood forest stand.

Figure 16-10.—An abundance of understory vegetation is available to white-tailed deer when loblolly pine-hardwood stands are frequently thinned.

Several examples illustrate how timber can be manipulated to meet the needs of various kinds of wildlife. Optimum habitats for ruffed grouse can be provided by a checkerboard pattern of different aged aspen stands with each aged block not over 100 acres in size and adjacent blocks differing by 10 to 15 years in age (Gullion 1977). Near Nacogdoches, Tex., pine-hardwood stands were generally similar to pine stands of comparable height in number of wintering bird species, but higher in species diversity, and lower in bird density (Dickson and Segelquist (1977), Dickson (1978) suggested that diversity in bottomland hardwood forests can be increased by planting, thinning, and harvesting practices that emphasize foliage layers beneath the canopy. He indicated that a basal area of about 90 square feet per acre would allow understory vegetation to develop but be dense enough to curtail epicormic branching. Alexander (1977) suggested that individual tree selection be used to attract birds that favor mature forests or that require vertical diversity in the timber stand.

On southern National Forests cutting practices are adjusted as needed to meet food and cover requirements for selected wildlife species (Gould 1977). In most cases white-tailed deer are featured, but harvest practices are often altered to meet the needs of other species such as wild turkey, grouse, bear, and squirrel. The presence of many birds and small mammals is clearly associated with the presence of snags and of living trees with cavities.

The timing of timber harvests can seriously affect wildlife productivity. Cuttings are less apt to be disruptive if made from late summer to late winter. If made in April and May, timber cuttings may seriously curtail breeding activity of ground nesting birds, such as eastern wild turkey and bobwhite quail.

The aftermath of logging has variable effects on wildlife. The debris from tree tops may help protect rodents from raptors, but exclude large animals from forage. Tree length logging may improve ruffed grouse habitat by removing the slash that provides concealment for predators and by favoring preferred herbaceous plants (Dolgaard et al. 1976). Log landings, roads, and trails may provide permanent openings favorable to wildlife.

RECREATION VALUES[6]

Forests provide many recreational opportunities: hunting, fishing, hiking, camping, picknicking, wilderness travel, and just looking at the scenery are but a few of the enjoyments. With the exception of wilderness and natural areas where logging is seldom allowed, the forests' attractiveness to the recreationist is largely fashioned by the kind, degree, and frequency of timber harvests, and unless special care is taken, the harvests are likely to detract from recreational values. Negative effects are most obvious immediately after a timber cut.

Clearcutting has the greatest visual impact of any cutting practice and in its early stages, at least, creates an objectionable appearance to most forest users, whether hikers, bird watchers, or hunters. Since most commercial pine-site forests will be clearcut, some efforts are justified in ameliorating the cutting impact. Usually this can be done by arranging the size, shape, and distribution of clearcuts so that they conform to natural shapes, break up the monotony of large expanses, and create balanced and unified patterns that complement the landscape (Barnes 1971). Even though small irregularly-shaped cuts may be more expensive and difficult to administer, they usually leave a better impression than cuts with symmetrical straight-line patterns. Edges of clearcuts can be made more appealing by thinning adjoining stands back from the edge and by retaining small trees and shrubs along the edges (Brush 1976). Clearcuts along skylines and at higher elevations should be avoided as they impair scenic values more than cuttings at lower elevations (Litton 1974). Nyland et al. (1977) questioned whether lopping hardwood tops was worth the cost ($17 to $147 per acre), since viewers still found the slash objectionable. In the South, rapid decay may reduce slash to unobjectionable remnants within 2 or 3 years. Alexander (1977) suggests that a modified shelterwood or individual tree selection be used to retain a landscape in the foreground of areas where openings have been made in the middleground and background.

---

[6]Text under this heading is taken from Halls (1978).

When even-aged stands mature they can be quite attractive to the recreationist. Brush (1976) says that mature and overmature even-aged stands without understory may be awesome and majestic, almost cathedral-like in quality. Shafer (1967) suggests that even-aged stands, in the long run, may be as desirable from an aesthetic standpoint in some timber types as single tree or group selection cuttings. Even-aged stands can provide park-like conditions which enhance outdoor activities such as hiking or other recreation. One distinct advantage of even-aged stands is that they are harvested less frequently than stands managed by the selection system.

To screen the viewer from unsightly clearcuts, many National Forests and timber companies retain strips of uncut timber along highways. The Natchez Trace Parkway in Mississippi and the Blue Ridge Parkway in the Appalachians are excellent examples of how a relatively narrow strip of timberland can be restricted from cutting and retained to satisfy the viewer. Anyone driving the Parkways is certain to be pleasantly impressed with the forest environment.

The selection system of harvest has less immediate visual impact than a clearcut, and is usually applied in areas where recreation is the dominant forest use. In these areas cuttings are usually designed to enhance recreational qualities rather than to achieve maximum timber growth. Trees with high scenic value, such as crooked trees with full crown, are likely to be favored at the expense of straight, smaller crowned trees of high timber value. Intermediate cuttings can provide an intermingling of dense and open stands (Rudolf 1967). They can also be used to create desirable sunlight and shade effects (James and Cordell 1970). Because large, old trees are scenically attractive they may be retained beyond the accepted rotation ages if they do not create unusual hazards. So that recreationists can easily see these "special" trees, felling patterns can be designed to increase the visibility of the entire bole and crowns (Shafer 1967). Too, log decks and felling patterns that surround the large trees can be arranged to provide for campsites and micro-vistas within the managed stand. To avoid interference with recreationists, harvest cuttings should be scheduled during the offseason.

Removing trees from recreation areas is likely to be complicated and costly. Ordinary logging practices will rarely be satisfactory because structures and trees selected to remain on the site must be protected from damage (Sieker 1955). Cull logs, sound logs, and all slash must be removed.

In commercial forest stands, whether evenaged or uneven-aged, the forester must keep in mind several practices that can enhance recreational values. A light and open stand is more attractive than a dark and dense stand (Clawson 1974); thus thinning to a relatively low tree basal area is best from the recreational standpoint. Recreational potential is increased when stumps are cut low or flush with the ground and the slash is removed (Hodges 1959). Logging along roadsides is particularly vulnerable to criticism, but with carefully planned tree selection, logging techniques, and location of landings, the negative impacts can be minimized (McDonald and Whiteley 1972).

Small openings in uneven-aged stands are desirable in forest landscapes, and strongly contrasting spatial effects can be attained by harvesting trees in groups (Brush 1976). In gently rolling terrain, clearings on woody hillsides often provide the only opportunity for viewing the surrounding countryside.

Harvesting

Large old trees are especially desirable, so long rotations generally increase the aesthetic appeal of a stand. Because tall trees increase scenic beauty, McGee (1975) suggested that the overstory be retained as long as possible in a series of hardwood cuts designed to regenerate the stand.

Logging roads, skid trails, and log decks can be quite unsightly when located strictly to facilitate timber harvests. But if arranged to be unobtrusive (fig. 16-11), the construction can be used to form networks of hiking trails, or access roads and paths to camping spots, springs, waterfalls, lakes, and scenic vistas (Shafer 1967). The unlimited use of forest roads was the major factor contributing to increased recreational use of vast forest tracts in Canada (Pulp and Paper Magazine of Canada 1969). Residual timber may be useless for recreational purposes because of cull logs and slash, but with well-placed timber access roads and complete utilization of slash, areas can be used for recreational purposes soon after timber harvest (Sieker 1955).

Figure 16-11.—Old logging road now leading to a scenic vista on the Kisatchie National Forest in Louisiana.

16-5 HARVESTING SYSTEMS IN WIDE USE[7]

Today's logging systems range all the way from horses to helicopters. It is beyond the scope of this book to present a treatise on all aspects of these harvesting systems. For such general texts the reader is referred to Conway (1976) or Simmons (1979). Additionally, useful summary reports of research programs on harvesting small wood, thinnings, and residual wood include Hakkila et al. (1979), McMillin (1978), Silversides (1974), and The Norwegian Forest Research Institute (1978). Especially useful are the symposia proceedings describing harvesting technology published periodically by the Forest Products Research Society, Madison, Wisconsin. Petro's (1975) manual on felling and bucking is explicit and thoroughly illustrated. Also, the U.S. Department of Agriculture, Forest Service (1976) provided a handbook designed to reduce the level of forest residues by improving felling and bucking practices through computer analysis of these practices. Whitmore and Jackson (1956) explain how correct bucking practice can significantly increase the value of black oak logs.

This text will focus on the uniquely difficult problems of harvesting the hardwoods that grow on southern pine sites.

In the South, three logging systems are in widespread use. These three systems—shortwood, longwood, and whole-tree chipping—with many variations of each system, were developed to suit timber and terrain characteristics, local markets, available equipment, and capabilities of local labor. Each system, with its variations, evolved slowly, after much trial and error. Most new concepts in methods or logging equipment successfully introduced have been adaptable to one of the accepted systems without upsetting overall operation or causing unusual personnel or operational changes.

Introduction of new systems, methods, or machinery is a slow and difficult process. Much good logging equipment has not found acceptance because it is not readily adaptable to local systems and labor. Experimental logging operations are, therefore, conducted by many large timber companies and logging equipment manufacturers. These are used to test and evaluate new methods, systems, and equipment to determine their usefulness in the field.

SHORTWOOD SYSTEM

By far the most prevalent harvesting method for small timber during the years since World War II has been the shortwood system which produces bolts 8 feet and shorter in length for pulpwood. The system is compatible with the production of other products such as veneer logs, sawlogs, and firewood. A typical shortwood operation consists of a producer-entrepreneur, who is also the truck driver, and his two helpers. The producer typically has an investment of about $12,000 to $18,000 (1979) which includes a bobtail truck equipped with a winch and cable loader, two chainsaws, and an axe. (See chapter 18 for data on chainsaw selection). In this typical operation, the producer acts as trucker, and while he is hauling a load to market the helpers are felling trees, limbing,

---

[7]The text under this heading is revised and expanded from: Altman, J. A. 1978. Unpublished Final Report FS-S0-3201-5.19, U.S. Dep. Agric., For. Serv., South. For. Exp. Stn., Pineville, La. 20 p.

topping, and measuring to proper length and bucking bolts from the stem as it lies by the stump. When the trucker returns to the woods, he is assisted by the two helpers as he drives from stump to stump and operates the loader (fig. 16-6).

The small capital investment required permits easy entry into this business. The mobility and small size of the equipment allows it to operate in even the smallest tracts, and no expensive roads are required. In the years immediately following World War II, the system was well suited to the small landowner of the South who was frequently a part-time logger, and part-time farmer or factory worker.

Shortwood to be moved long distances to market is usually trucked to a woodyard and transferred to rail cars for cheaper transport. A network of field woodyards or buying points has been established throughout the South; trucking distances from the woods to woodyards average 18 to 20 miles. When trucking direct to mill, hauling distances of 36 to 42 miles are common.

At the woodyards, mechanical unloading of trucks and weight-scaling reduce transfer time to a few minutes. Railroads provide cars designed for ease in mechanically loading and unloading shortwood; such pulpwood moves at a rate lower than that applied to other forest products.

Typical shortwood operations produce about three cords per man-day and at the lowest cost per cord of all widely used logging systems for pulpwood. The system, adaptable throughout most of the southern pine region, is also least capital intensive. It is, however, the most labor intensive, and the number of people willing to do work of this kind is decreasing. Insufficient labor will be available to supply industry's needs by this system. Moreover, the system is not profitable on stands of small trees.

Time to fell, limb, top, buck, and load a hardwood tree by the shortwood system increases with tree dbh (tables 16-36 and 16-37). Time to fell, buck, and load a cord of shortwood, however, varies inversely with tree dbh; for example, 5.6 man-hours are needed to cut and load a cord from 4-inch trees, but only 2.0 from 16-inch trees.

In hardwood stands cut for pulpwood by the shortwood system, cutters do little limbing; for the most part, they take only the stem section that is relatively free of limbs. Thus, a large proportion of hardwood tree biomass—the crown—is customarily left in the woods as logging residue.

The cost of logging by the shortwood system can be estimated through data provided in tables 16-36 and 16-37, knowledge of the number of trees required to yield a cord of pulpwood (table 27-7), and knowledge of the following factors (1979 prices, stands with trees averaging 6 inches dbh):

Factor	Value
Labor rate for chainsaw or truck operator	$5.00/hour
Availability of sawhand labor, i.e., hours of productive work as a decimal fraction of hours paid	.5
Availability of truck-winch loader labor	.9
Machine rate (cost) per hour of operation of a chainsaw	$1.50
Machine rate per hour of operation of the truck-mounted loader	$3.00
Number of 6-inch southern red oak trees required to yield a cord of pulpwood	28 to 45
Production rate on trees 6 inches in dbh:	
Walking to tree, felling, limbing, topping, and bucking it (a one-man job)	1.32 man-hours/cord
Loading shortwood with a winch onto a bobtail truck (a three-man job)	1.88 man-hours/cord

TABLE 16-36—*Tree diameter related to time per tree and per cord for one man to fell and buck southern red oak pulpwood with a chainsaw* (Data from field experience of D. M. Tufts, Pineville, La.)[1]

Dbh (inches)	Walk to tree	Fell	Limb and buck	Rest and delay	Total per tree	Trees per cord	Total time per cord	Cords per hour
	------------------------Seconds-----------------------						Minutes	
4	31	13	28	23	95	111.0	175.7	0.34
5	31	15	46	34	126	50.0	105.0	.54
6	31	18	71	49	169	28.2	79.5	.76
7	31	21	98	66	216	18.3	66.0	.91
8	31	25	133	88	277	12.8	59.0	1.08
9	31	34	173	114	352	9.2	54.0	1.14
10	31	46	219	145	441	7.0	51.4	1.17
11	31	58	267	184	540	5.3	47.7	1.26
12	31	70	320	219	640	4.3	46.0	1.28
13	31	82	373	252	738	3.6	44.3	1.32
14	31	92	427	286	836	3.1	43.2	1.39
15	31	103	482	320	936	2.6	40.6	1.48
16	31	115	539	347	1,032	2.2	37.8	1.59

[1]Based on a cut of five to ten cords per acre.

TABLE 16-37—*Tree diameter related to time per tree and per cord for a three-man crew to winch-load a bobtail truck with southern red oak pulpwood* (Data from field experience of D. M. Tufts, Pineville, La.)[1]

Dbh (inches)	Walk or drive to tree[2]	Load tree[2]	Total per tree[2]	Trees per cord	Time per cord[2]	Cords per crew-hour[3]
	-------------Crew minutes-------------				Crew-minutes	
4	0.35	0.13	0.48	111.0	53.3	1.13
5	.35	.48	.83	50.0	41.5	1.45
6	.35	.98	1.33	28.2	37.5	1.60
7	.35	1.65	2.00	18.3	36.6	1.64
8	.35	2.40	2.75	12.8	35.2	1.70
9	.35	3.32	3.67	9.2	33.8	1.78
10	.35	4.32	4.67	7.0	32.7	1.83
11	.35	5.43	5.78	5.3	30.6	1.96
12	.35	6.65	7.00	4.3	30.1	1.99
13	.35	7.90	8.25	3.6	29.7	2.02
14	.35	9.15	9.50	3.1	29.4	2.04
15	.35	10.43	10.83	2.6	28.2	2.13
16	.35	11.88	12.23	2.2	26.9	2.23

[1]Based on a cut of five to ten cords per acre.
[2]Multiply these values by 3 to convert them to man-minutes.
[3]Divide these values by 3 to convert to cords/man-hour.

Figure 16-12.—This pre-hauler loads wood at the stump and forwards it to roadside for transfer to truck. It can climb or descend slopes as steep as 45 percent and traverse slopes up to 15 percent, carries a 20,000 pound payload, has a loading bed measuring 80 inches wide by 174 inches long, and cost about $52,000 in 1979. Bearing pressure when loaded exceeds 10 psi. (Photo from Franklin Equipment Company.)

Use of these data in the computation outlined in table 16-38 indicates that the cost of southern red oak pulpwood from 6-inch-diameter trees logged from a site with five cords per acre is $38 to $61 per cord (depending on wage rate) delivered to mill or woodyard. These high costs, which are currently (1979) unacceptable to many mill procurement agents, are the central problem inhibiting utilization of pine-site hardwoods in small diameter classes.

Where terrain prevents access of trucks to the stump, **pre-haulers**—sometimes called **forwarders**—equipped with grapple loaders can be used to transport pulpwood or short logs to roadside for transfer to trucks (fig. 16-12). Loading time with the grapple varies inversely with log diameter, e.g., about 9 minutes per cord for 16-inch logs and 13 minutes for 10-inch logs. Haul distances may occasionally exceed 1 mile, but average about ¼-mile (American Pulpwood Association 1975). Travel time for a round trip from stump to roadside may be 30 minutes for a loaded-haul distance of 1 mile, and perhaps 10 minutes for a haul of ¼-mile (Timson 1975).

TABLE 16-38—*Computation of cost per cord to cut, load, and transport to mill pulpwood from pine-site southern red oaks averaging 6 inches in dbh and five cords per acre*

Function	Labor rate[1] ÷ availability	Machine rate	Total cost/hour	Production rate	Cost/ cord
	----------------Dollars/hour----------------			Hours/cord	Dollars
Fell..................	5/.5 = 10.00	1.50	11.50		
Limb and top...........	5/.5 = 10.00	1.50	11.50	1.32	15.18
Buck..................	5/.5 = 10.00	1.50	11.50		
Load (three men)........	15/ .9 = 16.67	3.00	19.67	.63	12.39
Cutting and loading cost per cord					27.57
Hauling cost per cord					5.00
Insurance and workman's compensation per cord..........................					1.00
Stumpage cost					4.00
Total cost per cord delivered to mill or woodyard......................					37.57

[1] In this tabulation a wage rate of $5.00/hour has been assumed. A $10.00/hour wage rate would increase total cost per cord from $37.57 to $61.27.

There are many variations of shortwood systems, but few if any produce wood at less cost than the simple system first described, when it is used by an aggressive crew. Readers desiring information on more mechanized shortwood harvesting equipment will find the following publications useful:

Subject	Citation
Koehring shortwood harvester	Boyd (1975)
Hahn shortwood harvester	Larson (1978a)
BM Volvo SM 880 processor	Heidersdorf (1974)
TH-100 thinner harvester	Curtin and Bunker (1971)
TH-210 thinner harvester	Bergeron (1978)
Propst thinner harvester	Stuart (1972)

All of the foregoing equipment is designed for use on coniferous trees; the spreading crowns of most southern hardwoods makes their use for hardwood harvest difficult and impractical.

LONGWOOD SYSTEM

Longwood logging is a system appropriate only for full-time professionals. Skilled operators for heavy equipment and trucks are needed; large tracts of timber are required to distribute over large volumes of production the nonproductive time needed to move equipment. Usually landings must be cleared and woods roads constructed. Since the system, as usually practiced, does not lend itself to thinning, landowners usually must consent to clear cutting. Equipment costs (about $150,000 in 1979) are about 10 times that required for shortwood operators using bobtailed trucks. Man-day production of from 5 to 10 cords for the longwood system is more than double that of shortwood operators.

Harvesting

Procedures in longwood logging vary substantially among operators, but the following operations are common to most (a 5-man woods crew is typical):

- Felling with a chainsaw when butt sections are to be used for lumber or veneer; shears can be used if butt splits are not objectionable.
- Limbing with a chainsaw; in hardwoods with heavy crowns little limbing is done above the main clear section of the bole.
- Topping, usually with a chainsaw, infrequently with shears. Entire tops of broad-crowned trees are frequently left in the woods. The merchantable top diameter of pine-site hardwoods is usually about 2 inches larger than that for southern pine. For hardwoods it may be 5 to 7 inches, or even larger.
- Bunching of long logs or stems into groups of appropriate weight for skidding; this is usually done by a skidder preparatory to dragging the assembly of stems to a landing for loading. For this purpose the **skidder** (a wheeled or tracked vehicle for rough terrain—fig. 16-13) is equipped at the rear with a winch drum holding 75 to 100 feet of cable called the **winchline**. The winchline may run from the drum directly to the logs or through an attached arch carrying a fairlead with rolling guides to reduce friction and wear on the cable. Attached to the end of the winchline by a shackle are 3 to 10 **chokers** several feet in length. The chokers are chains or cables somewhat smaller than the winchline, terminating in nooses usually secured around the butt ends of stems to be skidded. The skidder operator, or a choker setter, attaches the chokers to the group of logs in preparation for skidding. Usually two sets of chokers are used—one being left in the woods for attachment while the skidder pulls a load of logs to the landing at roadside. In pine-site hardwoods the skidder may move several times to accumulate a full load, after which the winch pulls the logs to the skidder. Arch skidders elevate the leading end of the logs to reduce hangups while skidding. Skidders without an arch drag logs small end first.
- Skidding stems or logs from the stump to a landing area typically accounts for one-third to one-half of the total cost of logging. From 1,000 to 10,000 pounds may be skidded per trip. Drawbar pull may range from less than 500 pounds for arch skidding of light loads on level ground to 12,000 pounds for ground-skidding heavy loads up 30-percent grades.
- Loading of stems or logs onto log trailers for highway transport, usually with a knuckleboom loader or front-end loader.

A number of driving forces have moved the logging industry toward longwood systems. On southern pine sites, artificial regeneration of plantations to insure prompt regrowth has favored clear cutting, to which the longwood system is well adapted. Labor shortages, whether real or perceived, have stimulated design of a great array of equipment to increase output of wood per man-day while easing the hard labor required in the shortwood system. Most of these have involved longwood systems. Finally, the high market value obtainable for veneer and solidwood products coupled with new sawmilling and veneer cutting

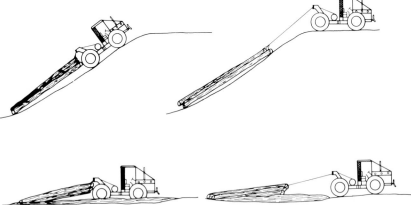

Figure 16-13.—Wheeled skidder with integral arch. (Top) Winchline attached to three chokers elevates three hardwood logs into position for skidding at about 3 miles per hour over level terrain. (Photo from Caterpillar Tractor Co.) (Bottom) Steep slopes and muddy areas often necessitate moving the wheeled skidder forward and then winching the logs to the skidder (Drawing after Hartman and Gibson 1970.)

techniques for small logs has impelled tree harvesters to divert each tree portion to the end use of highest value. This **merchandising** is best accomplished at central locations, requiring transport of entire stems to the mill.

These arguments have been widely accepted for southern pines, which are typified by long central stems and small crowns. The case is less compelling with pine-site hardwoods, however, because branch-free stems are typically short, crowns are heavy, and tree diameters are small. Even though man-day production may be relatively high with the longwood system, too great a portion of hardwood tree biomass is left in the woods and too little is taken to the mill (usually less than 50 percent). Moreover, it is a single-tree system; that is, trees are felled one at a time. Limbing in heavy crowns is both expensive and dangerous; hence little above the limb-free stem is harvested. Finally, in hardwood stands, the longwood system leaves a heavy residue of standing trees too small in diameter for economic removal.

In the longwood system, tree diameter has greater effect on logging cost than any other factor. Studies of hardwood longwood operations in Arkansas, Tennessee, and Mississippi by Tufts (1977b) indicated that harvesting cost per cord for 6-inch trees was three times that for 16-inch trees.

Skidding distance is the next most important factor. Some successful operators try to limit skidding distances to less than 600 feet; the average southwide is about 900 feet (American Pulpwood Association 1975). Readers interested in analyses of costs of skidder operations will find the following references useful: Anonymous (1977a); Donnelly (1978); Wimble (1976); McCraw and Hallet (1970); and McIntosh and Johnson (1974). These references all describe softwood operations.

Descriptions of skidder operations in pine-site hardwoods are scarce. Stevens and Smith (1978) studied the economics of thinning hardwood pole stands in Vermont by skidding trees suitable for firewood with very small 4-wheel-drive tractors (8 to 48 horsepower) equipped with winches. They concluded that when removing about 100 small trees per acre from terrain with slopes up to 20 percent, with skidding distances from 200 to 800 feet, skidding costs were about $13 per cord; this cost did not include stumpage, felling, bucking, or highway transport cost. Skidder production rate was about 0.6 cord per hour. For commercial production, minimum tractor size of 30 to 40 horsepower was recommended. In 1979, a 16-horsepower winch-equipped tractor cost $5,200 and one rated at 48 horsepower cost $25,000.

To aid loggers in comparing machines, Hensel (1976a) compiled data sheets describing specifications of rubber tired skidders manufactured in North America. The citations listed at the end of this section provide much additional performance data.

Longwood system using a feller-buncher and grapple skidder.—A longwood system finding much favor in the South is based on a feller-buncher teamed with a grapple skidder. Equipment investment for the system is about $250,000 (1979). The shearing head of the feller-buncher is carried on the front end of a prime mover (fig. 16-14) in such manner that trees can be severed near ground level, usually with shear blades but sometimes with mechanically actuated chainsaws (fig. 16-15) or with rotary cutters (fig. 18-49)—and bunched in assemblies convenient for pickup by a skidder equipped with a rear-mounted grapple (fig. 16-16). Some of these machines can shear and accumulate several stems before swinging to deposit them in a bunch—an efficient arrangement when harvesting small trees from coppice stands. The feller-buncher teamed with grapple skidder has much to recommend it in that chainsaw felling by men on the ground is avoided and stems are bunched with little manual labor and no choker setting.

This system can be used for thinning as well as clear cutting because the trees can be strategically bunched to be accessible to the skidder. Unless spreading crowns are removed, however, skidding will damage residual stems. Topping and limbing of bunched hardwoods such as the oaks is difficult with this system. Neither gate delimbers nor flail delimbers that work well with bundled stems of southern pine will do the job on the large, strong, spreading branches of most pine-site hardwoods. Therefore, if limb-free logs are to be loaded on highway trucks, crowns and some limbs must be severed with chainsaws from the bunched hardwood trees—a difficult and dangerous job for a man on the ground. This problem is a major deterrent to low-cost harvest of small hardwood stems having broadly spreading crowns.

Longwood system using a shear with carrier.—The spreading crowns of pine-site hardwoods prevent use of many mechanized longwood systems. While not yet in wide use for hardwoods, one promising approach employs a shear that severs stems, mechanically delimbs and tops them, and carries them to accumulate a grapple load for transfer by grapple skidder to a landing for forwarding or loading directly onto a truck. This procedure eliminates the danger and hard labor of limbing and topping. Disadvantages are degrade of valuable butt logs from shear-caused splitting, and loss of fiber in limbs and tops which remain in the woods.

Longwood thinning system using an inverted grapple skidder.—Pine-site hardwoods scattered among pines or in need of thinning, call for logging methods able to extract stems without damage to the residual stand. To do this, tops must be removed before skidding commences. Figure 16-17 diagrams a European longwood system in which trees are felled, limbed, and topped with a chainsaw, and are then mechanically loaded at the stump into a small inverted grapple for skidding to a landing where they are bucked to desired log lengths before forwarding to highway transport. Alternatively, a small dragging winch with chokers does the skidding (fig. 16-58). Much like the once-common horse logging, this system minimizes damage to residual trees, but it leaves in the woods the large crown portion of hardwood biomass.

Harvesting 1527

Figure 16-14.—(Top) Rubber-tired feller-buncher featuring frame-steered chassis with shear mounted in front; the shear accumulator can hold one to several trees. The machine drives up to each tree to be felled, severs it near ground level, and bunches the stems for pickup by a grapple skidder. (Photo from Clark Equipment Company.) (Bottom) Tracked feller-buncher with shear and tree accumulator on the end of a rotatable hydraulic boom. It does not have to closely approach each tree, but can extend its boom and shear while maintaining a fairly straight path through the woods. (Photo from Drott Division, J. I. Case.)

Figure 16-15.—Feller-buncher featuring a chainsaw felling head (top right) which severs stems without causing damaging splits in valuable butt logs. Extendable rotating boom (bottom left) reduces need to frequently reposition the machine. (Photos from Al Allbright Inc.)

Figure 16-16.—Grapple skidder grasping a load of hardwood stems before skidding them to roadside for loading on a truck. These oak stems have been topped and limbed by chainsaw after being bunched—a difficult task. (Photos from John Deere.)

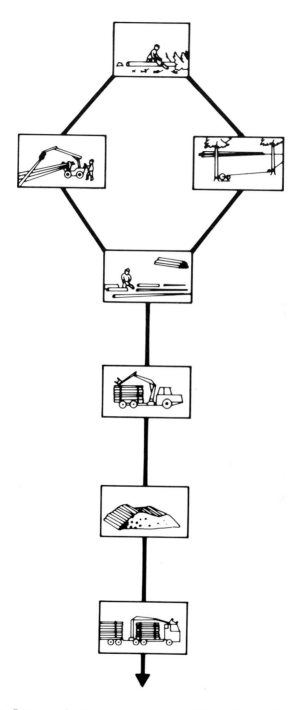

Figure 16-17.—European thinning system using small inverted grapples or dragging winches to skid longwood to a landing accessible to forwarder. (Drawing after Nilsson 1978.)

WHOLE-TREE CHIPPING

Past wood shortages plus the cost of supplying uniform-size chips free of fines, oversize chips, and bark have stimulated many unbleached kraft pulp mills and fiberboard plants to develop methods for using chips of lower quality. This relaxation of chip standards now permits these companies to accept whole-tree chips as part of their raw material supply (McKee[4], Conner 1978, Mills 1980). Whole-tree chips are also increasingly used by southern mills for fuel to replace natural gas and oil. Significant use of whole-tree chips for fiber and fuel opens the door to many opportunities in forest management heretofore not available. Chipping all above-ground portions, including limbs and tops, of nearly all trees in a stand may double the harvest recoverable by shortwood or longwood systems. Removal of limbs and crowns from the woods also eases site preparation and planting of the following forest.

In a typical whole-tree chipping operation, an investment of about $600,000 is required (1979 basis). Equipment needed includes a mobile roadside chipper capable of chipping whole hardwood trees with crowns attached (fig. 16-18), two feller-bunchers (fig. 16-14), and two grapple skidders (fig. 16-15). Also needed are a fifth-wheel tractor for spotting setout trailers in the woods, two fifth-wheel mainhaul tractors for highway transport, seven chip vans, two long-log setout trailers for recovery of valuable saw logs or veneer bolts, and support equipment for maintenance. Such operations require eight to ten-man crews, with weekly production of about 400 cords, or eight cords per man-day (Warren and Kluender 1976).

Massey et al. (1981) studied whole-tree chipping of residual hardwoods and pines in southeastern Louisiana and concluded that a minimum price of $15/green ton of whole-tree chips, delivered mill, was required to recover costs; for economic operation machine downtime must be minimal.

Sawlogs and veneer bolts are removed prior to some whole-tree chipping operations in which the unmerchantable tops and all remaining standing stems larger than 3 or 4 inches are brought to a landing for chipping. In others, sawlogs or veneer bolts are separated at the landing with the remainder being chipped. Studies in the North Central states indicate that a minimum of about 10 tons (green basis) of hardwood sawlogs must be obtainable per acre to justify sawlog sorting prior to whole-tree chipping (U.S. Department of Agriculture, Forest Service 1978, p. 55). Graves et al. (1977) concluded that sawlog and veneer log sorting was economical if the value of such roundwood exceeds $26 to $40 per cord at the landing, depending on the system used.

As with other logging systems, the widespread crowns of pine-site hardwoods damage residual trees; this system is best used for clear cuts in most hardwood stands on pine sites. The heavier limbs of many species are so inflexible that they must be notched by chainsaw before they can be compressed for chipping by the feed rolls of the roadside chipper (fig. 16-18). Such notching is hard work and dangerous.

---

[4]McKee, J. C. 1973. The pulping of whole-tree chips in the Southeast. Presented at the November 1973 meeting of the Southeastern Section of TAPPI, Charleston, S.C.

Figure 16-18.—(Top) Self-feeding, whole-tree chipper at roadside landing processing pine-site hardwoods. The machine carries a 575 horsepower diesel engine and can continuously chip white oak stems 11 inches in diameter. Stems to 22 inches in diameter can be chipped with intermittent feed. Chips are blown into vans for transport to the mill. (Photo from Nicholson Manufacturing Company.) (Bottom) Typical whole-tree chipping system (Drawing after Biltonen et al. 1976.)

A mechanized alternative procedure was proposed by Mattson et al. (1978). By this system, crowns too massive for the whole-tree chipper would be severed from the felled and bunched tree. After the stemwood is skidded away, a small, highly maneuverable, hydraulically actuated shear mounted on a knuckle boom of a vehicle (fig. 16-19) would sever large protruding limbs and align them with the butt of the main stem of the top. Thus compacted, the top is grapple skidded to the whole-tree chipper. In 1978, the cost per green ton to shear and compact the tops, skid them to roadside, and chip them was estimated at $6.32 per ton, green basis.

SUMMARY AND DISCUSSION

The shortwood system is applicable throughout most of the southern pine region and requires small capital investment, but it is labor intensive. The longwood system requires full-time professionals, lends itself to the widespread practice of clearcutting followed by planting, and is capital intensive. The whole-tree chipping system also requires full-time professional operators. It is an extension of the longwood system but requires relaxed chip standards in the using mills; it is the most capital intensive of the three systems.

Costs to produce a cord of wood by the three systems can be estimated by the method shown in table 16-38 for a shortwood operation. Some factors needed for use of this computational method are shown in tables 16-39 and 16-40.

Labor intensity, costs per cord, and return on initial equipment investment when harvesting mixed hardwood-pine stands and upland hardwood stands by shortwood, longwood, and whole-tree chipping systems, were compared by Stuart et al. (1978). They concluded that the shortwood system harvested wood at least cost per cord, and gave highest return on investment, but was somewhat more labor intensive than the other systems (table 16-41).

Readers interested in further analysis of harvesting costs in the South should consult publications of the LSU/MSU Logging and Forestry Operation Center, MSU/LSU Research Center, NSTL Station, Miss. 39529.

Harvesting 1533

Figure 16-19.—(Top) Experimental topwood harvester shearing and aligning limbs of hardwood tops. (Center) Typical hardwood top (Bottom) Compacted top. (Photo from Mattson et al. 1978.)

TABLE 16-39—*Examples of production rates for various operations during the harvest of hardwood trees measuring 4 to 10 inches in dbh* (Data from J. A. Altman, American Pulpwood Association, Jackson, Miss.)

Function	Production per man-hour
	Cords
Felling with a chainsaw	1.4
Limbing and topping with a chainsaw	.5
Bucking with a chainsaw	.6
Piling shortwood manually	2.5
Skidding with a cable skidder	2.0
Skidding with a grapple skidder	4.0
Loading onto a forwarder or truck with a mechanical grapple	8.0
Prehauling on forwarder	2.5
Felling with a feller-buncher	
Single-stem at a time	2.3
With accumulator shear head	2.7
Roadside whole-tree chipper (assumes a 2-man crew)	4.5

TABLE 16-40—*Examples of costs per scheduled hour, hauling costs per cord-mile, and percent availability of machines during hours scheduled for operation* (Data from J. A. Altman, American Pulpwood Association, Jackson, Miss., adjusted for 1979)

Machine	Cost per scheduled hour or mile of transport (including operator)	Availability
	Dollars	*Percent*
Chainsaw for felling, bucking, or limbing	9.20	50
Felling shear on a small crawler tractor	14.67	65
Feller-buncher (small)	15.37	60
Cable skidder on wheels	13.64	88
Cable skidder on tracks	14.30	85
Grapple skidder on wheels	13.67	81
Grapple skidder on tracks	14.80	71
Shortwood prehaul forwarder	11.22	64
Longwood prehaul forwarder	13.82	64
Hauling costs (per cord[1]/mile)[2]		
Tractor trailer (20-27 ton payload)	0.17–0.21	
Live[3] tandem (9-ton payload)	.19– .23	
Dead[4] tandem (9-ton payload)	.19– .23	
Single-axle bobtail truck (7-ton payload)	.21– .25	

[1] A cord of pine-site hardwood weighs about 6,000 pounds when green.
[2] Assumes one-way haul of 50 miles and return empty.
[3] Driven tandem axle for increased pulling power.
[4] Tandem axle (not driven) for better load distribution.

TABLE 16–41—*Labor intensity, cost per cord harvested, and return on initial investment when harvesting mixed hardwood-pine stands and upland stands by shortwood, longwood, and whole-tree chipping systems* (Stuart et al. 1978)

Harvesting system and statistic	Mixed hardwood-pine stand	Upland hardwood stand
Shortwood system for pulpwood only		
Cords per worker day index[1]	0.83	0.74
Cost index per cord[2]	.66	.64
Revenue required per cord to return 20 percent on initial investment[3]	$28.11	$27.80
Longwood system for pulpwood and roundwood grade products[4]		
Cords per worker index[1]	.95	.89
Cost index per cord[2]	1.06	.90
Revenue required per cord to return 20 percent on initial investment[3]	$44.67	$39.04
Whole-tree chipping system delivering pulp chips only		
Cords per worker day index[1]	.84	.79
Cost index per cord[2]	1.22	1.10
Revenue required per cord to return 20 percent on initial investment[3]	$57.44	$53.72
Whole-tree chipping system delivering pulp chips and roundwood grade products[4]		
Cords per worker day index[1]	.78	.73
Cost index per cord[2]	1.34	1.18
Revenue required per cord to return 20 percent on initial investment[3]	$62.45	$57.62

[1] Base value (index = 1) is 5.6 cords per worker day, including support workers.
[2] Base value (index = 1) is $36.49 per cord including stumpage and hauling, 1975 basis.
[3] Before income taxes; 1975 conditions.
[4] Roundwood products included and tallied as cords.

To conclude this brief discussion of harvesting systems that can be used for pine-site hardwoods, the following references are listed as useful to readers desiring more information on machine characteristics and performance.

Chainsaws
    Chapter 18
    Hensel (1976b)
    Morner (1976)
    Sarna (1979)
Tree shears
    Chapter 18
    Coughran (1978)
    Logging Research Foundation, Sweden (1975)
    Hakkila et al. (1979)
    White et al. (1969)

Feller-bunchers
    Finnish National Fund for Research (1979)
    Folkema (1977a)
    Folkema and Novak (1976)
    Gordon (1977)
    Logging Research Foundation, Sweden (1975)
    Powell (1975)
    Sarna (1976ab)
    Sondell and Svenson (1975)
    Sturos (1971)

Tufts (1976a)
World Wood (1978)
Skidders
  Anonymous (1977a)
  Biller (1970, 1971, 1972)
  Biller and Hartman (1971)
  Bredberg (1970)
  Darwin (1965)
  Donnelly (1977, 1978)
  Fiske and Fridley (1975)
  Garlicki and Calvert (1968)
  Gibson et al. (1969)
  Hartman and Gibson (1970)
  Hartman and Phillips (1970)
  Hassan (1977a)
  Hensel (1976a)
  Herrick (1955)
  Legault and Powell (1975)
  McCraw and Hallet (1970)
  McCullar (1976)
  McIntosh and Johnson (1974)
  Nilsson (1978)
  Stevens and Smith (1978)
  Vidrine et al. (1979)
  Wimble (1976)
Delimbers
  Dunfield (1971)
  Gordon (1978)
  Griffin (1977)
  Helgesson (1978)
  Logging Research Foundation, Sweden (1975)
  Nilsson (1978)
  Samset (1978)
  Upton (1976)
Shortwood and log forwarders
  Axelson (1972)
  Boyd (1975)
  Logging Research Foundation, Sweden (1975)
  Nickels (1978)
  Nilsson (1978)
Whole-tree forwarders
  Legault (1976)

Myles (1976)
Powell (1975)
Sarna (1976c)
Multiple-function machines
  Anonymous (1971, 1977b)
  Heidersdorf (1974)
  Kurelek (1975)
  Pickard (1972)
  Powell (1974a,b)
  Routhier (1971)
Whole-tree chippers
  Altman (1975)
  Anonymous (1968, 1974, 1975)
  Arend et al. (1955)
  Auchter (1972, 1975)
  Biltonen et al. (1979)
  Bradley and Stevens (1978)
  Bradley and Winsauer (1978)
  Bryan (1978)
  Conner (1978)
  Deal and Huff (1977)
  Erickson (1974)
  Folkema (1977b)
  Hakkila (1978)
  Hakkila et al. (1979)
  Harris (1977)
  Hillstrom and Steinhilb (1976)
  McCloud (1977)
  Martin (1977)
  Matics (1975, 1978)
  Morey (1975, 1978)
  Mulcahey (1974)
  Nelson (1976)
  Nelson (n.d.)
  Overby (1976)
  Pease (1972)
  Plummer (1976, 1978)
  Powell et al. (1975)
  Sterle (1971)
  Stuart et al. (1978)
  The Norwegian Forest Research Institute (1978)
  Tufts (1976b)
  Williams (1977ab)

16-6 MAXIMIZING VALUE OF TREES AND STEMS

There is great variation in value among individual trees and among parts of a single tree. For example, a sound 14-inch-diameter white oak tree containing two upper-grade veneer logs has several times the value of a red maple of equal size but crooked and infested with Columbian timber beetles. Within the white oak tree, the two veneer logs are several times more valuable than an equal weight of its topwood, bark, or rootwood. The problem of dismembering pine-site hardwoods to maximize their value is considerably more complex than with pine, because the hardwood species differ so markedly and the solidwood products they can yield are so diverse in purpose and value.

A partial listing of potential products illustrates the problem of achieving maximum value from each portion of a white oak tree, as follows (late 1979 values):

Tree portion and product	Value per ton—delivered (ovendry weight basis) *Dollars* (1979)
Stemwood	
Veneer log	162
No. 1 sawlog	71
No. 2 sawlog	40
No. 3 sawlog	27
No. 1 stave log	100
No. 2 stave log	87
No. 3 stave log	75
6½- to 8-foot post	32
Crosstie log	31
Pallet log	26
Pulpwood	16
Fuel	15
Topwood	
Bolter log for furniture dimension	30
Bolt for pallet lumber	25
Pulpwood	16
Fuel	15
Bark	
Architectural mulch	35
Horticultural mulch	35
Compost	70
Fuel	15
Polyflavanoid source	Not established
Rootwood	
Pulpwood	Not established
Fuel	15
Foliage for	
Fuel	15
Animal food supplement	Not established
Chlorophyll-carotene paste	Not established
Adhesive extender	Not established

The market price for hardwood pulp chips varies from less than that of roundwood pulpwood to as much as $64 per ton (ovendry basis with up to 7 percent bark content, 1979) loaded on ship at Gulf of Mexico ports. Among-species variations in product value further complicate the pattern.

The process of dismembering a tree to obtain maximum product value is termed tree **merchandising,** and the system of machines used for this purpose is called a **tree merchandiser.** In time, roots, branches, and foliage—as well as stems—will find economic use through this process. Figure 16-20 illustrates a system in which above-ground tree parts can be brought to the mill, and figures 16-32, 33, 34, and 35 show trees harvested with central root systems attached to stems. Thus, hardwood trees of species with compact crowns may enter a tree merchandiser that removes foliage, limbs, and bark before the central stem passes to a **stem merchandiser** for crosscutting into piles, logs, bolts, and pulpwood in combination to yield maximum value.

Fürstenberg Principality, Donaueshingen, West Germany, operates such a tree merchandiser on conifers. Kwasnitschka (1978) has described the process of whole tree transport with centralized delimbing, debarking, and stem bucking to maximize value. The largest trees processed have butt diameters of 15 inches and are 60 feet long; the smallest are 4 inches in dbh and about 18 feet long. Maximum transportation distance to the tree merchandiser is 40 miles. No major modification of Kwasnitschka's system (fig. 16-20) would be required to accommodate complete trees with central root mass attached.

By other systems (e.g., those shown in figs. 16-28, 29, 30, 31, 48, and 61), stumps, branches, and foliage may be cleaned from central stems and shipped separately to the millyard. Chapter 17 discusses recovery of clean wood, foliage, and fuel from tops and stumps so harvested. For description of a machine designed to strip foliage from trees, see Nedashkovskii et al. (1978).

However roots, foliage, and limbs are handled, stem merchandisers are needed to get maximum value from tree boles. Two systems have been developed. In one the stems move longitudinally (linearly) past one more cutoff saws; in the other they move transversely past a bank of saws spaced to cut them into selected lengths. A combination of both systems has been used in several large-scale complexes.

LINEAR STEM MERCHANDISERS

The linear system is by far the most common and requires only an infeed deck, linear conveyor, single or multiple cutoff saws, a debarker, and log pockets or decks from which segments are conveyed to process centers such as sawmills, veneer plants, bolter mills, or chippers. Linear systems can process 1½ to 2 stems per minute while making six to eight cuts per minute. They can yield 80 to 100 linear feet of logs and bolts per minute plus topwood to be chipped. To double output, two cutup systems can be advantageously arranged in parallel with topwood from both lines going to a single chipper. Such a dual system costs $850,000 to $1,000,000, requires four to six operators with one full-time cleanup man, and availability of millwright and electrician help. Manpower for log yard systems, such as rubber-tired log loader or boom crane, is additional (Murphy 1978).

The linear system allows great flexibility in selecting product length, but production is low on small stems. The stem of a 10-inch oak weighs about 1,000 pounds when green. A dual line might process three trees (1.5 tons) per minute—possibly 600 tons (200 cords) per shift. Stems of 6-inch oaks weigh perhaps 285 pounds (green basis); at four trees per minute, per-shift production would be only 228 tons (76 cords). Production is increased if entire stems destined for the chipper bypass the merchandiser and go directly through a barker to the chipper.

Blackford (1975) described a linear system particularly suited to small-volume central merchandising of hardwood stems of mixed species. The system (fig. 16-21) was installed at the sawmill of Inland Container Corporation, New Johnsonville, Tenn., but was later removed because its production was insufficient to satisfy sawmill demand. The system is described because of its simplicity, although throughput is only about 350 stems per shift. In this system, woods-run stems from 8 to 40 feet long and averaging 20 feet, were placed on the infeed deck by a front-end loader. Butt diameters were mostly 8 to 18 inches; 30 inches was the maximum admissible. Diameter at the small end of the logs was 4 to 5 inches. By means of a log stop and loader, stems were fed one at a time (**singulated**) onto the linear log haul which was controlled by the operator to position the log properly for crosscutting. The circular cutoff saw was driven by a 150-horsepower motor and hydraulically stroked (a chainsaw is cheaper and would serve, but more slowly and less reliably). Next a 50-inch-diameter, 72-inch-long drum chipper powered by a 300-horsepower motor chipped oversized cull-log sections ejected sideways into the chipper drum. Resulting pulp chips were conveyed to a van for transport to a pulpmill.

Next, logs going directly to the sawmill were ejected onto an infeed log deck leading through a log debarker to the sawmill. Beyond this kickoff station, small cull logs 8 to 25 feet long were ejected into a storage bin for transport to the pulpmill and conversion to chips.

The linear log haul terminated in a series of bins for sorting selected logs and bolts by lengths. Normally, two-thirds of the output went directly into the sawmill, but this varied with log quality. Cost of the system in 1975 (excluding the debarker) was about $150,000, of which about $60,000 was for the drum chipper and associated equipment. Production per 8-hour shift averaged 350 stems totaling about 250 tons (green basis). About 40 percent of the total tonnage ended as logs or bolts and the balance as chips or pulpwood.

To avoid the high cost of large front-end loaders and associated yard costs, a system has been developed whereby a small travelling grapple loader straddles the linear log haul conveyor and travels along it to grapple-feed logs from adjacent log piles. The travelling grapple loader also unloads logs from incoming trucks and piles them. Nerbonne (1974) has described several such systems in which a mobile hydraulic loader can travel along the combination loader track-log haul at 150 feet per minute.

Figure 16-20.—(A) Whole trees arriving at a tree merchandiser. (B) Whole trees being grapple-loaded onto the infeed rolls of the delimbing machine. (C) Rotating rolls equipped with knives delimb stems at 120 to 150 feet per minute; limbs are used for fuel or fiber. (D) Delimbed stems, after passing through a mechanical ring debarker, are crosscut for maximum value into poles, logs, and bolts. (Photos from Kwasnitschka 1978.)

Harvesting

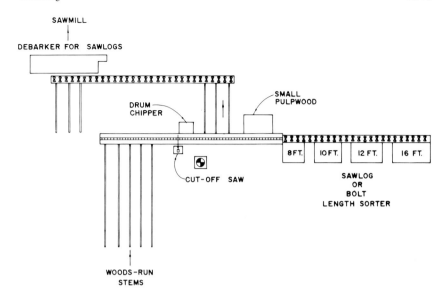

Figure 16-21.—Linear stem merchandiser for woods-run hardwood. Large-diameter cull log sections to 72 inches in length are delivered sideways to the drum chipper where most bark is segregated from pulp chips, and used for fuel. Pulpwood bolts are debarked and chipped at a nearby pulpmill. (Drawing after Blackford 1975.)

Figure 16-22 illustrates an installation at the Container Corp. plant in Georgiana, Ala., in which tree-length stems are crosscut by a 72-inch circular cutoff saw; sawlogs and veneer bolts can be ejected sideways into temporary storage and the balance of the wood is debarked and chipped. The chipper will accept roundwood up to 22 inches in diameter. In 1974, machinery for such a system cost about $150,000. Production per shift at that time was 200 to 225 tons of chips, 16 to 18 tons of sawlogs and veneer bolts, and 13 to 15 tons of bark. When operated to produce only chips, output per 8-hour shift is 250 to 275 tons of chips plus 16 to 18 tons of bark.

Because fireplace wood is selling for $140 or more a cord in urban centers, it will be manufactured in some areas in preference to pulpwood which sells for about one-fourth this price. Hardwood stem portions of low quality can be diverted from chippers through a low-cost linear merchandiser (fig. 16-23) to produce split firewood. With such a system, about four 2-foot-long bolts can be crosscut per minute, and split (but not piled) by three men. If bolts average 7 inches in diameter, production should be about 2¼ cords per hour. Cost of the equipment illustrated in figure 16-23 was about $38,000 in 1979.

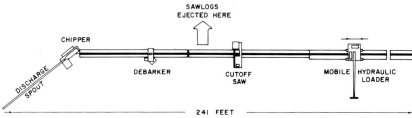

Figure 16-22.—(Top) Mobile hydraulic loader travelling along the log haul of a linear stem merchandiser can unload trucks along the track and then feed the log haul from this inventory. (Photo from Morbark Industries, Inc.) (Bottom) Plan view of linear stem merchandiser featuring such a travelling loader. (Drawing after Nerbonne 1974.)

Figure 16-23.—Linear merchandiser to convert low-grade hardwoods into split stove or fireplace wood 12 to 25 inches in length. Bolts are split by a hydraulic splitter exerting 25 tons force with 6-second ram cycle time. (Photo from La Font Corporation.)

TRANSVERSE STEM MERCHANDISERS[8]

Production through a linear stem merchandiser is limited to 1½ to 2 stems per minute. By moving stems transversely on lugged chains through banks of saws (fig. 16-24), production rates up to 10 stems per minute are possible, with total lineal input of 300 to 400 feet per minute. This productivity has stimulated the building of numerous installations for both hardwood and pine in the South.

The complexities of singulating individual logs on each lug, indexing the butt of each log to a given line, and spacing the saws for optimum combinations of cutting patterns severely complicate the design and increase capital costs for this system. While capable of operating at 8 to 10 stems per minute for short periods, such systems may be available only 50 to 85 percent of the time; the effective feed rate of most transverse merchandisers operating in 1978 was four to seven stems per minute. Operation of transverse stem merchandisers and their various complexities are further discussed in the following paragraphs.

Feeding infeed decks.—Delivery of stems to the infeed deck of a transverse tree merchandiser is crucial to its successful operation. Needed is a continuous single layer (not a double layer) of logs whose butts are reasonably aligned. Elevated decks serviced by a front-end loader (fig. 16-67 top) have been frequently used. Preferably the loader should be able to pass between the chains to deposit its load close to the headshaft or the trailing end of the log layer. With suitable topography, rubber-tired front-end loaders can drive directly onto the deck and

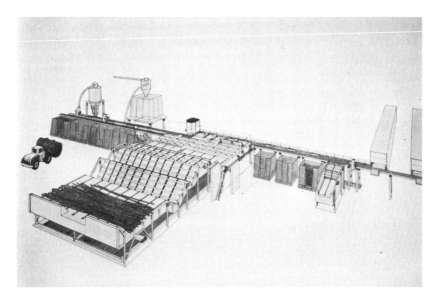

Figure 16-24.—Transverse stem merchandiser including infeed deck, unscrambler cascades, staggered crosscutting saws, and conveyors to bins for veneer logs, sawlogs, posts, and pulpwood. (Drawing after MoDoMekan.)

---

[8]Text under this heading is condensed from Murphy (1978).

thereby approach the headshaft area easily to deposit loads where needed. If ramp construction is necessary its slope should not exceed 8 degrees. Crane systems, cheaper than front-end loaders and needing less costly yard surfaces, can do a reasonably good job of spreading logs on log decks, but rubber-tired, front-end loaders are generally preferred. One arrangement for getting grapple loads of logs into a single orderly layer is the cascade, in which the first log deck is 18 to 24 inches higher than the successive deck; this works well on large logs but is ineffective on long small-diameter wood. Breaking the drive systems of the infeed decks into two tandem units permits logs to be straightened before they go over the headshaft in the cascade.

Singulating stems.—Separation of stems to reach the saws singly calls for either a lug-type unscrambler or a bar-type unscrambler (fig. 16-25 top) with 4- to 8-foot-long bars connecting two or more chains. The bar unscrambler is more successful, but is more difficult to keep operational because chains get loose and jump their sprockets; the large deck openings needed for the bars to pass around the sprockets present design problems. If lugs rather than bars are used, they must be ruggedly constructed and well maintained; constant and effective trash cleanup is required.

Butt indexing stems.—To obtain a reference from which to measure lengths for crosscutting, the butt end of stems must be aligned. This is accomplished with powered runover rolls, 4 to 8 feet long operating between lugged chains (fig. 16-25 bottom). To move stems longitudinally toward the reference line, they must be momentarily ahead of the lugs and resting directly on the rolls so they are free to shift. This can be accomplished by feeding directly from the headshaft of the unscrambler on to these rolls or by a slope in the chain table to allow the stem to coast or slide downhill away from the lug momentarily.

Crosscutting stems.—As the singulated stems, with butt ends aligned, move transversely through banks of circular saws, they are cut to desired lengths to maximum stem value. Saw arrangements commonly used are:

- Under-table saws (fig. 16-25 bottom)
- Overhead chop saws, shiftable (fig. 16-26 top)
- Staggered saws (fig. 16-26 bottom)
- Rotary-feed saws (fig. 16-27)

Saw spacing for optimum usable lengths can present problems. The solution is simple if the sawlog portion of the log is simply cut accurately to length with the balance going to the chipper. The problem becomes complicated, however, if peeler bolts (8.5 feet long minus 0 to plus 2 inches) and pallet or dimension bolts (nominal length, plus 1 or 2 inches) are removed as well as upper grade sawlogs (nominal length, plus 4 inches). The answer to the problem lies in selecting a reasonable number of combinations of cutting patterns that will maximize value of 80 to 90 percent of the volume to be handled. A lesser degree of optimization is accepted for the remaining 10 or 20 percent. Alternatively, transverse movement for most of the cutting, followed by a few cuts made with movement in the linear mode can give great flexibility.

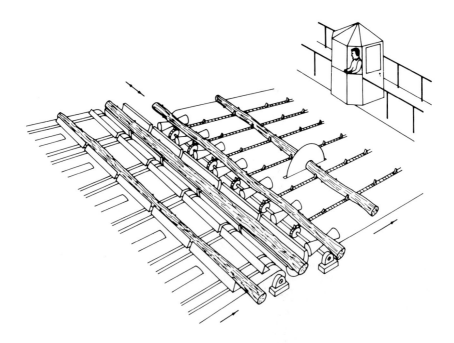

Figure 16-25.—Bar-type unscrambler singulating small hardwood stems in a transverse merchandiser. (Photo from MoDoMekan.) (Bottom) Butt indexing rolls to move each stem so that its butt end is aligned with those of the other stems. Note the under-table saw arrangement.

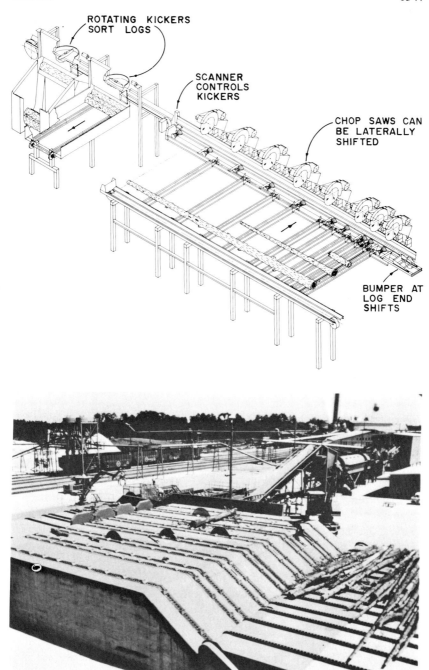

Figure 16-26.—Saw arrangements for transverse stem merchandisers. (Top) Laterally shiftable, 78-inch-diameter chop saws each driven by a 55-horsepower motor. (Drawing after Kockums Industries Inc.) (Bottom) Staggered under-table saws cutting hardwood stems. (Photo from Murphy 1978.)

Figure 16-27.—Rotary-feed device poised to carry a tree stem upward and through selectively lowered saws—in this instance to yield a sawlog from the visible part of the stem. (Photo from Murphy 1978.)

Product separation.—Separation of sawlogs, peeler logs, short bolts, posts, and pulpwood is accomplished by a number of systems including multiple tipple gates, lineal runout arrangements, lift-out systems, or multiple kickout pockets from lineal conveyors.

Cleaning up the trash.—Accumulations of bark, sawdust, and broken pieces cause as much downtime as crossed-up logs or equipment failures. The infeed deck generates large volumes of bark knocked or torn from logs, but the unscrambler is the major trash generator. An underneath sweep system featuring long bars can handle granular material, short cutoffs, grapevines, and limbs; it is probably the most successful system in use. The transfer from the transverse sweep bars to a linear conveyor going to fuel hog or barking drum must be handled so that it is non-binding and non-choking.

Costs and productivity.—Transverse stem merchandisers are expensive—costing $3 to $5 million per installation, but require only three or four men to operate them plus services of a millwright or electrician on call. Because of downtime, average productivity is four to seven stems per minute. The time required to handle a stem measuring 6 inches in dbh with 3 cubic foot content is the same as that for a 16-inch stem with 35 cubic-foot content. Merchandising is therefore highly sensitive to stem size and distribution. Costs for southern pines may range from about $5.80 per ton (green-weight basis) of input (6-inch stems) to only $0.33 per ton on 22-inch stems (table 16-42). Processing hardwood stems, which are shorter than pine stems of comparable dbh, is more expensive. Productivity can be raised and overall cost lowered if entire stems destined for the chipper can bypass the merchandiser.

TABLE 16-42—*Estimated unit costs to operate a transverse stem merchandiser on southern pine* (Data from Murphy 1978)[1]

Stem dbh (inches)	Per 100 cubic feet of solidwood	Per ton (green)	Per cord	Per stem	Volume per stem
	----------Dollars----------				Cubic feet
6	17.40	5.80	14.50	0.53	3.1
8	8.80	2.93	7.33	.67	7.6
10	5.20	1.73	4.33	.67	12.9
12	3.40	1.13	2.83	.69	20.2
14	2.50	.83	2.08	.72	28.9
16	1.80	.60	1.50	.67	35.4
18	1.50	.50	1.25	.63	50.3
20	1.20	.40	1.00	.77	64.1
22	1.00	.33	.83	.65	77.9

[1]No yard costs included; yard costs are estimated at $0.50/ton.

16-7 STUMP AND TREE PULLER—BUNCHERS

Proponents of complete-tree utilization view the central stump-root system of pine-site hardwoods as a major source of wood for energy or fiber. These central stump-root systems, with lateral roots severed at a 1-foot radius, comprise about 15 percent of the weight of above- and below-ground tree biomass (ovendry basis) and 25 to 30 percent of the weight of bark-free merchantable stems of small pine-site hardwoods (tables 14-2B and 16-2 and figs. 14-4 through 14-26). Ready availability of an unused wood resource of such magnitude is a strong incentive to devise practical stump harvesting methods. Moreover, removal of stumps in a commercial harvesting operation substantially reduces the cost of preparing the site for subsequent planting. Stump removal is not a new art. There were about 500 kinds of stump pullers on the American market in the 1800's, and stump pulling became a common American profession; two men with a yoke of oxen and a stump puller could travel indefinitely across the country pulling 20 to 50 stumps per day at $0.25 per stump—the standard price in 1850 (Sloan 1958).

Today, many entomologists, hydrologists, and soil scientists in the South view harvest of central stump-root systems with favor, because of diminished insect attraction to freshly exposed stump surfaces, improved percolation of water into many soils, and improvement of texture of some soils by the same mechanism as plowing of agricultural land. Uprooting is not unique to manmade forests; as noted by Stephens (1956), trees of the "primeval forest" had two general destinies: uprooting, or piece-by-piece disintegration in place.

STUMP PULLER-BUNCHERS

Andersson et al. (1978) and Walker (1976) have reviewed equipment available worldwide to pull stumps after trees have been felled. Major machines for this purpose include:

Machine	Reference
Pallari stump harvester	Hakkila and Makela (1973, 1974)
Cranab stump harvester	Andersson et al. (1978)
Dynapac stump harvester	Andersson et al. (1978)
ÖSA stump harvester	Anderson et al. (1978)
Wick-Bartlet stump harvester	Harrison (1975)
Rockland roto lifter	Walker (1976)
Cavaceppi stump auger	FAO (1962)
STFI oscillating saw	Andersson (1975)
FLECO stump blade	U.S. Department of Agriculture, Forest Service (1971)
Bulldozer-mounted root rake	U.S. Department of Agriculture, Forest Service (1971)
Foster vibro stump extractor	Anonymous (1978a)

The Pallari stump harvester appears well adapted to harvesting stumps of pine site hardwoods. Several versions of the machine were designed in Finland in 1975–1977 (fig. 16-28). The unit is generally mounted on an excavator-type crawler tractor, but tests have also been made on a feller-buncher chassis. Andersson et al. (1978) note that the machine, which is simple and reliable, can pull individual stumps and split them to the desired degree of fragmentation. Blomqvist (1978) reported that in commercial practice, the Pallari stump harvester produces 6.4 cubic meters (loose-wood basis) per hour and that cost of rootwood so harvested is $8.40 per cubic meter (loose-wood basis) delivered to the mill; transport distance to this mill averages about 25 miles.

The ÖSA 635 prototype stump harvester also appears to have some potential for pine-site hardwoods. The machine consists of a frame, a falling weight, and four movable knives. The falling weight drives the knives into the stump, splitting it in a cruciform pattern (fig. 16-29 top). This operations also frees the stump. When mounted on a feller-buncher or excavator chassis, grapple claws (fig. 16-29 bottom) fitted to the splitter can split, lift, and load stumps at a rate of about 5 cubic meters of solid wood per hour (Andersson et al. 1978). These authors propose three systems to deliver stumpwood at three levels of cleanliness into trucks for highway transport to the mill; each is based on initial splitting, harvesting, and limited fragmentation of stumps, with subsequent processing as follows:

System description	Figure	Cost, delivered 60 miles to mill (solid wood basis)
		Dollars/cubic meter
Uncleaned wood	16–30, top	28.52
Partial cleaning of wood accomplished in the forwarder in transit to roadside	16–30, center	28.75
Intensive cleaning by hammermilling and screening, at the landing before loading into trucks	16–30, bottom	32.43

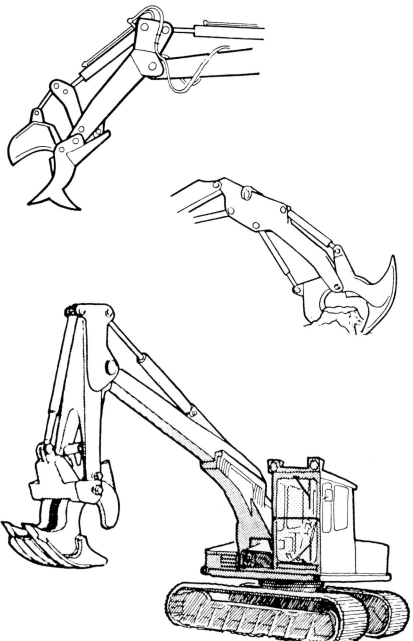

Figure 16-28.—The Pallari stump harvester. (Top) Two designs of splitter-grapple head. (Bottom) Mounted on an excavator tracked chassis. (Drawing after Andersson et al. 1978.)

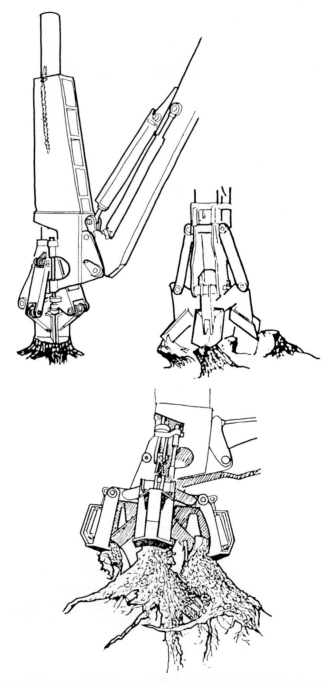

Figure 16-29.—ÖSA stump harvester. (Top) Splitter. (Bottom) Splitter combined with a grapple for mounting on a feller-buncher chassis. (Drawing after Andersson et al. 1978.)

Harvesting 1553

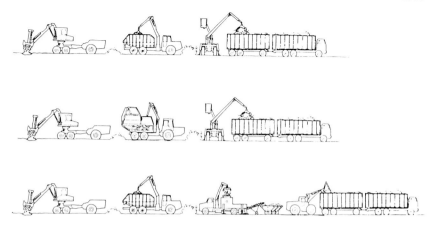

Figure 16-30.—Systems for utilizing the ÖSA stump harvester providing three degrees of stumpwood cleaning. (Top) Forwarded direct to roadside truck with no cleaning. (Center) Partially cleaned by tumbling in a forwarder-cleaner en route to highway truck. (Bottom) Forwarded to roadside landing for extensive cleaning with hammermills and screens before loading into truck. (Drawing after Andersson et al. 1978.)

Users of rootwood prefer it cleaned of dirt and rocks at time of harvest. To satisfy this need, the L. B. Foster Company, Pittsburgh, Pa., combined a vibrating grapple head with a hydraulic self-propelled loader in a prototype stump harvester (fig. 16-31). To pull and bunch a stump, the operator lowers the open grapple over the stump, with prongs rotated to avoid visible outstretched roots. Hydraulic pistons close the grapple prongs under the root system and the vibrator is started. The loader crane raises the grapple and extracts the stump-root system. Excess dirt is shaken off as the operator swings the vibrating stump-root to the bunching area, where it is dropped when clean. This cycle, as observed in experimental runs, takes 90 seconds or less. A grapple sized to pull 24-inch stumps weighs 3,000 pounds (Anonymous 1978a).

Procedures for cleaning stump-root systems at centralized crushing and chipping centers adjacent to pulp mills are described in chapter 17.

Figure 16-31.—Stump puller incorporating vibration device—just above grapple—to free stump from ground and loosen dirt from it. (Left) Grapple closing over stump. (Right) Extracted stump-root system. (Photo from L. B. Foster Company.)

TREE PULLER-BUNCHERS

Most pine-site hardwood trees are small in diameter and not very heavy. A 6-inch southern red oak, for example, complete with crown and lateral roots to a 1-foot radius, weighs about 430 pounds—of which about 16 percent is in the stump-root system (Table 16-2). Costs per ton for conventional felling and bunching are inversely proportional to the weight of the tree sections comprising the bunch. Harvesting the central stump-root system with the stem maximizes weight per section, and should improve efficiency.

Two tree pullers potentially practical for southern forest operations have been invented. One pulls trees including entire lateral root systems, which are subsequently sheared and left on the ground. It has the advantage of an adjustable-diameter shear to accommodate trees of varying diameters (Sederholm 1976). This machine is not further discussed because no data are available on its operation on hardwoods, for which pulling forces with lateral roots intact are very large.

The second machine, jointly developed by the Southern Forest Experiment Station of the USDA Forest Service and Rome Industries (Koch 1976), leaves lateral roots in place in the soil; only the central stump-root system is harvested. Machines of this design are in steady commercial operation in Florida pulling and bunching 45 to 90 small southern pines (to 12 inches dbh) per hour. Harvesting costs, including profit on equipment investment, total about $8.61 per green ton delivered in tree lengths to the kraft mill where both rootwood and stemwood are drum debarked and converted to pulp chips (Koch 1977). En route to the mill, trucks carrying the tree-length limb-free stems (with taproots attached) pass through a washing station consisting of two fire-hoze nozzles swivel-mounted atop 6-foot stands, one on each side of a concrete slab. As each truck passes through the station, the driver washes his load; total delay time is 25 to 30 minutes per load (Davis and Hurley 1978).

By 1983 several of these tree pullers were also employed in Mississippi to pull pines with central stump-root resin enriched through injection at the root collar with paraquat; the resin-rich wood is used in the production of naval stores, while the upper tree sections are pulped.

Grillot and McDermid (1977) found that in pine stands in west Florida, productivity of the machine when making clear cuts was 16 cords per machine-hour; when thinning it was less productive, averaging 7 cords per hour, because of the smaller average size of trees harvested.

The tree puller, manufactured by Rome Industries, Cedartown, Ga., can be fitted through a quick-hitch mechanisms on a number of prime movers—for example, a Caterpillar 920 or a John Deere 544B (fig. 16-32). Two elements are essential to the design (fig. 16-33). The first is a scissors-type grip achieved with a pair of stout horizontal knife blades that close at groundline and bite several inches into the stem from opposite sides.

With this grip anchor, the second element in the design comes into action. It is a clamshell-hinged tubular shear, 22 inches in diameter and made of ¾-inch-thick steel. The shear, sharpened on the lower edge, is forced vertically into the ground to a depth of 11 inches, thereby severing lateral roots all around the tree. At this point, broad steps on opposite sides of the shear strike the ground and limit further penetration. An additional stroke of the hydraulic cylinder raises the grips 9 inches while the steps remain pressed against the soil surface. The effect is to jack the stem and break it free of the ground. Finally, the complete tree is lifted into the air and bunched for skidding. Since shearing takes only a few seconds, a tree can be harvested and bunched in about 45 seconds.

Dirt that adheres to the stump root system can present problems. Such adherence is more severe in clay soils than in sandy soils. Hardwood stump root

Figure 16-32.—(Top) Tree puller-buncher poised to grip 8-inch southern red oak at groundline. (Bottom) Hickory tree pulled from ground after lateral roots near surface were severed, leaving central root mass intact. (Photos from D. Sirois, Southern Forest Experiment Station, Forest Service—USDA, Auburn, Ala.)

Figure 16-33.—(Top left) Rome TX-1600 lateral root shear and tree puller mounted on a conventional four-wheel-drive articulated loader, with hinged grip closed on tree to be harvested. (Top right) View of gripping knives in open position with shear retracted. (Bottom left) Grip closed; in operation the knives grip the tree stem at ground level; no other grip is needed. (Bottom right) Grip closed and tubular shear fully extended 20 inches. Steps that bear against soil surfaces during last 9 inches of extension are visible on opposite outer sides of the shear tube. Each side of the tube is independently driven through its 20-inch stroke by a 6-inch hydraulic cylinder housed in the vertical column.

Figure 16-34.—(Top) Grapple skidder dragging loblolly pines with taproots attached; trees were previously bunched by the tree puller. (Center) After extraction of all trees, sites are free of stumps and debris. (Bottom) Longleaf and slash pines with taproots attached, ready to load on trucks near Panama City, Fla. (Photos from Koch 1976).

Harvesting

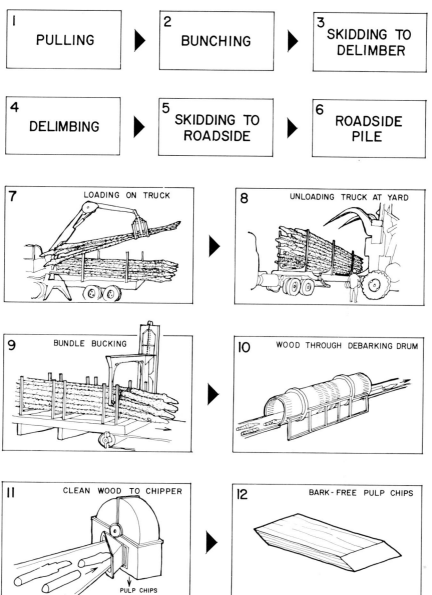

Figure 16-35.—Harvesting and utilization system for trees pulled with central stump-root system attached to stems. Omitted between steps 7 and 8 in this diagram is the truck-load washing station. (Drawing after Koch 1977.)

systems (fig. 14-4 through 14-26) are more difficult to clean than the more compact roots of the southern pines. (See chapter 17 for alternative procedure.) With pines, much dirt is shaken free during bunching, skidding, and stacking (fig. 16-34); washing en route to the mill dislodges more dirt, and finally a drum debarker can remove most of what remains so that bark-free (and remarkably dirt-free) root wood emerges from the drum debarker to pass through disk chippers and hence into the pulp mill. The procedure is summarized in figure 16-35.

Sites harvested by the tree puller show few holes (fig. 16-34); dirt falls back into the cavities and leaves a hole only a few inches to a foot in depth. These holes are filled up after rains and movement of machines over them.

DATA ON SOUTHERN HARDWOODS

To provide design information for modification of the Rome lateral root shear and tree puller to adapt it for use on southern hardwoods, Sirois (1977) collected data on stump-root biomass and on forces when shearing and pulling white oak, southern red oak, hickory sp., and sweetgum. In a 75-tree sample, green weight of the central stump-root system averaged 18 percent of the complete-tree weight of above-and below-ground parts. The harvested stump-root system weighed 22 percent as much as the total above-ground biomass. Harvested portions of the stump-root system had average green weight (when cleaned) of 104 pounds, with range from 79 to 164 pounds in trees that measured from 4 to 12 inches in dbh and averaged about 6.8 inches.

Hickories were the most difficult of the four species to extract because most have taproots with deep laterals (fig. 14-8). Because the lateral root shear penetrated only 10 to 12 inches into the ground, deeper laterals were not severed and pulling forces sometimes exceeded machine capability. Forces to shear laterals were greatest for the hickories, regression relationships indicating that more than 200,000 pounds are required to drive the shear to a 10- or 12-inch in clay-loam soils; lifting forces were less than 70,000 pounds (table 16-43).

The tree puller tested has shear blades 16 inches long that grip the tree at its base. Trees 12 inches in dbh of all four species may have butt diameters in excess of 16 inches (fig. 16-36), thus preventing closure of the tubular shear with resultant inability to shear laterals around the complete tree circumference; when lateral roots are not completely severed, pulling forces may be excessive. Tables 27-138 through 27-140 give additional data on the relationship between stump diameter and diameter at breast height.

The machine as tested weighed about 30,000 pound when mounted on a Caterpillar 930 wheeled loader, and ground bearing pressure was about 10 psi. In soft ground the machine sometimes penetrated the vegetative mat and became stuck.

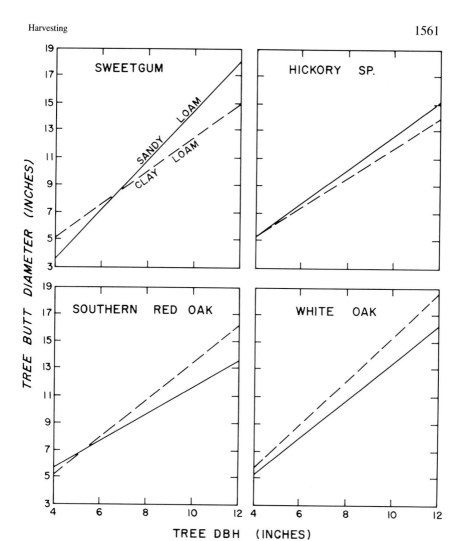

Figure 16-36.—Relationship of tree dbh to diameter at groundline in four species of pine-site hardwoods. (Drawing after Sirois 1977.)

TABLE 16-43—*Forces required by the Rome tree puller to shear lateral roots and pull small hardwood trees, of four species growing on two soil types, free of the ground with central stump-root system attached to stem* (Data from Sirois 1977)[1]

Species and tree dbh (inches)	Force to shear lateral roots		Force to pull trees	
	Sandy loam	Clay loam	Sandy loam	Clay loam
	------------------------Thousands of pounds------------------------			
Sweetgum				
4	68.3	37.0	3.1	18.0
8	90.8	95.3	35.4	29.4
12	113.3	153.6	67.7	40.8
Southern red oak				
4	39.3	46.4	14.4	17.2
8	87.1	97.3	25.4	26.1
12	134.8	148.2	36.5	35.0
White oak				
4	86.4	56.6	27.0	21.9
8	79.2	94.8	29.8	32.2
12	71.9	132.9	32.5	42.5
Hickory, sp.				
4	52.0	82.6	27.1	24.1
8	83.5	149.2	35.0	29.4
12	114.9	215.7	42.9	34.7

[1]Values based on regression equations relating dbh to force derived from data on about eight trees in each classification. R^2 values were generally 0.6 or above except for white oak in sandy loam, where R^2 values were less than 0.1.

From this test series, Sirois (1977) concluded that the equipment could be made to pull pine-site hardwoods on a production basis if modified and given operational limits as follows:

- The prime mover carrying the shear head should exert less ground pressure (about 6 psi) and be capable of exerting higher drawbar forces than normal wheeled loaders.
- The depth of penetration of the lateral root shear should be increased (mainly for sweetgum and hickory).
- Additional force is needed to close the scissor blades that grip the tree at groundline, thus insuring that the tubular shear is closed around the complete circumference of the tree so that all lateral roots will be severed during its cutting stroke.
- To keep machine cost and size within reason, an upper limit on tree size should be established at about 9 inches dbh.

16-8 CABLE YARDING

About 20 percent of pine sites are in highlands that may have steep slopes; another signficiant percentage, though level, is too soft when wet for loaded vehicles. Timber on steep hillsides and soft ground cannot be reached by wheel- or track-mounted feller-bunchers; trees are felled by chainsaw. After trees on such sites are felled, what are the options for getting them to roadside?

In these cases skidding or yarding cannot be avoided, but wheeled or tracked skidders must stay on roads built for them, and the roads must be closely spaced as skidder winches carry only 75 to 100 feet of cable. In many areas such close spacing of roads is excessively expensive or is aesthetically unacceptable. **Cable yarding** (fig. 16-37) offers possibilities for reducing road frequency because yarding distances can be as much as 1,000 feet on appropriate terrain.

In cable yarding, as in skidding with wheeled or tracked skidders, hardwood tree crowns present a problem. If trees are topped with a chainsaw, about half the tree weight is left at the stump. If the top is not severed from the stem, the spreading limbs are hard to drag and they cause great damage to residual trees in a partial cut. Moreover, once at roadside, the limbs and tops are difficult to reduce to useful form. This latter aspect will be discussed in sections 16-10, 11, and 12.

Ground lead cable yarding (figs. 16-17 and 16-37 top) will be further discussed and illustrated in section 16-11. The highlead system will be briefly explained here, but primary attention will be given to light equipment for skyline yarding. Readers needing a more extensive knowledge of cable yarding techniques are referred to Conway (1976, p. 186–305) and Studier and Binkley (1974). Larsen (1978) has assembled a compendium comparing major cable logging systems. Hawkes (1979) has described more than 40 systems for cable harvesting small timber, and The Norwegian Forest Research Institute's (1978) Report on Forest Operations for Research contains 63 pages describing such systems. Wendel and Kochenderfer (1978), Kochenderfer and Wendel (1979), and Patric (1980) describe environmental effects of cable logging in Appalachian forests.

Falk (1981) developed three programs for hand-held calculators to predict payload capability of cable logging systems including the effect of partial suspensions. The programs allow for downhill yarding and for yarding away from the yarder.

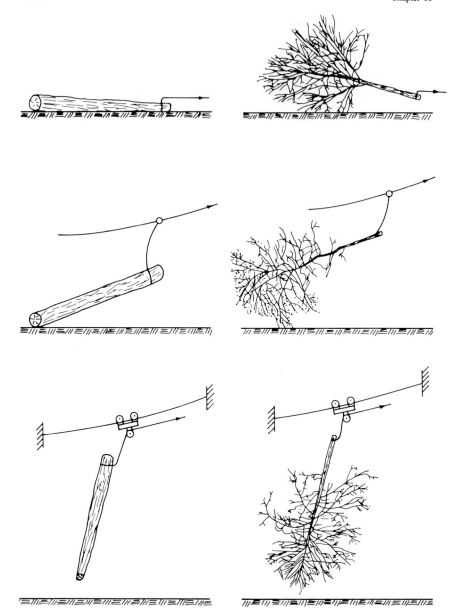

Figure 16-37.—Cable yarding systems for pine-site hardwoods. (Top) Ground lead, i.e., with no vertical lift capability. (Middle) Highlead with limited lift capability. (Bottom) Skyline which can lift logs or trees clear of the ground.

HIGHLEAD YARDING

Highlead yarding is a short-reach, primarily uphill system limited to clearcut operations. In this system (fig. 16-38), one drum reels in or pays out the **mainline,** while the second drum acts in concert to pay out or reel in the **haulback line.** A highlead system is not as mechanically efficient as a running skyline system with interlock (to be described next), because the haulback must be braked against the pull of the mainline to achieve log lift (Sirois and Iff 1978). To minimize capital investment required for cable yarding equipment—historically more costly than wheeled or tracked skidders—some loggers have converted two-drum shovels or cranes into highlead cable yarders.

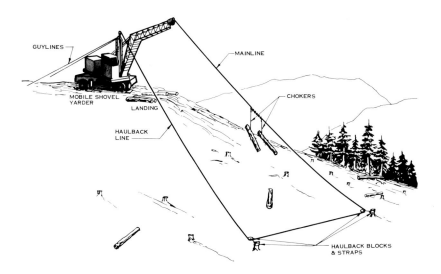

Figure 16-38.—Highlead system. (Drawing after Studier and Binkley 1974).

An experiment (Anonymous 1976a) with highlead cable yarding in the mountains of North Georgia is instructive in that less than a mile of road was required to log one 49-acre block; had it been harvested with wheeled or tracked skidders, 4 to 5 miles of road would have been required. Trees in the stand averaged 22 inches in dbh and 65 percent of the volume was yellow-poplar. The terrain harvested was nearly free of rocks and contained many slopes of 50 to 60 percent. The system featured a 25-foot tower, two drums, and ⅝-inch mainline and haulback cables. Maximum effective haul distance was 900 feet, although efficiency dwindled somewhat beyond 700 feet. A heelboom loader adjacent to the yarder fished logs from the jackstrawed pile at the yarder landing and loaded them on trucks for transport to the mill.

All trees 12 inches in dbh and up were harvested. The trees, averaging 215 board feet each, were topped at minimum merchantable diameter and bucked into two pieces. Daily production averaged 15,000 to 20,000 board feet (Scribner log scale). The five-man crew consisted of one yarder operator, two choker setters, one loader operator, and one general utility man. Two chainsaw operators were also needed. In this experiment, trees as small as 2 inches in diameter at groundline were felled to promote regeneration for even-aged management. The small timber was not harvested because of lack of market for hardwood pulpwood.

SKYLINE YARDING

Standing skyline systems (fig. 16-39) employ an elevated taut cable, anchored at one or both ends, along which a carriage travels. Additional operating lines move the carriage along the skyline. Capabilities of these systems range from those limited to uphill clearcut operation to those capable of partial-cut operation in either direction. Intermediate supports and special carriages (fig. 16-40 top) are used in multi-span systems to harvest timber on convex terrain (Mann[9]).

---

[9]Mann, C. N. 1979. A running skyline system for harvesting small logs. Presented at Amer. Soc. of Agric. Eng. West Coast Meeting, Aug. 6–9. Portland, Ore.

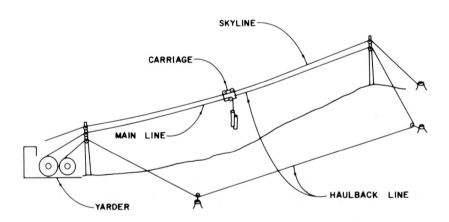

Figure 16-39.—Standing skyline system fitted with a carriage carrying choker cables.

Harvesting 1567

Carriages may be simple and non-locking (fig. 16-40 bottom). Others have locks that can be actuated to hold the carriage in position on the skyline while simultaneously releasing a lock on the lifting line to allow the choker hook to be lowered to the ground and carried out for lateral yarding.

Experience with a mobile standing skyline system in West Virginia indicated a logging cost for hardwoods of about $72 per thousand board feet (Scribner log scale) when harvesting an average of 5,000 board feet per acre (Smith 1978a).

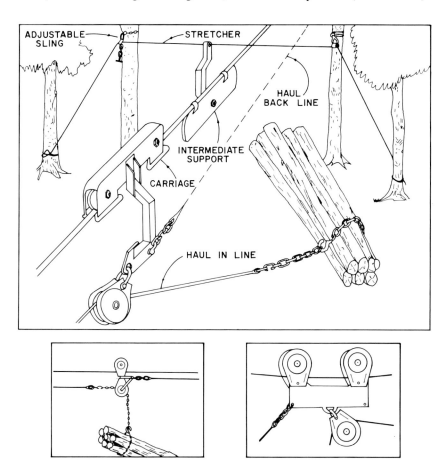

Figure 16-40.—Carriages with one or more pulleys that ride the skyline, below which is mounted a pulley over which the haul-in main line passes. (Top) Multi-span carriage slotted to pass over an intermediate support for the skyline. (Bottom) Single-span carriages. (Drawings after Hawkes 1979.)

Yarding small trees with a standing skyline system would be cheaper if such trees could first be bunched under the skyline (fig. 16-41 top) for group handling with a suspended grapple. To this end, the Equipment Development Center and the Forestry Sciences Laboratory in Missoula, Montana have developed a two-stage cable yarding system. In the first stage a suspended, gasoline-powered, radio-controlled vehicle is positioned by self-powered sheaves along the skyline, where it remains fixed while bunching stems (fig. 16-41 bottom). Movable at will, the buncher can be maneuvered along the skyline to eliminate hangups and minimize damage to residual trees in partial cuts. This maneuverability also permits stems to be aligned in neat piles for later cable yarding when, in the second stage, the system is re-rigged with a grapple suspended from the skyline (Taylor 1978).

Many engineers conclude that the **running skyline** is most likely to fill requirements for hardwood logging in the East and South. In this system, the carriage rides the haulback cable and both haulback and mainline support the load (fig. 16-42).

Kochenderfer and Wendel (1978) concluded, after 2 years of experience with a skyline yarder in the Appalachians, that a successful design should have the following characteristics:

- Mobility sufficient to negotiate steep roads of 12- to 14-foot width.
- Transportable on short moves over public roads without a lowboy trailer.
- Weight not to exceed 70,000 pounds.
- Running skyline design with carriage that can yard logs laterally from 10 feet on either side of the skyline.
- Capability to yard a 10,000-pound load.
- Capability to yard 1,000 feet with most work done in the range from 400 to 600 feet.
- Mainline speed of 1,000 feet per minute.
- Capability to log downhill as well as uphill (a feature of the running skyline).
- Equipped with a swinging boom so that yarded logs can be decked parallel to the road or loaded on trucks behind the yarder.
- Equipped with hydraulic guyline winches and hydraulic outriggers for levelling, to speed setup and teardown times.
- Operable with a four-man crew on partial cuts of 4,000 to 6,000 board feet per acre (Scribner log scale).
- Cost not to exceed $100,000.

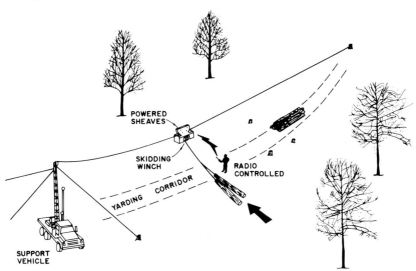

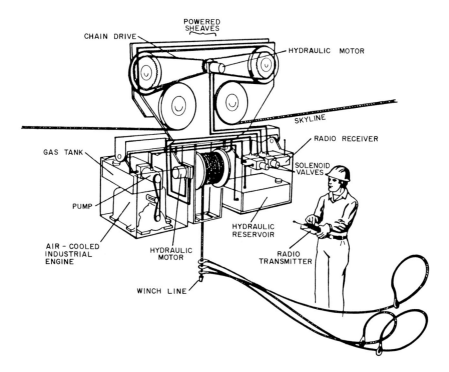

Figure 16-41.—Stage one of a two-stage skyline cable yarding system. (Top) A self-propelled radio-controlled vehicle rides the skyline cable to drag in and bunch small timber under the skyline for later yarding by a grapple fitted to the skyline system. (Bottom) Bunching vehicle. (Drawing after Taylor 1978.)

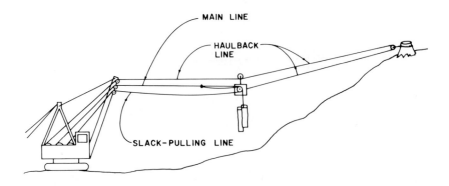
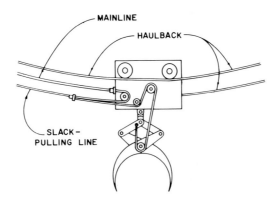

Figure 16-42.—(Top) Running skyline system fitted with choker cables. (Bottom) Fitted with a remotely operated grapple. (Drawing after Mann 1969.)

Harvesting

Such a machine could harvest hardwoods from terrain of varying contour (fig. 16-43); it remains to be proven, however, that such a machine is an economically viable alternative to existing ground skidding systems in the East and South (Kochenderfer and Wendel 1978).

Mann[9] described a 110-horsepower running skyline mounted on a wheeled skidder (fig. 16-44) that approaches the specifications recommended by Kochenderfer and Wendel, except that it requires a separate loader to keep the landing clear, sort logs, and load trucks. During a 262-hour operational test thinning and clearcutting small timber, about 71 percent of the test time was spent in actual yarding, about 16 percent in moving, 4 percent in miscellaneous delays, and the balance in scheduled and unscheduled maintenance; additionally, 98 man-hours were spent in rigging outside of normal working hours. During actual yarding time, production averaged 11 turns (loads) per hour and 37 pieces per hour. The designers believe that under the Appalachian conditions studied by Kochenderfer and Wendel the machine (in conjunction with a loader) might harvest about 120 logs per 8-hour shift. The operating crew consists of four men, as follows: one yarder operator, two choker setters, and one loader operator. If hauling is done by a self-loading truck, then a skidder is generally used in place of a loader to keep the landing clear. Additionally, two chainsaw operators are needed to fell and top trees.

Fisher et al. (1980) studied performance of a small running skyline yarder mounted on a 130-hp skidder and found that costs of a Virginia operation harvesting hardwood were about as follows:

Operation	Cost per thousand board feet, Doyle log scale
	Dollars
Road construction	12.34
Yarding (including changing roads)	59.78
Moving yarder	2.39
Loading	16.33
Hauling	25.00
Total	113.74

In this operation, average mid-point diameter of harvested logs was 15 inches with minimum top diameter of 8 inches. Average slope was 56 percent and maximum yarding distance was 700 feet. Haul distance from yarder to sawmill was 52 miles.

On a much larger scale, Washington Iron Works Model 78 Skylok running skyline yarders are teamed with whole-tree chippers in a major operation harvesting hardwoods from the steep hills of the Virginias (fig. 16-45). With this efficient system, whole trees are deposited on the landing, and high grade logs are removed by chainsaw before the loader feeds remaining tops and stems into the whole-tree chipper, thus clearing the landing. Landings are bulldozed large enough to accommodate the yarder, the chipper, two chip vans, and two skidders.

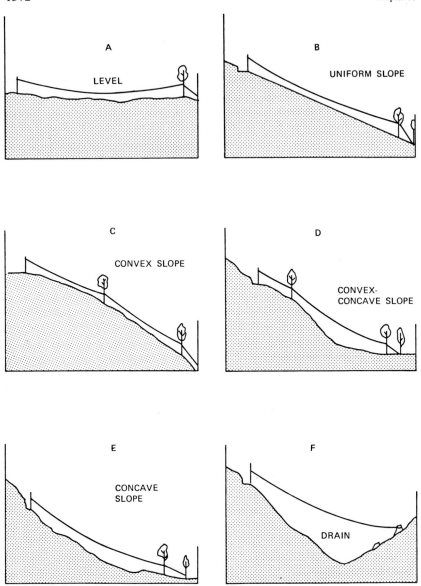

Figure 16-43.—A running skyline system rigged for various terrain contours. (Drawing after Kochenderfer and Wendel 1978.)

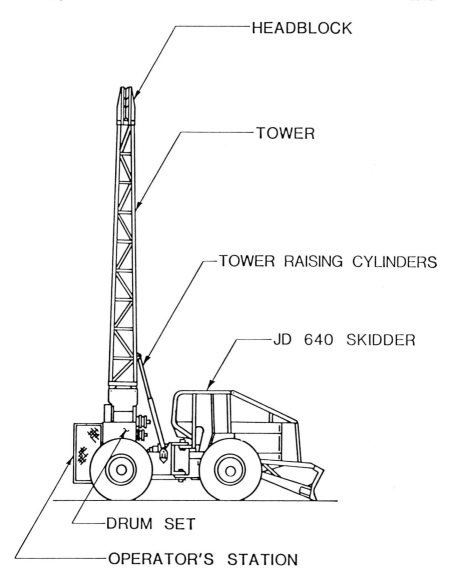

Figure 16-44.—Pewee yarder mounted on a wheeled skidder and featuring an interlock mechanism between mainline and haulback drums to conserve power and provide smooth control. All operating cables measure ½-inch in diameter. It has a 110-horsepower engine, can yard 1,200 feet with lateral yarding up to 150 feet to either side, and has cable speed of 750 feet per minute with line pull of 6,000 pounds. The tower, when erected, is 37 feet high and the machine weighs 44,000 pounds. Guylines are hydraulically tightened. (Drawing after Mann[9])

1574   Chapter 16

Figure 16-45.—In this West Virginia operation on steep terrain, a running skyline yards entire hardwood trees to the landing, where chokers are released. Sawlogs of high quality are extracted by chainsaw. The loader feeds tops to the whole-tree chipper which blows resultant chips into a waiting van. (Drawing after photo from Westvaco Corporation.)

Ordinarily clearcuts are 40 acres or less. Typically, the yarder is moved twice in a 40-acre set while the tractor that anchors the tail of the skyline (downhill, and about 800 feet away from the yarder) is moved 12 to 15 times. Virtually every tree, regardless of size, is dropped and skidded from the set. Production averages 200 tons of whole-tree chips per 8-hour day; another 15 tons leave the landing daily in the form of sawlogs. Excluding the logging supervisor, 11 men make up the crew. Three work under the carriage (a rigger and two choker setters), two man chainsaws, one runs the yarder and one operates the chipper. One man switches chip trailers, another unhooks chokers at the landing, and two men operate rubber-tired skidders. Additionally, a mechanic is available on call. Saw logs diverted from the chipper and chip vans are hauled by an independent operator (Pulpwood Production and Saw Mill Logging 1973).

The Westvaco experience has kindled renewed interest in cable logging in the South. Taylor Machine Works, Inc., of Louisville, Mississippi, has recognized this interest and is preparing to market a running skyline system with mechanically interlocked drums. The yarder has a 55-foot portable tower, four cable drums, and is self-propelled. Skyline, mainline, and haulback lines measure $5/8$-inch in diameter. Maximum line speed is 1,350 feet per minute and maximum line pull is 13,000 pounds. Overall width of the yarder is 11 feet 6 inches; weight is 75,000 pounds.

In any skyline operation, security of the tailhold anchor is crucial. Stout trees or stumps are customarily used; heavy tractors are both convenient and mobile. In the absence of other convenient tailholds, it is sometimes necessary to secure man-made anchors in soil or rock. Richardson (1979) has described anchors constructed of buried-log deadmen, multi-picket systems, plates, flukes, and screws; under development is an early-warning failure-detection system for such anchors.

16-9 BALLOON AND HELICOPTER YARDING

Thoughts of lifting whole trees and flying them from stump to landing are stimulated by difficult access and high road costs in mountainous or soft terrain, and site disturbance from ground skidding whole hardwood trees. Two methods of such aerial logging—balloon yarding and helicopter yarding—are in limited commercial operation on the West Coast. Weights of pine-site hardwoods are well within the lifting capacity of balloons and helicopters. The above-ground portion of a 6-inch dbh white oak weighs about 300 pounds when green; a 15-inch white oak weighs about 3,000 pounds (table 16-18).

BALLOON YARDING

Balloon yarding is a lift-augmented cable yarding system; the lift augmentation extends reach far beyond the capabilities of normal cable systems—e.g., up to 5,000 feet. Balloons may be employed in highlead, skyline, and running skyline modes (fig. 16-46). The balloons are filled with helium and have lifting capacities of 15,000 to 23,000 pounds. Yarders for the conventional balloon system have two continuously interlocked drums, with line speeds from 1,200 to 1,600 feet per minute. Productivity is related to volume per turn and turns per hour, with hourly production least at extreme yarding distance and maximum at about 1,200 yards. Current systems carry about 9,000 pounds of logs per turn and may yard 15 to 45 tons of wood per hour under West Coast conditions. Only 1 or 2 miles of road are required to log 640 acres by the balloon system, but investment cost in equipment is high. The yarder, balloon, and support equipment cost about $600,000 in 1976, with operating cost (including a 7-man crew) of $190 to $255 per operating hour.[10]

If pine-site white oak trees were yarded with stems and crowns intact, it would take 7 to 10 trees measuring 9 or 10 inches in dbh to make up a 9,000-pound load for the balloon yarding system. The problem of first bunching and then grappling or setting chokers on such a load seems substantial. If lesser loads are carried per turn, operating costs of $190 to $255 per hour make yarding costs per ton excessive.

---

[10] This discussion of balloon logging is condensed from Conway's (1976) report on work of J. E. Jorgensen, University of Washington, Seattle.

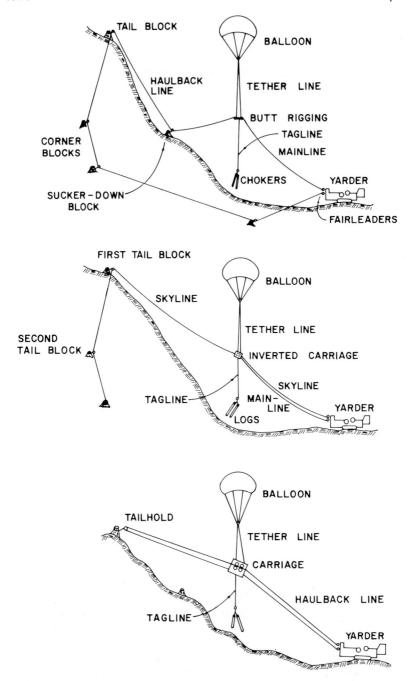

Figure 16-46.—Balloon yarding systems. (Top) Conventional or highlead. (Center) Inverted skyline. (Bottom) Running skyline. (Drawing after Conway 1976.)

HELICOPTER YARDING

Helicopters have instant appeal to the loggers' imagination because they are unincumbered by many of the obstacles that limit use of ground skidding and cable yarding systems. Cost analysis of helicopter yarding of low-value pine-site hardwoods is not favorable, however. This becomes evident from Conway's (1976) tabulation of important characteristics of four helicopters, as follows:

Maximum payload	Operating cost (with crew)	Purchase price
Pounds	*Dollars/hour*	*Dollars*
5,000	347	825,000
8,500	636	2,650,000
19,200	1,504	2,700,000
23,500	2,297	6,100,000

High-value timber has been commercially extracted by helicopter from boggy terrain in the South, however. In 1979, Georgia-Pacific Corp. yarded about 2 million board feet of cypress and black tupelo by helicopter from the Pee Dee River bottom in South Carolina. Daily, about 80,000 board feet of tree-length logs were transported ½-mile by a Boeing Vertol 107. The helicopter had a payload of about 9,000 pounds and averaged 23 trips per hour. It was equipped with 150 feet of cable terminating in an electric hook which enabled automatic log release at the landing. Three chokers were attached to the hook, each carrying up to three logs. Trees averaged about 14 inches in dbh (Bryan 1979).

Most pine-site hardwoods have low value compared to sawlog-size cypress and black tupelo from good hardwood sites. It seems doubtful that helicopter yarding can be seriously considered as a viable option for yarding such low-grade wood.

16-10 CLEARCUT HARVESTING ABOVE-GROUND BIOMASS OF SMALL HARDWOOD TREES

To minimize waste of logging residue, crowns on pine-site hardwoods should be harvested as well as the limb-free stems. Acceptable solutions to the problem have eluded loggers for generations. To date, the most practical system is the whole-tree chipper teamed with a feller-buncher as described in section 16-4 and figure 16-18. In clearcuts made by this method, stems and brush smaller than $3\frac{1}{2}$ or 4 inches in dbh are left standing. Also the product is, in most cases, a low-value whole-tree chip rather than higher-value logs, bolts, and bark-free chips. Alternatives discussed in the following paragraphs are somewhat tentative reflecting what is thought possible, rather than proved state of the art.

SWATHE-CUTTING FELLER-BUNCHER

Feller-bunchers in wide use shear one stem at a time and accumulate several stems before dropping them to the ground. For small stems, however, this is slow (i.e., up to about 100 stems per hour) because the shearing head must approach each tree individually. Needed is a practical machine for pine-site hardwoods and brush that will cut a swathe about 8 feet wide, severing at ground level everything in its path, while holding the severed trees until a suitable load for the grapple skidder is accumulated. For conifers with small crowns, solution to the problem is easier than it is for wide-crowned hardwoods. At least two prototypes have been built to test this idea on conifers—the Prince Albert Pulpwood machine (Stock 1978) and the Hydro-Ax swathcutter (Davidson 1978). Both use thick circular saws to sever all stems at groundline in a swathe several feet wide while the machine travels at 1 mile per hour or faster and accumulates a dozen or more severed trees before dropping them in a bunch. The Hydro-Ax has cut and bunched more than four trees per minute; in dense stands of very small spruce and fir the Prince Albert Pulpwood machine has averaged 25 trees per minute.

The rationale for development of a swathe-cutting feller-buncher is clear (fig. 16-47); its practical execution in a form appropriate for heavy-crowned small hardwoods is difficult, but probably not impossible (Säll 1981ab; Bryan 1980). Assuming that resulting bunches can be ground skidded (or skyline yarded free of the ground) to roadside, they could then be: whole-tree chipped, separated at roadside into components to maximize value, or transported entire to the mill (fig. 16-57) for such dismemberment. Highly mechanized roadsize machines for dismemberment of small hardwoods are not yet available, but something like the Hahn harvester (Larson 1978a), with added provision to sever limbs into chunks, is envisioned (fig. 16-48). The chunks would later be chipped for fuel or fiber at a central location.

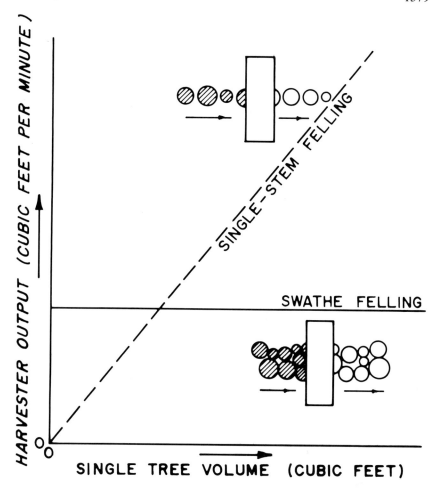

Figure 16-47.—Influence of single tree volume on harvester output when stems are severed singly and by a swathe cutter. (Drawing after Kerruish 1977.)

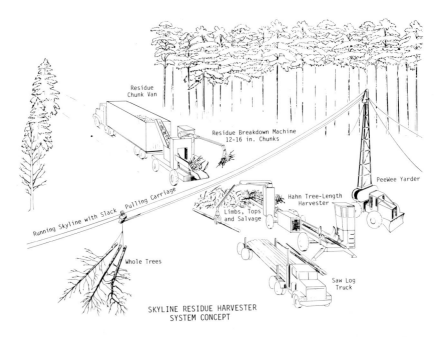

Figure 16-48.—(Top) Hahn harvester shear-delimbing aspen (*Populus tremuloides* Michx.) and sawing stems into logs or bolts. Machine functions are powered by a 160-hp diesel engine. The harvester is roadable and can process about 840 stems per 8-hour shift. (Photo from Hahn Machinery, Inc., Two Harbors, Minn.). (Bottom) Layout combining the Hahn harvester with the PeeWee yarder (fig. 16-37) and in a secondary operation reducing limbs to chunks for highway transport. These chunks would later be chipped for fuel or fiber. (Drawing from a proposal by the Pacific Northwest Forest and Range Experiment Station, U.S. Forest Service.)

As another alternative, a mobile multiple-purpose machine might be designed to swathe-fell small trees of all diameters and then, in a multiple-stem operation, remove and bale the limbs before depositing the limb-free stems in bunches ready for skidding to roadside. Removal of stiff, strong limbs continuously and simultaneously from multiple stems of whole hardwood trees is a very difficult job. The wide-span chain flails that work reasonably well on conifers are not likely to delimb oaks and hickories. Walbridge and Stuart (1978) concluded that it should be possible to bale limbs and tops, and have proposed some prototype machinery (Stuart and Walbridge 1976; Stuart et al.[11]; Stuart et al. 1980; Walbridge and Stuart 1981). Säll (1981ab) proposed mobile equipment to continuously bale whole trees of small diameter.

SWATHE-FELLING MOBILE CHIPPERS

The problem of accumulating pine-site hardwoods, with their wide-spread strong limbs, into compact bunches for skidding or cable yarding is so difficult that other alternatives have been examined. Koch and McKenzie (1976) proposed harvesting logging residue and standing trees with a swathe-cutting mobile chipper—the chipper to be trailed by a vehicle in which to accumulate the chips.

Felling-bar harvester.—Five timber companies with southern operations and Nicholson Manufacturing Company, cooperated with the Southern Forest Experiment Station of the U.S. Forest Service (with substantial financial assistance from the Energy Research and Development Administration) to develop a commercial prototype of such a harvester. After consideration of many designs, a mobile chipper with a ground-level cylindrical felling bar feeding a drum chipper was adopted (fig. 16-49). It was felt that chips from a drum chipper would be easier to handle and have more value than chunks from a hammer-type hogging head.

Performance goals were as follows:
- Operates primarily on terrain that is stone-free, has a slope of 30 percent or less, and supports 9 psi in track pressure. In follow-up designs, a track pressure of 8 psi will be goal; 6 psi would be preferable.
- Harvests 1 acre per hour at 1 mile per hour on land averaging 25 tons (green weight) of logging residue and standing culls per acre.
- Fells and chips standing stems of southern hardwoods and softwoods up to 12 inches in diameter (measured 6 inches above ground level) while moving at 1 mile per hour.
- Mills off the tops of 12-inch-diameter stumps to 6-inch height while travelling at 1 mile per hour (larger stumps at lower speeds).

---

[11]Stuart, W. B., R. G. Oderwald, J. N. Perumpral, and R. Williams 1979. The development of engineering criteria for an in-woods baler. Final Report, Forest Service, U.S. Dep. Agric. Contract 333927-1, March 16, 1979.

- Picks up and feeds into the drum chipper tops, branches, and cull stem sections residual from logging operations.
- Chips felled stems up to 19 inches in diameter if properly oriented to the infeeding hopper and with heavy lateral branches severed or notched to ease crown compaction.

A prototype felling bar equipped with four knives set at 38.5° rake angle on a 16.5-inch cutting circle was tested on Oregon ash (*Fraxinus latifolia* Benth.) and true hickory (*Carya* sp.). It was concluded that the felling bar on the commercial prototype should be arranged to use energy stored in the chipper drum, and should belted to turn at 600 rpm, with capability to slow to 100 rpm when picking up slash rather than felling trees. Peak power demands were brief but substantial and were related to felling-bar rpm, tree diameter, and tree species (fig. 16-50).

Down-thrust on the felling bar varied widely with tree size and species, but forces up to 16,000 pounds were frequently observed. Horizontal thrust (feed force parallel to the ground) was generally minimal and if felling-bar stalls were avoided did not exceed 3,000 pounds.

The felling bar worked reasonably well in that felled trees and high stumps were delivered butt-first toward the chipper. Under the right circumstances tops and branches could be picked up and delivered into the chipper hopper; slow rpm of the felling bar (e.g., 100 rpm) was better than 600 rpm in such pick-up operations.

Chips produced in felling were, for the most part, directed toward the drum chipper; chips were card-like in shape. In the commercial machine, such chips are re-chipped while passing through the drum chipper enroute to the chip carrier. Stumps of felled trees were smoothly severed, not shattered.

A prototype drum chipper, with two knives set at a rake angle of 52.5° on a cutting-circle diameter of 48 inches, and rotated at 544 rpm required an average of 464 horsepower-seconds per cubic foot of hickory chipped; Douglas-fir (*Pseudotsuga menziesii* Mirb.), which is much softer than hickory, averaged 261 horsepower seconds per cubic foot chipped. Power demand of the drum chipper varied with both species and log diameter, but was generally under 500 horsepower (fig. 16-51). The two-knife drum design was inadequate because it was not self-feeding. To improve feeding characteristics of the drum, the commercial prototype was equipped with three knives. Addition of a third knife does not alter specific cutting energy, but increases the horsepower demands graphed in figure 16-51 by about 50 percent.

Harvesting

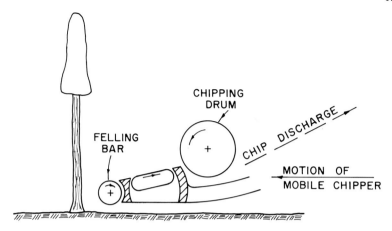

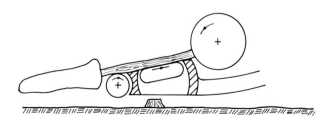

Figure 16-49.—Concept of a mobile machine with a felling bar arranged to feed a drum chipper. (Drawing after Koch and Savage 1980.)

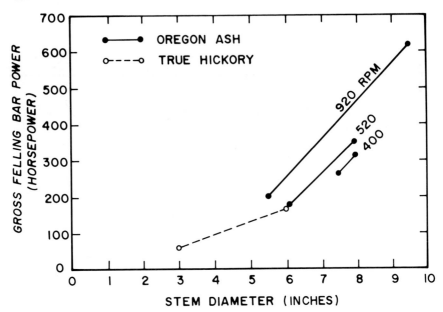

Figure 16-50.—Relationship between stem diameter and power required at 1-mile-per-hour traverse speed to fell single partially dry stems of two hardwood species at three felling-bar rotational speeds. (Data from Koch and Nicholson 1978.)

The commercial prototype was assembled on the hull of an FMC forwarder equipped with extended tracks (fig. 16-52). The machine has a 575-horsepower diesel engine which powers all functions including propulsion. Specifications are as follows:

Gross vehicle weight	70,000 pounds
Approximate ground contact area with 2-inch penetration of tracks	6,740 square inches
Approximate ground pressure	10.7 psi
Drum chipper characteristics	
Cutting circle diameter	48.0 inches
Spout width	47.5 inches
Number of knives	3
Rake angle of knives	52.5°
Drum speed	544 rpm
Nominal feed speed	136 feet per minute
Felling-bar characteristics	
Cutting-circle diameter	16.5 inches
Length	93.5 inches
Number of knives	4
Rake angle of knives	38.5°
Rotational speed	0–600 rpm
Clearance above ground	2 to 7 inches
Diameter of side feed rolls	24 inches
Machine ground speed	Creeping to 3 miles per hour

Harvesting

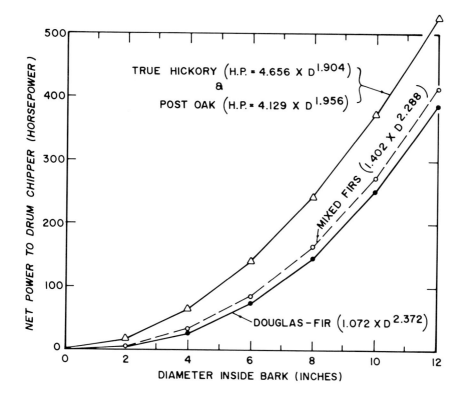

Figure 16-51.—Net power demand (idling power deducted) of two-knife, 48-inch-diameter, 550-RPM drum chipper fed at 95 feet per minute with single green logs of four species in a range of diameters. Rake angle of chipper knives was 52.5°.

In operation, chips from the drum chipper are blown to the rear of the moving machine into one of a pair of self-powered tracked vehicles each carrying a quick-dump chip bin with 10-ton holding capacity (fig. 16-53). Average speed of the mobile chipper should be 1 mile per hour over rock-free terrain of less than 30-percent slope that will support 10 psi ground pressure. At this speed the harvester will cover about 1 acre per hour on land averaging 25 tons (green weight) per acre of logging residues in the form of tops and limbs, standing cull trees, and stumps. About 85 percent of such residue should be recovered as chips and delivered into roadside piles at about $11.85 per green ton including 30 percent pretax profit on equipment investment (1977) of $470,000. When scheduled 7 days a week and 9.5 hours a day, the machine should harvest about 1500 acres per year (Koch and Nicholson 1978).

If this system can be put into successful operation, it will provide mills with wood for fuel and fiber that would otherwise be wasted; the system also has numerous other benefits:

- Changes some of the capital investment for site preparation to a harvesting expense.

- Should improve public reaction to harvesting because it eliminates waste wood and unsightly slash.
- Avoids the smoke of windrow and burn operations.
- Compared with the windrow and burn system, increases plantable area (by perhaps 10 percent)—because not all windrows are completely burnt.
- Increases land productivity, because scalping inherent in pile and burn operations is eliminated (fig. 16-8).
- Hastens replanting by several months because harvesting accomplishes site preparation.
- Wood harvested is forest residual chips (rather than chunks or shreds) which have high potential for fiber products more valuable than energy.
- Because no wood is skidded over the ground, wood delivered via mobile chipper and chip forwarding bin should be relatively free of dirt.

Field tests of this swathe-felling mobile chipper in Alabama and Mississippi during 1981 were promising, and industrial trials of the chipper teamed with two chip forwarders began in 1982. Readers interested in current test data should consult D.L. Sirois, Southern Experiment Station, Auburn, Alabama.

Circular-saw harvester.—An alternative design for a swathe-felling mobile chipper severs stems and feeds a disk chipper with dual ground-level circular saws counter-rotating toward each other (Smith and O'Dair 1980). NFI, Inc., Alexandria, La., cooperated with Georgia-Pacific Corp. to build a brush harvester on this principle (fig. 16-54). All functions of the harvester are driven by a 430-horsepower diesel engine. The tracked machine weights 48,000 pounds when carrying a 6,000-pound chip load and exerts a ground pressure of 8.3 psi. The 16-tooth cutter disks remove a 2¼-inch kerf and cuts a 7.5-foot swathe. Trees and brush 5 inches in dbh and smaller are easily felled and efficiently fed into the chipper. Design ground speed is about 1 mile per hour. The machine as originally built was priced (1978) at about $245,000, and had hourly production of about 6 tons (green basis) of whole-tree chips.[12]

The machine illustrated in figure 16-54 must cease harvesting each time the chip bin is filled, travel to roadside to empty its load onto a roadside chip pile, and then return to where harvesting was discontinued.

To avoid such discontinuous harvesting, the Pallari harvester, which operates on a similar felling and chipping principle (fig. 16-55), discharges chips onto portable chip sacks which are thrown from the harvester when filled. Use of sacks gives the harvester maneuverability around obstacles and enables it to reverse; also it permits formation of a small buffer storage in the forest. Disadvantages include additional original investment in sacks and costs of sack repair; also the sacking system requires at least one extra hand for the harvester and another for unloading sacks at roadside. Readers interested in the system are referred to Hakkila and Kalaja (1980), who have provided operational data on the machine.

---

[12] Koch, P. 1977. Correspondence relative to Study FS SO-3201-2.81, Southern Forest Experiment Station, U.S. Forest Service, Pineville, La.

Harvesting

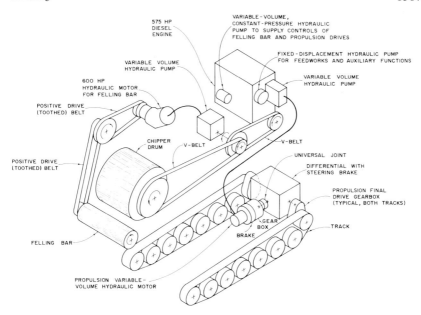

Figure 16-52.—Prototype commercial swathe-felling mobile chipper. (Top) Drive train from 575-hp diesel engine. (Drawing after Koch and Savage 1980.) (Bottom) Swathe-felling and chipping pine-site cull hardwoods; see fig. 17-10 (top) for resulting chips.

Figure 16-53.—Mobile chipper and companion chip forwarders retrieve standing trees and logging slash as chips and deposit them in roadside inventory piles. (Drawing after Koch and Nicholson 1978.) See also figure 28-11.

Harvesting 1589

Figure 16-54.—NFI/GP brush harvester. (Top left) Twin counter-rotating cutting disks that sever stems just above ground level and direct them into feed rolls leading to a drum chipper. (Top right) Front view from cleared swathe. (Lower left) Side view; grillwork on the front directs falling trees forward. The chip bin is directly behind operator's cab and the power unit trails. (Lower right) Chip bin self-dumps at roadside. (Photos from J. O'Dair, NFI.)

LOWER COST MACHINES TO HARVEST WHOLE TREES

Previously in this chapter four methods of harveting whole trees, ie., the above-ground biomass, have been described as follows:

Method	Comment
Large whole-tree chipper at roadside teamed with feller-bunchers and grapple skidders (fig. 16-18)	Not appropriate for stands with very small stems.
Skyline cable logging to processing center at roadside landing (figs. 16-42, 45, and 48)	Probably not appropriate for very small stems.
Swathe feller-buncher.	Hardwoods with spreading crowns are hard to bunch.
Swathe-felling mobile chipper, or chunker (figs. 16-53, 54, and 55)	Still in early stage of development; large trees must be circumvented.

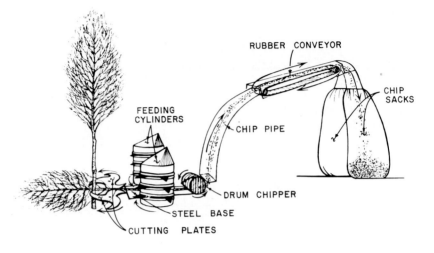

Figure 16-55.—Pallari swathe harvester. (Top) Principle of operation. (Bottom) Harvester in action, temporary chip storage in sacks, sack forwarder, and roadside discharge of sacks into a chip trailer. (Drawings from Hakkila and Kalaja 1980.)

All of these systems require substantial investments of $250,000 to $750,000.

If, as seems likely, a strong market for whole-tree fuel chips can be developed, alternatives are available. Some do not require large capital investment and are appropriate for seasonal activity. For example, using light farm machinery (fig. 16-56) a man can harvest during 1 working day about 4.3 tons of green whole-tree chips equal in fuel value to 1 ton (286 gallons) of oil. (A gallon of number 2 heating oil weighs 7.0 pounds and has a heat content—135,000 Btu—equal to about 30 pounds of green whole-tree chips of pine-site hardwoods.) At $1.00 per gallon, 286 gallons of heating oil (1 ton) will cost $286.

The poorer burning efficiency of wood, the high cost of wood-fueled burners compared to oil burners, and the inconvenience of handling wood (see chapter 26) make it doubtful that a seasonal worker could sell his day's output (4.3 tons of green chips) for $286, but he might very well get $143 or $33 per ton delivered to the user's fuel pile.

The low-investment harvesting system shown in figure 16-56 would be ecologically acceptable to many southern farmers and woodlot owners. Each worker in the two-man team might earn $143 per day, less his share of stumpage, daily cost of the chainsaw and simple harvester, and cost of trucking 8.6 tons to the customer's fuel pile. Thus, when fuel oil costs $1.00 per gallon, a farm woods worker cutting fuel chips might pocket $100 per day with very low capital investment.

Intermediate between the low-investment concept of figure 16-56 and the high-investment system shown in figure 16-18 are several arrangements whereby tractor-powered chippers operate along strip roads. Trees are dragged from distances of about 150 feet to the chippers by light winches, or small tractors. If strip roads can be spaced closely (e.g., 110 feet apart), long-reach sliding-boom cranes (fig. 16-60) can grip the trees after felling, draw them to the strip road, and feed them into moderately powered chippers. Chips can be blown into pallets or quick-dump bins for forwarding to highway trucks. These Scandinavian systems are further described by Nilsson (1978), Kalaja (1978), Hakkila (1978), and Hakkila et al. (1979).

If not chipped at the stump (fig. 16-56) or at strip roads or truck load landing (fig. 16-18), small trees could be forwarded to roadside, transferred to trucks, and transported entire to a centralized chipping center (fig. 16-57). Transporting undelimbed trees on public roads and compacting full-weight truckloads of such whole trees present problems. Danish and Finnish engineers increase load weights by bucking trees to truck bed length, alternating direction of butts and tops when loading, and compressing the load. The load space is enlarged during loading by tilting the bunk posts outward (fig. 16-57 bottom). Once the load is complete, the bunk posts are raised to vertical by a steel wire tightened by a hydraulic motor. The upper ends of the posts, which are jointed, are hinged 90° inward by the wire, thus compressing the load from above as well as from the sides. To avoid cost of removing extending branches, truck sides are sheathed with steel netting (fig. 16-57 bottom) (Hakkila et al. 1979).

Figure 16-56.—(Top) Two-man team harvesting whole-tree chips for fuel using a chain-saw equipped with a felling frame (to avoid stoop labor), and a light farm tractor powering a brush chipper. Daily output is 8 to 9 tons (green basis) for the two-man team. (Drawing after Hakkila et al. 1979.). (Bottom left) Felling a small tree with a frame-mounted chainsaw. (Bottom right) Manual bunching of trees. (Photos from files of P. Hakkila).

Harvesting

Figure 16-57.—Harvest and transport of small whole trees to centralized chipping plant. (Top) Trees chain-saw felled and bunched as in figure 16-56, grapple loaded on forwarder, and reloaded at roadside on truck with bunk posts rigged to compress the load. (Drawing after Hakkila et al. 1979.) (Bottom) Whole-tree portions of speckled alder (*Alnus incana* (L.) Moench) loaded onto a trailer rigged to compress the load and contain it within legal space. (Photo from Yhteisostot Oy Jyki Tehtaat, Finland.)

16-11 LIGHT MACHINERY TO SKID AND BUNCH LOGS FROM THINNED STANDS

In partial cuts, trees can be bunched and dragged to strip roads for chipping there or, in the case of roundwood, for forwarding to landings for later transfer to highway transport trucks. Biltonen et al. (1976) has described the economics of such thinning by chainsaw or shear felling (fig. 16-14) followed by skidding with heavy wheeled skidders (fig. 16-13) or grapple skidder (fig. 16-16). The wheeled skidder can reach about 150 feet off a skid trail to retrieve logs, while the grapple skidder must go to the bunch.

Other less expensive machines more suitable for small trees are being used in the Scandinavian countries to thin stands without damaging residual trees. Some of the equipment used may have a potential for pine-site hardwoods. Radio-controlled dragging winches are inexpensive, portable, and can drag logs 150 feet and bunch them for forwarding; winches attached to farm tractors are more expensive but easier to relocate (fig. 16-58).

A disadvantage of winch ground skidding is the need for time consuming changes of winch position or corner blocks (Nilsson 1978). A tractor-mounted grapple gives greater freedom of movement. For thinning, a dragging tractor must be narrow and small; it is therefore difficult to mount a normal cab for the driver. One solution is a machine guided by a driver who walks alongside (fig. 16-59).

Another system undergoing trials in Scandinavia uses long-reach, slide-boom cranes (fig. 16-60 top). They have a reach of about 55 feet, so the distance between strip roads for forwarders can be about 110 feet. Because the slide-boom crane also loads the forwarder, need for dragging by winch or tractor is eliminated and the harvesting system is simplified. Readers needing specifications and prices for such cranes are referred to Åkerman (1976) and Brunberg (1977).

An alternative thinning system for small woodland ownerships on hilly terrain is a light-weight cable yarder attached to a crawler tractor (fig. 16-60 bottom). With such equipment Biller and Peters (1981) thinned second-growth, 70-year-old even-aged stands of Appalachian hardwoods having basal areas of 125 to 140 ft^2 per acre. They made sanitation thinning cuts removing 0.9 Mbf of sawlogs per acre (International ¼-inch log scale) and 1.2 cords of pulpwood per acre. At other locations they thinned to a 9-inch-diameter limit, removing 13 Mbf of sawlogs of acre. Ground slope varied from 40 to 70 percent. All yarding was done uphill with ⅜-inch-diameter cable; the winch carried 1,500 feet of cable and could pay it out as well as pull it in under power. To yard, the tractor and yarder were oriented perpendicular to the logging road, the tractor blade lowered into the ground with down pressure, and the yarder boom positioned in line with the tractor, making the equipment stable. When a stem reached roadside, it was swung onto the road. Heavy stems required unhooking the tongs from log end and rehooking at stem center of gravity before swinging to roadside. Stems were bucked to length (10, 12, 14, or 16 feet) at roadside and stacked. Average log length when thinning to 9-inch-diameter limit was 12.4 feet. Minimum top diameter was 5 inches for the sanitation thinning and 7 inches for the diameter-limit thinning.

Figure 16-58.—Light dragging winches made in Sweden for thinning. (Top) Radio-controlled winch dragging logs about 50 meters to strip road. (Bottom) Dragging winch attached to tractor permits strip roads to be spaced about 100 meters apart. (Drawings after Nilsson 1978.)

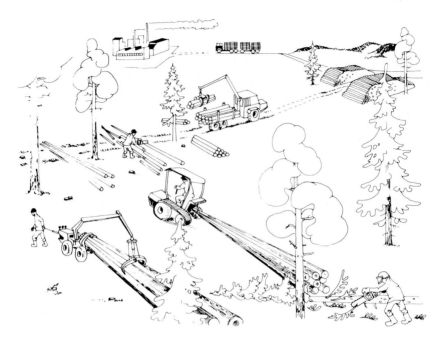

Figure 16-59.—Small dragging tractors used in Norway, Finland, and Sweden. (Drawing after Nilsson 1978.)

Biller and Peters used a 5-man crew, i.e., yarder operator, feller, tongsetter, and two chasers at the landing. Their logging cycle begins with a chaser carrying the tongs halfway down the hill where he hands them to the tongsetter and then returns to the landing. The tongsetter carries the tongs down the hill, hooks the stem at the top end, signals the yarder operator by radio to yard in the stem, and then follows the stem halfway up the hill to prepare for the next cycle. Tree felling is concurrent with yarding.

In these trials by Biller and Peters, yarding of single stems predominated. The largest volume yarded in a turn was 525 bd ft (130 cu ft) and maximum line pull was 8,000 lb. Tongs had maximum opening of 18 in. Butt diameter of the largest log was 25 in. Usual yarding distances up the slope was 134 to 170 ft., with 347 ft maximum.

Average cycle times, excluding delays, were 12.7 min for sawtimber cuts and 7.5 min for thinning cuts. When thinning to a 9-in-diameter limit under the test conditions, an experienced crew should be able to yard, buck, and stack about 40 stems per 8-hour shift. Average moving time from one setting to another, including rigging time, was only 5.4 min because no guylines were required. About 1 gal of diesel is expended per Mbf of log production (International ¼-inch log scale) Total daily yarding costs, including cost of ownership, operation, and labor, was $277. When thinning to a 9-inch-diameter limit, total cost (not including profit on investment) to fell, yard, buck, and stack logs at roadside, was $34.26 per Mbf of sawlogs, International ¼-inch log scale (Biller and Peters 1981).

Figure 16-60.—Two methods of thinning. (Top) Long-reach, sliding-boom cranes used for thinning eliminate need for dragging winches or small tractors. (Drawing after Nilsson 1978.) (Bottom) Cable yarder with boom over tractor blade for stability while yarding thinnings uphill from maximum distance of about 350 ft.—average about 150 ft. (Photo from Biller and Peters 1981.)

16-12 RESIDUE HARVESTING

It seems likely that integrated systems for clearcutting and retrieval of whole or complete trees will be practical on some southern ownerships. Many more acres, however, will be conventionally harvested for sawlogs and pulpwood only, leaving large amounts of logging residue, e.g., 20 to 30 tons per acre (green basis). These residues consist of standing culls of all species, standing trees of less than merchantable diameter, trees broken during road building or skidding, high stumps, unutilized stem sections, and tops and limbs of trees from which merchantable stems have been removed (Chappell and Beltz 1973, Craft 1978).

Some southern companies are retrieving part of this material for fuel, using whole-tree chipping systems (fig. 16-18). Harris (1977) reported that by this method Boise Southern Company is producing 150 tons of green fuel chips daily from logging residues. The swathe-felling mobile chipper (fig. 16-53) is another method for recovering 100 to 150 tons of similar fuel per day.

The large crowns of pine-site hardwoods present a substantial obstacle to residue harvesting. They can be arranged for convenient skidding if limbs are first lopped to permit manual bunching. A study by Gabriel et al. (1974) in New York State indicated that time to lop such tops—but not to bunch the limbs for skidding—varied with species as follows (trees averaged about 15 inches in dbh):

Species	Time to lop top
	Minutes/tree
Ash, white	1.2
Maple, red	1.0
Maple, sugar	1.5

Schick and Maxey (1978) found that an average of 9.3 minutes per tree was required to manually lop the tops of old-growth oak, yellow-poplar, and maple in West Virginia, whereas only 5.5 minutes was required to fell, limb, and buck the merchantable part of each tree. All operations were performed with a chainsaw. Time to manually bunch the limbs for skidding would be additional to the 9.3 minutes per tree.

To lessen the hard work of lopping and bunching tops from hardwood sawtimber, Mattson et al. (1978) proposed a mechanical system incorporating a shear on a maneuverable arm (fig. 16-19). During trials of the prototype equipment, 10.2 tons (green basis) were lopped and bunched per hour. Tops had average green weight of about 0.8 ton; therefore this was the average skidder load. Such small skidder loads increased costs above normal costs per ton for wood moved by grapple skidders. They concluded that costs of lopping and bunching the tops with the topwood processor, skidding the bunches to roadside, and chipping them there should total about $6.32 per green ton (1978 basis).

Stimulated by a computer analysis of the residue harvesting problem (Schmidt et al. 1970), Walbridge and Stuart (1978, 1981), and Stuart et al. (1980) proposed in-woods baling of both pine and hardwood logging residues (fig. 16-61). Whole trees would be skidded to the landing, and left alongside the loader-

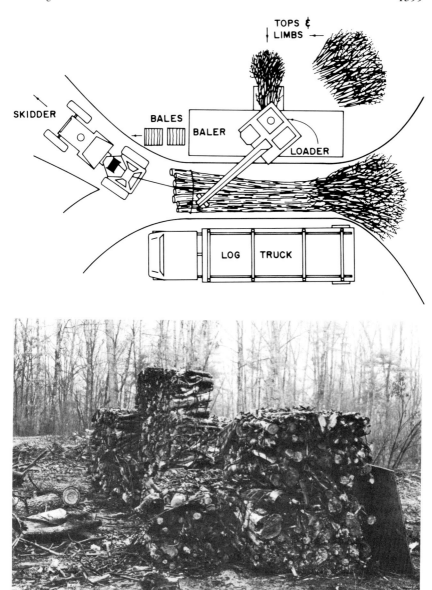

Figure 16-61.—Concept for baling tops and limbs at the landing. (Top) Layout of landing where stems are delimbed, topped, and loaded on trucks; limbs are loader-fed, sheared to short lengths, and baled. (Bottom) Cubical bales are strapped with 10-gauge annealed wire, measure about 3 feet on a side, and weight 1,500 to 1,800 pounds when green. (Drawing after Walbridge and Stuart 1978; photo from T. A. Walbridge, Jr.) See also figure 28-8.

baler machine where they would be delimbed and topped. The loader would not only load merchantable stems onto logging trucks, but would feed limbs and tops into a baler located beneath the loader.

Bales would measure about 3 feet square in end section and 3 or 4 feet long; with bulk density of 40 to 45 pounds per cubic foot, bales should weigh 1,500 to 1,800 pounds. The baler is estimated to cost $30,000 to $50,000 (1979 basis).

With increasing oil prices, it seems almost certain that light-weight mobile woods equipment will be developed to cut and bundle small packages of hardwood for stoves and fireplaces. Such equipment should be able to process limbwood, topwood, and culls in a range of diameters down to perhaps 3 inches. Residue volumes harvested for this purpose may be substantial when heating oil is priced at $1 per gallon or higher.

16-13 TRANSPORT, LOADING, AND OFF-LOADING

Costs of harvesting pine-site hardwoods include transport from roadside landing to mill, any intermediate handling at woodyards, and allocated costs of necessary woods roads. Discussion of these subjects is beyond the scope of this text. Data presented here are limited to piece weights and bulk densities of some major products, and tabulation of a few sources of pertinent information.

Variations in weights and volumes of sawlogs carried by trucks can be substantial. In a study of seven Appalachian mills (Timson 1974), a load range of 10,000 pounds, or 1,000 board feet, was commonplace for the same truck, driver, and cutting site. Differences in log size, shape, weight, and species caused a major share of this variation. In Timson's study, the loads of logs delivered to the seven mills by tandem-axle truck ranged from 17,710 to 37,520 pounds. Load volumes ranged from 1,199 to 3,759 board feet by the Doyle log rule and from 1,655 to 4,030 board feet by the International ¼-inch log rule, and from 342 to 689 cubic feet.

Average chip hauling distance to pulpmills in the South is about 50 miles; in North Carolina 1976 cost for chip transport by rail was $0.0222 per ton-mile, by truck about $0.060 (Hassan 1977). West of the Mississippi River rail freight for chips was about $0.041 per ton-mile in 1979, while truck transport in 1979 cost about $0.062 per ton-mile. Shortwood hauls to woodyards average about 20 miles and to mills about 40; average transport costs per cord in 1979 were $6.53 and $9.86 respectively, with an average of $8.74. Long logs or tree-length logs may be hauled up to 100 miles with transport cost in 1979 of about $0.065 per ton-mile.

WEIGHT OF GREEN WOOD IN SOLID FORM

Tables 7-2 and 7-2A show weights per cubic foot of freshly-cut green bark-free wood. Species descriptions in section 16-1 of sweetgum, yellow-poplar, and chestnut, scarlet, southern red, and white oaks contain data on weights per cubic foot of larger trees. Section 16-1 also gives weights of complete trees including above- and below-ground parts, and weights of above-ground whole trees and components. Tables 27-70 through 27-80 show log weights according to species, length, and diameter measured inside bark at the small end. Table 27-97 lists all of the weight tables and prediction equations in this text for stems and whole trees, by species.

Timson and Church (1972) reported that over 90 percent of all Appalachian hardwood logs sampled weighed less than 1 ton. In dense species like oak, from 50 to 60 percent of the logs weighed less than 1,000 pounds; for low-density species like yellow-poplar, about two-thirds of the logs weighed less than 1,000 pounds. Range in log weights for some of the species were as follows (sampled in West Virginia, southwestern Virginia, and western Maryland):

Species	Number of logs in sample	Weight range of logs
		Pounds
Ash, white	164	160-2,120
Hickory, *Carya* sp.	406	400-5,750
Maple, red	90	290-3,040
Oak, chestnut	313	290-3,390
Oak, northern red	1,366	270-6,440
Oak, white	209	320-4,170
Tupelo, black	74	490-4,190
Yellow-poplar	477	270-2,860

WEIGHT OF STACKED ROUNDWOOD

A cord of freshly cut hardwood pulpwood weighs about 5,800 pounds including bark, but varies considerably with species and bolt diameter and straightness (see discussion page 3268).

BULK DENSITY OF BALED CHIPS, BRANCHES, CHUNKS, AND FIREWOOD

Chips.—Hassan and Reeves (1978) reported that southern pine chips baled with a pressure of 300 psi had bulk density of 52.4 pounds per cubic foot; hardwood chips subjected to similar baling pressure might have a bulk density of 55 to 60 pounds per cubic foot. If baled at lower pressures, hardwood chips have bulk density of about 30 pounds per cubic foot (table 16-44). Uncompacted dry pulp chips have bulk density of 8 to 14 pounds per cubic foot, depending on species (table 16-44A). Readers interested in variables affecting chip bulk density

TABLE 16-44—*Bulk densities of green and air-dry hardwood residues when loose, settled by vibration, and compacted at five pressures* (Vidrine and Woodson 1982)[1,2]

Product and degree of packing	Sweetgum		Hickory, white and southern red oaks	
	Green	Airdry	Green	Airdry
	------------------Lb/cu ft------------------			
	-------(Percent moisture content)-------			
WHOLE-TREE CHIPS[3]				
Chipper-blown into van	29.6	—	30.8	—
	(96)		(58)	
Settled by vibration	26.2	—	26.2	—
Baled at pressures of (psi)				
0.0	22.9	—	23.3	—
12.5	27.1	—	27.7	—
25.0	28.8	—	29.0	—
50.0	30.8	—	30.7	—
75.0	30.8	—	31.2	—
100.0	30.8	—	31.2	—
BARK-FREE CHIPS[4]				
Air-blown into van	28.5	—	29.1	—
	(107)		(74)	
First loaded into hopper car	24.4	—	24.5	—
Baled at pressures of (psi)				
0.0	21.8	—	22.6	—
12.5	25.8	—	26.1	—
25.0	27.1	—	27.6	—
50.0	28.8	—	29.6	—
75.0	28.8	—	29.6	—
100.0	28.8	—	29.6	—
SAWDUST[5]				
Loose	18.9	10.8	19.7	16.2
	(108)	(11)	(84)	(44)
Settled by vibration	21.2	14.5	22.0	20.2
			(84)	
Blown into van or car	23.8	12.7	21.1	17.4
			(75)	
Baled at pressures of (psi)				
0.0	18.5	10.7	19.8	15.9
	(83)	(11)	(84)	(44)
12.5	28.2	15.1	30.2	20.5
25.0	29.3	15.8	32.2	21.9
50.0	29.3	15.9	32.6	22.4
75.0	29.3	15.9	32.6	22.4
100.0	29.3	15.9	32.6	22.4

[1] Residues were collected in North Louisiana.
[2] The first value is bulk density, lb/cu ft; the second value (in parentheses) is moisture content, percent of ovendry weight.
[3] From 75-in, three-knife disk chipper (mobile) with built-in trash separator.
[4] Unscreened chips from debarked sweetgum and oak logs.

TABLE 16-44—*Bulk densities of green and air-dry hardwood residues when loose, settled by vibration, and compacted at five pressures* (Vidrine and Woodson 1982)[1,2]
—Continued

Product and degree of packing	Sweetgum		Hickory, white and southern red oaks	
	Green	Airdry	Green	Airdry
	------------------Lb/cu ft------------------			
	-------(Percent moisture content)-------			
WHOLE-TREE CHIPS[3]				
PLANER SHAVINGS[6]				
Loose	12.6	5.6	12.4	6.1
	(93)	(11)	(54)	(12)
Blown into van or car	17.0	7.1	16.4	8.5
Settled by vibration	15.6	6.2	15.5	6.8
Baled at pressures of (psi)				
0.0	1.3	4.6	1.4	6.3
	(30)	(11)	(42)	(12)
12.5	5.7	8.4	5.8	12.7
25.0	6.4	9.4	6.5	13.5
50.0	7.0	11.1	7.2	14.4
75.0	8.1	12.6	7.5	15.4
100.0	8.6	13.3	8.0	16.6
FLAKES[7]				
Loose	14.3	5.8	7.7	5.7
	(119)	(9)	(76)	(8)
Blown into van or car	6.2	2.0	5.3	3.4
Settled by transport	18.4	—	10.5	—
Settled by vibration	10.8	4.2	7.7	2.8
Baled with pressures of (psi)				
0.0	15.3	6.4	10.9	4.6
12.5	21.9	10.4	14.5	6.0
25.0	22.4	10.4	14.5	6.0
50.0	22.4	10.4	14.5	6.0
75.0	22.4	10.4	14.5	6.0
100.0	22.4	10.4	14.5	6.0

[5]The oak sawdust was from a 50-in-diameter, inserted-tooth circular headsaw; the sweetgum sawdust was from a pneumatic conveyor that contained all sawdust from circle saws and bandsaws in a hardwood mill.

[6]Green shavings obtained from a pallet mill; dry shavings obtained from a manufacturer of kiln-dried dimension lumber.

[7]Produced from green wood on a shaping-lathe headrig (see figs. 18-98 top and 18-104C); flakes measured 0.02 in thick and 3 in long.

TABLE 16-44A—*Bulk density of whole-tree pulp chips and fibers of 14 pine-site hardwoods* (Woodson 1976)[1,2,3]

Species	Bulk density, gravity loaded	
	Pulp chips	Fibers
	------Pounds, ovendry/cubic foot, green------	
Ash, white	10.81	1.88
Elm, winged	12.22	2.13
Hackberry	10.55	1.84
Hickory, sp.	13.31	2.51
Maple, red	11.11	1.53
Oak, blackjack	13.24	2.11
Oak, post	14.10	2.28
Oak, southern red	12.24	1.39
Oak, water	12.18	1.92
Oak, white	13.55	1.78
Sweetbay	9.09	1.56
Sweetgum	9.52	1.48
Tupelo, black	10.71	1.29
Yellow-poplar	7.61	1.55

[1] Pulp chips were cut ¾-inch long on a disk chipper from three trees 6 inches in dbh of each species sampled in central Louisiana.
[2] The fibers were refined from green chips in a Bauer 418 pressurized disk mill at a steam pressure of 82 to 95 psi with retention time of 5 minutes and plate clearance of 0.050 inch.
[3] Bulk densities based on volume of green chips and fibers and ovendry weights; volume measurement was made by gravity-filling a 1-cu ft box.

will find useful the work by Edburg et al. (1973) who studied the influence of softwood chip dimensions on bulk density.

Branches.—Under extreme pressure, green hardwood residues comprised of tops and limbs can be bound into bales with bulk density of 55 to 65 pounds per cubic foot (fig. 16-61). With lower pressures, achievable bulk density is 20 to 50 pounds per cubic foot (table 16-45).

Logging-residue chunks.—Logging residues can be transported from forest to mill in chunk form to avoid expending energy to chip in the woods. Such chunks averaging 6 inches in length and up to 4 inches in diameter have bulk density of 20 to 37 pounds per cubic foot (table 16-45).

Firewood.—Steel-strapped firewood has a bulk density of 28 to 58 pounds per cubic foot depending on species, size of wood, and moisture content (table 16-45).

TABLE 16-45—*Bulk densities of green and air-dry baled branchwood, logging residue chunks, and steel strapped firewood* (Vidrine and Woodson 1982)[1,2]

Product and degree of packing	Sweetgum		Hickory, white and southern red oaks	
	Green	Airdry	Green	Airdry
	------------------lb/cu ft------------------			
	-------(Percent moisture content)-------			
Branches[3] baled at pressures of				
0.0 psi	35.6	21.9	41.9	33.9
	(66)	(22)	(67)	(18)
2.0	37.4	22.5	43.2	34.9
8.0	37.9	23.2	44.6	35.5
17.0	39.2	23.9	46.0	36.7
25.0	39.2	24.0	49.4	37.4
73.0	39.2	24.1	49.4	37.6
0.0 (Load removed)	37.3	23.1	44.0	35.9
Logging residue chunks[4]				
Loose	33.3	19.5	36.7	24.0
	(135)	(27)	(50)	(18)
Settled by vibration	5	5	5	5
Steel-strapped firewood				
Small roundwood[6]	50.0	31.4	56.7	45.9
	(117)	(38)	(67)	(26)
Split wood[7]	46.6	28.2	58.4	45.7
	(104)	(23)	(70)	(26)

[1]Hardwood residues collected in north Louisiana.
[2]The first value is bulk density, lb/cu ft; the second value (in parentheses) is moisture content, percent of ovendry weight.
[3]Before packing, 2-foot-long branchwood was piled about 20 inches high under a hydraulic press.
[4]Wood chunks about 6 inches long and up to 4 inches in diameter.
[5]No measurable change in bulk density.
[6]Four to 7 inches in diameter and 2 feet long.
[7]Split from wood 7 inches in diameter and larger; 2 feet long.

BULK DENSITY OF OTHER RESIDUES

For bulk density of hardwood **bark residues** see table 13-42 and related discussion. **Foliage** bulk density is tabulated in section 15-2.

Vidrine and Woodson (1982) found that bulk density of whole-tree chips, bark-free chips, sawdust, planer shavings, and flakes diminished in the order named, that airdry residues had lower bulk density than green residues, and that some densification of residues resulted by compaction under a platen exerting 100 psi pressure (table 16-44). Hardwood residues for their study were collected in north Louisiana.

Bois (1968) measured the bulk density of red maple and red oak sp. sawdust from southern Appalachian logs cut on a circular headrig fitted with inserted teeth cutting a 9/32-in-wide kerf (table 16-46). He found that bulk density of such sawdust loaded in a 2.43-cu ft steel drum increased linearly with moisture content. Sawdust packed by tamping it with a mall while filling the drum had about 50 percent greater bulk density than loose sawdust; if the drum was shaken after conveyor filling to simulate light compaction in a top-loaded van or car, bulk density was intermediate between that of loose sawdust and packed.

Chow et al. (1973) found that red oak sawdust at 5-percent moisture content, collected in Illinois, had bulk density of 11.7 lb/cu ft; hammermilling the sawdust through a ½-inch screen did not alter bulk density, but disk refining reduced it to 8.4 lb/cu ft.

White oak chips made from cooperage residue in Illinois were found by Chow et al. (1973) to have bulk densities as follows, at 6 percent moisture content:

Condition of chips	Bulk density
	Lb/cu ft
Unrefined	15.1
Hammermilled through ½-inch screen	20.0
Disk-refined	2.8

TABLE 16-46—*Bulk density of red maple and red oak sp. sawdust related to moisture content and degree of compaction* (Bois 1968)[1]

Moisture content ovendry-weight basis (percent)	Red oak sp.			Red maple	
	Loose[2]	Shaken[3]	Packed[4]	Loose[2]	Shaken[3]
	------------------------Lb/cu ft ------------------------				
0	11.0	13.9	16.8	8.9	12.2
5	11.5	14.6	17.3	9.3	12.8
10	12.1	15.3	17.7	9.8	13.4
15	12.6	16.0	18.3	10.2	14.0
20	13.2	16.7	18.9	10.7	14.6
25	13.7	17.4	19.5	11.1	15.2
30	14.3	18.1	20.0	11.6	15.9
50	16.5	20.8	22.8	13.3	18.3
75	19.2	24.3	26.2	15.6	21.3
100	22.0	27.8	31.0	17.8	24.4
125	24.7	31.3	36.0	20.0	27.4
140	26.4	33.3	40.0	21.4	29.3

[1]The sawdust was cut in a southern Appalachian sawmill equipped with an inserted-tooth circular headsaw with kerf of about 9/32-in.
[2]Drum with 2.43-cu ft capacity filled by allowing sawdust to fall into it.
[3]After filling, the drum was shaken and then dropped a short distance.
[4]To simulate packing from high-velocity blowing, the sawdust was tamped at regular intervals with a mall.

Woodson (1976) found that bulk density of dry fibers disk-refined from green hardwood chips varied with species from about 1.3 to 2.5 pounds per cubic foot (table 16-44A).

REFERENCES PERTINENT TO TRANSPORT

Following are some references that should be useful to readers seeking least-cost solutions to the problem of transporting wood from southern forests to market.

Topic and subject	Reference
Logging roads	
Proceedings of workshop on forest roads	West Virginia University (1970)
Spacing of roads and landings to minimize cost	Peters (1978)
Areas in skidroads, truck roads, and landings	Kochenderfer (1977)
Road layout to minimize erosion	Witter (1975)
Wood logging mats—An alternative to gravel	Boyer (1977)
Logging roads and the Federal Income tax	Siegel (1971)
Effect of wood form on transport cost	
Should wood be brought from forest to pulpmill in the form of whole-tree chips, bark-free chips, shortwood, or longwood?	Smith (1978b); Tufts (1977b)
Loading roundwood on trucks, and unloading	
Flexible system of pulpwood loaders and knuckleboom loaders to transfer shortwood, log lengths, and tree lengths in pulpwood yards	Jarck (1977)
Front-end log loaders—requirements and factors affecting their production	Phillips (1970); Phillips and Hartman (1970)
Review of use and performance of all kinds of loaders	Forestry and British Timber (1978)
Securing systems for truck loads of tree lengths	Lavoie (1981)
Costs to truck roundwood	
Long-distance pulpwood hauling; continued deterioration of rail services makes longer road hauls more economic	Fixmer (1976)
Factors that determine log hauling costs	Martin (1971)
How mountains influence trucking speeds and costs	Wysor and Goho (1969)
Causes of log-truck delays	Baumgras (1978)
Effect of load limits (in Louisiana) on cost of wood transport by truck	Larson (1978b)
Chip transport	
Cost analysis of rail and truck transport of chips to a 50-megawatt wood-fired power plant (3,150 tons/day)	Adler et al. (1978)
Transport and handling in the pulp and paper industry	Haas and Kalish (1975); Kalish and Haas (1977)
Transport outside the pulpmill	Oswald (1979)
Transport inside the pulpmill	Strucka (1979)
Rail transport of pulpwood chips	White (1970)
Blue Ox railcar to transport and automatically side-discharge longwood, shortwood, and chips	Altman (1976)
Chip unloading with rotary railcar dumps	Anonymous (1976b)
Union Pacific open-top railcar	Hensel (1970)
Transporting chips by hydraulic pipeline	Gardner (1977)

Topic and subject	Reference
Marine transport	
Trends in marine transport of forest products	Herbert (1977)
Ocean transport of wood chips	Hanaya (1975)
High-volume chip handling in ocean vessel loading	McClure and Sargent (1975)
Vehicular roll-on roll-off and lift-unit-frame systems in ocean transport	Olszowski (1977)
Chip compaction................................	Hassan (1977b, 1978); Oswald (1979)

16-14 ENERGY EXPENDED DURING HARVESTING

A cord of green pine-site hardwood weighs about 3 tons, of which about 2,570 pounds is water; its wood content of about 3,430 pounds has a heat content of about 26.8 million Btu. Of this total heat content, what percentage is expended during harvest,

The American Pulpwood Association (1975), after surveying operations in four national regions and analyzing 11 types of harvesting systems, concluded that in 1974 an average of 5.16 gallons of fuel—either gasoline or diesel—was required to harvest 1 cord of pulpwood and move it from stump to a concentration yard or mill by truck. Since diesel fuel contains about 135,000 Btu per gallon, energy expenditure per cord harvested averages about 0.70 million Btu/cord.

By totalling the energy invested in manufacture of a feller-buncher, wheeled skidder, loader, and truck and then allocating these energies over their useful lives, expressed in cords of production, it has been computed that an additional .05 million Btu of machinery are consumed during harvest of 1 cord (Tillman 1978, p. 91).

Energy expended per cord is therefore about 0.75 million Btu per cord or about 2.8 percent of the energy contained in the cord of wood.

Fuel consumption during harvesting operations from stump to roadside varied substantially by system. The bobtail-truck shortwood system with big-stick loader used 1.17 gallons of fuel per cord; the whole-tree chipper system consumed 2.97 gallons per cord, as follows (American Pulpwood Association 1975):

System	Fuel consumption
	Gallons per cord
• Manual shortwood system—fell, limb, top, and buck with chainsaw and load manually on board bobtail truck at the stump area.	0.41
• "Big-stick" loader—fell, limb, top, and buck with chainsaw and load on board bobtail truck at the stump area with "big-stick" loader winch.	1.17
• Prehaul shortwood—fell, limb, top, and buck with chainsaw, prehaul and load on board truck at the landing with hydraulic loader.	1.72
• Chainsaw and skid longwood, haul longwood—fell, limb, and top with chainsaw, skid and load long lengths with hydraulic loader at the landing.	1.83
• Buck at landing—fell, limb, and top with chainsaw, skid tree length to landing, buck with chainsaw and load on board truck with hydraulic loader at the landing.	2.22
• Shear and skid longwood, haul shortwood—shear, fell, limb, and top with chainsaw, skid long lengths, buck at landing with chainsaw and load shortwood on truck with hydraulic loader.	2.39
• Mechanical slasher—fell with chainsaw, skid long lengths, slash mechanically and load on board truck with hydraulic loader at the landing.	2.41
• Feller-buncher, 8-foot—feller-buncher, limb, top and buck with chainsaw to 8-foot lengths, skid or prehaul and load on board truck with hydraulic loader at the landing.	2.45
• Feller-buncher with mechanical slasher—feller-buncher, chainsaw limb and top, skid long lengths, slash mechanically, and load on board truck with hydraulic loader at the landing.	2.64
• Feller-buncher, buck at landing—feller-buncher, limb and top with chainsaw, skid long lengths, buck into shortwood at landing with chainsaw and load on board truck with hydraulic loader.	2.88
• Whole-tree chipping—feller-buncher, skid and chip whole trees.	2.97

For pine-site hardwoods, energy consumption per cord would be somewhat higher than that shown in the foregoing tabulation, because machines and men must work longer to harvest wood of small diameter from short trees in sparse stands. The values tabulated are industry averages based largely on coniferous stands, although many operations sampled cut hardwoods.

Data useful in computing energy consumed per cord harvested include engine fuel consumption rates for logging machinery (table 16-47), average productivity in cords per hour per function (table 16-48), average haul distances (table 16-49), and fuel consumption rate per function expressed in gallons per cord produced (table 16-50).

In addition to fuel used by harvesting and transport machines, about 0.72 gallons of fuel per cord produced is consumed by auxiliary and service equipment such as supervisors' cars, crew transport and maintenance vehicles, maintenance and repair equipment, and equipment to construct roads and landings.

Major advances in lowering the cost of harvesting low-grade hardwoods are likely to occur through development of energy-effective methods to reduce hardwood forest residues to chip or chunk form for easy transport. Readers needing an introduction to specific energy requirements during such comminution are referred to Jones (1981ab).

TABLE 16-47—*Engine fuel consumption rates* Data from American Pulpwood Association 1975)

Units and machine	Gasoline	Diesel
Gallons per hour		
Chainsaw at stump area	0.97	—
Chainsaw at landing	.74	—
Feller-buncher	—	5.36
Wheeled skidder	—	2.93
Prehauler or forwarder	—	2.63
Mechanical slasher	—	2.44
Hydraulic loader	3.87	2.83
Concentration-yard loader	3.50	3.09
Miles per gallon		
Wheeled skidder	1.07	1.94
Prehauler or forwarder	1.71	2.04
Two-axle truck	5.58	—
Three-axle truck	4.57	4.67
Tractor-trailer	4.10	4.92

TABLE 16-48—*Average roundwood production (cord/hour) by function* (Data from American Pulpwood Association 1975)[1]

Function	Cords/hour
Chainsaw (fell, limb, top)	2.60
Chainsaw (fell, limb, top, buck)	1.52
Hydraulic shear	6.97
Mechanical feller-buncher	8.38
Haul truck (manual loading)	2.15
Haul truck (mechanical loading)	3.32
Wheeled skidder	3.08
Crawler tractor (in-woods)	1.76
Prehauler or forwarder	3.13
Chainsaw (on landing)	2.75
Slasher (on landing)	5.43
Loader (on landing)	10.78
Concentration yard loader	18.00

[1] These are values based mostly on softwood production; machine productivity in pine-site hardwoods is significantly less. (See also table 16-34).

Harvesting

TABLE 16-49—*Average pulpwood transport distances* (Data from American Pulpwood Association 1975)

Location, function and distance unit	Round-trip distance
Round trip in woods (feet)	
Wheeled skidder	1,831
Prehauler or forwarder	2,579
Crawler tractor	1,084
Highway haul (miles)[1]	
Woods direct to mill	73-85
Woods to concentration yard	37-39
Concentration yard to mill	114-137

[1]The first value tabulated is for the south-central region; the second is for the southeast.

TABLE 16-50—*Fuel consumed per cord produced, by machine* (Data from American Pulpwood Association 1975.)

Machine	Gasoline	Diesel
	Gallons/cord	
Chainsaw (at stump area)	0.41	—
Hydraulic shear	—	0.58
Mechanical feller-buncher	—	.64
"Big stick" loader (while loading)	.76	—
Wheeled skidder	—	.95
Crawler tractor	—	1.06
Prehauler or forwarder	—	.84
Chainsaw (on landing)	.39	—
Slasher (on landing)	—	.58
Hydraulic loader (on landing)	.48	.47
Chipper (on landing)	—	1.11
Loader (at concentration yard)	.27	.27
Chipper (at concentration yard)	—	1.29

16-15 SYSTEM SELECTION BY MATHEMATICAL MODEL

It is clear that the great variability in woods conditions, harvesting and transport systems, and in product form and value make selection of optimum harvesting systems very difficult. Mathematical models and computer simulation should be useful in the selection process, but application of these techniques has not been widespread. Such techniques may afford eventual solutions to some hardwood harvesting problems. It is beyond the scope of this text, however, to discuss the subject in depth.

References describing a few of the available simulations are listed, as follows:

Simulation Name	Subject	References
T-H-A-T-S ...	Timber harvesting and transport simulator with subroutines for Appalachian logging	Martin (1975b)
LOGPLAN ...	Program to develop an optimum 1-year logging operations plan	Newnham (1976)
CANLOG	Harvesting machine simulator	Newnham (1971)
ECHO	Economic harvest optimization	Tedder et al. (1978)
———.......	Development of flexible timber harvesting models	Webster (1974)
———.......	Models appropriate for Appalachia	Biller and Johnson (1973)
———.......	Assessment of addition of small chippers or residue balers to three common harvesting systems	Stuart et al. (1981)
———.......	Pulpwood production simulation analysis	Hoole et al. (1972)
———.......	Simulation analysis of timber harvesting systems	Johnson et al. (1972)
HSS	The harvesting system simulator	O'Hearn et al. (1976)
SAPLOS	Simulation applied to logging systems	Goulet et al. (1978; 1980ab)
———.......	Eight views on harvesting simulation	Silversides (1974, p. 399–526)
———.......	Logging in Sweden	Logging Research Foundation, Sweden (1975)
———.......	Simulation of whole-tree chipping and trucking	Bradley et al. (1976)
———.......	Survey and comparison of eight simulation models	Goulet et al. (1979b)

While not computer simulations, many other cost analyses of logging systems are available that should be useful to logging managers in the South, e.g., Warren and Kluender (1976), Boyd and Novak (1977), Donnelly (1977), and Tufts (1977b).

16-16 STORING OF PULP CHIPS

Widespread adoption of chipping headrigs and satellite chip mills during the 1970s resulted in a significant shift toward storage of wood in outside chip piles, rather than as roundwood. Advantages include better chip measurement, lower handling costs, reduced space requirements, and smoother operation in the woodyard and woodroom. In addition, chip piles solve storage problems arising from establishment of chip mills at the wood source. Finally they simplify procedures for handling mixed species and sawmill residues.

CHIP DETERIORATION DURING STORAGE

Unless a hardwood chip pile is managed carefully, the pulping qualities of the chips may decline. Pulp yields may be lowered because the chips have lost weight, require extended pulping, or because more of the chip dissolves when pulped. Pulp strength may be lowered because cellulose in the chips has been degraded, or because the chips have to be overcooked to obtain desired grades of pulp. Pulp brightness may be lowered because chips have darkened. More pulping and bleaching chemicals may be required, or chemical recovery (in the kraft process) may be slower. Pulping equipment may corrode or wear more rapidly. In event of a fire in the chip pile, all or part of the chip inventory may be useless for pulp (Hulme 1979).

Section 11-6 (see subsection PULPWOOD CHIPS) describes agents (stain and decay fungi, bacteria, heat, and fire) that may deteriorate stored hardwood pulp chips, the effect of these agents on pulp yield and strength, and approaches to prevent chip degradation. Not previously discussed are subterranean termites, the only insects that have been noted in hardwood chip piles. Their damage is very limited and can be controlled by applying one of the recommended chemical treatments to the soil where the chip piles are to be located (Smith and Johnston 1970).

Unpublished (1980) data of M. S. White, Virginia Polytechnic Institute and State University, indicate that in Blacksburg, West Virginia, piles of whole-tree hardwood chips gain moisture content during exterior storage, as follows:

	Initial moisture	After 5 months
	--Percent of ovendry weight --	
Inner zones.	73	53
Outer zones	73	134
Average	73	110

White concluded that high piles (e.g., 20 feet high) gain less moisture during several months storage than low piles (e.g., those 10 feet high).

See figures 20-18AB for time vs. moisture content of chips spread 4 to 12 inches deep—roofed and with no roof.

The Bois et al. (1962) study of chip piles of oak (*Quercus* sp.) and gum (*Nyssa* sp.) is abstracted in section 11-6, but it is of interest here that moisture content and temperature in lower central portions of the three piles constructed in Brunswick, Ga., varied about as follows:

Date, species, and condition	Temperature	Moisture content
	°F	Percent
June (pile just built)		
Oak, green	126	41
Gum, green	111	49
Gum, wetted with water as pile was built	75	56
August		
Oak, green	106	52
Gum, green	96	59
Gum, wetted with water as pile was built	85	62
November		
Oak, green	97	53
Gum, green	91	62
Gum, wetted with water as pile was built	90	62

Bois et al. (1962) found higher wood losses in gum that in oak species during outdoor summer storage of chips. Specific gravity losses in chip pile storage may be reduced by compaction—the greater the proportion of the pile compacted, the less the deterioration. Oak in chip piles deteriorated less than oak in roundwood form; gum roundwood, however, deteriorated less than gum chips. Findings of Crane and Fassnacht (1960) were similar. Saturating a pile of gum chips with water and wetting it continuously thereafter was ineffective in controlling deterioration. Total pulp yield diminished after 60 days of chip storage. Pulp strength losses were confined primarily to loss of tearing resistance, which showed a drop of 10 percent after 90 days of storage; this was not considered excessive compared to expectations from roundwood storage.

Small, well-compacted piles on clean bases with inventory control to insure short storage time will minimize chip deterioration. The effectiveness of chemical treatment is discussed in section 11-6.

Readers wishing additional information on reducing deterioration of chips during outside storage, will find the treatise on the subject by Bergman and Nilsson (1979) useful.

CHIP HANDLING

Chips destined for outside storage adjacent to southern pulpmills may arrive by truck, or by rail, or may be processed from roundwood in the mill woodroom. If arriving by truck or rail, they must be discharged into conveyors leading to the chip pile (figs. 16-62 and 16-63). Systems of conveying chips into chip piles include wheeled front end loaders assisted by bulldozers to shape the piles, high-pressure blowers, and belt conveyors—sometimes terminating in chip slingers. Chips may be reclaimed by front-end loaders, under-the-pile screws, reciprocating stoker-type feeders, and mechanized rakes.

Good management of chip storage usually calls for species separation, and complete inventory turnover in about 30 days on a first-in, first-out basis. In addition to these basic needs, other requirements are minimal mechanical damage to the chips and low horsepower and manpower demands for high-volume handling.

These requirements are met with an overhead conveyorized stacker-reclaimer system (fig. 16-64) at International Paper Company's 1,700 ton per day pulpmill at Georgetown, S.C. Chips are unloaded at a drive-through hydraulic truck dumper (see fig. 16-62 for systems), capable of handling six trucks per hour, and a railcar rollover dumper (fig. 16-63 top). They are belt-conveyed first to a screening station and then to the 125 by 1,000-foot chip pile with storage capacity of 14,000 cords, segregated by species. Two remotely controlled belt-conveyor stackers reaching out 85 feet and adjustable for pile height speed handling and species sorting. Power demand of each stacker, when delivering 750 tons of chips per hour, is about 100 hp. The reclaimer is an overhead scraper type, on which scraper buckets attached to a chain conveyor rake chips from the top of pile through a transfer chute to the reclaim belt conveyor. Reclaimer capacity is 450 tons per hour with power demand of about 150 hp. To control turnover of chip inventory, the reclaimer is positioned along the chip pile relative to the stackers so that chips longest in storage are the first to be reclaimed. The reclaimer feeds chip silos which in turn supply the mill (Schlegel and Jepsen 1978).

The Owens-Illinois kraft pulp mill in Valdosta, Georgia, supplies its nine batch digesters with chips delivered by a live-bottom traveling-screw reclaim system (fig. 16-65). Active length of the screw (pile width) may be 20 to 30 feet and the length of travel (pile length) may be 400 or 500 feet. Several screws can travel on the same track to reclaim different species separated within the length of the pile. Storage height can be as much as 100 feet with reclaimable volume of 15,000 to 20,000 cords. The Valdosta installation normally delivers 65 to 67 cords of chips per hour, but has capacity to deliver 125 cords per hour. Availability of the reclaim screw has been greater than 99 percent since its installation in 1975 (Nilsson and Richter[13]).

---

[13]Nilsson, B. and R. Richter. 1978. Technological improvements in woodchips and waste handling. Unpublished paper presented at the June 25-30, 1978, 32nd Annual Meeting of the Forest Products Research Society, Atlanta, Georgia.

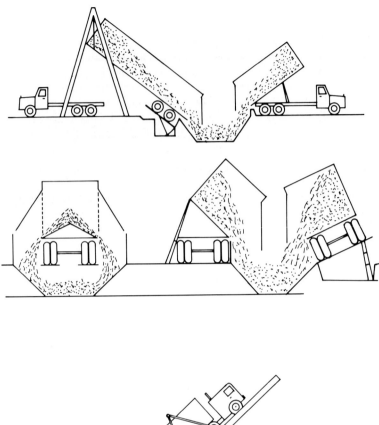
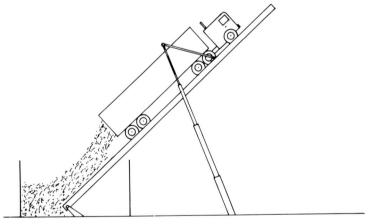

Figure 16-62.—Six methods of unloading chip trucks. (Drawing after Oswald 1979.)

Harvesting

Figure 16-63.—(Top) Rotary railcar dump for pulp chips. (Photo from Weyerhaeuser Co., Valliant, Okla.) (Bottom) Open-top side-discharge railcar for shortwood, longwood, or pulp chips. (Photo from Pullman, Inc.)

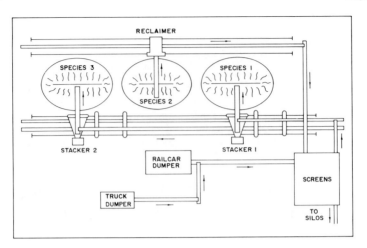

Figure 16-64.—(Top) Arrangement of truck and rail dumps, stackers, chip storage pile, and reclaim conveyor to supply a 1,700-ton-per-day pulpmill. (Center) Belt-conveyor stacker. (Bottom) Scraper-type reclaimer. (Drawings and photos from Schlegel and Jepsen 1978.)

Harvesting 1619

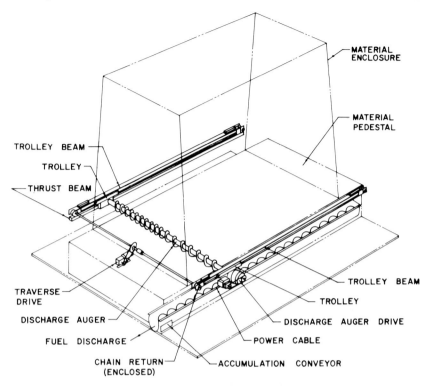

Figure 16-65.—Chip reclaimer system in which a screw conveyor travels back and forth on rails under the storage pile.

Pulpmills go to considerable expense to purchase or manufacture pulpchips according to specifications that are optimum for their processes. Inclusion of fines that pass a 3/16-inch screen and pin chips that pass a ⅜-inch screen lessen the efficiency of most pulping processes and generally increase operating costs. Chip handling systems both outside and inside the mill may mechanically degrade specification chips to form such fines and pin chips. A mill survey by Strucka (1979) indicated that pneumatic blowing systems, bulldozer manipulation of piles, and wheeled reclaim vehicles all cause appreciable chip damage. Belt conveyors did not cause significant mechanical chip degradation.

16-17 STORING OF WHOLE-TREE CHIPS AND FUEL

Incorporation of significant amounts of whole-tree chips in paper and fiber products, and the increasing use for fuel of whole-tree chips and hogged wood containing bark have stimulated interest in storage of such wood. Weight loss, and changes in pile temperature, moisture content, heat of combustion, and pH in piles of whole-tree chips are described in section 11-6 (under the heading HOGGED FUEL). See also opening paragraphs of section 16-16 for discussion of changes in moisture content.

One southern mill producing bleached paperboard plans to increase its consumption of whole-tree chips until 50 percent of the wood coming from company-owned lands is in this form. In summer, these whole-tree chips (75 percent hardwood and 25 percent pine) cannot be stored for any appreciable length of time. Even if stored a short time in the chip van in which they are transported, the chips start to blacken. For this reason, this mill unloads all whole-tree chips on arrival for direct delivery to the mill (Anonymous 1978b). Removal of fines during field chipping slows chip deterioration while in storage.

In a study of storage of whole-tree chips of southern pine, chip immersion in a dilute aqueous solution of formaldehyde and P-nitrophenol effectively preserved them during 6 months of storage in pile simulators. The treatment may be cost effective for pulpchip storage, but is probably not for fuel chips (Springer et al. 1978). Field testing is required to determine the effectiveness of this treatment in outside piles of whole-tree hardwood chips.

For use as fuel, slight degradation of whole-tree chips or hogged wood and bark in storage is unimportant. Most industrial wood-burning furnaces accept considerable amounts of fines, so there is more latitude in selecting chip handling equipment for fuel than for pulpchip storage.

Layouts of systems to store and reclaim hogged fuel are much like those for pulp chips except that a hog to reduce and control particle size replaces the screening function shown in figure 16-64 (top). Removal of contaminants such as rocks and metal fragments precedes the hogging operation. Species are not usually separated in fuel piles. (Fuel preparation and properties, and furnace feeding devices are further discussed in chapter 26.) Fuel reclaim systems may differ from pulpchip reclaim systems in that rather precise metering of fuel feed is required and mechanical damage to fuel particles is usually not a serious problem. One popular fuel reclaim mechanism is the stoker feeder (fig. 16-66) in which piston-actuated pusher plates slide under the pile base to eject metered volumes of fuel from the underside of the pile with each piston reciprocation. A single large wood-fired boiler for the wood industry may require a 20-ton truckload of fuel every 5 to 10 minutes, so the logistical problems involved are substantial, requiring rather large land areas for receiving wood, sizing it through hogs, and storing and reclaiming it.

Harvesting

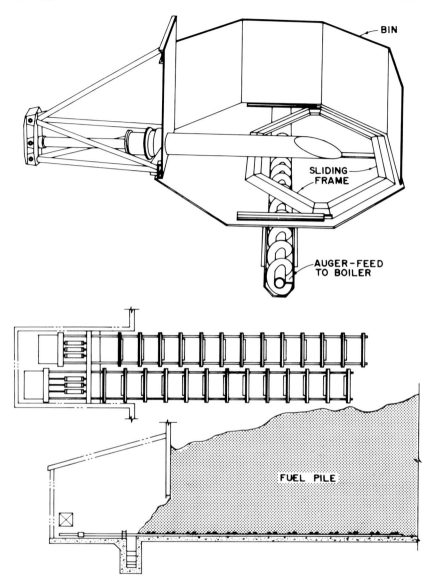

Figure 16-66.—Stoker feeders for reclaiming particulate wood from fuel piles. (Top) Sliding frame delivers fuel from bin bottom to central auger. (Drawing after Saxlund design.) (Bottom) Piston-actuated oscillating bars move fuel from bottom of pile to a conveyor feeding the burner. Bars, placed like rungs on a ladder, have beveled back edges to slide under the pile, but square front edges to scrape fuel toward the conveyor trough. (Drawing after Mill Supply Company.)

16-18 STORING OF ROUNDWOOD

Attacks by stain and decay fungi and bacteria in hardwood pulpwood and sawlogs are described in section 11-6 (under the heading ROUNDWOOD). If held under continuous water spray, hardwood logs of most species can be safely stored 6 to 18 months. Although the bark becomes darkened, the wood of most species, e.g., the oaks and sweetgum, remains sound and clean. Hackberry, however, may lose value because of stain. Wagner (1978) concluded that the annual cost of operating a water sprinking system for log storage is about $3 per thousand board feet of logs (Scribner log rule). A description of equipment required can be found in Koch (1972, p. 738–739).

Chemical treatments to protect logs against both insect and fungus attack are described in section 11-9 (under the heading PROTECTION OF LOGS AND PULPWOOD). Such treatments are less widely used than water sprays.

HANDLING OF PULPWOOD IN ROUNDWOOD FORM

Some wood procurement men in the southern pulp industry foresee a movement away from chip procurement toward more roundwood deliveries (both hardwoods and pine) to millyards. There are several driving forces behind this perceived move to roundwood. First, the fuel value of bark from millyard drum debarkers has increased with petroleum prices; no fuel component is purchased with bark-free chips. Second, the southern forest today has large volumes of hardwood (and small-diameter pine) in which the bark-to-wood ratio is very high (table 16-10). Third, technology is available to drum debark, at the millyard, many species of these hardwoods (and pine) of all diameters at low unit cost, with low fiber loss, and with clean bark/wood separation (see chapter 17 for debarking procedures). Finally, millyards handling short roundwood—perhaps 8 or 10 feet long instead of the traditional 5-foot-3-inch length—can be substantially mechanized for control by a few operators centrally located to give them direct line-of-sight or television-assisted views of the entire operation (Holekamp 1978). The shortwood can originate from shortwood loggers, longwood slashers, or tree-length merchandising systems (section 16-6).

With this concept, millyard roundwood storage is only temporary or for species separation. Stored roundwood is reclaimed by pushing or raking it into the system leading to the drum debarkers. Loading techniques, therefore, must build piles that permit such easy reclaim.

For a description of pulpwood handling at woodyards, readers are referred to Jarck (1977).

LOG HANDLING

Log inventory systems are numerous. The most frequently used calls for log stacking and unstacking by front end loaders (fig. 16-67). Phillips (1970) and Phillips and Hartman (1970) have described the requirements of front end loaders and factors affecting their production. With such equipment species and size sorts are possible; log piles may be built to a height of about 10 feet. The system is simple, but first-in first-out inventory control is difficult and heavy traffic over unsurfaced log yards may create substantial mud problems with dirt transferred to log surfaces.

Mud problems can be reduced and loader productivity increased by surfacing log yards with asphalt. Forest Industries (1976) estimated the cost of installing 12-inch-thick asphalt-concrete paving on a 20-acre log yard at about $1.5 million and annual operating costs, including loaders, at $342,000. Such a paved yard, serviced by four large front-end loaders (with operating costs of $41/hour each; $243,000/year) might unload 60,000 tons of logs annually from trucks or rail cars, and sort, deck, reload, or move them to the mill (Forest Industries 1976).

Tyre and Screpetis (1978) have described a roundwood inventory system appropriate for control of roundwood volumes handled by front-end loaders. As each load arrives, it is weighed and the logs counted. From this procedure volumes are determined. Each load is then assigned either directly to live decks or to one of several storage decks, the deck assignment being marked on the weight ticket. Tyre and Screpetis explain that by their system loads need only be scaled once and no scaling is needed at the live deck; they describe the flow of logs from weight scales to mill and provide sample inventory forms.

Many crane systems are in use. Two crane types that gained popularity during the 1970's are the jib crane and the log boom crane (fig. 16-68). Both can store logs in an annular pattern with piles 45 feet high or higher, and with minimum site preparation. Grapples on these cranes have capacities of 20 to 40 tons with which a truck can be unloaded in one or two movements requiring perhaps 2½ minutes per cycle.

Storage capability with the systems may range from 3,000 to 12,000 cords with a 120-foot boom radius and single-row storage, to 18,000 to 20,000 cords with 175-foot boom radius and multi-row storage. These systems have the capability to rotate inventory on a first-in first-out basis, and to inventory by species if incoming truck loads are of one species. Readers desiring illustration and discussion of various types of log handling cranes will find Lee and van Soest's (1978) paper useful.

For illustrations of small grapple loaders see figure 16-6, 16-12, 16-20, and 16-22.

Figure 16-67.—(Top) 120 horsepower front-end loader weighing 23,600 pounds and capable of lifting 10,000 pounds; the machine cost about $60,000 in 1979. (Bottom) 216-horsepower log stacker weighing 92,000 pounds with lift capacity of 80,000 pounds; machine cost in 1979 was $123,000. (Center) 107-horsepower sling loader for shortwood can lift 15,000 pounds and cost about $50,000 in 1979. (Photo from Taylor Machine Works, Inc.)

Figure 16-68.—Boom crane with 120-foot radius can build or reclaim an inventory of about 12,000 cords. The 15-ton grapple can unload a log truck in two bites. Front-end loader in the foreground is transferring hardwood logs, from a pallet previously loaded by the grapple, to the mill log deck. (Photo from LeTourneau Sales & Service, Inc.)

16-19 LITERATURE CITED

Adler, T. J., M. Blakey, and T. Meyer. 1978. The direct and indirect costs of transporting wood chips to supply a wood-fired power plant. TID-28737. Prepared by Dartmouth Coll., Thayer Sch. of Eng., Hanover, N.H., for U.S. Dep. Energy, Solar Energy, Washington, D.C. 75p.

Akerman, R. 1976. (Cranes—comparative data). Teknik. Forskningsstiftelsen Skogsarbeten 4: 8 p.

Alban, D. H. 1977. The effects of timber harvesting on the soil chemical resource. Proc., Soc. Amer. For., Oct. 2-6, Albuquerque, New Mexico, p. 238-243.

Alexander, R. R. 1977. Cutting methods in relation to resource use in central Rocky Mountain spruce-fir forests. J. For. 75: 395-400.

Altman, J. A. 1975. Precision whole tree chipping operation. Amer. Pulpw. Assoc., Tech. Release 75-R-8. 4 p.

Altman, J. A. 1976. Blue Ox pulpwood chip or long log rail car. Am. Pulpw. Assoc., Tech. Release 77-R-5. 2 p.

American Pulpwood Association. 1975. Fuel survey. 15 p. Amer. Pulpwood Assoc., Washington, D.C.

Andersson, S. 1975. (Utilization of smallwood, slash, and stumps.) Forskningsstiftelsen Skogsarbeten. Ekonomi No. 1E 1975. 4 p. Stockholm, Sweden.

Andersson, S., R. Hansen, Y. Jonsson, and M. Nylinder. 1978. Harvesting system for stumps and roots—a review of Scandinavian techniques. *In* Complete tree utilization of southern pine: symp. proc., New Orleans, La., April 17-1 C. W. McMillin, ed. For. Prod. Res. Soc., Madison, Wis., p. 130-146.

Anonymous. 1968. First turbine-powered mobile debarker-chipper. South. Lumberman 217(2695):27-28.

Anonymous. 1971. Tree-lengths at tree a minute clip. Can. For. Ind. 91(3):48-51, 53.

Anonymous. 1974. Mobile chippers enable Owens-Illinois to boost per-acre yield 10% to 100%. Pap. Trade J. 158(35):26-27.

Anonymous. 1975. Status of whole-tree chipping: spreading, but has its problems. Pulp & Pap. 49(6):126-127.

Anonymous. 1976a. Highlead cable yarding: is it coming back east? Pulpwood Prod. & Timber Harvesting 24(10):14-16.

Anonymous. 1976b. Jumbo railcar chip-unloading system. Pulp & Pap. 50(6):115.

Anonymous. 1977a. How to cope with smaller logs and their higher harvest costs. For. Ind. 104(10):34-36.

Anonymous. 1977b. Deere introduces new tree harvesting system. Dixie Logger and Lumberman 16(5):14-15.

Anonymous. 1978a. Vibrating stump puller undergoes field testing. For. Ind. 105(7):94.

Anonymous. 1978b. How Westvaco does it. Tappi 61(3):23-24.

Arend, J. L., R. N. Smith, and R. A. Ralston. 1955. Tests of a portable wood chipper. For. Prod. J. 5:156-160.

Auchter, R. J. 1972. Pulping without barking increases fiber yield. Pulp and Pap. 46(6):6-7.

Auchter, R. J. 1975. Whole-tree utilization: fact or fantasy. Tappi 58(11):4-5.

Axelsson, S. A. 1972. Repair statistics and performance of new logging machines: Koehring short-wood harvester/Report 1. LRR/47, Pulp and Pap. Res. Inst. Can., Pointe Claire, P.Q. 59 p.

Baker, J. B. 1978. Nutrient drain associated with hardwood plantation culture. *In* Proc., Second Symposium on Southeastern Hardwoods, p. 48-53. (Dothan, Ala., Ap. 20-22.)

Barnes, R. 1971. Patterned tree harvest proposed. West. Conserv. J. 28:44-47.

Barrett, L. I. and J. H. Buell. 1941. Oak topwood volume tables. USDA For. Serv. Tech. Note No. 48, 7 p. Appalachian For. Exp. Stn., Asheville, N.C.

Baumgras, J. E. 1978. The causes of logging truck delays on two West Virginia logging operations. U.S. Dep. Agric. For. Serv. Res. Pap. NE-421. 4 p.

Beckwith, S. L. 1964. Effect of site preparation on wildlife and vegetation in the sandhills of central Florida. Proc. Annu. Conf. Southeastern Assoc. Game and Fish Commissioners 18:39-48.

Beltz, R. C. 1976. Comparison of harvested volume to inventory and slash volumes in Midsouth logging operations. Ph.D. thesis. La. State Univ., Baton Rouge, La. 107 p.

Bergman, O., and T. Nilsson. 1979. Outside chip storage—Methods of reducing deterioration during OCS. Ch. 12 in Pulp and paper technology series no. 5, chip quality monograph (J. V. Hatton, ed.), TAPPI Joint Textbook Committee of the Paper Industry. p. 245-272.

Bergeron, G. A. 1978. Factors affecting the productivity of the TH-210 Thinner-Harvester in southern pine stands. McIntire-Stennis Proj. #1444, La. State Univ. Sch. For. and Wildl. Manage. and La. Agric. Exp. Stn. 117 p.

Bertelson, D. F. 1977. Pulpwood price trends in South. Pulp & Pap. 51(6):71.

Biller, C. J. 1970. Tables for estimating skidder tire wear. USDA For. Serv. Res. Pap. NE-168, 12 p. Northeast. For. Exp. Stn., Upper Darby, Pa.

Biller, C. J. 1971. Prediction of skidder tire wear in Appalachia. Amer. Soc. Agric. Eng. Trans. 14(2):350-352.

Biller, C. J. 1972. Are you overloading your skidder Tires? USDA For. Serv. Res. Pap. NE-247, 7 p. Northeast. For. Exp. Stn., Upper Darby, Pa.

Biller, C. J. and R. L. Hartman. 1971. Tractive characteristics of wheeled-skidder tires. Amer. Soc. Agric. Eng. Trans. 14(6):1024-1026, 1033.

Biller, C. J. and L. R. Johnson. 1973. Comparing logging systems through simulation. Amer. Soc. Agric. Eng. Trans. 16(6):1060-1063, 1071.

Biller, C. J., and P. A. Peters. 1981. Testing of the prototype Appalachian thinner. Pap. No. 81-1074, presented at 1981 Summer Meet., Amer. Soc. Agric. Eng., Orlando, Fla., June 21-24.

Biltonen, F. E., W. A. Hillstrom, H. M. Steinhilb, and R. M. Godman. 1976. Mechanized thinning of northern hardwood pole stands—methods and economics. USDA For. Serv. Res. Pap. NC-137, 17 p. North Cent. Exp. Stn., St. Paul, Minn.

Biltonen, F. E., J. A. Mattson, and E. D. Matson. 1979. The cost of debarked whole-tree chips stump to digester. U.S. Dep. Agric. For. Res. Pap. NC-174. 8 p.

Blackford, J. M. 1975. Low cost merchandiser log decks for hardwoods. FPRS Sep. No. MS-75-S59, 10 p. For. Prod. Res. Soc., Madison, Wis.

Blackmon, B. G. 1979. Estimates of nutrient drain by dormant-season harvests of coppice American sycamore. U.S. Dep. Agric. For. Serv. Res. Note SO-245. 5 p.

Blomqvist, L. 1978. An industrial plant for chipping stump and rootwood for chemical pulping. In Complete tree utilization of southern pine: symp. proc., New Orleans, La., April 17-19. C. W. McMillin, ed. For. Prod. Res. Soc., Madison, Wis., p. 294-298.

Blymyer, M. J. 1976. Impact of clearcutting on indigenous mammals of southwest Virginia. M.S. thesis. Virginia Polytechnic Institute and State Univ., Blacksburg. 218 pp.

Bois, P. J. 1968. Weight of sawdust from several southern Appalachian wood species. For. Prod. J. 18(10):52-54.

Bois, P. J., R. A. Flick, and W. D. Gilmer. 1962. A study of outside storage of hardwood pulp chips in the southeast. Tappi 45:609-618.

Boyd, J. H. 1975. Repair statistics and performance of new logging machines: Koehring short-wood harvester/Report 2. LRR/61, Pulp and Pap. Res. Inst. Can., Pointe Claire, P.Q. 57 p.

Boyd, C. W., W. W. Carson, and J. E. Jorgensen. 1977. Harvesting the forest resource—are we prepared? J. of For. 75:401-403.

Boyd, J. and W. Novak. 1977. Estimate wood cost, productivity and investment needs for 84 logging system combinations. Pulp & Pap. Can. 78(5):69-78.

Boyer, R. L. 1977. Logging mats—an alternative to high-cost gravel. Pap. presented at Spring Meet. of the Am. Pulpw. Assoc., Appalachian Tech. Div. (Williamsburg, Va. May 18-19.) 2 p.

Bradley, D. P., F. E. Biltonen, and S. A. Winsauer. 1976. A computer simulation of full-tree field chipping and trucking. USDA For. Serv. Res. Pap. NC-129, 14 p. North Cent. For. Exp. Stn., St. Paul, Minn.

Bradley, D., and D. Stevens. 1978. Wood for energy. An interim report on the whole-tree harvesting experiment, South Duxbury, Vermont. Dept. For., Parks, Recreation, Vermont Agency Environ. Conserv., and Vermont Nat. Resourc. Counc., Montpelier, Vermont. 40 p.

Bradley, D. P., and S. A. Winsauer. 1978. Simulated full-tree chipping: model compares favorably to the real world. For. Prod. J. 28(1):85-88.

Bredberg, C.-J. 1970. Evaluation of logging-machine prototypes: Timberjack 360 Grapple Skidder. WR/24, Pulp and Pap. Res. Inst. Can., Pointe Claire, P.Q. 22 p.

Brunberg, T. 1977. (Long-arm cranes—A comparison.) Teknik Forskningsstiftelsen Skogsarbeten 11: 1-4. (Stockholm)

Brush, R. O. 1976. Spaces within the woods: managing forests for visual enjoyment. J. For. 74(11):744-747.

Bryan, R. W. 1978. High chip production attributed to steady crew, equipment care. For. Ind. 105(11):36, 37.

Bryan, R. W. 1979. Helicopter called in to extract hardwood from boggy terrain. For. Ind. 106(3):64-65.

Bryan, R. W. 1980. Mobile shear-chipper delivers clean pulp chips to forwarder. For. Ind. 107(2):28-29.

Chappell, T. W. and R. C. Beltz. 1973. Southern logging residues: an opportunity. J. of For. 71:688-691.

Chow, P., C. S. Walters and J. K. Guiher. 1973. Specific gravity, bulk density, and screen analysis of midwestern plant-fiber residues. For. Prod. J. 23(2):57-60.

Clark, A., III. 1978. Total tree and its utilization in the southern United States. For. Prod. J. 28(10):47-52.

Clark, A. III, D. R. Phillips, and J. G. Schroeder. 1980a. Predicted weights and volumes of northern red oak trees in western North Carolina. Res. Pap. SE-209, U.S. Dep. Agric., For. Serv. 22 p.

Clark, A. III, D. R. Phillips, and H. C. Hitchcock III. 1980b. Predicted weights and volumes of scarlet oak trees on the Tennessee Cumberland Plateau. Res. Pap. SE-214, U.S. Dep. Agric., For. Serv. 23 p.

Clark, A. III, D. R. Phillips, and H. C. Hitchcock III. 1980c. Predicted weights and volumes of southern red oak trees on the Highland Rim in Tennessee. Res. Pap. SE-208, U.S. Dep. Agric., For. Serv. 23 p.

Clark, A. III, and J. G. Schroeder. 1977. Biomass of yellow-poplar in natural stands in western North Carolina. U.S. Dep. Agric. For. Serv. Res. Pap. SE-165. 41 p.

Clawson, M. 1974. Economic trade-offs in multiple-use management of forest lands. Am. J. Agric. Economics 56(5):919-926.

Colannino, A. 1976. Weight equations for white oak and chestnut oak in northern West Virginia. MS Thesis, West Va. University, Morgantown. 75 p.

Conner, A. L. 1978. Utilization of whole-tree chips in an unbleached Kraft linerboard mill. In Complete tree utilization of southern pine: symp. proc., New Orleans, La., April 17-19. C. W. McMillin, ed. For. Prod. Res. Soc., Madison, Wis., p. 268-274.

Conway, S. 1976. Logging practices—principles of timber harvesting systems. 416 p. San Francisco, Ca.: Miller Freeman Publ., Inc.

Corbett, E. W., J. A. Lynch, and W. E. Sopper. 1978. Timber harvesting practices and water quality in the eastern United States. J. For. 76:484-488.

Cost, N. D. 1978. Aboveground volume of hardwoods in the mountain region of North Carolina. U.S. Dep. Agric. For. Serv. Res. Note SE-266. Southeast. For. Exp. Stn., Asheville, N.C. 4 p.

Cottell, P. L., and R. T. Barth. 1974. Operator performance in logging. In Forest harvesting mechanization and automation. (C. R. Silversides, ed.) Publ. 5, IUFRO Div. 3, Can. For. Serv., Ottawa, p. 205-223.

Coughran, S. 1978. Tree shears: A technical approach. Amer. Logger and Lumberman 3(5):18-21.

Craft, E. P. 1976. Utilizing hardwood logging residue: A case study in the Appalachians. U.S. Dep. Agric. For. Serv. Res. Note NE-230. 7 p.

Craft, E. P. 1978. Harvesting and processing logging residue and poletimber thinnings in the Appalachians. Paper presented at the 1978 Sawmill Clinic and Machinery Show, Atlanta, Ga., October 11-13. 5 p.

Crane, T. P., Jr. and D. L. Fassnacht. 1960. Summer storage of hardwood chips at Pensacola, Fla. Tappi 43:188A-194A.

Curtin, D. T., and A. G. Bunker. 1971. Cooperative testing of a TH-100 plantation thinning system owned and operated by International Paper Company. Amer. Pulpw. Assoc., Atlanta, Ga. 20 p.

Darwin, W. N., Jr. 1965. Skidding coefficients on an alluvial soil. For. Prod. J. 15:302.

Davidson, D. A. 1978. Development of new feller buncher concepts. *In* Complete tree utilization of southern pine: symp. proc., New Orleans, La., April 17-19. C. W. McMillin, ed. For. Prod. Res. Soc., Madison, Wis., p. 115-120.

Davis, B. M., and Hurley, D. W. 1978. Fiber from a southern pine stump-root system? *In* Complete tree utilization of southern pine: symp. proc., New Orleans, La., C. W. McMillin, ed. For. Prod. Res. Soc., Madison, Wis., p. 274-276.

Deal, E. L., and J. L. Huff. 1977. Whole-tree chipper downtime and cost study. APA Tech. Rel. 77-R-15. 5 p. Amer. Plywood Assoc., Washington, D.C.

Dickerson, B. P. 1976. Soil compaction after tree-length skidding in northern Mississippi. Soil Sci. Soc. Am. J. 40:965-968.

Dickson, J. G. 1978. Forest bird communities of the bottomland hardwoods. *In* Proc. of the Workshop Management of Southern Forests for Nongame Birds. (Atlanta, Ga., Jan. 24-26.) U.S. Dep. Agric. For. Serv., Gen. Tech. Rep. SE-14. p. 66-73.

Dickson, J. G., and C. A. Segelquist. 1977. Winter bird populations in pine and pine-hardwood forest stands in East Texas. *In* Proc., Annu. Conf. Southeast. Assoc. Fish and Wildl. Agencies 31:134-137.

Dolgaard, S. J., G. W. Gullion, and J. C. Haas. 1976. Mechanized timber harvesting to improve ruffed grouse habitat. Univ. Minn. Agric. Exp. Stn., Univ. Minn. Tech. Bull. 308, For. Series 23. 11 pp.

Donnelly, D. M. 1977. Estimating the least cost combination of cable yarding and tractor skidding for a timber sale area. U.S. Dep. Agric. For. Serv. Res. Note RM-341. 8 p.

Donnelly, D. M. 1978. Computing average skidding distance for logging areas with irregular boundaries and variable log density. Gen. Tech. Rep. RM-58, U.S. Dep. Agric. For. Serv., Rocky Mountain For. Exp. Stn., Ft. Collins, Colo. 10 p.

Duffy, P. D., and D. C. McClurkin. 1974. Difficult eroded planting sites in north Mississippi evaluated by discriminant analysis. Soil Sci. Amer. Proc. 38:676-678.

Dunfield, J. D. 1971. Annotated bibliography on delimbing of trees. Can. For. Serv. Inf. Rep. FMR-X-31, 184 p. For. Manage. Inst., Ottawa, Ontario, Can.

Edberg, U., L. Engstrom, and N. Hartler. 1973. The influence of chip dimensions on chip bulk density. Svensk Papperstidning 76:529-533.

Erickson, J. R. 1974. Whole-tree chipping and residue utilization. *In* Int. Union For. Res. Org. Biomass Studies. Papers presented during meeting of S4.01. (Nancy, France, June 25-29, 1973.) p. 443-449. Univ. Maine, Orono.

Falk, G. D. 1981. Predicting the payload capability of cable logging systems including the effect of partial suspension. Res. Pap. NE-479, U.S. Dep. Agric., For. Serv. 29 p.

FAO 1962. Mechanical stump extractor and stump-extracting carriage. For. Equip. Notes. No. B.33.62. Food and Agric. Org. United Nations, Rome. 2 p.

Fisher, E. L., H. G. Gibson, and C. J. Biller. 1980. Production and cost of a live skyline cable yarder tested in Appalachia. Res. Pap. NE-465, U.S. Dep. Agric., For. Serv. 8 p.

Fiske, P. M. and R. B. Fridley. 1975. Some aspects of selecting log skidding tractors. Trans. of the ASAE 18(3):497-502.

Fixmer, F. N. 1976. Long distance pulpwood trucking—pro's and con's. Tech. Pap. 76-P-10. Amer. Pulpw. Assoc., Washington, D.C. 5 p.

Folkema, M. P., and W. P. Novak. 1976. Evaluation of Timmins "Fel-Del" harvester head. FERIC Tech. Rep. No. 5, For. Eng. Res. Inst. Can., Vancouver. 29 p.

Folkema, M. P. 1977a. Evaluation of Kockums 880 "Tree King" feller-buncher. Tech. Rep. TR-13, For. Eng. Res. Inst. Can., Vancouver. 16 p.

Folkema, M. P. 1977b. Whole-tree chipping with the Morbark Model 22 Chiparvestor. For. Eng. Res. Inst. Can., FERIC Tech. Note TN-16. 14 p.

Forest Industries. 1976. An analysis of some figures involved in paving the log yard. For. Ind. 103(11):52-53.

Forestry and British Timber. 1978. Loaders review. For. and Brit. Timb. 7(6):20-31.

Frederick, D. J., W. E. Gardner, R. C. Kellison, B. R. Brenneman, and P. L. Marsh. 1979. Predicting weight yields of West Virginia mountain hardwoods. J. For. 77: 762-764, 776.

Froehlich, H. A. 1978. Soil compaction from low ground-pressure, torsion-suspension logging vehicles on three forest soils. Res. Pap. 36, Sch. of For., Oregon State Univ., Corvallis. 13 p.

Gabriel, W. J., R. L. Nissen, Jr., and K. Burns. 1974. A time study of top-lopping northern hardwoods in New York State. AFRI Res. Rep. 23, Applied For. Res. Inst., State Univ. New York, 27 p.

Gardner, R. B. 1977. Transporting woodchips by hydraulic pipeline. Ch. 21 *in* Transport and handling in the pulp and paper industry, Kalish, J., and L. E. Haas, eds., 2:261-273. Miller-Freeman Publications, Inc., San Francisco.

Garlicki, A. M. and W. W. Calvert. 1968. Effect of tree-length orientation on skidding forces. For. Prod. J. 18(7):37-38.

Gibbons, B. W. 1977. Logging residue study. II. Tech. Rel. 77-R-18, Am. Pulpwood Assoc. 5 p.

Gibson, H. G., R. L. Hartman, D. L. Gochenour, Jr., and H. W. Parker. 1969. Wheeled skidder performance on sloping terrain. Agric. Eng. 50(3):152-154.

Goho, C. D. 1976. Study of logging residue at woods landings in Appalachia. U.S. Dep. Agric. For. Serv., Res. Note NE-219. 4 p.

Gordon, R. D. 1977. Tree shears and feller bunchers. Proj. Rep. No. 2, New Zealand Logging Ind. Res. Assoc., Inc., Rotorua. 43 p.

Gordon, R. D. 1978. Delimbing studies. Proj. Rep. No. 4, New Zealand Logging Ind. Res. Assoc., Inc., Rotorua. 48 p.

Gordon, R., J. H. Miller, and C. Brewer. 1981. Site preparation treatments and nutrient loss following complete harvest using the Nicholson-Koch mobile chipper. *In* Proc., 1st biennial southern silvicultural res. conf., Nov. 6-7, 1980, Atlanta, Ga. U.S. Dep. Agric. For. Serv., Gen. Tech. Rep. SO-34. p. 79-84.

Gould, N. E. 1977. Featured species planning for wildlife on southern national forests. Trans. N. Am. Wildl. Nat. Resour. Conf. 42:435-437.

Goulet, D. V., D. L. Sirois, R. H. Iff., C. A. Stutts, and M. P. Bailey. 1978. Harvesting simulation models—an evaluation and review for use in the southern United States. *In* Proc. of the IUFRO Workshop/Symposium on Simulation Techniques in Forest Operational Planning and Control. Wageningen, the Netherlands, Oct. 3-6. p. 127-136.

Goulet, D. V., R. H. Iff, and D. L. Sirois. 1979a. Tree-to-mill forest harvesting simulation models: Where are we? For. Prod. J. 29(10):50-55.

Goulet, D. V., D. L. Sirois, and R. H. Iff. 1979b. A survey of timber harvesting simulation models for use in the South. U.S. Dep. Agric. For. Serv. Gen. Tech. Rep. SO-25. 15 p.

Goulet, D. V., R. H. Iff, and D. L. Sirois. 1980a. Five forest harvesting simulation models. Part I: Modeling characteristics. For. Prod. J. 30(7):17-20.

Goulet, D. V., R. H. Iff, and D. L. Sirois. 1980b. Analysis of five forest harvesting simultion models. Part II. Paths, pitfalls, and other considerations. For. Prod. J. 30(8):18-22.

Graves, G. A., J. L. Bowyer, and D. P. Bradley. 1977. Economics of log separation in whole-tree chipping. Tappi 60(4):94-96.

Griffin, G. 1977. Simple delimber gains popularity across South. Pulpwood Prod. and Timber Harvesting 25(5):72-74.

Grillot, S. L., and R. W. McDermid. 1977. Productivity of the TX-1600 tree extractor in southern pine plantations. LSU For. Note No. 122. Louisiana State Univ., Baton Rouge. 2 p.

Gullion, G. W. 1977. Forest manipulation for ruffed grouse. Trans. N. Am. Wildl. Nat. Resour. Conf. 42:449-458.

Guttenberg, S. 1974. Woods labor problems in America's South. *In* Forest harvesting mechanization and automation. (C. R. Silversides, ed.) Publ. 5, IUFRO Div. 3, Can. For. Serv., Ottawa, p. 135-142.

Haas, L. E., and J. E. Kalish, eds. 1975. Transport and handling in the pulp and paper industry. Vol. 1. Miller-Freeman Publications, Inc., San Francisco. 336 p.

Hakkila, P. 1978. Whole-tree chipping systems in Europe. *In* Complete tree utilization of southern pine: symp. proc., New Orleans, La., April 17-20. C. W. McMillin, ed. For. Prod. Res. Soc., Madison, Wis., p. 107-202.

Hakkila, P., and H. Kalaja. 1980. Harvesting fuel chips with the Pallari swathe harvester. Folia Forestalia 418:1-24. Metsäntutkimuslaitos, Institutum Forestale Fenniae, Helsinki.

Hakkila, P., M. Leikola, and M. Salakari. 1979. Production, harvesting, and utilization of short-rotation wood. Fin. Rep. SARJA B N:o 46b. The Finnish National Fund for Res. and Develop. Helsinki, 163 p.

Hakkila, P., and M. Makela. 1973. Harvesting of stump and root wood by the Pallari Stumparvester. Commun. Inst. For. Fenn. 77.5 57 p.

Hakkila, P, and M. Makela. 1974. Further studies of the Pallari Stumparvester. Fol. For. Inst. For. Fenn. 200. 15 p.

Hakkila, P., L. Matti, and M. Salakari. 1979. Production, harvesting, and utilization of small-sized trees. SARJA B, N:o 46b, The Finnish National Fund for Res. and Develop., Helsinki. 163 p.

Halls, L. K. 1978. Effect of timber harvesting on wildlife, wildlife habitat and recreation values. *In* Complete tree utilization of southern pine: symp. proc., New Orleans, La., April 17-19. C. W. McMillin, ed. For. Prod. Res. Soc., Madison, Wis., p. 108-114.

Halls, L. K., and J. J. Stransky. 1971. Atlas of southern forest game. U.S. Dep. Agric. For. Serv. Southern For. Exp. Stn. 24 pp.

Hamilton, H. R., R. G. Bowman, Jr., R. W. Gardner, and J. J. Grimm. 1961. Phase report on factors affecting pulpwood production costs and technology in the southeastern United States to American Pulpwood Association. Battelle Memorial Institute, Columbus, Ohio. 44 p. + sep. book of appendices.

Hanaya, M. 1975. Ocean transport of woodchips—specialized ship design and handling facilities. Ch. 3 *in* Transportation and handling in the pulp and paper industry, Haas, L. E., and J. E. Kalish, eds., 1:60-95. Miller-Freeman Publications, Inc., San Francisco.

Harris, T. 1977. Boise Southern fiber/fuel operations. Tech. Rel. 77-R-31 Am. Pulpw. Assoc. 2 p.

Harris, W. F., R. A. Goldstein, and G. S. Henderson. 1973. Analysis of forest biomass pools, annual primary production and turnover of biomass for a mixed deciduous forest watershed. *In* IUFRO Biomass Studies, H. E. Young, ed. Univ. Maine, Orono, p. 43-64.

Harrison, R. T. 1975. Slash—equipment and methods for treatment and utilization. U.S. Dep. Agric. For. Serv., Equip. Develop. and Test. Rep. 7120-7. 47 p. Equip. Develop. Cent., San Dimas, Calif.

Hartman, R. L. and H. G. Gibson. 1970. Techniques for the wheeled-skidder operator. U.S. Dep. Agric. For. Serv., Res. Pap. NE-170. 18 p.

Hartman, R. L., and R. A. Phillips. 1970. Bunching with the wheeled skidder. The North. Logger and Timber Prod. 19(4):32-33.

Hassan, A. E. 1977a. Trafficability study of a cable skidder. ASAE Pap. No. 75-1508, 13 p. Amer. Soc. Agric. Eng., St. Joseph, Mich.

Hassan, A. E. 1977b. Compaction of wood chips—energy cost. Trans. ASAE 20(5):839-842. Amer. Soc. Agric. Eng., St. Joseph, Mich.

Hassan, A. E. 1978. Effect of mechanization on soils and forest regeneration. I. Coastal plain organic soil. Trans. ASAE 21(6):1107-1111. Amer. Soc. Agric. Eng., St. Joseph, Mich.

Hassan, A. E.-D., and R. H. Reeves. 1978. Compaction of wood chips—physical and pulping characteristics. Pap. No. 78-1575, presented at 1978 Winter Meet., Amer. Soc. Agric., Eng., Chicago, Dec. 18-20. 17 p.

Haussman, R. F. 1960. Permanent logging roads for better woodlot management. U.S. Dep. Agric. For. Serv., Div. State and Priv. For., Upper Darby, Pa. 38 p.

Hawkes, E. G. 1979. An introduction to cable harvesting systems for small timber, including descriptions of more than 40 systems. The Vermont Dept. of Forests, Parks, and Recreation. 119 p.

Heidersdorf, E. 1974. Evaluation of new logging machines: BM Volvo SM-880 Processor. LRR/55, Pulp and Pap. Res. Inst. Can., Pointe Claire, P.Q., 17 p.

Helgesson, T. 1978. (Flail delimbing—A bundle delimbing method for wood removed in thinning and cleaning processes.) Svenska Träforskningsinstitutet, No. 472. 21 p. (Sweden)

Hensel, J. S. 1976a. Wheeled skidder data. Am. Pulpwood Assoc. Tech. Rel. 76-R-44, 19 p. Washington D.C.

Hensel, J. S. 1976b. Chain saw data. Am. Pulpwood Assoc. Tech. Rel. 76-R-38, 32 p. Washington, D.C.

Hensel, J. S. 1970. Union Pacific Railroad opentop chips car. Tech. Rel. 70-R-26, Amer. Pulpwood Assoc., New York. 3 p.

Herbert, R. N. 1977. Trends in marine transport of forest products. Ch. 14 *in* Transport and handling in the pulp and paper industry, Kalish, J., and L. E. Haas, eds., 2:173-190. Miller-Freeman Publications, Inc., San Francisco.

Herrick, D. E. 1955. Tractive effort required to skid hardwood logs. For. Prod. J. 5:250-255.

Herrick, O. W. 1976. Key indicators of successful logging jobs in the Northeast. U.S. Dep. Agric. For. Serv., Res. Pap. NE-352. 5 p.

Hillstrom, W. A., and H. M. Steinhilb. 1976. Mechanized thinning shows promise in northern hardwoods. North. Logger and Timber Process. 24(12):12-13, 32.

Hitchcock, H. C. III. 1978. Aboveground tree weight equations for hardwood seedlings and saplings. Tappi 61(10):119-120.

Hitchcock, H. C. III. 1979. Biomass of southern hardwood regeneration estimated by vertical line sampling. J. For. 77:474-477.

Hodges, R. D., Jr. 1959. Logging in recreational areas. Proc. Soc. Am. For. 58:21-22.

Holekamp, J. A. 1978. Modern millyards for roundwood pulpwood and forest fuels. *In* Complete tree utilization of southern pine: symp. proc., New Orleans, La., April 17-19. C. W. McMillin, ed. For. Prod. Res. Soc., Madison, Wis., p. 224-234.

Hoole, J. N., W. H. Bussell, A. M. Leppert, and G. R. Harmon. 1972. Pulpwood production systems analysis—A simulation approach. J. For. 70:214-215.

Hulme, M. A. 1979. Outside chip storage—Deterioration during OCS. Ch. 11 *in* Pulp and paper technology series no. 5, chip quality monograph (J. V. Hatton, ed.), TAPPI Joint Textbook Committee of the Paper Industry. p. 213-244.

James, G. A., and H. K. Cordell. 1970. Importance of shading to visitors selecting a campsite at Indian Boundary Campground in Tennessee. U.S. Dep. Agric. For. Serv. Res. Note SE-130. 5 pp.

Jarck, W. 1977. A fast wood transfer technique for multiple length loads. Am. Pulpw. Assoc., Tech. Release 77-R-16. 5 p.

Johnson, L. R., D. L. Gochenour, Jr., and C. J. Biller. 1972. Simulation analysis of timber harvesting systems. Tech. Pap., 23rd An. Conf. and Conven. of the Amer. Inst. Indus. Eng., Anaheim, Calif., May 31-June 3. p. 353-361.

Johnson, F. L., and P. G. Risser. 1974. Biomass, annual net primary production, and dynamics of six mineral elements in a post oak-blackjack oak forest. Ecol. 55:1246-1258.

Jones, K. C. 1981a. Energy requirements to produce fuel from harvesting residues. *In* Proc., International Conf., "Harvesting and utilization of wood for energy purposes," ELMIA, Jonkoping, Sweden, Sept. 29-30, 1980. (J. E. Mattsson and P. O. Nilsson, eds.) No. 19, Swedish Univ. Agric. Sci., Dept. Operational Efficiency, Garpenberg, Sweden. p. 184-196.

Jones, K. C. 1981b. Fuel preparation from harvesting residues. Rep. No. P-28, FERIC, For. Eng. Res. Inst. of Canada, Pointe Claire, P.Q.

Jorgensen, J. R., C. G. Wells, and L. J. Metz. 1976. The nutrient cycle: key to continuous forest production. J. of For. 73:400-403.

Kalaja, H. 1978. Harvesting small-sized trees with terrain chipper TT 1000 F. Folia Forestalia 374:1-27. The Finnish For. Res. Inst., Helsinki.

Kalish, J., and L. E. Haas, eds. 1977. Transport and handling in the Pulp and Paper Industry. Vol. 2. Miller-Freeman Publications, Inc., San Francisco. 304 p.

Keays, J. L. 1971. Complete-tree utilization. An analysis of the literature. Part V: Stump, roots, and stump-root system. Inf. Rep. VP-X-79, For. Prod. Lab., Can. For. Serv., Dep. Fisheries and For. 62 p.

Kerruish, C. M. 1977. Developments in harvesting technology relevant to short rotation crops. Appita 31(1):41-48.

King, T., and S. Haines. 1979. Soil compaction absent in plantation thinning. Res. Note SO-251, U.S. Dep. Agric., For. Serv. 4 p.

Knight, H. A. 1978. Sizes of timber stands in the Piedmont of South Carolina. U.S. Dep. Agric. For. Serv. Res. Note SE-267. Southeast. For. Exp. Stn., Asheville, N.C. 5 p.

Koch, P. 1972. Changing raw material supplies and their effect upon wood processing technology. Seventh world for. congr., 9 p. Buenos Aires, Argentina.

Koch, P. 1976. New technique for harvesting southern pines with taproot attached can extend pulpwood resources significantly. Appl. Polym. Symp. 28:403-420.

Koch, P. 1977. Harvesting southern pines with taproots is economic way to boost tonnage per acre 20 percent. South. Pulp Pap. Manuf. 40(5): 17, 18, 20, 22.

Koch, P. 1980. Concept for southern pine plantation operation in the year 2020. J. For. 78:78-82.

Koch, P., and C. W. Boyd. 1978. Biomass removal from southern pine sites as affected by harvesting method. In Symp. on Principles of Maintaining Productivity on Prepared Sites. (Tippin T., ed.) p. 54-59. Mississippi State Univ.

Koch, P., and D. W. McKenzie. 1976. Machine to harvest slash, brush, and thinnings for fuel and fiber—a concept. J. For. 74:809-812.

Koch, P., and T. W. Nicholson, 1978. Harvesting residual biomass and swathe-felling with a mobile chipper. In Complete tree utilization of southern pine: symp. proc., New Orleans, La., April 17-19, C. W. McMillin, ed. For. Prod. Res. Soc., Madison, Wis. p. 146-154.

Koch, P., and T. E. Savage. 1980. Development of the swathe-felling mobile chipper. J. For. 78:17-21.

Kochenderfer, J. N. 1977. Area in skidroads, truck roads, and landings in the Central Appalachians. J. For. 75(8):507-508.

Kochenderfer, J. N., and G. W. Wendel. 1978. Skyline harvesting in Appalachia. U.S. Dep. Agric. For. Serv. Res. Pap. NE-400. Northeast. For. Exp. Stn., Broomall, Pa. 9 p.

Kochenderfer, J. N., and G. W. Wendel. 1980. Costs and environmental impacts of harvesting timber in Appalachia with a truck-mounted crane. Res. Pap. NE-456, U.S. Dep. Agric., For. Serv. 9 p.

Krinard, R. M., R. L. Johnson, and H. E. Kennedy, Jr. 1979. Volume, weight, and pulping properties of 5-year-old hardwoods. For. Prod. J. 29(8):52-55.

Kurelek, J. 1975. Multi-function forest harvesting machines—economics & productivity. Amer. Soc. of Agric. Eng. Pap. No. 75-1509, 15 p.

Kwasnitschka, K. 1978. Whole-tree utilization of Norway spruce in centralized timberyards. In Complete tree utilization of southern pine: symp. proc., New Orleans, La., April 17-19. C. W. McMillin, ed. For. Prod. Res. Soc., Madison, Wis., p. 211-216.

Larsen, R. S. 1978. Compendium of major cable logging systems. Interforest AB, Lidingo (Stockholm), Sweden. 112 p.

Larson, G. 1978a. Is revolution due in harvesting shortwood, Logging Manage. 1(5):4OE, 4OF.

Larson, G. 1978b. Tough Louisiana load limits may foreshadow national problem. Logging Manage. 1(5):29-31.

Lavoie, J.-M. 1981. Securing systems for truck loads of tree lengths. FERIC Tech. Note No. TN-41. For. Eng. Res. Inst. of Canada. 13 p.

Leaf, A. L. (ed.). 1979. Impact of intensive harvesting on forest nutrient cycling. Symp. Proc., Syracuse, N.Y., August 13-16. U.S. Dep. Agric. For. Serv. and U.S. Dep. Energy, Fuels from Biomass Systems. 421 p.

Lee, B., and C. J. M. van Soest. 1978. Log handling cranes. In Modern Sawmill Techniques. R. D. French, Ed. Proc., 8th Sawmill Clinic. (Portland, Ore., March.)

Legault, R. 1976. (Evaluation of cable logging systems in interior B.C. and Alberta. FERIC Tech. Note No. 9, For. Eng. Res. Inst. Can., Vancouver. 20 p.

Legault, R. and L. H. Powell. 1975. Evaluation of FMC 200 BG grapple skidder. FERIC Tech. Rep. No. 1, 23 p. For. Eng. Res. Inst. of Can.

Litton, R. B., Jr. 1974. Visual vulnerability of forest landscapes. J. For. 72:392-397.

Logging Research Foundation, Sweden. 1975. Logging in Sweden—design and operation of systems. 27 p. Stockholm. Sweden: Forskningsstiftelsen Skogsarbeten.

Lytle, S. A. 1959. Physiography and properties of southern forest soils. In Southern Forest Soils, p. 1-8, 8th Ann. For. Symp., La. State Univ., P. Y. Burns, ed. LSU Press, Baton Rouge.

McCloud, J. D. 1977. Masonite's whole-tree chipping. In Extending the resource through upgrading and increasing its value, proc. of Mid-South Sect. meet., pp. 25-26. Madison, Wis.: For. Prod. Res. Soc.

McClure, J. A. and E. L. Sargent. 1975. High-volume chip handling in ocean vessel loading. Tappi 58(2):106-109.

McClure, J. P., J. R. Saucier, and R. C. Biesterfeldt. 1981. Biomass in southeastern forests. U.S. Dep. Agric. For. Serv., Res. Pap. SE-227. 38 p.

McGraw, W. E. and R. M. Hallett. 1970. Studies on the productivity of skidding tractors. Can. For. Serv., Dep. of Fish. and For. Publ. No. 1282, 38 p.

McCullar, D. C. 1976. The logma clam skidder. Rev. Am. Pulpwood Assoc. Tech. Rel. 76-R-21, 2 p. Washington, D.C.: Am. Pulpwood Assoc.

McDonald, P. M., and R. V. Whiteley. 1972. Logging a roadside stand to protect scenic values. J. For. 70(2):80-83.

McGee, C. E. 1975. Regeneration alternatives in mixed oak stands. U.S. Dep. Agric. For. Serv. Res. Pap. SE-125. 8 pp.

McIntosh, J. A., and L. W. Johnson. 1974. Comparative skidding performances. British Columbia Logging News 1(11): 13-18.

McMillin, C. W. (ed.). 1978. Complete-tree utilization of southern pine. 484 p. For. Prod. Res. Soc., Madison, Wis.

Malkonen, E. 1976. The effect of fuller biomass harvesting on soil fertility, pp. 67-75. *In* Proc. of symp. on the harvesting of a larger part of the forest biomass, vol. 1. Food and Agric. Organ. of the United Nations.

Mann, C. N. 1969. Mechanics of running skylines. Res. Pap. PNW-75. U.S. Dep. Agric. For. Serv. 11 p.

Martin, A. J. 1971. The relative importance of factors that determine log-hauling costs. U.S. Dep. Agric. For. Serv., Res. Pap. NE-197. 15 p.

Martin, A. J. 1975a. REST: A computer system for estimating logging residue by using the line-intersect method. U.S. Dep. Agric. For. Serv., Res. Note NE-212. 4 p.

Martin, A. J. 1975b. Predicting logging residues: an interim equation for Appalachian oak sawtimber. U.S. Dep. Agric. For. Serv., Res. Note NE-203. 4 p.

Martin, A. J. 1975c. A first look at logging residue characteristics in West Virginia. U.S. Dep. Agric. For. Serv., Res. Note NE-200. 3 p.

Martin, A. J. 1976a. Suitability of the line intersect method for sampling hardwood logging residues. U.S. Dep. Agric. For. Serv., Res. Pap. NE-339. 6 p.

Martin, A. J. 1976b. A logging residue "yield" table for Appalachian hardwoods. U.S. Dep. Agric. For. Serv. Res. Note NE-227. 3 p.

Martin, A. J. 1977. Increased yields from near-complete multiproduct harvesting of a cove-hardwood stand. North. Logger and Timber Proc. 26(4):6, 7, 37.

Massey, J. G., M. P. McCollum, and W. C. Anderson. 1981. Cost of whole-tree chips for energy—Louisiana case study. For. Prod. J. 31(2):34-38.

Matics, H. E. 1975. Whole tree chipping—its effect on saw log production. *In* Impact of inflation on the management and utilization of hardwoods, proc. of the third annu. hardwood symp. of the Hardwood Res. Counc., pp. 60-63.

Matics, H. E. 1978. Dividends and problems of whole-tree utilization. TAPPI Ann. Mtg. Proc. 1978:85-87. (March 6-8, Chicago.)

Mattson, J. A., R. A. Arola and W. A. Hillstrom. 1978. Recovering and chipping hardwood cull trees having heavy limbs. *In* Complete tree utilization of southern pine: symp. proc., New Orleans, La., April 17-19, C. W. McMillin, ed. For. Res. Soc., Madison, Wis., p. 120-130.

Maxwell, W. G. and F. R. Ward. 1976a. Photo series for quantifying forest residues in the: coastal Douglas-fir-hemlock type, coastal Douglas-fir-hardwood type. U.S. Dep. Agric. For. Serv., Gen. Tech. Rep. PNW-51. 103 p.

Maxwell, W. G. and F. R. Ward. 1976b. Photo series for quantifying forest residues in the: ponderosa pine type, ponderosa pine and associated species type, and lodgepole pine type. U.S. Dep. Agric. For. Serv., Gen. Tech. Rep. PNW-52. 73 p.

Merrick, A. M. and M. Martinelli, Jr. 1952. Calculation of volume of hardwood logging waste. J. of For. 50:824.

Mills, C. F. 1980. What ever happened to whole tree chipping? *In* Proc., Ann. Meet., Tech. Assoc. Pulp and Pap. Ind., Atlanta 1980:365-372.

Morey, J. 1975. Conservation and economical harvesting of wood fiber by using the whole tree. Tappi 58(5):94-97.

Morey, J. M. 1978. Separated whole tree chips are changing pulpmill superintendents' minds about whole tree chips as a raw material source. TAPPI Ann. Mtg. Proc. (Chicago) :99-107, March 6-8.

Morner, B. 1976. Survey of chainsaws for logging. World Wood 17(8):28-40.

Mulcahey, J. 1974. Better utilization and better ecology—the Nicholson Complete Tree Utilizer and the Nicholson Ecolo Chipper. Amer. Soc. of Agric. Eng. Pap. No. 74-1566, 7 p.

Murphy, R. L. 1978. Log merchandising—state of the art and trends for the future. *In* Complete tree utilization of southern pine: symp. proc., New Orleans, La., April 17-19. C. W. McMillin, ed. For. Prod. Res. Soc., Madison, Wis., p. 234-245.

Myers, C., D. J. Polak, D. Raisanen, R. C. Schlesinger, and L. Stortz. 1980. Weight and volume equations and tables for six upland hardwoods in southern Illinois. U.S. Dep. Agric., For. Serv., Gen. Tech. Rep. NC-60. 17 p.

Myles, D. V. 1976. Development of a tree-length forwarder. Can. For. Serv. Inf. Rep. FMR-X-88, 71 p. For. Manage. Inst., Ottawa, Can.

Nedashkovskii, A. N., V. Skylar, and N. A. Oleinik. 1978. (Machine for stripping off tree foliage.) Lesnoe Khozvaistvo 1978(8):63-64.

Nelson, B. (n.d.) Total chips cost analysis. 12 p. Morbark Indus., Inc. Winn, Michigan.

Nelson, A. W., Jr. 1976. Whole-tree chipping as viewed by the tree farmer and the industrial landowner. Tappi 59(7):85-86.

Nerbonne, K. B. 1974. Handling and merchandising tree length material. In Modern sawmill techniques, vol. 4, proc. of the fourth sawmill clinic, pp. 188-197. San Francisco, Ca: Miller Freeman Publ. Inc.

Newnham, R. M. 1971. Canlog—the new CFS harvesting machine simulator. Pulp and Pap. Mag. of Can. 72(3):107-112.

Newnham, R. M. 1976. Computer may cut costs $1-2/cord. Pulp and Pap. Can. 77(9):73-79.

Nickels, J. 1978. John Deere forwarders. APA Tech. Res. 78-R-27. Amer. Pulpw. Assoc., Tacoma, Wash. 2 p.

Nilsson, P. O. 1978. Thinning systems used in northern European countries. In Complete tree utilization of southern pine: symp. proc., New Orleans, La., April 17-19. C. W. McMillin, ed. For. Prod. Res. Soc., Madison, Wis., p. 162-184.

Nyland, R. D. 1973. Center of gravity for tree-length hardwood sawlogs. Res. Note 9, 2 p. Appl. For. Res. Inst., State Univ. of N.Y., Syracuse, N.Y.

Nyland, R. D. 1977. Cubic volume tables for second-growth northern hardwoods in New York including English and metric units. AFRI Res. Rep. No. 35, 30 p. State Univ. of New York, Syracuse, N.Y.

Nyland, R. D., D. F. Behrend, P. J. Craul, and H. E. Echelberger. 1977. Effects of logging in northern hardwood forests. Tappi 60(6):58-61.

Oderwald, R. G., W. B. Stuart, and E. C. Ford III. 1979. Location of full-tree centers of gravity for Appalachian hardwoods. South. J. Appl. For. 3(3):123-125.

O'Hearn, S., W. B. Stuart, and T. A. Walbridge. 1976. Using computer simulation for comparing performance criteria between harvesting systems. Pap. No. 76-1567, Winter Meeting of Amer. Soc. Agric. Eng.

Olszowski, A. 1977. Large-unit ro-ro traffic. Ch. 19 in Transport and handling in the pulp and paper industry, Kalish, J., and L. E. Haas, eds., 2:237-246. Miller-Freeman Publications, Inc., San Francisco.

Oswald, D. 1979. Chip transportation—Outside the pulp mill. Ch. 9 in Pulp and paper technology series no. 5, chip quality monograph (J. V. Hatton, ed.), TAPPI Joint Textbook Committee of the Paper Industry. p. 171-190.

Overby, R. C. 1976. Drum chopping cost comparison: Whole tree chipping vs. conventional logging sites. Am. Pulpwood Assoc. Tech. Rel. 76-R-33. 2 p. Washington, D.C.

Patric, J. H. 1976. Soil erosion in the eastern forest. J. For. 74(1):671-677.

Patric, J. H. 1978. Harvesting effects on soil and water in the eastern hardwood forest. South. J. Appl. For. 2(3):66-73.

Patric, J. H. 1980. Some Environmental effects of cable logging in Appalachian forests. Gen. Tech. Rep. NE-55, U.S. Dep. Agric. For. Serv. 29 p.

Patric, J. H. and D. W. Smith. 1975. Forest management and nutrient cycling in eastern hardwoods. U.S. Dep. Agric. For. Serv., Res. Pap. NE-324. 12 p.

Pease, D. A. 1972. Hardwood chips produced in woods. For. Ind. 99(3):60-61.

Pennock, R., Jr., G. W. Wood, P. W. Shogren, Jr., K. Reinhart, V. C. Miles, and P. Younkin. 1975. The forest soil. Ch. 3 in Clearcutting in Pennsylvania. p. 21-32. Pa. State Univ. Coll. Agric. Sch. For. Resourc.

Peters, P. A. 1978. Spacing of roads and landings to minimize timber harvest cost. For. Sci. 24:209-217.

Petro, F. J. 1975. Felling and bucking hardwoods—how to improve your profit. Can. For. Serv. Publ. No. 1291, 141 p. Ottawa, Can.

Phillips, R. A. 1970. Potential of front end log loaders. ASAE Pap. No. 70-618, 12 p. St. Joseph, Mich.: Am. Soc. Agric. Eng.

Phillips, D. R. 1977. Total-tree weights and volumes for understory hardwoods. Tappi 60(6):68-71.

Phillips, D. R. 1981. Predicted total-tree biomass of understory hardwoods. Res. Pap. SE-223. U.S. Dep. Agric., For. Serv. 22 p.

Phillips, D. R., and N. D. Cost. 1979. Estimating the volume of hardwood crowns, stems, and the total tree. Res. Note SE-276, U.S. Dep. Agric. For. Serv. 6 p.

Phillips, R. A. and R. L. Hartman. 1970. Log loading with a rubber-tired front-end loader. The North. Logger and Timber Proc. 19(5):16, 17, 26, 27.

Phillips, D. R. and J. P. McClure. 1976. Composition, volume, and physical properties of the hardwood understory. Proc. fourth annu. hardwood symp. Hardwood Res. Counc., pp. 22-29.

Pickard, D. 1972. Cat tree-lengths harvester producing 4½ cords an hour. Can. For. Ind. 92(7):27-29.

Pierce, R. S., C. W. Martin, C. C. Reeves, G. E. Likens, and F. H. Bormann. 1972. Nutrient loss from clearcuttings in New Hampshire. In Watersheds in transition, symp. proc., pp. 285-295. Fort Collins, Colo. Urbana, Ill.: Amer. Water Resourc. Assoc.

Plummer, G. M. 1971. Increasing productivity and decreasing logging cost through supervision and goal setting. Tech. Rel. 71-R-28, Amer. Pulpw. Assoc. 5 p.

Plummer, G. M. 1972. Increasing productivity through job performance records and goal setting. APA Tech. Rel. 72-R-21. Amer. Pulpw. Assoc. 3p.

Plummer, G. M. 1976. Uses and potential productivity of whole-tree chippers. Tappi 59(7):64-65.

Plummer, G. M. 1978. The APA-TAPPI Whole-tree utilization committee. A review of achievements. Tappi 61(3):45-47.

Porterfield, R. L. and W. W. von Segen. 1976. Improved utilization: approaching the goal in Arkansas. J. For. 74:352-355.

Powell, L. H. 1974a. Evaluation of new logging machines: Arbomatik Roughwood Processor. LRR/58, Pulp and Pap. Res. Inst. Can., Pointe Claire, P.Q., 20 p.

Powell, L. H. 1974b. Evaluation of new logging machines: Tanguay Tree-Length Harvester. LRR/56, Pulp and Pap. Res. Inst. Can., Pointe Claire, P.Q., 24 p.

Powell, L. H. 1975. Evaluation of new logging machines: Forano BJ-20 Feller-Buncher. LRR/62, Pulp and Pap. Res. Inst. Can., Pointe Claire, P.Q., 22 p.

Powell, L. N., J. D. Shoemaker, R. Lazer, and R. G. Barker. 1975. Whole-tree chips as a source of papermaking fiber. Tappi 58(7):150-155.

Pritchett, W. L. 1977. The role of fertilizers in forest production. Tappi 60(6):74-77.

Pritchett, W. L., and C. G. Wells. 1978. Harvesting and site preparation increase nutrient mobilization. In Proc., Symp. on Principles of Maintaining Productivity on Prepared Sites. (Tippin, T., ed.) p. 98-110. Mississippi State Univ.

Pulp and Paper Magazine of Canada. 1969. Multiple-use: Can loggers and recreationists co-exist in the forest? Pulp and Pap. Mag. Can. 70(16):47-49.

Pulpwood Production and Saw Mill Logging. 1973. Westvaco weds yarders to whole tree chippers. Pulpwood Prod. and Saw Mill Logging 21(12):10-12.

Richardson, B. Y. 1979. Substitute anchors for cable logging systems. Field Notes 11(8):1-6. Engineering Tech. Info. Sys., U.S. Dep. Agric. For. Serv.

Routhier, J.-G. 1971. Evaluation of logging-machine prototypes: Tanguay Limber-Slasher. WR/36, Pulp and Pap. Res. Inst. Can., Pointe Claire, P.Q. 26 p.

Row, C. 1974. Effect of tract size on financial returns in forestry. In Forest harvesting mechanization and automation. (C. R. Silversides, ed.) Publ. 5, IUFRO Div. 3, Can. For. Serv., Ottawa, p. 105-134.

Rudolf, P. O. 1967. Silviculture for recreation area management. J. For. 65(6):385-390.

Säll, H.-O. 1981a. Development of harvesters for energy forest plantations. In Proc., International Conf., "Harvesting and utilization of wood for energy purposes" ELMIA, Jonkoping, Sweden, Sept. 29-30, 1980. (J. E. Mattson and P. O. Nilsson, eds.) No. 19, Swedish Univ. Agric. Sci., Dept. Operational Efficiency, Garpenberg, Sweden. p. 118-131.

Säll, H.-O. 1981b. Optimizing establishment, management, and harvesting energy crops from a technical standpoint. In Proc., Joint IEA/IUFRO Forestry Energy Workshop, Garpenberg, Sweden, Oct. 2, 1980. (J. E. Mattson and P. O. Nilsson, eds.) No. 20, Swedish Univ. Agric. Sci., Dept. Operational Efficiency Garpenberg, Sweden. p. 64-73.

Samset, I. 1978. The NISK-delimber in thinnings. Reports of the Norwegian Forest Research Institute, No. 34.4.

Sarna, R. P. 1976a. Caterpillar 225 with feller-buncher in Maine. Tech. Rel. 76-R-40, 8 p. Am. Pulpwood Assoc.

Sarna, R. P. 1976b. Kockums 880 on trial at Great Northern in Maine. Tech. Rel. 76-R-22, Am. Pulpwood Assoc. 8 p.

Sarna, R. P. 1976c. Koehring feller-forwarder—model KFF. Tech. Rel. 76-R-28, 4 p. Am. Pulpwood Assoc.

Sarna, R. P. 1979. Chain saw manual. Interstate Printers & Publishers, Inc., Danville, Ill. 118 p.

Schick, B. A., and W. R. Maxey. 1978. Costs of top lopping old-growth hardwoods: What price beauty? South J. Appl. For 2(3):94-95.

Schlegel, H., and A. Jepsen. 1978. Automatic chip piling, reclaiming system permits true 100% rotation. Pulp and Pap. 52(6):65-67.

Schmidt, J. W., Jr., W. D. Torlone, J. Byrd, and M. R. Fedorko. 1970. Feasibility study on the retrieval and use of primary wood residue. W. Va. Univ. Bull. Ser. 70, No. 8-3, Rep. No. 11, 250 p. Morgantown, W. Va.: W. Va. Univ.

Schnell, R. L. 1978. Biomass estimates of hickory tree components. Tech. Note B30, Div. For., Fish., and Wildl. Dev., Tenn. Val. Auth., Norris, Tenn. 39 p.

Sederholm, J. 1976. How should the stump's wood raw material best be utilized—Is the machine manufactured by Elektro-diesel, Sweden, a solution? Swedish For. Prod. Res. Lab., Stockholm, STFI Series A, No. 296, 31 p.

Shafer, E. L., Jr. 1967. Forest aesthetics—a focal point in multiple-use management and research. Reprinted from 14th IUFRO Congress Papers 7, Sec. 26, pp. 47-71.

Siegel, W. C. 1971. Logging roads and the Federal income tax. For. Prod. J. 21(10):12-14.

Sieker, J. 1955. Management of timber stands on recreation areas. Proc. Soc. Am. For. 54:106-109.

Silversides, C. R., Ed. 1974. Forest harvesting mechanization and automation. Publ. No. 5, IUFRO Div. 3. Can. For. Serv., Ottawa. 555 p.

Simmons, F. C. 1979. Handbook for eastern timber harvesting. U.S. Dep. Agric. For. Serv., Northeast. Area, State and Private For., Broomall, Pa. 180 p.

Sirois, D. L. 1977. Feasibility of harvesting southern hardwood trees by extraction. Amer. Soc. Agric. Eng. Pap. 77-1567, 16 p.

Sirois, D. L. 1983. Biomass of four hardwoods from lower Piedmont pine-hardwood stands in Alabama. U.S. Dep. Agric. For. Serv., Gen. Tech. Rep. S0-46.

Sirois, D. L., and R. H. Iff. 1978. Is there a potential for skyline logging of southern pines. *In* Complete tree utilization of southern pine: symp. proc., New Orleans, La., April 17-19. C. W. McMillin, ed. For. Prod. Res. Soc., Madison, Wis., p. 203-210.

Sloan, E. 1958. The seasons of America past. Wilfred Funk, Inc.: New York. 150 p.

Smith, C. 1978a. Skyline cable logging on the Fernow Experimental Forest. North. Logger and Timber Process. 27(5):20, 21, 32.

Smith, J. R. 1978b. Cost considerations in the transportation of raw materials from the woods to the mill. Pap. Trade J. 162(8):42-44, 46.

Smith, F. K., Jr., and H. R. Johnston. 1970. Eastern subterranean termite. For. Pest Leafl. 68. U.S. Dep. Agric. For. Serv. 8 p.

Smith, H. D., and N. I. Lamson. 1975. Grapevines in 12- to 15-year-old even-aged central Appalachian hardwood stands. *In* Impact on inflation on the management and utilization of hardwoods. Proceedings of third ann. hardwood symp. of the Hardwood Res. Counc. (Cashiers, N.C., May 1-3) p. 145-150.

Smith, F. M., and J. R. O'Dair. 1980. A biomass harvester. *In* Proc., 1980 Ann. Meet., Tech. Assoc. Pulp and Pap. Ind., Atlanta. p. 391-400.

Sondell, J., and A. Svensson. 1975. ÖSA 670 feller-buncher. No. 1 E 1975, Forskningsstiftelsen Skogsarbeten Teknik, Stockholm, Sweden. 6 p.

Springer, E. L., L. L. Zoch, Jr., W. C. Feist, C. J. Hajny, and R. W. Hemingway. 1978. Storage characteristics of southern pine whole-tree chips. *In* Complete tree utilization of southern pine: symposium proceedings, New Orleans, April 17-19, C. W. McMillin, ed., p. 216-223. For. Prod. Res. Soc.

State University of New York College of Environmental Science and Forestry. 1979. Impact of intensive harvesting on forest nutrient cycling. Symp. Proc., Aug. 13-16, 1979. State Univ. of New York Coll. Environ. Sci. and For., Syracuse. 421 p.

Stephens, E. P. 1956. The uprooting of trees: A forest process. Soil Sci. Soc. Amer. Proc. 20:113-116.

Sterle, J. R. 1971. Morbark metro chiparvestors. Amer. Pulpw. Assoc. Tech. Rel. 71-R-19, 5 p.

Sternitzke, H. S. 1978. Coastal Plain hardwood problem. J. For. 76:152-153.

Stevens, D. C., and N. R. Smith. 1978. Wood for energy: Skidding firewood with small tractors. A report on two field trials. Vermont Dept. For., Parks, and Recreation. 15 p.

Stock, S. 1978. Swather reaping benefits at Prince Albert. Pulp Pap. Mag. Can. 79(6):14-17.

Stransky, J. J. 1981. Site preparation effects on soil bulk density and pine seedling growth. South. J. Appl. For. 5:176-180.

Strucka, A. 1979. Chip transportation—Inside the pulp mill. Ch. 10 in Pulp and paper technology series no. 5, chip quality monograph (J. V. Hatton, ed.), TAPPI Joint Textbook Committee of the Paper Industry. p. 191-112.

Stuart, W. B. 1972. An evaluation of the first prototype of the Propst Timber Harvester. Amer. Pulpw. Assoc., Harvesting Res. Proj., Atlanta, Ga. 30 p.

Stuart, W. B., C. D. Porter, T. A. Walbridge, and R. G. Oderwald. 1980. The potential of modifying conventional harvesting systems for recovering logging residues as an energy source. In Proc., Tappi Ann. Meet. 1980: 377-390.

Stuart, W. B., C. D. Porter, T. A. Walbridge, and R. G. Oderwald. 1981. Economics of modifying harvesting systems to recover energy wood. For. Prod. J. 31(8):37-42.

Stuart, W. B., and T. A. Walbridge, Jr. 1974. The harvesting analysis technique—A simulation approach to the evaluation and analysis of harvesting systems. In Forest harvesting mechanization and automation. (C. R. Silversides, ed.) Publ. 5, IUFRO Div. 3. Can. For. Serv., Ottawa, p. 417-431.

Stuart, W. B. and T. A. Walbridge, Jr. 1976. Timber harvesting for the pallet and container industries. Va. Polytech. Inst. Div. Wood Res. & Wood Constr. Lab. Misc. Bull. No. 144, 12 p.

Stuart, W. B., T. A. Walbridge, Jr., S. E. O'Hearn. 1978. Whole-tree chipping and conventional harvesting systems—economic and productivity comparisons of a variety of stand types. Tappi 61(6):60-64.

Studier, D. D., and V. W. Binkley. 1974. Cable logging systems. U.S. Dep. Agric. For. Serv., Div. Timber Manage., Region 6, Portland, Ore. 205 p.

Sturos, J. A. 1971. The dynamic forces & moments required in handling tree-length logs. U.S. Dep. Agric. For. Serv., Res. Pap. NC-55. 8 p.

Switzer, G. L., L. E. Nelson, and L. E. Hinesley. 1978. Effects of utilization on nutrient regimes and site productivity. In Complete tree utilization of southern pine: symp. proc., New Orleans, La., April 17-19. C. W. McMillin, ed. For. Prod. Res. Soc., Madison, Wis., p. 91-102.

Taylor, H. 1978. Low-value timber harvesting. Field Notes 10(12):27-31. U.S. Dep. Agric. For. Serv. Eng. Tech. Inf. System.

Tedder, P. L., G. H. Weaver, and D. D. Kletke. 1978. Timber Ram and ECHO: Alternative harvest planning models for a southern forest. South. J. Appl. For. 2(3):98-103.

Tennessee Valley Authority. 1976. Wood ashes as fertilizer. Selected references. TVA Bibliography No. 1610. Tennessee Valley Authority, Muscle Shoals, Ala. 1 p.

The Norwegian Forest Research Institute. 1978. Report on Forest Operations Research. No. 17, Norwegian For. Res. Inst., Ås/NLH, Norway. 524 p.

Tillman, D. A. 1978. Wood as an energy source. New York: Academic Press. 252 p.

Timson, F. G. 1974. Weight and volume variation in truckloads of logs hauled in the central Appalachians. USDA For. Serv. Res. Pap. NE-300, 9 p. Northeast. For. Exp. Stn., Upper Darby, Pa.

Timson, F. G. 1975. Forwarders come to the Appalachian Mountains. North. Logger and Timber Proc. 24(2):14-15.

Timson, F. G. 1980. The quality and availabilty of hardwood logging residue based on developed quality levels. Res. Pap. NE-459. U.S. Dep. Agric., For. Serv. 10 p.

Timson, F. G., and T. W. Church, Jr. 1972. Appalachian loggers wrestle very few heavyweights! The North. Logger and Timber Process. 20(10):26, 28.

Trousdell, K. B. 1954. Peak population of seed-eating rodents and shrews occurs 1 year after loblolly stands are cut. U.S. Dep. Agric. For. Serv. Res. Note SE-68. 2 pp.

Tufts, D. M. 1976a. Effect of tree size on felling and bunching with Rome Industries' accumulator shear. Am. Pulpw. Assoc., Tech. Release 77-R-4. 6 p.

Tufts, D. M. 1976b. Whole-tree chipping. Tappi 59(7):60-62.

Tufts, D. M. 1977a. Factors influencing the economics of thinning pine plantations. *In* Harvesting, site preparation, and regeneration in southern pine plantations. Timber Harvesting Rep. No. 3, LSU/MSU Logging and For. Oper. Cent., MSU-NSTL Res. Cent., NSTL Station, MS. p. 21-74.

Tufts, D. M. 1977b. Planning a logging operation with analyses of some pulpwood logging systems. *In* Logging cost and production analysis. Timber harvesting Rep. No. 4, ed. 1. LSU/MSU Logging and Forestry Operations Center, Long Beach, Miss. p. 1-33.

Tyre, G. L. and G. D. Screpetis. 1978. System for roundwood inventory and control of veneer, sawtimber, and pulpwood volumes. For. Prod. J. 28(1):40-41.

U.S. Department of Agriculture, Forest Service. 1971. Tractor attachments for brush, slash, and root removal. U.S. Dep. Agric. For. Serv., Equip. Develop. Cen., San Dimas, Calif., ED&T Rep. 7120-3. 77 p.

U.S. Department of Agriculture, Forest Service. 1976. Improved harvesting program. FAB field instructions. U.S. Dep. Agric. For. Serv. S&PF. Washington, D.C. 43 p.

U.S. Department of Agriculture, Forest Service. 1977. Silvicultural activities and nonpoint pollution abatement: a cost-effectiveness analysis procedure. Rep. No. EPA-600/8-77-018, 121 p. Cincinnati, Ohio: U.S. Environ. Protection Agency.

U.S. Department of Agriculture, Forest Service. 1978. Final report—Forest residues energy program. U.S. Dep. Agric. For. Serv., North Central For. Exp. Stn., St. Paul, Minn., 297 p. (Prepared under sponsorship of ERDA. Contract number: E-(49-26)-1045.)

Upton, B. 1976. CIP beats delimbing costs with chains. Pulp and Pap. Can. 76(9):47-50.

Ursic, S. J. 1978. Effect of harvesting systems on water quality and supply. *In* Complete tree utilization of southern pine: symp. proc., New Orleans, La., April 17-19. C. W. McMillin, ed. For. Prod. Res. Soc., Madison, Wis., p. 103-108.

Vidrine, C. G., J. E. Carothers, and J. W. D. Robbins. 1979. "Downtime" in the use of four-wheeled drive rubber-tired logging skidders. Trans. of the ASAE 22(1):1-6, 12.

Vidrine, C. G., and G. E. Woodson. 1982. Bulk densities of selected materials from pine-site hardwoods. (For. Prod. J.32(7):21-24.)

Wagner, F. G., Jr. 1978. Preventing degrade in stored southern logs. For. Prod. Util. Bull., 4 p. U.S. Dep. Agric. For. Serv., State and Priv. For., Atlanta, Ga.

Waide, J. B., and W. T. Swank. 1976. Nutrient recycling and the stability of ecosystems: implications for forest management in the southeastern U.S. *In* America's renewable resource potential—1975: The turning point. Proc., 1975 National Conf. of the Soc. Amer. For., Washington, D.C. p. 404-424.

Walbridge, T. A., and W. B. Stuart. 1978. Future trends and research in timber harvesting. *In* Energy and the Southern Forest. E. T. Choong and J. L. Chambers, Ed. p. 166-170. Proc., 27th Ann. For. Symp., Louisiana State Univ., Baton Rouge.

Walbridge, T. A., Jr., and W. B. Stuart. 1981. An alternative to whole tree chipping for the recovery of logging residues. *In* Proc., International Conf., "Harvesting and utilization of wood for energy purposes", ELMIA, Jonkoping, Sweden, Sept. 29-30, 1980. (J. E. Mattsson and P. O. Nilsson, eds.). No. 19, Swedish Univ. Agric. Sci., Dept. Operational Efficiency, Garpenberg, Sweden, p. 132-148.

Walker, R. F. 1976. Stump extraction and utilization—a literary review with regard to conditions in New Zealand. For. Establ. Rep. No. 93, Proj. No. FE 32, 39. p. Rotorua, New Zealand: Prod. For. Div., For. Res. Inst.

Warren, B. J., and R. A. Kluender. 1976. Capital requirements for logging equipment used in thinnings. Am. Pulpwood Assoc. Tech. Rel. 76-R-45, 2 p. Wash., D.C.: Am. Pulpwood Assoc.

Wartluft, J. L. 1977. Weights of small Appalachian hardwood trees and components. Res. Pap. NE-366, U.S. Dep. Agric., For. Serv. 4 p.

Wartluft, J. L. 1978. Estimating top weights of hardwood sawtimber. Res. Pap. NE-427. U.S. Dep. Agric., For. Serv. 7 p.

Webster, D. B. 1974. Development of a flexible timber harvesting simulation model. FPRS Sep. No. MS-73-S4, 25 p. For. Prod. Res. Soc., Madison, Wis.

Wells, C. G., R. E. Campbell, L. F. DeBano, C. E. Lewis, R. L. Fredriksen, E. C. Franklin, R. C. Froelich, and P. H. Dunn. 1979. Effects of fire on soil. A state-of-knowledge review. U.S. Dep. Agric. For. Serv. Gen. Tech. Rep. WO-7. 34 p.

Wendel, G. W., and J. N. Kochenderfer. 1978. Damage to residual hardwood stands caused by cable yarding with a standing skyline. South. J. Appl. For. 2(4):121-125.

West Virginia University. 1970. Proc. For. Eng. Workshop on Forest Roads. W. Va. Univ. Ser. 71, No. 1-4, 40 p. Morgantown, W. Va. (In cooperation with U.S. Dep. Agric., For. Serv., Northeast. For. Exp. Stn., Upper Darby, Pa.)

White, W. B. 1970. Rail transportation of pulpwood and chips. Pulp and Pap. Mag. Canada 71(11-12):111-118.

White, E. H. 1974. Whole-tree harvesting depletes soil nutrients. Can. J. For. Res. 4:530-535.

White, M. C., R. R. Foil, and R. W. McDermid. 1969. Tree volume affects productivity of hydraulic tree-shears in southern logging. LSU For. Notes No. 85, 2 p. La. State Univ., Baton Rouge, La.

Whitmore, R. A., Jr. and W. L. Jackson. 1956. Increase your profit in the woods. U.S. Dep. Agric. For. Serv., Tech. Pap. 151. 11 p. Cent. States For. Exp. Stn., Columbus, Ohio.

Whittaker, R. H., N. Cohen, and J. S. Olson. 1963. Net production relations of three tree species at Oak Ridge, Tennessee. Ecol. 44:806-810.

Wiant, H. V., Jr., C. E. Sheetz, A. Colaninno, J. C. DeMoss, and F. Castaneda. 1977. Tables and procedures for estimating weights of some Appalachian hardwoods. Bull. 659T, Agric. and For. Exp. Stn., West V. Univ., Morgantown. 36 p.

Williams, J. 1977a. Masonite's whole tree/chip thinning program. Am. Pulpwood Assoc. Tech. Rel. 77-R-45, 3 p. Washington, D.C.: Am. Pulpwood Assoc.

Williams, J. 1977b. Morbark Model 12 in plantation thinning application. APA Tech. Rel. 77-R-44, 2 p. Washington, D.C.: Am. Pulpwood Assoc.

Wimble, A. W. 1976. Mini skidder for thinning. Tech. Rel. 76-R-47, 4 p. Am. Pulpw. Assoc., Washington, D.C.

Witter, C. J. 1975. Good logging roads do make a difference. Va. For. 29(4):12-13.

Woodson, G. E. 1976. Properties of medium-density fiberboad related to hardwood specific gravity. In Proc., 10th Particleboard Symp., Wash. State Univ., Pullman, Wash. p. 175-192.

World Wood. 1978. Feller-buncher can store stems and save swing cycles. World Wood 19(12):28.

Wysor, P. S., and C. D. Goho. 1969. How mountains influence highway trucking speeds and costs in Appalachia. South. Lumberman 219(2728):135-137.

Young, H. E. 1978. Forest Biomass inventory. In Complete tree utilization of southern pine: symp. proc., New Orleans, La., April 17-19. C. W. McMillin, ed. For. Prod. Res. Soc., Madison, Wis., p. 24-28.

Young, H. E., and A. J. Chase. 1965. Fiber weight and pulping characteristics of the logging residue of seven tree species in Maine. Tech. Bull. 17, Maine Agric. Exp. Stn., Univ. Maine, Orono. 44 p.

17

Debarking

Major portions of data drawn from research by:

R. A. Arola	J. L. Keays	G. M. Plummer
R. W. Berlyn	Kockums Industries, Inc.	L. N. Powell
J. M. Blackford	J. F. Lutz	T. Richardson
W. W. Calvert	Manitowoc Shipbuilding Co.	R. J. Smiltneek
Y. S. Chao	J. A. Mattson	J. A. Sturos
A. L. Conner	G. Nakamura	G. B. Tyler
R. P. Derrick	A. W. Nelson, Jr.	S. C. Uhr
J. R. Erickson	R. E. O'Brien	A. Wawer
J. S. Hensel	B. H. Paul	H. Wilcox
S. W. Hooper	E. Perem	G. E. Woodson
P. Koch		

CHAPTER 17
Debarking

CONTENTS

		Page
17-1	FACTORS AFFECTING DEBARKING	1644
	SPECIES	1644
	SEASON	1645
	PRETREATMENT TO EASE BARK REMOVAL	1647
17-2	DRUM DEBARKERS	1649
	WOOD RETENTION TIME	1650
	DRY VERSUS WET DEBARKING	1650
	SHORTWOOD VERSUS LONGWOOD	1651
	POWER TO TURN DRUM	1652
	DRUM PRODUCTIVITY AND SELECTION	1653
	ARRANGEMENT OF DRUM DEBARKERS	1655
17-3	IMPACT DEBARKERS	1655
17-4	RING DEBARKERS	1657
17-5	ROSSER-HEAD DEBARKERS AND POLE SHAVERS ..	1663
17-6	CHEMICAL DEBARKING	1667
17-7	BENEFICIATION OF WHOLE-TREE CHIPS	1667
	NEED FOR BENEFICIATION	1669
	Pulp mills..	1669
	Fiberboard plants	1671
	Particleboard plants	1671
	BENEFICIATION WITHIN THE CHIPPER	1673
	Modified disk chipper	1673
	Arasmith drum chipper..........................	1674
	CENTRALIZED SYSTEMS FOR WHOLE-TREE CHIPS	1675
	PROCESSING OF STUMP-ROOT SYSTEMS	1682
17-8	LITERATURE CITED	1684

CHAPTER 17
Debarking

Bark removal is the first step in most processes for converting pine-site hardwoods to products. Whole-tree chipping is a notable exception; separating bark from the wood in whole-tree chips is discussed in section 17-7.

Pulpmills prefer bark-free wood because most papers can contain little dirt and few bark fibers. Bark removal standards of the pulp and paper industry vary with pulping processes and finished products. In general, a progressively greater degree of bark removal is needed for the following pulping processes: unbleached kraft, bleached kraft, bleached sulfite, groundwood, and unbleached sulfite. Degree of bark removal also varies with end use; semi-chemical pulps used for corrugating stock need not be as clean as bleached chemical pulps used for fine papers. Among similar mills making unbleached kraft pulp, tolerance of bark entering digesters differs with pulp digestion and cleaning procedures. Most mills control bark content of chips to less than 3 percent, but some manufacturers of unbleached kraft pulp may accept chips with 4 percent bark. Mills using other pulping processes allow less (Erickson 1979). For further discussion of the effects of bark inclusion on pulp properties, see section 17-7.

The wood-treating industry removes bark from posts and poles to accelerate drying and to facilitate treatment with chemicals. Sawmills and veneer plants remove the bark from logs so that their wood residue will be bark-free and suitable when chipped for fiber or particle products. Cutting bark-free logs extends service life of saws and veneer knives; also bark-free logs are advantageously sawn because defects are visible. Finally, hardwood bark—8 to 19 percent of the dry weight of aboveground tree portions (table 16-10)—has economic value as fuel and for conversion into various agricultural, fiber, and chemical products.

The many and varied designs of debarkers have been reviewed extensively (e.g., Fobes 1957, Koch 1964 pp 169-178, Holzhey 1969, Weiner and Pollock 1972, Williston 1976, and Erickson 1979). Although pine-site hardwood species differ widely in the ease with which they are debarked, only a few major types of debarkers are in common use in the South. All of them remove bark from logs by one of three methods: rubbing or abrasion in a drum debarker, shear at the cambium layer with tools traveling circumferentially around the log, and cutting with knives mounted in rotating cutterheads to remove all of the bark plus a thin layer of wood. Bark is removed from whole-tree chips in beneficiating plants incorporating steaming, agitation in water, chip compression, air flotation, and screening.

Hydraulic log debarkers developed for use on the West Coast of North America are not used in the South and East, and are therefore not further discussed.

17-1 FACTORS AFFECTING DEBARKING

Completeness of bark removal depends not only on the equipment used, but also on log form and variability in adhesion between bark and wood. Straight, smooth, cylindrical logs are easier to debark than crooked logs with branch stubs, fissures, and stem forks. Adhesion between bark and wood varies among species, with season, and according to time and conditions of storage.

SPECIES

Lutz (1978) classified 25 of the hardwood species that are found on southern pine sites according to the ease with which machines can remove their bark, as follows (based on general industrial experience):

Easy to debark	Intermediate	Difficult to debark
Hackberry and sugarberry	Ash, green	Hickory, bitternut
Sweetbay	Ash, white	Hickory, mockernut
Sweetgum	Elm, American	Hickory, pignut
Tupelo, black	Elm, winged	Hickory, shagbark
Yellow-poplar	Maple, red	
	Oak, black	
	Oak, cherrybark	
	Oak, chestnut	
	Oak, laurel	
	Oak, northern red	
	Oak, post	
	Oak, scarlet	
	Oak, Shumard	
	Oak, southern red	
	Oak, water	
	Oak, white	

For comparison, southern pine is classified as easy to debark.

To help study variations among species, Berlyn (1964, 1965a) devised an apparatus for measuring the strength of bond between bark and wood. His data—taken when bark was tight in December and March—showed strengths of bark-wood bonds in softwoods one-third to one-half those in hardwoods. Values for unfrozen wood of four species were as follows (bond strength is essentially constant from butt to top in a log):

Species	Bond strength between bark and wood in winter
	----------------psi----------------
Elm, American	150
Maple, red and sugar	190
Hickory, bitternut	240
Ash, green	250

Berlyn found that bark was most readily removed from thin-barked species having low bark-wood bond strengths and proposed a mathematical model governing the relationship. His model showed good correlation between predicted and observed bark removal in drum debarkers; exceptions that did not follow the model were ash and hickory.

SEASON

The season of the year in which the wood is cut is the most important single factor affecting ease of debarking. Both hardwoods and softwoods are most readily debarked in the growing season. All pine-site hardwoods are much more difficult to debark in the dormant season; the tightly adhering bark of dormant hickories is particularly difficult to remove.

Easy peeling prevails only during the period of cambial activity, beginning with the swelling of the cambial cells preliminary to cell division, and continuing through the growing season when the cambial initials and the mother cells divide actively. In the beginning of the growing season, bark becomes peelable in a tree from the crown down, progressing downward about 8 feet per day but varying with the weather. Trees of northern species cut and topped during the dormant period and left in the forest may, to some extent, reach a peelable state by the next summer; this easy peeling season on winter cut wood begins a month later and ends earlier than in growing trees (Perem 1958). No comparable data are published on southern hardwoods, but summer dry storage in the South is impractical.

Pine-site hardwoods can be debarked more readily when wet than when their bark has dried; for this reason and to prevent losses from stain, beetle attack, and shrinkage checks, southern mills do not permit hardwood logs to dry in the yard.

Substantial increases in bark-wood bond strength occur below 45°F. On frozen hardwoods, bond strength may be several times that of unfrozen wood (O'Brien 1977).

Wilcox et al. (1956) developed an apparatus to measure the force required to shear bark from live standing trees. In determining the optimum season of the year for debarking they defined a **best peeling condition** to coincide with a 2000 g or less force on an 8-inch lever arm shearing a 1¼-inch isolated disk of bark from a standing tree. They called this measure **resistance to peeling** and made observations weekly on four to six trees of eight species from April 16 to August 31 in the central Adirondacks. The period of best peeling varied from 46 days for beech (*Fagus grandifolia* Ehrh.) to 110 days for red spruce (*Picea abies* Sarg.). Periods of best peeling for two species that occur on southern pine sites were as follows:

Species	Dates	Number of days
White ash	April 29—August 7	96
Red maple	May 4—August 12	96

Woodson[1] slightly modified the technique of Wilcox et al. (1956) to determine seasonal variation during 1980 and 1981 in bark peeling resistance of 13.

---

[1] Woodson, G. E. Unpublished data in Final Report FS-SO-3201-4, dated November 18, 1981. U.S. Dep. Agric., For. Serv., South. For. Exp. Stn., Pineville, La.

species of 6- to 10-inch-diameter pine-site hardwoods in central Louisiana. By measuring the torque required to shear 2-inch-diameter isolated disks of bark at breast height from growing trees, he found that bark was most difficult to remove from November through February and easiest to remove in June. Peeling resistance, as measured by torque, averaged 341 inch-pounds in February and only 67 inch-pounds in June (fig. 17-1). Hickory and water oak had greatest resistance to peeling in winter (593 and 430 inch-pounds, respectively), and white and post oaks had least (275 and 268 inch-pounds). In June, all had peeling resistance less than 80 inch-pounds torque, except blackjack oak which required 153 inch-pounds. Variation in moisture content of the disks of bark removed was substantially greater among species than with season. Average February moisture content was 57 percent of ovendry weight; the average for June was 63 percent (table 17-1).

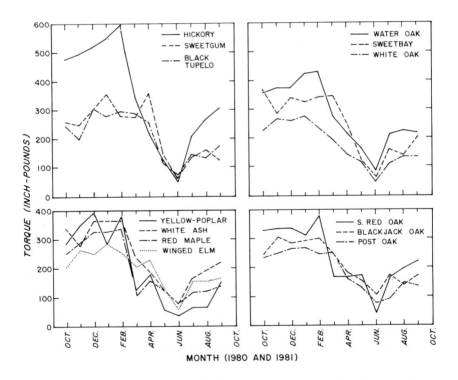

Figure 17-1.—Seasonal variation in resistance to peeling of bark from 13 species of hardwoods measuring 6 to 10 inches in dbh growing on southern pine sites in central Louisiana. Each curve is an average for three trees evaluated at breast height. (Drawing after Woodson; see text footnote[1].)

TABLE 17-1.—*Seasonal variation in moisture content (ovendry weight basis) in bark at breast height on 6- to 10-inch hardwoods of 13 species sampled during 1980 and 1981 in central Louisiana* (Data from G. E. Woodson[1])[2,3]

Species	Dec.	Jan.	Feb.	Mar.	Apr.	May[4]	June	July	Aug.	Sept.
	----------Percent moisture content----------									
Ash, white	42	36	43	58	57	81	60	50	38	39
Elm, winged	52	44	55	57	62	82	56	49	44	41
Hickory, mockernut	59	60	59	65	66	72	73	70	55	55
Maple, red	61	51	67	55	55	69	58	60	55	49
Oak, blackjack	40	39	43	43	48	52	46	40	38	43
Oak, post	35	31	41	43	41	56	41	41	37	31
Oak, southern red	42	42	47	49	46	57	49	47	38	40
Oak, water	43	38	37	39	44	58	46	44	42	40
Oak, white	48	41	46	45	49	53	47	51	41	39
Sweetbay	99	86	88	91	94	105	98	91	90	89
Sweetgum	—	43	54	57	47	77	51	55	47	45
Tupelo, black	55	49	48	58	66	117	69	68	46	45
Yellow-poplar	122	107	116	123	126	140	130	116	109	103
Average	—	51	57	60	62	78	63	60	52	51

[1]Final Report FS-SO-3201-4, dated November 18, 1981, Southern Forest Experiment Station, U.S. Dep. Agric., For. Serv., Pineville, La.
[2]Hardwood trees sampled were growing among southern pines. Each value is an average of three trees.
[3]No data taken in October and November.
[4]Data of May 19, 1981 were taken following a 4- to 5-inch rainfall.

PRETREATMENT TO EASE BARK REMOVAL

Paul[2] found that bark of winter-cut ash could be readily removed if logs were stored 3 or 4 weeks at 70° to 80°F in air of 100-percent relative humidity. He also noted that heating logs in hot water (190° to 200°F) for two hours loosened bark of white spruce (*Picea glauca* (Moench) Voss); only one hour was required to loosen bark of Douglas-fir (*Pseudotsuga menziesii* (Mirb.) Franco). Steaming for two hours at atmospheric pressure also loosened bark of white spruce.

Greaves (1968) found that steaming *Eucalyptus* sp. at 100 to 105°C for 2½ hours loosened bark sufficiently for easy removal. Vogt (1965) noted similar results with birch (*Betula* sp.) and spruce, and further observed that longitudinally scored cuts through the bark of birch logs facilitated bark removal.

Calvert and L'Ecuyer (1972) found that brief application of mechanical pressure normal to the bark loosened it somewhat. The pressure, applied to frozen softwood bolts with a roller, varied from 2,100 to 2,800 psi.

Woodson[1] found that bark on logs of southern red oak, red maple, sweetgum, and hickory trees 6 to 10 inches in diameter cut in February 1981 from among southern pines in central Louisiana could be loosened by steaming the logs at 212°F for 2 to 6 hours (table 17-2). Heating in water at 120°F for these periods did not loosen the bark, and 6 hours in water at 160°F loosened the bark only slightly.

[2]Paul, B. H. 1931. Peeling bark from winter-cut timber. Unpublished manuscript, U.S. Dep. Agric. For. Serv., For. Prod. Lab., Madison, Wis.

He found that steaming at 212°F or extended water soaking at 160°F more effectively loosened the bark if preceded by brief mechanical application of 1,000 psi pressure normal to the bark surface. When such pressure was followed by 6 hours of steaming, bark of all four species was removable without measurable shear force; 4 hours of steaming was equally effective on three of the species, but hickory bark still had slight resistance to peeling (table 17-2).

TABLE 17-2.—*Effect of type, intensity, and duration of heat—with and without a momentary mechanical pressure (1,000 psi) exerted normal to the bark—on resistance to peeling bark[1] from winter-cut 6- to 10-inch-diameter logs of four species of hardwoods grown on southern pine sites* (Data from G. E. Woodson[2])

Species and duration of heating (hours)	Control[3]	Heated in water at 120°F	Heated in water at 160°F	Heated in steam at 212°F
Inch-pounds torque				
WITHOUT PRELIMINARY APPLICATION OF PRESSURE				
Hickory, mockernut				
2	407	390	383	157
4	—	317	360	167
6	—	397	357	97
Maple, red				
2	210	220	230	93
4	—	173	217	97
6	—	220	180	53
Oak, southern red				
2	220	233	243	137
4	—	180	233	113
6	—	247	177	90
Sweetgum				
2	220	197	143	133
4	—	210	163	103
6	—	193	173	50
AFTER APPLICATION FOR 10 SECONDS OF A FORCE EXERTING 1,000 PSI PRESSURE NORMAL TO BARK SURFACE				
Hickory, mockernut				
2	340	367	383	153
4	—	390	293	70
6	—	373	330	0
Maple, red				
2	183	193	207	67
4	—	150	173	0
6	—	260	140	0
Oak, southern red				
2	207	243	183	160
4	—	167	200	0
6	—	207	160	0
Sweetgum				
2	210	207	183	133
4	—	170	180	0
6	—	197	170	0

[1] Measured by modifying the technique of Wilcox et al. (1956) to shear bark from a 2-inch-diameter isolated disk of bark. Each value is the average of three replications.
[2] Final Report FS-SO-3201-4, dated November 18, 1981, Southern Forest Experiment Station, For. Serv., U.S. Dep. Agric., Pineville, La.
[3] Freshly cut wood heated in air at 70°F; February 1981.

17-2 DRUM DEBARKERS

Rotating-drum debarkers (fig. 17-2) remove bark from pulpwood bolts and logs by abrasion against each other and against the corrugated interior of the drum as they are tumbled or rolled by the drum's rotation. Some drums are not corrugated but have steel staves welded to their interior. Debarking drums may be as short as 22 feet or as long as 200 feet; diameters are from 7 to 15 feet (O'Brien 1977).

Drum debarkers find favor in the southern pulp industry for several reasons. They are easy to feed with multiple sticks of pulpwood. They can—by suitable design of infeed and outfeed conveyors—accommodate shortwood (e.g., 4-foot lengths) or longwood (e.g., 24-foot lengths). One operator can tend several drums. Horsepower requirements are reasonable. Maintenance needs are not excessive, and productive capacity is in the range needed by the industry. A disadvantage of drum debarkers is their tendency to waste wood by breaking logs and bolts and "brooming" their ends (Tyler[3]).

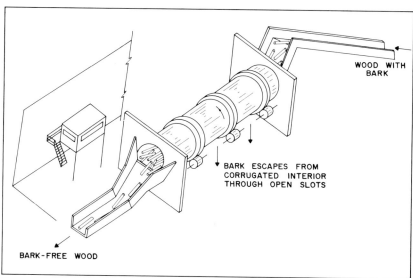

Figure 17-2.—Drum debarker. (Drawing after Erickson 1979.)

Drum debarked southern pine bolts typically display fiber damage around the periphery of the log extending 2 inches from both ends along the grain and penetrating 1/4- to 1/2-inch radially. Hardwood species suffer less damage from "brooming" than southern pine. Some proponents of drum debarkers contend that wood losses in drums are less than those from mechanical ring debarkers.

---

[3]Tyler, G. Billy. 1969. Capacity of large vs. small barking drums. TAPPI 24th Engineering Conference Paper dated September 16, 1969. 29 pp.

Berlyn (1965b) found that bark removal from a pulpwood stick in a drum debarker starts at the ends of the stick and advances toward midlength. Edges of intact bark areas are more vulnerable to drum action than the centers of bark areas. Areas of bare wood surrounded by intact bark seldom occur. Bark is removed in two stages: first the bark-to-wood bond is broken for some distance along and inwards from an exposed edge of bark; then, the edge bond is broken between the loosened patch and the area still intact. Berlyn observed no overall loosening of bark during passage of a pulp stick through a drum.

The drums are rotated by girth sprockets, gearing, or friction drives to achieve surface speeds on the drum shell of from 150 to 300 feet per minute for applications where short bolts tumble, to 370 to 590 feet per minute where longer logs roll in parallel mode. To drive the drum, 75 to 500 hp is usually required. Riding rings, encircling drum shells, are supported as they revolve on rigid or shock-absorbing trunnions or on chain suspension systems (O'Brien 1977).

The drum is constructed with a steel shell contoured or corrugated on its interior surface to promote agitation of logs or bolts. The corrugations, or lifters, are smooth with relatively large radii to minimize wood loss. Normal wood loss during drum debarking is about 1 percent. Loosened bark falls through slots in the shell of the drum, and is conveyed to fuel storage. The drum shell is subjected to severe shocks, particularly when debarking hardwoods that yield little on impact with the drum. Shell maintenance is reduced if wood loads are about 50 percent of drum volume, thereby permitting wood to be cushioned against wood, absorbing some of the impacts (O'Brien 1977).

WOOD RETENTION TIME

Bark is removed primarily by friction between bolts or logs. Wood depth within the drum and drum diameter are therefore positively correlated with percentage of bark removed.

Retention time in the drum varies directly with drum length. It is reduced as drum diameter and speed of revolution increase. A drum is normally set so that its centerline is level; feed rate through the drum is dependent on the slope of the wood-load gradient from infeed to outfeed. Intermittent feed results in a fluctuating discharge rate (O'Brien 1977).

Because of these relationships between drum dimensions and bark removal efficiency and wood retention time, the ratio of drum diameter to length is important. Wood depth and gradient must be maintained at the drum discharge point. Should depth be sacrificed for increased flow, part of the drum becomes a conveyor and not a bark removal device. Wood depth at discharge is controlled by a gate or dam facing the outlet end of the drum.

DRY VERSUS WET DEBARKING

In the southern pine region most drum debarking is performed dry. In the North, however, the first sections of the drum often thaw frozen bark in a hot water bath to facilitate its removal. Hooper's (1973) cost analysis indicates that wet drum debarkers out-produce dry ones of the same size, but costs per cord debarked are higher than for a dry system sized for comparable output because of equipment and effluent treatment costs. Substituting steam for hot water reduces pollution problems and produces drier bark, more valuable for fuel. Brief steaming probably loosens bark more effectively than soaking in hot water for the same period (table 17-2); data are not published, however, for the short dwell times common in a drum barker.

Erickson (1979) concluded that dry drum debarking, aided by steam in northern regions, will be preferred in future installations. For equivalent production, drum length and diameter must be increased, but dry debarking avoids pollution problems inherent in the wet system. Beak Consultants Ltd. (1978) noted, however, that wet drum debarking of hardwoods removes a greater percentage of bark than does dry debarking. With the exception of the hickories, bark of dormant pine-site hardwoods (even when frozen) can be removed to an acceptable degree by the dry system if retention times in the drum are appropriately extended. Most operators of drum debarkers are reluctant to accept hickory; in spring the bark of some hickories is released in sheets that foul conveyers, and in winter the bark is too tight to remove.

SHORTWOOD VERSUS LONGWOOD

If length of wood introduced to the drum approaches or exceeds the drum diameter, the wood tends to align parallel to the drum's central axis. For species with tightly adhering bark, and for crooked wood, the tumbling action of shortwood removes bark more effectively than the rubbing action obtained in the parallel mode (O'Brien 1977).

Even short logs can be debarked in the parallel mode if drum rotational speed is increased. Such increase in speed diminishes wood retention time and increases throughput, but reduces percentage of bark removed. Tight-barked shortwood is therefore generally debarked in the tumbling mode.

Longwood can be operated in the parallel mode more readily than shortwood but to equal the retention time possible in the shortwood tumbling mode, drums designed for longwood parallel mode must be larger in diameter and longer. Many of the longwood systems discharge logs from the lower portion of the drum end to avoid log whipping. Successful drum debarking requires careful design of infeed and outfeed chutes (fig. 17-4ABCD).

Mills that debark 24-foot lengths using the parallel mode have opportunity to obtain greater value from logs by crosscutting to yield bark-free sawlogs as well as pulpwood. Chips from longwood are of higher quality and require less rechipping than those from tumbled shortwood (O'Brien 1977).

East and Engelgau (1978) have described a system in which southern hardwood logs 12 to 22 feet long are fed directly from incoming trucks without slashing to shorter lengths; the system has performed well.

POWER TO TURN DRUM[4]

Data on debarking drums operated in the parallel mode are incomplete, but Derrick (1978) suggests that they be supplied with 15 percent more power than those operating in the tumbling mode. Wet-operation of debarking drums requires about one-third more power than dry-operation. Because dry debarking drums operated to tumble shortwood predominate in the South, discussion is confined to this type.

Debarking drums require power to lift bolts about half the diameter of the drum as they tumble; power is thus proportional to wood density and to percentage of solid wood in the mass of bolts being lifted. Debarking drums operate in the tumbling mode at from 5.75 to 7.5 rpm, one-fourth to one-third the speed at which centrifugal force causes wood to remain on the drum shell instead of tumbling. Power requirement is proportional to the cube of the drum diameter, and directly proportional to drum length and rpm.

For wood density of 50 pounds per cubic foot (green basis) recommended drive capacity horsepower (RDC) is as follows:

$$RDC = 0.00030 \ D^3LN + 25.0 \qquad (17\text{-}1)$$

Normal running load horsepower (NRL) is as follows:

$$NRL = 0.00025 \ D^3LN + 20.0 \qquad (17\text{-}2)$$

where:
- D = drum diameter, feet
- L = drum length, feet
- N = drum speed, rpm

Equations 17-1 and 17-2 are based on a dry drum operating in the tumbling mode in which the drum is filled to half the drum diameter, with solid wood component amounting to 15 to 20 percent of the total volume occupied by the tumbling wood. Pine-site hardwoods weigh significantly more (52 to 69 pounds per cubic feet, green basis) than the 50 pounds per cubic foot assumed by Derrick in deriving equations 17-1 and 17-2; therefore, debarking drums for pine-site hardwoods require proportionately more horsepower than the equations indicate.

Use of equations 17-1 and 17-2 to select the drive for a three-section, 14.5-foot-diameter, 40-foot-per-section dry debarking drum to operate at 6.25 rpm on 60-pound-per cubic foot logs yields values as follows:

$$RDC = 253.6 \times \frac{60}{50} = 304.3 \text{ horsepower per section}$$
$$NLR = 210.5 \times \frac{60}{50} = 252.6 \text{ horsepower per section}$$

A 300-horsepower motor per section is therefore suggested. The drive system should have at least 210 to 230 percent breakaway and starting torque available.

The most common drive for a debarking drum is a severe-duty squirrel-cage induction motor (1.15 service factor) connected by a fluid coupling which permits the motor to attain full speed very rapidly. The speed-torque characteristics of the motor must be carefully matched to the load.

---

[4]Text under this heading is condensed from Derrick (1978).

Wound-rotor motors have also been applied; they provide limited speed control, high starting torques, smooth acceleration during frequent starts without necessity of fluid couplings, eddy-current couplings, or mechanical clutches in the power train. Such motors are favored for larger drums even though initial costs are greater than for fluid-coupled, squirrel-cage induction motors (Derrick 1978).

DRUM PRODUCTIVITY AND SELECTION[5]

Wood handling operations at mill sites can be functional only about 18 hours per day; thus a 1500-cord-per-day mill must be able to debark about 83.5 cords per hour.

Production rate of debarkers is directly proportional to drum length and to the square of drum diameter. Dry drums operating on mixed southern hardwoods in the tumbling mode debark from 17 to 36 cords per hour depending on drum size and season of the year (table 17-3).

During winter, debarking power needed for the small drum was 13.2 horsepower per cord/hour; the large drum required 14.5 horsepower per cord/hour. Maintenance costs for the small drum should be about $4,000 per year (1969) and about 25 percent more for the large drum. In 1969, installed cost of the small drum was $169,400, while that of the large drum was $240,500.

Normal wood loss during drum debarking is about 1 percent, but if a drum is operated at 20 cords per hour when it is capable of 40 cords per hour, wood loss of 2 percent can be expected. Also, a 2-percent additional loss of wood can be expected for each rpm of drum speed above 6.5 rpm.

Debarking to supply a 1,500-cord-per-day mill with mixed southern hardwoods, 95-percent bark-free, during winter, could be accomplished with five small drums or three large drums. Although most efficient for winter conditions, the three large drums would cause substantial wood loss in summer because they would operate at 25 percent below capacity. Therefore, five small drums afford a better solution, since one could be shut down during the easy-barking summer season to avoid such wood loss.

Figure 17-3 shows the linear relationship between debarker productivity and drum volume during winter and summer in the South. During both seasons, output of drum debarkers is significantly greater on pine than on hardwoods.

Data provided by manufacturers (table 17-4) may vary considerably from those in figure 17-3. Drum debarkers are expensive; the smallest drum tabulated in table 17-4 was estimated to cost about $450,000 installed (1979). A 9-foot by 65-foot mill-made economy model described by Coleman and Evert (1977) cost about $150,000.

---

[5]Text under this heading is condensed from: Tyler, G. Billy. 1969. Capacity of large vs. small barking drums. TAPPI 24th Engineering Conf. Paper dated September 16, 1969. 29. p.

TABLE 17-3.—*Productivity of two sizes of dry drum debarkers operated in the tumbling mode on summer- and winter-cut southern hardwoods* (Tyler[1])

Drum class	Drum diameter	Drum length	Productivity	
			Summer	Winter
	------------Feet------------		------------Cords/hour------------	
Small (225 hp)............	12.0	67.5	22	17
Large (420 hp)............	14.5	80.0	36	29

[1]Tyler, G. Billy. 1969. Capacity of large vs. small barking drums. TAPPI 24th Engineering Conf. Paper dated September 16, 1969. 29 p.

TABLE 17-4.—*Debarking rates for 8-foot-long hardwoods in dry drums of various sizes operating in summer and winter*[1/2]

Drum diameter and length (feet) and season or condition	85-percent bark removal	95-percent bark removal
	----------------------Cords/hour----------------------	
7 by 45		
Summer........................	20	15
Winter (unfrozen)................	18	13
Winter (frozen)..................	13	10
12 by 45		
Summer........................	40	30
Winter (unfrozen)................	30	22
Winter (frozen)..................	22	17
12 by 55		
Summer........................	50	40
Winter (unfrozen)................	45	40
Winter (frozen)..................	30	25
12 by 68		
Summer........................	65	55
Winter (unfrozen)................	60	45
Winter (frozen)..................	40	30
12 by 75		
Summer........................	75	65
Winter (unfrozen)................	65	50
Winter (frozen)..................	45	35
12 by 80		
Summer........................	80	75
Winter (unfrozen)................	75	60
Winter (frozen	50	40
14.5 by 80		
Summer........................	105	90
Winter (unfrozen)................	90	70
Winter (frozen)..................	60	50

[1]1979 data from Manitowoc Shipbuilding Co.
[2]Rates are for 100 percent operating time with optimum conditions of feed, load, drum speed, and discharge. Rates would be lower for very-difficult-to-debark woods such as hickory, birch, and frozen aspen.

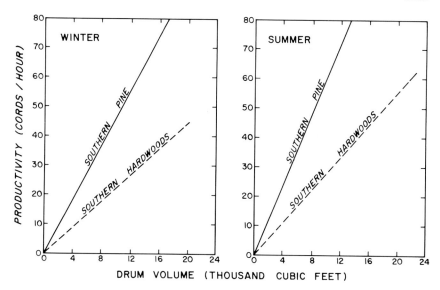

Figure 17-3.—Productivity of dry drum debarkers operating in tumbling mode (95-percent bark removal) on southern hardwoods and southern pine, related to volume of the drum. Both graphs depict Midsouth conditions. (Drawing after Tyler 1969[5].)

ARRANGEMENT OF DRUM DEBARKERS

Drum debarkers for pulpwood bolts can be arranged in multiples with infeed conveyor parallel (fig. 17-4A) or perpendicular to drum axis. Longwood parallel-mode operations require different conveyor arrangements (fig. 17-4BCD). Readers interested in the advantages and disadvantages of recovering high-grade logs from whole stems before or after drum debarking will find Harmon's (1977) discussion useful. Gilmer (1974), Nordquist (1975, p. 45), Tuuha and Sjogren (1976), and Koskinen and Sjogren (1977) discuss the shift toward parallel-mode debarking of longwood and mill layouts appropriate for this system.

17-3 IMPACT DEBARKERS OF DRUM TYPE

Hardwoods such as hickory, which are very difficult to debark when dormant, have stimulated efforts to further increase effectiveness of drum debarkers. Nakamura and Ohira (1964) equipped a drum debarker with a central drive shaft carrying spiral cutter knives; while this arrangement removed 90 percent of the bark in one-fourth the time required by an ordinary drum debarker, the knives caused too much wood loss. Knives that remove tight bark also notched some logs. Possibly this concept deserves re-examination for application to winter-cut hickory.

Increased utilization of logging residues calls for methods to remove bark from irregularly shaped pieces. Smiltneek (1974) found that by adding a quanti-

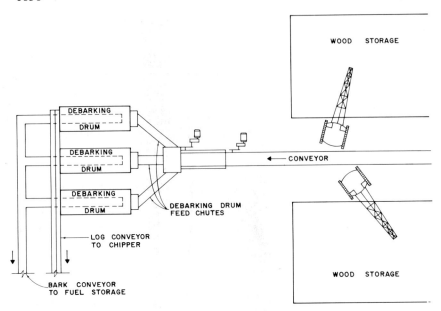

Figure 17-4A.—Multiple shortwood drum debarkers with infeed conveyor parallel to drum axes. (Drawing after Tyler [5].)

Figure 17-4B.—Single-line, long-log debarking system. Logs up to 24 feet in length enter drum via conveyor and infeed hopper at right, are parallel-tumbled in the drum (center), and discharged at left. (Photo from Fibre Making Processes, Inc.)

Debarking

Figure 17-4C.—(Top left) Long logs in-line conveyed to drum debarker. (Top right) Infeed hopper with long logs entering debarker drum. (Photos from Fibre Making Processes, Inc.)

ty of steel balls to the charge of wood in a batch-type drum debarker, extremely irregular chunks of wood (e.g., root portions) could be cleanly debarked quickly. Because separation of the steel balls from the cleaned wood was deemed a problem, the U.S. Forest Service Equipment Development Center at San Dimas, Calif. designed and tested a prototype debarker in which chain flails were attached to a rotating (300 rpm) central shaft; this drum, which rotated at 7 rpm, cleaned a charge of irregular softwood logging residues of bark, rot, dirt, and char in 2 minutes (Hensel 1975). While not tested on hardwoods, the chain-flail machine should work on them, but more slowly. Practicality of the chain-flail method appeared questionable, however, because of rapid chain wear and problems in modifying the machine for continuous rather than batch operation.

17-4 RING DEBARKERS

Capital costs for ring debarkers are considerably lower than for drum debarkers and power requirements per cubic foot of wood debarked are only about one-sixth as large (Beak Consultants Limited 1978). They are widely used to remove bark from southern hardwood sawlogs and veneer logs. In these debarkers, curved, scraping-tipped tools, mounted on a rotating ring apply radial and tangential pressure to shear bark at the cambium layer as the log passes through the ring (figs. 17-5 and 17-6 top). The barking tools are pivoted at the ring and forced against the log by springs, rubber bands, or hydraulic or air

Figure 17-4D.—Discharge orifice and log conveyor from long-log drum debarker. (Photo from Fibre Making Processes, Inc.)

pressure. The climbing edge of each debarking tool is projected forward somewhat and also ground sharp. As the log is driven against the closed tools by the debarker's feed mechanism, it meets these sharp edges first and the edges cut into the end of the log. Since the tools are rotating, they follow the cut in the log until the bark-removal edge of the tool reaches the bark periphery. The velocity of the tool as it opens causes it to open about 3/4-inch past the periphery of the bark, after which it plunges through the bark to the cambium layer. This amount of penetration is accomplished by adjusting the tensioning devices on the debarking tools. Rotation speed of the ring carrying the tools varies from about 90 to about 440 rpm, the higher speeds being suitable for small logs (table 17-5). Feed speeds of a mechanical ring debarker are determined by rotor speed, the number of tools mounted on the rotor, and the amount of overlap desired on tool paths. Feed speeds practical on dense hardwoods, such as oaks, are about one-half to two-thirds those appropriate for southern pine.

Figure 17-5.—Mechanical ring debarker. (Top) Outfeed side of a 26-inch, five-knife, rotating-ring mechanical debarker. (Bottom right) Southern red oak log entering the debarker, centered by spiked infeed rolls. (Bottom left) Southern red oak log on emergence from debarker; roughened surface is evidence that bark adhered tightly to wood. (Photos from Kockums Industries, Inc.)

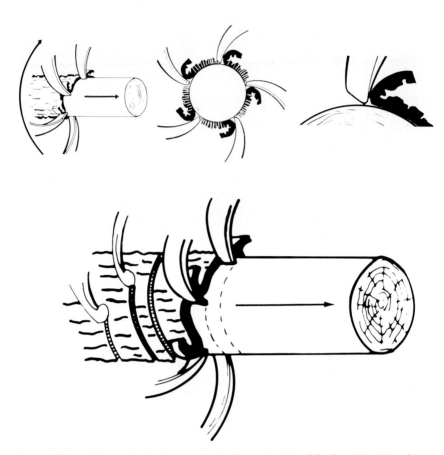

Figure 17-6.—Action of tools in a mechanical rotating-ring debarker. (Top) Usual arrangement of five debarking tools for easily debarked logs such as southern pine. (Bottom) Scoring knives in separate ring scribe the bark in a spiral pattern before debarking tools shear the cambium layer; fibrous barks are thereby reduced to manageable dimensions. In both drawings, knives and debarking tools are mounted in single transverse planes, but trace spiral paths relative to the advancing log. (Drawings from Kockums Industries, Inc.)

TABLE 17-5.—*Characteristics and feed speeds of five-knife mechanical ring debarkers suitable for southern oaks[1]* (Data from Kockums Industries, Inc.)

| Ring | | | Feed | | Average log |
Diameter	Speed	Power	Speed[2]	Power	diameter
Inches	Rpm	Hp	Fpm	Hp	Inches
14	440	30	100	5.0	5
18	361	50	150	7.5	9
21	242	40	80	5.0	9
26	222	50	95	5.0	11
30	221	75	150	15.0	11
35	90	75	70	7.5	15
40	90	75	70	7.5	15

[1]Hickories, elms, and hackberry may require special debarking techniques; see text description and figure 17-6.
[2]Appropriate for summer-cut wood; winter-cut oak would be fed more slowly.

On summer-cut wood of most hardwoods, the tangential force of each tool as it follows a spiral path over the surface of the wood, causes shear stresses in the cambium layer sufficient to remove the bark. The working edges of the debarking tools are slightly rounded so that a minimum of wood is removed. For tight-barked, winter-cut birch (*Betula* sp.) and aspen (*Populus* sp.) Frid (1967) found that optimum radius of tool edge was 1.5 mm; he noted that for these species the tool edge should be about 5 cm long and that two or three passages of a tool were required to remove the bark. Richardson[6] recommended that paths of succeeding tools should overlap at least 50 percent when debarking pine-site hardwoods. Berlyn (1970) found that frozen northern softwood logs peeled about the same, whether fed butt first or top first into the debarker. No comparable study has been made on southern hardwoods.

Richardson[6] advised that for hardwoods both infeed and outfeed feedworks should be as close to the debarking ring as possible for best centering of crooked logs. Branches should be smoothly trimmed, as hardwood branch stubs interfere with debarking tools.

Soft hardwoods such as sweetgum, yellow-poplar, black tupelo, and sweet-bay debark rather easily at high speeds with the same tool pressures used for southern pine. Summer-cut southern red, water, white, black, and post oaks can be debarked at fairly high speed when green; when winter cut, these oaks require 25 to 50 percent slower feed speeds.

---

[6]Richardson, T. 1975. Experience with ring debarkers on hardwoods from southern pine sites and comments on new developments to cope with difficult species. Paper presented at a symposium, "Utilization of hardwoods growing on southern pine sites," Alexandria, La., March 10-14.

Some winter-cut trees—particularly hickories—have bark tightly bonded to the wood at the cambium layer. Sharpened knives and increased tool forces cause substantial wood losses and risk machine damage and unacceptable maintenance costs. Hot water soaking or steaming of logs helps weaken the cambium layer (table 17-2), thereby facilitating bark removal—but even with such pretreatment, bark of winter-cut hickory is hard to remove. Double-ring machines are sometimes employed to get tool pressure applied twice to all points on log surfaces. Another pre-treatment that facilitates bark removal from tight-barked winter-cut frozen wood is application of either a circumferentially-directed or a longitudinally-directed rolling load perpendicular to the log surface just prior to its passage through the debarking ring (Calvert and L'Ecuyer 1972).

A 10-second static load, applied by pressing a plate perpendicular to the surface of the bark to achieve pressure of 1,000 psi, significantly reduces resistance to peeling if logs are subsequently steamed for 4 to 6 hours (table 17-2).

Hickory also presents problems in summer, when its bark may peel from logs in large sheets. These strong fibrous sheets cause delays by clogging tools and conveyors. Many operators of ring debarkers will not undertake to debark either winter-cut or summer-cut hickory. If hickory is needed in the mill, it may be passed through the debarker conveyor without engaging the tools.

Fibrous bark, such as that of the elms, hackberry, sugarberry, and the hickories, may clog conveyors because of its string-like nature. With these species, scoring knives (i.e., ringing tools) can be alternated with shear tools in a rotor carrying an even number of tools. The scoring knives ring the log in barber-pole pattern and cut the bark into short lengths to prevent it from tangling like a rope in conveyors. Alternatively, a double-rotor machine can be utilized (fig. 17-6 bottom) in which the first rotor carries ringing tools and the second finishes the job with debarking tools.

In contrast to southern pine, southern hardwoods may have flared butt sections that are difficult to debark and troublesome for subsequent manufacturing processes. One solution is to machine flared butts to cylindrical shape with a "reducer ring" independent of the debarking ring. For best results, logs should be presorted by diameter, and reducer rings of appropriate diameter employed with each diameter class. A 24-inch debarker fitted with the largest reducer ring can handle a flared butt up to 31.5 inches in diameter (Anonymous 1976). None are in operation in the United States, but Finnish and Swedish machines for this purpose are popular in Europe.

17-5 ROSSER-HEAD DEBARKERS AND POLE SHAVERS

Because some hardwoods cut in winter are so difficult to debark with mechanical ring debarkers, many mills use a **rosser head** to debark sawlogs and veneer logs. Machines with peripheral-milling cutterheads are also used to shave bark from hardwood posts and poles. In machines of this type, the log is revolved as it is fed longitudinally past the rotating rosser head. If knives are sharp and the cutterhead fixed in position, the log will be turned to a relatively constant diameter and its taper eliminated, with considerable loss of wood. If the knives are somewhat dull and the cutterhead is arranged to take a constant-depth cut and to float over branch knots and other irregularities, the loss of wood is minimized and natural taper is retained.

Small machines of this type are used to peel fence posts. Some are fixed in place with highly mechanized handling equipment; more common are portable machines small enough to be towed by a light truck. Favored hardwood species such as post oak are much heavier and harder than southern pine; machines to process them must be sturdily constructed. If well supplied with small-diameter post oak, a portable machine carrying the cutterhead shown in figure 17-7 can peel about 1,500 6-foot posts during an 8-hour day. Because of the weight and hardness of dense southern hardwoods, these machines should not be fed with posts larger than 12 inches in diameter. Typically the machines are driven by a 50-hp engine when operating on post oak.

Hardwood sawlogs and veneer logs may be debarked on heavier machines using the same principle (fig. 17-8). On some designs the logs are side delivered with a log-stop and loader into trunnions that revolve the log under the rosser head. The rotating head is traversed the length of the log by a cable-driven trolley, the head being held in contact with the log by an air cylinder. The operator stands on a platform beside the machine, from which point he controls both infeed and outfeed. On such a machine sized for hardwoods, a 30-horsepower, four-blade cutterhead is driven at about 1,750 rpm. On logs 8 to 16 feet in length and averaging 12 inches in diameter, this type of debarker will produce per 8-hour shift about as follows:

Species	Number of logs per hour	
	Summer	Winter
Oaks	70	50
Hickory	70	50
Southern pine	80	60

Relatively sharp knives are used for winter-cut logs and duller knives for logs cut during the growing season.

The heavy job of peeling poles and piling requires larger equipment with mechanized infeeding and outfeeding equipment (fig. 17-9). The pole is rotated and driven past the rosser heads by angled, air-filled, rotating tires. Production depends on species as well as efficiency of the handling system and capabilities of the machine.

Figure 17-7.—Trailer-mounted post peeler driven by gasoline engine. (Top) Angled wheels rotate post over fixed shoes, and advance it to the left past rosser head. (Center) Rosser head; carbide teeth remove bark and branch stubs; planer knives then smooth the wood. Fixed infeed and outfeed shoes control the amount of wood removed. (Bottom) The oak post will revolve as toothed feed wheels advance it over the rosser head. (Photos from Morbark Industries, Inc.)

Debarking

Figure 17-8.—Rosser-head debarker (sometimes termed a floating-head debarker) for hardwood logs. (Top) Tilted spiked feed rolls revolve the log while advancing it endwise. (Bottom) Rosser head beneath revolving log removes bark. Depth of cut is controlled by the operator. (Photos from Morbark Industries, Inc.)

Figure 17-9.—Pole shaver. (Top) The pole rotates as it passes under the rosser heads. (Middle) Feed wheels are mounted on turrets that control angularity of feedworks. (Bottom) Roughing head on left; finishing head on right. Depth of cut is controlled by shoes that ride on the rotating pole. (Photos from Nelson Electric; such machines are sold by S. O. Jones and Company, Tulsa, Oklahoma.)

Because of the weight and hardness of the hickories and white oaks and their unacceptability for utility poles, most southern pine pole manufacturers do not process these species. A pole shaver appropriate for white oak would have a pair of four-knife rosser heads that revolve at 3,550 rpm, one a 25-hp roughing cutter, the other a 15-hp finishing head. On southern pine, production of the machine illustrated in figure 17-9 should be at least 1,000 lineal feet per hour; on oak or hickory it would produce less—perhaps 600 lineal feet per hour.

17-6 CHEMICAL DEBARKING

Since about 1942 experiments have been conducted in which soluble arsenic compounds are applied on basal girdles of live trees when the cambium is actively growing, thereby killing the trees and causing subsequent loosening of the bark (Berkland 1957). Under southern conditions, wood stains rapidly and is attacked by insects soon after such treatment, so chemical debarking has proven impractical for southern hardwoods. Readers interested in literature on the subject will find the following references useful:

Reference and species	Reference and species
Schmidt (1956)	Thompson and Birdsall (1957)
Oak, northern red	Hickory, pignut
Oak, scarlet	Hickory, shagbark
Berkland (1957)	Oak, blackjack
Ash, white	Oak, post
Maple, red	Oak, southern red
Gammage and Furnival (1957)	Oak, white
Ash, green	Godman (1962)
Elm, American	Oak, black
Oak, water	Oak, northern red
Sugarberry	Oak, white
Sweetgum	Baldwin (1963)
Peterson et al. (1958)	Oak, northern red
Oak, black	

17-7 BENEFICIATION OF WHOLE-TREE CHIPS

Beneficiation is a term borrowed from the vocabulary of the metallurgical industry where it means preliminary conditioning of ore for refinement. As applied to the forest products industry, it has come to mean separation of mixtures of wood, bark, foliage, and contaminants (e.g., dirt, stones, or metal) into usable components such as clean wood for fiber, bark for fuel, and foliage for an animal food supplement.

As noted in chapter 16, increased utilization of pine-site hardwoods will likely be achieved in part by wider use of roadside whole-tree chippers (fig. 16-18) and through development of practical swathe felling equipment (figs. 16-49 through 16-56) yielding whole-tree chips of mixed species (fig. 17-10). Increasing demands for fiber and fuel will also stimulate harvest of stump-root systems (figs. 16-28 through 16-36).

Figure 17-10.—Whole-tree chips. (Top) Oak and hickory whole-tree chips harvested by the swathe-felling mobile chipper illustrated in figure 16-52. (Bottom) Whole-tree chips resulting from corridor thinning a young southern pine stand with the swathe-cutting mobile chipper illustrated in figure 16-54.

Plummer (1976) reported that 114 mills in the wood fiber industry were using whole-tree chips in 1975, as follows:

Type of plant	Number of mills
Unbleached kraft	29
Bleached kraft	30
Corrugating medium	18
Fiberboard	16
Hardboard	7
Metallurgical chips	4
Roofing	1
Fuel chips	1
Export chips	8
	114

Many industrialists and researchers in the forest products industry, anticipating increased use of whole-tree and complete-tree chips, give high priority to development of beneficiation systems. Accordingly, the literature reports much research on the subject—mostly applying to softwoods or northern hardwoods. Few extensive beneficiation systems for southern hardwood fiber were operational by 1980. Almost all managers of wood burning power plants routinely separate scrap metal and stones from their fuel, however.

It is not yet clear whether beneficiation of whole-tree chips destined for pulp mills—potentially the major users of such chips—should take place before or during the pulping process. Faced with a somewhat similar problem—where to dry wet wood fuel—most power plant operators combine drying with the combustion process. In the case of whole-tree chips for pulp, however, evidence suggests that dirt, foliage, fines, and most of the bark should be removed before admitting chips to the digester.

Readers needing a cost analysis of pulp chip beneficiation will find useful the economic study by Biltonen et al. (1979) of a proposed compression debarking plant.

NEED FOR BENEFICIATION

To clarify the need for beneficiation it is useful to review experiences of fiber using industries processing such whole-tree chips.

Pulp mills.—In the next few decades, more than half of the world's pulp will be produced by the kraft process, as follows (Keays 1974):

Pulping process	1970	2000
	Percent of total	
Kraft	50	55
Sulphite	10+	5
Mechanical	30	35
Neutral sulphite semichemical	5	5
Miscellaneous	5	0
	100	100

It is evident, therefore, that the needs of kraft mills will dominate industry action (see also figs. 25-4 and 25-5).

Mills pulping whole-tree chips of hardwood species find operations affected in several ways and have developed remedial strategies as follows:

- Excessive wear on digester feeding equipment, hot-stock refiners, pumps, paper mill refiners, fourdrinier fabrics, deflectors, and covers on paper mill rolls. These problems arise from sand and dirt content in whole-tree chips; this dirt content (up to 1.2 percent) can be significantly lessened by eliminating ground skidding during harvesting and by partial delimbing in the woods. As will be noted later, field chippers can be designed to reject significant proportions of whole-tree grit content (figs. 17-11 and 17-12). Most of the remaining grit can be removed by chip washers and sand separators installed on the digesters (Chao 1976; Conner 1978).
- Twigs, foliage, and stringy chips cause bridging in chip silos and hangups in silo outlets and in continuous digester feeders. These problems can be reduced by woods separation techniques (fig. 17-11 and 17-12) or by pre-screening at the mill.
- As noted in chapter 16, deterioration of chips in storage piles is more rapid if needle and bark content is high. Close control of chip storage to assure rapid first-in first-out rotation of inventory solves this problem (figs. 16-64 and 16-65).
- Fuel balance is altered. Bark from conventional drum debarkers is burned in boilers to generate steam. If bark is mingled with wood when chipped—as in whole-tree chipping—the bark is pulped along with the wood and its dissolved portion ends in the black liquor. This results in less bark and more black liquor to be burned, disrupting the fuel balance. Effective beneficiation systems at the mill could recover bark, foliage, and fines from whole-tree chips to fuel bark-fired boilers. These components have more fuel value than black liquor.
- Increased chemical consumption. Bleached kraft from whole-tree chips of northern hardwood species required more alkali (0.35 percent increase for each percent increase in bark and fines) and more chlorine and chlorine dioxide for bleaching (0.3 percent increase for each percent increase in bark content) than kraft from debarked chips (Wawer 1975). Powell et al. (1975) studied conditions required to produce bleached kraft pulp from 100-percent whole-tree chips of black oak, southern red oak, and sweetgum equivalent to that from debarked roundwood. No additional alkali was required to reach given permanganate numbers (a measure of lignin content), but the caustic requirement per ton of pulp from whole-tree chips was substantially higher because of lower yields and lower-than-normal permanganate number required to meet color requirements. Mill bleach requirements were the same for whole-tree chips as for standard chips. Chao (1976) reported that in a southern mill there was no significant difference in chemical consumption at 10- to 30-percent whole-tree chip levels compared to a furnish of 100-percent regular hardwood chips.

- Increased digester space needed. Powell et al. (1975) reported that chip packing in digesters is more variable with whole-tree chips than with normal chips—probably because chip size is less uniform; more elaborate chip screening procedures could overcome this problem. Because of lower yields from whole-tree chips, digesters need to be somewhat larger than if sized for normal chips.
- Increase in black liquor solids overloads recovery system. Powell et al. (1975) found that solids in the recovery system were 22 percent higher for operation on unscreened whole-tree chips than on normal chips; when screened whole-tree chips were used, the recovery system solids content was 15 percent greater than with normal chips. Wawer (1975) reported that 60 to 65 percent of digested bark ends in black liquor.
- Pulp yields are decreased. If whole-tree chips are cooked excessively, lower-than-desired pulp yields and heavier-than-needed loads on causticization and recovery processes result; if chips are not cooked enough, excessive color will remain in the final bleached product (Powell et al. 1975).
- Pulp properties. Data from Conner (1978), Chao (1976), Powell et al. (1975), and Wawer (1975) indicate that a mill with sufficient screen, digester, washer, bleach, and recovery capacity can use whole-tree chips without quality loss.

Fiberboard plants.—Plummer (1976) noted that 23 fiberboard and hardboard plants were using whole-tree chips in 1975. Unlike pulping which dissolves and diverts to fuel about half its input, fiberboard making is a high-yield process, and most of the material that enters the mill ends as board.

While most fiberboard products admit considerable amounts of bark and foliage, these components are deleterious to some fiberboard properties; beneficiation may be required if high proportions of whole-tree chips are to be used.

Particleboard plants.—Like fiberboard, particleboard is produced by a high-yield process, more than 80 percent of incoming furnish typically emerging as saleable board; in many plants product yield is between 95 and 98 percent. Excessive bark, foliage, and dirt in whole-tree chips, unless removed by prior beneficiation, will end in the board where they have a deleterious effect on most board properties.

Figure 7-11.—(Top) Roadside whole-tree chipper equipped with chip separating partition showing accumulation of bark, folige, soil and twigs culled from the chip stream. (Bottom) Roadside whole-tree chipper discharging into a portable screen from whence screened chips are blown into a chip van. (Photos from Morbark Industries.)

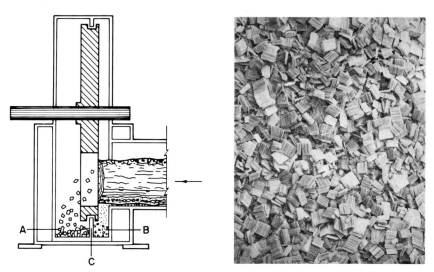

Figure 17-12.—(Left) Chip separating partition (C) in disk chipper follows the circumference of the chipper disk; acceptable chips flow into area A, while bark, foliage, and dirt are mostly drawn off from area B. (Right) Hardwood chips produced on a roadside chipper equipped with separating partition and subsequently screened on portable equipment at roadside. (Drawing from Morbark Industries.)

BENEFICIATION WITHIN THE CHIPPER

When wood with bark is chipped, those particles comprised mostly of bark, fines and foliage differ in form, weight, and specific surface from the particles that are mostly wood. Bark, fines, and foliage therefore tend to have different trajectories than wood chips when discharged from the chipping knives. This difference has been utilized in both disk and drum chippers to achieve beneficiation.

Modified disk chippers.—A field disk chipper provided with a chip separating partition (figs. 17-11 and 17-12) yields two streams of material—partially cleaned wood directed through screens into a chip van, and a stream of bark, foliage, and fines which is deposited on the ground at the chipping site for later use as fuel. Effectiveness of the system has not been extensively evaluated and probably varies with season of the year and with species.

A test of such a disk chipper in the tight bark season (January) and in the loose bark season (May) did not indicate significant reduction of bark percentage in unscreened chips blown directly into a van (table 17-6). Foliage content of the chips blown into the van in May was only about two-thirds that growing on the standing tree.

TABLE 17-6.—*Bark and foliage content (ovendry-weight basis) remaining in whole-tree chips of six species field chipped by the process illustrated in figures 17-11 (top) and 17-12 during February and May*[1]

Species	Chipped January 26, 1982			Chipped May 26, 1982		
	Bark	Fines[2]	Foliage	Bark	Fines[2]	Foliage
	------------------------------Percent------------------------------					
Hickory sp.	19.3	[3]	0.0	26.2	0.1	1.0
Maple, red	17.0	[3]	.0	11.1	.2	6.1
Oak, southern red	15.9	4.6	.0	13.9	.5	2.8
Oak, white	15.2	[3]	.0	16.9	1.6	2.4
Sweetgum	10.3	5.1	.0	13.6	1.3	2.3
Tupelo, black	8.7	[3]	.0	10.8	.4	3.6
Average	14.4	1.6	.0	15.4	.7	3.0

[1]Each value is an average representing all above-ground portions of two trees measuring 6 inches in dbh freshly cut among southern pines in central Louisiana.
[2]These fines, which were comprised of particles about the size of coarse sand, were predominantly bark.
[3]Quantity of sand-size fines was very small (near 0 percent).

Bark and foliage proportions of all above-ground parts of 6-inch trees of these species in mid-summer are about as follows (bark data derived from table 16-10; foliage data derived from table 16-2):

Species	Bark	Foliage
	------------Percent of ovendry weight------------	
Hickory sp.	19	3
Maple, red	12	6
Oak, southern red	20	3
Oak, white	12	4
Sweetgum	13	6
Tupelo, black	14	5

Arasmith drum chipper.—Since 1967, Hudson Pulp and Paper Corp. has processed a large percentage of hardwood removals from Florida lands through a tree-length merchandizing system. After removal of peeler logs, sawlogs and pulpwood there remained a quantity of large-diameter, unbarked, crooked, often defective, boltwood which could not be debarked and utilized, even as chips. The disposal and utilization problem was solved by installation of a low-power drum debarker-chipper manufactured by Arasmith Manufacturing Co., Rome, Ga. This chipper consists of a bolt-length shaft-mounted horizontal cylinder on the face of which are affixed, in a spiral arrangement, small knives set over slots in the cylinder. Bolts of wood are aligned parallel to and above the cylinder shaft and are fed broadside into the knives.

Hardwood bark, the inner portion of which is often fibrous and stringy, is torn from the surface of the bolt in relatively large pieces as the stick of wood makes contact with the knives. These bark shreds and some attached wood splinters thus do not go through the small slot at the base of the knife, but instead stay on the outside of the drum where they are conveyed off. Solid wood, on the other hand, is precisely cut by the small knives into narrow veneer and ribboned through the slot into the center of the drum. If the speed of the drum is sufficient, these narrow ribbons easily break up into acceptable-size chips having length along the grain equal to knife width.

Studies made by Hudson personnel indicate that approximately 20 percent of the original weight of an undebarked bolt ends outside the drum in relatively large pieces, nearly 80 percent inside the drum as uniform-size chips. Less than 3 percent of the material found inside the drum is estimated to be bark, while over 25 percent of the material outside the drum is bark. To recover some of the wood fiber from the outside-drum portion, this material is hammermilled and screened (fig. 17-13). The outer bark is separated to some degree from the inner bark and wood by the hammers and most of the bark is finely pulverized. The wood and inner bark are less affected by the hammermill action and a high percentage is retained on a 1/4-inch screen. Total recovery of solid wood fiber in the Cross City, Fla. installation is over 90 percent, and the usable chips normally contain less than 4 percent bark after screening (Uhr and Upchurch 1974).

CENTRALIZED SYSTEMS FOR WHOLE-TREE CHIPS

Nelson (1976) reviewed research sponsored by member companies of the American Pulpwood Association at the Battelle Institute in Columbus, Ohio on bark removal from chipped slabs after chipping. Battelle proposed a separation process which they named **vac-sink**, in which the mixture of chips and bark was floated in water under vacuum. Removal of air from the wood chips caused them to settle to the bottom of the tank, while the bark continued to float because its waxes resisted extraction of air.

A major disadvantage was incomplete separation of bark from wood. Many chips carried bark locked to them and either enriched the "float" (bark) with wood or contaminated the "sink" (wood) with bark. The process was practical enough so that one vac-sink plant was built and operated by Union Camp at its Savannah mill.

From these beginnings much additional work has been done on beneficiation of whole-tree chips. Section 17-1 describes variations among species in bark-wood bond strengths and some pre-treatments to loosen bark from wood. During the 1960's and 1970's, many researchers investigated methods to employ this knowledge to develop processes to separate bark and other contaminants, including foliage, from wood chips. Avenues of research on hardwoods and associated references are summarized as follows:

Method and reference	Hardwood species
Water flotation	
Liiri (1960)	*Betula* sp.
Womeldorf (1965)	*Betula papyrifera* Marsh.
	Fagus grandifolia Ehrh.
	Quercus rubra L.
Einspahr et al. (1969)	*Populus tremuloides* Michx.
Lea and Brawn (1970)	(not specified)
Julien et al. (1971, 1972)	*Populus tremuloides* Michx.
Air flotation	
Erickson (1972)	*Populus tremuloides* Michx.
	Acer saccharum Marsh.
Sturos (1972)	*Populus tremuloides* Michx.
Sturos (1973a)	*Populus tremuloides* Michx.
	Acer saccharum Marsh.
Sturos (1978)	*Populus tremuloides* Michx.
Winnowing by transverse airflow	
Sturos (1973b)	*Populus tremuloides* Mich.
	Acer saccharum Marsh.
Conditioning (by storing or steaming), agitating in water, and screening	
Berlyn et al. (1979)	Eastern Canadian hardwoods
Screening	
Snow (1952)	*Quercus* sp.
Christensen (1976)	Eastern hardwoods
Sturos (1978)	*Populus tremuloides* Michx.
Emanuel (1978)	Eastern hardwoods
Photosorting	
Sturos and Brumm (1978)	*Populus tremuloides* Michx.
	Acer saccharum Marsh.
Compression between nip rolls—with and without pre-steaming	
Blackford (1961)	Eastern and northern hardwoods
Erickson (1972)	*Populus tremuloides* Michx.
	Acer saccharum Marsh.
Arola and Erickson (1973, 1974)	*Populus tremuloides* Michx.
	Acer saccharum Marsh.
	Quercus sp.
	Carya sp.
	Liquidambar styraciflua L.
Arola (1974)	*Populus tremuloides* Michx.
	Acer saccharum Marsh.
Erickson and Hillstrom (1974)	(A patent)
Mattson (1974ab, 1975)	*Populus tremuloides* Michx.
	Acer saccharum Marsh.
Erickson (1976ab)	*Populus tremuloides* Michx.
	Acer saccharum Marsh.
Ball milling	
Mattson (1974a)	*Populus tremuloides* Michx.
	Acer saccharum Marsh.

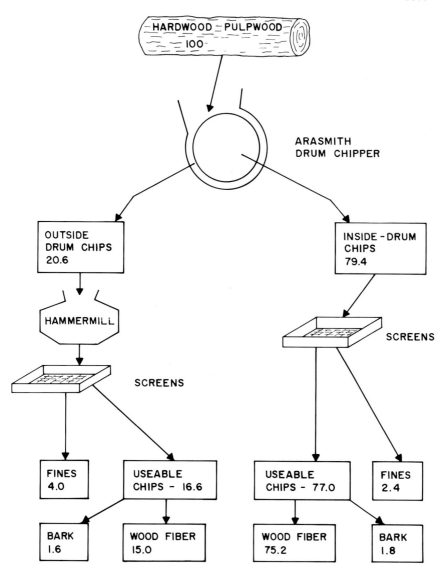

Figure 17-13.—Flow analysis indicating wood yield of 90.2 percent, by weight, from undebarked southern hardwoods chipped by Arasmith drum chipper. (Drawing after Uhr and Upchurch 1974.)

It is apparent that effective beneficiation of whole-tree chips of pine-site hardwoods requires several steps. Most approaches include compression debarking by passing a continuous single-layer flow of undebarked wood chips between two rotating steel rolls with a nip spacing much smaller than chip thickness. The rolls measure about 2 feet in diameter and revolve to yield a surface speed of about 740 feet per minute. Due to the nip action, bark particles separate from the wood, either adhering to the rolls for removal by roll scrapers, or fragmenting into smaller particles separable by screens (Blackford 1961); Arola and Erickson 1973). Rolls may be knurled or smooth depending on species. Steaming of chips at 2 to 30 psi gauge pressure for 1 to 10 minutes before compression debarking improves bark removal, but slightly increases wood loss (Erickson 1976a). Optimum conditions of roll knurling and presteaming vary substantially among species and with season of the year; data on most pine-site hardwoods are scarce.

Erickson's (1976b) data on seven species of southern hardwoods show that steaming of whole-tree chips, compression between nip rolls, and rejection of particles passing through a 3/16-inch screen yielded acceptable chips with bark content from 1.5 percent to 8.4 percent. Wood loss ranged from 7.7 percent to 35.1 percent, but was below 18 percent if both rolls were smooth; bark content under these conditions did not exceed 7.4 percent (table 17-7).

Subsequent study (Arola et al. 1976; Sturos and Erickson 1977; Sturos 1978; Sturos and Marvin 1978) indicated that vacuum-airlift separation before the process just described removes about 75 percent of the included foliage as a separate sort without loss of bark removal efficiency (fig. 17-14). Capital and operating costs of a chip debarking plant should be substantially lower with this combined system. Bark removal tests of the combined system on sweetgum and hickory bolewood chips give 78 and 71 percent bark removal and 85 and 86 percent wood recovery, respectively (table 17-8). This means that with sweetgum and hickory, 79 and 76 pounds of acceptable chips, containing 3 and 7 percent or less bark, respectively, can be recovered from 100 pounds of undebarked bolewood chips with an input bark content of 10 and 18 percent, respectively. Cost of beneficiation per ton of acceptable chips (ovendry basis) is estimated at $5.60. The fuel value of the separated bark more than covers the beneficiation costs.

The process diagrammed in figure 17-14 is designed for a pulp mill requiring maximum wood recovery. In the decades to come, many power plants will be fueled with whole-tree or complete-tree chips of pine-site hardwoods. It is likely that some receiving yards for such fuel chips will want to extract some clean pulp chips from the incoming wood because of their higher value compared to that of fuel chips. As suggested by Sturos and Dickson (1980), and indicated in table 17-8, about 40 pounds of clean wood chips (half the wood content) can be recovered from each 100 pounds of whole-tree pine-site hardwood chips through use of the vacuum-airlift system alone. (See left side of fig. 17-14.) This simple process will permit dirt and the majority of fines to fall through the screen belt and be rejected. About 75 percent of the foliage can be recovered at the first

TABLE 17-7.—*Bark content and wood loss in whole-tree chips of seven southern hardwood species after steaming[1], compression debarking, and screening* (Erickson 1976b)

					Total output		Output less fraction passing 3/16-inch screen	
Species	Date cut	Roll code[2]	Number of tests	Bark in[3]	Bark out[4]	Wood loss[5]	Bark out[4]	Wood loss[5]
					---------Percent---------			
Sweetgum.........	7/74	S	3	11.4	3.6	1.6	2.2	6.8
Sweetgum.........	1/75	S	1	17.1	7.9	3.8	3.2	11.4
Hickory	10/73	K	2	30.0[6]	12.1	15.5	8.4	27.5
Hickory *Carya ovata*	7/74	S	1	9.9	8.2	1.7	7.4	4.1
Black tupelo.......	7/74	S	3	5.2	1.9	0.9	0.9	3.1
Black tupelo.......	1/75	S	1	11.0	3.4	3.4	0.9	7.3
Southern red oak...	8/74	K	2	19.1	10.0	7.2	6.0	14.8
Southern red oak...	7/74	S	4	9.8	4.7	2.3	2.2	7.4
Water oak.........	7/74	S	6	10.9	6.0	1.7	3.0	8.5
Water oak.........	1/75	S	1	14.0	6.9	2.8	2.7	12.6
White oak.........	8/74	K	2	14.5	9.7	8.8	7.4	16.6
White oak.........	7/74	S	4	5.5	2.0	2.0	1.5	7.7
White oak.........	1/75	S	1	18.1	8.3	3.9	3.3	17.3
Yellow-poplar	10/73	K	4	23.0[6]	5.0	6.3	2.1	35.1

[1] Best results were obtained when steamed for 5 minutes at 30 psi gage pressure.
[2] S indicates both compression rolls smooth; K indicates one roll smooth and the other knurled.
[3] Bark, as percent of total input.
[4] Bark in output as a percent of total weight.
[5] Wood lost in bark fraction as a percent of total wood.
[6] Chips from slabs and edgings.

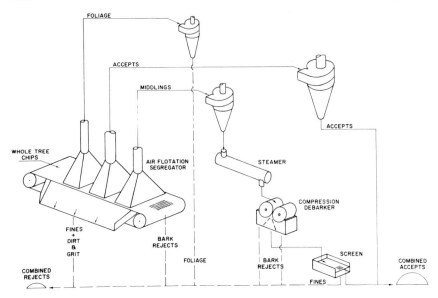

Figure 17-14.—The Forest Service system for beneficiating whole-tree chips through combination of vacuum airlift, steaming, compression nip rolls, and screening. (Drawing after Sturos and Marvin 1978.)

TABLE 17-8.—*Bark removal accomplished by vacuum-airlift segregation alone and in combination with steaming and compression debarking* [1,2,3]

	Input	Vacuum airlift segregation			Vacuum-airlift and compression debarking		
Species	bark content	Bark content of accepts	Wood recovery	Bark removal	Bark content of accepts	Wood recovery	Bark removal
	------------------------------Percent (ovendry weight basis)------------------------						
Sweetgum...	10.0	3.7	45.2	45.6	2.8	85.3	78.0
Hickory	17.7	5.7	49.3	50.3	6.6	86.3	71.7

[1] Average results of three tests on chips produced from bolewood.
[2] Data provided in personal communication from J. Sturos to P. Koch January 16, 1980.
[3] See text discussion for fuller explanation of percentages.

vacuum head for use as an animal food supplement. Acceptable pulp chips are drawn off at the second head. The third and final vacuum head is adjusted to remove the entire fuel fraction remaining, but to reject stones and stray metal.

The beneficiation procedure described by figure 17-14 is, except for the steaming stage, a dry process. Wet processes are also available. Berlyn et al. (1979) found that softwood and hardwood whole-tree chips can be effectively debarked if conditioned by piled storage for a few weeks (or steaming for 6 to 10 minutes) to loosen the bark-wood bond, then agitated in water to detach and fragment bark and foliage, and then screened to remove water. By this process, bark content of eastern Canadian hardwood whole-tree chips can readily be reduced to 4 percent. Effectiveness of the technique on oak and hickory chips has not yet been determined. Bark residue from the process is thoroughly wetted, diminishing its value for fuel.

In addition to removing bark and foliage from whole-tree chips, it is usually necessary to capture stones and metal from streams of pulp chips or fuel chips. A machine for this purpose is shown in figure 17-15.

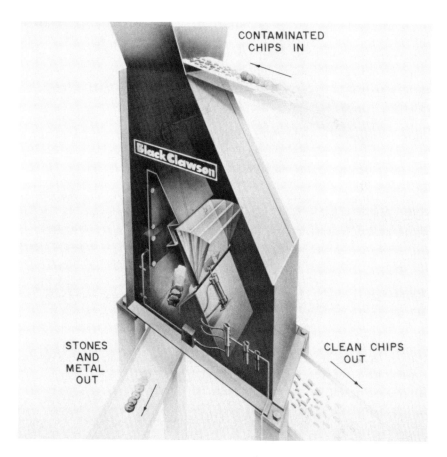

Figure 17-15.—Device for separating stones and metal from a moving stream of chips. Impact of hard, heavy objects actuates the diversion door by sonic differentiation. (Drawing after Black Clawson, Inc.)

PROCESSING OF STUMP-ROOT SYSTEMS

Stump-root systems including lateral roots to a 1-foot radius (figs. 14-3 through 14-26) comprise 12 to 21 percent of the dry weight of complete pine-site hardwood trees of small diameter (table 16-2). This potentially valuable wood source can be harvested together with above-ground tree portions (figs. 16-32 and 16-33) or pulled from the ground after stems have been harvested (figs. 16-28 through 16-31).

As with whole-tree chips, stumpwood can be partially cleaned at the harvesting site (fig. 16-30). Root systems still attached to stems can be water washed enroute to the mill (Davis and Hurley 1978). Although still experimental, vibrating devices attached to either type of stump harvesting equipment show promise of freeing the roots from most dirt and stones (Anonymous 1978; fig. 16-31).

Processes have been developed in Scandinavia (Blomqvist 1978) to crush, clean, and chip the spreading stump-root systems of northern conifers for use in a sulphate mill manufacturing unbleached and semi-bleached pulp. Because the root configuration of these conifers resembles that of pine-site hardwoods, it is likely that the Scandinavian system can be adapted to southern conditions.

A stump beneficiation plant of Joutseno Pulp Co. operated from 1975 to 1979, during which period it provided up to 20 percent of the chip mix without detrimental effect on the properties of the pulp. By this process, stumps are harvested by a tractor-mounted hydraulic hook (fig. 16-28) which splits them into pieces measuring about 1 foot by 2.3 feet at a rate of about 226 cubic feet per hour (loose basis). Haul distance to the mill is 10 to 40 miles. During 6- to 12-month storage on site before transport to the mill, bark, sand, and humus are loosened from the wood and a portion remains at the harvesting site.

At the mill the stump pieces undergo preliminary screening, crushing, washing, and flotation of crushed pieces to separate stones, further screening, and chipping of oversize pieces (fig. 17-16). Capacity of the beneficiation plant is about 1,400 cubic feet of stumps (loose basis) per hour. Power demand is 400 kW, and two operators are required per shift. Water circulation in the closed system is about 3,200 gallons per hour. Total cost of the stump chips in 1978 was $10.80 per cubic meter (35.3 cubic feet), loose basis, including harvesting (Blomqvist 1978). Development of the crushing equipment is described by Nisula (1975).

This process (fig. 17-16) is also used by a pulp mill in Mackmyra, Sweden.

In the United States is is likely that hardwood stump-root systems harvested simultaneously with above-ground tree portions will be severed at the mill for specialized beneficiation. While southern pine taproots destined for kraft pulping need only to pass through a drum debarker (Davis and Hurley 1978), hardwood stump-roots are more difficult to clean because of their numerous laterals (figs. 14-3 through 14-26). Specialized drum debarkers (Viklund 1978), impact drum debarkers (section 17-3), or crushing systems (fig. 17-16) may therefore be required.

Debarking

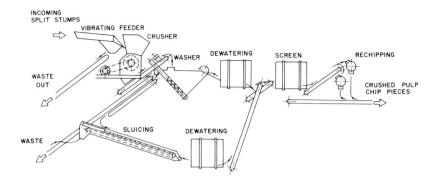

Figure 17-16.—Flow diagram of beneficiation plant to clean and chip rootwood for chemical pulping. (Drawing after Blomqvist 1978.)

17-8 LITERATURE CITED

Anonymous. 1976. A barker that removes flare butts. For. Ind. 103(11):69.

Anonymous. 1978. Vibrating stump puller undergoes field testing. For. Ind. 105(7):94.

Arola, R. A. 1974. Comparisons of wet and ovendry analyses of compression debarking tests on wood chips. U.S. Dep. Agric. For. Serv., Res. Note NC-178. 4 p.

Arola, R. A. and J. R. Erickson. 1973. Compression debarking of wood chips. U.S. Dep. Agric. For. Serv., Res. Pap. NC-85. 11 p.

Arola, R. A. and J. R. Erickson. 1974. Debarking of hardwood chips. FPRS Sep. No. MS-73-S5, 10 p. For. Prod. Res. Soc., Madison, Wis.

Arola, R. A., J. A. Sturos, and J. A. Mattson. 1976. Research in quality improvement of whole-tree chips. Tappi 59(7):66-70.

Baldwin, H. I. 1963. Reserving sawtimber in multiple-stem red oak during chemical debarking. Fox For. Notes No. 98, 1 p.

Beak Consultants Limited. 1978. Technical, economic, and environmental aspects of wet and dry debarking. Rep. No. EPS 3-WP-78-3, Water Pollution Control Directorate, Environ. Protect. Serv., Fish. and Environ. Canada. 189 p.

Berklund, B. L. 1957. Chemical debarking of pulpwood. Tappi 40(3):180A-182A.

Berlyn, R. W. 1964. A method for measuring the strength of the bond between bark and wood: the bark-wood bond-meter. Pulp Pap. Res. Inst. Can. Res. Note No. 43.

Berlyn, R. W. 1965a. The effect of variations in the strength of the bond between bark and wood in mechanical barking. Pulp and Pap. Res. Inst. Can. Res. Note No. 54. 22 p.

Berlyn, R. W. 1965b. The effect of some variables on the performance of a drum barker. Pulp Pap. Res. Inst. Can. Tech. Rep. No. 431.

Berlyn, R. W. 1970. The effect of some variables on the performance of a ring barker. Pulp and Pap. Res. Inst. Can. Woodland Pap. 15, 22 p.

Berlyn, R., J. Hutchinson, and R. Gooding. 1979. A method for upgrading chips from full trees. Pulp Pap. Canada 80(8):56-60.

Biltonen, F. E., J. A. Mattson, and E. D. Matson. 1979. The cost of debarked whole-tree chips stump to digestor. Res. Pap. NC 174, U.S. Dep. Agric., For. Serv. 8 p.

Blackford, J. M. 1961. Separating bark from wood chips. For. Prod. J. 11:515-519.

Blomqvist, L. 1978. An industrial plant for chipping stump and rootwood for chemical pulping. In Complete tree utilization of southern pine: symp. proc., New Orleans, La., April 17-19. C. W. McMillin, ed. For. Prod. Res. Soc., Madison, Wis., p. 294-298.

Calvert, W. W., and A. L'Ecuyer. 1972. Pretreatment allows debarking at below-zero temperatures. Can. Pulp Pap. Ind. 25(2):28-29.

Chao, Y. S. 1976. Hardwood whole-tree chip utilization at one southern kraft mill. FPRS Sep. No. MS-75-S69, 4 p. For. Prod. Res. Soc., Madison, Wis.

Christensen, E. 1976. Advancing the state-of-the-art in screening bark-free and non-bark-free chips. Tappi 59(5):93-95.

Coleman, S., and W. Evert. 1977. Economy model drum debarker. Am. Pulpwood Assoc. Tech. Rel. 77-R-11. 3 p. Washington, D.C.

Conner, A. L. 1978. Utilization of whole-tree chips in an unbleached Kraft linerboard mill. In Complete tree utilization of southern pine: symp. proc., New Orleans, La., April 17-19. C. W. McMillin, ed. For. Prod. Res. Soc., Madison, Wis., p. 268-274.

Davis, B. M., and Hurley, D. W. 1978. Fiber from a southern pine stump-root system? In Complete tree utilization of southern pine: symp. proc., New Orleans, La., April 17-19. C.W. McMillin, ed. For. Prod. Res. Soc., Madison, Wis., p. 274-276.

Derrick, R. P. 1978. Drive power requirements for barking drums. Tappi 61(7):53-57.

East, J. D., and W. G. Engelgau. 1978. Longwood barking and chipping. In Tappi Pulping Conf. Proc. 1978: 101-106.

Einspahr, D. W., M. K. Benson, and J. R. Peckham. 1969. Observations on a bark and wood chip separation procedure for aspen. For. Prod. J. 19(7):33-36.

Emanuel, D. M. 1978. Processing hardwood bark residues by screening. U.S. Dep. Agric. For. Serv. Res. Note NE-260. 3 p.

Erickson, J. R. 1972. The status of methods for debarking wood chips. Tappi 55:1216-1220.

Erickson, J. R. 1976a. Steaming chips facilitates bark removal. U.S. Dep. Agric. For. Serv., Res. Note NC-216. 4 p.

Erickson, J. R. 1976b. Removing bark from southern hardwood whole-tree chips. For. Prod. J. 26(2):45-48.

Erickson, J. R. 1979. Separation of bark from wood. Ch. 8 *in* Pulp and paper technology series no. 5, Chip quality monograph. (J. V. Hatton, ed.) TAPPI Joint Textbook Committee of the Paper Industry, p. 145-170.

Erickson, J. R., and W. A. Hillstrom. 1974. Process for removing bark from wood chips. U.S. Pat. No. 3,826,433. U.S. Pat. Off., Washington, D.C. 5 p.

Fobes, E. W. 1957. Bark-peeling machines and methods. U.S. Dep. Agric. For. Serv., For. Prod., Lab. Rep. 1730. Madison, Wis.

Frid, L. D. 1967. Investigation of the process of barking broadleaved species. Lesn. Z., Arhangel'sk 10(4):83-89.

Gammage, J. L., and G. M. Furnival. 1957. Chemical debarking in bottomland hardwoods. South. Pulp Pap. Manuf. 20(9):78-80.

Gilmer, W. D. 1974. Smallwood harvesting, debarking, and chipping. Tappi 57(6):54-57.

Godman, R. M. 1962. Sprouting of northern oaks reduced after debarking with sodium arsenite. U.S. Dep. Agric. For. Serv., Tech. Note No. 617. 2 p. Lake States For. Exp. Stn., St. Paul, Minn.

Greaves, H. 1968. Preliminary investigation of steam debarking. CSIRO Australian For. Prod. Newsl. 353:1-3.

Harmon, D. G. 1977. Long-log merchandising—before and after drum barking. Tappi 60(2):107-109.

Hensel, J. S. 1975. Developmental model chain-flail barker/cleaner. Am. Pulpwood Assoc. Tech. Rel. 75-R-54, 3 p. Washington, D.C.

Holzhey, G. 1969. Modern debarkers—their technological and economical aspects. Holz als Roh-und Werkstoff 27:81-102.

Hooper, S. W. 1973. Dry debarking of frozen wood. Pulp and Pap. Mag. Can. 74(7):105-107.

Julien, L. M., J. C. Edgar, and T. M. Conder. 1971. A density-gradient technique for obtaining wood and bark chip density. U.S. Dep. Agric. For. Serv., Res. Note NC-112. 4 p.

Julien, L. M., J. C. Edgar, and T. M. Conder. 1972. Segregation of aspen, balsam, and spruce wood and bark chips based on density differences. For. Prod. J. 22(6):56-59.

Keays, J. L. 1974. Full-tree and complete-tree utilization for pulp and paper. For. Prod. J. 24(11):13-16.

Koch, P. 1964. Wood machining processes. 530 p. The Ronald Press Co.: New York.

Koskinen, R. and N. V. Sjogren. 1977. Scandinavian millyard design. Two updated Finnish millyards. Tappi 60(2):98-100.

Lea, N. S., and J. S. Brawn. 1970. A new wood reclamation process. Tappi 53:622-624.

Liiri, O. 1960. Investigations on the debarking of birch chips by the soaking method. Paperi ja Puu 43:711-715.

Lutz, J. F. 1978. Wood veneer: log selection, cutting, and drying. U.S. Dep. Agric. For. Serv. Tech. Bull. No. 1577. 137 p. U.S. Govt. Print. Off., Washington, D.C.

Mattson, J. A. 1974a. Beneficiation of compression debarked wood chips. U.S. Dep. Agric. For. Serv., Res. Note NC-180. 4 p.

Mattson, J. A. 1974b. Compression debarking of stored wood chips. U.S. Dep. Agric. For. Serv., Res. Note NC-161. 4 p.

Mattson, J. A. 1975. Debarking chips from whole trees in the Lake States. U.S. Dep. Agric. For. Serv., Res. Pap. NC-115. 9 p.

Nakamura, G., and Y. Ohira. 1964. Industrial trials of bark removal. (III)—On the drum barker equipped with a central driving shaft having a spiral cutting edge. Bull. For. Exp. Stn. (Meguro, Tokyo) 171:155-168.

Nelson, A. W., Jr. 1976. Whole-tree chipping as viewed by the tree farmer and the industrial landowner. Tappi 59(7):85-86.

Nisula, P. 1975. (Stump crusher.) Folia Forestalia 245, 29 p.

Nordquist, H. 1975. Transversal system for treelength log processing. *In* Modern sawmill techniques, vol. 5: proc. of the fifth sawmill clinic. pp. 37-54. San Francisco, Ca.: Miller Freeman Publ., Inc.

O'Brien, R. E. 1977. Barking drum—state of the art. Tappi Alkaline Pulping/Secondary Fibers Conf. (Washington, D.C.) Papers: 109-115.

Perem, E. 1958. Review of literature on bark adhesion and methods of facilitating bark removal. Pulp Pap. Mag. Can. 59(9):109-114.

Peterson, K. R., C. S. Walters, and W. L. Meek. 1958. Barking black oak and jack pine fence posts with sodium arsenite. Bull. 626, Univ. Ill., Agric. Exp. Stn., Urbana. 30 p.

Plummer, G. M. 1976. Uses and potential productivity of whole-tree chippers. Tappi 59(7):64-65.

Powell, L. N., J. D. Shoemaker, R. Lazer, and R. G. Barker. 1975. Whole-tree chips as a source of papermaking fiber. Tappi 58(7):150-155.

Schmidt, O. A. 1956. A study of lumber grade recovery from chemically girdled red oaks. J. of For. 54: 243-245.

Smiltneek, R. J. 1974. Bark removal by the use of steel impactors. ASAE Pap. No. 74-1514, 6 p. Am. Soc. of Agric. Eng., St. Joseph, Mich.

Snow, E. A. 1952. Oak slabs and cordwood as a source of tannin and pulp. J. Amer. Leather Chem. Assoc. 47:563-577.

Sturos, J. A. 1972. Determining the terminal velocity of wood and bark chips. U.S. Dep. Agric. For. Serv., Res. Note NC-131. 4 p.

Sturos, J. A. 1973a. Predicting segregation of wood and bark chips by differences in terminal velocities. U.S. Dep. Agric. For. Serv., Res. Pap. NC-90. 8 p.

Sturos, J. A. 1973b. Segregation of foliage from chipped tree tops and limbs. U.S. Dep. Agric. For. Serv., Res. Note NC-146. 4 p.

Sturos, J. A. 1978. Bark, foliage, and grit removal from whole-tree chips—results and economics. *In* Proc., TAPPI Pulping Conf., Nov. 6-8, New Orleans. p. 121-134.

Sturos, J. S., and D. B. Brumm. 1978. Segregating wood and bark chips by photosorting. Res. Pap. NC-164, U.S. Dep. Agric. For. Serv. 5 p.

Sturos, J. A., and R. E. Dickson. 1980. Fiber, fuel, and food from whole-tree chips. Trans. of the ASAE 23(6):1353-1358.

Sturos, J. A. and J. R. Erickson. 1977. Proposed systems to remove bark and foliage from whole-tree chips. Trans. of the ASAE 20(2):206-209.

Sturos, J. A., and J. L. Marvin. 1978. A process for removing bark, foliage, and grit from southern pine whole-tree chips. *In* Complete tree utilization of southern pine: symp. proc., New Orleans, La., April 17-19. C.W. McMillin, ed. For. Prod. Res. Soc., Madison, Wis. p. 277-294.

Thompson, W. S., and K. C. Birdsall. 1957. Chemical debarking of southern trees. Agric. Exp. Stn. Tech. Bull. 42, 28 p. Miss. State Coll., State Coll., Miss.

Tuuha, R. and N. V. Sjogren. 1976. Scandinavian drumbarking and conveying equipment. Tappi 59(9):101-103.

Uhr., S. C. and J. R. Upchurch. 1974. Experiences with drum chipping of undebarked southern hardwoods. FPRS Sep. No. MS-73-S6, 8 p. For. Prod. Res. Soc., Madison, Wis.

Viklund, G. 1978. Cleaning raw material better in the forest must really reckon with stumpwood chips. Svensk Papperstidning 81(13):421-433.

Vogt, H. 1965. The bark's adhesion to wood. Norsk Skogindustri 19(5):181-186.

Wawer, A. 1975. Bark in hardwood chips—effect on mill operations. Pulp and Pap. Can. 76(7):51-54.

Weiner, J. and V. Pollock. 1972. Constitution and pulping of hardwoods. IV. Bibliogr. Ser. No. 233 (Suppl. I), 237 p. Appleton, Wis.: The Inst. of Pap. Chem.

Wilcox, H., F. J. Czabator, G. Girolami, D. E. Moreland, and R. F. Smith. 1956. Chemical debarking of some pulpwood species. State Univ. New York Coll. For., Syracuse, Tech. Pub. 77. 43 p.

Williston, E. 1976. State of the art—hardwood lumber manufacture. *In* Hardwood sawmill techniques, proc. of the first hardwood sawmill clinic program, pp. 1-19. San Francisco, Ca.: Miller Freeman Publ., Inc.

Womeldorff, F. M. 1965. Separation of heartwood from sapwood pulp chips by flotation methods. For. Prod. J. 15:407-408.

18

Machining

In section 18-2. G. W. Woodson provided the photographs of orthogonal cutting and the cutting force data specific to pine-site hardwoods. Information on inserted-tooth ripsaws for circular-saw headrigs is condensed from a publication by S. J. Lunstrom, and the discussion of over-arbor ripsaws compared to under-arbor ripsaws is from a paper by T. A. McLauchlan. Section 18-23, VENEER CUTTING is largely condensed from a summary publication by J. F. Lutz. G. H. Kyanka's review is the basis for section 18-29, dulling of cutting tools; and section 18-30 describing computer control of woodworking machines in secondary manufacture is from the introduction of C. W. McMillins' review of this subject.

Other data came from sources too numerous to completely enumerate, but major portions were drawn from research by:

M. Applefield	S. P. Hall	W. T. Nearn
P. A. Araman	H. Hallock	R. H. Page
R. A. Arola	L. F. Hanks	G. Pahlitzsch
R. D. Behm	R. A. Hann	R. L. Papworth
S. A. Bingham	N. Hartler	D. W. Patterson
R. Birkeland	J. V. Hatton	C. C. Peters
T. Bonac	D. Hayashi	I. L. Plough
G. Brown	B. G. Heebink	E. W. Price
H. W. Burry	R. W. Hemingway	P. S. Quelch
W. W. Calvert	K. Hirst	E. D. Rast
T. W. Church, Jr.	R. J. Hoyle, Jr.	G. P. Redman
L. Classen	H. A. Huber	H. W. Reynolds
B. E. Coleman	J. B. Huffman	B. H. River
B. M. Collet	J. S. Johnston	H. W. Rogers
H. H. Connelly	R. W. Jokerst	R. Sage
G. A. Cooper	C. W. Jones	R. Sarna
E. P. Craft	K. C. Jones	H. W. Saunders
E. E. Dargan	K. Kato	J. Schmied
E. M. Davis	E. D. Kirbach	J. Schroeder
R. K. Detjen	E. Kivimaa	N. C. Springate
W. R. DeVries	C. B. Koch	P. H. Steele
S. A. Dowdell	P. Koch	A. R. Stern
K. W. Duff	M. Komatsu	H. A. Stewart
J. Ekwall	F. E. Landt	A. St.-Laurent
J. E. Erickson	J. L. Lubkin	R. Szymani
J. R. Erickson	E. L. Lucas	C. M. Theien
D. Evans	W. M. McKenzie	A. Thrasher
A. O. Feihl	C. W. McMillin	U.S. Dept. of Agriculture,
I. B. Flann	F. B. Malcolm	Forest Service
E. W. Fobes	M. E. Martellotti	M. Wahlman
J. Fondronnier	J. A. Mattson	D. C. Walser
N. C. Franz	R. L. Miller	O. M. Walstad
F. Freese	J. T. Morgan	D. Ward
S. C. Gambrell, Jr.	C. D. Mote, Jr.	G. E. Woodson
C. J. Gatchell	G. Nakamura	

1687

CHAPTER 18
Machining

CONTENTS

	Page
18-1 MACHINABILITY—AMONG-SPECIES VARIATION	1701
18-2 ORTHOGONAL CUTTING	1702
DEFINITIONS	1703
EFFECTS OF CUTTING VELOCITY	1704
PLANING: 90-0 DIRECTION	1705
Chip formation	1705
Effects of width of cut	1709
Effects of knife angles	1709
Effects of depth of cut	1710
Effects of wood temperature	1712
Effects of moisture content	1712
Effects of specific gravity	1712
Summary of Woodson's data on cutting forces	1712
Scraping in the 90-0 direction	1783
Cutting excelsior	1783
PLANING: 90-0 to 45 DIRECTION	1783
VENEER CUTTING: 0-90 DIRECTION	1783
Chip formation	1784
Cutting forces	1785
CROSSCUTTING: 90-90 DIRECTION	1788
Chip formation	1789
Cutting forces	1789
OBLIQUE CUTTING	1790
INCLINED AND VIBRATORY CUTTING	1792
Vibratory cutting with lateral oscillation	1792
Vibratory cutting with perpendicular oscillation	1793
18-3 SHEARING AND CLEAVING	1794
SHEARING	1795
Factors affecting shearing force	1795
Estimating force and power requirements	
for crosscut shearing of roundwood	1795
Wood damage from shearing	1801
Bucking of pulpwood with continuous-feed shears	1803
Spiral-head shear to produce long chips	1803
Delimbing shears	1803

	CLEAVING 1804
	Pulpwood splitters 1806
	Firewood splitters 1807
	Cleaving transversely; kerfless cutting 1808
	Shearing veneer panels. 1808

18-4 PERIPHERAL MILLING PARALLEL TO
GRAIN (90-0 DIRECTION)......................... 1810
NOMENCLATURE................................. 1810
KINEMATICS 1810
CHIP FORMATION................................ 1813
FACTORS AFFECTING POWER 1815
Workpiece factors 1816
Cutterhead factors. 1816
Feed factors... 1817
Cutterhead power to up-mill six southern hardwoods..... 1818
SURFACE QUALITY................................ 1819
Definition of defects and their occurrence 1820
Defect control 1820
DOWN MILLING 1828

18-5 PERIPHERAL MILLING ACROSS THE GRAIN
(0-90 DIRECTION)................................. 1831

18-6 PERIPHERAL MILLING ACROSS THE GRAIN
(90-90 DIRECTION)................................ 1832

18-7 PERIPHERAL MILLING (PLANING) WITH
HELICAL CUTTERS................................ 1840

18-8 SAWING... 1842
FACTORS INFLUENCING SAWING PROCEDURES ... 1843
BANDSAWING 1844
Swaged-tooth wide bandsaws for primary manufacture ... 1844
Swaged-tooth band resaws for dry lumber 1860
Spring-set narrow bandsaws 1861
Portable horizontal band headrig for hardwoods.......... 1863
SASH GANGSAWING 1863
CIRCULAR SAWING 1865
Nomenclature 1865
Kinematics and fundamentals 1867
Inserted-tooth ripsaws for circular saw headrigs.......... 1869
Inserted-tooth ripsaws for edgers...................... 1875
Over-arbor ripsaws compared to under-arbor ripsaws 1876
Single- and double-arbor gangsaws 1878
Saw selection for circular gang ripsaws 1878

General-purpose swage-set ripsaws 1881
Spring-set ripsaws 1884
Glue-joint ripsaws 1884
Log cutoff saws 1886
Cutoff saws for rough trimmers 1888
Smooth-trim saws 1889
Hollow-ground combination planer saws 1891
Miter saws 1891
Combination saws 1892
Flat-ground combination saws 1893
Dado heads 1896
Ring saws 1897
Tubular core saws 1898
Portable circular-saw headrigs for hardwoods 1898
CARBIDE-TOOTHED CIRCULAR SAWS 1901
Gang ripsaws for green oak and hickory 1902
Gang ripsaws for dry hardwoods 1903
Smooth trimsaws for dry hardwoods 1903
Spiral rough trimsaw for hardwoods in linear motion 1903
CHAINSAWING 1904
Portable chainsaw headrigs for hardwoods 1907

18-9 CONVERTING WITH CHIPPING HEADRIGS 1908
CUTTING MODES 1909
POWER REQUIREMENTS 1909
Shaping-lathe headrig 1910
End-milling headrig chipper 1913
Down-milling chipping head 1914
CHIP FORM 1914
SURFACE QUALITY 1915
CHIPPING HEADRIGS PARTICULARLY SUITED
TO HARDWOODS 1916
Shaping-lathe headrig 1917
End-milling chipping headrig 1923
Climb-milling chipping headrig 1925

18-10 SAWMILLS FOR HIGH-QUALITY, LONG,
HARDWOOD LOGS 1931
LOG CHARACTERISTICS 1931
STATE-OF-ART SAWMILLS 1932
SAWING PATTERNS 1932
COMPUTERIZED SAWMILLS 1934
SAWING TIME 1934
SAWMILL DOWN TIME 1937
SAWING ACCURACY AND BOARD THICKNESS 1939
YIELDS AND RESIDUES 1940

18-11 SAWMILL LAYOUTS FOR SHORT, SMALL,
    HARDWOOD LOGS. 1954
    LOG AND LUMBER LENGTHS 1954
    BOLT SAWING PATTERNS 1956
    SAWMILL OUTPUT CORRELATED
       WITH FEED RATES 1961
    TWO-SAW SCRAG MILLS WITH
       CONTINUOUS FEED. 1964
    TWO-SAW SCRAG MILLS WITH
       RECIPROCATING FEED 1966
    LIVE SAWING 1969
    BOLTERS... 1970

18-12 RIPPING AND CROSSCUTTING LUMBER TO YIELD
    FURNITURE CUTTINGS 1975
    ROUGH MILL PROCEDURES 1975
       The gangsaw-rip-first system 1976
       The STUB system of crosscutting low-grade lumber first . 1978
    YIELD OF CUTTINGS 1980
    COMPUTER-ASSISTED OPTIMIZATION
       OF CUTTINGS 1988

18-13 JOINTING, PLANING, MOULDING, AND SHAPING..... 1989
    JOINTING 1989
       Purpose of jointing 1989
       Machine types 1989
    PLANING... 1992
       Single surfacer 1994
       Double surfacer. 1997
       Planer-matcher 2001
       Knives for top and bottom cylinders 2003
       Knives for sideheads.............................. 2005
       knives for profile cutterheads....................... 2007
    MOULDING....................................... 2008
       Productive capacity.............................. 2008
       Machine types 2010
       Moulder spindle types............................ 2012
       Knife grinding technique 2012
    SHAPING 2012
       Spindle design................................... 2013
       Cutterhead design 2014
       Single-spindle shaper 2015
       Double-spindle shaper 2016
       Double-head automatic shaper 2016

Double-table, fixed-spindle shaper 2019
Contour profiler 2020
Shaping attachment on a double-end tenoner 2021
Tape-controlled shaper 2021

18-14 MACHINING WITH ABRASIVES 2022
SURFACE QUALITY 2022
Effect on strength of glued joints 2022
Perceptibility of machining marks on painted surfaces.... 2022
Color variation of furniture related to sanding
  procedure. 2023
Surface quality measurement. 2023
TYPES OF COATED ABRASIVES 2023
Abrading agent 2023
Backing material..................................... 2027
Bond.. 2029
Distribution and orientation of abrasive particles 2029
Flexure pattern 2030
Product form 2031
FUNDAMENTAL ASPECTS......................... 2031
Scratch pattern 2031
Rate of wood removal................................ 2033
Power requirement 2034
Wood surface temperature 2038
ABRASIVE PLANING WITH WIDE-BELT SANDERS... 2039
Configuration of abrasive planers..................... 2039
Dry hardwood lumber................................ 2041
Dry laminated oak squares 2042
Green hardwood lumber............................. 2042
Glued-up lumber cores.............................. 2042
Particleboard 2042
Plywood 2043
Belt life 2044
Belt maintenance 2045
OTHER MACHINES TO SAND FLAT SURFACES 2045
Disk sander 2045
Drum sander 2046
Stroke sander....................................... 2046
Edge sander.. 2047
Wide-belt sander 2048
Polisher with drum of abrasive-impregnated,
  non-woven nylon................................. 2048
MACHINES TO SAND BROAD CURVED SURFACES .. 2050
MACHINES TO SAND CONTOURED EDGES 2050
Edge sander.. 2050
Spindle sander...................................... 2050

  Variety sander 2051
  Scroll sander 2051
  Open-drum sander 2051
 MACHINES TO SAND MOULDED EDGES 2052
  Formed-block sander 2052
  Formed-wheel sander 2052
  Abrasive impregnated non-woven nylon wheel 2052
  Brush-backed sanding wheel 2053
  Multiple-head moulding sander 2055
 MACHINES TO SAND TURNINGS 2055
  Brush-backed, automatic turning sander 2055
  Centerless spindle sander 2056
  Centerless spindle grinder 2057
 MACHINABILITY 2058
 ABRASIVE TUMBLING 2059

18-15 TURNING... 2060
 TOOL-POINT TYPE 2060
 LONG KNIFE TYPE 2060
  Bail-wood lathe 2061
  Back-knife lathe 2061
  Pringle and Brodie lathe 2063
  Variety lathe 2066
 PERIPHERAL-MILLING TYPE 2066
  Shaping-lathe headrig............................. 2067
  Automatic shaping lathe........................... 2067
  Copying lathe 2072
 CHUCKING TYPE 2074
  Dowel lathe 2074
  Sawing, chucking, and boring machine 2076
 MACHINABILITY 2077

18-16 BORING 2078
 BIT TYPES AND NOMENCLATURE................... 2078
 FUNDAMENTAL ASPECTS......................... 2083
 BORING DIRECTION AND CHIP FORMATION 2084
 FACTORS AFFECTING TORQUE AND THRUST
  BORING ACROSS THE GRAIN 2085
  Wood moisture content............................ 2085
  Specific gravity 2085
  Bit style 2086
  Hole depth...................................... 2086
  Bit diameter..................................... 2086
  Chip thickness................................... 2086
  Bit rpm... 2086

TORQUE AND THRUST BORING ALONG THE GRAIN . 2087
HOLE QUALITY 2087
    Effect of hole quality on strength of glued dowel joints ... 2088
BORING DEEP HOLES............................. 2089
MACHINE TYPES 2089
    Vertical multiple-spindle borer...................... 2089
    Horizontal multiple-spindle borer 2091

18-17 ROUTING.. 2092
    SPINDLE TYPES 2094
    MACHINE TYPES 2095
        Overhead router with fixed spindle 2095
        Overhead router with floating spindle................. 2095
        Radial-arm router 2099
        Inverted router 2099
        Computer-controlled routers 2099
        Dovetail routers.................................. 2103
        Special routers................................... 2103
        Portable routers.................................. 2103
    MACHINABILITY 2104

18-18 CARVING.. 2104
    MULTIPLE-SPINDLE CARVER 2104
    SINGLE-SPINDLE CARVER 2106
    MACHINABILITY 2106
    WOOD EMBOSSING MACHINES 2108

18-19 MORTISING 2110
    OSCILLATING ROUTER 2110
    HOLLOW-CHISEL MORTISER 2110
    CHAIN MORTISER 2114
    RECIPROCATING CHISEL MORTISER 2116
    MACHINABILITY 2116

18-20 TENONING 2119
    TENON CUTTERS 2119
    MACHINE TYPES 2122
        Single-end tenoner with manual carriage 2122
        Finger jointer 2122
        End matchers 2123
        Double-end tenoner................................ 2123
    CAPABILITIES OF DOUBLE-END TENONERS........ 2126
        Double cut-off (trim saws).......................... 2126
        Scoring and tenon stations.......................... 2127
        Relishing 2127
        Angular dado.................................... 2129

Sill horning 2129
Dovetailing. 2129
Cam-generated shaping 2129
Dado and jump dado 2129
Sanding 2131
Two-pass tenoners. 2132
FEED AND CONTROL OPTIONS FOR
    DOUBLE-END TENONERS 2132
    Traverse adjustment 2132
    Feed chain. 2132
    Numerical controls 2132
SOUND ENCLOSURES 2133

18-21 MACHINING WITH HIGH-VELOCITY LIQUID JETS.... 2133
    INDUSTRIAL APPLICATIONS 2134

18-22 MACHINING WITH LASERS 2137

18-23 VENEER CUTTING. 2143
    LOG SELECTION 2144
    Log conformation 2144
    Intrinsic wood properties 2144
    Log characteristics 2145
    LOG STORAGE 2145
    EFFECTS OF HEATING WOOD PRIOR TO
        VENEER CUTTING............................ 2148
    Plasticity and hardness. 2148
    Dimensional changes 2148
    Color changes 2148
    Torque to turn bolts 2148
    Shrinkage. 2149
    Drying time 2149
    Conclusions 2149
    TIME REQUIRED TO HEAT BOLTS AND
        FLITCHES 2149
    Effect of diameter. 2149
    Effect of temperature gradient. 2149
    Effect of bolt length 2149
    Effect of moisture content and specific gravity 2150
    Effect of heating medium. 2150
    Examples 2150
    FAVORABLE TEMPERATURES FOR VENEER
        CUTTING. 2151
    CHOICE OF CUTTING DIRECTION 2151
    FUNDAMENTAL ASPECTS OF VENEER CUTTING ... 2155
    Cutting forces 2155

LATHE VERSUS SLICER........................... 2158
   Lathe advantages 2159
   Slicer advantages 2159
   Stay-log lathe................................... 2159
   Back-roll lathe.................................. 2161
LATHE RIGIDITY AND CONTROL OF
   UNDESIRABLE MOVEMENT.................... 2161
   Bolt movement................................... 2161
   Machine movement................................ 2163
   Heat distortion of lathe 2163
SLICER RIGIDITY AND CONTROL OF
   UNDESIRABLE MOVEMENT.................... 2165
   Flitch movement 2165
   Machine movement................................ 2165
   Heat distortion.................................. 2166
CUTTING VELOCITY............................. 2166
VENEER KNIFE SPECIFICATION 2166
   Knife hardness 2167
   Angle of ground bevel 2167
GRINDING VENEER KNIVES 2167
   Grinding on the veneer side of the knife 2168
   Honing the knife................................. 2168
   Secondary knife bevels 2169
KNIFE SETTING 2169
   Setting the lathe knife level 2170
   Setting knife angle on a veneer lathe................. 2170
   Setting the slicer knife 2171
NOSEBARS 2172
   Fixed nosebar 2173
   Roller nosebar................................... 2173
   Comparison of fixed and roller nosebars 2173
SETTING NOSEBARS............................ 2174
   Setting a fixed nosebar on a lathe (by lead and gap) 2175
   Setting a fixed nosebar on a slicer (by lead and gap)..... 2178
   Setting a roller nosebar on a lathe (by lead and gap) 2179
   Setting a roller nosebar (by gap and exit gap).......... 2179
   Setting a fixed nosebar (by lead and exit gap).......... 2179
   Setting gap by pressure rather than to fixed stops........ 2180
POSSIBLE WAYS TO GENERALIZE SETTING
   OF LATHE AND SLICER....................... 2180
   Generalized knife settings.......................... 2180
   Generalized setting of a fixed nosebar 2180
   Generalized setting of a roller nosebar................ 2180
   Alternate generalized setting of a roller nosebar......... 2181
   Generalized setting of the gap by pressure............. 2181
   Summary of generalized lathe and slicer settings 2182

PANEL MANUFACTURE WITH VERY THIN FACES .. 2183
CUTTING THICK VENEER......................... 2183
   Thick sliced veneer 2183
   Thick rotary-peeled veneer 2185
STEAM-INJECTION KNIVES AND HEATED
   NOSEBARS 2188
POSITIONING BOLTS AND FLITCHES............... 2189
CONVEYING VENEER FROM LATHE AND SLICER .. 2189
   Conveying veneer from a lathe 2189
   Conveying veneer from a slicer 2191
CLIPPING 2191
   Clipping green veneer.............................. 2191
   Clipping dry veneer................................ 2191
VENEER YIELD...................................... 2192
   Lathe yield.. 2192
   Slicer yield.. 2194
MONOGRAPHS ON CUTTING VENEER OF
   PARTICULAR SOUTHERN SPECIES 2194

18-24 CHIPPING 2195
   CHIP DIMENSIONS................................ 2195
      Chips for kraft pulping............................. 2195
      Chips for very rapid impregnation and vapor-phase digestion 2196
      Chips for refiner mechanical pulp 2196
      Chips to be sliced into flakes 2196
      Fuel chips 2197
      Metallurgical chips 2197
   CHIP FORMATION AND POWER REQUIRED 2197
      Twin-disk chippers 2202
      Drum chipper fed with logs approaching endwise........ 2202
      Drum chipper with logs approaching sideways 2202
   DRIVE SELECTION 2203
   CHIP THICKNESS AND QUALITY................... 2203
   CHIPPER TYPES 2203
      Cordwood chippers at pulp mills..................... 2204
      Longwood chippers................................ 2204
      Residue chippers.................................. 2204
      Rechippers.. 2205
      Chip shredders 2205
      Portable chippers 2205
      Swathe-felling mobile chippers 2207
      Chipping headrigs................................. 2207
   CHIP SPECIFICATIONS............................. 2208

18-25 FLAKING 2208
    TYPES OF FLAKING MACHINES 2209
        Disk flakers 2210
        Drum flakers 2211
        Shaping-lathe headrig 2214
        Ring and cone flakers 2214
    POWER REQUIREMENTS 2216
    FLAKE QUALITY 2219

18-26 HOGGING 2222
    HOGGED WOOD PRODUCTS 2223
        Forest mulch 2223
        Fuel 2223
        Feedstock for chemical, fiber, charcoal, and briquette plants 2223
        Wood flour 2223
        Bark for fuel and charcoal 2224
    MACHINE TYPES 2224
        Knife hogs 2224
        Hammer hogs with horizontal axes 2224
        Bark shredder with vertical axis 2227
        Punch-and-die hogs 2227
        Double-rotor hogs 2230
        High-shear screw mixers 2231
        Disk mills 2232
        Ring mills 2232
        Ball and rod mills 2232
        Crushing rolls 2233
    GRINDING WOOD FLOUR 2233
        Coarse flour 2234
        Medium and fine flour 2236
    SCREENING 2237

18-27 DEFIBRATING 2238
    STONE GRINDING OF ROUNDWOOD 2238
    STEAM-GUN EXPLOSION TECHNIQUE OF
        DEFIBRATING CHIPS 2239
    DISK REFINING OF CHIPS 2239
    DEFIBRATING WITH CRUSHING ROLLS 2242

18-28 NOISE 2242

18-29 DULLING OF CUTTING TOOLS 2243
    TOOL MATERIALS AND HARDENING PROCEDURES 2244
        High-speed steel 2245
        Tungsten carbide 2245
        Stellites 2246
    TOOL WEAR 2247

**18-30 COMPUTER CONTROL OF WOODWORKING MACHINES
IN SECONDARY MANUFACTURE** 2251
    COMPUTER NUMERICAL CONTROLLERS 2252
    PROGRAMMING 2254
    COMPUTER CONTROLLED MACHINES 2256

18-31 LITERATURE CITED 2257

18
Machining

Since information on machining southern hardwoods is available in forms requiring some understanding of basic woodworking processes, essential explanations are provided. Broader coverage of wood machining processes is available in Koch (1964a). In addition, a number of comprehensive literature reviews have been made (Forest Products Research Society 1959, 1960, 1961; Koch and McMillin 1966; Koch 1968a, 1973; Kollmann and Côté 1968, p. 475-541; McMillin 1970, 1975). Historical reviews of the development of wood machining technology are also available (Mansfield 1952; Goodman 1964; Koch 1964a—p. 3-6, 1967, 1972—p. 758-759; Prokes 1966; Simons 1966; Wilkins 1966; Thunell 1967).

This chapter is concerned with industrial machining processes. Readers interested in craftwork are referred to the following texts:

Fine woodworking techniques, general
    Fine Woodworking Magazine (1978)
    Hoadley (1980)
Woodcarving and sculpture
    Rood (1950)
    Burk (1972)
    Hanna (1975)
    Butler (1974)
    Johnstone et al. (1971)
Joinery
    Frid (1979)
Chair making
    Dunbar (1976)
    Alexander (1978)

Wood turning
    Nish (1932)
Cabinet making and millwork
    Feirer (1970)
Small boat building
    Gardner (1977)
Cabin building
    Dunfield (1974)
    Leitch (1976)
    Rowell et al. (1977)
    Carlson (1977)
    Park (1978)

Pine site hardwoods require different machining techniques than those used for southern pine. Many of these hardwoods are extremely dense and therefore hard, heavy, strong, and stiff—characteristics that increase energy requirements and accelerate wear of cutting tools, machines, and conveyors. The wood of a few species, such as black tupelo, has interlocked grain which is difficult to plane without chipping. Tension wood in some species causes fuzzy grain on machined surfaces. Others, such as hickory, have internal stresses which distort log halves after center ripping. Such internal stresses also cause log ends of some oaks and hickories to check when softened for veneer cutting by hot water soaking. In many hardwood species complex warp in drying makes manufacture of flat, straight, dry lumber difficult.

Yield of long, wide defect-free cuttings from pine-site hardwoods is low because many logs are small and crooked, and often blemished by insect holes, stain, and large knots (see chapters 11 and 12). Plants utilizing pine-site hardwoods employ several strategies to increase yield, including:

- In some commodities, such as pallet boards and crossties, many defects common in pine-site hardwoods are acceptable; these products are therefore principal products of some mills.
- Acceptable yields of defect-free solid-wood cuttings can sometimes be obtained by reducing log lengths to 3 to 6 feet, instead of the usual 12 to 16 feet, and by special sawing techniques (see sec. 18-11).
- Another approach cuts logs into thin, short, or narrow parts and reassembles them. For example, furniture frame stock can be laminated from partly defective veneers; randomization of defects diminishes their effect on strength.
- Some commodities made of fibers, flakes, or particles may tolerate inclusion of many species and much defective wood. Pine-site hardwoods can be the principal raw material for such products.

Woods from pine-site hardwoods vary significantly in machinability. Before discussing the various machining processes, it is useful to have some knowledge of this variability.

18-1 MACHINABILITY—AMONG-SPECIES VARIATION

Behm (1974) rated from 1 (worst) to 10 (best) the overall machinability of 40 hardwoods and softwoods commonly used by furniture manufacturers. Rating 10 on his scale were mahogany from Central America (*Swietenia macrophylla* King.) and teak (*Tectona grandis* L. F.); southern pine rated 8. Highest rated domestic hardwood, at 9.8, was butternut (*Juglans cinerea* L.), followed—at 9.6—by black cherry (*Prunus serotina* Ehrh.), yellow birch (*Betula alleghaniensis* Britton), and black walnut (*Juglans nigra* L.). Pine-site hardwoods so classified were rated as follows:

Species	Rating
Ash, white	9.4
Oak, white	9.3
Oak, northern red	9.2
Sweetgum	9.2
Oak, southern red	9.0
Hickory	9.0
Tupelo, black	8.5
Hackberry	8.0
Elm, American	7.4

Thus, the predominating species on southern pine sites, i.e., the oaks, hickories, and sweetgum, number among this country's most machinable hardwoods.

Davis (1962) evaluated the machinability of 34 species of domestic hardwoods, including white ash, American elm, hackberry, hickory, red maple, northern red oak, white oak, sweetgum, black tupelo, and yellow-poplar (table 18-1). Among these 10, the most readily machined were as follows:

Machining operation	Species
Planing	Oaks
Shaping	White ash
Turning	Sweetgum, white oak, northern red oak, and hickory
Boring	Hickory, hackberry, the oaks, white ash, American elm, and sweetgum
Hollow-chisel mortising	White oak, hickory, and northern red oak
Sanding	White oak, northern red oak, and hickory

Cantin (1965) found that red maple and northern red oak, as well as white ash, were readily shaped.

TABLE 18-1.—*Percentage of fair to perfect pieces yielded by 10 hardwoods machined by six processes* (Data from Davis 1962, p. 66)

Species	Planing[1]– perfect pieces	Shaping[2]— good to excellent pieces	Turning[3]— fair to excellent pieces	Boring[4]— Good to excellent pieces	Mortising[4,5]— fair to excellent pieces	Sanding[4]— good to excellent pieces
	---Percent---					
Ash, white	75	55	79	94	58	75
Elm, American	33	13	65	94	75	66
Hackberry	74	10	77	99	72	—
Hickory	76	20	84	100	98	80
Maple, red	41	25	76	80	34	37
Oak, northern red	91	28	84	99	95	81
Oak, white[6]	87	35	85	95	99	83
Sweetgum	51	28	86	92	58	23
Tupelo, black	48	32	75	82	24	21
Yellow-poplar	70	13	81	87	63	19

[1]Wood at 6 percent moisture content; 1/16-inch depth of cut, rake angles of 15°, 20°, and 25°— data averaged.
[2]Moisture content 6 and 12 percent—data averaged.
[3]Moisture content 6, 12, and 20 percent—data averaged.
[4]Moisture content 6 percent.
[5]Hollow-chisel mortise.
[6]Includes chestnut oak.

18-2 ORTHOGONAL CUTTING[1]

Wood is machined by removing chips that range in size from sanderdust to pulp chips or larger. There are two basic machining processes. In the first, known as orthogonal cutting, the cutting edge is perpendicular to the direction of the relative motion of tool and workpiece. A carpenter's hand plane cuts ortho-

---

[1]In sec. 18-2, the photographs and cutting-force data specific to pine-site hardwoods are all taken from Woodson (1979), unless otherwise noted.

gonally, as does a bandsaw. Rotary peeling of veneer approximates orthogonal cutting.

The second is a rotary-cutting process (peripheral milling) in which single chips are formed and removed by the intermittent engagement of knives carried on the periphery of a rotating cutterhead or saw. A rotary planer machines wood by the peripheral milling process.

To separate a chip from the workpiece during any wood machining process, it is first necessary to cause a structural failure at the juncture of chip and workpiece. Since the strength of wood varies with grain direction, chip configuration, cutting power, and surface quality are all strongly affected by the direction of cut (fig. 18-1), as well as the knife geometry.

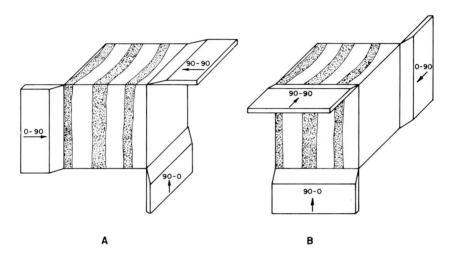

Figure 18-1.—Designation of the three major machining directions. The first number is the angle the cutting edge makes with the grain; the second is the angle between cutter movement and grain.

A two-number notation used by McKenzie (1961) is useful in describing the orthogonal machining situation. With this system the first figure given is the angle the cutting edge makes with the grain of the wood; the second figure is the angle between direction of tool motion and grain. Thus, hand planing has the simple notation of 90-0. The first number indicates the cutter edge is perpendicular to the grain and the second indicates that cutter movement is parallel to the grain. Similarly, veneer cutting is 0-90 and crosscutting is 90-90 (fig. 18-1).

DEFINITIONS

Figure 18-2 illustrates standard nomenclature of wood machining terms applicable to orthogonal cutting.

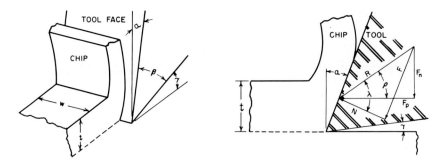

Figure 18-2.—Nomenclature in orthogonal cutting.

α—*Rake angle:* angle between the tool face and a plane perpendicular to the direction of tool travel.
β—*Sharpness angle:* angle between the tool face and back.
γ—*Clearance angle:* angle between the back of the tool and the work surface behind the tool.
t—*Nominal chip thickness; depth of cut.*
w—*Width of undeformed chip.*
F_n—*Normal tool force:* force component acting perpendicular to parallel tool force and perpendicular to the surface generated.
F_p—*Parallel tool force:* force component acting parallel to tool motion in workpiece, i.e., parallel to cut surface.
R—*Resultant tool force:* the resultant of normal and parallel tool force components.
ρ—*Angle of tool force resultant:* the angle whose tangent is equal to the normal tool force divided by the parallel tool force.
F—*Friction force:* force component acting along the interface between tool and chip.
N—*Normal to the friction force:* force component acting normal to tool face.
λ—*Angle between resultant tool force and the normal frictional force;* the angle whose tangent is equal to the friction force divided by the normal friction force.

EFFECTS OF CUTTING VELOCITY

In the experimental data that follow in this chapter, cutting velocity is always stated because the effect of cutting velocity on cutting forces is not well established. Endersby (1965), when cutting in a near-orthogonal mode, found that in the range from 1,000 to 9,000 feet per minute (f.p.m.), cutting velocity had little effect on cutting forces; however, when velocity was reduced from 1,000 to 7 f.p.m., cutting force increased about 2½ times. Inoue and Mori (1979) found that parallel cutting force changed only slightly with cutting velocity at small depths of cut or with large rake angles; when deep cuts (0.3 mm) were combined with small rake angles (15°), however, parallel cutting force decreased significantly as velocity was increased from 0.1 to 1.0 meter per second.

Factors that may alter the cutting resistance of wood as cutting velocity is increased include the following:

- More force is required to accelerate the chip at high cutting velocity than at low.
- Strength of wood increases with increasing rate of deformation.

- Strength of wood decreases as temperature increases; there may be localized changes in workpiece temperature near the juncture of chip and workpiece.
- The coefficient of friction between tool and chip may change as cutting velocity is varied (Ivanovskij and Goronok 1978).
- When cutting wet wood, hydraulic action of water in proximity to the knife may alter cutting forces as velocity is changed.

It may be that in most situations these several factors are mutually counteracting so that the net effect of changing cutting velocity is minor.

As a final comment, high cutting velocity may sometimes assist in accomplishing clean severance of fibers because of chip inertia. This effect can be observed when cutting grass with a scythe or rotary power mower.

PLANING: 90-0 DIRECTION

General discussions of orthogonal cutting in the planing mode (visualize a carpenter's hand plane) essentially parallel to the grain are available (Franz 1958; Koch 1964a, p. 35-87; Stewart 1971a; Huang and Hayashi 1973). This text will be restricted to data specific to hardwood species found on southern pine sites.

Chip formation.—As described by Koch (1955, p. 261; 1956, p. 397) and enumerated by Franz (1958), three basic chip types (figs. 18-3, 18-4, and 18-5) may result when wood is machined parallel to the grain in the 90-0 direction. **Type I chips** are broken splinters formed by cleavage along the grain; **Type II chips** fail in shear and tend to form continuous spirals; **Type III chips** are severely compressed parallel to the grain and are more or less formless. Type II chips leave the best surface.

Type I chips are formed when cutting conditions are such that the wood splits ahead of the tool by cleavage until it fails in bending as a cantilever beam, as illustrated by dry wood in figure 18-3 top. Type I chips sometimes peel off in segmented spirals without breaking abruptly (fig. 18-3 bottom). Factors leading to formation of Type I chips are:

- Low resistance in cleavage combined with high stiffness and strength in bending.
- Deep cuts (Type I chips can form with any depth of cut, depending on other factors.)
- Large rake angles (25° and more).
- Low coefficient of friction between chip and tool face.
- Low moisture content in the wood.

Type I chips leave **chipped grain,** i.e., the split ahead of the cutting edge frequently runs below the planed surface. The amount of roughness depends upon the depth to which the cleavage runs into the wood. Power consumed by a knife forming Type I chips is low because wood fails relatively easily in tension perpendicular to the grain, and the knife severs few fibers. Beause it is seldom cutting, the knife edge dulls slowly.

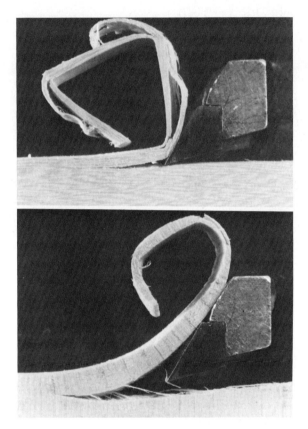

Figure 18-3.—Franz Type I chips from cuts in 90-0 direction. (Top) Broken chips formed when making a .060-inch cut in winged elm at 10.9-percent moisture content with a 30° rake angle and 15° clearance angle. (Bottom) Tearing ahead of the knife produced when making .030-inch cut in black tupelo with interlocked grain at 10.9-percent moisture content with a 30° rake angle and 15° clearance angle. (Photos from Woodson 1979.)

Rake angles of 25 and 35° tend to cause Type I chips because the normal cutting force (F_n) is negative at all depths of cut for all moisture contents.

Type II chips occur when under limited conditions (the mechanics are explained by Franz 1958 and Koch 1964a, p. 79-87) which induce continuous wood failures extending from the cutting edge to the work surface ahead of the tool (fig. 18-4). The movement of the tool strains the wood ahead of the tool in compression parallel to the grain and causes diagonal shearing stresses; as the wood fails, it forms a continuous, smooth spiral chip. The radius of the spiral increases as chip thickness increases. Chips may display laminae or layers. The resultant surface is excellent. Thin chips, intermediate to high moisture content, and 5 to 20° rake angles favor formation of Type II chips in hardwoods. The cutting edge is in intimate contact with the wood at all times, and dulling may be rapid. Power demand is intermediate between that for Type I and Type III chips.

Machining 1707

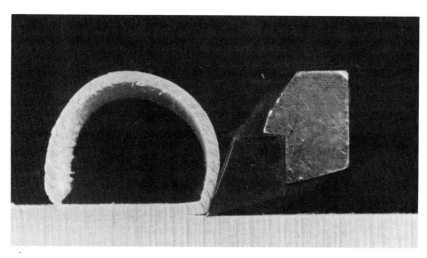

Figure 18-4.—Franz Type II chip (approximate) from an 0.045-inch cut (90-0 direction) in black tupelo at 10.9 percent moisture content with 20° rake angel and 15° clearance angle. In a true Type II chip, no split should be visible ahead of the knife tip. (Photo from Woodson 1979.)

Figure 18-5.—Franz Type III chip from an 0.060-inch cut in black tupelo at 10.9 percent moisture content with a 10° rake angle and a 15° clearance angle. (Photo from Woodson 1979.)

Because Type II chips yield excellent surfaces, it is useful to be able to predict rake angles that promote formation of Type II chips. Stewart (1977) advanced the simplified concept that if the normal force (F_n) is zero, Type II chips should form. This assumption permits quick calculation of optimum rake angle (α) if F_n and F_p are known, as follows:

The **cutting coefficient of friction** is expressed

$$\mu = \tan(\arctan F_n/F_p + \alpha) \quad (18\text{-}1)$$

If $F_n = 0$, then

$$\mu = \tan \alpha \quad (18\text{-}2)$$

From values of F_n and F_p in tables 18-5A through 18-26A, the cutting coefficient of friction (μ) can be calculated. For example, table 18-6A gives the following values of F_n and F_p for white ash cut at 11 percent moisture content in the 90-0 direction with 20-degree rake angle and 0.030-inch chip thickness:

F_n = 3.2 pounds per 0.1-inch knife length
F_p = 25.6 pounds per 0.1-inch knife length

Therefore,

μ = Tan (arctan 3.2/25.6 + 20 degrees)
= tan 27.1 degrees = 0.51

Thus, Stewart's (1977) method predicts that Type II chips should form if rake angle is 27.1 degrees when cutting the chip 0.030-inch thick from white ash at 11 percent moisture content.

The cutting coefficient of friction calculated from equation 18-1 is significatnly greater than static and dynamic coefficients of friction observed when blocks of wood are moved over smooth steel surfaces (fig. 9-22 and table 9-13). This disparity arises because cutting friction includes plastic deformation of the forming chip and resistance at the cutting edge and along the clearance face—as well as along the rake face.

Type III chips tend to form in cycles. Wood ahead of the tool is stressed in compression parallel to the grain and ruptures in shear parallel to the grain and compression parallel to the grain. The chip does not escape freely up the tool, and the deformed wood is compacted against the tool face (fig. 8-5). Stresses are then transferred to undeformed areas that fail in turn. When the accumulation of compressed material becomes critical, the chip buckles and escapes upward, and the cycle begins again. Factors favorable to the formation of Type III chips include:

- Small or negative rake angles.
- Dull cutting edges (the rounded edge presents a negative rake angle at tool edge extremity).
- High coefficient of friction between chip and tool face.

Wood failures ahead of the tool establish the surface, frequently extending below the plane of the cut or leaving incompletely severed wood elements prominent on the surface. This machining defect is termed **fuzzy grain.** Power consumption is high, and dulling may be rapid.

Readers interested in mathematical models to predict chip types when cutting in the 90-0 direction will find useful Franz (1958), Koch (1964a, p. 79-87),

Stewart (1971a, 1977), Huang and Hayashi (1973), and Klamecki (1979).

Effects of the width of cut.—If the tool is wider than the workpiece, cutting forces are directly proportional to width of the cut; doubling width doubles cutting forces.

Effects of knife angles.—During all orthogonal cutting (all directions), the angle between the tool face and a plane perpendicular to the direction of tool travel (rake angle) strongly affects tool forces as well as chip type and smoothness of cut; forces are negatively correlated with rake angle (table 18-2 and 18-3, and figures 18-8 and 18-60 through 18-63).

TABLE 18-2.—*Average parallel (F_p) and normal (F_n) tool forces per 0.1 inch of knife for orthogonally cutting southern red oak and yellow-poplar in three directions at 0.030 inch depth of cut, with various rake angles, and at three moisture contents* (Data from Woodson 1979)[1]

Species and Rake angle (degrees)	10% M.C.		20% M.C.		Saturated	
	F_p	F_n	F_p	F_n	F_p	F_n
	----------------------Pounds----------------------					
PLANING: 90-0 DIRECTION						
Southern red oak[2]						
10	37.97	10.15	22.46	5.54	18.90	4.55
20	32.78	3.05	19.11	1.74	14.23	1.18
30	21.47	−1.29	11.38	−.71	8.91	−.37
Yellow-poplar[3]						
10	20.12	4.66	15.36	4.81	12.81	5.67
20	18.59	2.34	13.90	1.81	10.24	1.61
30	10.74	−.66	11.69	−.08	7.53	.81
VENEER CUTTING: 0-90 DIRECTION						
Southern red oak[2]						
50	6.12	.19	4.57	−.48	3.76	−.26
60	6.40	.35	3.83	−1.16	3.19	−.82
70	4.72	.29	3.39	−.08	2.49	.09
Yellow-poplar[3]						
50	2.20	.05	2.12	−.08	1.84	.10
60	1.93	−.28	2.02	−.38	1.70	−.12
70	1.89	.18	1.98	.12	1.49	.28
CROSSCUTTING: 90-90 DIRECTION						
Southern red oak[2]						
20	43.01	2.89	31.01	1.04	25.30	.52
30	37.03	−3.03	25.74	−2.57	17.13	−2.02
40	26.86	−6.74	18.86	−4.78	14.05	−3.93
Yellow-poplar[3]						
20	18.73	4.07	22.12	2.24	15.47	1.34
30	16.74	1.34	15.04	−.30	12.78	−.10
40	12.27	−1.03	12.27	−1.80	8.08	−1.10

[1]Each value is the average for five replications; cutting velocity was 5 inches per minute.
[2]Specific gravity averaged 0.618 based on ovendry weight and green volume.
[3]Specific gravity averaged 0.376 based on ovendry weight and green volume.

TABLE 18-3.—*Tool forces per 0.1 inch of knife when orthogonally cutting in the 90-0 direction with various depths of cut, moisture contents, and rake angles. Values are averaged over all variables and over 22 pine site species of hardwoods* (Data from Woodson 1979)[1]

Principal factor	Parallel force		Normal force		
	Average	Maximum	Average	Maximum	Minimum
Depth of cut, inches	---------------------- Pounds ----------------------				
.015	9.9	12.0	1.7	2.3	1.0
.030	18.5	23.1	2.5	3.5	1.5
.045	24.2	31.4	3.1	4.4	1.7
.060	30.0	40.1	3.8	5.4	2.0
Moisture content, percent					
10.9	28.2	38.4	4.0	5.5	2.2
18.9	20.0	24.5	2.6	3.5	1.5
104.3	13.8	17.1	1.8	2.6	1.0
Rake angle, degrees					
10	26.3	33.2	6.7	8.6	4.6
20	23.1	27.0	2.3	2.9	1.6
30	12.5	19.8	− .6	.2	− 1.5

[1]Cutting velocity was 5 inches per minute.

The angle between the tool face and back (sharpness angle) strongly affects the rate at which the cutting edge dulls. Minute fracturing of a freshly sharpened and honed knife edge occurs as the very first few chips are cut and continues until equilibrium is reached between the cutting edge—which grows thicker and more rigid as dulling proceeds—and the cutting forces; from this time, wear proceeds at a slower rate. Effective rake angle is decreased as wear proceeds (fig. 18-9); cutting forces rise, and chip formation is altered.

The angle between the back of the tool and the work surface behind the tool, i.e., clearance angle, does not have a critical effect on cutting force or chip formation; 15° is usual. As it is reduced below 15°, tool forces rise moderately. Dulling of the tool reduces the effective clearance angle, which may in fact become negative; a negative clearance angle increases the cutting forces exerted by the knife and usually adversely affects surface quality by causing **raised grain**—a roughened condition in which dense latewood, after being depressed by the dull knife, swells subsequent to planing so that it is raised above the less dense (and therefore less swollen) earlywood. At the other extreme, if the clearance angle is very large and rake angle is kept constant, the cutting edge becomes thin, and resulting rapid dulling increases cutting forces. In the cutting force data presented in this section (18-2), the clearance angle is 15° unless otherwise stated.

Effects of depth of cut.—In all orthogonal cutting, depth of cut is synonymous with the thickness of the undeformed chip. As Lubkin (1957) and others have observed, in a given cutting situation two types of parallel-force curves

Machining 1711

may develop with changing chip thickness. When chips are very thin, the parallel force varies according to a power curve, and F_p becomes zero at zero chip thickness.

$$F_p = Kt^m w \qquad (18\text{-}3)$$

where:
- F_p = parallel tool force
- K = a constant
- t = chip thickness
- m = a constant between 1 and 0 (generally observed to be from 0.25 to 0.67)
- w = width of chip

Beyond the region of very thin chips it is possible, with suitably chosen constants A and B, to approximate considerable portions of this curve with a straight-line function of t:

$$F_p = (A + Bt)w \qquad (18\text{-}4)$$

In some situations the experimentally determined parallel cutting force defined by equation 18-3 holds for the entire practical range of chip thicknesses. In other situations, however, the curve straightens beyond a certain chip thickness and continues linearly as described by equation 18-4.

Pooled data from 22 pine-site hardwoods show that both parallel and normal forces during cutting in the 90-0 direction are positively correlated with specific gravity and chip thickness (fig. 18-10). Figures 18-6 through 18-8 show how chip thickness, rake angle, and moisture content interact to influence cutting forces in white ash.

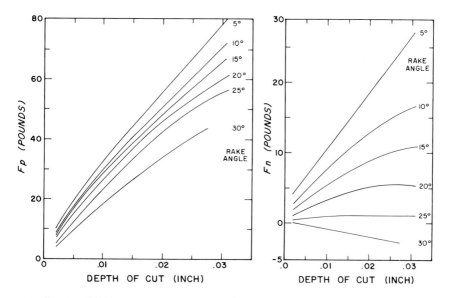

Figure 18-6.—White ash at 1.5-percent moisture content; relationship of cutting forces per 0.25-inch knife length to depth of cut and rake angle when orthogonally cutting in the 90-0 direction. (Left) Parallel cutting force, F_p. (Right) Normal cutting force, F_n. (Drawings after Franz 1958.)

Effects of wood temperature.—Kivimaa (1950), working with Finnish birch (*Betula* sp.), found that force required for cutting parallel to the grain, as well as for the other modes, decreases as wood temperature increases (fig. 18-11). It is probable that workpiece temperature interacts strongly with moisture content, chip thickness, and rake angle to affect chip formation and cutting forces. Wood is weakened by heat; steamed wood is softened more than wood subjected to dry heat, due mainly to more softening of the middle lamella. Even at temperatures somewhat below 100°C, wood may be somewhat plastic; when wood has cooled completely, however, most of the original strong bond between the cell walls is restored (Necesany 1965). Figure 10-15 and related discussion give the immediate effects of elevated temperatures on the mechanical properties of pine-site hardwoods.

Effects of moisture content.—Cutting forces are usually negatively correlated with moisture content, being greatest in dry wood and least in green wood (tables 18-2 and 18-3, and figures 18-6 through 18-8).

Effects of specific gravity.—Cutting forces are proportional to wood specific gravity (fig. 18-10). Thus, yellow-poplar requires much less force to cut than southern red oak (table 18-2).

Summary of Woodson's data on cutting forces.—Woodson (1979) orthogonally machined specimens (1/8- to 1/4-inch thick) from 5 trees of each of 22 hardwood species grown on southern pine sites. He measured parallel and normal cutting forces when cutting in the 90-0 direction at a cutting velocity of 5 inches per minute, as affected by selected rake angles (10, 20, and 30 degrees), depths of cut (0.015, 0.030, 0.045, and 0.060 inch), and moisture contents (about 10 percent, about 20 percent, and saturated). Specific gravities and saturated moisture contents of the specimens are given in table 18-4).

When dry wood was planed with 10° rake angle (fig. 18-10) parallel and normal forces increased with increasing depth of cut and increasing specific gravity. At low rake angle, chips were continuous (fig. 18-4) and normal forces were positive (i.e., the knife was pushing the workpiece). As rake angle increased, the normal forces were reduced and produced broken chips (i.e., broken as a cantilever beam as in figure 18-3 top). With further increase in rake angle to about 20°, the normal forces changed from positive to negative.

Consequently, the relationship between average normal force and specific gravity was not significant with a 20° rake angle. The 30° rake angle produced increasingly negative cutting forces as the specific gravity increased. A combination of low rake angle and low wood specific gravity produced extensive crushing ahead of the knife and a chip similar to that shown in figure 18-5. In species exhibiting interlocked grain, such as black tupelo and the elms, high rake angles produced tearing ahead of the knife and left a rough surface (fig. 18-3, bottom).

Maximum parallel forces appeared to be about 30 percent greater than the average parallel forces (table 18-3). Similarly, the maximum normal forces were 30 to 40 percent greater than the average normal forces.

Tables 18-5A, 18-6A, etc. through 18-26A contain Woodson's data on tool forces when planing orthogonally in the 90-0 direction, as affected by species,

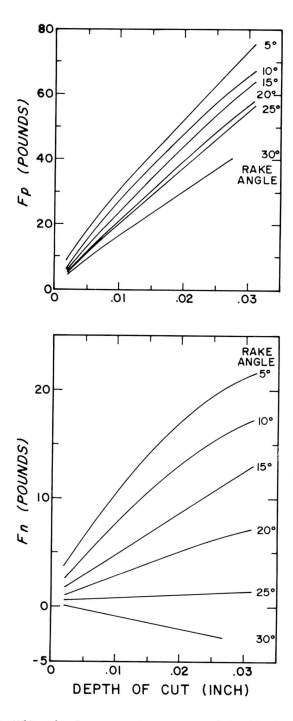

Table 18-7.—White ash at 8-percent moisture content; relationship of cutting forces per 0.25-inch knife length to depth of cut and rake angle when orthogonally cutting in the 90-0 direction. (Top) Parallel cutting force, F_p. (Bottom) Normal cutting force, F_n. (Drawings after Franz 1958.)

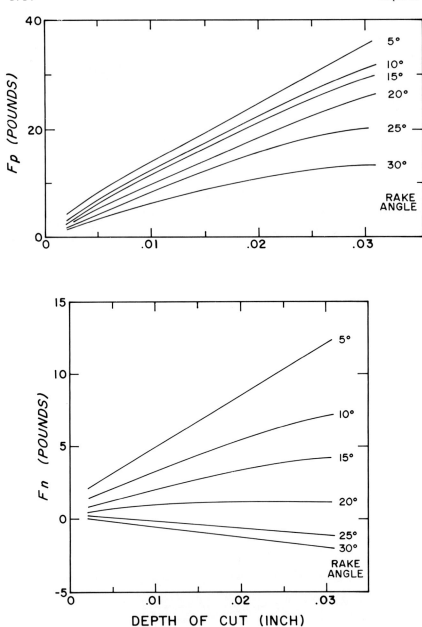

Figure 18-8.—White ash saturated with water; relationship of cutting forces per 0.25-inch knife length to depth of cut and rake angle when orthogonally cutting in the 90-0 direction. (Top) Parallel cutting force, F_p. (Bottom) Normal cutting force, F_n. (Drawings after Franz 1958.)

Machining 1715

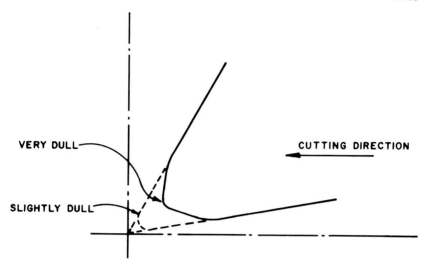

Figure 18-9.—Profile of a cutting edge as dulling proceeds. (Drawing after McKenzie and Cowling 1971.)

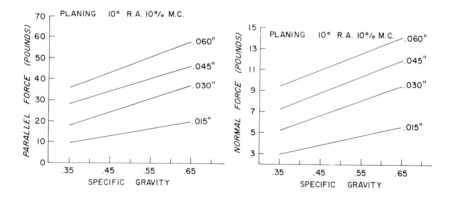

Figure 18-10.—Relationship of parallel and normal cutting force per 0.1-inch knife length with specific gravity (ovendry weight and green volume) at four depths of cut for 90-0 planing. Average of 22 species, moisture content 10 percent, rake angle 10°, clearance angle 15°. All regressions are significant at .01 level. (Drawing after Woodson 1979.)

depth of cut, moisture content, and rake angle. Table 18-4 provides a species index to these tables.

The reader is cautioned that at higher feed speeds (e.g., 10,000 feet per minute), parallel cutting forces may be significantly lower than these tabulated values observed during slow-speed cutting (5 inches per minute).

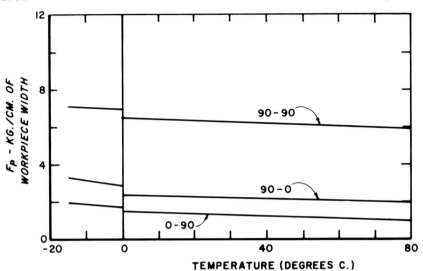

Figure 18-11.—Effect of temperature on parallel tool force. See figure 18-1 for explanation of cutting directions. Species, Finnish birch; chip thickness, 0.1 mm; moisture content, approximately 80 percent; rake angle, 35°; clearance angle, 10°. (Drawing after Kivimaa 1950.)

TABLE 18-4.—*Specific gravity (ovendry weight and green volume basis) and average saturated moisture contents of machining specimens cut from 22 southern hardwoods grown on southern pine sites; right hand column is an index to tabular data on cutting forces presented in tables 18-5A through 26C.* (Data from Woodson 1979)

Species	Saturated moisture content	Specific gravity	Table number[1]
	Percent		
Ash, green	107	0.562	18-5A,B,C
Ash, white	104	.577	18-6A,B,C
Elm, American	106	.551	18-7A,B,C
Elm, winged	89	.642	18-8A,B,C
Hackberry	115	.541	18-9A,B,C
Hickory	92	.640	18-10A,B,C
Maple, red	132	.489	18-11A,B,C
Oak, black	94	.598	18-12A,B,C
Oak, blackjack	118	.626	18-13A,B,C
Oak, cherrybark	92	.609	18-14A,B,C
Oak, laurel	92	.597	18-15A,B,C
Oak, northern red	88	.626	18-16A,B,C
Oak, post	80	.636	18-17A,B,C
Oak, scarlet	92	.618	18-18A,B,C
Oak, Shumard	90	.620	18-19A,B,C
Oak, southern red	91	.618	18-20A,B,C
Oak, water	101	.573	18-21A,B,C
Oak, white	81	.644	18-22A,B,C
Sweetbay	140	.442	18-23A,B,C
Sweetgum	145	.439	18-24A,B,C
Tupelo, black	121	.501	18-25A,B,C
Yellow-poplar	157	.376	18-26A,B,C

[1]*A* indicates planing direction (90-0); *B* indicates veneer direction (0-90); *C* indicates crosscut direction (90-90).

TABLE 18-5A.—*Tool forces when orthogonally cutting green ash parallel to the grain in the 90-0 direction[1]* (Data from Woodson 1979)

Depth of cut and moisture content (percent)	Rake angle, degrees		
	10	20	30
	----------------Pounds per 0.1 inch of knife----------------		
	PARALLEL FORCE[2]		
0.015 inch			
10.7.................	17.7 (21.7)	16.8 (18.6)	10.7 (14.2)
18.8.................	11.6 (14.6)	9.5 (10.6)	5.9 (8.2)
100.7.................	8.4 (9.2)	7.1 (7.9)	4.5 (5.4)
0.030 inch			
10.7.................	32.4 (41.1)	34.7 (39.0)	12.1 (29.5)
18.8.................	21.9 (25.7)	20.3 (22.8)	12.7 (17.2)
100.7.................	15.6 (17.4)	13.2 (15.1)	8.0 (9.8)
0.045 inch			
10.7.................	43.2 (54.5)	43.4 (48.1)	15.4 (38.3)
18.8.................	30.2 (36.5)	26.8 (29.3)	17.4 (23.2)
100.7.................	21.3 (25.0)	18.5 (21.6)	11.1 (13.7)
0.060 inch			
10.7.................	57.0 (72.7)	57.4 (63.2)	16.2 (48.8)
18.8.................	38.1 (46.4)	34.7 (38.8)	18.1 (28.1)
100.7.................	28.0 (33.8)	23.1 (26.9)	12.5 (16.6)
	NORMAL FORCE[3,4]		
0.015 inch			
10.7.................	5.3 (3.9 to 6.5)	2.2 (1.8 to 2.7)	−0.3 (−0.7 to 0.1)
18.8.................	3.4 (2.4 to 4.3)	1.4 (1.1 to 1.8)	.0 (−.4 to .7)
100.7.................	2.5 (2.1 to 2.9)	.9 (.7 to 1.3)	.2 (−.2 to .6)
0.030 inch			
10.7.................	8.8 (6.3 to 11.2)	4.3 (3.8 to 4.8)	−.5 (−1.6 to .6)
18.8.................	5.9 (4.3 to 7.1)	2.3 (1.6 to 2.7)	−.6 (−1.2 to .2)
100.7.................	3.7 (3.2 to 4.2)	1.2 (.7 to 1.7)	−.3 (−.8 to .1)
0.045 inch			
10.7.................	11.3 (7.1 to 14.3)	5.5 (4.7 to 6.1)	−1.1 (−3.1 to .2)
18.8.................	8.0 (6.3 to 9.5)	2.7 (2.3 to 3.2)	−.9 (−1.8 to −.3)
100.7.................	4.8 (3.8 to 5.6)	1.2 (.6 to 1.7)	−.9 (−1.4 to −.1)
0.060 inch			
10.7.................	15.2 (10.3 to 18.5)	6.8 (6.0 to 7.7)	−.8 (−3.2 to 1.0)
18.8.................	9.7 (7.7 to 12.1)	3.6 (2.8 to 4.1)	−1.1 (−2.2 to −.3)
100.7.................	5.9 (4.6 to 6.8)	1.1 (.5 to 1.7)	−.8 (−1.8 to −.2)

[1]Clearance angle 15°; cutting velocity 5 inches per minute.

[2]The first number in each entry is the average cutting force; the numbers following in parentheses are maximum forces (average of five).

[3]The first number in each entry is the average cutting force; the numbers following in parentheses are minimum and maximum forces (average of five).

[4]A negative normal force indicates that the knife tended to lift the workpiece; force was positive when the knife pushed the workpiece.

TABLE 18-5B.—*Tool forces when orthogonally cutting green ash veneer in the 0-90 direction[1]* (Data from Woodson 1979)

Depth of cut and moisture content (percent)	Rake angle, degrees		
	50	60	70
	----------------Pounds per 0.1 inch of knife----------------		
PARALLEL FORCE[2]			
0.015 inch			
11.0	2.9 (6.8)	3.4 (7.1)	3.3 (5.8)
19.1	2.7 (4.6)	2.5 (5.0)	4.6 (9.6)
109.4	1.6 (3.2)	1.7 (2.7)	3.9 (6.9)
0.030 inch			
11.0	4.1 (10.7)	4.3 (9.2)	3.7 (7.0)
19.1	3.6 (7.5)	3.4 (7.7)	3.1 (5.3)
109.4	2.4 (5.2)	2.3 (4.1)	2.1 (3.0)
0.045 inch			
11.0	4.6 (13.5)	4.6 (11.2)	4.1 (9.0)
19.1	4.5 (8.7)	4.6 (9.8)	3.8 (7.2)
109.4	2.9 (6.8)	2.9 (5.7)	2.4 (4.1)
0.060 inch			
11.0	5.9 (15.7)	5.2 (12.5)	5.8 (11.4)
19.1	5.5 (11.0)	4.6 (10.5)	4.7 (8.7)
109.4	3.4 (7.4)	3.5 (7.0)	3.0 (5.2)
NORMAL FORCE[3,4]			
0.015 inch			
11.0	0.3 (−0.8 to 1.9)	0.4 (−0.7 to 1.6)	−0.1 (−1.2 to 1.0)
19.1	−.2 (−.8 to .6)	−.1 (−.9 to .9)	.0 (−1.5 to 1.3)
109.4	−.2 (−.6 to .2)	−.2 (−.6 to .2)	−.2 (−1.4 to .8)
0.030 inch			
11.0	.0 (−1.3 to 1.5)	−.2 (−1.6 to .8)	.3 (−1.0 to 1.5)
19.1	−.6 (−1.5 to .1)	−.7 (−2.0 to .3)	.0 (−.9 to 1.1)
109.4	−.4 (−1.2 to .1)	−.6 (−1.3 to .2)	−.2 (−.6 to .3)
0.045 inch			
11.0	−.1 (−1.5 to 1.4)	−.2 (−1.9 to .9)	.2 (−1.6 to 1.6)
19.1	−.7 (−1.8 to .4)	−1.1 (−2.9 to .3)	−.3 (−1.4 to .8)
109.4	−.5 (−1.5 to .1)	−.9 (−1.8 to .3)	−.7 (−1.5 to .0)
0.060 inch			
11.0	−.1 (−1.8 to 1.6)	−.4 (−2.2 to .9)	−.2 (−1.9 to 1.4)
19.1	−.9 (−2.4 to .3)	−1.0 (−2.7 to .2)	−.5 (−2.0 to .9)
109.4	−.6 (−1.9 to .3)	−.9 (−2.2 to .3)	−.8 (−1.9 to .1)

[1]Clearance angle 0°; cutting velocity 5 inches per minute.

[2]The first number in each entry is the average cutting force; the numbers following in parentheses are maximum forces (average of five).

[3]The first number in each entry is the average cutting force; the numbers following in parentheses are minimum and maximum forces (average of five).

[4]A negative normal force indicates that the knife tended to lift the workpiece; force was positive when the knife pushed the workpiece.

TABLE 18-5C.—*Tool forces when orthogonally cutting green ash across the grain in the 90-90 direction[1]* (Data from Woodson 1979)

Depth of cut and moisture content (percent)	Rake angle, degrees		
	20	30	40
	---------Pounds per 0.1 inch of knife---------		
	PARALLEL FORCE[2]		
0.015 inch			
10.5	23.4 (28.5)	18.7 (22.4)	15.2 (17.5)
18.4	16.6 (19.5)	13.7 (16.8)	11.5 (13.9)
111.7	12.7 (15.1)	10.8 (12.2)	8.6 (10.3)
0.030 inch			
10.5	35.9 (45.0)	32.4 (39.5)	25.3 (29.7)
18.4	30.3 (35.4)	22.3 (26.4)	18.3 (22.0)
111.7	21.2 (23.7)	16.2 (18.4)	12.2 (14.7)
0.045 inch			
10.5	45.7 (56.6)	46.5 (57.9)	34.0 (39.3)
18.4	37.2 (42.3)	30.2 (35.8)	24.2 (28.7)
111.7	28.3 (33.1)	21.2 (24.5)	16.5 (20.6)
0.060 inch			
10.5	44.5 (59.1)	54.6 (66.8)	38.4 (46.4)
18.4	45.5 (53.5)	36.4 (43.6)	28.7 (34.7)
111.7	30.2 (35.7)	25.4 (29.2)	20.5 (25.2)
	NORMAL FORCE[3,4]		
0.015 inch			
10.5	2.5 (1.8 to 4.5)	0.2 (−0.9 to 2.2)	−2.5 (−3.4 to −1.1)
18.4	1.9 (1.1 to 3.3)	−.2 (−1.1 to .7)	−1.6 (−2.6 to −.6)
111.7	.6 (−.1 to 1.2)	−.5 (−1.2 to .1)	−1.8 (−2.4 to −1.1)
0.030 inch			
10.5	4.1 (2.5 to 7.1)	−1.3 (−3.2 to 1.3)	−5.5 (−7.2 to −2.3)
18.4	2.4 (1.2 to 4.3)	−1.7 (−2.8 to −.2)	−4.0 (−5.1 to −2.0)
111.7	.5 (−.1 to 1.2)	−1.8 (−2.6 to −.9)	−3.6 (−4.8 to −2.2)
0.045 inch			
10.5	5.4 (2.5 to 8.4)	−1.9 (−4.7 to 3.1)	−7.9 (−10.0 to −3.5)
18.4	2.5 (1.0 to 4.3)	−3.0 (−4.7 to −.4)	−5.9 (−8.0 to −3.0)
111.7	.5 (−.6 to 1.5)	−2.8 (−3.9 to −1.7)	−5.1 (−6.8 to −3.1)
0.060 inch			
10.5	6.3 (2.1 to 10.3)	.9 (−6.5 to 11.9)	−9.1 (−11.8 to −5.1)
18.4	2.6 (.8 to 5.6)	−3.8 (−6.3 to −.4)	−7.3 (−9.7 to −4.3)
111.7	.6 (−.8 to 2.3)	−3.7 (−4.9 to −2.5)	−6.4 (−8.3 to −4.1)

[1] Clearance angle 15°; cutting velocity 5 inches per minute.

[2] The first number in each entry is the average cutting force; the numbers following in parentheses are maximum forces (average of five).

[3] The first number in each entry is the average cutting force; the numbers following in parentheses are minimum and maximum forces (average of five).

[4] A negative normal force indicates that the knife tended to lift the workpiece; force was positive when the knife pushed the workpiece.

TABLE 18-6A.—*Tool forces when orthogonally cutting white ash parallel to the grain in the 90-0 direction*[1] (Data from Woodson 1979)

Depth of cut and moisture content (percent)	Rake angle, degrees		
	10	20	30

----------------Pounds per 0.1 inch of knife-----------------

PARALLEL FORCE[2]

0.015 inch			
11.0	17.3 (21.9)	14.3 (15.8)	9.7 (12.8)
19.2	12.3 (14.2)	8.5 (9.9)	7.0 (8.0)
98.5	7.5 (8.7)	6.2 (7.4)	4.4 (5.1)
0.030 inch			
11.0	29.3 (39.6)	25.6 (28.3)	18.3 (23.6)
19.2	22.6 (26.7)	17.2 (19.6)	13.1 (14.4)
98.5	14.1 (16.4)	11.0 (13.1)	8.0 (9.0)
0.045 inch			
11.0	40.0 (52.7)	37.5 (41.3)	21.9 (32.3)
19.2	28.4 (37.2)	23.1 (26.2)	17.0 (18.9)
98.5	20.8 (25.8)	16.5 (18.1)	10.5 (11.7)
0.060 inch			
11.0	49.0 (64.5)	53.3 (58.3)	22.1 (37.8)
19.2	34.3 (45.1)	32.7 (36.4)	21.0 (24.0)
98.5	26.3 (32.8)	21.8 (24.7)	11.3 (13.6)

NORMAL FORCE[3,4]

0.015 inch			
11.0	5.0 (3.7 to 6.6)	1.9 (1.5 to 2.3)	−0.1 (−0.4 to 0.4)
19.2	3.4 (2.8 to 4.1)	1.1 (.8 to 1.6)	−.1 (−.4 to .3)
98.5	2.3 (1.8 to 2.9)	.9 (.6 to 1.2)	.1 (−.1 to .4)
0.030 inch			
11.0	8.4 (3.9 to 10.8)	3.2 (2.8 to 3.8)	−.7 (−1.4 to .0)
19.2	5.8 (3.9 to 7.1)	2.0 (1.5 to 2.5)	−.6 (−1.0 to −.2)
98.5	3.5 (2.6 to 4.4)	.9 (.5 to 1.4)	−.3 (−.7 to .2)
0.045 inch			
11.0	11.4 (6.4 to 14.3)	4.4 (3.9 to 5.1)	−1.0 (−2.0 to .0)
19.2	7.3 (4.8 to 9.6)	2.5 (2.0 to 3.0)	−1.0 (−1.4 to −.3)
98.5	4.8 (3.8 to 6.1)	1.0 (.6 to 1.5)	−.6 (−1.1 to −.2)
0.060 inch			
11.0	12.8 (7.9 to 16.7)	5.8 (4.8 to 6.7)	−1.1 (−2.6 to .3)
19.2	8.7 (5.6 to 10.6)	3.4 (2.7 to 3.9)	−1.1 (−1.4 to −.5)
98.5	5.5 (3.6 to 7.2)	.8 (.3 to 1.4)	−1.2 (−1.7 to −.6)

[1]Clearance angle 15°; cutting velocity 5 inches per minute.
[2]The first number in each entry is the average cutting force; the numbers following in parentheses are maximum forces (average of five).
[3]The first number in each entry is the average cutting force; the numbers following in parentheses are minimum and maximum forces (average of five).
[4]A negative normal force indicates that the knife tended to lift the workpiece; force was positive when the knife pushed the workpiece.

TABLE 18-6B.—*Tool forces when orthogonally cutting white ash veneer in the 0-90 direction[1] (Data from Woodson 1979)*

Depth of cut and moisture content (percent)	Rake angle, degrees		
	50	60	70
	----------------Pounds per 0.1 inch of knife----------------		
PARALLEL FORCE[2]			
0.015 inch			
11.2	3.7 (7.4)	3.7 (6.7)	4.2 (7.5)
19.1	2.9 (5.5)	2.5 (4.7)	5.0 (11.5)
106.0	2.1 (3.6)	2.0 (3.1)	4.7 (8.0)
0.030 inch			
11.2	4.8 (12.9)	4.8 (10.4)	4.7 (7.9)
19.1	4.5 (8.9)	4.6 (8.2)	3.3 (5.7)
106.0	2.9 (6.0)	2.8 (4.5)	2.2 (3.8)
0.045 inch			
11.2	5.8 (15.0)	5.8 (13.9)	5.8 (10.9)
19.1	4.7 (10.9)	5.3 (10.2)	4.4 (7.8)
106.0	3.5 (7.3)	3.4 (5.9)	2.8 (4.7)
0.060 inch			
11.2	6.5 (15.8)	6.6 (15.6)	6.8 (13.9)
19.1	5.9 (13.0)	5.9 (10.4)	5.5 (11.0)
106.0	4.1 (8.2)	4.1 (7.3)	3.3 (5.5)
NORMAL FORCE[3,4]			
0.015 inch			
11.2	−0.1 (−1.0 to 1.2)	0.2 (−1.1 to 1.2)	0.1 (−2.2 to 1.4)
19.1	−.2 (−.9 to 1.0)	−.2 (−1.0 to .6)	.3 (−1.2 to 2.5)
106.0	−.2 (−.6 to .4)	−.3 (−.8 to .2)	−.3 (−1.4 to .7)
0.030 inch			
11.2	−.2 (−1.8 to 1.6)	−.1 (−1.4 to 1.3)	.7 (−1.3 to 2.2)
19.1	−.7 (−1.8 to .1)	−1.1 (−2.5 to .2)	−.4 (−1.2 to .9)
106.0	−.4 (−1.2 to .5)	−.8 (−1.8 to .1)	.3 (−.2 to 1.1)
0.045 inch			
11.2	−.1 (−1.9 to 1.5)	−.1 (−1.9 to 1.8)	.4 (−1.0 to 2.4)
19.1	−.7 (−2.0 to .8)	−1.1 (−3.2 to .2)	−.9 (−2.2 to .6)
106.0	−.6 (−1.6 to .3)	−1.1 (−2.5 to .1)	−.2 (−1.1 to .6)
0.060 inch			
11.2	−.0 (−2.7 to 1.8)	.0 (−1.8 to 1.7)	.6 (−1.1 to 2.8)
19.1	−1.0 (−2.4 to .3)	−1.4 (−3.3 to .1)	−1.0 (−2.4 to 1.3)
106.0	−.7 (−1.6 to .2)	−1.1 (−2.7 to .2)	−.3 (−1.5 to .6)

[1] Clearance angle 0°; cutting velocity 5 inches per minute.

[2] The first number in each entry is the average cutting force; the numbers following in parentheses are maximum forces (average of five).

[3] The first number in each entry is the average cutting force; the numbers following in parentheses are minimum and maximum forces (average of five).

[4] A negative normal force indicates that the knife tended to lift the workpiece; force was positive when the knife pushed the workpiece.

TABLE 18-6C.—*Tool forces when orthogonally cutting white ash across the grain in the 90-90 direction[1]* (Data from Woodson 1979)

Depth of cut and moisture content (percent)	Rake angle, degrees		
	20	30	40
	----------Pounds per 0.1 inch of knife----------		
PARALLEL FORCE[2]			
0.015 inch			
10.8	19.9 (24.0)	22.6 (26.2)	17.1 (20.6)
18.3	22.3 (25.0)	12.6 (14.4)	11.2 (13.2)
106.2	13.5 (15.3)	9.9 (11.7)	9.1 (10.5)
0.030 inch			
10.8	42.1 (49.2)	33.7 (39.2)	27.2 (31.5)
18.3	33.8 (37.8)	23.2 (26.6)	20.3 (23.2)
106.2	21.2 (23.8)	15.6 (17.8)	13.7 (15.3)
0.045 inch			
10.8	51.3 (60.7)	44.8 (50.7)	35.0 (43.4)
18.3	41.8 (46.9)	31.1 (35.2)	24.7 (28.0)
106.2	27.5 (31.5)	20.5 (23.3)	17.4 (20.3)
0.060 inch			
10.8	62.7 (76.2)	51.2 (60.2)	41.8 (49.7)
18.3	49.4 (57.8)	38.1 (43.6)	30.3 (35.8)
106.2	35.5 (41.3)	25.0 (28.2)	21.7 (24.6)
NORMAL FORCE[3,4]			
0.015 inch			
10.8	2.3 (1.3 to 3.6)	−0.8 (−1.9 to 0.4)	−3.2 (−4.4 to −1.9)
18.3	1.6 (1.0 to 2.5)	−.3 (−1.0 to .3)	−2.0 (−2.6 to −1.2)
106.2	.4 (.0 to 1.0)	−.7 (−1.3 to −.2)	−1.8 (−2.3 to −1.2)
0.030 inch			
10.8	3.1 (1.7 to 4.9)	−2.3 (−3.8 to −.4)	−6.1 (−8.2 to −4.2)
18.3	1.5 (.7 to 3.0)	−2.3 (−3.4 to −1.3)	−4.6 (−5.9 to −2.9)
106.2	.2 (−.5 to .8)	−2.1 (−2.9 to −1.3)	−3.8 (−4.8 to −2.7)
0.045 inch			
10.8	4.1 (1.5 to 6.1)	−3.8 (−6.1 to −1.1)	−9.1 (−11.8 to −6.3)
18.3	1.6 (.1 to 3.5)	−3.7 (−5.1 to −2.0)	−6.8 (−8.2 to −4.2)
106.2	−.1 (−.9 to 1.0)	−3.3 (−4.4 to −2.2)	−5.3 (−6.6 to −4.0)
0.060 inch			
10.8	4.8 (1.6 to 8.0)	−4.4 (−7.7 to −.7)	−11.3 (−14.0 to −7.9)
18.3	1.9 (−.2 to 4.1)	−5.1 (−7.5 to −2.8)	−8.5 (−11.5 to −4.9)
106.2	−.2 (−1.4 to 1.1)	−4.3 (−5.5 to −2.9)	−6.9 (−8.7 to −5.2)

[1]Clearance angle 15°; cutting velocity 5 inches per minute.

[2]The first number in each entry is the average cutting force; the numbers following in parentheses are maximum forces (average of five).

[3]The first number in each entry is the average cutting force; the numbers following in parentheses are minimum and maximum forces (average of five).

[4]A negative normal force indicates that the knife tended to lift the workpiece; force was positive when the knife pushed the workpiece.

TABLE 18-7A.—*Tool forces when orthogonally cutting American elm parallel to the grain in the 90-0 direction[1]* (Data from Woodson 1979)

Depth of cut and moisture content (percent)	Rake angle, degrees		
	10	20	30

----------------*Pounds per 0.1 inch of knife*----------------

PARALLEL FORCE[2]

0.015 inch			
10.8	15.8 (18.7)	14.9 (17.1)	9.3 (12.0)
18.8	12.0 (13.6)	9.2 (10.2)	6.7 (8.7)
102.8	8.7 (10.4)	7.5 (8.0)	5.0 (6.1)
0.030 inch			
10.8	30.3 (38.2)	31.0 (35.0)	11.9 (21.1)
18.8	22.5 (27.9)	19.1 (21.4)	13.5 (17.5)
102.8	14.1 (17.2)	14.1 (15.9)	8.0 (9.6)
0.045 inch			
10.8	40.5 (50.8)	38.7 (46.6)	13.0 (35.1)
18.8	30.7 (37.7)	26.4 (30.5)	14.7 (23.7)
102.8	25.4 (31.9)	19.6 (22.9)	10.7 (15.8)
0.060 inch			
10.8	50.1 (64.6)	50.6 (60.4)	13.9 (43.3)
18.8	37.7 (46.3)	33.4 (37.6)	16.4 (32.3)
102.8	30.0 (38.5)	23.0 (27.4)	12.7 (18.6)

NORMAL FORCE[3,4]

0.015 inch			
10.8	4.6 (3.8 to 5.2)	1.8 (1.5 to 2.1)	−0.2 (−0.5 to 0.2)
18.8	3.5 (2.7 to 4.0)	1.3 (.9 to 1.6)	−.2 (−.5 to .4)
102.8	2.6 (2.0 to 3.2)	.8 (.6 to 1.1)	−.1 (−.4 to .2)
0.030 inch			
10.8	8.2 (6.0 to 9.7)	3.6 (3.0 to 4.0)	−.6 (−1.5 to .1)
18.8	5.9 (3.8 to 7.3)	2.0 (1.6 to 2.4)	−1.0 (−1.3 to −.5)
102.8	4.1 (3.3 to 5.0)	1.1 (.8 to 1.5)	−.5 (−.8 to −.1)
0.045 inch			
10.8	10.5 (7.5 to 12.8)	3.8 (2.7 to 4.6)	−.6 (−2.4 to .2)
18.8	7.6 (5.0 to 9.4)	2.5 (2.0 to 2.9)	−1.2 (−2.0 to −.1)
102.8	5.7 (4.6 to 6.9)	1.2 (.8 to 1.7)	−.8 (−1.6 to −.2)
0.060 inch			
10.8	12.7 (8.9 to 16.5)	4.9 (3.6 to 6.0)	−.8 (−3.1 to .0)
18.8	9.2 (6.1 to 11.4)	2.9 (2.3 to 3.5)	−1.3 (−2.7 to −.3)
102.8	6.8 (5.0 to 8.4)	1.2 (.6 to 1.7)	−1.1 (−2.1 to −.3)

[1] Clearance angle 15°; cutting velocity 5 inches per minute.

[2] The first number in each entry is the average cutting force; the numbers following in parentheses are maximum forces (average of five).

[3] The first number in each entry is the average cutting force; the numbers following in parentheses are minimum and maximum forces (average of five).

[4] A negative normal force indicates that the knife tended to lift the workpiece; force was positive when the knife pushed the workpiece.

TABLE 18-7B.—*Tool forces when orthogonally cutting American elm veneer in the 0-90 direction*[1] *(Data from Woodson 1979)*

Depth of cut and moisture content (percent)	Rake angle, degrees		
	50	60	70
	----------------Pounds per 0.1 inch of knife-----------------		
PARALLEL FORCE[2]			
0.015 inch			
11.7..................	2.7 (5.5)	2.9 (5.0)	2.8 (4.5)
18.9..................	2.2 (4.4)	2.0 (3.3)	3.6 (7.4)
108.0.................	1.8 (2.9)	1.5 (2.5)	3.3 (5.3)
0.030 inch			
11.7..................	4.0 (8.5)	3.8 (7.5)	3.3 (5.6)
18.9..................	3.3 (6.4)	3.0 (5.5)	2.5 (4.2)
108.0.................	2.7 (4.8)	2.2 (3.6)	2.1 (3.1)
0.045 inch			
11.7..................	4.9 (10.7)	4.8 (9.1)	3.7 (7.1)
18.9..................	4.3 (8.8)	3.8 (7.0)	3.3 (5.3)
108.0.................	3.1 (5.8)	3.0 (4.8)	2.7 (4.1)
0.060 inch			
11.7..................	5.8 (12.1)	5.1 (11.1)	4.4 (8.1)
18.9..................	5.2 (10.0)	4.3 (8.4)	3.8 (6.5)
108.0.................	3.6 (7.0)	3.5 (5.8)	3.0 (4.8)
NORMAL FORCE[3,4]			
0.015 inch			
11.7..................	0.1 (−0.6 to .6)	0.4 (−0.4 to 1.1)	0.2 (−1.0 to 0.6)
18.9..................	−.2 (−.8 to .7)	−.3 (−.8 to .2)	−.4 (−1.8 to .9)
108.0.................	−.2 (−.5 to .2)	−.3 (−.6 to .2)	−.4 (−1.6 to .7)
0.030 inch			
11.7..................	−.2 (−1.3 to .6)	−.1 (−1.3 to .8)	.3 (−.6 to 1.2)
18.9..................	−.5 (−1.5 to .8)	−.9 (−1.8 to .1)	−.2 (−.8 to .6)
108.0.................	−.5 (−1.1 to .2)	−.7 (−1.3 to .0)	.1 (−.4 to .6)
0.045 inch			
11.7..................	−.5 (−1.9 to .8)	−.3 (−1.8 to 1.2)	−.2 (−1.6 to .9)
18.9..................	−.7 (−2.1 to .9)	−1.2 (−2.7 to .2)	−.7 (−1.7 to .3)
108.0.................	−.7 (−1.6 to .3)	−.9 (−1.8 to .1)	−.5 (−1.1 to .3)
0.060 inch			
11.7..................	−.7 (−2.5 to 1.1)	−.6 (−2.6 to 1.2)	−.4 (−1.9 to .9)
18.9..................	−.9 (−2.4 to .5)	−1.4 (−3.1 to .1)	−1.0 (−2.2 to .4)
108.0.................	−.9 (−2.2 to .2)	−1.2 (−2.5 to −.2)	−.8 (−1.9 to .3)

[1] Clearance angle 0°; cutting velocity 5 inches per minute.

[2] The first number in each entry is the average cutting force; the numbers following in parentheses are maximum forces (average of five).

[3] The first number in each entry is the average cutting force; the numbers following in parentheses are minimum and maximum forces (average of five).

[4] A negative normal force indicates that the knife tended to lift the workpiece; force was positive when the knife pushed the workpiece.

TABLE 18-7C.—*Tool forces when orthogonally cutting American elm across the grain in the 90-90 direction[1]* (Data from Woodson 1979)

Depth of cut and moisture content (percent)	Rake angle, degrees		
	20	30	40

----------------------Pounds per 0.1 inch of knife----------------------

PARALLEL FORCE[2]

0.015 inch			
10.4............	20.1 (25.0)	18.9 (23.0)	13.8 (18.3)
18.4............	17.2 (19.3)	13.2 (15.4)	11.0 (13.3)
107.9............	12.0 (14.0)	10.4 (11.8)	7.8 (9.0)
0.030 inch			
10.4............	38.3 (44.0)	30.5 (35.3)	23.1 (28.7)
18.4............	30.1 (33.0)	22.9 (26.4)	17.3 (21.1)
107.9............	22.3 (25.1)	17.1 (19.5)	12.6 (14.6)
0.045 inch			
10.4............	42.9 (50.0)	38.0 (44.6)	30.6 (36.6)
18.4............	38.5 (43.4)	29.1 (33.2)	22.7 (26.6)
107.9............	30.4 (33.3)	21.8 (24.0)	15.3 (18.7)
0.060 inch			
10.4............	51.3 (60.5)	42.0 (51.5)	35.5 (43.7)
18.4............	48.4 (54.8)	35.5 (39.7)	28.0 (36.0)
107.9............	38.2 (42.9)	26.7 (30.7)	19.4 (22.5)

NORMAL FORCE[3,4]

0.015 inch			
10.4............	1.7 (0.8 to 2.3)	−0.6 (−1.8 to 0.8)	−2.6 (−3.8 to −1.6)
18.4............	1.5 (1.0 to 1.8)	−.7 (−1.2 to −.1)	−1.6 (−2.5 to −1.1)
107.9............	.6 (.3 to .9)	−.5 (−.9 to −.1)	−1.7 (−2.4 to −.9)
0.030 inch			
10.4............	2.3 (1.0 to 4.3)	−2.3 (−3.9 to −.2)	−5.3 (−7.3 to −3.7)
18.4............	1.6 (1.0 to 2.8)	−2.0 (−3.2 to −1.1)	−4.1 (−5.9 to −3.2)
107.9............	.5 (−.1 to 1.0)	−1.7 (−2.6 to −.9)	−3.6 (−4.4 to −2.7)
0.045 inch			
10.4............	3.0 (.1 to 4.7)	−3.1 (−5.3 to −.3)	−7.6 (−10.2 to −5.1)
18.4............	1.8 (.8 to 3.6)	−3.0 (−4.3 to −1.9)	−6.2 (−7.9 to −4.3)
107.9............	.6 (−.3 to 1.4)	−2.6 (−3.7 to −1.6)	−4.9 (−6.8 to −3.7)
0.060 inch			
10.4............	3.9 (1.1 to 6.7)	−3.6 (−6.9 to −1.3)	−9.8 (−13.1 to −6.5)
18.4............	2.3 (.8 to 4.4)	−4.4 (−5.9 to −2.6)	−7.8 (−10.6 to −3.9)
107.9............	.9 (−.5 to 2.0)	−3.4 (−4.9 to −2.2)	−6.6 (−8.0 to −4.7)

[1]Clearance angle 15°; cutting velocity 5 inches per minute.
[2]The first number in each entry is the average cutting force; the numbers following in parentheses are maximum forces (average of five).
[3]The first number in each entry is the average cutting force; the numbers following in parentheses are minimum and maximum forces (average of five).
[4]A negative normal force indicates that the knife tended to lift the workpiece; force was positive when the knife pushed the workpiece.

TABLE 18-8A.—*Tool forces when orthogonally cutting winged elm parallel to the grain in the 90-0 direction[1]* (Data from Woodson 1979)

Depth of cut and moisture content (percent)	Rake angle, degrees		
	10	20	30
	----------------Pounds per 0.1 inch of knife----------------		
	PARALLEL FORCE[2]		
0.015 inch			
10.9	15.5 (18.9)	13.7 (15.7)	8.4 (10.9)
19.2	13.0 (15.3)	9.4 (10.4)	6.2 (7.8)
87.7	9.7 (11.1)	7.2 (8.8)	4.8 (6.7)
0.030 inch			
10.9	33.0 (41.8)	30.4 (34.7)	14.1 (20.3)
19.2	24.4 (29.0)	20.2 (22.8)	12.6 (15.7)
87.7	15.8 (20.5)	13.2 (18.3)	7.8 (10.5)
0.045 inch			
10.9	39.6 (55.4)	41.0 (46.0)	16.0 (28.5)
19.2	29.8 (38.2)	25.1 (29.3)	16.4 (21.3)
87.7	21.3 (31.5)	15.8 (22.9)	9.5 (13.6)
0.060 inch			
10.9	51.8 (71.3)	50.3 (56.2)	19.7 (34.6)
19.2	36.4 (47.7)	29.2 (35.0)	18.7 (25.4)
87.7	28.6 (40.9)	18.6 (30.6)	12.0 (15.9)
	NORMAL FORCE[3,4]		
0.015 inch			
10.9	4.7 (3.9 to 5.7)	2.0 (1.6 to 2.4)	−0.1 (−0.5 to 0.5)
19.2	3.8 (2.6 to 4.7)	1.2 (1.0 to 1.5)	.1 (−.2 to .4)
87.7	2.6 (2.1 to 3.0)	.8 (.6 to 1.1)	.0 (−.3 to .3)
0.030 inch			
10.9	9.0 (5.9 to 10.8)	3.7 (3.0 to 4.4)	−.5 (−1.3 to .3)
19.2	6.1 (4.8 to 7.1)	1.9 (1.5 to 2.2)	−.5 (−.9 to .1)
87.7	3.9 (2.6 to 5.0)	1.0 (.5 to 1.4)	−.4 (−.8 to .0)
0.045 inch			
10.9	11.1 (6.5 to 14.4)	4.7 (4.1 to 5.3)	−.7 (−2.2 to .1)
19.2	7.6 (5.3 to 9.1)	2.2 (1.9 to 2.6)	−.8 (−1.4 to −.1)
87.7	5.1 (2.8 to 7.3)	1.1 (.5 to 1.9)	−.6 (−1.2 to −.1)
0.060 inch			
10.9	13.9 (7.6 to 18.0)	5.6 (4.6 to 6.4)	−1.1 (−2.7 to .2)
19.2	9.0 (5.7 to 11.0)	2.5 (1.7 to 3.0)	−1.0 (−1.8 to −.1)
87.7	6.2 (3.9 to 8.9)	1.3 (.6 to 1.9)	−.7 (−1.4 to .1)

[1] Clearance angle 15°; cutting velocity 5 inches per minute.

[2] The first number in each entry is the average cutting force; the numbers following in parentheses are maximum forces (average of five).

[3] The first number in each entry is the average cutting force: the numbers following in parentheses are minimum and maximum forces (average of five).

[4] A negative normal force indicates that the knife tended to lift the workpiece; force was positive when the knife pushed the workpiece.

TABLE 18-8B.—*Tool forces when orthogonally cutting winged elm veneer in the 0-90 direction[1]* (Data from Woodson 1979)

Depth of cut and moisture content (percent)	Rake angle, degrees		
	50	60	70
	----------------Pounds per 0.1 inch of knife----------------		
	PARALLEL FORCE[2]		
0.015 inch			
11.6	3.2 (6.3)	3.9 (6.2)	3.3 (5.4)
19.1	2.6 (4.5)	2.7 (4.4)	5.3 (10.0)
88.8	2.2 (3.6)	2.3 (3.3)	4.4 (8.6)
0.030 inch			
11.6	5.3 (10.3)	5.4 (9.8)	4.3 (6.6)
19.1	4.1 (8.0)	3.9 (6.2)	3.3 (5.6)
88.8	3.3 (6.7)	3.3 (5.3)	2.6 (3.9)
0.045 inch			
11.6	6.8 (13.8)	6.5 (11.9)	5.0 (9.1)
19.1	5.2 (10.4)	4.7 (9.0)	4.4 (7.2)
88.8	4.3 (9.0)	3.8 (6.8)	3.1 (5.3)
0.060 inch			
11.6	7.5 (16.2)	7.0 (15.2)	5.9 (11.3)
19.1	6.1 (13.8)	5.8 (10.5)	5.2 (9.3)
88.8	5.0 (10.8)	4.4 (9.1)	3.8 (6.5)
	NORMAL FORCE[3,4]		
0.015 inch			
11.6	−0.3 (−0.9 to .5)	0.7 (0.2 to 1.4)	0.0 (−1.0 to 1.1)
19.1	−.5 (−1.0 to .1)	−.7 (−1.1 to .1)	−.5 (−2.0 to 1.2)
88.8	−.3 (−.7 to .2)	−.2 (−.6 to .3)	−.9 (−2.2 to .7)
0.030 inch			
11.6	−.7 (−1.9 to .6)	−.3 (−1.6 to 1.2)	.1 (−.8 to 1.2)
19.1	−.9 (−1.9 to .1)	−1.3 (−2.3 to −.2)	−.4 (−1.2 to .8)
88.8	−.7 (−1.7 to .5)	−1.0 (−1.9 to .0)	−.1 (−.8 to .4)
0.045 inch			
11.6	−.8 (−2.8 to .8)	−.7 (−2.4 to .9)	−.7 (−2.0 to .7)
19.1	−1.0 (−2.5 to .5)	−1.5 (−2.9 to .2)	−1.3 (−2.2 to .3)
88.8	−.8 (−2.4 to .7)	−1.1 (−2.7 to .1)	−.9 (−1.7 to .0)
0.060 inch			
11.6	−.9 (−3.2 to 1.2)	−.6 (−2.6 to 1.2)	−.6 (−2.4 to .6)
19.1	−1.3 (−3.2 to .5)	−1.7 (−3.5 to .1)	−1.4 (−2.8 to .5)
88.8	−.9 (−2.8 to .6)	−1.2 (−3.0 to .6)	−1.2 (−2.5 to .3)

[1] Clearance angle 0°; cutting velocity 5 inches per minute.

[2] The first number in each entry is the average cutting force; the numbers following in parentheses are maximum forces (average of five).

[3] The first number in each entry is the average cutting force; the numbers following in parentheses are minimum and maximum forces (average of five).

[4] A negative normal force indicates that the knife tended to lift the workpiece; force was positive when the knife pushed the workpiece.

TABLE 18-8C.—*Tool forces when orthogonally cutting winged elm across the grain in the 90-90 direction[1]* (Data from Woodson 1979)

Depth of cut and moisture content (percent)	Rake angle, degrees		
	20	30	40

----------------------Pounds per 0.1 inch of knife----------------------

PARALLEL FORCE[2]

0.015 inch			
11.6............	22.0 (28.2)	19.7 (23.2)	15.8 (18.6)
18.6............	17.4 (20.7)	14.2 (16.9)	11.9 (14.0)
91.2............	14.5 (16.7)	10.2 (12.3)	8.3 (9.4)
0.030 inch			
11.6............	42.5 (48.8)	35.5 (42.1)	25.6 (30.1)
18.6............	34.5 (38.8)	24.9 (28.7)	19.3 (21.2)
91.2............	26.2 (28.8)	17.7 (19.8)	14.0 (16.6)
0.045 inch			
11.6............	54.3 (60.3)	40.7 (46.9)	34.2 (41.0)
18.6............	41.9 (46.9)	32.8 (38.9)	24.8 (28.4)
91.2............	35.4 (39.1)	24.6 (28.4)	19.0 (21.7)
0.060 inch			
11.6............	60.2 (68.1)	53.7 (62.3)	39.9 (48.4)
18.6............	52.6 (61.6)	39.0 (45.4)	29.7 (35.6)
91.2............	44.0 (48.8)	30.7 (34.2)	23.3 (27.7)

NORMAL FORCE[3,4]

0.015 inch			
11.6............	1.6 (0.9 to 2.5)	−1.0 (−2.0 to −0.2)	−3.3 (−4.3 to −2.4)
18.6............	1.3 (1.0 to 2.0)	−1.0 (−1.6 to − .4)	−2.4 (−2.9 to −1.7)
91.2............	.5 (.1 to 1.0)	− .7 (−1.1 to − .3)	−1.9 (−2.4 to −1.3)
0.030 inch			
11.6............	2.0 (.8 to 4.9)	−3.3 (−5.1 to − .7)	−6.7 (−8.3 to −4.2)
18.6............	1.3 (.4 to 2.4)	−2.9 (−4.1 to −1.8)	−5.2 (−6.4 to −4.1)
91.2............	.3 (−.5 to .9)	−2.4 (−3.3 to −1.7)	−4.3 (−5.1 to −3.5)
0.045 inch			
11.6............	2.9 (−.1 to 5.8)	−4.2 (−6.3 to −1.6)	−9.2 (−11.5 to −6.7)
18.6............	1.5 (.4 to 2.9)	−4.3 (−6.1 to 2.7)	−7.3 (−9.1 to −5.3)
91.2............	.3 (−.9 to 1.2)	−3.6 (−4.5 to −2.4)	−5.6 (−6.9 to −4.3)
0.060 inch			
11.6............	3.4 (−.6 to 6.8)	−6.3 (−9.1 to −1.0)	−11.7 (−14.2 to −8.1)
18.6............	1.8 (.3 to 3.6)	−5.6 (−7.5 to −3.4)	−9.7 (−12.1 to −5.8)
91.2............	.5 (−.8 to 1.8)	−4.7 (−6.0 to −3.5)	−7.5 (−9.4 to −4.6)

[1]Clearance angle 15°; cutting velocity 5 inches per minute.
[2]The first number in each entry is the average cutting force; the numbers following in parentheses are maximum forces (average of five).
[3]The first number in each entry is the average cutting force: the numbers following in parentheses are minimum and maximum forces (average of five).
[4]A negative normal force indicates that the knife tended to lift the workpiece; force was positive when the knife pushed the workpiece.

TABLE 18-9A.—*Tool forces when orthogonally cutting hackberry parallel to the grain in the 90-0 direction[1]* (Data from Woodson 1979)

Depth of cut and moisture content (percent)	Rake angle, degrees		
	10	20	30
	---------------- *Pounds per 0.1 inch of knife* ----------------		
	PARALLEL FORCE[2]		
0.015 inch			
10.7.................	14.6 (18.7)	13.8 (16.2)	8.3 (12.9)
19.6.................	9.3 (11.9)	7.8 (8.9)	6.2 (7.3)
117.6................	8.5 (11.1)	6.2 (7.7)	3.9 (4.8)
0.030 inch			
10.7.................	30.0 (37.8)	29.5 (34.4)	15.1 (24.7)
19.6.................	18.7 (25.0)	15.4 (18.6)	11.5 (13.4)
117.6................	15.1 (18.4)	11.1 (12.6)	7.0 (8.7)
0.045 inch			
10.7.................	40.7 (52.3)	37.7 (40.9)	14.2 (33.9)
19.6.................	24.2 (33.9)	21.8 (24.8)	15.7 (19.1)
117.6................	21.2 (26.6)	15.7 (18.6)	9.1 (11.5)
0.060 inch			
10.7.................	52.3 (66.0)	50.4 (56.4)	17.6 (43.0)
19.6.................	31.8 (43.7)	28.2 (32.4)	17.6 (25.3)
117.6................	25.9 (31.9)	18.0 (22.0)	11.6 (14.1)
	NORMAL FORCE[3,4]		
0.015 inch			
10.7.................	4.2 (2.7 to 5.6)	1.7 (1.4 to 2.3)	−0.2 (−0.7 to 0.5)
19.6.................	2.9 (1.7 to 4.1)	1.1 (.7 to 1.6)	.0 (−.4 to .3)
117.6................	2.7 (1.9 to 3.6)	1.2 (.6 to 1.8)	.5 (−.1 to 1.2)
0.030 inch			
10.7.................	8.1 (5.0 to 10.2)	3.4 (2.7 to 4.1)	−.9 (−1.6 to −.1)
19.6.................	4.9 (2.9 to 6.1)	1.7 (1.2 to 2.1)	−.5 (−.8 to .1)
117.6................	3.9 (2.5 to 5.0)	1.2 (.7 to 1.8)	−.1 (−.7 to .9)
0.045 inch			
10.7.................	11.0 (7.2 to 13.9)	4.1 (3.4 to 4.7)	−.8 (−2.6 to .4)
19.6.................	6.3 (3.2 to 8.3)	2.1 (1.5 to 2.7)	−.8 (−1.3 to .0)
117.6................	5.1 (3.4 to 6.4)	1.3 (.6 to 2.1)	−.2 (− .8 to .5)
0.060 inch			
10.7.................	13.5 (8.7 to 16.9)	5.6 (5.1 to 6.2)	−1.2 (−3.4 to .1)
19.6.................	7.9 (3.7 to 9.9)	2.6 (1.8 to 3.3)	−1.0 (−2.0 to .2)
117.6................	5.6 (3.3 to 7.4)	1.4 (.5 to 2.4)	−.5 (−1.1 to .4)

[1]Clearance angle 15°; cutting velocity 5 inches per minute.

[2]The first number in each entry is the average cutting force; the numbers following in parentheses are maximum forces (average of five).

[3]The first number in each entry is the average cutting force; the numbers following in parentheses are minimum and maximum forces (average of five).

[4]A negative normal force indicates that the knife tended to lift the workpiece; force was positive when the knife pushed the workpiece.

TABLE 18-9B.—*Tool forces when orthogonally cutting hackberry veneer in the 0-90 direction[1]* (Data from Woodson 1979)

Depth of cut and moisture content (percent)	Rake angle, degrees		
	50	60	70
	----------------Pounds per 0.1 inch of knife----------------		
PARALLEL FORCE[2]			
0.015 inch			
11.1	2.8 (5.7)	2.8 (4.8)	3.9 (7.0)
19.5	2.3 (3.8)	1.9 (3.3)	3.8 (7.8)
115.0	1.8 (3.0)	1.7 (2.9)	3.7 (5.4)
0.030 inch			
11.1	4.1 (9.5)	3.6 (7.4)	3.6 (6.1)
19.5	3.6 (6.0)	3.5 (5.7)	2.9 (4.2)
115.0	2.7 (5.1)	2.6 (4.4)	2.4 (3.4)
0.045 inch			
11.1	5.8 (12.7)	4.4 (9.4)	4.3 (7.8)
19.5	4.2 (7.9)	4.0 (7.7)	3.6 (5.3)
115.0	3.2 (6.5)	3.3 (5.7)	3.0 (4.3)
0.060 inch			
11.1	5.1 (14.9)	5.3 (11.3)	5.3 (10.4)
19.5	4.8 (9.6)	4.8 (8.6)	4.2 (6.9)
115.0	3.6 (7.6)	3.5 (6.8)	3.4 (5.2)
NORMAL FORCE[3,4]			
0.015 inch			
11.1	0.1 (−0.6 to 1.3)	0.2 (−0.6 to 1.1)	−0.1 (−1.7 to 1.2)
19.5	−.1 (−.8 to .7)	−.4 (−.9 to .1)	.0 (−1.2 to 1.3)
115.0	−.3 (−.6 to .2)	−.3 (−.6 to .2)	−.7 (−1.5 to .4)
0.030 inch			
11.1	−.1 (−1.4 to 1.3)	−.6 (−1.5 to .9)	.3 (−.7 to 1.2)
19.5	−.7 (−1.3 to .2)	−1.0 (−1.9 to .2)	−.6 (−1.2 to .2)
115.0	−.5 (−1.1 to .1)	−.7 (−1.3 to −.1)	−.1 (−.5 to .6)
0.045 inch			
11.1	−.1 (−1.5 to 2.0)	−.4 (−2.0 to 1.1)	.1 (−1.2 to 1.5)
19.5	−.8 (−1.8 to .4)	−1.1 (−2.5 to .0)	−1.1 (−1.9 to .0)
115.0	−.6 (−1.6 to .2)	−.9 (−2.1 to .1)	−.6 (−1.3 to .2)
0.060 inch			
11.1	−.2 (−2.0 to 1.8)	−.3 (−1.8 to 1.8)	.2 (−1.5 to 1.9)
19.5	−.9 (−2.3 to .3)	−1.3 (−2.9 to .0)	−1.3 (−2.4 to −.1)
115.0	−.6 (−2.1 to .6)	−.8 (−2.2 to .3)	−1.0 (−1.7 to .2)

[1] Clearance angle 0°; cutting velocity 5 inches per minute.
[2] The first number in each entry is the average cutting force; the numbers following in parentheses are maximum forces (average of five).
[3] The first number in each entry is the average cutting force; the numbers following in parentheses are minimum and maximum forces (average of five).
[4] A negative normal force indicates that the knife tended to lift the workpiece; force was positive when the knife pushed the workpiece.

TABLE 18-9C.—*Tool forces when orthogonally cutting hackberry across the grain in the 90-90 direction[1]* (Data from Woodson 1979)

Depth of cut and moisture content (percent)	Rake angle, degrees		
	20	30	40
	----------Pounds per 0.1 inch of knife----------		
	PARALLEL FORCE[2]		
0.015 inch			
10.6............	20.2 (25.7)	17.7 (20.9)	15.1 (18.3)
18.9............	18.1 (21.8)	13.8 (16.4)	11.1 (13.3)
111.7............	10.9 (12.3)	9.2 (10.5)	7.7 (8.8)
0.030 inch			
10.6............	36.6 (43.7)	32.3 (37.7)	25.6 (29.6)
18.9............	29.7 (35.9)	20.5 (24.4)	17.1 (20.0)
111.7............	21.0 (22.6)	15.1 (17.2)	12.1 (14.9)
0.045 inch			
10.6............	43.7 (53.1)	38.1 (45.1)	31.9 (38.3)
18.9............	37.5 (42.9)	27.6 (32.4)	22.0 (26.4)
111.7............	26.6 (29.7)	19.9 (22.3)	15.6 (18.4)
0.060 inch			
10.6............	53.1 (67.0)	46.4 (55.0)	39.1 (48.6)
18.9............	45.0 (51.1)	33.8 (38.5)	26.6 (32.2)
111.7............	33.0 (37.2)	23.6 (27.6)	19.2 (22.0)
	NORMAL FORCE[3,4]		
0.015 inch			
10.6............	2.3 (0.9 to 3.8)	0.0 (−1.0 to 1.3)	−2.5 (−3.5 to −1.3)
18.9............	1.4 (.8 to 2.0)	−.7 (−1.4 to .2)	−1.8 (−2.4 to −1.2)
111.7............	.9 (.5 to 1.1)	−.1 (−.8 to .4)	−1.4 (−2.1 to −.9)
0.030 inch			
10.6............	3.4 (.8 to 6.5)	−1.8 (−3.9 to .2)	−5.1 (−6.7 to −3.1)
18.9............	1.4 (.5 to 2.8)	−1.9 (−3.2 to −.9)	−4.1 (−5.2 to −3.1)
111.7............	.5 (−.1 to 1.2)	−1.3 (−2.3 to −.6)	−3.3 (−4.1 to −2.5)
0.045 inch			
10.6............	4.1 (1.4 to 8.4)	−2.1 (−4.9 to 2.2)	−7.0 (−9.3 to −3.2)
18.9............	1.5 (.3 to 3.2)	−3.1 (−4.7 to −1.7)	−6.0 (−7.8 to −4.1)
111.7............	.6 (−.2 to 1.5)	−2.3 (−3.2 to −1.3)	−4.6 (−6.0 to −3.4)
0.060 inch			
10.6............	5.1 (1.7 to 10.0)	−2.8 (−6.2 to 2.2)	−8.8 (−12.2 to −2.7)
18.9............	2.1 (.1 to 3.8)	−4.1 (−6.1 to −2.3)	−7.6 (−9.9 to −4.6)
111.7............	.5 (−.6 to 1.7)	−3.3 (−5.2 to −1.8)	−6.2 (−7.6 to −4.7)

[1] Clearance angle 15°; cutting velocity 5 inches per minute.

[2] The first number in each entry is the average cutting force; the numbers following in parentheses are maximum forces (average of five).

[3] The first number in each entry is the average cutting force; the numbers following in parentheses are minimum and maximum forces (average of five).

[4] A negative normal force indicates that the knife tended to lift the workpiece; force was positive when the knife pushed the workpiece.

TABLE 18-10A.—*Tool forces when orthogonally cutting hickory parallel to the grain in the 90-0 direction[1]* (Data from Woodson 1979)

Depth of cut and moisture content (percent)	Rake angle, degrees		
	10	20	30
	----------------Pounds per 0.1 inch of knife----------------		
PARALLEL FORCE[2]			
0.015 inch			
11.0	18.2 (21.9)	17.8 (19.7)	9.5 (14.5)
19.6	10.7 (12.4)	8.3 (10.0)	6.3 (7.7)
91.2	7.2 (9.1)	6.0 (7.5)	3.9 (4.7)
0.030 inch			
11.0	37.2 (45.2)	40.9 (45.2)	16.7 (26.7)
19.6	21.4 (24.5)	16.0 (18.3)	13.1 (15.9)
91.2	14.4 (18.0)	10.9 (12.9)	6.9 (8.4)
0.045 inch			
11.0	47.6 (62.2)	47.5 (52.7)	14.7 (35.1)
19.6	29.1 (33.0)	21.9 (26.5)	17.4 (19.8)
91.2	20.1 (24.0)	14.6 (17.1)	8.4 (11.9)
0.060 inch			
11.0	59.5 (78.0)	58.5 (71.7)	19.8 (49.5)
19.6	36.1 (42.9)	28.5 (34.3)	19.1 (25.5)
91.2	26.9 (31.7)	18.6 (22.5)	10.0 (14.4)
NORMAL FORCE[3,4]			
0.015 inch			
11.0	5.2 (4.4 to 6.3)	2.0 (1.5 to 2.5)	−0.3 (−0.9 to 0.3)
19.6	3.0 (2.3 to 3.4)	1.1 (.8 to 1.4)	−.2 (−.4 to .2)
91.2	2.3 (1.7 to 2.9)	.9 (.5 to 1.4)	.3 (.0 to .6)
0.030 inch			
11.0	9.7 (7.2 to 11.5)	3.7 (2.9 to 4.5)	−.9 (−2.2 to .1)
19.6	5.5 (4.3 to 6.3)	1.8 (1.4 to 2.1)	−.8 (−1.2 to −.4)
91.2	3.6 (2.6 to 4.5)	1.2 (.9 to 1.9)	−.1 (−.4 to .3)
0.045 inch			
11.0	12.2 (7.2 to 16.5)	4.3 (2.9 to 5.5)	−1.1 (−3.1 to .5)
19.6	6.9 (5.1 to 7.9)	2.5 (2.0 to 3.0)	−1.1 (−1.6 to −.5)
91.2	4.7 (3.0 to 6.0)	1.3 (1.0 to 1.7)	−.2 (−.6 to .3)
0.060 inch			
11.0	14.4 (8.4 to 20.0)	5.8 (3.0 to 7.3)	−1.4 (−4.0 to .0)
19.6	8.5 (6.9 to 9.9)	2.9 (1.9 to 3.7)	−1.2 (−2.2 to −.5)
91.2	5.8 (4.1 to 7.2)	1.5 (1.0 to 2.0)	−.6 (−1.2 to −.1)

[1] Clearance angle 15°; cutting velocity 5 inches per minute.
[2] The first number in each entry is the average cutting force; the numbers following in parentheses are maximum forces (average of five).
[3] The first number in each entry is the average cutting force: the numbers following in parentheses are minimum and maximum forces (average of five).
[4] A negative normal force indicates that the knife tended to lift the workpiece; force was positive when the knife pushed the workpiece.

TABLE 18-10B.—*Tool forces when orthogonally cutting hickory veneer in the 0-90 direction[1] (Data from Woodson 1979)*

Depth of cut and moisture content (percent)	Rake angle, degrees		
	50	60	70
	----------------Pounds per 0.1 inch of knife----------------		
PARALLEL FORCE[2]			
0.015 inch			
12.1	3.6 (8.1)	4.1 (7.6)	3.5 (5.4)
20.5	2.9 (5.6)	2.6 (4.5)	6.1 (11.3)
98.2	1.9 (3.4)	1.7 (3.0)	4.3 (7.7)
0.030 inch			
12.1	5.9 (14.6)	5.9 (12.4)	4.4 (8.5)
20.5	4.5 (9.0)	4.1 (7.6)	3.9 (5.6)
98.2	3.2 (6.2)	2.7 (4.7)	2.7 (3.7)
0.045 inch			
12.1	7.3 (18.3)	6.6 (15.0)	5.9 (11.5)
20.5	5.6 (11.1)	5.0 (10.2)	4.7 (7.6)
98.2	3.9 (7.9)	3.3 (6.0)	3.2 (4.6)
0.060 inch			
12.1	8.5 (22.2)	7.5 (18.1)	7.2 (14.2)
20.5	6.3 (13.9)	5.6 (11.5)	5.5 (9.5)
98.2	4.6 (10.1)	3.9 (7.3)	3.9 (5.9)
NORMAL FORCE[3,4]			
0.015 inch			
12.1	−0.3 (−1.4 to 0.8)	0.8 (−0.3 to 2.0)	0.1 (−0.7 to 1.1)
20.5	−.2 (−1.0 to 1.3)	−.3 (−1.0 to .4)	.3 (−1.8 to 2.2)
98.2	−.1 (−.6 to .6)	−.1 (−.7 to .4)	−.5 (−2.0 to 1.2)
0.030 inch			
12.1	−.4 (−2.4 to 1.7)	.5 (−1.6 to 2.4)	.6 (−.8 to 1.8)
20.5	−.4 (−1.8 to 1.1)	−1.1 (−2.5 to .3)	−.4 (−1.4 to .7)
98.2	−.4 (−1.3 to .5)	−.7 (−1.5 to .4)	−.1 (−.6 to .6)
0.045 inch			
12.1	−.3 (−3.3 to 2.2)	−.1 (−2.6 to 2.1)	.2 (−2.2 to 1.6)
20.5	−.8 (−2.6 to 1.3)	−1.4 (−3.4 to .5)	−.8 (−2.5 to .8)
98.2	−.5 (−1.6 to .5)	−.8 (−1.9 to .2)	−.6 (−1.2 to .2)
0.060 inch			
12.1	−.5 (−2.8 to 1.8)	−.4 (−2.7 to 2.5)	−.3 (−2.6 to 1.4)
20.5	−.6 (−3.0 to 2.0)	−1.8 (−4.1 to .4)	−1.6 (−3.5 to .6)
98.2	−.6 (−1.8 to .7)	−1.0 (−1.8 to .7)	−.8 (−1.8 to .3)

[1] Clearance angle 0°; cutting velocity 5 inches per minute.

[2] The first number in each entry is the average cutting force; the numbers following in parentheses are maximum forces (average of five).

[3] The first number in each entry is the average cutting force; the numbers following in parentheses are minimum and maximum forces (average of five).

[4] A negative normal force indicates that the knife tended to lift the workpiece; force was positive when the knife pushed the workpiece.

TABLE 18-10C.—*Tool forces when orthogonally cutting hickory across the grain in the 90-90 direction[1]* (Data from Woodson 1979)

Depth of cut and moisture content (percent)	Rake angle, degrees		
	20	30	40

----------Pounds per 0.1 inch of knife----------

PARALLEL FORCE[2]

0.015 inch			
10.7............	25.8 (32.2)	21.8 (28.6)	16.3 (20.2)
20.0............	19.7 (22.6)	15.1 (17.8)	14.8 (17.9)
85.9............	12.8 (15.1)	13.0 (14.5)	9.0 (10.6)
0.030 inch			
10.7............	42.6 (52.4)	36.8 (45.2)	29.0 (35.3)
20.0............	34.1 (39.8)	25.9 (30.3)	22.4 (27.1)
85.9............	23.9 (26.3)	19.3 (22.2)	14.7 (16.5)
0.045 inch			
10.7............	49.1 (61.2)	49.2 (61.9)	35.3 (44.2)
20.0............	43.2 (48.2)	33.6 (39.5)	29.1 (35.4)
85.9............	31.9 (36.6)	23.5 (27.2)	18.6 (21.0)
0.060 inch			
10.7............	56.0 (71.1)	52.1 (68.8)	46.1 (55.1)
20.0............	52.6 (59.8)	40.5 (49.1)	34.6 (41.7)
85.9............	38.8 (43.5)	29.3 (35.5)	22.0 (25.3)

NORMAL FORCE[3,4]

0.015 inch			
10.7............	0.9 (−0.4 to 2.4)	−1.8 (−3.5 to −0.8)	−4.3 (−5.7 to −2.6)
20.0............	1.2 (.6 to 1.8)	−1.4 (−2.1 to −.5)	−2.9 (−4.0 to −1.9)
85.9............	.5 (.0 to 1.1)	−1.1 (−1.9 to −.3)	−2.2 (−2.8 to −1.4)
0.030 inch			
10.7............	.2 (−2.1 to 2.4)	−4.8 (−7.1 to −2.9)	−8.7 (−11.4 to −5.4)
20.0............	.8 (−.1 to 2.0)	−3.5 (−4.8 to −2.0)	−6.0 (−8.1 to −4.1)
85.9............	.0 (−1.0 to .8)	−2.6 (−3.8 to −1.3)	−4.3 (−5.4 to −3.1)
0.045 inch			
10.7............	.2 (−3.3 to 3.4)	−7.0 (−10.2 to −3.7)	−11.6 (−14.8 to −6.7)
20.0............	1.0 (−.7 to 2.6)	−4.8 (−7.1 to −2.8)	−8.4 (−10.7 to −5.1)
85.9............	−.3 (−1.5 to 1.2)	−3.7 (−5.5 to −2.0)	−6.0 (−7.7 to −3.7)
0.060 inch			
10.7............	.2 (−4.0 to 5.6)	−8.3 (−11.9 to −3.2)	−15.1 (−19.1 to −9.7)
20.0............	1.3 (−.3 to 2.9)	−6.4 (−9.2 to −3.5)	−10.9 (−14.5 to −6.1)
85.9............	.1 (−1.3 to 1.8)	−4.8 (−7.5 to −2.7)	−7.8 (−10.0 to −4.8)

[1] Clearance angle 15°; cutting velocity 5 inches per minute.
[2] The first number in each entry is the average cutting force; the numbers following in parentheses are maximum forces (average of five).
[3] The first number in each entry is the average cutting force; the numbers following in parentheses are minimum and maximum forces (average of five).
[4] A negative normal force indicates that the knife tended to lift the workpiece; force was positive when the knife pushed the workpiece.

TABLE 18-11A.—*Tool forces when orthogonally cutting red maple parallel to the grain in the 90-0 direction*[1] (Data from Woodson 1979)

Depth of cut and moisture content (percent)	Rake angle, degrees		
	10	20	30
	----------------Pounds per 0.1 inch of knife----------------		
	PARALLEL FORCE[2]		
0.015 inch			
10.8	13.6 (19.1)	13.2 (14.4)	10.5 (13.2)
18.7	10.2 (13.2)	7.5 (8.7)	7.4 (8.4)
124.8	7.7 (10.0)	5.4 (7.2)	4.7 (5.7)
0.030 inch			
10.8	25.2 (36.1)	27.1 (29.2)	13.5 (25.4)
18.7	17.3 (23.8)	16.9 (18.7)	14.2 (17.0)
124.8	14.8 (17.8)	11.2 (13.4)	8.5 (10.2)
0.045 inch			
10.8	34.3 (48.5)	37.4 (40.8)	11.6 (35.6)
18.7	24.3 (33.6)	25.5 (28.0)	16.7 (22.7)
124.8	21.2 (25.1)	16.7 (18.9)	10.6 (13.1)
0.060 inch			
10.8	43.7 (62.4)	47.3 (51.6)	12.8 (45.0)
18.7	31.5 (44.3)	31.9 (34.4)	18.3 (29.6)
124.8	27.0 (32.3)	22.0 (25.0)	13.2 (18.7)
	NORMAL FORCE[3,4]		
0.015 inch			
10.8	4.1 (2.5 to 6.0)	2.0 (1.6 to 2.5)	−0.1 (−0.5 to 0.5)
18.7	3.3 (2.1 to 4.8)	1.4 (1.0 to 1.7)	.2 (−.2 to .6)
124.8	2.8 (1.6 to 4.2)	.9 (.4 to 2.1)	.6 (.0 to 1.1)
0.030 inch			
10.8	7.0 (4.2 to 9.3)	3.6 (3.3 to 4.1)	−.6 (−1.5 to .5)
18.7	4.7 (2.9 to 6.4)	2.1 (1.7 to 2.5)	−.4 (−1.0 to .2)
124.8	3.5 (2.7 to 4.7)	.9 (.5 to 1.4)	.0 (−.6 to .4)
0.045 inch			
10.8	8.8 (4.9 to 11.8)	4.8 (4.2 to 5.2)	−.6 (−2.2 to .4)
18.7	6.4 (3.4 to 8.4)	2.9 (2.3 to 3.3)	−.7 (−1.5 to .2)
124.8	4.8 (3.4 to 6.1)	1.1 (.5 to 1.9)	−.3 (−1.1 to .5)
0.060 inch			
10.8	11.3 (6.4 to 15.4)	5.7 (4.6 to 6.5)	−.6 (−2.8 to .3)
18.7	7.6 (4.5 to 12.3)	3.1 (2.8 to 3.5)	−8 (−2.0 to .4)
124.8	5.1 (3.1 to 6.6)	1.2 (.5 to 2.1)	−.9 (−2.1 to .3)

[1] Clearance angle 15°; cutting velocity 5 inches per minute.
[2] The first number in each entry is the average cutting force; the numbers following in parentheses are maximum forces (average of five).
[3] The first number in each entry is the average cutting force: the numbers following in parentheses are minimum and maximum forces (average of five).
[4] A negative normal force indicates that the knife tended to lift the workpiece; force was positive when the knife pushed the workpiece.

TABLE 18-11B.—*Tool forces when orthogonally cutting red maple veneer in the 0-90 direction[1]* (Data from Woodson 1979)

Depth of cut and moisture content (percent)	Rake angle, degrees		
	50	60	70
	----------------*Pounds per 0.1 inch of knife*----------------		
PARALLEL FORCE[2]			
0.015 inch			
11.7	2.6 (4.4)	2.6 (4.1)	3.2 (5.2)
18.8	2.3 (3.5)	1.8 (2.7)	3.4 (7.3)
136.2	1.9 (2.8)	1.6 (2.1)	2.9 (5.2)
0.030 inch			
11.7	3.4 (7.4)	3.5 (6.2)	3.0 (4.7)
18.8	3.2 (5.0)	2.8 (4.2)	2.4 (3.7)
136.2	2.6 (4.0)	2.3 (3.2)	1.9 (2.5)
0.045 inch			
11.7	4.2 (10.0)	3.8 (7.8)	3.6 (6.3)
18.8	3.8 (7.2)	3.4 (5.5)	3.0 (4.5)
136.2	2.8 (5.1)	2.7 (4.1)	2.3 (3.4)
0.060 inch			
11.7	4.5 (11.4)	4.1 (9.4)	4.1 (7.7)
18.8	4.1 (7.7)	3.7 (7.0)	3.4 (5.3)
136.2	3.5 (6.2)	2.9 (4.6)	2.7 (4.1)
NORMAL FORCE[3,4]			
0.015 inch			
11.7	0.1 (−0.3 to 1.0)	0.0 (−0.4 to 0.5)	−0.5 (−1.4 to 0.3)
18.8	.0 (−.5 to .7)	−.3 (−.7 to .0)	−.4 (−1.4 to 1.0)
136.2	.1 (−.3 to .5)	−.1 (−.4 to .3)	−.5 (−1.4 to 1.0)
0.030 inch			
11.7	−.3 (−1.2 to .6)	−.4 (−1.3 to .4)	−.2 (−.8 to .4)
18.8	−.7 (−1.2 to .0)	−.9 (−1.5 to −.1)	−.5 (−.9 to −.1)
136.2	−.3 (−.8 to .3)	−.5 (−1.0 to .0)	.0 (−.3 to .5)
0.045 inch			
11.7	−.5 (−1.7 to .7)	−.6 (−1.9 to .7)	−.6 (−1.4 to .2)
18.8	−.7 (−1.7 to .1)	−1.0 (−2.1 to −.2)	−1.1 (−1.7 to −.2)
136.2	−.4 (−1.1 to .4)	−.7 (−1.5 to −.1)	−.5 (−.9 to .2)
0.060 inch			
11.7	−.6 (−2.1 to .7)	−.7 (−2.2 to .7)	−.6 (−1.9 to .5)
18.8	−1.0 (−2.2 to .1)	−1.4 (−2.6 to .0)	−1.3 (−2.4 to .1)
136.2	−.6 (−1.5 to .3)	−.8 (−1.9 to .0)	−.8 (−1.4 to .0)

[1] Clearance angle 0°; cutting velocity 5 inches per minute.

[2] The first number in each entry is the average cutting force; the numbers following in parentheses are maximum forces (average of five).

[3] The first number in each entry is the average cutting force; the numbers following in parentheses are minimum and maximum forces (average of five).

[4] A negative normal force indicates that the knife tended to lift the workpiece; force was positive when the knife pushed the workpiece.

TABLE 18-11C.—*Tool forces when orthogonally cutting red maple across the grain in the 90-90 direction[1]* (Data from Woodson 1979)

Depth of cut and moisture content (percent)	Rake angle, degrees		
	20	30	40

----------Pounds per 0.1 inch of knife----------

PARALLEL FORCE[2]

0.015 inch			
11.3............	18.3 (22.7)	16.1 (20.6)	12.6 (14.4)
18.3............	14.9 (18.1)	12.6 (14.5)	9.5 (11.4)
135.1...........	12.0 (13.7)	9.5 (10.5)	8.1 (9.0)
0.030 inch			
11.3............	29.1 (36.3)	25.6 (30.9)	19.1 (22.7)
18.3............	26.1 (30.9)	20.1 (23.2)	15.1 (17.9)
135.1...........	19.3 (21.5)	15.1 (17.2)	11.2 (13.3)
0.045 inch			
11.3............	34.6 (46.9)	36.5 (43.5)	26.7 (32.4)
18.3............	33.1 (38.1)	26.7 (30.2)	19.6 (24.1)
135.1...........	26.3 (30.5)	19.4 (22.7)	14.1 (17.1)
0.060 inch			
11.3............	32.0 (48.1)	40.4 (53.1)	28.3 (36.2)
18.3............	41.3 (46.1)	32.1 (35.9)	24.1 (29.6)
135.1...........	31.9 (36.9)	24.5 (27.6)	17.5 (20.7)

NORMAL FORCE[3,4]

0.015 inch			
11.3............	2.8 (1.7 to 4.8)	1.1 (-0.7 to 2.7)	-0.9 (-2.0 to 0.4)
18.3............	1.7 (1.1 to 2.6)	.3 ($-.4$ to 1.1)	-1.1 (-1.7 to $-.5$)
135.1...........	.8 (.5 to 1.3)	.0 ($-.3$ to .5)	-1.2 (-1.6 to $-.9$)
0.030 inch			
11.3............	4.6 (2.5 to 7.8)	1.0 (-1.4 to 4.5)	-2.8 (-4.5 to $-.6$)
18.3............	2.2 (1.2 to 3.4)	$-.7$ (-2.0 to .4)	-2.9 (-3.7 to -1.8)
135.1...........	1.0 (.4 to 1.4)	$-.9$ (-1.6 to .0)	-2.4 (-3.3 to -1.7)
0.045 inch			
11.3............	6.7 (2.3 to 12.6)	1.8 (-1.9 to 6.8)	-4.2 (-6.3 to .0)
18.3............	2.4 (1.3 to 4.2)	-1.7 (-3.0 to .1)	-4.4 (-6.4 to -2.9)
135.1...........	.8 (.1 to 1.9)	-1.6 (-2.6 to $-.5$)	-3.5 (-5.1 to -2.6)
0.060 inch			
11.3............	8.4 (1.1 to 16.5)	1.5 (-3.6 to 7.6)	-4.1 (-7.9 to 1.4)
18.3............	2.4 (.9 to 4.9)	-2.4 (-4.2 to $-.1$)	-6.2 (-8.6 to -3.4)
135.1...........	1.0 ($-.3$ to 2.2)	-2.5 (-3.8 to -1.1)	-4.7 (-6.4 to -3.1)

[1]Clearance angle 15°; cutting velocity 5 inches per minute.

[2]The first number in each entry is the average cutting force; the numbers following in parentheses are maximum forces (average of five).

[3]The first number in each entry is the average cutting force; the numbers following in parentheses are minimum and maximum forces (average of five).

[4]A negative normal force indicates that the knife tended to lift the workpiece; force was positive when the knife pushed the workpiece.

TABLE 18-12A.—*Tool forces when orthogonally cutting black oak parallel to the grain in the 90-0 direction[1]* (Data from Woodson 1979)

Depth of cut and moisture content (percent)	Rake angle, degrees		
	10	20	30

----------------*Pounds per 0.1 inch of knife*----------------

PARALLEL FORCE[2]

Depth/MC	10	20	30
0.015 inch			
10.6	19.2 (23.4)	15.2 (18.3)	9.8 (14.7)
18.3	13.2 (15.3)	10.5 (12.3)	8.3 (9.7)
90.5	8.4 (10.1)	7.2 (8.8)	4.9 (6.2)
0.030 inch			
10.6	36.3 (42.0)	32.8 (39.0)	13.4 (26.1)
18.3	27.1 (31.9)	23.7 (26.6)	15.0 (18.8)
90.5	17.7 (20.7)	12.6 (15.8)	9.0 (12.3)
0.045 inch			
10.6	47.9 (58.0)	41.7 (52.7)	16.8 (40.2)
18.3	36.9 (43.4)	31.7 (35.0)	19.4 (25.5)
90.5	25.6 (31.3)	20.0 (23.4)	9.3 (15.2)
0.060 inch			
10.6	57.6 (73.2)	52.0 (63.8)	17.7 (44.3)
18.3	45.1 (54.6)	39.3 (44.8)	21.7 (33.6)
90.5	33.1 (41.3)	23.6 (30.4)	10.9 (18.2)

NORMAL FORCE[3,4]

Depth/MC	10	20	30
0.015 inch			
10.6	5.2 (3.7 to 6.2)	1.8 (1.3 to 2.3)	−0.5 (−1.0 to 0.1)
18.3	3.4 (2.6 to 4.2)	1.5 (1.1 to 1.9)	−.2 (−.7 to .3)
90.5	2.4 (1.8 to 2.9)	.9 (.5 to 1.2)	.2 (−.4 to .9)
0.030 inch			
10.6	8.6 (5.8 to 10.6)	3.3 (2.1 to 4.5)	−1.0 (−2.1 to .1)
18.3	6.6 (5.3 to 7.7)	2.1 (1.5 to 2.5)	−1.0 (−1.7 to −.2)
90.5	4.2 (3.6 to 5.1)	1.0 (.6 to 1.4)	−.6 (−1.2 to .0)
0.045 inch			
10.6	11.6 (8.4 to 14.7)	4.0 (2.3 to 5.2)	−1.1 (−2.9 to .4)
18.3	9.1 (7.2 to 10.5)	2.6 (1.8 to 3.1)	−1.0 (−2.0 to −.1)
90.5	5.8 (3.7 to 6.7)	1.2 (.7 to 1.8)	−.5 (−1.5 to .2)
0.060 inch			
10.6	12.8 (7.1 to 17.6)	5.1 (3.3 to 6.9)	−1.3 (−3.4 to .0)
18.3	10.5 (8.2 to 12.8)	2.9 (2.1 to 3.8)	−2.0 (−3.3 to −.9)
90.5	7.7 (5.4 to 9.2)	1.3 (.7 to 2.2)	−.8 (−1.7 to −.1)

[1] Clearance angle 15°; cutting velocity 5 inches per minute.

[2] The first number in each entry is the average cutting force; the numbers following in parentheses are maximum forces (average of five).

[3] The first number in each entry is the average cutting force; the numbers following in parentheses are minimum and maximum forces (average of five).

[4] A negative normal force indicates that the knife tended to lift the workpiece; force was positive when the knife pushed the workpiece.

TABLE 18-12B.—*Tool forces when orthogonally cutting black oak veneer in the 0-90 direction[1] (Data from Woodson 1979)*

Depth of cut and moisture content (percent)	Rake angle, degrees		
	50	60	70

---------------- Pounds per 0.1 inch of knife ----------------

PARALLEL FORCE[2]

0.015 inch			
10.6	3.5 (9.3)	3.3 (8.0)	3.4 (6.3)
18.5	2.8 (6.8)	2.5 (5.6)	5.5 (12.0)
92.3	2.0 (5.1)	1.9 (4.8)	4.3 (8.5)
0.030 inch			
10.6	4.9 (13.4)	4.7 (11.3)	5.0 (10.4)
18.5	3.8 (10.6)	4.4 (8.8)	3.6 (6.8)
92.3	3.4 (7.7)	2.9 (6.1)	2.6 (5.7)
0.045 inch			
10.6	5.6 (17.7)	5.5 (14.6)	5.3 (12.1)
18.5	4.5 (10.8)	5.2 (11.6)	4.4 (8.3)
92.3	3.9 (9.0)	3.6 (7.8)	3.5 (6.9)
0.060 inch			
10.6	6.6 (19.4)	5.9 (16.0)	6.6 (14.2)
18.5	5.8 (14.4)	5.5 (12.2)	5.2 (11.0)
92.3	4.5 (10.4)	4.0 (8.6)	3.8 (8.0)

NORMAL FORCE[3,4]

0.015 inch			
10.6	0.6 (−0.8 to 3.8)	0.9 (−0.2 to 3.1)	0.3 (−0.7 to 2.0)
18.5	.7 (−.8 to 2.7)	.3 (−.9 to 2.0)	.5 (−1.9 to 2.7)
92.3	.4 (−.5 to 1.6)	.2 (−.7 to 1.3)	.2 (−1.4 to 1.9)
0.030 inch			
10.6	.4 (−1.7 to 3.9)	.6 (−1.4 to 3.1)	1.5 (−.4 to 3.5)
18.5	.0 (−1.5 to 2.1)	.0 (−1.7 to 1.7)	.3 (−.7 to 1.8)
92.3	.0 (−1.0 to 1.4)	−.3 (−1.6 to .9)	.4 (−.7 to 1.5)
0.045 inch			
10.6	.8 (−1.6 to 6.3)	.6 (−1.5 to 3.7)	.8 (−1.8 to 3.8)
18.5	.0 (−1.7 to 1.7)	−.1 (−1.9 to 1.9)	−.3 (−1.6 to 1.8)
92.3	−.1 (−1.2 to 1.3)	−.4 (−2.4 to 1.3)	.0 (−1.0 to 1.2)
0.060 inch			
10.6	.3 (−2.1 to 4.8)	.3 (−2.1 to 3.9)	.8 (−2.2 to 4.1)
18.5	−.4 (−2.1 to 2.1)	−.1 (−2.4 to 2.6)	−.6 (−2.3 to 1.6)
92.3	−.3 (−1.8 to 1.1)	−.7 (−2.5 to 1.3)	−.2 (−1.5 to 1.1)

[1]Clearance angle 0°; cutting velocity 5 inches per minute.

[2]The first number in each entry is the average cutting force; the numbers following in parentheses are maximum forces (average of five).

[3]The first number in each entry is the average cutting force; the numbers following in parentheses are minimum and maximum forces (average of five).

[4]A negative normal force indicates that the knife tended to lift the workpiece; force was positive when the knife pushed the workpiece.

TABLE 18-12C.—*Tool forces when orthogonally cutting black oak across the grain in the 90-90 direction[1]* (Data from Woodson 1979)

Depth of cut and moisture content (percent)	Rake angle, degrees		
	20	30	40

----------Pounds per 0.1 inch of knife----------

PARALLEL FORCE[2]

Depth of cut and moisture content (percent)	20	30	40
0.015 inch			
10.7	21.6 (28.6)	17.7 (24.1)	14.5 (19.3)
18.3	16.4 (20.6)	13.2 (16.9)	11.9 (15.5)
99.4	14.8 (17.9)	11.1 (13.9)	8.9 (11.7)
0.030 inch			
10.7	37.9 (50.2)	32.0 (42.2)	24.2 (31.2)
18.3	30.0 (36.5)	23.6 (28.6)	19.8 (25.1)
99.4	24.1 (28.6)	17.6 (20.9)	14.4 (18.4)
0.045 inch			
10.7	44.1 (55.3)	37.4 (47.5)	30.7 (38.2)
18.3	37.0 (46.5)	31.0 (38.9)	24.4 (31.8)
99.4	32.1 (37.7)	22.2 (26.6)	18.2 (22.9)
0.060 inch			
10.7	49.8 (63.2)	45.9 (60.7)	34.9 (45.5)
18.3	46.6 (58.5)	37.2 (45.1)	29.4 (36.5)
99.4	38.9 (46.8)	27.2 (32.4)	22.6 (28.3)

NORMAL FORCE[3,4]

Depth of cut and moisture content (percent)	20	30	40
0.015 inch			
10.7	2.0 (0.7 to 3.6)	−0.7 (−2.1 to 1.2)	−2.7 (−4.4 to −1.0)
18.3	1.3 (.5 to 2.1)	−.7 (−1.7 to 1.0)	−2.0 (−3.3 to −.8)
99.4	.8 (.2 to 1.2)	−.7 (−1.5 to .3)	−1.9 (−3.0 to −.8)
0.030 inch			
10.7	2.6 (.4 to 5.4)	−3.0 (−5.6 to .8)	−5.4 (−8.6 to −2.7)
18.3	1.2 (.2 to 3.3)	−2.4 (−4.4 to −.6)	−4.8 (−6.9 to −3.1)
99.4	.7 (−.2 to 1.4)	−1.7 (−2.7 to −.5)	−4.0 (−5.4 to −2.0)
0.045 inch			
10.7	3.5 (.1 to 7.1)	−4.0 (−7.1 to .9)	−7.5 (−11.2 to −3.1)
18.3	1.8 (−.3 to 3.6)	−3.5 (−6.1 to −.4)	−6.8 (−10.0 to −4.0)
99.4	.9 (−.7 to 2.7)	−2.7 (−4.1 to −1.3)	−5.2 (−7.0 to −3.1)
0.060 inch			
10.7	4.6 (.5 to 8.9)	−4.4 (−8.9 to 1.7)	−9.1 (−12.6 to −4.3)
18.3	2.0 (−.4 to 4.9)	−5.3 (−7.7 to −1.2)	−8.7 (−11.6 to −4.2)
99.4	1.0 (−.9 to 2.6)	−3.5 (−5.0 to −1.8)	−6.8 (−10.0 to −3.2)

[1]Clearance angle 15°; cutting velocity 5 inches per minute.

[2]The first number in each entry is the average cutting force; the numbers following in parentheses are maximum forces (average of five).

[3]The first number in each entry is the average cutting force; the numbers following in parentheses are minimum and maximum forces (average of five).

[4]A negative normal force indicates that the knife tended to lift the workpiece; force was positive when the knife pushed the workpiece.

TABLE 18-13A.—*Tool forces when orthogonally cutting blackjack oak parallel to the grain in the 90 0 direction*[1] (Data from Woodson 1979)

Depth of cut and moisture content (percent)	Rake angle, degrees		
	10	20	30
	----------------Pounds per 0.1 inch of knife----------------		
	PARALLEL FORCE[2]		
0.015 inch			
11.6	19.0 (22.7)	15.2 (18.2)	10.1 (14.6)
19.2	12.1 (13.8)	10.6 (12.2)	7.3 (9.0)
80.8	9.2 (10.7)	6.9 (8.4)	4.5 (6.0)
0.030 inch			
11.6	33.2 (41.9)	31.3 (36.0)	17.8 (27.6)
19.2	24.9 (29.1)	20.8 (24.1)	12.2 (16.5)
80.8	17.2 (20.1)	12.3 (15.3)	7.4 (9.8)
0.045 inch			
11.6	48.1 (56.4)	40.0 (46.2)	19.9 (38.2)
19.2	31.7 (39.3)	27.1 (30.1)	15.5 (19.9)
80.8	23.8 (27.6)	17.2 (21.4)	10.1 (13.0)
0.060 inch			
11.6	57.7 (73.4)	53.7 (63.0)	20.4 (50.7)
19.2	42.3 (48.6)	35.7 (40.2)	16.8 (25.5)
80.8	30.6 (36.1)	22.7 (27.7)	11.9 (15.7)
	NORMAL FORCE[3,4]		
0.015 inch			
11.6	5.3 (4.0 to 6.4)	1.7 (1.1 to 2.3)	−0.3 (−0.8 to 0.2)
19.2	3.4 (2.5 to 4.1)	1.3 (.8 to 1.5)	−.3 (−.6 to .1)
80.8	2.5 (1.8 to 3.2)	.8 (.5 to 1.1)	−.2 (−.5 to .4)
0.030 inch			
11.6	9.0 (6.6 to 10.7)	2.9 (2.2 to 3.7)	−1.1 (−2.6 to .8)
19.2	5.9 (4.3 to 7.1)	1.7 (1.1 to 2.3)	−.9 (−1.4 to −.3)
80.8	4.0 (3.0 to 4.8)	1.0 (.5 to 1.4)	−.6 (−1.0 to −.2)
0.045 inch			
11.6	11.5 (8.4 to 14.1)	3.3 (2.0 to 4.6)	−1.4 (−3.8 to .9)
19.2	7.5 (5.7 to 9.2)	2.1 (1.4 to 2.7)	−1.2 (−1.8 to −.6)
80.8	5.3 (4.0 to 6.2)	1.1 (.6 to 1.7)	−1.0 (−1.5 to −.5)
0.060 inch			
11.6	13.7 (10.2 to 17.1)	4.2 (2.4 to 5.8)	−1.4 (−5.1 to 1.2)
19.2	9.5 (7.1 to 11.0)	2.6 (1.9 to 3.1)	−1.5 (−2.8 to −.3)
80.8	6.3 (4.7 to 7.6)	1.2 (.5 to 1.8)	−1.2 (−1.9 to −.6)

[1] Clearance angle 15°; cutting velocity 5 inches per minute.

[2] The first number in each entry is the average cutting force; the numbers following in parentheses are maximum forces (average of five).

[3] The first number in each entry is the average cutting force; the numbers following in parentheses are minimum and maximum forces (average of five).

[4] A negative normal force indicates that the knife tended to lift the workpiece; force was positive when the knife pushed the workpiece.

TABLE 18-13B.—*Tool forces when orthogonally cutting blackjack oak veneer in the 0-90 direction*[1] (Data from Woodson 1979)

Depth of cut and moisture content (percent)	Rake angle, degrees		
	50	60	70

---------------Pounds per 0.1 inch of knife----------------

PARALLEL FORCE[2]

Depth of cut and moisture content	50	60	70
0.015 inch			
11.7	3.9 (8.9)	4.4 (8.9)	3.7 (7.1)
19.2	2.9 (6.2)	5.8 (11.9)	2.9 (5.5)
82.5	2.1 (5.1)	4.2 (8.4)	2.4 (4.6)
0.030 inch			
11.7	5.8 (14.5)	4.6 (9.9)	4.9 (10.6)
19.2	4.4 (9.3)	3.6 (6.9)	3.9 (7.4)
82.5	4.0 (7.4)	2.8 (5.4)	3.2 (5.9)
0.045 inch			
11.7	6.7 (15.9)	5.4 (12.6)	6.0 (14.0)
19.2	5.2 (10.1)	4.3 (8.6)	4.8 (8.3)
82.5	4.6 (9.1)	3.6 (7.6)	3.9 (7.3)
0.060 inch			
11.7	7.9 (19.3)	6.4 (14.3)	7.2 (15.5)
19.2	6.0 (12.0)	5.6 (10.6)	5.8 (10.8)
82.5	5.3 (11.2)	4.6 (8.8)	4.6 (7.6)

NORMAL FORCE[3,4]

Depth of cut and moisture content	50	60	70
0.015 inch			
11.7	−0.2 (−1.3 to 1.5)	−0.5 (−2.3 to .9)	0.5 (−0.6 to 1.6)
19.2	.0 (−.9 to 1.7)	−.6 (−2.3 to 1.1)	−.2 (−.8 to .8)
82.5	−.1 (−.8 to .6)	−1.0 (−3.2 to .3)	.1 (−.5 to .8)
0.030 inch			
11.7	−.5 (−2.2 to 1.6)	−.4 (−2.0 to .9)	−.5 (−2.1 to .8)
19.2	−.9 (−2.0 to .3)	−.8 (−1.8 to .5)	−1.2 (−2.4 to .2)
82.5	−.8 (−2.0 to .3)	−.8 (−2.0 to −.1)	−.9 (−1.8 to .1)
0.045 inch			
11.7	−.5 (−2.9 to 2.1)	−.9 (−2.8 to .5)	−.8 (−2.9 to 1.2)
19.2	−1.0 (−2.6 to .4)	−1.3 (−3.0 to .2)	−1.5 (−3.0 to .2)
82.5	−1.1 (−2.6 to .2)	−1.3 (−2.9 to .3)	−1.4 (−2.6 to .1)
0.060 inch			
11.7	−.9 (−4.2 to 1.4)	−1.0 (−3.0 to .8)	−1.4 (−4.0 to .7)
19.2	−1.5 (−3.2 to .1)	−1.8 (−3.9 to −.1)	−2.3 (−3.8 to .0)
82.5	−1.3 (−2.7 to .0)	−1.7 (−3.7 to .0)	−2.2 (−3.4 to −.4)

[1]Clearance angle 0°; cutting velocity 5 inches per minute.

[2]The first number in each entry is the average cutting force; the numbers following in parentheses are maximum forces (average of five).

[3]The first number in each entry is the average cutting force; the numbers following in parentheses are minimum and maximum forces (average of five).

[4]A negative normal force indicates that the knife tended to lift the workpiece; force was positive when the knife pushed the workpiece.

TABLE 18-13C.—*Tool forces when orthogonally cutting blackjack oak across the grain in the 90-90 direction*[1] (Data from Woodson 1979)

Depth of cut and moisture content (percent)	Rake angle, degrees		
	20	30	40

----------Pounds per 0.1 inch of knife----------

PARALLEL FORCE[2]

0.015 inch			
10.9............	17.3 (25.6)	15.4 (21.8)	12.2 (17.1)
17.7............	16.5 (21.5)	12.9 (16.9)	11.7 (15.0)
89.2............	11.9 (15.4)	10.1 (12.4)	8.4 (11.1)
0.030 inch			
10.9............	31.5 (41.9)	27.7 (35.4)	20.6 (28.2)
17.7............	31.7 (38.5)	22.2 (26.9)	17.7 (21.9)
89.2............	21.3 (25.4)	16.4 (19.3)	14.3 (18.2)
0.045 inch			
10.9............	33.8 (49.2)	33.3 (46.6)	25.5 (35.7)
17.7............	36.7 (43.9)	29.9 (36.3)	23.8 (30.0)
89.2............	28.0 (34.9)	21.1 (25.9)	18.3 (25.2)
0.060 inch			
10.9............	40.5 (55.4)	40.9 (51.7)	34.1 (45.4)
17.7............	43.1 (52.9)	35.4 (43.3)	28.2 (36.1)
89.2............	33.7 (42.2)	26.9 (32.8)	22.3 (28.6)

NORMAL FORCE[3,4]

0.015 inch			
10.9............	1.4 (−0.2 to 4.1)	−0.4 (−2.0 to 1.8)	−2.4 (−4.3 to −0.1)
17.7............	1.3 (.3 to 2.8)	−.9 (−2.0 to .1)	−2.5 (−3.8 to −1.1)
89.2............	.5 (−.1 to 1.3)	−.9 (−1.8 to .0)	−1.7 (−2.6 to −.7)
0.030 inch			
10.9............	2.1 (−1.5 to 7.9)	−1.7 (−4.8 to 3.7)	−5.2 (−8.5 to −1.3)
17.7............	1.0 (−.7 to 4.8)	−2.4 (−4.0 to −.6)	−4.7 (−6.4 to −2.6)
89.2............	.3 (−.7 to 1.8)	−1.9 (−3.1 to −.3)	−3.9 (−5.7 to −1.7)
0.045 inch			
10.9............	2.0 (−2.2 to 8.9)	−2.8 (−7.2 to 3.7)	−7.2 (−11.3 to −3.2)
17.7............	1.2 (−.8 to 4.4)	−4.2 (−6.7 to −1.8)	−6.8 (−9.2 to −4.1)
89.2............	.6 (−.9 to 3.3)	−2.8 (−4.4 to −.6)	−5.2 (−7.7 to −3.3)
0.060 inch			
10.9............	2.6 (−2.6 to 11.6)	−3.7 (−8.1 to 4.2)	−9.1 (−14.0 to −2.7)
17.7............	1.2 (−2.1 to 5.0)	−5.2 (−9.0 to −2.4)	−8.6 (−12.1 to −4.2)
89.2............	.7 (−1.5 to 4.2)	−3.8 (−5.9 to −.8)	−6.7 (−9.2 to −3.7)

[1]Clearance angle 15°; cutting velocity 5 inches per minute.
[2]The first number in each entry is the average cutting force; the numbers following in parentheses are maximum forces (average of five).
[3]The first number in each entry is the average cutting force; the numbers following in parentheses are minimum and maximum forces (average of five).
[4]A negative normal force indicates that the knife tended to lift the workpiece; force was positive when the knife pushed the workpiece.

TABLE 18-14A.—*Tool forces when orthogonally cutting cherrybark oak parallel to the grain in the 90-0 direction[1]* (Data from Woodson 1979)

Depth of cut and moisture content (percent)	Rake angle, degrees		
	10	20	30
	----------------Pounds per 0.1 inch of knife----------------		
PARALLEL FORCE[2]			
0.015 inch			
10.4	17.8 (21.5)	16.7 (19.3)	12.9 (16.6)
18.7	11.7 (13.7)	10.1 (12.0)	7.4 (8.4)
87.9	8.5 (10.4)	6.0 (7.3)	4.7 (6.0)
0.030 inch			
10.4	37.0 (45.0)	37.6 (41.8)	21.9 (29.7)
18.7	23.6 (27.0)	19.8 (22.9)	14.6 (17.8)
87.9	16.9 (20.1)	11.9 (13.5)	7.2 (9.1)
0.045 inch			
10.4	47.9 (59.0)	45.8 (52.3)	19.5 (42.6)
18.7	30.0 (38.1)	27.4 (30.5)	17.9 (22.1)
87.9	24.3 (30.0)	17.9 (20.2)	9.3 (11.6)
0.060 inch			
10.4	58.0 (72.9)	55.3 (66.8)	23.9 (49.9)
18.7	38.1 (47.5)	34.3 (36.8)	23.0 (29.5)
87.9	30.0 (34.6)	22.4 (25.7)	13.3 (16.7)
NORMAL FORCE[3,4]			
0.015 inch			
10.4	5.1 (4.1 to 6.2)	2.0 (1.4 to 2.6)	−0.7 (−1.1 to −0.1)
18.7	3.2 (2.6 to 3.8)	1.2 (.9 to 1.6)	−.1 (−.5 to .3)
87.9	2.4 (2.0 to 3.3)	.8 (.4 to 1.4)	.0 (−.4 to .5)
0.030 inch			
10.4	8.9 (6.1 to 11.3)	3.6 (2.6 to 4.5)	−1.7 (−2.5 to −.7)
18.7	6.0 (4.4 to 7.0)	1.9 (1.5 to 2.3)	−.9 (−1.4 to −.4)
87.9	4.0 (3.2 to 5.0)	1.0 (.6 to 1.6)	−.2 (−.8 to .5)
0.045 inch			
10.4	11.3 (8.0 to 15.2)	4.1 (2.4 to 5.4)	−1.5 (−3.5 to −.1)
18.7	7.4 (5.4 to 8.9)	2.3 (1.8 to 2.9)	−1.3 (−2.1 to −.4)
87.9	5.8 (4.6 to 7.3)	1.1 (.5 to 1.8)	−.5 (−1.2 to .4)
0.060 inch			
10.4	14.7 (10.7 to 18.3)	4.5 (2.5 to 6.3)	−1.6 (−4.2 to .1)
18.7	9.2 (7.0 to 10.9)	3.0 (2.3 to 3.5)	−1.8 (−2.8 to −.7)
87.9	6.7 (5.2 to 8.1)	1.2 (.6 to 1.8)	−1.2 (−2.0 to −.2)

[1] Clearance angle 15°; cutting velocity 5 inches per minute.

[2] The first number in each entry is the average cutting force; the numbers following in parentheses are maximum forces (average of five).

[3] The first number in each entry is the average cutting force; the numbers following in parentheses are minimum and maximum forces (average of five).

[4] A negative normal force indicates that the knife tended to lift the workpiece; force was positive when the knife pushed the workpiece.

TABLE 18-14B.—*Tool forces when orthogonally cutting cherrybark oak veneer in the 0-90 direction[1]* (Data from Woodson 1979)

Depth of cut and moisture content (percent)	Rake angle, degrees		
	50	60	70
	----------------Pounds per 0.1 inch of knife----------------		
PARALLEL FORCE[2]			
0.015 inch			
11.4	3.8 (9.3)	4.6 (11.6)	3.5 (7.8)
18.8	3.3 (8.1)	2.9 (5.9)	6.5 (14.0)
93.4	2.1 (5.0)	2.4 (4.5)	4.7 (9.6)
0.030 inch			
11.4	5.5 (15.3)	5.5 (14.3)	4.5 (10.6)
18.8	4.4 (9.9)	4.1 (8.0)	3.5 (6.8)
93.4	3.0 (7.3)	3.0 (6.6)	2.5 (5.4)
0.045 inch			
11.4	6.1 (17.8)	6.7 (18.3)	6.2 (14.7)
18.8	5.2 (12.6)	4.7 (9.9)	4.7 (8.7)
93.4	3.8 (8.8)	3.8 (8.8)	3.2 (7.0)
0.060 inch			
11.4	7.9 (21.7)	7.9 (20.4)	6.8 (15.9)
18.8	6.7 (15.0)	5.1 (12.0)	5.0 (9.5)
93.4	4.6 (11.1)	4.3 (9.5)	3.8 (7.9)
NORMAL FORCE[3,4]			
0.015 inch			
11.4	0.5 (−0.9 to 3.1)	1.2 (−0.4 to 3.2)	−0.1 (−1.6 to 1.6)
18.8	.5 (−1.0 to 2.9)	−.1 (−.9 to 1.6)	.0 (−2.5 to 2.4)
93.4	.3 (−.4 to 1.5)	.0 (−.8 to .9)	−1.0 (−2.4 to 1.3)
0.030 inch			
11.4	.6 (−1.9 to 5.2)	.8 (−1.1 to 3.6)	1.4 (−.3 to 3.6)
18.8	.3 (−1.6 to 3.1)	−.9 (−2.5 to .9)	−.1 (−1.5 to 2.0)
93.4	−.1 (−1.1 to 1.3)	−.5 (−1.6 to .6)	.4 (−.6 to 1.8)
0.045 inch			
11.4	.7 (−1.9 to 5.1)	.7 (−1.5 to 3.9)	.2 (−2.0 to 3.0)
18.8	−.1 (−1.9 to 2.4)	−1.1 (−2.8 to .7)	−.7 (−2.7 to 1.8)
93.4	−.2 (−1.3 to 1.3)	−.7 (−2.2 to 1.3)	−.3 (−1.5 to 1.3)
0.060 inch			
11.4	.8 (−1.9 to 5.9)	.9 (−2.3 to 4.9)	.2 (−2.5 to 3.3)
18.8	−.2 (−2.2 to 2.7)	−1.3 (−3.8 to .8)	−1.3 (−3.6 to 1.1)
93.4	−.5 (−2.1 to 1.6)	−.9 (−2.8 to .7)	−.5 (−1.9 to 1.2)

[1] Clearance angle 0°; cutting velocity 5 inches per minute.

[2] The first number in each entry is the average cutting force; the numbers following in parentheses are maximum forces (average of five).

[3] The first number in each entry is the average cutting force; the numbers following in parentheses are minimum and maximum forces (average of five).

[4] A negative normal force indicates that the knife tended to lift the workpiece; force was positive when the knife pushed the workpiece.

TABLE 18-14C.—*Tool forces when orthogonally cutting cherrybark oak across the grain in the 90-90 direction[1]* (Data from Woodson 1979)

Depth of cut and moisture content (percent)	Rake angle, degrees		
	20	30	40

----------Pounds per 0.1 inch of knife----------

PARALLEL FORCE[2]

0.015 inch			
9.9............	25.0 (32.6)	18.1 (24.1)	15.5 (20.7)
18.6............	18.0 (22.6)	13.7 (17.3)	11.0 (14.2)
96.0............	13.3 (15.6)	10.6 (12.8)	8.7 (11.4)
0.030 inch			
9.9............	40.6 (52.3)	35.4 (44.4)	27.3 (34.2)
18.6............	30.7 (36.9)	23.4 (28.0)	18.4 (23.3)
96.0............	24.7 (28.6)	17.5 (21.9)	14.1 (17.9)
0.045 inch			
9.9............	48.5 (61.7)	43.2 (54.2)	32.8 (43.9)
18.6............	37.9 (45.1)	29.2 (34.5)	24.2 (29.5)
96.0............	32.5 (37.5)	22.5 (27.0)	18.0 (22.1)
0.060 inch			
9.9............	63.6 (79.3)	48.2 (63.9)	40.6 (53.5)
18.6............	46.2 (54.1)	35.9 (42.9)	28.3 (37.2)
96.0............	39.3 (46.2)	27.3 (33.8)	21.2 (27.0)

NORMAL FORCE[3,4]

0.015 inch			
9.9............	2.3 (1.0 to 3.9)	−0.4 (−1.9 to 1.8)	−2.9 (−5.0 to −1.2)
18.6............	1.5 (.6 to 2.6)	−.5 (−1.3 to .5)	−1.9 (−3.0 to −.3)
96.0............	.7 (−.1 to 1.6)	−.5 (−1.2 to .0)	−1.9 (−2.7 to −.8)
0.030 inch			
9.9............	3.3 (.8 to 7.2)	−2.5 (−5.1 to 1.0)	−6.2 (−8.9 to −2.6)
18.6............	1.5 (.0 to 3.6)	−2.2 (−3.7 to −.3)	−4.7 (−6.6 to −2.6)
96.0............	.4 (−.7 to 1.7)	−2.2 (−3.4 to −.9)	−3.9 (−5.3 to −2.4)
0.045 inch			
9.9............	4.1 (.8 to 7.9)	−3.2 (−6.9 to 2.2)	−8.4 (−12.2 to −3.6)
18.6............	1.6 (−.4 to 4.0)	−3.1 (−5.2 to .2)	−6.6 (−8.7 to −3.7)
96.0............	.4 (−1.1 to 1.9)	−3.2 (−4.9 to −1.5)	−5.3 (−7.2 to −3.2)
0.060 inch			
9.9............	4.9 (.1 to 9.7)	−3.9 (−8.8 to 3.2)	−10.9 (−15.7 to −4.0)
18.6............	2.2 (−.5 to 4.8)	−4.5 (−7.3 to −1.6)	−8.7 (−12.1 to −4.2)
96.0............	.4 (−1.3 to 2.7)	−4.2 (−6.2 to −1.9)	−6.8 (−9.6 to −3.7)

[1] Clearance angle 15°; cutting velocity 5 inches per minute.

[2] The first number in each entry is the average cutting force; the numbers following in parentheses are maximum forces (average of five).

[3] The first number in each entry is the average cutting force; the numbers following in parentheses are minimum and maximum forces (average of five).

[4] A negative normal force indicates that the knife tended to lift the workpiece; force was positive when the knife pushed the workpiece.

TABLE 18-15A.—*Tool forces when orthogonally cutting laurel oak parallel to the grain in the 90-0 direction[1]* (Data from Woodson 1979)

Depth of cut and moisture content (percent)	Rake angle, degrees		
	10	20	30
	----------------*Pounds per 0.1 inch of knife*----------------		
	PARALLEL FORCE[2]		
0.015 inch			
10.9	19.0 (22.3)	16.8 (19.3)	11.4 (14.3)
19.0	12.0 (13.5)	10.1 (11.5)	6.6 (8.3)
89.4	8.2 (10.5)	7.0 (8.3)	4.7 (6.1)
0.030 inch			
10.9	36.3 (43.2)	31.6 (35.4)	17.6 (28.9)
19.0	23.7 (29.1)	20.2 (22.2)	12.9 (17.1)
89.4	15.9 (19.7)	12.8 (15.7)	7.8 (10.1)
0.045 inch			
10.9	48.9 (58.0)	44.4 (48.3)	19.3 (40.6)
19.0	30.8 (39.1)	28.3 (32.0)	16.9 (23.2)
89.4	20.5 (27.9)	18.1 (21.8)	10.7 (14.6)
0.060 inch			
10.9	61.3 (77.5)	61.2 (64.1)	18.5 (50.1)
19.0	39.5 (50.7)	35.3 (39.9)	18.7 (31.6)
89.4	27.4 (34.7)	23.4 (28.3)	14.4 (18.0)
	NORMAL FORCE[3,4]		
0.015 inch			
10.9	5.8 (4.9 to 6.8)	2.1 (1.7 to 2.5)	−0.6 (−1.1 to 0.0)
19.0	3.5 (2.8 to 4.1)	1.2 (.9 to 1.6)	−.1 (−.4 to .5)
89.4	2.6 (1.6 to 4.0)	.8 (.4 to 1.2)	−.1 (−.5 to .4)
0.030 inch			
10.9	9.7 (7.6 to 11.5)	3.6 (3.1 to 4.2)	−1.3 (−2.4 to −.1)
19.0	6.1 (3.7 to 7.7)	1.9 (1.5 to 2.3)	−.7 (−1.4 to −.1)
89.4	3.7 (2.3 to 4.9)	1.1 (.6 to 1.7)	−.4 (−1.1 to .5)
0.045 inch			
10.9	12.7 (10.5 to 15.2)	4.5 (3.9 to 5.2)	−1.5 (−3.4 to .3)
19.0	7.6 (5.1 to 9.4)	2.4 (1.8 to 2.9)	−1.2 (−2.0 to −.5)
89.4	4.6 (3.0 to 6.3)	1.2 (.7 to 1.8)	−.8 (−1.4 to −.1)
0.060 inch			
10.9	15.4 (11.6 to 18.9)	6.2 (5.4 to 7.0)	−1.5 (−4.4 to .1)
19.0	9.9 (6.8 to 12.4)	2.8 (2.1 to 3.5)	−1.4 (−2.6 to −.7)
89.4	6.0 (4.4 to 7.8)	1.4 (.8 to 2.1)	−1.4 (−2.2 to −.7)

[1]Clearance angle 15°; cutting velocity 5 inches per minute.
[2]The first number in each entry is the average cutting force; the numbers following in parentheses are maximum forces (average of five).
[3]The first number in each entry is the average cutting force; the numbers following in parentheses are minimum and maximum forces (average of five).
[4]A negative normal force indicates that the knife tended to lift the workpiece; force was positive when the knife pushed the workpiece.

TABLE 18-15B.—*Tool forces when orthogonally cutting laurel oak veneer in the 0-90 direction[1]* (Data from Woodson 1979)

Depth of cut and moisture content (percent)	Rake angle, degrees		
	50	60	70
	----------------Pounds per 0.1 inch of knife----------------		
PARALLEL FORCE[2]			
0.015 inch			
11.1.....................	4.2 (10.0)	4.7 (10.8)	4.1 (8.7)
19.0.....................	2.9 (6.5)	4.9 (12.2)	2.9 (6.3)
91.2.....................	2.5 (5.9)	4.4 (8.9)	2.1 (4.5)
0.030 inch			
11.1.....................	5.0 (13.7)	4.3 (9.2)	5.2 (12.0)
19.0.....................	4.3 (9.3)	3.6 (7.4)	3.8 (8.0)
91.2.....................	3.2 (8.5)	2.8 (5.9)	2.9 (6.1)
0.045 inch			
11.1.....................	5.2 (15.9)	5.2 (13.4)	5.9 (13.6)
19.0.....................	5.0 (10.9)	4.3 (9.3)	4.5 (9.1)
91.2.....................	3.6 (10.1)	3.4 (7.3)	3.5 (6.9)
0.060 inch			
11.1.....................	6.7 (18.5)	6.5 (16.2)	6.8 (15.4)
19.0.....................	6.1 (13.5)	5.8 (11.4)	5.3 (11.5)
91.2.....................	4.1 (10.3)	4.3 (9.4)	4.0 (8.1)
NORMAL FORCE[3,4]			
0.015 inch			
11.1.....................	0.1 (−1.2 to 2.6)	−0.1 (−1.7 to 2.2)	0.8 (−0.6 to 2.4)
19.0.....................	.2 (−.7 to 2.0)	−.1 (−1.8 to 2.1)	.3 (−.7 to 1.8)
91.2.....................	.2 (−.7 to 1.9)	−.6 (−2.1 to 1.3)	.3 (−.5 to 1.2)
0.030 inch			
11.1.....................	.4 (−1.4 to 3.1)	.7 (−.9 to 2.6)	.3 (−2.1 to 3.6)
19.0.....................	−.4 (−1.9 to 2.0)	−.3 (−1.5 to 1.8)	−.5 (−1.7 to 1.1)
91.2.....................	.2 (−1.2 to 1.8)	−.6 (−1.6 to .5)	−.1 (−1.3 to 1.3)
0.045 inch			
11.1.....................	.1 (−1.8 to 3.0)	.1 (−1.7 to 1.9)	.1 (−2.6 to 3.5)
19.0.....................	−.3 (−2.1 to 1.9)	−.8 (−2.5 to 1.7)	−.8 (−2.4 to 1.2)
91.2.....................	−.1 (−1.8 to 1.9)	−.7 (−2.3 to .6)	−.1 (−1.8 to 2.3)
0.060 inch			
11.1.....................	.1 (−2.0 to 3.3)	.4 (−2.0 to 3.3)	.1 (−2.8 to 3.6)
19.0.....................	−.6 (−2.8 to 1.8)	−1.2 (−3.3 to 1.2)	−1.5 (−3.3 to .3)
91.2.....................	.0 (−1.6 to 2.3)	−1.1 (−3.3 to .9)	−.2 (−1.8 to 1.1)

[1] Clearance angle 0°; cutting velocity 5 inches per minute.

[2] The first number in each entry is the average cutting force; the numbers following in parentheses are maximum forces (average of five).

[3] The first number in each entry is the average cutting force; the numbers following in parentheses are minimum and maximum forces (average of five).

[4] A negative normal force indicates that the knife tended to lift the workpiece; force was positive when the knife pushed the workpiece.

TABLE 18-15C.—*Tool forces when orthogonally cutting laurel oak across the grain in the 90-90 direction[1]* (Data from Woodson 1979)

Depth of cut and moisture content (percent)	Rake angle, degrees		
	20	30	40

------------------Pounds per 0.1 inch of knife------------------

PARALLEL FORCE[2]

0.015 inch			
10.7	23.4 (29.2)	21.2 (27.6)	16.0 (19.3)
18.6	18.3 (21.8)	12.1 (15.6)	10.8 (14.2)
94.4	14.1 (17.6)	12.1 (15.0)	8.3 (11.0)
0.030 inch			
10.7	36.5 (48.6)	33.4 (41.6)	28.1 (35.1)
18.6	33.6 (39.9)	23.9 (29.5)	18.5 (22.2)
94.4	24.8 (30.3)	16.9 (21.1)	13.5 (16.9)
0.045 inch			
10.7	47.9 (59.1)	43.0 (53.7)	34.2 (40.5)
18.6	42.1 (47.7)	30.8 (37.3)	23.5 (29.4)
94.4	32.3 (39.7)	22.9 (29.1)	17.2 (22.7)
0.060 inch			
10.7	49.5 (70.5)	48.1 (63.0)	43.8 (54.7)
18.6	50.9 (58.9)	39.3 (46.3)	30.2 (38.0)
94.4	39.4 (49.4)	28.1 (35.1)	20.8 (27.0)

NORMAL FORCE[3,4]

0.015 inch			
10.7	2.2 (0.6 to 4.7)	−0.5 (−2.0 to 1.6)	−2.5 (−3.9 to −0.7)
18.6	1.4 (.8 to 2.6)	−.2 (−1.2 to .8)	−1.9 (−3.0 to −.7)
94.4	.8 (.1 to 1.9)	−.8 (−1.7 to .0)	−1.7 (−2.6 to −.6)
0.030 inch			
10.7	3.2 (.7 to 7.2)	−1.8 (−4.6 to 2.6)	−5.9 (−8.3 to −3.0)
18.6	1.4 (.2 to 3.1)	−2.4 (−4.1 to −.6)	−4.6 (−6.5 to −2.4)
94.4	.5 (−.9 to 1.6)	−2.0 (−3.1 to −.6)	−3.7 (−5.2 to −1.7)
0.045 inch			
10.7	4.5 (.8 to 9.2)	−2.9 (−6.0 to 2.8)	−7.9 (−10.7 to −3.8)
18.6	1.5 (.0 to 3.9)	−3.8 (−5.7 to 1.6)	−6.7 (−8.9 to −3.1)
94.4	.5 (−1.1 to 2.1)	−3.2 (−4.7 to −1.1)	−5.1 (−7.7 to −2.6)
0.060 inch			
10.7	5.7 (1.5 to 10.2)	−2.8 (−7.7 to 4.6)	−10.4 (−14.2 to −4.6)
18.6	1.4 (−.7 to 3.9)	−5.2 (−8.1 to −1.3)	−8.7 (−12.0 to −4.2)
94.4	.3 (−1.9 to 2.2)	−4.3 (−6.3 to −1.6)	−6.6 (−10.0 to −3.2)

[1] Clearance angle 15°; cutting velocity 5 inches per minute.

[2] The first number in each entry is the average cutting force; the numbers following in parentheses are maximum forces (average of five).

[3] The first number in each entry is the average cutting force; the numbers following in parentheses are minimum and maximum forces (average of five).

[4] A negative normal force indicates that the knife tended to lift the workpiece; force was positive when the knife pushed the workpiece.

TABLE 18-16A.—*Tool forces when orthogonally cutting northern red oak parallel to the grain in the 90-0 direction[1]* (Data from Woodson 1979)

Depth of cut and moisture content (percent)	Rake angle, degrees		
	10	20	30
	------------------Pounds per 0.1 inch of knife------------------		
PARALLEL FORCE[2]			
0.015 inch			
10.4	20.1 (23.3)	17.6 (19.7)	11.9 (16.0)
17.8	13.6 (15.2)	11.9 (13.4)	8.8 (10.8)
78.9	8.8 (10.1)	8.0 (9.0)	5.8 (7.1)
0.030 inch			
10.4	37.2 (44.9)	31.9 (38.4)	18.9 (31.8)
17.8	25.6 (31.5)	23.4 (25.5)	15.9 (21.0)
78.9	19.9 (22.5)	15.5 (17.0)	10.5 (12.3)
0.045 inch			
10.4	45.2 (58.1)	40.4 (51.7)	21.4 (38.7)
17.8	32.3 (38.2)	32.9 (37.1)	17.8 (27.5)
78.9	27.1 (30.3)	21.1 (23.9)	15.4 (18.1)
0.060 inch			
10.4	58.9 (77.7)	47.3 (70.6)	24.9 (52.4)
17.8	44.0 (54.5)	40.9 (43.8)	20.4 (34.7)
78.9	36.4 (40.9)	27.9 (33.1)	17.5 (22.8)
NORMAL FORCE[3,4]			
0.015 inch			
10.4	5.6 (4.5 to 6.7)	2.0 (1.4 to 2.5)	−.4 (−0.9 to 0.3)
17.8	3.5 (3.0 to 4.2)	1.3 (.9 to 1.7)	−.4 (−.8 to .1)
78.9	2.6 (1.9 to 3.1)	.9 (.6 to 1.3)	.1 (−.3 to .5)
0.030 inch			
10.4	9.1 (6.8 to 11.8)	3.1 (2.2 to 4.0)	−1.2 (−2.8 to −.1)
17.8	6.4 (4.9 to 7.6)	1.8 (1.1 to 2.4)	−1.3 (−2.0 to −.5)
78.9	4.2 (3.5 to 5.2)	1.0 (.6 to 1.6)	−.8 (−1.2 to −.2)
0.045 inch			
10.4	10.8 (8.1 to 13.5)	3.8 (2.4 to 5.2)	−1.6 (−3.5 to −.1)
17.8	7.7 (5.7 to 8.9)	2.3 (1.6 to 2.9)	−1.5 (−2.6 to −.4)
78.9	6.0 (4.8 to 7.0)	1.1 (.6 to 1.9)	−1.4 (−2.0 to −.6)
0.060 inch			
10.4	13.6 (9.5 to 17.8)	3.7 (1.5 to 5.7)	−2.0 (−4.7 to −.3)
17.8	10.4 (7.7 to 12.7)	2.8 (1.9 to 3.4)	−1.9 (−3.6 to −.5)
78.9	7.3 (5.8 to 8.5)	1.1 (.5 to 1.7)	−1.6 (−2.6 to −.7)

[1] Clearance angle 15°; cutting velocity 5 inches per minute.

[2] The first number in each entry is the average cutting force; the numbers following in parentheses are maximum forces (average of five).

[3] The first number in each entry is the average cutting force; the numbers following in parentheses are minimum and maximum forces (average of five).

[4] A negative normal force indicates that the knife tended to lift the workpiece; force was positive when the knife pushed the workpiece.

Machining

TABLE 18-16B.—*Tool forces when orthogonally cutting northern red oak veneer in the 0-90 direction[1]* (Data from Woodson 1979)

Depth of cut and moisture content (percent)	Rake angle, degrees		
	50	60	70
	----------------Pounds per 0.1 inch of knife----------------		
	PARALLEL FORCE[2]		
0.015 inch			
10.9................	2.9 (9.0)	3.5 (8.6)	3.5 (6.9)
18.1................	2.8 (6.6)	2.7 (5.8)	4.8 (11.9)
91.9................	2.2 (4.8)	2.1 (4.4)	3.9 (7.6)
0.030 inch			
10.9................	4.2 (13.1)	4.2 (12.0)	3.8 (8.6)
18.1................	4.2 (9.8)	3.6 (8.3)	3.3 (6.3)
91.9................	3.3 (8.3)	3.0 (6.5)	2.6 (5.3)
0.045 inch			
10.9................	4.6 (15.0)	5.4 (14.7)	5.3 (10.6)
18.1................	4.2 (11.6)	4.4 (10.3)	3.9 (8.3)
91.9................	3.6 (9.5)	3.6 (7.3)	3.1 (6.1)
0.060 inch			
10.9................	6.2 (18.4)	5.9 (16.4)	5.9 (14.1)
18.1................	4.6 (12.6)	5.0 (11.4)	4.6 (9.6)
91.9................	4.7 (12.1)	4.1 (8.8)	3.8 (7.3)
	NORMAL FORCE[3,4]		
0.015 inch			
10.9................	0.5 (−0.7 to 2.7)	0.9 (−0.4 to 3.0)	0.6 (−0.6 to 1.6)
18.1................	.3 (−.7 to 1.9)	.3 (−.8 to 2.0)	1.0 (−1.2 to 4.3)
91.9................	.2 (−.5 to 1.3)	.2 (−.4 to .8)	.4 (−1.2 to 1.9)
0.030 inch			
10.9................	.5 (−1.1 to 3.8)	.6 (−1.0 to 3.3)	1.4 (−.4 to 3.1)
18.1................	.1 (−1.3 to 2.3)	−.1 (−1.5 to 2.0)	.4 (−1.1 to 2.0)
91.9................	.0 (−.8 to 1.5)	−.3 (−1.4 to .8)	.4 (−.5 to 1.3)
0.045 inch			
10.9................	.8 (−1.1 to 4.2)	.4 (−1.7 to 3.1)	1.7 (−.1 to 4.0)
18.1................	.2 (−1.5 to 2.3)	.1 (−2.0 to 2.5)	.1 (−1.7 to 2.3)
91.9................	−.1 (−1.2 to 1.7)	−.4 (−1.9 to .8)	−.1 (−1.2 to 1.0)
0.060 inch			
10.9................	.6 (−1.4 to 4.6)	.8 (−1.2 to 4.0)	1.8 (−.6 to 5.1)
18.1................	−.1 (−1.6 to 2.1)	.1 (−2.4 to 2.9)	−.3 (−1.8 to 1.8)
91.9................	.0 (−1.5 to 1.7)	−.7 (−2.7 to .7)	−.4 (−1.7 to 1.1)

[1] Clearance angle 0°; cutting velocity 5 inches per minute.

[2] The first number in each entry is the average cutting force; the numbers following in parentheses are maximum forces (average of five).

[3] The first number in each entry is the average cutting force; the numbers following in parentheses are minimum and maximum forces (average of five).

[4] A negative normal force indicates that the knife tended to lift the workpiece; force was positive when the knife pushed the workpiece.

TABLE 18-16C.—*Tool forces when orthogonally cutting northern red oak across the grain in the 90-90 direction[1]* (Data from Woodson 1979)

Depth of cut and moisture content (percent)	Rake angle, degrees		
	20	30	40

----------------------Pounds per 0.1 inch of knife----------------------

PARALLEL FORCE[2]

0.015 inch			
10.6............	21.8 (27.4)	17.1 (21.7)	17.4 (22.0)
17.8............	17.2 (21.3)	14.8 (17.5)	13.0 (16.2)
92.3............	12.8 (15.0)	10.6 (13.5)	8.6 (10.9)
0.030 inch			
10.6............	34.5 (42.5)	29.4 (40.9)	24.4 (31.3)
17.8............	29.9 (36.3)	21.9 (26.3)	18.3 (23.6)
92.3............	23.1 (27.4)	18.3 (22.0)	14.0 (17.0)
0.045 inch			
10.6............	37.0 (49.8)	35.0 (46.6)	29.3 (36.8)
17.8............	41.4 (47.8)	29.6 (36.2)	24.5 (32.7)
92.3............	31.3 (37.6)	22.7 (29.1)	17.4 (21.4)
0.060 inch			
10.6............	45.0 (56.9)	40.8 (52.9)	36.9 (46.7)
17.8............	46.1 (53.3)	36.1 (43.0)	29.2 (35.9)
92.3............	36.7 (43.8)	27.6 (32.7)	20.9 (25.5)

NORMAL FORCE[3,4]

0.015 inch			
10.6............	1.7 (0.6 to 2.9)	−0.3 (−1.5 to 1.2)	−3.5 (−5.2 to −1.7)
17.8............	1.3 (.5 to 2.1)	−1.0 (−1.9 to −.1)	−2.6 (−3.8 to −1.2)
92.3............	.7 (.2 to 1.3)	−.7 (−1.7 to .1)	−2.0 (−2.8 to −1.0)
0.030 inch			
10.6............	2.3 (.5 to 4.2)	−2.0 (−4.1 to 1.2)	−5.6 (−8.0 to −2.7)
17.8............	1.4 (−.2 to 2.9)	−2.5 (−3.8 to −1.0)	−4.7 (−6.5 to −2.4)
92.3............	.4 (−.5 to 1.4)	−2.2 (−3.3 to −.8)	−4.0 (−5.2 to −2.5)
0.045 inch			
10.6............	2.4 (−.2 to 5.4)	−3.0 (−5.7 to −.5)	−7.4 (−10.4 to −3.7)
17.8............	1.3 (−.7 to 3.5)	−3.4 (−4.9 to −2.0)	−7.1 (−10.1 to −4.4)
92.3............	.7 (−.6 to 1.8)	−3.1 (−4.6 to −1.6)	−5.2 (−7.3 to −3.3)
0.060 inch			
10.6............	3.6 (−.1 to 8.2)	−3.6 (−6.6 to −.2)	−9.9 (−13.5 to −6.4)
17.8............	1.7 (−.7 to 4.7)	−4.3 (−7.0 to −1.6)	−8.3 (−11.5 to −4.9)
92.3............	1.3 (−.5 to 2.9)	−4.0 (−6.0 to −2.2)	−6.8 (−9.0 to −3.8)

[1] Clearance angle 15°; cutting velocity 5 inches per minute.

[2] The first number in each entry is the average cutting force; the numbers following in parentheses are maximum forces (average of five).

[3] The first number in each entry is the average cutting force; the numbers following in parentheses are minimum and maximum forces (average of five).

[4] A negative normal force indicates that the knife tended to lift the workpiece; force was positive when the knife pushed the workpiece.

TABLE 18-17A.—*Tool forces when orthogonally cutting post oak parallel to the grain in the 90-0 direction[1]* (Data from Woodson 1979)

Depth of cut and moisture content (percent)	Rake angle, degrees		
	10	20	30
	............Pounds per 0.1 inch of knife............		
	PARALLEL FORCE[2]		
0.015 inch			
11.3	18.0 (20.3)	16.7 (20.4)	10.2 (13.6)
19.7	12.1 (14.1)	9.9 (11.2)	5.9 (7.4)
79.6	8.8 (10.8)	7.4 (8.7)	4.9 (5.9)
0.030 inch			
11.3	32.6 (40.4)	31.8 (36.1)	15.3 (22.1)
19.7	22.3 (26.8)	17.2 (20.1)	10.7 (14.3)
79.6	17.4 (21.4)	11.6 (14.1)	7.0 (9.7)
0.045 inch			
11.3	40.8 (55.9)	38.9 (46.9)	19.4 (36.5)
19.7	31.6 (38.0)	24.0 (27.5)	13.0 (15.2)
79.6	23.8 (31.5)	15.8 (19.3)	9.5 (12.3)
0.060 inch			
11.3	52.3 (73.2)	45.9 (57.7)	21.2 (40.7)
19.7	38.3 (50.1)	28.8 (35.0)	15.5 (21.0)
79.6	29.8 (38.0)	18.0 (23.6)	12.4 (15.4)
	NORMAL FORCE[3,4]		
0.015 inch			
11.3	4.8 (3.5 to 5.8)	1.8 (1.3 to 2.4)	−0.6 (−1.1 to 0.0)
19.7	3.5 (2.6 to 4.1)	1.2 (.8 to 1.6)	.1 (−.6 to .8)
79.6	2.4 (1.7 to 3.1)	.8 (.5 to 1.3)	.0 (−.3 to .4)
0.030 inch			
11.3	8.2 (5.8 to 10.7)	2.9 (2.0 to 3.7)	−1.1 (−2.0 to .1)
19.7	5.4 (4.0 to 6.6)	1.6 (1.0 to 2.0)	−.8 (−1.4 to −.2)
79.6	4.1 (3.0 to 5.3)	1.1 (.6 to 1.6)	−.4 (−.8 to .3)
0.045 inch			
11.3	10.3 (6.9 to 13.5)	3.4 (2.0 to 4.6)	−1.5 (−3.2 to −.4)
19.7	7.4 (5.4 to 9.2)	2.0 (1.3 to 2.6)	−1.1 (−1.5 to −.4)
79.6	5.0 (3.4 to 6.1)	1.3 (.6 to 1.9)	−.5 (−1.0 to .2)
0.060 inch			
11.3	11.8 (7.6 to 18.1)	3.9 (1.9 to 5.6)	−1.7 (−4.1 to .1)
19.7	8.8 (5.8 to 11.5)	2.0 (1.3 to 2.8)	−1.3 (−1.9 to −.6)
79.6	6.1 (4.2 to 7.5)	1.4 (.9 to 2.1)	−.6 (−1.3 to .1)

[1]Clearance angle 15°; cutting velocity 5 inches per minute.

[2]The first number in each entry is the average cutting force; the numbers following in parentheses are maximum forces (average of five).

[3]The first number in each entry is the average cutting force; the numbers following in parentheses are minimum and maximum forces (average of five).

[4]A negative normal force indicates that the knife tended to lift the workpiece; force was positive when the knife pushed the workpiece.

TABLE 18-17B.—*Tool forces when orthogonally cutting post oak veneer in the 0-90 direction[1]* (Data from Woodson 1979)

Depth of cut and moisture content (percent)	Rake angle, degrees		
	50	60	70
	............Pounds per 0.1 inch of knife............		
PARALLEL FORCE[2]			
0.015 inch			
11.6	3.8 (9.6)	4.0 (8.0)	3.6 (5.9)
19.5	3.0 (6.0)	2.5 (5.1)	5.9 (11.0)
78.5	2.4 (4.3)	2.1 (4.3)	4.8 (8.8)
0.030 inch			
11.6	5.6 (13.4)	5.1 (11.9)	4.6 (9.0)
19.5	4.5 (8.9)	3.8 (7.6)	3.2 (5.4)
78.5	3.4 (6.4)	3.1 (6.5)	2.8 (4.9)
0.045 inch			
11.6	6.2 (17.4)	6.3 (13.7)	5.4 (10.9)
19.5	5.5 (11.1)	4.8 (10.2)	4.4 (7.2)
78.5	4.0 (8.5)	3.7 (8.1)	3.5 (6.2)
0.060 inch			
11.6	7.7 (18.4)	6.6 (17.6)	6.9 (14.4)
19.5	6.5 (12.3)	5.6 (10.7)	5.5 (9.1)
78.5	5.0 (9.5)	4.4 (9.4)	4.0 (7.3)
NORMAL FORCE[3,4]			
0.015 inch			
11.6	−0.3 (−1.4 to 1.2)	0.2 (−1.0 to 1.3)	−0.6 (−1.6 to 0.8)
19.5	−.3 (−1.2 to 1.4)	−.5 (−1.3 to .9)	−.7 (−2.7 to 1.7)
78.5	−.1 (−.7 to .8)	−.2 (−1.0 to .7)	.0 (−2.0 to 1.9)
0.030 inch			
11.6	−.3 (−2.1 to 1.7)	−.4 (−2.2 to 1.1)	.3 (−1.2 to 1.9)
19.5	−.8 (−2.0 to .5)	−1.1 (−2.2 to .4)	−.5 (−1.4 to .5)
78.5	−.6 (−1.4 to .4)	−.8 (−1.8 to .2)	.1 (−.6 to .9)
0.045 inch			
11.6	−.7 (−2.6 to 1.8)	−.9 (−3.0 to 1.2)	−.3 (−2.3 to 1.3)
19.5	−1.1 (−2.5 to .6)	−1.6 (−3.2 to .3)	−1.4 (−2.7 to .3)
78.5	−.8 (−2.1 to .4)	−1.1 (−2.5 to .3)	−.6 (−1.5 to 1.0)
0.060 inch			
11.6	−.9 (−3.4 to 1.4)	−1.0 (−3.9 to 1.9)	−.7 (−2.8 to 2.1)
19.5	−1.2 (−3.2 to .7)	−1.9 (−3.6 to .0)	−1.6 (−3.6 to .8)
78.5	−1.1 (−2.9 to .4)	−1.2 (−2.9 to .5)	−.9 (−2.2 to .7)

[1] Clearance angle 0°; cutting velocity 5 inches per minute.

[2] The first number in each entry is the average cutting force; the numbers following in parentheses are maximum forces (average of five).

[3] The first number in each entry is the average cutting force; the numbers following in parentheses are minimum and maximum forces (average of five).

[4] A negative normal force indicates that the knife tended to lift the workpiece; force was positive when the knife pushed the workpiece.

TABLE 18-17C.—*Tool forces when orthogonally cutting post oak across the grain in the 90-90 direction[1]* (Data from Woodson 1979)

Depth of cut and moisture content (percent)	Rake angle, degrees		
	20	30	40

..................Pounds per 0.1 inch of knife......................

PARALLEL FORCE[2]

0.015 inch			
10.0............	24.5 (32.1)	18.0 (23.3)	17.1 (22.1)
19.5............	16.4 (20.6)	14.1 (18.1)	9.7 (12.5)
83.1............	15.4 (18.5)	11.4 (14.1)	8.5 (11.2)
0.030 inch			
10.0............	40.7 (50.1)	35.2 (44.5)	28.5 (38.1)
19.5............	31.2 (38.0)	23.0 (29.1)	17.2 (22.1)
83.1............	23.6 (27.5)	18.3 (21.9)	13.7 (17.3)
0.045 inch			
10.0............	48.3 (61.1)	43.4 (56.4)	36.2 (48.0)
19.5............	39.2 (48.0)	31.4 (39.1)	23.0 (30.1)
83.1............	34.7 (41.4)	23.1 (29.3)	18.4 (22.9)
0.060 inch			
10.0............	51.2 (65.2)	51.3 (67.6)	45.0 (60.1)
19.5............	45.9 (55.0)	37.7 (45.4)	28.5 (36.4)
83.1............	41.6 (50.6)	29.2 (34.6)	22.1 (29.0)

NORMAL FORCE[3,4]

0.015 inch			
10.0............	1.6 (0.2 to 4.3)	−0.8 (−2.0 to 0.8)	−3.2 (−5.0 to −1.3)
19.5............	1.1 (.3 to 1.8)	−.9 (−1.9 to .8)	−2.0 (−3.0 to −.9)
83.1............	.5 (−.1 to 1.2)	−.9 (−1.6 to −.2)	−2.1 (−3.0 to −.9)
0.030 inch			
10.0............	1.3 (−.9 to 4.7)	−4.2 (−6.5 to −.9)	−7.6 (−11.4 to −4.6)
19.5............	.8 (.6 to 2.6)	−2.6 (−4.7 to −.3)	−4.6 (−6.3 to −2.3)
83.1............	.3 (−.5 to 1.2)	−2.4 (−3.5 to −.9)	−4.1 (−5.6 to −2.4)
0.045 inch			
10.0............	1.5 (−1.5 to 7.5)	−5.6 (−8.4 to −.9)	−10.7 (−15.0 to −6.6)
19.5............	.9 (−.9 to 3.4)	−4.0 (−6.5 to −.7)	−6.8 (−9.6 to −3.8)
83.1............	.2 (−1.7 to 1.7)	−3.4 (−5.2 to −1.8)	−5.7 (−7.6 to −3.3)
0.060 inch			
10.0............	2.0 (−2.6 to 7.8)	−7.4 (−11.9 to −2.5)	−13.7 (−19.4 to −7.1)
19.5............	1.3 (−1.4 to 4.3)	−5.3 (−8.3 to −2.0)	−8.6 (−12.4 to −4.7)
83.1............	.3 (−2.1 to 2.7)	−4.5 (−6.6 to −2.5)	−6.8 (−9.6 to −4.0)

[1] Clearance angle 15°; cutting velocity 5 inches per minute.

[2] The first number in each entry is the average cutting force; the numbers following in parentheses are maximum forces (average of five).

[3] The first number in each entry is the average cutting force; the numbers following in parentheses are minimum and maximum forces (average of five).

[4] A negative normal force indicates that the knife tended to lift the workpiece; force was positive when the knife pushed the workpiece.

TABLE 18-18A.—*Tool forces when orthogonally cutting scarlet oak parallel to the grain in the 90-0 direction*[1] (Data from Woodson 1979)

Depth of cut and moisture content (percent)	Rake angle, degrees		
	10	20	30
	----------------Pounds per 0.1 inch of knife----------------		
PARALLEL FORCE[2]			
0.015 inch			
10.3	19.8 (23.8)	15.4 (18.8)	10.4 (16.0)
18.2	11.4 (13.1)	9.1 (10.8)	7.4 (9.2)
88.7	8.9 (11.2)	7.1 (8.6)	4.6 (5.7)
0.030 inch			
10.3	34.6 (41.9)	28.5 (36.2)	15.2 (27.0)
18.2	24.5 (29.0)	20.1 (23.2)	13.9 (18.1)
88.7	17.0 (19.8)	12.7 (15.6)	7.8 (10.5)
0.045 inch			
10.3	44.5 (59.2)	38.8 (50.5)	18.2 (38.7)
18.2	29.8 (36.3)	22.3 (30.6)	19.7 (23.1)
88.7	22.4 (27.9)	17.6 (21.7)	10.3 (15.4)
0.060 inch			
10.3	54.4 (74.3)	50.9 (64.7)	25.7 (51.7)
18.2	42.8 (49.7)	35.8 (38.7)	21.8 (26.7)
88.7	28.6 (36.9)	21.6 (26.2)	11.9 (11.5)
NORMAL FORCE[3,4]			
0.015 inch			
10.3	5.7 (4.3 to 7.3)	1.9 (1.2 to 3.4)	−0.4 (−1.0 to 0.6)
18.2	3.1 (2.3 to 4.1)	1.1 (.6 to 1.5)	−.1 (−.6 to .4)
88.7	2.7 (1.8 to 4.1)	1.0 (.5 to 1.5)	.4 (−.3 to .8)
0.030 inch			
10.3	8.4 (6.0 to 10.9)	2.9 (1.6 to 4.1)	−1.1 (−2.1 to .2)
18.2	6.2 (5.0 to 7.4)	1.9 (1.2 to 2.4)	−.9 (−1.5 to − .2)
88.7	4.1 (3.1 to 5.4)	1.2 (.6 to 2.0)	−.3 (−.8 to .3)
0.045 inch			
10.3	11.0 (7.3 to 14.3)	4.0 (2.6 to 5.3)	−1.3 (−3.3 to .2)
18.2	7.4 (5.4 to 9.0)	1.8 (1.1 to 2.9)	−1.8 (−2.3 to −1.0)
88.7	5.0 (3.7 to 6.8)	1.3 (.6 to 2.1)	−.8 (−1.4 to .0)
0.060 inch			
10.3	12.4 (8.2 to 19.1)	5.0 (3.3 to 6.6)	−1.6 (−4.4 to .9)
18.2	9.8 (7.0 to 11.5)	2.7 (1.8 to 3.2)	−2.0 (−2.6 to −1.2)
88.7	6.4 (3.8 to 8.5)	1.3 (.6 to 2.4)	−1.0 (−1.9 to − .2)

[1] Clearance angle 15°; cutting velocity 5 inches per minute.

[2] The first number in each entry is the average cutting force; the numbers following in parentheses are maximum forces (average of five).

[3] The first number in each entry is the average cutting force; the numbers following in parentheses are minimum and maximum forces (average of five).

[4] A negative normal force indicates that the knife tended to lift the workpiece; force was positive when the knife pushed the workpiece.

TABLE 18-18B.—*Tool forces when orthogonally cutting scarlet oak veneer in the 0-90 direction[1]* (Data from Woodson 1979)

Depth of cut and moisture content (percent)	Rake angle, degrees		
	50	60	70
	----------------Pounds per 0.1 inch of knife----------------		
	PARALLEL FORCE[2]		
0.015 inch			
11.0..................	3.8 (11.9)	4.8 (10.3)	4.2 (8.5)
18.8..................	3.1 (6.8)	3.0 (5.5)	5.6 (11.2)
92.3..................	2.2 (5.0)	2.4 (5.3)	2.3 (4.5)
0.030 inch			
11.0..................	5.1 (15.1)	5.4 (14.1)	5.1 (12.5)
18.8..................	4.3 (9.5)	4.1 (8.7)	3.7 (6.8)
92.3..................	3.3 (7.2)	3.4 (7.3)	2.9 (5.7)
0.045 inch			
11.0..................	5.7 (16.5)	6.3 (17.7)	5.9 (13.0)
18.8..................	4.9 (11.5)	5.0 (10.7)	4.5 (9.2)
92.3..................	3.9 (7.9)	3.5 (7.6)	3.5 (7.1)
0.060 inch			
11.0..................	6.8 (19.2)	7.3 (17.2)	7.4 (16.9)
18.8..................	6.0 (13.4)	5.6 (12.7)	5.2 (10.5)
92.3..................	4.5 (9.7)	3.9 (8.6)	3.9 (7.7)
	NORMAL FORCE[3,4]		
0.015 inch			
11.0..................	1.1 (−1.2 to 4.5)	1.5 (−0.2 to 3.7)	0.7 (−0.6 to 2.0)
18.8..................	.4 (−.8 to 2.3)	.5 (−1.0 to 2.8)	.2 (−1.8 to 2.8)
92.3..................	.1 (−.6 to 1.0)	−.2 (−1.1 to 1.0)	.3 (−.4 to 1.6)
0.030 inch			
11.0..................	.3 (−1.7 to 5.5)	.6 (−1.9 to 5.0)	.9 (−1.5 to 3.8)
18.8..................	−.1 (−1.7 to 2.5)	−.4 (−2.0 to 1.3)	−.1 (−1.4 to 1.8)
92.3..................	−.3 (−1.4 to 1.3)	−.8 (−2.0 to .8)	−.4 (−1.4 to .8)
0.045 inch			
11.0..................	.5 (−1.6 to 5.6)	.7 (−1.8 to 5.2)	.9 (−1.6 to 4.1)
18.8..................	−.2 (−1.9 to 2.0)	−.8 (−2.9 to 1.3)	−1.1 (−2.9 to 1.3)
92.3..................	−.4 (−1.9 to 1.2)	−1.1 (−2.7 to .5)	−.8 (−2.2 to .5)
0.060 inch			
11.0..................	.7 (−1.9 to 5.4)	1.2 (−2.4 to 5.2)	.7 (−2.1 to 4.9)
18.8..................	−.2 (−2.2 to 2.8)	−1.2 (−3.8 to 1.5)	−1.5 (−3.5 to 1.0)
92.3..................	−.4 (−2.3 to 1.5)	−1.0 (−2.9 to .8)	−.9 (−2.6 to .8)

[1]Clearance angle 0°; cutting velocity 5 inches per minute.

[2]The first number in each entry is the average cutting force; the numbers following in parentheses are maximum forces (average of five).

[3]The first number in each entry is the average cutting force; the numbers following in parentheses are minimum and maximum forces (average of five).

[4]A negative normal force indicates that the knife tended to lift the workpiece; force was positive when the knife pushed the workpiece.

TABLE 18-18C.—*Tool forces when orthogonally cutting scarlet oak across the grain in the 90-90 direction[1]* (Data from Woodson 1979)

Depth of cut and moisture content (percent)	Rake angle, degrees		
	20	30	40
	----------Pounds per 0.1 inch of knife----------		
PARALLEL FORCE[2]			
0.015 inch			
10.6	23.0 (31.3)	21.4 (28.5)	17.2 (21.8)
18.2	17.3 (20.8)	14.6 (17.8)	11.8 (15.8)
94.5	14.1 (17.5)	10.1 (12.2)	9.6 (12.0)
0.030 inch			
10.6	38.1 (49.2)	35.4 (44.5)	27.1 (34.9)
18.2	29.8 (36.8)	21.5 (25.8)	18.6 (23.3)
94.5	23.7 (27.7)	17.0 (21.0)	14.2 (18.3)
0.045 inch			
10.6	41.3 (54.0)	39.8 (50.2)	34.5 (43.5)
18.2	39.5 (44.8)	29.3 (35.3)	23.0 (29.7)
94.5	30.4 (36.7)	21.1 (26.4)	18.1 (23.3)
0.060 inch			
10.6	47.5 (64.1)	46.9 (60.7)	43.1 (53.5)
18.2	47.1 (57.0)	32.8 (43.6)	27.9 (39.1)
94.5	34.3 (40.3)	26.9 (33.1)	22.0 (28.5)
NORMAL FORCE[3,4]			
0.015 inch			
10.6	1.9 (0.7 to 4.3)	−1.1 (−2.6 to 1.5)	−3.3 (−4.8 to −1.6)
18.2	1.2 (.2 to 2.1)	−1.2 (−2.0 to .3)	−2.4 (−3.6 to −1.0)
94.5	.5 (−.2 to 1.1)	−.6 (−1.4 to .0)	−2.2 (−3.3 to −1.2)
0.030 inch			
10.6	2.9 (.9 to 5.4)	−3.1 (−5.0 to −.4)	−6.4 (−8.9 to −3.0)
18.2	1.0 (−.2 to 2.2)	−2.7 (−4.2 to −.9)	−5.0 (−6.6 to −3.0)
94.5	.3 (−1.0 to 1.3)	−2.1 (−3.2 to −.8)	−4.0 (−5.9 to −2.3)
0.045 inch			
10.6	3.9 (−.5 to 8.3)	−3.2 (−6.2 to .3)	−8.8 (−11.8 to −4.9)
18.2	1.1 (−.6 to 2.7)	−4.2 (−6.6 to −1.6)	−6.9 (−9.5 to −3.5)
94.5	.5 (−1.0 to 1.8)	−2.9 (−4.6 to −1.5)	−5.6 (−8.1 to −3.5)
0.060 inch			
10.6	4.7 (.6 to 9.7)	−4.0 (−8.2 to 3.3)	−11.7 (−16.0 to −5.2)
18.2	1.4 (−1.2 to 3.7)	−5.5 (−9.1 to −2.2)	−9.2 (−12.9 to −3.5)
94.5	.8 (−.8 to 2.3)	−4.2 (−6.3 to −2.2)	−7.8 (−10.3 to −4.5)

[1] Clearance angle 15°; cutting velocity 5 inches per minute.

[2] The first number in each entry is the average cutting force; the numbers following in parentheses are maximum forces (average of five).

[3] The first number in each entry is the average cutting force; the numbers following in parentheses are minimum and maximum forces (average of five).

[4] A negative normal force indicates that the knife tended to lift the workpiece; force was positive when the knife pushed the workpiece.

TABLE 18-19A.—*Tool forces when orthogonally cutting Shumard oak parallel to the grain in the 90-0 direction[1]* (Data from Woodson 1979)

Depth of cut and moisture content (percent)	Rake angle, degrees		
	10	20	30
	---------------- Pounds per 0.1 inch of knife ----------------		
PARALLEL FORCE[2]			
0.015 inch			
10.4	19.7 (23.1)	17.4 (21.5)	11.6 (16.6)
18.8	12.3 (14.2)	8.8 (11.5)	7.3 (8.9)
88.0	9.3 (10.7)	7.0 (7.9)	4.6 (6.3)
0.030 inch			
10.4	35.8 (45.0)	29.1 (40.9)	16.5 (31.0)
18.8	24.5 (28.3)	20.5 (23.7)	13.8 (17.1)
88.0	17.7 (20.6)	13.3 (15.5)	8.3 (10.5)
0.045 inch			
10.4	43.7 (55.6)	40.1 (52.7)	20.8 (39.9)
18.8	32.5 (36.6)	25.5 (29.0)	18.4 (23.0)
88.0	25.5 (30.3)	19.2 (22.0)	11.4 (14.2)
0.060 inch			
10.4	51.2 (72.3)	47.2 (70.1)	20.3 (47.4)
18.8	40.3 (49.1)	34.3 (38.1)	21.0 (27.6)
88.0	31.1 (37.1)	24.8 (28.5)	14.8 (18.5)
NORMAL FORCE[3,4]			
0.015 inch			
10.4	5.6 (4.3 to 6.5)	2.3 (1.6 to 2.8)	−0.2 (−0.8 to 0.3)
18.8	3.7 (2.5 to 4.7)	1.1 (.7 to 1.6)	−.1 (−.5 to .3)
88.0	2.6 (2.2 to 3.1)	1.1 (.7 to 1.4)	.3 (−.1 to .7)
0.030 inch			
10.4	9.6 (7.4 to 11.6)	3.5 (2.0 to 4.6)	−.9 (−2.0 to .0)
18.8	6.3 (5.4 to 7.3)	1.9 (1.3 to 2.7)	−.9 (−1.4 to −.4)
88.0	4.3 (3.5 to 4.9)	1.4 (.8 to 1.9)	−.2 (−.6 to .3)
0.045 inch			
10.4	11.2 (8.9 to 13.7)	4.1 (2.6 to 5.4)	−1.1 (−3.2 to .2)
18.8	7.9 (6.4 to 8.9)	2.3 (1.5 to 3.0)	−1.4 (−2.0 to −.5)
88.0	5.9 (4.7 to 7.0)	1.5 (1.0 to 1.8)	−.6 (−1.1 to .1)
0.060 inch			
10.4	12.5 (8.4 to 17.3)	4.4 (2.3 to 6.5)	−1.2 (−3.6 to .3)
18.8	9.8 (6.7 to 11.5)	3.0 (2.2 to 3.7)	−1.7 (−2.6 to −.7)
88.0	7.1 (5.2 to 8.4)	1.6 (.9 to 2.2)	−.7 (−1.6 to .4)

[1] Clearance angle 15°; cutting velocity 5 inches per minute.

[2] The first number in each entry is the average cutting force; the numbers following in parentheses are maximum forces (average of five).

[3] The first number in each entry is the average cutting force; the numbers following in parentheses are minimum and maximum forces (average of five).

[4] A negative normal force indicates that the knife tended to lift the workpiece; force was positive when the knife pushed the workpiece.

TABLE 18-19B.—*Tool forces when orthogonally cutting Shumard oak veneer in the 90-0 direction[1]* (Data from Woodson 1979)

Depth of cut and moisture content (percent)	Rake angle, degrees		
	50	60	70
	----------------Pounds per 0.1 inch of knife----------------		
	PARALLEL FORCE[2]		
0.015 inch			
11.0	4.0 (9.1)	3.9 (8.7)	4.4 (9.2)
18.4	3.1 (5.6)	5.5 (12.5)	3.0 (5.7)
92.5	2.4 (5.4)	4.9 (9.3)	2.6 (4.5)
0.030 inch			
11.0	5.2 (15.0)	4.5 (8.8)	5.3 (10.8)
18.4	4.2 (9.1)	3.3 (6.4)	3.9 (7.6)
92.5	3.5 (8.4)	2.8 (5.9)	3.3 (6.3)
0.045 inch			
11.0	5.8 (16.8)	5.7 (12.9)	5.9 (13.3)
18.4	5.6 (11.4)	4.1 (8.6)	4.8 (9.0)
92.5	4.0 (9.6)	3.9 (8.4)	3.5 (7.4)
0.060 inch			
11.0	6.7 (18.5)	6.0 (15.2)	6.5 (15.8)
18.4	6.3 (13.0)	4.8 (10.0)	5.4 (10.3)
92.5	4.7 (12.1)	4.3 (9.2)	4.4 (8.4)
	NORMAL FORCE[3,4]		
0.015 inch			
11.0	0.8 (−1.1 to 3.6)	−0.5 (−2.4 to 1.4)	1.5 (−0.4 to 3.8)
18.4	.3 (−.7 to 1.8)	.4 (−1.5 to 3.0)	.6 (−.8 to 2.0)
92.5	.6 (−.4 to 2.3)	.1 (−1.4 to 1.9)	.8 (−.2 to 1.8)
0.030 inch			
11.0	.3 (−1.8 to 4.0)	1.2 (−.6 to 3.1)	.9 (−1.0 to 3.5)
18.4	−.4 (−1.5 to 2.1)	−.4 (−1.6 to 1.2)	.0 (−1.7 to 2.1)
92.5	.3 (−1.2 to 2.0)	.1 (−1.2 to 1.5)	.4 (−.7 to 1.6)
0.045 inch			
11.0	.4 (−1.8 to 5.0)	.6 (−1.3 to 3.1)	.4 (−1.4 to 3.0)
18.4	−.7 (−1.9 to 1.7)	−.4 (−2.1 to 1.8)	−.6 (−2.2 to 1.4)
92.5	.1 (−1.1 to 2.4)	−.2 (−1.9 to 2.0)	.1 (−1.4 to 1.7)
0.060 inch			
11.0	.5 (−2.1 to 4.6)	.3 (−1.7 to 3.0)	.5 (−1.9 to 3.8)
18.4	−.7 (−2.3 to 1.9)	−.6 (−2.6 to 1.6)	−.4 (−2.3 to 1.7)
92.5	.2 (−1.1 to 2.3)	−.5 (−2.6 to 1.8)	−.2 (−1.9 to 1.4)

[1] Clearance angle 0°; cutting velocity 5 inches per minute.

[2] The first number in each entry is the average cutting force; the numbers following in parentheses are maximum forces (average of five).

[3] The first number in each entry is the average cutting force; the numbers following in parentheses are minimum and maximum forces (average of five).

[4] A negative normal force indicates that the knife tended to lift the workpiece; force was positive when the knife pushed the workpiece.

TABLE 18-19C.—*Tool forces when orthogonally cutting Shumard oak across the grain in the 90-90 direction[1]* (Data from Woodson 1979)

Depth of cut and moisture content (percent)	Rake angle, degrees		
	20	30	40
	----------Pounds per 0.1 inch of knife----------		
PARALLEL FORCE[2]			
0.015 inch			
10.4	28.6 (36.8)	20.0 (25.7)	16.6 (21.9)
17.5	18.2 (21.7)	14.0 (18.7)	11.3 (14.2)
90.8	13.6 (15.8)	11.8 (13.8)	9.5 (11.3)
0.030 inch			
10.4	41.0 (51.2)	36.8 (43.0)	26.5 (33.2)
17.5	30.6 (36.7)	23.0 (27.6)	17.3 (22.4)
90.8	24.5 (29.3)	18.0 (20.8)	14.7 (17.3)
0.045 inch			
10.4	48.8 (59.1)	42.4 (53.0)	33.9 (42.5)
17.5	38.6 (44.7)	29.9 (36.2)	23.3 (30.2)
90.8	31.2 (36.5)	23.7 (27.4)	18.6 (21.9)
0.060 inch			
10.4	59.1 (75.8)	52.2 (68.5)	40.2 (53.2)
17.5	45.5 (53.5)	35.3 (43.9)	26.9 (35.9)
90.8	36.9 (42.1)	28.8 (33.1)	22.3 (26.8)
NORMAL FORCE[3,4]			
0.015 inch			
10.4	1.9 (0.4 to 3.4)	−0.6 (−1.8 to 0.7)	−3.3 (−4.6 to −2.1)
17.5	1.4 (.8 to 2.0)	−.9 (−2.2 to .0)	−2.1 (−3.1 to −1.2)
90.8	.7 (.1 to 1.2)	−.8 (−1.4 to −.2)	−2.0 (−2.9 to −1.2)
0.030 inch			
10.4	2.4 (.5 to 4.5)	−3.4 (−5.2 to −.6)	−6.3 (−8.8 to −3.6)
17.5	1.4 (.2 to 2.8)	−2.3 (−4.1 to −.9)	−4.5 (−6.3 to −2.7)
90.8	.5 (−.4 to 1.6)	−2.0 (−3.0 to −1.0)	−4.0 (−5.4 to −2.6)
0.045 inch			
10.4	3.0 (.5 to 6.1)	−4.0 (−7.0 to −1.3)	−8.9 (−12.1 to −5.1)
17.5	1.2 (−.3 to 3.2)	−3.7 (−5.9 to −2.0)	−6.7 (−9.9 to −3.4)
90.8	.6 (−.8 to 1.8)	−3.1 (−4.5 to −1.6)	−5.7 (−7.4 to −3.9)
0.060 inch			
10.4	3.7 (.9 to 7.4)	−5.5 (−9.4 to −1.4)	−11.5 (−14.7 to −6.4)
17.5	1.5 (−.2 to 4.1)	−4.7 (−7.1 to −2.4)	−8.4 (−12.5 to −3.9)
90.8	1.1 (−.5 to 2.6)	−4.3 (−6.3 to −2.1)	−7.2 (−9.7 to −4.7)

[1] Clearance angle 15°; cutting velocity 5 inches per minute.

[2] The first number in each entry is the average cutting force; the numbers following in parentheses are maximum forces (average of five).

[3] The first number in each entry is the average cutting force; the numbers following in parentheses are minimum and maximum forces (average of five).

[4] A negative normal force indicates that the knife tended to lift the workpiece; force was positive when the knife pushed the workpiece.

TABLE 18-20A.—*Tool forces when orthogonally cutting southern red oak parallel to the grain in the 90-0 direction*[1] (Data from Woodson 1979)

Depth of cut and moisture content (percent)	Rake angle, degrees		
	10	20	30
	----------------Pounds per 0.1 inch of knife----------------		
PARALLEL FORCE[2]			
0.015 inch			
10.9	18.1 (23.1)	14.8 (18.4)	12.6 (16.0)
19.5	12.1 (14.6)	9.6 (11.2)	7.1 (9.1)
89.1	10.1 (11.9)	8.1 (9.5)	5.5 (7.3)
0.030 inch			
10.9	38.0 (47.8)	32.8 (38.3)	21.5 (28.9)
19.5	22.5 (27.4)	19.1 (21.6)	11.4 (13.7)
89.1	18.9 (22.6)	14.2 (17.1)	8.9 (13.5)
0.045 inch			
10.9	47.7 (61.4)	42.4 (48.4)	23.3 (38.2)
19.5	31.9 (38.2)	25.1 (29.5)	13.4 (17.0)
89.1	25.4 (30.7)	18.3 (21.0)	11.8 (17.6)
0.060 inch			
10.9	54.8 (70.2)	53.3 (63.7)	24.3 (49.3)
19.5	40.5 (48.6)	30.4 (36.6)	15.4 (20.9)
89.1	31.6 (38.9)	22.9 (30.5)	13.4 (20.2)
NORMAL FORCE[3,4]			
0.015 inch			
10.9	5.7 (4.3 to 6.7)	1.8 (1.4 to 2.3)	−0.5 (−1.1 to 0.2)
19.5	3.1 (2.2 to 4.0)	1.1 (.8 to 1.6)	−.3 (−.6 to .1)
89.1	2.9 (2.0 to 3.9)	1.0 (.6 to 1.4)	.1 (−.3 to .6)
0.030 inch			
10.9	10.2 (7.7 to 13.4)	3.1 (1.9 to 3.8)	−1.3 (−2.3 to −.3)
19.5	5.5 (3.8 to 6.6)	1.7 (1.2 to 2.2)	−.7 (−1.1 to .0)
89.1	4.5 (3.2 to 5.5)	1.2 (.7 to 1.7)	−.4 (−.9 to .6)
0.045 inch			
10.9	12.1 (8.3 to 15.6)	3.9 (2.7 to 4.9)	−1.5 (−3.1 to −.4)
19.5	7.2 (5.1 to 8.6)	2.0 (1.3 to 2.8)	−.8 (−1.4 to −.1)
89.1	6.0 (4.4 to 7.5)	1.4 (.8 to 2.1)	−.7 (−1.7 to .4)
0.060 inch			
10.9	14.1 (8.7 to 17.9)	4.7 (2.4 to 6.4)	−1.4 (−3.4 to .3)
19.5	9.3 (7.2 to 10.7)	2.5 (1.7 to 3.2)	−1.1 (−2.0 to −.2)
89.1	7.3 (5.4 to 9.7)	1.5 (1.0 to 2.3)	−.9 (−1.9 to .0)

[1] Clearance angle 15°; cutting velocity 5 inches per minute.
[2] The first number in each entry is the average cutting force; the numbers following in parentheses are maximum forces (average of five).
[3] The first number in each entry is the average cutting force; the numbers following in parentheses are minimum and maximum forces (average of five).
[4] A negative normal force indicates that the knife tended to lift the workpiece; force was positive when the knife pushed the workpiece.

TABLE 18-20B.—*Tool forces when orthogonally cutting southern red oak veneer in the 0-90 direction[1]* (Data from Woodson 1979)

Depth of cut and moisture content (percent)	Rake angle, degrees		
	50	60	70
	----------------Pounds per 0.1 inch of knife----------------		
	PARALLEL FORCE[2]		
0.015 inch			
11.2	4.4 (11.0)	4.7 (10.6)	4.1 (7.5)
19.4	3.1 (7.6)	2.6 (5.0)	6.1 (12.9)
91.3	2.4 (5.2)	2.2 (4.5)	4.6 (7.9)
0.030 inch			
11.2	6.1 (16.3)	6.4 (14.5)	4.7 (10.0)
19.4	4.6 (10.2)	3.8 (7.5)	3.4 (6.1)
91.3	3.8 (8.2)	3.4 (7.6)	2.5 (5.1)
0.045 inch			
11.2	7.1 (18.8)	7.2 (15.2)	6.1 (12.3)
19.4	5.5 (11.8)	5.3 (11.1)	4.2 (7.4)
91.3	4.2 (10.1)	4.0 (8.6)	3.3 (7.1)
0.060 inch			
11.2	7.8 (21.9)	8.1 (18.1)	6.4 (14.1)
19.4	6.3 (13.2)	6.0 (11.9)	4.9 (8.4)
91.3	4.8 (11.8)	4.7 (9.7)	4.0 (7.8)
	NORMAL FORCE[3,4]		
0.015 inch			
11.2	0.4 (−1.1 to 3.0)	0.8 (−0.9 to 2.3)	−0.4 (−2.2 to 0.9)
19.4	.0 (−1.1 to 2.0)	−.3 (−1.1 to .8)	.5 (−2.1 to 3.9)
91.3	.1 (−.5 to 1.2)	−.1 (−1.0 to .9)	−.5 (−2.1 to .8)
0.030 inch			
11.2	.2 (−1.5 to 2.7)	.3 (−1.9 to 3.1)	.3 (−1.5 to 1.7)
19.4	−.5 (−1.8 to 1.6)	−1.2 (−2.4 to .1)	−.1 (−1.3 to 1.1)
91.3	−.3 (−1.6 to 1.3)	−.8 (−2.1 to .8)	.1 (−1.0 to 1.1)
0.045 inch			
11.2	.3 (−3.0 to 4.9)	−.1 (−2.7 to 3.1)	.2 (−2.0 to 2.5)
19.4	−.9 (−2.5 to 1.1)	−1.2 (−3.3 to 1.1)	−1.1 (−2.3 to .4)
91.3	−.4 (−1.6 to 1.1)	−.7 (−2.9 to 1.2)	−.3 (−1.5 to .9)
0.060 inch			
11.2	.8 (−2.3 to 6.5)	−.2 (−3.7 to 3.0)	−.8 (−3.4 to 1.8)
19.4	−1.0 (−3.1 to .7)	−1.5 (−4.1 to 1.2)	−1.6 (−2.8 to .1)
91.3	−.2 (−1.7 to 1.5)	−1.2 (−3.3 to 1.2)	−1.1 (−2.6 to .4)

[1] Clearance angle 0°; cutting velocity 5 inches per minute.

[2] The first number in each entry is the average cutting force; the numbers following in parentheses are maximum forces (average of five).

[3] The first number in each entry is the average cutting force; the numbers following in parentheses are minimum and maximum forces (average of five).

[4] A negative normal force indicates that the knife tended to lift the workpiece; force was positive when the knife pushed the workpiece.

TABLE 18-20C.—*Tool forces when orthogonally cutting southern red oak across the grain in the 90-90 direction[1]* (Data from Woodson 1979)

Depth of cut and moisture content (percent)	Rake angle, degrees		
	20	30	40

----------Pounds per 0.1 inch of knife----------

PARALLEL FORCE[2]

0.015 inch			
10.6	25.4 (33.0)	21.0 (26.3)	17.5 (23.7)
18.7	13.7 (18.4)	13.2 (16.7)	10.8 (13.8)
91.9	14.5 (17.5)	11.0 (14.0)	8.9 (11.5)
0.030 inch			
10.6	43.0 (52.5)	37.0 (46.6)	26.9 (33.8)
18.7	31.0 (36.7)	25.7 (30.7)	18.9 (22.5)
91.9	25.3 (30.2)	17.1 (21.9)	14.0 (17.7)
0.045 inch			
10.6	54.5 (65.4)	43.1 (52.3)	36.0 (47.2)
18.7	39.6 (46.0)	30.2 (37.6)	23.1 (30.5)
91.9	36.0 (43.3)	23.6 (29.2)	18.2 (23.4)
0.060 inch			
10.6	59.3 (73.0)	50.9 (64.2)	42.3 (53.6)
18.7	44.1 (52.8)	39.4 (50.9)	28.4 (37.3)
91.9	39.2 (49.6)	29.5 (40.1)	21.1 (28.5)

NORMAL FORCE[3,4]

0.015 inch			
10.6	2.2 (0.9 to 3.8)	−1.1 (−2.3 to 0.6)	−3.7 (−5.5 to −1.5)
18.7	1.1 (.4 to 2.6)	−.8 (−1.6 to .4)	−2.0 (−2.9 to −1.1)
91.9	.5 (.0 to 1.1)	−.7 (−1.6 to .2)	−2.0 (−2.8 to −1.0)
0.030 inch			
10.6	2.9 (1.3 to 6.1)	−3.0 (−5.6 to −.6)	−6.7 (−9.2 to −3.3)
18.7	1.0 (−.1 to 2.8)	−2.6 (−4.6 to −.9)	−4.8 (−6.5 to −2.9)
91.9	.5 (−.4 to 1.5)	−2.0 (−3.3 to −.7)	−3.9 (−5.5 to −2.2)
0.045 inch			
.10.6	3.4 (.0 to 7.2)	−4.1 (−6.5 to −.2)	−9.4 (−13.4 to −4.1)
18.7	1.3 (−.6 to 4.0)	−3.9 (−6.4 to .6)	−6.6 (−9.6 to −3.2)
91.9	.6 (−1.2 to 2.2)	−3.2 (−5.0 to −1.4)	−5.4 (−7.9 to −3.0)
0.060 inch			
10.6	4.8 (.0 to 11.6)	−5.0 (−9.1 to −1.3)	−11.0 (−15.5 to −2.1)
18.7	1.4 (−1.0 to 4.9)	−5.5 (−9.4 to −1.0)	−8.5 (−13.0 to −3.6)
91.9	1.3 (−.8 to 3.1)	−4.7 (−6.7 to −1.8)	−7.0 (−9.9 to −3.2)

[1]Clearance angle 15°; cutting velocity 5 inches per minute.

[2]The first number in each entry is the average cutting force; the numbers following in parentheses are maximum forces (average of five).

[3]The first number in each entry is the average cutting force; the numbers following in parentheses are minimum and maximum forces (average of five).

[4]A negative normal force indicates that the knife tended to lift the workpiece; force was positive when the knife pushed the workpiece.

TABLE 18-21A.—*Tool forces when orthogonally cutting water oak parallel to the grain in the 90-0 direction[1]* (Data from Woodson 1979)

Depth of cut and moisture content (percent)	Rake angle, degrees		
	10	20	30
	---------------- Pounds per 0.1 inch of knife ----------------		
PARALLEL FORCE[2]			
0.015 inch			
10.4	20.6 (25.7)	17.7 (20.1)	11.6 (15.6)
18.4	13.6 (14.9)	12.2 (14.0)	8.3 (10.0)
95.5	9.6 (11.0)	7.0 (8.7)	5.8 (7.5)
0.030 inch			
10.4	34.9 (41.8)	35.7 (40.1)	17.0 (32.1)
18.4	25.2 (29.4)	22.6 (25.2)	16.1 (19.7)
95.5	17.7 (20.8)	14.0 (16.3)	10.0 (12.1)
0.045 inch			
10.4	48.8 (59.9)	44.8 (52.3)	17.9 (43.5)
18.4	29.9 (38.2)	32.5 (37.1)	22.0 (28.7)
95.5	25.2 (29.7)	20.3 (22.9)	13.3 (16.7)
0.060 inch			
10.4	61.3 (80.0)	58.4 (67.7)	19.4 (58.3)
18.4	39.6 (51.2)	40.3 (44.4)	17.9 (36.8)
95.5	32.9 (38.3)	23.6 (27.2)	17.0 (23.2)
NORMAL FORCE[3,4]			
0.015 inch			
10.4	6.1 (4.6 to 7.6)	2.1 (1.6 to 2.5)	−0.1 (−0.8 to 0.4)
18.4	4.0 (3.2 to 4.7)	1.5 (1.1 to 1.8)	.0 (−.3 to .4)
95.5	2.6 (2.0 to 3.0)	1.0 (.5 to 1.8)	.0 (−.5 to .5)
0.030 inch			
10.4	9.3 (7.7 to 10.9)	3.3 (2.3 to 4.2)	−1.1 (−2.5 to .3)
18.4	6.1 (4.5 to 6.9)	2.0 (1.5 to 2.6)	−.8 (−1.4 to .2)
95.5	3.9 (2.9 to 5.0)	1.0 (.4 to 1.8)	−.7 (−1.1 to .0)
0.045 inch			
10.4	12.1 (8.5 to 15.7)	3.8 (2.6 to 5.1)	−1.2 (−3.3 to .6)
18.4	8.1 (5.5 to 9.8)	2.5 (1.8 to 3.1)	−1.3 (−2.4 to −.1)
95.5	5.4 (4.0 to 6.5)	1.0 (.4 to 2.0)	−1.0 (−1.8 to −.1)
0.060 inch			
10.4	14.6 (10.5 to 19.5)	4.6 (2.1 to 6.5)	−1.0 (−4.6 to 1.1)
18.4	9.8 (6.7 to 12.5)	3.1 (2.3 to 3.8)	−1.3 (−3.4 to .3)
95.5	6.0 (5.1 to 7.2)	1.0 (.3 to 1.9)	−1.4 (−2.8 to −.2)

[1] Clearance angle 15°; cutting velocity 5 inches per minute.

[2] The first number in each entry is the average cutting force; the numbers following in parentheses are maximum forces (average of five).

[3] The first number in each entry is the average cutting force; the numbers following in parentheses are minimum and maximum forces (average of five).

[4] A negative normal force indicates that the knife tended to lift the workpiece; force was positive when the knife pushed the workpiece.

TABLE 18-21B.—*Tool forces when orthogonally cutting water oak veneer in the 0-90 direction[1] (Data from Woodson 1979)*

Depth of cut and moisture content (percent)	Rake angle, degrees		
	50	60	70
	---------------- Pounds per 0.1 inch of knife ----------------		
	PARALLEL FORCE[2]		
0.015 inch			
10.7	3.7 (9.5)	3.3 (7.3)	4.2 (7.9)
18.7	2.8 (6.9)	2.5 (5.5)	4.5 (10.6)
102.6	1.9 (4.8)	1.8 (4.0)	3.7 (6.9)
0.030 inch			
10.7	4.5 (13.4)	4.1 (9.5)	3.7 (8.6)
18.7	3.9 (9.1)	3.3 (7.4)	2.8 (6.0)
102.6	3.0 (6.8)	2.9 (5.7)	2.2 (4.6)
0.045 inch			
10.7	6.0 (16.6)	4.7 (11.6)	4.7 (11.1)
18.7	4.8 (10.4)	4.6 (9.6)	3.4 (7.1)
102.6	3.5 (8.1)	3.3 (6.9)	3.0 (5.5)
0.060 inch			
10.7	6.7 (18.0)	5.6 (13.7)	5.0 (12.3)
18.7	5.2 (11.7)	5.1 (10.2)	4.3 (8.7)
102.6	3.9 (9.6)	3.7 (8.5)	3.4 (6.5)
	NORMAL FORCE[3,4]		
0.015 inch			
10.7	0.8 (−0.7 to 3.2)	0.4 (−0.8 to 1.3)	0.0 (−1.9 to 2.0)
18.7	.5 (−.7 to 2.5)	−.1 (−1.0 to 1.2)	−.1 (−1.9 to 1.6)
102.6	.2 (−.5 to 1.7)	−.1 (−.8 to .4)	−.4 (−2.0 to 1.1)
0.030 inch			
10.7	.4 (−1.1 to 3.5)	−.2 (−1.6 to 1.4)	.8 (−.9 to 2.6)
18.7	−.1 (−1.4 to 2.0)	−.4 (−1.9 to 1.3)	−.2 (−1.2 to 1.6)
102.6	−.1 (−1.1 to 1.7)	−.6 (−1.6 to .6)	.1 (−.6 to 1.2)
0.045 inch			
10.7	.4 (−1.6 to 3.0)	−.3 (−2.3 to 1.5)	.0 (−1.8 to 2.4)
18.7	−.4 (−2.0 to 1.8)	−.7 (−2.5 to 1.2)	−.5 (−1.8 to 1.1)
102.6	−.1 (−1.5 to 1.9)	−.7 (−2.1 to .7)	−.3 (−1.6 to .7)
0.060 inch			
10.7	.5 (−2.3 to 4.1)	−.4 (−2.3 to 1.4)	−.3 (−2.6 to 2.4)
18.7	−.8 (−2.6 to 2.0)	−1.2 (−3.3 to .6)	−.7 (−2.2 to 1.5)
102.6	−.2 (−1.8 to 1.6)	−.8 (−2.8 to 1.2)	−.4 (−1.8 to 1.1)

[1]Clearance angle 0°; cutting velocity 5 inches per minute.

[2]The first number in each entry is the average cutting force; the numbers following in parentheses are maximum forces (average of five).

[3]The first number in each entry is the average cutting force; the numbers following in parentheses are minimum and maximum forces (average of five).

[4]A negative normal force indicates that the knife tended to lift the workpiece; force was positive when the knife pushed the workpiece.

TABLE 18-21C.—*Tool forces when orthogonally cutting water oak across the grain in the 90-90 direction[1]* (Data from Woodson 1979)

Depth of cut and moisture content (percent)	Rake angle, degrees		
	20	30	40
	----------Pounds per 0.1 inch of knife----------		
PARALLEL FORCE[2]			
0.015 inch			
10.7	18.8 (24.3)	19.8 (25.7)	16.6 (23.0)
18.0	15.2 (18.9)	14.2 (18.0)	10.9 (13.6)
105.3	13.1 (16.2)	9.9 (12.0)	8.2 (10.4)
0.030 inch			
10.7	35.0 (42.6)	30.8 (39.2)	25.1 (32.9)
18.0	29.0 (35.5)	22.0 (27.7)	17.4 (21.5)
105.3	21.6 (26.3)	16.7 (20.7)	13.4 (17.1)
0.045 inch			
10.7	38.8 (50.4)	39.0 (52.3)	32.3 (40.1)
18.0	35.8 (43.8)	28.7 (35.0)	21.4 (28.1)
105.3	28.8 (33.5)	20.5 (25.6)	16.9 (21.2)
0.060 inch			
10.7	47.9 (59.8)	45.0 (58.0)	39.4 (50.6)
18.0	39.5 (48.9)	36.2 (46.6)	29.1 (37.9)
105.3	35.4 (41.8)	26.0 (31.1)	20.8 (26.1)
NORMAL FORCE[3,4]			
0.015 inch			
10.7	2.5 (0.4 to 4.3)	−0.4 (−1.9 to 2.2)	−2.8 (−5.0 to −0.6)
18.0	1.6 (.8 to 2.7)	−.5 (−1.6 to 1.0)	−1.7 (−2.4 to −.8)
105.3	.7 (−.2 to 1.6)	−.3 (−.9 to .2)	−1.7 (−2.4 to −.7)
0.030 inch			
10.7	3.4 (.4 to 6.2)	−1.8 (−3.8 to 1.7)	−5.0 (−7.6 to −2.3)
18.0	1.7 (.2 to 3.7)	−1.7 (−3.2 to .2)	−3.7 (−5.4 to −2.3)
105.3	.8 (−.4 to 1.8)	−1.5 (−2.7 to −.3)	−3.5 (−4.9 to −1.9)
0.045 inch			
10.7	3.9 (.5 to 7.4)	−2.7 (−5.5 to .6)	−7.0 (−9.9 to −3.8)
18.0	2.2 (.1 to 4.8)	−2.7 (−5.2 to −.6)	−5.4 (−8.1 to −2.9)
105.3	.8 (−.7 to 2.4)	−2.3 (−4.1 to −.8)	−4.8 (−6.6 to −2.6)
0.060 inch			
10.7	4.2 (.0 to 8.8)	−3.0 (−6.7 to 1.1)	−9.9 (−13.4 to −4.5)
18.0	3.3 (.7 to 6.8)	−3.9 (−6.9 to −.8)	−7.8 (−11.5 to −3.0)
105.3	1.0 (−1.5 to 2.9)	−3.2 (−5.2 to−1.0)	−6.4 (−9.0 to −3.4)

[1]Clearance angle 15°; cutting velocity 5 inches per minute.

[2]The first number in each entry is the average cutting force; the numbers following in parentheses are maximum forces (average of five).

[3]The first number in each entry is the average cutting force; the numbers following in parentheses are minimum and maximum forces (average of five).

[4]A negative normal force indicates that the knife tended to lift the workpiece; force was positive when the knife pushed the workpiece.

TABLE 18-22A.—*Tool forces when orthogonally cutting white oak parallel to the grain in the 90-0 direction*[1] (Data from Woodson 1979)

Depth of cut and moisture content (percent)	Rake angle, degrees		
	10	20	30
	----------------Pounds per 0.1 inch of knife----------------		
PARALLEL FORCE[2]			
0.015 inch			
10.9	19.3 (22.7)	16.2 (19.2)	5.4 (9.3)
19.6	12.0 (14.0)	8.5 (10.1)	6.4 (7.6)
75.1	9.4 (11.2)	6.5 (8.9)	4.5 (5.8)
0.030 inch			
10.9	33.3 (42.7)	31.2 (34.6)	10.4 (15.9)
19.6	21.1 (27.7)	18.6 (21.1)	11.2 (12.8)
75.1	16.4 (21.1)	10.5 (12.7)	7.4 (9.6)
0.045 inch			
10.9	41.0 (52.1)	40.8 (45.1)	12.0 (23.5)
19.6	28.0 (36.5)	24.4 (28.0)	15.5 (21.0)
75.1	24.4 (30.2)	14.9 (16.8)	9.5 (11.5)
0.060 inch			
10.9	50.2 (70.3)	54.8 (61.9)	12.7 (22.2)
19.6	35.9 (49.8)	28.2 (32.3)	18.7 (25.7)
75.1	30.8 (38.4)	18.1 (22.3)	11.4 (13.7)
NORMAL FORCE[3,4]			
0.015 inch			
10.9	5.2 (3.4 to 6.3)	1.9 (1.3 to 2.6)	−0.4 (−0.7 to 0.2)
19.6	3.8 (2.6 to 4.3)	1.3 (.8 to 1.7)	−.1 (−.3 to .3)
75.1	2.8 (1.8 to 3.5)	1.1 (.5 to 1.9)	.2 (−.2 to .8)
0.030 inch			
10.9	8.5 (5.7 to 10.8)	3.0 (2.3 to 4.2)	−1.0 (−1.8 to −.3)
19.6	5.9 (3.9 to 7.6)	2.0 (1.5 to 2.4)	−.6 (−1.0 to −.1)
75.1	4.2 (3.0 to 5.3)	1.3 (.6 to 1.9)	−.1 (−.6 to .6)
0.045 inch			
10.9	9.5 (5.5 to 13.7)	3.9 (2.9 to 4.8)	−1.2 (−2.0 to −.4)
19.6	7.2 (4.5 to 8.9)	2.4 (1.7 to 3.2)	−1.0 (−1.3 to −.6)
75.1	6.1 (4.4 to 7.3)	1.5 (.6 to 2.0)	−.3 (−1.1 to .2)
0.060 inch			
10.9	11.6 (7.0 to 16.2)	5.3 (4.2 to 6.4)	−1.3 (−2.4 to −.5)
19.6	9.1 (6.1 to 11.8)	2.9 (2.3 to 3.6)	−1.5 (−2.4 to −.6)
75.1	7.5 (4.1 to 9.6)	1.8 (.7 to 2.8)	−.5 (−1.1 to .2)

[1] Clearance angle 15°; cutting velocity 5 inches per minute.

[2] The first number in each entry is the average cutting force; the numbers following in parentheses are maximum forces (average of five).

[3] The first number in each entry is the average cutting force; the numbers following in parentheses are minimum and maximum forces (average of five).

[4] A negative normal force indicates that the knife tended to lift the workpiece; force was positive when the knife pushed the workpiece.

TABLE 18-22B.—*Tool forces when orthogonally cutting white oak veneer in the 0-90 direction[1]* (Data from Woodson 1979)

Depth of cut and moisture content (percent)	Rake angle, degrees		
	50	60	70
	----------------Pounds per 0.1 inch of knife----------------		
PARALLEL FORCE[2]			
0.015 inch			
11.5	4.0 (8.9)	4.1 (9.0)	4.0 (7.4)
19.3	3.2 (7.4)	2.9 (5.5)	5.5 (12.4)
82.1	2.0 (5.0)	2.0 (4.5)	5.3 (9.6)
0.030 inch			
11.5	5.4 (13.4)	5.1 (12.3)	4.3 (10.4)
19.3	4.5 (10.2)	4.3 (8.9)	3.7 (6.5)
82.1	3.4 (7.0)	2.9 (6.1)	2.8 (4.9)
0.045 inch			
11.5	6.6 (17.6)	6.4 (15.0)	5.7 (12.5)
19.3	5.5 (13.3)	5.3 (11.0)	5.2 (9.8)
82.1	3.9 (9.8)	3.4 (8.2)	3.3 (6.1)
0.060 inch			
11.5	7.7 (20.5)	7.1 (18.4)	6.6 (15.6)
19.3	6.5 (14.5)	6.4 (13.2)	6.4 (12.0)
82.1	4.1 (9.9)	4.4 (8.3)	4.1 (8.5)
NORMAL FORCE[3,4]			
0.015 inch			
11.5	0.2 (−1.0 to 2.1)	0.4 (−0.9 to 1.8)	−0.3 (−2.3 to 2.6)
19.3	−.1 (−1.2 to 2.0)	−.3 (−1.5 to 1.2)	−.5 (−2.9 to 1.4)
82.1	.1 (−.6 to 1.2)	−.2 (−1.2 to .7)	−.9 (−3.2 to 1.4)
0.030 inch			
11.5	.1 (−2.1 to 2.9)	−.1 (−1.5 to 1.9)	1.3 (−1.0 to 3.2)
19.3	−.3 (−2.5 to 2.9)	−1.2 (−3.0 to .8)	.3 (−1.0 to 1.4)
82.1	−.2 (−1.2 to 1.4)	−.6 (−1.9 to .9)	.2 (−.5 to 1.4)
0.045 inch			
11.5	−.4 (−3.2 to 3.3)	−.1 (−2.3 to 2.2)	.5 (−1.6 to 3.2)
19.3	−.5 (−2.3 to 2.3)	−1.3 (−4.2 to .6)	−1.0 (−2.9 to 1.0)
82.1	−.2 (−1.6 to 1.6)	−.8 (−2.7 to .7)	−.2 (−1.5 to 1.2)
0.060 inch			
11.5	−.0 (−2.8 to 4.2)	.0 (−2.7 to 3.0)	−.2 (−3.3 to 3.3)
19.3	−.6 (−3.0 to 2.4)	−1.5 (−4.3 to 1.3)	−1.5 (−3.7 to 1.7)
82.1	−.5 (−2.0 to 1.1)	−1.0 (−2.9 to .7)	.1 (−1.8 to 2.0)

[1]Clearance angle 0°; cutting velocity 5 inches per minute.

[2]The first number in each entry is the average cutting force; the numbers following in parentheses are maximum forces (average of five).

[3]The first number in each entry is the average cutting force; the numbers following in parentheses are minimum and maximum forces (average of five).

[4]A negative normal force indicates that the knife tended to lift the workpiece; force was positive when the knife pushed the workpiece.

TABLE 18-22C.—*Tool forces when orthogonally cutting white oak across the grain in the 90-90 direction[1]* (Data from Woodson 1979)

Depth of cut and moisture content (percent)	Rake angle, degrees		
	20	30	40

----------Pounds per 0.1 inch of knife----------

PARALLEL FORCE[2]

0.015 inch			
11.2	24.6 (32.1)	20.9 (26.9)	16.1 (20.9)
18.8	16.2 (20.3)	14.1 (17.7)	10.0 (14.0)
86.3	13.9 (16.5)	11.6 (13.8)	8.8 (11.1)
0.030 inch			
11.2	39.6 (50.9)	31.6 (43.8)	26.7 (34.0)
18.8	29.9 (37.7)	21.4 (27.8)	17.2 (21.8)
86.3	23.4 (27.3)	17.4 (20.8)	13.2 (16.5)
0.045 inch			
11.2	48.7 (57.7)	43.1 (53.2)	34.3 (44.8)
18.8	34.5 (43.9)	31.0 (37.7)	20.7 (27.5)
86.3	32.8 (38.1)	23.3 (27.9)	16.9 (20.8)
0.060 inch			
11.2	54.6 (69.5)	48.0 (61.2)	41.1 (51.1)
18.8	42.3 (54.1)	36.3 (47.6)	26.8 (35.1)
86.3	38.6 (47.6)	28.0 (33.1)	21.0 (26.5)

NORMAL FORCE[3,4]

0.015 inch			
11.2	1.3 (−0.2 to 3.7)	−1.7 (−3.0 to 0.1)	−3.7 (−5.3 to −1.6)
18.8	1.0 (.3 to 1.8)	−.7 (−1.8 to .4)	−2.1 (−3.4 to −1.1)
83.3	.5 (.0 to 1.1)	−.8 (−1.6 to −.1)	−1.9 (−2.5 to −.9)
0.030 inch			
11.2	.9 (−1.3 to 4.6)	−3.6 (−6.2 to −.7)	−7.2 (−10.0 to −4.4)
18.8	1.1 (−.3 to 3.2)	−2.3 (−4.1 to −.7)	−4.6 (−6.4 to −2.4)
83.1	.4 (−.6 to 1.3)	−1.9 (−3.2 to −.7)	−3.7 (−5.0 to −1.9)
0.045 inch			
11.2	1.0 (−2.4 to 3.8)	−5.5 (−8.5 to −1.2)	−9.9 (−13.4 to −4.7)
18.8	1.1 (−.8 to 4.2)	−3.7 (−6.4 to −.8)	−6.0 (−9.0 to −3.2)
86.3	.4 (−1.0 to 1.6)	−3.2 (−4.9 to −1.4)	−5.6 (−7.2 to −3.1)
0.060 inch			
11.2	1.0 (−2.7 to 6.2)	−6.3 (−10.8 to −.5)	−12.2 (−17.4 to −5.7)
18.8	.9 (−1.0 to 4.5)	−4.1 (−7.5 to −.1)	−8.1 (−12.3 to −3.7)
86.3	.6 (−1.4 to 2.7)	−4.1 (−6.3 to −2.2)	−6.8 (−9.0 to −3.3)

[1] Clearance angle 15°; cutting velocity 5 inches per minute.

[2] The first number in each entry is the average cutting force; the numbers following in parentheses are maximum forces (average of five).

[3] The first number in each entry is the average cutting force; the numbers following in parentheses are minimum and maximum forces (average of five).

[4] A negative normal force indicates that the knife tended to lift the workpiece; force was positive when the knife pushed the workpiece.

TABLE 18-23A.—*Tool forces when orthogonally cutting sweetbay parallel to the grain in the 90-0 direction*[1] (Data from Woodson 1979)

Depth of cut and moisture content (percent)	Rake angle, degrees		
	10	20	30
	----------------*Pounds per 0.1 inch of knife*----------------		
	PARALLEL FORCE[2]		
0.015 inch			
10.7..................	12.1 (15.6)	12.3 (13.2)	8.6 (11.3)
19.2..................	9.9 (11.9)	8.4 (9.6)	6.9 (7.8)
102.7.................	6.3 (8.5)	4.0 (5.7)	3.4 (4.5)
0.030 inch			
10.7..................	20.6 (27.7)	23.4 (25.8)	12.2 (22.1)
19.2..................	18.3 (24.4)	16.0 (17.8)	13.0 (15.7)
102.7.................	12.6 (15.9)	8.8 (11.5)	5.2 (7.3)
0.045 inch			
10.7..................	31.7 (41.1)	36.0 (39.5)	11.0 (29.5)
19.2..................	25.4 (32.7)	23.9 (25.6)	16.1 (21.3)
102.7.................	16.3 (23.0)	12.6 (17.0)	7.0 (10.3)
0.060 inch			
10.7..................	40.8 (50.9)	43.4 (48.2)	10.5 (34.8)
19.2..................	31.3 (43.7)	30.7 (33.5)	19.9 (26.1)
102.7.................	20.6 (27.9)	17.7 (22.3)	8.1 (11.7)
	NORMAL FORCE[3,4]		
0.015 inch			
10.7..................	3.6 (2.2 to 4.9)	1.7 (1.4 to 1.9)	−0.3 (−0.6 to 0.1)
19.2..................	3.4 (2.2 to 4.3)	1.4 (1.0 to 1.7)	.3 (−.1 to .6)
102.7.................	2.7 (1.3 to 4.8)	1.1 (.5 to 2.1)	.4 (−.1 to 1.2)
0.030 inch			
10.7..................	5.7 (3.1 to 7.6)	2.8 (2.5 to 3.2)	−.8 (−1.5 to .1)
19.2..................	5.6 (3.2 to 8.0)	2.1 (1.7 to 2.6)	−.4 (−.9 to .1)
102.7.................	3.3 (2.0 to 5.0)	.9 (.5 to 1.7)	.1 (−.4 to .7)
0.045 inch			
10.7..................	8.2 (5.0 to 11.1)	3.7 (2.7 to 4.5)	−.7 (−2.0 to .2)
19.2..................	6.7 (3.8 to 10.8)	2.7 (2.3 to 3.1)	−.9 (−1.5 to −.3)
102.7.................	4.3 (2.2 to 6.0)	1.1 (.5 to 1.8)	−.3 (−1.0 to .2)
0.060 inch			
10.7..................	10.3 (4.9 to 13.4)	5.1 (4.6 to 5.7)	−.6 (−2.6 to .3)
19.2..................	8.0 (5.0 to 12.2)	3.4 (2.6 to 3.8)	−1.1 (−1.9 to .0)
102.7.................	5.6 (3.0 to 8.6)	1.1 (.5 to 1.8)	−.5 (−1.1 to .3)

[1] Clearance angle 15°; cutting velocity 5 inches per minute.

[2] The first number in each entry is the average cutting force; the numbers following in parentheses are maximum forces (average of five).

[3] The first number in each entry is the average cutting force; the numbers following in parentheses are minimum and maximum forces (average of five).

[4] A negative normal force indicates that the knife tended to lift the workpiece; force was positive when the knife pushed the workpiece.

TABLE 18-23B.—*Tool forces when orthogonally cutting sweetbay veneer in the 0-90 direction[1]* (Data from Woodson 1979)

Depth of cut and moisture content (percent)	Rake angle, degrees		
	50	60	70
	----------------Pounds per 0.1 inch of knife----------------		
PARALLEL FORCE[2]			
0.015 inch			
11.3	2.2 (4.3)	2.0 (3.5)	2.1 (3.2)
18.7	2.1 (4.3)	1.9 (3.1)	2.9 (5.4)
150.8	1.6 (2.8)	1.5 (2.0)	2.5 (3.9)
0.030 inch			
11.3	3.3 (7.0)	2.7 (5.2)	2.5 (4.0)
18.7	2.8 (5.4)	2.5 (4.0)	2.3 (3.7)
150.8	2.1 (3.8)	1.9 (2.9)	1.7 (2.5)
0.045 inch			
11.3	3.8 (8.3)	3.3 (6.4)	3.0 (6.0)
18.7	2.9 (5.5)	3.0 (5.2)	2.9 (4.7)
150.8	2.2 (4.7)	2.3 (3.6)	2.0 (3.0)
0.060 inch			
11.3	4.0 (9.4)	3.5 (7.5)	3.4 (6.7)
18.7	3.5 (6.5)	3.4 (6.2)	3.0 (4.8)
150.8	2.6 (5.4)	2.5 (4.2)	2.3 (3.3)
NORMAL FORCE[3,4]			
0.015 inch			
11.3	0.1 (−0.4 to 1.1)	−0.1 (−0.5 to .5)	−0.1 (−0.7 to 0.6)
18.7	.6 (−.2 to 2.0)	−.2 (−.6 to .5)	−.4 (−1.4 to .6)
150.8	.2 (−.2 to 1.1)	.0 (−.3 to .4)	−.5 (−1.3 to .5)
0.030 inch			
11.3	−.1 (−.8 to 1.2)	−.5 (−1.3 to .2)	.2 (−.4 to 1.1)
18.7	−.3 (−.9 to .8)	−.5 (−1.2 to .2)	.1 (−.5 to .9)
150.8	−.1 (−.7 to .7)	−.3 (−.8 to .3)	.2 (−.3 to .5)
0.045 inch			
11.3	−.2 (−1.2 to 1.0)	−.8 (−1.9 to .1)	−.2 (−1.2 to .6)
18.7	−.4 (−1.1 to .6)	−.9 (−1.8 to .1)	−.4 (−1.2 to .9)
150.8	−.2 (−.8 to .5)	−.6 (−1.2 to .1)	−.4 (−.8 to .1)
0.060 inch			
11.3	−.2 (−1.5 to 1.0)	−.6 (−2.0 to .6)	−.2 (−1.5 to 1.1)
18.7	−.4 (−1.5 to .6)	−1.0 (−2.2 to .2)	−1.0 (−1.9 to .1)
150.8	−.4 (−1.2 to .5)	−.7 (−1.7 to .0)	−.7 (−1.4 to −.1)

[1] Clearance angle 0°; cutting velocity 5 inches per minute.

[2] The first number in each entry is the average cutting force; the numbers following in parentheses are maximum forces (average of five).

[3] The first number in each entry is the average cutting force; the numbers following in parentheses are minimum and maximum forces (average of five).

[4] A negative normal force indicates that the knife tended to lift the workpiece; force was positive when the knife pushed the workpiece.

TABLE 18-23C.—*Tool forces when orthogonally cutting sweetbay across the grain in the 90-90 direction[1]* (Data from Woodson 1979)

Depth of cut and moisture content (percent)	Rake angle, degrees		
	20	30	40
	----------Pounds per 0.1 inch of knife----------		
PARALLEL FORCE[2]			
0.015 inch			
11.0............	16.9 (22.1)	14.5 (19.6)	11.1 (15.1)
17.6............	14.5 (17.4)	11.8 (15.4)	9.4 (11.4)
146.5...........	9.8 (11.6)	9.5 (11.0)	6.8 (8.2)
0.030 inch			
11.0............	26.1 (33.1)	22.7 (30.9)	18.6 (22.0)
17.6............	25.3 (29.4)	19.0 (22.8)	14.7 (18.0)
146.5...........	19.0 (21.4)	15.7 (19.3)	10.1 (12.4)
0.045 inch			
11.0............	28.6 (37.4)	27.4 (37.0)	24.5 (31.2)
17.6............	29.4 (36.0)	25.7 (30.2)	19.0 (23.2)
146.5...........	24.6 (28.5)	19.1 (23.9)	13.2 (15.9)
0.060 inch			
11.0............	33.6 (49.0)	29.5 (38.5)	27.6 (40.4)
17.6............	31.5 (40.0)	30.6 (37.1)	22.7 (28.9)
146.5...........	29.2 (35.8)	22.8 (28.3)	17.4 (22.7)
NORMAL FORCE[3,4]			
0.015 inch			
11.0............	3.2 (1.4 to 6.9)	1.7 (0.0 to 4.3)	−0.2 (−2.2 to 1.9)
17.6............	1.8 (1.1 to 3.5)	.4 (−.4 to 1.4)	−.8 (−1.3 to .1)
146.5...........	.8 (.6 to 1.3)	.1 (−.4 to .6)	−.7 (−1.2 to −.1)
0.030 inch			
11.0............	5.1 (2.1 to 9.4)	2.0 (−1.0 to 6.9)	−2.2 (−3.4 to .9)
17.6............	2.8 (1.3 to 4.9)	−.2 (−1.5 to 1.2)	−2.1 (−3.5 to −.6)
146.5...........	1.1 (.5 to 1.8)	−.8 (−2.0 to −.2)	−1.8 (−2.8 to −1.0)
0.045 inch			
11.0............	7.2 (2.6 to 13.9)	2.8 (−1.3 to 7.9)	−2.4 (−5.6 to 3.9)
17.6............	3.3 (.5 to 5.5)	−1.2 (−3.1 to 1.2)	−3.3 (−5.2 to −.6)
146.5...........	1.2 (.1 to 2.3)	−1.6 (−3.0 to .0)	−2.9 (−4.5 to −1.2)
0.060 inch			
11.0............	7.3 (1.6 to 17.9)	4.5 (−2.1 to 12.2)	−1.9 (−7.5 to 4.8)
17.6............	3.0 (.4 to 6.1)	−1.6 (−4.3 to 1.7)	−4.6 (−7.2 to −.8)
146.5...........	1.3 (−.8 to 3.6)	−2.3 (−4.3 to −.2)	−4.7 (−6.7 to −2.1)

[1]Clearance angle 15°; cutting velocity 5 inches per minute.

[2]The first number in each entry is the average cutting force; the numbers following in parentheses are maximum forces (average of five).

[3]The first number in each entry is the average cutting force; the numbers following in parentheses are minimum and maximum forces (average of five).

[4]A negative normal force indicates that the knife tended to lift the workpiece; force was positive when the knife pushed the workpiece.

TABLE 18-24A.—*Tool forces when orthogonally cutting sweetgum parallel to the grain in the 90-0 direction[1]* (Data from Woodson 1979)

Depth of cut and moisture content (percent)	Rake angle, degrees		
	10	20	30
	----------------Pounds per 0.1 inch of knife-----------------		
PARALLEL FORCE[2]			
0.015 inch			
10.9	13.5 (18.2)	12.6 (13.9)	9.2 (11.8)
20.0	8.3 (9.7)	7.4 (7.9)	6.1 (6.9)
130.0	5.8 (7.3)	4.9 (6.0)	3.5 (4.2)
0.030 inch			
10.9	27.7 (36.2)	24.2 (26.8)	14.1 (23.3)
20.0	15.3 (19.6)	14.1 (15.7)	11.2 (13.2)
130.0	10.7 (14.0)	8.7 (10.6)	6.4 (8.0)
0.045 inch			
10.9	33.7 (45.6)	32.6 (36.7)	12.7 (34.4)
20.0	21.4 (28.6)	20.8 (22.8)	15.2 (19.0)
130.0	14.8 (20.3)	12.2 (15.4)	8.0 (10.7)
0.060 inch			
10.9	43.2 (56.8)	42.0 (49.9)	11.3 (40.5)
20.0	28.2 (37.3)	27.4 (31.5)	16.3 (24.9)
130.0	17.7 (25.5)	16.5 (19.6)	9.1 (12.4)
NORMAL FORCE[3,4]			
0.015 inch			
10.9	4.3 (2.2 to 6.2)	2.1 (1.7 to 2.4)	−0.2 (−0.6 to 0.2)
20.0	2.6 (1.8 to 3.3)	1.1 (.9 to 1.5)	.2 (−.1 to .3)
130.0	2.2 (1.2 to 2.7)	.9 (.8 to 1.1)	.5 (.2 to .8)
0.030 inch			
10.9	7.7 (4.6 to 10.7)	3.3 (2.9 to 3.8)	−.7 (−1.5 to .0)
20.0	4.3 (2.8 to 5.3)	1.7 (1.4 to 2.1)	−.4 (−.7 to .0)
130.0	3.2 (1.9 to 4.4)	1.2 (.9 to 1.5)	.3 (−.2 to .6)
0.045 inch			
10.9	9.3 (5.5 to 12.5)	4.2 (3.6 to 4.8)	−.7 (−2.3 to .2)
20.0	6.0 (3.9 to 8.6)	2.3 (1.9 to 2.6)	−.7 (−1.3 to −.1)
130.0	4.0 (2.3 to 5.3)	1.3 (.9 to 1.8)	.0 (−.5 to .7)
0.060 inch			
10.9	11.6 (7.4 to 15.5)	5.1 (4.1 to 5.9)	−.9 (−2.6 to .0)
20.0	7.4 (4.3 to 10.3)	2.4 (1.9 to 2.9)	−.8 (−1.8 to .1)
130.0	4.8 (2.7 to 6.4)	1.5 (1.2 to 2.0)	.0 (−.6 to .9)

[1] Clearance angle 15°; cutting velocity 5 inches per minute.

[2] The first number in each entry is the average cutting force; the numbers following in parentheses are maximum forces (average of five).

[3] The first number in each entry is the average cutting force; the numbers following in parentheses are minimum and maximum forces (average of five).

[4] A negative normal force indicates that the knife tended to lift the workpiece; force was positive when the knife pushed the workpiece.

TABLE 18-24B.—*Tool forces when orthogonally cutting sweetgum veneer in the 0-90 direction[1]* (Data from Woodson 1979)

Depth of cut and moisture content (percent)	Rake angle, degrees		
	50	60	70
	----------------Pounds per 0.1 inch of knife----------------		
PARALLEL FORCE[2]			
0.015 inch			
11.4	2.2 (3.5)	2.4 (3.3)	3.8 (7.2)
20.4	2.0 (2.8)	1.7 (2.0)	2.9 (4.9)
157.0	1.6 (2.0)	1.3 (1.6)	2.9 (4.1)
0.030 inch			
11.4	3.3 (5.8)	3.3 (5.1)	2.5 (3.6)
20.4	3.0 (4.6)	2.3 (3.4)	2.1 (2.6)
157.0	2.0 (3.2)	1.9 (2.4)	1.7 (2.0)
0.045 inch			
11.4	3.6 (7.5)	3.6 (6.7)	2.8 (4.9)
20.4	3.4 (5.5)	3.1 (4.3)	2.7 (3.6)
157.0	2.6 (4.3)	2.1 (3.0)	2.1 (2.6)
0.060 inch			
11.4	4.0 (9.6)	3.6 (8.0)	3.2 (6.2)
20.4	3.8 (6.8)	3.3 (5.5)	3.3 (4.4)
157.0	3.0 (5.0)	2.5 (4.3)	2.4 (3.5)
NORMAL FORCE[3,4]			
0.015 inch			
11.4	0.0 (−0.3 to 0.5)	−0.2 (−0.5 to 0.2)	−1.2 (−2.5 to 0.1)
20.4	−.1 (−.4 to .3)	−.4 (−.7 to −.1)	−.5 (−1.6 to .6)
157.0	.1 (−.1 to .3)	.0 (−.3 to .3)	−.5 (−1.3 to .3)
0.030 inch			
11.4	−.4 (−1.1 to .3)	−.8 (−1.4 to .0)	−.3 (−.7 to .3)
20.4	−.6 (−1.0 to −.1)	−.9 (−1.3 to −.5)	−.4 (−.6 to −.1)
157.0	−.3 (−.6 to .1)	−.4 (−.8 to .0)	.2 (−.2 to .4)
0.045 inch			
11.4	−.7 (−1.7 to .2)	−.8 (−2.1 to .2)	−.7 (−1.4 to .0)
20.4	−.8 (−1.4 to .0)	−1.3 (−1.9 to −.5)	−.9 (−1.3 to −.2)
157.0	−.4 (−1.0 to .2)	−.7 (−1.2 to −.1)	−.3 (−.7 to .2)
0.060 inch			
11.4	−.7 (−2.0 to .4)	−.9 (−2.4 to .3)	−.8 (−1.9 to .3)
20.4	−1.0 (−1.9 to .1)	−1.5 (−2.4 to −.2)	−1.3 (−2.0 to .1)
157.0	−.6 (−1.3 to .2)	−.9 (−1.6 to .0)	−.8 (−1.2 to .1)

[1] Clearance angle 0°; cutting velocity 5 inches per minute.

[2] The first number in each entry is the average cutting force; the numbers following in parentheses are maximum forces (average of five).

[3] The first number in each entry is the average cutting force; the numbers following in parentheses are minimum and maximum forces (average of five).

[4] A negative normal force indicates that the knife tended to lift the workpiece; force was positive when the knife pushed the workpiece.

TABLE 18-24C.—*Tool forces when orthogonally cutting sweetgum across the grain in the 90-90 direction[1]* (Data from Woodson 1979)

Depth of cut and moisture content (percent)	Rake angle, degrees		
	20	30	40

----------Pounds per 0.1 inch of knife----------

PARALLEL FORCE[2]

0.015 inch			
10.6	15.9 (20.3)	13.6 (16.7)	10.0 (12.1)
19.2	14.0 (16.6)	10.6 (12.6)	7.8 (9.7)
147.9	9.5 (10.8)	8.6 (10.2)	6.9 (7.9)
0.030 inch			
10.6	28.0 (33.8)	21.6 (26.0)	16.4 (21.2)
19.2	24.9 (27.9)	16.8 (18.9)	12.8 (14.8)
147.9	17.4 (20.0)	14.9 (17.6)	10.3 (12.5)
0.045 inch			
10.6	36.1 (44.7)	31.8 (39.0)	20.7 (26.8)
19.2	32.9 (37.0)	23.2 (26.7)	17.1 (21.5)
147.9	24.2 (27.8)	18.3 (22.3)	13.0 (16.9)
0.060 inch			
10.6	40.7 (53.8)	36.3 (45.4)	27.5 (37.8)
19.2	36.1 (42.7)	28.7 (33.6)	21.0 (28.5)
147.9	30.0 (53.9)	22.4 (27.1)	15.3 (20.5)

NORMAL FORCE[3,4]

0.015 inch			
10.6	1.8 (0.7 to 2.8)	0.5 (−0.2 to 1.8)	−1.3 (−2.0 to −0.4)
19.2	1.8 (1.1 to 2.6)	.2 (−.5 to .6)	−1.0 (−1.4 to −.4)
147.9	.9 (.6 to 1.3)	.0 (−.4 to .4)	−.7 (−1.0 to −.1)
0.030 inch			
10.6	4.0 (1.7 to 6.0)	.2 (−1.0 to 1.7)	−3.1 (−4.4 to −1.6)
19.2	2.2 (1.2 to 3.5)	−.7 (−1.8 to .4)	−2.4 (−3.2 to −1.4)
147.9	.9 (.6 to 1.6)	−.8 (−1.8 to .0)	−2.0 (−2.9 to −1.0)
0.045 inch			
10.6	5.6 (2.5 to 9.3)	−.5 (−2.7 to 2.0)	−4.3 (−6.2 to −1.8)
19.2	3.1 (1.6 to 5.0)	−1.7 (−3.2 to −.2)	−3.7 (−5.3 to −1.7)
147.9	.8 (−.2 to 1.9)	−1.5 (−2.9 to −.3)	−3.3 (−4.9 to −1.4)
0.060 inch			
10.6	5.2 (1.6 to 9.4)	−.1 (−2.5 to 3.4)	−5.7 (−8.9 to −1.8)
19.2	2.6 (.2 to 4.8)	−2.6 (−4.7 to −.6)	−5.2 (−7.6 to −1.7)
147.9	.6 (−1.2 to 2.2)	−2.3 (−4.5 to −.5)	−4.7 (−7.0 to −.8)

[1]Clearance angle 15°; cutting velocity 5 inches per minute.

[2]The first number in each entry is the average cutting force; the numbers following in parentheses are maximum forces (average of five).

[3]The first number in each entry is the average cutting force; the numbers following in parentheses are minimum and maximum forces (average of five).

[4]A negative normal force indicates that the knife tended to lift the workpiece; force was positive when the knife pushed the workpiece.

TABLE 18-25A.—*Tool forces when orthogonally cutting black tupelo parallel to the grain in the 90-0 direction[1]* (Data from Woodson 1979)

Depth of cut and moisture content (percent)	Rake angle, degrees		
	10	20	30
	---------------- Pounds per 0.1 inch of knife ----------------		
PARALLEL FORCE[2]			
0.015 inch			
10.9................	15.0 (20.2)	11.6 (13.4)	9.0 (12.3)
19.1................	9.9 (14.8)	7.5 (9.0)	7.5 (8.9)
113.0................	7.3 (9.6)	5.5 (6.5)	4.1 (4.6)
0.030 inch			
10.9................	26.8 (35.3)	26.7 (30.6)	14.7 (23.1)
19.1................	17.8 (24.8)	15.6 (18.4)	12.7 (14.7)
113.0................	14.0 (18.9)	9.1 (10.7)	6.8 (8.0)
0.045 inch			
10.9................	35.3 (47.7)	37.1 (41.4)	14.7 (31.2)
19.1................	25.8 (34.4)	21.2 (24.2)	14.7 (18.8)
113.0................	17.2 (25.2)	11.2 (14.3)	9.3 (11.6)
0.060 inch			
10.9................	43.3 (58.7)	48.5 (52.8)	15.6 (40.7)
19.1................	34.8 (43.8)	28.3 (31.0)	18.0 (22.9)
113.0................	21.2 (27.4)	14.8 (19.9)	13.3 (16.3)
NORMAL FORCE[3,4]			
0.015 inch			
10.9................	4.9 (2.9 to 7.0)	1.7 (1.4 to 2.2)	−0.1 (−0.5 to 0.3)
19.1................	4.2 (1.8 to 6.5)	1.3 (.9 to 2.1)	.1 (−.3 to .5)
113.0................	3.0 (1.5 to 5.2)	1.2 (.8 to 1.4)	.5 (.2 to .7)
0.030 inch			
10.9................	7.8 (4.1 to 10.3)	3.3 (2.8 to 3.8)	−.5 (−1.4 to .2)
19.1................	6.4 (2.8 to 10.1)	2.0 (1.4 to 2.7)	−.3 (−.8 to .1)
113.0................	4.7 (2.3 to 6.5)	1.2 (.7 to 1.6)	.2 (−.1 to .4)
0.045 inch			
10.9................	9.9 (6.0 to 13.0)	4.4 (3.7 to 5.0)	−.7 (−2.3 to .5)
19.1................	8.0 (4.4 to 11.4)	2.4 (2.0 to 3.2)	−.4 (−1.1 to .2)
113.0................	5.4 (2.7 to 8.0)	1.2 (.7 to 1.6)	.0 (−.3 to .4)
0.060 inch			
10.9................	11.3 (7.2 to 15.5)	5.7 (5.1 to 6.1)	−1.0 (−3.0 to .3)
19.1................	9.9 (5.2 to 13.0)	3.2 (2.5 to 3.9)	−.9 (−1.7 to −.1)
113.0................	6.3 (3.3 to 8.9)	1.2 (.5 to 2.2)	−.6 (−1.0 to −.1)

[1]Clearance angle 15°; cutting velocity 5 inches per minute.

[2]The first number in each entry is the average cutting force; the numbers following in parentheses are maximum forces (average of five).

[3]The first number in each entry is the average cutting force; the numbers following in parentheses are minimum and maximum forces (average of five).

[4]A negative normal force indicates that the knife tended to lift the workpiece; force was positive when the knife pushed the workpiece.

TABLE 18-25B.—*Tool forces when orthogonally cutting black tupelo veneer in the 0-90 direction[1]* (Data from Woodson 1979)

Depth of cut and moisture content (percent)	Rake angle, degrees		
	50	60	70

----------------Pounds per 0.1 inch of knife----------------

PARALLEL FORCE[2]

0.015 inch			
11.4	3.1 (5.4)	2.8 (4.2)	3.7 (5.7)
19.5	2.7 (3.8)	2.3 (3.4)	3.8 (7.5)
124.7	1.8 (2.5)	1.5 (2.1)	3.7 (5.4)
0.030 inch			
11.4	4.5 (7.8)	3.9 (6.6)	3.3 (4.9)
19.5	3.8 (5.5)	3.4 (4.8)	2.8 (3.7)
124.7	2.7 (3.9)	2.2 (3.0)	2.0 (2.4)
0.045 inch			
11.4	5.1 (10.3)	4.6 (9.2)	4.4 (7.0)
19.5	4.8 (7.5)	4.1 (6.3)	3.3 (4.7)
124.7	3.0 (5.2)	2.8 (4.2)	2.6 (3.4)
0.060 inch			
11.4	5.7 (12.4)	5.4 (10.4)	5.2 (10.0)
19.5	5.3 (8.8)	4.8 (7.5)	4.1 (6.0)
124.7	3.5 (6.4)	3.1 (5.2)	3.0 (4.1)

NORMAL FORCE[3,4]

0.015 inch			
11.4	−0.4 (−1.0 to 0.3)	−0.5 (−0.9 to .0)	−0.5 (−1.6 to 0.7)
19.5	−.3 (−.7 to .0)	−.7 (−1.0 to .0)	−.8 (−2.6 to .5)
124.7	−.1 (−.3 to .2)	−.3 (−.6 to .1)	−1.3 (−2.4 to −.1)
0.030 inch			
11.4	−.9 (−1.9 to .2)	−1.1 (−2.2 to −.1)	−.6 (−1.5 to .3)
19.5	−1.0 (−1.6 to −.1)	−1.4 (−2.1 to −.6)	−.8 (−1.2 to −.1)
124.7	−.5 (−.9 to .1)	−.8 (−1.3 to −.3)	−.1 (−.5 to .4)
0.045 inch			
11.4	−1.1 (−2.3 to .0)	−1.2 (−2.6 to .1)	−1.1 (−2.3 to .0)
19.5	−1.1 (−2.3 to −.2)	−1.8 (−2.7 to −.5)	−1.5 (−2.2 to −.5)
124.7	−.7 (−1.3 to .2)	−1.1 (−1.8 to −.3)	−.9 (−1.3 to −.3)
0.060 inch			
11.4	−1.2 (−3.0 to .3)	−1.3 (−2.9 to .3)	−1.2 (−2.9 to .7)
19.5	−1.5 (−2.8 to −.2)	−2.0 (−3.3 to −.5)	−2.0 (−3.0 to −.7)
124.7	−.8 (−1.7 to .2)	−1.5 (−2.5 to −.3)	−1.3 (−2.0 to −.3)

[1] Clearance angle 0°; cutting velocity 5 inches per minute.

[2] The first number in each entry is the average cutting force; the numbers following in parentheses are maximum forces (average of five).

[3] The first number in each entry is the average cutting force; the numbers following in parentheses are minimum and maximum forces (average of five).

[4] A negative normal force indicates that the knife tended to lift the workpiece; force was positive when the knife pushed the workpiece.

TABLE 18-25C.—*Tool forces when orthogonally cutting black tupelo across the grain in the 90-90 direction[1]* (Data from Woodson 1979)

Depth of cut and moisture content (percent)	Rake angle, degrees		
	20	30	40
	----------Pounds per 0.1 inch of knife----------		
PARALLEL FORCE[2]			
0.015 inch			
10.4............	20.1 (27.8)	17.0 (20.7)	10.9 (13.7)
19.0............	15.6 (17.7)	12.2 (14.2)	9.9 (11.6)
125.2............	11.5 (13.5)	9.5 (11.0)	7.5 (8.7)
0.030 inch			
10.4............	34.4 (44.6)	27.4 (34.4)	20.5 (25.5)
19.0............	30.0 (35.1)	20.5 (24.3)	15.3 (18.2)
125.2............	19.8 (22.7)	15.0 (18.3)	10.9 (12.4)
0.045 inch			
10.4............	45.9 (61.4)	37.3 (45.2)	28.3 (34.4)
19.0............	37.8 (45.2)	25.6 (30.3)	20.1 (24.2)
125.2............	28.1 (32.3)	19.5 (23.4)	14.1 (17.0)
0.060 inch			
10.4............	59.5 (71.8)	46.0 (56.7)	35.9 (45.0)
19.0............	45.9 (56.6)	34.4 (41.9)	25.0 (31.4)
125.2............	34.2 (39.5)	24.7 (31.5)	17.5 (22.2)
NORMAL FORCE[3,4]			
0.015 inch			
10.4............	2.6 (0.8 to 7.8)	0.4 (−0.6 to 2.9)	−1.5 (−2.3 to −0.4)
19.0............	1.6 (1.1 to 2.1)	−.1 (−.6 to .5)	−1.3 (−1.9 to −.7)
125.2............	.8 (.6 to 1.3)	−.1 (−.5 to .6)	−1.2 (−1.8 to −.8)
0.030 inch			
10.4............	3.5 (.8 to 7.2)	−.7 (−2.6 to 3.3)	−4.6 (−6.2 to −1.4)
19.0............	2.1 (1.3 to 3.7)	−1.3 (−2.2 to −.2)	−3.2 (−4.3 to −2.2)
125.2............	.9 (.4 to 1.8)	−1.0 (−2.1 to .1)	−2.6 (−3.4 to −1.7)
0.045 inch			
10.4............	4.6 (1.3 to 9.1)	−1.5 (−4.6 to 3.9)	−6.1 (−8.7 to −1.6)
19.0............	2.3 (1.2 to 4.0)	−2.4 (−4.4 to −.5)	−4.8 (−6.7 to −2.9)
125.2............	.8 (−.4 to 1.7)	−2.3 (−3.4 to −.7)	−3.9 (−5.6 to −2.6)
0.060 inch			
10.4............	4.9 (−.2 to 12.3)	−3.1 (−6.3 to 5.4)	−8.4 (−12.7 to −3.5)
19.0............	1.7 (.1 to 4.7)	−3.6 (−6.4 to −.7)	−6.9 (−9.7 to −3.3)
125.2............	.6 (−.5 to 2.6)	−3.2 (−5.9 to −1.5)	−5.0 (−7.1 to −2.7)

[1] Clearance angle 15°; cutting velocity 5 inches per minute.

[2] The first number in each entry is the average cutting force; the numbers following in parentheses are maximum forces (average of five).

[3] The first number in each entry is the average cutting force; the numbers following in parentheses are minimum and maximum forces (average of five).

[4] A negative normal force indicates that the knife tended to lift the workpiece; force was positive when the knife pushed the workpiece.

TABLE 18-26A.—*Tool forces when orthogonally cutting yellow-poplar parallel to the grain in the 90-0 direction[1]* (Data from Woodson 1979)

Depth of cut and moisture content (percent)	Rake angle, degrees		
	10	20	30
	----------------Pounds per 0.1 inch of knife----------------		
PARALLEL FORCE[2]			
0.015 inch			
10.4................	9.8 (12.9)	10.1 (12.2)	6.9 (10.3)
19.2................	8.6 (10.7)	6.9 (7.8)	6.4 (7.4)
124.4................	6.5 (7.9)	5.6 (7.1)	4.2 (5.4)
0.030 inch			
10.4................	20.1 (25.5)	18.6 (25.2)	10.7 (19.8)
19.2................	15.4 (20.2)	13.9 (15.9)	11.7 (15.0)
124.4................	12.8 (16.8)	10.2 (11.8)	7.5 (9.7)
0.045 inch			
10.4................	28.4 (36.4)	25.8 (35.3)	6.9 (24.6)
19.2................	21.0 (26.5)	20.2 (23.1)	14.3 (21.8)
124.4................	16.6 (20.4)	14.7 (17.6)	8.9 (12.0)
0.060 inch			
10.4................	37.4 (45.7)	30.5 (43.3)	6.6 (30.2)
19.2................	27.3 (34.2)	26.1 (30.7)	16.9 (29.2)
124.4................	20.3 (26.6)	19.4 (22.7)	10.6 (17.4)
NORMAL FORCE[3,4]			
0.015 inch			
10.4................	2.6 (1.3 to 4.0)	1.4 (1.0 to 2.0)	−0.3 (−0.7 to 0.0)
19.2................	3.2 (2.3 to 4.2)	1.3 (.9 to 1.6)	.4 (.0 to 1.0)
124.4................	3.6 (2.3 to 4.9)	1.3 (.7 to 2.0)	1.0 (.3 to 1.7)
0.030 inch			
10.4................	4.7 (2.6 to 7.0)	2.3 (1.6 to 2.9)	−.7 (−1.7 to .0)
19.2................	4.8 (3.0 to 6.4)	1.8 (1.3 to 2.4)	−.1 (−.6 to .7)
124.4................	5.7 (2.4 to 8.6)	1.6 (1.0 to 2.5)	.8 (−.0 to 2.1)
0.045 inch			
10.4................	6.0 (2.6 to 8.8)	3.4 (2.6 to 4.2)	−.4 (−2.2 to .3)
19.2................	6.6 (3.2 to 9.5)	2.2 (1.8 to 2.7)	−.3 (−1.3 to .5)
124.4................	5.8 (2.6 to 8.7)	1.6 (.9 to 2.3)	.5 (−.2 to 1.4)
0.060 inch			
10.4................	8.1 (3.0 to 11.7)	3.4 (2.5 to 4.1)	−.5 (−2.6 to .3)
19.2................	7.7 (3.9 to 10.9)	2.9 (2.2 to 3.8)	−.4 (−1.8 to .8)
124.4................	7.2 (4.0 to 8.9)	1.7 (.9 to 2.5)	.7 (−.4 to 1.6)

[1]Clearance angle 15°; cutting velocity 5 inches per minute.

[2]The first number in each entry is the average cutting force; the numbers following in parentheses are maximum forces (average of five).

[3]The first number in each entry is the average cutting force; the numbers following in parentheses are minimum and maximum forces (average of five).

[4]A negative normal force indicates that the knife tended to lift the workpiece; force was positive when the knife pushed the workpiece.

TABLE 18-26B.—*Tool forces when orthogonally cutting yellow-poplar veneer in the 0-90 direction[1]* (Data from Woodson 1979)

Depth of cut and moisture content (percent)	Rake angle, degrees		
	50	60	70

----------------Pounds per 0.1 inch of knife----------------

	PARALLEL FORCE[2]		
0.015 inch			
10.6	1.7 (3.5)	1.6 (2.8)	3.4 (5.4)
18.8	1.6 (3.4)	1.4 (2.5)	2.4 (5.0)
180.9	1.6 (2.3)	1.2 (1.8)	2.2 (3.7)
0.030 inch			
10.6	2.2 (4.7)	1.9 (3.9)	1.9 (3.3)
18.8	2.1 (4.0)	2.0 (3.5)	2.0 (3.0)
180.9	1.8 (3.1)	1.7 (2.3)	1.5 (2.2)
0.045 inch			
10.6	2.6 (6.1)	2.2 (4.9)	2.3 (4.4)
18.8	2.7 (5.4)	2.3 (4.2)	2.1 (3.4)
180.9	2.1 (3.7)	1.7 (2.7)	1.8 (2.4)
0.060 inch			
10.6	2.7 (6.8)	2.2 (5.7)	2.4 (4.7)
18.8	3.1 (5.8)	2.9 (5.3)	2.5 (4.1)
180.9	2.2 (4.3)	2.0 (3.4)	2.1 (3.0)
	NORMAL FORCE[3,4]		
0.015 inch			
10.6	0.2 (−0.2 to 0.8)	0.1 (−0.4 to .6)	−0.3 (−1.8 to 1.2)
18.8	.3 (−.1 to 1.1)	.3 (−.3 to 1.0)	.3 (−.6 to 1.4)
180.9	.5 (.0 to 1.0)	.3 (−.2 to .8)	.0 (−1.0 to .9)
0.030 inch			
10.6	.0 (−.6 to .5)	−.3 (−.8 to .2)	.2 (−.5 to .9)
18.8	−.1 (−.5 to .6)	−.4 (−.9 to .2)	.1 (−.4 to .7)
180.9	.1 (−.4 to .7)	−.1 (−.6 to .2)	.3 (.0 to .7)
0.045 inch			
10.6	−.1 (−.9 to .6)	−.3 (−1.0 to .3)	−.3 (−1.0 to .5)
18.8	−.3 (−1.0 to .5)	−.5 (−1.2 to .0)	−.4 (−.9 to .4)
180.9	.1 (−.5 to .7)	−.3 (−.9 to .2)	−.1 (−.5 to .3)
0.060 inch			
10.6	−.1 (−1.4 to 1.2)	−.5 (−1.6 to .4)	−.4 (−1.3 to .7)
18.8	−.4 (−1.2 to .5)	−.8 (−1.6 to .3)	−.6 (−1.2 to .1)
180.9	−.1 (−.7 to .6)	−.6 (−1.2 to .0)	−.3 (−.8 to .4)

[1]Clearance angle 0°; cutting velocity 5 inches per minute.

[2]The first number in each entry is the average cutting force; the numbers following in parentheses are maximum forces (average of five).

[3]The first number in each entry is the average cutting force; the numbers following in parentheses are minimum and maximum forces (average of five).

[4]A negative normal force indicates that the knife tended to lift the workpiece; force was positive when the knife pushed the workpiece.

TABLE 18-26C.—*Tool forces when orthogonally cutting yellow-poplar across the grain in the 90-90 direction[1]* (Data from Woodson 1979)

Depth of cut and moisture content (percent)	Rake angle, degrees		
	20	30	40

----------Pounds per 0.1 inch of knife----------

PARALLEL FORCE[2]

0.015 inch			
10.8..........	12.1 (16.7)	19.8 (23.3)	7.7 (10.8)
17.7..........	12.3 (14.7)	9.6 (13.4)	8.2 (10.4)
165.5..........	8.5 (10.2)	7.8 (9.0)	5.8 (7.0)
0.030 inch			
10.8..........	18.7 (26.4)	16.7 (22.9)	12.3 (15.8)
17.7..........	22.1 (27.4)	15.0 (20.1)	12.3 (15.1)
165.5..........	15.5 (18.2)	12.8 (15.2)	8.1 (9.8)
0.045 inch			
10.8..........	24.1 (37.2)	19.1 (24.2)	17.2 (22.5)
17.7..........	26.2 (32.7)	20.3 (25.0)	14.7 (19.0)
165.5..........	21.1 (24.3)	16.3 (19.5)	10.6 (14.0)
0.060 inch			
10.8..........	28.2 (46.3)	18.9 (28.8)	20.5 (25.1)
17.7..........	27.4 (35.6)	24.9 (29.2)	18.8 (24.3)
165.5..........	25.7 (29.2)	20.4 (25.4)	12.9 (16.6)

NORMAL FORCE[3,4]

0.015 inch			
10.8..........	3.3 (1.2 to 6.9)	1.2 (−0.2 to 3.0)	−0.3 (−1.5 to 1.4)
17.7..........	1.6 (.9 to 2.6)	.2 (−.3 to .8)	−.6 (−1.3 to .1)
165.5..........	.8 (.6 to 1.2)	.3 (.0 to .8)	−.5 (−.9 to .0)
0.030 inch			
10.8..........	4.1 (1.4 to 7.6)	1.3 (−.6 to 4.0)	−1.0 (−2.4 to .4)
17.7..........	2.2 (1.1 to 3.8)	−.3 (−1.1 to .6)	−1.8 (−2.8 to −.9)
165.5..........	1.3 (.7 to 1.7)	−.1 (−.7 to .6)	−1.1 (−2.1 to −.5)
0.045 inch			
10.8..........	5.1 (1.9 to 9.4)	2.0 (−.4 to 5.4)	−2.1 (−4.2 to .6)
17.7..........	2.7 (.8 to 4.8)	−.8 (−2.3 to .3)	−2.7 (−4.5 to −1.2)
165.5..........	1.4 (.5 to 2.1)	−.5 (−1.6 to .3)	−2.1 (−3.3 to −1.0)
0.060 inch			
10.8..........	5.0 (.4 to 10.7)	3.4 (−.7 to 10.4)	−2.8 (−5.9 to 1.1)
17.7..........	2.9 (.2 to 4.9)	−1.4 (−2.7 to −.1)	−3.8 (−6.0 to −1.5)
165.5..........	1.3 (.1 to 2.6)	−.9 (−2.1 to .3)	−3.0 (−4.9 to −1.2)

[1]Clearance angle 15°; cutting velocity 5 inches per minute.

[2]The first number in each entry is the average cutting force; the numbers following in parentheses are maximum forces (average of five).

[3]The first number in each entry is the average cutting force; the numbers following in parentheses are minimum and maximum forces (average of five).

[4]A negative normal force indicates that the knife tended to lift the workpiece; force was positive when the knife pushed the workpiece.

Scraping in the 90-0 direction.—Varbanov (1972) employed an electrically heated knife with negative rake angle to scrape surfaces of oak (*Quercus* sp.) and beech (*Fagus* sp.); surface quality obtained in a single pass was as good as that obtained by sanding with abrasive belts. Optimum conditions were about as follows: rake angle, -10 to -14 degrees; depth of cut, 0.1 to 0.5 mm; cutting velocity, 12 to 30m/min.' knife temperature 300°C.

Cutting excelsior.—Excelsior consists of thin, narrow, ribbon-like strands of wood. It is usually cut from ***Populus*** sp. but can be made from any wood that is light in weight and color, soft, tough, straight grained, absorbent, and free from odor. Among the pine-site hardwoods, yellow-poplar should be suitable. Excelsior is orthogonally cut on simple machines in which a number of small knives are mounted in parallel in a heavy frame to score the face of bolts to the ribbon width desired; a planing knife cutting in the 90-0 direction shaves the surface to the depth of scoring, thus producing excelsior.

PLANING: 90-0 TO 45 DIRECTION

Ideally, the carpenter planes parallel to the grain in true 90-0 direction (fig. 18-1). In practice, however, he frequently must plane against or with the slope of the grain. Stewart (1969) showed that on white ash at 7 percent moisture content, parallel cutting forces increase as grain angle varies from 0°—regardless of whether the tool moves with or against its slope. He additionally found that when cutting against the grain the cutting coefficient of friction increased significantly with grain slope at angles up to about 45°; when cutting with the grain, however, the slope of the grain had little effect on cutting coefficient of friction. A change in cutting coefficient of friction alters the relationship between normal and parallel cutting forces (eq. 18-1).

When orthogonally planing white ash against the grain at 7-percent moisture content, Stewart (1971a) concluded that chipped grain does not occur at any grain angle when cutting with a rake angle of 20° or less. A rake angle of 25° produced best surfaces over the widest range of grain angles.

VENEER CUTTING: 0-90 DIRECTION[2]

To study chip formation in pine-site hardwoods and cutting forces required Woodson (1979) orthogonally cut room-temperature specimens ¼-inch thick at 5 inches per minute in the 0-90 direction as shown in figure 18-1 (left). His experiments simulated slicing veneer across growth rings rather than rotary peeling veneer concentric with growth rings. Kivimaa (1950) found that Finnish birch (*Betula* sp.) at 12-percent moisture content cut in this mode with 35° rake angle required 12 percent less parallel cutting force for 0.1 mm-thick veneer than if cut as shown in figure 18-1 (right); he concluded that the influence of wood rays on birch mechanical properties accounted for the difference.

---

[2]This subsection is based on data from Woodson (1979).

Woodson's (1979) specimens were taken from five 6-inch trees of each of 22 species grown on southern pine sites. He measured cutting forces as affected by various rake angles (50, 60, and 70°), depths of cut (0.015, .030, .045, and .060 inch), and moisture contents (about 10 percent, about 20 percent, and saturated). Saturated moisture contents and specific gravities of the specimens are given in table 18-4. Clearance angle on all knives was 0 degrees. These tests differed from commercial veneer peeling because the wood was cut at low temperature (unsteamed) and no restraint was applied above the knife (see section 18-23).

Chip formation.—Under favorable conditions, chips formed by cutting in the 0-90 mode emerge as **continuous veneer**, defined by Leney (1960) as an unbroken sheet in which the original wood structure is essentially unchanged by the cutting process. With a suitably sharp knife and a thin cut, continuous veneer with relatively smooth unbroken surfaces on both sides can be cut (fig. 18-12).

For 0-90 cutting, McMillin (1958) has accounted for the formation of veneer of various types in terms of the mechanical properties of the wood.

Following initial incision, the cells above the cutting edge must move upward along the face of the knife. Being restrained by the wood above them they are compressed, developing a force normal to the knife face, together with a frictional force along the face of the knife. As the cut proceeds, the forces reach a maximum when the veneer begins to bend as a cantilever beam. This bending deformation creates a zone of maximum tension above the cutting edge and causes compression near the top surface of the forming veneer.

As depth of cut is increased, rake angle decreased, or cutting edge dulled, critical zones of stress develop as depicted in figure 18-13. In Zone 1 maximum tension due to bending develops close to the cutting edge and at right angles to the long axis of the zone as drawn; failure occurs as a **tension check**. In Zone 2 the frictional force along the face of the knife may cause a more or less horizontal shear plane to develop between the compressed cells at the knife chip interface and the resisting wood above them. In Zone 3, the cutting edge deflects the wood elements into a slight bulge preceding the edge, and the somewhat compacted cell walls may fail in tension either above or below the cutting plane (**compression tearing**).

Stewart (1979a) concluded that although the complexity of the 0-90 cutting situation precludes complete analysis, chip type can be approximately determined through knowledge of tool forces and strength properties of the wood.

Woodson (1979) found that thinner cuts and cuts in saturated wood generally produced continuous chips with the knife continuously in the wood (fig. 18-12). Conversely, the thickest cuts and cuts in dry wood generally produced discontinuous chips (fig. 18-14 top) with the knife disengaged from the wood much of the time. Thus, discontinuous chips could produce higher maximum parallel forces and yet have lower average parallel forces. Averaged over all depths of cut, diffuse-porous hardwoods generally yielded chips that were slightly more continuous than ring-porous hardwoods.

Figure 18-12.—Continuous veneer in saturated red maple cut 0.030 inch thick 70° rake angle and 0° clearance angle. (Photo from Woodson 1979.)

When cut radially, ring-porous hardwoods exhibit zones of weakness in thin-walled earlywood cells. As the knife edge approaches the earlywood zone, stresses build until the chip fails as a cantilever beam (18-14 top). These zones of weakness are not present if the cuts are made tangentially (fig. 18-14 bottom). For consistent sampling in his study of 0-90 cutting, therefore, Woodson cut across the grain as in veneer slicing, not tangentially, as in rotary veneer cutting.

Cutting forces.—Woodson (1979) found that average parallel cutting force was consistently greatest for the thinnest cuts, i.e., 0.015 inch (fig. 18-15), but at the other three depths of cut studied, average parallel force increased with

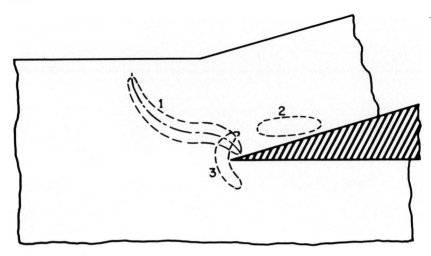

Figure 18-13.—Critical zones of stress in veneer cut without a nosebar. 1, tension; 2, shear; 3, compression tearing.

deeper cuts. Average parallel force also increased with specific gravity, but decreased with increased rake angle and moisture content (table 18-2).

Among the species studied, average normal forces were not correlated with specific gravity, moisture content, or rake angle, but increased with deeper cuts.

To summarize the effects of the principal factors (depth of cut, moisture content, and rake angle), the tool forces were averaged over all other variables (table 18-27). In general, maximum parallel forces for veneer cutting were about twice as large as the average forces. Maximum and minimum forces generally were similar in magnitude but opposite in sign.

TABLE 18-27.—*Tool forces per 0.1 inch of knife when orthogonally cutting veneer in the 0-90 direction at selected depths of cut, moisture contents, and rake angles. Values are averaged over all variables and over 22 pine-site species of hardwoods[1] Data from Woodson 1979)*

Principal factor	Parallel force		Normal force		
	Average	Maximum	Average	Maximum	Minimum
	--------Pounds--------				
Depth of cut, inches					
.015	3.1	6.2	0.0	1.3	−1.0
.030	3.5	7.2	−.2	1.2	−1.3
.045	4.2	9.0	−.5	1.2	−1.9
.060	4.9	10.6	−.7	1.3	−2.4
Moisture content, percent					
10.9	4.7	10.8	0.0	2.0	−1.6
18.9	4.0	8.0	−.6	1.0	−1.9
104.3	3.1	6.0	−.4	.7	−1.4
Rake, angle, degrees					
50	4.1	9.3	−0.2	1.5	−1.5
60	3.9	8.0	−.5	1.0	−1.9
70	3.9	7.4	−.3	1.2	−1.6

[1]Cutting velocity was 5 inches per minute.

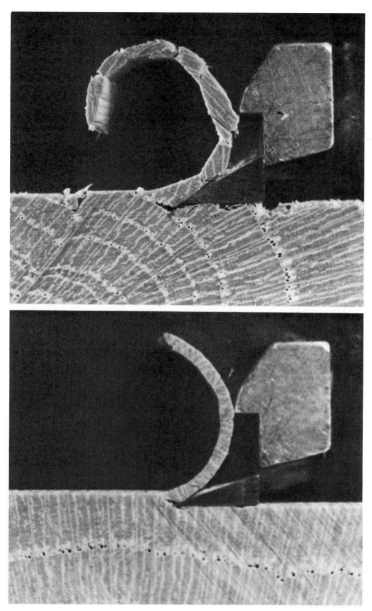

Figure 18-14.—Cutting orthogonally in the 0-90 direction. (Top) Cantilever beam type failure in dry (11.4 percent) cherrybark oak cut across the growth rings 0.045-inch deep with 60° rake angle and 0° clearance angle. (Bottom) Continuous veneer in dry (11.4 percent) cherrybark oak cut tangentially parallel to growth rings 0.060-inch deep with 70° rake angle and 0° clearance angle. (Photos from Woodson 1979.)

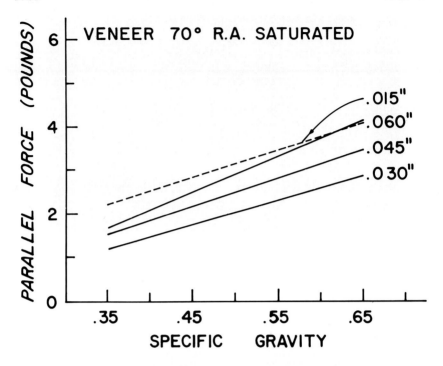

Figure 18-15.—Relationship of average parallel cutting force per 0.1-inch knife length with specific gravity (ovendry weight and green volume) for veneer cutting, in the 0-90 direction, saturated wood of 22 species with a 70° rake angle (0° clearance angle) at four depths of cut. The dotted line indicates regression significant at .05 level; others are significant at .01 level. (Drawing after Woodson 1979.)

Tables 18-5B, 18-6B, etc., through 18-26B contain Woodson't data on tool forces when cutting veneer in the 0-90 direction, as affected by species, depth of cut, moisture content, and rake angle. Table 18-4 provides a species index to these tables.

CROSSCUTTING: 90-90 DIRECTION[2]

Because gangsaws, bandsaws, tenoners, and end-grain shapers cut across the grain in the 90-90 direction (fig. 18-1), this mode of cutting is of practical interest to woodworkers.

McKenzie (1961, 1962) employed a tool dynamometer, slow-velocity cutting, and microscopic examination of forming chips to develop a mathematical model to predict patterns of chip formation. His method is summarized in Koch (1964a, p. 88-109).

To study chip formation in pine-site hardwoods and cutting forces required, Woodson (1979) orthogonally cut room-temperature specimens ⅛-inch thick at 5 inches per minute in the 90-90 direction as shown in figure 18-1 (left). His specimens were taken from five 6-inch trees from each of 22 species grown on southern pine sites. He measured cutting forces as affected by various rake angles (20, 30, and 40°) depths of cut (0.015, .030, .045, and .060 inch), and moisture contents (about 10 percent, about 20 percent, and saturated). Saturated moisture contents and specific gravities of the specimens are given in table 18-4. Clearance angle on all knives was 15 degrees.

Chip formation.—Uniform chips and smooth surfaces were produced with high rake angles in all types of wood (fig. 18-16); sharp knives promote good surfaces. Low rake angles in dry wood produced very rough surfaces with considerable failure beneath the plane of the knife cut (fig. 18-17 top). Low-density species, of all moisture contents, cut with low rake angles produced chips less than the desired thickness because the knife skimmed over the surface and did little cutting (fig. 18-17 bottom). Dull knives yield poor surfaces.

Cutting forces.—Average parallel force increased with increasing depth of cut and specific gravity (fig. 18-18, tables 18-2 and 18-28). Generally, average parallel forces decreased with increasing moisture content. Maximum parallel forces were about 23 percent greater than average parallel forces.

Usually, average normal forces were positive for wood specific gravities below 0.50 and negative for specific gravities above 0.50. For all specific gravities, normal force increased with depth of cut. Average normal force changed from positive to negative as rake angle increased from 20 to 30 degrees (table 18-28).

Tables 18-5C, 18-6C, etc. through 18-26C contain Woodson's data on tool forces when crosscutting orthogonally in the 90-90 direction, as affected by species, depth of cut, moisture content, and rake angle. Table 18-4 provides a species index to these tables.

TABLE 18-28.—*Tool forces per 0.1 inch of knife when orthogonally crosscutting in the 90-90 direction with various depths of cut, moisture contents, and rake angles. Values are averaged over all variables and over 22 pine-site species of hardwoods[1] Data from Woodson 1979)*

Principal factor	Parallel force		Normal force		
	Average	Maximum	Average	Maximum	Minimum
	------------------------------Pounds------------------------------				
Depth of cut, inches					
.015....................	14.1	17.5	−0.4	0.8	−1.3
.030....................	23.5	28.6	−1.5	.4	−3.0
.045....................	29.9	36.5	−2.3	.3	−4.5
.060....................	35.7	44.2	−3.1	.5	−6.0
Moisture content, percent					
10.9....................	32.5	41.1	−1.6	1.9	−4.4
18.9....................	25.5	30.8	−1.9	.1	−3.6
104.3....................	19.4	23.2	−1.9	−.5	−3.1
Rake, angle, degrees					
20.....................	31.5	38.3	1.9	4.1	−0.1
30.....................	25.6	31.4	−2.2	.1	−4.1
40.....................	20.2	25.3	−5.1	−2.6	−7.1

[1]Cutting velocity was 5 inches per minute.

Figure 18-16.—When crosscutting in the 90-90 direction, uniform chips and smooth surfaces were formed when cutting 0.030 inch deep in black oak at 10.7 percent moisture content with a 40° rake angle and 15° clearance angle. (Photo from Woodson 1979.)

OBLIQUE CUTTING

In strictly orthogonal cutting, the cutting edge is perpendicular to the motion of the tool over the workpiece. In some applications, the cutting edge is set obliquely to its direction of movement; the **deviation angle** between the edge and a line normal to the motion measures the degree of obliquity. Kivimaa (1950) has shown that when wood is cut parallel to the grain, parallel cutting force is negatively correlated with deviation angle; however, when cutting

Machining 1791

Figure 18-17.—Undesirable chip formation when crosscutting in the 90-90 direction. (Top) Broken chips and rough surfaces formed when making a 0.060-inch cut in cherrybark oak at 9.9 percent moisture content with a 20° rake angle and 15° clearance angle. (Bottom) Undersized chips formed when making a 0.045-inch cut in saturated sweetbay with a 20° rake angle and 15° clearance angle. (Photo from Woodson 1979.)

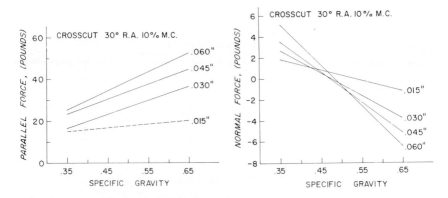

Figure 18-18.—Relationship of average parallel and normal cutting force per 0.1-inch knife length with specific gravity (ovendry weight and green volume) for crosscutting in the 90-90 direction dry wood (10 percent moisture content) of 22 species with a 30° rake angle (15° clearance angle) at four depths of cut. Dotted line indicates regression significant at .05 level; others significant at .01 level. (Drawing after Woodson 1979.)

tangentially across the grain (as in veneer slicing), parallel cutting force stays the same or rises as deviation angle is increased.

Readers interested in additional data on oblique cutting should find work by Kinoshita (1980ab) useful.

INCLINED AND VIBRATORY CUTTING

If a cutting edge is drawn transversely during an otherwise orthogonal cutting operation (visualize bread being sliced with a long knife), the process is termed **inclined cutting**. If the knife oscillates laterally across the direction of cut (as when halving a large grapefruit with a knife not much longer than the diameter of the grapefruit), or vibrates perpendicular to the knife edge in the direction of the cut (visualize a jackhammer), the process is termed **vibratory cutting**.

Vibratory cutting with lateral oscillation.—Cuts made by a knife having deviation angle are sometimes erroneously equated with cuts made by a laterally oscillating knife; a moment's thought about the pattern cut by a nicked knife should clarify the difference between the two situations. A knife given deviation angle has a slightly decreased effective sharpness angle, and therefore cutting forces are reduced when cutting orthotropic materials. A laterally oscillating knife, however, has a substantially reduced effective sharpness angle. Also, the oscillating knife, because of slight imperfections in the cutting edge, exerts a toothed cutting action that is more effective than the simple pressure of a knife cutting with deviation angle (visualize drawing a toothed knife across a tomato skin compared to simply pressing the knife against the skin).

Numerous workers have established that lateral oscillation of the cutting edge substantially reduces parallel cutting forces, slightly reduces normal cutting forces, and—particularly when cutting in the 90-90 direction (fig. 18-19)—

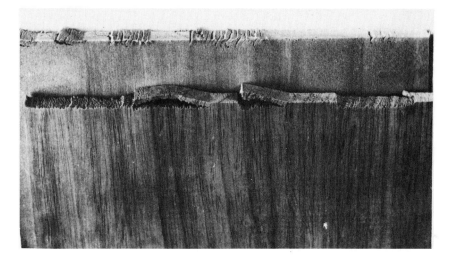

Figure 18-19.—Lateral vibration of the knife at 120 cycles per second improved surfaces of wood crosscut in the 90-90 direction at 0.5 inch per minute. (Top) Yellow-poplar. (Bottom) Common persimmon (*Diospyros virginiana* L.). The poorer surfaces shown resulted when lateral vibration was stopped. Rake angle 25°; normal chip thickness, 0.03 inch. (Photo from McKenzie 1961.)

greatly improves surface quality. The force reductions occur because lateral vibration of the knife not only reduces effective sharpness and rake angles but also alters cutting action through lateral shearing and reduction of the cutting coefficient of friction. The surface improvements occur because chip types are altered in a favorable way. Researchers have used vibration frequencies from 20 to 25,000 Hz with amplitudes from 10 μm to 25 mm. The principle has been successfully used in microtomes to section tissues. In the forest products industry, vibratory cutting with lateral oscillation has had limited application to veneer lathes in the United States. In Germany, an effort to cut 8 mm-thick parquet flooring with vibrating knives was not commercially successful, however.

Few data have been published on pine-site hardwoods, but readers interested in the history of research on inclined cutting and vibratory cutting with lateral oscillation will find useful reviews by Szymani and Dickinson (1975) and Shreve (1972). Additional later references include Kato and Asano (1974ab) and Fujiwara et al. (1976).

Vibratory cutting with perpendicular oscillation.—Kato and Asano (1974c) orthogonally cut parallel to the grain in the 90-0 direction with a knife ultrasonically vibrating in the direction of the cut (like a jackhammer). In their study of cypress (*Chamaecyparis obtusa* Endl.) parallel cutting forces were reduced to about one-fourth of those with a non-vibrating knife, and normal cutting forces were near zero. Rake angles used were 38.5° and 50°; frequencies were 18.12 and 18.56 kHz; amplitudes were 24 and 28 μm; feed speeds were 1.3 and 5.2 mm/second; clearance angle was 5° and depth of cut 0.2 mm. Hamamoto and Mori (1971) observed parallel cutting force reductions only when cutting in the 90-90 direction with such a vibrating knife. Parallel cutting forces in the 0-90 and 90-0 direction were about the same as with a non-vibrating knife. When cutting in the 90-90 direction, a vibrating knife lessened fiber deformation ahead of the cutting edge; chips formed by shearing along the grain, and surfaces were better than those produced with a non-vibrating knife.

18-3 SHEARING AND CLEAVING

These processes are distinguished from conventional orthogonal cutting because of the extreme depth of cut, i.e., the chip and the workpiece are equally massive, stiff, and difficult to deform (fig. 18-20).

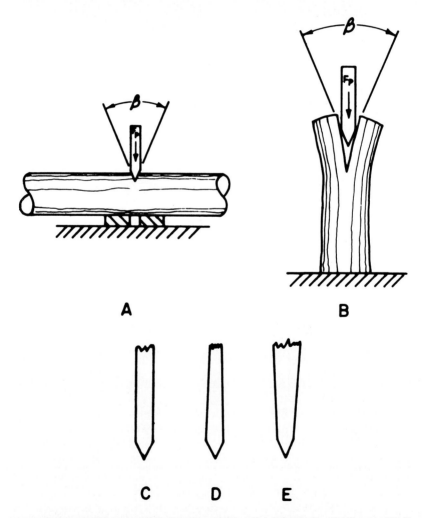

Figure 18-20.—(A) Shearing. (B) Cleaving. (C) Parallel-sided knife. (D) Tapered knife. (E) Knife with thick root.

SHEARING

When very great depths of cut are taken by a knife crosscutting in the 90-90 mode against an anvil or opposing knife (visualize rose stems cut with pruning shears), the process is described as shearing (fig. 18-20A). Golob's (1976) illustrated review of the patent literature on shears for tree felling chronicles more than 200 years of evolution. Today, tree felling shears (fig. 16-14 and 18-28), delimbing shears, and shears to reduce long logs to shorter pulpwood lengths are in common use in the South. Surprisingly, however, none of the published research describes work on hardwoods commonly found on southern pine sites, and data on oaks and hickories are conspicuously lacking.

Some generalizations are possible, however, from work done on other speices by Erickson (1967); Johnston (1967, 1968abcd); Kempe (1967); Wiklund (1967, 1971); McIntosh and Kerbes (1969); Johnston and St. Laurent (1970, 1971, 1974ab); Arola (1971, 1972); Koch (1971); Arola and Grimm (1974); McIntosh and McLauchlan (1974); McLauchlan and Kusec (1975).

Factors affecting shearing force.—Shear forces are less in warm than in frozen wood, in clear than in knotty wood, in heartwood than in sapwood, in low-density wood than in dense wood, and where the shearing direction is perpendicular to the annual rings (fig. 18-1 left) than parallel to them (fig. 18-1 right). Above the fiber saturation point, moisture content apparently makes little difference in the force required to shear, but dry wood requires more force than saturated wood.

Cutting velocity has little effect on shearing force. Shear force is least when the specimen is cut between opposing knives; if cut by a single knife against an anvil, a narrow anvil requires less force than a wide one.

The friction coefficient between a steel knife and green wood is approximately 0.2. Grease lubrication between knife and wood does not greatly reduce shearing forces; Teflon surfaces on the knife are more effective. Axial loads (simulating the weight of a standing tree) do not appreciably increase shearing forces. Lateral vibration of the cutter reduces shear forces required, as does tapering the cutter plate (fig. 18-20D) to give clearance between the plate and the wood. In a review of Russian work, Kubler (1960) reported that vibration in the feed direction also reduces shear forces required.

For parallel-sided cutters (fig. 18-20C), thin blades shear with less force than thick blades. Blades tapered so that the plate near the cutting edge is thin and the root thick (fig. 18-20E) require forces intermediate to thin and thick plates without taper. Sharp blades shear with less force than dull blades. Cutting-edge profiles in the shape of an open V do not appreciably lower forces required to shear logs fit inside the V. Guillotine shear blades, with cutting edge obliquely inclined to the direction of cut, require less force than a blade of equal thickness and bevel angle with cutting edge orthogonal to the direction of cut.

Estimating force and power requirements for crosscut shearing of roundwood.—Based on his previous work (Arola 1971), Arola (1972) devised a useful procedure for estimating force and power requirements. Although based on northern forest species, the method should also be valid for southern hardwoods. Arola's study included wood with specific gravities between 0.30 and 0.65, log diameters from 5 to 10 inches, shear blade thicknesses from 1/8- to 1/2-inch with sharpness angles from 20° to 50° and shearing speeds from 2 to 12 inches per second. Effects of shear blade dulling and internal wood temperature are incorporated in the procedure, which is accomplished in steps as follows:

Step. 1—Determine the following.—
- G = specific gravity (based on ovendry weight, green volume) of the species to be crosscut sheared. (See tables 7-6 and 7-7.)
- t = shear-blade thickness (inches).
- D = maximum log diameter to be sheared (inches).
- P = desired hydraulic pressure (psi).
- V = desired shearing speed (ips). (Recommended range is 2 to 12 inches per second.)
- T = estimated lowest temperature of operation (°F.).

Step. 2—Determine the total shear force requirement. F_m—Enter nomograph 1 (fig. 18-21) at the appropriate wood specific gravity G and follow this value up to the selected blade thickness t (see example). At this intercept follow a horizontal line to intercept the maximum log diameter D to be sheared and then along a vertical line to determine the total shearing force F_m required to effect the cut. Based on data from Arola (1971), shearing force values in nomograph 1 have been adjusted to approximate the 95-percent upper confidence limits on a single estimate plus a 15-percent increase in force due to moderate blade dulling (approximated by a 1/32-inch flat along the entire cutting edge).

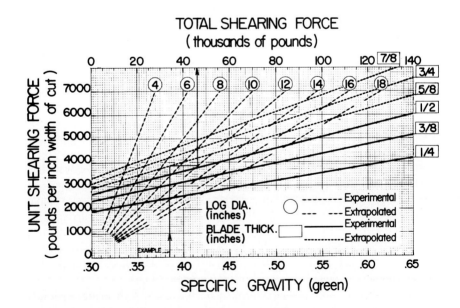

Figure 18-21.—Nomograph number 1 for determining force to shear green logs, according to wood specific gravity, shear-blade thickness, and maximum log diameter. (Drawing after Arola 1972.)

Machining 1797

Step. 3.—Correct the shear force for temperature $F_{c/t}$. The total shearing force as determined in Step 2 is for an approximate temperature of 60°F. This value must be corrected for shearing at lower temperatures—particularly for frozen wood. Increases in force for northern woods sheared at temperatures down to 0°F range from 10 to 32 lb/in width of cut per °F drop in temperature (Arola 1971). An average value of 20 lb/in width of cut per °F is recommended as a multiplier of the anticipated temperature drop and maximum log diameter. Thus,

$$F_{c/t} = F_m + 20(60-T)D \qquad (18\text{-}5)$$

Step 4.—Determine the cylinder size d.—Locate the maximum shearing force $F_{c/t}$ from Step 3 on nomograph 2 (fig. 18-22) and the desired operating pressure P. Place a straight edge along values to find the intercept that determines the piston diameter d. If this is not a standard cylinder size, select the next higher diameter. This selection procedure assumes a direct relationship between the maximum force requirement and the hydraulic cylinder. If a mechanical linkage is provided that voids this assumption, the force on the cylinder must be adjusted accordingly.

Step 5.—Determine the hydraulic oil flow rate Q.—The intercept for hydraulic oil flow rate Q is determined from nomograph 3 (fig. 18-23) by placing a straight edge on the piston diameter d determined from Step 4 and the desired shearing speed V.

Step 6.—Determine the theoretical motor horsepower requirements HP_t.—To determine the theoretical horsepower of the motor to drive the hydraulic pump (assuming 100-percent overall efficiency of the pump), the straight line intercept on the horsepower scale (nomograph 4, fig. 18-24) is found by connecting the flow rate Q from Step 5 and the desired hydraulic pressure P from Step 1.

Step 7.—Adjust the motor horsepower for the overall efficiency of the pump HP.—The horsepower requirement (Step 6) must be adjusted for the overall efficiency of the hydraulic pump (product of volumetric and mechanical efficiencies). The manufacturer's performance curves for the particular pump should be consulted. However, in lieu of the actual performance curves, the approximate values of overall pump efficiency from table 18-29 can be used for estimating purposes. For mobile equipment applications the balanced vane pump having fixed displacement has become the most universally accepted because of high operational speeds and pressures. The adjusted motor horsepower is determined as follows:

$$HP_e = \frac{HP_t}{e} \qquad (18\text{-}6$$

The estimator should be aware that line losses from the pump to the cylinder may decrease the usable hydraulic pressure in the cylinder by 10 to 25 percent.

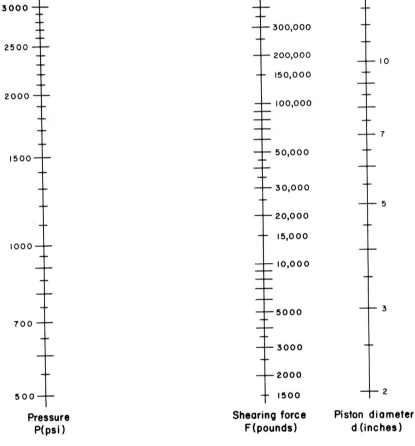

Figure 18-22.—Nomograph number 2 for determining hydraulic cylinder diameter to deliver required shearing force with selected hydraulic pressure. (Drawing after Arola 1972.)

FLOW RATE

SOLVES: $Q = 0.204 \, V d^2$

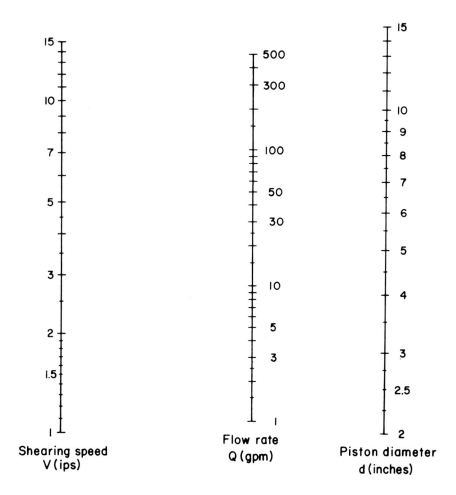

Figure 18-23.—Nomograph number 3 for determining hydraulic oil flow rate as affected by cylinder diameter, and shearing speed.

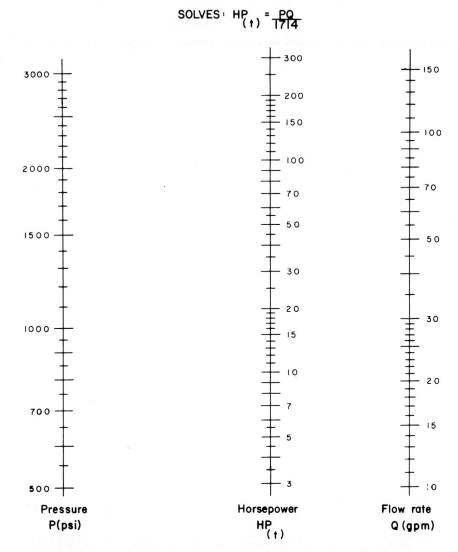

Figure 18-24.—Nomograph number 4 to determine theoretical motor horsepower required to drive a hydraulic pump as affected by hydraulic pressure and flow rate. (Drawing after Arola 1972.)

TABLE 18-29.—*Efficiencies, pressure ratings, weight, and speed of five types of hydraulic pumps* (Data from Arola 1972)

Pump type	Pressure rating (P)	Overall efficiency (e)	Weight	Rated speed
	psi	*Percent*	*Lb/HP*	*rpm*
External gear............	2,000– 3,000	80–90	0.5	1,200–2,500
Internal gear............	500– 2,000	60–85	.5	1,200–2,500
Vane.................	1,000– 2,000	80–95	.5	1,200–1,800
Axial piston............	2,000–10,000	90–98	.25	1,200–3,600
Radial piston............	3,000–10,000	85–95	.35	1,200–1,800

Wood damage from shearing.—Wood adjacent to a shear blade is damaged by passage of the blade (fig. 18-25). The damage—2- to 10-inch splits along the grain and crushed fiber ends—increases with increase in tree diameter. Such damage does not greatly concern pulpwood users. Few operators of southern sawmills and veneer mills, however, will accept logs or bolts severed by shears. Procedures to reduce damage caused by shears are therefore of interest.

Literature previously cited indicates that shear damage is less in warm than in frozen wood, in small than in large trees, from thin than from thick blades, from blades of 50° than of 20° sharpness angle (fig. 18-26), and with double shears than with a single shear acting against a flat anvil. McLauchlan and Kusec (1975) reported signifcantly less wood damage from a ribbed shear having grooves in the upper surface; apparently the ribs localize damage and reduce incidence of long splits.

Golob (1976) proposed that splitting damage should be reduced if a pair of opposed shear blades were arranged to cut in the 90-45 direction. (Visualize driving sharp spades at a 45° angle downward into the stump from opposite sides until they meet at the pith below groundline and sever the tree.) Trials of the concept indicated some diminution of splitting damage.

There is great need for mechanized felling without damage to the valuable butt log; alternatives to shears that provide the needed degree of mechanization include chainsaw feller-bunchers (fig. 16-15) and fellor-bunchers equipped with rotary cutting bars (fig. 18-49).

Bucking of pulpwood with continuous-feed shears.—Stock (1976) described a 140-foot-long longitudinal conveyor whereby tree-length pulpwood stems are sheared into pulpwood lengths while in continuous endwise motion. As the trees move through an 18-foot section, pairs of guillotine-like knives travel with them to cut off pulpwood lengths. The shear, driven by a 75-hp motor, can handle trees up to 20 inches in diameter and process more than 5,000 cu ft of wood per hour. Unlike saws, shears make no sawdust—thereby increasing yield of usable wood fiber by about 1 percent.

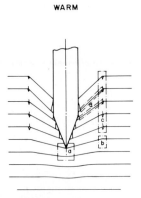
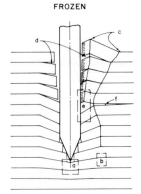

WARM

a. Compression perpendicular to the grain, and flextural and tensile stresses
b. Flexural stress in conjunction with tension parallel to the grain
c. Bending fracture in conjunction with compression parallel to the grain
d. Shear stress

FROZEN

a. Compression perpendicular to the grain, and flextural and tensile stresses
b. Flexural stress in conjunction with tension parallel to the grain
c. Bending fracture in conjunction with compression parallel to the grain
d. Shear fracture
e. Compression damage
f. Splitting

Figure 18-25.—Cutting sequence and wood damage caused by shearing warm and frozen wood across the grain. (Drawing after Wiklund 1967.)

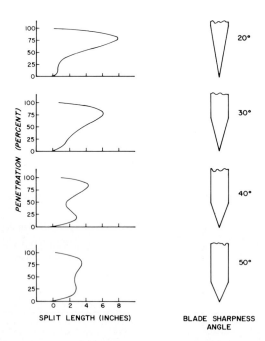

Figure 18-26.—Location and length of splits in yellow birch (*Betula alleghaniensis* Britton) roundwood, related to sharpness angle of a 3/8-inch-thick blade used to shear across the grain at 4.7 inches per second. (Drawing after Arola 1971.)

Spiral-head shear to produce long chips.—The expanding structural flakeboard industry will build plants in the South in coming decades. Flakes for cores in such structural boards can be produced by a ring flaker (fig. 18-27I) from finger-sized chips. Such **fingerling chips** can be manufactured with a spiral-head shear (Barwise et al. 1977). The tapered spiral shear severs small hardwood stems into finger-length disks and disk portions that are subsequently hammer-milled to yield fingerling chips (fig. 18-27). Erickson (1976) found that red maple and northern red oak required 9.8 and 9.3 hp min per cubic foot of wood sheared, respectively. In 1980 the equipment was still in the experimental and demonstration stage. By 1981, however, fingerling or block chips were being commercially cut in Finland, by a farm-tractor-driven KOPO PH-10 spiral-screw chipper, from small trees—delimbed or with limbs; chipping energy requirement was about 1.1 kWhr/m^3. Chips produced by the equipment measure about 50-60 mm long, 43 mm wide, and 8.7 mm thick (Hakkila and Kalaja 1981). This concept shows promise for producing not only fingerling chips for ring flaking, but also chunks of wood much larger than pulp chips which can be used as an industrial fuel.

More recently, a prototype involuted-disk chunker—a variation of the spiral-head shear—showed sufficient promise to warrant additional research; see Arola et al. (1982)*, and Mattson (1985).**

Delimbing shears.—The large, strong, stiff branches of most pine site hardwoods are a major obstacle to mechanized harvesting. Southern pine, most other conifers, and a few northern hardwoods (e.g., *Betula* sp.) have smaller limbs of less strength that can be readily sheared from the stem (fig. 18-28). Johnston and St. Laurent (1974b) found that a force of about 1,000 pounds per inch of branch diameter is required to orthogonally shear limbs from northern softwoods with a ¼-inch-thick knife having a sharpness angle of 45°. Force data on southern hardwoods are lacking but 1,500 to 2,000 pounds per inch of branch diameter might be appropriate for oaks and hickories. Delimbing shears that vibrate in the fashion of jackhammers (visualize a carpenter tapping a chisel with a mallet) might significantly reduce shear forces required (Kato et al. 1971; Kato and Asano 1974c).

Chain flailing devices are available (Gauthier 1975; Folkema and Giguere 1979) that knock limbs from northern conifers and birch (*Betula* sp.); they are ineffective in removing the strong, thick limbs of most southern hardwoods, however.

---

*Arola, Rodger A., Robert C. Radcliffe, Sharon A. Winsauer, and Edsel D. Matson. 1982. A new machine for producing chunkwood. U.S. Department of Agriculture Forest Service, Research Paper NC-211, 8 p. U.S. Department of Agriculture Forest Service, North Central Forest Experiment Station, St. Paul, Minnesota.

**Mattson, J.A. 1985. Chunking forest residues for energy wood. Comminution of wood and bark symposium, Forest Products Research Society, held October 1-3, 1984, Chicago, Illinois.

Figure 18-27.—Rotating spiral shear to produce fingerling chips from small hardwoods.(Top) Infeed at right, shear in center. (Bottom left) Infeed end of spiral shear in action. (Bottom right) Disks and disk portions produced by the spiral-head shear; the disks can be hammermilled to yield fingerlings. (Photo from Erickson 1976.)

CLEAVING

Longitudinal splitting of a short log in the 90-0 direction (as with a hatchet) is termed **cleaving** (fig. 18-20B). Industrial processes that call for knowledge of cleaving include splitting of pulpwood too large in diameter for accommodation

Figure 18-28.—JD743 harvester. (Top) Clamshell shear on boom can sever conifers 18 inches in diameter at the groundline; delimbing shears, on left, are shown open—ready to receive tree stem. (Bottom) Sheared tree in foreground; tree stem in background being propelled—left to right—butt first by spiked feed rolls through delimbing shears closed to conform to stem contour. (Photos from John Deere Co.)

in chipper infeed spouts, splitting of firewood (figs. 16-23 and 18-29), and splitting of stumps (figs. 16-28 and 16-29).

Figure 18-29.—Trailer mounted firewood splitter driven by a 5-hp gasoline engine with pump and cylinder system capable of hydraulically pushing a 27-inch-long bolt, with 12 tons force, against the splitting blade at left.

Maximum parallel cutting force (F_p) during cleaving is less in warm than in frozen wood, in clear than in knotty wood, in low-density wood than in dense wood, in small-diameter wood than in large wood, if split close to the log periphery than if split through the pith, in straight-grained wood than in wood with interlocked or crooked grain, and in green wood than in dry wood. Data on southern pine indicate that bolt length in the range from 14 to 42 inches does not greatly affect maximum force to cleave, but long bolts do require more work to cleave than short bolts (Koch 1972, p. 805).

Pulpwood splitters.—Southern woodyards occasionally receive wood that is too large, or has too much sweep or crook to allow it to pass through the spout of the mill chipper. To utilize such logs, some woodyards employ stationary or portable hydraulic splitters that quarter the log in one pass. A typical portable unit is mounted on a 2-ton truck and arranged to quarter 60 logs per hour measuring up to 8 feet long and 28 inches in diameter. Logs are placed on the splitter by a knuckleboom loader installed behind the truck cab. Both loader and splitter are operated hydraulically from an auxiliary 60-hp engine mounted behind the cab of the truck. Quartered pieces fall to either side of the truck.

De Vries (1973) reviewed equipment used in Australia to split hardwood pulpwood. Although the installations processed *Eucalyptus* species, the equipment would be equally applicable to pine-site hardwoods of size comparable to *Eucalyptus*. Splitter investment varied from $15,000 to $87,500 (1973), hourly production from 7.5 to 42 tons (green), and costs per ton from $0.63 to $2.65 (table 18-30).

Manual splitting with axes, crowbars, wedges, and saws was also studied, and costs per ton split were less than with mechanized equipment—but in 1973 prevailing Australian wage rates were low in comparison to those of the United States. In Australia, bigger logs and knotty logs are often split with a splitting gun. With this method, two cleaving axe cuts are made on one end of the log, forming a cross through the heart; this is to ease entry of the splitting gun and to start the initial splits. About 1/10 of a pound of black powder is put into the tubular splitting gun and it is hammered 4 to 6 inches into the end of the log. A fuse 6 inches long is then inserted into the gun and lit. The blown logs often have to be further split with axes or chainsaws. Although costs are low (about $0.77/ton, green in 1973), fire hazard may be substantial (De Vries 1973).

For North American experience mechanically splitting oversize hemlock logs for chipping, see: (Nichols, A.F. 1985. Splitting and chipping oversize logs for pulp chips—operator's perspective. Comminution of wood and bark symposium, Forest Products Research Society, held October 1-3, 1984, Chicago, Illinois).

TABLE 18-30.—*Some specifications of pulpwood splitters used in Australia to split Eucalyptus species* (Data from de Vries 1973)

Type and power source	Maximum splitting force	Maximum log size Length Diameter		Average size of split billets	Hourly production	Installed cost	Cost per ton split
	Tons	*Feet*	--------*Inches*--------		*Tons, greeen*	--*Dollars, 1973*--	
Stationary splitters							
Steam cylinder	12.1	4.0	42	5 × 7	22	26,000	1.11
Hydraulic, driven by 240-hp diesel engine .	50.3	6.5	72	11 × 11	22	87,500	2.00
Hydraulic, driven by 50-hp, electric motor .	37.7	8.0	24	7 × 7	26-38[1]	60,000	.63-.92[1]
Mobile splitters							
Caterpillar D7E tractor carrying a splitting knife attached to the dozer blade	—	40.0	No limit	8 × 12	42	62,400	1.09
Hydraulic, driven by 34-hp gasoline engine; trailer-mounted	—	8.3	48	6 × 8	7.5	15,000	2.65

[1]Species with interlocked grain caused lowest hourly production and highest cost per ton.

Firewood splitters.—Increased use of split wood for home heating has created a market for portable firewood splitters. Most are trailer mounted, hydraulically actuated to deliver 10 to 25 tons force, and powered either by small gas engines (5 to 20 horsepower) or arranged for connection to the power takeoff of a farm tractor (fig. 18-29). More mechanized firewood splitters are also available (fig. 16-23).

Cleaving transversely; kerfless cutting.—Wielders of hatchets are familiar with the technique of splitting kindling by placing the stick to be split flat on a chopping block and striking it a cleaving blow. Attempts have been made to use this principle as a method for kerfless cutting of wood.

Johnston and St. Laurent (1975a) proposed to reduce splitting ahead of the cutting edge by compressing the wood to at least 25 percent of its yield strength, between clamps, then cutting it with a knife along a path between clamps (fig. 18-30 top), releasing the clamps to remove the compression, and then separating the wood from the knife. A prototype longitudinal compression slicer designed to cut green softwood 6-inch-deep cants at 300 fpm has been constructed (Johnston and St. Laurent 1979).

Murphey and Schneider (1956) developed a prototype machine to slice (transversely split) wood with a series of round knives mounted on mandrels and spaced by disks as wide as the desired lumber thickness. A pair of these roller assemblies constituted a roller set which cut the upper and lower surface of the wood and fed it into the next set of rollers. The knives of one mandrel were arranged to slice the wood to a depth of about ¼-inch; thus two sets of rollers were needed to slice 1-inch lumber. To prevent cleavage ahead of the knives, a set of pressure plates were placed on the mandrel and spaced the same width as the lumber being sliced—which had to be of uniform width. The knife action compresses wood in the vicinity of the cut, but this compressed wood is later removed when the wood is planed. With feed speed of 20 feet per minute, the cutting disks rotated at 10 rpm. One-inch-thick oak and maple were successfully cut—both green and dry, but the idea has yet to be applied on a commercial scale.

E. Anderson and W. Maier patented a green-wood splitter resembling a circular saw with a number of teeth equally spaced around the periphery; unlike a fast-rotating saw, however, the slitter's edge is honed to a sharp knife edge and the slitter rotates slowly with the feed at the same peripheral speed as the wood feed speed. Since the slitter cuts like a knife, it produces no sawdust. Lumber slit with the experimental equipment must be planed to smooth the cut (Forest Industries 1981).

Shearing veneer and panels.—Veneer guillotines are in widespread commercial use to true veneer edges prior to edge gluing the veneers. These guillotines can also trim packs (a bundle perhaps 2 inches thick) of veneer to length.

Shears to produce sized and punched panels from fiberboards, flakeboards, and plywood are also commercially available (fig. 18-30 bottom). Also in commercial use are shear dies arranged to cut finger-joints in short strips of veneer so they can be rapidly assembled into long strips.

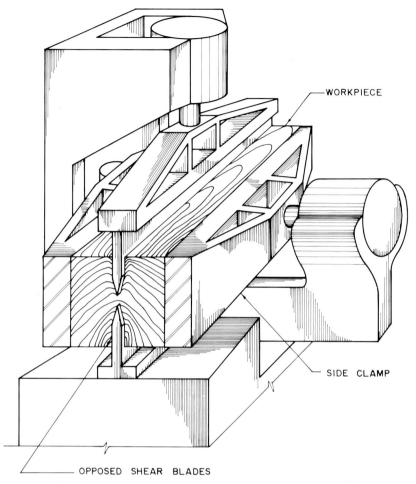

Figure 18-30.—(Top) Orthogonal cleaving in the 0-90 direction. (Bottom) Cutting die fixed under a moving top pressure platen (not visible) arranged to shear panels up to 5/8-inch thick to semi-elipsoid pattern; spring-loaded ejection system inside of the shear blade retracts to expose machine punches that produced the two holes visible in the finished part. (Photo from Ontario Die Company of America.)

18-4 PERIPHERAL MILLING PARALLEL TO GRAIN (90-0 DIRECTION)

Peripheral milling or planing may be defined as the removal of excess wood in the form of single chips formed by intermittent engagement with the workpiece of knives carried on the periphery of a rotating cutterhead. The cutterhead usually carries several knives, removable for sharpening, which are precisely adjusted to cut on a common cutting circle. Final adjustment involves **jointing** the knives with an abrasive hone while the cutterhead revolves at operating speed. The machined surface therefore consists of individual knife traces generated by the successive engagements of each knife.

For detailed treatment of kinematics, force systems, power requirements, chip formation, and chip severance phenomena, the reader should first read Koch (1972, p. 808-827), then Koch (1964a, p. 111-166), and finally—for photographs of forming chips—Koch (1954).

This text will explain needed nomenclature, briefly outline kinematics of the process, illustrate principal modes of chip formation, and summarize factors affecting power requirements. Major emphasis will be given findings related to severity of planing defects and recommendations to minimize these defects.

NOMENCLATURE

Figures 18-31 and 18-32 illustrate nomenclature in peripheral milling.

KINEMATICS

In conventional planers, the engaged knives move counter to the movement of the workpiece—an action termed up-milling (fig. 18-33A, 18-34CD); if the engaged knives move in the same direction as the workpiece, the process is called down-milling (fig. 18-33B, 18-34AB).

The trochoidal path taken by each knife tip can be represented (fig. 18-33C) by considering the workpiece fixed in space and allowing the cutterhead to rotate about a roll circle of a diameter that gives a relative translatory velocity equal to the desired feed speed (Martellotti 1941, 1945):

Feed per jointed knife (F_t) can be stated:

$$F_t = \frac{12F}{Tn} \qquad (18\text{-}7)$$

The distance between knife marks can be reduced by decreasing the feed speed, increasing the number of jointed knives in the cutterhead, or by increasing the cutterhead speed.

Machining

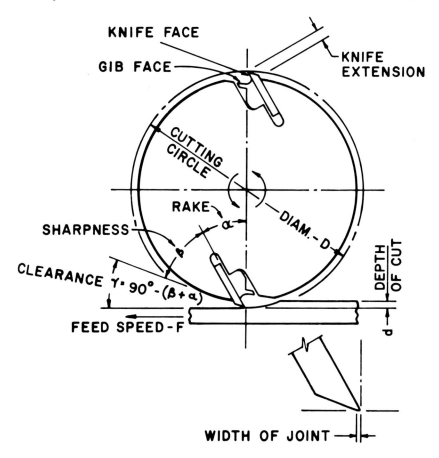

Figure 18-31.—Terminology for peripheral milling cutterhead. Up-milling illustrated.
α   Rake angle, degrees
β   Sharpness angle, degrees
λ   Clearance angle, degrees
d   Depth of cut, inches
D   Cutting-circle diameter, inches
F   Feed speed of workpiece, feet per minute
(Drawing after Koch 1955.)

If the knives are accurately jointed, for up-milling the wave height can be expressed:

$$h = \frac{F_t^2}{8\left(R + \dfrac{F_t T}{\pi}\right)} \quad (18\text{-}8)$$

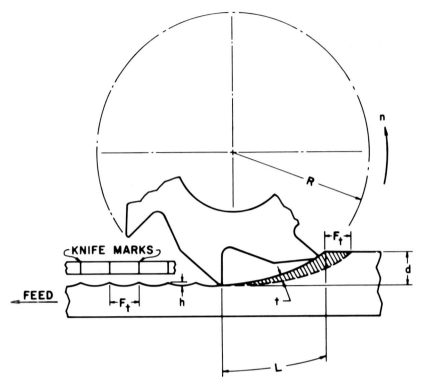

Figure 18-32.—Path generated by up-milling knife.
F_t Feed per knife; translation of the workpiece between engagement of successive knifes, inch
h   Height of knife marks above point of lowest level, inch
L   Length of knife path during each engagement with the workpiece, inches
n   Revolutions per minute of cutterhead, rpm
R   Cutting-circle radius, inches
t_{avg} Average thickness of undeformed chip, inch
T   Number of jointed knives in the cutterhead
d   Depth of cut, inch.
(Drawing after Koch 1964a, p. 113.)

Wave height can be reduced by increasing the radius of the cutterhead or by decreasing the feed per knife.

For up-milling the length of knife path engagement is as follows:

$$L = R \text{ arc cos} \left(1 + \frac{d}{R}\right) + \frac{F_t T}{\pi D} (Dd - d^2)^{1/2} \qquad (18\text{-}9)$$

When up-milling, the average thickness of the undeformed chip can be stated:

$$t_{ang} = \frac{F_t d}{L} \qquad (18\text{-}10)$$

The significance of these formulae can be visualized from a tabulation of the dimensions of chips and surfaces produced by two commonly used up-milling

Machining

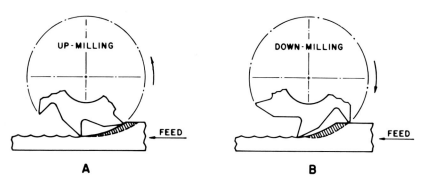

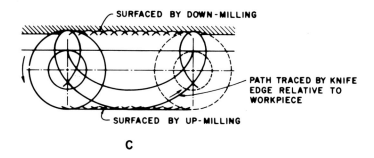

Figure 18-33.—(A) Up-milling. (B) Down-milling. (C) Knife path, relative to workpiece, is a curtate trochoid. (Drawing after Martellotti 1941, 1945.)

cutterheads. For purposes of comparison, it is assumed that both heads rotate at 3,450 revolutions per minute (rpm) and take a ⅛-inch-deep cut.

Dimension	T = eight knives D = nine inches F = 300 f.p.m.	T = 16 knives D = 11 inches F = 1,000 f.p.m.
	----------------------------Inch----------------------------	
F_t	0.130	0.217
h	0.441×10^{-3}	0.891×10^{-3}
L	1.100	1.289
t_{avg}	0.015	0.021

CHIP FORMATION

Koch's (1954, 1955, 1956) high-speed photographs of up-milling knives cutting Douglas fir, *Pseudotsuga menziesii* (Mirb.) Franco (none for southern hardwoods have been published) show the same chip formations observed in orthogonal cutting of pine-site hardwoods in the 90-0 or planing direction (e.g., figs. 18-3, 18-4, 18-5). Peripheral up-milling differs in one important respect from 90-0 orthogonal cutting; while the initial up-milling cut is essentially parallel to the grain, the emerging cut may be at a considerable angle to the grain

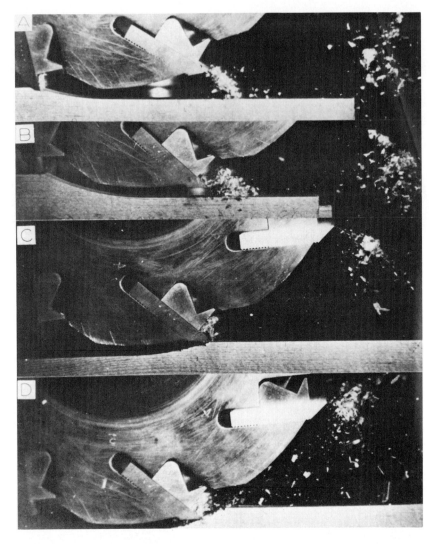

Figure 18-34.—Comparison of up-milling and down-milling at 175 fpm with eight-knife cutterhead rotating at 3,600 rpm and carrying knives with 30° rake. (A) Down-milling, ¼-inch cut. (B) Down-milling, ½-inch cut. (C) Up-milling, ¼-inch cut. (D) Up-milling, ½-inch cut. (Photos from Koch 1956.)

(fig. 18-34CD). Furthermore, in peripheral up-milling the undeformed chip thickness constantly changes from a minute value at contact to a maximum just prior to emergence; chips cut at emergence are frequently Franze Type I (fig. 18-3).

Fortunately, the part of the knife path that remains visible on the machined surface is the initial portion where chip thickness is minute; a Franz Type II failure can therefore be induced (fig. 18-35). With very low rake angles, Type III chips are common.

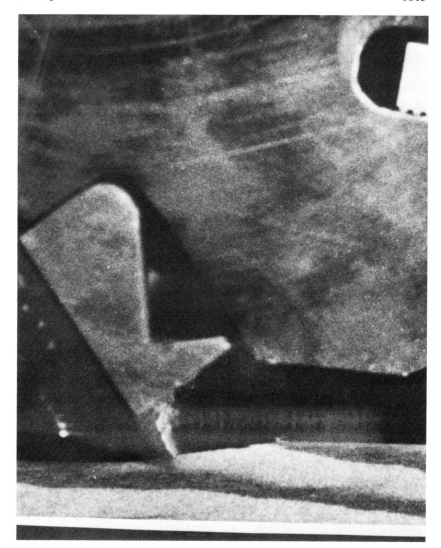

Figure 18-35.—A type II chip formed during peripheral up-milling with an eight-knife, 9-inch cutterhead turning at 3,450 rpm taking a 3/16-inch depth of cut in dry Douglas-fir. Rake angle 30°; clearance angle 20°. The flat-faced gib illustrated interferes with chip flow; gibs with faces curved to ease chip flow and to meet relief provided in cutterhead body would be preferable. (Photo from Koch 1955.)

FACTORS AFFECTING POWER

Trends apparent in Koch's (1972, p. 819-825) data on Douglas-fir should be generally applicable to pine-site hardwoods. Detailed discussions of the causes behind the effect of each factor are available (Koch 1964a, p. 121), and may be summarized as follows.

Workpiece factors.—Power required is directly proportional to width of cut. With commonly used rake angles, more cutterhead power is required to plane wet wood than dry because of the power consumed accelerating heavy wet chips; with very low or negative rake angles, however, dry wood takes more power than wet (fig. 18-36). This is reasonable inasmuch as knives with low rake angles form Type III chips by compressing the wood parallel to the grain, and dry wood is much stronger than wet when so loaded.

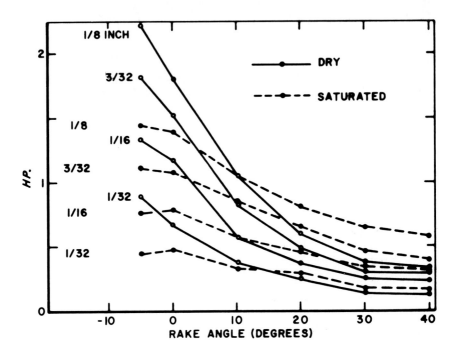

Figure 18-36.—Effect of rake angle and depth of cut on net cutterhead power requirement (per inch of workpiece width) for planing dry and saturated Douglas-fir by upmilling. F_t, 0.127 inch; clearance angle, 20°; cutting circle diameter, 9.05 inches; nominal speed, 3,600 rpm. (Drawing after Koch 1964a, p. 149.)

Flat-grain wood without prominent rays has more of a tendency to split ahead of the knife than does edge-grain wood; therefore, such flat-grain wood may require less power to mill. In woods with prominent rays, e.g., oaks, this relationship may not apply.

Because wood of high density is strong and the heavy chips require substantial energy for acceleration to cutting velocity, high-density wood requires more power to plane than wood of low density (fig. 18-37).

Cutterhead factors.—Net power required by an up-milling cutterhead is less with low cutting velocity than with high, less with small-diameter cutterheads than with large, less with large rake angles than with small (fig. 18-36), slightly less with large clearance angles than with small, less with sharp knives than with dull or heavily jointed knives, and less with knives well extended (0.3 inch or more) beyond the gib face than with knives extended less than 0.3 inch.

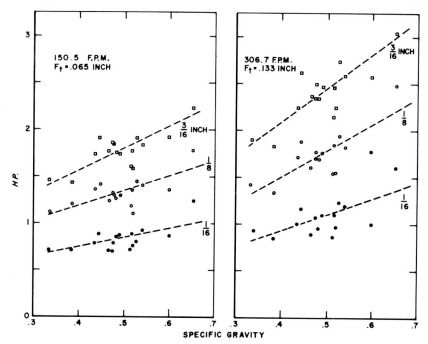

Figure 18-37.—Effect of specific gravity (basis of ovendry weight and volume) of Douglas-fir on net cutterhead horsepower per inch of workpiece width. Data are for upmilling at two feed speeds (150.5 and 306.7 feet per minute) and three depths of cut with an eight-knife, 9-inch cutterhead turning at 3,600 rpm. Rake angle, 30°. Moisture content, 8.5 percent. (Drawing after Koch 1956.)

At feed speeds below $F_t = 0.3$ inch and cuts less than ⅛-inch deep, horsepower demand increases with number of jointed knives cutting, although in no case does a doubling of the number of knives cause a doubling of power demand. With deep cuts and values of F_t over 0.3, power demand may be negatively correlated with number of knives cutting because the large chips formed with fewer knives cannot readily escape from the knife.

If the cutterhead is skewed so that its rotational axis makes an angle other than 90° with the direction of feed, each element of each cutting edge cuts at an angle to the grain direction. The effective knife extension and rake angle are increased by an amount dependent on the angle through which skewed, the feed speed, and the cutterhead peripheral velocity; also, the knives cut at an angle to the grain and less energy is required for chip formation. Therefore, cutterhead power is slightly reduced.

Feed factors.—In the special case where F_t is held constant while feed speed is increased (i.e., the number of knives in the cutterhead is doubled each time the feed speed is doubled), the horsepower requirement per knife is approximately constant within the feed speed range from 100 to 1,000 fpm.

In the more general case in which all other factors remain constant, an increase in feed speed increases the height of the individual knife marks and increases the distance between them, lowering surface quality and raising cutterhead power demand (fig. 18-38). With cuts less than ⅛-inch deep, the horsepower requirement does not double when F_t is doubled; in other words, it takes less energy to cut a given volume of wood into long chips than into short.

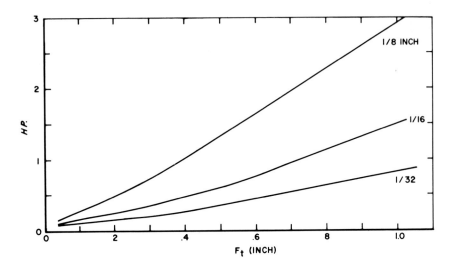

Figure 18-38.—Effect of feed per knife (F_t) on net cutterhead power (per knife per inch of workpiece width). Data for upmilling at three depths of cut in Douglas-fir at 7.3-percent moisture content. Rake angle, 27½°; cutting circle diameter, 9.44 inches; nominal speed, 3,600 rpm. (Drawing after Koch 1964a, p. 157.)

Depth of cut is positively correlated with cutterhead power demand. With conventional feeds per knife (F_t of less than 0.2 inch) and relatively shallow cuts, cutterhead power demand falls short of doubling when depth of cut is doubled; under these conditions it generally requires less energy to remove a given volume of wood in a single cut than in two shallow cuts (fig. 18-36). Where F_t exceeds 0.5 inch, however, some tests have shown that horsepower requirement per knife is approximately proportional to depth of cut, i.e., a ⅛-inch cut takes nearly twice as much power as a 1/16-inch cut and nearly four times as much as 1/32-inch cut (fig. 18-38); in this range of F_t, capacious and smoothly rounded gullets are required to ease chip flow and minimize cutterhead power (see figs. 18-98 bottom, 18-99 bottom).

Cutterhead power to up-mill six southern hardwoods.—Davis (1962) provided data on planing yellow-poplar, sweetgum, and white oak. He found that a 1/16-inch depth of cut in white oak at 6 percent moisture content required about 50 percent more cutterhead power than a similar cut in sweetgum at the same moisture content, and that a rake angle of 0° required nearly three times the cutterhead power of a 40° angle (table 18-31). At all five rake angles tested, sweetgum and Douglas-fir required about the same cutterhead power.

When planing hickory, northern red oak, yellow-poplar, and basswood (*Tilia americana* L.), Stewart (1974c) found that 80 to 90 percent of the variation in power requirement could be accounted for in regression equations involving depth of cut, feed rate, moisture content, and specific gravity (table 18-70); power required was positively correlated with all these factors.

SURFACE QUALITY

Quality of the machined surface primarily results from the cutting geometry and type of chip formed. In general, machined surfaces are improved by maintaining low values for F_t (feed per knife) and h (height of knife marks); this is accomplished by increasing the cutting-circle diameter and the number of jointed knives in the cutterhead and by reducing the feed speed.

TABLE 18-31.—*Cutterhead power requirement to make 1/16-inch depth upmilling cuts[1] in three hardwoods (6-percent moisture content) using high-speed steel knives ground to five rake angles* (Data from Davis 1962, p. 20)

Rake angle (degrees)	Species[2]		
	Sweetgum	Yellow-poplar	White oak
	----------Kilowatts----------		
0	6.3	7.0	7.4
10	4.5	4.9	5.0
20	3.0	3.3	3.5
30	2.7	3.1	3.5
40	2.1	2.4	2.6
Average	3.7	4.1	4.4

[1] Boards were 4 inches wide; feed speed was set to yield 20 knife marks per inch (i.e., F_t = 0.05 inch). Cutting-circle diameter was about 6 inches.
[2] Specific gravity of these species based on weight and volume at test were as follows:

Yellow-poplar 0.48
Sweetgum 52
White oak 65

When planing hardwood flooring, cutterhead speed is commonly fixed at a nominal 3,600 rpm because this is the synchronous speed of the usual motor mounted direct on the spindle; 3,600 rpm is also the highest speed at which knives can be consistently jointed so all track in the same path and share equally in the cut. Moulders, however, are frequently equipped with cutterheads to be operated at speeds of 4,800, 5,400, 6,000, and 7,200 rpm as well as 3,600 rpm. At the higher speeds, it is difficult to joint effectively so that usually the trace of only a single knife shows on the workpiece, in which case knife marks per minute (regardless of feed speed) are numerically equal to the speed of the cutterhead. Therefore, to get 20 knife marks per inch at these speeds, feed rates are as follows:

Cutterhead speed	Feed speed
Rpm	Feet per minute
3,600	15.0
4,800	20.0
5,400	22.5
6,000	25.0
7,200	30.0

In contrast, a jointed 16-knife flooring head rotating at 3,600 rpm will yield 20 knife marks per inch if fed at 240 feet per minute. (See equation 18-7.)

Definition of defects and their occurrence.—Davis (1962) found that among southern hardwoods, the oaks yielded planed wood with fewest defects. American elm, red maple, black tupelo, and sweetgum had most defects when planed; white ash, hackberry, hickory, and yellow-poplar were intermediate (table 18-1).

The best surfaces result when Type II chips are formed during the early part of knife engagement (fig. 18-35). For dry southern hardwoods, rake angles from 5 to 25° in combination with a clearance angle of about 20° (not less than 15°) are appropriate. Type II chips are more likely to occur with shallow cuts (less than ⅛-inch) than with deep cuts, but are difficult to form if wood is planed against a sloping grain.

Type I chips (fig. 18-3)—if the splits run below the cutting plane—cause **chipped grain** (fig. 18-39). Among planing defects in pine-site hardwoods, chipped grain is the most damaging to surface quality. It is most prevalent in red maple and hickory and least prevalent in yellow-poplar.

Type III chips (fig 18-5) tend to cause incomplete fiber severence and **fuzzy grain** (fig. 18-39 bottom); the defect is a particular problem when cutting wet wood with knives of low rake angle. Pine-site hardwoods do not commonly display this defect on planed surfaces; it is negligible in ash, oak, and hickory.

Dull knives, knives that have been jointed too many times between sharpenings, and knives with insufficient clearance angle cause **raised grain**—a roughened condition in which hard portions of the annual ring are raised above the softer portion but not torn loose from it (fig. 18-39). The defect may show up subsequent to machining, as the wood swells when exposed to high humidity. Raised grain is most common in American elm and hackberry, intermediate in the oaks, and rare in the ashes and hickories.

Chip marks are caused by shavings or fiber bundles that fold over and adhere to the cutting edge so that they are carried around and indented into the surface of the wood (fig. 18-39). Davis (1962) noted that chipped grain and chip marks can be easily distinguished by applying a few drops of water and waiting a few minutes. Chipped grain, which is caused by broken-out particles, will not be affected; chip marks, which are dents where wood is compressed, will swell as they absorb water and become conspicuous. Red maple and yellow-poplar frequently display chip marks, but oaks seldom do; hickory is intermediate. While a remedy is difficult to find, wiping each knife edge with a solvent-soaked rag sometimes helps. Reducing depth of cut or feed rate may alleviate the problem when inadequate blowpipe suction is a contributing factor.

Defect control.—The defect most commonly observed in planed hardwoods is a washboard effect caused by poor jointing (fig. 18-40). Properly jointed knives display a fine land (jointing plane) on each cutting edge (fig. 18-31 bottom). This land widens with repeated jointing; knives should be reground when the land becomes about 1/32-inch wide. For effective jointing, spindle bearings must be precise and cutterheads must be in dynamic balance at operating speed.

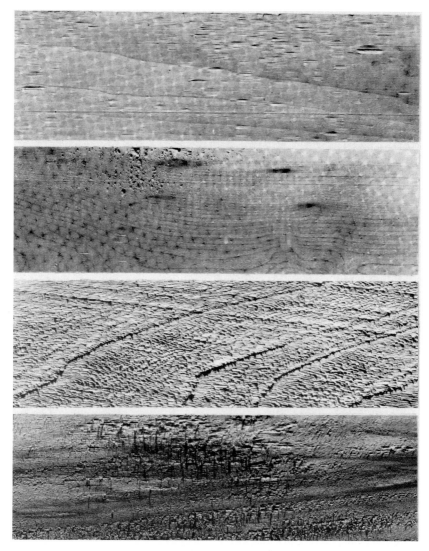

M-93419-F, M-93420-F, M-116452-F, M-116453-F

Figure 18.39.—Typical machining defects in hardwoods. (Top) Chip marks in yellow-poplar. (Upper center) Chipped grain in sugar maple (*Acer saccharum* Marsh.). (Lower center) Raised grain in American elm. (Bottom) Fuzzy grain in willow. (*Salix* sp.). (Photos from Davis 1962.)

Specific gravity and number of rings per inch of pine-site hardwoods seem to have little effect on frequency of defects in planed wood.

Cutting velocity is not of great importance in controlling planing defects, but the number of knife cuts per inch has strong negative correlation with number of planing defects, i.e., planing defects are minimum with 20 or more knife marks per inch for most pine-site hardwoods. In 1/16-inch deep cuts on three pine-site hardwoods plus yellow birch (*Betula alleghaniensis* Britton) and hard maple

M 34182 F

Figure 18-40.—(A) One knife mark per cutterhead revolution before jointing of planer knives. (B) Four knife marks per revolution after jointing. (Photo from Davis 1962.)

(*Acer saccharum* Marsh.), Davis (1962) found that for the five hardwoods as a group, 12 knife cuts per inch yielded 3½ times as many defect-free specimens as 6 knife cuts, while 20 knife cuts per inch yielded 4½ times as many (fig. 18-41). Cantin (1965) made a similar evaluation of northern red oak and white ash; the curve for northern red oak resembled that of white oak, while that for white ash resembled that of sweetgum in figure 18-41.

Shallow cuts leave fewer defects than deep cuts, with most difference observable as depth of cut in increased from 1/32- to 3/32-inch (table 18-32).

TABLE 18-32.—*Percent defect-free pieces after planing 10 southern hardwoods with four depths of cut[1]* (Data from Davis 1962)

Species	Depth of cut, inch			
	1/32	2/32	3/32	4/32
	---------------Percent---------------			
Ash, sp.	58	38	32	26
Elm, sp.	6	4	0	0
Hackberry	28	10	6	4
Hickory	46	6	14	6
Maple, red	40	28	30	14
Oak, northern red	74	56	36	28
Oak, white	58	34	22	24
Sweetgum	36	22	14	16
Tupelo, black	62	38	40	34
Yellow-poplar	64	36	44	34

[1] Upmilling; 30° rake angle; 8 knife marks per inch; wood at 6 percent moisture content.

Machining

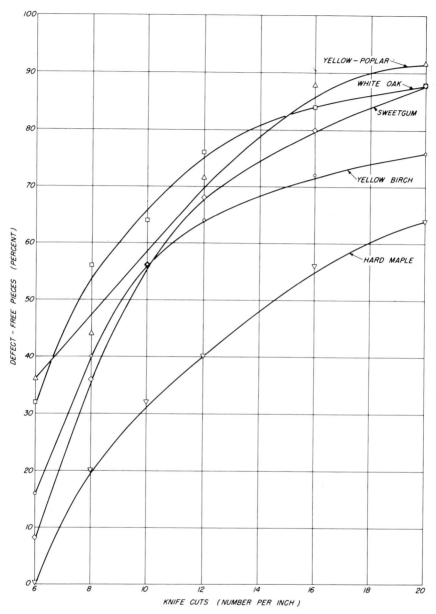

Figure 18-41.—Effects of knife marks per inch on quality of machined surface when planing 1/16-inch depth of cut from five hardwoods at 6-percent moisture content. Rake angle was 20°. (Drawing after Davis 1962.)

Davis (1962) found that planing defects in southern hardwoods were less serious when lumber was machined at 6 percent moisture content than at 20 percent; wood at 12 percent moisture content was intermediate (table 18-33). Stewart (1979b) concluded that optimum moisture content for machining hardwoods is between 8 and 12 percent.

Rake angle is a major determinant of surface quality obtained when planing, but sensitivity to rake angle varies with species. The oaks are not much affected and plane well through a wide range of angles. Hackberry, American elm, and red maple, however, may yield twice as many defect-free pieces at the optimum rake angle compared to the poorest one (table 18-34). Hardwood flooring plants that plane only oak may use rake angles of 10 or 15°. Factories that must plane numerous hardwood species frequently use a rake angle of 20°. Although angles smaller than 20° may give good results, power required is high and cutting edges dull rapidly.

TABLE 18-33.—*Percent defect-free pieces in wood of 10 hardwood species planed at three moisture contents[1]* (Data from Davis 1962)

Species	Moisture content		
	6 percent	12 percent	20 percent
	------Percent------		
Ash, sp.	53	39	35
Elm, sp.	18	5	6
Hackberry	20	8	7
Hickory	35	27	16
Maple, red	17	12	15
Oak, northern red	65	54	48
Oak, white	37	37	26
Sweetgum	46	38	36
Tupelo, black	36	40	30
Yellow-poplar	47	37	40

[1]Upmilling; 30° rake angle; about 33 knife cuts per inch; 1/16-inch depth of cut.

TABLE 18-34.—*Percent defect-free pieces in wood of 10 hardwood species planed with knives ground at six rake angles[1]* (Data from Davis 1962)

Species	Rake angle					
	5°	10°	15°	20°	25°	30°
	------Percent------					
Ash, sp.	69	70	72	73	79	53
Elm, American	24	24	48	33	19	18
Hackberry	37	47	75	93	54	20
Hickory[2]	—	74	—	81	—	74
Maple, red	43	61	57	33	34	18
Oak, northern red	66	96	95	92	87	65
Oak, white	74	98	95	93	74	37
Sweetgum	35	66	54	51	49	44
Tupelo, black	42	52	47	53	43	37
Yellow-poplar	66	75	75	67	67	48

[1]Cutting conditions not explicitly stated but believed to be as follows: upmilling, 33 knife cuts per inch, 1/16-inch depth of cut, 6-percent moisture content. Machined on a 4-knife cabinet planer.
[2]20 knife cuts per inch; 1/16-inch depth of cut; 6 percent moisture content. Machined on a 4-knife moulder.

A major manufacturer of moulders recommends rake angles from 5 to 25° depending on species and moisture content (table 18-35).

Cantin (1965) found that planer knives with 15° rake angle yielded most defect-free pieces, as follows:

Rake angle	Percent defect-free pieces		
	Red maple	Northern red oak	White ash
Degrees	------------------------*Percent*------------------------		
15	72	100	96
20	52	92	86
25	50	88	74
30	42	92	74

In these tests, depth of cut was 1/16-inch with 20 knife marks per inch; lumber was kiln dried.

TABLE 18-35.—*Rake angles recommended by Mattison Machine Works for planing (upmilling) airdry and kiln-dry wood of seven species of southern hardwoods* (Ekwall 1976)

Species and grain type	Air-dried wood	Kiln-dried wood
	---------------*Rake angle, degrees*---------------	
Ash..................................	20	15
Elm, American	25	20
Elm, winged	20	15
Hickory	15	10
Maple (curly)........................	10	5
Maple (plain)........................	20	15
Oak (plain)..........................	20	15
Oak (quartered)	15	10
Sweetgum.............................	20	15

Chipped grain is a major defect observable in planed pine-site hardwoods. Stewart (1970a, 1971b) found that the depth of the defect below the surface and not the frequency of occurrence determines the required amount of subsequent processing to remove chipped grain.

In a study of hard maple (*Acer saccharum* Marsh.) Stewart (1971b) found that the most severe chipped grain occurs at approximately 10° slope of grain which is frequently found near knots and areas of cross grain (fig. 18-42). The shallowest chipped grain was produced at 30 knife marks per inch and 1/16-inch depth of cut, and at moderate rake angles of 20 and 25° (fig. 18-43). At this feed per knife, about 1/64-inch would have been sanded away to remove the chipped grain when slope of grain does not exceed 5°; at greater slopes of grain, approximately 1/32-inch would have to be removed. When planing with a rake angle of 25°, the depth of defect increased more rapidly as the number of knife marks per inch decreased, and at a slope of grain of approximately 10° compared to slopes of grain less than 5° (fig. 18-44).

Stewart (1971b) also found that the severest defects occurred at pockets of short grain, and that the depth of the defect associated with these areas appeared

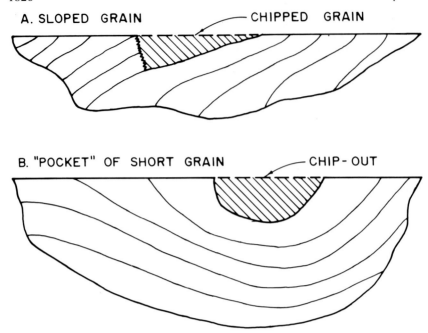

Figure 18-42.—(A) Sloped grain associated with chipped grain. (B) A pocket of short grain associated with chip-out. (Drawing after Stewart 1971b.)

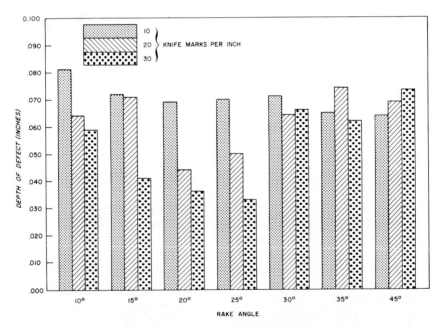

Figure 18-43.—Maximum depth of chipped grain in kiln-dried hard maple having 10° slope of grain and planed 1/16-inch deep at three feed speeds with knives ground at seven rake angles. (Drawing after Stewart 1971b.)

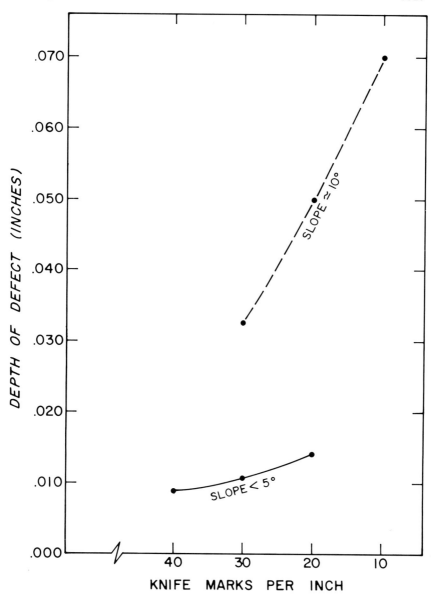

Figure 18-44.—Effect of slope of grain and knife marks per inch on maximum depth of chipped grain in kiln-dried hard maple planed 1/16-inch deep with rake angle of 25° (Drawing after Stewart 1971b.)

to be independent of the rake angles and feed rates studied. The chip-outs occur most frequently on the heart side of boards and often near an overgrown knot; they are also associated with wavy grain. To reduce defect occurrence, such areas are often removed during crosscutting and ripping operations; when only one side of a board will show, the sap side should be used.

Stewart and Crist (1982) found that knife planing slightly crushed red oak, yellow-poplar, hard maple, and aspen below the machined surface; such subsurface wood deformation was less in knife-planed than in abrasive-planed wood.

DOWN MILLING

Most conventional planing is accomplished by the up-milling process as described in the previous portion of this section. Some machining, however, is performed in the down-milling mode (fig. 18-45). Development of the **chipping headrig**, which reduces logs to square or rectangular timbers plus pulp chips (no sawdust), has further stimulated interest in down-milling (Koch 1964b).

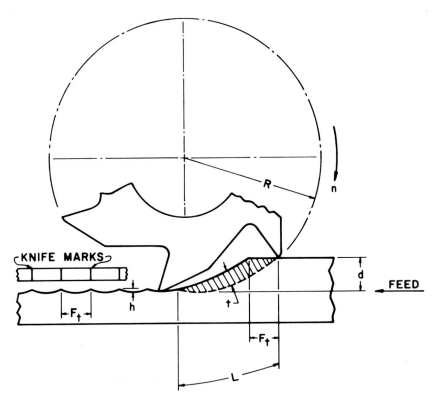

Figure 18-45.—Path generated by down-milling knife. The path of knife engagement is shorter than in up-milling because feed of workpiece shortens time knife is cutting.

    d   Depth of cut, inch
    F_t Feed per knife; translation of the workpiece between engagement of successive knives, inch
    h   Height of knife marks above point of lowest level, inch
    L   Length of knife path during each engagement with the workpiece, inches
    n   Revolutions per minute of cutterhead, r.p.m.
    R   Cutting-circle radius, inches
    $t_{avg.}$ Average thickness of undeformed chip, inch
    T   Number of jointed knives in the cutterhead
(Drawing after Koch 1964a.)

As with up-milling, the feed per jointed knife is given by the equation:

$$F_t = \frac{12F}{Tn} \qquad (18\text{-}7)$$

The distance between knife marks is reduced by decreasing the feed speed, increasing the cutterhead speed, or by increasing the number of jointed knives carried in the cutterhead.

If the knives have been jointed to a common cutting circle, the wave height can be expressed (Martellotti 1941, 1945):

$$h = \frac{R}{8} \left[\frac{F_t}{R - \frac{F_t T}{2\pi}} \right]^2 \qquad (18\text{-}11)$$

As in up-milling, wave height can be reduced by increasing the radius of the cutterhead and by decreasing the feed per knife.

For down-milling, the length of knife path engagement is as follows (Martellotti 1941, 1945):

$$L = R \text{ arc cos} \frac{R-d}{R} - \frac{F_t T}{2\pi R} (2Rd - d^2)^{1/2} \qquad (18\text{-}12)$$

As with up-milling, the average thickness of the undeformed chip can be stated:

$$t_{avg} = \frac{F_t d}{L} \qquad (18\text{-}10)$$

Down-milling at 300 fpm (⅛-inch-deep cut) with an eight-knife, 9-inch-diameter cutterhead turning at 3,600 rpm can be compared with up-milling in similar circumstances:

Dimension	Down-milling	Up-milling
	----------------- Inch -----------------	
F_t	0.130	0.130
h	$.551 \times 10^{-3}$	$.441 \times 10^{-3}$
L	1.022	1.100
t_{avg}	.016	.015

In down-milling, wave height, average chip thickness, and maximum chip thickness are greater than in up-milling; however, length of tool path is shorter for down-milling.

Because down-milling cuts thicker chips than up-milling, it requires substantially more cutterhead power (fig. 18-46). The attitude of the knife at engagement in down-milling is less conducive to advance splitting (Type 1 chip formation).

With sharp knives and in the conventional range of rake angles, the up-milling knife exerts very little pressure on the workpiece or tends to lift it away from the bed plate. Down-milling, however, holds the workpiece strongly against the bed plate due to the manner of knife engagement (fig. 18-34A). While this force may be of some advantage, the accompanying horizontal force vector, tending to uncontrollably accelerate the rate of feed, is usually a disadvantage. Since less feed power is required, however, the down-milling process frequently requires less total power than up-milling.

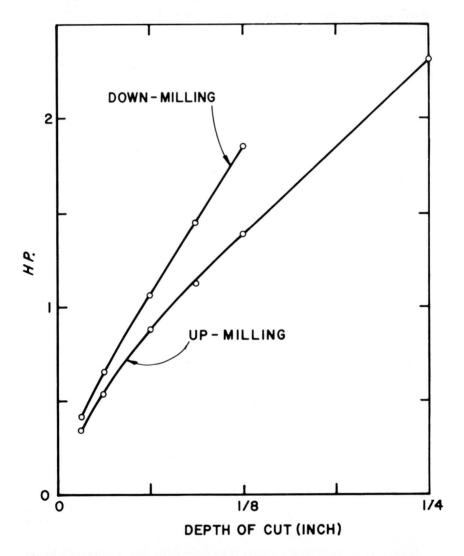

Figure 18-46.—Net cutterhead power per inch of workpiece width in up-milling, as compared to down-milling. F_t, 0.0895 inch; rake angle, 30°; feed speed, 206 fpm; cutting-circle diameter, 9.08 inches; cutterhead speed, 3,600 rpm; number of knives, eight; species, Douglas-fir at 8.9 percent moisture content. (Drawing after Koch 1956.)

Down-milling differs from up-milling in two additional respects. In down-milling the knife enters at the rough surface, thus wiping adhering bundles of fibers from the cutting edge before chip marks can be indented into the finished surface. Also, as figure 18-34 shows, down-milling chips are discharged horizontally along the workpiece.

18-5 PERIPHERAL MILLING ACROSS THE GRAIN (0-90 DIRECTION)

As one considers peripheral milling procedures that might plane lumber without chipping defects in areas where grain is sloped, two methods come to mind. The first, which was used in some early hardwood flooring machines, is not true peripheral milling but calls for mounting knives on the face of a horizontally disposed doughnut-shaped disk with diameter large enough so that a sweeping curved line is produced as the blades pass over the material being planed; this method is termed **face milling.** By this method the knives cut essentially in the 0-90 direction and chips removed are flake-like and suitable for use in structural flakeboards. Lutz et al. (1969a) and Heebink (1975), after tests of such a ring-head planer on Douglas-fir, found that better results were obtained from green wood than from wood with 16-percent moisture content, that quality of flakes produced on a ring-head planer is equal to that produced on a disk flaker, but that planed surfaces produced were rougher and had more slivers than those produced by a conventional cylinder head. Flakes cut by a ring-head planer made acceptable structural flakeboard. Stewart (1974a) milled basswood (*Tilia americana* L.) with a ring-head planer and concluded that a face milling operation could be designed to surface hardwoods and manufacture flakes simultaneously. Readers interested in the geometry and power requirement of face milling should find useful the work of Yokochi and Fukui (1978ab).

A second method, which has been extensively studied by Stewart (1970b, 1971c, 1974b, 1975a), Stewart and Lehmann (1973), and Stewart and Parks (1980), calls for a conventional peripheral-milling planer head to machine hardwood panels of short cuttings across the grain(fig. 18-47). Stewart (1975a) found that knife planing across the grain of hardwoods produces a shallower maximum-depth of defect than planing along the grain, and also yields a flake more useable in structural flakeboards than the spiral shavings typically produced when planing along the grain.

When planing kiln dried hardwoods across the grain:
- Surface quality generally improves as specific gravity of the wood increases.
- Larger rake angles yield better surfaces than small rake angles, but the effect is more pronounced on low-density woods.
- To yield satisfactory surfaces, feed speed should be slowed to yield 10 knife marks an inch or more.
- At usual feed speeds (more than five knife marks per inch), surface

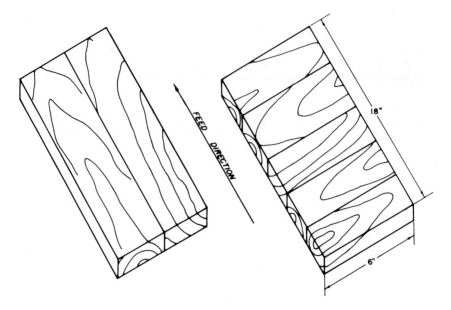

Figure 18-47.—Panels illustrating planing parallel to the grain (left), and across the grain (right).

quality is unaffected by depths of cut up to ⅛-inch.
- Depths of defect were less than 1/64-inch with rake angles of 30° or more.
- Slightly less power is required than when planing along the grain.
- The surface quality of panels knife planed across the grain equals or exceeds the quality of surfaces machined with a 36-grit abrasive belt (fig. 18-48).

18-6 PERIPHERAL MILLING ACROSS THE GRAIN (90-90 DIRECTION)

Wood—usually kiln-dried—is peripherally milled across the grain in the 90-90 direction when patterns are shaped on the end-grain of furniture panels (sec. 18-11), when routing and carving (sec. 18-15), and when mortising and tenoning (sec. 18-16).

Green wood may be peripherally milled across the grain in the 90-90 direction on certain classes of chipping headrigs (sec. 18-9), and in drum chippers and tree felling devices that use a rotary cutting bar (figs. 16-49, 16-52, and 18-49).

At 95 feet per minute feed speed, net power requirements of a drum chipper equipped with two knives set at 52.5° rake angle on a 48-inch cutting circle vary from less than 10 horsepower to about 600 horsepower depending on tree species and diameter (fig. 16-51).

At 1-mile-per-hour feed speed, net power requirements of a tree felling bar equipped with four straight knives set at 38.5° rake angle on 16.5-inch cutting

Machining

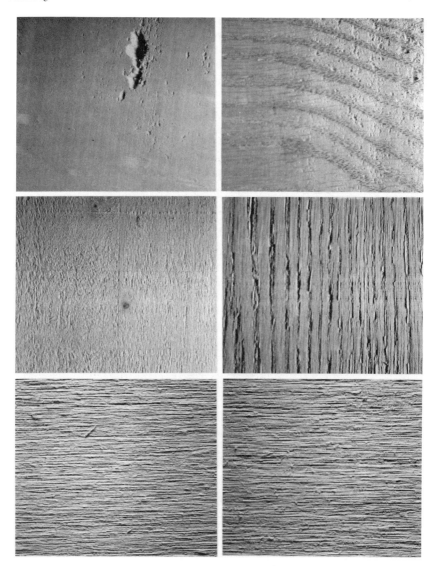

Figure 18-48.—Kiln-dried yellow-poplar (left) and northern red oak (right) planed 1/16-inch deep by various systems. (Top) Knife planed with 25° rake angle at 20 feet per minute; yellow-poplar at left was planed across the grain while the oak at right was planed parallel to the grain. Chipped grain is evident in areas of short grain; see figure 18-42. (Center) Knife planed across the grain with 52° rake angle; yellow-poplar at left has 20 knife marks per inch and oak at right has 10. (Bottom) Abrasive planed parallel to grain with 36-grit belt. (Photos from Stewart and Parks 1980).

circle vary from 100 to 600 horsepower, depending on tree diameter, species, and felling bar rpm (fig. 16-50).

As noted in sec. 18-3, feller-bunchers are available that utilize a small-diameter rotary cutting bar equipped with helically disposed knives in place of

Figure 18-49.—Feller-buncher felling head equipment with 24-inch-long 2-1/2-inch-diameter rotary cutting bar. The bar turns at 2,400 rpm and can cut through a 14-inch northern red oak in 8 to 10 seconds. (Photo from Drott Manufacturing Division of J. I. Case Co.)

shear blades (figs. 18-49 and 18-50). Such cutting bars sever tree stems without the objectionable splits caused by shears; and are reported to harvest about 70 softwood trees per hour (Skory 1979); productivity data on hardwoods are not available.

To determine force and power requirements to fell trees, Mattson (1980) extensively studied rotary cutting across the grain of bigtooth aspen (*Poplus grandidentata* Michx.), sugar maple (*Acer saccharum* Marsh.), and red pine (*Pinus resinosa* Ait.) from Michigan. In his experiments he used twelve two-knife helical cutterheads (fig. 18-50) with cutting-circle diameter of 2.5 inches, clearance angle of 10°, and rake and helix angles as follows:

> Normal rake angle: 15°, 30°, and 45°
> Helix angle: 0°, 20°, 40°, and 60°

His experimental work was accomplished on a milling machine with table speeds arranged to yield feeds per knife in the range from 0.002 to 0.02 inch.

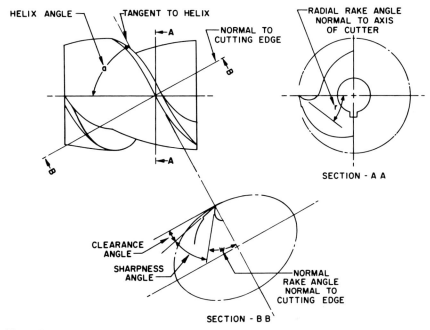

Figure 18-50. Rotary cutting bar for tree felling with definition of helix angle and rake angle. (Drawing after Mattson 1980.)

Mattson evaluated the rake and helix angles that produced the lowest axial force (figs. 18-51A) and torque. He found that a cutter with 40° helix angle and 30° rake angle was an optimum compromise. Increasing helix angles up to 40° reduced cutting torque, but larger angles caused increased torque (fig. 18-51B). Cutting forces decreased with increasing rake angle, but cutters dulled rapidly at rake angles larger than 30°. Mattson also found that torque on cutterheads increased linearly with feed per knife (fig. 18-52).

Power requirement of the cutter depended on both cutter rotational speed and the speed with which the cutter moved through the wood; when the cutter moved through the wood at constant speed, horsepower requirement increased with an increase in cutter rotational speed (fig. 18-53).

Mattson (1980) found that forces acting on the cutterheads, and power required, were proportional to wood specific gravity, i.e., dense woods exerted more force on the knives and required more horsepower to cut than woods of low density. Frozen wood required 20 to 25 percent more power to cut than unfrozen green wood.

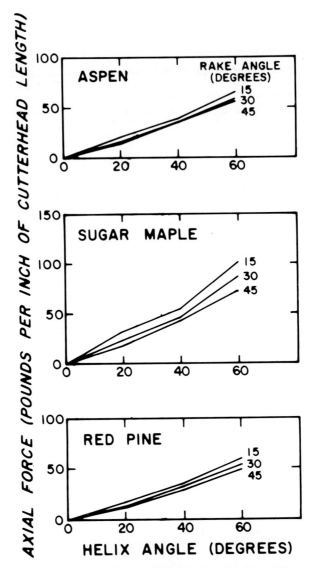

Figure 18-51A.—Axial force on 2.5-inch-diameter, two-knife cutterheads related to normal rake angle and helix angle when cutting green wood of three species across end grain. Graphs represent pooled data for feeds per knife in the range from 0.020 to 0.002 inch. (Drawing after Mattson 1980.)

Christopherson[*] obtained experimental data on a 2½-inch diameter rotary cutting bar used to delimb hardwoods, and concluded such cutting bars would make clean cuts with little or no branch stub remaining, and that they would be rugged, self cleaning, and easily sharpened. He found that horsepower requirements were low enough to allow the use of hydraulic motors.

---

[*]Christopherson, N.S. 1984. Removing single limbs using a rotary auger cutter. Res. Note NC-311, North Central Forest Experiment Station, USDA Forest Service. 5p.

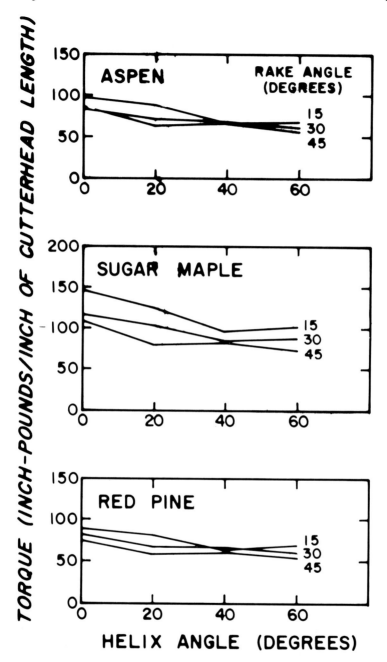

Figure 18-51B.—Maximum torque observed on 2.5-inch-diameter, two-knife cutterheads related to normal rake angle and helix angle when cutting green wood of three species across end grain. Graphs represent pooled data for feeds per knife in the range from 0.020 to 0.002 inch. (Drawing after Mattson 1980.)

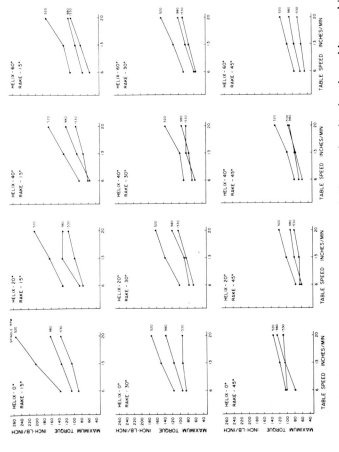

Figure 18-52.—Maximum torque observed on 2.5-inch-diameter, two-knife cutterheads related to table speed (i.e., .002 to .020 inch feed per knife), normal rake angle, and helix angle when cutting green unfrozen sugar maple across end grain. (Drawing after Mattson 1980.)

Machining

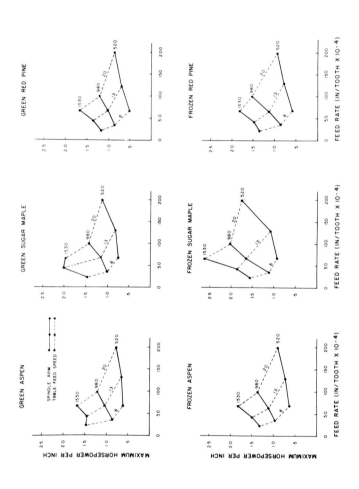

Figure 18-53.—Maximum horsepower required by 2.5-inch-diameter, two-knife cutterhead with 30° rake angle and 40° helix angle related to feed per knife (tooth), table feed speed, and cutterhead speed when cutting wood of three species across the grain. The graph represents pooled data for all rake angles and helix angles evaluated. (Drawing after Mattson 1980.)

18-7 PERIPHERAL MILLING (PLANING) WITH HELICAL CUTTERS

The previous section described the process of cutting 90-90 across the grain with a rotary cutting bar equipped with helical knives. In such applications interest is focused on the quality of the severed surface, i.e., avoidance of splits.

Helical cutters are being used increasingly for planing kiln-dried hardwoods (fig. 18-143) because they are quieter than regular cutters (Stewart 1975b, 1978a; Boles and Flanigan 1976). Furthermore, surface quality is generally better when machining parallel to the grain with a helical cutting edge (fig. 18-54) than with a straight cutting edge (Stewart 1971d), and cutterhead torque is less (Kimura 1980). For further discussion of helical planing heads see text related to figure 18-143.

Stewart and Lehmann (1974) found that cross-grain planing with a segmented helical cutter (fig. 18-55 top) yielded good surfaces (18-55 bottom) and resultant flakes could be manufactured into strong and linearly stable particleboard. Northern red oak was one of the species they evaluated.

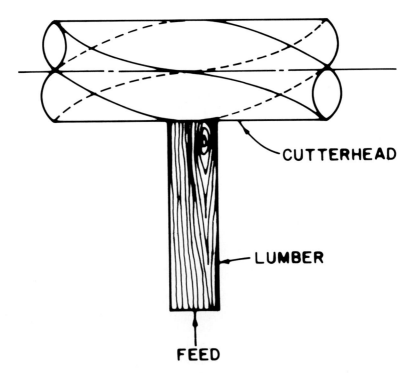

Figure 18-54.—Arrangement to use a helical cutter to plane lumber parallel to the grain.

Headrig and edger chippers cutting in the 90-0 and 90-90 modes are in widespread use in mills throughout North America (see sec. 18-9), and to a lesser extent throughout the world. They have been adopted because of their

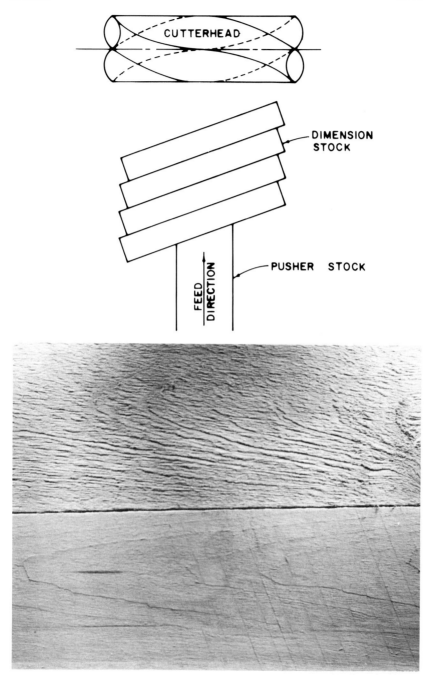

Figure 18-55.—(Top) Arrangement to use a helical cutter to plane lumber across the grain. (Bottom) A uniformly textured basswood (*Tilia Americana* L.) surface machined when cuttings were aligned parallel to a tangent of the helix angle at high feed rates (upper photo) compared with a smooth surface when cuttings were parallel to the rotational axis of the cutterhead at slower feed rates (bottom photo). (Drawings and photos from Stewart and Lehmann 1974.)

productivity per man-hour and because they make no sawdust. The chippers are not without disadvantages, however, in that these peripheral- and end-milling cutterheads take large bites per tooth to make pulp chips ⅝- to ⅞-inch long and therefore tear out grain around knots and tend to splinter board edges. These rough surfaces require smoothing by planers adjusted to remove about ⅛-inch from board faces and edges. Because of the low value of planer shavings relative to lumber, however, mill operators are reluctant to take this remedial action.

Koch (1976a) tested and described a system in which flaking heads arranged to follow headrig and edger chipper heads would smooth machined surfaces and produce high-value flakes of near optimum dimensions for structural flakeboard. In the proposed concept, eight knives are closely grouped in a 45° helix on a cutterhead tipped at 45° angle to the direction of workpiece feed. Each knife is set out in cutting radius the thickness of one flake (about 0.015 inch) from the immediately preceding knife. Thus an eight-knife head would, in each revolution, remove eight flakes 0.015 inch thick and perhaps 1 inch in length while making a cut about ⅛-inch deep, thereby removing traces of torn grain caused by chipper knives. In the time interval following engagement of the eight closely grouped knives, and their re-engagement during the next revolution, the workpiece advances a selected distance (e.g., 1 inch) thus determining flake length.

18-8 SAWING

The kinematics and cutting forces in sawing have been described by Koch (1964a, p. 179-284); techniques of saw fitting and sharpening were described by Hanchett (1946) and Quelch (1970). Here the sawing process is briefly described as necessary to present the available information specific to sawing pine-site hardwoods.

Common to all saws are multiple teeth arranged to cut in sequence. The teeth may be formed into, or attached to, the periphery of round flat plates (circular saws), chains, (chainsaws), thin continuous metal bands (bandsaws), or long flat rectangular plates or webs arranged in multiples (sash gangsaws).

Saw teeth are designed to make ripping cuts parallel to the grain for headrigs, gangsaws, resaws, and edgers; they are made to cut across the grain for chainsaws, log cut-off saws, and trim saws.

Headrigs designed to saw logs into timbers, cants, thick lumber, and boards may have log carriages with reciprocating motion; with this arrangement lumber is ripped from each log one board at a time with a single circular saw or bandsaw. Frequently used are headrigs that saw logs into wany-edged cants or lumber in a single pass through the machine, e.g., with multiple circular saws (scrag mills), or multiple-band saws (twin or quad bandmills). Less frequently used on pine-site hardwoods are oscillating web saws (sash gangsaws).

Hardwood cants produced on a headrig are resawn into lumber of smaller dimension by multisaw ripsaws (gangsaws carrying bandsaws or circular saws, or infrequently, sash gangsaws); alternatively, a band resaw (usually a vertical linebar resaw) is used to give greater flexibility in resawing. By adjustment of the distance between linebar (guide) and bandsaw, lumber of any thickness can be ripped from a cant; cants that will yield several boards are conveyed back to the infeed end of the linebar resaw for additional cuts as required.

Except for chainsaws, crosscut saws in mills are virtually all circular. The design of ripping and crosscutting teeth is controlled by the size of timber to be sawn, feed speed, moisture content and density of the wood, and smoothness of surface desired.

FACTORS INFLUENCING SAWING PROCEDURES

Sawing procedures for hardwoods growing on southern pine sites are strongly influenced by the density, form, and quality of wood obtainable from the trees. The oaks listed in table 2-18 comprise 47.8 percent of the hardwood volume on pine sites and hickory amounts to 8.5 percent of the volume. These, and most other pine-site hardwoods, are dense, heavy woods. Only six of the species have specific gravity equal to, or less than, the specific gravity of slash and longleaf pine, i.e., 0.53 based on ovendry weight and green volume (table 7-7). These six species (hackberry, black tupelo, red maple, sweetgum, sweetbay, and yellow-poplar) comprise only 23.6 percent of the volume of pine-site hardwoods.

Sawing techniques are, to a large extent, dictated by the small size of pine-site hardwood trees. Of those 5 inches and larger in dbh, about 46 percent measure 10.9 inches or less in dbh outside bark and 75 percent measure 14.9 inches or less in dbh (table 2-4). If one could get a 16-foot log from such trees, the top diameter inside bark of this first log would be 12 or 13 inches maximum, would average about 9 inches, and in many cases would be less than 5 inches.

Moreover, because of shortness of bole and crookedness of stem, logs must frequently be reduced in length to 10, 8, 6, or even 4 feet, rather than the 16 feet that is standard for good southern pine or bottomland hardwood sawlogs. To yield upper-grade lumber from its outer portions a log should be 12 to 16 feet long and over 12 inches in diameter inside bark at the small end. Because logs from pine-site hardwoods are generally smaller and shorter than these minimums, sawing patterns aim at high production rather than standard lumber grades. This generally means that these short, small-diameter logs are **live-sawn,** i.e., sawn through and through without turning, or are slabbed on two sides in one pass through twin saws or cutterheads, turned flat, and resawn into desired cants or boards with multiple circular saws or bandsaws.

These considerations focus the sawmiller's attention on two-side chipping headrigs (see sec. 18-9), single or multiple bandsaws, and single or twin circular saws for ripping cuts during primary log breakdown, bandsaws and circular saws for resawing cants parallel to the grain, circular saws for edging, and circular saws for end trimming or defect trimming green wood across the grain. Additionally, a variety of circular saws are required to saw dry hardwoods in ripping cuts parallel to the grain and in trim cuts across the grain. Also, chainsaws and circular saws are used to crosscut tree stems into logs and bolts. Although more or less obsolete, manual crosscut saws are sometimes used to cut tree stems into shorter lengths. The technology of manual crosscut saws is not discussed here; readers needing such information should consult Miller (1978).

The remainder of this section will describe the cutting action of bandsaws, sash gangsaws, and circular saws. The following sections will discuss chipping headrigs, mill layout for primary breakdown of low-grade, short, small-diameter logs; some comment will be addressed to yield and sawing time.

BANDSAWING

Bandsaw teeth cut orthogonally (fig. 18-56B). When rip sawing the length of a log or timber, the teeth cut across the grain in the 90-90 mode (fig. 18-1). In general, **swage-set** wide bandsaws (fig. 18-56C, 18-57A,D) are used for longitudinal cutting in primary manufacturing because they make less sawdust (have a narrower kerf), and can cut wider boards than circular saws. In the woodworking shop, **spring-set** narrow bandsaws (fig. 18-57E) are used because they can cut curves and irregular shapes that are difficult to cut with other tools. All teeth require clearance between kerf wall and saw plate. The swage tooth (fig. 18-57D) is formed by upsetting the tooth tip against an anvil; teeth are spring set (fig. 18-57E) by bending alternate teeth sideways.

Nomenclature for bandsaws is indicated in figures 18-56 and 18-57.

Cutting edge velocity is:

$$v = \sqrt{c^2 + f^2} \qquad (18\text{-}13)$$

Feed per tooth, or bite, is:

$$t = \frac{pf}{c} \qquad (18\text{-}14)$$

The volume of wood V (cubic inches) removed by each swaged tooth as it travels with kerf k (inch) through the workpiece of depth d (inches) is:

$$V = tdk = \frac{pfdk}{c} \qquad (18\text{-}15)$$

Swaged-tooth wide bandsaws for primary manufacture.—To resist feeding forces and stay on the saw whelels (fig. 18-58) bandsaws require tension making them tightest on the cutting edge and stiff throughout their width. Hanchett (1946) and Quelch (1970) give instructions for tensioning bandsaws with power-driven stretcher rolls and by application of heat.

In ripping, cutting force on a bandsaw tooth is comprised of several elements: severing fibers in the 90-90 direction, shearing the chips from the kerf boundary, dragging the expanded chips past the kerf boundary, chambering the sawdust in the gullet space, and accelerating the sawdust to ejection velocity (fig. 18-59). Power consumed to accelerate bandsaw sawdust to not great—perhaps 5 hp on a headrig, and the accelerational force on each bandsaw tooth is also low—about 1 pound per tooth in the cut (see Koch 1964a, p. 141-144 and 227 for computational method). Forces to shear chips from the kerf boundary, drug them past the boundary, and chamber them are not easy to determine; in aggregate they might

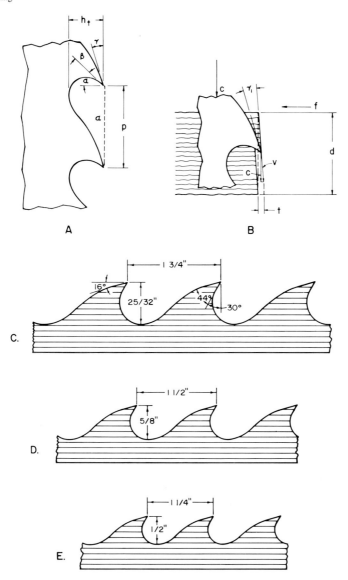

Figure 18-56.—(A and B) Bandsaw nomenclature.

- α Rake angle, degrees
- β Sharpness angle, degrees
- λ Clearance angle, degrees
- c   Saw velocity, fpm
- f   Feed speed, fpm
- v   Resultant velocity of cutting edge with respect to workpiece, fpm
- p   Tooth pitch, i.e., distance between teeth, inches
- h_t Depth of gullet, inches
- a   Area of tooth gullet, square inches
- d   Depth of workpiece, inches
- t   Tooth bite, i.e., depth of cut per tooth, inch

(C) Tooth form suitable for headrigs or linebar resaws cutting dense hardwoods; a tooth pitch of 2 inches with depth of 7/8-inch is also in common use. (D) Tooth form for resawing green hardwoods. (E) Commonly used tooth for resawing dry pine-site hardwoods.

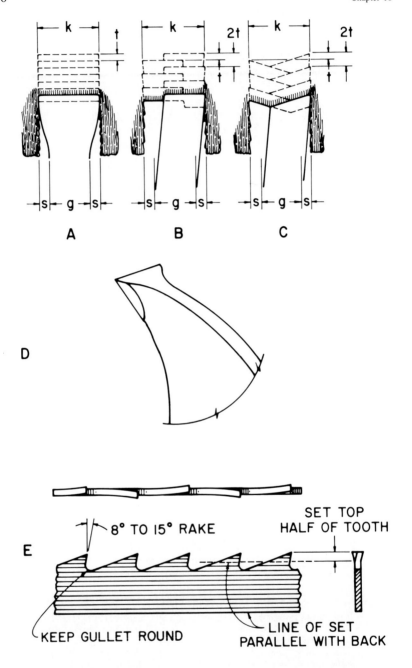

Figure 18-57.—Side clearance and shape of cut of bandsaw teeth. Width of kerf in inches = k; thickness of saw plate in inches = g; set or swage to each side of saw plate in inches = s; for swaged teeth, k is approximately equal to width of swage. (A) Swage set. (B) Spring set with no top bevel. (C) Spring set with top bevel. (D) Swaged tooth for wide bandsaw cutting dense hardwoods. (E) Tooth contours for spring-set narrow bandsaws. On dense hardwoods tooth styles B and C frequently perform well in cases where style A performs poorly.

Figure 18-58.—Eight-foot hand mill. (Photo from Filer and Stonewell Company.)

approximately equal the severance force. The primary tooth load is from severing fibers in the 90-90 direction. Tables 18-5C through 18-25C provide data on parallel cutting forces (F_p) and normal cutting force (F_n) for orthogonally cutting knives 0.1 inch long. These data are for slow-speed cutting (5 inches per minute); at normal cutting velocities of 8,000 to 10,000 fpm, parallel cutting forces—but probably not normal forces—may be significantly lower.

Cutting forces are directly proportional to the length of the cutting edge. Also, they are proportional to the specific gravity of the wood being cut (figs. 18-60 and 18-61). Additionally, the parallel cutting force is proportional to the thickness of the undeformed chip, i.e., feed per tooth (figs. 18-62 and 18-63).

These average cutting forces can be sizeable. For example, when cutting kiln-dry oak or hickory at a feed per tooth of 0.060 inch with 20° rake angle, the severance load on a tooth with cutting edge 0.1 inch long will average nearly 60 pounds (fig. 18-63); peak loads will be substantially higher. Green oak or hickory cut with 30° rake angle would offer less resistance—averaging about 35 pounds per tooth (fig. 18-62).

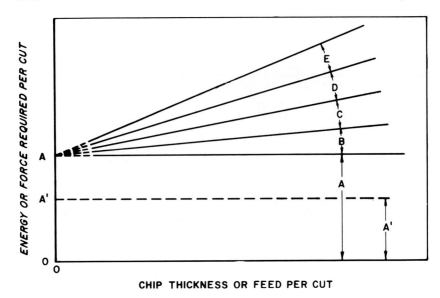

Figure 18-59.—Graphical analysis of cutting resistance in the ripsawing process. A, fiber indentation and incision phase of cutting; A^1, lower indentation and incision energy for sharper cutting edge; B, chip formation; C, chip breakage and associated friction; D, chip removal and associated friction; E, chip acceleration. (Drawing after Koch 1964a, p. 186.)

Cutting energy per unit volume of wood removed, i.e., specific cutting energy, is also proportional to wood specific gravity, but is inversely proportional to chip thickness or bite per tooth. Thus, white oak requires substantially more energy to cut than cottonwood (*Populus deltoides* Bartr. ex Marsh.), but energy expenditure when cutting both species can be reduced by increasing bite per tooth (fig. 18-64).

Because of these relationships, underfeeding can substantially increase energy consumed during sawing. For example, a narrow-kerf bandsaw cutting cottonwood might be limited (by gullet area and tooth strength) to a feed per tooth of 0.044 inch; should a feed per tooth of only 0.025 be used, energy consumed will be 50 percent greater than if the saw were fed at 0.044 inch, as follows (see fig. 18-64):

$$\frac{2.78 \times 0.025^{-0.72026}}{2.78 \times 0.044^{-0.72026}} = 1.5$$

The maximum feed per tooth is limited by tooth strength and gullet capacity; as a rule of thumb it should not exceed saw plate thickness plus 10 percent (Allen 1975). See table 18-36 for saw gauges.

Quelch (1964), in his treatise on sawmill feeds and speeds appropriate for timber of western Canada, recommends that feed per tooth on bandsaws should equal saw blade thickness less 4 percent for each inch less than 10-inch saw width, or plus 4 percent for each inch over 10-inch saw width; on saws 10 inches

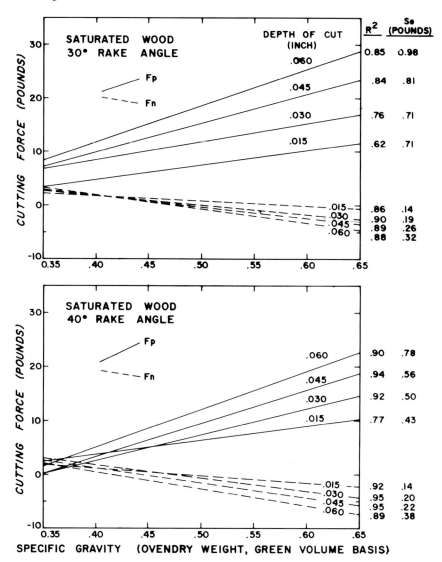

Figure 18-60.—Effect of wood specific gravity on parallel (F_p) and normal (F_n) cutting forces per 0.1 inch of knife when orthogonally cutting saturated pine-site hardwoods across the grain in the 90-90 direction with two rake angles and four depths of cut. (Regression lines derived from tables 18-5C through 18-26C); coefficients of determination (R^2) and standard error of estimates (S_e) are indicated.)

wide, feed per tooth should equal blade thickness. For oaks and hickories, with specific gravities up to 50 percent greater than western woods, such feeds per tooth should probably be considered near maximums because tooth loads when cutting dense hardwoods can be large (figs. 18-61 and 18-62).

The data in figure 18-64 permit ready computation of net horsepower requirement to make a bandsaw cut in green hardwood logs or cants. Needed informa-

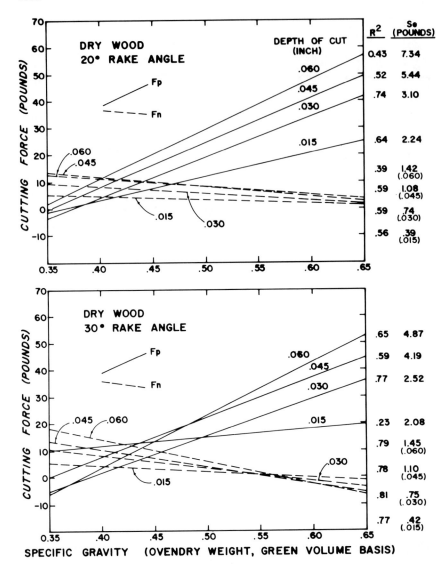

Figure 18-61.—Effect of wood specific gravity on parallel (F_p) and normal (F_n) cutting forces per 0.1 inch of knife when cutting dry pine-site hardwoods across the grain in the 90-90 direction with two rake angles and at four depths of cut. (Regression lines derived from tables 18-56C through 18-26C; coefficients of determination (R^2) and standard error of estimates (S_e) are indicated.)

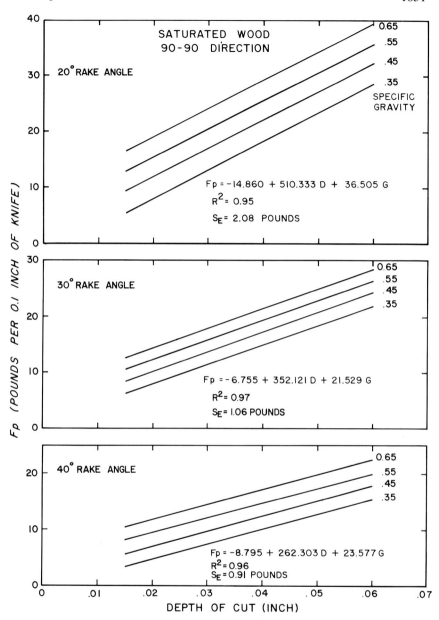

Figure 18-62.—Effect of depth of cut on parallel cutting force (F_p) when orthogonally cutting saturated pine-site hardwoods of four specific gravities (green volume, ovendry weight basis) across the grain in the 90-90 direction with a knife having 20, 30, or 40° rake angle. (Based on regression analysis of data in tables 18-5C through 18-26C.)

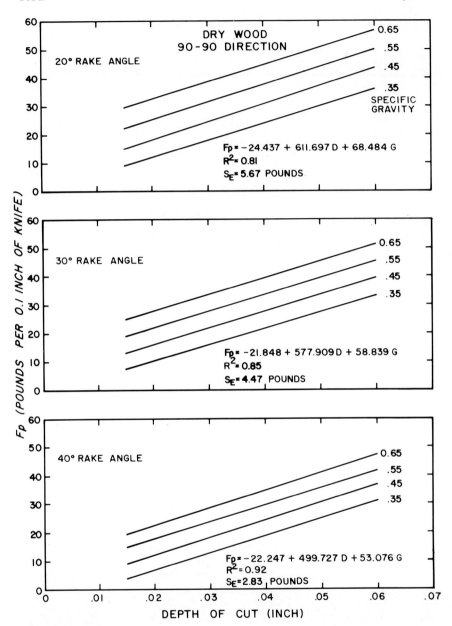

Figure 18-63.—Effect of depth of cut on parallel cutting force (F_p) when orthogonally cutting dry (10% MC) pine-site hardwoods of four specific gravities (green volume, ovendry weight basis) across the grain in the 90-90 direction with a knife having 20, 30, or 40° rake angle. (based on regression analysis of data in tables 18-5C through 18-26C.)

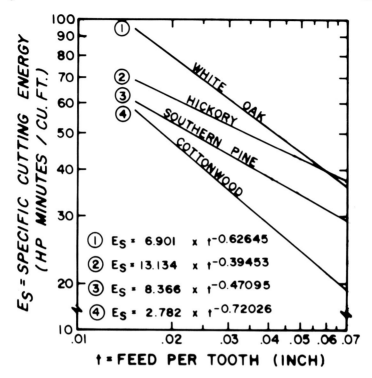

Figure 18-64.—Specific cutting energy to remove kerf when bandsawing green logs and cants of four species at various chip thicknesses, i.e., feeds per tooth. (Drawing after Allen 1975.)

tion includes feed per tooth (equation 18-14), log feed speed (f), and width of kerf (k). With a 10-inch-wide, 16-gauge bandsaw swaged to twice the blade thickness plus one gauge, width of kerf will be about 0.145 inch (table 18-36); feed per tooth might be about equal to blade thickness or 0.065 inch. With saw velocity of 10,000 feet per minute and tooth spacing of 1.75 inches, feed speed of the log would be 371 fpm. A single saw cutting through a 10-inch-deep cut would therefore remove 3.74 cubic feet of wood per minute. From figure 18-64 it is seen that, at a feed per tooth of 0.065 inch, specific cutting energy for oak or hickory is about 40 hp minutes per cubic food removed. Therefore, net sawing power required for this single bandsaw cutting 10 inches deep in oak or hickory at 371 feet per minute is 3.74 x 40 or 150 hp. To this must be added the idling power consumed by the saw when it is not in a cut.

TABLE 18-36.—*Saw gauge (Birmingham wire gauge) equivalent in inches and millimeters.*

Gauge of blade	Fraction of inch	Inch (decimal)	Millimeters
0	11/32 scant	0.340	8.63
1	5/16 scant	0.300	7.62
2	9/32	0.284	7.21
3	¼ full	0.259	6.57
4	15/64	0.238	6.04
5	7/32	0.220	5.59
6	13/64	0.203	5.18
7	3/16 scant	0.180	4.57
8	5/32 full	0.165	4.19
9	5/32 scant	0.148	3.76
10	1/8 full	0.134	3.40
11	1/8 scant	0.120	3.05
12	7/64	0.109	2.77
13	3/32	0.095	2.41
14	5/64 full	0.083	2.10
15	5/64 scant	0.072	1.82
16	1/16 full	0.065	1.65
17	1/16 scant	0.058	1.47
18	3/64	0.049	1.24
19		0.042	1.06
20		0.035	0.89
21	1/32	0.032	0.81
22		0.028	0.71
23		0.025	0.64
24		0.022	0.56
25		0.020	0.51
26		0.018	0.46
27	1/64	0.016	0.41
28		0.014	0.36
29		0.013	0.33
30		0.012	0.30

Feed per tooth is limited by gullet capacity as well as by cutting force exerted on each tooth. Quelch (1964) has provided a procedure to compute maximum depth of cut permissible before gullets become overloaded. He found that for tooth shapes in common use (fig. 18-56C) gullet area is about as follows:

$$\text{Gullet area} = a = \frac{p \times h_t}{1.75} \qquad (18\text{-}16)$$

For example, a bandsaw with 1.75-inch pitch and 25/32-inch gullet depth would have a gullet area of about 0.78 square inches. The capacity of the gullet to chamber sawdust is only about 70 percent of this gullet area; i.e., 0.55 square inch in the example given. The depth of cut that can be made with this tooth without overloading the gullet is dependent on the feed per tooth; at a feed per tooth of 0.065 inch a conservative estimate of a maximum depth of cut is therefore 0.55/0.065 = 8.46 inches. Because sawdust starts to spill from the bottom of the cut just before the tooth emerges from the cut, 75 percent of one tooth pitch can be added to this previously computed value to yield a maximum depth of cut of 8.46 + .75 x 1.75 = 9.8 inches.

Readers interested in graphically presented relationships that ease determination of optimum cutting conditions should find Suchsland's (1980) publication useful.

Swage widths for bandsaws cutting green hardwoods are variously computed. Some bandsaw manufacturers suggest sawblade thickness plus five gauges. Another rule of thumb calls for teeth to be swaged to twice blade thickness plus one gauge. In either case, swage width increases as thicker and wider blades carried on larger wheels are employed to increase feed per tooth and depth of cut (table 18-37).

TABLE 18-37.—*Relationship of wheel diameter, saw gauge, saw width, and saw power for swage-set bandsaws cutting pine-site hardwood logs and timbers*

Wheel diameter (inches)	Saw thickness[1]		Saw width	Tooth[2] pitch	h_t/p^3	Saw[4] power
	BWG[5]	Inch	----------Inches----------			Horsepower
54............	17	0.058	7 to 10	1-1/4 to 2	0.38	100
60............	16	.065	8 to 10	1-1/2 to 2	.40	125
66............	16	.065	10	1-1/2 to 2	.42	150
72............	15	.072	10 or 12	1-3/4 to 2	.44	175
84............	14	.083	12	1-3/4 to 2	.46	200
96............	13	.095	15	2-1/4 to 2-3/4	.48	250

[1]Saws with very high strain (force between wheels) may use thinner saws.
[2]For saw velocities from 7,000 to 11,000 fpm; smaller pitches are for slower velocities.
[3]Ratio of depth of gullet to tooth pitch.
[4]With saw velocity of 10,000 fpm and medium-pitch teeth; slower saw speeds with smaller tooth pitches require less power.
[5]Birmingham wire gauge.

Practical strength considerations make a 44° sharpness angle fairly standard; sharpening, swaging, and shaping tools are designed with this in mind. The tooth proportions shown in figure 18-56C are in general use; gullets are rounded, and backs of teeth are full to give maximum strength. Rake angle is commonly 30°, clearance angle 16°. For band headrigs, tooth pitches of 1½ to 2 inches matched with gullet depths of 5/8- and 7/8-inch are in common use. In the South, where frozen wood is uncommon, some researchers favor a tooth with a shorter rake face and smaller gullet than shown in figure 18-56. This more pyramidal tooth is stiffer and cuts more smoothly, but must be fed more slowly because of its limited gullet (Jones1965; Thrasher 1977).

Teeth are first swaged, next lightly faced on a grinding machine, then side-dressed with a pressure-type swage shaper, and finally ground on the face. Teeth should be straight, and swaged with 4° clearance behind the side-cutting edges.

Cutting corners should be perfectly formed, sharp, and shaped to have 6° side clearance (fig. 18-57A,D).

Kirbach and Bonac (1979ab) found that the top clearance angle for industrial ripsaws should not be less than 6°; they concluded that an 8° clearance angle would avoid interference and afford greater tooth strength than the 16° clearance angle customarily used.

The force (**strain**) between wheels (fig. 18-58) required to properly stretch the saw can be expressed in terms of blade width and thickness, as follows:

$$F = (1000)QWg \qquad (18\text{-}17)$$

where:
- F = total upward acting force applied to the spindle of the upper wheel, i.e., strain, pounds (minimum practical operating level)
- W = width of blade (gullet to back), inches
- g = thickness of blade, inches
- Q = a constant; generally 10 for headsaws and 8 for resaws

For example, a 14-inch, 14-gauge headsaw requires a minimum of 11,620 pounds strain. To this force must be added the weight of the top wheel which depends on machine design but averages about as follows (Quelch 1964):

Wheel diameter	Wheel weight
Inches	*Pounds*
54	950
60	1,450
72	2,500
84	3,050
96	3,600

Bandsaws carrying very much higher strains are in successful operation. Since saws with high strain may cut a straight line even though thinner than normal, kerf can be reduced proportionally. It is likely, therefore, that high-strain bandsaws will be increasingly used. Readers interested in further information on high-strain bandsaws will find Clark (1969), Cumming (1969), Foschi and Porter (1970), Porter (1970, 1971), and Allen (1975) useful.

Power to drive a bandsaw is proportional to the amount of sawdust removed per minute and to feed per tooth (fig. 18-64); since large wheels carry thicker and wider saws than small wheels, horsepower to drive them is proportional to wheel diameter (table 18-37).

Power to drive a sawmill carriage can be applied by electric, hydraulic, or steam mechanisms. Average carriage speed on the cutting stroke can be calculated from equation 18-14; maximum speed when returning to pick up a new log may exceed 1,000 f.p.m. Power required is a function of the combined weight of log and carriage and the maximum acceleration required. Normally, the heaviest carriages are used in conjunction with the largest band mills cutting the heaviest logs. If the carriage is driven by a **steam gun,** the diameter of the gun (cylinder) is related to the carriage weight.

Gun diameter	Weight of carriage alone
Inches	*Pounds*
14	over 20,000
12	to 20,000
10	to 15,000
8	to 10,000

When hydraulic power is used to drive the carriage, the size of the electric motor required to turn the pump is related to carriage weight.

Carriage weight	Motor size
Pounds	*Horsepower*
28,000	400
18,000	150-200
10,000	100
6,000	60
3,000	25

In many hardwood mills cants or heavy slabs cut with the headrig are resawn into lumber of thinner dimension on a smaller bandsaw called a resaw. Such resaws can cut horizontally or vertically (fig. 18-65). From a study of four eastern Canadian sawmills using each type of saw to cut hard maple and yellow birch, Robichaud (1975) concluded that vertical resaws sawed more accurately, handled sweepy timber better, and cost less. Horizontal saws permit slabs to be fed face-down on the feed table and are thus easy to feed; also, a split infeed table is sometimes provided so that two cants or slabs can be fed simultaneously and a different board thickness cut from each. All things considered, however, vertical band resaws were favored over horizontal.

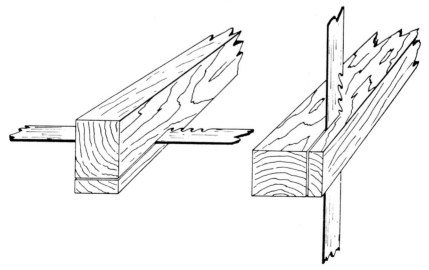

Figure 18-65.—Band resaws can cut horizontally (left) or vertically (right).

The vertical band resaw most frequently employed is called a **linebar resaw** (fig. 18-66). In an efficient installation, random-length slabs and cants flow nearly continuously through these heavy machines as a single board is ripped from each. The linebar, or guide, is equipped with setworks so that distance between guide and saw can be quickly and accurately set. Vertical press rolls, or horizontal spiral rolls, align the stock against the guide before the saw enters the cut. Since only one board is removed with each pass of a cant, each cant is repeatedly returned on a merry-go-round conveyor to the infeed side of the resaw until reduced to lumber of the desired size. Detjen (1975), in his discussion of linebar resaws for small hardwoods, favors linebars short enough to follow sweep in hardwood cants; also favored is independent air pressure on each press roll so the press roll closest to the saw can dominate when cants have sweep.

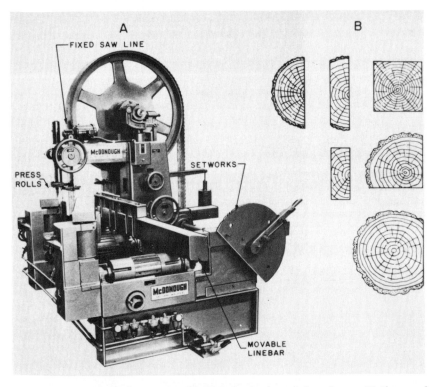

Figure 18-66.—(A) Linebar resaw with 54-inch wheels and short frame. (B) Shapes of cants and slabs that can be resawn on the linebar resaw. Since only one board is removed with each pass, each cant is repeatedly returned on a merry-go-round conveyor to the infeed of the resaw until reduced to lumber of the desired size. The short frame permits cants to approach the infeed from either saw or linebar side. (Photo and drawings from McDonough Manufacturing Company.)

One layout for small hardwood logs calls for an end-dogging carriage on a twin-band headrig to produce two-sided cants which then proceed to a twin combination edger and linebar resaw for removal of the remaining two slabs and recirculation of the cant for resawing into lumber.

All bandsaws are sized by wheel diameter, which on linebar resaws range from 46 to 72 inches (Detjen 1975); the machines can carry single, twin, or quad saws and can have fixed saws with shifting fence or fixed fence with shifting saws. As noted in table 18-37, sawblade thickness for bandsaws is about 0.001 inch per inch of wheel diameter, and larger wheels carry wider saws. Opinion varies on the width of swage required to cut green hardwood on a band resaw. Johnston and St. Laurent (1975b) found when cutting hard maple that a minimum side clearance of 0.015 inch on each side of the saw plate allowed clearance from wood springback in the kerf under most conditions. Rules of thumb for determining swage widths suggest twice blade thickness plus one gauge, sawblade thickness plus five gauges, or blade thickness plus 0.007 inch for each inch of sawblade width.

For a 10-inch-wide, 16-gauge bandsaw, these various rules would work out as follows:

Rule	Swage width
	Inch
Twice blade thickness plus one gauge	0.134
Blade thickness plus five gauges	.120
Blade thickness plus 0.007 inch per inch of blade width	.135

The illustrated 54-inch resaw (fig. 18-66), if cutting oak, might carry a 26-foot-long, 8-inch-wide, 17-gauge saw swaged to 11 or 12 gauge with 1-½-inch tooth spacing. The saw normally carries a "strain" of 4,640 pounds. The saw itself is driven by a 100-hp motor. Saw speed is 7,400 fpm. The four 8-inch horizontal feed rolls and two 10-inch vertical rolls are driven by a 3- or 5-hp electric variable-speed drive to give feed speeds infinitely variable between 100 and 300 lineal fpm. The maximum feed speed of 300 fpm yields a bite per tooth of 0.061 inch. Normal high speed on a 10-inch-thick cant is 180 fpm, giving a bite per tooth of 0.036 inch. Maximum cant size that can be split in the center is 20 inches thick by 24 inches wide. The setworks on the linebar can be air or hydraulically operated and remotely controlled by the sawyer. While one cant is being run, it is possible to preset the linebar for the next cant. Production per 8-hour shift on this machine is determined by stock size and cutting program but can reach 50,000 bd ft. Automatic off-bearing is provided so that the operating crew consists of the sawyer only, unless the machine is used to salvage stock out of thin irregular slabs. Slab resawing usually requires a tail sawyer because the waste slab may break up as it leaves the feedworks.

It is possible to orient a band resaw at an angle intermediate between vertical and horizontal. The feedworks of such a slant resaw enjoy some of the advantages of a horizontal resaw in that gravity keeps the cant positioned against both bottom and side feed rolls. Filer and Stowell (1977) have illustrated and described the virtues of such an arrangement.

A discussion of sawing accuracy and thickness variation in bandsawn hardwood lumber is beyond the scope of this text, but readers needing information on the subject will find useful articles by Birkeland (1968), Chardin (1973, 1979), Allen (1975), and Robichaud (1975). Forces acting on bandsaws have been discussed by Cumming (1969), Sugihara (1970), Porter (1971), and Pahlitzsch and Puttkammer (1975, 1976). Tanaka et al. (1981), Kirbach and Bonac (1978ab), and Ulsoy and Mote (1978) have discussed vibrational characteristics of bandsaws; the latter reference is a review of the literature concerning the effects on saw vibration and stability of bandsaw velocity, tension, exitation to vibrate, and loads normal to the cutting edge. Unanswered are many fundamental questions regarding the effect of guides, thermally induced and initial stresses, cutting forces, propagations of disturbances, and blade workpiece lateral interactions. To provide a theoretical basis, Ulsoy and Mote (1980) studied two methods for analyzing bandsaw transverse and torsional vibration and stability, and developed computer programs based on the methods.

Swaged-tooth band resaws for dry lumber.—Because remanufacturing plants commonly keep only a few thicknesses of dry lumber in inventory, they require resaws to convert these thicknesses to a multiplicity of thinner sizes. For maximum lumber recovery, bandsaws for remanufacturing have much thinner kerf than primary bandsaws. Saw gauge varies according to wheel diameter (table 18-38A). For pine-site hardwoods the tooth shape illustrated in figure 18-56E is widely used with rake angle of 30° and sharpness angle of 44°.

Because efficient feeds per tooth are a function of saw gauge, gullet depths can be expressed as a function of blade thickness and tooth pitch. Thin saws take small bites per tooth and therefore need less gullet space (table 18-38B). Tooth spacing on wide saws is greater than on narrow saws because they are capable of deeper cuts and need more gullet space.

Saw velocities of 7,000 to 9,000 fpm are usual for cutting dry hardwoods. Saw strain is given by equation 18-17. Practical feed speeds lie in a range defined by equation 18-14, with feed per tooth restricted to values one-half to three-quarters of sawblade thickness. Swage width for cutting dry hardwoods should be 0.016 to 0.024 inch wider than saw thickness; for example, a 19-gauge saw is commonly swaged to about 16 gauge.

TABLE 18-38A.—*Proportions of band resaws for dry pine-site hardwoods*

Wheel diameter (inches)	Saw thickness[1]		Saw width	Saw power
	BWG	Inch	Inches	Horsepower
36	21-19	0.032-.042	3-4	15-25
44	20-19	.035-.042	4-5	30-40
48	19-18	.042-.049	4-6	40-50
54	19-17	.042-.058	5-7	60-75
60	18-16	.049-.065	6-8	100-125

[1]Saw thickness should not exceed wheel diameter (in inches) divided by 1,000 plus 1 saw gauge.

TABLE 18-38B.—*Tooth proportions for wide bandsaws ripping dry pine-site hardwoods*[1]

Saw width (inches)	Saw gauge	Pitch	Gullet depth
	BWG	*Inches*	*Inch*
4	20 (18)	1¾ (1)	7/16 (1/4)
5	19 (17)	1¾ (1-½)	½ (7/16)
6	18 (16)	1¾ (—)	⅝—
7	17 (15)	1¾ (—)	1-11/16 —
8	16 (14)	2 (1¾)	15/16 (¾)

[1]The first number in each entry is for hardwood of low density; the second number, in parentheses, is for oaks and hickory.

Most band resaws for dry hardwoods are vertical; they may be in single, twin, or tandem arrangement depending on the production required. Figure 18-67 (top) shows a single vertical resaw on which the rolls tilt so that standard boards or cants, rectangular in cross section, can be center-resawn into two bevelled pieces of equal size and shape. Much moulding stock is manufactured in this manner, as are triangular corner blocks for crates and pallet boxes. When equipped to resaw dry oak moulding stock, this 36-inch band resaw might carry a 17-foot-long, 4-inch-wide, 18-gauge saw swaged to 15 or 16 gauge with a 1-inch tooth pitch and 9/32-inch gullet depth. Normal "strain" for this saw is 1,470 pounds. A 20-hp motor drives the saw at 7,800 fpm. When splitting a 4-inch dry board, the saw would typically be fed at a bite per tooth of 0.02 inch, yielding a feed speed of 156 fpm. The four feed rolls are each 6 inches in diameter and 6 inches tall. They are driven by a pair of ½-hp variable-speed motors that are mounted on, and tilt with, the feed roll mechanism. The feed speed is variable from 25 to 250 fpm. The rolls can be tilted up to 45°. The maximum-size workpiece that can be center-split on the saw is 8 inches wide and 8 inches thick.

Spring-set narrow bandsaws.—Saws more than ½-inch wide are primarily used to rip. Narrower saws can cut curves; for example, a ½-inch, 21-gauge saw can easily cut a curve of 6-inch radius (fig. 18-67 bottom), and a ⅛-inch, 25-gauge saw can cut a curve of ⅜-inch radius.

Proportions of narrow bandsaws are shown in figure 18-57E. Teeth on ¼-inch-wide saws are commonly set 0.005 inch to each side; those on saws 1 inch wide are set approximately 0.010 to each side, with other saw widths set proportionately. The thinnest and narrowest saws are run on the smallest wheels (table 18-39).

Wheels for saws less than 2 inches wide are commonly rubber covered, and the saw teeth do not project over the edge. For wide saws, easier-to-clean, hard-faced wheels are used; the spring-set teeth must project beyond the front rim of these wheels.

1862   Chapter 18

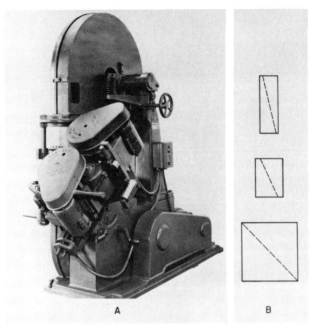

Figure 18-67.—Thirty-six-inch bandsaws for dry lumber. (Top) Resaw with tilting rolls; typical cuts indicated at B; photo from Tri-State Machinery Company. (Bottom) Bandsaw for template-controlled sawing. (Photo from Tannewitz.)

TABLE 18-39.—*Narrow bandsaw dimensions* (Koch 1964a, p. 200)

Wheel diameter (inches)	Saw gauge	Saw width	Saw speed	Points per inch[1]	Saw length
	BWG	Inches	Fpm		Feet
12-18	25	3/16 to ½	3000	7-5	6-10
20-30	22	3/16 to 1½	3500 (to 7500)[2]	7-4	10-15
36	21	¼ to 2	4000-4500 (to 8500)[2]	6-3	18-20
42	20	¼ to 2	5000-5500 (to 8000)[2]	5-2	over 21

[1]Counts teeth in a 1-inch space, i.e., includes the tooth at each end of a measured inch; thus, if tooth points are spaced 1/10-inch apart, there are considered to be 11 teeth per inch.
[2]On some current designs.

Portable horizontal band headrig for hardwoods.—In the South, where about 75 percent of commercial forest land is held by non-industrial private owners, average landholding is small (less than 100 acres), and markets are nearby, there should be demand for a low-price, low-power portable sawmill. Because saw power is proportional to kerf width, bandsaws need less power than circular saws; moreover, even thin-kerf bandsaws can cut lumber from sizeable logs if feed is appropriately regulated. Figure 18-68 (top) illustrates an arrangement whereby hardwood logs up to 20 inches in diameter and 12 feet long can be live-sawn, and placed on sticks for drying and subsequent edging. Figure 18-68 (bottom) shows the hydraulically actuated parallelogram mechanism whereby the bandsaw is lowered with each successive cut to yield the thickness of lumber desired. The bandsaw wheels which carry the air-tensioned saw are 28 inches in diameter; the 19 gauge saw is 3½ inches wide and swage set. As saw kerf is only about 1/16-inch, the air cooled gasoline engine which drives the saw is rated at only 25 horsepower. In 1980, such a trailer-mounted mill was available from the manufacturer, complete with 14 saws, saw sharpening device, and swaging tools, for less than $10,000. Production under favorable circumstances, can approach 5,000 board feed of wany-edged planks and boards per 8-hour shift. Production of 3,500 board feet would require sawing an 8-foot, 10-inch log about every 5 minutes.

Other portable band sawmills are described by Mills (1976).

SASH GANGSAWING

The **sash-gangsaw** carries several straight, short sawblades clamped in a reciprocating frame; the log or cant is fed continuously through this frame. The action is similar to that of a handsaw in the hands of a carpenter ripping a board. The saws cut on the downstroke only.

The sash gangsaw is best suited for long, smooth, softwood logs of uniform diameter and minimum taper, or long cants previously slabbed on two sides.

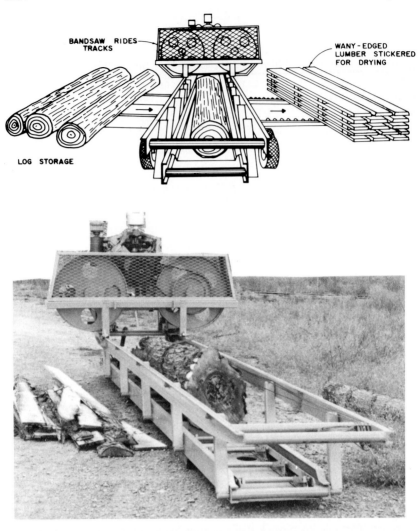

Figure 18-68.—Trailer-mounted portable band sawmill carrying a 19-gauge saw on 28-inch wheels driven by a 25-horsepower gasoline engine. (Top) Logs are rolled between the tracks and the saw indexed downward before each cut; carriage is pushed through the cut and lumber manually offloaded and stacked. (Bottom) Hand-pumped hydraulic cylinders adjust level of saw through a parallelogram mechanism; the gasoline engine powers only the bandsaw blade. (Drawing and photo from Paul Bunyan, Jr. Enterprises, Eagle Creek, Ore.)

Sash gangsaws have therefore not been particularly successful on the dense, hard, short, small-diameter, rough, tapered logs so typical of pine-site hardwoods.

Readers interested in a discussion of kinematics, power requirements, and productivity of sash gangsaws are referred to Koch (1964a, 1972).

CIRCULAR SAWING

The circular saw, in numerous variations, is used in all stages of hardwood manufacture. This text will discuss only a few of the most important applications. For discussions in greater depth, see Endersby (1953), Lubkin (1957), Harris (1960), Koch (1964a, p. 217-273), and Kollmann and Côté (1968, p. 490).

Nomenclature.—Figures 18-69, 18-70, and 18-71 illustrate nomenclature for circular saws.

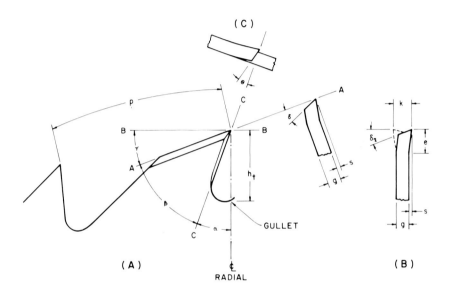

Figure 18-69.—Geometry of spring-set circular saw teeth. (A) Side view of tooth. (B) Front view of tooth. (C) Projection along rake plane.

α Rake angle, degrees
γ Clearance angle, degrees
β Sharpness angle, degrees
δ Top bevel angle, degrees (measured from clearance plane A-A, Fig. 18-69A)
θ Front bevel angle, degrees (measured from the rake plane, fig. 18-69A)
g   Thickness of saw blade, inch (may be variable from saw center to tooth extremity)
s   Amount of set (or swage) to each side of saw plate, inch
e   Length of tooth affected by set (measured from tooth extremity in a radial direction to the line of set, i.e., the line of set falls along a circle concentric with the cutting circle but of slightly smaller diameter)
k   Width of kerf, inch (nominally 2s + g; actual kerf may vary from nominal width because of vibration, runout, or other factors)
h_t Gullet depth measured radially, inches
N   Number of teeth in saw
P   Tooth pitch, inches
a   Area of tooth gullet, square inches.

(Drawing after Koch 1964a, p. 219.)

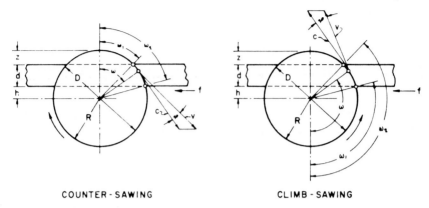

COUNTER-SAWING   CLIMB-SAWING

Figure 18-70.—Angles, dimensions, and velocity vectors for circular saws.
- z Saw projection beyond the workpiece, inches
- d Depth of workpiece, inches
- h Distance between workpiece and axis of saw rotation, inches
- ω Instantaneous position angle of tooth under consideration
- ω_1 Angle through which tooth edge had rotated from reference line to entry in wood
- ω_2 Angle through which tooth edge has rotated from reference line to exit from wood
- f Feed speed, feet per minute
- n Saw blade, revolutions per minute
- c Cutting velocity, feet per minute (i.e., peripheral velocity of cutting edge)
- v Velocity of cutting edge relative to workpiece, feet per minute
- f_r Feed per revolution of saw blade, inches
- b Length of arc of tooth engagement with workpiece, inches
- R Blade radius, inches

(Drawing after Lubkin 1957.)

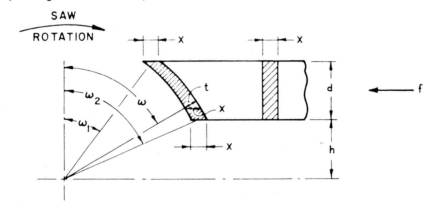

Figure 18-71.—Approximate chip geometry for circular saws.
- t Instantaneous chip thickness, inch (measured in a direction perpendicular to a tangent to the tooth trace, i.e., in an approximately radial direction)
- t_a Average chip thickness, inch
- x Feed per tooth, or "bite" per tooth, inch
- V Volume of wood removed by a single tooth as it travels through the workpiece, cubic inches
- E_s Specific cutting energy, kilowatthours per cubic inch kerf removed

(Drawing after Lubkin 1957.)

Kinematics and fundamentals.—Depending on the direction of cut through the workpiece, circular saws may cut by **counter-sawing** or by **climb sawing** (fig. 18-70). Counter-sawing resembles up-milling in that the cutting edge emerges from the workpiece more nearly at right angles to the direction of feed than where it entered. In climb-sawing, the cut is more nearly parallel to the feed where it leaves the workpiece.

In order to rationally specify circular saws and matching drive motors, it is helpful to understand some basic relationships.

In planing, where the angle ω is small, the cumbersome expressions for the true trochoidal cutting path are necessary for accuracy. In circular sawing, however, the tooth traces can be closely approximated as arcs of circles. For counter-sawing (fig. 18-70):

$$\omega_1 = \text{arc cos}\left(\frac{d+h}{R}\right) \tag{18-18}$$

$$\omega_2 = \text{arc cos}\ \frac{h}{r} \tag{18-19}$$

For climb-sawing (fig. 18-70):

$$\omega_1 = \text{arc cos}\ \frac{h}{R} \tag{18-20}$$

$$\omega_2 = \text{arc cos}\left(\frac{d+h}{R}\right) \tag{18-21}$$

The angle at which instantaneous chip thickness is the average chip thickness can be approximated as follows:

$$\omega_a \approx \text{arc cos}\ \frac{\left(h+\dfrac{d}{2}\right)}{R} \tag{18-22}$$

The length of path of tooth engagement for both counter-sawing and climb-sawing can be stated:

$$b = R(\omega_2 - \omega_1) \tag{18-23}$$

The tooth pitch for uniformly spaced teeth is

$$p = \frac{2\pi R}{N} \tag{18-24}$$

Feed per revolution of blade is:

$$f_r = \frac{f}{n} \tag{18-25}$$

The feed per tooth is:

$$x = \frac{2\pi R f}{Nc} = \frac{pf}{c} = \frac{12f}{nN} \tag{18-26}$$

Chip thickness at any instant (fig. 18-71) can be approximately stated for swage-set teeth.

$$t = x \sin \omega = \frac{f \sin \omega}{nN} = \frac{pf}{c} \sin \omega \qquad (18\text{-}27)$$

For swage-set teeth the average chip thickness is:

$$ta = \frac{xd}{b} = \frac{pfd}{bc} \qquad (18\text{-}28)$$

As shown in figures 18-57B and C, spring-set teeth penetrate to twice the depth that swage-set teeth penetrate for a given feed per tooth because the points of alternate teeth cut on opposite sides of the kerf; for spring-set teeth, the average chip thickness is:

$$t_{\text{avg spring-set}} = \frac{gt_a + (g + 2s - g)t_a}{g} = \frac{k}{g} t_a \qquad (18\text{-}29)$$

Rim speed, i.e., peripheral velocity of the saw teeth, can be expressed:

$$c = \frac{2\pi Rn}{12} = \frac{\pi Rn}{6} \qquad (18\text{-}30)$$

For both counter-sawing and climb-sawing, resultant tooth velocity (fig. 18-70) with respect to the workpiece is:

$$v = \sqrt{c^2 + f^2 + 2cf \cos \omega} \qquad (18\text{-}31)$$

The actual number of teeth engaged will alternate between the two integral numbers closest in value to the ratio b:p.

Obviously total cutting power required is positively correlated with width of kerf and with length of cutting path (thickness of workpiece). When evaluating the effect of feed per tooth on power required, it is convenient to use the concept of specific cutting energy, i.e., the energy to remove a unit volume of wood. From equations 18-3 and 18-4 it was observed that the parallel cutting force could be expressed in terms of chip thickness; curvilinearly (kt^m) or linearly ($A + Bt$). If the experimentally determined constants K and m (or A and B) are known, the specific cutting energy of a circular saw (E_s, kilowatt hours per cubic inch of kerf removed) can be calculated for any value of average chip thickness (t_a, equation 18-28 for swage-set saws, and 18-29 for spring-set saws); see Koch (1964a, p. 225) for development of the equations.

$$E_g = \frac{(0.377)(10^{-6})}{12} \left[\frac{K}{t_a^{1-m}} \right] \qquad (18\text{-}32)$$

$$E_g = \frac{(0.377)(10^{-6})}{12} \left(B + \frac{A}{t_a} \right) \qquad (18\text{-}32)$$

The shape of these curves is shown in figure 18-72. Rim speed for circular saws cutting hardwoods is usually 8,000 to 10,000 fpm. It is, however, practical to increase the chip thickness and thereby reduce specific cutting energy by reducing saw revolutions per minute, increasing tooth spacing, or increasing

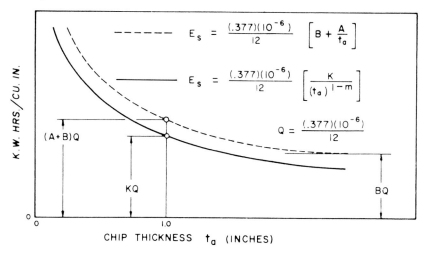

Figure 18-72.—Forms of specific cutting energy curves. While most bandsaws cut a chip thinner than ¼-inch, chipping headrigs commonly feed ½-inch to 1 inch per knife. (Drawing after Lubkin 1957.)

feed speed. Gullet capacity, tooth strength, plate strength, and surface quality limit the feed per tooth to values considerably less than ¼-inch for log saws and much less for most other circular saws. A detailed discussion of these limiting factors, the effects of size and orientation of the sawblade, and the effects of tooth angles, can be found in Koch (1964a, p. 226-271). See also Suchsland (1980).

As with bandsaws, cutting resistance to teeth of circular ripsaws is comprised of fiber indentation and severance, chip formation, chip breakage and associated friction, chip removal and associated friction, and chip acceleration. Teeth on circular saws used for ripping approximate 90-90 cutting when depth of cut is shallow and saw projection is substantial. With deep cuts, cutting mode changes from near 90-0 at one surface to near 90-90 at the other surface (fig. 18-70).

Severance forces for pine-site hardwoods orthogonally cut in the 90-90 and 90-0 directions with 0.1-inch knife length are given in tables 18-5AC through 18-26AC. Force trends for the 90-90 direction—which predominates in ripping cuts with circular saws—are shown in figures 18-60 through 18-63.

Inserted-tooth ripsaws for circular-saw headrigs[3].—Circle headrigs used on pine-site hardwoods commonly are equipped with two-piece teeth having easily replaceable bits and shanks (fig. 18-73). The bit is drop-forged and has clearance angles factory-ground on top and sides. It is sharpened by grinding on the rake face only so that the cutting edge is perpendicular to the plane of the saw plate. The assembly, consisting of bit and shank, is friction fitted into a grooved seat in the saw plate. The shank forms a rounded and capacious gullet, and the end nearest the bit is thickened to chamber the sawdust (fig. 18-73).

---

[3]Data under this heading are, for the most part, condensed from Lunstrum (1972); readers needing fuller information on operation of circle headrigs should consult this reference.

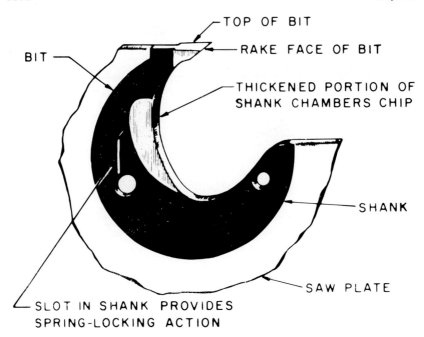

Figure 18-73.—Two piece inserted tooth for a circular ripsaw.

Because the bits are replaceable when worn, the diameter of the saw remains constant regardless of time the saw plate has been in service. To suit saws of all diameters and plate thicknesses, bits are available in a range of sizes and styles (table 18-40).

A rake angle of 30-40° is commonly employed for all insert-tooth ripsaws cutting pine-site hardwoods. The maximum width of the forged bit is generally slightly less than twice the plate thickness, and may range from 7/32- to 3/8-inch but is commonly ¼- to 5/16-inch in width.

Diameter of a headsaw should be the smallest possible that can saw the largest logs expected, as follows:

Diameter of largest log	Saw diameter
Inches	Inches
18	40
20	44
22	48
26	52
30	56
34	60

Saw plate thickness should match expected load; large logs, hard logs, and fast saw and carriage speeds require thick plates to withstand large stresses (table 18-40). Tooth size and style, designated by an arbitrary system involving both letters and numerals, must afford the gullet area needed to chamber the sawdust. Gullet sawdust capacity is about 70 percent of gullet area. Depth of cut (cant face width) without overloading the gullet is therefore conservatively computed as equal to 0.7 × gullet area/feed per tooth; because sawdust spillage occurs just before each tooth leaves the cut, about 3 inches can be added to the foregoing value (Quelch 1964). Gullet areas of some standard tooth styles are as follows:

Tooth style	Gullet area
	Square inches
2½	1.5
F	2.0
3, B	2.5
3½	3.0
D, 4½	4.0

TABLE 18-40.—*Gauges[1] of circular saw plates recommended for various tooth styles, saw sizes, and load demands* (Data from Lunstrum 1972)

Tooth style and saw diameter (inches)	Light load	Medium load	Heavy load
	----------BWG gauges----------		
2½, F			
40 to 54	9/10	8/9	7/8
56 to 60	8/9	7/8	—
B, 3			
40 to 54	8/9	7/8	6/7
56 to 68	7/8	6/7	6/7
3½			
40 to 54	—	7/8	5/6
56 to 60	7/8	6/7	5/6
D, 4½			
40 to 54	—	6/7	5/6
56 to 60	6/7	5/6	5/6

[1] The first gauge given is for the center; the second is the gauge of the rim.

Teeth with the larger gullet areas are carried in the thicker saw plates (table 18-40). Recommended tooth styles for four classes of hardwood logs are shown in table 18-41; teeth with large gullets are appropriate for large logs, but not for small logs. Tooth pitch is commonly 3 to 4 inches but may be as large as 6 inches.

The cutting edge of saw teeth must provide a path (saw kerf) wide enough to prevent excessive rubbing of the saw plate against kerf walls. Pine site hardwoods do not require as much clearance as softwoods. Hardwoods when not frozen require ¼-inch to ⅜-inch width of cutting edge (table 18-42). Frozen hardwoods cut cleanly and require about 1/64-inch (on thinner saw plates) to 2/64-inch (on thicker saw plates) less cutting edge width than unfrozen hardwoods.

Readers interested in the distribution and magnitude of stresses causing tooth deflection and breakage in inserted-tooth circular saws will find the work by Malcolm and Koster (1970) useful.

TABLE 18-41.—*Recommended tooth styles[1] for circular headsaws of logs of various classes and diameter* (Data from Lunstrum 1972)

Class of log	Diameter range of log, inches		
	10 to 19	13 to 21	20 to 33
	------------------------Tooth style----------------------		
Knotty	F, 2½	B	—
Frozen	F, 2½	B, 3½	D, 4½
Hard hardwoods	F	B	D
Soft hardwoods	2½	3, 3½	4½

[1]Bits in F, B, and D tooth styles are interchangeable, but shanks are not.

TABLE 18-42.—*Recommended cutting edge widths of inserted saw teeth in circular headsaws related to saw gauge and tooth style* (Data from Lunstrum 1972)

Tooth style	Saw gauge				
	9/10	8/9	7/8	6/7	5/6
	--------------------64ths of an inch--------------------				
2½, F	16	17	18		
B, 3		17	18	20	
3½			18	20	22
D, 4½				22	24

Saw power required for ripping with an inserted tooth circular saw can be computed by assuming a standard situation and then adjusting power needed for this standard situation by factors reflecting deviation from the standard. The amount of power required is determined by:

- Feed per tooth (positively correlated, fig. 18-74)
- Specific gravity (positively correlated, fig. 18-74, table 18-43)
- Kerf width (positively correlated, footnote 2 of table 18-43)
- Cant face width to be sawn (positively correlated, fig. 18-75 and table 18-43)
- Saw speed (positively correlated with rpm of saw, footnote 1 of table 18-43)
- Tooth style (table 18-43)

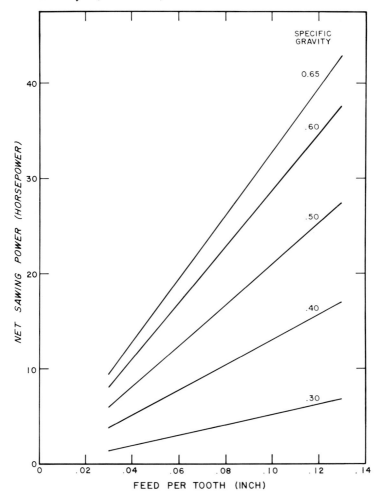

Figure 18-74.—Net cutting horsepower to ripsaw 10-inch-deep cants of five specific gravities (ovendry-weight, green-volume basis) with a 48-inch, 7-8 gauge circular saw fitted with 40 B-style inserted teeth having 9/32-inch kerf and rake angle of 43°; saw speed was held constant at 300 rpm. (Data from Telford 1949.)

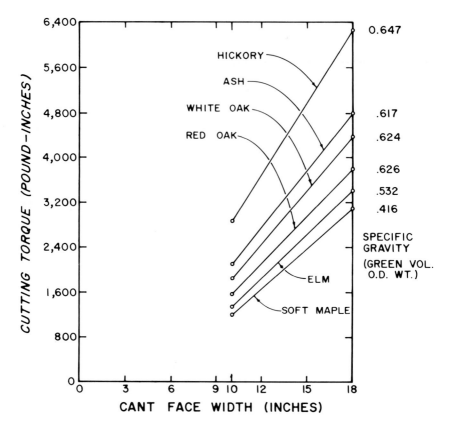

Figure 18-75.—Effect of cant face width on torque required at saw mandrel to ripsaw boards from six species of hardwood cants. The 48-inch circular saw turned at 300 rpm and carried 48 inserted teeth ground with 43° rake angle to cut a kerf 9/32-inch wide. Feed per tooth held constant at 0.065 inch. (Drawing after Telford 1949.)

Malcolm (1967) found that a 3/16-inch feed per tooth on northern red oak yielded as good a surface as an ⅛-inch feed, but tearout at the edge of each board was greater and necessitated a larger allowance for planing. For this, and other reasons, manufacturers of sawteeth have advocated standardizing feed per tooth of inserted-tooth headsaws at ⅛-inch for softwoods and 1/10-inch for hardwoods. A feed per tooth of 1/9-inch (0.11) was used in preparation of table 18-43. Saw feed can be expressed as inches per revolution of the saw. For example, to maintain a feed per tooth of 0.11 inch with a 50-tooth saw requires a feed per revolution of 5.5 inches.

Table 18-43 defines power used by the saw only. Additional power is needed to drive the carriage, as outlined in the preceding subsection on bandsaws.

TABLE 18-43.—*Basic cutting power required for an inserted-tooth, 550-rpm[1] circular saw operating on pine-site hardwoods at a feed per tooth of 0.11 inch with kerf of 9/32-inch[2]*
(Data from Lunstrum 1972)

Species group and tooth style	Maximum cant face width	Feed rate, inches/revolution of saw					
		4.0	4.5	5.0	5.5	6.0	6.5
	Inches	----------------Horsepower----------------					
All pine-site hardwoods[3] except oaks and hickories							
2½............................	12			97	105	118	126
F	16		118	129	140	157	168
B, 3...........................	19	133	147	161	175	196	
3½............................	22	160	176	193	210		
D, 4½.........................	28	213	235	257			
Oaks and hickories[4]							
2½............................	12			117	127	142	152
F	16		142	156	169	189	203
B, 3...........................	19	161	178	195	211	237	
3½............................	22	193	213	233	254		
D, 4½.........................	28	257	284	311			

[1]For a saw speed other than 550 rpm multiply basic horsepower values by the appropriate correction value, as follows:

Saw speed	Correction factor
rpm	
400	.75
450	.82
500	.91
600	1.09
650	1.18
700	1.28
750	1.37

[2]For each 1/32-inch increase or decrease in kerf width, adjust basic horsepower values by 11 percent. For further data, see: Cuppett, D.G. 1982. Power consumption and lumber yields for reduced-kerf circular saws cutting hardwoods. U.S. Dept. Agric. For. Serv. Res. Pap. NE-505, 6p.

[3]Assumes 95 percent of production is in species with specific gravity in the range from 0.46 to 0.55 (basis of green volume and ovendry weight).

[4]Assumes 95 percent of production is in species with specific gravities of 0.56 and greater (basis of green volume and ovendry weight).

Inserted-tooth ripsaws for edgers.—Hardwood sawmills cutting 1-inch- and 2-inch-thick lumber usually have an edger carrying two or three inserted-tooth saws. These saws are generally 9/10 gauge, 14 inches in diameter, rotate at 2,700 rpm, carry 14 style 2½ teeth and are driven by a 30 horsepower motor which provides feed power as well as saw power. Typically, feed speed is 175 fpm yielding a feed per tooth of about 0.056 inch. Feed per tooth on an edger is usually less than on a headsaw because edger feed mechanisms and hold-down devices are less secure than those on headsaws.

The accuracy with which an edger cuts is greatly influenced by the linearity of feed into and through the saws. Schliewe's (1973) study of roll-feed transports indicated substantial lateral movement of cants when fed through an edger without saws. A lag-bed (endless-bed) feed system can significantly reduce lateral movement, but there is still room for improvement.

Over-arbor ripsaws compared to under-arbor ripsaws[4].—The cutting action of circular ripsaws is more complex than that of bandsaws because of constantly changing chip thickness and grain direction. Until recently, circular ripsaws commonly cut in **mode I,** i.e., counter sawed with over-arbor feed systems (fig. 18-76). With the development of guided saws and floating saw collars, it became evident that there is merit in **mode II,** i.e., climb sawing with under-arbor feed systems (fig. 18-77). A floating saw collar permits free movement of the collar, and the saw which it grips, along an arbor which is keyed or splined to transmit torque to the saw; with this arrangement, saws are held in desired positions by saw guides, i.e., pairs of anti-friction pads applied near saw peripheries.

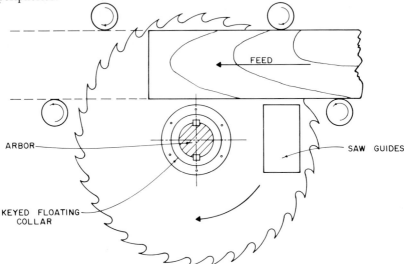

Figure 18-76.—Mode I sawing: keyed floating collar with over-arbor feed and counter-cutting saw. (Drawing after McLauchlan 1974.)

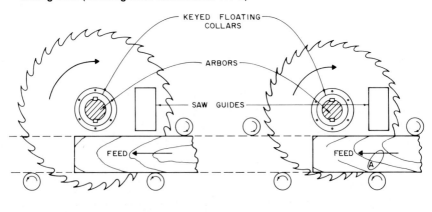

Figure 18-77.—Mode II sawing; keyed floating collars with under-arbor feeds and climb-cutting saws. (Drawing after McLauchlan 1974.)

---

[4]Discussion under this heading is condensed from McLauchlan (1974).

In theory and in the laboratory climb sawing in mode II yields smoother cuts and requires less power than counter sawing in mode I; on production machines, however, evidence for these advantages is slight.

In counter sawing in mode I, the feed force exerted by the saw is always in the same direction (left to right in fig. 18-76) so that there is no slack in the feed system; simple mechanical systems are needed to prevent lumber from being kicked back toward the operator. With climb cutting, the feed force exerted by the feedworks must first be positive (right to left in fig. 18-77) to feed a workpiece to the saw at a specified rate, but thereafter the workpiece must be held back against the sawing force to limit the size of bites per tooth; feed system requirements for mode II climb cutting are therefore rigorous.

For under-arbor mode II feed, the bottom fixed feed rolls can be much closer together than in mode I feed, providing more stability—particularly when feeding short, thin cants. Mode II cutting also permits use of a dip-chain endless bed feed in place of lower fixed rolls, thus providing exceptionally good control over lumber in the cut.

Saw guides are most effective when close to a cut. For over-arbor mode I feed, the bottom side of the wood always passes very close to the guides, and thus the guides approach maximum effectiveness. In mode II under-arbor feed, guides are close to the workpiece only for maximum depths of cut, thereby losing effectiveness on thinner cants.

With over-arbor feed systems, there is more freedom in the maximum depth of wood that can be passed through a machine because top feed rolls usually can be raised. With an under-arbor feed, maximum depth is very little more than the rated size and it is desirable, as noted above, to be sawing close to the maximum. Hence, with an under-arbor feed, there is a greater probability that some pieces being fed to a machine will be too deep to pass through it.

Guides always drag slightly on a saw, tending to make the periphery of the saw slack before entering the guide and taut (in tension) after leaving the guides. For mode I (fig. 18-76) the saw will tend to be undesirably slack in the cut, whereas in mode II (fig. 18-77) it will be taut and should saw more accurately. Moreover, in the mode I arrangement guides are more frequently fouled with knots and slivers than those of mode II.

Offsetting the advantages of saw tautness and anti-fouling due to guide position in mode II, is the disadvantage of greater guide distance from the saw teeth that have loaded gullets. In mode I, the teeth leaving the cut with full gullets need maximum lateral restraint and get this restraint from the immediately adjacent guides. In mode II, however, maximum lateral restraint is applied where the saw least needs it, i.e., adjacent to the entering cut where gullets are empty of sawdust.

Small saws are generally more desirable than large saws; they can be thinner in plate and kerf because they are stiffer and deviate less in the cut. Smaller saws can be used in mode I cutting because teeth need only protrude slightly above the top cant surface. In mode II cutting, however, teeth should fully clear the bottom surface of the cant to easily discharge loaded gullets. If too small a saw is used (fig. 18-77 right), sawdust from overfilled gullets spills and is dragged between the saw plate and sides of the cut (see area marked A), roughening cut surfaces and causing excessive friction, heating, and power consumption.

McLaughlin (1974) concludes his comparison by observing that there is merit in a third arrangement mode in commercial operation. **Mode III** combines an underfeed arbor with a counter-cutting saw. This would be depicted by figure 18-77 if the direction of saw rotation were reversed, so that motion of the teeth in the cut opposes the feed. A basic advantage of this mode of sawing is that the filled gullets move only a short distance (almost vertically upward) in a cut. The surface of this cut is not roughened by dragging sawdust along it, feed force is always positive as in Mode I, and power consumption is sometimes greatly reduced compared with Mode II. Although in Mode III operation the saw ahead of the guides is slack in a cut, this has almost no effect on accuracy or roughness of cut because sawdust drag in gullets is minimal except very close to the guides.

Single- and double-arbor gangsaws.—The pallet industry requires very large amounts of hardwood lumber in thicknesses from $\frac{3}{8}$-inch to 2 inches and in widths of 4 and 6 inches. Pallet boards are usually short—mostly in lengths from 36 to 60 inches. This need for low-cost pallet deckboards and stringers has resulted in widespread use of gangsaws mounting circular saws to ripsaw 4- and 6-inch-thick cants.

Machines designed for cants 4 inches and less in thickness usually carry only one arbor, arranged as shown in figures 18-76 or 18-77. For 6- and 8-inch cants, double-arbor saws are usual; such saws have a lower arbor arranged as shown in figure 18-76, and above it—but slightly offset—is an arbor arranged in mode II (fig. 18-77) or in Mode III described above. Saws on the lower arbor cut up into the cant a fixed distance—perhaps 3 inches; saws on the upper arbor are exactly aligned with, and cut into the kerf of, those on the lower arbor. The cut of the top saws is therefore dependent on cant thickness, being about 1 inch on a 4-inch cant and 3 inches on a 6-inch cant.

Sage and Olson (1975) have described a typical double-arbor circular gang ripsaw for pallet cants up to $6\frac{3}{8}$ inches thick and 12 inches wide. The roll feed machine is designed to continuously feed cants as short as 36 inches, but 26-inch-long cants can be fed if they are continuously butted. Lumber thickness (as thin as $\frac{1}{2}$-inch) is determined by saw spacer sleeves on the arbors. The infeed table is provided with three pockets for three different sawing patterns, and the operator positions each cant in the appropriate pocket from whence it feeds itself through the sawing zone. Ten horsepower is adequate to drive the feedworks. Saw arbors are individually powered for a rim speed of 10,000 to 12,500 rpm on 16-inch saws. The 0.120-inch saw plates carry carbide-tipped teeth to cut an 0.190-inch kerf. Feed speed is 80 to 100 feet per minute. Average feed per tooth is 0.050 inch (0.030 to 0.065 inch is in the range recommended by both machine manufacturer and user). For eastern hardwoods—including red maple, ash, oak, and hickory—15 to 20 horsepower should be provided per saw, preferably the latter.

Saw selection for circular gang ripsaws.—The technique of high-speed ripping of thick cants with circular saws is developing rapidly and numerous designs for saws, saw collars, and guides have evolved since 1965. Readers interested in thin-kerf circular saws for gang ripsaws, should find the following references useful:

Jones (1963, 1965)	Betts (1969)	Thrasher (1972)
Demsky (1967)	Salemme (1969)	Campbell (1973)
DuClos (1967)	Schliewe (1969)	Ratliff (1973)
Hallock (1968)	Pearson (1972)	Mote and Szymani (1978)
		Szymani (1978)

Mote and Szymani (1978), summarizing research on vibration in circular saws, concluded that vibration reduction and control may require improved saw design and manufacturing specifications, as well as feedback control of saw response. Readers interested in understanding vibration in circular saws should consult this review. See also Radcliffe and Mote (1981). Das (1979) summarized a systems approach to circular saw design, and Anderson (1979) presented the arguments for purchasing saw blades to satisfy prescribed standard specifications.

Mote and Szymani (1978) listed important practical observations on saw noise, as follows:

- During idle running, a high frequency whistle or self-excited vibration may occur that is more intense than the noise during sawing. The noise intensity increases with increasing rpm, saw diameter, blade thickness, and number of teeth.
- During sawing a linear relationship exists between noise intensity, cutting velocity, and feed rate.
- The noise generation mechanism is sensitive to the design of the teeth and their number and arrangement (pitch).

They noted four methods of noise reduction: (1) a second disk rotating with the saw blade to provide damping; (2) fixed damping plates near the saw blade; (3) lamination of the blade with a viscoelastic layer; and (4) use of dissipative saw guides. The best results were obtained with laminated (compound) blades—now commercially available; noise reductions ranged from 10 to 20 decibels on the A scale (db(A)).

They noted that saw vibration damping with magnetic fields (eddy current damping) reduced noise up to 15 dB(A) in saws 2 to 3.6 mm thick, 300 to 500 mm in diameter, and rotating at speeds of 1,000 to 4,000 rpm. Irregular pitch (tooth spacing) distributing the exitation energy of cutting over a broader spectrum, reduces blade response where an excitation harmonic becomes centered on a saw blade resonance. A reduction exceeding 15 dB(A) has been reported from this approach.

Seemingly subtle modifications of the tooth profile itself are significant factors in saw blade excitation. Noise reduction of 6 to 9 dB(A) during cutting and 23 to 25 dB(A) during idling has been achieved by tapering both sides of the teeth (i.e., bevelling tops and backs of teeth on both sides (Kimura and Fukui 1976). The aerodynamics of the gullets also significantly affect noise, but inadequate understanding of noise source mechanisms hinders effective remedies.

The literature on circular sawing with thin saws is complex. To simplify selection of circular saws for single and double arbor edgers, Lutz Classen developed, during the mid-1970's, a method of saw selection that he found practical for industrial use. Szymani (1978) reviewed Classen's[5] experience in writings and summarized his conclusion that saw stiffness and cutting force are important factors in predicting accuracy of sawing. Thus, combining the values for saw stiffness and the cutting force yields an index or number indicative of saw deformation during sawing. Classen defined this load index (LI) as follows:

$$LI = W \times F_t \qquad (18\text{-}32)$$

where:
- W = The saw deflection, inch, per pound force (applied as shown in figure 18-78); deflection is inversely related to stiffness.
- F_t = The tangential component of the cutting force exerted on the saw periphery, pounds.

Saws with the lowest load index are likely to saw most accurately in the sawing situation for which the load index is computed.

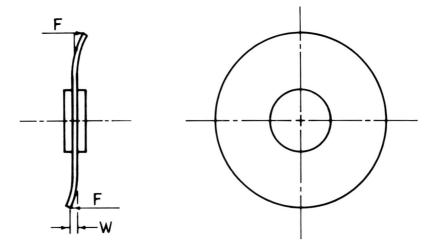

Figure 18-78.—Procedure for applying one-pound loads, F, to a saw to measure deflection W for use in equation 18-32. (Drawing after Classen, as reviewed by Szymani 1978.)

Deflection (W, inches) can be readily determined experimentally (fig. 18-78). The tangential component of cutting force (F_t, pounds) is computed from knowledge of the net cutting horsepower needed to drive a single saw making the cut under study, saw rpm, and saw radius (R, inches), as follows:

$$F_t = \frac{Hp \times 33{,}000 \times 12}{2\pi R \times rpm} \qquad (18\text{-}32)$$

---

[5]Lutz Classen's death in 1977 had left much of his valuable information unpublished.

Classen's index considers the feed speed, depth of cut, tooth type, saw blade geometry, degree of tensioning, clamping method, and saw guide system. Some of Classen's representative load index values for a particular sawing situation (table 18-44) indicated a range from 0.33 (best) to 0.76 (poorest).

Classen found that each circular saw has one speed at which it operates best, free of harmonic vibration. Szymani and Mote (1977) recommended that circular saws be operated at a speed 10 to 15 percent below the **critical speed** at which they become unstable.

Tensioning or prestressing thin circular saws increases their stability by introducing favorable in-plane residual stresses—either by local plastic deformation or by heating. Deforming the saw slightly between opposed rollers is the standard procedure for introducing such stresses. Interested readers will find useful the analysis of Szymani and Mote (1979) which describes a method of evaluating tension and analyzes the relationship between rolling load and resulting tension stresses.

TABLE 18-44.—*Classen's load index for a particular sawing situation[1] related to circular saw type and diameter, guidance, and type of collar* (Classen's data reviewed by Szymani 1978)

Saw type, arrangement and clamping method[2]	Saw diameter, inches	
	20	26
	------------Load index------------	
Strob Saw[3]		
Edge-clamped collar	0.76	0.85
Face-clamped collar	0.61	0.63
Floating Saw with collar		
(3 point guides at saw entry and at		
90° and 180° from mid-point of cut)		
Edge-clamped collar	0.65	0.67
Face-clamped collar	0.61	0.56
Saw with Fixed Collars and No Guide		
or Guide with Large Clearance		
Edge-clamped collar	0.55	0.65
Face-clamped collar	0.37	0.36
Saw with Fixed Collars and One		
Point Guide at Saw Entry		
Edge-clamped collar	0.45	0.46
Face-clamped collar	0.33	0.31

[1] Softwood, ⅛-inch kerf, feed rate 160 fpm, depth of cut 6 inches.

[2] An edge-clamped collar clamps the saw only at the outside radius of the collar; it is relieved elsewhere.

[3] In addition to the normal gullets, a strob saw has two very deep gullets of constant width that run nearly to the saw collar but at a trailing angle; cutters the width of the kerf are attached to the face of each long gullet to cut themselves free whenever saw deflection occurs and eject sawdust forcibly from the tops of the long gullets.

General-purpose swage-set ripsaws.—For application where change in saw diameter with wear is not critical, swage-set steel ripsaws may be more economical than insert-tooth saws. As with bandsaws, swaging tools are designed for a limited range of tooth shapes, commonly a rake angle of 30°, with sharpness angle of 44°, resulting in a clearange angle of 16°. Kirbach and Bonac (1979ab) have proposed reducing clearance angle to a minimum of 8° and increasing sharpness angle to 60°, thereby stiffening the tooth and making the cutting edge less vulnerable to mechanical damage. Rim speeds are in the range from 8,000 to 14,000 rpm. Table 18-45 shows the proportions of small swage-set ripsaws. For thin feed stock use the maximum number of teeth. Saws in the heavier gauges are suitable for power-fed machines or when cutting green or thick hardwood. For saws 15 gauge and thicker the tooth height should be approximately (0.43)(pitch); for saws less than 15 gauge, tooth height can be somewhat less. Width of swage is commonly twice plate thickness.

Figure 18-79 (top) illustrates a heavy-duty power ripping machine that employs swage-set steel saws for work such as multiple ripping of dry oak plank into flooring stock. This machine has four power-driven feed rolls, the top rolls being 7½ inches in diameter. It will rip stock up to 3-5/16 inches thick and 16 inches wide into as many strips as saw power permits. The spindle is driven at 3,600 rpm by V-belt from a 40-60-hp motor. The feed rolls are powered by a 5-hp motor. In a typical operation ripping 1-inch, rough, kiln-dry oak into multiple strips at 200 fpm, the machine might carry four, 14-inch, 30-tooth, 12-gauge saws swaged to 6 gauge. Thus at 200 fpm feed and 3,600 rpm spindle speed, the feed per tooth is 0.022 inch and the rim speed is 13,000 fpm. Under these conditions each saw in the cut demands approximately 12½ hp, for a total of 50 hp with four saws engaged. Similar machines are available that will feed up to 300 fpm and with as much as 125 hp on the saw spindle for multiple ripping of wider stock.

TABLE 18-45.—*Proportions of general-purpose, solid-tooth swaged ripsaws* (Koch 1964a, p. 253)

Diameter (inches)	Gauge	Number of teeth	Diameter (inches)	Gauge	Number of teeth
	BWG		Inches	BWG	
6	18	36-40	22	8-12	30-36
7	18	36-40	24	7-11	30-36
8	18	36-40	26	7-11	30-36
9	16	36-38	28	10	36
10	12-16	24-36	30	10	36
12	10-14	24-36	32	10	36
14	10-14	24-44	34	9	36
16	10-14	24-40	36	9	36
18	9-13	24-40	38	8	36
20	8-13	24-36	40	8	36

During the 1970's, industry largely converted from all-steel saws to carbide-tipped blades, and machines were produced to accommodate the change. Figure 18-79 (bottom) illustrates a gang ripsaw equipped with six top press rolls and an

Figure 18-79.—Multiple ripsaws. (Top) Over-arbor-feed machine designed with roll feed for all-steel saws; photo from Yates-American Machine Co. (Bottom) Under-arbor-feed machine designed for carbide-tipped saws. Top feed rolls press lumber against lower feed chain which dips slightly to accommodate saws; outboard return table has a removable center section to ease sawblade changes. (Photo from Stetson-Ross.)

under-arbor chain feed that dips ⅛- to 3/16-inch to provide saw clearance in the cutting zone. Carbide-tipped saws, that vary little in diameter with wear, are appropriate for this machine which carries 50 to 125 hp on the 3,600- to 5,000-rpm arbor and 3 to 7½ hp on the feed chain. On a typical operation ripping 1-inch rough, kiln-dry oak into multiple strips at 200 fpm, the saw arbor might turn at 3,600 rpm and carry 12-inch, 24-tooth, carbide-tipped saws with plate thickness of 0.109 inch and kerf of 0.155 inch. When ripping turning squares from 1-¾-inch-thick green hardwood planks, this dip-chain ripsaw—if equipped with five 12-inch saws—will handle about 10,000 board feet of lumber per 8-hour shift.

Spring-set ripsaws.—Spring-set ripsaws, while not common on gang ripsaws in North America, are widely used elsewhere. Because spring-set teeth use less rake and the tooth configuration is not limited by swaging, there is more freedom for selection of tooth angles. Spring-set ripsaws are used for the same purpose as swage-set ripsaws, but on the spring-set saw tooth penetration is twice that of the swage-set tooth on a saw of otherwise equal specification (fig. 18-57BC). British specifications for spring-set ripsaws are summarized in Koch (1964a, p. 254-257). Sharpening procedures for carbide teeth that resemble spring-set teeth are given in Hewitt (1978).

Glue-joint ripsaws.—Glue-joint ripsaws (fig. 18-80) are important in machining pine-site hardwoods because the furniture industry uses large quantities of wood panels glued up of narrow hardwood strips from which unacceptable defects have been removed by ripping. A glue-joint ripsaw simultaneously rips and prepares the edge surface of lumber for edge gluing. Thus the saw must rip in a straight line and generate a smooth kerf wall surface.

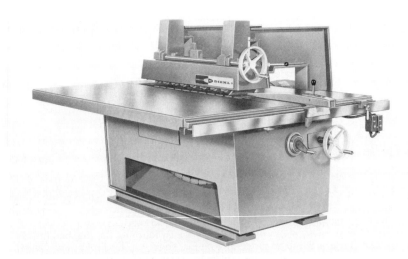

Figure 18-80.—Glue-joint, straight-line, counter-cutting ripsaw in which the saw projects upward through a split endless-bed feed; top feed rolls press lumber against the lower endless bed. (Photo from Diehl Machines, Inc.)

Machining

From 1930 to 1960, both spring-set and swaged steel saws (table 18-46) were used for this purpose, but swaged saws predominated. Both types were jointed for both roundness and trueness of tracking (fig. 18-81) while they rotated at operating speed.

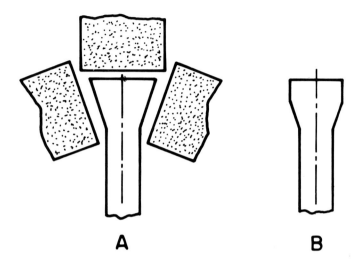

Figure 18-81.—Jointing of steel saw teeth on a glue-joint ripsaw. (A) Correct. (B) Incorrect. Swaged teeth commonly have 30° rake angle, 16° top clearance angle, and no front bevel. On spring-set saws only, a top bevel of 10° is sometimes used.

TABLE 18-46.—*Steel saw blades for operation at 3,600 rpm on glue-joint ripsaws* (Koch, 1964a, p. 259)

Diameter (inches)	Gauge	Spring set	Swage set
	BWG	----------------Number of teeth----------------	
12[1]	12	36	24
14[1]	12	36	24
16	12	36	30
18	11	36	30

[1]With style A carbide-tipped teeth (straight face and back) the specifications would call for:
  12-inch, 11-gauge, 24-teeth (for stock up to 2 inches thick)
  14-inch, 10-gauge, 24-teeth (for stock up to 2 inches thick)
  14-inch, 9-gauge, 20-teeth (for stock over 2 inches thick)

Adoption by the industry of carbide-tipped saws for glue jointers about 1970 eliminated need for on-machine saw truing. Regardless of saw design, however, saw collars of maximum size are used to stiffen the blade. Glue-joint ripsaws may be mounted either above or below the saw table, but counter-sawing is the usual mode. Smoothest surfaces are obtained at low feed speeds and with the countercutting saw protruding a minimum distance beyond the workpiece surface (St. Laurent 1973).

Figure 18-80 illustrates a counter-cutting ripsaw distinguished by a bottom, split, **endless bed** (chain feed) through which the saw protrudes and eight idle top rolls, all of which serve to keep the workpiece moving in a straight line. In a typical application on low-density hardwood 1¾ inches thick, the machine would be powered by 2-hp, variable-speed feed motor set to feed at 120 fpm with a single 14-inch, 24-tooth, swage-set saw direct driven at 3,600 rpm by a 15-hp motor. Under these conditions the feed per tooth is 0.017 inch. Machines of this type can cut stock ranging from ⅛-inch to 4½ inches thick at feed speeds ranging up to 250 fpm with saw power up to 20 hp.

Overcutting machines that carry two or more adjustable saws cutting against a lower endless bed are also available. On these machines the endless bed has a controlled sag immediately beneath the saws, to permit the saws to be shifted along the saw arbor (fig. 18-79 bottom).

Log cutoff saws.—The trend toward tree-length logging and diversion of each portion of the stem to its most appropriate use has stimulated new interest in log cutoff saws for central log decks (figs. 16-20 through 16-27). Figure 18-82A shows solid-tooth styles and bevels. Typically, rake angle is negative (e.g., -20^0), sharpness angle 45^0, clearance angle 65^0, and tooth height 0.76 times the pitch. Bevel is equally divided between front and back of the tooth, i.e., both front and top carry a 12 to 15^0 bevel. The point only is bevelled; the remainder of the tooth and gullet are ground straight across. The teeth are spring set about 3/64-inch to each side of the saw plate.

The inserted tooth shown in figure 18-82B is a style commonly used for log cutoff saws ranging from 66 to 108 inches in diameter. It is spring-set (1/8-inch to each side is common), 4-9/32 inches long, and is riveted into a radially oriented V-milled socket in the saw plate that is 2-13/16 inches deep.

Specifications for log cutoff saws from 60 to 108 inches in diameter are shown in table 18-47. Smaller circular saws for bucking small ash and oak logs for firewood (up to 12 inches in diameter) might have the following specifications:

Diameter	36 inches
Gauge	7
Number of teeth (solid)	80
rpm	1100
Horsepower	25
Tooth style	
for overhead saws	see figure 18-82A
for underslung saws	see figure 18-83

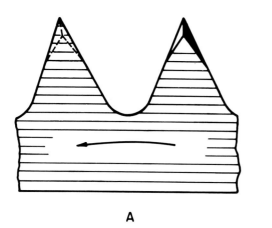
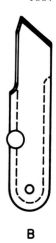
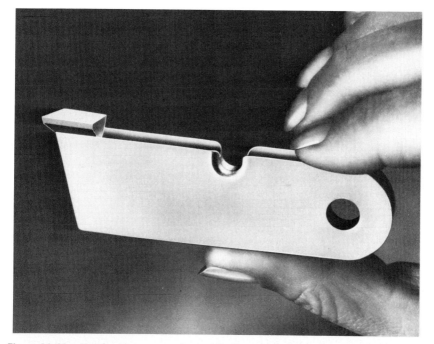

Figure 18-82.—Tooth styles commonly used for log cutoff saws. (Top) A. Solid tooth. B. Inserted tooth. (Bottom) Carbide-tipped inserted tooth. (Photo from Hannaco Knives and Saws, Div. of IKS, Inc.)

TABLE 18-47.—*Specifications for log cutoff saws* (Koch 1964a, p. 261)

Diameter (inches)	Gauge	Maximum number of teeth[1]
	BWG	
60	6	96
62	6	100
64	6	104
66	5	108
96	4	158
108	3	164

[1]Usual number is significantly fewer; often equal to the saw diameter in inches.

Cutoff saws for rough trimmers.—These crosscutting saws are used to trim green lumber to length. They may be mounted above or below the workpiece and arranged to either counter-saw or climb-saw. For underslung trimmers, the rake angle is typically 0 and the clearance angle 45. Face bevel should not exceed top bevel and is commonly 12 to 15 (see fig. 18-82A for overhead trimmers and 18-83 for underslung). The face bevel should not extend into the gullet. Teeth are spring-set; a set of 2½ BWG gauges to each side of the saw is common. Table 18-48 shows common specifications.

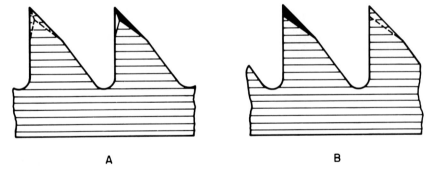

Figure 18-83.—Tooth styles for rough trimmer saws.

A typical sawmill trimmer chain might travel at 70 fpm, which, with a 24-inch, 70-tooth saw rotating at 1,915 rpm, would produce a rim speed of 12,000 fpm and a feed per tooth of 0.063 inch.

A swing saw designed to cut 4- and 6-inch-thick oak pallet cants to exact length prior to ripping into deck boards could have the following specifications:

```
Diameter .......................  30 inches
Number of teeth ................  120
Gauge ..........................  8
rpm ............................  1,300
Horsepower .....................  20
Tooth style
    for overhead saws ..........  See figure 18-82A
    for underslung saws ........  See figure 18-83
```

TABLE 18-48.—*Specifications for rough trimmer saws* (Koch 1964a, p. 263)

Diameter (inches)	Gauge	Number of teeth	Diameter (inches)	Gauge	Number of teeth
	BWG			BWG	
6	18	100	24	10-11	70
8	18	100	26	10	70
10	16	100-150	28	10	70
12	12-14	70-150	30	9-10	70
14	12-14	60-150	32	10	70
16	12-14	60-150	34	9	70
18	10-13	60-100	36	8-9	70-80
20	10-13	70-80	38	8	80
22	10-12	70	40	6-7	80

Smooth-trim saws.—Dry hardwood glued-up lumber panels, oak stair stepping, and a wide variety of hardwood parts for furniture must be given an accurate and smooth length trim. Usually this is accomplished on a double-end trim saw or by saws on horizontal spindles of a double-end tenoner. Saws employed for this purpose are hollow-ground, run at a rim speed of 20,000 fpm or higher, and are commonly arranged for climb-sawing. For use on dry lumber, tooth pitch is approximately ⅝-inch and tooth height is 7/16-inch (fig. 18-84). Saws are hollow ground to reduce the plate thickness 4 to 5 gauges in an area concentric to the rim (table 18-49). The 45 face bevel, lack of set, high rim speed, and numerous teeth result in smoothly machined kerf walls.

For applications where the durability of carbide cutting edges is required, specifications could be as follows:

    Diameter...................... 20 or 22 inches
    Number of teeth................ 120
    Tooth style .... 25 alternate top bevel and 15 alternate face bevel
    Saw gauge .... 9-gauge plate; 0.200-inch kerf

Saws with this configuration operating at a rim speed near 20,000 fpm are noisy. Intensity of noise decreases with decreasing rpm, saw diameter, blade thickness, number of teeth, and feed rate. Laminated (compound) blades make less noise than one-piece blades.

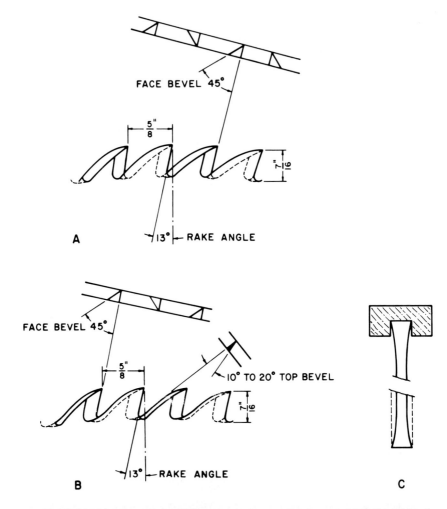

Figure 18-84.—Tooth styles for hollow-ground, smooth-trim saws. (A) Round-back tooth (approximately 45° top bevel). (B) Skew-back tooth (10° to 20° top bevel); this configuration gives more strength to the point of the tooth and is an exception to the rule of equally dividing face and top bevel. (C) Hollow-ground saw plate gives clearance to cutting corners. (Drawing after Koch 1964a, p. 247, 264.)

TABLE 18-49.—*Hollow-ground smooth-trim steel saws for trimming kiln-dried southern hardwoods*

Feed method and saw diameter (inches)	Gauge[1]	Number of teeth
	BWG	
Hand-fed machines		
8	14-18-14	100
10	13-17-13	100
12	12-16-12	120-150
14	11-15-11	120-170
16	10-14-10	130
Power-fed machines		
18	8-12-8	100
20	7-11-7	120
22	6-11-6	140
24	5-10-5	140

[1]Birmingham wire gauge. The first and third values are the gauge of the saw plate, the second is the gauge to which the plate is hollow-ground in an area concentric to the rim. (See fig. 18-84C.)

In laboratory work it is sometimes necessary to crosscut or ripsaw very thin sections without damaging wood structure at the sawn surface. Bramhall and McLauchlan (1970) have described a technique using a 4.5-inch-diameter, 34-tooth, hollow-ground saw revolved on precision bearings at a rim speed of 6,000 fpm to cut transverse sections 900 m thick and longitudinal sections only 100 m thick. The saw recommended has neither set nor swage, but is hollow ground from 0.05-inch thickness at the rim to 0.04-inch thickness at a distance 1 inch inward from the rim. Rake angle is 15 with alternate front bevel angle of 15.

Hollow-ground combination planer saws.—These versatile concave-ground saws are designed for hand-fed machines to produce smooth surfaces on crosscutting and miter cuts as well as on ripping cuts. Their teeth are neither swaged nor spring-set; clearance is provided by reducing (tapering) the saw plate thickness three gauges in an area concentric to the rim. They are not well suited to radial arm saws and they will not stand up to the fast feeds required on tenoners because feed per tooth is limited by the small clearance between saw plate and kerf wall, and because saw projection should be sizeable to avoid having too much of the thick rim in the cut.

The grouped teeth (fig. 18-85) include raker teeth that clean out kerf material left by scoring teeth; rakers should be 1/64-inch shorter than scoring teeth to produce the kerf contour shown in figure 18-85 (bottom). These saws are available in diameters from 6 to 24 inches with 16 to 9 gauge rim thickness (table 18-50).

Miter saws.—Miter saws, used principally for smooth crosscutting of thin cabinet wood or making miter cuts in small pieces such as picture-frame mouldings, are hollow ground and have no raker teeth. They are finely and evenly toothed with a tooth style similar to the scoring teeth in figure 18-85, but with rake angle, front bevel, and top bevel all ground at about 13. Diameters range from 6 to 16 inches with gauges at the rim from 16 to 12 gauge (table 18-51).

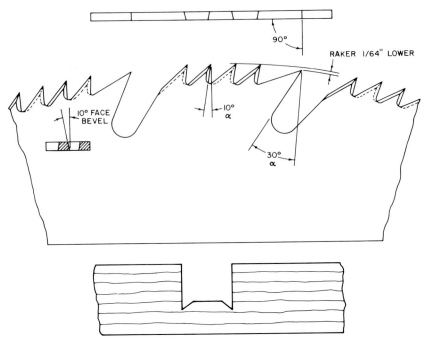

Figure 18-85.—Hollow-ground combination planer saw. (Top) Tooth style. (Bottom) Correct kerf profile made by having raker tooth shorter than scoring teeth. (Drawing after Koch 1964a, p. 267.)

Combination saws.—These saws are designed to do faster, but rougher, work than hollow-ground combination planer saws. They are suitable for use on electric hand-saws. Teeth are spring-set 2½ gauges to each side and are ground with both top and front bevel and with intermediate rake angle (fig. 18-86), thus suiting them to crosscutting and mitering, as well as ripping cuts. These saws typically have 44 teeth and are available in diameters from 6 to 18 inches with saw plates 18 to 12 gauge in thickness (table 18-52).

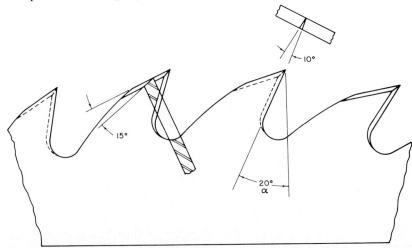

Figure 18-86.—Tooth style for combination saw.

TABLE 18-50.—*Specifications for hollow-ground combination planer saws*

Diameter (inches)	Gauge[1]	Number of sections	Collar size
	BWG		Inches
6	16-19-16	10-12	2½
8	14-17-14	12-14	3½
10	13-16-13	16	4
12	12-15-12	18	5
14	11-14-11	18-20	5
16	11-14-11	20-22	6
18	10-13-10	22-24	7
20	10-13-10	26	7
24	9-12-9	30	7

[1]The first and third values are the gauge of the saw plate, the second is the gauge to which the plate is hollow-ground in an area concentric to the rim.

TABLE 18-51.—*Specifications for miter saws*

Diameter (inches)	Gauge[1]	Number of teeth
	BWG	
6	16-19-16	150
8	15-18-15	150
10	14-17-14	150
12	13-16-13	150
14	12-15-12	200
16	12-15-12	200

[1]The first and third values are the gauge of the saw plate; the second is the gauge to which the plate is hollow-ground in an area concentric to the rim.

TABLE 18-52.—*Specifications for combination saws*

Diameter (inches)	Gauge	Number of teeth
	BWG	
6	18	44
8	18	44
10	16	44
12	14	44
14	13	44
16	13	44
18	12	44

Flat-ground combination saws.—These saws have tooth styles resembling those of hollow-ground combination saws but the saw plates are flat (not hollow ground) and side clearance is achieved by spring setting the teeth. Because of this additional clearance, the saw need project very little beyond the workpiece. Therefore, with two scoring teeth and one raker in each tooth grouping (fig. 18-87 top), the saw is much used on radial-arm saws to crosscut, miter, and rip.

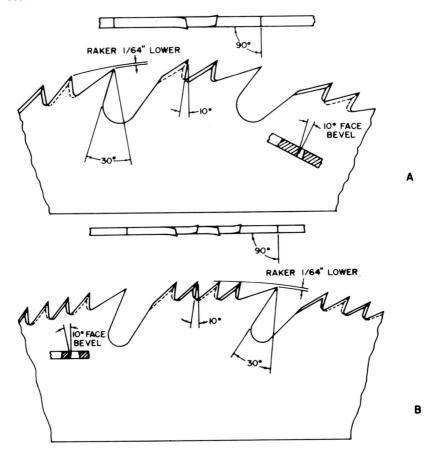

Figure 18-87.—Tooth styles for flat-ground combination saws having grouped teeth. Scoring teeth are spring-set with top and front bevel; raker teeth have no set and neither top nor front bevel. (Top) Two scoring teeth and one raker tooth per section. (Bottom) Four scoring teeth and one raker tooth per section.

With four scoring teeth per group (fig. 18-87 bottom), the saw requires more power, but can withstand the feeds employed on double-end tenoners. The kerf made by this saw appears the same as that of hollow-ground combination saws (fig. 18-85).

The saws are available in diameters from 8 to 22 inches with saw-plate thicknesses from 18 to 8 gauge (table 18-53). They are usually operated at 3600 rpm, mounted directly on the arbor of a synchronous motor.

TABLE 18-53.—*Specifications for flat-ground combination saws with grouped teeth*[1]

Number of scoring teeth and diameter (inches)	Gauge	Number of sections
	BWG	
Two scoring teeth per group		
8	18	16
10	16	18
12	14	20
14	14	22
16	13	24
18	12	26
20	12	28
22	8	30
Four scoring teeth per group		
6	18	12
8	18	14
10	16	16
12	14	18
14	13	18
16	13	20

[1]Scoring teeth are spring-set; raker teeth are not set.

Power requirements when ripsawing with swage-set teeth are described in table 18-43 and in figures 18-74 and 18-75. Data on crosscuts and miter cuts have yet to be presented, however. Orthogonal cutting experiments (tables 18-5 through 18-26) have shown that it takes a greater parallel cutting force to sever wood chips perpendicular to the grain (as in band ripsawing) than it does tangential to the grain (as in veneer cutting). It is therefore reasonable that ripping with a combination circular saw requires more cutting power than crosscutting with the same saw.

McMillin and Lubkin (1959) developed an equation to predict power demand for a variety of cutting conditions encountered when using a climb-cutting, overcutting, radial-arm saw; the equation includes the effect of grain orientation.

The 16-inch, flat-ground combination saw had two spring-set scoring teeth for each raker tooth; specifications were as follows:

Thickness, 0.097 inch	Front and top bevel angle of scorers, 10°
Set, 0.025 inch per side	Clearance angle of scorers, 30°
Actual kerf, 0.16 inch	Clearance angle of rakers, 20°
Number of teeth, 72	Rake angle of scorers, 10°
Number of groups of teeth, 24	Rake angle of rakers, 27°
Number of scorers, 48	Rakers had no additional bevel angles.
Collar diameter, 4 inches	

The workpiece was flat-sawn and knot-free hard maple at 11 percent moisture content. Except for those cuts involving thickness as a variable, the workpiece was 1⅝ inches thick. Nominal saw speed was 3600 rpm.

Their equation (derivation summarized in Koch 1964a, p. 240-241) was expressed as follows:

$$P = d(C_3 + C_4 f)(1 + E \sin^2 \theta) \qquad (18\text{-}33)$$

where:
- P = net cutting horsepower
- d = depth of workpiece (depth of cut), inches
- θ = the angle the sawblade makes with the crosscutting direction, i.e.:
  - θ = 0° when crosscutting
  - θ = 45° in a miter cut
  - θ = 90° when ripsawing

and where C_3, C_4, and E are constants for a cutting situation; for these particular saw and climb-cutting conditions:

$C_3 = 0.284$
$C_4 = .0548$
$E = 1.343$

When the foregoing constants are inserted in equation 18-33, the curves drawn in figure 18-88 result; experimentally determined points are indicated to illustrate the good fit of this equation.

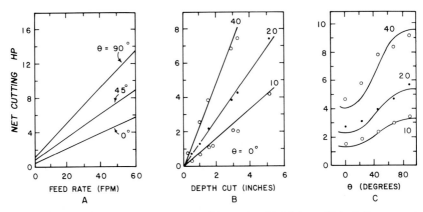

Figure 18-88.—Net cutting power required (equation 18-33) by a 16-inch flat-ground combination saw, climb sawing dry hard maple, related to the angle of cut (θ), depth of cut, and feed rate. Numbers adjacent to graphs B and C show feed speeds in feet per minute. Graphs A and C represent cuts through 1-5/8-inch-thick lumber. (Drawings after McMillin and Lubkin 1959.)

Dado heads.—Dado heads are designed to cut smooth-surfaced grooves from ⅛ to 4 inches wide in any grain direction. A dado head assembly is comprised of two hollow-ground outside saws (fig. 18-89 left) in combination with one or more inside cutters (fig. 18-89 right), depending on the width of groove desired. A single outer saw cuts a groove ⅛-inch wide and when the two outer saws are used without inside cutters they cut a ¼-inch groove. By a suitable selection of

Machining 1897

Figure 18-89.—Elements of a dado head assembly. (Left) Outside saw. (Right) Inside cutter.

inside cutters, widths in 1/16-inch increments from ¼ to 4 inches can be cut. Standard diameters are 5, 6, 7, 8, 9, 10, 11, 12, 14, 16, 18, and 20 inches. The swaged inside cutters are available in plate thicknesses of 1/16- 1/8-, and 1/4-inch.

The following points should be observed when assembling dado heads:

- The rakers of one outside saw should be opposite the scoring teeth of the other.
- Inside cutters should be arranged evenly around the circumference with their swaged cutting edges in line with the gullets of the outside cutters.
- The swaged tips of the inside cutters must not touch the tips of other inside cutters nor the outside teeth.
- When grooving across the grain only, use of inside cutters no wider than 1/16-inch minimizes fraying at the end of the cut.
- As width and depth of cut are increased, feed rate should be reduced.

To sharpen the elements of a dado head, the entire assembly, including all inside cutters in the set, must be first jointed or rounded on the periphery. Filing should reduce the resulting "land" to a barely visible minimum but not beyond that point. The raker teeth on the outside saws and on the inside cutters should be filed 1/64-inch below the scoring teeth of the outside saws, and should have neither top nor front bevel. The inside cutters are filed on the cutting faces and then jointed on the tops.

Ring saws.—A portable gas-engine-driven saw similar in function to a conventional chain saw but having a circular, ring-shaped cutting blade and no bar is manufactured by Ring Saws, Inc., Waltham, Mass. Used for crosscutting small logs, the saw has minimal tendency to kick back uncontrollably, and most of the blade is guarded. A splitter blade located in the lower half of the ring is thicker than the inside of the blade, yet thinner than the kerf, so much binding is eliminated. A friction drive allows the blade to slip at a predetermined load (Swan 1971a).

Tubular core saws.—Hardwood logs must frequently be reduced to short lengths because of crook and other defects. E. E. Dargan and E. M. Ervin devised a unique sawing operation to convert such short lengths into 2¼- to 3⅛-inch dowels (Dargan 1969). The short logs are chucked on centers, and a long, tubular core saw makes an initial cut for escape of sawdust parallel to the grain and adjacent to the periphery of the log. Successive cuts are made as the log is indexed (visualize holes in the cylinder of a revolver) so that each bore overlaps slightly into the kerf of a previous bore (fig. 18-90). If the log is large enough a second ring of bores is made closer to the log pith. The machine will handle logs from 8 to 40 inches in diameter and from 36 to 52 inches long. Its cyclic operating rate is 12 cuts per minute, and it is capable of cutting dowels as large as 3⅞ inches in diameter. The tubes (fig. 18-91), which carry three swage-set teeth, cut a ¼-inch kerf, and rotate at 4,000 rpm. Tubes used are a scant ¼-inch thick with inside diameters of 2¼, 2¾, 3⅝, and 3⅞ inches. To cut dowels of maximum length and diameter requires about 75 connected horsepower (sometimes operating at 300-percent overload). Plunge speed is 67 fpm, resulting in a 1½-second duration of cut.

Portable circular-saw headrigs for hardwoods.—The simplicity of inserted-tooth circular saws makes them suitable for portable sawmills that can be easily and quickly moved from woodlot to woodlot. Mills (1976) reviewed designs of portable sawmills available, and the reader is directed to this review for a summary of available designs. Bryan (1978) described a hardwood operation in which a mobile circular-saw headrig is combined with a two-saw vertical edger on a trailer-mounted frame; output when working in oak and red maple timber measuring about 18 inches in dbh is about 10,000 board feet per day with a seven-man crew that includes logging force.

Among several designs offering extreme simplicity and portability is a one-man rig that moves a circular sawblade through a stationary log (fig. 18-92). It can be dismantled and carried by hand, truck, or helicopter to areas where log haulage is not practical. For use on 6- to 16-foot-long pine-site hardwood logs not exceeding 60 inches in diameter, the machine would carry a 30-inch diameter, 9 gauge saw with 12 inserted teeth of 5/16B style. The saw would be revolved at 1,750 rpm by a 53 hp gasoline or 44 hp diesel engine. On such logs, production of hardwood planks, pallet cants, mine timbers, and crossties should average 4 to 8 thousand board feet per 8-hour shift. The machine requires about 1 hour to set up on a new location. It sells, complete with engine, trailer, and two saws for about $10,000 (1980).

Figure 18-90.—Tubular core saw. (Top) The short log has been chucked and vertical-spindle cutter at right center has made a 1-inch cut from which initial sawdust will escape. The tubular saw has begun its first cut. (Bottom) Yield from 10-inch-diameter, 52-inch-long bolt. Taper in the log prevented cuts closer to the log surface. Residual wood is chipped for pulp.

Figure 18-91.—Three-tooth tubular core saw, which turns at 4,000 rpm, cuts 2-1/4- to 4-inch diameter dowels at rate of about 12 per minute. Tubes are changed to suit desired dowel diameter.

Figure 18-92.—One man portable sawmill transportable to the stump by pickup truck or trailer. Vertical and horizontal saws move along tracks; the log is stationary. (Photo from International Enterprises of America, Portland, Oregon.)

CARBIDE-TOOTHED CIRCULAR SAWS

Knowledge of the tooth forms described in preceding sub-sections is needed to understand the cutting action of circular saws. The hardwood-using industry has, however, in many—perhaps most—situations switched from these all steel saws to steel saw blades carrying carbide cutting edges. Thus, carbide-toothed saws are widely used on edgers, trimmers, and panel saws. Of the many variations of carbide-toothed saws, three important types will be described, as well as a new saw that crosscuts lumber while it is in motion. Readers needing information on techniques for fitting and sharpening carbide-toothed saws will find Hewitt's (1978) handbook useful.

Because swaging tools are not used on carbide saws, tooth sharpness angle is not standardized at 44°, as on steel saws. To provide solid support for the carbide inserts, which are brazed in place, top clearance angles are kept low—about 8° (fig. 18-93). At the sides of the teeth radial clearance angles are generally about 2° and tangential clearance about 3°

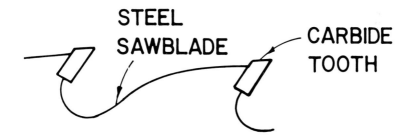

Figure 18-93.—Carbide teeth brazed onto a steel saw blade. Tooth form should give maximum support to the cutting tip and at the same time provide adequate gullet area. (Drawing after Hewitt 1978.)

In Kirbach's (n.d.) discussion of carbide-tipped ripsaws he noted the following three types:

Type	Comment
Solid, unslotted	Particularly favored for thin-kerf sawing; kerf can be as little as 0.125 inch. Successful operation of these saws requires highly effective and well maintained guide systems. Saw blades must be manufactured accurately to uniform thickness.
Slotted, without carbide bars mounted in the slots.	Not commonly used in sawmill ripsawing operations. The saws are slotted to make them more stable when heated by cutting.
Slotted, with carbide bars mounted in the slots.	Have found wide acceptance in sawmill resaw operations. In one type the slots are open to the rim; in another only some of the slots are open. The saws are fairly trouble-free because the bar-carrying slots virtually eliminate problems arising from slivers and sawdust that settle between the cut wood surface and the sawblade. Saws with open slots are generally thicker than saws with closed slots or no slots, resulting in thicker kerf (about 0.200 inch).

Gang ripsaws for green oak and hickory.—Counter-cutting saws (fig. 18-76) are widely used to gang rip 4-inch green hardwood cants into pallet deck boards. Specifications for a typical carbide-tipped saw for this purpose are as follows for a feed per tooth of 0.05 inch at 160, 240, and 300 fpm:

Statistic	160 fpm[6]	240 fpm[7]	300 fpm[8]
Diameter (inches)	16	16	16
Saw plate thickness (inches)	0.120	0.120	0.120
Number of teeth	16	24	24
rpm	2,400	2,400	3,000
Carbide grade	OM-1 or OM-2	W-2	C-2
Kerf (inch)	0.190	0.190	0.190
Rake angle	30°	28°	30°
Top clearance angle	8°	12°	7°
Radial clearance angle (each side)	1.5°	1°	2°
Tangential clearance angle (each side)	3°	4°	3°
Front bevel	0°	0°	0°
Top bevel	0°	0°	0°

Sage and Olson (1975) found that slotted "strob" plates performed better than standard unslotted saw plates. Opinions vary among saw specialists, however.

---

[6]Personal correspondence 14 May 1980 with Roy Sage, Schurman Machine Works, Inc., Woodland, Wash.
[7]Personal correspondence 12 May 1980 with Kenneth W. Duff, Disston, Inc., Seattle, Wash.
[8]Personal correspondence 13 July 1980 with George Brown, Industrial Carbide Tooling, Inc., Eugene, Ore.

Machining

Gang ripsaws for dry hardwoods.—Top-arbor, dip-chain, climb-cutting gang ripsaws (fig. 18-79 bottom) are effectively used to straight-line rip wide hardwood boards and planks into narrower strips. (See Ward 1975ab for description of a highly mechanized operation.) Saw specifications for such ripsaws, cutting in the mode shown in figure 18-77A, might be as follows for feed speeds of 150 and 200 fpm on 2-inch-thick stock:

Statistic	150 fpm[7]	200 fpm[8]
Diameter (inches)	16	12
Gauge (inch)	0.120	0.110
Number of teeth	30	40
rpm	2,400	3,000
Carbide grade	W-2	C-2
Kerf (inch)	0.180	0.160
Rake angle	28°	15°
Top clearance angle	12°	12°
Radial clearance angle (each side)	1°	½°
Tangential clearance angle	4°	1°
Front bevel	0°	0°
Top bevel	0°	0°
Maximum depth of cut (inches)	—	4

Smooth trimsaws for dry hardwoods.—Hardwood furniture or cabinet manufacture requires smooth crosscuts, done typically on a double-end trimmer or tenoner, which might carry carbide-tipped saws with one of the following specifications:

Statistic	Reference text footnote 7	Reference text footnote 8
Diameter (inches)	16	14
Gauge (inch)	0.148	0.134
Number of teeth	100	100
rpm	3,600	3,600
Carbide grade	W-2	C-2
Kerf (inch)	0.188	0.194
Rake angle	10°	10°
Top clearance angle	8°	10°
Radial clearance angle (each side)	1°	2°
Tangential clearance angle	4°	3°
Front bevel (alternate)	15°	15°
Top bevel (alternate)	20°	15°
Maximum depth of cut (inches)	3	2
Normal feed speed on 1-inch-thick lumber (fpm)	—	180

Spiral rough trimsaw for hardwoods in linear motion.—Conversion of hardwood lumber to parts for furniture or fixtures usually requires that boards or strips be crosscut into random lengths to remove defects. Such defect removal and trimming to length is slow and labor intensive because, by usual practice, lumber motion is stopped during positioning and cutting. The Drexel-Heritage plant at Whittier, N.C., in cooperation with Black Clawson, Inc., has put into successful operation a spiral-helix chop saw (fig. 18-94) that cuts without interrupting the lineal flow of lumber moving endwise past the saw continuously at 100 fpm.

To accomplish this, a spiral-shaped sawblade with a helix lead of 1½ inches is canted to make the cutting edge square with the workpiece. To accomplish a crosscut, it is triggered to make a single, high-speed revolution (1,500 rpm) and stop instantly poised for the next cut; the 1- by 3½-inch lumber is thereby crosscut in motion and flows through a missing quadrant of the sawblade (fig. 18-94). Sharpening is complex, but is required only two or three times a year.

Ward (1975a) has described the computer logic and layout that permits the operator to do nothing more than tap a control (like a telegraph key) to mark ends of stock and defects as they pass before him under a shadow-line marker.

Figure 18-94.—Spiral-helix chop saw that crosscuts linearly moving strips of hardwood to remove defects and trim to lengths called for by cutting orders. Teeth start close to the hub and progress radially outward in a spiral profile to maximum radius 270° later. The root of the blade also follows a screw-auger path where it joins the blade. (Photo from Woodworking and Furniture Digest; description from Ward 1975a.)

CHAINSAWING

Since the 1950's, chain saws have affored a practical, light-weight, one-man logging tool that can fell or buck trees in a fraction of the time required by manual crosscut saws. Prior to the oil embargo of 1973 chainsaws were primarily used by professional loggers. By 1980, however, a significant proportion of

Machining

home owners had purchased small chainsaws to cut wood for residential stoves and fireplaces.

Sarna (1976), in his review of chainsaw developments, noted that they have been made safer and more productive. Engine housings have been redesigned to improve weight distribution and eliminate sharp edges. Handles and throttles have been redesigned to allow use of heavy gloves. New air intake filters allow easier cleaning. Better design of mufflers, chain guides, and air intake baffles have reduced noise (currently 104-106 dB(A)). Saw vibration has been reduced by weighting the crank shaft and isolating the piston from other engine components with rubber bushings. Heated handlebars have been provided for winter use. The chisel tooth has resulted in faster cutting and less maintenance. Narrower kerf, improved drive sprockets, bar guides, and nose guide sprockets have improved cutting efficiency, and new chain brake systems have reduced hazards of kickbacks.

In chain crosscutting, the main cutting action approaches orthogonal cutting perpendicular to the grain with cutting edge parallel to the grain (0-90 direction); the cutting action is complicated because kerf boundaries are established by cutting (90-90 mode) in the planes of the kerf walls simultaneously with advance of the cutting edge (figs. 18-95, 18-96).

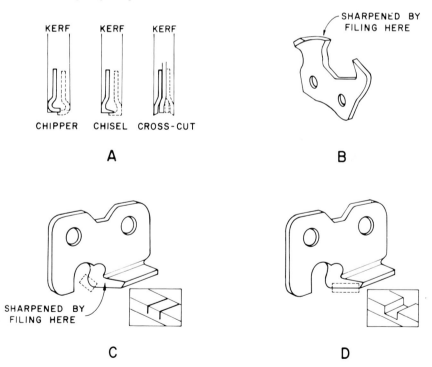

Figure 18-95.—Chain saw teeth. (A) Three types of chain saw teeth. (B) Saw tooth, intermediate between chipper and chisel type that can be sharpened on the back with a flat file. (C) Chipper-type chain; inset indicates that the corner cuts kerf boundary. (D) Chipper-type chain; inset indicates that the cutting edge lifts out the chip. (Drawings after Penburthy 1968.)

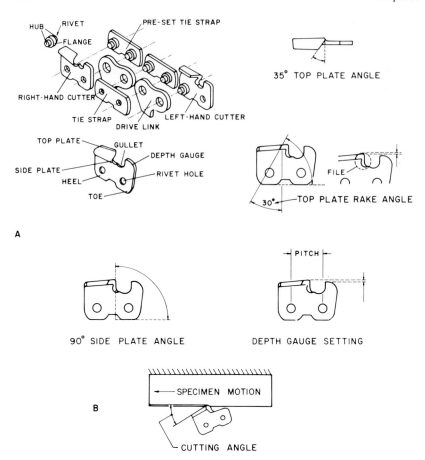

Figure 18-96.—(A) Filing angles and nomenclature for chainsaws. (B) Definition of cutting angle in chainsaws. (Drawing A from Omark Industries, Inc.; drawing B after Gambrell and Byars 1966.)

Gambrell and Byars (1966) reported data collected while chainsawing green northern red oak. In the range of chain speeds from 500 to 3,640 fpm in cuts from 0.010 to 0.060 inch deep per toothed link, cutting forces in the three principal directions increased appreciably with increased depth of cut and decreased slightly with increased cutting speed; energy consumption was relatively independent of cutting angle (fig. 18-96B). Reynolds et al. (1970) and Reynolds and Soedel (1972) developed relationships that predict power requirements of cutting chains; saw and engine designers should find the information useful.

Most southern hardwoods are cut with one-man saws of less than 6 hp that carry teeth similar to those illustrated in figure 18-95BCD, although the chisel-type tooth is gaining in popularity. These one-man saws may have straight bars 14 to 16 inches long, or they may be equipped with curved rims or bows. The bowsaw is used primarily when bucking small roundwood on the ground. The bow enables the operator to make plunging cuts and thus avoid unnecessary stooping; because of its narrow rim, it is pinched less in such cuts than a saw equipped with a straight bar. The trend is toward shorter bars and higher chain speeds. Factors determining chain selection include the following.

Saw chains are primarily classified according to the distance, or pitch, between articulations. Large-pitch chains with ½- to ⅞-inch between articulations are stronger because individual links, pins, and cutters are heavier. They carry longer, wider cutters that produce a wider kerf, take bigger chips, and require more power, but have a longer service life. Since there is more space between individual cutting edges they are less suited for removing small limbs or cutting small trees than chains of smaller pitch.

Small-pitch chains (less than ½-inch between articulations) operate better on direct-drive saws because at their high chain speed (2,400 to 3,800 fpm) the lighter links, smaller kerf, and lighter feed per tooth minimize shock load on the engine and pounding on bar entry, rails, and sprockets. Only small-pitch chains, whether direct-drive or gear-drive, are satisfactory on low-horsepower saws.

Medium- to large-pitch chains are more suitable for the slower chain speeds of gear-drive saws (800 to 2,000 fpm), where the greater weight of the links has less adverse effects on the mechanism and power is adequate for heavier feeds per tooth. Specific cutting energy is lower with thick chips than with thin chips.

Chainsaws are manufactured in many countries and offered in hundreds of models. To aid loggers in comparisons of the models, Hensel (1976) tabulated specifications and performance data on chainsaws manufactured for use by the logging industry of the United States and Canada. Morner (1976) also surveyed chainsaws used in harvesting.

Manuals on the use and care of chainsaws have been published by Johnson (1977) and Sarna (1979). While most chainsaws are designed to be hand-held, some mechanized versions are designed into feller bunchers for hardwoods (fig. 16-15), log-deck cutoff saws, and log-bundle bucking machines.

There is a substantial literature on vibrations of chainsaws that may be transmitted to the hands of operators with deleterious effect. A sampling of this literature published since 1969 includes the following:

Sorensson (1969)   Firth (1972)
Keighley (1970, 1973)   Thompson (1973)
Welch (1971)

Portable chainsaw headrigs for hardwoods.—The most portable of all headrigs is a chainsaw rigged to make ripping cuts (rather than the usual cross-grain cuts) to yield planks or cants (fig. 18-97). Machines of this class may be equipped with one or two engines. With a single engine, one man might produce 500 board feet of cants and planks in 8 hours from oak logs averaging 12 inches in diameter and 8 to 10 feet long. A double-engine machine with a two-man crew will produce more (Sperber 1977; Sperber Tool Works 1977). For another design of chainsaw headrig see The Logger and Lumberman (1980).

Figure 18-97.—Portable chainsaw headrig weighing 75 pounds. The 0.404-inch pitch chain cuts a 3/8-inch kerf and is driven by two 7-horsepower engines. It is capable of milling southern hardwood logs up to 50 inches in diameter. On a 12-inch log, it can ripsaw at about 12 fpm with manual feed. On oak, the chain must be sharpened after about 1 hour of operation. (Photo from Sperber Tool Works, Inc., West Caldwell, N.J.)

18-9 CONVERTING WITH CHIPPING HEADRIGS

Widespread adoption of chipping headrigs and edgers has greatly changed lumber manufacturing methods and chip procurement procedures throughout North America. These machines—designed to convert small logs into timbers (cants or flitches) without simultaneously producing sawdust—are probably the most important innovations in mechanical conversion of logs since invention of the mechanical ring debarker. In most applications the resulting squared (or partially squared) material is resawn into lumber. The machines, invented to eliminate sawdust from slabbing cuts on conventional headrigs, are still under development. Events contributing to the development of various configurations of chipping headrigs were reviewed by Koch (1967, 1968b).

Chipping headrigs remove wood in the form of single chips formed by the intermittent engagement with the workpiece of knives carried on the periphery (or on the end or face) of a rotating cutterhead. Each cutterhead usually carries several knives, removable for sharpening, which are precisely set to cut on a common cutting circle (or in a common plane in the case of end mills). The finished surface therefore consists of individual knife traces generated by the successive engagement of each knife.

This text will briefly discuss information essential to understanding the function of chipping headrigs. Readers desiring more complete data are referred to Koch and Dobie's (1979) profusely illustrated description of production of chips by chipping headrigs.

CUTTING MODES

For the three basic directions in which a knife can be oriented to cut wood (fig. 18-1), kinematics, force systems, and chip severence phenomenon cutting are described in sections 18-2, 18-4, 18-5, and 18-6. To exploit these three possibilities, three modes of cutting have been developed for chipping headrigs:

- Cutting with a shaping-lathe configuration (fig. 18-98A), in which the knife edge is parallel to the grain but moves perpendicular to the grain, i.e., 0-90 mode.
- End milling (fig. 18-98B), with chip severence accomplished by cutting across the grain with the knife edge at right angles to the grain, i.e., 90-90 mode.
- Peripheral down-milling (planing) with the knife edge perpendicular to the grain, but travelling more-or-less parallel to the grain, i.e., approximately in the 90-0 mode (fig. 19-98C).

Cutterheads operating in all three modes are manufactured commercially (fig. 18-99). The shaping-lathe configuration is employed only on headrigs. End-milling cutterheads are widely used on two-head edgers as well as on two-head canting headrigs. Downmilling heads, widely used on two- and four-head headrigs and on two-head edgers, are also used to convert heavy slabs into rectangular boards by thicknessing the face-down slab from the top and simultaneously milling both edges with profile-cutting heads mounted on one or more arbors.

POWER REQUIREMENTS

The high density, hardness, taper, butt swells, and surface bumps so typical of pine-site hardwood logs are obstacles to use of chipping headrigs because these attributes cause extraordinary power demands on the cutterheads and problems in maintaining linearity of feed. Linearity of feed is difficult for two reasons: (1) unbalanced cutting forces buffeting the log in the cutting zone, and (2) inability to advance a knobby irregular log in a controlled manner.

The power required by any head of a chipping headrig is determined by properties of the wood, chip dimensions, volume of wood chipped per unit of time, cutter configuration, and cutting mode. As a common base for comparison, it is convenient to use the concept of specific cutting energy, i.e., the amount of energy required to remove a unit volume of wood.

Figure 18-98.—Three configurations of chipping headrigs and resulting chips. (A) Shaping-lathe, 0-90 mode. (B) End-milling, 90-90 mode. (C) Down-milling, approximately 90-0 mode. Arrow in (A) points to cam and cam follower controlling shape of cant. The saw in the end-milling configuration (B) proved unnecessary and was later removed. (Photos from U.S. Department of Agriculture, Forest Service 1964, p. 60.)

Shaping-lathe headrig.—Specific cutting energy for the shaping-lathe configuration (figs. 18-98A, 18-99 top, 0-90 cutting direction) is less than for the other two modes if chip thickness is held constant at the usual pulp chip dimension of ⅛-inch to 3/16-inch; a specific cutting energy of less than 5 hp min/ft^3 is probable under such conditions when cutting oak. The lathe is best adapted to cutting flakes rather than chips, however. Net specific cutting energy for removal of flakes cut in the 0-90 mode is inversely correlated with flake thickness and positively correlated with wood specific gravity; water-soaked wood cut at

Machining 1911

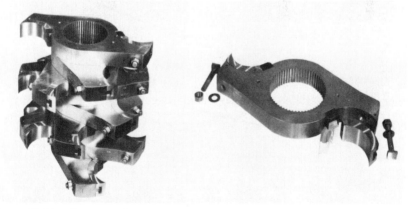

Figure 18-99.—Commercial cutterheads for chipping headrigs designed for cutting in three modes. (Top) Cutterhead of shaping-lathe headrig designed to cut in 0-90 mode; photo from Stetson-Ross. (Center) End-milling headrig chipper cutting approximately 90-90; bandsaw is at left; photo from Chipper Machines and Engineering Company. (Bottom) Down-milling head for cutting approximately 90-0; photo from Stetson-Ross.

160°F required 5.5 percent less specific energy than green wood cut at 72°F (fig. 18-100). More complete information on cutterhead power required by the shaping-lathe headrig is contained in Koch (1974).

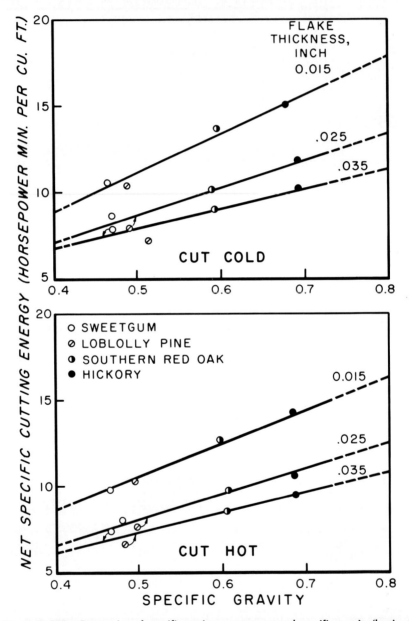

Figure 18-100.—Regression of specific cutting energy on wood specific gravity (basis of ovendry weight and green volume) for wet wood cut in the 0-90 mode on a shaping-lathe headrig at 72°F (top) and 160°F (bottom). Plotted points are averages for the species named. Data are for a 10-knife, 18-inch-diameter cutterhead, with knives having rake angle of 43° and clearance angle of 5°, cutting flakes measuring 3-1/4 inches along the grain. (Drawing after Koch 1974.)

End-milling headrig chipper.—Miller and Roppel (1973) provided a chart (fig. 18-101) whereby horsepower requirements can be determined for an end-milling headrig chipper (fig. 18-99 center) slabbing softwood logs; for oak and hickory these values should be increased by 30 to 40 percent. To use this chart,

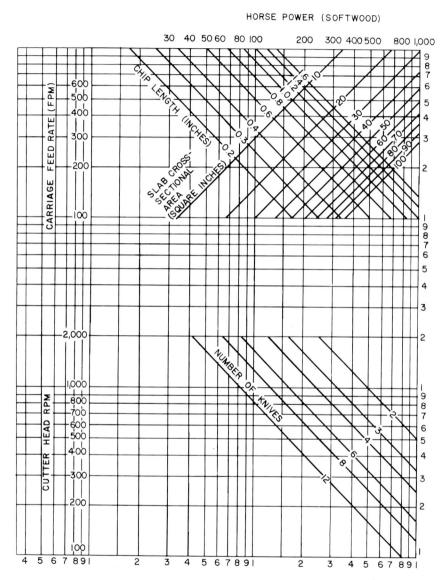

Figure 18-101.—Chart for determination of power requirement of an end-milling (90-90 mode) cutterhead chipping slabs from softwood logs. (Drawing after Miller and Roppel 1973.)

carriage feed rate, cross-sectional area of slab to be removed, and desired chip length must be known. For a carriage feed rate of 375 fpm, cross-sectional slab area to be removed of 20 square inches, and a chip length of 0.8 inch, procedure is as follows.

- To estimate cutterhead horsepower (softwoods): Locate feed rate of 375 fpm at upper left and proceed horizontally to the right until cross-sectional area of 20 square inches is intersected; horsepower requirement is found directly **above** this intersection (hp = 250).
- To determine the number of knives and rpm of chipping head: Locate feed rate of 375 fpm at upper left, and proceed horizontally to the right to intersection with chip length of 0.8 inch; then proceed **down** to intersect number of knives and horizontally to left to rpm. Thus a chip length of 0.8 inch can be accomplished with a six-knife chipper at 900 rpm, an eight-knife chipper at 700 rpm, and a 12-knife chipper at 450 rpm.

According to these determinations, specific cutting energy for the cutterhead illustrated (fig. 18-99 center) is 4.8 hp min/ft^3 of softwood chipped. On oak or hickory 6.2 to 6.7 hp min/ft^3 would likely be needed.

Down-milling chipping head.—When cutting in the mode of figure 18-98C with a cutterhead having rake angle of 45°, clearance angle of 7½°, and cutting-circle diameter of 11-1/16-inch, specific cutting energy milling ¾-inch-long chips from slash pine was 4 hp min/ft^3 (Koch 1964b). No comparable data are available for oak and hickory, but 5 to 5½ hp min/ft^3 is likely required.

CHIP FORM

Chips cut by the three cutterhead styles are distinctly different (fig. 18-98). In commercial practice, chips vary with species, cutterhead design, and production and maintenance procedures.

The shaping-lathe headrig cuts flakes suitable for the manufacture of structural flakeboard. Hardwoods with small vessels yield cohesive flakes whereas those with large vessels, such as southern red oak, tend to yield strands. Cohesive flake formation is more likely if green wood is heated in steam or water before machining (fig. 18-102). Through use of scoring knives, high-quality chips measuring about 3/16-thick and of precise length (e.g. 1-inch) can also be cut on the shaping lathe (Plough and Koch, 1983).

Chips (fig. 18-103) cut with end-milling cutterheads such as that illustrated in figure 18-99 (center), resemble pulp chips cut with a disk chipper.

Down-milled chips cut with cutterheads similar to that shown in figure 18-99 (bottom) also resemble pulp chips with a disk chipper, but part of each chip is thin because of the cutting geometry (fig. 18-98 bottom).

Figure 18-102.—Up-milled flakes 0.015-inch thick and 3 inches long cut with a shaping-lathe headrig from sweetgum (top), southern red oak (center), and hickory (bottom), soaked in water at 72°F (left) and 160°F (right). (Photo from Koch 1974.)

SURFACE QUALITY

The shaping-lathe headrig, because it cuts thin flakes in the veneer direction, leaves a very smooth surface with no tearout around knots, in areas of short grain, or at cant corners. Cants or cylinders cut on the shaping-lathe headrig appear to have been planed.

Surfaces cut with end-milling heads or with down-milling heads tend to be rough and may exhibit torn grain around knots in areas of short grain, and at cant corners. This torn grain is caused by the large feed per knife required for pulp chip production—5/8 to 1 inch. Addition of scoring knives to end-mills and careful sharpening and fitting practices can generally control roughness on end-

Figure 18-103.—White oak chips cut with an end-milling chipping headrig. (See fig. 18-99.)

milled cants to acceptable surfaces resembling those on lumber sawn at large feeds per tooth.

When climb-milling at ⅝-inch to 1-inch feed per knife, height of knife marks (see h in fig. 18-45) is considerable—0.005 to 0.014 inch depending on feed speed and number of knives cutting. The usual rake angle for climb-milling softwood logs is 55°; this high rake angle promotes good chip formation and is favorable to low power consumption. On some southern hardwoods with interlocked grain, however, rake angle should be reduced to about 40° to reduce incidence of torn grain; power consumption is therefore increased on hardwoods for two reasons—the greater wood density and the diminished rake angle.

Woodson and Rigby (1978) outlined measures to minimize milling defects on end-milling and down-milling headrig chippers and improve surface quality.

CHIPPING HEADRIGS PARTICULARLY SUITED TO HARDWOODS

Prior to 1981 chipping headrigs were not applied widely to the conversion of hardwood logs in North America. For examples of early experiences of mills using chipping headrigs on hardwoods the reader is referred to Rajala (1974), Rajala Timber Co. (1978), and Griffin (1977). Deterring application were the mechanical problems of feeding short, crooked, bumpy, tapered, dense, hard logs in a straight line through clusters of opposed pairs of chipping heads. Additionally, the hardwood lumber industry has traditionally thrived on large-diameter upper-grade logs which are carefully turned on sawmill carriages to achieve maximum recovery of clear lumber in standard grades that will yield clear cuttings (Malcolm 1965; Stumbo 1974; Vick 1979).

It appears to many mill operators that logs from small hardwood trees growing on pine sites contain very little clear lumber and that high production is called for at the possible expense of some loss of upper-grade lumber. A substantial body of literature advances the argument that low-grade logs should be live-sawn, i.e., not turned on the carriage (Bousquet 1972; Kersavage 1972; Pnevmaticos and Bousquet 1972; Flann 1974a, 1976; Huyler 1974; Robichaud et al. 1974; Richards 1977, 1978).

At the risk of omitting numerous praiseworthy designs, the following paragraphs will describe three chipping headrig designs suited to pine-site hardwoods, one cutting in each of the three modes.

Shaping-lathe headrig.—The shaping-lathe headrig (figs. 18-104ABCD) is designed to machine short logs into crossties, or into cants for resawing, or into cylinders for veneer peeling, with residue in the form of postage-stamp-size flakes (fig. 18-102) well suited for use in structural composite panels and flakeboard (fig. 24-2) or molded-flake products (figs. 24-60 and 24-62). Alternatively, the residue can be pulp chips (fig. 28-20). Firmly chucking the short logs on both ends securely controls workpiece movement and feedrate during machining. Thus, both cants and flakes (or chips) are precisely cut to desired dimensions and shape, and surface quality of cants is superior.

The machine, as originally designed, functioned as follows (fig. 18-104AB). Bolts are clamped in the chucks of the workpiece spindle, which turns at about 15 rpm. Attached to the spindle is a replaceable cam having the shape and dimensions of the desired cant. The cam rotates and moves with the workpiece until it strikes a follower aligned with the cutterhead. As the workpiece makes a single revolution, the center distance between cutterhead and workpiece changes in response to the cam, and the workpiece (log) is machined to the shape and dimensions of the cam. Since the log makes only a single revolution while being sized, machining time is brief, approximately 4 seconds. Feed rate is six bolts per minute.

In 1981 the shaping-lathe headrig was redesigned to incoroporate an automatic mechanical log centering device, a log scanner, and automatic setworks for pattern selection; thus need for diameter sort of logs was eliminated. Typical layouts of this machine are shown in figures 18-104CD and 18-252.

Widespread use of the shaping-lathe headrig may not occur until there is a strong demand for the flakes it produces as a residue. It is believed that before the year 2000, consumption of structural flakeboard (a competitor of sheathing grades of softwood structural plywood) will create this demand. It is pertinent to note that during 1979, structural flakeboard (100 pounds of dry wood flakes yields about 100 pounds of panel) sold for $200 an ovendry ton; unbleached kraft linerboard (100 pounds of dry pulp chips yields only 50 pounds of kraft linerboard) sold for less than $200/ton f.o.b. mill. It would appear that precision-cut flakes should be substantially more valuable than pulp chips.

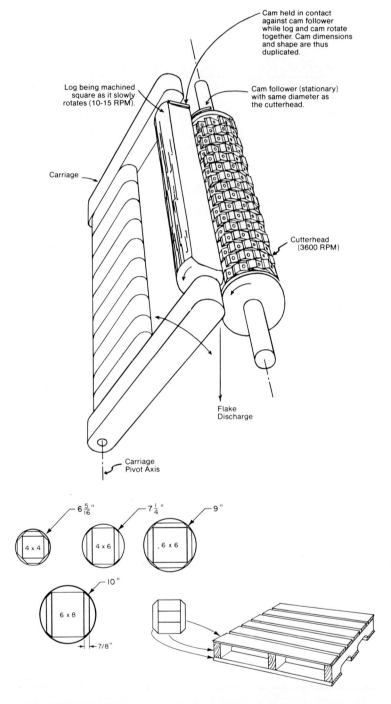

Figure 18-104A.—(Top) Principle of operation of shaping-lathe headrig. A log scanner coupled with automatic control of carriage oscillation eliminates need for diameter sort of logs. (See fig. 18-104 D top.) (Bottom) Typical cant shapes and resaw patterns when the machine is employed to make pallet parts.

Figure 18-104B.—Prototype shaping-lathe headrig to accept logs to 40 to 53 inches in length and 4 to 12 inches in diameter. Log unscrambler and infeed deck in foreground deliver logs to centering vee's. Machined cant shown end-gripped in chucks, will be discharged with flakes onto conveyor belt below centering vee's. Design feed rate is 6 logs per minute. Smoothly machined cants have shape and dimensions of replaceable cams (see left end of cant) mounted on workpiece spindle. (Photo from Stetson-Ross, Seattle.)

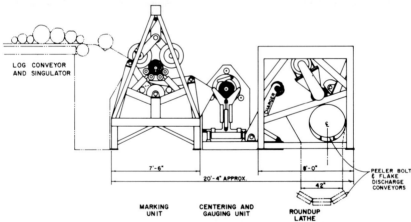

Figure 18-104C.—Production version of shaping-lathe headrig designed to machine 4-foot hardwood logs measuring 6 to 24 inches in diameter into the largest cylinders they contain for subsequent peeling into veneer; production rate is 6 to 7 logs per minute. Flakes from roundup are used for structural flakeboard. In the top photo, cutterhead with knife setting bar is in foreground at left of control station. See figure 18-252 for machine layout. (Photo and drawing from Stetson-Ross.)

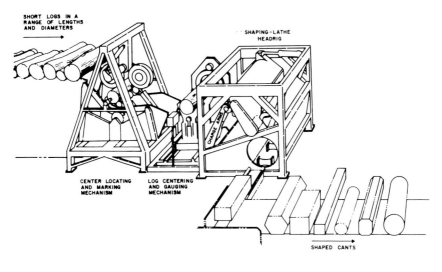
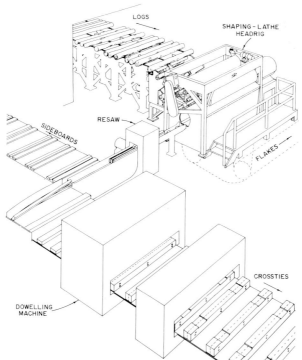

Figure 18-104D.—(Top) Shaping-lathe headrig arranged to produce cants for resawing into pallet parts, or perfect cylinders for subsequent peeling into veneer; incoming logs measure 36 to 52 inches in length and 6 to 24 inches in diameter, and can be fed in random order. (Bottom) In a 9-foot version, the shaping-lathe headrig can yield crossties plus sideboards. The diagram shows 4.5- by 7-inch cants being produced to assemble with six steel dowels (no glue) into laminated 7- by 9-inch crossties. The headrig will also manufacture 6- to 9-foot logs into posts, mine timbers, and cants for industrial lumber.

When projected demand for flakes does develop, the shaping-lathe headrig can be a major key to the problem of utilizing hardwoods that grow on southern pine sites. The literature contains several descriptions of enterprises employing the shaping-lathe headrig as follows (see also chapter 28):

Description of enterprise	Citation
Manufacture of pallet cants, pallets, and flakeboard	Koch (1975)
Manufacture of crossties, light timbers, lumber pallets, and moulded-flake pallets	Koch and Caughey (1978)
Manufacture of composite panels with veneer faces and flake cores	Springate et al. (1978), Springate (1978a)
Manufacture of flakeboard, decorative plywood, and long fabricated joists	Koch (1982); sect. 28-31
Manufacture of crossties, mine timbers, pallet cants, 2 by 4 studs, lumber laminated from veneer, and structural flakeboard	Koch (1978a), Anderson (1981); sect. 28-32
Manufacture of pallet deck boards	Koch and Gruenhut (1981); sect. 28-17
Manufacture of lumber pallet parts and composite structural panels from oak	Roubicek and Koch (1981); sect. 28-26
Manufacture of hardwood flakeboard in four southern locations	Koch (1978bc); sect. 28-27
Manufacture of dowel-laminated crossties plus pulpchips (or flakes)	Sect. 28-28, fig. 20-12, Koch (1976b), Howe and Koch (1976)
The shaping-lathe headrig (developments in 1984).	Koch (in press); *in* Proceedings of a symposium "Comminution of Wood and Bark", Chicago, Ill., Oct. 1-3, 1984. Published by the Forest Products Research Society, Madison, Wis.

A mill combining a shaping-lathe headrig with a veneer lathe (figs. 18-104C and 18-252) to manufacture pallet deckboards could give extremely high product yield from bolts bucked from tree-length pine-site hardwoods. Veneer for composite panels of veneer over flakes would be peeled from deck-board-length bolts 10 inches and larger. Cores would be ejected from the veneer lathe when they reach 7.6 inches in diameter, and converted on the shaping-lathe headrig to octagons containing a 4- by 6-inch central cant plus two bevel-edged sideboards for resawing into deck boards. Bolts smaller than 10 inches in diameter would be converted to smaller octagons or squares on the shaping-lathe headrig, and resawn into pallet parts. (See sect. 28-26 for an economic feasibility analysis of such an operation.)

Another operation that appears promising would use a roundup shaping-lathe in the manufacture of structural flakeboard, decorative plywood, and fabricated long trusses. Sect. 28-31 summarizes an economic feasibility analysis of such an enterprise (Koch 1982).

Still another would employ 9-foot and 4-foot-long shaping lathes equipped to convert hardwood stems into crossties, highway posts, short light timbers, and pallet cants; residue from the operation would be pulp chips. See sect. 28-18 for an economic feasibility analysis.

End-milling chipping headrig.—When a dense, hard, knotty, tapered, eccentric, bumpy, swell-butted log passes between two opposed cutterheads, each exerts a force on the log proportional to its depth of cut. Because the two cutterheads are never removing the same amount of wood, unbalanced forces act on the log and urge it to shift position on the feedworks. To solve the problem, Conveyor Machinery Corporation, Montgomery, Ala., has designed a reciprocating vertical rectangular carriage, into which the log is clamped at 2-foot intervals from both top and bottom; the carriage is guided firmly along both top and bottom edges as it is advanced through the cutting zone (fig. 18-105). The log, firmly clamped within the carriage, is slabbed on two sides by a pair of opposed end-milling cutterheads each looking like the one illustrated in figure 18-99 (center).

Figure 18-105.—Infeed of end-milling chipping headrig that slabs logs on two sides simultaneously. The log is clamped firmly from both top and bottom in a vertical reciprocating carriage that is guided from both top and bottom as it passes through the cutting zone. The log can be accurately centered; chipping heads are equipped with a setworks accommodating logs of varying sizes. After passing through the cutting zone, the log is discharged. (Photo from Conveyor Machinery Corporation.)

The Roberson Lumber Company, Simsboro, La., operates such a machine on small pine-site hardwoods to produce crossties, cants, and lumber from logs 8 to 10 feet long. The headrig will process about 800 logs during an 8-hour shift yielding about 25,000 board feet of solid wood products. Each cutterhead is driven by a 125-hp motor at 1,750 rpm, with feed rate of about 300 fpm. Resulting pulp chips are 0.8 inch long (fig. 18-103 top). With their layout (fig. 18-106), lumber recovery factor is about 5.2 board feet per cubic foot of log. Log lengths are 8, 8½, 9, and 10 feet, and diameter range is from 8 to 14 inches at the small end. Oak and hickory logs are held firmly during the slabbing cuts.

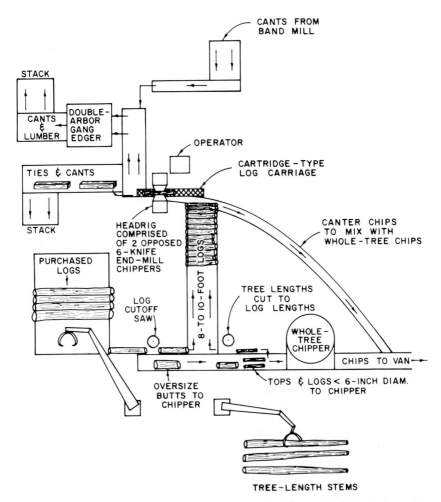

Figure 18-106.—Mill layout featuring end-milling chipping headrig shown in figure 18-105. Logs processed through the headrig are 8 to 14 inches in small-end diameter and 8 to 10 feet long. The double-arbor gang edger (top left) is equipped with a conveyor— not shown— to return to it miscut cants for remanufacture. (Drawing after Robertson Lumber Company, Simsboro, La.)

A major trend toward use of "sharp-chain" log transport systems in conjunction with two slabbing heads and twin band saws has developed for production of two-sided softwood cants; such systems frequently have no log hold-down devices, but rely on the gripping action of the sharp-chain transporter. Because tapered, rough knotty hardwood logs are buffeted by unequal slabbing forces on the two sides of the log, hold-down devices appear to be more necessary on hardwood than on softwood logs; vertical hold downs seem to be more effective than angular holddowns.

End-milling slabbing chippers used in conjunction with a carriage-type single bandmill have also proven viable, especially when coupled with log scanners, ball screw or linear positioning setworks, digital readout devices in the sawyer's cab, and some method for a second line on small logs downstream in the mill layout (so that only one line per log—establishing the best opening face—is required on the bandmill carriage).

Climb-milling chipping headrig.—Jones (1976) presented a candid view of the problems of processing southern hardwoods with a climb-milling, continuous-feed, chipping headrig designed to employ the double-taper method of converting logs to cants (fig. 18-107). By this method, a fixed bottom head and a fixed side head chip full-length faces on the log. These opening cuts are oriented 90° from each other (as indicated by the heavy dotted lines in the lower right hand corner of fig. 18-107 bottom). Taper can then accumulate across the horizontal axis from the fixed side head as well as along the vertical axis from the fixed bottom head of the machine. A top chipperhead is set to chip a face parallel to that opened by the bottom head, and the other side head is set to open a face parallel to that chipped by the fixed side head. If the log has significant taper, these selectively-set faces will be less than full length (fig. 18-107 bottom). All four cutterheads are down-milling heads.

The canter is preceded by a log-orienting device consisting of three rolls that can be skewed to rotate a log to best advantage before the log is advanced into a buckhorn chain (fig. 18-108) that provides a vertical and horizontal index for the fixed heads to follow. Powered outfeed rolls pull the cant through the chipper-

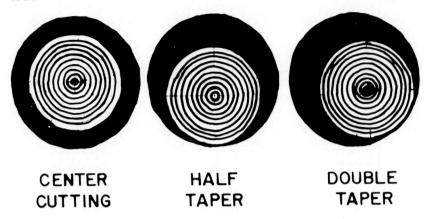

CENTER CUTTING   HALF TAPER   DOUBLE TAPER

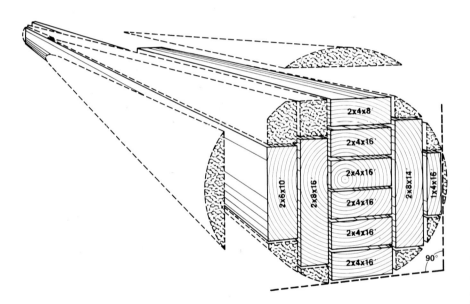

Figure 18-107.—(Top) Comparison of three basic cutting plans for chipping headrigs. (Bottom) Diagram of yield from a 16-foot log of 10-1/4-inch top and 12-1/4-inch bottom diameter based on two full-length double-tapered opening faces—indicated by dashed lines, 90° apart. (Drawings after Jones 1976.)

heads and advance it for further breakdown. Full discussions of the rationale for double-taper canting are given in Anderson and Ailport (1973).

In commenting on the 1975 installation of a double-taper down-milling chipper canter in the Waynesboro, Georgia plant of Kimberly-Clark, Jones (1976) indicated that the mill processes 40 percent southern hardwoods and 60 percent southern pine. The canter machines all four log faces in one pass, as logs flow more-or-less continuously through it. A band resaw converts the cants to lumber. Jones noted that several requirements must be satisfied if a continuous-feed chipping headrig is to be successful on hardwoods, as follows:

- A market for hardwood chips is essential.
- Logs must be straight; sweep should not exceed 1 inch.
- Logs must be 8 feet long or longer.
- Some positive means is required to remove stub branches.
- Logs must be green (not dry).
- The headrig must have power enough on the cutterheads and feedworks to feed at a rate sufficient to yield pulp chips at least ⅝-inch long when cutting oak and hickory.
- To be economic, at least 1,500 logs in diameter classes from 5 to 12 inches, and in lengths averaging at least 12 feet must be processed per 8-hour shift. (Some authorities question this requirement).
- Rake angle of knives should be 40° for hardwoods (rathern than the 55° angle common on softwood) to reduce severity of torn grain.

Even if logs are sorted to prescribed diameter limits, these conditions are difficult to meet; woods-run southern hardwood logs usually have sweep exceeding 1 inch, butt swells, bumps, branch stubs, and substantial taper.

The chipping headrig shown in figure 18-108 is a single-pass machine through which logs flow at feed speeds generally between 100 and 250 lineal feet per minute. For a given delay time per shift and interval time between logs, lineal throughput increases with increased feed speed. The numbers of logs processed per shift can be estimated by the following equation:

$$n = \frac{mr(1 - d)}{l + tr} \qquad (18\text{-}34)$$

where:
- n = number of logs per shift
- m = minutes per shift
- r = machine feed speed, fpm
- l = average log length, feet
- t = average interval of time between logs, minutes
- d = delay time as a proportion of m (e.g., delay from mechanical breakdown, rest periods, or lack of logs)

For example, 1,770 logs will be run per 480-minute shift if machine feed speed is 100 fpm, delay time is one-quarter of the shift, average log length is 12 feet, and the time interval between logs averages 5 seconds.

Figure 18-108.—Double-taper chipping headrig. This four-head canter is viewed from the infeed end to show vertical rolls of the log turner; the log is oriented for double-taper presentation to the cutterheads while held in a buckhorn feed chain and five hold-down wheels (shown retracted). The canter opens bottom and one side face with fixed heads, and top and other side with moveable heads. (Photo from Stetson-Ross.)

Climb-milling headrigs to not have to be continuous-feed, single-pass machines. To feed short hardwood logs in a straight line through the buffeting forces of opposed climb-milling heads, Koch (1964b) proposed that 100-inch-long logs be chucked from the ends and securely guided through clusters of opposed climb-milling heads, rotated 90°, and carried back through the cutterheads for discharge as squared (or partially squared) cants, as follows (fig. 18-109 top):

- A. Charging jaws close symmetrically and center bolt in two planes.
- B. Traverse ways for charging jaws permit centered bolt to be moved into line with the end dogs of the carriage.
- C. Overhead track (with matching track below) guides carriage, with end dogs gripping the bolt at both ends, at constant speed according to desired chip length by motor-driven ball screw.
- D. Opposed pair of two-knife, climb-milling, 20-inch-high, 14-inch cutting circle, 75-hp chipping heads turning at 3,450 rpm. This pair cuts only on the first pass.
- E. Opposed pair of 12-knife, 15-hp jointer heads set 1/64-inch inside chipping heads D and F cuts on both in and out pass for a well-planed final cant surface.

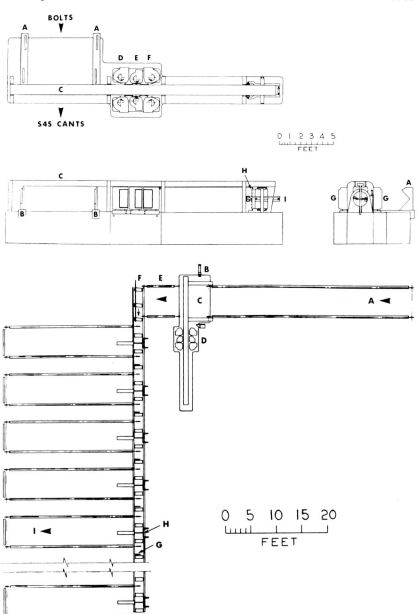

Figure 18-109.—(Top) Design concept of a climb-milling headrig in which short logs are carried through the cutting zone while end-chucked firmly in a vertical carriage guided from top and bottom, revolved 90°, machined again on the return travel and discharged as S4S cants. (Bottom) Flow plan. See text for discussion of sequence of operations. (Drawing after Koch 1964b.)

F. Chipping heads same as D but rotating in the opposite direction to climb-cut on the return pass. (Note: Cutterheads D, E, and F are mounted in two opposed assemblies; automatic setworks control cant size.
G. Chip outlet hoods. The few shavings made by jointer heads E must be separately exhausted.
H. End dogs (narrow enough to pass between cutterheads when in their most closed position).
I. Turning tongs to rotate partially machined bolt after initial pass. Only center pins of end dogs remain engaged during 90° rotation.

Notes: Bolt length, 100 ± 2 inches; bolt diameter, 4½-inch minimum and 24-inch maximum; S4S cant size, 3- by 4-inch through 12- by 12-inch; cycle rate, 5 cants per minute; chip length, ½-, ⅝-, or ¾-inch as specified.

Major sequences for this headrig, with layout shown in figure 18-109 (bottom) would be as follows:

Charger cycle, after passing bolt spacer and indexer (not shown):

1. Bark-free bolt proceeds down log deck (A) into charger, and is positioned endwise by cylinder (B).
2. Clamps of bolt charger (C) close symmetrically and center the bolt in two planes. Degree of closure of charging clamps, with program instructions relating S4S cant size to bolt diameter, directs a memory system to control setworks and sorter mechanisms.
3. The charger assembly (with bolt) advances until its centerline coincides with the centerline of the carriage. End clamps on carriage engage and automatic setworks positions cutterhead pairs (D) and charger clamps release.

Headrig cycle:

4. As carriage advances at constant speed past heads (D) and (E) the charger retracts and picks up another bolt.
5. At the end of the forward stroke, the turner grips the finished surfaces on one end of the bolt, carriage end clamps release the outermost dogs, but not the center pin from the bolt.
6. Turner then rotates the bolt 90° (by rack and pinion) while setworks repositions cutterheads (F) and (E)—cutterhead (D) moves simultaneously with (F). The end dogs are reset, the carriage completes its return stroke, discharges the S4S cant, a new bolt is centered, and the cycle repeats.

Sorter cycle:

7. The finished S4S cant is discharged onto the sorter receiving chains (E), proceeds to roll case (F), where the proper sorter stop—(G) for example—is energized by the computer, and the cant is pushed off onto accumulation chains (I).
8. Cants are accumulated at end of chains (I) and picked up in tiers to be stickered and stacked by forklift truck.

On all chipping headrigs—whatever their cutting mode—cutting patterns should be controlled to maximize lumber recovery, because lumber is more valuable than chips or flakes. Investment in equipment to maximize lumber yield—such as log scanners, precise log positioners, and computer controlled setworks—is justified by provable increase in production and lumber yield obtainable, and also on the ratio of pulp chip (or flake) value to lumber value. Under some circumstances, flakes from a shaping-lathe headrig might approach the value of crossties or mine timbers on a weight basis. Pulp chips, however, have only a fraction of the value of lumber.

Mills seeking high lumber yield with low labor expenditure, can combine headrig chippers with twin or quad bandsaws—all under computer control; Sohn (1973) described the mill that pioneered this concept.

18-10 SAWMILLS FOR HIGH-QUALITY, LONG, HARDWOOD LOGS

This text is concerned with hardwoods that grow on southern pine sites, where stemwood is predominantly of small diameter, and of such form that short logs must predominate if stem crook and major defects are removed during log bucking. The sawing of large, long, high-quality logs—while a complex process requiring skill and good equipment—is a problem that industry manages well with existing technology. Much of this section is therefore primarily restricted to a listing of references for readers who wish to study the subject.

LOG CHARACTERISTICS

Chapter 11 describes log defects arising from rot and insect attack. Chapter 12 defines log and tree grades. Girard form classes, a measure of taper in tree stem (See chapter 11 in Belyea 1931, and Larson 1963), are discussed in the following references:

Reference	Subject
Minckler and Green (1958)	Upland hardwoods in Illinois
Harris and Nash (1960)	White oak in Missouri
Hilt and Dale (1979)	Upland oaks in Kentucky, after thinning

Frequency data on Appalachian sawlogs classified by length, diameter, volume, species, and log grade were tabulated by Goho and Wysor (1970). Data for five regions on frequency distributions of diameter and length of sawlogs from eight Appalachian species were graphed by Goho and Martin (1973).

Church (1973) found sweep of 2 inches or more on 17 percent of 4,510 logs measured at Appalachian sawmills; volume deductions for sweep scaled at least 10 percent in 1 of every 7 logs, and at least 15 percent in 1 of every 9 logs measured. Garrett (1970) studied the physical suitability of Appalachian hardwood sawlogs for manufacture of sawed timbers and found that most grade 1 logs can be used to saw timbers as large as 8 by 9 inches in end dimension; most grade 2 logs are suited for manufacture of timbers 6 by 8 inches and smaller; and most grade 3 logs are suited for manufacture of timbers 5 by 7 inches and smaller.

Quantitative data explaining the superior profitability of large hardwood logs of good grade were published by Martens (1965), Pnevmaticos et al. (1971), Bousquet et al. (1972), and Smith et al. (1979).

STATE-OF-ART SAWMILLS

Some guidelines for state-of-art mills particularly designed for grade recovery from long logs of eastern hardwoods have been provided by Allen (1975), Hartman (1975), Seffens (1975), Worley (1975), Little (1976), Williston (1976, 1979), and White (1976).

Major sawmill machines are illustrated as follows:

Bandsaw headrig (fig. 18-58)
Linebar resaw (fig. 18-65)
Circular saws for headrigs (figs. 18-73 through 18-75)
Circular gang ripsaws (figs. 18-76, 18-77)
End-milling two-head canter (fig. 18-105, 18-106)
Double-taper, climb-milling, four-head chipping headrig (figs. 18-107, 18-108)

Lunstrum (1972) has provided a practical manual for circular sawmill operation in primary log breakdown. His manual discusses sawing accuracy, equipment selection, set-up, operation, and maintenance; it also contains a troubleshooting guide to help locate causes of more common problems found in circle-sawmill operations. Timber Processing Industry (1980) illustrated and briefly described 24 different sawmill edgers available in North America.

Brief discussions of portable sawmills are illustrated in section 18-8 (bandsaw, fig. 18-68; circular saw, fig. 18-92; chainsaw, fig. 18-97).

SAWING PATTERNS

Sawmillers cutting high-quality logs are primarily interested in patterns that recover all possible high-grade lumber; they must also be concerned with total lumber yield per log, and with fast production, however. The method most frequently used to maximize grade is that of turning the log whenever there is higher grade on another face. Patterson (1979) described the system. Faces may be cut parallel to the bark or parallel to the pith. In sawing around, all defects are positioned if possible in one face which is turned toward the sawyer. This face is sawed shallow, removing knots and taper, the next face is sawed deep until grade drops below the grade expected from the next adjacent face. This is continued until no grade change is expected, then the center is live-sawn. If the defects are scattered, it is best to position them on the corners of the faces; again the worst face should be sawn shallow first and the other faces sawn deep for grade. Logs with a sweep or crook are positioned with the belly toward the sawyer. Short boards should be sawn until a good bearing surface is developed, then the log turned 180° and heavy slabbed to remove the horns. Grade lumber is sawn from the remaining two faces. Further descriptions of the virtues and methodology of grade sawing have been provided by Malcolm (1965), Stumbo (1974), Vick (1976, 1979), and Richards et al. (1980).

Patterson (1979) also summarized procedures in **live-sawing**. He recommends that if possible all defects be grouped in one face which is positioned straight up or straight down. The log is sawn into boards to the center then the log is turned 180° and again sawn through-and-through until only the dog board remains. If the defects are scattered, they should be positioned to give the best faces opposite each other; the best face should be put against the carriage knees so that taper is removed from a poorer face. Descriptions of the virtues and methodology of live-sawing hardwoods have been provided by King (1956), Burry (1969), Bousquet (1972), Kersavage (1972), Pnevmaticos and Bousquet (1972), Richards (1973, 1977, 1978), Flann (1974ab, 1976), Flann and Bousquet (1974), Huyler (1974), Robichaud et al. (1974), Bousquet and Flann (1975), Richards and Newman (1979), and Richards et al. (1980). Patterson (1979) further observes that with either live-sawing or sawing around for grade, small logs should be slabbed heavy enough to produce a No. 1 common face, and large logs should be slabbed for a 6-½-inch-wide face for at least 8 feet (minimum for FAS grade).

Most researchers have found that live-sawing with grade ripping of the central boards (to remove low-grade central portions) produces more value from a log than grade sawing around the log (Richards et al. 1980); if, however, a log has a poor center and the outside 5 inches is clear, lumber of greater value may result from sawing around the poor center, rathern than live-sawing through it. Live sawing generally produces fewer pieces of lumber, but in aggregate these pieces contain more volume than lumber produced when the log is sawn around. Switching from grade sawing to live sawing increases headrig production, but since all boards must be edged, the edger may become a bottleneck (Patterson 1979).

To obtain some of the advantages of live sawing, plus better stability of the log in on the carriage, Peter (1967) and Murphey and Cochran (1972) suggested that a three-turn method of sawing was practical. By this method, the log rests on a flattened surface and can be dogged securely while live sawn.

Headrig output per shift can be increased by producing hardwood lumber in combination with ties and timbers cut from central log portions. The advantages of this system have been explained by Church and Niskala (1966), Niskala and Church (1966), Garrett (1969, 1970), and Church and Garrett (1970).

Systems for computer simulations of sawing patterns for hardwoods have been explained by Reynolds and Gatchell (1969), Reynolds (1970), Pnevmaticos and Mouland (1978), Richards et al. (1979), and Richards et al. (1980). Adkins et al. (1980) provided four computer programs for simulation of hardwood log sawing. McMillin (1982) proposed that computerized axial tomography be used to assess the location and extent of defects within logs prior to breakdown to facilitate optimum value recovery at the headrig.

With a view toward maximizing lumber yield without consideration for grade, Hallock et al. (1976, n.d.), by computerized analysis, studied eight of the most commonly used sawing patterns and concluded that for each class of log there is a best method. Hallock et al. (1979) compared centered versus offset sawing, and Hallock and Lewis (1979) reported on individual log yields by four centered sawing systems. In general, it was found that offset methods are best (offset patterns are not laterally centered on the geometric center of the log). While the comparisons were made for softwood logs, the results should be useful to those interested in hardwood yield only—ignoring grade recovery. Steele et al. (1981) presented a geometric model with which to calculate the volumes of green lumber, dry lumber, green pulp chips, green sawdust, and dry planer shavings produced by the sawmilling process; they note that a minimum of data gathering is required, and illustrate the mathematical theory comprising the model with equations and drawings.

COMPUTERIZED SAWMILLS

Sohn's (1973) computerized softwood sawmill pioneered use of log scanners and positioners and computers on setworks controlling headrig chippers coupled with thin-kerf bandsaws. Research has concentrated on softwood sawing patterns (Hallock and Lewis 1971; Hallock 1973), and applications in softwood mills have greatly outnumbered those in hardwood mills. Worley (1975) and Moen and Seffens (1975) have described, however, two hardwood mills employing computerized setworks.

SAWING TIME

Rast (1974a) related sawing time on conventional headrigs per log (table 18-54), per thousand board feet (table 18-55) and per tree (table 18-56) to log and tree size, resaw capability, and percent scaling defect. His data were based on 18 species of hardwoods sawn at numerous eastern mills, some of which had circular-saw headrigs and some of which had bandsaw headrigs. Nearly half of the mills had resaws.

Log length does not greatly affect sawing time per log, but time is about proportional to log diameter (table 18-54). Sawing time per thousand board feet is inversely proportional to log diameter and log length, and defective logs take significantly more time to saw than defect-free logs (table 18-55). Sawing time per tree is proportional to tree dbh and to merchantable height (table 18-56).

In all cases, addition of a resaw to the mill layout substantially reduces headsaw time per log or tree, and the time to saw a thousand board feet of lumber.

TABLE 18-54.—*Time[1] to saw a hardwood log on a sawmill heading with (and without) a resaw, related to log diameter and length* (Data from Rast 1974)[2]

Equipment and log length (feet)	Log diameter inside bark, inches					
	8	10	12	14	16	18
	---------------------Minutes--------------------					
No resaw						
8	1.39	1.66	1.99	2.38	2.83	3.34
10	1.50	1.82	2.21	2.67	3.20	3.80
12	1.62	1.99	2.44	2.97	3.58	4.27
14	1.75	2.16	2.67	3.27	3.96	4.75
16	1.88	2.35	2.91	3.58	4.36	5.23
With resaw						
8	1.02	1.22	1.46	1.74	2.06	2.43
10	1.04	1.26	1.54	1.86	2.23	2.66
12	1.06	1.31	1.62	1.99	2.41	2.89
14	1.08	1.37	1.71	2.12	2.59	3.13
16	1.11	1.42	1.80	2.26	2.78	3.37

[1]Sawing time commences when the log is rolled onto the carriage and continues until the log is completely sawn, the dog board or cant is released, and the carriage returns and stops in front of the log deck ready for the next log; delay time in excess of 20 sec. during this cycle not included.

[2]Based on 6,850 logs of 18 species sawn at 20 different sawmills, of which seven used circular saws and 13 used bandsaws; eight of the mills had resaws.

TABLE 18-55—*Time[1] to saw a thousand board feet of hardwood lumber on a sawmill headrig with (and without) a resaw, related to log diameter and length, and percent of defect scaled* (Data from Rast 1974)[2]

Equipment and log length (feet)	Log diameter inside bark, inches					
	8	10	12	14	16	18
	---------------------Minutes--------------------					
LESS THAN 25 PERCENT SCALING DEFECT						
No resaw						
8	83.9	63.8	52.9	46.3	42.0	39.1
10	69.9	53.1	44.0	38.5	35.0	32.5
12	61.2	46.6	38.7	33.9	30.9	28.7
14	55.2	42.3	35.2	31.0	28.2	26.3
16	50.9	39.1	32.7	28.9	26.4	24.6
With resaw						
8	58.9	44.9	37.2	32.6	29.6	27.6
10	48.3	36.4	29.9	26.0	23.4	21.7
12	41.8	31.3	25.6	22.1	19.9	18.3
14	37.4	27.9	22.8	19.6	17.6	16.2
16	34.3	25.6	20.8	17.9	16.1	14.8

[1]Sawing time commences when the log is rolled onto the carriage and continues until the log is completely sawn, the dog board or cant is released, and the carriage returns and stops in front of the log deck ready for the next log; delay time in excess of 20 sec. during this cycle not included.

[2]Based on 6,850 logs of 18 species sawn at 20 different sawmills, of which seven used circular saws and 13 used bandsaws; eight of the mills had resaws.
Table continued on next page.

TABLE 18-55.—*Time¹ to saw a thousand board feet of hardwood lumber on a sawmill heading with (and without) a resaw, related to log diameter and length and percent of defect scaled* (Data from Rast 1974)²—Continued

Equipment and log length (feet)	Log diameter inside bark, inches					
	8	10	12	14	16	18
	------------------Minutes------------------					
MORE THAN 25 PERCENT SCALING DEFECT						
No resaw						
8	99.8	80.7	68.8	62.2	57.9	55.0
10	85.8	69.0	59.9	54.4	50.8	48.4
12	77.1	62.5	54.6	49.8	46.7	44.6
14	71.1	58.1	51.1	46.8	44.1	42.2
16	66.8	55.0	48.6	44.7	42.2	40.5
With resaw						
8	74.8	60.7	53.1	48.5	45.5	43.5
10	64.2	52.2	45.7	41.8	39.3	37.6
12	57.7	47.1	41.4	38.0	35.7	34.2
14	53.3	43.8	38.6	35.5	33.5	32.1
16	50.2	41.5	36.7	33.8	32.0	30.7

[1]Sawing time commences when the log is rolled onto the carriage and continues until the log is completely sawn, the dog board or cant is released, and the carriage returns and stops in front of the log deck ready for the next log; delay time in excess of 20 sec. during this cycle not included.

[2]Based on 6,850 logs of 18 species sawn at 20 different sawmills, of which seven used circular saws and 13 used bandsaws; eight of the mills had resaws.

TABLE 18-56—*Time¹ to saw a hardwood tree on a sawmill headrig with (and without) a resaw, related to tree dbh and merchantable height and percent of defect scaled* (Data from Rast 1974)²

Equipment and tree merchantable height (number of 16-foot logs)	Tree dbh, inches					
	10	12	14	16	18	20
	------------------Minutes------------------					
No resaw						
1	2.3	3.7	4.8	5.9	6.9	8.0
2	4.2	5.3	6.5	7.6	8.8	10.2
3	5.1	6.3	7.6	9.0	10.5	12.2
4	5.8	7.2	8.6	10.3	12.1	14.1
With resaw						
1	0.7	1.9	2.7	3.5	4.2	4.8
2	1.5	2.4	3.2	4.1	5.0	5.9
3	—	—	—	4.4	5.5	6.8
4	—	—	—	4.5	6.0	7.6

[1]Sawing time commences when the log is rolled onto the carriage and continues until the log is completely sawn, the dog board or cant is released, and the carriage returns and stops in front of the log deck ready for the next log; delay time in excess of 20 sec. during this cycle not included.

[2]Based on 1,181 trees of eight hardwood species sawn at 11 different sawmills, of which three used circular saws and eight bandsaws. Four of the mills with bandsaws. Four of the mills with bandsaws had resaws.

SAWMILL DOWN TIME

Rast (1974b) defined sawmill down time as any delay of 20 seconds or more during the sawing shift. While collecting data on 6,083 logs sawn at 20 sawmills, he observed that causes for down time were as follows (figs. 18-110, 18-111):

- Repairs
- Change or sharpen saw
- Jam-up on log deck
- Trouble with log (getting it on carriage; positioning it; cleaning it of mud, rocks, or metal)
- Talk (sawyer to millwright, owner, salesman, etc.)
- Miscellaneous (no logs on deck; changing sawyers; cleaning carriage, sawyer's cage, around saw, and carriage tracks; lumber jam on live rolls that remove lumber from the saw)

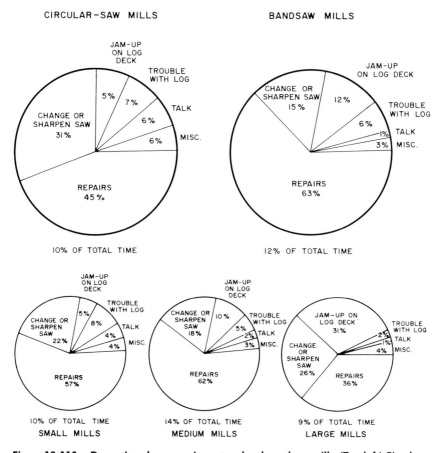

Figure 18-110.—Down time, by cause, in eastern hardwood sawmills. (Top left) Circular saw mills. (Top right) Bandsaw mills. (Bottom left) Small mills. (Bottom center) Medium mills. (Bottom right) Large mills. (Drawings after Rast 1974b.)

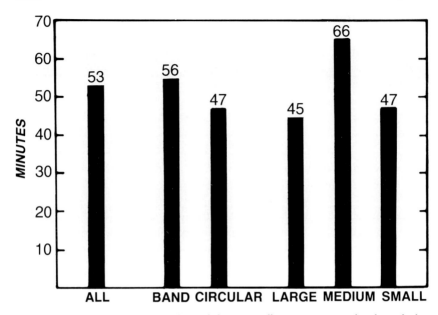

Figure 18-111.—Down time per 8-hour shift in sawmills cutting eastern hardwoods, by mill categories. (Drawing after Rast 1974b.)

Rast summarized his conclusions as follows.

Average down time for all mills combined was 11% of total operating time—10% for circular saws and 12% for bandsaws. Band mills have a larger percentage of their down time caused by repairs (63%) than do circular mills (45%) because they are usually larger mills and tend to have more auxiliary equipment supporting the head-saw. Sharpening the (circular) saw took longer (an average of 10 minutes) than changing a band saw (8½ minutes), and the band mills sawed an average of 138 logs before changing saws, as compared to 75 for circular mills; hence, the difference between 31% and 14% for changing or sharpening saws. The difference in jam-up on the log deck (circular saws 5%; band mills 12%) occurred because the band mills tended to push harder for production. The time taken up by talk was higher for circular mills (6%) than for band mills (1%). The lower percentage of down time for repairs for the large mills (36%) was due mainly to their greater emphasis on preventive maintenance, their employment of professional millwrights, and their stocking of some parts. The large percentage of down time caused by jam-up on log deck (31%) in the large mills was mainly due to the operator's haste or carelessness. On the other hand, in the larger mills, both trouble with logs and talk time took smaller percentages of total down time (5%, 2%, and 2%, 1%, respectively for medium and large mills).

Analysts seeking ways to minimize sawmill down time and maximize production have resorted to various recording methods and computer analyses. A sampling of the literature on the subject includes the following; Holemo and Dyson (1972), Bennett (1973), Dutrow and Granskog (1973), Stephens (1975), Aune (1977), Kersavage (1978), and Carino and Bowyer (1979).

SAWING ACCURACY AND BOARD THICKNESS

A review of the substantial body of literature related to sawing accuracy and board thickness is beyond the scope of this text, but abstracts of a few pertinent references seem useful.

Freese et al. (1976) studied thickness losses encountered in kiln-drying and planing 4/4 red oak lumber. Their results suggest that lumber slightly less than 1 inch thick when rough dry can produce moderately long, wide cuttings planed on two sides to a thickness of 13/16-inch. Rough dry thickness for panels should be slightly over 1 inch to plane to 13/16-inch. Green 4/4 red oak lumber should be ⅛-inch thicker than required rough dry size to allow for shrinkage. Sawing variation and warp during drying (see chapter 20) are two major problems deterring reduction of thicknesses of green rough hardwood lumber.

Thickness variations found in hardwood lumber are substantial. Applefield and Bois (1966) measured lumber in 16 furniture plants at 12 locations in North Carolina. They found that 37.4% of the lumber was mismanufactured, as follows:

Types of mismanufacture	Percent
Miscuts	19.7
Too thin	16.0
Oversize	1.7
Total	37.4

Stern et al. (1979) studied the influence of sawing accuracy on the amount of planed dry lumber recovered from softwood logs; although the study was not specific to hardwood, observed trends should be applicable. Their graphical data show the yield a mill operator can expect at all levels of sawing accuracy. For example, if the mill has a headsaw kerf of 0.375 inch, a cant resaw kerf of 0.375, and a sawing accuracy of 0.3 inch, it can expect to obtain 850 board feet from the same logs that would yield 1,000 board feet if there were perfect sawing accuracy. With kerfs of 0.125 and 0.125 inch and sawing accuracy of 0.3 inch, expected yield would be 830 board feet from logs that would yield 1,000 board feet if sawn with perfect accuracy. Of course, the initial unit of logs from which 1,000 board feet lumber tally could be obtained is larger for the wider kerfs.

Stern et al. (1979) concluded that exponentially larger increases in lumber yield can be expected as sawing accuracy increases. This increase in yield, without additional raw material consumed, tends to compensate for the exponentially increasing costs of improving sawing accuracy.

Mason (1975) studied thickness variations in 50 sawmills in the Northwest and South and concluded that hardwood mills must cut 20 to 25 million board feet per year to justify the investment required for the logic systems, control systems, positioning systems, and sawing equipment needed to saw as accurately, and with as thin kerf, as the state of the art allows.

Mill managers wishing to develop a lumber-size control program will find procedures outlined by Whitehead (1978) useful.

YIELDS AND RESIDUES

Sawmills traditionally have measured yield in board feet. A convenient term expressing a mill's converting efficiency, therefore, is **lumber recovery factor,** usually abbreviated to LRF. LRF is the ratio of nominal board feet of rough green lumber recovered per green cubic foot of bark-free log input to a mill. LRF varies among mills in the United States from a low of 3 to 4 to a high of about 12; the average for the Nation is 7 to 7.5. The national LRF average is weighted heavily by large efficient western mills; since they produce over half of the Nation's softwood lumber, they tend to raise the overall average (Lunstrum 1975). Hardwood lumber is usually sawn thicker than nominal thickness (as explained earlier in this section), while softwood is sawn less than nominal thickness; thus, average LRF in hardwood mills is less than in softwood mills. Burry's (1976) survey of 10 hardwood mills in New York State indicated average LRF of 6.7 for bandsaw mills and 6.4 for circular-saw mills; the hard maple logs analyzed in his study averaged 13.4 inches in diameter and 11.5 feet in length.

Data relating LRF to volumetric recovery of rough green and dry planed hardwood lumber are not available, but information gathered on softwood mills during the sawmill improvement program of the U.S. Forest Service is instructive (fig. 18-112). With LRF of 7, about average, volumetric recovery of green rough lumber is approximately 53 percent, and dry planed lumber about 41 percent of green log volume. In the transition from rough green softwood lumber to dry planed lumber, 17 to 24 percent of board volume is lost to shrinkage and planer shavings, as follows:

LRF	Computation	Percent volume loss
5	(100)(38-29)/38	24
7	(100)(53-41)/53	21
9	(100)(64-53)/64	17

The mills achieving an LRF of 9 saw more accurately than those with LRF of 5, and therefore are able to reduce their allowance for wood removal during planing.

Log size, grade, species, and sawing method are related in complex ways with sawing time, and quantity and grade of lumber yield. To help hardwood mill managers constructively manipulate these relationships, Adams and Dunmire (1978) developed a computer analysis technique they term SOLVE II. The technique is used to determine:

- Mill efficiency in terms of LRF.
- Grade recovery for comparison with usually achieved recovery from particular log grades.
- Prices the mill manager can pay for logs, if he is to maintain a stated profit level.
- Break-even log sizes.
- Pulp chip yield from slabs and edgings.

Hardwood sawmill managers should find the technique useful.

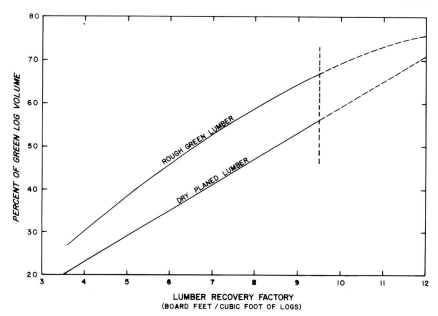

Figure 18-112.—Average volumetric recovery of rough green and dry planed softwood 4/4 and 8/4 lumber, as a percentage of green log volume by mill conversion efficiency. Data reliable to left of vertical dotted line. (Data from S. J. Lunstrum, U.S. Forest Service, Forest Products Laboratory, Madison, Wis., private communication, January 25, 1979.)

To guide resource decisions, woodland and sawmill managers need data on the yield of lumber, sawdust, and bark-free residue obtainable from bark-free stems of trees of different species, dbh, and merchantable height to a top not smaller than 8 inches in diameter inside bark. Hanks (1977) tabulated such data for eight eastern hardwood species of which six are germane to this text (black, chestnut, white, and northern red oaks plus yellow-poplar and red maple). Of the six, red maple was sampled north of the southern pine region, but data on the species are included with the rest which were sampled in the eastern range of southern pine. Each of the tables (18-57 through 18-62) represent data from 100 trees. For red maple and yellow-poplar the tables show data representative of mills with circular-saw headrigs as well as those with bandsaw headrigs. On yellow-poplar the circular mills get more lumber than the band mills, because the logs that went to the bandmills contained some rot. Data for the other four species were obtained from band mills only. The volume of lumber (rough green) is expressed in cubic feet and includes 2 inches of overlength beyond nominal board length. Sawdust volume is expressed in cubic feet of solid wood, as is the residue volume.

TABLE 18-57—*Red maple; predicted yields of sawmill products in bandsaw mills and circular-saw mills from bark-free merchantable stems to an 8-inch top* (Hands 1977)[1]

DBH (inches)	Product	Merchantable height, feet					
		24	32	40	48	56	64
		Cubic feet					
	BANDSAW MILLS (10/64-INCH KERF)						
10	Lumber	5.0	6.5	8.0	9.6	11.1	12.6
	Sawdust	1.0	1.4	1.8	2.2	2.6	2.9
	Residue	3.3	5.2	7.0	8.8	10.7	12.5
	Gross tree	9.3	12.9	16.5	20.1	23.7	27.3
11	Lumber	6.5	8.2	9.9	11.6	13.3	15.0
	Sawdust	1.3	1.7	2.1	2.5	2.9	3.3
	Residue	4.1	5.9	7.7	9.4	11.2	13.0
	Gross Tree	11.5	15.3	19.2	23.1	26.9	30.8
12	Lumber	8.0	10.0	11.9	13.9	15.8	17.7
	Sawdust	1.5	1.9	2.3	2.8	3.2	3.6
	Residue	4.9	6.7	8.4	10.1	11.9	13.6
	Gross Tree	13.9	18.0	22.2	26.4	30.5	34.7
13	Lumber	9.7	11.9	14.1	16.3	18.5	20.7
	Sawdust	1.7	2.2	2.6	3.1	3.5	4.0
	Residue	5.8	7.5	9.2	10.9	12.5	14.2
	Gross Tree	16.5	21.0	25.5	29.9	34.4	38.9
14	Lumber	11.6	14.0	16.5	18.9	21.4	23.9
	Sawdust	2.0	2.5	3.0	3.4	3.9	4.4
	Residue	6.8	8.4	10.0	11.6	13.3	14.9
	Gross Tree	19.4	24.2	29.0	33.8	38.6	43.4
15	Lumber	13.5	16.3	19.0	21.8	24.5	27.3
	Sawdust	2.3	2.8	3.3	3.8	4.3	4.8
	Residue	7.9	9.4	11.0	12.5	14.0	15.6
	Gross Tree	22.4	27.6	32.8	38.0	43.1	48.3
16	Lumber	15.6	18.7	21.8	24.8	27.9	30.9
	Sawdust	2.6	3.2	3.7	4.2	4.8	5.3
	Residue	9.0	10.5	11.9	13.4	14.9	16.4
	Gross Tree	25.7	31.2	36.8	42.4	48.0	53.5
17	Lumber	17.9	21.3	24.7	28.0	31.4	34.8
	Sawdust	3.0	3.5	4.1	4.7	5.2	5.8
	Residue	10.2	11.6	13.0	14.4	15.8	17.2
	Gross Tree	29.1	35.1	41.1	47.1	53.1	59.1
18	Lumber	20.3	24.0	27.7	31.5	35.2	38.9
	Sawdust	3.3	3.9	4.5	5.1	5.7	6.3
	Residue	11.5	12.8	14.1	15.4	16.7	18.0
	Gross Tree	32.8	39.3	45.7	52.1	58.6	65.0
19	Lumber	22.8	26.9	31.0	35.1	39.2	43.3
	Sawdust	3.7	4.4	5.0	5.6	6.2	6.9
	Residue	12.8	14.0	15.3	16.5	17.7	18.9
	Gross Tree	36.7	43.6	50.5	57.4	64.3	71.2

[1] Bold lines indicate range of observed data.

TABLE 18-57—*Red maple; predicted yields of sawmill products in bandsaw mills and circular-saw mills from bark-free merchantable stems to an 8-inch top* (Hands 1977)[1]
—Continued

DBH (inches)	Product	Merchantable height, feet					
		24	32	40	48	56	64
		----------------- Cubic feet -----------------					
20	Lumber	25.4	29.9	34.4	38.9	43.4	47.8
	Sawdust	4.1	4.8	5.5	6.1	6.8	7.4
	Residue	14.2	15.4	16.5	17.6	18.8	19.9
	Gross Tree	40.8	48.2	55.6	63.0	70.4	77.8
21	Lumber	28.2	33.1	38.0	42.9	47.8	52.7
	Sawdust	4.6	5.3	6.0	6.7	7.4	8.1
	Residue	15.7	16.8	17.8	18.8	19.9	20.9
	Gross Tree	45.1	53.1	61.0	68.9	76.8	84.7
22	Lumber	31.2	36.5	41.8	47.1	52.4	57.7
	Sawdust	5.0	5.7	6.5	7.2	8.0	8.7
	Residue	17.3	18.2	19.2	20.1	21.0	22.0
	Gross Tree	49.7	58.1	66.6	75.0	83.5	92.0
23	Lumber	34.2	40.0	45.7	51.5	57.3	63.0
	Sawdust	5.5	6.2	7.0	7.8	8.6	9.4
	Residue	18.9	19.8	20.6	21.4	22.3	23.1
	Gross Tree	54.4	63.4	72.5	81.5	90.5	99.5
24	Lumber	37.4	43.6	49.9	56.1	62.3	68.5
	Sawdust	5.9	6.8	7.6	8.4	9.2	10.1
	Residue	20.6	21.4	22.1	22.8	23.5	24.3
	Gross Tree	59.3	69.0	78.6	88.2	97.8	107.5
	CIRCULAR-SAW MILLS (17/64-INCH KERF)						
10	Lumber	2.1	3.9	5.6	7.4	9.2	11.0
	Sawdust	0.9	1.4	1.9	2.5	3.0	3.5
	Residue	3.3	5.2	7.0	8.8	10.7	12.5
	Gross Tree	9.3	12.9	16.5	20.1	23.7	27.3
11	Lumber	3.4	5.4	7.4	9.3	11.3	13.3
	Sawdust	1.2	1.8	2.4	3.0	3.6	4.2
	Residue	4.1	5.9	7.7	9.4	11.2	13.0
	Gross Tree	11.5	15.3	19.2	23.1	26.9	30.8
12	Lumber	4.8	7.0	9.2	11.4	13.6	15.9
	Sawdust	1.6	2.3	2.9	3.6	4.2	4.9
	Residue	4.9	6.7	8.4	10.1	11.9	13.6
	Gross Tree	13.9	18.0	22.2	26.4	30.5	34.7
13	Lumber	6.4	8.8	11.3	13.7	16.2	18.6
	Sawdust	2.0	2.8	3.5	4.2	4.9	5.7
	Residue	5.8	7.5	9.2	10.9	12.5	14.2
	Gross Tree	16.5	21.0	25.5	29.9	34.4	38.9
14	Lumber	8.1	10.8	13.5	16.2	18.9	21.6
	Sawdust	2.5	3.3	4.1	4.9	5.7	6.5
	Residue	6.8	8.4	10.0	11.6	13.3	14.9
	Gross Tree	19.4	24.2	29.0	33.8	38.6	43.4

Table continued next page.

TABLE 18-57—*Red maple; predicted yields of sawmill products in bandsaw mills and circular-saw mills from bark-free merchantable stems to an 8-inch top* (Hands 1977)[1]
—Continued

DBH (inches)	Product	Merchantable height, feet					
		24	32	40	48	56	64
		---------------- Cubic feet ----------------					
15	Lumber	9.9	12.9	15.8	18.8	21.8	24.8
	Sawdust	3.0	3.9	4.7	5.6	6.5	7.4
	Residue	7.9	9.4	11.0	12.5	14.0	15.6
	Gross Tree	22.4	27.6	32.8	38.0	43.1	48.3
16	Lumber	11.8	15.1	18.4	21.6	24.9	28.2
	Sawdust	3.5	4.5	5.4	6.4	7.4	8.3
	Residue	9.0	10.5	11.9	13.4	14.9	16.4
	Gross Tree	25.7	31.2	36.8	42.4	48.0	53.5
17	Lumber	13.9	17.5	21.1	24.7	28.3	31.9
	Sawdust	4.1	5.1	6.2	7.2	8.3	9.3
	Residue	10.2	11.6	13.0	14.4	15.8	17.2
	Gross Tree	29.1	35.1	41.1	47.1	53.1	59.1
18	Lumber	16.0	20.0	23.9	27.8	31.8	35.7
	Sawdust	4.6	5.8	6.9	8.1	9.2	10.4
	Residue	11.5	12.8	14.1	15.4	16.7	18.0
	Gross Tree	32.8	39.3	45.7	52.1	58.6	65.0
19	Lumber	18.4	22.6	26.9	31.2	35.5	39.8
	Sawdust	5.3	6.5	7.8	9.0	10.3	11.5
	Residue	12.8	14.0	15.3	16.5	17.7	18.9
	Gross Tree	36.7	43.6	50.5	57.4	64.3	71.2
20	Lumber	20.8	25.5	30.1	34.8	39.4	44.1
	Sawdust	5.9	7.3	8.6	10.0	11.4	12.7
	Residue	14.2	15.4	16.5	17.6	18.8	19.9
	Gross Tree	40.8	48.2	55.6	63.0	70.4	77.8
21	Lumber	23.3	28.4	33.5	38.5	43.6	48.6
	Sawdust	6.6	8.1	9.6	11.0	12.5	14.0
	Residue	15.7	16.8	17.8	18.8	19.9	20.9
	Gross Tree	45.1	53.1	61.0	68.9	76.8	84.7
22	Lumber	26.0	31.5	37.0	42.4	47.9	53.4
	Sawdust	7.3	8.9	10.5	12.1	13.7	15.3
	Residue	17.3	18.2	19.2	20.1	21.0	22.0
	Gross Tree	49.7	58.1	66.6	75.0	83.5	92.0
23	Lumber	28.8	34.7	40.6	46.5	52.4	58.3
	Sawdust	8.1	9.8	11.5	13.2	15.0	16.7
	Residue	18.9	19.8	20.6	21.4	22.3	23.1
	Gross Tree	54.4	63.4	72.5	81.5	90.5	99.5
24	Lumber	31.8	38.1	44.5	50.8	57.2	63.5
	Sawdust	8.9	10.7	12.6	14.4	16.3	18.1
	Residue	20.6	21.4	22.1	22.8	23.5	24.3
	Gross Tree	59.3	69.0	78.6	88.2	97.8	107.5

[1] Bold lines indicate range of observed data.

TABLE 18-58—*Black oak; predicted yields of sawmill products in bandsaw mills (10/64-inch kerf) from bark-free, merchantable stems to an 8-inch top* (Hanks 1977)[1]

DBH (inches)	Product	Merchantable height, feet					
		24	32	40	48	56	64
		---------------- Cubic feet ----------------					
10	Lumber	4.0	4.8	5.6	6.4	7.1	7.9
	Sawdust	0.7	1.0	1.3	1.5	1.8	2.1
	Residue	5.5	6.0	6.5	7.0	7.4	7.9
	Gross Tree	10.5	12.2	13.8	15.5	17.1	18.8
11	Lumber	5.4	6.5	7.6	8.7	9.8	10.9
	Sawdust	0.9	1.2	1.6	1.9	2.2	2.5
	Residue	6.0	6.6	7.1	7.7	8.3	8.9
	Gross Tree	12.6	14.7	16.8	18.9	21.0	23.1
12	Lumber	6.9	8.4	9.8	11.2	12.7	14.1
	Sawdust	1.1	1.5	1.9	2.2	2.6	2.9
	Residue	6.5	7.2	7.9	8.5	9.2	9.9
	Gross Tree	14.8	17.4	20.0	22.6	25.2	27.8
13	Lumber	8.6	10.4	12.2	14.0	15.8	17.6
	Sawdust	1.4	1.8	2.2	2.6	3.0	3.4
	Residue	7.1	7.9	8.6	9.4	10.2	11.0
	Gross Tree	17.3	20.4	23.5	26.7	29.8	32.9
14	Lumber	10.4	12.6	14.8	17.0	19.2	21.4
	Sawdust	1.7	2.1	2.6	3.0	3.5	3.9
	Residue	7.7	8.6	9.5	10.4	11.3	12.2
	Gross Tree	19.9	23.6	27.3	31.0	34.7	38.4
15	Lumber	12.3	14.9	17.6	20.2	22.8	25.5
	Sawdust	1.9	2.4	3.0	3.5	4.0	4.5
	Residue	8.3	9.4	10.4	11.4	12.5	13.5
	Gross Tree	22.8	27.1	31.4	35.7	40.0	44.4
16	Lumber	14.4	17.5	20.5	23.6	26.7	29.8
	Sawdust	2.2	2.8	3.4	4.0	4.5	5.1
	Residue	9.0	10.2	11.4	12.5	13.7	14.9
	Gross Tree	25.8	30.8	35.8	40.8	45.7	50.7
17	Lumber	16.6	20.1	23.7	27.3	30.9	34.4
	Sawdust	2.6	3.2	3.8	4.5	5.1	5.7
	Residue	9.8	11.1	12.4	13.7	15.0	16.4
	Gross Tree	29.1	34.8	40.4	46.1	51.8	57.5
18	Lumber	18.9	23.0	27.1	31.2	35.3	39.4
	Sawdust	2.9	3.6	4.3	5.0	5.7	6.4
	Residue	10.6	12.0	13.5	15.0	16.5	17.9
	Gross Tree	32.5	39.0	45.4	51.8	58.2	64.6
19	Lumber	21.3	26.0	30.6	35.3	39.9	44.6
	Sawdust	3.2	4.0	4.8	5.6	6.3	7.1
	Residue	11.4	13.0	14.7	16.3	17.9	19.6
	Gross Tree	36.2	43.4	50.6	57.8	65.0	72.2

Table continued next page.

TABLE 18-58—*Black oak; predicted yields of sawmill products in bandsaw mills (10/64-inch kerf) from bark-free, merchantable stems to an 8-inch top* (Hanks 1977)[1]
—Continued

DBH (inches)	Product	Merchantable height, feet					
		24	32	40	48	56	64
		.... Cubic feet					
20	Lumber	23.9	29.1	34.4	39.6	44.8	50.0
	Sawdust	3.6	4.5	5.3	6.2	7.0	7.9
	Residue	12.3	14.1	15.9	17.7	19.5	21.3
	Gross Tree	40.0	48.0	56.1	64.1	72.2	80.2
21	Lumber	26.6	32.5	38.3	44.1	50.0	55.8
	Sawdust	4.0	5.0	5.9	6.8	7.7	8.7
	Residue	13.2	15.2	17.2	19.2	21.2	23.2
	Gross Tree	44.0	52.9	61.9	70.8	79.7	88.6
22	Lumber	29.5	36.0	42.4	48.9	55.4	61.8
	Sawdust	4.4	5.5	6.5	7.5	8.5	9.5
	Residue	14.2	16.4	18.5	20.7	22.9	25.1
	Gross Tree	48.3	58.1	67.9	77.8	87.6	97.4
23	Lumber	32.5	39.6	46.7	53.9	61.0	68.1
	Sawdust	4.9	6.0	7.1	8.2	9.3	10.4
	Residue	15.2	17.6	20.0	22.3	24.7	27.1
	Gross Tree	52.7	63.5	74.3	85.1	95.8	106.6
24	Lumber	35.6	43.4	51.3	59.1	66.9	74.7
	Sawdust	5.3	6.5	7.7	8.9	10.1	11.3
	Residue	16.3	18.9	21.4	24.0	26.6	29.2
	Gross Tree	57.3	69.1	80.9	92.7	104.5	116.3

[1] Bold lines indicate range of observed data.

TABLE 18-59—*Chestnut oak; predicted yields of sawmill products in bandsaw mills (10/64-inch kerf) from bark-free merchantable stems to an 8-inch top* (Hanks 1977)[1]

DBH (inches)	Product	Merchantable height, feet					
		24	32	40	48	56	64
		-----------------Cubic feet----------------					
10	Lumber	5.3	4.4	3.5	2.6	1.7	0.8
	Sawdust	1.0	1.1	1.1	1.2	1.3	1.3
	Residue	3.6	5.9	8.2	10.5	12.8	15.1
	Gross Tree	9.3	11.0	12.7	14.3	16.0	17.7
11	Lumber	6.5	6.0	5.5	5.0	4.5	4.0
	Sawdust	1.2	1.3	1.4	1.5	1.6	1.7
	Residue	4.1	6.5	8.8	11.1	13.4	15.7
	Gross Tree	11.3	13.4	15.5	17.6	19.7	21.9
12	Lumber	7.7	7.6	7.6	7.5	7.5	7.4
	Sawdust	1.4	1.6	1.7	1.9	2.1	2.2
	Residue	4.8	7.1	9.4	11.7	14.0	16.3
	Gross Tree	13.4	16.0	18.6	21.2	23.8	26.4

13	Lumber	9.0	9.5	9.9	10.3	10.7	11.2
	Sawdust	1.6	1.8	2.1	2.3	2.5	2.8
	Residue	5.4	7.7	10.1	12.4	14.7	17.0
	Gross Tree	15.8	18.9	22.0	25.1	28.3	31.4
14	Lumber	10.5	11.4	12.4	13.3	14.3	15.2
	Sawdust	1.8	2.1	2.4	2.7	3.0	3.3
	Residue	6.1	8.5	10.8	13.1	15.5	17.8
	Gross Tree	18.3	22.0	25.7	29.4	33.1	36.8
15	Lumber	12.0	13.5	15.0	16.5	18.1	19.6
	Sawdust	2.1	2.5	2.8	3.2	3.6	4.0
	Residue	6.9	9.2	11.6	13.9	16.3	18.6
	Gross Tree	21.0	25.3	29.6	33.9	38.2	42.5
16	Lumber	13.7	15.8	17.9	20.0	22.1	24.2
	Sawdust	2.3	2.8	3.2	3.7	4.2	4.6
	Residue	7.7	10.1	12.4	14.8	17.1	19.5
	Gross Tree	23.9	28.9	33.8	38.8	43.7	48.7
17	Lumber	15.4	18.2	20.9	23.7	26.4	29.2
	Sawdust	2.6	3.1	3.7	4.2	4.8	5.3
	Residue	8.6	11.0	13.3	15.7	18.0	20.4
	Gross Tree	27.0	32.7	38.3	44.0	49.6	55.3
18	Lumber	17.3	20.7	24.1	27.6	31.0	34.4
	Sawdust	2.9	3.5	4.2	4.8	5.4	6.1
	Residue	9.6	11.9	14.3	16.6	19.0	21.4
	Gross Tree	30.3	36.7	43.1	49.5	55.8	62.2
19	Lumber	19.3	23.4	27.6	31.7	35.8	40.0
	Sawdust	3.2	3.9	4.7	5.4	6.1	6.9
	Residue	10.5	12.9	15.3	17.6	20.0	22.4
	Gross Tree	33.8	40.9	48.1	55.3	62.4	69.6
20	Lumber	21.3	26.2	31.1	36.0	40.9	45.8
	Sawdust	3.5	4.3	5.2	6.0	6.8	7.7
	Residue	11.6	14.0	16.3	18.7	21.1	23.5
	Gross Tree	37.4	45.4	53.4	61.4	69.4	77.3
21	Lumber	23.5	29.2	34.9	40.6	46.3	52.0
	Sawdust	3.9	4.8	5.7	6.7	7.6	8.6
	Residue	12.7	15.1	17.4	19.8	22.2	24.6
	Gross Tree	41.3	50.1	59.0	67.8	76.7	85.5
22	Lumber	25.8	32.4	38.9	45.4	51.9	58.4
	Sawdust	4.2	5.3	6.3	7.4	8.4	9.5
	Residue	13.8	16.2	18.6	21.0	23.4	25.8
	Gross Tree	45.3	55.1	64.8	74.6	84.3	94.0
23	Lumber	28.2	35.6	43.0	50.4	57.8	65.2
	Sawdust	4.6	5.7	6.9	8.1	9.3	10.4
	Residue	15.0	17.4	19.8	22.3	24.7	27.1
	Gross Tree	49.5	60.2	70.9	81.6	92.3	103.0
24	Lumber	30.7	39.0	47.3	55.6	63.9	72.3
	Sawdust	5.0	6.3	7.5	8.8	10.1	11.4
	Residue	16.3	18.7	21.1	23.5	26.0	28.4
	Gross Tree	53.9	65.6	77.3	89.0	100.7	112.3

[1]Bold lines indicate range of observed data.

TABLE 18-60—*Northern red oak; predicted yields of sawmill products in bandsaw mills (10/64-inch kerf) from bark-free merchantable stems to an 8-inch top* (Hanks 1977)[1]

DBH (inches)	Product	Merchantable height, feet					
		24	32	40	48	56	64
		------Cubic feet------					
10	Lumber	4.9	4.8	4.7	4.6	4.5	4.4
	Sawdust	1.2	1.3	1.5	1.7	1.9	2.1
	Residue	4.8	7.7	10.6	13.6	16.5	19.4
	Gross Tree	11.2	13.8	16.3	18.9	21.5	24.0
11	Lumber	6.1	6.4	6.6	6.9	7.2	7.5
	Sawdust	1.3	1.6	1.8	2.0	2.3	2.5
	Residue	5.4	8.2	11.1	14.0	16.9	19.7
	Gross Tree	13.2	16.2	19.1	22.1	25.0	28.0
12	Lumber	7.4	8.1	8.8	9.5	10.2	10.9
	Sawdust	1.5	1.8	2.1	2.4	2.7	3.0
	Residue	6.0	8.8	11.6	14.5	17.3	20.1
	Gross Tree	15.4	18.8	22.2	25.5	28.9	32.3
13	Lumber	8.8	10.0	11.1	12.3	13.4	14.6
	Sawdust	1.7	2.1	2.4	2.8	3.1	3.5
	Residue	6.7	9.4	12.2	15.0	17.7	20.5
	Gross Tree	17.8	21.6	25.5	29.3	33.2	37.0
14	Lumber	10.4	12.0	13.6	15.3	16.9	18.5
	Sawdust	2.0	2.4	2.8	3.2	3.6	4.1
	Residue	7.4	10.1	12.8	15.5	18.2	20.9
	Gross Tree	20.4	24.7	29.1	33.4	37.8	42.1
15	Lumber	12.0	14.2	16.4	18.5	20.7	22.8
	Sawdust	2.2	2.7	3.2	3.7	4.2	4.6
	Residue	8.2	10.8	13.5	16.1	18.8	21.4
	Gross Tree	23.1	28.0	32.9	37.8	42.7	47.6
16	Lumber	13.8	16.5	19.3	22.0	24.7	27.4
	Sawdust	2.5	3.0	3.6	4.2	4.7	5.3
	Residue	9.0	11.6	14.2	16.7	19.3	21.9
	Gross Tree	26.1	31.6	37.0	42.5	47.9	53.4
17	Lumber	15.7	19.0	22.4	25.7	29.0	32.3
	Sawdust	2.7	3.4	4.0	4.7	5.3	6.0
	Residue	9.9	12.4	14.9	17.4	19.9	22.4
	Gross Tree	29.2	35.3	41.4	47.5	53.5	59.6
18	Lumber	17.7	21.7	25.6	29.6	33.5	37.5
	Sawdust	3.0	3.8	4.5	5.2	5.9	6.7
	Residue	10.9	13.3	15.7	18.1	20.5	23.0
	Gross Tree	32.6	39.3	46.0	52.8	59.5	66.2
19	Lumber	19.9	24.5	29.1	33.7	38.3	42.9
	Sawdust	3.3	4.2	5.0	5.8	6.6	7.4
	Residue	11.9	14.2	16.6	18.9	21.2	23.6
	Gross Tree	36.1	43.5	50.9	58.3	65.8	73.2

DBH	Product					
20	Lumber	22.1	27.4	32.7	38.1	43.4	48.7
	Sawdust	3.7	4.6	5.5	6.4	7.3	8.2
	Residue	13.0	15.2	17.4	19.7	21.9	24.2
	Gross Tree	39.8	48.0	56.1	64.2	72.4	80.5
21	Lumber	24.4	30.5	36.6	42.6	48.7	54.8
	Sawdust	4.0	5.0	6.0	7.0	8.1	9.1
	Residue	14.1	16.2	18.4	20.5	22.7	24.8
	Gross Tree	43.7	52.6	61.5	70.4	79.3	88.3
22	Lumber	26.9	33.8	40.6	47.4	54.3	61.1
	Sawdust	4.4	5.5	6.6	7.7	8.8	10.0
	Residue	15.3	17.3	19.4	21.4	23.5	25.5
	Gross Tree	47.8	57.5	67.2	76.9	86.6	96.4
23	Lumber	29.5	37.2	44.8	52.5	60.1	67.8
	Sawdust	4.7	6.0	7.2	8.4	9.6	10.9
	Residue	16.5	18.4	20.4	22.3	24.3	26.2
	Gross Tree	52.1	62.6	73.2	83.7	94.3	104.8
24	Lumber	32.2	40.7	49.2	57.7	66.2	74.7
	Sawdust	5.1	6.5	7.8	9.2	10.5	11.8
	Residue	17.8	19.6	21.4	23.3	25.1	27.0
	Gross Tree	56.6	68.0	79.4	90.8	102.3	113.7

[1] Bold lines indicate range of observed data.

TABLE 18-61—*White oak; predicted yields of sawmill products in bandsaw mills (10/64-inch kerf) from bark-free merchantable stems to an 8-inch top* (Hanks 1977)[1]

DBH (inches)	Product	Merchantable height, feet					
		24	32	40	48	56	64
		---------------- Cubic feet ----------------					
10	Lumber	3.9	3.9	4.0	4.0	4.1	4.1
	Sawdust	0.8	1.0	1.2	1.4	1.6	1.8
	Residue	3.3	6.2	9.0	11.9	14.8	17.7
	Gross Tree	8.2	10.9	13.6	16.3	19.1	21.8
11	Lumber	5.3	5.6	6.0	6.3	6.7	7.0
	Sawdust	1.0	1.2	1.5	1.7	2.0	2.2
	Residue	3.9	6.8	9.6	12.5	15.4	18.2
	Gross Tree	10.4	13.5	16.6	19.7	22.8	25.9
12	Lumber	6.8	7.5	8.2	8.9	9.6	10.3
	Sawdust	1.2	1.5	1.8	2.1	2.4	2.7
	Residue	4.6	7.5	10.3	13.1	16.0	18.8
	Gross Tree	12.9	16.4	19.8	23.3	26.8	30.3
13	Lumber	8.5	9.5	10.6	11.6	12.7	13.7
	Sawdust	1.5	1.8	2.2	2.5	2.8	3.2
	Residue	5.4	8.2	11.0	13.8	16.6	19.5
	Gross tree	15.5	19.5	23.4	27.3	31.2	35.2
14	Lumber	10.3	11.7	13.2	14.6	16.1	17.5
	Sawdust	1.8	2.1	2.5	2.9	3.3	3.7
	Residue	6.2	9.0	11.8	14.6	17.4	20.2
	Gross tree	18.4	22.8	27.2	31.6	36.0	40.4

Table continued next page.

TABLE 18-61—*White oak; predicted yields of sawmill products in bandsaw mills (10/64-inch kerf) from bark-free, merchantable stems to an 8-inch top* (Hanks 1977)[1]
—Continued

DBH (inches)	Product	Merchantable height, feet					
		24	32	40	48	56	64
		Cubic feet					
15	Lumber	12.2	14.1	16.0	17.8	19.7	21.6
	Sawdust	2.1	2.5	3.0	3.4	3.8	4.3
	Residue	7.1	9.8	12.6	15.4	18.1	20.9
	Gross tree	21.5	26.4	31.3	36.2	41.1	46.0
16	Lumber	14.3	16.6	18.9	21.3	23.6	25.9
	Sawdust	2.4	2.9	3.4	3.9	4.4	4.9
	Residue	8.0	10.7	13.5	16.2	19.0	21.7
	Gross tree	24.8	30.2	35.7	41.1	46.6	52.0
17	Lumber	16.5	19.3	22.1	24.9	27.7	30.5
	Sawdust	2.7	3.3	3.9	4.4	5.0	5.6
	Residue	9.0	11.7	14.4	17.1	19.9	22.6
	Gross tree	28.3	34.3	40.4	46.4	52.4	58.4
18	Lumber	18.9	22.2	25.5	28.8	32.1	35.4
	Sawdust	3.1	3.7	4.4	5.0	5.6	6.3
	Residue	10.1	12.7	15.4	18.1	20.8	23.5
	Gross tree	32.0	38.7	45.3	51.9	58.6	65.2
19	Lumber	21.4	25.2	29.0	32.9	36.7	40.5
	Sawdust	3.5	4.2	4.9	5.6	6.3	7.0
	Residue	11.2	13.8	16.5	19.1	21.8	24.5
	Gross tree	36.0	43.3	50.5	57.8	65.1	72.4
20	Lumber	24.0	28.4	32.8	37.2	41.6	46.0
	Sawdust	3.9	4.7	5.4	6.2	7.0	7.8
	Residue	12.4	15.0	17.6	20.2	22.8	25.5
	Gross tree	40.1	48.1	56.1	64.0	72.0	79.9
21	Lumber	26.7	31.7	36.7	41.7	46.7	51.7
	Sawdust	4.3	5.2	6.0	6.9	7.8	8.6
	Residue	13.6	16.2	18.8	21.4	23.9	26.5
	Gross tree	44.5	53.2	61.9	70.5	79.2	87.9
22	Lumber	29.6	35.2	40.8	46.5	52.1	57.7
	Sawdust	4.8	5.7	6.6	7.6	8.5	9.5
	Residue	14.9	17.5	20.0	22.6	25.1	27.7
	Gross tree	49.1	58.5	67.9	77.4	86.8	96.2
23	Lumber	32.6	38.9	45.2	51.4	57.7	64.0
	Sawdust	5.2	6.3	7.3	8.3	9.3	10.4
	Residue	16.3	18.8	21.3	23.8	26.3	28.8
	Gross tree	53.9	64.1	74.3	84.5	94.7	104.9
24	Lumber	35.8	42.7	49.7	56.6	63.6	70.6
	Sawdust	5.7	6.8	8.0	9.1	10.2	11.3
	Residue	17.7	20.2	22.6	25.1	27.6	30.0
	Gross Tree	58.9	69.9	81.0	92.0	103.0	114.1

[1]Bold lines indicate range of observed data.

TABLE 18-62—*Yellow-poplar; predicted yields of sawmill products in bandsaw mills and circular-saw mills from bark-free merchantable stems to an 8-inch top* (Hanks 1977)[1]

DBH (inches)	Product	Merchantable height, feet					
		24	32	40	48	56	64
		---------------Cubic feet---------------					
	BANDSAW MILLS (10/64-INCH KERF)						
10	Lumber	3.8	3.9	3.9	4.0	4.0	4.1
	Sawdust	0.8	0.9	1.0	1.1	1.2	1.3
	Residue	5.0	6.1	7.1	8.2	9.3	10.3
	Gross tree	10.3	11.7	13.0	14.3	15.7	17.0
11	Lumber	5.5	5.9	6.2	6.6	6.9	7.3
	Sawdust	1.1	1.2	1.4	1.5	1.7	1.8
	Residue	5.3	6.4	7.5	8.6	9.8	10.9
	Gross tree	12.4	14.2	16.0	17.8	19.6	21.3
12	Lumber	7.3	8.0	8.7	9.4	10.1	10.7
	Sawdust	1.3	1.5	1.8	2.0	2.2	2.4
	Residue	5.6	6.8	8.0	9.2	10.3	11.5
	Gross tree	14.8	17.0	19.3	21.5	23.8	26.0
13	Lumber	9.3	10.3	11.4	12.4	13.5	14.5
	Sawdust	1.6	1.9	2.2	2.4	2.7	3.0
	Residue	6.0	7.2	8.5	9.7	10.9	12.2
	Gross tree	17.3	20.1	22.9	25.6	28.4	31.1
14	Lumber	11.4	12.9	14.3	15.7	17.2	18.6
	Sawdust	1.9	2.3	2.6	2.9	3.3	3.6
	Residue	6.4	7.7	9.0	10.3	11.6	12.9
	Gross tree	20.1	23.4	26.7	30.0	33.3	36.6
15	Lumber	13.7	15.6	17.4	19.3	21.1	23.0
	Sawdust	2.3	2.7	3.1	3.5	3.9	4.3
	Residue	6.8	8.2	9.5	10.9	12.3	13.7
	Gross tree	23.1	27.0	30.9	34.7	38.6	42.5
16	Lumber	16.2	18.5	20.8	23.1	25.3	27.6
	Sawdust	2.6	3.1	3.6	4.1	4.6	5.0
	Residue	7.2	8.7	10.1	11.6	13.1	14.5
	Gross tree	26.2	30.8	35.3	39.8	44.3	48.8
17	Lumber	18.8	21.6	24.3	27.1	29.8	32.6
	Sawdust	3.0	3.6	4.1	4.7	5.3	5.8
	Residue	7.7	9.2	10.8	12.3	13.9	15.4
	Gross tree	29.6	34.8	40.0	45.2	50.4	55.6
18	Lumber	21.6	24.8	28.1	31.4	34.6	37.9
	Sawdust	3.4	4.1	4.7	5.4	6.0	6.7
	Residue	8.2	9.8	11.5	13.1	14.7	16.4
	Gross tree	33.2	39.1	45.0	50.9	56.8	62.7
19	Lumber	24.5	28.3	32.1	35.9	39.7	43.5
	Sawdust	3.9	4.6	5.3	6.1	6.8	7.5
	Residue	8.7	10.4	12.2	13.9	15.6	17.4
	Gross tree	37.0	43.6	50.3	56.9	63.6	70.2

Table continued next page.

TABLE 18-62—*Yellow-poplar; predicted yields of sawmill products in bandsaw mills and circular-saw mills from bark-free, merchantable stems to an 8-inch top* (Hanks 1977)[1]
—Continued

DBH (inches)	Product	Merchantable height, feet				
		24	32	40	48	56	64
		.... Cubic feet					
20	Lumber	27.6	32.0	36.3	40.7	45.0	49.3
	Sawdust	4.3	5.1	6.0	6.8	7.6	8.5
	Residue	9.3	11.1	12.9	14.8	16.6	18.4
	Gross tree	40.9	48.4	55.8	63.3	70.7	78.2
21	Lumber	30.9	35.8	40.7	45.7	50.6	55.5
	Sawdust	4.8	5.7	6.6	7.6	8.5	9.4
	Residue	9.9	11.8	13.7	15.7	17.6	19.5
	Gross tree	45.1	53.4	61.7	70.0	78.3	86.5
22	Lumber	34.3	39.8	45.4	50.9	56.5	62.0
	Sawdust	5.3	6.3	7.3	8.4	9.4	10.4
	Residue	10.5	12.5	14.6	16.6	18.6	20.7
	Gross tree	49.5	58.7	67.8	77.0	86.1	95.3
23	Lumber	37.8	44.0	50.2	56.4	62.6	68.8
	Sawdust	5.8	6.9	8.1	9.2	10.4	11.5
	Residue	11.5	13.3	15.4	17.6	19.7	21.9
	Gross tree	54.1	64.2	74.3	84.3	94.4	104.5
24	Lumber	41.6	48.4	55.3	62.2	69.0	75.9
	Sawdust	6.3	7.6	8.9	10.1	11.4	12.6
	Residue	11.8	14.1	16.3	18.6	20.9	23.2
	Gross tree	58.9	70.0	81.0	92.0	103.0	114.0
	CIRCULAR-SAW MILLS (17/64-INCH KERF)						
10	Lumber	4.0	4.6	5.2	5.8	6.4	7.0
	Sawdust	1.4	1.8	2.1	2.4	2.7	3.1
	Residue	5.0	6.1	7.1	8.2	9.3	10.3
	Gross tree	10.3	11.7	13.0	14.3	15.7	17.0
11	Lumber	5.4	6.3	7.2	8.1	8.9	9.8
	Sawdust	1.8	2.2	2.6	3.0	3.4	3.8
	Residue	5.3	6.4	7.5	8.6	9.8	10.9
	Gross tree	12.4	14.2	16.0	17.8	19.6	21.3
12	Lumber	6.9	8.1	9.3	10.5	11.7	12.9
	Sawdust	2.2	2.7	3.2	3.7	4.2	4.7
	Residue	5.6	6.8	8.0	9.2	10.3	11.5
	Gross tree	14.8	17.0	19.3	21.5	23.8	26.0
13	Lumber	8.6	10.1	11.6	13.1	14.7	16.2
	Sawdust	2.6	3.2	3.8	4.4	5.0	5.6
	Residue	6.0	7.2	8.5	9.7	10.9	12.2
	Gross tree	17.3	20.1	22.9	25.6	28.4	31.1
14	Lumber	10.4	12.3	14.1	16.0	17.9	19.8
	Sawdust	3.1	3.8	4.5	5.2	5.9	6.6
	Residue	6.4	7.7	9.0	10.3	11.6	12.9
	Gross tree	20.1	23.4	26.7	30.0	33.3	36.6

TABLE 18-62—*Yellow-poplar; predicted yields of sawmill products in bandsaw mills and circular-saw mills from bark-free, merchantable stems to an 8-inch top* (Hanks 1977)[1]
—Continued

DBH (inches)	Product	Merchantable height, feet					
		24	32	40	48	56	64
		.... Cubic feet					
15	Lumber	12.3	14.6	16.8	19.1	21.3	23.6
	Sawdust	3.6	4.4	5.2	6.0	6.8	7.6
	Residue	6.8	8.2	9.5	10.9	12.3	13.7
	Gross tree	23.1	27.0	30.9	34.7	38.6	42.5
16	Lumber	14.3	17.0	19.7	22.4	25.0	27.7
	Sawdust	4.1	5.0	6.0	6.9	7.8	8.8
	Residue	7.2	8.7	10.1	11.6	13.1	14.5
	Gross tree	26.2	30.8	35.3	39.8	44.3	48.8
17	Lumber	16.5	19.6	22.7	25.9	29.0	32.1
	Sawdust	4.7	5.7	6.8	7.9	8.9	10.0
	Residue	7.7	9.2	10.8	12.3	13.9	15.4
	Gross tree	29.6	34.8	40.0	45.2	50.4	55.6
18	Lumber	18.8	22.4	26.0	29.6	33.1	36.7
	Sawdust	5.3	6.5	7.7	8.9	10.1	11.2
	Residue	8.2	9.8	11.5	13.1	14.7	16.4
	Gross tree	33.2	39.1	45.0	50.9	56.8	62.7
19	Lumber	21.3	25.4	29.4	33.5	37.6	41.6
	Sawdust	5.9	7.3	8.6	9.9	11.3	12.6
	Residue	8.7	10.4	12.2	13.9	15.6	17.4
	Gross tree	37.0	43.6	50.3	56.9	63.6	70.2
20	Lumber	23.9	28.4	33.0	37.6	42.2	46.8
	Sawdust	6.6	8.1	9.6	11.0	12.5	14.0
	Residue	9.3	11.1	12.9	14.8	16.6	18.4
	Gross tree	40.9	48.4	55.8	63.3	70.7	78.2
21	Lumber	26.6	31.7	36.8	42.0	47.1	52.2
	Sawdust	7.3	8.9	10.6	12.2	13.9	15.5
	Residue	9.9	11.8	13.7	15.7	17.6	19.5
	Gross tree	45.1	53.4	61.7	70.0	78.3	86.5
22	Lumber	29.4	35.1	40.8	46.5	52.2	57.9
	Sawdust	8.0	9.8	11.7	13.5	15.3	17.1
	Residue	10.5	12.5	14.6	16.6	18.6	20.7
	Gross tree	49.5	58.7	67.8	77.0	86.1	95.3
23	Lumber	32.4	38.7	45.0	51.3	57.6	63.9
	Sawdust	8.8	10.8	12.8	14.8	16.8	18.7
	Residue	11.1	13.3	15.4	17.6	19.7	21.9
	Gross tree	54.1	64.2	74.3	84.3	94.4	104.5
24	Lumber	35.5	42.4	49.3	56.3	63.2	70.1
	Sawdust	9.6	11.8	13.9	16.1	18.3	20.5
	Residue	11.8	14.1	16.3	18.6	20.9	23.2
	Gross tree	58.9	70.0	81.0	92.0	103.0	114.0

[1]Bold lines indicate range of observed data.

A portion of the residue tabulated is not usable because of rot; the remainder is usable for pulp chips. Averages for rot in Hank's sample trees were:

Species	Average rot
	Percent
Maple, red	3
Oak, black	2
Oak, chestnut	4
Oak, northern red	4
Oak, white	2
Yellow-poplar	2

Hank's (1977) tables included trees to 30 inches in dbh and to 72 feet of merchantable height; the portions reproduced here are truncated at 24 inches in dbh and 64 feet in merchantable height.

Other yield and residue studies of mills sawing southern hardwoods include Lane (1954), Massengale (1971), Clark et al. (1974), Phillips et al. (1974), Phillips (1975), and Clark (1976). These studies are summarized in chapter 27, mostly in sect. 27-3. Steele et al. (1981) presented a geometric model with which to calculate volumes of green sawdust, green pulp chips, dry planer shavings, and green or dry lumber produced by the sawmilling process.

18-11 SAWMILL LAYOUTS FOR SHORT, SMALL, HARDWOOD LOGS

Section 18-9 described chipping headrigs appropriate for short, small, hardwood logs of low grade; such headrigs can process 1,600 to 2,000 logs or bolts per 8-hour shift. In this section, headrigs using saws are discussed. Obviously cutterheads chipping slabs from log exteriors can be teamed with multiple bandsaws or circular saws making interior cuts.

Chapter 28 contains abstracts of financial operating results (capital required and profit potential) of manufacturing enterprises based on a variety of headrig types.

In 1984, T.D. McKay and H.W. Wilson (Forest Products Journal 34(6):43-48) analyzed nine alternative small mills for processing low-grade hardwoods in southwest Virginia, and concluded that only SHOLO (fig. 18-114) and scragg mills would show a 15 percent return on investment when operating at 80 percent of capacity. The scragg mill evaluated consisted of two shifting circular saws mounted on a common arbor; logs pass through such a saw on a feed chain.

LOG AND LUMBER LENGTHS

Pine-site hardwoods are short of bole, crooked, and have defective sections. To obtain straight logs with an acceptable degree of defect, tree-length stems must be judiciously crosscut into shorter lengths. Standard sawmills producing grade lumber seek logs 12 to 16 feet long, but such logs are not plentiful in stems of pine-site hardwoods. Bolts from 3 to 8 feet in length are readily obtainable, however. Huffman's (1973) data on red maple logs from Florida showed that yield of both lumber tally and furniture cuttings were increased significantly when long logs were reduced to short lengths before sawing (fig. 18-113). Other studies (Reynolds and Schroeder 1977, 1978) support this conclusion. Lumber from such short logs should have high utility because the two major consumers

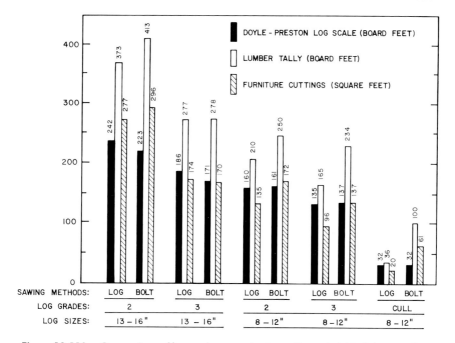

Figure 18-113.—Comparison of log scale, green lumber tally, and yield of clear cuttings from standard-length red maple logs and from bolts 4, 5, and 6 feet long. (Drawing after Huffman 1973.)

of hardwood lumber in the United States are the furniture industry and the pallet industry; both use short cuttings.

Bingham and Schroeder (1976, 1977) found that most furniture parts are quite short. An analysis of one set of cutting bills showed that 96 percent of the pieces could be cut from 42-inch lumber, and that average cutting length was 24 inches; chair parts could be cut from the 42-inch lumber with yield of about 80 percent. Another analysis of cutting bills for a full line of furniture showed that average cutting length was 23 inches. In a third study of an entire year's production of a major manufacturer of a full line of furniture (including chairs but not upholstered furniture), Bingham and Schroeder found that the average cutting length required by this manufacturer was 31 inches, as follows:

Cutting length	Proportion of cuttings shorter than this length
Inches	*Percent*
36	66.3
42	77.4
48	85.4
54	92.0
60	95.4
66	98.9
72	99.1
78	99.9
84	100.0

For another analysis, see: Araman, Philip A. Rough-part sizes needed from lumber for manufacturing furniture and kitchen cabinets. Broomall, PA: Northeast. For. Exp. Stn.; 1982; USDA For. Serv. Res. Pap. NE-503. 8 p.

Most pallet parts are 30 to 48 inches in length; only a minor proportion exceeds 72 inches in length.

Craft and Emanuel (1981) found that woods-run 4- and 6-foot-long bolts from hardwood poletimber thinnings are well suited for commercial pallet production. Bolts containing unsound heart defects and those containing sweep exceeding 1.5 inches should be eliminated. Resulting bolts (6 to 10 inches in scaling diameter) need not be segregated and should yield about 55 percent of cubic-foot volume in acceptable pallet cants and boards. When only 4- by 4- and 4- by 6-inch cants are produced (no side lumber), product yield is approximately 45 percent of the cubic foot volume. The quality of cants from such sound woods-run bolts is adequate for production of permanent or returnable warehouse pallets (Craft and Emanuel 1981). For additional information see subsection PALLETS in section 27-2, and table 27-122A. See also: Craft, E. Paul; Whitenack, Kenneth R., Jr. A classification system for predicting pallet part quality from hardwood cants. Broomall, PA: Northeast. For. Exp. Stn.; 1982; USDA For. Serv. Res. Pap. NE-515. 7 p.

BOLT SAWING PATTERNS

As noted in section 18-10, there is a substantial body of literature that advocates live sawing of 12- to 16-foot eastern hardwood logs, particularly those of low grade; large logs with low-grade centers and clear outer wood yield more value if sawn around, however. The same arguments hold true for logs 3 to 8 feet long, but the need to process the short bolts rapidly favors live sawing patterns producing cants to be resawn and planks to be ripped.

On a few timber species it is practical to center-rip a small log and then convert the resulting log halves to lumber on a horizontal bandsaw or linebar vertical resaw. This live-sawing procedure gives high yield but requires edging of every board—a labor intensive operation. Additionally, warp of halves of center-ripped logs of some southern species (hickory and post oak, for example), due to internal log stresses, deters the use of this system.

Arrangements which simultaneously remove slabs from two sides of logs are more practical. For the smallest logs, sawing a narrow face on two or four sides followed by gang ripping of the resultant cant is efficient. On larger logs, boards may be cut from two or four sides of the log before releasing the central cant from the headrig for sale as a cant, or for resawing. Alternatively, slabs can be ripped from the log for diversion to a resaw for recovery of boards.

Coleman and Reynolds (1973) compared three of these methods of sawing 44- and 52-inch-long factory-grade oak logs, as follows:

- Selective method (fig. 18-114 left), based on live sawing logs into boards and selectively ripping and end-trimming these boards into pallet parts. The log was placed on the sawmill carriage so that the poorest face of the log faced the saw blade.
- Gang method (fig. 18-114 right), in which each log was canted on four sides, resawed into boards, and the boards end-trimmed to yield pallet parts.
- Combination method (fig. 18-115), in which a cant was produced by slabbing and the cant was resawn into boards; the slabs were also resawn to yield boards which were then selectively ripped to desired width. Boards from both cants and slabs were end-trimmed to yield pallet parts.

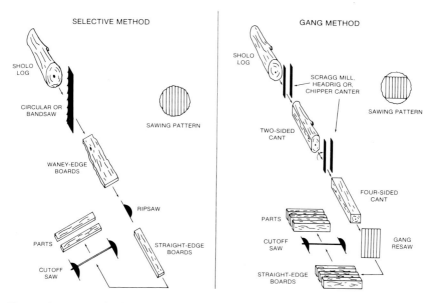

Figure 18-114.—Methods for sawing short hardwood logs, here termed SHOLO. (Left) Selective method. (Right) Gang method. (Drawings after Coleman and Reynolds 1973.)

The gang method gave the highest production rate, but the yield of pallet parts was about 40 percent less than from the other two methods, which showed about equal yield. For very low grade logs, the selective method was deemed most appropriate. For better logs, the combination method seemed best.

Reynolds (1974) reviewed industrial experience using circular saws to convert low-grade hardwood short logs by the selective and combination methods; yields were as follows:

Method, log source and product	Proportion of green log weight
	Percent
Selective method applied to very low grade logs from eastern Tennessee	
Pallet parts	20
Pulp chips	50
Sawdust and bark	30
Combination method applied to low-grade West Virginia timber of better grade than that from eastern Tennessee	
Pallet parts	40
Pulp chips	30
Sawdust and bark	30

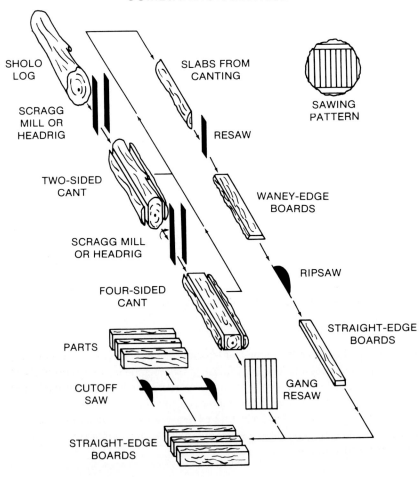

Figure 18-115.—Combination method of sawing short hardwood logs. (Drawing after Coleman and Reynolds 1973.)

Hardwoods from Arkansas and New York State were too poor in quality or too expensive for either method to be economic.

Small short southern hardwood logs can furnish lumber for both the pallet and furniture industries. Pallet parts do not require clear lumber, but furniture cuttings generally call for at least one clear face, and sometimes require that both faces be clear. Moreover, sizes of pallet parts are different than sizes of lumber from which furniture cuttings are made. Therefore, a method of separating short logs intended for furniture parts from pallet logs is useful. Reynolds and Schroeder (1978) described a method of making the separation.

By their method, stems and residue are bucked into 6-, 7-, and 8-foot lengths with a 2-inch trim allowance, to remove big knots, and minimize sweep. Very low-grade wood is bucked into 4-foot pulpwood lengths. At the sawmill, bolts that do not have one or more clear faces are sawed to pallet cants (4 by 4, 4 by 6, or 6 by 6 inches). From the remaining bolts a light slabbing cut is taken from the best face to expose a face 3 inches wide. If this face is poor, the bolt is sawed to pallet cants. If the face is good, the bolt is sawn as shown in figure 18-116.

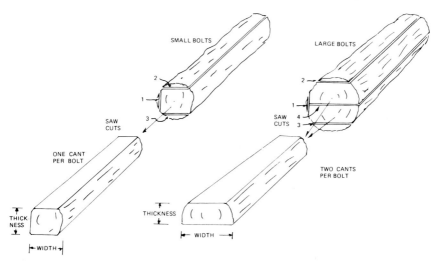

Figure 18-116.—Sawing three-sided cants from short bolts for resawing into furniture cuttings. (Drawing after Reynolds and Schroeder 1978.)

Bolts sawn by Reynolds and Schroeder were small; of 114 in their study, only three were large enough to yield two cants per bolt; average cant size was as follows:

Bolt length	Average cant size	
	Height	Width
Feet	*Inches*	
6	5.6	6.6
7	5.8	6.8
8	5.6	8.1

The cants were resawed into boards and dried. The dried boards were gang ripped and defects removed to obtain clear-one-face cuttings of random width and length. Logs 6-, 7-, and 8-feet-long gave good yields by their system; 4- and 5-foot logs did not.

The percentage of cuttings from the kiln-dried boards after gang ripping and defecting followed by salvage ripping was as follows:

Board length	Percent
Feet	
6	57.5
7	66.8
8	65.6

About 36 percent of the cuttings were 40 inches in length or longer; the remainder were 12 to 40 inches long. Widths were 1 inch to 3 inches in half-inch increments.

In another system proposed by Reynolds and Schroeder (1979) and more fully described by Reynolds and Gatchell (1982), felled trees are bucked primarily into "System 6 bolts" for use in producing furniture stock. These System 6 bolts are sound, 6 feet 3 inches in length, and have no more than 1.5 inches sweep; diameter is from 7.5 to 12.5 inches at the small end. Bucking to this length permits sweep and crook reduction and avoidance of defects. The remainder of the tree is bucked into pallet bolts, pulpwood bolts, and firewood. The System 6 bolts are sawed into two slabs and a 3¾-inch-thick cant on a two-saw scrag headrig. The cant is sawed into 1-inch boards on a single-arbor circular gang resaw in a single pass. Only 1- by 4-inch boards are made from the cant. The slabs are sawn into 1-inch boards which are edged to 4- and 6-inch widths. End-trimming to 72-inch length completes the sawing system. (See sect. 28-13 for an economic analysis of System 6.) See also Northeastern Forest Experiment Station 1983 reports RP-NE-520 and RP-NE-525 for discussions of cutting furniture frames and kitchen cabinet blanks by System 6. For a System 6 rough-mill operating manual see Res. Pap. NE-542 (1984); and for a sample plant design see GTR-NE-87 (1984)—both available from the Northeastern Forest Experiment Station.

Using computer simulation to determine the values that could be obtained by sawing low-grade oak and hickory logs in many different ways, the USDA Forest Service's Forestry Sciences Laboratory at Princeton, West Virginia (Reynolds 1969; Reynolds and Gatchell 1970, 1971) developed the SHOLO (SHOrt LOg) process for manufacturing parts for 40- by 48-inch standard warehouse pallets. The steps in the process are: debarking tree-length logs, bucking into lengths of 44 or 52 inches, live-sawing into 1- or 2-inch boards, ripping into needed widths, planing to finished thickness, and end-trimming. Selective bucking of tree-lengths yielded 57 percent acceptable short logs and 43 percent random-length round residue suitable for chipping. Sixteen percent of the original long-log volume was converted into pallet parts, 55 percent into pulp chips, and the remaining 29 percent into bark, sawdust, and planer shavings.

Cost for a mill of this type (including yard, mill, and tools) was estimated at $550,000. With a 16-man crew, such a mill could utilize approximately 70 cords of tree-length logs in one 8-hour shift. Annual sales were estimated at $640,000, based on 1969 prices.

Among the products of short hardwood logs and bolts are small dowels and turnings, the blanks for which are ½-inch-thick clear boards 4 inches in width. Saunders (1969) concluded after much mill experience that optimum recovery of these short clear boards is accomplished by live sawing the bolts with saws spaced to yield 4-inch-thick cants. These cants are then resawn into ½-inch boards.

Larger turned products requiring squares measuring 1.25 inches on a side are also products of short hardwood logs. Saunders (1979) concluded that yield of clear squares is maximized if short logs (10 feet or less in length) are live sawn with a thin-kerf bandsaw headrig; resulting 1.25-inch-thick round-edge boards are then ripped into squares with a merry-go-round thin band resaw fed to commence cutting at the sapwood edge until unusable heartwood is encountered, at which point the plank is turned and squares are ripped from the opposite sapwood edge until only the unusable heartwood in the center of the plank remains as a residue.

An alternative method of producing turning blanks of this size with tube saws, practiced by Dargan (1969), is described in connection with figures 18-90 and 18-91. Fasick and Lawrence (1971) graphed yields of rounds and chippable residues for a range of bolt lengths and diameters; they concluded that 9- to 15-inch-diameter logs in lengths from 4 to 5 feet are best suited to Dargan's technique.

In addition to more conventional bolt sawing patterns, it is possible to cut radially into the pith to yield tapered boards (visualize tangerine segments viewed in cross section); such a sawing pattern was widely used around 1900 to produce most of the white-pine siding in New England. More recently Hasenwinkle (1974) has elaborated on this concept by rejoining such tapered segments into glued assemblies. A variation of the radial-sawing procedure yields square timbers with hollow centers (fig. 18-117).

SAWMILL OUTPUT CORRELATED WITH FEED RATES

Some sawmill engineers feel that 20,000 board feet of cants, or more, must be produced on a headrig per 8-hour shift to be economic. Machinery constraints, however, may limit the number of short, low-grade logs or bolts processed per shift to about 2,000, i.e., 6 logs per minute for 360 minutes. This assumes 25 percent delay time ($\frac{480-360}{480}$ x 100) from mechanical breakdown, relief periods, or waiting on logs. The processing of 2,000 logs per shift will yield 6,000 to 60,000 board feet of cants depending on cant size and length (table 18-63); side boards from the larger logs will yield additional footage.

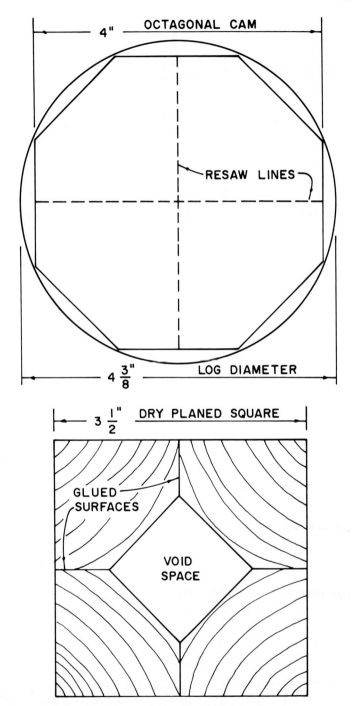

Figure 18-117.—Method whereby square hollow posts or rafters can be assembled from small logs turned to octagonal shape and then quartered and reassembled.

TABLE 18-63—*Board feet contained in 2,000 cants of five lengths and five sizes*

Cant length in size (inches)	Board feet in 2,000 cants
Three feet	
3 x 4	6,000
4 x 4	8,000
4 x 5	10,000
5 x 5	12,500
6 x 6	18,000
Four feet	
3 x 4	8,000
4 x 4	10,667
4 x 5	13,333
5 x 5	16,667
6 x 6	24,000
Six feet	
3 x 4	12,000
4 x 4	16,000
4 x 5	20,000
5 x 5	25,000
6 x 6	36,000
Eight feet	
3 x 4	16,000
4 x 4	21,333
4 x 5	26,667
5 x 5	33,333
6 x 6	48,000
Ten feet	
3 x 4	20,000
4 x 4	26,667
4 x 5	33,333
5 x 5	41,667
6 x 6	60,000

If loaded on a continuously running chain infeed so that logs occupy half the space on the chain, i.e., the space between logs is equal to the log length, then feed speed to process 2,000 logs in 360 minutes with no delay time is as follows:

Log length	Lineal feed speed of chain
Feet	*Feet/min*
3	33.3
4	44.4
6	66.7
8	88.9
10	111.1

If resulting cants are ripsawn on a roll-feed gang ripsaw, one at a time with space interval between cants equal to cant length, the foregoing feed speeds are also required to resaw 2,000 cants in 360 minutes. If logs could be processed in lengths that yield cants of twice the desired length (for example, 6-foot logs to yield cants that could be cut into two 3-foot lengths from which 36-inch pallet deck boards could be ripped), a 3-saw precision trimmer to make two end cuts and a center crosscut would need process only about one cant every 11 seconds—a leisurely rate. Even with small logs, 2,000 6-foot cants per shift would yield about 20,000 board feet and speed of infeed chains on the headrig would be an easily attainable 67 feet per minute. In the layout of mills manufacturing short lumber, care must be taken not to overload edgers with an excess of wany boards, and trim saws with an excess of trim-backs due to barky ends.

The procedures for producing 2,000 short logs per 8-hour shift and getting bark removed from them are discussed in section 16-6 and chapter 17. How are they to be sawn? The following paragraphs describe twin-saw headrigs; some are designed for continuous flow of logs, and some—with lower log output per shift—have reciprocating feedworks. Also, single-saw headrigs for live-sawing of lesser numbers of short logs into thick planks for subsequent ripping into squares will be briefly described.

TWO-SAW SCRAG MILLS WITH CONTINUOUS FEED

This approach would seem to solve the problem of achieving production of about 20,000 board feet per day on small logs averaging perhaps 6 feet in length. The infeed chain could run at less than 100 feet per minute if space between logs did not exceed log length. A major problem is that of quickly (in 11 seconds or less) positioning each log accurately, holding it firmly, and quickly setting the twin saws to appropriate spacing with appropriate offset to maximize lumber yield. Detjen (1978) reviewed the problem and outlined a solution in which a bottom feed chain with pusher dogs and a top chain with hold-back dogs were coupled with a series of opposed, short, log-turning wheels mounted on a carriage having capability to travel 4 feet downstream and back on command (fig. 18-118). Each pair of turning wheels has the capability of opening or closing to suit log diameter and side shifting through the use of programmable setworks. This wheel and carriage concept permits the use of a series of overhead camera-type scanners to achieve spot readings while the log is turning and linear readings during downstream travel—thereby providing all the data needed for computer determination of the best opening face to maximize lumber yield.

According to Detjen's (1978) layout of a twin-bandsaw headrig, logs would enter from the log deck through a scanning curtain for diameter, taper, and length resolution. The logs would be deposited into the turning system on the twin end-dogging log feeder provided with a variable fence to give one excellent slab in the double-taper mode (fig. 18-107 top), one straight, accurate heart flitch, and one slab having the total log taper. All slabs would pass to a trough-fed linebar resaw having a slat-bed bottom feed chain, a fixed fence, a shifting vertical band resaw, and a shifting end-mill chipping head to yield one or two optimum-thickness pieces from each slab on a one-pass basis. Heart flitches (cants) would be resawn on a gang edger equipped with thin-kerf circular saws. A simple board edger and a trimmer would round out the equipment requirements.

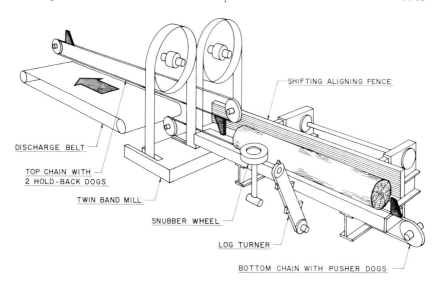

Figure 18-118.—Feed chains and turning devices to accurately position and firmly hold small, short, low-grade hardwood logs while they are sawn on multiple-saw headrigs. (Top) Twin bandsaw arrangement with an aligning fence to permit sawing parallel to the bark on one side. (Bottom) Equipped with turning roll, aligning fence, bottom chipping head, two banks of four edging saws (spindles vertical), and two 48-inch shifting circular headsaws. (Drawing and photo from McDonough Manufacturing Co.)

A simpler and cheaper headrig (fig. 18-119) employing twin circular saws can also achieve production of 20,000 board feet per 8-hour shift on small short hardwood logs, but at the expense of some loss in lumber yield due to: (1) less than optimum log positioning because of lack of scanners and computer logic; (2) lesser degree of feed linearity because logs are less firmly clamped; and (3) wider kerf on circular saws compared to bandsaws. Such a mill, with the layout shown in figure 18-120 and staffed with 10 men plus a foreman, might produce 20,000 board feet in 8 hours from 42- to 64-inch-long logs 9 to 13 inches in diameter at the small end (5-inch minimum). A cord of wood (6,000 pounds or 128 cubic feet) processed in this manner yields about 550 board feet of lumber. Mill cost in 1974, including a 6,000 square foot building was about $500,000.

Figure 18-119.—Twin circular-saw, continuous-feed scrag mill suitable for short, small-diameter hardwood logs. Each saw is driven at 750 rpm by a 75- to 100-hp motor; setworks permit sawing cants from 3-1/2 to 12 inches in width. Rolls adjacent to the infeed chain permit log to be rotated for best presentation to the saws. Three overhead disks hold logs against the lower feed chain while the log is being sawn. The feed chain travels at 150 fpm through the 48-inch-diameter, 20-tooth circular saws, which have 9/32-inch kerf. (Photo from Corley Manufacturing Co.)

TWO-SAW SCRAG MILLS WITH RECIPROCATING FEED

To eliminate the need for a resaw to recover boards from heavy slabs, some twin-circular-saw scrag mills have reciprocating feedworks featuring a vertical carriage into which the log is securely clamped from top and bottom (fig. 18-121 top). By this method, the sawyer can bring the log back through the saws to repetitively reset them and thus remove side boards until the desired cant thickness is achieved.

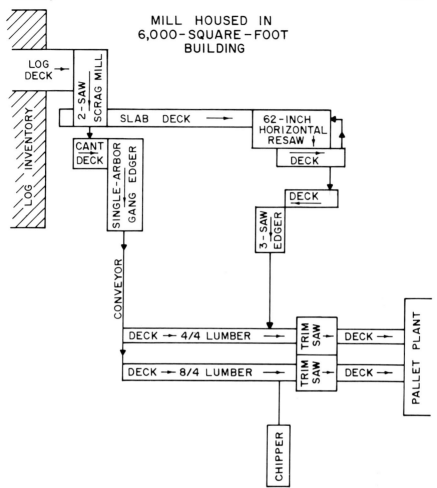

Figure 18-120.—Sawmill layout for small, short, low-grade hardwood logs featuring a continuously feeding twin-circular-saw headrig, a horizontal band resaw for slabs, and a gang edger with circular saws for cants. (Drawing after Abell Lumber Corp., Lawrenceville, Vir.)

In a variation of this idea, Hartzell (1976) built a short-log, two-man scrag mill with reciprocating overhead carriage and rotating end dogs (fig. 18-121 bottom). By this method side boards can be cut from all four sides. The headrig will handle logs as small as 5 inches in diameter but is best suited for logs larger than 10 inches in diameter. It will handle logs 4 to 16 feet in length, but was designed primarily for 12-foot logs not exceeding 18 inches in diameter. Hartzell found that the mill can process 433 logs in 7 hours with potential for some increase. Logs with heart rot cannot be revolved on the carriage. This mill is also available in a portable version.

An Australian design for 6-foot-long logs features twin bandsaws and a reciprocating, recumbent-J-shaped vertical carriage that dogs logs on their lower

Figure 18-121.—Two-saw scrag mills with reciprocating feedworks for short, small hardwood logs. (Top) Mill with vertical carriage into which the log is clamped from top and bottom, permitting boards to be sawn from two sides. The log is then returned, undogged, repositioned, dogged, and squared on the second pass. (Photo from Frick Forest Products, Inc.) (Bottom) Mill with reciprocating overhead carriage and rotating end dogs which permit rapid log rotation so that slabs and boards can be quickly cut from all four sides. (Photo from Carthage Machine Co., Inc.)

surface and for a couple of feet on the top surface of the log end nearest the saw (fig. 18-122). By having an open-ended vertical carriage, a log can be brought into position and appropriately rotated while the last pair of cuts is being made on the previous log. Such a headrig efficiently live-saws small logs into planks—particularly if equipped with a pair of chipping heads to remove two thin slabs in the first pass.

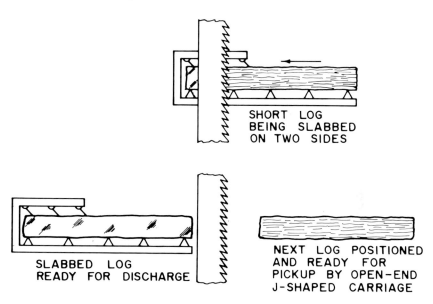

Figure 18-122.—(Top) Recumbent-J-shaped carriage grips logs on one end. (Bottom) The open rear end of this carriage permits the incoming log to be rolled into position for instant pickup after the last cut it made on the preceding log.

LIVE SAWING

In the manufacture of some hardwood products—white ash handles, for example—logs 40 to 66 inches long are live-sawn through-and-through into 1-¾-inch-thick planks. These planks are then selectively ripped (fig. 18-80) or gang-ripped (fig. 18-79) into 1¾-inch squares. The primary plank production can be accomplished on the reciprocating carriage of a conventional sawmill (Saunders 1979), or on the reciprocating carriage of a two-saw scrag mill (figs. 18-121 top, 18-122) if the log clamping device is less than 1¾ inches thick.

If log value warrants, it would be possible to use the continuous feed chain illustrated in figure 18-118 running through a series of twin bandsaws arranged to live-saw 1¾-inch-thick plank. A cheaper solution might call for a continuous-feed, twin-bandsaw scrag mill equipped with a pair of end-milling chipper heads, plus a vertical linebar resaw (fig. 18-66); with this arrangement the linebar resaw would require a feed speed three to four times the feed speed of the scrag mill.

In the short-log hardwood sawmill described by Saunders (1979), a conventional double-cut bandsaw headrig and carriage are tilted 15° so that gravity holds logs cradled aginst the inclined knees of the carriage. In this configuration logs are easily rotated. Saunders concluded that logs 10 feet and shorter can be live-sawn into 1.25-inch-thick planks at two per minute.

BOLTERS

The **bolter** is a table-type ripsawing machine (fig. 18-123) capable of sawing logs 2 to 8 feet in length, or more commonly 3 to 6 feet. Most employ circular saws whose action tends to push the log down on the table and back against the rear stop. When a bandsaw is used, some type of hold-down device is generally needed. The saw table (carriage) advances and recedes at high speed, actuated either by hydraulic rams or by cable drive controlled through foot pedals. The sawing guide is a fence, beyond and parallel to the saw; this guide can be rapidly and easily adjusted to rip slabs or boards of selected thicknesses. Logs are manually pressed against the guide, opening cuts usually being parallel to the bark. The operator manually turns the log to saw the log face desired, and can even turn short logs end for end if desired. The log is not dogged, but lies freely on the table, secured only by the toothed log stop at the rear end of the table. Lumber thickness is surprisingly uniform.

In New England, bolters are much used to saw birch (*Betula* sp.) and maple for the furniture and turnery industries. A two-man crew (sawyer and offbearer) sawing 4- to 5-foot-long birch bolts might produce, per 8-hour shift, about 3,500 board feet of live-sawn, wany-edged 4/4 lumber; when sawing around the log, production would be reduced to about 2,500 board feet of 4/4 lumber. A live deck to deliver logs conveniently over the top of the guide increases production somewhat, as does sawing of thicker boards or cants.

When textile shuttles made of flowering dogwood (*Cornus florida* L.) were in demand, many bolters were operated in the South to provide the clear cuttings needed. Hickory handle blanks are also frequently produced on bolter saws. During the 1970's, interest in bolters for manufacture of furniture parts was renewed. See sect. 28-4 for an economic analysis of bolter operation to produce short lumber for upholstered furniture.

Calvert's (1955) study of bolter sawing of white ash indicated sawing times per thousand board feet of 4/4 lumber and volume recovery as follows:

Bolt diameter inside bark at small end	Sawing time per thousand board feet of lumber	Percent of bolt volume (inside bark) recovered as lumber
Inches	*Minutes*	*Percent*
4	375	42
6	290	47
8	225	52
10	180	56
12	160	57
14	155	58
16	170	57
18	200	54

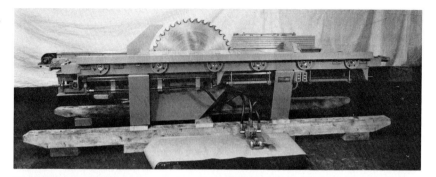

Figure 18-123.—(Top) Circular-saw short log bolter with automatic offbear belt, cable-driven table, and 40-inch-diameter, 9-gauge saw with style 2-1/2 inserted teeth, driven by a 75-hp motor. The bolter can be equipped with a 20-hp, 24-inch top saw. (Bottom) Continuous resaw system to complement the bolter. Feed speed is about 160 fpm. Maximum size of cant is 14 inches wide by 6 feet long. (Photo from J. W. Penney and Sons Co.)

Percentage of white ash tree volume converted to clear cuttings was independent of tree dbh, in the range from 8 to 20 inches, and averaged about 34 percent of volume inside bark to a 4-inch top. The proportions of clear cuttings in sawn lumber were 69 percent for 8-inch trees, 65 percent for 14-inch trees, and 63 percent for 20-inch trees.

The value of lumber sawn from a bolt is proportional to bolt diameter; Flann (1978) found that yellow birch (*Betula alleghaniensis* Britton) and sugar maple

(*Acer saccharum* Marsh.) bolts 10 inches in diameter yielded lumber worth about $6, those 12.5 inches in diameter about $10, and those 18 inches in diameter $20 to $25.

Redman (1957) produced clear-one-face cuttings from oak, sweetgum, and yellow-poplar bolts cut in North Carolina. His data (table 18-64) were based on bolter-sawing the rough green boards, kiln-drying them to 6 percent moisture content, and then glue-joint ripping them to develop the clear-one-face cuttings. Yields per cord of bolts and per thousand board feet of bolts, Doyle scale, were positively correlated with bolt diameter; upper grade bolts yielded significantly more cuttings than lower-grade bolts. For reasons not clear, bolts rift-sawn yielded significantly greater quantities of clear-one-face cuttings than bolts cut by a modified live-sawing technique (table 18-64).

TABLE 18-64.—*Board feet of 4/4 kiln-dried clear-one-face cuttings from bolts of five grades, six diameters, and three species* (Data from Redman 1957)[1/2/3]

Bolt species, sawing method, and grade[4]	Bolt diameter inside bark at small end, inches					
	8	10	12	14	16	18
	Board feet of clear-one-face cuttings					
	PER CORD OF BOLTS					
Oak sp., live-sawn[5]						
0	164	173	180	184	188	191
1	151	160	166	170	173	176
2	138	146	152	155	159	161
3	130	137	142	146	149	151
4	117	123	128	131	134	136
Oak, sp., rift-sawn[6]						
0	221	237	250	257	259	—
1	213	229	241	248	250	—
2	205	220	232	238	240	—
3	186	211	222	229	230	—
4	184	198	209	214	216	—
Sweetgum, live-sawn						
0	177	187	194	198	203	206
1	164	173	180	184	188	191
2	152	160	166	170	173	175
3	138	146	151	155	158	161
4	121	128	132	136	139	141
Sweetgum, rift-sawn						
0	234	251	264	272	274	—
1	230	246	260	267	269	—
2	221	238	250	257	259	—
3	213	229	241	248	250	—
4	209	224	236	243	245	—
Yellow-poplar, live-sawn						
0	298	315	327	335	341	346
1	259	274	284	291	297	301
2	237	251	260	267	272	276
3	194	205	213	218	223	226
4	134	141	147	150	153	156

Flitches cut 1½ or 2 inches thick from hardwood bolts are sometimes purchased by manufacturers of clear furniture squares and random-width stock; yield of cuttings from such flitches can be predicted by the technique of Rast and Chebetar (1980), diagrammed in figure 27-19.

TABLE 18-64.—*Board feet of 4/4 kiln-dried clear-one-face cuttings from bolts of five grades, six diameters, and three species* (Data from Redman 1957)[1,2,3]
—Continued

Bolt species, sawing method, and grade[4]	Bolt diameter inside bark at small end, inches					
	8	10	12	14	16	18
	------------- Board feet of clear-one-face cuttings -------------					
Oak, sp., live-sawn[5]	PER 1,000 BOARD FEET OF BOLTS, DOYLE SCALE					
0	840	620	510	460	430	410
1	770	580	470	430	400	380
2	700	530	430	390	360	350
3	660	490	400	370	340	320
4	580	450	360	330	310	290
Oak, sp., rift-sawn[6]						
0	1,120	830	730	640	590	—
1	1,080	800	700	610	560	—
2	1,040	770	670	590	540	—
3	1,000	740	650	560	520	—
4	940	690	590	530	490	—
Sweetgum, live-sawn						
0	900	680	550	500	460	440
1	840	630	510	460	430	410
2	770	580	470	430	400	380
3	700	530	430	390	360	350
4	620	460	380	340	320	300
Sweetgum, rift-sawn						
0	1,190	880	770	670	620	—
1	1,160	860	760	660	610	—
2	1,120	830	730	640	590	—
3	1,080	800	690	610	570	—
4	1,060	780	670	600	550	—
Yellow-poplar, live-sawn						
0	1,520	1,140	930	840	780	750
1	1,320	990	810	730	680	650
2	1,210	910	740	670	620	590
3	990	740	610	550	510	490
4	680	510	420	380	350	320

[1] Bolts were 26, 36, 43, and 72 inches long, and end-coated as soon as cut from the tree to inhibit checking.

[2] Lumber was cut 1-3/16 inches thick with a bolter saw having 5/16-inch kerf.

[3] The product was clear-one-face cuttings, 2 inches shorter than the bolt length, kiln-dried to 6-percent moisture content, random width (1-3/4 inches and wider), ready to glue into panels.

[4] Bolt grade 0 had no visible defects; No. 1 grade had visible defects confined to one quarter of the bolt, i.e., to one face; No. 2 grade had defects on two faces; grade No. 3 had defects on three faces; grade No. 4 had defects on all four faces.

[5] Sawn through-and-through, except a board was first sawn from one face and that face then turned down.

[6] Sawn to expose as many boards as possible showing edge grain.

Data on volume, and percent of total bolt volume, of clear-one-face and better furniture cuttings from graded white oak bolts of three lengths and several diameters are given in tables 12-32 and 12-33.

Koch et al. (1968) converted black cherry (*Prunus serotina* Ehrh.) bolts to 4/4 lumber and recovered about 42 percent of bolt cubic volume as lumber from 6-inch bolts, 50 percent from 10-inch bolts, and 57 percent from 14-inch bolts.

Huffman's (1973) study of red maple indicated that yields of lumber and furniture parts could be significantly increased by bucking long logs into short bolts for conversion on a bolter rather than on a conventional long-log mill (fig. 18-113).

Cooper and Schlesinger (1974) found that yellow-poplar bolts 4 to 6 feet long and 6 to 7 inches in diameter could be live sawn into 1-1/8-inch boards at the rate of 1.7 bolts per minute; with such bolts, sawing time per 1000 board feet of lumber was about 95 minutes including idle saw time while bolts were placed on the saw table.

In western North Carolina Dowdell and Plotkin (1975) compared a short-log bolter with a regular hardwood sawmill. The bolter-sawn lumber compared very favorably with conventionally sawn lumber in quantity and quality yield of dimension stock and furniture parts. They concluded that the operation must start at the tree-length log deck. High-quality sawlogs and veneer logs should be handled in the customary manner. Low-quality logs and pulpwood-size logs should be cut to rough dimension length (or multiples of this length) up to the maximum length that the bolter mill can handle. The diameter of the bolts should not exceed 14 inches. Ten to 14 inches was the optimum sawing size. The bolter mill should have a live infeed deck, a green chain to take away the sawn lumber, and an adequate sawdust removal system. This pilot project, and its logs and product values, suggested that a fully mechanized setup and minimum labor would break even at approximately 10,000 board feet of lumber production per day. Most of the lumber bolter-sawn was yellow-poplar. However, pine, oak, and maple were also produced. No difficulties were experienced in air-drying and kiln-drying the short lengths of lumber. To maximize the yield of furniture parts, there should be coordination between the lengths of these parts and the length of the bolts.

A subsequent study[9] with this bolter, in which 4- to 6-foot-long red oak bolts were sawn into 4/4 lumber, indicated that bolts for conversion by this method should have at least two clear faces and that cubic yield of lumber can be about 49 percent of cubic bolt volume (inside bark).

Morgan (1979) found that 5,535 cords of mixed hardwoods converted to 5/4 lumber on a bolter saw at Marimont Furniture Company, Marion, North Carolina, yielded an average of 498 board feet per cord. The lumber, when kiln-dried and cut into frame stock for upholstered furniture, yielded 50.9 percent frame stock—slightly more than the yield of frame stock from long-length No. 2 Common oak lumber (48.1 percent). In this study, only about half the bolts had

---

[9]Short log system. Unpublished undated report of a pilot project to determine the economic feasibility of operating a short log system in western North Carolina hardwoods. The project was a cooperative effort of the North Carolina Forest Service, the Appalachian Regional Commission, North Carolina State University, and U.S. Forest Service.

one or more clear faces. In a sample of 65 bolts—all but two red oak—which averaged 9.7 inches in diameter, 7 feet 2 inches long, and 319 pounds in weight (including bark), average yield per bolt was 71 pounds of slabs, 61 pounds of sawdust, and 187 pounds (25 board feet) of lumber; average moisture content of the bolts was 61.5 percent. With an experienced crew, production averaged about 1,000 board feet of lumber per operational hour.

Productivity of bolters can be increased by incorporating a continuous resaw system (fig. 18-123 bottom) to minimize the number of saw lines cut on the bolter.

18-12 RIPPING AND CROSSCUTTING LUMBER TO YIELD FURNITURE CUTTINGS

Furniture parts, i.e., cuttings of desired size that may be clear on one or both sides, are usually cut from rough or planed, kiln-dry, random-length, random-width lumber of specified standard lumber grades (see chapter 12). The portion of a furniture plant where this operation is conducted is termed a **rough mill**. Such lumber is graded visually although much work has been done to mechanize the process by developing computerized non-destructive methods of sensing, locating, and plotting defects (Hallock and Galiger 1971; Bulgrin 1974; McDonald 1975, 1978; Leslie 1976).

Lumber in standard grades is useful to wholesalers, procurement agents, and most others who engage in lumber commerce and use. Short logs obtainable from pine-site hardwoods yield lumber too short to qualify for the upper grades of wood as defined in these standard grading rules. It has been necessary to consider other ways to classify such short lumber. (See chapter 12; also Bingham and Schroeder's 1976 article on bolt and lumber grading.)

Although the need for standard grades of square-edged lumber is recognized, it would seem more direct to produce parts from wany-edged lumber live-sawn from bolts; Reynolds and Gatchell (1979) summarized the economics of this approach. It appears wasteful that the last sawmill operation performed on each board to be graded by standard rules is edging, and one of the first operations in the furniture plant's rough mill is re-edging each board while ripping it to desired widths.

It might also be argued that for certain products, such as oak flooring, cuttings should be made from green rather than kiln-dry lumber; this would greatly reduce kiln capacity needed and energy consumption during drying. Reynolds (1968) concluded from test data that it is possible to convert green, planed, oak lumber into flooring blanks and pallet parts whose grade when dry can be predicted with about 95 percent accuracy.

ROUGH MILL PROCEDURES

There are two distinct methods of converting lumber into usable parts or cuttings: crosscutting the board, followed by ripping to width; and gang ripping the board to width, followed by crosscutting to length. Many other techniques, such as gang crosscutting and selective ripping are also available. The first two methods predominate, however, crosscutting first most frequently—particularly on lumber that is twisted, bowed, or not flat (Raymond 1975).

Researchers are not in complete agreement on the merits of the two systems. Hall et al. (1980) compared yields of 5/4 northern red oak stair parts from rip-first and crosscut-first production lines. They found that No. 1 Common lumber could be processed through either line without affecting total yield of parts; there was a significantly higher (7 percent) yield of the longest cutting (49 inches) on the rip-first line, however.

Hallock and Giese (1980) mathematically modelled hardwood cutting yields by the gang-rip method and by the crosscut-first method. They found that gang ripping produced higher total yields; the gain tended to be in medium and shorter lengths. Their study involved very narrow cuttings (1 and 1-1/2 inches in width), thereby obtaining higher yields; but use of such narrow cuttings requires more adhesive in panel manufacture than the 2-1/2-inch-wide cuttings sawn by Hall et al. This difference in width of cuttings probably accounted for the differing conclusions of the two teams of researchers.

If lumber is flat and straight, the gang ripping method can be highly efficient; about 24,000 board feet of 4/4 lumber can be gang ripped per 8-hour shift at about 1 man-day per 1,000 board feet of cuttings. A crosscut-first mill may average about 600 board feet of cuttings per man-day. Percentage yield of parts by the two methods is considered about equal. Boards may be planed before or after gang ripping. Advocates feel that machine jams are avoided and handling is eased by planing first. Those who rip first believe they get more yield through more uniform planing of the ripped pieces. The flatness and straightness of the rough lumber is important to the decision on when to plane. In many operations, the gang ripsaw must produce a square smooth edge suitable for edge gluing without further smoothing. Gang ripping of uniform widths makes it possible to fingerjoint short unusable lengths into longer usable pieces. In conjunction with one or more crosscut systems, it is best suited to operations requiring 75,000 to 100,000 board feet of lumber daily. Its relative inflexibility makes it less useful to the manufacturer who requires solid parts in a wide range of widths and shapes (Raymond 1975).

The gangsaw-rip-first system.—Lucas and Araman (1975) compared the advantages and disadvantages of four manufacturing sequences for **interior furniture parts**, i.e., those structural parts of a furniture frame that are hidden from view. Because their primary function is structural, they admit sound defect except in the outer inch of either end where machining for joining is necessary. For a given line of furniture, most interior parts have the same width and thickness, varying only in length. In three of the sequences evaluated by Lucas and Araman, gang ripping was the first step; in the fourth, the lumber was crosscut first. Though the grade of lumber used affected the percentage yield of parts, the manufacturing sequence used did not—but it could effect the cost per part. They concluded that the selection of the best method should be based on factors other than parts yield. In an automated mill described by Araman and

Lucas (1975) and Araman (1977), lumber is planed on both sides and gangsaw ripped into full-length strips of standard width; the strips are marked for defect location with an electrolytic solution, then crosscut to remove defects and to desired lengths (fig. 18-124). Requiring only three workers, the mill was designed to reduce the need for human decisions and to process nearly 6,000 board

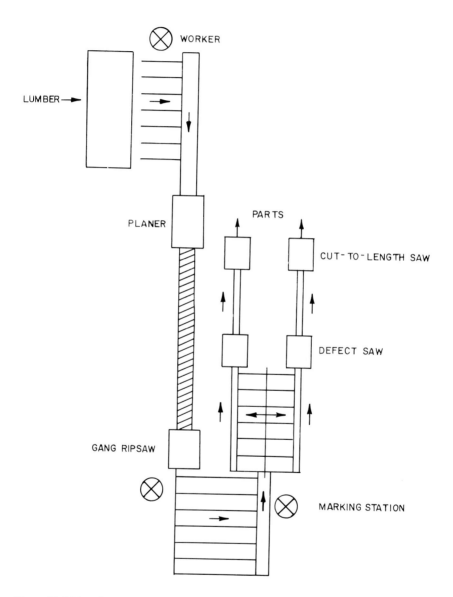

Figure 18-124.—Flow chart of a gangsaw-rip-first system for producing interior hardwood furniture parts. (Drawing after Araman 1977.)

feet of No. 1 Common or 2a Common yellow-poplar lumber per day while yielding about 4,000 board feet of finished parts per day, as follows.

Lumber grade	Rough lumber processed	Yield of finished parts
	----------Board feet----------	
No. 1 common................	5,700	4,400
No. 2A common	5,500	3,900

In this analysis, all strips were 1-11/16 inches wide, with part lengths from 13 to 73 inches.

Ward (1975ab) described and illustrated a highly automated industrial operation resulting from the above work featuring a spiral cutoff saw for crosscutting strips while they are in motion. A planer used in rough mills of this type is illustrated in figure 18-133 (or 18-138), the gang ripsaw in figure 18-79, and the defect saw in figure 18-94.

Defect marking with an electrolytic solution and saws that crosscut automatically to fill cutting schedules are described by Ward (1972).

The STUB system of crosscutting low-grade lumber first.—To yield a variety of widths of cuttings that are clear on one face, Lucas and Gatchell (1976) devised a manufacturing system for low-grade lumber they designated *STUB* (*S*hort *T*emporarily *U*pdated *B*oards). The STUB system starts by crosscutting long boards into short boards, but here the similarity between the STUB rough mill and the conventional roughmill ends. Instead of producing cut-to-length dimension, the crosscut operation produces three random-length products: STUB boards, residual boards, and waste (fig. 18-125). STUB boards differ from standard hardwood lumber boards in that each STUB board has at least one clear-one-face cutting *of a specified minimum width* that runs its full length. The crosscut operator has only three restrictions to observe:

1. The minimum width of clear area extending the full length of the STUB board.
2. The minimum length of the STUB board.
3. The maximum length of the STUB board.

The crosscut operator has no specific cutting lengths to be concerned about; he need only develop a mental image of the width and length of the minimum clear area. The crosscut operation consists of a station where the operator marks the board with a conducting type of electrolytic solution at the places where he wants the board to be crosscut; following this is an automatic mark-sensing crosscut saw.

Resulting STUB boards are inventoried by length classes—perhaps 2- to 4-inch classes. The rough-mill foreman will draw from this STUB inventory to satisfy his requirements, matching the STUB boards with the specific cutting lengths. The selected STUB boards will then be sent to the ripsaw where the full-length cuttings will be ripped out. Since the ripsaw operator is concerned only about recovering the full-length cuttings, he should be able to perform efficiently and consistently.

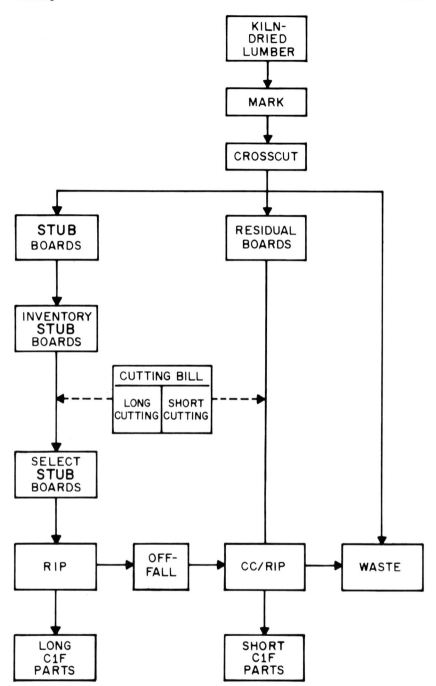

Figure 18-125.—Flow diagram for the STUB rough-mill line. CIF means clear one face; CC means crosscut. (Drawing after Lucas and Gatchell 1976.)

Processing of the residual boards—a salvage operation—occurs on another line and operates much the same as a conventional rough mill; that is, specified part lengths are crosscut directly.

Hafley and Hanson (1973) describe how to determine the optimum sequence for filling cutting bills in a conventional rough mill.

Mullin (1975) concluded that cut-off saw operators should produce 5,000 to 8,000 board feet of 4/4 lumber per day and that crosscutting efficiency can be improved by reducing complexity of decisions required; color-coded back gauges are a help and a color-coded gauge on the infeed table showing combinations of lengths needed is also useful. Mullin found that use of such color-coded tapes increased cutting yield by 4-1/2 percent. For further descriptions of systems using computer-generated gauge tapes, see Anonymous (1977a) and Helmers (1975).

YIELD OF CUTTINGS

It is beyond the scope of this text to present comprehensive yield data for various cutting methods and schedules as affected by lumber grades, but a sampling of the literature is abstracted here, in chapter 12, and in chapter 27.

For nomograms to predict furniture clear cutting yields from standard National Hardwood Lumber Association lumber grades *when the lumber is processed by gang ripping first* see figures 27-14AB through 27-18AB.

While pine-site hardwoods yield much No. 2 common lumber (see ch. 12), this grade is little used by the furniture industry when they are seeking cuttings longer than 40 inches. Lucas (1973) concluded, however, that if cuttings sought were in 10 widths (1.5- to 6-inch width by half-inch increments) and 10 lengths, No. 2 Common oak lumber could yield substantial quantities of cuttings 40 to 60 inches long, as follows:

Cutting length	Yield of clear-one-face cuttings
Inches	*Percent of total board volume*
60	17.4
48	7.7
44	4.3
40	4.2
36	4.2
32	3.5
30	2.3
24	4.7
22	2.7
18	4.0
Total	55.0

Pine-site hardwoods are much used for pallet parts, and Pepke et al. (1977) found that many pieces of high quality wood could be selected from cut stock in pallet plants. Such clear material when kiln-dried yielded 60 to 87 percent cuttings for drawer sides.

Cooper and Schlesinger (1974) thinned 59 yellow-poplar trees from a 30-year-old plantation in southern Illinois. All were at least 6 inches in dbh and of low quality; of 349 bolts measuring at least 5.5 inches dib at the small end and 2 to 6.5 feet long, only 17 had at least one face clear. All bolts were live-sawn into 1-1/8-inch-thick boards, dried, and skip-planed to 15/16-inch thickness. These dry boards were then diagrammed into clear-one-face cuttings 1 to 6 iches wide and 12 to 75 inches long that could be obtained by ripping, then crosscutting, and then re-ripping if required. The square feet of clear-one-face cuttings per tree (Y) were as follows (the average was 74 percent of volume computed by the International 1/4-inch log scale):

$$Y = 0.00565 \times dbh^{2.6810} \times height^{0.7397} \quad (18\text{-}35)$$

where dbh is measured in inches and height to a 5.5-inch top dib in feet. For example, a 10-inch tree measuring 30 feet high to a 6-inch top dib, yielded 33.5 square feet of clear-one-face cuttings.

Dimension recovery factor (Y_1), which equals the square feet of clear-one-face cuttings divided by total cubic feet in the tree to a 5.5-inch top, was as follows:

$$Y_1 = \frac{Y}{0.01444 \times dbh^{1.9568} \times height^{0.6076}} \quad (18\text{-}36)$$

For example, dimension recovery factor for a 10-inch tree measuring 30 feet high to a 5.5-inch dib top was 3.3.

Individual bolts from these trees yielded square feet of clear-one-side cuttings (Y_2), as follows:

$$Y_2 = 0.0001919 \times diameter^{2.4687} \times length^{1.3123} \quad (18\text{-}37)$$

For example, a bolt 8 inches in diameter inside bark at the small end and 48 inches long yielded 5.2 square feet of clear-one-face cuttings.

Nearly half the cuttings from these 59 yellow-poplar trees were 37 to 75 inches long; predominant widths were 1.6 to 3.5 inches. Yield of clear-two-face cuttings from these trees (2,451 square feet) was only slightly less than that of clear-one-face cuttings (2,474 square feet).

Landt (1974) felled 42 small low-quality red maple trees in an improvement cut of an 18- to 44-year-old stand in southern Illinois and bucked them into 302 short bolts 2 to 6.5 feet long measuring at least 6 inches in diameter inside bark at the small end. The bolts were live-sawn into 1-1/8-inch-thick lumber, dried, skip-planed to 15/16-inch thickness, and measured for yield of clear-one-face cuttings 1 to 6 inches wide and 12 to 72 inches long that could be obtained by ripping, then crosscutting, and then re-ripping if necessary. The square feet of clear-one-face cuttings per tree (Y) were as follows:

$$Y = 0.009326 \times dbh^{2.09589} \times height^{1.03289} \quad (18\text{-}38)$$

For example, a 10-inch tree measuring 30 feet high to a 6-inch top dib, yielded 39.0 square feet of clear-one-face cuttings.

Dimension recovery factor (Y_1), which equals the square feet of clear-one-face cuttings divided by the total cubic feet in the tree to a 6-inch top, was as follows:

$$Y_1 = \frac{Y}{0.008293 \times dbh^{1.81788} \times height^{0.86923}} \quad (18\text{-}39)$$

For example, dimension recovery factor for a 10-inch tree measuring 30 feet high to a 6-inch top was 3.7.

Yield of clear-one-face cuttings from individual red maple bolts (Y_2, square feet), was as follows:

$$Y_2 = 0.000163 \times diameter^{2.66086} \times length^{1.262316} \quad (18\text{-}40)$$

For example, a bolt measuring 8 inches in diameter at the small end and 48 inches long had a yield of 5.5 square feet of clear-one-face cuttings. Over half (61 percent) of the cuttings from these 42 red maple trees were in the 2- to 3-inch width class. Major length classes were as follows:

Length	Percent
24 inches and less	41
36 to 60 inches	52

Use of the data collected by Cooper and Schlesinger (1974) on yellow-poplar, and by Landt (1974) on red maple has been simplified through development (Rosen et al., 1980) of nomographs that predict yields of cuttings, as follows (based on first ripping, then crosscutting, and then re-ripping if required):

Species and type of cutting	Figure number
Yellow-poplar	
Clear two-sides	18-126
Clear one-side	18-127
Sound character-marked	18-128
Red maple	
Clear two-sides	18-129

The three yellow-poplar nomographs predict yields of three grades of cuttings from bolts 2 to 7 feet in length and 5 to 18 inches in diameter. The red maple nomograph is for clear-two-side cuttings from bolts 2 to 7 feet in length and 5 to 18 inches in diameter; the same nomograph predicts yield of clear one-side and sound character-marked cuttings without much error.

To use the charts, first locate the maximum cutting length required, sometimes called the primary length, on the right-hand side of the chart. The percent yield, in surface area of a 1-inch-wide cutting, is found by moving horizontally to the left until the point of intersection with the percent yield scale on the far left. This is the percent yield of a 1-inch-wide cutting of the primary length. To find the percent yield of the second longest cutting needed, start at the point on the right corresponding to the primary cutting length and proceed vertically to the intersection with the line corresponding to the length of the second cutting. Now move horizontally to the percent yield scale. The percent yield of the second longest cutting, given the removal of the longest cuttings, is obtained by

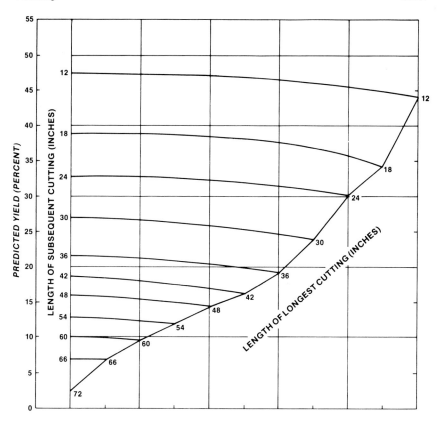
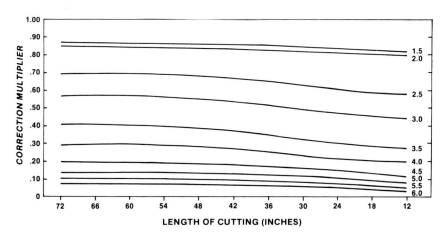

Figure 18-126.—Nomograph to predict **clear-two-side cuttings** from 2- to 7-foot-long **yellow-poplar** bolts 5 to 18 inches in diameter from southern Illinois. (Drawing after Rosen et al. (1980).)

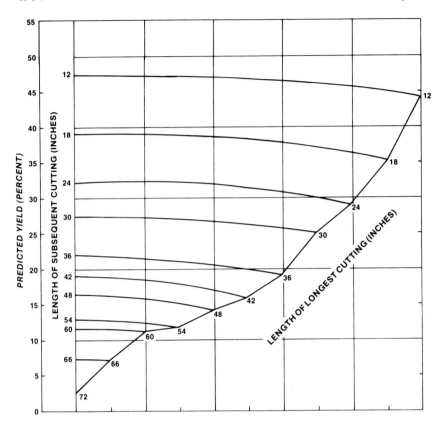

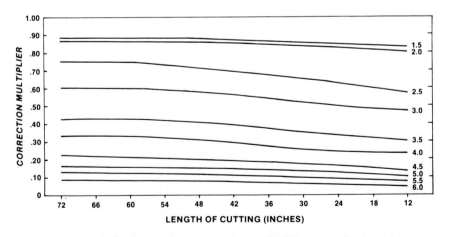

Figure 18-127.—Nomograph to predict **clear-one-side cuttings** from 2- to 7-foot-long **yellow-poplar bolts** 5 to 18 inches in diameter from southern Illinois. (Drawing after Rosen et al. 1980.)

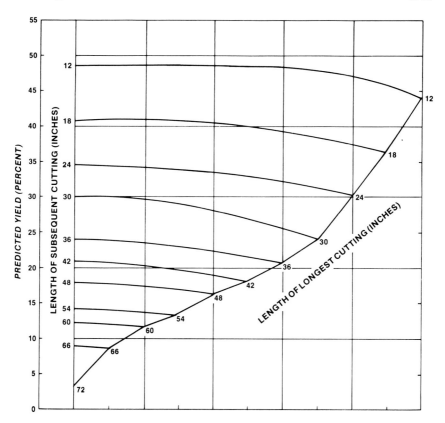
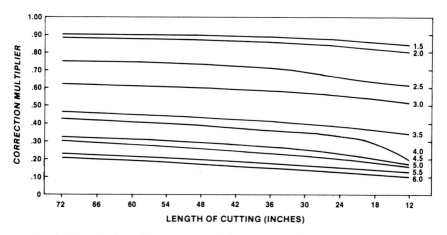

Figure 18-128.—Nomograph to predict sound **character-marked** cuttings from 2- to 7-foot-long **yellow-poplar** bolts 5 to 18 inches in diameter from southern Illinois. (Drawing after Rosen et al. 1980.)

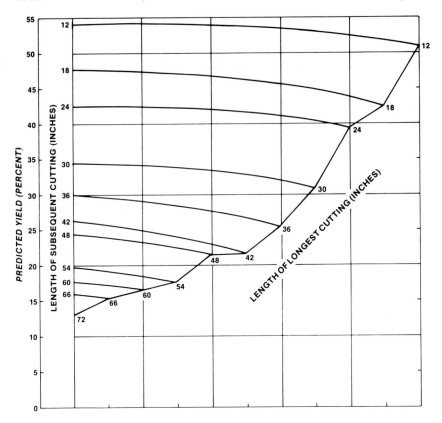
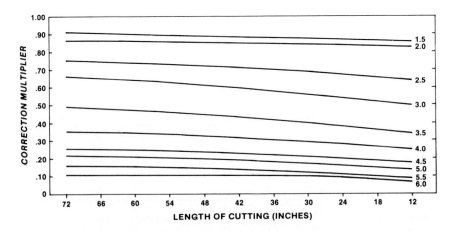

Figure 18-129.—Nomograph to predict **clear-two-side** cuttings from 2- to 7-foot-long **red maple** bolts 5 to 18 inches in diameter from southern Illinois. (Drawing after Rosen et al. 1980.)

subtracting the percent yield for the longest cutting from the percent yield for the second longest cutting. Other yields are obtained in a similar fashion always starting at the point on the right corresponding to the primary cutting length proceeding vertically to the line corresponding to the next desired length and then horizontally to the percent yield scale. The percent yield for each new length is found by subtracting the unadjusted percent yield of the previous cutting length from the percent yield of the cutting length in question.

For example, if the cutting bill called for red maple, clear-two-sides, 2-foot minimum bolt length, 1-inch-wide cuttings of lengths 48, 24, and 12 inches, the percent yield of 48-inch cuttings would be 21 percent (fig. 18-129). For 24-inch cuttings, the percent yield would be 21 percent (42 percent - 21 percent), and finally the percent yield of 12-inch cuttings would be 12 percent (54 percent - 42 percent). The total percent yield of all cuttings would be 54 percent (which is also the accumulated percentage of the shortest length) or 540 square feet of surface area per 1,000 board feet of bolts, scaled according to the International 1/4-inch Rule, brought into the mill.

As mentioned earlier, the yield charts represent 1-inch cutting widths. To determine the percent yield of cuttings other than 1-inch-wide, a correction multiplier chart is given. The percent yields for cuttings greater than 1 inch are found by locating the length of the desired cutting at the base of the correction multiplier chart. Then move vertically to the intersection with the line for needed width and then horizontally to the correction multiplier scale. Multiply the correction multiplier by the present yield, for a 1-inch-wide cutting of the required length, to obtain the percent yield for the needed width.

For example, given a cutting bill as before, but with a varying width requirement of 2.5 inches for 48-inch cuttings, 3 inches for 24-inch cuttings, and 2 inches for 12-inch cuttings, take the percent yield before subtraction for previous lengths and multiply it by the width correction multiplier of the length needed (fig. 18-129). Now the percent yield of 48-inch cuttings is 15 percent, the original percent yield of 21 percent times the correction multiplier for a 2.5-inch-wide, 48-inch-long cutting of 0.72. For the 3-inch-wide, 24-inch cutting, the percent yield is now 8 percent, the original 42 percent times the correction of 0.54 and finally minus 15 percent for the 48-inch cuttings already removed. Similarly, the percent yield for the 2-inch-wide, 12-inch cuttings is 21 percent, 54 percent times 0.82 minus 23 for the 48- and 24-inch cuttings already removed. The total percent yield of all desired cuttings is 44 percent or 440 square feet per 1,000 board feet of bolts, as scaled by the International 1/4-inch Rule, brought into the mill.

Yields are easily converted from percentages to number of cuttings per thousand board feet by dividing the percent yield by 100, multiplying by 1,000, and dividing by the area of the cutting surface in square feet.

In North Carolina, Page and Rodenbach (1962) measured the yield of clear cuttings for exposed furniture tops, end panels, and drawer fronts from red maple and hard maple. Also measured were yields of usable interior cuttings for drawer sides, bottom rails, and partition rails. Results were as follows:

Lumber grade	Clear-two-face cuttings	Usable interior cuttings	Total yield
	---------Percent of total lumber--------		
No. 1 Common	56	15	71
No. 2 Common	15	46	61

COMPUTER-ASSISTED OPTIMIZATION OF CUTTINGS

The rapidly developing technology of applying scanners and computers to help saw operators optimize cutting cannot be fully described here. Early work included that by Dunmire and Englerth (1967), Huber (1971), Hallet (1972), and Hanover et al. (1973). Applications of computer-generated paper gauge tapes are described by Helmers (1975). Stern and McDonald (1978) provided a computer program optimizing yield from multiple-ripped boards; solutions include rip cut and crosscut locations as well as percent yield of clear cuttings. Huber and Harsh (1977) described a rough mill improvement program which utilizes computer analysis to select lumber grades and sizes that will most cost-efficiently yield desired cutting bills. Lucas and Catron (1973) provided a computerized data bank describing the defects in No. 2 Common oak boards according to location, size, and type; also recorded are the length and width of the 637 boards analyzed. Such data are useful in developing computerized cutting schedules for No. 2 Common lumber; over half the hardwood lumber produced in sawmills of the Appalachian Region is No. 2 Common and below (Lucas 1973).

It seems likely that the trend toward computerized least-cost solutions to cutting bills in rough mills will continue. The capital costs of such sophisticated rough mills may, however, result in their removal from furniture plants into separate enterprises. It seems entirely possible that an optical image analyzer—developed for space exploration—could digitize defect size and location in a board. This digitized information could drive a machining table along X-Y coordinates (fig. 18-234) so that a laser (fig. 18-232) could saw out and optimize yield of clear-one-face parts called for in a cutting bill. The ability of a laser to make blind cuts without completely traversing board width or length should significantly increase achievable yield. Moreover, operator decisions would be minimized. McMillin (1982) further described such a system, and Huber et al. (1982) concluded that it should be economically feasible.

Readers needing a status report on other systems for detection of defects in lumber should find Szymani's (1979) review useful.

18-13 JOINTING, PLANING, MOULDING, AND SHAPING

Wood is smoothed on one surface by jointing, thicknessed by planing, machined into patterns by molding, and edge-contoured by shaping. All of these processes are accomplished by peripherally milling cutterheads (see sec. 18-4 through 18-7).

Among-species variation in machinability is discussed in section 18-1. While some species are more easily machined under a wider range of conditions than others, virtually any wood can be machined without serious defect if cutting conditions are made optimum for that species. (See *defect-control* discussion related to tables 18-32 through 18-35.)

JOINTING

As used in this text section, **jointing** is defined as the smoothing or trueing of one surface of a board, timber or veneer by means of a single peripheral milling head; almost without exception these cutterheads are revolved against the feed in up-milling mode. Thicknessing is not the function of a jointer.

Purpose of jointing.—Primary functions performed on jointers are as follows:

- To smooth a surface into a true flat plane that requires no further machining.
- To machine a surface relatively flat, i.e., to remove warp, twist, and cup from a distorted surface by machining off the high spots. When a jointer is used for this purpose, the wood is usually resurfaced on a thicknessing planer prior to some subsequent operation.
- To reduce the thickness of a previously thicknessed workpiece.

In addition to these production functions, a simple jointer (fig. 18-130) can be used in a small shop to perform a variety of jobs, such as squaring stock, end planing, taper planing, rabbeting, or beveling.

Machine types.—To accomplish the foregoing functions, four types of jointers have been developed.

The **hand-fed jointer** (fig. 18-130) carries a cutterhead, generally direct-motor driven at 3,600 rpm, mounting two, three, or four knives. The depth of cut required to smoothly joint a flat surface is determined by its roughness. On well-sawn surfaces of headrig-cut lumber about 3/32-inch depth of cut is required to remove sawtooth witness marks and evidence of torn grain. Poorly sawn lumber with deep torn grain may require twice this depth of cut. Edges of dry lumber ripped on a planer-type saw (fig. 18-80) may need no jointing.

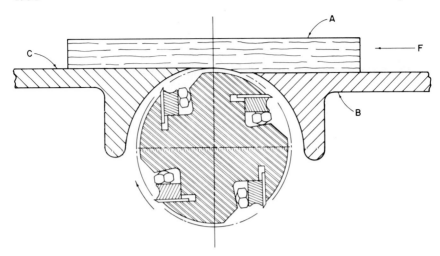

Figure 18-130.—Jointer smooths but doesn't plane to thickness. (A) Workpiece. (B) Infeed platen vertically adjustable to define the depth of the cut. (C) Outfeed platen with surface tangent to cutting circle.

A **glue jointer** machine edges of dry lumber to prepare them for glue-assembly into panels. As pieces going into the machine may be tapered in width, the two edges must be machined in two separate operations. Wood emerging from the machine will still be tapered in width but with edges planed. This is accomplished by feeding the lumber on edge across a jointing head, turning the other edge down, and then returning it across a second jointer head. On both passes the lumber is guided in a straight line and firmly secured between yielding pressure rolls and an endless chain that acts as a vertical reference plane. The same endless chain is utilized on both passes (fig. 18-131). In production operations, the machine is equipped with a hopper feed at the infeed and an automatic board turner and feeder at the other end of the machine to reintroduce the piece on its second pass.

The machine illustrated in figure 18-131 will handle stock up to 4-1/2 inches thick and has a speed range from 33 to 100 fpm. The feed works is powered by a 5-hp motor. The two jointer heads are direct-connected to separate 5-hp, 3600-rpm motors. When glue-jointing 1-1/4-inch white oak, the machine might typically carry two 7-1/4-inch-cutting-circle round cutterheads, each mounting six jointed[10] knives having a 20° rake angle and a 20° clearance angle, and might be fed at 67 fpm to yield approximately 26 knife cuts per inch. Thus, at 75 percent efficiency, 24,000 lineal feet of stock could be fed through the machine in one 8-hour shift.

While this type of machine is principally used to make straight glue joints, it is also capable of cutting rabbets or moulded edge patterns. The two cutterhead spindles may each be tilted as much as 15° in either direction to cut bevels.

Many peripheral-milling glue jointers are in operation, but development of straight-line ripsaws equipped with carbide teeth (fig. 18-80) have made them largely obsolete.

---

[10]Here the term "joint" is used in the sense defined in figure 18-31.

Machining 1991

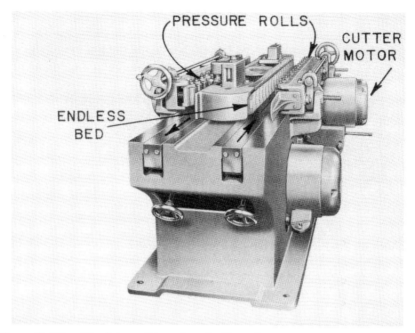

Figure 18-131.—Glue jointer. (Photo from Diehl Machines, Inc.)

The **traveling-head veneer jointer** (fig. 18-132) machines a glue joint on the edges of long veneer sheets. To assure superior jointing on veneers over 6 feet long, the veneer is held stationary on edge, and the cutterhead travels. To accomplish this, a clamp secures vertical bundles of veneer up to 3 inches thick, after which the traveling cutterhead joints the exposed lower edge of the bundle in a perfectly straight line parallel to veneer grain; depth of cut is usually about 0.03 inch. Standard clamp lengths are 102, 126, and 150 inches. The veneer to be jointed is placed between the clamps and rests on pins which extend across the gap under the clamps. After closing the clamps on the vertically oriented bundle of veneer, the operator retracts the supporting pins to clear a path for the cutterhead. The traveling head assembly incorporates a 3,600 rpm, 7-1/2-hp, 8-knife, 3-inch-wide cutterhead with 7-1/2-inch cutting circle. The 1-1/2-hp feed motor advances the traveling head assembly at about 40 fpm on the cutting stroke, producing a surface with nearly 60 knife marks per inch. An automatic motor-driven, glue-applicator wheel can be attached to the traveling head assembly to spread glue as the cutterhead assembly returns to start position.

A **facing jointer** machines one surface of a board relatively flat by smoothing the high spots to reduce warp, twist, and cup. It is accomplished by mechanical feeding fingers that more or less duplicate the action of hand feeding the workpiece across a simple jointing head. The most simple configuration of this machine consists of an upper endless bed carrying flexible-rubber or spring-tensioned steel fingers mounted over a simple jointer as illustrated in figure 18-130. This face-jointing operation is normally followed by thicknessing on a top-head planer prior to other remanufacturing operations. The facing jointer is

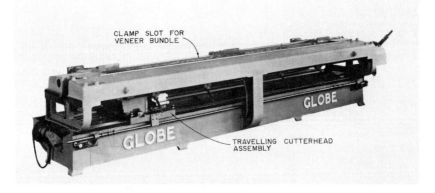

Figure 18-132.—Travelling-head veneer jointer, operator's side. Veneer bundles—visualize a deck of cards on edge—are inserted in the clamp slot where gravity brings lower edges flush. (Photo from Globe Machine Mfg. Co., Inc.)

frequently modified to incorporate a top thicknessing head following the bottom jointer head (fig. 18-133) to bring the jointed board to oversize thickness prior to double surfacing in a second operation. The board must not be forcibly flattened, but should be fed with a minimum of flexing so that the cutterhead alone accomplishes the flattening effect. A maximum of 1/4-inch warp or twist can be eliminated by this process. Facing jointers equipped with thicknessing heads are available in widths from 18 to 52 inches and will accept lumber from 1/2-inch to 8 inches thick. The feeding mechanism will handle minimum lengths of 16 inches if the stock is fed intermittently or 12 inches if the feed is continuous and each short piece is followed by a longer piece. The cutting-circle diameter of both top and bottom heads is 6 inches. Each 4-knife head (not jointed) is belt driven at 3,600 rpm from a separate 40- to 75-hp motor. The 5-hp feedworks is driven by a variable-speed transmission to provide feed speeds up to 120 fpm. Increased production can be obtained by feeding the stock in multiple widths. In a typical rough mill operation handling southern hardwoods, the machine might be fitted with knives having a 20° rake angle, and a 20° clearance angle. Feed speed is usually 85 to 100 fpm. Thus, if stock averaged 8 inches in width and was fed continuously at 75 percent efficiency, the output per 8-hour shift on 1-inch stock would be approximately 22,000 board feet. Since 1970 facing jointers equipped with thicknessing heads have been routinely fitted with carbide cutting edges.

PLANING

The peripheral milling of lumber to smooth one or more surfaces with simultaneous sizing to some predetermined thickness or profile pattern is defined as **planing**. Planing cutterheads, often called **cylinders**, usually cut in the up-milling direction (fig. 18-31).

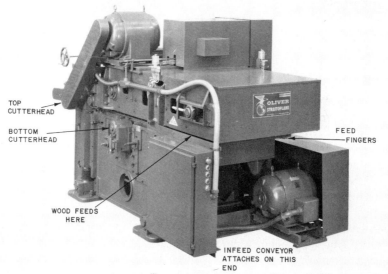

Figure 18-133.—Facing jointer followed by, and integrally combined with, a top thicknessing head. (Top) Spring-loaded fingers designed to feed lumber over direct-driven bottom head visible at right. (Bottom) Infeed end of machine; bottom cutterhead is at mid-length of the machine; the top cutterhead is near the outfeed end and is belt driven. (Photos from Oliver Machinery Company.)

Single surfacer.—This most elementary planer smooths one side of a board while removing enough wood to reduce it to a predetermined thickness (fig. 18-134). The depth of cut that must be removed to smoothly plane a flat board is determined by the roughness of the sawn surface or, when planing glued-up panels, by the amount of mis-match between adjacent boards. Shallow cuts cause fewer defects than deep cuts (table 18-32).

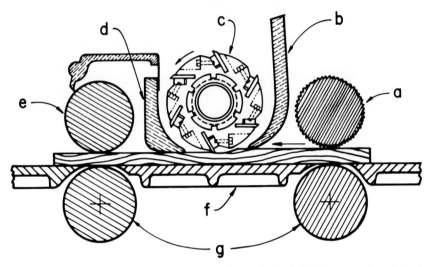

Figure 18-134.—Single surfacer planes to thickness desired. (a) Corrugated top infeed roll. (b) Chipbreaker. (c) Cutterhead. (d) Pressure bar. (e) Top outfeed roll. (f) Lower platen. (g) Bottom feed rolls.

All industrial planers are power-fed to maintain a uniform feed rate, reduce breakage, and avoid stoppages that cause burns from cutterheads and skidding rolls. Effectiveness of the feed is improved if both top and bottom rolls are power driven, and if the rolls are large in diameter, corrugated, and mounted in multiples—that is, two pairs of infeeding rolls are more effective than one pair. Ordinarily, the top and bottom rolls are both solid. If, however, the lumber is very uneven in thickness, or if it is fed in multiples across the width of the machine, the top roll may carry independent narrow sections mounted on a common arbor but spring-supported on an internal arbor; in this way each section can yield as much as 3/8-inch independently (fig. 18-135). To attain positive feed of rough or wet lumber, the bottom infeed roll should be corrugated and set a fraction of an inch above the platen; with dry, smooth, flat lumber the infeed roll should be smooth and raised barely above the platen. Elastomeric top feed rolls are sometimes used to increase roll traction and to accommodate variations in board thickness.

In all planers a **chipbreaker** precedes the cutterhead (fig. 18-136). It holds the lumber firmly against the opposite **platen** limiting the cut to yield an accurate board thickness. Pressure of the tips of properly designed chipbreaker shoes

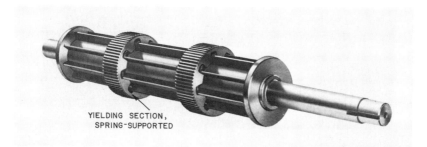

Figure 18-135.—Sectional top infeed roll. (Photo from Newman Machine Company, Inc.)

helps reduce advance splitting (fig. 18-3), permitting knives with higher rake angles and accompanying lower power consumption. The front face directs chips into the shavings collector pipe. Chipbreakers also minimize gouged ends or **snipes** when board-ends enter or leave the cutting zone out of control. In cruder designs the chipbreaker is simply a weighted bar resting on the lumber. In more sophisticated machines the bar is divided into counterbalanced hinged sections, permitting each chipbreaker tip to hug the cutting circle as it rises and falls with varying stock thickness (in some designs these thickness variations are sensed by movement of the infeed rolls.) Shoes also rock to follow the longitudinal undulations of the lumber. On very fast planers, pressure on the individual shoes may be regulated by air cylinders to suit lumber conditions. This arrangement also permits remote control of pressure and quick lifting of the chipbreaker assembly in the event of a breakup.

The opening between the cutting circle and the platen can be adjusted to yield lumber of desired thickness by lifting the top assembly (fig. 18-134) on multiple hoist screws or, in a preferable design, by raising or lowering the bottom platen assembly on inclined ways. In the latter design, however, it is difficult to utilize infeed conveyors effectively because the height of the bottom platen of the planer varies in relation to the infeed conveyor forcing boards to flex as they pass into the planer.

Rotating cutterheads in planing machines may be as short as 4 inches or as long as 54 inches depending on the work to be done. Cutterhead diameter depends on the number of knives to be fitted into the cutting circle. The 4- to 8-knife cabinet planers used on southern hardwoods use cutting circles from 6 to 9-1/4 inches in diameter. High-speed planers for oak flooring carry 8 to 20 knives and require cutting circles from 9 to 12 inches in diameter. As explained in section 18-4, the number of knife marks per inch substantially defines the quality of a planed surface (table 18-65 and equation 18-7). Large cutting circles improve surface quality (equation 18-8); however, the desirability of compact planer designs and inability to handle hardwood flowing at speeds much above 500 fpm have, so far, limited cutterhead diameters to about 12 inches.

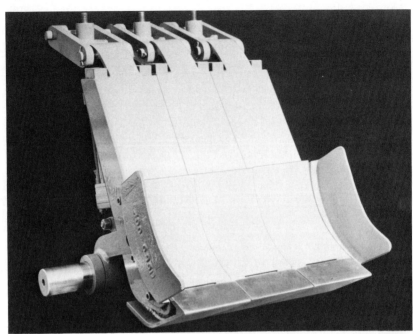
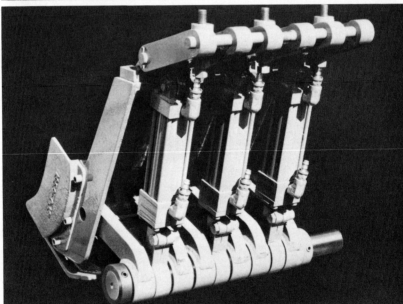

Figure 18-136.—Chipbreaker assembly from heavy-duty planer. (Top) Sectional rocking shoes and shaving deflector. (Bottom) Air cylinders attached to each shoe permit pressure adjustment to suit lumber being planed. Air lift of entire assembly eases cleanout of the machine. (Photos from Stetson-Ross.)

A **pressure bar** behind the cutterhead is common to all planers (fig. 18-134). The lower surface of this bar is adjusted parallel to the opposite platen and 0.001 to 0.003 inch above tangency to the cutting circle. It holds the lumber down as it passes through and leaves the cutting zone. Some designs feature a quick-release mechanism to assist in clearing jams or breakups in the machine.

Depending on the machine design, the opening between cutterhead and lower platen is set to yield the desired board thickness by adjusting either the cutterhead or the lower platen.

Thin flexible material such as tempered hardboard is planed on a specialized type of single surfacer designed to prevent fluttering of such material in the cutting zone (fig. 18-137). The sheet material to be planed is introduced to and removed from the machine on belts running over the infeed and outfeed tables. In the cutting zone, the workpiece is driven by two pairs of powered rolls. The lower rolls are chrome plated and smooth, the upper ones sectional and rubber-faced. The cutterhead is equipped with as many as six carbide knives and rotates at 4500 rpm with the feed (down-milling). Guide shoes on both sides of the cutterhead are not in contact with the workpiece. The thin and flexible workpiece is held down during the sizing operation by means of a perforated vacuum table immediately under the cutterhead. If the carbides knives are jointed, a feed speed of 150 fpm would produce 15 knife marks per inch; if unjointed, the feed rate would drop to 25 fpm for the same number of knife marks per inch.

Double surfacer.—The double surfacer is a key machine in most factories processing hardwood furniture parts and glued-up panels; it smooths both sides of a board or panel and simultaneously reduces it to a predetermined finished thickness. The arrangement shown in figure 18-138A is used on hardwoods virtually to the exclusion of other systems in spite of its obvious shortcomings. The board is forced against the lower bed first, and the top cylinder removes the excess thickness, if any, to finish the top. The lower cylinder then takes a fixed cut from the lower side of the lumber regardless of whether the top has been surfaced or not. If the board is too thin to be surfaced on the top, it will be made still thinner by the cut from the lower cylinder. The top cylinder machines the top of the lumber, while the lumber is still rough on the bottom. With this arrangement, boards are usually fed best-face-down.

The arrangement in figure 18-138B is used on facing planers, where it is desirable to plane off some degree of cup and twist. Because of the yielding hold-down device over the bottom cutterhead, however, it is not used on production planers where a good surface must be established in one pass through the machine; sudden excessive forces normal to the surface moving the board away from the cutterhead leave the surface uneven.

TABLE 18-65.—*Relationship of planer feed rate and number of knives to knife cuts per inch; spindle speed 3,450 rpm*

Lineal feed rate (f.p.m.)	Number of knives in cutterhead[1]									
	1	2	4	6	8	10	12	14	16	18	20
	----Knife marks per inch----									
10	28.8									
15	19.2									
20	14.4	28.8								
25	11.5	23.0								
30	9.6	19.2								
35	8.2	16.4								
40	7.2	14.4	28.8							
45	6.4	12.8	25.6							
50	5.8	11.5	23.0							
60	4.8	9.6	19.2	28.8						
70	4.1	8.2	16.4	24.6						
80		7.2	14.4	21.6	28.8					
90		6.4	12.8	19.2	25.6					
100		5.8	11.5	17.3	23.0	28.8				
110		5.2	10.5	15.7	20.9	26.1				
120		4.8	9.6	14.4	19.2	24.0	28.8			
130		4.4	8.8	13.3	17.7	22.1	26.5			
140		4.1	8.2	12.3	16.4	20.5	24.6	28.8		
150			7.7	11.5	15.3	19.2	23.0	26.8		
160			7.2	10.8	14.4	18.0	21.6	25.2	28.8	
170			6.8	10.1	13.5	16.9	20.3	23.7	27.1	
180			6.4	9.6	12.8	16.0	19.2	22.4	25.6	28.8
190			6.1	9.1	12.1	15.1	18.2	21.2	24.2	27.2
200			5.8	8.6	11.5	14.4	17.3	20.1	23.0	25.9	28.8
220			5.2	7.8	10.5	13.1	15.7	18.3	20.9	23.5	26.1
240			4.8	7.2	9.6	12.0	14.4	16.8	19.2	21.6	24.0
260			4.4	6.6	8.8	11.1	13.3	15.5	17.7	19.9	22.1
280			4.1	6.2	8.2	10.3	12.3	14.4	16.4	18.5	20.5
300				5.8	7.7	9.6	11.5	13.4	15.3	17.3	19.2
325				5.3	7.1	8.8	10.6	12.4	14.2	15.9	17.7
350				4.9	6.6	8.2	9.9	11.5	13.1	14.8	16.4
375				4.6	6.1	7.7	9.2	10.7	12.3	13.8	15.3
400				4.3	5.8	7.2	8.6	10.1	11.5	12.9	14.4
425				4.1	5.4	6.8	8.1	9.5	10.8	12.2	13.5
450					5.1	6.4	7.7	8.9	10.2	11.5	12.8
475					4.8	6.1	7.3	8.5	9.7	10.9	12.1
500					4.6	5.8	6.9	8.1	9.2	10.4	11.5
550					4.2	5.2	6.3	7.3	8.4	9.4	10.5
600						4.8	5.8	6.7	7.7	8.6	9.6
700						4.1	4.9	5.8	6.6	7.4	8.2
800							4.3	5.0	5.8	6.5	7.2
900								4.5	5.1	5.8	6.4
1,000								4.0	4.6	5.2	5.8
1,100									4.2	4.7	5.2
1,200										4.3	4.8
1,300										4.0	4.4
1,400											4.1

[1] Assumes knives are jointed.

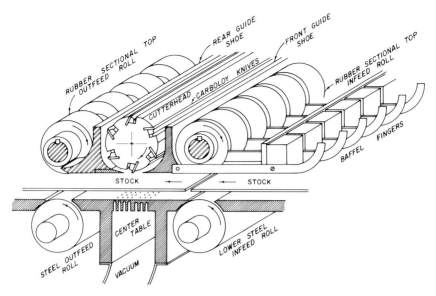

Figure 18-137.—A 50-inch single surfacer for very thin and flexible materials. (Diagram after Buss Machine Works, Inc.)

Although expensive, the arrangement in figure 18-138C has much to recommend it. Lumber is fed face up. The bottom cylinder cuts first, with the workpiece forced upward against the rigid top platen, cutting it to final thickness plus the thickness necessary for surfacing the top face. Thus, a varying cut is taken on the back of the workpiece by the bottom cylinder while it cuts against the solid overplate. Excess thickness is thereby removed from the back or the low-grade side. If the stock is too thin to allow for planing fully on both sides, only a light cut, or no cut, will be removed from the back. After the lumber passes the bottom cylinder, it is flexed or pressed downward by the top chipbreakers and accurately thicknessed by the top cylinder. Thus a measured and predetermined cut is taken from the face of the board. The face surfacing is accomplished against a previously surfaced back. A rough board that is less than the desired thickness will emerge from this planer still rough on both sides.

The arrangement shown in figure 18-138C can, in some plants, advantageously double-surface hardwood lumber at very high speeds before (or after) it is kiln-dried; the resulting smooth, uniformly-thick lumber can then be accurately ripped and planed to final pattern and width on conventional planers (fig. 18-138A).

Conventional double surfacers (fig. 18-138A) for furniture and other woodworking plants are available in widths up to 50 inches and with feed roll diameters up to 6 inches. Feed and cutterhead horsepower are related to the number of knives in the head (seldom more than 6), class of work involved, width of machine and feed speed desired. Generally speaking, the top cylinder carries less than 40 hp, the bottom cylinder less than 20 hp, and the feed less than 15 hp. Feed speeds are generally below 125 fpm.

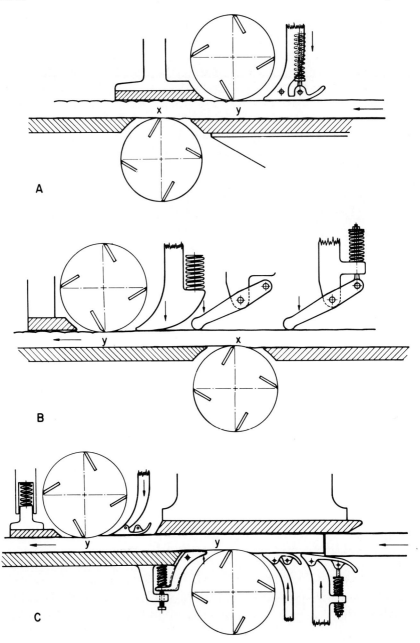

Figure 18-138.—Cutterhead arrangements for double surfacers; x indicates cuts of fixed depth, and y indicates cuts to controlled thickness. (A) Conventional double surfacer. (B) Facing head followed by thicknessing head. (C) Two-way thicknessing planer. If lumber had portions where thickness allowance for planing is scant, arrangements (A) and (B) will yield rough surfaces on upper side. (Drawings after Koch 1948.)

Machining

Planer-matcher.—A planer-matcher is a double surfacer equipped with two opposed sideheads that can simultaneously machine both edges of a board. The machine usually has two additional horizontal spindles carrying **profile heads** to machine patterns on the top and/or bottom of the lumber or to rip wide boards into narrower strips. Normally a planer-matcher is used only for lumber 6 feet in length or longer, although it is possible to arrange them for feeding shorter material.

Figure 18-139 (top) illustrates a six-head machine; each cutterhead carries 12 jointed knives. The cylinders (arrangement of 18-138A) are shown surfacing top and bottom, the opposed sideheads are cutting a shiplap pattern on both edges, and the top profile head is cutting a drop-siding pattern. The bottom profiler is idle and not visible. Most planers and matchers for southern hardwoods have cylinders 15 inches long and can carry sideheads to machine timbers 6 to 8 inches thick. Proportions of machines suitable for southern hardwoods are given in table 18-66. In southern hardwood mills such machines might typically be used to plane dry sweetgum and black tupelo boards on both surfaces and both edges for use as crating lumber, to plane yellow-poplar 2 by 4s, or to run tongue-and-groove white oak wall panelling with an edge or center pattern machined by one of the profile heads.

TABLE 18-66.—*Typical proportions for six-head planer-matchers with double profilers*[1,2,3]

	Number of jointed knives per cutterhead			
	8 or 10	12	14	16
Cutting-circle diameter, inches	9	$9^{3/8}$	10	$12^{3/16}$
Feed-roll diameter, inches	10	12	14	14 to 16
Maximum feed speed, fpm[4]		425	500	575
Horsepower:				
Feed table (variable speed)	15	20	25	30
Planer feed (variable speed)	25	40	50	60 to 75
Top cylinder	50	80	130	150
Bottom cylinder	25	40	65	75
Outside sidehead	25	40	65	65 to 75
Inside sidehead	15	25	25	35
Top profile head	25	40	65	75
Bottom profile head	15	25	25	35

[1]The usual machine for southern hardwoods will accept lumber 15 inches wide and 6 or 8 inches thick.
[2]Planer-matchers usually have six powered rolls; i.e., two pair infeeding and one pair outfeeding.
[3]For southern hardwoods, a rake angle of 20° is commonly used; some profile knives for dry lumber have 15° rake angles.
[4]To yield about 8 knife marks per inch.

Planer-matchers for tongue-and-groove random length oak flooring are specialized machines (fig. 18-139 bottom) that normally need not accept lumber more than 7 inches wide, and which usually operate on lumber less than 3 inches wide. Hardwood flooring is run face down on 5-head machines having the configuration shown in figures 18-138A and 18-139, but with the bottom profile

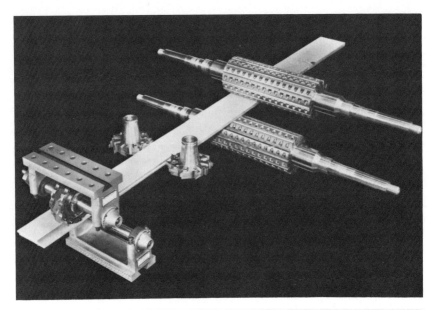

Figure 18-139.—Planer-matchers. (Top) Arrangement with double profilers; no cutterhead is mounted on lower profile spindle. Photo from Stetson-Ross. (Bottom) Hardwood flooring planer-matcher; the flooring is run face down, groove to the inside (right) guide. A hold-down roll operates between the sideheads, and a final top cutterhead (between outfeed rolls) machines a hollow in the back of the flooring. (Photo from Yates American Machine Company.)

head omitted. The top profiler is used to machine a hollow back on the flooring so that the flooring strips, when nailed in place, will rest only on the edges of their lower surfaces. A heavy pressure roll mounted between the matcher heads holds the flooring strips firmly to the lower matcher shoes so that even if the lumber is under thickness the face is planed and the tongue and groove are positioned relative to the face. Some flooring machines feature top and bottom cylinders (cutterheads) that can be traversed laterally 3 inches at a time to periodically bring sharp knife segments into action on the 3-inch-wide flooring strips.

Machining

Of the many knife types found on planers and matchers, five are so widely used that they require description. Sharpening practices vary according to knife type and location on cylinder, sidehead, or profiler.

Knives for top and bottom cylinders.—Top and bottom cylinders are designed to smooth flat surfaces and have historically carried 5/32-inch-thick, smooth-back knives. These long thin knives are periodically sharpened (in balanced pairs) to produce a keen, straight cutting edge. The knives are set accurately in the cylinder by means of a dial gauge indicator or, preferably, a roller knife gauge (fig. 18-140). Jointing, i.e., the process of bringing all knives to a common cutting circle, is performed by traversing a suitable stone along the periphery of the rotating cutterhead (fig. 18-141). The knives are "heel ground" (an interim grinding operation that removes all but a hairline trace of the "land" produced by the jointing operation) with the attachment shown in figure 18-142.

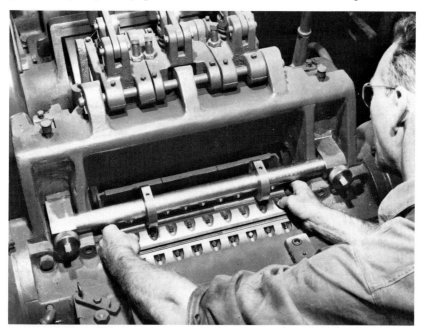

Figure 18-140.—Roller knife setting gauge used to precisely position each knife before the knife is firmly clamped in place. (Photo from Stetson-Ross.)

When planing green or kiln dried white oak, steel knives in the cylinders may require jointing two to four times per shift, and heel grinding after two or three 8-hour shifts. Normally, steel knives are removed from each cylinder and resharpened weekly. Time required to joint is perhaps 2 minutes. To heel grind steel knives in one cylinder requires about an hour, as does changing the knives in one cylinder.

During the 1970's, legislation limiting noise levels in United States woodworking factories stimulated research to reduce planer noise traceable largely to lumber vibration caused by impact of knives (Stewart 1972a). To reduce impact and vibration, a cutterhead with helically mounted blades was developed (Stew-

Figure 18-141.—Jointer for the top cylinder of a double surfacer equipped with carbide-tipped helical knives. The jointing device is fitted with a diamond hone. Jointing is accomplished while the cutterhead cylinder rotates at 3,600 rpm. (Photo from Newman-Whitney.)

art and Hart 1976; Stewart 1978a). In this cylinder (fig. 18-143), the helix angle is near 45°. A 6-knife design with 5-inch-diameter cutting circle is in wide use on hardwood double-surfacers. The cutterhead body is solid, with helical grooves

Machining

Figure 18-142.—Heel-grinding knives in the bottom cutterhead of a planer-matcher. Mounted ready to use on the top cutterhead are jointer bar and jointer. (Photo from Stetson-Ross.)

that accept carbide-tipped knife blocks fitting precisely into the slot and secured by a screw; desired rake and clearance angles are pre-ground. Final tip grinding is done after the cutterhead has been installed (Stewart 1975b). York (1975) found that such cutterheads work well on kiln-dried lumber but green oak rapidly dulled the carbide cutting edges.

Carbide knives can be lightly jointed several times before excessive land width causes surface defects in the workpiece. At this time, usually after several operational months, knives are resharpened with a power-driven wheel rigidly mounted and precisely traversed to grind just enough material from the clearance surface of each knife to remove the land caused by jointing. Carbide knives in planer heads can usually be ground five to ten times yielding tip life of several years.

When cutting dry white oak, carbide-tipped helical knives in cylinders may run 60 to 80 8-hour shifts before rejointing is required. On green (wet) oak boards carbide knives dull much more rapidly requiring jointing every 3 to 4 weeks. Readers interested in more detailed instructions on jointing and sharpening carbide-tipped helical knives in planer cylinders are referred to the maintenance manual on this subject published by Newman Machine Co., Inc., Greensboro, N.C. See also sect. 18-29.

Knives for sideheads.—For machining flat square edges on boards, sideheads usually carry 5/16-inch-thick corrugated-back steel knives (fig. 18-144B). To machine an edge contour, this type of knife can be **ground-to-pattern** (fig.

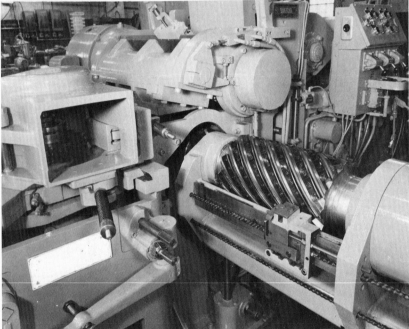

Figure 18-143.—Helical cutterheads for planers. (Top) 30-inch-long, cabinet-planer cutterhead fitted with six knives in a 5-inch-diameter cutting circle. (Bottom) 15-inch-long cutterhead for a planer and matcher fitted with 14 knives in a 10-inch-diameter cutting circle. (Photos from Newman Machine Company, Inc.)

18-145). Standard patterns, whether applicable to sideheads or profile heads are usually **milled-to-pattern** knives (see sideheads and top profile head in figure 18-139); these knives are sharpened easily by grinding the face of the knife so that the pattern milled on the knife back remains unchanged (fig. 18-144AC).

Sideheads are slipped onto the side spindles and centered either by self-centering sleeves or by tapered spindle tops matched to taper-bored cutterheads. They must usually be given a final joint after being secured on the sidehead spindle. Therefore, the procedure for straight sidehead knives and ''milled-to-

Machining 2007

Figure 18-144.—Combination grinder and jointer. (A) Grinding sideheads. (B) Jointing sideheads. (C) Grinding profile heads (in yoke assembly). (D) Jointing profile heads (in yoke assembly). (Photo from Stetson-Ross.)

pattern" knives is to sharpen them without removing them from the head (fig. 18-144A), then joint them (fig. 18-144B), regrind them to the joint, and finally transfer them to the sidehead spindle. **Ground-to-pattern** knives are handled somewhat differently as explained in the next paragraph.

Knives for profile cutterheads.—The profile arbors on some planers are fixed and the cutterheads are secured in place with self-centering collars. In this situation, the procedure is the same as with the sideheads described above. In another design, the cutterheads are carried in self-contained units (fig. 18-139 top) that eliminate the necessity for jointing in place on the planer and matcher. Figure 18-144CD illustrates the manner in which cutterheads mounted in these yokes are ground and finish jointed in the grinding room.

Ground-to-pattern knives must be removed from the cutterhead in order to sharpen them. They can be hand sharpened individually, but preferably are sharpened in a jig (fig. 18-145A). Subsequently the knives are repositioned in the cutterhead (fig. 18-145B) jointed, and finally the head is slipped into position in top or bottom profiler.

Figure 18-145.—Ground-to-pattern knives. (A) Fixture for gang grinding. (B) Sharpened knives mounted in profile head (yoke assembly). Photo from Stetson-Ross.)

MOULDING

The purpose of the moulder is to machine complex shapes on the surfaces or edges of long or short lumber (fig. 18-146 top). Moulding, like planing, is a peripheral milling process. A simple moulder has a top cutterhead followed by two or three sideheads followed by a bottom cutterhead (fig. 18-146 bottom). Instead of being directly opposed as in a planer and matcher, moulder sideheads are staggered to permit their spindle-mounted motors to clear each other when one of the spindles is tilted. Tilting sideheads permit angled saw cuts and varied bevels on the edges of the workpiece without changing knives. While the moulder can machine a very broad range of shapes, it cannot make tongue and groove flooring as accurately as a planer and matcher because the staggered sideheads do not rigidly control workpiece width and position of tongue and groove.

Productive capacity.—Characteristically moulders are designed for relatively short runs of any pattern. In some plants as many as 20 different shapes may be run in a day. In others the moulders setup may be unchanged for several days.

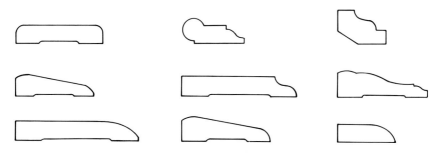

Figure 18-146.—(Top) Cross sections of typical mouldings. (Bottom) A 6-inch, four-head moulder. Cutterheads can run jointed at 3,600 rpm or unjointed at 6,000 or 7,200 rpm. Hopper feed is at infeed end. (Photo from Mattison Machine Works.)

The productivity of a moulder may be expressed by the following formula:

$$P = V(60T - CX)(Y)(K) \qquad (18\text{-}41)$$

where:

- P = lineal footage of mouldings produced per shift
- V = feed speed, feet per minute
- T = length of shift, hours
- C = idle machine time due to each pattern change, minutes
- X = number of pattern changes per shift
- Y = pattern multiples
- K = continuity of feed, percent efficiency expressed as a decimal fraction. This factor must include all nonmachining time due to all causes other than pattern change

Obviously, quick pattern change is important. If, for example, four different patterns were required during an 8-hour shift, and each change took 60 minutes, production would be half that with no pattern change. Also, production is directly proportional to feed rate, which is in turn dictated by the desired surface quality as expressed in knife marks per inch.

There are two ways to increase the lineal rate of feed without reducing surface quality (12 to 16 knife marks per inch for most mouldings):

- Increasing the cutterhead spindle speed.
- Increasing the number of jointed knives in the cutterhead.

Of the several synchronous spindle speeds in common use, i.e., 3,600 rpm, 6,000 rpm, and 7,200 rpm, none but the 3,600-rpm speed can be consistently jointed. For this reason moulders with spindle speeds of 6,000 or 7,200 rpm are limited to single-knife operation, that is, one cutting knife and one balancing knife per head. On the other hand, the feeding rate of moulders equipped with spindles operating at 3,600 rpm may be increased in direct proportion to the number of jointed knives in each cutterhead. To maintain 20 knife marks per inch, a 7,200-rpm, one-knife machine must feed at 30 fpm, while a 3,600 rpm machine with six jointed knives in each cutterhead can feed at 90 fpm.

Machine types.—The controls of many machines are so designed that synchronous spindle speeds can be selected at 3,600, 6,000, or 7,200 rpm. This arrangement permits one-knife operation at high spindle speed on short runs of nonstandard patterns as well as multiknife, 3,600-rpm operation with jointed knives at high feed speeds on long runs of standard patterns.

In the cutterhead arrangement most frequently used on a four-head moulder, the top cutterhead cuts first and the bottom one last. Since the sides are machined while the bottom of the workpiece is still rough, pattern registration on the edges may not be accurate. Prior surfacing would, of course, eliminate this difficulty. The usual five-head arrangement places a second top head between the side-heads and the final bottom head. Designs are also available in which the fifth head cuts first and is on the bottom in order to smooth the bottom surface of the stock as an initial machining step. Six-head machines are also available which surface the bottom and inside (guide side) edge of stock prior to machining by the conventionally arranged four heads following.

In one arrangement, an initial bottom cutterhead machines a series of narrow shallow grooves in the lower surface of the lumber. Fins on the lower platen fit these grooves and provide guidance through the cutting zone. A final bottom cutterhead surfaces the bottom of the lumber to remove the grooves. (See Ward 1975ab for description of an operation using this technique.)

While 5-head and 6-head moulders are readily available and are in widespread use, it is sometimes desirable to attach cutting units to the outfeed end of a 4-head moulder to increase its versatility (fig. 18-147). These additional spindles can be arranged to surface, gang rip, or machine supplementary grooves that are not easily cut with the other heads. **Jump spindles** activated by the stock to make interrupted cuts, have been designed for both horizontal and vertical attachment to the outfeed end of a standard moulder.

Figure 18-147.—(Top) Three-hp vertical and horizontal motorized spindles attached to the outfeed end of a four-head moulder. These spindles permit some special patterns to be machined with low-cost cutterheads. (Bottom) A 7-1/2-hp tiltable saw arbor mounted on outfeed end of four-head moulder. Guards and shaving heads, here removed for clarity, must be in place during operation. (Photos from Mattison Machine Works.)

Moulders are commonly equipped with two pairs of power-driven infeeding rolls, or with two top infeeding rolls over a lower endless bed. Figure 18-146 illustrates a machine with two 6-inch-diameter, corrugated, sectional, top infeed rolls and a bottom endless bed, all powered by a 5-hp variable-speed drive. Moulders do not have outfeed rolls, but some have feed rolls between cutting zones to facilitate feeding of short stock. A hopper feeding device permits feeding short stock at speeds above 50 fpm (fig. 18-146).

Moulders are made in 4-, 6-, 8-, and 12-inch widths and usually accept a workpiece up to 4 inches thick. Cutting circles range in diameter from 5 to 9 inches. Sideheads are provided with as much as 45° inward tilt and 15° outward. Cutterhead power depends entirely on the class of work to be done. The 6-inch moulder illustrated in figure 18-146 carries 7-1/2 hp on the top and bottom heads, and 5 hp on each sidehead.

Moulder spindle types.—Moulder spindle types can be classified as follows:

- Fixed arbor with cutterhead fixed on spindle (illustrated by top and bottom planing cylinders in fig. 18-139).
- Fixed, straight spindle with slip-on cutterheads—cutterheads equipped with self-centering sleeves (fig. 18-148 top).
- Fixed, taper-top spindle with slip-on, taper-bore cutterheads (illustrated by sideheads in figs. 18-139 and 18-144AB).
- Removable, tapered-end horizontal spindle (fig. 18-148 bottom).
- Yoke or "cartridge" cutterhead unit consisting of removable aluminum yoke complete with ball bearings, spindle, and cutterhead (figs. 18-144CD and 18-145).

Knife grinding technique—Considerable skill is required to make ground-to-pattern moulding knives that will reproduce the types of contours shown in figure 18-146 top. The projection necessary to properly form a knife blank to reproduce the pattern desired is shown in figure 18-149. The construction is based on the minimum allowable cutting circle. The thickest portion of the moulding lies on ST, the tangent to this minimum cutting circle. Note that the rake angle is different at each of the various depths of cut along the profile. Having obtained one knife blank by this projection, other knives can be ground to match it by means of a suitable jig (fig. 18-145).

For long runs of standard patterns, milled-to-pattern steel knives (fig. 18-144ABC) are practical and require much less grinding skill. Carbide-tipped moulding knives require special expertise in brazing and grinding but dull less rapidly than steel milled-to-pattern knives.

SHAPING

It is the purpose of the shaper to cut an edge profile or edge pattern on the side, end, or periphery of a workpiece. Some designs also permit inside contours to be shaped. In addition, scooping cuts such as those required in the shaping of bowl interiors or chair seats may be accomplished on some specialized equipment.

Machining

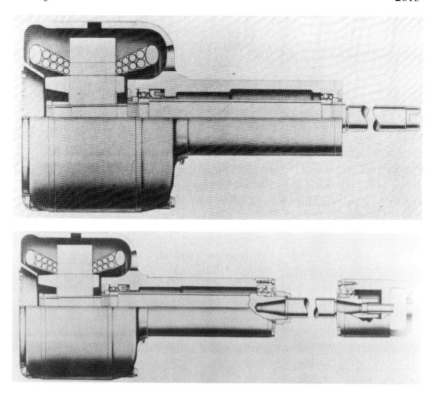

Figure 18-148.—(Top) Straight moulder spindle for slip-on cutterheads equipped with self-centering sleeves. (Bottom) Removable, tapered-end spindle. (Photo from Diehl Machines, Inc.)

Shaping is a peripheral milling process, as are jointing, planing, and moulding. Sections 18-4 through 18-7 describe fundamental aspects.

Davis (1962) evaluated 10 of the species that are the subject of this text and concluded that white ash, black tupelo, white and northern red oaks, sweetgum, and red maple were most easily shaped, while hickory, yellow-poplar, American elm, and hackberry yielded fewer pieces of good quality (table 18-1). He found that cuts made parallel to the grain or in a diagonal direction were consistently better than cuts perpendicular to the grain. In parallel and diagonal cuts, raised grain was the worst defect. In end-grain cuts surface roughness was the worst defect; dense woods were shaped more readily across the grain than woods of low density. Wood moisture content, in the range from 6 to 10 percent, did not significantly affect quality of shaped surfaces.

Spindle design.—The shaping spindle may carry several knives mounted on a cutterhead but, as the knives are not usually jointed, only one knife leaves a trace on the workpiece. To minimize centrifugal forces and to permit shaping of

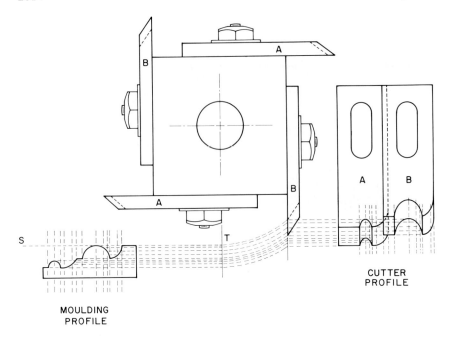

Figure 18-149.—Method of laying out moulding knives.

small radii, the cutting circle employed on shaper heads is comparatively small, usually 5 inches or less. Shaping the periphery of a workpiece involves cutting at all grain orientations. These several factors make high spindle rpm desirable. Spindle speeds from 7,200 to 10,000 rpm are common, with some designs providing 15,000 or even 20,000 rpm. Power on a shaper spindle may be 3 to 20 hp depending on the application. Shaper spindles are commonly belt driven from 1,800- or 3,600-rpm motors. In some designs the drive motor is two-speed to give a choice of spindle speeds (10,000 or 20,000 rpm, for instance) without a pulley change. Figure 18-150 illustrates the design of a belted spindle for a table shaper.

If 7,200 rpm is the maximum spindle speed required, it is possible to mount the motor directly on the cutter spindle. This assumes motor operation on 120-cycle current.

Cutterhead design.—Figure 18-151 illustrates some typical shaper cutters. When the length of run justifies the expense, formed cutters of small diameter are frequently made so that cutting edge and head body are integral. For short runs, shaper heads that carry ground-to-pattern knives are commonly used. The projection technique to develop a knife profile from an edge-shape profile is identical to that used for a ground-to-pattern moulding knife (fig. 18-149).

In many shaping operations better quality work can be obtained by employing down-milling rather than conventional up-milling. A discussion of down-milling compared to up-milling can be found in section 18-4 in connection with figures 18-32, 33, 34, and 45. Many of the machine types described in the following paragraphs provide the positive feed control needed for down-milling.

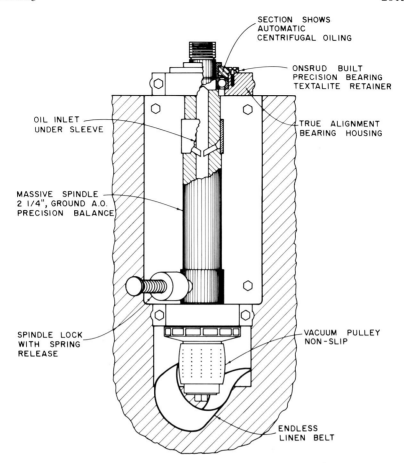

Figure 18-150.—Belted shaper spindle. The high rotational speeds of such spindles require that they be very rigid and in dynamic balance; ball bearings carrying the spindle must be selected for precision. Improved belting material developed since 1960 makes unnecessary application of a vacuum to the lower pulley. (Drawing from Onsrud Division, Danly Machine Corporation.)

Single-spindle shaper.—The shaper in its most simple configuration carries a 3-, 5-, or 7-1/2-hp motor belted to a single vertical spindle rotated at 8,500 rpm. Control and position of the workpiece as it is guided past the shaping head is achieved by one of the following techniques:

- The workpiece can be held against smooth collars on the spindle itself.
- Straight stock can be held against a straight guide bar.
- Stock to be shaped can be secured to a template that is, in turn, held against the shaper collars. (See Ekwall 1979 for a discussion of template design.)
- The stock can be held against special forms, pins, or jigs. An example would be the rotation of the workpiece around a pin fixed to the table in such a way that the shaper head generates a perfect circle as the workpiece turns.

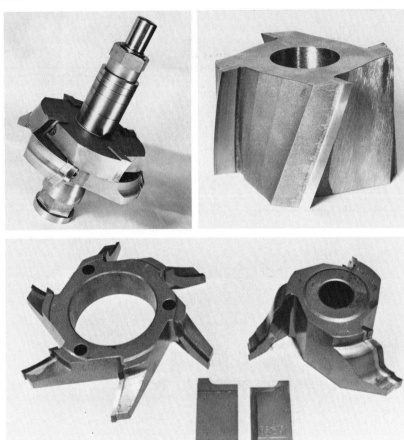

Figure 18-151.—Shaper heads. Most carry carbide cutting edges.

Double-spindle shaper.—On a double spindle shaper, the two spindles rotate in opposite directions so that a workpiece can always be presented to the cutter with favorable grain orientation (fig. 18-152).

Double-head automatic shaper.—This machine consists of a revolving table which feeds work to two cutting spindles (fig. 18-153). Each workpiece is manually placed on top of a template that is bolted to the table. As the table revolves and the work approaches the cutting stations, cam-actuated air clamps grip the work and hold it under pressure as it is being shaped. After passing the cutterheads, the clamps automatically release the shaped workpiece for removal.

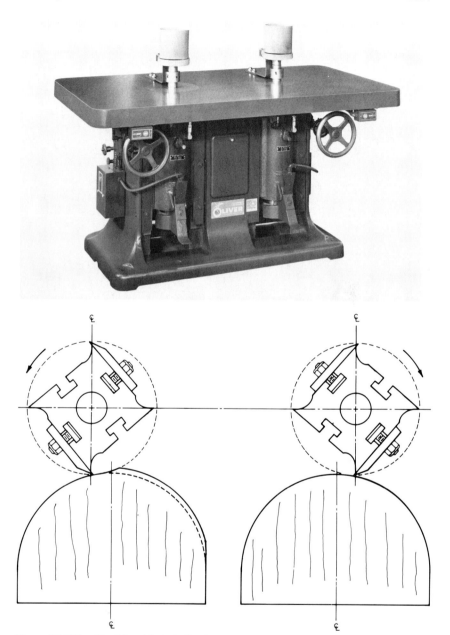

Figure 18-152.—(Top) Double-spindle shaper. (Bottom) Plan view of a double-spindle shaper showing spindle selection for favorable grain orientation. (Photo from Oliver Machinery Co.)

The cutterheads follow the pattern on the template by means of idle guide rollers located below the cutterheads. The two moving cutterhead spindles are individually driven by separate arbor-mounted 20-hp motors that operate on

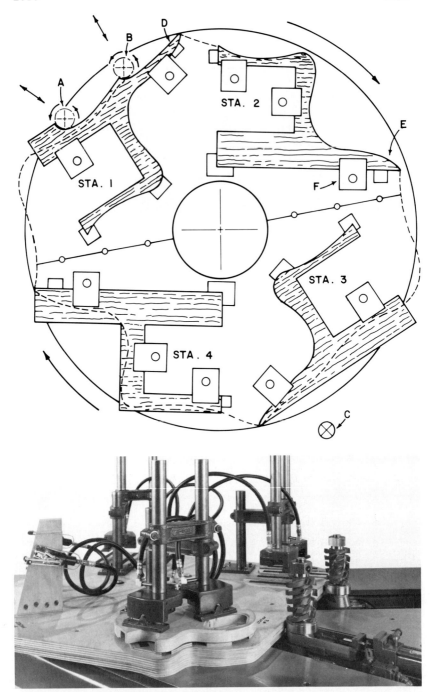

Figure 18-153.—Double head automatic shaper. (Top) Set-up with a four-station template. (A) roughing head; (B) finishing head; (C) operator's position; (D) profile cut at stations 1 and 3; (E) profile cut at stations 2 and 4; (F) clamps. (Bottom) Machine set-up to show template and air clamps. (Photo from Danly Machine Corporation, Onsrud Division.)

120-cycle current and turn at 7,200 rpm. The guide rollers are kept in contact with the template by spring-cushioned air cylinders. The pressure of the guide roller against the template may be regulated according to the requirements of the pattern. Template patterns are made of plywood, band sawn and sanded to the desired shape, and bolted to the table. They are classified as single-station or multiple-station templates. Single-station templates are used for shaping around the entire periphery of a single piece in one pass. The principle of the multiple-station pattern is illustrated by figure 18-153 (top).

The table, hydraulically driven by a 7-1/2-hp motor, can accommodate a round workpiece up to 86 inches in diameter. Table rotational speed is variable (0 to 4 rpm) and is controlled by cams so that feed speed is regulated to conform to depth of cut and grain orientation. It can be rotated in either direction so that up- or down-milling can be used, depending on the cut to be made.

Because the machine is equipped with two cutterheads, one roughing and one finishing, it is frequently unnecessary to mark out and bandsaw the pattern on the workpiece prior to shaping.

Double-table, fixed-spindle shaper.—This shaper is particularly designed for a very rapid shaping of pieces that measure less than 28 inches in diameter and 3 inches thick (fig. 18-154). Dual feed tables permit the operator to unload and load one table while work is being shaped on the other table. Rates of production range from 4 to 20 pieces per minute. Hairbrush backs, for example, might be turned out at the rate of 8 per minute.

Figure 18-154.—Dual table shaper. (Photo from Danly Machine Corporation, Onsrud Division.)

The machine has a single overhead 5- to 10-hp cutterhead spindle which rotates at 5,000 to 10,000 rpm. During the machining cycle, the workpiece is held on the table-mounted pattern by a center air clamp. The overhead cutter, together with a roller pattern follower below and aligned with it, is mounted on an arm which pivots between the two tables and is held in position by an air cylinder. Contact of the pattern follower with the pattern controls position of the cutter in relation to the workpiece and provides a machined part with contours duplicated from the pattern. At the end of each cutting cycle, the operator actuates a foot switch which initiates a cutting cycle on the second table.

Contour profiler.—This machine (fig. 18-155) shapes straight-lines and contours quickly, accurately, and safely. It is practical for both long and short runs. The machine has a rectangular table (to which the workpiece is clamped) with a long reciprocating stroke, thus carrying the work past the cutters and returning it to the starting position. The cutterheads move in and out at right angles to the table travel, controlled by a template or pattern, that is part of the work set-up. Air clamps grip the workpiece firmly while it is in the cutting zone but release it at the unloading position.

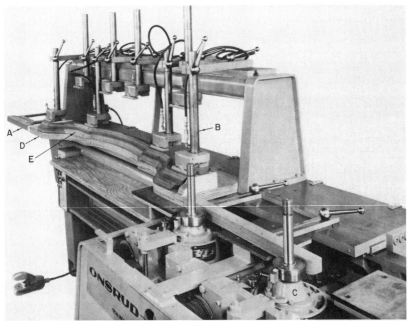

Figure 18-155.—Dual-head contour profiler. (A) Table; traversed by a combination of air and hydraulic action; lengths available for work up to 84 inches long. (B) Air clamps to secure workpiece. (C) Each spindle carries an arbor-mounted, 20-hp, 7,200-rpm, 120-cycle motor; spindle accommodates up to a 10-inch-high cutterhead having a cutting circle 2-1/2 to 4 inches in diameter; maximum in-and-out spindle movement is 6 inches. Motion is controlled by a guide roller following a template under the workpiece; the smallest workpiece radius equals the cutting-circle radius plus one-third; contours having a steepness up to 60° between line of table travel and the tangent of the curve can be accomplished. (D) Template. (E) Workpiece. (Photo from Danly Machine Corporation, Onsrud Division.)

The speed at which the table traverses can be automatically cam controlled from 0 to 80 fpm on the cutting stroke to accommodate areas of difficult cutting geometry; return speed is 84 fpm. Cutter spindles—one roughing and one finishing—rotate clockwise to accomplish down-milling. About 300 long parts can be machined per hour. If multiple-station templates are used for short parts, the production rate can be substantially increased. Machines are available that have shaper spindles on both sides of the table to simultaneously contour two edges of a workpiece.

Shaping attachment on a double-end tenoner.—Refer to figure 18-227 (bottom) with accompanying discussion of *Cam generated shaping.*

Tape-controlled shaper.—Through the mid 1970's, numerically controlled and computer-numerically controlled shapers and routers (fig. 18-156) were slow to gain acceptance due to the durability of manual machines, inexperience with electronic devices, initial high cost, and to conservatism in industry. During the late 1970s advances in technology significantly reduced the costs of computer control, and use of computer-numerically controlled shapers and routers should increase substantially. Smith (1978) described the cost and function of equipment made by four major manufacturers; C. W. McMillin's review of computer control of woodworking machines in secondary manufacture is summarized in section 18-30.

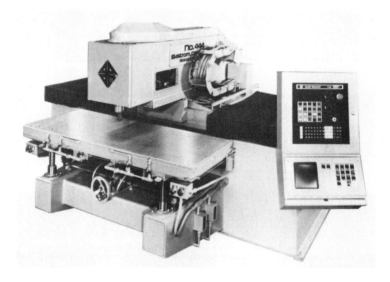

Figure 18-156.—Computer-numerically controlled shaper-router with fixed table and moving ram carrying a 15-hp overhead cutter. Vertical adjustment is manually set; longitudinal and transverse feed directions and rates are computer controlled. (Photo from Ekstrom, Carlson and Co.)

18-14 MACHINING WITH ABRASIVES

Flat furniture parts and panels may be sanded in an abrasive planer to reduce them to desired thickness. Nearly all exposed surfaces of furniture must be smoothed with abrasives before high-quality decorative and protective finishes can be applied. Such abrasive machining before applying finishes, is usually done in two stages. Rough sanding levels the wood surface to remove marks of previous machining operations. The second stage—polishing—reduces and smooths coarse scratches left from rough sanding.

SURFACE QUALITY

Effect on strength of glued joints.—Knife-cut surfaces produced by peripheral milling heads or by precision saws may be adequately level and smooth for glue joints, but the strength of the joint is significantly affected by the surface preparation; the effect is species dependent, as follows (River and Miniutti 1975):

Species and surface preparation	Strength of glued joint as a proportion of strength of solid wood
	Percent
Yellow-poplar	
Sawn..........................	54
Planed.........................	63
Glued-jointed (fig. 18-131)..........	93
Red oak sp.	
Sawn..........................	70
Planed.........................	108
Glue-jointed.....................	96

The presence of surface cells crushed and torn during machining weaken glue joints. Because abrasive-planed wood has significantly more crushed and torn surface cells than knife-planed wood, glue joints made from wood surfaces prepared by abrasive planing have less resistance to glue line separation during accelerated aging than those prepared by knife planing; shear strengths of glued joints prepared by the two systems are about equal, however (Jokerst and Stewart 1976.)

Through use of scanning electron micrographs, Stewart and Crist (1982) found that subsurface crushing was more severe in abrasive-planed than knife-planed wood; red oak, yellow-poplar, hard maple, and aspen were analyzed in their experiments.

Perceptibility of machining marks on painted surfaces.—Hirst (1971) found that machining marks from sanding and planing hardwoods and softwoods were generally less perceptible on flat- and satin-painted surfaces than on gloss-painted. Black flat paints hid machining marks better than white or clear coatings; with gloss coatings, however, paint color had little effect on perceptibility of marks. A second coat of paint reduced the visibility of marks on sanded but not on planed surfaces.

With gloss coatings, 32 knife marks per inch were barely perceived; with flat or satin coatings, however, 20 marks per inch could not be noticed. Under most finishes, 4 to 8 knife marks per inch were readily visible.

When sanded along the grain, score marks of 120-grit paper were nearly invisible under three coats of paint. Score marks made by sanding across the grain were much more difficult to hide.

Color variation of furniture related to sanding procedure.—The color of hardwood furniture made on a production basis sometimes varies from that of sample furniture shown to get an order. Connelly (1975a) attributes the problem to departure from standard grit sizes for each sanding operation, using dull abrasives, and inadequately inspecting the furniture just before finishes are applied. He recommends that all parts be lightly sanded to remove raised grain just before this inspection, and that burn marks and dents be repaired following inspection. Finishes should be applied within 8 hours of inspection and surface repair to promote even and predictable penetration of stain.

Surface quality measurement.—Non-destructive measurement of surface roughness of wood and reconstituted wood products such as particleboard and fiberboard is difficult. Methods include two-dimensional analyses by profile meters, which may be optical, mechanical, or electronic (Hann 1957; Pahlitzsch and Dziobek 1965ab; Hefty and Brooks 1968; Peters and Mergen 1971; Umetsu et al. 1979). Pneumatic systems that monitor air escape from a pressurized tube or cup placed against the surface, give a measure of three-dimensional texture (Porter et al. 1971; Bonac 1975, 1979). Other three-dimensional systems include measurement of contact angle of water drops on surfaces (Suzuki 1958) and incident light reflection (Nakamura and Takachio 1960). Potrebic (1969) developed a technique for measuring four parameters that permit numerical expression of surface smoothness. Hse (1972) developed equations for determining roughness factor (ratio of true surface to apparent area) of rotary-cut veneer based on an assumed fiber model.

Reviews of methods of measuring surface roughness have been provided by Elmendorf and Vaughan (1958), Davis (1958), American Society for Testing Materials (1959), Pahlitzsch and Dziobek (1961a), Sieminski (1966), and Peters and Cumming (1970).

TYPES OF COATED ABRASIVES

Six elements contribute to the structure and performance of coated abrasives (fig. 18-157 top) as follows: abrading agent, backing material, bond, distribution and orientation of abrasive particles, flexure pattern, and product form.

Abrading agent.—Of the five minerals (fig. 18-158 AB) commonly used to sand wood, two occur in nature (flint and garnet), and three are man-made (aluminum oxide, silicon carbide, and zirconium-based grit).

Flint, a grayish white to faint pink silicon dioxide quartz found in large natural deposits in many areas of North America, is the abrasive on common sheet sandpaper for hand sanding. Although flint breaks up into sharp fragments

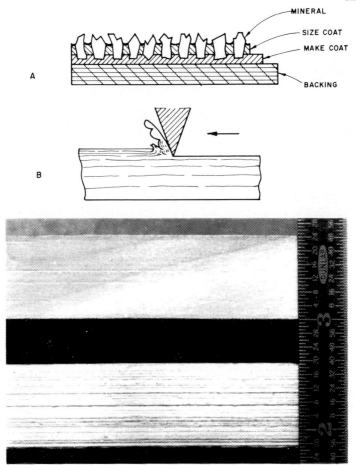

Figure 18-157.—(Top) Construction of a belt coated with abrasives. (Center) Cutting action of an abrasive particle. (Bottom) Surfaces on dry (7-percent moisture content) yellow-poplar produced with 36-grit, aluminum oxide, open-coat paper. The rougher of the two surfaces was made with commercial paper in which the mineral was imbedded on the backing without regard to the location of the cutting points; the smooth surface was made with the same grit carefully arranged so that the cutting points were all in a common plane. (Photo from R. Birkeland.)

and is low in cost, it is little used for industrial applications because it is neither as hard nor durable as other available abrasives.

Garnet is a mixed orthosilicate of iron, aluminum, calcium, and magnesium. It is red in color. Of the several known forms in which it naturally occurs, Almandite ($Al_2O_3FeO.3SiO_2$) is the principal one used for coated abrasives. The northeastern part of the United States is one of the important sources of these crystals. Garnet crystals, when crushed, provide light wedge-shaped grits that are harder and sharper than flint. Having a hardness of approximately 7.5 on the Moh hardness scale, garnet is harder than glass and is a widely used natural abrasive. It is second in toughness only to aluminum oxide. The specific gravity of garnet ranges from 3.4 to 4.3.

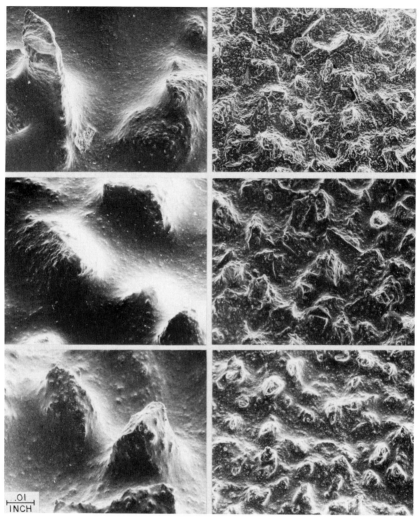

Figure 18-158A.—Thirty-six-grit (left) and 100-grit (right), closed-coat, man-made abrasives from Norton Co. Scale mark shows 0.01 inch. (Top) Silicon carbide. (Center) Aluminum oxide. (Bottom) Zirconium-based grit. (Scanning electron micrographs from files of C. W. McMillin, Southern Forest Experiment Station, U.S. Dep. Agric. For. Serv., Pineville, La.)

According to Gillson et al. (1960), practically all garnet used today is heat treated before marketing. Originally all garnet used in the coated abrasive industry was heat treated at 200-300°C to improve its capillarity and thus improve its adhesion to the bond used. Early in the 1930's, heat treatment temperatures were increased to 700-800°C to improve hardness, toughness, fracture properties, and color.

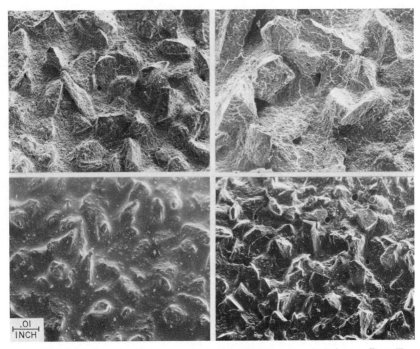

Figure 18-158B.—Natural abrasives. (Top left) Medium-grit, closed-coat flint. (Top right) Coarse-grit, closed-coat flint. (Bottom left) Open-coat, 100-grit garnet. (Bottom right) Closed-coat, 100-grit garnet. Scale mark shows 0.01 inch. (Scanning electron micrographs from files of C. W. McMillin, Southern Forest Experiment Station, U.S. Dep. Agric. For. Serv., Pineville, LA.)

Aluminum oxide, first created synthetically about 1900, is a reddish brown smelted derivative of bauxite ore. This ore, heated in an electric furnace to approximately 3500°F, together with a small amount of coke and iron filings, produces a "pig" that may contain as much as 50 percent aluminum oxide. When crushed, aluminum oxide forms a grit having heavy wedge-shaped particles with a hardness of 9.5 on the Moh scale and a specific gravity of 3.96. Aluminum oxide particles are the toughest of the abrasives under discussion and are exceeded in sharpness and hardness only by silicon carbide. In combination with a resin bond, aluminum oxide is very resistant to breakdown, and is widely accepted for sanding applications requiring high pressures. When dull, cutting points become rounded, and heat induced burnishing of surfaces may result.

Silicon carbide, blue black in color, was first experimentally produced in the early 1890's. It is manufactured commercially by combining a mixture of sand (silicon dioxide), powdered coke (carbon), and a small quantity of sawdust and salt in an electric resistance furnace at about 4,000°F. The sawdust makes the mass porous and aids in the escape of carbon monoxide. The salt helps remove iron impurities by forming a volatile chloride. The crystals of silicon carbide that form around the electrode have a hardness of about 9.6 on the Moh scale and a specific gravity of 3.2. When these crystals are cooled and crushed, the resulting grit is chiefly sharp, wedge-shaped particles. Although silicon carbide is the hardest and sharpest of the minerals used in the manufacture of coated abrasives, it is the most readily crumbled because of its brittleness. It is excellent for light sanding operations, such as removing raised fibers from previously sanded wood. It is also an efficient abrasive for sanding hardboard and particleboard, which have resin binders.

Zirconium-based grit, an alloy of aluminum oxide and zirconium oxide, became commercially available about 1972 under the trade names *Cubacut* (3M) and *Norzon* (Norton Company). *Cubacut* grit is brownish-black, while *Norzon* grit is white or gray. The grit has a specific gravity of about 4.56. Its hardness cannot be easily measured on the Moh scale because fractures develop that make the Moh test invalid. The zirconium-based grit, i.e., **alumina zirconia**, is used for heavy stock removal and high-pressure grinding. It is suited for use on abrasive planers.

The sizes of particles used on coated abrasives are established by sifting the **grit** through screens of standard mesh. The fineness of the mesh is part of the trade designation. The finest screen, constructed of silk threads, has 220 openings to the linear inch or 48,400 openings per square inch. Grains finer than this are segregated by sedimentation or by air flotation. Grain sizes range from 12 (coarsest) to 600 (finest). Earlier nomenclature for the particle size of flint ranged from No. 4 (coarsest) to 10/0 (finest). The relationship of the two numbering systems is shown in table 18-67.

TABLE 18-67.—*Comparison of current and old systems of designating grit sizes for woodworking coated abrasives*[1]

Current designation based on screen mesh	Old system	Current designation based on screen mesh	Old system
12	—	120	3/0
16	4	150	4/0
20	3-1/2	180	5/0
24	3	220	6/0
30	2-1/2	240	7/0
36	2	280	8/0
40	1-1/2	320	9/0
50	1	360	—
60	1/2	400	10/0
80	0	500	—
100	2/0	600	—

[1]Silicon carbide is available in grit meshes 16 through 600, aluminum oxide in meshes 24 through 500, zirconium-based grit in meshes 20 through 220, garnet in meshes 36 through 320, and flint in coarse (36-50), medium (80-100), fine (150), and extra fine (180).

Backing material.—**Backing** is the foundation of coated abrasives (fig. 18-157 top). Paper, cloth, vulcanized fiber, and combinations of the foregoing are used as backing materials.

Paper backings (table 18-68) with weights of 90 pounds and less are used to make sheet goods for hand sanding or for reciprocating and orbital sanding machines. The 40-pound paper is best for sanding of curved surfaces where a flexible sheet is required to conform to the surface of the workpiece. The 70- and 90-pound weights are stiffer and cut faster on flat or nearly flat surfaces. The 110-pound paper is combined with a light-weight cloth to make a tear-resistant backing desirable in many drum sanding operations. Drum sander wraps and wide sander belts use 130-pound paper as the standard backing material. Papers of 100-pound weight and less are single-layer kraft papers made on a Fourdrinier machine from wood fibers. Papers heavier than 100-pound weight may be four or five ply and are made on a cylinder machine from jute or from hemp fibers reclaimed from old rope. The longitudinal strength of the cylinder papers make them particularly suitable for belts and drum wraps.

TABLE 18-68.—*Paper backings for coated abrasives*

Letter indicator	Weight per ream[1]	Product form
	Pounds	
A................................	40	Sheet goods
C................................	70	Sheet goods
D................................	90	Sheet goods
	110	In combination with cloth
E................................	130	Rolls—belts

[1] A standard papermaker's ream is defined as 480 sheets measuring 24 inches by 36 inches.

Cloth backings are summarized in table 18-69. The drill cloth has fewer but heavier threads than those in jeans cloth. Coarser grades of minerals are coated on drill cloth because it provides substantial support for greater effectiveness in the heavier abrading applications. Edge sanding, coarse stroke sanding, and wide-belt sanding are applications that might use drill backing material. Jeans cloth is used in situations requiring more flexibility, as when mould, spool, and ribbon sanding. Cloth is not recommended for drum sander wraps because it tends to stretch. Heavy polyester backings are used for abrasive planing or similar severe duty.

Combination paper (110-1b) and cloth backing is available in roll form for spirally wrapping sander drums. The material, which is 24 inches wide with 50 feet to the roll, has greater breakage resistance than E-weight paper provides. The paper and cloth combination is, however, being displaced by the increasingly popular vulcanized fiber.

Vulcanized fiber is made from cotton rag base paper treated with zinc chloride, which gelatinizes the cotton cellulose. Multiple sheets (5 to 7) of this treated stock are vulcanized and calendered to give the surface a smooth finish. Coated abrasives with 30-mil fiber backing are used in disk form to a limited degree and are well suited to flexible-shaft-type sanders. The principal application of vulcanized fiber backing is in 20-mil thickness for 24-inch-wide rolls of coated abrasive for drum-sander wraps.

TABLE 18-69.—*Cloth backings for coated abrasives*

Cloth designation and thread count (per lineal inch)	Product form
J (Jeans)[1] 96 by 64 84 by 56	Belts, rolls, disks, and sheets
X (Drills)[1] 76 by 48 72 by 48	Belts, rolls, and disks
H (heavy polyester for heavy-duty lumber sanding and abrasive machining)	Belts only

[1]Twill-weave long-staple cotton. Largest number of threads run longitudinally in the belt.

Backing material largely determines the cost of coated abrasives. Cloth is more expensive than paper, and the weight of the paper determines the price paid for paper-backed belts.

Bond.—Adhesives which bond grit to backing are applied in two layers (fig. 18-157 top), the **make coat** in which the grit is initially secured, and the **size** coat applied later, in combinations as follows:

Make coat	Size coat
Hide glue	Hide glue
Hide glue	Urea resin
Urea resin	Urea resin
Hide glue	Phenolic resin
Phenolic resin	Phenolic resin

An all-glue adhesive bond has the lowest cost and the lowest mineral-holding strength, in part because it softens with heat and high humidity. An all-resin bond has the highest cost and the greatest mineral-holding ability. Due to the relative freedom from tackiness of the resin bond, the tendency of the abrasive to fill or load is considerably reduced. Thermo-setting and waterproof resin bonds have significantly broadened the efficient application of coated abrasives.

Resins and glues can be used with all types of backings. Sheet goods are generally made with the cheaper hide glue because they are used in operations producing little heat. In wet sanding or where heat is produced or where particles receive mechanical shock, resin adhesives give better service.

Distribution and orientation of abrasive particles.—Coated abrasives are manufactured in a continuous multi-stage operation commencing with unreeling and imprinting the backing (fig. 18-159). Printed with product name, mineral, and grade, the continuous web receives a closely controlled "make coat" of adhesive and a carefully monitored and distributed coat of abrasive. Continuously dried in an oven, the strip then passes through a second machine that applies the "size" adhesive coat, and is finally dried in a second continuous oven.

Two methods are used to apply the mineral particles to the moving strip of backing material. When applied by gravity, good mineral orientation is achieved

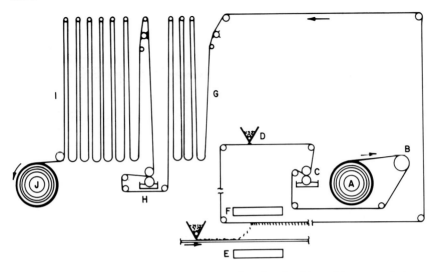

Figure 18-159.—Diagram of continuous operations involved in making coated abrasives. A. Backing material. B. Printing. C. Application of "make coat" of bond. D. Gravity coater. E. Electrostatic coater—negative coater. F. Electrostatic coater—positive electrode. G. Festoons of strip in predrying zone. H. Application of "size coat" of bond. I. Final drying zone. J. Completed coated abrasive.

by agitating the strip causing the particles to stand on their long axes. In the second method the mineral is spread on an endless belt passing between two electrodes (fig. 18-159EF) below and close to the adhesive-coated backing. As the particles of grit on the lower conveyor enter the electrostatic field, they become polarized and are attracted to, and become imbedded upright in, the adhesive bond of the backing.

On **closed-coat** adhesive strips, the abrasive particles completely cover the adhesive side of the backing. On **open-coat** strips spacing of particles on the backing applies about 40 percent less than that on closed-coat strips. Open-coated sandpaper is more flexible and has less of a tendency to clog than does closed-coat paper. Closed-coat paper, with more mineral exposed, may have a longer life than open-coated paper when sanding species that do not tend to clog the abrasive surface.

Birkeland (1971), of the National Institute of Technology in Oslo, Norway, has proposed that abrasive grains should be placed in the "make coat" in such a manner that their cutting tips fall on a common plane; He would accomplish this either by classifying the abrasive particles by length as well as screen size, or by imbedding the particles at variable depths in the "make coat." He found that coarse grits so arranged cut smoother surfaces than those made by commercial papers of the same grit (fig. 18-157 Bottom).

Flexure pattern.—Flexibility of coated adhesives can be varied by controlled breaking of the adhesive into a variety of flexure patterns (fig. 18-160). Most flexibility results from a full-flex pattern with break lines very close together; single-flexing with widely-spaced break lines yields least flexibility.

Machining

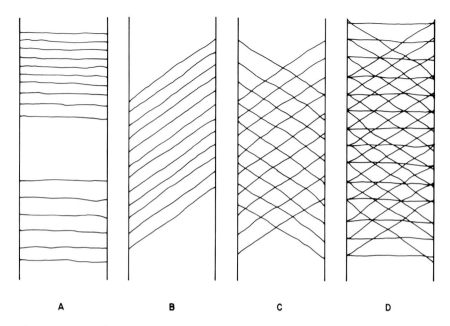

Figure 18-160.—Flexure patterns for coated abrasives. A. Single flex. B. Angle flex. C. Double flex. D. Full flex.

Product form.—Coated abrasives for woodworking are available in sheet, roll, belt, and specialty form. Standard roll length is 50 yards. The abrasive material can be obtained in practically any width, the only limitation being the width in which it was coated. Slitted stock is made into endless belts by means of 45-degree (more or less) pressure bonded joints. The abrasive in the area of the joint is ground off in a bevel skiving operation so that the joint thickness is the same as the thickness of the rest of the belt. Special cones, coils, and rolls are available, with contours preformed to conform to the curved or irregular surface to be sanded.

In addition to the forms described above, it is possible to obtain wheels and sheets of non-woven fabric impregnated throughout with grit and adhesive. These yielding materials can be used with light pressure and abrasive surface speeds on the order of 2,000 to 5,000 fpm to produce matte or satin finishes on wood. Non-woven fabric webs can be impregnated with any of the five mineral abrasives.

FUNDAMENTAL ASPECTS

Users of coated abrasives desire a good surface (scratch pattern) on the workpiece, fast wood removal, low power consumption, low wood temperatures at the abrasive-wood interface, and long belt life. All these desires can seldom be accommodated simultaneously.

Scratch pattern.—A belt sander produces a surface characterized by innumerable overlapping grooves (fig. 18-161) left by individual pieces of grit on the

Figure 18-161.—Scratch patterns of three different grit sizes on hard maple. A. 120 (3/0) grit. B. 80 (1/0) grit. C. 40 (1-1/2) grit. (Photo from U.S. Forest Products Laboratory.)

abrasive belt, usually cutting with negative rake angle (fig. 18-157 center). Patterns made are related to the shape of the cutting point of the grit, its penetration force, its proximity and similarity to, and alignment (fig. 18-157 bottom) with neighboring grit particles (Meyer 1970; Birkeland 1971; Dziobek 1971b; Umetsu et al. 1976ab; and Kato and Fukui 1977). Abrasives with small grit machine smoother surfaces than those with large (fig. 18-161).

Nakamura (1966) found no change in surface quality between up- and down-milling, that is, between operating the sanding belt with or against the feed. According to Ward (1963) and Seto and Nozaki (1966), however, it is preferable to feed panels against the direction of belt travel.

When kiln-dry hard maple was abrasive-planed across the grain, depth of the scratch pattern did not differ significantly from that obtained when abrasive-planing parallel to the grain (Stewart 1972b, 1976a). Kato and Fukui (1976a), however, found that sanding hardwoods across the grain yielded smoother surfaces than sanding parallel to the grain.

Nakamura (1966) found that feed speed through a belt sander was inversely correlated with surface quality. Pahlitzsch (1970) found that belt speed had little effect on surface quality, and quality of surface was almost independent of belt pressure within the range tested.

Rate of wood removal.—Coarse grits remove more wood per unit of time than fine grits (fig. 18-162, Pahlitzsch and Dziobek 1959, 1961b). Data comparing grits of different minerals are scarce; Stewart (1978c) belt sanded kiln-dried hard maple with garnet and aluminum-oxide grits of comparable sizes and distribution and found that initially garnet removed wood at a higher rate, but after 30 minutes aluminum oxide surpassed garnet (fig. 18-163). Franz and Hinken (1954) reached a similar conclusion; in their test on dry hard maple (fig. 18-164), at a belt speed of 5,000 fpm, the garnet and aluminum oxide belts initially removed wood at the same rate, but after 80 minutes the garnet belt was cutting at less than half the rate of the aluminum oxide.

Connelly (1976), however, concluded from 40-minute tests of 100-grit mineral on x-weight drill that garnet consistently removed dry hard maple at a faster rate than aluminum oxide.

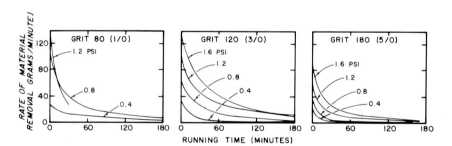

Figure 18-162.—Effect of platen pressure on the rate of wood removal with coated abrasive belts of various grit sizes. Data applies to a 6-inch-wide, 240-inch-long, closed-coat garnet belt, machining 6-inch-wide, by 4-inch-long hard maple (ovendry) parallel to the grain. The belt employed a hide glue bond and travelled at 5,000 fpm. Length of pressure shoe, 4 inches. Each curve is identified by its corresponding platen pressure in psi. (Drawings after Franz and Hinken 1954.)

Increasing the pressure of an abrasive belt against the workpiece increases rate of wood removal (fig. 18-162; Pahlitzsch and Dziobek 1959; Hayashi and Hara 1964; Pahlitzsch 1970; Kato and Fukui 1976ab; Stewart 1978c).

Rate of wood removed increases as belt velocity increases (fig. 18-164; Hayashi and Hara 1964; Pahlitzsch 1970), particularly at higher sanding pressures (Kato and Fukui 1976ab).

The effect of feed rate—other factors remaining constant—on rate of wood removal appears to vary with coarseness of grit. Researchers experienced in

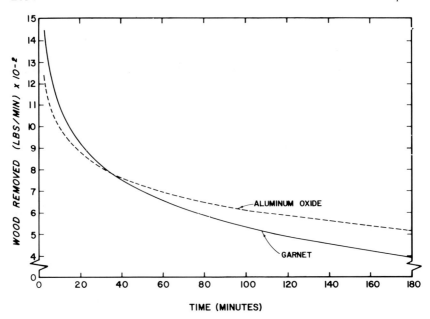

Figure 18-163—Wood removal rates related to running time for aluminum oxide and garnet 6-inch-wide sanding belts close-coated with 100-mesh grit and operated at a platen pressure of 0.75 psi. Belt speed was 5,000 fpm; the grit was glued to the backing with a urea resin sand-size coat over an animal glue make coat. After 30 minutes of running time, both belts removed about 0.08 pounds/minute; thereafter, aluminum oxide removed wood faster than garnet. (Drawing after Stewart 1978c.)

abrasive planing with coarse grits (e.g., 36-grit) conclude that wood removal rate is significantly increased when feed rate is doubled. With finer grits, i.e., 80 to 240, Nakamura (1966) found that rate of wood removal on a belt sander was inversely proportional to feed speed. In the range from 20 to 50 meters per minute the slope of the curve was steep; from 50 to 70 meters per minute the curved flattened somewhat. He found that rate of wood removal was less influenced by feed rate when belts were of fine grit than when they were coarse grit.

Lateral oscillation of a sander belt increases rate of wood removal, particularly with fine grit at low sanding pressures, large oscillation amplitude, and operation on dense wood (Hamamoto and Mori 1973; Kato and Fukui 1976b).

Rate of wood removal decreases as running time on the belt increases (fig. 18-164 top), and decreases as specific gravity of wood increases (fig. 18-164 bottom).

Franz and Hinken (1954) found that at constant sanding pressure rate of wood removal increased as wood moisture content increased because of reduced resistance to grit penetration.

Power requirement.—Cutting force and power required on an abrasive belt are determined by properties of the belt and its arrangement, by the manner in which wood is fed to the belt, and by properties of the wood.

In abrasive planing of hardwoods, grit size is inversely related to power consumption, i.e., power increases so that production through an abrasive

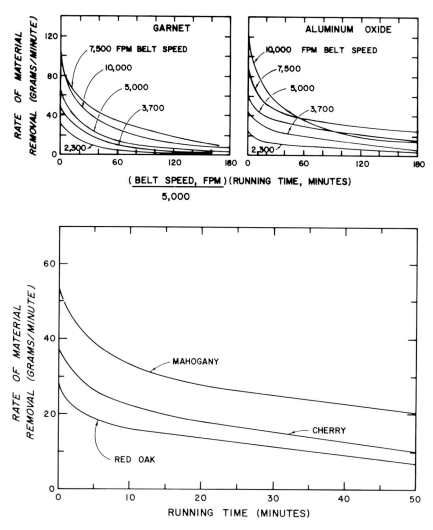

Figure 18-164.—Running time and belt speed related to rate of wood removal. (Top) With garnet and aluminum oxide 120-grit belts 6 inches wide machining dry hard maple parallel to the grain at a sanding pressure of 0.8 psi. (Bottom) Rate of wood removal on northern red oak, black cherry (*Prunus serotina* Ehrh.), and mahogany (*Swietinia macrophylla* King) at 6 percent moisture content; platen pressure 0.4 psi on a 6-inch-wide, closed-coat, 120-grit, garnet belt with hide-glue bond running parallel to the wood grain under a 4-inch-long platen. (Drawings after Franz and Hinken 1954.)

planer is greatly reduced with grit meshes of 60 or more; an abrupt increase in power occurs as stock removal capacity of a grit is reached. Grit meshes of 24 and 36 are used by industry for heavy stock removal (Stewart 1974c). Parallel cutting force is a function not only of grit size but also of grit spacing. Umetsu et al. (1976a) found that parallel cutting force on a single-point pyramid-shaped tool was maximum when the distance between grooves (scratches) was 0.7 to

0.8 of the distance between adjacent grooves at which adjacent groove surfaces would intersect at the workpiece surface. Nakamura (1966) measured power required to machine three hardwoods with 129-cm-wide aluminum-oxide paper-backed belts in grits from 80 to 150 and a silicon-carbide cloth-backed belt of 240 grit; with data on a number of feed arrangements pooled, belt specific power (energy to remove 1,000 cubic cm of wood) required to machine parallel to the grain was approximately linearly related to grit mesh in the range from 80 to 240 mesh, as follows:

Species	Grit size (mesh)	
	80	240
	----------kWh/1000 cubic cm----------	
Lauan (*Shorea* sp.)	0.10	0.20
Sen (*Kalopanax* sp.)	.15	.33
Makamba (*Betula* sp.)	.15	.57

Similarly, when abrasive planing across the grain, belt power required increases with grit mesh. Stewart (1976a) machined hard maple at 8 percent moisture content across the grain with width of cut of 12 inches, depth of cut of 0.040 inch, belt speed of 5,800 fpm, and feed rate of 36 fpm; power required was as follows:

Grit mesh	Maximum depth of scratch	Belt power requirement
	Inch	*kW*
36	0.012	18.8
60	.006	20.4
80	.005	24.0

Pressure between belt and workpiece is positively correlated with power required, i.e., as pressure increases, power requirement increases (Nakamura 1966; Pahlitzsch 1970; Kato and Fukui 1976ab).

With sanding pressure, depth of cut, and feed speed constant, increased **belt velocity** should require increased belt power, but the evidence in the literature is conflicting. Pahlitzsch (1970) notes that belt velocity has practically no influence on power requirement; Nakamura (1966) found that belt power required was proportional to belt velocity. Because of blunting of grit particles, belt power requirement increases with increased **belt running time**.

Stewart (1970c, 1972b) found that the power required by a belt sander is positively correlated with rate of wood removed and is therefore positively correlated with both **depth of cut** and **feed rate** of the wood being sanded (fig. 18-165). In further tests on air-dry and kiln-dry yellow-poplar, northern red oak, and hickory, with open-coat, aluminum-oxide, cloth-backed, resin-bonded belts having grit sizes from 36 to 100, Stewart (1974c) found that power required was linearly proportional to depth of cut at moderate depths of cut, but increased curvilinearly above a straight line at depths of cut greater than 1/8-inch. Nakamura's (1966) data on belt sanding hardwoods indicate nearly direct proportionality between rate of wood removal and belt power required; feed power was also shown linearly related to rate of wood removed. Stewart (1974c) concluded that the effect of depth of cut and feed rate on belt power required could best be

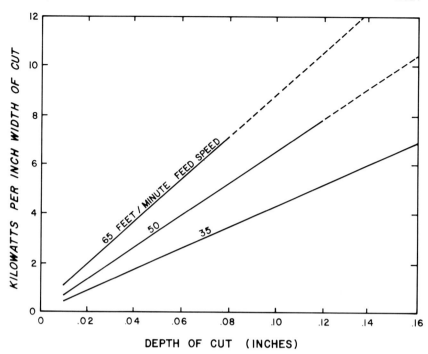

Figure 18-165.—Power requirement to drive sanding belt related to depth of cut and feed speed. The 36-grit, open-coat, aluminum oxide belt was driven at 5,800 fpm while machining hard maple at 9-percent moisture content. (Drawing after Stewart 1972b).

expressed as a function of the product of these two factors squared, and that this product accounted for more than half the observed variation in power requirement.

In abrasive planing grit is forced into the wood to produce grooves or scratches; resistance to such penetration is proportional to **specific gravity**. Stewart's (1974c) analysis of belt power requirement shows such a positive correlation. The same study indicates that **moisture content** of yellow-poplar, northern red oak, and hickory (in the range from 8 to 20 percent) is positively and linearly correlated with belt power required. **Grain direction** strongly influences requirements for belt power. Stewart (1974c) found that only three-fourths the power needed to sand parallel to the grain was required to sand dry hard maple across the grain (in the veneer cutting direction), when depth of cut, feed speed, and grit size were held constant. He concluded that hard maple could be sanded across the grain with an 80-grit belt yielding scratches only 0.005 inch deep with the same power expenditure as sanding along the grain with a 36-grit belt yielding scratches more than twice as deep (0.012 inch).

Stewart's (1974c) regression equations predicting power required to drive abrasive planer belts and knife-carrying planer heads summarize the effect of major variables when planing parallel to the grain (table 18-70). Abrasive planing requires significantly more power than knife planing, but machining defects may be less deep than when knife planing.

TABLE 18-70.—*Regression equations for power required (P, kilowatts) to abrasive-plane and knife-plane 4-inch-wide boards of four species parallel to the grain, related to depth of cut (DC, inch), feed rate (FR, inches/minute), grit size (GS, mesh number), moisture content (MC, percent), and specific gravity (SG)* (Stewart 1974c)[1]

Species and equation	Regression coefficient (R^2)	Standard error of the estimate (S_E)
ABRASIVE PLANING[2]		
Basswood (*Tilia americana* L.)		
$P = -8.51 + 4.71(DC \times FR) - 0.16(DC \times FR)^2 + 0.18MC + 0.07GS$	0.93	2.1
Yellow-poplar		
$P = -11.05 + 6.14(DC \times FR) - 0.23(DC \times FR)^2 + 0.33MC + 0.10GS$	.89	2.9
Northern red oak		
$P = -9.89 + 7.43(DC \times FR) - 0.41(DC \times FR)^2 + 0.21MC + 0.09GS$	.92	2.5
Hickory		
$P = -12.18 + 10.63(DC \times FR) - 0.94(DC \times FR)^2 + 0.23MC + 0.13GS$	.87	2.8
All four species, data pooled		
$P = -20.07 + 6.34(DC + FR) - 0.30(DC \times FR)^2 + 0.21MC + 0.09GS + 21.39SG$	.87	2.9
KNIFE PLANING[3]		
Basswood		
$P = -0.37 + 0.63(DC \times FR) - 0.04(DC \times FR)^2 + 0.06MC$	.88	.2
Yellow-poplar		
$P = -0.31 + 0.78(DC \times FR) - 0.09(DC \times FR)^2 + 0.05MC$	.79	.3
Northern red oak		
$P = -0.15 + 0.85(DC \times FR) - 0.10(DC \times FR)^2 + 0.05MC$	.89	.2
Hickory		
$P = -0.03 + 1.03(DC \times FR) - 0.13(DC \times FR)^2 + 0.07MC$	.90	.2
All four species, data pooled		
$P = -1.21 + 0.75(DC \times FR) - 0.07(DC \times FR)^2 + 0.05MC + 2.02SG$	.81	.3

[1] Cuts were 0.01 to 0.06 inch deep, feed rates were 35 to 55 fpm, grit sizes were 36 to 100 mesh, and moisture contents were 8 to 20 percent. Based on green volume and ovendry weight, specific gravities averaged: basswood 0.37, yellow-poplar 0.49, northern red oak 0.62, and hickory 0.67.

[2] The abrasive planing was done with 18-inch-wide, 103-inch-long, open-coat, aluminum oxide belts running at 5,800 fpm.

[3] The knife planing was done with four jointed knives having 25° rake angle, mounted in a cutterhead with 5-inch-diameter cutting circle, revolved at 3,750 rpm.

Wood surface temperature.—At all sanding pressures, coarse grits tend to produce higher temperatures on the surface of the wood than fine grits (fig. 18-166 left) because more machining energy is converted to heat. High belt speeds also elevate temperatures at the wood surface (fig. 18-166 right) because the belt

Machining

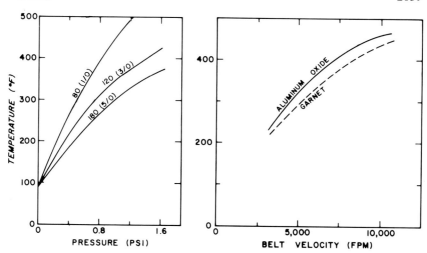

Figure 18-166.—Surface temperature of dry hard maple when belt sanded parallel to the grain. (Left) Temperature related to sanding pressure and grit size. (Right) Temperature related to belt velocity and grit mineral. (Drawings after Franz and Hinken 1954.)

has a shorter time to cool between successive engagements with the wood than at slow belt speed; also, high rubbing speeds promote high temperatures. Mineral cutting points dull in a process similar to knife dulling and surface temperatures may reach more than 400°F. (fig. 18-166 right); at such temperatures, belts may clog or load because the wood is somewhat plastic and also because grit bonds of animal glue may become tacky and soft enough to permit some movement of grit. Franz and Hinken (1954) found that short platens reduced contact time and surface temperatures, and decreased glazing of wood surfaces; platens studied were 4 to 10 inches long.

Frictional drag and accompanying temperature rise can be reduced by injecting an air cushion between sanding belt and platen (Dziobek 1971a; Pahlitzsch and Argyropoulos 1974).

ABRASIVE PLANING WITH WIDE-BELT SANDERS

Abrasive planers are distinguished from other belt sanders in that their primary function is thicknessing (while smoothing) oversize lumber, particleboard, or plywood. The need to thickness rough hardwood lumber—usually dry—calls for very aggressive cutting action, substantial cutting horsepower, and feedworks adaptable to warped lumber.

Configuration of abrasive planers.—Abrasive planers are usually **double-deck**, simultaneously machining both top and bottom of the wood in one pass. Feed speeds may be as fast as 700 fpm. Coated abrasive belts are commonly 50 to 52, 63, or 67 inches wide. Belts 103 or 142 inches long are widely used. Most machines have four heads, and eight heads are not uncommon; half the heads cut on the top and half on the bottom (figs. 18-167, 18-168). The belts run counter to the feed direction.

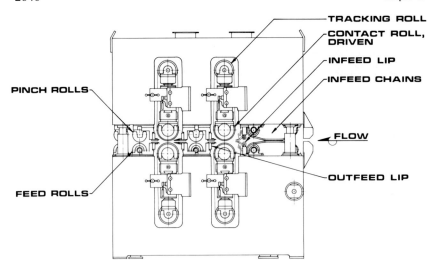

Figure 18-167.—Four-head, batch-feed abrasive planer for dry hardwood lumber and glued lumber panels. Abrasive belts travel counter to lumber flow. Both pairs of heads are directly opposed to insure thickness control and to remove an equal amount of stock from each side of the lumber. On a 53-inch-wide machine capable of feed speeds to 200 fpm, each head carries 250 hp. A total of 30 hp drives infeed chains and bottom feed rolls. Top pinch rolls are spring-loaded idlers. (Drawing from Kimwood Corp.)

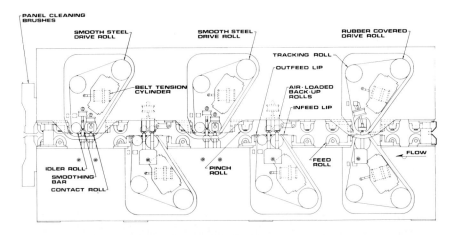

Figure 18-168.—Six-head particleboard sander. Abrasive belts travel counter to direction of panel flow. The first two heads are opposed for maximum stock removal and size control. The next two pairs of top and bottom heads are staggered and provided with both contact rolls and smoothing bars to yield smooth surfaces. On a 63-inch machine capable of feed speeds to 250 fpm, the first two heads each carry 250 hp, the second pair each 150 hp, and the third pair each 125 hp. A total of 40 hp drives the feed rolls. (Drawing from Kimwood Corp.)

The **primary** heads, i.e., the first top and bottom heads, carry coarse grits and do the major cutting job. Cuts of 0.03 inch per primary head on plywood, 0.04 inch on particleboard, 0.06 inch on hardwood glued-up lumber cores, and up to 0.07 inch on rough dry hardwood boards are common; cuts on any of these materials should not exceed about 0.12 inch, even when using coarsest grits and slowest feed speeds. The steel **contact roll** (fig. 18-168) on each primary head usually has a smooth surface. **Secondary** or finishing heads carry finer grits and also have smooth steel contact rolls. Contact rolls on secondary heads may be followed by smoothing bars.

Belt speeds of 5,000 to 9,500 fpm are common. Belts are assembled with a skived splice 3/8-inch wide made at an angle to the length of the belt. Cloth backing for the abrasive is the toughest possible X-weight drill or H-weight heavy twill, woven from long-staple cotton and filled to increase wear resistance and stiffness. A thin film of urethane polymer on the inside of belts for primary heads may increase life of belts and rolls, but on secondary heads tends to strip the graphite covers from smoother bars.

The abrasive grains are bonded to primary belts with phenolic resin on both the underlying **make coat** and the later applied **size coat. Open-coat** construction—that is, with abrasive grains spaced apart—is more flexible and preferred for sanding with fine grits on stock such as mouldings. The closed-coat belts are heavy and stiff and afford a maximum number of cutting points, but have a tendency to load up; closed-coat belts are commonly used on primary heads for aggressive stock removal from flat surfaces.

Dry hardwood lumber.—In furniture plants, dry hardwood lumber may be thicknessed by abrasive planing before or after it is ripped and prior to cross cutting to yield clear cuttings. Wide lumber may be fed in-line, one board following another at high speed (up to 700 fpm) as in a planer equipped with knives. Strips may be match-fed in multiple widths but at slower speed (about 120 fpm). In-line abrasive planers usually have four top and four bottom heads, whereas batch machines may have only two of each (fig. 18-167). For structural lumber, such as yellow-poplar studs, a four-head, in-line abrasive surfacer might incorporate a pair of opposed cutterheads to knife plane the edges. Abrasive planing requires significantly more power than knife planing, but has the following advantages:

- Splitting of cupped boards is reduced.
- Tearout around knots and in other areas of sloped grain is reduced.
- Chipped knots and sniped board ends are largely eliminated.
- Planing allowance can be reduced.

Belts carried on the four-head batch abrasive planer illustrated in figure 18-167 for machining kiln-dry white oak and black tupelo lumber at 120 fpm might be as follows:

Specification	Primary heads	Secondary heads
Belt width, inches	52	
Belt length, inches	108 or 142	
Mineral	Zirconium-based	
Grit mesh	24-36	50
Backing	H-weight	X-weight drill
Bond	resin	
Open- or closed-coat	closed	
Belt life, hours of operation	8 to 24	

Dry laminated oak squares.—In thicknessing laminated oak squares (fed so that glue lines are vertical), a heavy roughing cut is required to level the laminations, which vary slightly in width. To accomplish this with minimum power expenditure, a 24-inch-wide two-head thicknesser has been applied in which the first heavy cut is made by a 20-horsepower spirally segmented cutterhead carrying carbide knives. The second cut, 1/64-inch deep, is made by a 40-horsepower abrasive belt; both heads machine from the top to achieve thicknessing cuts. Feed speed is 65 fpm (Anonymous 1976a).

Green hardwood lumber.—Considerable hardwood is thicknessed green in knife planers, e.g., cants to be gang ripped in pallet plants. In such applications, however, chipped knots or torn grain are not critical defects, and the more expensive and energy intensive method of abrasive planing is not usually competitive.

Glued-up lumber cores.—Lumber cores to be overlaid with veneer must be smoothly surfaced with no chipped grain on surfaces or edges. If edge-banded, the edge-band lumber strip must not be chipped. Such glued-up panels are commonly surfaced by abrasive planing. The glue lines in the panels, which cause rapid dulling of planer knives, have less effect on abrasive belts, and chipped grain is avoided. Stewart (1975a) has shown that panels oriented to feed across the grain require less power than those fed parallel to the grain, and therefore finer grits can be used without exceeding power available. Use of the finer grits reduces scratch depth, thereby easing subsequent polishing operations. Belts for abrasive planing glued-up edge-banded hardwood lumber cores at 70 fpm might have the following specifications:

Specification	Primary heads	Secondary heads
Belt width, inches	52	
Mineral	Zirconium based	
Grit mesh	36	60
Backing	H-weight	
Bond	resin	
Open- or closed-coat	closed	
Belt life, lineal feet of panel	28,000-60,000	10,000-30,000

Particleboard.—Abrasive planers are well suited to thicknessing and smoothing particleboard, particularly when edge-banded with lumber that may chip if knife planed. Maximum depths of cut per head have been related to feed speed and grit size (table 18-71).

For particleboard Connelly (1975c) proposed use of three pairs of sanding heads, the first equipped with steel contact rolls, the second with rubber covered contact rolls, and the third with smoothing bars.

Kimwood Corporation also recommends six-head machines for particleboard but with slightly different arrangement (fig. 18-168) in which the first pair of heads have smooth steel contact rolls and no smoothing bars; the second and third pairs also have smooth steel contact rolls, but are followed by smoothing bars. The machine shown in figure 18-168, when thicknessing and smoothing edge-bonded particleboard core panels at 60 to 250 fpm, might carry belts as follows:

Specification	Primary heads	Secondary heads
Belt width, inches	---------------------52, 63, or 67-------------------	
Belt length, inches	-------------------------142-------------------------	
Mineral	------------------Zirconium based------------------	
Grit mesh	36	50
Backing	H-weight	X-drill
Bond	-------------------------resin------------------------	
Open or closed coat	-------------------------closed-----------------------	
Belt life, lineal feet of panel	28,000	60,000

TABLE 18-71.—*Maximum depths of cut when abrasive planing particleboard at four feed speeds with eight grit sizes*[1] (Data from Kimwood Corp)[2]

Grit mesh	Feed speed, feet per minute			
	50	100	150	200
	------------------------------Inch------------------------------			
24	0.100	0.060	0.032	0.025
36	.075	.040	.028	.020
40	.050	.025	.020	.015
50	.040	.020	.016	.012
60	.025	.015	.010	.009
80	.015	.010	.008	.006
100	.012	.008	.005	.004
120	.010	.005	.004	.003

[1] Data appropriate for a belt velocity of about 6,300 fpm.
[2] Personal correspondence dated May 2, 1980 with D. Evans.

Plywood.—Hardwood plywood or particleboard being smoothed and polished prior to application of finish requires finer grits than those used when sizing banded particleboard cores to be veneered. Plywood with a character-marked white oak veneer face being sanded at 60 fpm on the machine illustrated in figure 18-169 might carry the following belts:

Specification	Primary heads	Secondary heads
Belt width, inches	52	
Belt length, inches	108	
Mineral	garnet	
Grit mesh	80	100
Backing	X-drill	
Bond	resin	
Open- or closed-coat	Closed	Open
Belt life, lineal feet of panel	28,000-60,000	10,000-25,000

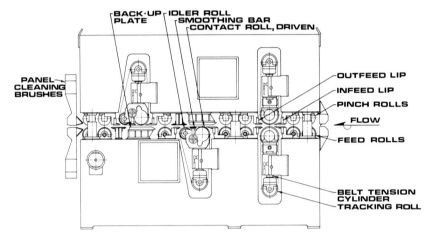

Figure 18-169.—Four-head plywood sander. Abrasive belts travel counter to panel feed. The first two heads are opposed for maximum stock removal and thickness control. The second pair are staggered and have smoothing bars. On a 53-inch-wide machine capable of feed speeds to 200 fpm, the first two heads each carry 150 hp, the second pair 125 hp each. A total of 20 hp drives the seven bottom feed rolls; the top rolls are spring-loaded idlers. (Drawing from Kimwood Corp.)

Thicknessing and polishing plywood panels is made difficult by variations in thickness of cross band veneers and face veneers, and by tape splices in face veneers. One approach to avoid sanding through the face veneer calls for thicknessing the completed panel by removing excess thickness only from the back of the panel, reserving a minimum cut for levelling and polishing the face. Appropriate grits are as follows:

	Dense hardwoods	Softer hardwoods
	Grit mesh	
First cuts	80	80
Finishing cuts	100	120
Polishing cuts	150	180

Ward (1975c) suggested somewhat coarser grits (50 and 60) on first cuts and finer grits (180 and 220) on polishing cuts.

Belt life.—Abrasive grits dull, and spaces between them become loaded with fine wood particles as running time is extended. The literature contains no data on the newer zirconium-based grits used for abrasive planing, but Stewart (1978b) has provided some information on northern red oak and yellow-poplar at 12-percent moisture content planed with a 36-grit, aluminum-oxide belt travelling at 5,800 fpm. He found that belt loading is reduced and belt life prolonged by using high stock removal rates, e.g., up to 90 fpm feed rate with depths of cut up to 0.12 inch; optimum combinations were not determined. These findings supported earlier conclusions drawn from belt-life data on abrasive planing of *Pinus ponderosa* Dougl. ex Laws. (Stewart 1976b).

Connelly (1975c), in discussing belt life versus belt cost, suggests that open-coat garnet with an all resin bond may be the most economical combination to use for abrasive planing. The literature offers no comprehensive data relating belt construction and cost to belt life.

Belt maintenance.—Belt backings and some of the bonding materials are hygroscopic so their length changes with moisture content. Belts heated by friction during use have lower moisture content (and are shorter) than after overnight cooling in a humid atmosphere. Belts should be stored in a room at about 70°F maintained at 50 percent relative humidity. Used belts, belts taken from the machine between shifts, or belts removed from the carton prior to use, should be hung on rack spindles in horizontal position; the spindles should be at least 4 inches in diameter. Belts exposed to high humidity can be reconditioned by storing 12 to 24 hours at 50 percent relative humidity. Belts in cartons should be stored on end, unopened, and insulated from damp floors. If belts have been stored a long time in the shipping carton, they should be removed and hung on the spindles for at least 24 hours to relieve stresses and normalize moisture content (Connelly 1975b).

OTHER MACHINES TO SAND FLAT SURFACES

The sander designed for mounting on a tenoner (fig. 18-229) for fine-finishing flat or formed edges, is typical of many arrangements which combine belt sanders with other machines.

In addition to abrasive-belt planers, there are numerous other machines designed to machine flat surfaces and edges.

Disk sander.—The disk sander consists of a work table set across the face of an 18- to 48-inch-diameter revolving plate to which coated abrasive disks are attached by adhesive or sometimes by mechanical clamp. The worktable usually can be tilted to permit angle or miter sanding. Mid-radius speed of a disk sander is usually about 4,000 fpm with peripheral speed of about 8,000 fpm; disks are commonly driven by a 5-hp motor.

The machine is especially adapted for bevel and angle sanding, end-grain sanding, and sanding of drawers and square portions of turned legs and posts. Its major disadvantage is its pattern of circular scratches which may have to be removed before applying finishes.

For heavy material removal with disk sanders, Behr-Manning (1967) recommended closed-coat 40-grit aluminum oxide or silicon carbide with an all resin bond on vulcanized fiber backing. For lighter work, closed-coat garnet or aluminum oxide on E-weight paper or X-weight cloth, with an all resin bond or resin on glue bond was recommended as follows:

Species and operation	Grit mesh
Medium density hardwoods	
Rough sanding	60
Finish sanding	100 or 120
Hard hardwoods	
Rough sanding	50
Finish sanding	100 or 120

Drum sander.—Multiple-drum sanders, described in considerable detail by Koch (1964a, p. 415-419), have been made obsolete by development of the wide-belt sander. Interested readers are referred to the 1964 publication.

Stroke sander.—A flat-belt stroke sander is used to remove scratches left by a previous wide-belt (or multiple-drum) sanding operation and to prepare a surface suitable for application of finishes. It is a horizontal-belt machine; some designs carry two 6-inch-wide belts of different grit size, each separately powered and running in opposite directions (fig. 18-170). In this case the belts are used one at a time, first the coarse and then the fine, and are brought into contact with the work by a pressure pad that reciprocates along the belt while the operator pushes the worktable under the belt to apply the abrasive action to the

Figure 18-170.—Double-belt, reversible, hydraulic stroke-sander. (Photo from Mattison Machine Works.)

workpiece as desired. The machine illustrated will handle a workpiece 8 feet long. The speed of the 5-hp hydraulic stroking mechanism is adjustable from a creeping speed to a maximum of 90 strokes per minute, the length of stroke being adjustable to conform to the length of stock. Pressure on the sanding pad is lever controlled by the operator. Each belt is powered by a 7-1/2-, 10-, or 15-hp motor. The belts are commonly driven at 5,000 or 7,500 fpm. The belts may be run in opposite directions so that those fibers that were bent in the first stage of sanding will be cut off during the second stage. Each belt on the 8-foot machine illustrated is 6 inches wide by 28 feet long.

To facilitate straight-line production, stroke sanders may be so designed that the stock to be machined passes continuously beneath the belts on an endless bed. In this case both belts, of a two-belt machine, remain in contact with the work.

For automatic stroke sanders, Behr Manning (1967, p. 55) recommended closed coat abrasives on E-weight paper or J-weight cloth, as follows (glue/resin bonds for less severe service and resin/resin bonds for particleboard and hardboard):

Function and material	Grit mesh	Mineral
Cutting down		
Particleboard..........................	80	Garnet or aluminum oxide
Hardboard............................	100	Silicon carbide
Hardwoods (solid)	100	Garnet or aluminum oxide
Straight-grain hardwood plywood	100	Aluminum oxide
Fancy-face hardwood plywood	100	Garnet or aluminum oxide
Butt, crotch, or swirl plywood	120 to 180	Garnet or aluminum oxide
Polishing		
Hard hardwoods (solid)	150	Aluminum oxide
Medium-density hardwoods (solid).........	180	Aluminum oxide
Fancy-face hardwood plywood	180	Aluminum oxide

In correspondence (May 1980), industry consultant H. H. Connelly suggested aluminum oxide to cut down hardboard and garnet to cut down straight-grain hardwood plywood; he also recommended garnet for polishing the three products tabulated above.

Edge sander. It is the primary purpose of the edge sander to machine a flat surface on the edge of a workpiece. It carries a 6- to 9-inch-wide abrasive belt on two pulleys having vertical axes; thus the plane of the belt is perpendicular to the horizontal worktable. The belt is backed by a platen. To minimize heating and loading of the belt, either the belt or the platen may oscillate. If a corrugated, moving, rubber, contact belt is interposed between the abrasive belt and the platen, heating of the abrasive belt is reduced and the efficiency of the cutting action is increased.

Most edge sanders are hand fed, but they may be equipped with power feed and a tilting belt to permit the sanding of edge bevels. With 5 hp on the sanding belt and 1/2-hp on the feedworks, feed speeds of 20 to 110 fpm are usual. Belts run at 3,000 to 5,000 fpm.

Abrasive belts for flat-surface edge sanders are usually glue-resin or resin/resin closed-coat bonded on E-weight paper or X-weight cloth; Behr-Manning (1967, p. 57) suggests abrasives, as follows:

Function and material	Grit mesh	Mineral
Cutting down		
Hard hardwoods	60	Garnet or aluminum oxide
Medium-density hardwoods (with grain or end-grain)	80	Garnet or aluminum oxide
Polishing		
Hard hardwoods	100	Aluminum oxide
Medium-density hardwoods (with grain or end-grain)	120	Garnet or aluminum oxide

To minimize overheating of end-grain pieces, many operators prefer open-coat garnet belts, particularly on those that run at slower speeds (Stevens 1977). Garnet 120-grit mineral, with all resin bond, on X-weight cloth backing has given good service when edge-sanding oak (Rea 1975).

Wide-belt sander.—Multi-head, double-deck, abrasive planers for thicknessing and smoothing both sides of the workpiece have been previously described (figs. 18-167, 18-168, and 18-169). Single-deck machines designed for cutting down and polishing may be arranged with upper heads to cut against a lower fixed or traveling bed. The belts may run over serrated rubber contact rolls, smooth steel contact rolls, smoothing bars, or a combination, depending on requirements. Machines are available in widths from 18 to 64 inches to accept stock up to 3 inches thick. For a 50-inch machine equipped with a 5-hp feed motor and 60-75 hp, 50 hp, and 40 hp on the first, second, and third heads, respectively, feed speeds from 20 to 60 fpm are practical. A machine of this type working on oak panels or frames might produce 16,000 lineal feet per 8-hour shift at a feed speed of 50 to 60 fpm. Grit progression might call for 100 mesh on the first belt, 140 mesh on the second, and 180 on the third. A glue/resin bond with closed-coat arrangement on E-weight paper or X-weight cloth is commonly used.

The nuances of polishing and burnishing prefinished plywood panels are complex and beyond the scope of this text. Readers interested in an introduction to the subject are referred to Carroll (1977) and Stevens (1978).

Exterior surfaces of assembled drawers are manually sanded to remove glue and to smooth joints. The drawer sander used for this purpose carries a single wide belt which runs over a horizontal bench-level platen with a 90° angle at one end to permit flush sanding of projecting joints. Belt speeds of 3,000 to 4,000 fpm are usual. Garnet grit of 50 to 60 mesh resin-bonded in closed-coat arrangement on X-weight cloth is commonly used. If considerable glue must be removed, aluminum oxide may give better service than garnet.

Polisher with drum of abrasive-impregnated, non-woven nylon.—A matte or satin surface on sheet material can be produced using a top-surfacer polishing machine employing a yielding abrasive drum of non-woven nylon impregnated throughout with grit and adhesive (fig. 18-171). It can be used to remove projecting fibers from white wood and to scuff sealer or filler coats. The drum may be given 3/8-inch oscillation at frequencies ranging from 0 to 200 cycles per minute. The drum is dressed, or trued as it becomes worn, so that a range of drum diameters from 12 inches to 6 inches may be accommodated. Typically, a 7½-hp, variable-speed motor is arranged to drive the drum at a

Figure 18-171.—Two-head, roller-bed, top-cutting polishing machine employing abrasive drums. (Top) Overall view. (Bottom) One abrasive drum partially removed. (Photos from Timesavers, Inc.)

constant peripheral speed of 3,000-5,000 fpm, regardless of drum diameter. Light abrasive pressure is used with this type of machine. The workpiece is backed up by a powered "billy roll" on the under side opposite the abrasive drum. Feed speeds from 20 to 60 fpm are practical. A 1-hp feed motor driving the two lower steel rolls, the two rubber-covered top rolls, and the "billy roll" provides the feed force and control necessary. On some machines of this type the drum rotates against the feed; on others, operators have found it advantageous to rotate the drum with the feed. An automatic sensing device is required to control the abrasive-to-stock contact so that the drum is lowered and raised at beginning and end of the cut to avoid rounding the leading and trailing edges of each workpiece. Non-woven nylon impregnated with garnet, aluminum oxide, or silicon-carbide is available in the wheel form required for this machine.

MACHINES TO SAND BROAD CURVED SURFACES

While it is possible to sand the surfaces of a ski, serpentine drawer front, curved slat, or other similar shapes on an open drum sander, a belt sander with suitable feed roll orientation can do the job much quicker and with greater precision. A typical machine will accept a workpiece of variable curvature as thin as 1/64-inch and as thick as 2½ inches. The 8-inch top feed roll is comprised of rubber segments separated by thin spacers to insure that the stock is fed positively at feed rates of 40-75 fpm. The abrasive belt on the bottom side travels at 3,800 fpm counter to the feed direction and runs over a 14-inch, serrated rubber contact roll. The resiliency and shape of the contact roll achieves a longer contact area than would a felt-covered steel roll. The machine requires approximately 2½ hp to drive the abrasive belt and feed rolls for every foot of width, i.e., a 24-inch machine requires 5 hp. Roughing jobs are done with grits from 36 to 120, whitewood polishing jobs are done with grits from 120 to 320, and finishing following the sealing coat may require grits from 280 to 400. Adjustment of the counterweights controlling top feed-roll pressure regulates the aggressiveness of the abrasive action. As an example of production capacity, curved chair backs for theater seats might be sanded on one side at a rate of 20 pieces per minute. Occasional workpiece kickback from these machines has limited their application.

MACHINES TO SAND CONTOURED EDGES

Contoured edges produced by a straight shaper head following a band-sawn edge contour can be sanded on a variety of machines.
Edge sander.—An edge sander can be used for this operation if one of the pulleys (spindle vertical) is arranged so that the work can be presented to the belt as it passes around the pulley. A very small idle pulley, or wood form, may be required to accommodate small radii.
Spindle sander.—A vertical, oscillating, rotating spindle that protrudes through a horizontal worktable can be used to sand edge contours in much the

Machining 2051

same way that a single-spindle shaper is used to shape curved contours. Surface speeds of 2500-3000 fpm are suitable. Close-coat garnet with J-weight cloth or E-weight paper backing is commonly used with grits in the range from 60-150.

Variety sander.—A variety sander (fig. 18-172) that carries a narrow abrasive belt in the same orientation as a band saw blade can be used to get into very sharp corners. It has the disadvantage of sanding across the grain when used for this purpose. Belt widths are 1/2-inch to 1½ inches. Abrasive speeds seldom exceed 900 fpm. Open-coat garnet on J-weight angle-flex cloth is generally preferred for this operation. Grit sizes may vary from 80 through 120.

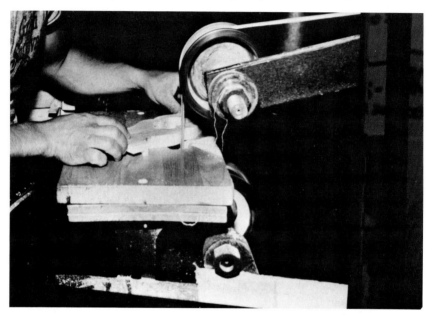

Figure 18-172.—Variety sander.

Scroll sander.—A scroll sander is used to sand interior curves of such items as pierced screens or grills. It consists of a flat worktable through which protrudes a square-ended metal blade. A narrow sanding belt runs over the top of the blade. The top of the blade serves as the idler, and the side of the blade serves as the platen. Abrasive speed and specifications are similar to those applicable to the variety sander described above.

Open-drum sander.—An open-drum sander is a cantilevered pulley or drum, sheathed on the outside with a coated abrasive, that rotates around a horizontal axis. It may be a hard cylinder with a felt pad cemented between cylinder and abrasive, or it may incorporate an inflatable sleeve to back up the belt. Drum diameter is commonly 8 inches with surface velocity of 1,000 to 3,000 rpm. On hardwoods both garnet and aluminum oxide closed-coat abrasives are used on J-weight cloth backing, as follows:

Species and operation	Grit mesh
Medium-density hardwoods	
Cutting down	100
Finishing	120
Hard hardwoods	
Cutting down	80
Finishing	120

MACHINES TO SAND MOULDED EDGES

Moulded edges machined by a formed shaper or moulder head (fig. 18-146 top) can be sanded on a variety of machines, of which the following are typical.

Formed-block sander.—A formed-block sander is a belt sander. In one configuration it looks similar to the stroke sander (fig. 18-170) except that the table is not moved during sanding, and the pressure shoe or block is formed to make the sanding belt conform to the desired workpiece shape. In usual practice the pressure and stroking of the formed block involves hand operation rather than mechanical stroking. A slightly resilient material is sometimes used to face the formed block.

Another configuration (fig. 18-173 top) resembles the edge sander except that no oscillation is provided and the back-up platen is made much shorter, given a resilient mounting, and formed to cause the belt to assume the desired shape. When manually fed, this configuration is useful because it permits the operator to use both hands on the workpiece. Furthermore, if the belt is pushed considerably out of line by the formed block, the edges of concave, straight, and convex workpieces can be sanded. A single-flex cloth belt driven at approximately 3,000 fpm is frequently used in this application. A double- or full-flex belt is sometimes too flexible to retain the shape of the moulding. A 3-hp drive is suitable. Power-fed machines are also available (fig. 18-173 bottom).

Formed-wheel sander.—If the moulded shape is not too deep or complex, a belt can be made to turn around a formed, idle contact wheel. The work to be sanded can then be presented to this wheel. In an alternative configuration, abrasive is cast into a formed wheel which in turn can be presented to the workpiece (fig. 18-174). The spindle may be adjusted to any degree of tilt from vertical to horizontal, and elevated or depressed to bring the motor completely below the table.

Abrasive-impregnated non-woven nylon wheel.—If a non-rigid, narrow wheel is dressed to conform to the workpiece shape desired, it is capable of doing a light polishing job on mouldings. One version of such wheels resembles a stiff nylon rug impregnated with silicon carbide, and is available in grits from 60 to 500 mesh.

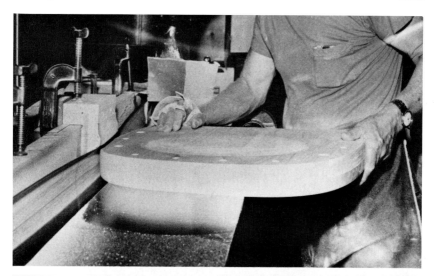

Figure 18-173.—(Top) Manually fed belt sander smoothing contoured edges; the belt rides against a formed block. (Bottom) Power-fed, edge-contour sander viewed from the infeed end; sander belt is at left of workpiece.

Brush-backed sanding wheel.—It is the purpose of the brush-backed wheel to sand contoured, carved, reeded, or turned work. It is suitable for light sanding and polishing, but is not too effective in removing moulder knife marks. Excessive abrasive pressure tends to distort the pattern and round whatever sharp corners the moulding may have. The assembly consists of a number of brushes in holders spaced around the periphery of the wheel. Multiple strips of abrasive are stored within the wheel and dispensed so that a brush backs up each strip of

Figure 18-174.—Formed-wheel sander. (Top) Arranged to contour-sand a curved moulding. (Bottom) Foam wheel loaded with abrasive grit.

abrasive. Centrifugal action and the stiffening effect of the brushes causes the shredded tips of the abrasive to reach into grooves and depressions in the workpiece.

Wheels of this type may be pedestal mounted, flexible-shaft mounted for portable use, or in-line mounted for power-feed, straight-line production. Width of wheel face ranges from 2 to 6 inches and rotational speeds are commonly 600 to 1800 rpm. The abrasive strips may be slashed into filaments if the pattern to be sanded has considerable detail. Garnet is most used, but both aluminum oxide and silicon carbide are applicable. Both closed and open coatings are used depending on the loading tendency. E-weight paper and X-weight cloth backing are commonly used for moderate contours. J-weight, angle-flex cloth backing is suitable for intricate patterns. Grit size ranges from 100 to 150 for most work.

Radial flap wheels operate on a principle similar to the brush-backed wheel. They substitute closely packed, radially oriented abrasive filaments for the action of the brushes. In some applications the shape of the contour to be sanded is dressed into the wheel. Peripheral speed of a radial flap wheel is about 4,000 fpm. Closed-coat garnet on J-weight cloth backing in grit sizes from 60 through 150 is commonly used.

Multiple-head moulding sander.—From the foregoing discussion it is evident that sequential, power-fed sanding operations on moulded shapes can be arranged if production warrants. Multiple-head machines that can sand more than one side of a workpiece in a single pass are available. For critical moulded patterns a formed sanding shoe is used to back up the belt, and for less critical areas a contact roll is used. Designs are on the market that will handle stock up to 12 inches wide and 4 inches thick. Feed speeds up to 60 fpm are practical on hardwood mouldings.

As an alternative to belts, formed wheels of polyurethane foam loaded with abrasive grit (fig. 18-174 bottom) can be used in series to sand mouldings at 30 to 40 lineal fpm; they perform best on the softer hardwoods (Ward 1976).

MACHINES TO SAND TURNINGS

Sanding machines for turnings fall into two classes, i.e., workpiece-centered and centerless. Many machine designs are available; descriptions of two illustrate the principles on which they operate.

Brush-backed, automatic turning sander.—On these workpiece-centered machines (fig. 18-175 top), each workpiece is rotated at high speed around its own axis and is at the same time conveyed on a reel past banks of oscillating brush-backed abrasive strips. Although reel motion is continuous, the sanded workpiece stops rotating about its own axis long enough to permit bottom discharge and to permit the operator to insert a new workpiece in its place. A 10-spindle machine typically has a capacity of 10-30 pieces per minute in sizes from 3/8-inch to 6 inches in diameter and 2 to 72 inches in length. Reel speed, and hence rate of production, is determined by species, pattern, and the relative roughness of the stock as it comes from the lathe. Spindle speeds are variable from 1,000 to 3,000 rpm. Spindle speed should be adjusted to produce a peripheral velocity on the workpiece ranging from 1,200 fpm on small diameters to 1,800 fpm on larger diameters. Pressure exerted by the several (five is typical) brush-backed sanding racks is adjustable. Each rack can carry a progressively finer grit to accomplish a complete roughing through polishing job in one pass of

Figure 18-175.—Workpiece-centered machines to sand turnings. (Top) Workpiece, rotating around its own axis, is carried in a reel past banks of oscillating brush-backed abrasive strips. (Bottom) Rapidly rotating turnings are conveyed under a pair of laterally shifting narrow sander belts.

the workpiece. Oscillation of the racks is variable in amplitude up to 2 inches, but is ordinarily held in the range from 3/8-inch to 1/2-inch. A 3-hp, variable-speed motor drives the spindles and a 2-hp, variable-speed motor drives the feed reel.

Centerless spindle sander.—The two-stage belt sander illustrated in figure 18-176A is designed for centerless sanding of straight, tapered, round, or oval turnings measuring 5/16- to 2½ inches in diameter. The first belt is used to cut down and smooth the turning and the second performs a polishing job. Feed rates

Figure 18-176A.—Two-stage centerless-spindle belt sander. The two tipped feed wheels advance stock from right to left. The two sander belts run downwards against stock and rotate it. (Photo from Critz Machinery.)

up to 45 fpm are attainable. The 1/3-hp, two-speed rotating discs impart peripheral velocity to the workpiece, and their degree of tilt controls the feed speed. The 6-inch by 120-inch sand belts are spaced on 54-inch centers and are independently driven by 3-hp motors at 4,350 fpm.

Centerless spindle grinder.—Through use of wet-grinding procedures with a formed-abrasive wheel it is possible—in a one-step process—to centerless-grind dowels into spindle-shaped patterns with contours as deep as half an inch. Typically, such spindle-shaped pieces measure 3/8-inch to 10 inches long and 1/16-inch to 2 inches in diameter (fig. 18-176B). According to the manufacturer (Von Arnauld Corp., Franklin Lakes, N.J.), perfect spheres can also be ground on the machine. The formed grinding wheel is powered by a 10-hp motor and turns at 2,400 rpm.

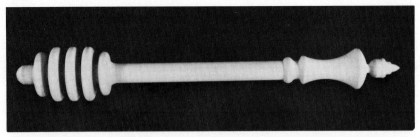

Figure 18-176B.—One-inch-diameter, 7-inch-long wood spindle wet-ground from a straight dowel pressed against an abrasive wheel in a centerless grinding machine.

MACHINABILITY

With suitable selection of abrasive mineral, grit size, abrasive speed and pressure, and in some cases special surface treatment, virtually all wood species can be given a good quality surface by means of coated abrasive machining. Some species are, however, more difficult to sand than others. Davis (1962) evaluated the sanding characteristics of a number of hardwood species according to their tendency to scratch and fuzz (table 18-72) under abrasive action. The tests were made with garnet 100 grit on wood at 6 percent moisture content. Results from both drum and belt sanding were evaluated. Abrasive speed on the drum sander was about 2,600 fpm and on the belt sander 4,200 fpm. The belt sander produced more scratch-free pieces and less fuzzing than the drum sander. Sanding scratches tend to be obscured by the structure of ring-porous woods of relatively coarse texture. Diffuse-porous, fine-textured, relatively hard woods require finer abrasives to avoid obvious scratches. The ring-porous woods, with the exception of elm, exhibited the least tendency to fuzz. Hard species fuzz less than soft species. Sanding characteristics are improved by reducing the moisture content of the workpiece.

Davis' data (table 18-1) indicated that white oak, northern red oak, and hickory yielded most defect-free pieces when sanded; white ash and American elm were intermediate, and red maple, sweetgum, black tupelo, and yellow-poplar were most difficult to sand without defect.

TABLE 18-72.—*Resistance of 10 southern hardwoods to scratching and fuzzing when sanded at 6-percent moisture content with 100 grit garnet* (Davis 1962)

Species	Scratch-free pieces	Fuzz-free pieces
	------Percent------	
Elm, American	70	62
Hickory	67	92
Oak, northern red	66	95
Oak, white	66	99
Ash, white	52	98
Oak, chestnut	50	100
Yellow-poplar	15	23
Sweetgum	8	37
Maple, red	8	66
Tupelo, black	4	38

ABRASIVE TUMBLING

Abrasive tumbling, sometimes termed barrel finishing, can produce a clean, well-sanded finish on all surfaces of small wood parts. For effective barrel finishing, parts should be small, have no sharp corners or large flat surfaces, and be designed with small radii on external corners and internal fillets. Without these radii, corners will be irregularly rounded, chipped and nicked by uneven abrasion. Parts suitable for abrasive tumbling include small turnings, such as screw-driver handles, utensil handles, and short dowel products.

In abrasive tumbling, a container more or less filled with parts mixed with an abrasive medium is rotated slowly. Barrel size may vary from 2 to 4 feet in diameter and from 2 to 6 feet long. The access door should be solid, but replaceable by a panel perforated to allow dust to escape. A pulley-driven shaft extending through a barrel provides the most simple configuration. The barrel, usually of wood or metal—sometimes rubber lined to protect the tumbling parts—may be round in cross section, or octagonal, to accentuate the tumbling action. A skewed shaft can add an eccentric motion to the normal rotational motion.

In some designs, longitudinal vibration of the barrel with variable amplitude and frequencies up to 2,700 vibrations per minute adds further motion. A vibrating barrel rotates at speeds from 5 to 10 rpm compared to 20 or 30 rpm for a non-vibrating barrel. Barrels are filled with parts to about 70 to 95 percent of capacity, slowly turning vibrating barrels being filled closer to capacity than those without vibration. Small barrels are used for small runs to maintain desired loading. A 42-inch by 72-inch barrel might handle 10,000 to 15,000 screwdriver handles in one batch.

Pumice is a commonly used abrasive. Wood parts are tumbled in a dry barrel, with one or two pounds of pumice powder per cubic yard of parts, more being needed if parts are intricately shaped. Coarse textured woods require more abrasive than do species with fine texture. Excess abrasive has no important adverse effects, but too little increases time to achieve required quality of finish.

The time required to produce a well-sanded surface may range from a couple of hours to 8 hours or more. After tumbling, pumice powder and wood meal adheres to the barrel and the parts. This is removed by continued tumbling with a perforated access door, or in a perforated barrel.

More complete removal results from tumbling with a small quantity of coarse sawdust which works like a sweeping compound. Final tumbling over perforations separates sawdust, wood meal, and abrasive—yielding smooth, clean parts.

18-5 TURNING

Hardwood turnings are produced in substantial volumes for use as handles, in furniture, housewares, tools, toys, novelties, and sports gear, and by industry. Essentially all are turned between centers or while gripped by end-chucks. Very little face-plate turning of flatware or bowls is done commercially. Because most turnings are short (1 to 36 inches in length) and small in cross section (usually less than 2-1/2 inches in diameter), blanks for their production are readily produced from pine-site-hardwoods.

A rich diversity of machine designs for turning has evolved since 1800, when Marc Isambard Brunel invented a series of machining operations to rapidly produce block pulleys for the rigging of English ships. These designs can be categorized by type as tool-point, long-knife, peripheral-milling, and chucking.

TOOL-POINT TYPE

In this family of designs the workpiece revolves between centers at 2,000 to 6,000 rpm, and a narrow tool is presented at an angle to achieve variation of orthogonal cutting perpendicular to the grain with cutting edge parallel to the grain (fig. 18-177). A hand lathe is the simplest configuration of this type, the tool being a hand-held chisel. In both hand and automatic lathes, primary tool motion is transverse, i.e., parallel to the axis of workpiece rotation. A secondary motion involves cross feed to bring the tool point closer or further away from the axis of workpiece rotation, decreasing or increasing workpiece diameter. By combining transverse and cross motions, any profile that is round in cross section can be machined.

LONG-KNIFE TYPE

The finishing cut on a lathe of this type is made by a milled-to-pattern knife having the same length as the workpiece. The formed, long knife is presented to the rotating workpiece somewhat as a veneer knife is presented to a rotating log.

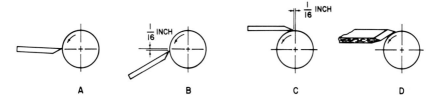

Figure 18-177.—Types of cutting action. A. Scraping (0° rake). B. Cutting (30° rake). C. Cutting (80° rake). D. Shearing.

Most, if not all, of the designs require that the blank workpiece, in the form of a rough square, be first turned to dowel shape; some also require contouring with a tracing knife to approximately 1/32-inch oversize before the long knife takes its finishing cut. This roughing operation is usually done by tool-point action, as described previously. The following lathes illustrate design variations within the type.

Bail-wood lathe.—Wooden handles for bail wires of buckets, round clothes pins, tapered furniture legs, file handles, and similar small turnings can be produced at rates up to 100 per minute on this machine. Maximum practical workpiece size is about 1-3/4 inches in diameter by 5 to 8 inches long. To eliminate roughing cuts, incoming stock is in the form of accurately sized dowels cut to precise length. The stock is hopper fed, one piece at a time, and automatically chucked between the rapidly rotating spindles. A long milled-to-pattern knife is briefly pivoted into the cut with a cam-controlled hinged action and then withdrawn as the finished part is ejected (fig. 18-178). Sequential boring or end-forming operations may be performed before or after turning by automatic transfer from or to a second machine.

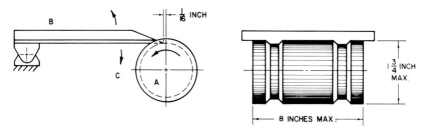

Figure 18-178.—Action of knife on bail-wood lathe. A. Rotating workpiece (2,000 to 6,000 rpm). B. Pivoted, long milled-to-pattern knife. C. Cam-controlled motion of knife (up to 100 cycles per minute).

Back-knife lathe.—The back-knife lathe is one of the most simple, efficient, and accurate lathes (fig. 18-179). It is best applied to produce slender turnings 6 to 30 inches long measuring less than 3 inches in diameter; for example, a 12-inch-long piece measuring 1-1/2 inches at the major diameter and 5/8-inch at the minor diameter could be machined at a rate of 10 to 14 per minute with expenditure of 10 to 15 hp. Good surface quality results from the shearing action of its long milled-to-pattern knife.

Figure 18-179—(Top) Automatic, hopper-fed, hydraulic back-knife lathe. A. Live center (12 speeds from 1,600 to 7,200 rpm). B. Transverse carriage carrying roughing gouge, support ring, and contouring chisel. C. Milled-to-pattern back knife mounted on guillotine slides (Photo from Goodspeed Machine Company.) (Bottom) Support ring and roughing gouge in action.

Stock, in the form of a square, or preferably a dowel, that has previously been accurately cut to length, is hopper fed and automatically centered between rotating spindle centers. Machining requires three knives, and proceeds from right to left (from tailstock to headstock) along the workpiece. The first two knives are tool points, both mounted on a single power-driven carriage that traverses the workpiece. The first tool is a gouge that reduces the square section to circular, the second a round-nose chisel that pivots toward and away from the axis of rotation as it traverses to produce the desired profile, but slightly oversize. The pivoting action is controlled by a follower and template. On the same traveling carriage, located between the two chisels just described, is a ring which passes over and supports the rough dowel produced by the roughing gouge. This support minimizes deflection of the workpiece, permitting slender turnings to be accurately and smoothly machined.

As the carriage carrying the roughing gouge, support ring, and contouring chisel traverses the length of the workpiece, the third knife comes into action following passage of the carriage. This back knife is milled-to-pattern and extends the full length of the workpiece. It is set in a plane parallel to the axis of workpiece rotation but with its edge at a 20° angle to horizontal (fig. 18-179 top) and is mounted on two vertical slides. As the back knife moves down, it engages the workpiece on the right-hand end immediately following the traveling carriage and this portion of the blade extends below the workpiece as it continues down. Thus at any instant the ring is supporting the workpiece a few inches ahead of engagement of the back knife. After passage of the carriage—with roughing gouge, support ring, and contouring chisel—and completion of the back knife stroke, a parting tool on the back knife may cut off the finished turning, or the centers may be withdrawn and the turning placed by picker fingers on a conveyor. The back knife then rises as the carriage returns for another cycle. Figures 18-177C and 18-179 (bottom) show the attitude of the roughing gouge and contouring chisel that move with the carriage. Figure 18-180 shows the attitude of the back knife.

Back-knife lathes are made in configurations other than that shown in figure 18-179 and 18-180. A very fast design is available in which the back knife moves on a horizontal slide under the turning. In still another configuration the back knife is divided into two sections, each section separately stroked. In these designs attachments can be provided that permit a boring tool to operate through the center of the tailstock. A production rate of 500-600 pieces per hour is readily available on turnings from 5 inches to 12 inches long and from 125 to 500 pieces per hour on longer turnings. Handling equipment between machines has been given special attention in this family of designs to permit rough squares or dowels to be cut to length, finish turned, and bored completely through—all with one operator—at a rate in excess of 500 pieces per hour (fig. 18-181).

Pringle and Brodie lathe.—This rotary-knife lathe (fig. 18-182) is designed to make extremely slender turnings such as hickory drum sticks. During an 8-hour shift, a lathe of this type might produce 2,000 drum sticks, each stick perhaps 16 inches long with minimum diameters ranging from 1/4-inch to 3/8-inch.

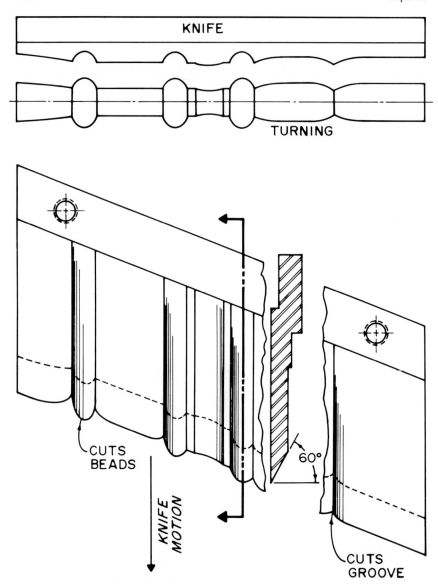

Figure 18-180.—Milled-to-pattern knife for back-knife lathe. (Top) Top view of knife and turning. (Bottom) Side view. Clearance angle is about 1°; the knife edge is ground so that cutting is initiated at the high spots on the turning.

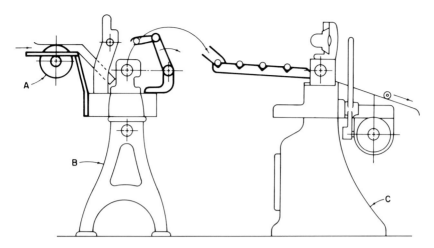

Figure 18-181.—Transfer machinery for sequential operations. A. Saw for cutting long squares into accurate short lengths. B. Hopper-fed, automatic, back-knife lathe (back-knife located on horizontal slides beneath workpiece). C. Double-end boring machine. (Drawing after Walter Hempel.)

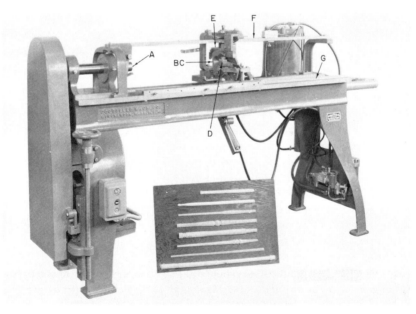

Figure 18-182.—Rotary-knife lathe. Motion of carriage during the machining stroke is from right to left. A. Live center. B. Roughing gouge. C. Support ring. D. Contouring chisel. E. Milled-to-pattern cylinder-shaped knife. F. Rack and pinion device to revolve cylindrical knife one revolution in length of workpiece. G. Tailstock (not shown) is clamped in these guideways. (Photo from Goodspeed Machine Company.)

The rotary-knife lathe operates on the same principle as a back-knife lathe. If the long milled-to-pattern back knife illustrated in figure 18-180 were oriented without any shear angle and then bent into a hoop shape so that the cutting edge was the bottom of the hoop, it would illustrate the cylindrical knife of the Pringle and Brodie lathe. This configuration permits the cylinder-shape back knife to closely approach the support ring without interference. The cylinder-shaped knife must make one complete revolution about a vertical axis as it cuts from one end of the workpiece to the other, in order to machine the complete pattern. Only a few of these machines remain in operation because modern back-knife lathes can produce equally slender turnings and are mechanically simpler.

Variety lathe.—The variety lathe is distinguished by the fact that the workpiece is not turned between centers but is periodically advanced through a hollow powered chuck which grips it tightly on one end only while it is being machined; thus the stock need not be cut to workpiece length prior to its introduction to the lathe. On larger variety lathes, long squares are rotated at high speed (2,000 to 6,000 rpm) and are axially fed past a roughing tool to form them into dowels of proper diameter to fit snugly into the adjacent support ring. The attitude of the roughing tool is illustrated in figure 18-177B. The emerging dowel is advanced beyond the support ring the length of the desired turning. The turning is then formed with a milled-to-pattern knife (similar to fig. 18-178), end-machined or bored as desired by a tool in the tail stock, and finally cut off from the dowel by a pivoted or slide-mounted parting tool or saw (fig. 18-183 top). The major disadvantage of a variety lathe is the waste of raw material caused by inability of the lathe to use the final few inches of each square. This raw material waste is 15 percent for a large variety lathe, but is less on lathes designed to make very small turnings.

Variety lathes are made in two styles. For large turnings, 3/4-inch to 3 inches in diameter and up to 6 inches long, an offset sliding headstock is used so that the long square is rotated by a socket-shaped spur in the sliding headstock and is guided by the support ring as discussed above. This style lathe requires approximately 5 to 7½ hp, and on larger turnings it can produce 300 to 600 pieces per hour. One operator can feed two machines.

The second style uses stock in the form of long dowels which are hopper-fed through the hollow headstock (fig. 18-183 bottom) to make small turnings such as golf tees, beads, and small knobs. A collect device within the headstock holds the dowel firmly as soon as the axial advance of the dowel is completed during each cycle. Rate of production is about 4,000 pieces per hour, with one person tending as many as six machines. Maximum workpiece diameter is approximately 3/4-inch. A 5-hp motor is adequate for most lathes of this style. Headstock speed is in the range from 2,000 to 5,000 rpm.

PERIPHERAL-MILLING TYPE

In contrast with the lathe types discussed in the foregoing paragraphs, wherein one or more tool points or wide knives are guided against a rapidly rotating workpiece, the peripheral milling type relies on a multi-knife, rapidly rotating cutterhead brought into action with the slowly rotating workpiece. The axis of

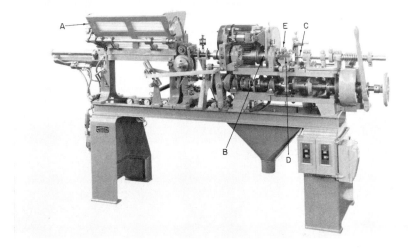

Figure 18-183.—Variety lathes. (Top) Three-inch-diameter bowl and lid at left shown being machined at right by sequentially acting forming, parting, and boring tools. (Bottom) Hopper-fed, hollow-headstock variety lathe for small-diameter turnings. A. Hopper-feeding device for long dowel stock. B. Hollow headstock through which dowels are intermittently advanced. Internal collet at right-hand end. C. Tailstock in which boring or end-machining tools can be mounted. D. Milled-to-pattern long knife with cam-controlled pivoted action. E. Parting tool. (Bottom photo from Goodspeed Machine Company.)

rotation of the cutterhead is parallel to the axis of rotation of the workpiece, and the distance between them is adjusted to bring the cutterhead and workpiece into contact.

Shaping-lathe headrig.—The shaping-lathe headrig is described in section 18-9 (figs. 18-104ABCD); it is designed to machine short logs into cants for resawing or into cylinders for veneer peeling, with residue in the form of postage-stamp size flakes well suited for use in structural flakeboards or composite panels (fig. 18-102). Alternatively, cutter heads can be fitted with knives to cut pulp chips.

Automatic shaping lathe.—The automatic shaping lathe used by the furniture industry (fig. 18-184) is applied to kiln-dry squares or rounds to make straight, tapered, or contoured turnings that are round, oval, or polygonal in

Figure 18-184.—A fully automatic, hopper-fed shaping-lathe that can accept stock up to 4 inches square and 30 inches long. Production of 12-inch-long turnings from the 2-inch by 2-inch hardwood squares illustrated is about 2,400 pieces per 8-hour shift. Machines with boring attachments are available. (Photo from Mattison Machine Works.)

cross-section. It finds its best application in the manufacture of the less slender turnings from 3/4-inch to 4 inches in diameter. Lathes of this type are available that will machine a continuous pattern up to 54 inches in length (fig. 18-185 bottom). In general, turnings larger than 3 inches in diameter do not require steady rests. As the slenderness ratio between diameter and length decreases to less than approximately 3/30, steady rests become necessary. Although ingenious support devices have been designed for this lathe, it is more difficult to get effective support close to the cutting action than it is on a backknife or Pringle

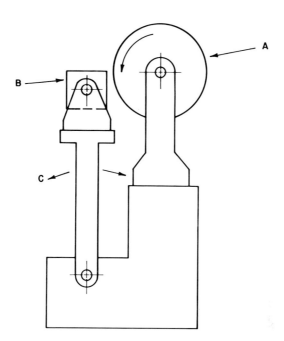

Figure 18-185.—(Top) Diagram of automatic shaping lathe. A. Rotating cutterhead acting full length of the pattern (2,700-3,600 rpm). B. Slowly rotating workpiece, which may turn clockwise or counterclockwise at 2 to 30 rpm. C. Carriage motion that controls diameter of completed workpiece. (Bottom) Timed oscillation of the carriage can produce turnings that are polygonal or oval in cross section—the one illustrated has a square cross section.

and Brodie lathe. However, the ability of the shaping lathe to turn polygonal shapes puts it in a class by itself. Turnings measuring 2 inches in diameter by 16 inches long might be produced on the shaping lathe at a rate of 4-6 per minute.

Figure 18-185 illustrates the shaping-lathe's principle of operation; through action of a cam mounted on the spindle of the workpiece and kept in contact with the follower shoe on the cutterhead spindle, the carriage moves in and out in proper sequence to machine the desired cross-sectional shape (see also fig. 18-104A top).

The 15- to 30-hp cutterhead rotates at 2,700 rpm, while the speed of the 1-1/2-hp workpiece spindle is variable from 2 to 30 rpm. The workpiece may be rotated in either direction to achieve up- or down-milling. The cutterhead extends the full length of the turned pattern (fig. 18-186), with the knives mounted at a rake angle of about 20° and a shear angle of about 45°. The cutting work is divided in such a way that when forming a bead, the knives start at the highest point and shear downward, toward the right and toward the left. In this way the knives can be oriented to shear with the grain. Because the knife body has six

Figure 18-186.—Typical cutterheads for an automatic shaping lathe. (Photo from Mattison Machine Works.)

clamping seats, and because the complete cylinder assembly is made up of short sections, it is possible to divide or arrange the knives around the periphery of the head so that there are rarely more than one or two knives cutting at any one time. The knives in the cutterhead are not jointed. Therefore each surface element on the workpiece is generated by a single knife. If the workpiece revolves at 10 rpm and has a diameter of 3 inches, a 2,700-rpm cutterhead will cut 28 knife marks per peripheral inch on the surface of the workpiece.

The skew angle of the knives makes determination of the knife outline somewhat difficult. In practice, a template corresponding to the profile of the desired workpiece is placed in a special marking and setting-up machine, and the cutter blanks are clamped in appropriate positions on the head. The projected profile is scribed on the knife blank by means of a stylus arranged to follow the template. After the knife blank is ground to the scribed line, the trueness of the cutter pattern is tested against the template, and a correction is made if necessary. The assembled cutterhead must, of course, be in dynamic balance.

Geometric development of a knife profile is illustrated in figure 18-187. A is the contour of the workpiece. The knife B is being developed to cut from the high point of the pattern to the low point. To accomplish this, the knife is skewed so that the high point of the bead is touched first. Knife and workpiece are shown

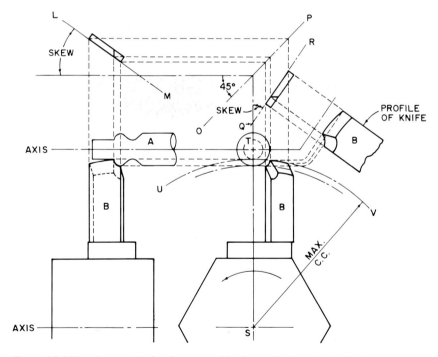

Figure 18-187.—Geometric development of knife profile for shaping lathe.

in a side view at the lower left. Above is the plan view of the head showing angle of skew and position of the cutting edge in relation to the axis of rotation of the head. On the lower right is an end view of the workpiece, head, and cutter. The true profile of the cutter is projected on the extreme right.

The steps involved in arriving at this projection are as follows:

1. Erect vertical lines from elements of the pattern A to intersect the line of the cutting edge LM, and from these intersection points draw horizontal lines to intersect line OP, which is oriented at an angle of 45° to the axis of rotation and hence vertically downward.
2. Draw horizontal lines from elements of the pattern A to intersect line ST.
3. Draw arcs of cutting circles, UV, corresponding to the intersections of line ST and the horizontal lines projected from A.
4. Draw line QR at the proper angle of skew to line ST and, where the vertical lines from OP intersect QR, draw a series of lines perpendicular to QR.
5. From the intersection points of the vertical lines from OP and the various corresponding cutting circles, project lines horizontally and then turn them through an angle equal to the complement of angle of skew so as to parallel QR. The intersection of these projections with the lines drawn perpendicular to QR establish points on the true profile of the cutter.

Copying lathe.—The copying lathe is used to machine furniture legs, gun stocks, shoe lasts, wooden legs, and other articles that are asymmetrical longitudinally as well as in cross-section. The principle of the machine is somewhat similar to the shaping lathe just described, in that a rapidly rotating peripheral milling head cuts against a slowly rotating workpiece. On the copying lathe, however, the cutterhead is narrow and small in diameter. The cutterhead is guided in its movement by a copyroller acting on a master model of the part to be reproduced. In this sense the copying lathe might be compared to a mechanized version of the multiple-spindle carver illustrated in figure 18-210.

Figure 18-188 (top) illustrates a four-spindle vertical copying lathe with one spindle in use. The master model or template, at the top, is mounted between centers and rotates slowly in time with the four workpiece spindles below. The copying ball that follows the contours of the master model is carried, with the cutterheads, in a light-weight oscillating frame. The three-knife cutterheads, which rotate at about 10,000 rpm, are the same diameter as the 3-inch copying ball and generate on each workpiece the same contours described by the copying ball as it passes over the surface of the master model. A single 5-hp motor drives all four cutterheads. A small feed motor traverses the cutterhead carriage and slowly rotates the master model and the four workpiece spindles at 15 to 30 rpm. The longitudinal feed rate of the carriage ranges from 1/32- to 5/8-inch per rotation of the workpiece spindles. Machines of the design illustrated usually accept stock up to about 32 inches long and might turn out as many as 100 to 200 pieces in an 8-hour shift, the rate depending on the species, length, and contour of the workpiece.

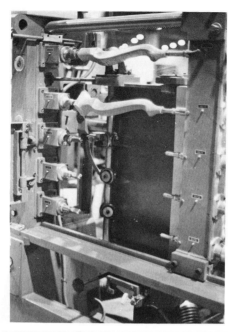
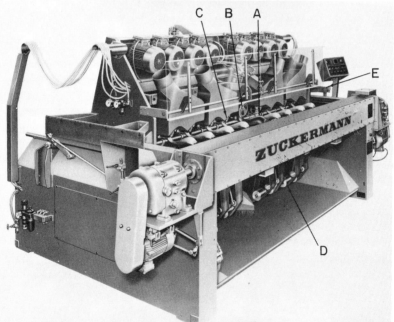

Figure 18-188.—Copying lathes. (Top) Vertical four-spindle machine. (Bottom) Horizontal eight-spindle machine. A. Master model. B. Copyroller. C. Cutterheads. D. Sanding system. E. Cams to control rate of rotational and transverse movement. (Bottom photo from Maschinenfabrik Zuckermann.)

The most advanced versions of copying lathes are also equipped with a longitudinal (with-the-grain) sanding system which accomplishes a uniform sanding finish while the part is being shaped. The sanding mechanism works on the same principle as the shaper in that a rapidly moving sanding belt is pressed against the rotating workpiece by a cylindrical contact roll. The movement of the contact roll, analogous to the movement of the cutterhead, is controlled by a sanding copyroller. The sanding system is usually positioned on the opposite side of the workpiece from the shaper. Since shaping must necessarily be performed prior to sanding, the cutterhead precedes the sanding contact roll in its traverse movement by about 1 inch.

Figure 18-188 (bottom) illustrates an 8-spindle high production copyshaper and sander. The master model (A) is visible in the middle of the machine, with four workpiece spindles to each side. The copyroller (B), just above the master form, controls the up and down movement of the cutterheads (C). Each cutterhead has a turning diameter of 160 mm and carries six circular knives, each 35 mm in diameter. Oversize master models facilitate smoother operation of the machine, and allow faster workpiece rotation, so a slightly undersize copyroller is needed for exact reproduction. Each cutterhead is driven at 6,000 rpm by a 5.5-hp motor.

The sanding system (D) is located just below the workpieces (partially obscured by the machine frame). Cylindrical sanding contact rolls, each 60 mm in diameter by 150 mm wide, act on the workpiece in accordance with the movement of a similarly sized copyroller traversing the master model. Here also, it is necessary for the sanding copyroller to be somewhat undersize to compensate for the oversize master model. Each two sanding heads are powered by a 3.0-hp motor which drives the sanding belts at a speed of approximately 15 m/sec.

The traverse movement of the copyroller and cutterheads is driven by a 2.0-hp motor, and can be varied in speed from 20 to 800 mm/min. Workpiece rotation, which ranges from 15 to 75 rpm, is controlled by a 4.0-hp motor. During the working cycle both the rotation and the traverse movement are independently variable within the foregoing parameters. Change in the rate of these movements is controlled by cams riding along templates located to either side of the machine (E).

The machine illustrated can accept stock up to 1,000 mm long and 200 mm wide. In an 8-hour shift it is capable of turning out as many as 500 parts, depending on the configuration of the part, the species of wood, and the surface quality required.

CHUCKING TYPE

On dowel lathes which produce cylindrical dowels, and on chucking machines that end-machine turnings and squares, knives revolve around the non-rotating workpiece.

Dowel lathe.—In the dowel lathe (fig. 18-189A), non-rotating square or oversize round stock is fed axially through the center of a four-knife revolving cutterhead (fig. 18-189B). The curved portion of each cutting edge makes the

Machining 2075

Figure 18-189A.—Dowel turning machine. From right to left: Hopper for rough squares. Infeed rolls. Housing and rotating cutterhead. Outfeed rolls. (Photo from Hawker-Dayton Corporation.)

Figure 18-189B.—Cutterheads from dowel machines. (Top) Rotor-mounted knives. Photo from J. S. Richardson Company, Inc. (Bottom) Cutting arrangement. Photo from Hawker-Dayton Corp.

roughing cut. The straight section of each knife accurately sizes and finishes the dowel. A projecting tip rides against the finished dowel and, in combination with the tips on the other three knives, acts as a supporting ring through which the finished dowel must pass.

Two standard sizes of dowel machines are available. The smaller will turn dowels from 5/16-inch to 2 inches in diameter, and uses 15 hp to drive the cutterhead at 3,600 rpm; it will accept squares as short as 14 inches. The larger will turn dowels from 1-1/4 to 3-1/4 inches in diameter, and needs 25 hp to drive the cutterhead at 3,600 rpm. Both have feed speeds from 40 to 275 fpm; 50 fpm is typical, yielding perhaps 16,000 lineal feet per shift.

The largest dowel lathes can turn dowels 6 inches in diameter. One operator, who also trims defects from the squares, tends each hopper-fed machine. These dowel machines are frequently used to convert squares to rounds preparatory to further turning operations. For alternative methods of obtaining rounds, see figures 18-90, 18-91, and 18-104CD.

Sawing, chucking, and boring machine.—This equipment double-end trims, double-end chucks (e.g., machines conical ends), end bores, and cross-bores dowels, turnings, squares, or rectangles. The machine illustrated in figure 18-190 will accept stock up to 52 inches long, 2-1/2 inches thick, and 3 inches wide. During chucking the workpiece has neither rotational or translatory motion. The cutterhead, however, both rotates and translates.

Figure 18-190.—A hopper-fed, sawing, chucking, and boring machine. The top boring rail permits crossboring. (Photo from J. S. Richardson Co.)

A 3-hp feed motor provides feed rates variable from 20 to 60 pieces per minute. Saw and chuck motors are each 2 hp at 3,600 rpm or 4 hp at 7,200 rpm. The action of the machine is as follows. The feed quadrants pick up the lowest piece in the hopper and feed it past the two trim saws where it is cut off and placed into position for chucking and boring; because the quadrants move rapidly, saw-cut quality may be poor if feed is too fast. The piece is clamped by rubber blocks, and the chucks and bits feed in. During infeeding the quadrants return to pick up another piece. As the second piece is brought past the saws, the chucking stroke on the first is completed, the clamps are released, the finished piece is ejected, and the next one is clamped. Fifty percent of the cycle is used during infeed of chucks and bits, 25 percent in their return to original position, and 25 percent in unclamping and ejecting the finished piece and in feeding the next piece past the trim saws into clamping position.

This distribution of time gives the greatest portion of time to the most difficult job—end-machining with knives mounted in a rotating chuck. At a feed rate of 40 pieces per minute on pieces with a 5/8-inch length to be machined on an end, the infeeding chuck moves at about 50 inches per minute, which, with a two-knife chuck rotating at 7,200 rpm, results in a chip thickness of approximately 0.004 inch.

MACHINABILITY

Davis (1962) evaluated turning machinability of a number of hardwoods found on southern pine sites (table 18-1). He used a milled-to-pattern knife presented to the workpiece in a manner similar to that employed on a back-knife lathe. The 5-inch-long turnings contained a bead, cove, and fillet; they were cut from 3/4-inch squares without benefit of a roughing cut. Knife angles and feed speeds were not specified. Tests made at four spindle speeds ranging from 950 to 3300 rpm showed that the higher spindle speeds gave better results. Six of the seven poorest turning woods, aspen, gumbo-limbo, cottonwood, basswood, willow, and buckeye—were the lightest woods tested; but turning qualities were not consistently related to specific gravity.

Except for hickory and red maple, the species turned equally well at 6 and 12 percent moisture content. At 20 percent moisture content the quality of turnings decreased markedly, as follows (spindle speed 3,300 rpm);

Species	At moisture content of		
	6 percent	12 percent	20 percent
	----------Percent fair to excellent pieces----------		
Ash, white	93	98	45
Elm, American	60	70	61
Hackberry	78	87	78
Hickory	99	88	83
Maple, red	96	82	54
Oak, chestnut	100	95	84
Oak, northern red	90	95	75
Oak, white	93	92	78
Sweetgum	95	92	77
Tupelo, black	88	94	67
Yellow-poplar	98	98	49

Cantin (1965) found that rings per inch and specific gravity were both positively correlated with turning machinability of 10 eastern Canadian hardwood species, three of which are found on southern pine sites. It is likely that adjustment of cutting variables permits most species to be machined sasisfactorily, although some are undoubtedly more difficult to handle than others.

18-16 BORING

Machine boring is a common operation whenever dowels, rungs, or screws are required in assembling wood components. Holes are also needed for bolted connections in containers, trusses, furniture frames, poles, structural timbers, and frames of boats and ships.

BIT TYPES AND NOMENCLATURE

Although there are many specialized bit designs, six types are most important. Figure 18-191A shows a double-spur, double-lip, single-twist, solid-center bit on which the spurs cut ahead of the lips. This bit may have a threaded or brad (plain) point.

Figure 18-191B illustrates a double-spur, double-lip, double-twist bit which may also have a threaded or brad point. The flat-cut, double-lip, double-twist bit (fig. 18-191C) is similar in design except that outlying spurs are not used. With the flat-cut bit, the side cutting spurs sever the end surface of the chip simultaneously with the cutting action of the lips.

Holes in excess of 6 inches deep—sometimes bored with double-spur, double-lip, double-twist bits—are more frequently bored with a ship auger (fig. 18-191D), a single-twist design with one cutting lip and one side cutting spur. The four types of bits described above are sharpened by filing on the rake face of the cutting lips and the inside surfaces of the spurs. The lips are commonly filed with a rake angle (α) of 30 to 35° and a clearance angle (γ) of 10 to 15°.

In contrast to the foregoing types, the spur machine bit illustrated in fig. 18-191E is sharpened by grinding on the clearance surface, or back side, of the lips. The rake angle therefore continuously varies along the cutting lip from about 0° near the axis of rotation to about 45° at the bit periphery.

The twist drill illustrated in figure 18-191F is also sharpened by grinding on the clearance surface of the lips. It has neither spurs nor point, and is most frequently used to drill in end grain and to bore dowel holes.

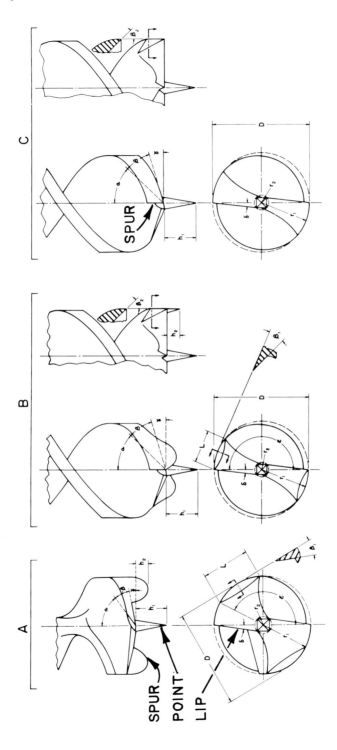

Figure 18-191ABC—Bit types. (A) Double-spur, double-lip, single-twist, solid-center bit. (B) Double-spur, double-lip, double-twist bit. (C) Flat-cut, double-lip, double-twist bit. (Drawings after McMillin and Woodson 1974.)

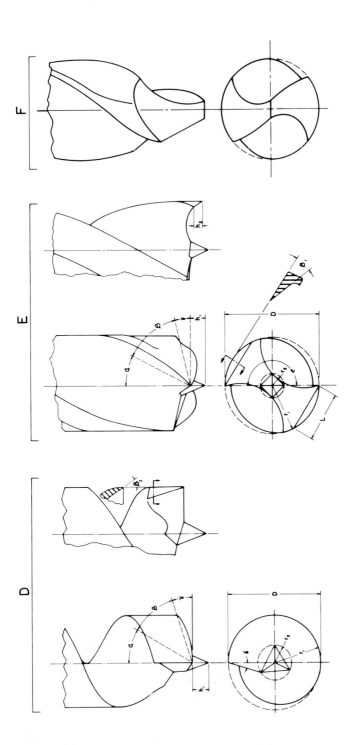

Figure 18-191DEF.—Bit types. (D) Ship auger. (E) Spur machine bit. (F) Twist drill. (Drawings after Woodson and McMillin in 1972; McMillin and Woodson 1974.)

The symbols used in subsequent text of this section are defined as shown in figure 18-191 and the following tabulation.

α	Rake angle, degrees	P	Power required at the spindle, horsepower
β	Sharpness angle of lips, degrees		
β_1	Sharpness angle of spurs, degrees	T	Torque on spindle, inch-pounds
β_2	Sharpness angle of side cutting spurs, degrees	n	Spindle speed, revolutions per minute
γ	Clearance angle of lips, degrees	f	Feed speed, inches per minute
δ	Skew angle of lips, degrees	E	Energy, kilowatthours
ϵ	Angle of lead (spur to lip measured at circumference), degrees	E_s	Specific cutting energy, kilowatt-hours per cubic inch
D	Diameter of bit, inches	t	Chip thickness, inches
h_1	Height of point above lip, inches	N	Number of cutting lips
h_2	Height of spur above lip, inches	v	Velocity of cutting edge of lip, feet per minute
L	Length of spur at root, inches		
r_1	Radius of bit, inches	d	Depth of hole, inches
r_2	Effective radius of point, inches	V	Volume of hole, cubic inches

Table 18-73 lists typical geometrical specifications for all types except the twist drill, and describes the bits used by McMillin and Woodson (1974) to bore southern pine.

TABLE 18-73.—*Geometrical specifications for five bit types*

Bit type and diameter (inches)	α	β	β₁	β₂	γ	δ	ε	h₁	h₂	L	r₁	r₂
	---Degrees---							---Inches---				
Spur machine bit[1]												
0.50	20.2	54.4	37.3	------	15.4	------	------	0.10	0.03	0.21	0.250	0.07
1.00	19.3	60.6	35.1	------	10.1	------	------	.20	.10	.53	.500	.09
1.25	18.6	61.4	36.3	------	10.0	------	------	.20	.11	.62	.625	.11
Double-spur, double-twist												
0.50	30.6	44.9	31.9	33.2	14.6	18.2	161.8	.11	.05	.23	.250	.08
1.00	33.5	45.6	29.6	32.4	10.9	14.8	164.9	.21	.11	.45	.500	.14
1.25	31.8	47.7	29.7	35.9	10.4	12.1	167.9	.22	.12	.55	.625	.14
Flat-cut, double-twist												
0.50	29.6	45.1	------	31.1	15.3	17.8	------	.13	------	------	.250	.08
1.00	35.7	40.9	------	35.2	13.4	13.2	------	.23	------	------	.500	.11
1.25	33.3	42.7	------	33.7	14.1	10.9	------	.24	------	------	.625	.12
Double-spur, single-twist, solid-center												
0.50	30.0	47.8	29.8	------	12.2	15.5	151.5	.10	.04	.25	.250	.09
1.00	27.4	51.6	28.2	------	11.0	12.8	148.3	.22	.12	.46	.500	.13
1.25	31.8	46.1	27.8	------	12.1	11.7	152.3	.23	.12	.58	.625	.15
Ship auger, 12-inch twist												
1.00	37.5	45.2	------	35.0	7.3	15.1	------	.29	------	------	.500	.15

[1]For the spur machine bit, α, β, and γ were measured at the midpoint of the bit radius.

Machining

FUNDAMENTAL ASPECTS

The velocity of the cutting edge varies with the spindle speed (n) and the distance (r_t) from the axis of rotation.

$$v = 2\pi r_t n/12 = 0.5236 r_t n \qquad (18\text{-}42)$$

A 1-inch-diameter bit rotating at 3,600 rpm has a maximum cutting velocity of 942 fpm.

The thickness of the undeformed chip (t) is directly proportional to the feed speed (f) and inversely proportional to the number of cutting lips (N) and the spindle speed.

$$t = f/nN \qquad (18\text{-}43)$$

The tabulation below gives feed speeds required to yield 0.010, 0.020, and 0.030-inch-thick chips at spindle speeds of 1,200, 2,400, and 3,600 rpm for bits having two cutting lips. For bits having only one cutting lip, feed speeds are one-half those shown.

Chip thickness	Spindle speed		
Inches	1,200 rpm	2,400 rpm	3,600 rpm
	---------Inches/minute---------		
0.010	24	48	72
.020	48	96	144
.030	72	144	216

The net horsepower (P) requirement at the spindle is a positive linear function of the torque (T) and the rotational speed of the spindle.

$$P = \frac{2\pi nT}{(33,000)(12)} = (1.587)(10^{-5})nT \qquad (18\text{-}44)$$

Since equation 18-44 neglects no-load idling losses of the motor and spindle assembly, actual power demand is somewhat higher than that indicated. Neither does the equation include power (normally only a fraction of a horsepower) to overcome thrust when advancing the bit.

Least energy is consumed boring a hole if bits cut thick chips. The net cutting energy (E) consumed in boring a hole can be calculated in kilowatthours from the following:

$$E = \frac{0.746 Pd}{60 f} = \frac{(12.43)(10^{-3})Pd}{tnN} \qquad (18\text{-}45)$$

Specific boring energy (E_s), an expression of efficiency of the cutting action, is defined as follows:

$$E_s = \frac{\text{Net cutting energy}}{\text{Volume removed}} = \frac{\text{Kilowatthours}}{\text{Cubic inches}} \qquad (18\text{-}46)$$

Since the volume of wood removed in boring a hole of diameter (D) and depth (d) is

$$V = \frac{d\pi D^2}{4} \qquad (18\text{-}47)$$

then:

$$E_s = \frac{15.83(10^{-3})P}{fD^2} = \frac{(15.83)(10^{-3})P}{tnND^2} \qquad (18\text{-}48)$$

BORING DIRECTION AND CHIP FORMATION

Holes may be bored in any of the three primary directions illustrated in figure 18-192, or at intermediate angles. Torque and thrust do not differ significantly for holes bored in the tangential and radial directions (Goodchild 1955; McMillin and Woodson 1974; Komatsu 1978a). Generally torque is greater and thrust is less when boring along the grain (longitudinal direction) than when boring across the grain. The tabulation following, although determined on dry southern pine averaging about 0.54 specific gravity (based on ovendry weight and volume at 10.4 percent moisture content), is probably also representative of most hardwood species; it compares the effect of direction for a double-spur, double-lip, single-twist, solid-center bit at 2,400 rpm removing 0.020-inch-thick chips (McMillin and Woodson 1974).

Bit diameter (inches) and parameter	Along the grain	Across the grain
0.50		
Torque, inch-pounds	16.6	13.7
Thrust, pounds	73.5	106.3
1.00		
Torque, inch-pounds	56.6	37.7
Thrust, pounds	137.3	183.1
1.25		
Torque, inch-pounds	71.3	53.6
Thrust, pounds	161.4	197.5

Goodchild (1955) and McMillin and Woodson (1974) found that thrust is related to the force required to advance the spurs and brad into the work and, to a lesser extent, to the normal force component exerted by the cutting lips. When drilling across the grain, the spurs and brad bend and sever fibers perpendicular to their long axes. When drilling along the grain, fibers are separated parallel to their long axes. Since fibers are stronger when deflected perpendicular to their axes, greater thrust forces would be expected when drilling across the grain than when drilling along the grain.

Torque is primarily related to the parallel tool force component of the lips and the spurs and brad. When drilling along the grain, the lips sever fibers perpendicular to their long axes. In contrast, when drilling across the grain, fibers are cut in a plane parallel to their axes. Since fibers offer greatest resistance when severed in the perpendicular direction, greater torque would be expected when boring along the grain than when boring across the grain.

When drilling in the longitudinal direction, the action of the lips (the cutting edges generating chips) approximates orthogonal cutting across the grain. For across the grain boring, cutting continuously alternates from the veneer cutting direction to the planing direction. (See sec. 18-2 for a discussion of the basic modes of chip formation in orthogonal cutting.)

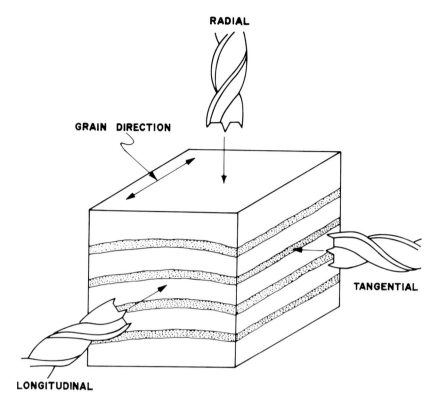

Figure 18-192.—Designation of the three main boring directions: feed longitudinal along the grain, and feed across the grain in radial or tangential direction. (Drawing after McMillin and Woodson 1974.)

FACTORS AFFECTING TORQUE AND THRUST BORING ACROSS THE GRAIN

Wood moisture content.—Komatsu (1978a) found that torque and thrust were maximum at wood moisture contents from 5 to 20 percent, decreased from this maximum with increasing moisture content to fiber saturation, and then remained about constant. McMillin and Woodson (1974) found that thrust when boring wood at 10 percent moisture content was less than at 73 percent; effect on torque varied with hole diameter and bit style.

McMillin and Woodson (1972) bored 3-1/2-inch-deep holes across and along the grain of southern pine with a 1-inch spur machine bit rotated at 2,400 rpm while removing chips 0.020-inch thick; wood moisture contents were varied from ovendry to saturation. They found that for both boring directions, torque and thrust increased with increasing moisture content to a maximum at about 5 to 10 percent, then decreased to a constant value at about the fiber-saturation point.

Specific gravity.—Torque, thrust, and power to bore are all positively correlated (fig. 18-193) with specific gravity (Davis 1962; McMillin and Woodson 1974; Komatsu 1975).

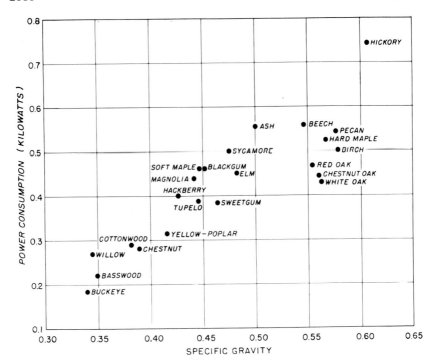

Figure 18-193.—Relationship of specific gravity to spindle power requirement when boring wood of 22 hardwood species across the grain with a 1-inch-diameter, double-spur, double-lip, brad-pointed bit rotated at 2,400 rpm. Moisture content was 6 percent. Feed rate was uniform but not specified. (Drawing after Davis 1962.)

Bit style.—Goodchild (1955) found that drill points (see fig. 18-191A) cause about 17 percent of the torque and 25 percent of the thrust needed; spurs cause 37 percent of the torque and 10 percent of the thrust; and lips cause about 46 percent of the torque and 65 percent of the thrust. McMillin and Woodson (1974) found that a bit with no spurs (fig. 18-191C) required less thrust at all bit diameters and at all moisture contents than bits with spurs.

Hole depth.—Torque required to bore increases with increasing hole depth; thrust, however, may not vary significantly with hole depth (McMillin and Woodson 1974).

Bit diameter.—Thrust, torque, and power required to bore all are functions of bit diameter; i.e., large-diameter bits require more thrust, torque, and power than small-diameter bits (Goodchild 1955; McMillin and Woodson 1974).

Chip thickness.—Thrust, torque, and power required to bore are all positively correlated with chip thickness; i.e., at constant rpm, feed rates yielding thick chips require more thrust, torque, and power than those yielding thin chips (Goodchild 1955; Komatsu 1975, 1976, 1977).

Bit rpm.—At constant feed speed but varying bit rpm, thrust, torque, and power are negatively correlated with bit rpm, because chip thickness is negatively correlated with bit rpm. McMillin and Woodson (1974) found, however, that at constant chip thickness, thrust is increased slightly with increase in bit rpm.

TORQUE AND THRUST BORING ALONG THE GRAIN

As previously noted, less thrust is required to bore along the grain than across the grain with most bits. Komatsu (1979a) observed that thrust is about constant from an across the grain direction to 45° to the grain; as the boring direction approaches parallelism with the grain, thrust required decreases. Twist drills (fig. 18-191F), however, require increasing thrust force as boring direction varies from across the grain to parallel to the grain (Komatsu 1979a).

When boring along the grain, torque is a positive curvilinear function of bit diameter, and linearly increases with increased chip thickness or increased specific gravity. For larger bit diameters, flat-cut, double-twist bits require less torque than other bit types. While boring along the grain, torque varies little with variation in spindle speed if chip thickness is held constant.

Thrust required in boring along the grain is positively correlated with bit diameter; it is less when boring wood of low density than when boring wood of high density and less for wet wood than dry wood (McMillin and Woodson 1974).

HOLE QUALITY

Davis (1962) rated 22 hardwood species, 10 of which can be found on southern pine sites, according to smoothness of cut and accuracy of finished holes bored across the grain (table 18-74). Holes were consistently larger across the grain than parallel to the grain.

Off-size or different-size holes in different woods bored with the same bit may explain why some woods split considerably more than others when dowelled. Davis found that holes in some pieces of beech (*Fagus grandifolia* Ehrh.) were 0.002 inch undersize and those in some pieces of *Magnolia* sp. 0.006 inch oversize.

The size of a hole, after it is bored, changes with variation in wood moisture content, increasing with increased moisture content. The increase across the grain is more marked than the increase parallel to the grain.

McMillin and Woodson (1974) found that smoother holes could be bored in southern pine across the grain than along the grain, and in dry wood compared to wet wood; when boring across the grain, bits equipped with spurs yielded smoother holes than flat-cut double-twist bits (fig. 18-191C). When boring along the grain, hole smoothness varied little with bit type. Hole smoothness improved with decreasing chip thickness, but was unaffected by spindle speed (for chips of constant thickness) and specific gravity. Trends are probably similar in pine-site hardwoods.

TABLE 18-74.—*Quality of machine-bored holes in 10 hardwoods as measured by percentage yield of good to perfect pieces and variation from nominal hole size* (Data from Davis 1962[1])

Species	Good to excellent pieces	Amount off size[2]
	Percent	*Ten-thousandths inch*
Hickory	100	3
Hackberry	99	6
Oak, northern red	99	2
Oak, white	95	6
Ash, white	94	5
Elm, American	94	4
Sweetgum	92	5
Yellow-poplar	87	11
Tupelo, black	82	10
Maple, red	80	22

[1] Each value is the average from 100 to 200 holes bored in flat-grain wood at 6-percent moisture content with a 1-inch-diameter, single-twist, solid-center, brad-point bit turned at 2,400 rpm; feed rate and cutting angles not specified.
[2] Values represent off-size either across the grain or parallel to it, whichever was greatest.

Komatsu (1976), in **across-the-grain** tests of Asian woods, found that bits equipped with spurs yielded smoother, more concentric holes than twist drills (fig. 18-191F). With the twist drill, roundness and surface quality were best at spindle speeds of 1,500 to 2,500 rpm; spindle speed did not affect quality of holes bored by the spur-equipped bits. For best hole quality (roundness and smoothness) twist drills should be ground with a point included angle of about 100° (Komatsu 1978b).

When boring **along the grain** or nearly along the grain, however, Komatsu (1979b) found that twist drills yielded rounder and less oversize holes than spur- and brad-equipped, double-lip, single-twist, solid-center bits.

Effect of hole quality on strength of glued dowel joints.—Nearn et al. (1953) determined the effect of hole size on joint strength; in their study 3/8-inch standard, compressed, and spiral dowels were inserted into 24/64-, 25/64-, and 27/64-inch holes in lumber of the red oak group. Liquid, cold, hide glue was used. Joints with spiral dowels were strongest; strength values were about the same regardless of the hole size. Average strength in tension was 842 psi. Standard and compressed dowels showed a well-defined response to hole size, superior joints resulting when holes were 1/64- to 2/64-inch larger than the dowel. Average strengths of joints with standard and compressed dowels were 694 and 685 psi, respectively. With holes 1/64- or 2/64-inch oversize, however, joints with the standard dowel were essentially equal in strength to those with spiral dowels. The compressed dowels did not perform as well as the other two in this study, partly because they did not expand fully. Standard dowels should be used for joints that will be severely stressed since, for a given size, they contain the full volume of unaltered wood.

Hoyle (1956) found that strength of glued dowel joints in northern red oak, using both plain dowels and spiral dowels, was little affected by thickness of chips cut by drill lips during boring. With chip thickness held constant in the range from 1/128- to 1/16-inch, a spindle speed of 2,880 rpm consistently yielded stronger dowelled joints than spindle speeds of 1,620 or 4,430 rpm. The double-spur machine bit with brad point measured 0.391 inch in diameter.

BORING DEEP HOLES

The literature contains no information on boring deep holes in the hardwoods that are the subject of this text. Data of Woodson and McMillin (1972) on deep-boring southern pine may have application also to hardwoods. When holes 10-1/2 inches deep and 1 inch in diameter were made with either a ship auger or a double-spur, double-twist machine bit, clogging (fig. 18-194) occurred at a shallower depth (avg. 6.5 inches) when boring across the grain than when boring along the grain (avg. 10.1 inches). In both boring directions, thrust force and torque demand for unclogged bits were less for the ship auger than for the machine bit. Generally, torque and thrust were positively correlated with chip thickness and specific gravity; they were unrelated to spindle speed when the thickness of chips was held constant. For the machine bit, thrust was less in wet than in dry wood. Although the ship auger required less horsepower than the machine bit, it was slightly less efficient; i.e., more energy was required to remove a unit volume of wood. In boring along the grain, the ship auger made better holes than the machine bit when the wood was dry; in wet wood hole quality did not differ between bit types. When boring across the grain, the machine bit made better holes in both wet and dry wood.

MACHINE TYPES

Boring machines range from the simple vertical, single-spindle model to complex transfer machines with multiple vertical, horizontal, and angular heads that are sequentially operated. A discussion of two machine types illustrates the design possibilities. As with shapers and routers, boring machines can be arranged for computer control.

Vertical multiple-spindle borer.—It is the purpose of this machine (fig. 18-195) to face bore patterns of holes in "flatwork"—all in a single stroke. Large, medium, or small holes can be bored on close or extended centers with regular and irregular spacing; spindles adjust 12 inches across-line and also laterally. Without special attachments, the needle-bearing-equipped boring heads can drill holes at a minimum center distance of 1-1/16 inches.

The table is moved hydraulically to present the work to the bits. Feed speeds from 10 to 140 inches per minute are possible. Table motion can be set to provide automatic-continuous or intermittent stroking. The length of table stroke is adjustable from 1/2-inch to 16 inches. A 1-1/2-hp motor drives the hydraulic pump controlling the feed-table movement.

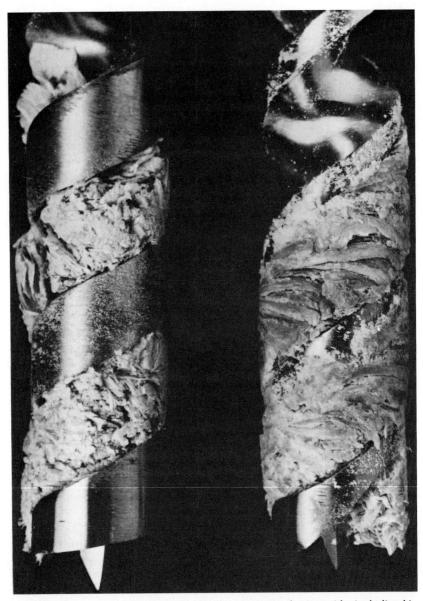

Figure 18-194.—Severe chip clogging during cross-grain boring with single-lip ship auger (left), and double-spur, double-twist bit (right). (Photo from Woodson and McMillin 1972.)

The boring spindles are gear driven at 1,800 rpm from a single 10-hp motor. A telescopic shaft, with universal joint at top and bottom, connects the upper and lower spindle units. Special boring heads are available that locate multiple bits at extremely close hole spacing in desired cluster patterns.

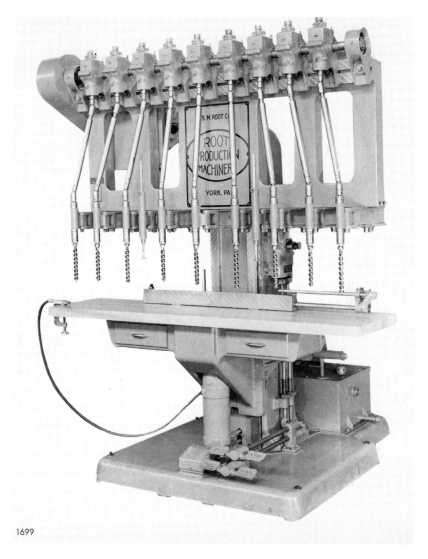

Figure 18-195.—A 6-foot, vertical, multiple-spindle borer with hydraulic feed table. (Photo from B. M. Root Company.)

Horizontal multiple-spindle borer.—It is the purpose of this machine (fig. 18-196) to edge and end-bore furniture parts to accept screws and dowels. Face boring on a workpiece of limited size is also practical. Each bit, or bit cluster, is driven at 3,600 rpm by an individual 1- to 2-hp motor that can be laterally adjusted to accomplish the desired hole spacing. Bits are advanced horizontally into the workpiece by means of a 1-1/2-hp motor driving a hydraulic stroking mechanism. Spindle feed rates are adjustable from 10 to 165 inches per minute. Stroke length is adjustable from 1/2-inch to 9 inches.

Figure 18-196.—A 6-foot, horizontal, multiple-spindle, boring machine with hydraulic feed. (Photograph courtesy B. M. Root Company.)

18-17 ROUTING

Routers employ cutterheads of small diameter designed for both side cutting and end cutting. They may simultaneously peripheral mill, bore, and face mill.

Router spindles typically carry 3/4 to 10 hp and rotate at 6,000 to 50,000 rpm. At these speeds the cutters cannot be effectively jointed, although the heads may carry 2, 3, or 4 cutting edges. Typical tools, and the cuts resulting from their application, are shown in figure 18-197.

Typical router operations (fig. 18-198) include:

- Slotting cuts for veining and inlaying
- Shaping cuts on small mouldings, or on cabinet components
- Internal shaping cuts
- Piercing cuts on screens, chairbacks, or cabinets
- Carving of panels, mouldings, rosettes, or flutes
- Boring
- Dovetailing, dadoing, and mortising
- Machining serpentine end joints (Gatchell et al. 1977). See figure 22-4.

Little data on fundamental aspects of routing have been published in the English language, perhaps because power requirements of routers are small. Readers interested in the non-English literature should consult Pahlitzsch and Sandvoss (1969), Kondo et al. (1971), and Mori (1971).

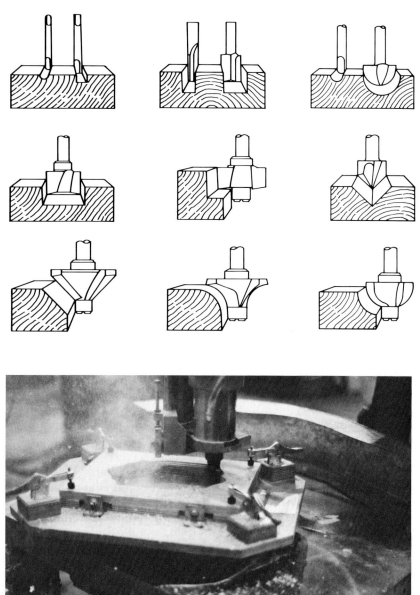

Figure 18-197.—(Top) Representative router cuts. (Bottom) Router in action.

Figure 18-198.—Typical work produced on routing machines. (Top) Flatwork. (Bottom) Spirals.

SPINDLE TYPES

Router spindle speeds are significantly higher than those for most other woodworking machines; four designs in common use span a range from 6,000 to 50,000 rpm.

Direct, synchronous-motor drives with spindle-mounted armatures are compact and permit flexibility in designing tilt, plunge, and traverse mechanisms that move the router head; common spindle speeds for two-pole motors are as follows:

Current characteristic	Spindle rpm
Cycles/sec.	
100	6,000
120	7,200
240	14,400
360	21,600

For even higher speeds, the Precise Corporation of Racine, Wis., manufactures 3/4- to 1-1/2-hp liquid-cooled electric spindles that rotate at speeds to 54,000 rpm, and fractional-hp spindles that rotate at speeds to 120,000 rpm. For these high speeds, single-lip cutters made from solid carbide cylinders measure only about 0.03 inch in diameter with flute length of less than 0.06 inch. They are used to machine cavities about 0.05 inch deep to receive fine inlays.

Since 1970 direct-driven, high-rpm router spindles have been largely replaced by belt-driven spindles (fig. 18-150); improved belt materials have greater pulling power than older materials and allow use of greater horsepower. Two-speed motors can yield spindle speeds of 10,000 and 20,000 rpm.

Jobs requiring little power may employ air-turbines (fig. 18-199) that develop ¼-hp to 2 hp and drive router spindles at 22,000 to 50,000 rpm.

MACHINE TYPES

Because a routing bit cuts in three-dimensional motion, the manner in which work is presented to the tool can be unusually flexible. To control the plunging direction of tool motion relative to the workpiece, it is possible to leave the cutterhead fixed and move the worktable (and workpiece) up and down by means of a counterbalanced, foot-controlled mechanism (fig. 18-200). Alternatively, the worktable can be fixed and the cutting spindle can be given the vertical motion by foot-controlled mechanical linkages or air or hydraulic mechanisms. In some designs the spindle is tiltable fore-and-aft; in others, the table can be tilted fore-and-aft as well as laterally. For most jobs, the workpiece must move horizontally in relation to the cutterhead. In many designs this is accomplished by manually sliding the workpiece over the worktable and under the spindle. Alternatively, the work can be power driven under the spindle, or the spindle assembly can be moved over the surface of the workpiece.

Overhead router with fixed spindle.—A typical fixed-spindle overhead router (fig. 18-200) has a vertically adjustable work table that can be dropped as much as 7-1/2 inches below the spindle chuck. A counter-balanced foot treadle quickly raises the table 4-1/4 inches to operating position and locks it; a foot-operated trip latch lowers the table when machining is complete. The spindle, which turns at 20,000 rpm, is driven by an endless silk belt from a 3-hp motor. Typical work set-ups for this machine are shown in figure 18-201.

Overhead router with floating spindle.—On this machine (fig. 18-202) the table is fixed in a vertical sense (except for initial elevation or tilt adjustment);

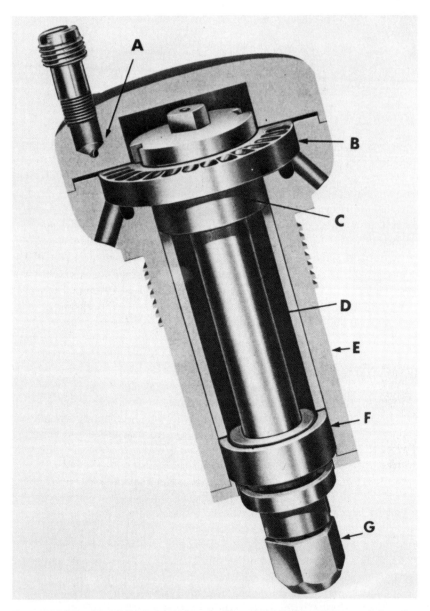

Figure 18-199.—Air turbine motor integral with router spindle. A. Inlet nozzle for air at 90 to 100 psi. B. Impeller wheel. C. Top ball bearing (grease packed). D. Spindle body. E. Quill. F. Bottom ball bearing (grease packed). G. Chuck nut.

depth of cut is determined by air-controlled plunging action of the spindle which is belt driven at 10,000 or 20,000 rpm by a two-speed constant horsepower motor available in sizes from 5 to 10 hp. The maximum distance from spindle to bare table is 13 inches. The table can, if so specified, tilt 45°. Maximum spindle stroke is 3-1/2 inches.

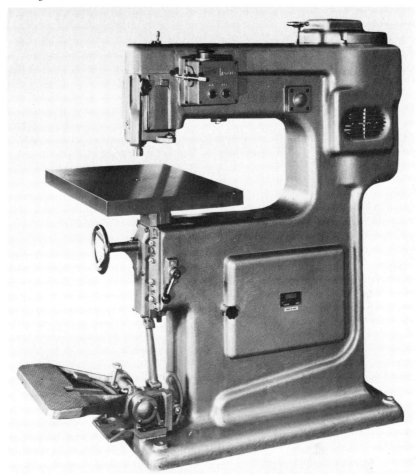

Figure 18-200.—Fixed-head router with vertically adjustable table equipped with removal guide pin. (Photo from Onsrud Division, Danly Machine Corporation.)

When routing a slot of uniform depth in a curved workpiece, the spindle can float to follow curved contours (fig. 18-202). Sensitive air actuation of spindle movement permits the cutter to be automatically held to a preset depth at both the tops of the humps and the bottoms of the valleys. The pressure of the sensing elements is light enough so that the operator can move the work easily without marring the workpiece.

Air actuation of the spindle movement permits, if so specified, automatic vertical cycling; for example, the spindle can be set for the rout position for 2 seconds and then retracted for 3 seconds, permitting the workpiece to be shifted before the router automatically returns for 2 seconds to the routing position. Alternatively, the spindle can be cycled to penetrate to two different depths of cuts, alternating each time the foot pedal is depressed.

The fixed bed can be equipped with a power-fed, numerically-controlled compound worktable capable of movement in both longitudinal and transverse directions (fig. 18-203).

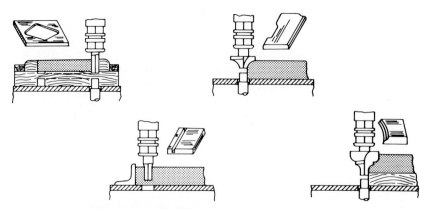

Figure 18-201.—Typical set-ups for router. (Upper left) Inside cut-out. Slot on under side of template controls position to produce desired cut-out on work. (Lower left) Inlet routing. Work is fed along fence for straight-line work. Template held against guide pin can be used for contour work. (Upper right) Top-edge routing. Work is fed along guide pin or fence. (Lower right) Contour edge routing. Template held against guide pin determines path generated by cutter.

Figure 18-202.—Overhead router with floating spindle. Vertical spindle movement is by pneumatic cylinder. (Photo from Onsrud Division, Danly Machine Corporation.)

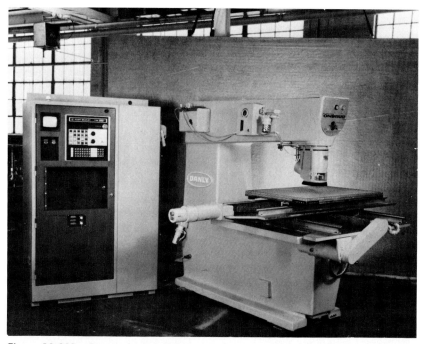

Figure 18-203.—Router equipped for computer-controlled spindle plunge movement and work table movement in both longitudinal and transverse directions. (Photo from Onsrud Division, Danly Machine Corporation.)

Radial-arm router.—For stock that is too heavy or awkward to conveniently move around a guide pin, heavy radial-arm machines are available that permit the router head to move freely over the workpiece within a radius of nearly 4 feet.

For light industrial use, smaller radial-arm routers are available in two designs, In one, the router head is radial-arm-mounted (fig. 18-204 top), and in the other, the cutterhead is stationary but the work table is movable on a radial arm (fig. 18-204 bottom). In either case, the router spindle may be belt-driven from a 2- or 3-hp motor to rotate at 18,000 to 23,000 rpm.

Inverted router.—Spiral router bits make less noise than those with straight cutting edges (Fukui et al. 1974), cut smoother, and if located below the table can pull the workpiece firmly to the table while ejecting shavings downward (fig. 18-205 top). On an inverted router, the guide pin is above the table and the routing template is placed on top of the workpiece; no clamping is necessary. The machine can be adapted to make curved cuts like a bandsaw by addition of a top chuck to hold a long slender helical router cutter in tension while it rotates at 20,000 rpm (fig. 18-205 bottom). Wood 10 times thicker than the bit diameter can be cut. Unlike bandsawn edges, routed edges are smooth (Ekwall 1978).

Computer controlled routers.—Building on experience of the metal machining industry, numerically (computer) controlled routing, shaping, and boring machines (figs. 18-203 and 18-206) for wood gained acceptance during the

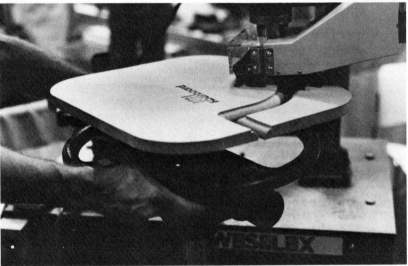

Figure 18-204.—Radial arm routers. (Top) Spindle movable, table stationary. (Bottom) Table movable, spindle stationary.

1970's (see sect. 18-30). Relative motion between bit and workpiece is controlled in X, Y, and Z directions by a multichannel (typically 8-channel) perforated tape. Most patterns are readily programmed for tape preparation.

Computer-controlled routers may carry single spindles or multiple spindles carried on a T-slotted mounting bar; cutterheads may be belt driven at 11,500 rpm or 32,000 rpm by 15-hp motors (Anonymous 1976b).

Some designs carry as many as eight heads (for routing, boring, or shaping), all mounted on a rotatable turret; under computer control, the heads can be brought into cutting position and programmed for machining patterns along X,

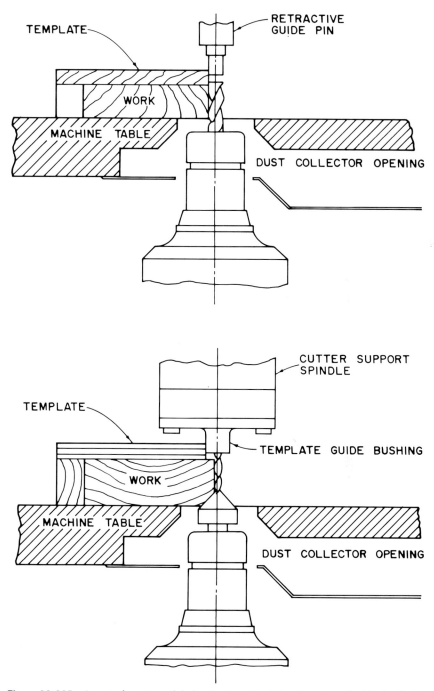

Figure 18-205.—Inverted router with helical cutter. (Top) Routing edge with bit chucked on one end; thrust on workpiece is downward, as is chip ejection. (Bottom) Slender bit chucked from both ends to put it in tension for bandsaw-like cuts. (Drawing after C. R. Onsrud, Inc., Troutman, N.C.)

Figure 18-206.—Multi-head computer-controlled routing and boring machine.

Y, and Z axes (Scully 1977a). Vacuum chucks are commonly used to grip the workpiece on such machines (Anonymous 1975; Thompson 1976).

Smith (1978) has described computer controlled equipment made by four manufacturers and provided cost data for the operations they perform. Further descriptions of operations using computer-controlled routers have been provided by Anonymous (1975, 1977b), Mattingly (1975), Ward (1975d), and Coleman (1977). Vacuum chucks for such equipment are described by Thompson (1976).

These fast, multi-head, power-fed machines have exposed some problem areas in the routing process. The literature indicates that diameters of small routing bits may vary substantially so that adjacent bits on a multi-head machine cut grooves of differing width even though bits are nominally the same size (Ward 1975d).

Theien (1970) found that power-fed routers should be capable of routing at least 1.5 inches deep at a feed rate of 300 to 400 inches per minute. He recommended a 3/4-inch tool shank to provide strength needed, and cutters with two carbide-tipped flutes, also 3/4-inch in diameter where possible. C-2 and C-3 grades of carbide are appropriate; cutting edges should have small clearance angles to lengthen service time before dulling. With such a cutter routing 1.5-inch-thick stock at 300 inches per minute with spindle speed of 23,000 rpm, horsepower required varied with material, as follows (Theien 1970):

Material	Horsepower required
Pine sp.	3
Hard maple	9
Particleboard @	
28 lb/cu ft	2
45 lb/cu ft	6
60 lb/cu ft	10

Dovetail routers.—Dovetail routers machine dovetail tenons in drawer sides and route grooves in drawer fronts and backs with standardized bits (fig. 18-207). Machines typically carry 15, 20, or 25 spindles operating at 6,000 rpm, and can be quickly set up to dovetail either drawer sides, or fronts and backs. Such machines produce up to 150 drawers per hour.

Special routers.—Compact electric or air operated router spindle assemblies can be designed into a great variety of automatic or semi-automatic machinery. An example would be addition of jump router assemblies to a double-end tenoner.

Portable routers.—Light-weight portable routers function like radial-arm routers except they are supported directly on the workpiece (or template pattern) and are freely guided by the operator. One-quarter to 2-hp, air-turbine models are available with spindle speeds up to 50,000 rpm. Portable electric routers can carry as much as 5 hp with spindle speeds of about 15,000 rpm.

Figure 18-207.—(Top left) Sliding dovetail. (Top right) Dovetail tenons in a drawer side, and dovetail grooves in a drawer front. (Bottom) Typical cutters to machine such joints.

MACHINABILITY

Woods that can be readily shaped and bored (table 18-1) also machine well when routed.

Flake orientation in particleboards has considerable influence on machinability by routing. More tearout is experienced when routing across the grain of such oriented boards than when routing parallel to the flakes. If a piercing cut or a slot is to be routed in hardboard, best results are obtained with helical cutters in which the helix is wound to exert a down pressure on the fibers at the top edge of the cut. This tends to minimize the fuzzing that otherwise takes place.

18-18 CARVING

The appeal of much fine wood furniture and millwork is attributable to the artistry of carved designs. It seems likely that significant numbers of consumers will continue to appreciate the beauty of carved wood. Machine carving is executed with small bits designed for both peripheral and face milling (fig. 18-208). Spindle speeds are commonly 10,000 rpm to 20,000 rpm.

MULTIPLE-SPINDLE CARVER

Carvings for furniture or industrial applications can be produced in quantity on multiple-spindle carving machines that produce as many as 30 carvings at a time. These machines are three-dimensional pantographs that permit a master template or model to be precisely copied. The model is mounted in the center of the machine, and the operator, by moving a somewhat blunt follower over the entire surface area of the model, causes the gang of motor-driven cutterheads to duplicate the contours of the model; all workpieces are carved simultaneously and accurately. The multiple-spindle carver completely finishes the carving except for undercutting, which is done individually on each workpiece, either by hand or on a single-spindle carving machine.

Multiple-spindle carvers can be arranged to carve centered work (fig. 18-209) or flat work (fig. 18-210). Figure 18-209 depicts a multiple-spindle machine, each spindle holding a part end-gripped between centers. The operator's right hand is guiding the follower over the surface of the model, while the carving spindles are duplicating the action of the follower. The centers carrying the workpiece, and the model, can be rotated or indexed in unison by the operator so that all surface contours on the periphery of the model can be reproduced on the workpieces. By suitable changes in the gearing and pantograph mechanisms of some carver designs, motion on one half of the apparatus can be reversed from that on the other half of the other half, permitting the operator to simultaneously machine right-hand and left-hand carvings with a single setup.

The blanks are usually turned or band-sawn to a rough shape before placing them between centers in the carving machine.

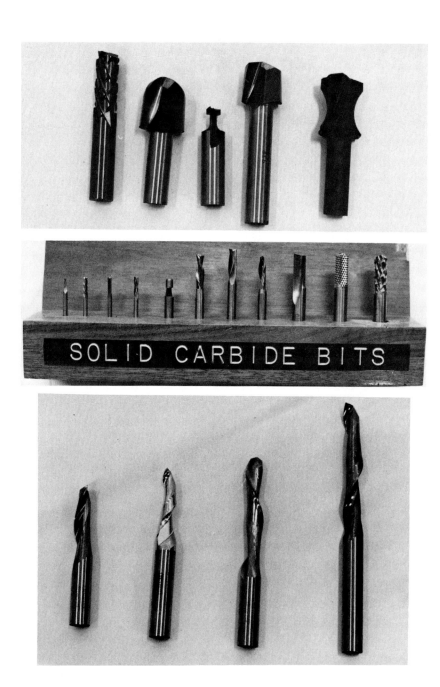

Figure 18-208.—(Top and center) Carving bits. (Bottom) Spiral routing bits.

Figure 18-209.—Pantograph multiple-spindle carving machine for flat and rotary carving work, here arranged to carve chair legs. Roughing cutters are engaged; finishing cutters can be indexed into action quickly from their location at upper end of each cutter-motor shaft. Two-speed motors rotate cutters at 12,000 or 18,000 rpm.

SINGLE-SPINDLE CARVER

Carving designs requiring undercuts can be executed on the single-spindle carver (fig. 18-211 top). It is also used to advantage on short runs of carving work that would not justify the set-up time or investment of a multiple-spindle machine. The operator works freehand, and no master model or template is required. To avoid self-feeding tendencies, with attendant danger to the operator's hands, the cutting tools are designed with rake angles close to zero degrees. The spindle of the machine illustrated is belt-driven at 10,000 rpm by a 3-hp motor. Single-spindle machines can also be arranged to duplicate a model (fig. 18-211 bottom).

MACHINABILITY

Among pine-site hardwoods, white oak, the red oaks, and the ashes are readily carved. Yellow-poplar is one of the more difficult species.

Figure 18-210.—Carved mouldings.

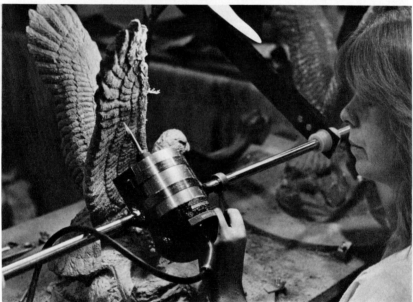

Figure 18-211.—(Top) Industrial single-spindle carver belt driven at 10,000 rpm by a 3-hp motor. (Photo from Ekstrom Carlson & Co.) (Bottom) The operator of this carver is duplicating a model barely visible at right.

WOOD EMBOSSING MACHINES

Embossing wheels (fig. 18-212) and embossing presses can inexpensively produce products that resemble carvings. Scully (1977b) has described the technique whereby a heated die (about 500°F) is pressed at about 500 psi specific pressure against wood of 6- to 8-percent moisture content until the wood is warmed (about 2 seconds); pressure is then increased to 1,500 to 2,000 psi over

Figure 18-212.—Heated embossing wheel. The workpiece was embossed with another pattern, but illustrates the concept.

the next 5 seconds—sometimes in two steps between which the press is backed off to let steam escape. Patterns deeper than 1/8-inch may be roughly pre-routed to remove some of the excess wood.

Fiberboards, particleboards, and plywood can also be embossed, but each material reacts differently.

18-19 MORTISING

The mortise and tenon joint has been used for millennia to join structural and decorative wood members. In the hewn timbers of colonial buildings, hand tools were employed to make the mortises; today's furniture factory cuts mortise cavities, principally of three shapes (fig. 18-213), with oscillating routers, hollow chisels, chain cutters, and reciprocating chisels.

OSCILLATING ROUTER

Tool action of a conventional oscillating router bit is evident from figures 18-213C and 18-214. The cavity thus formed has the disadvantage of rounded ends.

A new concept introduced in 1980 calls for a three-flute rotating router bit to move within a square-cornered die in such a way that the flutes are timed to roll into die corners and therefore execute square-cornered cuts (fig. 18-215).

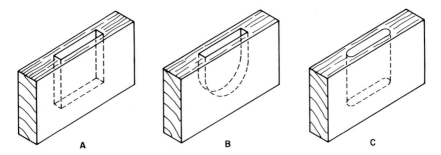

Figure 18-213.—Mortise shapes. A. Formed by hollow-chisel mortiser or by reciprocating chisel mortiser. B. Formed by saw-chain mortiser. C. Formed by oscillating router bit.

HOLLOW-CHISEL MORTISER

Square, rectangular, or even triangular mortises can be made by a hollow-chisel bit, i.e., a boring bit revolving inside a shell which terminates in cutting edges (fig. 18-216). The rotating inner bit projects slightly beyond the cutting end of the non-rotating external hollow chisel; both advance and withdraw together. When forced into the workpiece by the motion of the ram (fig. 18-217), the bit bores a hole and the sharp edges of the chisel broach the circular hole square. Thus the tool combines boring action with orthogonal cutting. Rectangular mortises are made by withdrawing the tool and moving the workpiece to form a succession of overlapping square holes. A single- or double-slotted opening in the hollow chisel, which permits the shavings to escape (fig. 18-216), must remain above the surface of the wood when the chisel is in its deepest cut.

Figure 18-214.—Oscillating router mortiser with multiple bits.

Figure 18-215.—Three-flute router bit, square-cornered die in which it rotates timed to sweep die corners, and resulting square-cornered, flat-bottomed mortise.

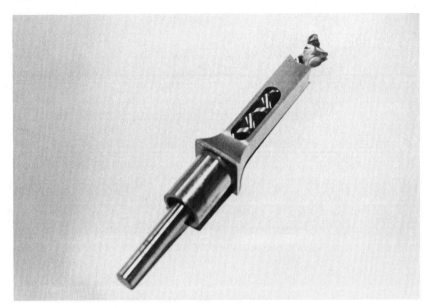

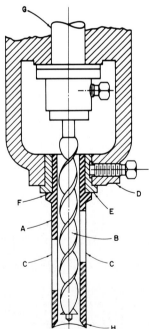

Figure 18-216.—(Top) Hollow mortising chisel and matching bit. (Bottom) Method of mounting hollow-chisel mortise bits. A. Fixed hollow chisel. B. Rotating boring bit. C. Openings in chisel for chip escape. D. Chisel socket with clamp screw. E. Bushing to accomodate size of each chisel. F. Shoulder of chisel insures concentric alignment with bit. G. Rotating spindle that drives the bit; no end-play permitted or proper clearance at H cannot be maintained (1/32-inch for chisels smaller than 3/4-inch; on larger chisels, as much as 1/16-inch.)

Hollow chisels are available in sizes from 1/4-inch to 1-1/2 inches square. Its sharp cutting edge has a clearance angle of 0°. The rake angle is cut on the interior face by means of a double-angle, conical reaming cutter (fig. 18-217 top), yielding a primary rake angle of 65° and a secondary rake angle of 45°. Chisels are ground in three steps as follows: (1) the end is ground square; (2) the rake angle is reamed; (3) corners are grooved with a square file.

Figure 18-217 (bottom) illustrates the elements of a horizontal, hollow-chisel

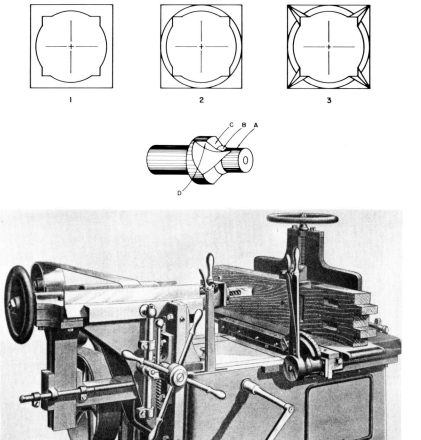

Figure 18-217.—(Top) Three steps in sharpening a chisel and the double-angle reamer to accomplish the second step. A. Pilot. B. Reams primary angle. C. Reams secondary angle. D. Reamer sharpened on this face. A separate reamer is required for each size of chisel. (Bottom) This horizontal hollow-chisel mortising machine of the 1890's handled 16-inch hardwood timber. (Drawing after Greenlee Bros. & Co. 1962.)

mortiser in its original simple configuration. In a typical modern machine, the chisel ram, or carriage, is power driven by means of a worm and worm gear. The clutch is foot actuated to stop the ram at any point in the stroke. When held engaged it actuates the carriage continuously at 15 to 52 strokes per minutes, as needed. One horsepower is sufficient for bits ranging in size from 1/4-inch to 1-1/2 inches square. Typically, the chisel stroke is adjustable from 2-1/2 to 4 inches. The compound worktable can be tilted, adjusted vertically, traversed laterally and traversed longitudinally, and carries side and top clamps to hold the workpiece in place during both plunge and withdrawal strokes.

Hollow-chisel mortisers are available in numerous designs incorporating gang-vertical, diagonal, and horizonal mountings of the chisels. Feed can be accomplished by moving either chisel or workpiece. In special-purpose production lines, mortising stations can be installed for sequential operations involving one or more double-end tenoners in series, with automatic transfer between stations.

CHAIN MORTISER

A chain mortiser (fig. 18-218) will plunge-cut cavities 1/4- to 1/2-inch wide (in certain applications up to 1-1/4 inches wide), from 3 to 5-1/2 inches deep, and from 1 to 6 inches long. The chain is simply constructed of left-hand, right-hand, and center links carrying teeth with rake angle of about 30°, but with neither top or front bevel. Clearance is provided by the swaged shape of teeth in the outer links. Rake angle should be kept equal on all teeth and front bevel avoided; teeth should be kept of equal height and pitch. Because the chain bars are straight sided (or wider at the bottom), the plunge cut of the chain mortiser can be compared to a circular ripsaw in a plunge cut with blade parallel to the grain and motion perpendicular to the grain.

The mortise chain assembly consists of a top-drive sprocket, chain bar with roller wheel at the bottom, and the chain. The chain bar for round-bottom slots illustrated in figure 18-218 (center) will cut a mortise 3/8-inch wide, 3 inches long, and 4 inches deep. Assuming that the drive sprocket turns at 3,600 rpm and that the chain pitch is 3/8-inch (giving an outside tooth pitch of 3/4-inch), a feed speed of 1 inch per second would give maximum tooth penetration of 0.002 inch per tooth. Under these circumstances the chain would run at 1,800 fpm, comparable to many hand-held chainsaws (800 to 3,800 fpm).

For relatively flat-bottomed blind mortises, a double-bearing guide bar can be used (fig. 18-218 bottom). In contrast with the chain saws described in section 18-8, the mortise chain is guided by the outside links that straddle a center guide rail. Lubrication of the mortise chain and the condition of the guide bar have a pronounced effect on performance.

A chain mortising machine in its most simple configuration is shown in figure 18-218 top. The driving sprocket of the mortise chain turns at 3,600 rpm and is direct connected to a 7-1/2-hp motor. This provides sufficient power to carry chains that will cut mortises up to 1-1/4 inches wide. If so specified, the machine

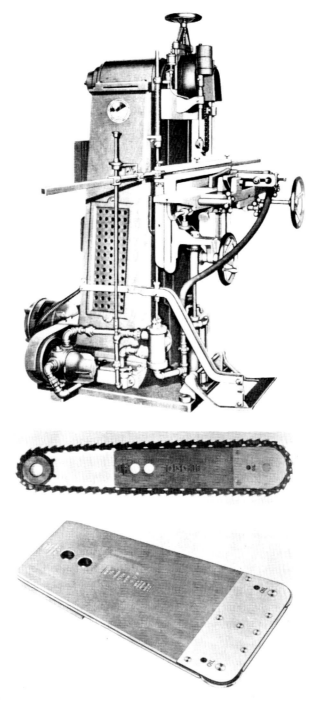

Figure 18-218.—(Top) Chain mortiser. Photo from Northfield Foundry and Machine Company. (Bottom) Mortise chain and bar assembly for round-bottom mortise slot, and double-bearing bar for blind mortise cavity with flat bottom. (Photos from William H. Field Company, Inc.)

can be arranged with an extended spindle to carry two sets of mortise chains in order to permit simultaneous cutting of two cavities.

The length of the vertical power feed stroke is adjustable from 0 to 7 inches. It is actuated by a hydraulic cylinder that is in turn powered by a 1-hp pump and motor combination. In contrast with the hollow-chisel mortiser (fig. 18-217), which has a moving head, the chain mortise assembly is rigid and the vertically moving table provides the feed motion. In addition to a maximum power stroke of 7 inches, the compound table may be tilted 45°, vertically adjusted 13 inches, laterally traversed 2-1/2 inches, and longitudinally traversed 11 inches. Mechanical or hydraulic clamps for the workpiece can be specified for the table, including a hold-down foot that rests on the stock, where the chain emerges from the workpiece, thereby reducing tearout.

More complex machines provide gang mounting or incorporate lateral, longitudinal, and swivel motions of the cutter.

RECIPROCATING CHISEL MORTISER

Early reciprocating chisel mortisers utilized a single-point tool reciprocating to more-or-less duplicate the hand operation of chiseling a mortise cavity. More recently, machines have been designed to reciprocate and oscillate multi-point chisels to excavate square-ended cavities at high speed (fig. 18-219). Typically, chisels are driven at 2,800 to 4,500 strokes per minute and require 2 to 5 hp.

MACHINABILITY

Smoothness and accuracy of mortise cuts vary significantly among hardwood species. No data are available for oscillating routers, chain mortisers, or reciprocating-chisel mortisers, but Davis (1962) provided data on a 1/2-inch square, hollow-chisel mortiser with spindle speed of 2,400 rpm (feed speed and rake angle not specified). Wood specimens were 3/4-inch thick and held at 6 percent moisture content.

Mortise cuts were made in side grain and arranged so that two of the four sides of each square mortise ran across the grain and two parallel to the grain. While cuts parallel to the grain were reasonably smooth in all woods, the cuts across the grain varied widely in quality, with some showing considerable crushing and tearing. On the more refractory species (soft maple, for example), the corners of the mortise tended to tear very badly in the area where the square chisel must broach out the excess material left by the boring bit. This tearout extended the complete depth of the mortise (fig. 18-220).

In most of the hardwoods tested, the mortises varied from the chisel size by amounts up to 0.006 inch parallel to the grain and 0.002 inch across the grain. In addition, the mortises tended to taper slightly, being largest on the entry side and smallest on the side where the chisel emerged. The taper was about 0.003 inch parallel to the grain and less than 0.001 across the grain. The holes in nearly all

Figure 18-219.—Reciprocating chisel mortiser. (Top) Cutting action. (Bottom) Resulting square-cornered mortise cavities. (Photos from MAKA.)

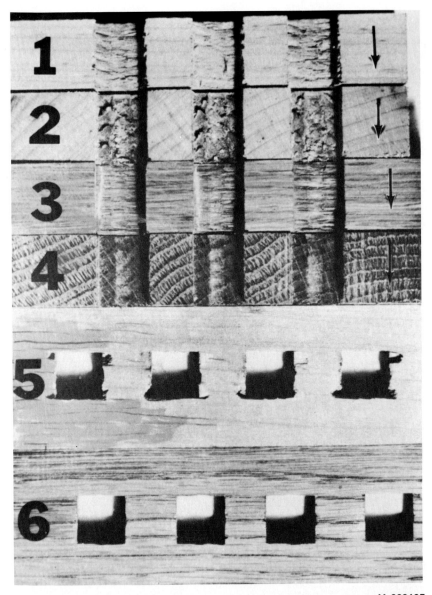

Figure 18-220.—Hollow-chisel mortises in side and end grain of two species: 1, 2, and 5, soft maple; 3, 4, and 6, red oak; 1 and 3, side grain; 2 and 4, end grain. Arrows indicate direction of cut in samples 1 to 4. (Photo from Davis 1962.)

cases, were smaller than the chisel. Heavier woods usually produced smoother cuts and more accurate mortises than lighter woods, although black tupelo gave poorer results than its specific gravity would indicate. Pore arangement had little influence on mortise quality. Davis' (1962) results on 11 hardwood species found on southern pine sites are summarized as follows:

Species	Fair to excellent pieces	Amount off size
	Percent	Thousandths of an inch
Oak, white	99	3.6
Hickory	98	2.6
Oak, chestnut	96	3.6
Oak, northern red	95	3.4
Elm, American	75	3.9
Hackberry	72	3.0
Yellow-poplar	63	4.6
Ash, white	58	4.1
Sweetgum	58	4.9
Maple, red	34	4.6
Tupelo, black	24	4.3

Black cherry and true mahogany, both important furniture woods, yielded 100 percent fair to excellent pieces in Davis's experiment.

18-20 TENONING

Tenoning machines cut patterns on one or both ends of a workpiece (fig. 18-221) by conveying it past a sequence of cut-off saws, slotting or notching saws, and cutterheads which peripheral mill and saw with cutting orientations of 90-0, 0-90, and 90-90. Horizontal spindles carry **tenon** heads, and vertical spindles carry **cope** heads. **Cutoff saws** have horizontal spindles which may be tilted for angular cuts. See figure 18-222A for illustration of these three principal machining units common to most tenoners. Many cope and tenon heads are fitted with saws only, to make required cuts. Other applications call for cutterheads with knives or for knives in combination with saws. Figure 18-222A illustrates groove-cutting knives on the two cope heads and knives combined with spurs (to make clean shoulder-cuts) on the tenon heads. In all tenoning machines the workpiece is conveyed continuously past saws and cutterheads to achieve patterns typified in figure 18-221.

TENON CUTTERS

The heads and knives that are carried on tenon spindles are distinct from most other cutterheads. In the tenoning operation the cut is perpendicular to the grain, and it is customary to mount the knives diagonally on the cutterhead to minimize the impact on an individual knife as it becomes engaged in the cut. Because of this geometry, the rake angle varies from one end of the knife to the other, being minimum where the face of the knife is closest to the axis of the cutterhead. In

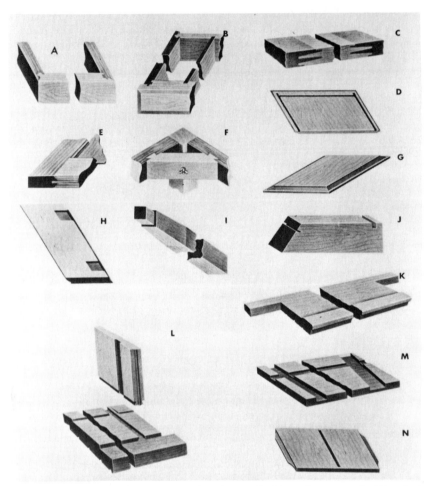

Figure 18-221.—Typical patterns produced on a double-end tenoner. A. Miter lock joint. B. Square lock joint. C. Finger joint. D. Coping and blind dado. E. Tenon and cope on door rail. F. Mitered furniture rail. G. Shaped panel. H. Recess for drop-leaf hinge. I. Taper and ferrule cut on leg. J. Miter, dado, and clamp nail groove. K. Sill-horning. L. Dovetail for sills and stiles. M. Straight and angular dado. N. Blind dado and relished tenon. (Photo from Greenlee Bros. & Co.)

addition, the cutting edge must be ground to convex curvature to enable it to generate a flat cut. Figure 18-222B illustrates the manner in which the curved profile of the knife must be developed in order to achieve a flat cut on the workpiece.

Tenons heads normally carry shoulder cutters (sometimes called **spurs** or **combs**) to accomplish a clean shoulder cut. They act as scoring cutters and generate the finished surface of the shoulder. The cutting action precedes that of the tenon knives. The cutting circle of the shoulder cutters should be about 1/64-inch larger than the cutting circle of the tenon knives, and their peripheral edges

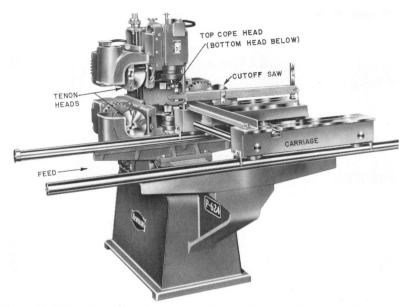

Figure 18-222A.—A single-end tenoner equipped with manually traversed work table. (Photo from Newman-Whitney.)

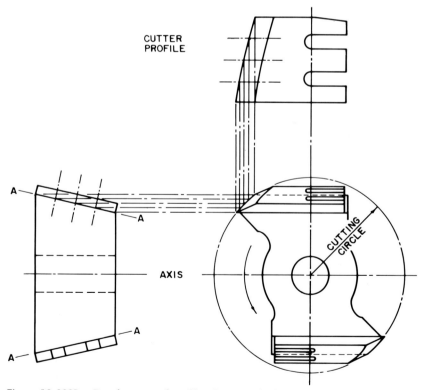

Figure 18-222B.—Development of profile of a tenon knife to cut a straight tenon; A-A indicates plane of cutting edge.

should be shaped to this cutting circle. As the shoulder cutters, rather than the tenon knives, generate the surface on the shoulder, the shoulder cutters should project approximately 1/32-inch beyond the ends of the tenon knives. In addition, the cutting edges on the faces of the shoulder cutters must have a few degrees of clearance.

MACHINE TYPES

Tenoners are made in infinite variety but principle machine types include single-end tenoners with manual carriage, finger jointers, end matchers, and double-end tenoners.

Single-end tenoner with manual carriage.—In this most simple configuration of a tenoner (fig. 18-222A), the workpiece is clamped to a manually traversed carriage that is pushed past horizontal top and bottom tenon heads, top and bottom vertical cope heads, and finally a cut-off saw in the order named. Only one end of the workpiece is machined at a time.

Finger jointer.—Machines to cut fingerjoints (fig. 18-221C) are specialized single-end tenoners usually provided with a cut-off saw, not tiltable, followed by a single cope head which acts in sequence to machine a fingerjoint (fig. 18-222C). Although machines are available that utilize manual feed, more productive designs utilize continuous, lugged feed chains with automatic transfers between right- and left-hand machines. With this arrangement random-length boards can be machined on both ends at rates up to 60 pieces per minute. For such an application the cut-off saw might carry a 3-hp, 3,600-rpm motor and the cope head a 10-hp, 3,600-rpm motor. Random-length stock up to 2 inches thick

Figure 18-222C.—Manually fed machine to cut finger joints; formed die at left applies glue to fingers.

and 12 inches wide can be accepted, but for best results stock should be 6 inches or less in width. Accurate fingerjoints are difficult to machine because they require deep cuts across the grain and permit no tearout on the back side. In addition, vertical positioning of right- and left-hand patterns, and angular positioning of the workpiece must be near perfect to avoid unsightly gaps in the end joints.

Readers interested in end-glued joints will find the following references useful (see also fig. 22-4):

Subject	Reference
Optimum geometry of structural and non-structural fingerjoints, butt joints, and scarf joints	Koch (1972, p. 1163-1175)
Fingerjointing green lumber	Dobie (1976)
Fingerjoints appropriate for hardwood furniture	Murphey and Rishel (1972)
Structural fingerjoints and methods of curing joint gluelines	Edlund (1975), Northcross (1979)

End matchers.—End matchers are specialized single-end tenoners installed in pairs (fig. 18-223) to cut a tongue on one end and a groove on the other end of random lengths of hardwood (or softwood) flooring. Each end matcher is fed by a lugged chain travelling at 60 to 90 lugs per minute. The two-machine system is usually provided with an automatic transfer between right- and left-hand units. The stock is fed face up, with the lugs pushing against the groove side of the flooring. Pressure is exerted on the under side of the workpiece so that it is forced against a solid overplate while passing through the machining zone. By this means the face of the flooring becomes the reference plane controlling the vertical location of the end-matched tongue and groove. The left-hand groove machine carries a 5-hp cut-off saw followed by a 5-hp cope head. The right-hand tongue machine carries a 5-hp cut-off saw followed by 2-hp top and 2-hp bottom tenon heads. The spindles normally rotate at 3,600 rpm but can be operated at 6,000 rpm if 120-cycle current is available. Each machine requires a 3-hp feed motor. On most end matchers the thinnest stock that can be end-matched is 3/8-inch and the thickest is 2-inch. Normal spacing between feed lugs is 6-1/4 inches, although wider spacing can be specified. It is much more difficult to end match accurately a wide board than a narrow one. The shortest workpiece that can be end matched is 9 inches.

Double-end tenoner.—The double-end tenoner, unlike the end matcher, is designed to handle stock that has been previously sorted to length. The workpiece is carried through the machine on two traveling chains and is held down in the cutting zone by two interlocking rubber-faced, overhead, traveling chains that are supported by two pressure bars. Thus the reference plane on a double-end tenoner is the bottom or underside of the stock. Figure 18-224 shows the two lower feed chains and the two upper hold down beam assemblies. One side of the machine (the near side in figure 18-224) is fixed on the floor-mounted base, while the other can be power traversed and precisely spaced (within plus or minus 0.005 inch) in relation to the fixed side for running material of various

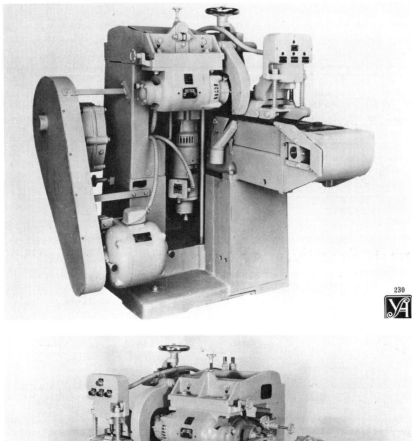

Figure 18-223.—End matchers for hardwood flooring. (Top) Left-hand groove machine. (Bottom) Right-hand tongue machine. (Photo from Yates-American Machine Co.)

Figure 18-224.—Infeed end of a double-end tenoner. (Photo from Mereen-Johnson Machine Co.)

lengths and widths. In some cases, it is necessary to make the right hand side of the machine adjustable or make both sides adjustable symmetrically about the center line of the machine for best flow of material.

The most common double end tenoner is capable of end trimming 96 inches and lug feeding 48-inch-wide material. Machines are available for material from 12 inches to 24 feet in length. The length of the lower infeed chains can be specified to accept panel sizes from 12 inches to over 120 inches ahead of the upper hold down beam assemblies. The normal maximum thickness of stock that can be accepted is about 4 inches but machines can be provided to accept 8-inch stock. The speed of the chains is commonly 20 to 60 fpm. Hydraulic and D/C electric feed drives are available providing a wider range of speed, e.g., 5 to 100 fpm.

Standard cutter-head arrangement on each side of a furniture type double end tenoner would include one tilting cut-off trim saw at the first station, top and bottom horizontal spindle scoring motors at the second station, top and bottom horizontal-spindle tenon motors at the third station, one high-speed tilting vertical-spindle cope motor mounted below at the fourth station and a full length dado arbor assembly mounted above the fourth station. Figure 18-225 illustrates this machine arrangement. Trim motors are normally 7-1/2 hp and are available to 20 hp; scoring motors are 3 hp or 5 hp; tenon motors are 5 hp or 7-1/2 hp; and cope motors are 7-1/2 hp up to 20 hp. All spindle speeds are 3,600 rpm, except the cope motors which can be provided with alternate spindle speeds up to 7,200 rpm. The variable speed feed motor is normally 3 hp.

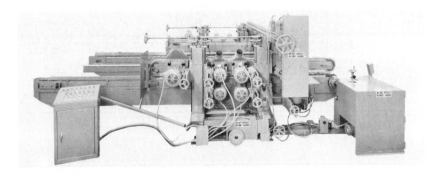

Figure 18-225.—Side view of double-end tenoner. First station (on left) tilting trim motors, second station top and bottom scoring motors, third station top and bottom tenon motors, fourth station above dado arbor assembly, fourth station below universal air-jump cope motor assembly. (Photo from Mereen-Johnson Machine Co.)

CAPABILITIES OF DOUBLE-END TENONERS

This most versatile of woodworking machines can perform a broad variety of work; the following paragraphs outline some of the possibilities. Readers interested in intensive machine utilization and reduction of setup time will find Ekwall's (1977) analysis of setup times useful.

Double cut-off (trim saws).—Each end of the machine carries a tilting cut-off saw. Therefore, stock can be smooth trimmed to precise length either with or without a bevel (fig. 18-221J). Each trim-saw motor is adjustable horizontally and vertically for use above or below the material. Combination trim/hog tooling is available which hogs all of the edge-trim waste so it can be pulled into the dust collection system. Trim/hog tooling, available up to 2 inches in width, is normally used below panels with saw teeth cutting counter to panel flow; this arrangmenet reduces panel vibration because cutting forces hold panels firmly against lower feed chains. To further stabilize panels during the cut, an anvil supports them as they are trimmed.

Trim motors are available with unit-mounted scoring motors (fig. 18-226) that adjust to position with the trim motor and include independent horizontal adjustment for aligning the scoring blade with the trim saw and vertical adjustment for setting the depth of cut of the scoring blade. The scoring motor can also be air actuated to jump up through the trailing edge of the panel as it passes through the machine to prevent any trailing edge tear-out by the trim saw. Scoring saws normally precede trim saws using combination trim/hog tooling to prevent tear-out on wood veneer, high pressure laminates and other decorative surfaced panels.

Figure 18-226.—Combination trim saw and jump scoring saw assembly for a double-end tenoner. Feed is from right to left. (Photo from Mereen-Johnson Machine Co.)

Scoring and tenon station.—Figure 18-225 illustrates the first station trim motor, second station top and bottom scoring motors and third station top and bottom tenon motors and fourth station high-speed tilting cope motor (behind guard) which can machine a multiplicity of shapes (fig. 18-221ABEFGI). Scoring saws are normally used ahead of the tenon heads to cut through veneer and thus eliminate tearing by the tenon heads or cope heads. Scoring saw blades run with the feed (climb-sawing) and are able to make clean, accurate cuts without tearing fine veeners. Scoring and tenon motors are normally not tiltable but can be provided with tiltability. Top and bottom scoring motors and the top and bottom tenon motors have individual vertical and horizontal adjustment and alternately can be provided with an adjustment which will move the top and bottom motors simultaneously for faster size changes. Detachable extended spindles are sometimes provided on the tenon motors; these spindles project through the hold down beam or feed chain beam to machine center dado grooves in panels.

Relishing.—In furniture manufacture tongues or tenons must sometimes be contoured as shown in figure 18-221N; such contouring is termed **relishing**. Cope motors with air-cylinder-actuated relishing attachments (fig. 18-227 top) are used for so trimming tenons to fit into blind dado grooves. The relishing cutters are driven at 7,200 rpm from the 3,600-rpm cope motor through positive timing belts. The offset high-speed spindle is air actuated to perform the relishing operation by a timing drum setting system which is driven from the feed

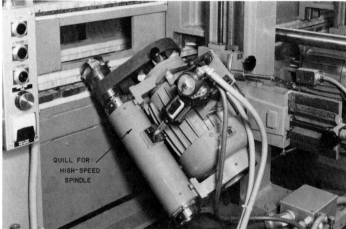

Figure 18-227.—Air-actuated jump cope motors. (Top) Assembly for relishing. See figure 18-221N for example of tenons trimmed by this equipment. (Bottom) Universal, tilting 7,200-rpm quill with shaping cutter. (Photos from Mereen-Johnson Co.)

drive shaft and is arranged for individual jumping of two of the cope assemblies and of a dado arbor assembly in and out of cutting position. The air jump cope motor assembly illustrated in figure 18-227 includes an air/hydraulic cylinder for controlled speed of the jump actuation.

Angular dado.—A tenon head attachment makes possible the cutting of dado grooves with angularity of 5 to 16° on window-frame stiles (fig. 18-221M). The attachment automatically moves the top and bottom tenoning motors diagonally across the work. With this arrangement, both ends of the stile can be trimmed square, angular dado cuts made on the bottom, and right-angle dado cuts made on the top—all in one pass through the machine. If desired, it is also possible to cut off the workpiece parallel to the angled dado by using combination heads that dado and cut off simultaneously.

Sill-horning.—The meaning of this phrase is illustrated by figure 18-221K. The operation is accomplished by the use of special clamp dogs that hold the stock at the correct angle for trimming and notching. The clamping and unclamping action of the dogs is actuated by cams on the chain beams. Sills up to 7-1/2 inches in width can be handled if the machine is properly specified. With extended spindles, notching saws on the lower tenon spindles cut to 5-1/4 inches from below, and saws on the cope units cut to 4-5/8 inches from the end of the piece.

Dovetailing.—Figure 18-221L illustrates a dovetail slot that can be cut with a vertical router attachment (fig. 18-228 top) driven at approximately 14,000 rpm. By tilting the attachment, it is also possible to make dovetail cuts at desired angles on the end of the workpiece. In either case, it is recommended that the workpiece first be grooved with a scoring or tenon cutter before dovetail slotting to increase the life of the router bit and to permit higher feed speeds.

Cam-generated shaping.—One or both ends of a workpiece can be contour shaped (without prior bandsawing) on a double-end tenoner by the application of cam-operated shaping attachments to the cope spindles. The offset shaping spindle, which is driven from the cope spindle, revolves at 7,200 rpm. A cam, cam roller, lever, and segment gear arrangement is used to transmit the cam pattern to the shaping spindle.

A simplified cam-forming unit is used in making the cam from the original pattern. To make the cam, the cam roller lever is removed and the cam-forming unit fastened in its place. When a collar of the same diameter as the shaping cutter is placed on the shaping spindle and the pattern is fed through the machine, the cutter on the cam former automatically generates the cam outline. This generated cam will, in turn, guide the shaper cutter to accurately reproduce the original pattern.

Since attachments on each end of the tenoner operate on individual cams, each end of the workpiece can be given a different contour. Tenoners equipped with these attachments will handle a shaped length up to 60 inches in one complete revolution of the cam; longer lengths can be obtained through different gear ratios to the feed shaft.

Dado and jump-dado.—Jump- and stationary-type dado attachments provide the means for cutting grooves between, and outside of, the pressure beams.

Figure 18-228.—(Top) Tilting jump router assembly at the third station of a double-end tenoner; jump actuation is triggered by a timing drum. (Bottom) Air jump dado saw; motion is controlled by a timing drum and limit switches. (Photos from Mereen-Johnson Machine Co.)

Typical dado work is illustrated in figure 18-221HN. The jump, or lift arrangement, is controlled through a timing drum integrated to the air actuated jump cylinders that raise the cutter head normally ¾-inch (fig. 18-228 bottom). Various jump lengths are obtainable from 10 to 120 inches. This arrangement permits the making of a groove open at one end or, if the cutter drops into the work and lifts out again, the forming of grooves that are blind at both ends. A single dado head might be driven by a 7½-hp motor at 2,100 to 3,600 rpm. Specifications for dado heads are discussed in connection with figure 18-89. Router assemblies can also be mounted on an overbeam to route slots.

Sanding.—Long-belt edge sanding for flat and profile sanding are available with belt lengths up to 200 inches (fig. 18-229). Multiple sanders can be

Figure 18-229.—Tilting, 5-hp, belt sanding attachment on outfeed end of a double-end tenoner can smooth flat or formed edges. (Photo from Mereen-Johnson Machine Co.)

provided on each side of the machine with 120° tiltability; various grits of sanding belts can be used at each station to obtain optimum sanding quality. Air actuated sanding platens are used to eliminate dubbing of leading and trailing ends of panels. Features include air cooling of the platen to dissipate sanding belt heat, striated heat sinks behind the sanding platen for additional heat dissipation, variable speed drive of the sanding belt so belt speed can be matched to panel feed speed, air tensioning of the sand belt and a belt breakage limit switch to stop the feed drive when a belt breaks. Powered vertical, horizontal and tilt adjustment with digital counter position reference is available on the sanders to minimize set-up time.

Two-pass tenoners.—When two double-end tenoners are coupled at 90° to each other it is possible to machine all four sides of a tile or panel in a single pass. For example, 12- by 12-inch ceiling tiles—a product requiring precise squareness—are being processed at 90 to 120 pieces per minute. Four- by 8-foot plywood panels are sized in width and length at speeds up to 30 panels per minute, 4- by 16-foot hardwood panels at 16 per minute, and thin (0.02 inch) decorative laminates in 5- by 12-foot size at 10 panels per minute. Hollow-core doors for houses can be trimmed, edge sanded, and the corners eased by sanding at a rate of 12 to 22 doors per minute.

Machines to size particleboard panels have been highly developed to operate under numeric control at very high production rates, e.g., 10 to 15 tons of board per hour. Readers interested in such an operation will find Scully's (1975) description useful.

Because double-end tenoners have such extraordinary flexibility, they are offered with increasingly complex machining and transfer stations. Readers interested in a state-of-art account of such machines installed in series will find useful Saul and Lagerquist's (1975) description of an installation at American Furniture Company, Martinsville, Virginia.

FEED AND CONTROL OPTIONS FOR DOUBLE-END TENONERS

Traverse adjustment.—The basic method of moving the adjustable chain beam assembly is with an A/C motor driving a screw attached to a nut on the movable beam; position reference is by a digital counter reading in hundredths of an inch. Push-button setting systems using D/C control are also available. They adjust the beam very rapidly then slow prior to coming to final position—always from the same direction to eliminate backlash.

Feed chain.—Smooth-top feed chain is used when panels will be fed by feed lugs and when excessive side forces from cutterheads are not anticipated. Corrugated feed chain is used for friction and lug feeding of solid wood panels and veneer-surfaced panels. Panels with decorative surfaces susceptible to scratching and marring are friction or lug fed on feed chains with replaceable friction pads.

Numerical controls.—Programmable logic controllers are available having the capability of moving all of the cutter head motors to position, traversing the machine to position, adjusting the feed speed, setting the hold downs to the proper thickness and controlling jump actuation of cope motors or the dado assembly. Input of setting information can be by manual push button, perforated tape or floppy disk. A CRT television tube display will provide production control information to the operator including a drawing of the panel illustrating the size, edge machining being accomplished, location of grooves and type of panel being processed. The automatic feeder and stacker can be controlled from this unit. Controller integration with a bar code optical reader will illustrate on the CRT to the operator if he is processing the correct panel and a laser scanning device will determine whether the machine is set up correctly for that specific panel. See also sect. 18-30.

SOUND ENCLOSURES

Machine-mounted sound enclosures are being provided as an integral part of double end tenoners. Acoustical insulation lines the enclosure and protective coverings are provided over the insulation. The enclosure doubles as a dust collection system, eliminating the need for dust hoods on cutting motors and decreasing time required for tooling changes. A refuse belt is provided under each trim motor to eject refuse from the enclosure. Hinged access doors are electrically integrated so they can not be raised unless the motors have been stopped; electrical brakes on motors stop rotation of the tooling in 5 to 10 seconds. Fluorescent lamps and impact-resistant windows are provided in the doors.

18-21 MACHINING WITH HIGH-VELOCITY LIQUID JETS

When a liquid jet impinges on wood, a force is exerted between jet and wood which is related to liquid flow rate. For water impingement perpendicular to the surface, the force (F, pounds) normal to the surface is as follows:

$$F = 1.56 d^2 p \qquad (18\text{-}49)$$

where:

d = jet diameter, inch
p = fluid pressure, psi

Water consumption (V, gallons per hour) can be expressed as follows:

$$V = 1{,}790 d^2 p^{1/2} \qquad (18\text{-}50)$$

Thus, at a water pressure of 35,000 psi, force for a 0.010-inch nozzle orifice is 5.5 pounds with water consumption of 33.5 gallons per hour.

For effective cutting fluid pressure must be high, flow continuous, and the liquid jet must be coherent (not dispersed); Franz (1970ab) found that addition of low-concentration polyethylene oxide or emulsion of other long-chain polymer to water significantly increases water jet coherence and cutting effectiveness.

McCartney Manufacturing Company, Inc., Baxter Springs, Kansas commercially manufactures a continuous-flow, high-pressure pumping system (fig. 18-230 top) that delivers water with additive at 40,000 to 55,000 psi through sapphire nozzles. The sapphire orifices measure 50 to 375 μm in diameter with a square-edge entry and a conical exit.

INDUSTRIAL APPLICATIONS

Liquid jets cut with very narrow kerf, have no tool wear, can make interrupted cuts, can be programmed to cut complex patterns, use little water, make little residue (which is carried off by the jet and can be removed by filtration), emit no dust to the atmosphere, operate at low noise levels and fairly low energy levels, and do not excessively wet cut surfaces. They are, however, limited in depth of cut; feed rates on thick material (e.g., 1-inch-thick) are slow and kerf walls may be tapered (narrow at jet entry and wider at jet exit). For crosscutting plywood or boards, power requirements are substantially greater for liquid jets than for circular saws. Also, the nozzle must travel in very close proximity to the surface being cut, so surfaces must be smooth and regular.

Chamberlin (1973) and Walstad (1974) described the operation of Alton Box Board Company in Jackson, Tenn., in which 1/2-inch-thick laminated paper tubes are contour cut at 150 to 180 inches per minute by coordinating vertical motion of a water jet with rotational movement of the tubes (fig. 18-230 bottom); a 30-hp motor drives the pump in this sytem. The water-polymer requirement is less than 50 gallons per hour with a jet orifice 0.01 inch in diameter and fluid pressure of about 45,000 psi.

In 1976 Flow Industries, Kent, Washington, installed a fluid-jet slitter on an 87-inch corrugator at Weyerhaeuser's Austin, Minn. plant. The computer-directed system carries six jets emitting pure water to slit wide corrugated board into desired widths without crushing (fig. 18-231 top), at lineal feed speeds up to 600 fpm. The 60-hp pumping system can deliver 90 gallons of water per hour at a continuous pressure of 55,000 psi; usually, however, only about 80 percent of this is used. Figure 18-231 (bottom) relates horsepower required to flow rate and fluid pressure (Paperboard Packaging 1977).

Szymani (1972), in a study of liquid-jet cutting of 11 types of corrugated board, found that edgewise compression strength of jet-cut board was about twice that of board cut with a conventional slitter; the principal failure mechanism during jet cutting is breaking of interfiber bonds resulting in fiber separation.

Wood products such as plywood and lumber can also be cut with fluid jets. A coherent water jet from a nozzle with 0.0079-inch (200 μm) diameter driven by a fluid pressure of 60,000 psi can cut 3/4-inch plywood at about 30 feet per minute with kerf width of less than 0.012 inch (300 μm).

If live-sawn, dry, pre-surfaced boards about 1-1/8 inches thick cut from pine-site hardwood short logs could be effectively cut by this method, the fluid jet

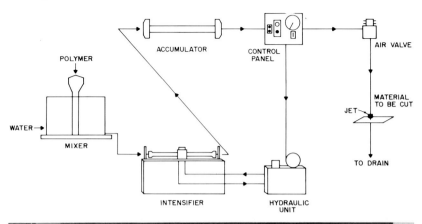

Figure 18-230.—(Top) Diagram of fluid jet cutting system. (Bottom) Upholstery frames being cut from laminated paperboard with a fluid jet. (Drawing and photo from McCartney Manufacturing Company, Inc.)

could be coupled to a defect-detecting image analyzer and computer to make blind cuts and effectively excise clear-one-face cuttings. At a cutting speed of about 50 fpm, energy requirement would not be excessive (less than 100 hp), water consumption would be low, and the water jet would not seriously wet kerf walls. It seems possible that required technology for pumps to deliver fluid continuously at required pressures, additives to attain needed jet coherence, and wear-resistant orifices will be developed to do this job by the year 2000.

Readers interested in the history and technology of fluid-jet cutting should find Szymani and Dickinson's (1975) review useful.

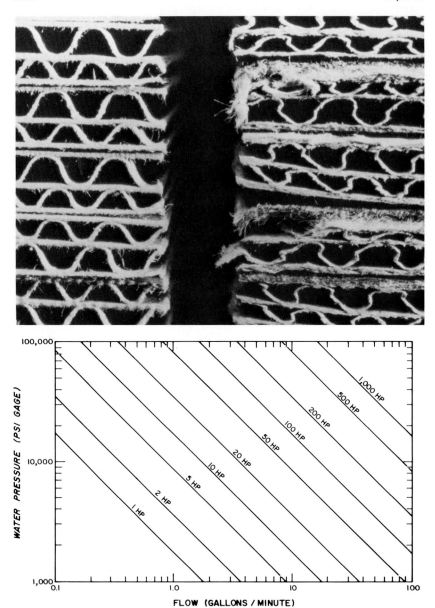

Figure 18-231.—(Top) Two piles of sized corrugated board; the pile on the right was cut by conventional mechanical slitters, that on the left by water jets. (Bottom) Nomogram relating pump horsepower to flow volume and water pressure. (Photo and drawing from Flow Industries, Inc.)

18-22 MACHINING WITH LASERS

Lasers hold considerable promise for making thin-kerf cuts of moderate depth in solid wood and wood composites. The history of their development through 1974 has been reviewed by Szymani and Dickinson (1975). Stimulated by defense needs and by needs of other industries, the technology continues to develop rapidly.

Lasers (acronym for light amplification by stimulated emission of radiation) generate photon energy through an internal amplification process in which stimulated emission plays a dominant role. They emit a coherent beam of highly collimated monochromatic light that when focused to minimum diameter can produce power densities sufficient to vaporize most materials. Lasers offer a number of advantages over conventional machining processes, as follows:

- No residue (sawdust) is formed.
- Narrow kerf reduces waste.
- Ability to cut complicated profiles.
- No tool wear.
- Produces a smooth surface.
- Little noise.
- No reaction force exerted on the workpiece.

Potential for laser cutting was increased with the development of the carbon-dioxide molecular gas laser. The collimated beam from this laser is continuous and output powers in excess of 5,000 watts are possible. The cutting action of the carbon dioxide laser can be further improved by using a co-axial jet of gas, usually air, to assist in removal of vapor and particles from the region and cool the top surface (fig. 18-232). With the gas jet, it is possible to produce deep, uniform cuts with square edges in a variety of materials.

Carbon-dioxide laser profile cutting machines have been developed to prepare steel-rule die blocks of the type used for cutting and/or creasing paper cartons, gaskets and cloth. In this application, an intricate and accurate pattern of narrow slots is required in 3/4-inch plywood (figs. 18-233, 18-234); steel rules are inserted into the slots. At cutting speeds of 8 to 20 inches per minute, laser preparation of the die blocks is much faster than conventional methods. Descriptions of this computer-controlled technique, which was introduced about 1970, include accounts in Paper, Film, and Foil Converter (1977) and by Rau and Bauer (1978/1979) and Anonymous (1976c).

In another commercial operation, Laser Craft of Santa Rosa, Calif., is laser-etching finely detailed carved patterns in hardwood flat stock.

McMillin and Harry (1971) demonstrated that southern pine lumber and wood composites can be cut with a carbon-dioxide laser and explored factors which

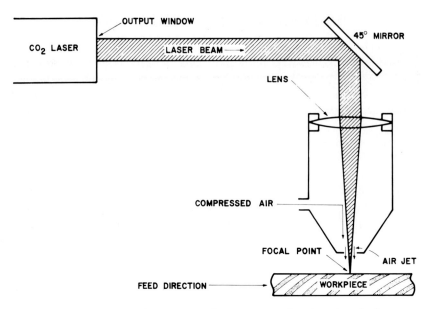

Figure 18-232.—Diagram of experimental air-jet-assisted carbon-dioxide, laser-cutting device. (Drawing after McMillin and Harry 1971.)

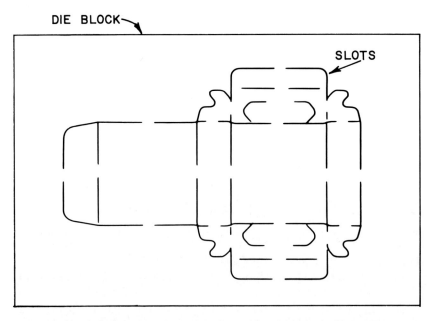

Figure 18-233.—Pattern of laser-cut kerf slots in steel-rule die block of 3/4-inch birch plywood. With steel rules inserted in the kerf slots, the die is used to cut and crease stock for a cardboard box. The pattern measures about 7 inches in length; the kerf, which penetrates through the block, measures about 0.028 inch wide. (Photo from British Oxygen Co., Ltd., London, England.)

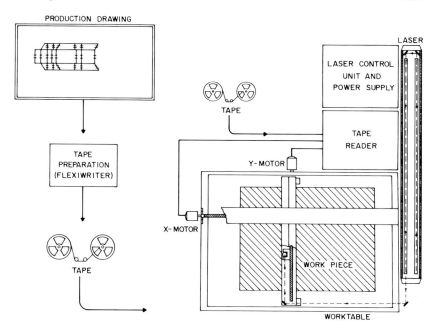

MODEL 600 SCHEMATIC

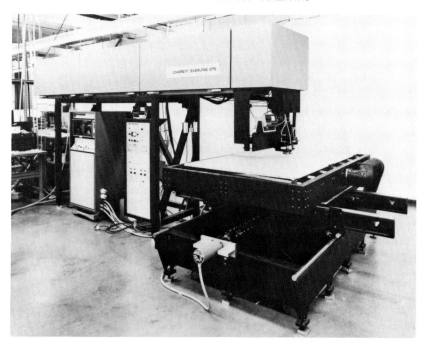

Figure 18-234.—Laser die cutting machine. (Top) Functional diagram. (Bottom) Numerically-controlled machine in operation. (Drawing and photo from Coherent Radiation.)

affect the maximum rate of cutting. The laser used contained a mixture of carbon dioxide, nitrogen, and helium and emitted radiation at a wavelength of 10.6 μm. The beam emerged horizontally from the laser tube and was deflected downwards by a 45° mirror (fig. 18-232). It was then focused by a lens which formed the upper sealing surface of an air-jet nozzle. The focused beam passed concentrically down the axis of the nozzle and was at minimum diameter about 2 mm outside the nozzle. An air-hydraulic, variable speed feed system was used to traverse the workpiece past the focused beam. All cuts were made at 240 watts of output power.

The maximum feed speed at full penetration of the laser beam differed wth workpiece thickness, wood specific gravity, and moisture content. There was significant difference in feed speed when cutting in a direction along the grain as compared to cutting across the grain.

Maximum feed speed at which wood could be cut decreased with increasing workpiece thickness in both wet and dry samples. For a given thickness, slower feed speeds were required for wet than for dry wood. The magnitude of the difference increased as the thickness of the workpiece decreased. Cutting speeds were in the range from 10 feet per minute for 1/4-inch-thick dry wood of low specific gravity, to about 1 foot per minute for 1-inch-thick green wood of high specific gravity.

The width of the kerf produced by the laser beam is extremely small (avg. 0.012 inch) compared to the kerfs produced by conventional saws. McMillin and Harry's data showed that kerf width was unrelated to cutting direction, moisture content, and specific gravity, but increased with increasing workpiece thickness. Kerf widths were 0.009, 0.012, and 0.015 inch for 0.25-, 0.50-, and 1.00-inch-thick samples, respectively.

Wood-based materials and paper products may also be cut with the carbon-dioxide laser. McMillin and Harry provided the following examples:

Material	Thickness	Power	Feed speed
	Inch	Watts	Inches/minute
Plywood, southern pine	0.50	240	20
Particleboard, southern pine	.50	240	16
Tempered hardboard	.25	240	13
Hardboard	.25	180	12
Fiber insulation board	.50	180	14
Corrugated boxboard	.17	180	236
Illustration board	.10	180	91
Kraft linerboard	.02	180	207

Scanning electron micrographs prepared by McMillin and Harry (1971) show that laser-cut surfaces—while blackened—are far smoother than conventionally cut surfaces. On laser-cut surfaces, there is little evident damage to wood structure (fig. 18-235); some carbon deposits, however, are evident on cell walls and in lumen cavities.

Peters and Marshall (1975) determined cutting rates on some eastern hardwoods using a 250-watt carbon dioxide laser and obtained results similar to those of McMillin and Harry, except they observed that the upper 1/16-inch of the kerf

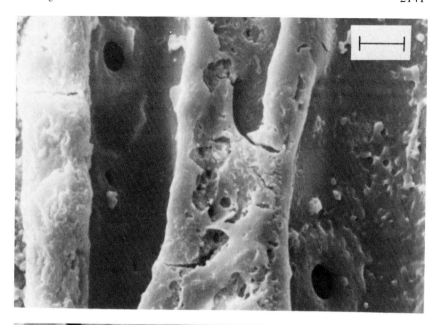
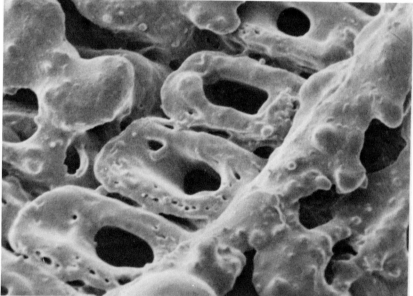

F-520 992

Figrue 18-235.—Scanning electron micrographs of southern pine surfaces cut with a carbon dioxide laser along the grain (top) and across the grain (bottom). Scale mark shows 10 μm. (Photos from McMillin and Harry 1971.)

wall was vaporized in such manner that charring was not evident; below the 1/16-inch depth, blackening of kerf walls was evident. Their results pertinent to hardwoods are summarized in table 18-75. Phenolic and silicate bonds in plywood or particleboard slowed cutting rates and increased specific cutting energy significantly compared to urea bonds. Knots required more specific cutting energy than knot-free wood.

TABLE 18-75.—*Kerf width, cutting speed, and specific cutting energy when laser cutting hardwoods and wood composites with a 250-watt carbon-dioxide laser* (Peters and Marshall 1975[1])

Material[2]	Specific gravity[3]	Depth of cut	Kerf width	Speed of cut[4]	Net specific cutting energy[5]
		----------Inch----------		Feet/minute	kWh/100 cu in
LUMBER					
Maple, red	0.54	0.506	0.025	3.00	0.92
Maple, red	.54	1.017	.024	1.10	1.29
Oak, white	.68	.866	.022	.85	2.15
Hickory	.72	.722	.021	.93	2.48
Oak, northern red	.63	.989	.030	1.1-1.1	1.06
Oak, northern red (80 percent M.C.)	.63	1.145	.030	0.3-0.4	2.52
KNOTS					
Oak, northern red	—	.830	.018	.90	2.59
Hickory	—	.735	.026	.50	3.62
PLYWOOD					
Urea-bonded birch	—	.591	.019	2.00	1.54
Phenolic-bonded birch	—	.603	.026	.50	4.43
PARTICLEBOARD[6]					
Urea-bonded softwood	—	.514	.019	2.00	1.78
Phenolic-bonded softwood	—	.522	.028	.15	16.02
Silicate-bonded softwood	—	.525	.024	.15	18.12

[1] With coaxial N_2 jet assist.
[2] All at 6-percent moisture content except as noted.
[3] Based on volume at 12 percent moisture content and ovendry weight.
[4] When two values are given, the second is for cutting across the grain.
[5] Net energy to produce 1 cubic inch of kerf volume.
[6] Density as cut was 40 pounds per cubic foot.

To evaluate cutting performance with a carbon dioxide laser of higher power, Peters and Banas (1977) cut a number of woods including hickory and northern red oak, at continuous power outputs ranging up to 6 kW; results on these two species at 3 kW output were as follows (all boards were 0.75 inch thick):

Machining

Parameter	Northern red oak		Hickory at 6-percent MC
	At 6-percent MC	At 80-percent MC	
Density as cut, lb/ft^3	43	63	50
Cutting speed, fpm	16.7	16.7	13.3
Kerf width, inch	0.022	0.018	0.020
Specific cutting energy (watt-hour/inch^3 of kerf	15.1	18.5	20.9

Cut surfaces of the green oak were less browned than those of dry oak; kerf walls in hickory were slightly charred. Cutting speed of the woods tested was inversely and linearly related to density and inversely related to thickness squared. Cutting speed increased rapidly with an increase in power (speed = constant × power$^{1.35}$). With 3 kW of power, green oak was cut at the same speed as dry oak, but with narrower kerf and therefore more specific cutting energy.

When cutting with more than 1 kW of power, Peters and Banas (1977) detected no different in cutting speeds between urea- and phenolic-bonded 1/2-inch particleboards; for both, cutting speeds were 10 fpm with 1 kW of power and 37.5 fpm at 4 kW.

18-23 VENEER CUTTING[11]

The economic competitiveness of products made of veneer cut from pine-site hardwoods is limited by the small diameter, short lengths, and low grade of available logs. In spite of these limitations, there is a potential for substantially increased sales of veneer products, so the veneer cutting process is described here in considerable detail. Chapters 22 and 24 discuss product uses, and chapter 28 some economic feasibility studies. This text section is concerned with the process of veneer cutting and selection and preparation of wood prior to veneer cutting.

While some veneer is produced by sawing (Truax 1939, see also sect. 28-2), this discussion is limited to rotary peeling and slicing. Peeled and sliced hardwood veneer is commonly produced in thicknesses from 1/50- to 1/4-inch. For special purposes it might be cut as thick as 1/2-inch or as thin as 1/110-inch. Veneer cutting closely approaches orthogonal cutting perpendicular to the grain with knife edge parallel to the grain (see sec. 18-2); it is treated separately because the rake angles used are very large (about 70°) and because the additional factor of nosebar pressure is introduced into the cutting situation. Moreover, the clearance angle commonly used ($0°\pm 1°$) distinguishes veneer cutting from most other orthogonal cutting situations.

Appropriate selection and pretreatment of wood is essential to successful peeling and slicing. These aspects are therefore discussed first.

---

[11]This section is largely condensed from Lutz (1977); some illustrations and data specific to pine site hardwoods and to cutting thick veneer have been added, however. Readers desiring fuller information on a wider range of species should consult the original work.

LOG SELECTION

Veneer can be cut from most woods, but the job is eased if logs have favorable conformation, intrinsic wood properties and internal characteristics.

Log conformation.—For rotary peeling logs should be cylindrical with pith centered in log ends. Tapered logs yield short veneer, and eccentric logs narrow veneer. Sweep in logs is adverse to both peeling and slicing as tension wood is often associated with sweep. Sweep generally diminishes yields of peeled veneer; slight sweep can be tolerated for slicing if flitches are sawn so that full-length veneers can be produced. Some experts at peeling low-grade logs believe that the shaping-lathe headrig (figs. 18-104ABCD and 18-252) can advantageously convert tapered eccentric bolts to the cylindrical cylinders favorable for high veneer production. (See sect. 28-31 for an economic analysis of such an operation.) Ideal minimum log diameter for peeling is 14 inches, for flat slicing 18 inches, and for quarter slicing 24 inches. Since pine-site hardwoods provide few logs of such large diameter, technologies have been developed that can competitively use much smaller logs, e.g., an average of 8 to 12 inches (fig. 18-252 and 18-254).

Intrinsic wood properties.—While all of the pine-site hardwoods can be cut into veneer by suitable manipulation of cutting conditions, wood of intermediate **specific gravity** cuts best. Dense species like hickory require more power to cut and tend to develop deep cracks as the veneer passes over the knife. Yellow-poplar and red maple, among the least dense of the pine-site species, cut well, but good face veneers are difficult to cut from some soft low-density woods. **Moisture content** in the tree is generally not a decisive factor, although wood with very high moisture content (e.g., 120 percent) is usually more difficult to cut into veneer than wood of moderate moisture content (e.g., 50 to 60 percent). It is difficult or impossible to cut good veneer from wood below fiber saturation point, or approximately 30 percent. **Permeable** wood, such as yellow-poplar, is easier to cut than impermeable wood because water is readily forced out by action of knife and nosebar, thereby reducing forces that could rupture the wood; moreover, plywood made from veneers of permeable woods is less subject to "blowout" in the hot press than impermeable woods such as white oak. Woods with little **shrinkage** are favored for veneer production. Crossbands in plywood must be stable in the longitudinal direction to avoid cracks in face veneers during service; longitudinal shrinkage in face veneers bows panels, and radial or tangential shrinkage stresses glue lines. The stable heartwood veneer of some species (yellow-poplar, for example) displays less buckling and end-wrinkling than sapwood veneer. Veneer cutting is eased if **wood structure** is uniform as in diffuse-porous yellow-poplar and sweetgum; uniform wood structure also minimizes "telegraphing" of grain from core veneers to face veneers. In some ring porous woods such as the oaks, there is a pronounced difference in density between earlywood and latewood; such woods selected for veneer cutting should have a minimum of six rings per inch with narrow earlywood bands and moderately dense latewood. Wood so selected cuts well, does not shell readily between rings, and performs well as furniture, panelling, or flooring.

Wood with **straight grain** is easier to cut than wood with irregular or interlocked grain such as black tupelo, and is more likely to remain flat. High market value of some irregularly grained wood, e.g., curly grain in black tupelo and maple, however, may warrant extra care in cutting and drying such veneer for specialty products. **Extraneous cell contents** such as deposits of calcium carbonate or magnesium carbonate in insect-caused wounds of hickory may be hard enough to nick a sharp veneer knife. Although **heartwood veneer** tends to be more stable than sapwood veneer, checks in rotary-cut heartwood veneer are measurably deeper than in sapwood cut under the same conditions, possibly because polyphenols in heartwood render it less plastic. Many polyphenols react with iron and steel in the presence of water to form a **blue-black stain** that can be very obvious and objectionable on oak face veneers if the wet wood is in contact with iron or steel for even a brief time; wet wood stains more readily when hot than cold.

Log characteristics.—Problems in veneer cutting can be eased by knowledge of log characteristics troublesome to the process. Some pine-site hardwoods such as hickory and post oak have internal growth stresses that promote **end splits** typically radiating from the pith like spokes of a wheel. When green wood is heated in preparation for veneer cutting, the splits enlarge. Such splits, particularly damaging in logs to be rotary cut, are not quite so serious when the wood is to be sliced because the log can often be sawed to eliminate the major split. Flitches with **growth stresses** tend to bow toward the bark side, particularly during heating. Logs with marked growth stresses will generally yield more veneer by flat-slicing than by quarter slicing (fig. 18-236). In oaks and hickories **bark pockets** may cause defects in veneer that exclude it from many uses; such bark pockets may show on log ends or as overgrowths on the bark. Extensive **fire scars** (fig. 11-10) make logs of doubtful use for veneer. **Bird peck** and insect galleries in wood generally limit veneer cut from it to use as cores, as does **stain** in standing trees (secs. 11-6 and 11-7). Sap stain during storage is particularly troublesome with sweetgum; oxidative stain causes objectionable discoloration of light-colored face veneers of red maple (see sec. 11-2). Lutz (1975) summarized log properties of 13 southern hardwood species that relate to their suitability for veneer (table 18-76).

LOG STORAGE

End drying and splits in log ends can occur in dense hardwoods like hickory and oak in one hot, dry, windy day when sunlight falls directly on log ends. Log quality loss from such cracking and from insect attack and stain can be reduced by keeping veneer log storage to a minimum; the first logs into storage should be the first out. For ideal storage, logs should be tree-length, with bark intact, end-coated, and completely submerged in cold water; the next best system is storage of logs under roof with all surfaces wetted by a water spray. (See sec. 16-18). End-splits in veneer bolts are minimized if tree-length logs are bucked to bolt length just prior to veneer cutting.

LATHE

A. ROTARY (YELLOW BIRCH)

B. ONE-HALF ROUND (RED OAK)

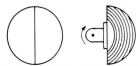

C. ONE-HALF ROUND (BLACK CHERRY)

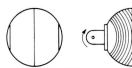

D. BACK CUT (ROSEWOOD)

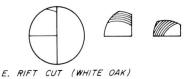

E. RIFT CUT (WHITE OAK)

SLICER

F. FLAT SLICED (WALNUT)

G. QUARTER SLICED (PRIMAVERA)

H. RIFT SLICED (WHITE OAK)

I. WHOLE LOG (FLAT SLICED) (ASPEN)

J. 1. FLAT SLICED
   2. BACK CUT
   3. QUARTER SLICED

M 140 660

Figure 18-236.—Some of the cutting directions used to obtain different grain patterns in veneer. The species in parentheses are typical of those cut by the method diagrammed. The wide dark lines under "slicer" represent the backboard left at the end of slicing. (Drawing after Lutz 1977.)

TABLE 18-76.—*Some properties of logs of 13 southern hardwoods of interest to veneer manufacturers* (Lutz 1975)

Species	Eccentricity	Crook	Taper	End splits	Relative freedom of logs from...[1] Shake	Decay	Knots	Insect attack	Bird peck	Stain
Ash, green...............	B	B	C	B	A	B	A	B	A	B
Ash, white...............	A	A	B	C	A	B	B	B-C	B	B
Hickory..................	A	A	A	C	C	B	B	C	C	C
Oak, black...............	B	B	B	B	C	B	C	B	B	C
Oak, cherrybark..........	A	A	B	B	A	A	A-B	B	A-B	B
Oak, post................	B	B-C	B	B	B	B	C	C	B	C
Oak, southern red........	A	A	B	B	A	B	B	B	B	B
Oak, water...............	A	B	B-C	B-C	B	B	B	B-C	B	C
Oak, white...............	B	A	B	B	A	A	B	B	A	B
Sweetbay.................	B	B	B	A	A	B	A	A	B	C
Sweetgum.................	A	A	A	A	A-B	B	A	B	B	B
Tupelo, black............	B	B	B	A	A	B	A-B	A	B	B
Yellow-poplar............	A	A	A	A	A	B	A	B	B	C

[1]A rating, species property very suitable for veneer; B rating, intermediate; C rating, less desirable for veneer.

EFFECTS OF HEATING WOOD PRIOR TO VENEER CUTTING

Temperature of the wood when it is cut into veneer is under control of the mill manager, and is a major factor affecting the cutting operation. The effects of heat on green wood are numerous.

Plasticity and hardness.—Heating green wood makes it more plastic; as soon as wood reaches a given temperature it is as plastic as it will get at that temperature. Veneer cut from heated bolts or flitches, particularly of dense species, can be cut thicker and with fewer cracks than from unheated wood. Heating wood makes it softer; hard knots, which if unheated may nick a sharp knife, will often be softened by heating so they can be cut. Heating does not, however, soften mineral deposits like calcium carbonate in hickory. Wood of low density may be oversoftened by heating; if cutting leaves such wood with a fuzzy surface, it is too hot.

Dimensional changes.—When green wood is heated, it expands tangentially and shinks radially; the rate of movement increases slowly with the increases in temperature up to about 150°F and then increases more rapidly. Consequently, species tending to develop shakes and splits through the pith and shake probably should not be heated above 150°F. Flitches that do not contain pith can be heated to higher temperature than bolts that do contain the pith. Fewer end splits develop in logs heated in long lengths and then cut to bolt lengths, than in bolts cut to length before heating. While slow heating may slightly reduce end splits, the maximum temperature is more important; the higher the heating temperature, the larger the end splits.

An idea of the variation of log splitting due to heating is provided by the following:

Splitting not a serious problem	Intermediate	Splitting is a problem
Hackberry	Ash, white	Hickory, sp.
Sweetbay	Elm, American	Oak, black
Sweetgum	Elm, winged	Oak, cherrybark
Tupelo, black	Maple, red	Oak, chestnut
Yellow-poplar		Oak, laurel
		Oak, northern red
		Oak, post
		Oak, scarlet
		Oak, southern red
		Oak, water
		Oak, white

Color changes.—Heating tends to darken sapwood of all species and may darken heartwood as well. To keep the wood as light in color as possible, minimum heating times and temperatures are recommended for ash, maple, and oak.

Torque to turn bolts.—The torque required to peel a bolt into veneer depends on wood density, veneer thickness, bolt diameter, setting of the knife and pressure bar, and wood temperature. In general, evenly heating bolts reduces cutting forces about as much as it reduces the wood's ability to resist spin-out. If bolts are heated at high temperature for a short time, however, both ends become hot and soft while the inner part of the bolt is cooler and requires relatively higher cutting forces. Best procedure is to heat the bolts at a lower temperature long enough so each is uniformly hot throughout; this procedure also minimizes splits at bolt ends that may contribute to break-out of bolts during rotary cutting.

Shrinkage.—Heating wood of some collapse-susceptible hardwoods (e.g., some ash species) may significantly increase shrinkage of the veneer when it is dried; the shrinkage increases with increased conditioning temperature and heating time.

Drying time.—When sound green logs with a high moisture content are heated in hot water or steam to 150°F, they generally lose 1 to 10 percent of their moisture. Thin veneers cut from hardwoods so heated require the same drying time as those cut from non-heated wood. If logs or flitches are heated and then cooled in water, the end grain will pick up water; if the veneer is not spurred (end-trimmed) at the lathe, this extra water may slow drying at the ends of the sheets.

Conclusions.—Heating veneer logs reduces cutting power and makes it possible to cut tight veneer sheets with improved strength perpendicular to the grain, less splitting during handling, and less checking in service. Also, knots are softened, reducing nicks in veneer knives and thereby promoting smooth veneer surfaces. Overheating may cause excessive splitting in hickory and oak, fuzzy surfaces on earlywood of some species, shelling (separation of earlywood and latewood), unwanted darkening of veneer, and increased spin-out. For hardwoods, the advantages outweigh the disadvantages, and almost all producers of hardwood face veneer heat bolts or flitches.

TIME REQUIRED TO HEAT BOLTS AND FLITCHES

The bolt or flitch should be heated long enough so that temperature of the wood from the start to the end of cutting varies no more than 10°F; to achieve this in minimum time, the heating medium (hot water or steam) must circulate freely to all surfaces of bolts and flitches. The time required is controlled by a number of factors, as follows.

Effect of diameter.—Much more time is required to heat a large-diameter bolt than a small one, i.e., heating time increases with the square of the diameter.

Effect of temperature gradient.—The greater the difference in temperature between the wood and the heating medium the faster the heating rate; as the wood temperature approaches that of the heating medium, rate of heating slows. It is generally practical to aim for a core temperature 10°F cooler than the heating medium. The colder the wood, the longer the heating time required; heating capacity should be based on the coldest winter conditions anticipated. Heating times for frozen wood of high moisture content like yellow-poplar, are longer than for frozen low-moisture-content woods like ash.

Effect of bolt length.—Rate of heating along end grain is about 2½ times as fast as for side grain; it is about equal for tangential and radial directions.

Because most bolts and flitches are long compared to their diameter, heating rate through side grain generally controls heating time. Faster end-grain heating probably causes knots to heat, and therefore soften, faster than the surrounding wood.

Effect of moisture content and specific gravity.—At moisture contents above 30 percent, differences in moisture content have little effect on heating rate. This rate varies inversely, however, with specific gravity; thus light woods like yellow-poplar take less time to heat than dense woods like hickory.

Effect of heating medium.—Hot water at temperatures just below the boiling point heats veneer bolts at about the same rate as a steam-air mixture at the same temperature. Some plants inject steam into the heating vat and at the same time spray hot water over the bolts or flitches; in addition to adding heat to the wood, the hot water prevents drying and checking. Modifying this method by blowing the steam through an alkaline water solution with slightly higher boiling point than pure water is questionable; these few degrees higher temperature would not seem important and the mildly alkaline treatment would penetrate only a fraction of an inch in most woods.

Heating with hot water is preferred by the hardwood industry because temperature throughout the vat is readily controlled and end-drying is never a problem. Two men are generally required to operate hot water vats, however, while only one man is needed to load and unload steam vats with a forklift. Steam chambers are probably safer; a fall into a hot water vat is generally fatal.

Temperatures in heating vats should be recorded at half-hour or shorter intervals and automatically regulated to within 2° to 4° of the desired temperature. A system that works well is to pump the water from one vat to another. After heating one vat, the hot water is pumped to a second vat containing unheated logs. As the hot water goes from tank to tank, only enough heat is added to maintain the water temperature.

If bolts or flitches are about the same diameter, they can be moved progressively through a hot water or steam tunnel, thus gaining the advantage of straight-line production. Core temperature should be checked on larger bolts by inserting a thermometer in a hole drilled 1 to 2 inches deep at mid-length of the core as it comes from the lathe. If cores of large-diameter bolts are within 10°F of the temperature in the vat, smaller bolts will also be adequately heated.

Examples.—For water or air-steam vat temperature of 150°F, some examples of heating times are given in the following paragraphs; more detailed information and tables are given by Feihl (1972) and Fleischer (1959).

- Heating time related to diameter (green 8-foot-long bolts with specific gravity of 0.50, an initial temperature of 70°F, and a final temperature at a 6-inch core of 140°F)

Bolt diameter	Required heating time
Inches	*Hours*
12	14
24	60

- Heating time in water at 150°F related to core temperature (12 inches from one end of green, 11.5-inch-diameter southern red oak bolts with a specific gravity of 0.57, an initial temperature of 63°F, and various final temperatures at a 6-inch core; unpublished data from R. C. Paul, Southern Forest Experiment Station, Pineville, La.)

Final core temperature	Required heating time
°F	Hours
100	3
120	5
140	11

- Heating time related to initial wood temperature (8-foot-long, 12-inch-diameter bolts with a specific gravity of 0.56, a green moisture content of 80 percent, various initial temperatures, and final temperatures at a 4-inch core of 140°F)

Initial wood temperature	Required heating time
°F	Hours
0	27
40	21
70	16

Frozen wood takes longer to heat than unfrozen wood. Steinhagen (1977) cooled red oak (*Quercus* sp.) bolts to a uniform temperature and then immersed them in agitated water maintained at 180°F. He found that considerable energy and time were required to melt ice in the interior of the bolt (fig. 18-237), indicating that heating curves suitable for unfrozen bolts are not appropriate for frozen bolts.

FAVORABLE TEMPERATURES FOR VENEER CUTTING

Lutz (1977) related specific gravity to favorable temperature for cutting veneer from hardwoods (fig. 18-238); woods of high specific gravity require higher temperatures. Unfortunately, the dense hickories and oaks are prone to end splits when heated too hot. His recommendations for hardwoods that grow on southern pine sites are shown in table 18-77.

CHOICE OF CUTTING DIRECTION

Some of the ways bolts or flitches are prepared and cut into veneer on a lathe or slicer are illustrated in figure 18-236. There are two main directions in which veneer can be cut—parallel to the annual rings (rotary-cut) or parallel to the wood rays (quarter-sliced). The other directions fall between these two extremes. Half-round, flat-slicing, and back cutting all result in cutting parallel to the rings in the middle of the veneer sheets and at angles to the rings at the two edges. **Rift slicing** (fig. 18-239) cuts midway between parallel to the rays and

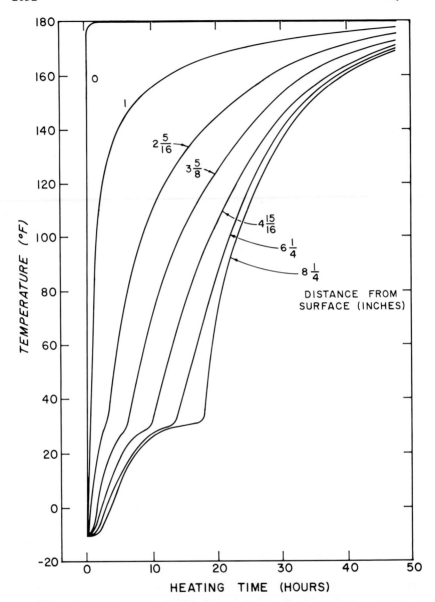

Figure 18-237.—Interior temperatures in red oak peeler logs (*Quercus* sp.) related to time immersed in an agitated water bath maintained at 180° F, and to distance from log surface. The logs were all heartwood and were uniformly cooled to −10° F prior to immersion in the hot water. (Drawing after Steinhagen 1977.)

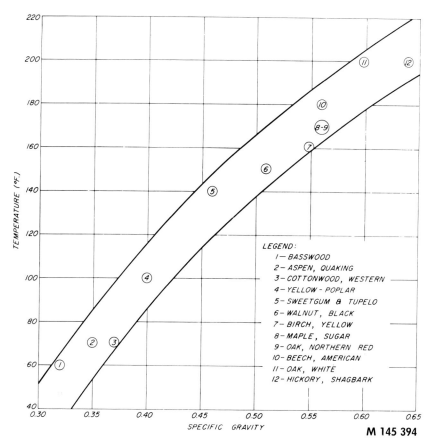

Figure 18-238.—Favorable temperature range (area between heavy lines) for cutting veneer of hardwood species of various specific gravities. Points show favorable temperatures for the individual hardwood species indicated. The data apply to the rotary cutting of veneer 1/8-inch thick, of straight-grained wood, free of defects such as knots or tension wood, i.e., "soft streaks." (Drawing after Lutz 1977.)

perpendicular to them. Rotary-cut (tangential) and quarter-sliced (radial) surfaces of pine-site hardwoods are illustrated in color in figure 5-4 through 5-15.

The lathe is used to cut practically all veneer used in construction plywood, some decorative face veneer, and most container, core, and crossband veneer. Slicing and staylog cutting is primarily for decorative face veneers. A **stay-log** is an attachment for a veneer lathe on which flitches may be mounted for cutting into half-round, back-cut, or rift veneer. Very high-quality core and crossband veneer is occasionally produced by quarter slicing. Small, fast slicers have been used to produce container veneer.

Figure 18-239.—Rift-sliced white oak. The pencil-stripe figure is obtained by cutting the wood rays at an angle of about 45°. (Photo from Lutz 1977.)

TABLE 18-77.—*Suggested temperatures to which 21 species of southern hardwoods should be heated prior to rotary-cutting and slicing veneer* (Lutz 1975, 1977)

Species	Rotary cutting	Slicing
	°F	
Ash, green	140-160	160-180
Ash, white	140-160	160-180
Elm, American	120-140	150-170
Hackberry (including sugarberry)	120-140	140-160
Hickory, true, sp.	160-180	190-200
Maple, red	100-140	130-150
Oak, black	140-160	180-200
Oak, cherrybark	140-160	180-200
Oak, chestnut	140-160	180-200
Oak, laurel	140-160	180-200
Oak, northern red	140-160	180-200
Oak, post	140-160	180-200
Oak, scarlet	140-160	180-200
Oak, Shumard	140-160	180-200
Oak, southern red	140-160	180-200
Oak, water	140-160	180-200
Oak, white	140-160	180-200
Sweetbay	70-120	120-140
Sweetgum	120-140	140-160
Tupelo, black	120-140	150-160
Yellow-poplar	70-120	120-140

FUNDAMENTAL ASPECTS OF VENEER CUTTING

Veneer peeling and slicing closely approximate orthogonal cutting in the 0-90 mode (figs. 18-1, 18-12 through 18-15) except that a **nosebar** is used to compress the wood ahead of the cutting edge (figs. 18-240 and 18-241). Peeled or sliced veneer has a **loose** side (the side with tension checks—see figs. 18-14 and 18-240DE) and a **tight** side.

Whether peeled or sliced, if the knife is sharp, the surface of hardwood veneer varies according to cell type being cut; vessels make substantial cavities, thin-walled cells are usually severed cleanly at the cutting plane, and thick-walled cells tend to separate at the middle lamella.

Figure 18-240 illustrates peeling; figure 18-241 shows slicing. In figure 18-240 (bottom), veneer cutting geometry is drawn to correspond to the terminology and diagrams of section 18-2 (fig. 18-2). Adjustment of the nosebar is sometimes stated in terms of nosebar compression:

$$\text{Percent nosebar compression} = \frac{(100)(t_1 - c)}{t_1} \qquad (18\text{-}51)$$

The **face** of the knife is the surface in contact with the veneer. (While this is not the terminology used by industry, it conforms to that used in fundamental machining studies.) The **back** of the knife is the ground bevel next to the bolt or flitch.

Cutting forces.—Parallel and normal cutting forces per 0.1 inch of knife length when cutting thin veneer (0.015, 0.030, 0.045, and 0.060-inch thick) from unheated green wood of 22 pine site hardwoods are given in tables 18-5B through 18-26B; the data tabulated for 70° rake angle are appropriate for veneer cutting without a nosebar. Normal cutting forces (F_n) are usually negative, i.e., the knife tends to draw the veneer bolt toward it, and small in magnitude—generally less than 1 pound per tenth inch of knife length. Parallel cutting forces (F_p are larger—up to 4 pounds per tenth inch of knife length when cutting 0.06 inch-thick veneer from green unheated oaks and hickories. On a 100-inch-long bolt, therefore, parallel cutting force on the knife (cutting at slow speed without a nosebar) may approach 4,000 pounds when peeling or slicing 1/16-inch oak or hickory from unheated green bolts.

When parallel cutting forces are plotted against depth of cut, the relationship is either a straight line function of the form

$$F_p = (A + Bt)w$$

or a power function of the form

$$F_p = Kt^m w$$

where A, B, K, and m are constants; t is the chip thickness; and w is the width of cut.

Parallel cutting forces have a positive linear correlation with specific gravity so that the dense oaks and hickories require more parallel cutting force than less dense species such as yellow-poplar, red maple, sweetbay, and sweetgum (fig. 18-15).

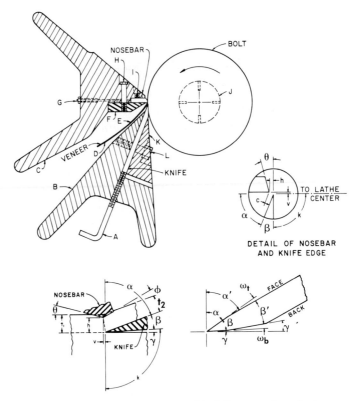

Figure 18-240.—Nomenclature in veneer cutting. (Top) Cross section of rotary veneer lathe. A, knife adjusting screw; B, knife bar; C, pressure bar; D, loose side of veneer; E, tight side of veneer; F, nosebar cap; G, nosebar adjusting screw (horizontal); H, nosebar locking screw; I, nosebar adjusting screw (vertical); J, chuck; K, knife cap; L, knife cap bolt; Inset, detail of cutting edge and nosebar. (Bottom) Cross sections through cutting edge and solid nosebar arranged in convention of orthogonal cutting diagrams.

α	Primary rake angle
α'	Secondary rake angle
β	Primary sharpness angle
β'	Grinding angle, angle of ground bevel
γ	Primary clearance angle
γ'	Secondary clearance angle
ω_t	Face honing angle
ω_b	Back honing angle
θ	Nosebar compression angle
ϕ	Nosebar clearance angle
k	Knife angle used in commercial practice (90° plus the clearance angle)
t_1	Depth of cut; undeformed veneer thickness
t_2	Actual veneer thickness
h	Horizontal nosebar opening (frequently called gap)
v	Vertical nosebar opening (frequently called lead)
c	Exit gap

when v/h is equal to or less than tan (90 − α)
$$c = [h + v \tan(90 - \alpha)] \cos(90 - \alpha)$$
when v/h is more than tan (90 − α)
$$c = \sqrt{v^2 + h^2}$$

(Drawing after Koch 1964a.)

Machining 2157

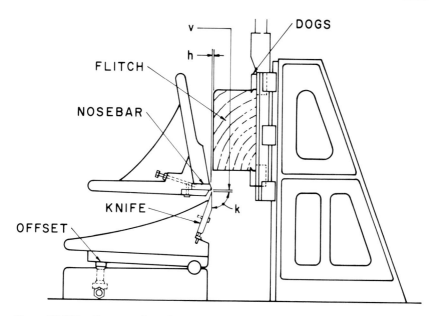

Figure 18-241.—Cross section of veneer slicer. Knife is stationary; dogs holding flitch move up and down in vertical (or inclined) guides.

In the range of rake angles from 70° to 50°, the normal force (F_n) changes little with rake angle, and the parallel cutting force (F_p) shows only modest increase as the rake angle is decreased from 70° to 50°.

The magnitude of cutting forces should be considered when selecting veneer cutting equipment. Lutz and Patzer (1966) found that loads while cutting thick veneer were as high as 200 pounds per inch of knife and 500 pounds per inch of nosebar. Because forces are substantial, wear on moving parts can be significant. To reduce slack and play caused by wear or heat distortion, pre-loaded anti-friction bearings, wear plates, and take up mechanisms are desirable. If knife and nosebar can be maintained at a uniform warm temperature during both setup and cutting, heat distortion will be minimized and condensation on the knife (and hence wood staining) will be reduced.

To counter any tendency for the wood being cut to come loose during cutting, retractable hydraulic dogs on slicers and hydraulic chucks on lathes are recommended; when dogs or chucks are retracted, they permit continuous cutting to thin backboards or small-diameter cores.

Lathes in high-production installations may be automatically fed at rates up to seven logs per minute. Slicer production is slower; vacuum lift devices can speed loading of slicers, however, if flitch backs are wide, flat, and smooth.

Slicer drives are powered with 40 to 75 hp eddy-current drives or by DC motors, the latter sometimes equipped with air clutches and air brakes. Lathe drives take several forms; options include steam engines, a.c. motors with speed changers, d.c. motors with motor generator sets, and hydraulic motors. In all cases, it is desirable to increase spindle speed as block diameter decreases to keep cutting speed constant. Four-foot lathes appropriate for cutting dense southern hardwood veneer up to 1/4-inch thick might have 100-hp drives; 8-foot lathes for southern hardwoods commonly have drives of 125 to 200 hp. Readers needing a review of lathe drives will find the discussion by Hancock and Hailey (1975) useful.

LATHE VERSUS SLICER

The heart of any lathe or slicer is the knife and the nosebar assembly or **pressure bar**, which are similar on both machines. The knife bevel angle is about the same for slicer and lathe, but lathe knives may be slightly more hollow-ground.

The nosebar on both the lathe and slicer compresses the wood, with maximum compression ideally just ahead of the knife edge. This compression reduces splitting of the wood ahead of the knife, reduces breaks into the veneer from the knife side, and forces the knife bar assembly against the feed mechanism, thereby helping control veneer thickness. For both the lathe and slicer, the pressure bar is, therefore, important in controlling the roughness, depth of checks, and thickness of the veneer. The slicer has a fixed nosebar, while the lathe may have a fixed nosebar or a rotating roller bar (fig. 18-242).

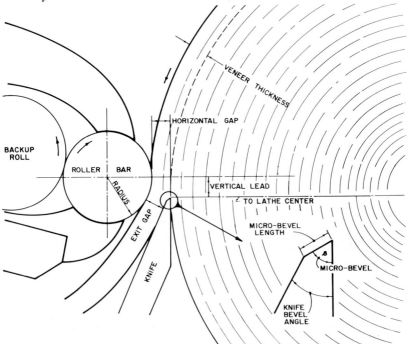

Figure 18-242.—Cross section of rotary veneer lathe equipped with a roller nosebar. Inset shows microvebel β which commonly is 30° and extends 0.020 inch in length. (Drawing after Forintek Canada Corporation 1979.)

In general, the greatest yield is obtained by rotary cutting. Half-round, flat-slicing, or back cutting provide intermediate yields; and the least yield is obtained by quarter- or rift-slicing.

The smoothest and tightest veneer can be produced by quarter- or rift-slicing, followed by rotary cutting; the roughest and loosest veneer is produced by flat slicing, half-round, or back-cutting. Differences in roughness are due to the effect of wood structure orientation (fig. 18-243).

While slicing and rotary cutting involve some differences and inherent advantages, good quality veneer can generally be produced by either method. The quality of the end product is determined more by the log quality, the heating of the bolts or flitches, and the setting of the knife and pressure bar than by differences in the cutting method.

Lathe advantages.—Logs to be cut into veneer on a lathe need to be crosscut to the desired bolt length, but they do not need to be processed through a sawmill prior to cutting veneer. After roundup of the bolt, the lathe cuts a continuous strip of veneer. Continuous cutting is advantageous because it means more production with a given cutting velocity, wider sheets of veneer, and a more uniform cutting condition. Full rotary cutting is approximately tangential to the annual rings and knots are exposed at their smallest cross section. In full rotary cutting, there is no impact at the start of cutting or tearoff at the end of cutting as may occur when slicing or cutting with a stay-log.

Slicer advantages.—The most decorative grain patterns are obtained by slicing flitches sawn from logs (fig. 18-236 right), rather than rotary peeling entire logs. As sliced veneer sheets are kept in consecutive order, figured veneer can be readily matched. Flitches can be heated with less danger of end splits developing than in comparable bolts being heated for rotary cutting. Sliced veneer is always cut from a flat surface, and most veneer is used on a flat surface. By contrast, rotary veneer cut from a curved surface must be flattened for most uses. The disadvantage of cutting from a curved surface becomes more pronounced with thicker veneers cut from small-diameter bolts.

Logs with sweep are more readily sliced than rotary peeled because flitches can be sawn and oriented so that full width sheets are obtained on the first cuts. This capability of a slicer contributes to the fact that veneer as long as 16 feet is produced on a slicer while most rotary-cut veneer is 10 feet or shorter.

Stay-log lathe.—A stay-log lathe carries a plate, eccentrically rotated between centers as shown in figure 18-236CD, to which half of a log or a flitch may be secured. The plate, or **stay-log**, makes it possible to produce veneer on a lathe, similar in appearance to sliced face veneer (fig. 18-236C). The advantages of stay-log cutting on the lathe are very similar to the advantages of slicing. Flitches can be selected for appearance of the grain and consecutive sheets can be matched for decorative faces. Sheets cut with the stay-log are generally wider than sheets cut on the slicer. Veneer cut with a stay-log is taken from a curved surface in comparison with veneer that is sliced from a flat surface. Veneer cut with stay-log may be up to 10 feet in length.

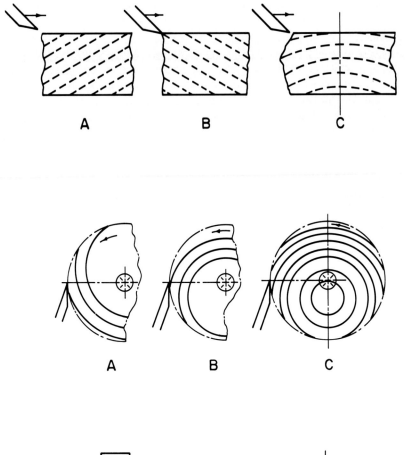

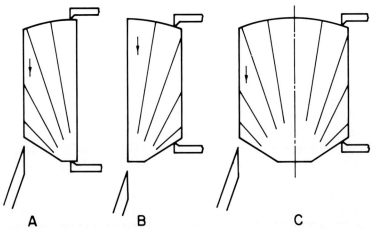

Figure 18-243.—Optimum orientation of wood structure to favor cutting of smooth veneer. A, favorable; B, unfavorable; C, combination of favorable and unfavorable. (Top) Slope of grain. (Center) Annual ring orientation. (Bottom) Ray orientation.

Machining 2161

Back-roll lathe.—To avoid the necessity of clipping veneer to width in a separate operation, a rotary lathe is sometimes equipped with a back-roll carrying radially mounted knives (fig. 18-244). The back roll is fed toward the log by feed screws at the same rate at which the knife is fed so that the back-roll knives make an impression into the veneer bolt slightly deeper than the thickness of the veneer being cut. As the veneer is cut, it therefore separates into pieces having the width of the spacing of the back-roll knives. Since the scoring knives generally leave a light score mark on the tight side of the next piece of veneer, the back-roll lathe is better suited for cutting thick container veneer than thin decorative veneer. All lathes are generally equipped with spur knives so veneer can be cut to one or more lengths while it is being peeled. To manufacture container veneer from short pine-site logs, the round-up shaping-lathe should be a good companion machine to a back-roll veneer lathe (fig. 18-252).

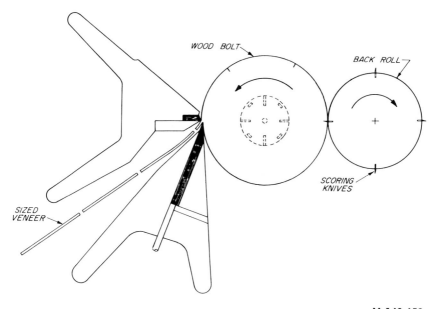

Figure 18-244.—Back roll lathe. (Drawing after Lutz 1977.)

LATHE RIGIDITY AND CONTROL OF UNDESIRABLE MOVEMENT

Knife and pressure bar settings are meaningful only if the wood is held securely in the lathe or slicer and if the machine parts have a minimum of play.

Bolt movement.—Bolts are held in a lathe by chucks. In general, the larger the chucks the more securely the bolt is held. The chucks transmit the torque needed to cut the veneer and also must resist the tendency of the bolts to ride up on the knife. The spurs on the chucks should, therefore, be designed not only to transmit power to turn the bolt but also to keep it from shifting from the spindle center. The best spur configuration is not well established. Some mills prefer half circles; others, star-shaped spurs and a ring around the circumference of the chuck. In practice, the spurs sometimes become battered and bent and may collect wood debris. For best performance, they should be tapered for a positive secure fit.

The pressure used to set the chucks in the bolt ends depends on the wood species, heating, and chuck size. Generally, enough pressure is used to indent the spurs at least three-fourths their length into the bolt ends. Square-cut bolt ends allow a more uniform grip than bolts that are end trimmed on a bias.

The wood in contact with the spurs receives fluctuating loads during cutting, which may cause the bolt to become loose in the chucks. On older lathes, the operator must watch for this and further indent the spurs if looseness develops. Newer lathes have hydraulic chucking. A relatively high pressure is used to set the chucks and a lower pressure insures that spurs remain seated during cutting. Too high end pressure may bend the wood bolt when it reaches a small diameter.

Another modern solution to holding the bolts more securely is the use of retractable chucks. Larger chucks and spindles hold the bolt at the start of peeling; they are retracted during peeling, allowing smaller innner chucks and spindles to hold and drive the bolt until the final core diameter is reached. A modification of this is sequentially retractable chucks such as 5-inch inner chucks, with an 8-inch outer chuck on one end and a 12-inch outer chuck at the other end. The bolt is first driven with the 12- and 8-inch chucks. At a bolt diameter of about 14 inches, the 12-inch chuck is withdrawn and the bolt is then driven with one 8- and one 5-inch chuck. At a diameter of about 10 inches, the 8-inch chuck is withdrawn, and cutting is completed with the two 5-inch chucks driving the bolt to the final core diameter.

To obtain maximum recovery, bolts are turned to as small a diameter as practical. The bolt is loaded as a beam by the knife and pressure bar. Its resistance to bending is directly related to the cube of the radius of the bolt. At small diameters, an unsupported bolt bends in the middle away from the knife. The bolt becomes barrel-shaped and the veneer ribbon wrinkles in the middle. To overcome this problem, backup rolls have been built to support the bolt during cutting.

Some early backup rolls operated with a fixed pressure against the bolt. But this caused problems. The cutting force fluctuates during peeling, and a fixed pressure against the bolt surface sometimes increased rather than reduced bowing of the bolt.

Improved backup rolls fix their position geometrically to keep the bolt cylindrical. One method of doing this is a servo-system with a follower at the end of the block that signals adjustments of pressure on the backup roll. Another method is to have this backup roll positioned mechanically by the feed mechanism so the bolt remains a cylinder. When properly made and operated, backup rolls permit cutting 8-foot bolts to a final core diameter of about 4 inches, and 4-foot bolts to a 3-inch core.

Machine movement.—Production lathes frequently wear excessively, developing play in spindle sleeves and bearings, feed screws, headblock or knife-angle trunnions, nosebar eccentric, and blocks under screws used to change the lead (vertical nosebar adjustment). These problems can be minimized by using preloaded bearings for the spindles, an air cylinder to keep the knife bar always against one side of the feed screw, a ball-feed screw drive to reduce feed screw wear, and replaceable wear surfaces for the ways.

Forces between bolt and knife carriage reverse depending on whether or not the nosebar is in contact with the bolt (fig. 18-245). In the absence of a nosebar, carriage and bolt are pulled together; when the nosebar is pressed against the bolt, carriage and bolt are forced apart. Normally veneer is peeled with the nosebar closed, but since many operators round up the log with nosebar open, and at intervals clear jams by opening the nosebar, forces are frequently reversed and the play in the lathe carriage causes variations in veneer thickness. In brief, variation in veneer thickness is least if the nosebar can be kept in contact with the bolt at all times.

To detect and correct for play, dial gages should be mounted at each end of the lathe with the gage on the knife frame and the sensing tip against a bracket on the bar frame. These gages should be zeroed after setting the gap or horizonal opening. Any play will show on the gages as a reading other than zero and the original gap or horizontal opening restored by adjusting the nosebar until the gages read zero. Walser (1975) describes a method to preload the pressure bar assembly to improve accuracy when setting the veneer lathe.

Other things being equal, the greater the overhang of the spindles the more spring in the cutting system. This is most noticeable when short bolts are cut on a long lathe. If both short and long bolts are to be cut on the same lathe, the lathe should be equipped with spindle steady rests.

Heat distortion of lathe.—Bolts that have been heat-conditioned prior to cutting may cause the knife and pressure bar to distort. Heating causes the knife to rise in the middle, decreasing the lead. Heat may also cause the nosebar to drop or move in a horizontal plane depending on the lathe. On some lathes, one method of correcting for these changes is to adjust the pull screws on the A-frame built over the pressure bar for this purpose. A better solution is to heat the knife and pressure bar to the expected operating condition prior to the final fitting (setting) of the knife and bar. Some lathes have had heating elements built in them to prevent heat distortion.

Another good practice is to store sharpened knives in a warm area so they are at the same temperature they attain during cutting. Feihl and Godin (1971) suggest heat distortion can also be controlled by continuous cooling of the knife bed and the pressure bar bed. However, they and others indicate heating the knife and bar works better than cooling, particularly for long lathes.

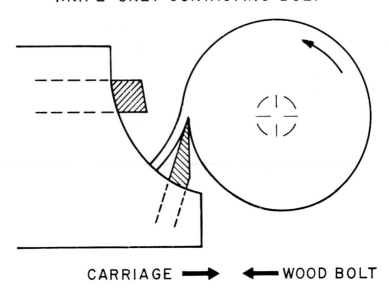

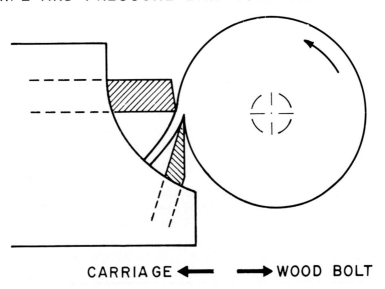

Figure 18-245.—Direction of forces acting on carriage and bolts when the knife only is contacting the bolt (top) and when both nosebar and knife are contacting the bolt (bottom). (Drawings after Lutz et al. 1969b.)

SLICER RIGIDITY AND CONTROL OF UNDESIRABLE MOVEMENT

Flitch movement.—In some vertical and all horizontal slicers, gravity helps hold the back of the flitch against the flitch bed. In the most common vertically operating face veneer slicers, however, the flitch is cantilevered from the bed. Regardless of design, the security with which the flitch is dogged to the flitch bed affects accuracy of slicing. Hydraulic cylinders actuating dogs maintain good contact and have check valves to prevent flitches from shifting during cutting; older slicers with screw-set dogs require periodic resetting of the dogs as they become loose during intermittent cutting.

Heated flitches may be bowed or twisted. Very often this bow or twist can be removed by forcing the flitch flat against the flitch table and dogging it securely. Here oversized dogs are used at the start of the cutting. Recently developed are retractable dogs, extended for maximum holding power at the start of slicing and then automatically retracted when the slicing cut approaches the dogs.

A recent practice is to glue valuable flitches to an inexpensive backboard and then slice to the glueline. Special glues and gluing techniques are used to bond the hot wet flitches to the backboards. Another innovation is to hold the flitch against the table with a pattern of vacuum cups. The flitch back should be wide, smooth, and flat or the flitch may break loose from the table during cutting.

Machine movement.—Most modern slicers incorporate take-up mechanisms to remove slack as wear develops in feed screws, offset mechanism, flitch table ways, and knife carriage ways. Some slicers advance the knife by a pawl and ratchet for each stroke. This is highly accurate providing the same number of teeth are advanced each stroke, there is little play in the feed mechanism, and there is no overtravel of the carriage. The number of teeth advanced each stroke should be checked several times before and during actual cutting. The brake on the shaft which advanced the knife each stroke should be adjusted so there is no overtravel.

Other slicers feed by moving the previously cut surface against a stop plate. The surface of the flitch and of the stop plate must be free of splinters or other debris and the flitch must be advanced flush to the stop to produce veneer of uniform thickness.

The offset mechanism on modern slicers (fig. 18-241) is hydraulically operated and does not generally require attention once the cam is set to retract the knife at the bottom of the stroke. The amount of offset is adjustable and should be large enough to insure clearance of the flitch on the upstroke. Excess offset should not be used as it may induce slight vibration to the knife. The knife and bar carriage pivot on half bearings for the offset. Since the half bearings are not held at the top, if the flitch fails to clear on the upstroke, the knife bar carriage may be lifted from the half bearings. Similarly, high nosebar pressure cannot be used without danger of unwanted movement of the knife carriage on the half bearings.

As with the lathe, it is desirable to have dial gages mounted at each end of the slicer with the gage on the knife frame and the sensing tip against a bracket on the bar frame. The gages are particularly useful for returning to the previous setting after the bar has been retracted to hone the knife.

Heat distortion.—Since face veneer slicers are generally longer than lathes, heat distortion of the knife and bar may be more of a problem. As on the lathe, the heated knife rises in the middle and the pressure bar drops. The pull screws on the A-frame on the casting holding the bar can compensate for movement due to heat. A better solution, and one that is built into modern slicers, is a means of heating the knife and bar prior to fitting them, and then keeping these parts continually warm. This greatly reduces changes in the knife-bar setting due to cutting hot flitches, and also the condensate and iron-tannate stain from contact of iron or steel particles with wet wood.

CUTTING VELOCITY

Cutting speed does not seem to be a critical controlling factor for most veneer production. If optimum veneer tightness and smoothness are important, however, a moderately slow cutting speed (100 to 500 feet per minute) is recommended for lathes. On a slicer, the number of cutting strokes per minute varies with knife length, as follows:

Knife length	Cutting strokes per minute
Inches	
118	80 to 100
165, 185, or 195	60 to 75
213 or 225	60 to 70

When slicing 1/8-inch and thicker veneer, there may be a slight vibration of the slicer due to impact at the start of the cut. Inclining the length of the flitch 3° to 5° from the long direction of the knife lessens this impact as the cutting starts at one corner of the flitch. Slower speed also reduces initial impact at each cut.

Lathes with efficient charging equipment may process four to six small veneer bolts per minute when cutting 1/8-inch-thick veneer. Slicers cutting veneer 1/32-inch thick may process three to five flitches per hour.

VENEER KNIFE SPECIFICATION

Because data on percent carbon and other components of steel in veneer knives are usually not published by knife manufacturers, knives are generally specified for length, thickness, depth, hardness, angle of ground bevel, and if hardened throughout or inlaid with a hard insert on mild steel backing. Also, the presence or absence of slots for clamping will be determined by the equipment used. An ideal knife should be stiff, tough, and resistant to corrosion and wear.

The most common knife thickness for lathes is 5/8-inch and for face veneer slicers, 3/4-inch. One-half-inch-thick knives are sometimes used on the lathe; they are less expensive but also less stiff. European horizontal slicers may use a knife 15 mm (19/32-inch) in thickness, supported with a blade holder. In general, the veneer knife should be thicker when cutting thick veneer. When cutting thin veneer, a thinner knife can be used if properly supported.

The choice of an inlaid knife or one hardened throughout may depend on the end product. Hardwood face veneer is generally cut with an inlaid knife. The mild steel used for backing is stable and easy to grind. It can be readily drilled so that the knife can be held firmly when back grinding. The highly refined hardened tool steel insert is generally of highest quality for cutting wood.

Knives that are hardened throughout reportedly may stand up better when cutting hard knots. They are sometimes, but not always, used in plants producing construction plywood.

Most veneer knives are supplied the full length of the lathe or slicer. However, two- and three-piece knives are sometimes used with a special clamping arrangement so they can be ground and set as a unit. If one section is damaged, it can be replaced without replacing the entire knife.

Knife hardness.—The hardness of the knife should be specified and can readily be tested. A soft knife can be easily honed and is tough but also wears rapidly. A hard knife is difficult to hone, is more likely to chip if it hits something hard, but holds a sharp edge much better. Most rotary veneer plants prefer a knife with a Rockwell hardness on the C scale of 56 to 58. Knives for face veneer slicers are often 58 to 60 on the Rockwell C scale. To keep as sharp an edge as possible when cutting low-density woods like basswood (*Tilia Americana* L.), a knife with a Rockwell hardness of 60 to 62 may be used.

Angle of ground bevel.—As supplied by the manufacturer, bevel angle β^1 (see fig. 18-240) may be specified in the range from about 18° to 23°. The smaller the angle the less the veneer is bent as it is cut and hence the tighter the veneer. The larger the bevel angle, the stiffer, more resistant the edge. More care must be taken when grinding the smaller bevel angles, as the edge is more likely to heat than when grinding a knife to a large bevel angle.

The ground surface is generally slightly concave to make the knife easier to hone. For the lathe, the recommended hollow grind is 0.002 to 0.004 inch, while slicer knives generally have a hollow of 0.001 to 0.002 inch. The flatter grind for a slicer knife means less chance for the flitch to rub against the heel of the knife and stain the wood. More hollow can be used on a lathe knife as the bolt surface curves away from the ground surface of the knife. However, the hollow should not exceed 0.004 inch, as this weakens the knife edge. While the details of knife bevel can be changed by grinding at the veneer producing plant, the knife should be ordered as it will be used to eliminate an extra grinding.

GRINDING VENEER KNIVES

The cutting of accurate veneer requires a straight, sharp, tough knife edge. To grind a straight edge, the grinder must be level and rigid. The most satisfactory veneer knife grinders have a fixed bed for mounting the knife and a travelling grinding wheel. The abrasive may be a solid cup wheel or a segmented wheel which some operators prefer because it requires less dressing, and replacement segments are less expensive than a new cup wheel. A magnetic chuck on the grinder speeds knife setup for grinding. A V-belt drive in place of gears is smooth and reduces chatter marks on the knife.

The knife bed as well as the ways on which the grinder moves must be straight, rigid, and parallel to each other. To maintain even wear of the ways, the grinding wheel should traverse the entire length of the grinder even when grinding short knives.

The knife surface facing the grinder bed must be checked for bumps or other rough spots that will prevent the knife from lying perfectly flat. If necessary, the back of the knife should also be ground to restore a plane surface. (See *Grinding on the veneer side of the knife.*)

Because heat causes metal to expand and deform, the grinder and knife should be kept at uniform temperature during grinding. Water used to cool the grinding wheel and knife should be at room temperature and recirculated. A stream of water with synthetic coolant should be directed against the grinding stone 1/2-inch ahead of where the stone contacts the knife edge during grinding.

Godin (1968) considers overheating of the knife tip the most serious problem in grinding and lists four main causes: (1) too heavy a cut; (2) inadequate cooling; (3) clogged grinding wheel; and (4) too hard a grade of grinding wheel. Heating is less likely to occur if the knife edge is pointed up and engages the grinding wheel first during grinding. A feed of 0.0003 to 0.0005 inch is suggested for each complete traverse of the wheel. Some operators like to dress the wheel and use a very fine feed for the last one or two traverses of the sharpening. This helps give a fine surface. Some manufacturers polish the knife by multiple passes without feeding. The smooth edge reportedly aids good veneer cutting. Care must be used with this technique or the grinding wheel may rub, heat, and weaken the knife tip.

Another cause of an irregular edge is dubbing at the two ends of the knife. The most likely causes are looseness in the grinding wheel spindle bearings, excessive end play, and slack in the feeding mechanism. However, even a grinder in good mechanical condition may slightly round the ends of the knife. If it is important to have the knife straight to the extreme ends, discarded knife sections 4 to 6 inches long can be attached to the knife bed at the two ends and in line with the knife being ground. These dummy sections absorb the heavier cut at the start of each traverse of the wheel and the main knife is not dubbed at the ends.

Grinding on the veneer side of the knife.—A knife in use may wear unevenly on the side where the veneer passes through the exit gap between the pressure bar and the knife. It may also be bent by excessive local pressure as from a knot or chip buildup. The remedy for such wear and bends is to grind a flat surface on the veneer side of the knife. The grinder bed is tilted 1/2° to 3° toward the knife and the knife is ground to produce a short microbevel (fig. 18-242). A magnetic chuck on the grinder facilitates this grinding. Otherwise, the knife body must be drilled and tapped not more than 12 inches apart so the knife can be mounted securely.

Some modern grinders are equipped with two grinding wheels so the face and back of the knife can be ground at the same time.

Honing the knife.—The knife should be ground only enough to obtain a thin wire edge the length of the knife. The wire edge is removed by careful honing with a kerosene-saturated, medium-grade, medium to soft stone on one side of the knife and then the other. Some operators use one stone and others use two, one on each side of the knife simultaneously. In either case, each pass of the stone cuts at the base of the wire edge and bends it away from the stone. Honing is continued until all of the wire edge is removed. If a badly nicked knife develops a heavy wire edge, the grinding wheel can be stopped and the wire edge removed while the knife is still clamped in the grinder. A few more passes of the wheel will create a new fine wire edge that can be easily honed away. After the wire edge is removed, the edge is finished by lightly honing with a fine-textured kerosene-soaked stone.

More detailed suggestions for grinding and honing veneer knives are given by Godin (1968).

Secondary knife bevels.—When a sharp knife ground to a bevel angle of about 21° is first put in the lathe or slicer, it is easily nicked by a knot or other hard substances. These nicks are removed by honing the knife in place on the lathe or slicer. After several bolts or flitches are cut, the knife edge wears slightly and this, plus the honing, makes the extreme edge more resistant to damage, a condition sometimes called work-sharp. When examined under a microscope, the edge is seen to be slightly rounded so it is probably closer to 30° to 35° than to 21° at the extreme tip. Such a knife will remain sharp and do a good job of cutting for several hours if no very hard material is hit.

Veneer knives, like planer knives, wear faster if the bevel or sharpness angle is less than 30° to 35°. Realizing this, some operators grind a short (0.020-inch) bevel on the veneer side of the knife (fig. 18-242). This strengthens the edge and is commonly used with knives in core lathes peeling unheated softwoods. For additional discussion of knife dulling, see sec. 18-29.

Kivimaa and Kovanen (1953), Feihl (1959), and others have studied the use of a precision microbevel on either side of the lathe knife. They report that a second bevel can be honed on either or both sides of the knife, and that a final included angle of 30° or 35° with a microbevel 0.010 to 0.020 inch in length greatly strengthens the knife edge. At least one commercial grinder has a separate grinding wheel that can grind a microbevel at the same time the main bevel is being ground.

Some slicer operators use a two-bevel knife. The main bevel is 19° and the second bevel is 21°. Grinding of the second bevel is continued until the length of the second bevel is about 1/2-inch. This is the only part of the knife that rubs against the flitch, and so the two-bevel knife reduces strain. Other operators prefer a single-bevel knife.

KNIFE SETTING

Before a lathe veneer knife is set, knife frame and pressure bar frame must be accurately aligned with the axis of rotation of the lathe spindles. Similarly, slicer knife and bar ways must be level and perpendicular to the flitch ways. Play in moving parts must be at a minimum, and machine parts should be at the temperature they attain in use. Feihl and Godin (1967) describe methods of checking alignment of lathes.

A knife must be correctly ground, flat and with a straight cutting edge. The knife bed must also be flat and clean, and if a knife holder is used, it also must be clean and flat. The knife or knife and knife holder is then set on the two end adjusting screws. The clamping screws are tightened by hand so that the knife is flat against the bed but free to move. To this point, the procedure is the same for the lathe and the slicer.

Setting the lathe knife level.—After the knife is resting on the two end adjusting screws on the lathe, the knife edge is raised until it is level with the center of the spindles. An accurately machined wood block cut out at one end to one-half the diameter of the spindle is useful as a template. The cutout end rests on the spindle and the other end on the knife edge. The height of the knife is then adjusted until a spirit level on the back of the template indicates level. The same adjustment is then made at the other end of the knife. If the span is short and the knife deep and stiff, the knife height should be the same across the lathe. However, longer knives or old ones may sag in the middle. One way of checking this is to level a transit with a telescope about 20 feet from the lathe and swing it from one end of the knife to the other. The knife edge should be in line with the crosshairs along its length. If the knife sags in the middle, it should be raised with the leveling screws near the center of the knife. Sag in the knife can also be checked with a tautly stretched fine wire. Once the knife edge is true, some operators make scribe marks on the lathe so they can reposition knives with precision. Another method is to measure the extension of the knife from the top of the knife bed.

To speed knife changes, some lathes have knife holders. After grinding, the knife is preset to the desired height in the holder, and the holder quickly bolted in place in the lathe. Some plants in effect preset the knife by pouring babbit metal at the bottom edge of the knife after each grind. The depth of the knife is thus kept constant and the knife can then be placed on the height-adjusting screws without changing them.

If there is wear in the spindle bearings, the bolt will ride up during cutting, taking up the play. To compensate for this, the knife edge is sometimes set above the spindle centers, a distance equal to the play in the spindles, making it level with the spindle centers during cutting.

Setting knife angle on a veneer lathe.—After the knife is set to the spindle centers, the knife angle is adjusted. In general, the side of the knife that contacts the bolt is approximately vertical (tangent to the surface of the bolt). Such a knife is said to have an angle of 90°. If the knife leads into the bolt 2°, the knife angle is 92° and the clearance angle 2°. A lathe knife can also be set with a negative clearance. A knife angle of 89° means the knife has 1° negative clearance.

Most lathes are built so the knife angle can be made to change automatically with the bolt diameter to keep a constant angle with bolts of small and large diameter. Without this adjustment, when cutting at a bolt diameter of 3 feet, the knife angle may be 91°; at 6 inches the angle may be 89°30'. Feihl and Godin (1967) describe several methods that can be used to properly set the pitch ways.

Hailey et al. (1980) described the use of curved pitch rails to enable smooth veneer to be cut from the part of a peeler block under 10 inches in diameter. Lathe manufacturers should be consulted for recommended procedure for use with their lathes.

In general, lathe operators use less lead into the bolt (lower knife angles) when cutting low-density woods than when cutting dense veneer. For example, Fleischer (1949) suggests a knife angle of 90°30' when cutting 1/32-inch yellow-poplar (low-density wood) and 90°45' when cutting 1/32-inch yellow birch (high density wood). Fleischer shows a pronounced effect of veneer thickness on the best knife setting. For 1/100-inch birch, he recommends a knife setting of 92°, for 1/32-inch 90°45', for 1/16-inch 90°15', and for ⅛-inch veneer 90°. These settings are for log diameters from 20 to 12 inches.

When the correct knife angle is used, the knife side next to the bolt will show 1/16- to 1/8-inch of bright rub below the knife edge. Too high an angle causes chatter and corrugation on the veneer and the bolt surfaces. The waves are closely spaced with three or more waves per inch of veneer width. Too low a knife angle results in too much bearing on the knife, forcing it out of the ideal spiral cutting line, with a tendency to plunge into the bolt, resulting in thick and thin veneer with waves a foot or more apart.

Some lathe operators use low knife angles, as the heavy bearing of the knife against the bolt tends to smooth the surface of the veneer. Manufacturers do not recommend this because pressures may bend the knife. Low knife angles also require more power, and cause more stains and wear to the lathe.

To prevent these problems, some lathe operators increase the angle of the knife until a corrugated veneer surface results. They then reduce the knife angle gradually until the corrugations disappear and use this knife angle for cutting.

For best results, satisfactory knife angles should be determined by instrument. Instruments for measuring the knife angle are described by Fleischer (1956), Feihl and Godin (1967), Fondronnier and Guillerm (1967), and Dokken and Godin (1975). While all are suitable, the French design (fig. 18-246) and the Canadian design (Dokken and Godin 1975) are easily read.

If the knives are all ground the same, they can be interchanged on a lathe or slicer without changing the knife angle or clearance angle. However, if the knives are ground so the bevel or sharpness angle is as little as 1/2° different, the cutting can be altered significantly. Consequently, knife angle should be checked with an instrument after each knife change.

Setting the slicer knife.—Setting the knife in the slicer is similar to setting the knife in a lathe except that the edge is positioned by the extension of the knife from the bed. The slicer knife edge should extend above the knife bed just enough so the ground face of the knife clears the bolts that hold it against the knife bed. In other words, the knife should extend as little as possible and still make certain the flitch will clear. On vertical face veneer slicers, this distance is about 2 inches (fig. 18-241).

Like the lathe knife, the slicer knife should rest on the two end adjusting screws. It is then brought against the bed and any sag in the middle is removed with the height-adjusting screws near the middle of the slicer. Since slicer knives

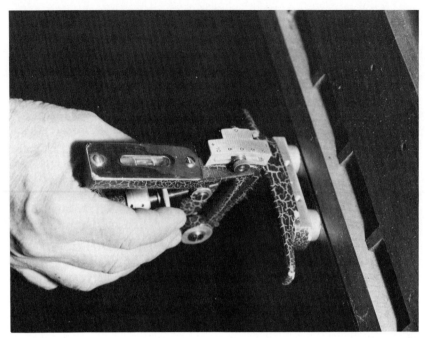

M 139 939

Figure 18-246.—Instrument of French design (Fondronnier and Guillerm 1975) for measuring knife angle. It is held by magnets to the knife, the bubble centered, and the knife angle read on the vernier. (Photo from Lutz 1977.)

are often longer than lathe knives, this adjustment is more critical on the slicer. A taut fine wire can be used as a guide to determine sag in the knife or, if the nosebar bed is known to be straight, it can be used as a guide. A nosebar ground to uniform thickness is brought up against the nosebar bed, and its bottom used as a reference to determine if there is a sag in the slicer knife.

Once the knife edge is determined to be straight, the knife is bolted firmly in place and all of the adjusting screws are brought in contact with the bottom of the knife.

The knife angle of the slicer is relatively easy to set compared to the lathe knife. Since all cutting is from a flat surface, the knife angle does not change with flitch size. The knife must lead into the flitch so the heel of the knife does not rub hard against the flitch. A slicer knife angle from 90°20' to 90°30' (about 1/2° clearance angle) can be used to slice wood from 1/100- to 1/4-inch in thickness from both low-density and high-density woods.

Like the lathe knife, the angle of the slicer knife should be checked with an instrument each time a knife is replaced.

NOSEBARS

The **nosebar** on a veneer lathe or slicer helps control veneer thickness, smoothness, and depth of checks into the veneer. It compresses the wood just ahead of the knife and so allows the knife to cut rather than split the veneer from the bolt or flitch.

By keeping a force between the knife carriage and the flitch or bolt, the pressure bar takes up slack in the machinery always in the same direction and so aids control of the veneer thickness (fig. 18-245).

There are two common types of nosebars—the **fixed bar** (figs. 18-240 and 18-241), and the **roller bar** (fig. 18-242).

Fixed nosebar.—Two factors to consider when selecting a fixed pressure bar are its stability and wear resistance. The most common metals are tool steel, stellite, and stainless steel. The tool steel bar is relatively inexpensive. A stellite bar is more expensive, harder to grind, and less stable. However, the stellite bar will wear many times longer than the tool steel. Stainless steel is easier to grind than stellite, and, like stellite, does not stain the veneer.

The fixed bar is generally ground to a bevel angle of about 74° to 78°. As the wood bolt or flitch approaches the fixed bar in the lathe or slicer, the wood is compressed along a plane 12° to 16° (θ in fig. 18-240) from the motion of the wood. When cutting 1/28-inch or thinner veneer from dense hardwoods, the bar should be ground to a sharp edge. The edge of the bar is generally slightly eased or rounded when cutting thicker veneer from low-density woods or woods subject to rupture on the tight side of the veneer from rubbing against the bar. Various researchers recommend an edge radius of about 0.015 inch.

Roller nosebar.—The roller bar, the second major type of pressure bar, is commonly of bronze, generally 5/8-inch in diameter if it is a single bar and 1/2-inch in diameter if it is a double roller bar. The single roller bar is driven directly while the double roller bar is driven by a backup roll (fig. 18-242). Two advantages of the double roller type stand out: (1) the drive roller can be larger so there is less breakage of the rollers, and (2) the knife and pressure bar can advance very close to the chucks, to permit peeling to smaller diameter cores than with a single roller bar. The drive chain for a single roller bar may protrude up to 1 inch beyond the surface of the roller bar. Roller bars are generally lubricated with 1 percent vegetable oil mixed in water and introduced through holes in the cap that holds the bar. Other suggestions for maintaining roller nosebars can be found in Hancock and Feihl (1976).

Comparison of fixed and roller nosebars.—The fixed bar is the simplest and most commonly used nosebar. It is used exclusively on slicers and is by far the most common bar used to cut hardwoods on a lathe. The roller bar is more common in the United States for cutting West Coast softwoods and hardwoods. The fixed bar can be used to cut veneer of any thickness. The 5/8-inch-diameter roller bar cannot be set to cut veneer much thinner than 1/16-inch. Most veneer peeled with the aid of a roller bar is used in construction plywood and is 1/12-inch or thicker. In general, it is easier to set a fixed bar precisely than a roller bar.

A major advantage of the driven roller bar is that it reduces torque for turning the bolt; this in turn means less spinout of the bolts at the chucks and less breakage at shake and splits in bolts. The roller bar also pushes through small splinters that jam between a fixed bar and the bolt, degrading the veneer.

SETTING NOSEBARS

Procedures for setting the nosebar, like those for setting the knife, are effective only if the lathe or slicer is in good mechanical condition with minimum looseness in moving parts and if knife, pressure bar, and surrounding metal parts are at about cutting temperature.

Cross sections of the knife with a conventional single-bevel fixed bar (figs. 18-240 center right and 18-241) and roller bar (fig. 18-242) show three openings between the knife and the bar—lead (v), gap (h), and exit gap (c). With any knife-bar combination, the position of the bar with respect to the knife is fixed if any two of the three openings are fixed, i.e., if the lead and gap are set, this also sets the exit gap. Which two are chosen for setting the knife and bar should depend on how the knife and bar can be adjusted on a specific lathe or slicer. Examples of how these three openings are interrelated for different veneer thicknesses and different settings are given in table 18-78.

TABLE 18-78.—*Lathe settings for fixed and roller nosebars with constant and variable lead* (Lutz 1977)[1]

Feed (veneer thickness, inch)	Lead	Gap	Exit gap
	------------------------------- *Inch* -------------------------------		
	FIXED BAR AND CONSTANT LEAD[2]		
0.010	0.030	0.009	0.019
.032	.030	.029	.038
.042	.030	.038	.046
.0625	.030	.056	.063
.100	.030	.090	.095
.125	.030	.112	.115
.1875	.030	.169	.168
.250	.030	.225	.221
	FIXED BAR AND VARIABLE LEAD[3]		
.010	.005	.009	.010
.032	.010	.029	.031
.042	.012	.038	.040
.0625	.017	.056	.058
.100	.024	.090	.093
.125	.030	.112	.115
.1875	.043	.169	.173
.250	.056	.225	.230
	ROLLER BAR AND FIXED LEAD[4]		
.0625	.085	.056	.062
.100	.085	.090	.094
.125	.085	.112	.114
.1875	.085	.169	.167
.250	.085	.225	.220
	ROLLER BAR AND VARIABLE LEAD[5]		
.0625	.068	.056	.056
.100	.075	.090	.090
.125	.079	.112	.112
.1875	.089	.169	.169
.250	.100	.225	.225

[1]Knife angle 90° (0° clearance angle) and gap 10 percent less than feed.
[2]Knife bevel 20°.
[3]Knife bevel 21°.
[4]Knife bevel 20°; 5/8-inch-diameter roller bar.
[5]Knife bevel 21°; 5/8-inch-diameter roller bar; gap equal to exit gap.

Setting a fixed nosebar on a lathe (by lead and gap).—For precision veneer cutting, both nosebar edge and knife edge must be straight and as perfectly aligned as possible. The bed for the bar and the nosebar cap should be clean and straight. The bar is inserted between the bed and the cap and the nosebar locking screw tightened just enough to hold the bar against the bed but loose enough so it can be moved without bending. The bar should extend from the supporting casting only a minimum amount so it is as rigid as practical. After the knife is set, the bar is moved toward the knife with adjusting screws at the two ends of the bar until the bar is about 1/32-inch behind the knife edge.

Adjusting screws at the two ends of the nosebar bed on most lathes raise or lower the entire bed to increase or decrease the lead between the nosebar edge and the knife edge.

All agree that the bar edge should be set above rather than at or below the knife edge, and the amount of lead must be the same at all points along the knife edge. The amount of **lead** (vertical opening) is adjusted primarily for the thickness of veneer being cut. Some lathe operators set the lead one-third of the thickness of veneer being cut. Fleischer (1949) suggests a straight-line relationship, with a lead of 0.005 inch when cutting 1/100-inch and 0.030 inch when cutting 1/8-inch veneer. Some settings using a variable lead that depend on veneer thickness are shown in table 18-78. Certain lathes made in Germany do not have a lead or vertical-opening adjustment. This distance is built in the lathe to be about 0.020 inch, coinciding with the lead suggested by Fleischer for 1/4-inch-thick veneer.

A common method of checking this opening is to insert a feeler gage of the proper thickness at a right angle to the ground surface of the knife in the lead (fig. 18-247) between the knife edge and the bar. After the bar is set this way at both ends, it should also be checked at other intervals along the knife. Some lathes have push-pulls so the bar can be warped locally to make the lead or vertical opening uniform across the lathe. If the knife and bar are ground straight and the knife bed and bar are also straight, however, any local adjustment of the lead should be minimal. Use of a feeler gage may slightly nick the blade. It is, therefore, good practice to lightly hone the knife after setting the lead.

The second bar adjustment is the **gap** or horizontal opening. This is the distance from the leading edge of the nosebar to a plane extended from the ground surface of the knife. Some experienced operators first bring the edge of the bar to the same plane as the knife edge, to check whether there are any spots where the bar is ahead or behind the knife edge. These local spots are brought in line with the push-pull screws in back of the bar. Once the bar is fitted to the knife, it is retracted to give the desired opening or gap and clamped.

Instruments to help make this critical setting are recommended, however. Two such instruments are described by Fleischer (1956) and Feihl and Godin (1967). Both are dial-micrometer depth gages that use the ground surface of the knife as a reference and measure to the edge of the bar. To automatically position the measuring pin, Fleischer (1956) suggests that the instrument rest on the top of the pressure bar and on the ground face of the knife (fig. 18-248).

While one man holds the instrument in contact with the knife and the movable sensing pin against the leading nosebar edge a second man advances the bar until the correct gap or horizontal opening is indicated. When advancing the bar, the adjustment should always be made to take the play out of the adjusting screws. First the two ends are checked. If they do not indicate the same opening, then they must be brought to the same position with the adjusting screws at each end of the pressure bar bed. Assuming the knife and bar were ground straight and were not warped when mounted on the lathe, the gap should now be the same

M 139 942

Figure 18-247.—Adjusting the lead of the nosebar with a feeler gage. The lead of the bar is moved until a feeler gage of the desired thickness is at a right angle to the ground surface of the knife when the gage is inserted in the opening between the knife and the bar. (Photo from Lutz 1977.)

across the lathe. However, since this is one of the critical lathe settings, the opening or gap should be checked at 4-inch intervals along the bar. The value of each reading is chalked on the casting holding the pressure bar. Any gradual bends or humps in the bar are then plainly evident. Local deviations are corrected by the push-pull screws at the back of the bar. For accurate cutting, the gap should be within ±0.001 inch at all positions.

The actual value of the gap will depend on the thickness of veneer and somewhat on the species being cut. A figure commonly quoted is for the gap to be 20 percent smaller than the thickness of veneer being cut. Experiments at the U.S. Forest Products Laboratory indicate this results in high compression of the wood by the nosebar. It would only be used when cutting thin veneer from an easily compressible species that is resistant to damage by scraping the nosebar over the wood surface.

When the pressure bar is set as described earlier, a compression of 10 to 15 percent has been found good for cutting veneer from 1/16- to 1/8-inch thick. Twenty percent compression may be satisfactory for thin veneer. Higher compression (smaller gap or horizontal opening) may result in tighter veneer, but may also cause the veneer to be thinner than the knife feed and cause damage to the tight side, such as shelling of the grain on susceptible species.

M 139 940

Figure 18-248.—Measuring the gap between the knife and nosebar edge. Measurements are chalked on the nosebar casting and any deviations greater than 0.001 inch removed with the push-pull adjustment of the bar. (Photo from Lutz 1977.)

Use of instruments to measure knife angle and pressure bar settings insures setups that can be readily duplicated. When experience shows that a certain setting is good for cutting a given thickness of veneer from a given species at a given temperature, then the information should be recorded and the conditions exactly duplicated when this item is produced again.

Setting a fixed nosebar on a slicer (by lead and gap).—The slicer bar is ground and set as described for the fixed bar on the lathe, but spacings are different. Instead of variable (0.010 to 0.030 inch) leads set on the lathe, on the slicer the lead or vertical opening is generally about 0.030 inch. Veneer of satisfactory quality from 1/100- to ¼-inch in thickness has been cut at the U.S. Forest Products Laboratory with this lead. A smaller lead such as 0.020 inch can be used for cutting 1/28-inch and thinner veneer, but may result in more splinters breaking off at the end of the cut and more chance for rub marks from splinters jammed between the knife and bar.

Pressure on the nosebar of a vertical-operating face-veneer slicer must be less than can be applied on a lathe. The knife and nosebar rest on half bearings which allow offsetting the knife and bar to clear the flitch on the upstroke. Excessive pressure against the flitch will rock the knife and bar carriage on the half bearing, causing poor veneer and possible damage to the slicer.

When slicing 1/28-inch (0.036) veneer, the range of satisfactory gap or horizontal openings between the knife and bar is between 0.029 and 0.032 inch. In effect, the bar is then compressing the wood just ahead of the knife edge 0.004 to 0.007 inch. Face veneer producers sometimes set the bar to compress the wood only 0.001 or 0.002 inch. When slicing thicker veneer such as 1/8 (0.125) inch, the bar may be set to leave a gap 0.010 inch less than the feed, i.e., 0.115 inch.

As with the lathe, more compression (slightly smaller openings) can be used when cutting low-density woods than when cutting high-density woods.

Setting a roller nosebar on a lathe (by lead and gap).—The roller bar is most commonly used when rotary-cutting western softwoods 1/12 to 3/16-inch in thickness, but is also used on some lathes cutting core veneer 1/16-inch or more in thickness from southern hardwoods. It cannot be used for cutting veneer thinner than 1/16-inch, because a roller bar would over-compress the veneer after it is cut by restricting the throat between the roller bar and the knife causing the veneer to jam and break.

In industry practice, 5/8-inch diameter roller bars are generally set with a lead of 1/16-inch or more. Lathe settings for several veneer thicknesses using a fixed lead are shown in table 18-78.

The gap is set much the same as with a fixed bar, compressing the wood ahead of the knife about 10 to 15 percent of the veneer thickness. This varies with species, wood density, and veneer thickness as discussed under the fixed nosebar. The gap or horizontal opening can be set and checked with a depth gage reading to 0.001 inch.

Setting a roller nosebar (by gap and exit gap).—Collett et al. (1971) suggest that lathes having a roller bar be set by gap and exit gap. For veneer thicknesses from 1/10- to 1/4-inch, the literature indicates that the gap and exit gap can be the same. This simplifies the recordkeeping as only one value needs to be recorded for each veneer thickness of each species. They recommend use of a depth gage to measure the gap and a feeler gage to measure the exit gap. They suggest compression at both the gap and exit gap of 10 to 20 percent of veneer thickness. Table 18-78 (bottom tabulation) shows some settings where the gap and exit gap are the same. Forintek Canada Corporation (1979) published handy nomographs permitting quick determination of gap and lead to maintain gap equal to exit gap.

Setting a fixed nosebar (by lead and exit gap).—Lead and exit gap are suggested by Fondronnier and Guillerm (1967) as the openings to be measured when setting a lathe with a fixed bar. They list the lead changing in a regular manner with veneer thickness as follows:

Veneer thickness		Lead or vertical opening	
Inch	Mm	Inch	Mm
0.039	1	0.020	0.5
.078	2	.024	.6
.118	3	.028	.7
.157	4	.031	.8
.197	5	.035	.9
.236	6	.039	1.0

They suggest the exit gap should be 10 to 20 percent less than the veneer thickness, and recommend feeler gages to measure both the lead and exit gap.

Setting gap by pressure rather than to fixed stops.—During rotary cutting of veneer, the force against the nosebar may vary from 10 to 500 pounds per lineal inch of contact with the wood. Feihl and Carroll (1969) adapted a research lathe to allow the bar to float and maintain the gap by pressure delivered by a cylinder and piston acting against the bar frame. Thus, they set the lead to stops but allowed the gap to be determined by the force against the bar. They reported that the method eliminates play in the horizontal mechanism, provides a direct measure of pressure against the bar and so gives the operator good control of the setting, and that the veneer produced was equal in quality to veneer produced with a bar set to fixed stops. The method is being tried commercially.

Readers needing information on the operation and maintenance of nosebars should find useful the manual on the subject by Hancock and Feihl (1980).

POSSIBLE WAYS TO GENERALIZE SETTING OF LATHE AND SLICER

Optimization of veneer peeling or slicing may require different knife and pressure bar settings for each specific cutting situation. However, it would be convenient to have one knife setting that could be used to cut veneer of any species into any thickness from 1/32- to 1/4-inch. Similarly, it would simplify pressure bar settings if one lead could be used for cutting all veneer. From an examination of the literature and experience at the U.S. Forest Products Laboratory, it is possible to do this.

Generalized knife settings.—The knife settings specified in figure 18-249 are broadly applicable, and may be particularly useful as starting points for cutting unfamiliar species. The knife should be ground to a 21° bevel with 0.002 inch hollow grind. The knife angle can be set to 90°30', or, stated another way, with 1/2° clearance angle. For lathes having an automatic change of knife angle with change in bolt diameter, the knife can be set at 90°30' when it is 12 inches from the spindle center. This knife setting can be used to cut veneer 1/32- to 1/4-inch in thickness from any species on the slicer or on the lathe from bolt diameters of 24 inches to a 6-inch core.

Generalized setting of a fixed nosebar.—The nosebar should be ground to have an included angle of 75°. This results in the wood being compressed along a plane approximately 15° from the cutting direction. The edge of the bar that contacts the wood should be rounded to an edge having a radius of about 0.015 inch.

The lead of the fixed nosebar ahead of the knife edge can be 0.03 inch for both the lathe and the slicer. The gap from the edge of the pressure bar to the plane of the ground surface of the knife can be 90 percent of the thickness of the veneer being cut. Veneer 1/32- to 1/4-inch in thickness and of various species can be cut with these fixed pressure bar settings (fig. 18-249).

Generalized setting of roller nosebar.—The generalized settings for lathes with a roller nosebar are for cutting veneer 1/16- to 1/4-inch in thickness. The center of the ⅝-inch-diameter roller bar should lead the knife edge by 0.085 inch. The gap should be 90 percent of veneer thickness (fig. 18-249).

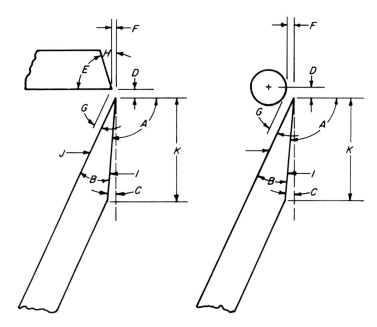

KNIFE AND FIXED BAR   KNIFE AND ROLLER BAR

M 144 168

Figure 18-249.—Knife and nosebar settings of general applicability are specified in terms of the diagram. These settings might be used to cut veneer from 1/32- to 1/4-inch in thickness.

Symbol	Generalized Settings	Symbol	Generalized Settings
A	Knife angle = 90° 30'	E	Nosebar bevel = 75° (θ = 15°)
B	Knife bevel = 21° with 0.002-inch hollow grind	F	Gap = 90 percent of veneer thickness (10 pct compression)
C	Clearance angle = 30' (½°)	G	Exit gap = Gap = 90 percent of veneer thickness (roller bar)
D	Lead = 0.030 inch for fixed bar or 0.085 for ⅝-inch-diameter roller bar	H	Nosebar compression angle = 15° (fixed bar)

(Drawing and data from Lutz 1977.)

Alternate generalized setting of roller nosebar.—Collett et al. (1971) describe setting a roller bar with the gap and exit gap equal. As with the rigid bar, a generalized setting would be to have the gap and exit gap both 90 percent of the thickness of the veneer being cut (fig. 18-249 and table 18-78, bottom tabulation).

Generalized setting of the gap by pressure.—Feihl and Carroll (1969) reported that pine veneer 1/10- to 1/6-inch in thickness can be cut satisfactorily with pressure on a floating roller bar of about 60 pounds per linear inch of bar contacting the wood bolt. They further conclude: "It is not impossible that in some mills (when all species are fairly similar and veneer thicknesses are in the same range) it would be practical to use one pressure setting." In the South, where oak species predominate among pine-site hardwoods, mills producing oak veneer might find this technique useful.

Summary of generalized lathe and slicer settings.—Suggested "universal" lathe and slicer settings—listed in figure 18-249—are not optimum settings, but they should permit cutting veneer of moderate quality from any species into any thickness from 1/32- to 1/4-inch. (The roller bar is not satisfactory for use when cutting veneer thinner than 1/16 inch.)

In general, excluding the extreme ranges of specific gravity, one species of wood acts much like another and the veneer cutting process does not change abruptly within the range of thickness from 1/32- to 1/4-inch.

The settings listed with figure 18-249 will generally result in a moderately tight cut. If tighter and smoother veneer is desired, smaller openings between the knife and pressure bar may be used. Lathes having automatic pitch adjustment could be set to have a knife angle of 91° at a bolt diameter of 36 inches and a knife angle of 89°30′ at 6 inches. Ideally, the rate of change of the knife pitch should be increased at the smaller diameters. A smaller fixed nosebar lead such as 0.020 or 0.015 inch can be used for cutting 1/16-inch and thinner veneer.

Ash, elm, most oaks, and hickory are intermediate in sensitivity to knife and nosebar settings, while hackberry, red maple, sweetbay, sweetgum, black tupelo and yellow-poplar are least sensitive to these settings. Laurel oak is among the species most sensitive to nosebar settings (table 18-79a).

TABLE 18-79a.—*Sensitivity of 21 hardwoods found on southern pine sites to settings of knife and nosebar* (Lutz 1977)[1]

Species	Knife	Nosebar
Ash, white	B	B
Elm, American	B	B
Elm, winged	B	B
Hackberry	A	A
Hickory, true	B	B
Maple, red	A	A
Oak, black	B	B
Oak, cherrybark	B	B
Oak, chestnut	B	B
Oak, laurel	B	C
Oak, northern red	B	B
Oak, post	B	B
Oak, scarlet	B	B
Oak, Shumard	B	B
Oak, southern red	B	B
Oak, water	B	B
Oak, white	B	B
Sweetbay	A	A
Sweetgum	A	B
Tupelo, black	A	A
Yellow-poplar	A	A

[1] A, species least sensitive to settings; B, species intermediate; and C, species most sensitive (least desirable).

PANEL MANUFACTURE WITH VERY THIN FACES

Peeler bolts that will yield face veneers are scarce in pine-site hardwoods. Demand for good quality face veneers is strong, however. One approach to solving the supply problem is to peel thinner face veneers. However, the use of very thin face veneer aggravates some problems such as sand-throughs, resin or glue bleedthroughs, core defects, telegraphing, and handling damage. To study the problems associated with use of very thin face veneers, Feihl et al. (1976) manufactured about 1,000 panels with yellow birch (*Betula alleghaniensis* Britton) faces ranging from 1/28-inch, the usual minimum thickness, to 1/50-inch. Their study showed no problem in facing particleboards with 1/50-inch veneer. The 1/36-inch faces were suitable for filled and sanded plywood blanks. The quality of some softwood core types presently available, however, is too low for application of thin faces. Carroll (1978) described methods for successfully sanding thin face veneers.

Yellow birch is a favored veneer species; manufacture of very thin face veneers from the oaks and hickories predominating on southern pine sites is more difficult and is not described in the literature.

CUTTING THICK VENEER

Most of the preceding discussion relates to cutting veneer from 1/32- to 1/4-inch thick. Because most peeler bolts available from pine-site hardwoods are short (e.g., 52 inches long), dense, and contain defects that preclude their use for face veneers, they are most likely to yield veneer for structural applications and cores. Such uses, if they are to be significantly expanded, may call for veneer cut 1/4- to 1/2-inch in thickness. Cutting such thick veneer requires special techniques, whether rotary peeled or sliced.

Thick sliced veneer.—Research on rotary peeling thick veneer was preceded by study of slicing thick veneer. Lutz et al. (1962) sliced 7/16-inch thick yellow-poplar and 3/8-inch oak at 40 strokes per minute on a slicer having a knife ground to a 22° bevel angle and set with a knife angle of 90°20′. The fixed bar was ground with a 12° nosebar compression angle and set with lead of 0.030 inch. When the slicer approached the top of the flitch, the wood sometimes splintered and broke instead of cutting smoothly to the edge. Favorable flitch temperatures for thick slicing were 190°F for yellow-poplar and 210° for oak.

Following construction of a heavy-duty experimental slicer, Peters et al. (1969a) published additional data specific to northern red oak and yellow-poplar. One-half and 1-inch-thick slices were cut at 5, 50, 200, and 500 fpm from flat-grain cants heated in water to 190°F. Average specific gravity values of the two northern red oak logs cut were 0.51 and 0.59; those of the yellow-poplar logs were 0.33 and 0.37. The oak had seven or eight rings per inch, while the yellow-poplar had four. The knife had 20° sharpness angle and 0.5° clearance angle. The 15° fixed nosebar was set to give lead of 0.060 inch and **restraint** (nominal veneer thickness minus c in fig. 18-240) of 0.057 inch for ½-inch slices of both species. When cutting 1-inch slices, lead was 0.200 inch and restraint 0.133 inch for oak; for 1-inch yellow-poplar lead was 0.275 inch and restraint was 0.153 inch.

Forces exerted on the knife and bar are shown in table 18-79b.

TABLE 18-79b.—*Summary of forces on knife and nosebar when northern red oak and yellow-poplar are sliced into thick veneer* (Peters et al. 1969a)

Feed thickness and cutting velocity (fpm)	Forces[1] and horsepower per inch of bolt length					Net cutting power[4]
	Knife		Bar		Combined	
	Parallel	Perpen-dicular[2]	Parallel	Perpen-dicular	Parallel[3]	
	----------------------------Pounds----------------------------					Hp
NORTHERN RED OAK						
½-inch						
5	203	260	91	458	294	0.04
50	196	284	72	454	268	.41
200	215	351	98	553	313	1.90
500	215	372	115	548	330	5.00
1 inch						
5	265	308	154	693	419	.06
50	275	391	165	760	440	.67
200	254	439	211	840	465	2.82
500	291	454	239	858	530	8.01
YELLOW-POPLAR						
½-inch						
5	150	191	51	289	201	.03
50	184	250	54	340	238	.36
200	178	281	72	393	250	1.52
500	179	305	74	404	253	3.83
1 inch						
5	163	155	130	461	293	.04
50	204	229	154	598	358	.54
200	181	284	195	676	376	2.28
500	216	340	168	734	384	5.82

[1]Parallel and perpendicular to cutting direction.
[2]Perpendicular knife forces are in the opposite direction to perpendicular bar forces.
[3]Perpendicular forces as measured in this study cannot be combined.
[4]Horsepower calculated from combined parallel force and velocity.

Depth of knife checks was positively correlated with cutting velocity as shown by the following tabulation of fracture depth expressed as a percent of veneer thickness.

Species and cutting velocity	½-inch thick	1 inch thick
fpm	----------Percent----------	
Northern red oak		
5	32	46
50	48	67
200	70	76
500	72	78
Yellow-poplar		
5	32	32
50	46	33
200	64	44
500	74	78

When slicing northern red oak and yellow-poplar veneer 1/2-inch thick from flitches heated to 140°F, Peters and Patzer (1976) also found that best results were obtained with very slow cutting speeds (5 fpm) with a fixed nosebar set for 20 percent compression. When slicing 1-inch-thick northern red oak veneer, even slower speeds improved quality; knife fracture depth decreased from 0.48 inch at 5 fpm to 0.06 inch at 0.05 fpm.

Peters et al. (1972) found that quality of thick-sliced veneer was best if the knife edge was maintained parallel to the grain of the flitch, so that cutting was truly in the 0-90 direction.

For readers desiring more detailed information, a tabulation of major publications related to thick slicing of veneer follows:

Lutz (1963)	Carlson (1968)
Peters (1968, 1976)	Peters and Patzer (1976)
Peters and Zenk (1968)	Johnston and St. Laurent (1978)
Peters et al. (1968, 1969ab, 1972, 1976)	Johnston and St. Laurent (1979)

While work on thick slicing continues, many experts in the field believe that rotary peeling offers greater potential for early commercialization.

Thick rotary-peeled veneer.—Hann et al. (1971) rotary peeled northern red oak to 5/16- and 7/16-inch thicknesses (fig. 18-250). For this thick cutting, 50-inch bolts were first heated at 170°F for 12 hours. The hot bolts were centered between 8-inch diameter lathe chucks and turned at 16 rpm. The lathe was equipped with a fixed nosebar having a 15° nosebar compression angle; the veneer knife had a ground bevel angle of 21°. The nosebar was set for 0.060-inch compression of veneer of both thicknesses. Thickness in the green condition ranged from 0.261 to 0.302 inch for the 5/16 inch veneer and 0.431 to 0.447 for the 7/16-inch veneer. With core diameter fixed at 8½ inches, veneer yield from 16-inch bolts was 70 to 80 percent. When clipped, press-dried and parallel-glue laminated into 2-ply and 3-ply pallet deckboards, average performance of the deckboards was equal to or better than similar pallets with solid oak or hickory deckboards.

Spin-out is a problem when cutting such thick veneer from pine-site hardwoods. To obtain acceptable yield from small-diameter bolts, core diameter should be substantially less than 8½ inches—perhaps 5¼ inches. It seems likely that such a small core diameter can be achieved when peeling 1/4-inch or 5/16-

Figure 18-250.—Sheet of 7/16-inch-thick, rotary-peeled oak veneer emerging from lathe. (Photo from Hann et al. 1971.)

inch oak veneer if a friction drive can be supplied to the peeler block surface to reduce torque application at the end chucks (Lutz and Patzer 1976). Development of such a device is in progress (fig. 18-251). Preliminary roundup of bolts (fig. 18-252) will assist such a friction drive. Cutting forces are high so a heavy-duty lathe is required. Core diameters can be smaller for 4-foot bolts than for 8-foot lengths.

Machining 2187

Figure 18-251.—(Top) Prototype of Hydraulically powered backup rolls which bear against peeler log surface and assist the lathe chucks in providing torque to the log. A hydraulic cylinder holds the pair of backup rolls in contact with the log; plastic-faced rolls have a coefficient of friction on green wood of 0.5 to 0.8. (Photo from Forest Products Laboratory, For. Serv., U.S. Dep. Agric.) (Bottom) Commercial application of the concept. (Drawing from Coe Manufacturing Company.)

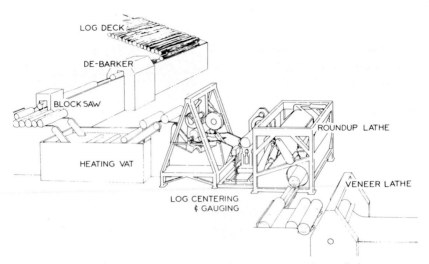

Figure 18-252.—Production line to automatically singulate, mechanically locate center axis, chuck, and machine 52-inch-long hardwood peeler bolts to maximum-diameter cylinders before passing them to a veneer lathe. The line will machine seven bolts per minute; residue is in the form of flakes for structural flakeboard. (Drawing from Stetson-Ross after Roubicek and Koch 1981.) See figure 18-104C for photograph of machine.

STEAM-INJECTION KNIVES AND HEATED NOSEBARS

On western softwoods, Walser (1974) found that injection of steam into the exit gap along the loose side of the veneer (fig. 18-253) resulted in smoother veneer and less degrade from roughness. In his laboratory studies Walser used steam pressure of 54 psi with knife tip temperatures from 185° to 220°F. Later commercial trials used steam superheated to yield knife tip temperatures up to 400°F. No data on southern hardwoods cut with the steam knife have been published.

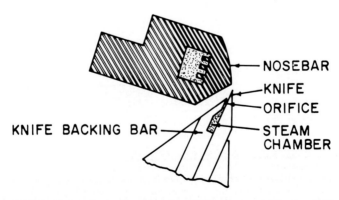

Figure 18-253.—Steam-heated nosebar coupled with steam-injection knife. (Drawing after Walser 1978.)

Walser (1978) also proposed that the nosebar be heated and given a double bevel with curved upper surface. The top curved contour allows slivers to be swept past the nosebar by the moving surface of the bolt. The nosebar is heated by steam, the steam being retained within the nosebar (fig. 18-253), or ejected onto the surface of the bolt. Walser concluded that this heating reduces friction between nosebar and bolt and therefore reduces torque needed to turn the bolt. No data are published on use of the heated nosebar on southern hardwoods.

POSITIONING BOLTS AND FLITCHES

To gain maximum yield of veneer, bolts must be chucked at their geometric center (Bolton 1975; Kuykendall 1975; Foschi 1976; and Washburn 1976). Some of the X-Y centering devices rely on sophisticated scanning equipment coupled to computers (Kuykendall 1975; Knokey 1979); others are simple and remarkably effective mechanical devices (see fig. 18-252) capable of centering as many as seven small peeler bolts per minute.

The way a flitch is mounted on the slicer table has little effect on yield, but it can affect the smoothness of the veneer. An eccentric flat-cut flitch should be dogged with the pith toward the start of the knife cut. A quartered flitch should be turned 180° when the cut approaches the true quarter (fig. 18-243 bottom). These, and related phenomena are discussed by Lutz (1956).

CONVEYING VENEER FROM LATHE AND SLICER

Conveying veneer from a lathe.—As veneer comes from the lathe, it may be manually pulled out on a table, but more generally it is moved to long trays in line with the clippers or is reeled.

The **tray system** (fig. 18-254 top) is most common in both hardwood and softwood plants. As the veneer comes from the lathe, a short tipple directs unusable veneer to a waste conveyor. Usable veneer is directed into one of the trays with belts synchronized to the lathe speed. After one tray is full, the veneer is broken or cut, and the veneer directed to another tray. This must be done carefully to prevent the veneer ribbon from being folded and split.

In the **close-coupled system** (fig. 18-254 bottom), the infeed tipple to the trays and the trays themselves are replaced by a conveyor between lathe and the machine which clips the veneer to width. The distance between the lathe spindles and clipper knife is approximately 60 feet. This belt conveyor is so arranged that veneer emerging from the lathe can be conveyed at high speed, while the clipper can finish clipping the end of the previous band of veneer at a slower speed. Springate (1978b) compared the two systems and reviewed techniques appropriate for small logs.

A third mechanical means of conveying veneer from the lathe is with a reel. The **reel system** works best with 1/8-inch and thinner hardwoods cut from sound bolts. Like the tray system, the first unusable veneer is directed to a waste

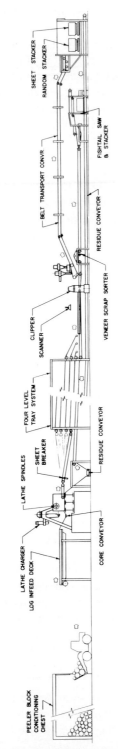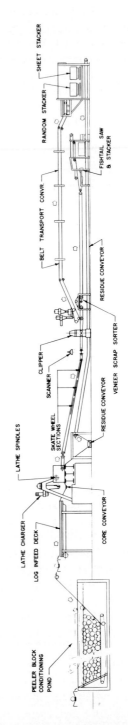

Figure 18-254.—Tray system and close-coupled system for rotary peeling, clipping, and stacking veneer from small logs. If preceded by the roundup lathe shown in figure 18-252, these lines can peel seven veneer bolts per minute. (Drawing after Springate 1978b.)

conveyor. Then the usable roundup is collected on a short table. Finally, when a sound ribbon of veneer comes from the lathe, it is tacked to a reel and the veneer reeled up as it is peeled. The speed of the reel is synchronized with the lathe. The veneer is reeled with the loose side out.

Tray and reeling are combined at some plants, the better grades being cut into thin face stock and reeled, and lower grades into thicker core stock conveyed on trays.

Conveying veneer from a slicer.—Sliced veneer sheets are usually kept in consecutive order. In many plants, two men turn the veneer as the sheets come from the slicer and stack them consecutively with the loose side up. In some cases, a short conveyor takes the veneer from the slicer to a convenient stacking position. Some European plants automatically convey the sliced veneer to a dryer sized for the wood species and thickness and production rate of the slicer.

A German machinery manufacturer recently announced a system to reel sliced veneer by first applying string to the ends of the veneer sheets as they come from the slicer. The string then "leads" the veneer onto the reel where it can then be stored before unreeling into a dryer.

CLIPPING

Clipping green veneer.—Veneer stored on trays is fed to one or more clippers (fig. 18-254). In a typical installation, with six trays from a lathe, three trays would feed to one clipper and the other three to a second clipper. A modern clipper has some sensing and measuring device so veneer can be clipped to nominal 4-foot, 2 foot, or random widths. Random widths may be generated when defects such as knots and splits are clipped from the veneer ribbon. An accurate sensing device coupled with the clipper soon pays for itself by greater yields of usable veneer. The green veneer is then sorted by widths, grades, and possibly by sapwood and heartwood in preparation for drying.

Reeled veneer is stored in racks and unreeled just ahead of the clipper. The clipping operation is much the same as that described for veneer stored on trays. One limitation of reeled veneer is that, if it is cut from hot bolts, it should be clipped before the veneer cools and sets in a curved shape.

Flitches of green sliced veneer sometimes have defects clipped out or are trimmed before drying. Packs about 1/4-inch deep are clipped together as a book. The green clipping saves drying of material that will not be used.

Clipping dry veneer.—Veneer on trays or on reels is sometimes fed to the dryer in a continuous ribbon. As the veneer comes from the dryer, it is clipped to size. This system reportedly results in less waste and split veneer. One dryer manufacturer states that drying of a continuous ribbon will result in at least a 4-percent increase in recovery of dry veneer.

VENEER YIELD

Lathe yield.—With knife-cut veneer one might assume that veneer recovery could equal the volume of log minus the volume of core. Unfortunately, this is not the case. Losses include those due to roundup (about 5½ percent), spurring veneer to length (about 2 percent), green-end clipper loss (about 22 percent), below-grade veneer (about 6 percent), core volume (about 9½ percent), and veneer shrinkage when dried (about 3 percent). Data are not published on losses when peeling southern hardwoods, but the foregoing data from Douglas-fir (*Pseudotsuga menziesii* (Mirb.) Franco) commercial operations (Woodfin 1973) result in an actual dry veneer recovery of only 52 percent of the total volume of the green peeler bolt.

Pine-site hardwood logs are crooked, and veneer yield can be increased by peeling 4-foot rather than 8-foot-long bolts. Comparative data for hardwoods are not published, but Fahey's (1978) data on western hemlock (*Tsuga heterophylla* (Raf.) Sarg.) are indicative of gains to be made by peeling 4-foot bolts (fig. 18-255).

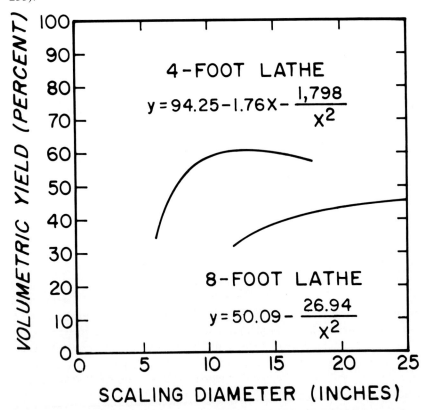

Figure 18-255.—Volume yield of dry western hemlock veneer (percent of bolt volume) from 4-foot and 8-foot lathes related to diameter inside bark at the small end of the peeler bolt. Core diameter on the 4-foot lathe was 3 inches, while that on the 8-foot lathe was 7 inches. Thickness of the veneer when dry was 0.205 inch. (Drawing after Fahey 1978.)

Springate (1978b) provided a nomograph (fig. 18-256) relating diameter of 8-foot peeler bolts to veneer yield expressed as square feet of 3/8-inch-thick panel per cubic foot of log, per board foot Doyle log scale, and per cord of input volume.

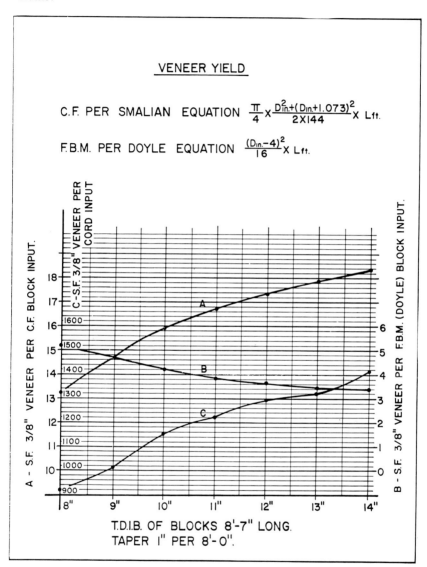

Figure 18-256.—Veneer yield expressed as square-foot area of 3/8-inch-thick panels from 103-inch-long veneer bolts, related to bolt diameter inside bark at the small end.
  A. Per cubic foot of log input.
  B. Per thousand board feet Doyle log scale input.
  C. Per standard cord input.
(Drawing after Springate 1978b.)

Slicer yield.—In general veneer recovery is highest by rotary cutting, less by flat-slicing, and least by quarter-slicing. Yields are less for slicing because of losses when sawing the flitches and when clipping straight edges on the relatively narrow sliced veneer.

Some commercial slicing operators have reported that, for logs 15 inches and larger in diameter, the yield of flat-sliced veneer is about equal in volume to the board foot value by the Scribner Decimal C log rule. For example, an 8-foot-long log with scaling diameter of 19 inches contains 120 board feet by the Scribner Decimal C log rule; one might expect, therefore, to get 10 cubic feet (120/12) of sliced veneer from such a log.

No data are published on slicer yields from low-grade pine-site hardwoods, but Rast (1978) provided yield data on 8- to 12-foot-long, veneer-quality logs of northern red oak from northern Pennsylvania. He found that volume of logs flitched and sliced was distributed as shown in figure 18-257; 56.7 percent of total log volume ended as dry veneer.

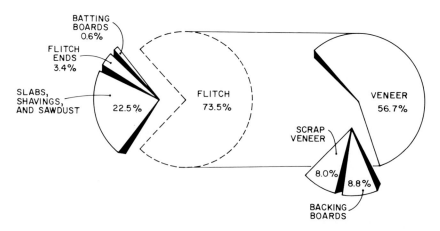

Figure 18-257.—Components of sliced northern red oak logs of veneer quality, as a percentage of total log volume. Batting boards are usable boards recovered when logs are sawn into flitches; they are used to crate the dry "books" of veneer. (Drawing after Rast 1978.)

MONOGRAPHS ON CUTTING VENEER OF PARTICULAR SOUTHERN SPECIES

The text under this section (18-23) has necessarily been general; available information on pine site hardwoods has been included, however. Readers interested in short monographs on cutting particular species will find the following reports useful:

Species	Citation
Elm, American	Feihl (1956)
Hickory	Lutz (1955)
Oak, chestnut	Lutz and McAlister (1963)
Oak, water	U.S. Department of Agriculture, Forest Service (1965)
Tupelo, black	U.S. Department of Agriculture, Forest Service (1965)
Yellow-poplar	Lutz and Patzer (1966)

18-24 CHIPPING

Discussion in this section is limited to a few important aspects of chipping hardwoods. Readers needing more detailed data on the chipping process—from history through production, transport, storage, and analysis—are referred to Hatton (1979a). See also proceedings of the symposium "Comminution of wood and bark" held October 1-3, 1984 in Chicago, Illinois; available from the Forest Products Research Society, Madison, Wisconsin.

To make chemical pulp or refiner mechanical pulp, hardwoods must first be reduced to chips of relatively uniform size. Even energy chips destined for combustion, gasification, pyrolysis, or conversion to liquid fuels and other chemicals serve best if optimally sized for processing.

CHIP DIMENSIONS

Chips for kraft pulp and for refiner groundwood are the dominant hardwood chip products, and will perhaps continue to dominate the market into the 21st Century. Other chip uses, however, will likely increase in importance.

Chips for kraft pulping.—Optimum size and proportions of pulp chips vary according to pulping process and equipment. Schmied (1964) studied the effects of chip size and shape on the uniformity of wood delignification. Size of chips affects the cook when the cooking is rapid and if the chips have a high moisture content. In large chips, long diffusion paths delay penetration of the pulping chemicals to the chip centers. Doubling of chip size requires a fourfold prolongation of the cooking time. In mixed sizes, excessive absorption and side reactions in the small chips may slow delignification of large ones by depleting chemicals in the liquor penetrating the large chips. This is one reason for avoiding mixtures of chips with sawdust. Initially, the diffusion front of the chemicals follows the edges of the chips, but in later stages of cooking the front is rounded and undercooked chips become cylindrical and ellipsoidal. Schmied concluded that these effects are considerably reduced if chips are dry, because the lumens of dry wood can be filled more easily.

Specific recommendations concerning chip thickness were made by Wahlman (1967); he found that maximum-strength alkaline pulps were made from chips 2 to 5 mm (0.08 to 0.20 inch) thick and that screenings, i.e., particles of undigested wood, became excessive with chip thickness exceeding 5 mm. Hatton (1977a, 1978ab) concluded that chips 2 mm thick achieved maximum screened yield of kraft pulp from hardwoods. In the southern pine region, pulp chip dimensions shown in figure 18-258 are commonly used, but preferences differ among mills. For example, some mills with continuous digesters prefer chips 0.75 to 1.00 inch long. For the interested reader, Borlew and Miller (1970) reviewed the literature on the effect of chip thickness on the kraft pulping process, and Sobieski (1980) provided a selected annotated bibliography on the effect of chip thickness on pulping.

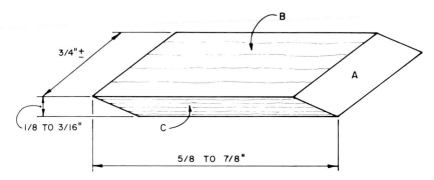

Figure 18-258.—Chip dimensions commonly observed in southern kraft mills. Surface A cut by chipping knife, surfaces B and C are split parallel to the grain. For maximum screened yield from hardwoods pulped by the kraft process, Hatton (1977a, 1978ab) recommends a chip thickness of 2 mm (0.08 inch).

Chips for very rapid impregnation and vapor-phase digestion.—Accurately cut flakes measuring 20 or 30 mm long and only 0.3 to 0.5 mm thick can be very rapidly impregnated with liquid and cooked quickly in a separate operation. While not presently used in the South, the process may become commercially important for pulping hardwoods (Reitter 1974).

Chips for refiner mechanical pulp.—No data descriptive of pine-site hardwoods are published, but thin chips (e.g., 0.10 inch thick) should require less specific refining energy than thicker chips. Preliminary efforts by the Southern Forest Experiment Station, Pineville, La., to disk refine very thin southern red oak flakes (0.015 to 0.025 inch thick) yielded unsatisfactory fiber. It seems probable, however, that further research could develop techniques for obtaining acceptable fiber from thin flakes with low expenditure of energy. At present, the chip dimensions shown in figure 18-258 are in common use.

Chips to be sliced into flakes.—Growth of the structural flakeboard industry in the South should increase demand for chips suitable for conversion to flakes (see sec. 18-25). Such chips, termed **fingerlings**, are longer and thicker than the normal pulp chip. Requirements vary, but a chip 1.5 to 2.5 inches long and 1/2-inch in width and thickness should be suitable for most machines designed to cut flakes from fingerling chips.

Fuel chips.—Users of stoker-type wood-burning furnaces (fig. 26-11) frequently describe optimum fuel as walnut-size. Operators of suspension burners (fig. 26-21), however, need much smaller particles—perhaps passing a screen with ⅜-inch holes. Charcoal briquet manufacturers (fig. 26-32) and operators of wood gasifiers (fig. 26-26) can use chips of the size illustrated in figure 18-258. Fluidized bed burners (fig. 26-19) will tolerate large chunks of wood as well as pulp-chip size pieces.

Metallurgical chips.—In the Ohio and Tennessee River Valleys there is a significant market for hardwood chips used in the manufacture of ferro-alloys. These chips measure 2.5 to 8 inches square and 0.5 to 1.0 inch thick. Hard hardwoods are preferred species. (See section 22-16 for further discussion of this product.)

CHIP FORMATION AND POWER REQUIRED

Pulp chips can be cut in any of three major modes (fig. 18-1, 18-98, 18-259). Energy consumed per cubic inch of wood chipped is least if chips are long and thick rather than short and thin, if rake angles are high, if knives are sharp, and if 0-90 or 90-0 cutting mode is used rather than the 90-90 mode.

Conventional disk-type chippers (fig. 18-259A) cut in a mode intermediate between 90-90 and 90-0. In these chippers, several straight knives are bolted in more or less radial disposition into a heavy disk that revolves in a vertical plane. Severed chips pass through a slot in the disk and may be discharged from top, bottom, or sides of the disk housing. Logs, slabs, or edgings are fed against the disk through the infeed spout (figs. 18-260 top and 18-261). The angle between the face of the disk and the axis of the spout is usually 37½° (fig. 18-262, top); this angle may be attained by attaching the spout at a horizontal angle (ω in fig. 18-261) only, or in combination with a vertical angle (not illustrated). Chippers with a vertical spout angle are usually gravity fed; a powered conveyor delivers wood into horizontally fed chippers.

Erickson (1964) and Papworth and Erickson (1966) on tests of a three-knife disk chipper cutting 4- by 4-inch by 8-foot wood found that vertical spout angle had no effect on specific cutting energy; however, a 30° horizontal spout angle (ω in fig. 18-261) required slightly less cutting energy than a 0° angle. Knives with small sharpness angle required less power than those with large sharpness angle, and sharp knives required less power than blunt. Long chips required less specific cutting energy than short chips.

On disk chippers, the angle θ (figure 18-262) appears to control the ratio of chip thickness to length, the chips becoming thicker as θ becomes larger; if θ is less than 90°, bristles are formed on the ends of the chips; if θ is more than 90°, the ends of the chips are compressed (Hartler 1962). Helically-formed surfaces following each knife (fig. 18-262) keep the workpiece in full contact with the face plate and help control chip size. Swept-back knives (fig. 18-261) diminish knife impact and provide an oblique cut that should reduce power and diminish bruising. Many manufacturers place the knives radially, but place the spout so that the bedknives and workpieces (see figs. 18-261 and 18-262B) are aligned to

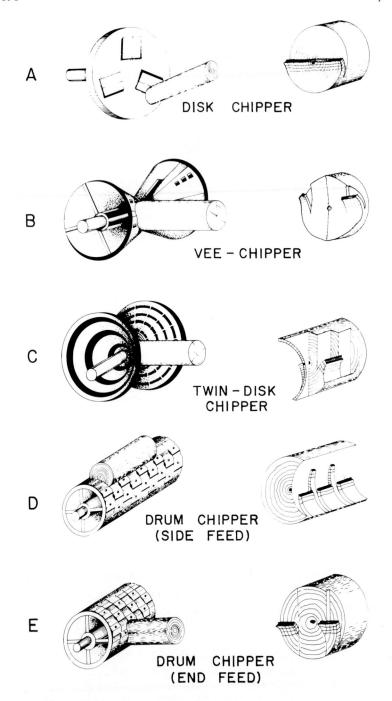

Figure 18-259.—Chipping modes. (Drawing after Erickson 1966.)

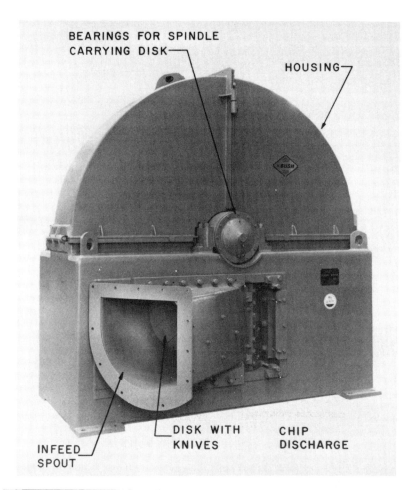

Figure 18-260.—(Top) Horizontal-feed, whole-log disk chipper designed for bottom discharge. Photo from Bush Manufacturing Company. (Bottom left) Infeed end of horizontal-feed drum chipper, and (bottom right) the chipping drum. (Photos from Pallman Pulverizer Co., Inc.)

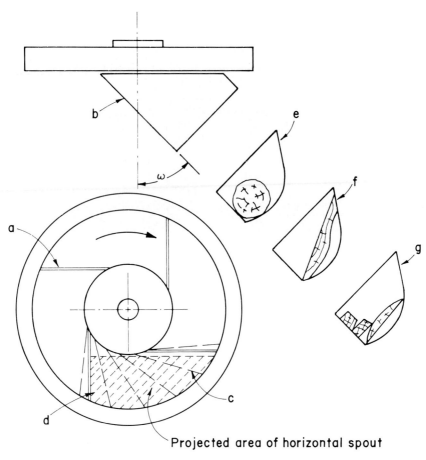

Figure 18-261.—Diagram of disk chipper. a, swept-back knives; b, spout position for horizontal feed; c, slicing action of knives across projected spout area; d, location of bedknife or anvil, against which the wood is pressed by the knives in passing. Infeed spout will admit wood in several forms; e, roundwood; f, wide slabs; g, slabs and edgings. ω = horizontal spout angle (commonly 90° − 37½° = 52½°).

prevent simultaneous impact of all parts of the knife edge across the full width of a rectangular piece of wood to be chipped.

The power demand of a chipper is proportional to the volume of wood it chips in a unit time. The number of cuts a machine makes per cubic foot of solid roundwood chipped is as follows:

$$X = 6912/L\pi D^2 \qquad (18\text{-}52)$$

where

X = number of cuts per cubic foot of solid wood chipped
L = chip length, inches
D = diameter of bolt, inches

Thus, when cutting 5/8-inch-long chips, about 35 cuts per cubic foot are required for a 10-inch log, and 880 cuts per cubic foot are required for a pulp stick measuring 2 inches in diameter, or 3.5 and 88 revolutions, respectively, for a 10-knife disk.

Machining

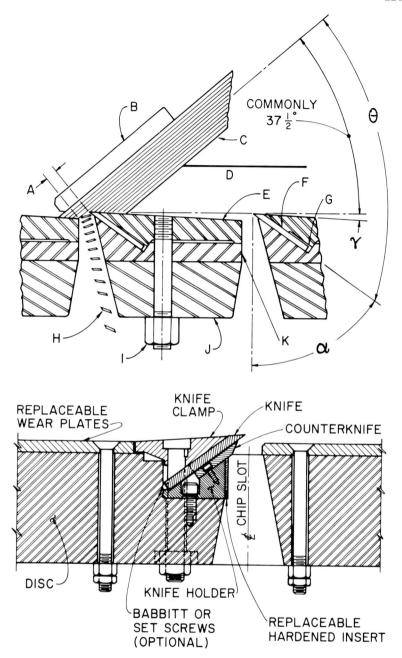

Figure 18-262.—Cutting action of knives in disk chipper. (Top) Cross section through one type of disk. A, chip length; B, side bedknife; C, workpiece; D, bottom bedknife; E, helical face plate; F, knife; G, shim; H, chips; I, face plate stud and nut; J, chipper disk; K, knife carrier; α, rake angle (approximately 50°); γ, clearance angle (2° to 8°), θ, angle between rake face of knife and grain direction of workpiece. (Bottom) Cross section through disk in common use on southern hardwoods. (Drawing at top from Sumner Iron Works; drawing at bottom from Bush Manufacturing Company.)

The productiveness of a chipper is determined by the size of workpiece it will admit and the number of cuts it makes per minute. Therefore, the following relationship expresses the output of a chipper.

$$V = nN/X \qquad (18\text{-}53)$$

where:
- V = cubic feet of solid wood chipped per minute
- n = revolutions per minute of chipper disk
- N = number of knives in the disk
- X = number of cuts per cubic foot (from equation 18-52)

According to Rogers (1948) and Fobes (1959), specific cutting energy for disk-type chippers is proportional to wood specific gravity as follows:

Specific gravity	Horsepower seconds per cubic foot of solid wood chipped
0.3	195
.4	300
.5	430
.6	570

The specific cutting energy, as computed by Rogers (1948) for disk chippers, is approximately 400 hp sec/ft^3 for hardwoods and 280 hp sec/ft^3 for softwoods. Values of 10 hp hr/cord to chip hardwoods and 7 hp hr/cord to chip softwoods have long been used. Specific cutting energy increases slightly with decreasing chip length, and increases with decreasing moisture content.

In selecting power units for disk chippers, many users rely on the formulae and nomographs published by TAPPI (1969) to predict power required.

Twin-disk chippers.—Papworth and Johnson (1968) found that a Soderhamn HP chipper comprised of two conical disks with small ends joined together to form a Vee-drum, and fed as shown in figure 18-259C, took less specific cutting energy than a disk chipper (fig. 18-259A), as follows (chips were 1-1/4 inches long and 1/4-inch thick):

Species	Twin-disk chipper	Disk chipper
	----------Hp sec/ft^3 of wood chipped----------	
Aspen (*Populus tremuloides* Michx.)	120	205
Maple, red	152	229
Maple, sugar (*Acer saccharum* Marsh.)	178	285

Drum chipper fed with logs approaching endwise.—Drum chippers (figs. 18-259E, 18-260 bottom, 16-52, and 16-53) operate in a different mode than disk chippers; horsepower requirements, graphed in figure 16-51, indicate a specific cutting energy for oak and hickory of about 400 hp sec/ft^3 of wood chipped.

Drum chipper with logs approaching sideways.—In some applications, wood is presented sideways to a drum chipper (figs. 17-13, 18-259D, and 18-104). Specific cutting energy when cutting thin chips (flakes) in this mode is quite high—about 700 hp sec/ft^3 for unheated southern red oak cut 0.015 inch thick (fig. 18-100). For chips measuring 1/8- or 1/4-inch in thickness, however, specific cutting energy should not be greater than that required for a disk chipper, and might be less. See Plough and Koch (1983) for a discussion of this chipping technology.

DRIVE SELECTION

Choice of a chipper drive is generally between synchronous motors that do not slow down if properly sized, and wound-rotor motors which may slow under heavy load but are cheaper than synchronous motors. Synchronous motors must be sized for the largest log to be chipped; they can be direct coupled without flywheel or belt drive, and because they do not lose speed can be used on chippers fitted with blower vanes to transport chips. For the same reason they are suitable for long logs.

Wound rotor motors lose speed under heavy load and are fitted with flywheels to store energy; they are suitable for short logs and gravity discharge.

CHIP THICKNESS AND QUALITY

Engelgau (1977) found that chip thickness cut on a disk chipper is positively correlated with chip length. Thinner chips are obtained if the inclination of the incoming wood can be made steeper in gravity-feed chippers. Altering knife sharpness angle will also alter chip thickness.

Uniform chips with low fractions of overthick, undersized, and pin chips are favored by closely controlling direction and feed speed of wood entering and passing through the chipper, exact positioning of all knives in relation to the bed plate and disk, as small a knife sharpness angle (β in fig. 18-2) as possible while maintaining knife sharpness during the operating cycle, and moderate cutting speed (Hartler and Stade 1977). Engelgau (1977) suggests that for a 112-diameter chipper, a 9,000 fpm rim speed yields an unbruised chip of good quality without sacrificing required disk momentum.

CHIPPER TYPES

Disk chippers predominate in the industry and most of the discussion under this heading pertains to disk chippers. For interested readers, McKenzie (1970) briefly reviewed the advantages and disadvantages of several types of chippers other than disk chippers.

Chippers are designed specifically for the wood to be chipped, e.g., pulpwood bolts, long logs, sawmill residues, or veneer residues. Spout shapes are tailored to the raw material and may be rectangular, square, V-shaped, round, or modified round.

Chippers may be designed so that fan blades throw (and blow) chips to discharge them in an upward direction, in which case they can be blown a vertical plus horizontal distance not to exceed 60 feet. Ductwork should be arranged to avoid transfer of the chips within the flow path from one side to the other to avoid chip damage from impact with the ductwork. Alternatively, chips can be gravity discharged downward onto a conveyor, thereby reducing power requirement by 25 to 50 hp. Impact damage on downwardly discharged chips can be lessened by ejecting the chips at a glancing angle to the chip conveyor and by locating the conveyor a considerable distance below the chipper (Veuger 1976).

Cordwood chippers at pulp mills.—The usual pulpwood chipper for pine-site hardwoods receives wood in short lengths as it comes from the drum debarker. These machines usually have vertically inclined spouts, are gravity-fed, will accommodate sticks up to 24 inches in diameter, and discharge from the bottom. A typical machine carries 10 to 12 knives in a 104-inch disk that rotates at 400 rpm and is driven by a direct-connected 1,250-hp wound-rotor (or synchronous) motor. When efficiently fed, output of the chipper is approximately 40 to 50 cords per hour.

Longwood chippers.—At many locations in the South, long-log chip mills have been installed in conjunction with a ring barker (commonly 26 to 30 inches). Disk chippers for these mills are chain fed horizontally; typically, the 75- or 84-inch disk carries eight knives and is rotated at 500 to 450 rpm by a synchronous, 500- to 1,000-hp motor (fig. 18-260). Assuming that operation is reasonably continuous, output should be about 30 cords per hour if wood averages 6 inches in diameter.

Vee-drum chippers (figs. 18-259B and 18-276) coupled to ring debarkers are also available in a range of capacities; Bryan (1975) and Hensel (1971) described factory-assembled chip mills of this type.

Standard disk chippers have been modified to yield long "fingerling" chips needed by the growing flakeboard industry, but the literature through 1982 contains no descriptions of such operations. Erickson (1976) described a spiral-head chipper to make hardwood fingerling chips 2-1/2 inches long with cross-sections 1 inch by 1 inch or less (see fig. 18-27). The technology of making "fingerling" chips on disk and helical chippers is more fully described in proceedings of the symposium "Comminution of wood and bark" held October 1-3 in Chicago, Illinois; available from the Forest Products Research Society, Madison, Wisconsin.

Residue chippers.—Disk chippers for slabs and edgings usually carry three to six knives; smaller disks turn at higher rotational speed than larger disks to achieve comparable outputs. A mill equipped with saws and edgers (as contrasted to a mill with chipping headrig and chipping edger) that produces 10,000 bd ft of lumber per hour might chip all its residues in one chipper. Typically the 58-inch disk would carry six knives, turn at 720 rpm, and be driven by a 150-hp squirrel-cage induction motor. Such a chipper would normally produce about 15 tons of green chips per hour of mill operation, and be arranged for horizontal feed. Herrick and Christensen (1967) provided a guide for evaluating the economic feasibility of installing residue chippers in hardwood sawmills.

A veneer plant with two lathes could have three chippers. Cores and lathe spinouts might be chipped on a horizontal-feed, top-discharge, 66-inch, eight-knife, 250-hp, 600-rpm disk chipper. Waste veneer requires a special horizontal feed with crushing rolls; a typical installation would discharge from top or bottom, carry eight knives in an 84-inch disk driven at 500 rpm by a 250-hp motor. The trim ends of the veneer bolts (**lily pads**) require a special chipper that cuts chips in the 0-90 mode by an action illustrated in figure 18-259D. The 40- to 60-inch drum that carries the knives rotates at 205 to 100 rpm and is driven by a 30- to 150-hp motor. Productivity is in the range from 10 to 20 tons per hour.

Stump and rootwood present special problems during preparation for pulping (see fig. 17-16 and related discussion).

Rechippers.—Pulp chips may come from many sources other than pulp mill wood rooms. Sawmill residue chippers, chipping headrigs, and whole-tree chippers all furnish significant quantities of chips to southern pulp mills. Characteristics of chips vary with source, but most contain some oversize chips which are objectionable to the pulp mills. They are screened out and either recycled through the primary chipper or rechipped on equipment specially designed for the job.

Screening to separate chips by thickness is a rapidly developing technology, the description of which is beyond the scope of this text. Interested readers are referred to Steffes (1978) and Hatton (1976, 1977b and 1979b).

Disk chippers and other types of conventional chippers are employed as rechippers. Chip slicers (fig. 18-263), long used by particleboard manufacturers, were introduced during the 1970's to rechip oversize pulp chips.

Chip shredders.—Miller and Rothrock (1963) and Nolan (1967) have shown that for kraft pulping there are some advantages to be gained from shredding chips prior to digestion. In Nolan's experiments, conventionally cut chips at 40- to 45-percent moisture content were passed through a 28-inch Vertiflex attrition mill. The plates had teeth 3/4-inch high in the inner zone and 3/8-inch high at the outer periphery. Plate clearance for shredding was 0.900 inch, corresponding to a clearance of 0.525 inch between the tips of the teeth on rotor and stator. Rotor speed was 1,800 rpm. Feed rate at the mill was 1.2 tons (air-dry) of chips per hour, which was less than 10 percent of the capacity of the 100-hp unit.

Shredding increased the exposed surface of chips by splitting them along natural lines of cleavage without breakage across the grain and without crushing or otherwise damaging the fibers. The chief gains were: (1) high-yield pulps produced more easily; (2) chip screens eliminated; (3) knot breakers eliminated or operated lightly; (4) washing improved; (5) fiberizing power reduced; (6) pulp made cleaner; (7) cooking time reduced; and (8) digestion production increased. These experiments were made on slash pine chips.

Eastern hardwoods, including red maple and American elm, were studied by Temler and Bryce (1979) who also concluded that shredding chips reduced production costs for bleached kraft.

Portable chippers.—Because it is becoming increasingly difficult to find labor to harvest southern pine pulpwood in the traditional cordwood lengths, much effort has been expended to improve the processing of tree-length material. Some mills do not have large consolidated timber holdings, but must rely on small woodyards that may be 100 to 150 miles away. Such chip users may find it

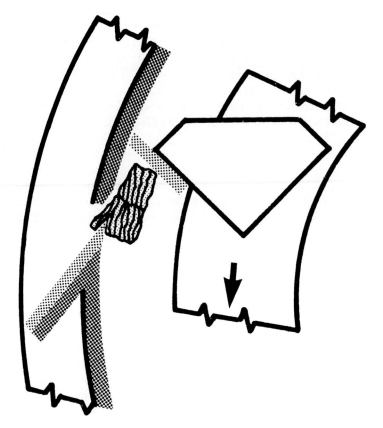

Figure 18-263.—Principle of operation of one design of chip slicer. The central rotor uses centrifugal force to orient and hold the chips before slicing. (Drawing from Rader Companies, Inc.)

economically feasible to use mobile chip mills to process tree-length wood.

Several designs have been developed and are in use in the South, including some that are self-mobile with all the components on one chassis. A hydraulic loader places long logs in the infeed conveyor, which carries them at speeds up to 100 fpm through an 18-inch ring barker and directly into a close-coupled chipper (disk fig. 18-259A, or vee-drum fig. 18-259B). The chipper is fan discharged to a chip truck. In locations where bark has high value, it may go to a pulverizer (bark hog)—also equipped with a fan discharge. Total power for the unit is commonly 300 to 600 hp. Production capacities vary, but 15 to 20 cords per hour on logs having an average diameter of 6 inches is considered attainable. Altman (1975) described a southern hardwood operation in which 200 to 240 tons of green chips are produced daily. The features, performance, and cost of seven different designs of mobile chip mills have been reviewed by Grant (1967).

As discussed in chapter 19, southern pulp mills are accepting increasing amounts of hardwood chipped with bark in place. In-woods, whole-tree chippers are discussed in text accompanying figures 16-18 and 16-45. Mills' (1980) review of whole-tree chipping indicated that use of whole tree chippers is increasing and that in 1979 a total of 603 such machines were supplying pulp mills of North America.

It is possible that demand for fuel chips will grow sufficiently to warrant use of small, hand-fed, trailer-type chippers suitable for hardwood brush and small stems (figs. 16-56, 18-275). Swan (1971b) described a typical machine with 10- by 12-inch, or 10- by 16-inch throat sized for towing by a pick-up truck and powered with a small gasoline engine. At least two men are required for its operation: one to feed the machine, a second to cut brush, and (sometimes) a third to assist in cutting and feeding. These machines will chip small brush about as fast as they can be hand fed.

Readers interested in production of fuel chips with tractor-mounted chippers will find useful Johansson's (1979) comparison of four such machines. For operating data on a tractor powered "fingerling or block chipper" see Hakkila and Kalaja (1981). For a profusely illustrated and comprehensive discussion of producing chips in the forest, see: (Hakkila, P. 1984. Forest chips as fuel for heating plants in Finland. Folia Forestalia 586. 62 p.).

Crossley and Sullivan (1980) provided specification sheets—most with illustrations—of 76 mobile chippers manufactured in Europe and North America, classified as follows:

Mobile chipper class	Designs described
	Number
Brush chipper	4
Three-point hitch mounted as a tractor implement	20
Trailer-mounted, light weight	9
Trailer-mounted, medium weight	8
Truck- or trailer-mounted heavy weight	30
Terrain chipper (incorporates a chip box or dumpable hopper on a forwarder)	5

Swathe-felling mobile chippers.—Mobile machines that chip as they move through the forest are a new development stimulated by the need for fuel chips. Operations and features of two such machines are described by the text accompanying figures 16-52 and 16-53.

Chipping headrigs.—The application of chipping headrigs to conversion of softwoods greatly changed chip procurement patterns in North America (see chapter 3 in Hatton 1979a). Hardwoods, however, are less easily converted with chipping headrigs than softwoods (see sec. 18-9). It seems likely, however, that the shaping-lathe headrig (figs. 18-104ABCD and 18-252) will find increasing application in the conversion of southern hardwoods; these machines yield flakes (fig. 18-102) that could be useful for pulp as well as in flakeboard. (See Plough and Koch, (1983) for a description of the technology of cutting pulp chips on a shaping-lathe headrig.) Also, end-milling chipping headrigs (fig. 18-105) will probably be increasingly used.

CHIP SPECIFICATIONS

Hatton (1978c) reviewed chip quality parameters whose determination could form a comprehensive new evaluation and specification procedure, the development of which he thinks is timely. He suggests that chip parameters routinely monitored should include moisture content, bulk density, bark content, and chip size classification including chip thickness.

N. Hartler and Y. Stade, in chapter 13 of Hatton (1979a), summarize chip specifications for various pulping processes, as follows:

> In **kraft pulping** chip thickness is the most important parameter determining chip quality. Instead of nominal thickness, the effective thickness should be used. The latter is markedly lower than the former due to lamellation, which results in the measurement of a falsely high thickness, and to cracks which result in a shortening of diffusion distances.
>
> In **sulphite pulping** the most important chip parameters are chip length, thickness, and chip damage. In order to counteract the negative effects of long chips, extensive impregnation-promoting measures have to be taken.
>
> In **refiner-mechanical pulping** an undisturbed constant flow of chips into the refiner is most important. In order to achieve undisturbed feeding, chip dimensions must be kept constant. Furthermore the chips must have as high a moisture content as possible, the ideal situation being completely moisture-saturated wood chips.
>
> Chips to be used in **Kamyr continuous digesters** must have a fairly low content of pin chips and undersize in order to secure a sufficiently high flow of cooking liquor through the circulation system. For pin chips and undersize, there are certain percentage-content levels which must not be exceeded.
>
> Regardless of the pulping process used the following chip requirements are recommended: a constancy in chip density, a low bark content and a very low content of hard impurities like sand. For chips purchased at the full contract price the tolerance limit is zero, but an economic optimization of the total production unit results in a bark content related to specific conditions in a given mill. Hard impurities like sand are particularly undesirable in the production of refiner-mechanical pulp and high yield pulps with in-line refining, due to the wear the impurities cause on the refiner plates.

18-25 FLAKING

Growth of the structural flakeboard industry in North America has stimulated interest in machines to cut flakes from roundwood, veneer cores, slabs and edgings, and chips. Flakes for hardwood structural flakeboard are generally 0.01 to 0.03 inch thick; those designed for board surfaces may be about 0.015 inch thick, while those in the core might average 0.025 inch thick. Length of the flakes is usually in the range from 3/4-inch to 3 inches; strength and stiffness of boards are enhanced by using long flakes in face layers. Width of flakes has much less effect on board mechanical properties than either thickness or length; flake width is difficult to control but is generally 0.2 to 1.0 inch.

From more than a decade of research at the Southern Forest Experiment Station in Pineville, La., it is concluded that the ideal face flake for structural flakeboard from pine-site hardwoods should be flat, cut smoothly in the 0-90 direction (fig. 18-1) and with no cross grain, cleanly cut to length without end-bruising (ends tapered in thickness are preferable), about 0.015 inch thick, and about ⅜-inch wide (fig. 18-264). Flakes of good quality are more readily cut from green wood than dry, and are more easily cut from heated wood (see table 18-77 for optimum temperatures) than cold.

Few, if any, flaking machines can duplicate the quality of a carefully cut veneer flake (fig. 18-264), but some approach it.

Figure 18-264.—These near-optimum flakes for use in face layers of hardwood structural flakeboard were smoothly rotary-peeled 0.015 inch thick from a water saturated and heated sweetgum log, scored to 3-inch length, and clipped to 3/8-inch width.

TYPES OF FLAKING MACHINES

Most flaking machines for roundwood, veneer cores, and slabs and edgings can be classified as disk flakers or drum flakers; the shaping-lathe flaker is a special category of drum flaker. Chips may also be sliced into flakes by a variety of ring flakers.

Less important types of flakers, not further discussed, include ring-head planers (Heebink 1975), helical cutterheads to smooth surfaces left by headrig or edger chipping heads (Koch 1976a; 1978d), knife-planers that cut across the grain with straight knives (Stewart 1970b; 1975a, fig. 18-47), or helical knives (fig. 18-55).

Disk flakers.—Disk flakers carry radially disposed castellated knives made with alternatively high and low sections arranged to cut flakes of controlled thickness and length from short roundwood bolts, slabs, or edgings presented sideways to the disk. The disk may be vertically oriented as in figure 18-265, or horizontally placed to permit hopper feed (fig. 18-266). A typical 24-knife vertical-disk flaker measures 114 inches in diameter, is arranged to accept 36-inch-long bolts, and is driven at 1,200 rpm by a 350-hp motor. The manufacturer estimates it will produce 65 tons of 0.020-inch flakes (ovendry basis) per 8-hour shift. The knives and feed are arranged so that cutting is in the 0-90 (veneer cutting) direction.

Figure 18-265.—Disk flaker measuring 93 inches in diameter. (Top) Hopper removed to show disposition of castellated knives mounted in the disk; flake length is determined by the length of cutting edge sections. (Bottom) Feed arrangement for 24-inch-long hardwood bolts. (Photo from CAE Machinery Ltd.)

Figure 18-266.—Horizontal-disk flaking machine. (Left) Door to feed pocket closed; wood is disposed horizontally. (Right) Feed pocket showing horizontal disk at bottom carrying removable knives and scorers; the scorers are normally spaced about 3/4-inch apart. Lugged feed chains form two sides of the feed pocket. (Photos from Albert Bezner Machinenfabrik.)

A typical hopper-fed machine with horizontal disk (fig. 18-266) carries a 15-knife, 90-inch-diameter disk and is driven at 374 rpm by a 160-hp motor. The feed chains are driven by a 5-hp motor at speeds from creeping to over 50 inches per minute, depending on wood type and flake thickness desired. Good hardwood flakes can be cut from round logs (to 16 inches in diameter), half-rounds, veneer cores, slabs and edgings, all cut to 15-inch lengths. Production data on southern hardwoods are not available, but output of 0.015-inch-thick flakes per 8-hour shift might approach 15 tons (ovendry basis). If flake thickness is increased to 0.025 inch, output per shift would increase proportionately to perhaps 25 tons. When flaking green hardwood, 3 to 5 percent of fines and 1 to 2 percent of splinters can be expected. Fines and splinters are automatically separated from the acceptable flakes and exhausted via a secondary outlet. One man can feed several machines if they are appropriately arranged.

For further discussion of disk flakers see: (Cramond, P.J. 1985. The CAE waferizer. In proceedings of the symposium "Comminution of wood and bark" held October 1-3, 1984 in Chicago, Illinois. Forest Products Research Society, Madison, Wisconsin).

Drum flakers.—The cutterhead on a drum flaker cuts flakes to length by one of two systems. Either the knives are castellated (made with short cutting-edge sections) and staggered in alternate knife slots (fig. 18-99 top), or continuous knives are used in conjunction with spaced scoring cutters. Figure 18-267 illustrates three configurations of drum flaker. In each case the grain of the workpiece is parallel to the drum axis, and the feed direction is radial to the drum.

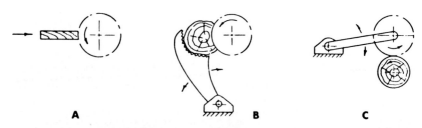

Figure 18-267.—Configurations for flake-cutting machines of the drum type. A. Utilizes slabs and edgings. Flake width is determined by the thickness of the workpiece. B. Utilizes roundwood. C. Utilizes roundwood. Because the workpiece is revolved slowly between centers as cutting proceeds, resulting flakes are tapered on both edges.

Hopper-fed drum flakers for cut-to-length wood operate in a manner similar to that shown in figure 18-266 except the wood is fed against a drum having short knives (sometimes equipped with spurs to fix flake length), helically disposed on its periphery (fig. 18-268). A typical machine admits green debarked roundwood bolts 4 feet long that have been heated in hot water to optimum flaking temperature. The manufacturer indicates that such a machine equipped with a 350-hp motor should produce at least 5 tons per hour (ovendry weight basis) of hardwood flakes 0.015 inch thick. The flaking drum measures about 2 feet in diameter and typically rotates at 1,000 or 1,250 rpm.

Figure 18-268.—Hopper-fed drum flaker for 4-foot-long roundwood. The mouth of the hopper is at the top of the machine at the rear, where wood can be conveyed sideways to it. The cutterhead at bottom, shown with cover removed, is equipped with built-in automatic wrenches to facilitate rapid knife change. Flakes are discharged downward. (Photo from Albert Bezner Machinenfabrik.)

Drum flaking of long random-length logs requires a different infeed procedure. Figure 18-269 illustrates a design in which groups of long random-length logs (or bundles of slabs and edgings) are fed endwise against a stop; the drum-type flake cutter is mounted on a travelling carriage which advances the cutterhead toward and through the bundle of wood. In a typical design the cutterhead,

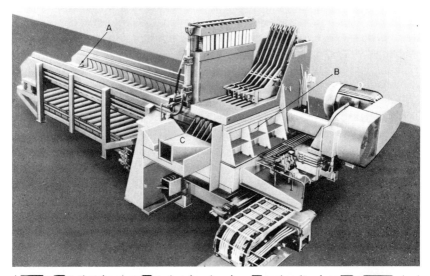

Figure 18-269.—Drum flaker for random-length long logs or bundles of slabs and edgings. (Top) Machine layout showing log infeed, A, flaking chamber, B, and flake discharge, C. (Bottom) View of log stop, D, terminating endwise movement of log bundle, travelling cutterhead, E, and fixed backstop, F. (Photos from Coe Manufacturing Company illustrating Hombak design.)

which requires 700 hp to drive it at 750 rpm, is 39½ inches in diameter and 58¼ inches long. In one traverse of the cutterhead, the log bundle is reduced in length by 58½ inches, whereupon the cutterhead is retracted and the bundle advances against the stop. The drum carries 96 serrated knives (usually with cutting edges ¾-inch to 1¼ inches long) disposed helically about the drum. The largest log that can be admitted is 29½ inches in diameter. When cutting hardwood flakes 0.020 inch thick from random-length hardwood logs, output of flakes per 8-hour shift is about 120 tons (ovendry basis).

Steiner's (1972) review should be useful to readers needing a summary of the technology of flaking long logs. Sybertz (1985)* provided another review of the technology of drum flaking.

Shaping-lathe headrig.—Using the principle shown in figs. 18-185 and 18-267C, the shaping-lathe headrig produces flakes of high quality (fig. 18-102) while shaping logs into cants to be resawn (fig. 18-104ABD), crossties (fig. 28-21 top), or cylinders to be rotary peeled (fig. 18-104C and 18-252).

A shaping-lathe headrig arranged to make hardwood cants 36 to 52 inches long should carry 300 hp on the cutterhead and will produce 16 to 26 tons of flakes per 8-hour shift (ovendry basis) plus substantially more than this weight of cants (see sect. 28-17 and 28-26).

A shaping-lathe roundup machine arranged to make 52-inch-long cylindrical oak peeler bolts might carry 200 to 250 hp on the cutterhead and will produce about 5 tons of flakes per 8-hour shift plus about 2,500 cylinders (see sect. 28-26 and 28-31).

Ring and cone flakers.—These machines are used to slice large chips into flakes; ideally the chips should be of "fingerling" size, i.e., 2 or 3 inches in length along the grain and about 1/2 to 1 inch in thickness and width. The cutting action is in the 0-90 direction (veneer cutting direction). They are made in several configurations.

In one arrangement, the incoming wood chips are centrifugally thrown by a high-speed rotor against an outer cutter cage that is stationary within a housing (fig. 18-270). Thickness of flakes is altered by adjusting the stationary knives in relation to the rotor and cage. A typical machine of this design might carry a 26-inch rotor turned at 1,850 rpm by a 45-hp motor and would produce approximately 1,760 to 2,200 pounds per hour (ovendry basis) of flakes suitable for flakeboard cores. A smaller machine of this type is also available for pilot plants. It has an output of 440-660 pounds per hour. The rotor is driven at 3,200 rpm by a 25-hp motor.

In a variation of the foregoing arrangement, both rotor impeller and cutter cage revolve, but in opposite directions—each driven by a separate motor. In this design (fig. 18-271), incoming coarse chips (green or dry) hit the rotating impeller and are thrown to the periphery where they are reduced to thin flat flakes by the action of the counter-rotating cutter cage. Adjustment of the knives in the cutter cage controls the flake thickness (0.012-inch minimum to 0.060-inch maximum). Cutting speed is approximately 15,000 fpm. On a large 42-

---

* Sybertz, H. 1985. Drum flakers for cut-to-length and random-length roundwood. In proceedings of the symposium "Comminution of wood and bark" held October 1-3, 1984 in Chicago, Illinois. Forest Products Research Society, Madison, Wisconsin.

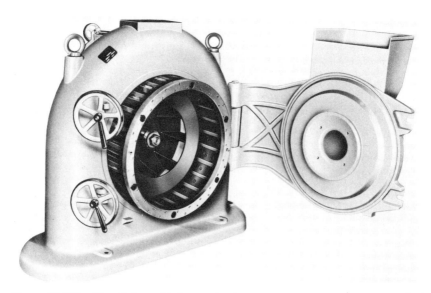

Figure 18-270.—Ring flaker with knives fixed in a cutter cage. Cutter cage pulled forward in this view. (Photo from Condux-Werk.)

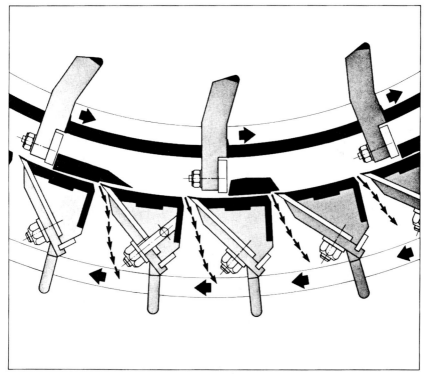

Figure 18-271.—Ring flaker with counter-rotating cutter cage. (Drawing after Pallmann Pulverizer Company, Inc.)

knife machine of this type, the impeller would be driven at 980 rpm by a 300-hp motor and the cutter cage would be counter-rotated at 50 rpm by a 30-hp motor. When fed with chunks or chips of wood measuring approximately 2 inches by 1 inch by 5/8-inch, the manufacturer states that the machine should turn out about 5 tons per hour of flakes measuring 0.015 inch thick or 6-1/2 tons per hour of 0.020-inch flakes. The foregoing production figures are based on softwoods, ovendry weight basis. Production rates on hardwoods should be somewhat greater. Readers interested in flakers of this type will find Fischer's (1974) description useful.

Ring flakers are made in conical configuration as well as in the cylindrical arrangements just described. As in all ring flakers, the tip of each rotating vane is equipped with a counter-knife that pushes the material past the cutting knives which project inwardly beyond the wearplates a distance approximately equal to the thickness of the flakes produced. The flakes flow outwardly through the knife slots and out into the surrounding casing from whence they drop into a negative-pressure conveyor system. The conical ring flaker (fig. 18-272) differs from the cylindrical ring flaker in that the inside of the knife ring and the outside of the rotor have the shape of a frustrum of a cone. This conical design permits ready adjustment of the clearance between knives and counter knives, by advancing or backing the rotor, to reduce or increase clearance. To produce thin flat flakes with minimum oversize, Plough (1974) found that knife-to-counter-knife clearance must be held as small as possible—preferably 0.030 inch or less. In this design, the rotor is driven by a 150-hp motor; the knife ring does not revolve. When cutting oak, flake production should be about 6,170 pounds per hour of 0.012-inch-thick flakes or 15,650 pounds per hour of 0.032-inch flakes (ovendry weight basis).

POWER REQUIREMENTS

Specific cutting energy, i.e., the energy to cut a cubic foot of wood into particles, is usually greater for flaking than for chipping.

Comparisons among machine types is difficult, but Sybertz (1974) reported comparative data for hardwoods and softwoods mixed. He found that when cutting core flakes 0.016 to 0.020 inch thick from wood of 0.45 specific gravity and moisture content of 30 to 50 percent (green weight basis), specific cutting energy differed according to machines selected for the job. Of the three examples given, specific cutting energy was least when random-length wood was cut into 8- or 9-foot lengths and flaked on a drum flaker (17 kWh per ton—ovendry basis—flaked). Next lowest was flaking of random-length wood with a reciprocating drum flaker head on a pocket flaker (fig. 18-269) with specific cutting energy of 21 kWh per ton (ovendry basis) flaked. Requiring most specific cutting energy was the system of chipping random-length logs followed by reducing the chips to flakes in a ring flaker (40 kWh per ton—ovendry basis—flaked); of this total 13.3 kWh was required to make the chips and 26.7 kWh to reduce them to flakes in a ring flaker.

Machining

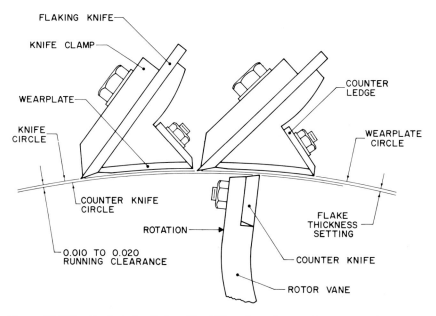

Figure 18-272.—Conical ring flaker. (Top) View through open front cover of rotor in place; the knife ring has been removed and is suspended on the crane hook. (Bottom) Cross-section through rotor and knife ring. (Photo from Black Clawson, Inc.; drawing after Plough 1974.)

The shaping-lathe headrig, which is a type of drum flaker in which the wood being cut is under precise position control (fig. 18-104ABCD), requires significantly less cutterhead power to flake than any of the three foregoing examples. From figure 18-100 it can be computed that only about 9 kWh are required to cut a ton (ovendry basis) of hardwood into flakes 0.016 to 0.020 inch thick from wood averaging 0.45 specific gravity:

i.e., $\dfrac{10 \times .746 \times 2{,}000}{60 \times .45 \times 62.4}$

= 8.9 kWh/ton).

Plough's (1974) data for a conical ring flaker with freshly sharpened knives and counter knives indicate very low specific cutting energy—about 6 kWh per ton (ovendry basis) of 0.016-inch-thick flakes cut from green Doulas-fir chips—when operating with a clearance of 0.030 inch between counter knives and cutting knives (fig. 18-273). To this value, the specific energy to chip the wood prior to its introduction into the flaker must be added, however. The total of chipping and flaking specific energy might therefore total about 20 kWh per ton of ovendry flakes—possibly more.

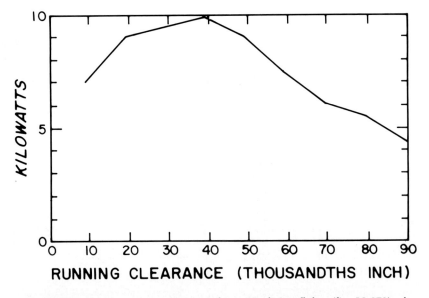

Figure 18-273.—Net power requirement of a conical ring flaker (fig. 18-273) when cutting 3 tons per hour of green Douglas-fir flakes 0.016 inch thick, with clearance between counter knives and cutting knives of 0.01 to 0.09 inch. (Drawing after Plough 1974.)

FLAKE QUALITY

Modulus of rupture (bending strength) and modulus of elasticity (stiffness) of hardwood structural flakeboards are very sensitive to the quality of flakes in the face layers. Flakes illustrated in figure 18-265 are near optimum for use in face layers.

Approaching this quality are flakes made on the shaping-lathe headrig (fig. 18-102). Properly adjusted disk flakers and drum flakers produce flakes resembling those shown in figure 18-102. Flakes made by slicing chips and "fingerlings" in ring flakers are not as well suited for face layers in structural boards, but are much used in cores of three-layer boards (fig. 18-274ABC).

Price and Lehmann (1978) compared three-layer structural flakeboards of sweetgum, southern red oak, and mockernut hickory made from 2-1/4-inch-long, 0.020-inch-thick flakes cut from green wood on a shaping-lathe headrig and on disk, drum, and ring flakers. The ring flaker produced 23.8 percent fines (from chips 2-1/2 inches long); the drum, lathe, and disk produced only 7.7, 3.5, and 2.2 percent fines (fig. 18-274ABC).

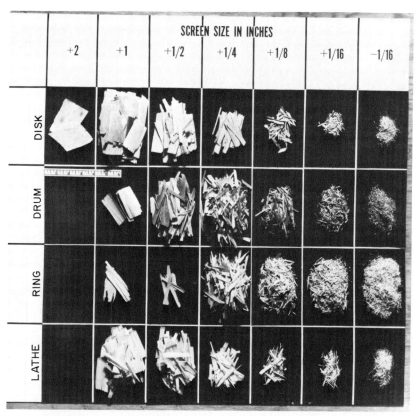

Figure 18-274A.—Sweetgum flakes from each of four flakers illustrating proportions retained on each screen. (Photo from Price and Lehmann 1978.)

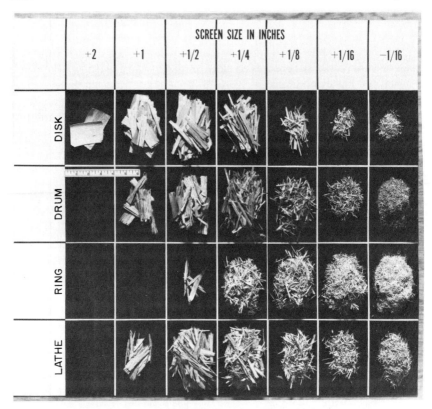

Figure 18-274B.—Southern red oak flakes from each of four flakers illustrating proportions retained on each screen. (Photo from Price and Lehmann 1978.)

When oak and hickory boards were evaluated, those made from shaping-lathe-cut flakes had highest bending strength and stiffness. Internal bond strength was highest in boards made of ring-cut flakes. Linear-expansion values were low in panels of drum-cut flakes pressed to high compression ratios and for boards of lathe-cut flakes at low compression ratios.

At high compression ratios, sweetgum panels had highest bending strength and stiffness when made from disk-cut flakes but highest internal bond strength when made from ring-cut flakes. Sweetgum boards were most stable when fabricated from lathe-cut flakes.

Figures 24-4 through 24-12 graphically portray these comparisons for oak, hickory, and sweetgum flakeboards.

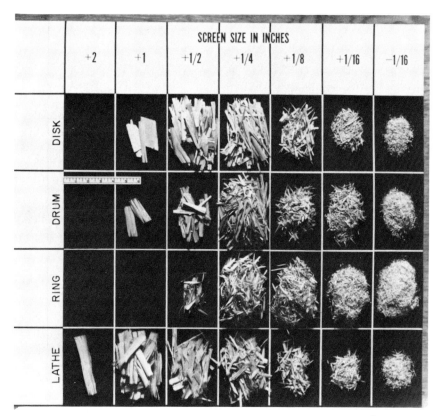

Figure 18-274C.—Mockernut hickory flakes from each of four flakers illustrating properties retained on each screen. (Photo from Price and Lehmann 1978.)

18-26 HOGGING

Comminution machines, i.e., **hogs**, for wood, bark, branches, and roots are designed to reduce pieces of random dimensions to particles of specified smaller size. In contrast with machines to make pulp chips or flakes, hogs require no special orientation of the workpiece other than that dictated by the infeed spout configuration. Most hogs damage some fibers by bruising them, and shorten others because of lack of control over the minimum-size particle produced.

Selection of a hog and assessment of power requirement depends not only on the throughput desired, but also on the material to be hogged (and its moisture content), the type of hog, and the end product desired. An array of some of these factors follows:

Material to be hogged	Type of machine	Product desired
Standing cull trees	Knife hog	Forest mulch
Logging residue	Hammer hog	Fuel for:
Branches	Punch-and-die hog	Grate-type burners
Roots	Disk mill	Suspension burners
Bark	Ring mill	Gasifiers
Mill-yard debris	Double-rotor hog	Fluidized-bed burners
Slabs and edgings	Horizontal-disk hammer hog	Feedstock for:
Chunks	Roll-crusher	Charcoal manufacture
Chips	Ball or rod mill	Liquid fuel production
Sawdust	Chain flail	Pulp mills
Shavings	High-shear screw mixer	Fiberboard plants
Veneer clippings		Particleboard plants
Particleboard trim		Briquette manufacture
Log ends		Wood flour
Baled material		Bark particles for:
		Drilling mud
		Fuel
		Mulch
		Growing medium

The literature affords little quantitative data on power requirements for comminution of wood and bark. In general, specific energy to comminute is inversely correlated with particle size, i.e., much more energy is required to produce a ton of wood flour than a ton of coarse hog fuel from forest residues. Energy consumption is also lessened if infeed systems are devised so that comminution machines operate steadily at near full capacities. Knife hogs require less specific comminution energy for a given particle size than hammer hogs. Readers needing an introduction to energy requirements of comminution machines will find useful Jones' (1981) analysis of the problem. See also: Proceedings of the symposium "Comminution of wood and bark" held October 1-3, 1984, in Chicago, Illinois; available from the Forest Products Research Society, Madison, Wisconsin.

Selection of a machine appropriate for both the material to be hogged and the product desired can be simplified by first considering some product characteristics and then discussing machine characteristics. Regardless of machine utilized it is usually prudent to eliminate metal and rocks from material before it is hogged. Waller and Halgrimson (1980) have described a device for this purpose. (See also fig. 17-15.)

HOGGED WOOD PRODUCTS

Forest mulch.—Standing cull trees, logging residues, and unwanted brush may be reduced to chips or chunks and spread as mulch on the forest floor. The size and shape of particles constituting such mulch is not critical. Mobile hammer hogs (both vertical and horizontal), knife hogs (figs. 16-52 and 16-53), and chain flails (Upton 1976; Helgesson 1978) have been used for on-site comminution of forest residues into mulch. Procedures will not be further discussed in this text section; interested readers can obtain additional information from the following references:

Reference	Subject
Cammack and Richardson (1980)	Mobile hammer-type flail with horizontal axis
Gallup (1976)	Review of clearing, grubbing, and slash disposal
Gardner and Gibson (1974)	Utilization and disposal of logging residue
Harrison (1975)	Treatment and utilization of logging residues
Lambert (1974a)	Power requirements and blade design for slash cutting machinery
Koch and McKenzie (1976)	Mobile mulcher
Koch and Nicholson (1978)	Swathe-felling mobile chipper
Koch and Savage (1980)	Swathe-felling mobile chipper
McLean (1974)	Techniques for forest residue reduction
Sickeler (1974)	Mobile hammer-type flail with horizontal axis of rotation
U.S. Department of Agriculture, Forest Service (1970)	Field trials of a tree and brush masticator

Fuel.—Industrial fuel wood is used in sizes ranging from brick-size chunks to finely-divided dust and at moisture contents from ovendry to half water by weight. Consequently, any of the machines listed may be used during its preparation, with the possible exception of chain flails, roll crushers, ball or rod mills, ring mills, and disk mills. Hammer hogs and punch-and-die hogs probably predominate. Systems for fuel preparation, storage, and reclaim are described in section 16-17 and by figures 16-66 and 17-14.

Feedstock for chemical, fiber, charcoal, and briquette plants.—Pulp chips are made primarily on machines equipped with knives. Particles for board plants generally are sized in ring or disk mills. Wood for productin of briquettes, charcoal, or liquid fuels is usually sized through a hammer hog or punch-and-die hog. Rootwood might be roll-crushed before comminution on a hammer or punch-and-die hog.

Wood flour.—The manufacture of wood flour may commence with punch-and-die hogs; final milling is achieved with hammer hogs and disk mills; for very fine grinds, ball or rod mills could be used. For a further discussion of this technology, see the sub-section heading GRINDING WOOD FLOUR, that follows in this section.

Bark for fuel and charcoal.—Bark that is not stringy is frequently sized through hammer hogs having vertical or horizontal axes, and through punch-and-die hogs. Stringy bark, such as that from spring-cut hickory and elm, can absorb much horsepower if hammered, and is best reduced through a knife hog; the dirt generally attached to bark causes rapid dulling of knives, however. An alternative appropriate for stringy bark is the double-rotor hog.

MACHINE TYPES*

Knife hogs.—Two categories of knife hogs are in common use. The first, a portable hog or chipper, is designed to dispose of brush, branches, and other woody residues measuring 6 inches or less in diameter (fig. 18-275). The four-knife cutterhead is driven at about 2,500 rpm and requires 20 to 110 hp, depending on the size of wood admitted.

Vee-drum knife hogs (figs. 18-259B and 18-276) are used to reduce green slabs and edgings to a particle size suitable for power plant fuel. The knives are staggered in the rotor so that the entire area under the infeed spout is swept by the knives. Particle size is altered by adjusting knife extension beyond the rotor (fig. 18-276 bottom). These machines are available with rotor diameters ranging from 24 to 60 inches. The smallest rotors carry 16 knives, are rotated at 1,500 rpm by a 25- to 50-hp motor, and can produce approximately 18 tons (green basis) of hogged fuel per hour. The largest rotors carry 36 knives, are rotated at 600 rpm by a 250-hp motor, and can produce approximately 60 tons (green basis) of hogged fuel per hour from oak slabs and edgings.

Hammer hogs with horizontal axes.—These hammer hogs (fig. 18-277) are probably the most widely used fuel comminution machines. Incoming wood or bark is reduced to smaller pieces by the impact of revolving hammers. Screens or spaced grate bars placed below the hammers limit the size of particles ejected from the machine. The hammer hog is particularly effective in reducing solid wood residues to fuel; it is less effective on fibrous material such as spring-cut hickory or elm bark. The machines are commonly available in a power range from 20 to 400 hp. On smaller models the rotor turns at 1,200 to 1,800 rpm, while on larger machines rotor speeds are generally about 750 to 900 rpm. Hammers and screens can be selected to reduce a variety of material (e.g., oversize chips, granular bark, pulp laps, or baled branches) to particles of desired size—from walnut size to coarse sawdust or finer. The literature contains little quantitative data on specific energy required by these machines to comminute wood and bark of various forms and moisture contents into particles of specified size.

---

* For a comprehensive review of hogging, and other comminution processes, see proceedings of the symposium "Comminution of wood and bark" held October 1-3, 1984 in Chicago, Illinois; available from the Forest Products Research Society, Madison, Wisconsin.

In one experiment (personal correspondence March 9, 1982), A. Wiley and W. K. Murphey found that bark freshly removed from sweetgum and red oak sp. logs by a rosserhead debarker required about 0.861 watt hour/ovendry pound (including idling energy) if reduced when green with a horizontal-axis hammer hog to particles sizes such that 80 to 90 percent would pass through a screen with 3/8-inch openings. Before hogging, about one-third of the bark was too coarse to pass through a 3/8-inch screen.

Hammer hogs with vertical infeed spouts predominate. For a discussion of horizontal-feed machines the reader is referred to Neubaumer (1971).

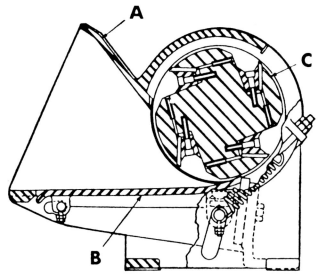

Figure 18-275.—Portable brush chipper in action, with cross-section. (A) Feed spout; (B) bed plate; (C) cutterhead. (Photo and drawing from M-B Company, Inc. of Wisconsin.)

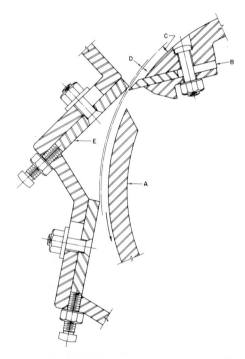

Figure 18-276.—Vee-drum knife hog. (Top) Arranged for top feed and bottom discharge; see figure 18-259 for perspective view of vee-cutterhead. (Bottom) Cross-sectional view showing relationship between knives, anvil, and rotor. (A) Rotor; (B) knife; (C) knife setting; (D) anvil clearance of about 1/16-inch; (E) anvil. (Photo and drawing from Black Clawson, Inc.)

Figure 18-277.—Cutaway model of hammer hog with horizontal axis. Hammers are pivoted. The breaker plate at left is top pivoted so clearance between it and the hammers can be adjusted. Milled material is bottom discharged. Compartment at upper right traps metal and rocks. (Diagram from Williams Patent Crusher and Pulverizer Company, Inc.)

Bark shredder with vertical axis.—These machines (fig. 18-278) are designed to reduce bark to a particle size suited for industrial boiler fuel; they can also pulverize occasional wood pieces. Bark to be shredded is introduced into a vertical hopper which directs the bark around the rotor into a zone where swinging knives fragment it by impact and shear (fig. 18-278 bottom). Size of emerging particles is determined by the number and size of the stationary anvils. Typically, bark is reduced to pieces that do not exceed 1 by 4 inches in size. The rotor turns at 750 to 950 rpm depending on model size with power and productivity reported by the manufacturer as follows:

Hopper diameter	Horsepower	Throughput of green bark
Feet		*Tons/hour*
		(green weight basis)
4	125	50
5	250	100

Throughput per hour depends on the material being shredded. A 125-hp machine arranged to produce particles that pass through a sizing screen having 2-inch openings might have a throughput of 50 tons of white oak bark per hour (green weight basis).

Punch-and-die hogs.—The general configuration of the punch-and-die hog is similar to the horizontal-axis hammer hog, but cutting action is distinctly

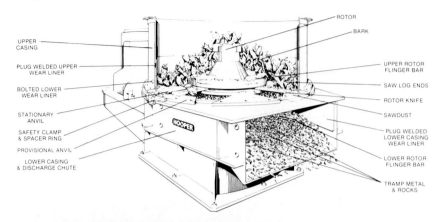

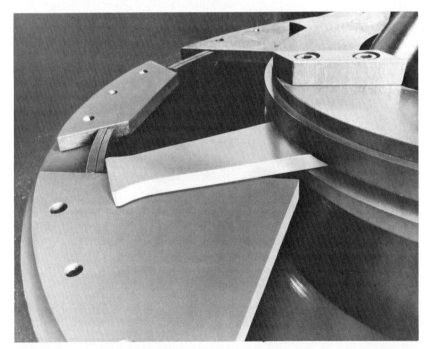

Figure 18-278.—Bark shredder with vertical axis designed to shred bark by impact and shear. (Top) Major features. (Bottom) Action of pivoted rotor knife swinging across stationary anvil. (Drawing and photo from S. W. Hooper Corp.)

different (fig. 18-279). High and low teeth (punches) alternate along the length of the rotor, closely meshing with notches (dies) in the anvil assembly. As material enters the hog the high teeth make first contact and punch out a piece the size of the tooth. The second stage action occurs when the material drops down between adjacent high teeth and the low teeth cut off the material left between the holes punched by the high teeth. Typically, a machine of this type having a 22-inch rotor turning at 1,175 rpm might be driven by a 100- to 250-hp motor. A

Figure 18-279.—Punch-and-die hog. (Top) Horizontal-rotor top-loaded machine used for light and medium duty grinding of bark, and lumber trim-ends. (Bottom) Rotor carrying alternating high and low teeth grinds material against corresponding stationary anvil parts. (Photos from Jacksonville Blow Pipe Company.)

similar machine with 36-inch rotor also turning at 1,175 rpm might be driven by a 250- to 500-hp motor. The punch-and-die hog is particularly adapted to the reduction of tough, stringy bark (e.g., hickory bark) as well as wood. Specific hogging energy is more for wood than for bark so that throughput when hogging wood is less than when hogging bark with equal power expenditure. In either case, hourly throughput is proportional to horsepower applied and inversely proportional to moisture content and size of screen openings.

Lambert (1974b) has described the flow plan and operation of a punch-and-die hog used to convert wood from construction and demolition sites, broken pallets, and mill residues into usable products.

Double-rotor hogs.—To accelerate the flow of bark through a hog, one design utilizes two opposed rotors revolving in opposite directions in the manner of a pair of meshing spur gears or a pair of squeeze rolls. Each rotor carries helically disposed short carbide-faced teeth and is independently driven at 1,800 rpm by a 20-hp motor (fig. 18-280). Bark enters the top of the hog through a 15- by 32-inch opening and is discharged from the bottom without passing through a screen. Specific hogging energy, throughput, and screen analysis varies with the species of bark pulverized (table 18-80a).

Very large double-rotor hogs with infeed hoppers measuring up to 12 by 15 feet are available for shredding logging residue, stumps, bark, and lumber from demolition sites. Such a machine driven by a 450-hp diesel engine might hog 40 to 90 tons of wood waste per hour. Resulting particles might typically measure 2 or 3 inches long.

Figure 18-280.—Double rotor bark shredder. One of the two toothed rotors is visible through the top opening. Bark enters at top and is discharged from the bottom without passing through screens. (Photo from Forest Tool Company.)

TABLE 18-80a.—*Specific hogging energy, throughput, and screen analysis of a double-rotor hog (fig. 18-280) when fed with fresh bark of summer-cut wood of three species*[1,2,3] (Data from Forest Tool Company, Birmingham, Ala.[4])

Statistic	Hickory	White oak	Sweetgum
Specific hogging energy, hp hr/ton, ovendry basis	----------2.7 to 4.0----------		
Throughput, tons/hr, fresh basis	------------15 to 20------------		
Throughput tons/hr, ovendry basis	------------10 to 15-----------		
Screen analysis, percent			
Retained on 2-inch screen	5	5	5
Retained on 1-inch screen	25	30	30
Retained on 1/2-inch screen	35	25	30
Through 1/2-inch screen	35	40	35

[1] Fresh bark from a mechanical ring debarker.
[2] Gap between teeth of opposed rotors set at 0.030 inch.
[3] Each rotor driven at 1,800 rpm by a 20-hp motor.
[4] Personal correspondence dated June 18, 1980.

High-shear screw mixers.—Screw mixers, much used in plastics manufacture, may be a key to achieving rapid reactions during chemical conversions of wood. These single-screw (fig. 18-280A) and double-screw (fig. 26-39) devices can convert wood chips to fiber on a continuous basis. Hemingway and Koch[12]

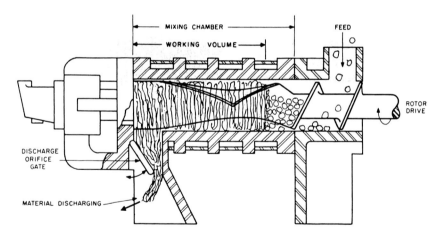

Figure 18-280A.—Cross section through a single-screw continous mixer. At 40 to 60 applied horsepower, and 500 to 800 rpm on the screw rotor, this size 4 mixer yielded about 400 pounds of fiber per hour from fresh whole-tree chips of southern red oak. (Drawing from Farrel Company, Ansonia, Conn.)

[12] Hemingway, R.W., and P. Koch. Comminution of whole-tree oak chips and southern pine bark with a Farrel continuous mixer and a Banbury mixer. Final Report FS-SO-3201-41 dated January 20, 1982. U.S. Dep. Agric. For. Serv., South. For. Exp. Stn., Pineville, La. 19 p. plus 20 figures. (See also Final Report FS-SO-3201-62 dated April 2, 1983.)

processed green (69.4 percent moisture content) southern red oak whole-tree chips with a single-screw mixer (fig. 18-280A) at an applied specific energy of 0.15 hp-hr/lb to produce a fibrous product of gently defibrated material with about 18 percent moisture content (ovendry-weight basis). Although a high proportion of large softly-pelletized particles were obtained from dry screening, Bauer-McNett fiber classification showed that the product was similar to some thermomechanical pulps.

Hardwood barks were not fed to the machine, but Hemingway and Koch found that processing southern pine bark through the mixer did not remove the large quantities of water present in wet bark obtained from outside fuel piles; the product contained about 40 percent coarse particles, primarily woody shives and some large clusters of sclerids.

For additional, and more recent, information see: Proceedings of the symposium "Comminution of wood and bark" held October 1-3, 1984 in Chicago, Illinois; available from the Forest Products Research Society, Madison, Wisconsin.

Disk mills.—Disk mills may operate with axes horizontal or vertical. The former are described in a following subsection GRINDING WOOD FLOUR and section 18-27 DEFIBERATING.

The toothed disk mill (fig. 18-281), with axes vertical, further reduces in size small wood particles such as shavings. A wide variety of particle forms, ranging from small flat flakes or strands for particleboard to wood flour, can be produced by varying the toothing and the spacing between disks. Incoming raw material is introduced into the top central hopper. The centrifugal action of the lower rotating disk causes the material to move outward through the progressively finer teeth until discharged at the periphery of the disk. This design permits wet material to be processed without clogging.

Wood meal can be produced from moist chips by using stone disks (fig. 18-281E) and allowing the heat generated in the process to convert the water content of the material to steam and thus soften the incoming chips. This softening action permits the production of wood meal suitable for filler in linoleum and plastics compounds. A somewhat longer fibered material can be similarly produced for use in loading cardboard or paper.

The mill illustrated (fig. 18-281 top) has a disk diameter of 1,000 mm. The lower disk is rotated at approximately 400 rpm by a 60-hp motor and can produce approximately 3,300 pounds per hour of flat chips for the center layer of particleboard. If driven by an 80-hp motor, the mill can produce 330-440 pounds of 100-mesh wood flour per hour. The foregoing figures are on an ovendry basis and are approximate; actual output depends on species, moisture content, and particle geometry of incoming material.

Ring mills.—These machines, sometimes called wing-beater mills or ring flakers are described by figures 18-270, 18-271, and 18-272 and accompanying discussion.

Ball and rod mills.—Ball and rod mills are little used in the comminution of wood. Readers interested in the effects of ball milling on wood properties are

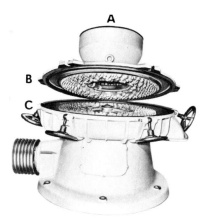
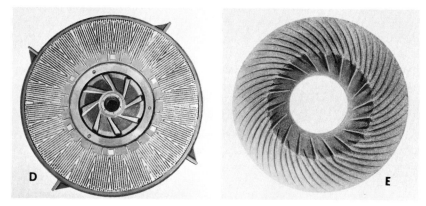

Figure 18-281.—Vertical-axes disk mill. (A) Infeed hopper; (B) top fixed toothed disk; (C) lower rotating toothed disk; (D) interchangeable bar-design disk; (E) interchangeable stone disk. (Photo from Condux-Werk.)

referred to Fujii (1964), Revilla-Perez (1972), Venkateswaran (1972), Mattson (1974), and Fengel et al. (1978).

Crushing rolls.—Harvey (1972) described how green hardwood slabs, edgings, or entire small stems can be converted to strand-like mats by crushing them in repeated passes between heavy opposed rolls (fig. 18-285). These crushed mats can then be knife hogged to yield strands of short length useful in reconstituted flat and molded products in which foamed urethane resin is the binder (fig. 24-63). Marra et al. (1975) further described the process. For an economic analysis, see sect. 28-14.

GRINDING WOOD FLOUR

Coarse wood flour (4 to 20 mesh), which closely resembles ordinary sawdust is commercially prepared and marketed as floor sweeping compound, animal bedding, packing medium, ground covering, roofing felts, and filler for some moulded items. Medium wood flour (finer than 20 but coarser than 80 mesh) is used for roofing felts, foundry moulding sands, and linoleum. Fine wood flour (in the range finer than 80 mesh but coarser than 200 mesh) is used as a filler in a variety of moulded plastics, as a filler for explosives, and to clean furs. Production of wood flour in the United States increased significantly in the years from 1958 to 1977 (fig. 22-70).

Unless it is important to retain a fibrous characteristic in the product, the wood waste admitted to the grinders should be relatively dry. A moisture content of 12 percent might be considered a desirable maximum as the heat generated in the grinding process can then, in all probability, bring the product to the 6 to 8 percent moisture content generally desired by the trade. Dry wood waste requires less power to grind to fine mesh sizes than green, and dry wood flour is more readily sifted. Flour from green wood cannot be sifted into fine mesh sizes as it plugs the screens.

Because of the explosive nature of a mixture of dry wood flour and air, wood flour systems must be very carefully designed for safety. Usual requirements would include explosion-proof motors, static-free V-belts, magnetic protection, explosion vents, and carefully engineered dust filters. Some processors use inert gas in the conveying systems as well as automatic devices to detect and suppress explosions.

Wood flour is made from both hardwoods and softwoods. Ponderosa pine (*Pinus ponderosa* Dougl. ex Laws) is most used in the United States, followed by eastern white pine (*Pinus strobus* L.), southern pine, and Douglas-fir. In general, the softwoods can be ground finer and require less grinding energy than hardwoods. To clean furs, wood flour comprised of a mixture of sugar maple (*Acer saccharum* Marsh.), beech (*Fagus grandifolia* Ehrh.), dogwood (*Cornus florida* L.), and birch (*Betula* sp.) is frequently used.

Southern hardwoods that find limited use in wood floor manufacture include sweetgum, black tupelo, red maple, beech, and cottonwood (*Populus deltoides* Bartr. ex Marsh.)—the latter ground to about 16 mesh for use in roofing felt.

Coarse flour.—A hammer hog (fig. 18-277) or a disk attrition mill (fig. 18-282) is suitable for making coarse wood flour of 4 to 20 mesh. Hammer mills

Figure 18-282.—Sectional view of double-disk attrition mill. (A) Infeed spout with metering device; (B) individually powered counter-rotating disks; (C) replaceable grinding plates; (D) adjustable gap between plates. Pressure on the plates is maintained by heavy coil springs that act as a safety device to allow plates to separate and discharge foreign material without damage to the plates. (Photo from The Bauer Bros. Co.)

produce a somewhat granular particle, while the product of attrition mills is softer and more fibrous. Hammer mills can tolerate a range in sizes of incoming wood waste; attrition mills require uniform material of small size, with no splinters exceeding 3 or 4 inches long.

A 60-hp, 3600-rpm hammer mill having a hammer velocity of 21,700 fpm and a 1/4-inch screen size might grind hardwood solid mill waste (11.8 percent moisture content) at the rate of 5,580 pounds per hour or 93 pounds per horsepower hour. The product of the mill might weigh 11.3 pounds per cubic foot and show the following screen analysis:

+20 mesh 53.5%	+ 80 mesh 1.0%
+35 mesh 29.5%	+100 mesh 2.3%
+48 mesh 6.2%	−100 mesh 2.7%
+65 mesh 4.8%	

By comparison, a 36-inch, double-disk attrition mill carrying two 75-hp, 1,800-rpm motors driving the plates shown in figure 18-283 might turn out 3,920 pounds of coarse flour per hour at a power demand of 92 hp. This amounts

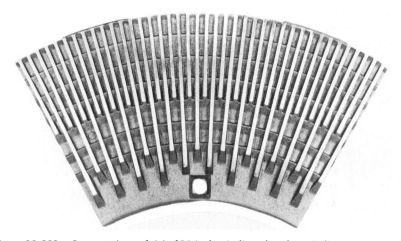

Figure 18-283.—Segment (one of six) of 36-inch grinding plate for grinding coarse wood flour on a double-disk attrition mill. (Photo from The Bauer Bros. Co.)

to a specific grinding energy of 47 hp hours per ton to coarse-grind previously reduced hardwood waste. The product of the attrition mill under these circumstances might show the following anlysis:

+ 8 mesh 2.9%	+ 48 mesh 12.8%
+14 mesh 16.2%	+100 mesh 15.7%
+20 mesh 5.7%	−100 mesh 9.6%
+35 mesh 37.1%	

Plates for each disc of a 36-inch mill are made in six identical sections (fig. 18-283) and are balanced in sets. As the ribs on the plates are radial, direction of rotation may be reversed to prolong plate life. Satisfactory results are obtained if the direction is reversed once each day.

The plates for each disc look very similar; however, the plate dams on the two sides do not match and result in a wavy line for material that passes between them. A plate clearance between the opposing discs of about 0.010 inch is quite commonly employed.

In contrast to a hammer mill with screen, an attrition mill has no built-in sizing device and is basically not a fiber shortening machine. For some coarse products the output of the attrition mill need not be classified, but for most medium grades of wood flour, classification of the product by screening is mandatory.

Medium and fine flour.—To make fine wood flour, it is advantageous to use two mills in series (fig. 18-284). Assuming that the first unit is a hammer mill, it might be equipped with a 1/8-inch or 1/4-inch screen. The second mill is fed from the cyclone collector of the first mill and discharges its product into another pneumatic collecting system, or into a screw conveyor with choke and dust-relief system. The latter design may be preferred because the tightly packed materials in the choke smother any sparks originating from passage of foreign material.

Figure 18-284 shows the recirculating path at the second mill. Since all oversized product is recycled, a given yield of a given size wood flour can be produced either by feeding the mill at a high rate, or at a slow rate with screens or plates set finer; power consumption is about the same either way. A higher feed

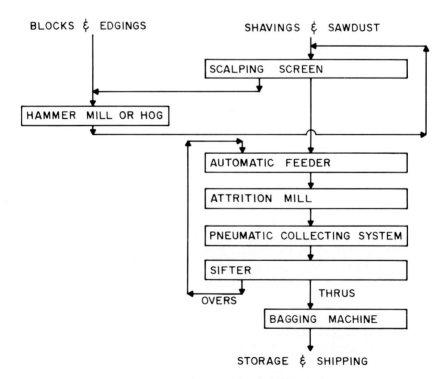

Figure 18-284.—Flow diagram for wood flour manufacture.

rate holds down temperature rise, but requires greater screening capacity to classify and return the oversize particles.

A typical operation producing an end product predominantly in the 60- to 70-mesh size might employ one or two men processing dry hardwood shavings through an initial 160-hp hammer mill and then through a 140-hp attrition mill, screened to turn out 10 to 12 tons of bagged flour per 8-hour shift.

For production of wood flours of 80 mesh and finer, attrition mills are frequently used as the second milling unit. In attrition milling the material is rolled and rubbed by the action of the plates while it has comparatively small lineal velocity. It is expelled primarily by centrifugal action. By suitable choice of speeds, disk diameter, grinding plates, and subsequent screens, the attrition mill can be made to produce flour in the quantity and quality desired.

Application of the attrition mill in the second stage of fine flour grinding may be illustrated by the following example. Dry hardwood mill waste, previously hammer milled to a fineness of 46.5 percent through a 20-mesh screen, when fed to a 36-inch double-disc attrition mill driven by two 75-hp motors at 1800 rpm used 102 hp to produce 1,950 pounds per hour; i.e., specific grinding energy was 105 hp hours per ton. Screen analysis of the product showed the following (in this example, retention on the 65-mesh screen was unusually high):

+20	1.4%	+ 80	6.4%
+35	6.7%	+100	11.6%
+48	9.3%	−100	47.3%
+65	17.3%	Total	100.0%

Most plants grind at higher rates than those indicated above. On a 36-inch double-disc machine it is possible to handle up to 8,000 pounds per hour through the mill, and power input may be as high as 200 hp. Output is, of course, dependent on the mesh grouping desired.

Grinding plate life varies with the type and cleanliness of the material being processed. In general, a set of 36-inch reversible plates will produce about 400-500 tons of 50-mesh flour before a change is necessary.

SCREENING

Screening of wood particles is a complex technology beyond the scope of this text. Readers needing an introduction to the subject are referred to figures 17-13, 17-14, and 17-16 and to:

Dilly (1975)	Steffes (1978)	Fuller*
Christensen (1976)	Hatton (1979a, Ch. 4)	Lynn*
Keating (1976)	Malito (1979)	Thireault*
Emanuel (1978)	Dessens*	

One method for removal of stones and metals from moving streams of chips is shown in figure 17-15.

---

* In proceedings of the symposium, "Comminution of wood and bark" held October 1-3, 1984 in Chicago, Illinois; available from the Forest Products Research Society, Madison, Wisconsin.

18-27 DEFIBRATING

The techniques by which wood is mechanically defibrated can be considered machining processes in which the characteristics of the separated fibers—rather than the contours of the wood substrate—are the properties of interest. Three mechanical defibration processes are in widespread use: stone grinding, steam-gun explosion, and disk refining.

STONE GRINDING OF ROUNDWOOD

Stone grinding, in which a bolt of freshly-cut wood is pressed against a revolving grindstone, was the first mechanical defibration process developed. It remains in widespread use, principally producing mechanical pulp from softwoods for mixture with chemical pulp in the manufacture of newsprint. See section 25-5, subsection STONE GRINDING, for a summary of data pertinent to pulping southern hardwoods by this process.

Because pulp yields by the stone groundwood process approach 100 percent (about double that for the kraft process), and because pine-site hardwood volume is increasing and hardwood cordwood is cheaper than softwood cordwood, efforts to develop acceptable stone groundwood pulps from dense hardwoods such as oak and hickory will likely continue. Although production of stone-ground pulp requires about twice the electrical energy needed for production of kraft pulp, it requires no heat. Readers seeking an introduction to the process should find the following references useful:

Texts describing machines and process parameters
    Perry (1950, chapter 3)   Gavelin (1966)
    Koch (1964a, p. 499-508)   Aario et al. (1980)
    White (1964, p. 231-251)
Experiments and reviews describing manufacture of stone groundwood from hardwoods
    Schafer and Pew (1942)   Keays (1975)
    Corbin and Slentz (1956)   Atack (1976)
    Swartz (1960)   McGovern and Auchter (1976)
    McCarty (1973)   Leask (1977)
    Namiki (1973)   Brecht (1979)
Other papers describing properties of stone groundwood and its manufacture
    TAPPI[12]   Paulapuro (1977ab)
    Höglund (1973)   DeVries et al. (1978)
    Rudstrom et al. (1973)   Dornfeld and Wu (1978)
    EUCEPA International   Dornfeld et al. (1978)
      Mechanical Pulping
      Conference, Helsinki (1977)

For a review of defibrating with pressurized stone grinders see: (Aario, M.I. 1985. Developments in stone grinding. In proceedings of the symposium "Comminution of wood and bark" held October 1-3, in Chicago, Illinois. Available from the Forest Products Research Society, Madison, Wisconsin).

STEAM-GUN EXPLOSION TECHNIQUE OF DEFIBRATING CHIPS

Developed by W. H. Mason in 1924, the steam-gun technique continues to be one of the most important processes by which mechanical fiber is manufactured for use in hardboard. The Masonite fiberboard plant at Laurel, Miss., is a major consumer of pine-site hardwoods, which it converts to fibers by this and other processes.

Each defibration cycle is initiated by introducing screened wet chips to a vertical cylindrical pressure vessel. The steel cylinder, or "gun," is then closed, and high pressure steam is admitted. This treatment causes the lignin in the intercellular substance to soften. Steam pressure is raised over a predetermined time interval to 300-600 psi and is then raised quickly to a terminal pressure of 600-1200 psi. In one example of a cycle, steam pressure is maintained at 350 psi for 30 to 40 seconds and is then increased to 1000-1200 psi for 5 seconds. During this brief period each chip develops high internal pressure, which is instantly released by opening a quick acting valve on the bottom of the cylinder. The chips emerge with explosive force and defibrate into fibers and fiber bundles. Steam and fibers are separated in a cyclone, and the fibers drop into a stock chest for subsequent further separation of the fibers and fiber bundles in a revolving-disk mill. Pollution of processing water can be reduced if time and pressure of steam treatment are reduced. Fibers separate at the intercellular layer or between the primary and secondary walls. Fibrillation of the individual fibers during refining is limited by presence of most of the original lignin content.

See section 23-6 for additional discussion of this process and resulting fiberboard products.

DISK REFINING OF CHIPS

Resource managers seeking commodity uses for pine-site hardwoods are frequently thwarted by the small size and low quality of the stems, and by the diversity of density, anatomy, and color of wood of these species. One method of homogonizing the raw material involves chipping the wood (sometimes with bark included) and defibrating the chips in a **disk refiner**. Metering wet chips of the major species groups to disk refiners in mixes with carefully controlled species proportions can produce mechanical fiber with predictable properties.

Even with control of species mixes, however, resulting fibers may be difficult to use in some major products. Pine-site hardwoods are poorly suited for paper-grade mechanical pulp made on disk refiners because of their short fibers, high proportion of vessels, and color. The vessel elements of hardwoods—especially the short, wide, barrel-shaped type present in oaks—do not bond readily into a fiber network, and when present on the surface of print-grade papers cause picks on the printing press from tacky inks (Byrd et al. 1967).

There is a large and rapidly growing literature, however, related to semichemical processes involving the use of disk refiners to make paper-grade pulp from hardwoods. Chapter 25 provides a review of manufacturing techniques, properties, and markets for mechanical and semichemical pulp from hardwoods that grow on southern pine sites. Keays and Leask (1973) reviewed the status of refiner groundwood for pulping and the EUCEPA International Technical Pulping Conference in Helsinki (1977) provided further review data. Stevens (1980) presented examples of refiner system selection. Leask (1981) reviewed the theory of chip refining.

For many fiberboard products, the pine-site hardwoods are well suited. For others—e.g., prefinished exterior house siding—inclusion of large proportions of oak in surface layers diminishes acceptability and durability of factory-applied exterior finishes. This, and other aspects of fiberboard manufacture are discussed in chapter 23. Section 28-30 summarizes an economic analysis of manufacture of hardboard siding from disk-refined southern hardwoods.

In spite of the problems involved in using mixed pine-site hardwoods for products made of mechanical fiber, the disk refining process has so much promise for solutions that essentials of the technology are briefly described in the following paragraphs and in section 23-6.

Disk refiners may incorporate either one or two revolving disks. In a double-disk refiner (fig. 18-282 and sect. 23-6), wet chips or other fibrous materials are fed in at the center of the mill and fiberizing is accomplished as the materials pass outwardly between the disks which revolve in opposite directions. Single-disk machines are similar except that only one disk revolves, the other being stationary. Clearance between the plates is adjustable in increments of 0.001 inch up to a maximum of 3 inches, but during operation is generally in the range from 0.001 to 0.050 inch.

In theory, there would appear to be a real difference in the performance of a single-disk machine compared to a mill in which both disks revolve. A ball placed between double-revolving plates will receive equal but oppositely directed forces from each plate; it will roll rapidly but will not travel radially outwardly except as affected by gravity and the forces incident to the feeding of additional material. The only centrifugal force acting on a ball under these conditions is that tending to explode it.

On the other hand, a ball in contact with both plates of a single-disk unit will be rolled by the revolving disk over the stationary plate, and under centrifugal force will travel in a spiral path toward the periphery. It will be subject to two centrifugal forces, i.e., the rapid rolling will tend to explode the ball, while the spiral travel accelerates discharge of the ball from the periphery of the disks.

Gravity feed is more effective in a double-disk mill than in a single-disk mill, and the material can be retained longer between the disks of a double revolving unit.

In addition to the differences inherent between the single- and double-disk designs, the following variables affect the performance of a revolving-disk mill.

- Rotational speed of disks (range: 400 to 1,800 rpm)
- Peripheral speed of disks (range of disk diameter: 24 to 48 inches)
- Rate of feed
- Proportion of water to dry fiber (**consistency** is the percentage by weight of dry fiber in a fiber-water mix)

- Pattern of fiberizing plates (see fig. 18-283 for an example)
- Temperature of incoming chips and defiberizing zone
- Power applied (range: 100 to 2,500 hp per revolving disk)
- Distance between plates
- Precision with which disks rotate

The effect of these variables is determined largely by trial and error. While it is not entirely clear what happens to wood chips as they progress through a revolving-disk mill, it is evident that liquids are violently agitated, fibers are compressed and lumens collapsed. Disk machines are effective in separating individual fibers, or fiber bundles if desired, from wood chips or other fibrous aggregates without substantially reducing their length. In this respect revolving-disk mills appear to have an advantage over pulp stone grinders.

From photographic analysis of chips and fibers within a single-disk refiner, Atack et al. (1982) made the following observations: Chips and dilution water fall to the base of the feeder housing and are fed along it to the refiner eye, where the chips are reduced to coarse pulp. This coarse pulp proceeds through the breaker bars into the refining zone. Some of the pulp in the inner part of the refining zone flows back to the breaker bars along grooves of the stationary plates. Pulp in the outer part of the refining zone moves radially outwards. There is considerable recirculation of coarse pulp around the inner part of the refining zone and the breaker bar section. For a short fraction of its passage through the refiner, most of the fibrous material is constrained to move in the direction of rotation of the moving plates. Some of this material is stapled momentarily in a tangential orientation across the bars of both sets of plates. The immobilized fibres are then subjected to refining action between the relatively moving bars before being disgorged into the adjacent grooves. The reject content of sampled material decreases gradually from the breaker bar section to the refiner periphery and the strength of screened pulp increases.

Readers interested in the hydraulic action and the influence of chip moisture content within disk refiners are referred to Leider and Rihs (1977) and Hartler (1977).

Softwood chips processed through revolving disk mills, at atmospheric pressure or at elevated pressure in thermo-mechanical pulping, yield pulps of higher strength than those obtained by stone grinding. Hardwoods, however, do not generally yield acceptable pulps unless given prior alkaline impregnation (e.g., with a mixture of NaOH and Na_2SO_3). Such pre-treated hardwood chips yield fibers having fibrillar fines on their surfaces, which enhance fiber to fiber bonds.

Disk mills may be used to defibrate, disperse, or to develop fiber quality. Defibration reduces aggregates of fibers into individual fibers. The aggregate involved may be raw chips, screen rejects, semichemically cooked chips, or old paper. The colloid-mill action of a revolving-disk machine causes thorough dispersal of extractives, pulverizable dirt, and fiber bundles so that they become less visible. In quality development, fiber bonding properties and strength are improved.

These three functions of a disk mill are not sharply defined. For example, all fiberizing treatments are accompanied by some dispersion and by some quality development, but the type of mill and the variables involved can be adjusted to particularly accentuate the function desired. Fiberizing is assisted by high speeds of about 900 to 1,800 rpm and consistencies of 8 to 12 percent. Dispersal is most effectively accomplished at high rpm with consistencies of 3 to 4 percent. Where fiber strength and reduction of power consumption are primary objectives, low speeds of perhaps 450 to 900 rpm may be preferred in conjunction with consistencies of 20 to 30 percent (Keays and Leask 1973).

Optimum plate patterns are designed more by trial and error than by theory. Different wood species require different plate patterns, and chemically pretreated chips require patterns different from those used with untreated chips.

There is some indication from mill trials that double-disk refiners equipped with thermoplastic plates, having a modulus of elasticity near that of wood fibers, may require less power than conventional steel plates, and yield pulp of improved strength (Basile and Matthew 1978).

Section 23-6 describes a variety of defibrators, including Asplund, pressurized, and rapid-cycle disk refiners. Section 25-5 defines the disk refining processes of RMP, TMP, and CCMP. For further reading see Franzen (1985).

DEFIBRATING WITH CRUSHING ROLLS

Harvey (1972) described a system of defibrating hardwood logs, cants, or slabs in which the green wood is crushed between slowly turning, but powerful, nip rolls (fig. 18-285). Resultant strand-like aggregates of hardwood fibers have potential for use in reconstituted wood as described in figure 24-63, in section 28-14, and by Marra et al. (1975). The machine illustrated is powered by a 50-hp motor on the top roll and a 10-hp motor on the bottom roll. When fed at 25 fpm on 6-inch northern red oak to achieve the degree of fiberization illustrated, the machine might produce 50 tons (green-weight basis) of fiber per 8-hour shift in the laboratory.

Industrial-scale applications of the machine have not been made in North America; in Russia, however, a 75-hp machine operating at a log deck crushed 64 tons of wood per hour (Smerdov and Sotonin 1978). For a review of experiments in Canada see DuSault (1985).

18-28 NOISE

There is a large and growing body of literature on reduction of noise emanating from woodworking machines. (The Abstract Information Digest Service of the Forest Products Research Society — currently termed "Forest" — is a good bibliographic source on the subject.)

Circular saws and planers are frequently among the excessively noisy machines in woodworking plants. Two approaches to noise reduction have been widely used. In the first, offending machines are isolated in sound enclosures. The technology of constructing such enclosures has been described by Greenwood (1968), Schmutzler (1970), Pease (1972), Stewart (1974d), Martyr (1976) Saljé and Bartsch (1976), Heydt and Schwarz (1977), and Woodworking Industry (1978).

Figure 18-285.—TVA fiberizer capable of crushing a log or cant to yield a fibrous mat. (Top left) Feed is from right to left. (Top right) Crushing rolls. (Center) Fibrous mat emerging from crushing rolls. (Bottom) Southern red oak fibers about 2 inches long resulting from hammermilling and screening the mat shown at center. (Photos from A. A. Marra.)

The second approach calls for modification of saws and planers to reduce the sound they generate. For circular saws, this technology is summarized in section 18-8 under the subheading CIRCULAR SAWING, *Saw selection for circular gang ripsaws*. Planer noise can be significantly reduced by use of cutterheads having helically disposed knives, as described in connection with figure 18-143.

18-29 DULLING OF CUTTING TOOLS

Craftsmen agree unanimously that sharp tools are essential to achievement of smoothly machined surfaces with minimum energy expenditure. It is beyond the scope of this text to discuss in depth techniques for sharpening cutting edges, but a few conclusions of Kirbach and Bonac (1981) are instructive. They found that edge breakage during grinding increases with increasing rake angle and with increase in grit size. In their study, grinding depth did not influence the degree of edge breakage. Kirbach and Bonac found that all grinding directions that can be employed in sharpening cemented tungsten-carbide-tipped tools involve edge breakage, but the breakage differs considerably with direction. Grinding parallel to the cutting edge should be avoided; they found that the most suitable grinding direction is at an angle of 45° or larger to the cutting edge. Also, edge breakage varies with the manner in which the grinding wheel is withdrawn from the cut. Retraction of the wheel in the plane of grinding provides the poorest edge; a better practice is retraction perpendicular to the plane of grinding. For a discussion of grinding sharp cutting edges with large rake angles on high-speed steel knives see section 18-23, subsection GRINDING VENEER KNIVES.

In use cutting edges dull (fig. 18-9). Scollard (1980) suggested that in some applications it is economically more feasible to discard dull knives than it is to sharpen them; in most circumstances they must be resharpened, however. Users of cutting tools are therefore interested in slowing the dulling process. One approach to prolonging edge sharpness is coating the sharpened tool with a wear-resistant smooth hard material; for example, Stirling (1979) described a procedure in which cutting edges are honed with a fine abrasive, submerged in a chemical bath at a temperature not exceeding 140°F and electrically coated with about 0.0002 inch of chromium having a hardness of up to 72 Rockwell C; Stirling found that such treatment significantly extended the life of chipper knives and teeth on circular saws.

Most cutting tools in the woodworking industry are designed for repeated sharpening, however, and hard coatings have limited use in the United States. The balance of this section is therefore concerned with tool materials and the mechanism of dulling.

TOOL MATERIALS AND HARDENING PROCEDURES[13]

The American Society of Tool and Manufacturing Engineers defines **tool steel** as "either carbon or alloy steels capable of being hardened and tempered." These steels are utilized for machining materials at ordinary and elevated temperatures. All steel is composed of a combination of iron and carbon with varying amounts of other elements. Carbon added to iron at levels as low as 0.01 percent forms an intersticial solid solution. That is, the carbon atoms move into spaces between the iron atoms and remain their when the solution solidifies from the liquid form. These carbon atoms act to lock up the matrix and transform normally soft, maleable iron into the harder, stronger material we know as steel.

Other materials present as **alloys** play an important part in developing desired properties. Certain metallic elements can replace iron atoms in the matrix and enhance performance. Other combinations can re-form the carbon into very hard structural configurations—the structures known as carbides. The most useful elements used in tool steel are chromium, tungsten, molybdenum, vanadium, and cobalt. Combinations of these elements must be hardenable. To harden steel, a heat treatment is necessary.

---

[13]Text under this heading is condensed from a review prepared by G. H. Kyanka, State University of New York, Syracuse, 1982. (See sect. 18-14, subsection TYPES OF COATED ABRASIVES and figures 18-58AB for description and illustration of materials used for coated abrasives.)

All steels are hardened by causing an unnatural formation of the basic elements to occur. **Surface hardening** can be accomplished by causing carbon or nitrogen atoms to migrate into the matrix. The terms **carbonizing** or **nitriding** describe such processes. The hardening process requires heating above a critical temperature and then controlling the cooling process to lock certain crystalline structures into the resulting metallic system. Hence the term **heat treating**.

The most typical hardening procedure involves heating steel above a transformation temperature where a structure known as austenite occurs and then quickly cooling (**quenching**) the material into a phase known as martensite. **Martensite** is a very hard metastable system characterized by a superheated solid solution of carbon in a body-centered tetragonal crystal with a highly distorted structure; the carbon in the system does not diffuse into the structure during cooling. Martensite is extremely brittle and hard to machine, so is not useful as formed, but must be tempered. **Tempering** is a gentler reheating process which allows some carbon diffusion and results in a hard, yet more flexible structure. The steel is now **tough**; it can be deformed without crack formation and can absorb considerable energy from impact or deflection due to applied stress.

Hardness, which promotes long wear, and toughness, which promotes resistance to fracture, are contradictory properties in most tool materials; i.e., hard steels are usually not tough. The tool designer must compromise to insure that the cutting edge can withstand cutting impacts as well as resist dulling.

High speed steel.—**High-speed steel** is a tool steel originally designed to machine other metals at high rates of removal. The term is now also part of the nomenclature in wood machining. The most important properties of a high-speed steel are its resistance to crack propagation and hardness sufficient for the cutting task. Table 18-80 shows properties of two high-speed steels, together with their American Iron and Steel Institute designations. They have modulus of elasticity of about 30,000,000 psi and Rockwell C hardness of 62-66. The **Rockwell C hardness** number is determined by measuring a ball indent area under a prescribed load.

Figure 18-286 depicts a badly fractured and dulled cutting edge of high-speed steel.

Tungsten carbide.—Carbides are compressed assemblies of hard crystals held together by a binder which gives flexibility and toughness to the composite. Most carbide is produced by reducing a tungstick oxide or ammonium paratungstate in hydrogen to tungsten powder; the tungsten powder is carburized by heating with pure carbon and then ball milling with cobalt to produce the final compound to be pressed into the shape of a cutting tool. This process of hard metal pressing requires careful cutting-tip design to insure an accurate, non-stressed part. Sintering (agglomerating by heating) is employed to control hardness and carbon content of the material. A controlled atmosphere of hydrogen or carbon monoxide is used, as well as vacuum sintering, to regulate the diffusion process which produces the tungsten carbide configuration desired. Both systems have their proponents and the choice of process is limited mainly by cost.

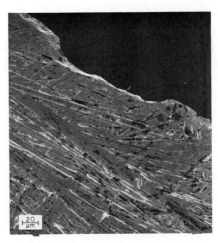

Figure 18-286.—Clearance face of a high-speed steel knife showing fracture and dulling of the cutting edge. (Micrograph from G. Kyanka.)

Vacuum sintering can produce superior material, but is more subject to process variables and requires sophisticated equipment of high cost and complexity.

Hard metals are porous. They shrink upon pressing into a tool shape, but voids are left and can cause wear. Also, tungsten carbide and cobalt have different expansion coefficients; the carbide grains in a tool are therefore in compression and the cobalt in tension as the tool is heated. Metallurgists work constantly to minimize void space and stress raisers to keep fracture strength high without sacrificing hardness. Tungsten carbides are more dense than high-speed steel, harder, and have much higher modulus of elasticity (table 18-80b).

Stellites.—**Stellites** encompass a family of cobalt-chromium alloys noted for hardness, stability and wear resistance, even at high temperature. They have found wide use for valves and gears in internal combustion engines. Their use as tool materials evolved from their inherent mechanical properties and from their corrosion resistance and low friction coefficients.

Stellite is used most commonly as a coating or tipping material; it is deposited on a clean cutting edge in a process similar to conventional welding. The edge is then ground to desired tool configuration, a slow and labor intensive process. Tipped tools may, however, be resharpened a few times before the coating needs replacement. Recently developed machinery can deposit stellite in a preformed shape on a cutting surface, reducing preparation costs and grinding losses.

A typical stellite comprised of 50 percent cobalt and 30 percent chromium (with carbon, nickel, molybdenum, iron, and other alloying ingredients making up the remainder) will have a modulus of elasticity comparable to steel and other properties as noted in table 18-81.

TABLE 18-80b.—*Properties of three classes of cutting tool materials* (Data from Kyanka; see text footnote[13])

Material	Hardness R_c	Modulus of elasticity	Density	Thermal conductivity	Coefficient of thermal expansion up to 650°C $\times 10^6$/°C
		Psi	G/cm^3	$Cal/°C/m/s$	
High-speed steel					
Medium-carbon, heat treated	66 (max)	32,500,000	8.6	0.61	12.6
High-carbon (1.10 percent C), heat treated	62 (max)	29,500,000	7.8	.41	14.7
Tungsten carbides	67-82	61,000,000-94,300,000	11.1-15.2	0.068-0.290	4.5-7.2
Stellites	63-67	30,000,000	8.4-8.8	low	14.5-16.9

TOOL WEAR[13]

Wear between two sliding surfaces can occur by plastic deformation, diffusion, viscous flow, and fracture (fig. 18-286). Viscous flow, and perhaps other wear mechanisms, is accelerated by chemical/electrical reactions at the interface which soften tool material, e.g., the cobalt in carbide cutters (fig. 18-287).

All commercial cutting tools are crystalline materials and have planes of strength and weakness. Because it is impossible to orient the cutting face so that only strong planes are exposed to the applied forces, it is necessary to examine how weaker planes behave. The micro structures of high-speed steel and carbide tools expose planes of atoms which can slide more readily than others. These planes (defined by physicists by the Miller Index) can be shifted by high stresses at impact during cutting, and atoms are dislodged to be carried away with the chips formed by the workpiece. Even diamond has these weak planes which are exploited by the diamond cutter to cause easy fracture. After these planes are fractured by the workpiece, the more rigid planes remaining must continue the work of material removal.

To minimize dulling, then, the tool designer attempts to place the hardest yet most energy absorbant and chemically inert cutting surface possible at the wood-tool interface. It is a difficult task with no easy solutions. Researchers across the world have attempted to model the systems involved in tool wear and have developed some interesting and potentially useful concepts.

The most common approach is to separate tool wear into mechanical and electro-mechanical categories. These are then looked at separately and interactive schemes are modelled. Another approach is to measure heat-friction wear and that due to silica and other abrasive inclusions. These paramenters can then be weighted to develop a relative wear scale for species of varying specific gravity (surface wear from heat) and crystal inclusions (wear from abrasion).

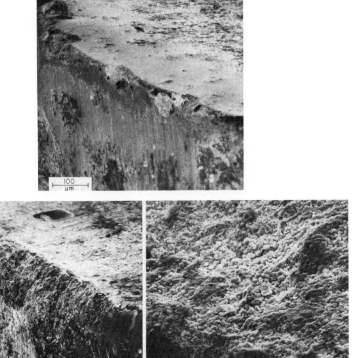

Figure 18-287.—(Top) Tungsten carbide cutting edge showing chemical corrosive pitting in and adjacent to chipped areas. (Bottom left) Same cutting edge at higher magnification. (Bottom right) At still higher magnification, nearly loose grains of tungsten carbide can be seen; the cobalt binder has been eroded. (Micrographs from G. Kyanka.)

To understand tool wear when machining southern hardwoods, chemical factors must be considered. For example, green oak is especially wearing on any type of tool material, and appears to cause dissolution of the cobalt binder in tungsten carbide tools.

Galvanic corrosion and cathodic-anodic ion transfer can develop and be accelerated by certain combinations of species, moisture content, and extractives. Some work has indicated that keeping workpiece-tool voltages balanced or over-balanced can decrease wear rates significantly. Whether such systems have commercial value for mass production remains to be shown. Possibly ceramic cutting edges could better resist such wear.

It has been shown that surface hardening or treatment can improve wear significantly. Coating techniques involving chromium, stellite, and other hard surfaces show some value, but increased temperature from heavy use can soon negate the benefit. Once coating on the treated zone is materially reduced in depth, overall wear rate may not be greatly reduced. The success of chromium suggests that smoothness can be as important as hardness. Some hard-cased tools do not perform as well as chromium plated tools in some applications.

The data in table 18-81 suggest how wear on cutting edges varies with tool material and cutting situation. Wear is least on tungsten carbide edges cutting dry wood with low extractives and silicon content, and high pH. Wear is most rapid on steel cutting green oak with high silicon content and low pH.

TABLE 18-81.—*Severity of cutting edge wear (loss of sharpness) related to tool material and to wood pH and moisture, extractives, and silicon content (Data from Kyanka; see text footnote 13)[1]*

Material	Hardness	Dry wood (<15 percent moisture content)	Green wood (above fiber saturation)	Effect of high extractives content, e.g., oak or cedar	Effect of high silicon content	Effect of low pH
	Rockwell C	---------- *Wear factor* ----------				
High-speed steel (reference material)	42-48	1.0	1.0	severe loss	severe loss	moderate loss
Carbide tip (tungsten-C)	65-84	0.05-0.20	0.10-0.30	severe loss	moderate loss	mild loss
Stellite tips	60-62	.15- .20	.20- .25	moderate loss	moderate loss	mild loss
Hard chromium plating	66-70	.10- .50	.20- .75	moderate loss	moderate loss	mild loss
High frequency induction hardened tip	60-70	.30- .50	.60+	severe loss	severe loss	moderate loss

[1] Guideline data tabulated are from a variety of sources in the literature, and from personal communications. Each combination of wood and tool material requires tests to determine severity of cutting edge wear.

18-30 COMPUTER CONTROL OF WOODWORKING MACHINES IN SECONDARY MANUFACTURE[14]

In the late 1800's innovative automatic machines were developed rapidly, and by the turn of the century, woodworking machines could turn out thousands of identical parts with little human interaction. While productivity improved, such automatic machines were limited in that they could only perform a series of sequential operations. That is, the "program" was created by a series of cams and ratchets that actuated tools needed to manufacture the part. Little change could be made in the variety of sequences and the cost of creating new "hardware" programs was high.

It was clear that more versatile machines were needed. Much more desirable would be a machine that could be guided by a set of written instructions, then on command carry out operations described in the instructions—e.g., position the workpiece, route grooves, or drill holes in specified locations—with no human intervention. Refined concepts of feedback, and the modern digital computer have made such machines possible.

In 1966, a numerically controlled routing and shaping machine was introduced into the wood industry. The machine could mill, drill, and bore at any angle on the workpiece using a punched-paper-tape program read by an electronic controller. The program positioned the tool and dictated its depth and rate of cut. To alter the shape of the part or create an entirely different one, it was necessary only to amend the old punched tape or create a new one.

These early machines were expensive due to the cost of the computer-controller, and the industry was slow to respond. By 1982 the cost of micro-processor-based controllers no larger than desktop calculators has decreased dramatically, magnetic media and solid state bubble memory were replacing punched paper tape, and microprocessor controllers added relatively little to the cost of numerically controlled machines.

Aside from reduced machine cost, other factors favor increased introduction of computer-numerically-controlled equipment for secondary wood conversion. Costs associated with labor are such that returns on investments for programmable machinery are attractive. But perhaps more important is the potential for improved productivity. In many applications, computer numerical control of machines can materially increase production, reduce rejects and waste, minimize material handling, and improve dimensional accuracy.

---

[14]Text in this section is from the introductory pages of McMillin, C. W. 1983. Computer numerical control of woodworking machines in secondary manufacture. Gen. Tech. Rep. SO-42, New Orleans, La. U.S. Dep. Agric. For. Serv., South. For. Exp. Stn. 24 p. Readers needing a fuller description of the subject should consult the source document.

COMPUTER NUMERICAL CONTROLLERS

At the heart of the computerized numerical control (CNC) is a low cost, small microprocessor that is the central arithmetic and logic unit of the system. Such microprocessors are miniaturized to fit on a single silicon chip and frequently hold thousands of transistors, resistors, and related circuit elements. By adding additional chips to provide timing, program memory interfaces for input/output signals, random-access memory, and other ancillary functions, it is possible to assemble a numerical controller on boards no larger than 8 by 10-inch pieces of paper.

It is the function of the microprocessor to accept data in the form of binary digits (0's and 1's), to store the data, and to perform arithmetic and logic operations in accordance with a previously programmed set of instructions. After processing, the microprocessor must then deliver the results to a user output mechanism. Typically, a microprocessor contains the following components—a decode and code control unit to interpret instructions from programs, an arithmetic and logic unit, registers for manipulating data, an accumulation register, address buffers to provide access to sequential instructions, and input-output buffers to read instructions or data into the microprocessor or to send them out.

A computer numerical controller consists of an operator control panel, part program data reader, and peripheral device connecter panel. Typically the components are also available for individual mounting according to the machine tool builder's requirements. Such free-standing cabinet configurations measure about $30 \times 30 \times 70$ inches high, weigh about 250 pounds and are air cooled during operation.

The controller's power supply usually operates at 60 HZ, 120 volts AC, and is normally provided with over and under voltage protection as well as overtemperature detection display circuitry and automatic shutdown if excessive internal temperatures are reached. Some power supplies feature battery backup to maintain part program data storage during power failures as well as diagnostic indicators and external voltage test points.

The operators' control panel usually consists of a CRT video display, various push buttons, indicators, and selector switches used to initiate, monitor, and govern control operations. Typical operations include control on/off, manual axis jog, machine home, and part program load/execute functions. A serial communication link is usually provided between the operator control panel and the controller circuit boards so that the control panel may be on or near the machine tool with the controller and power supply placed in a more desirable location.

CRT displays range in size from 5 to 12 inches measured diagonally. Some are capable of displaying up to 15 lines with 32 alphanumeric characters per line. Most feature selectable character size to ease reading of position coordinates and other data at a distance. Typical readouts include stored part program, current position coordinates, distance to go, command blocks, cutter compensation and diagnostic data messages from the programmable controller interface, alarm messages, and total and part machining operation time.

A typewriter-like alphanumeric keyboard enables the operator to manually enter part program data and tool offsets as well as edit programs and initiate various control operations. Some control panels contain space for additional hardware used in specialized applications; they may also provide for emergency termination of machine operation and interlocks to prevent unauthorized modification of stored part programs.

The electronic circuitry needed to implement machine operations are sometimes contained on individual printed circuit modules assigned to specific slots within the controller chassis. In other controllers, the functions of individual digital logic boards are combined and incorporated on a single board using very large scale integrated (VLSI) circuitry. Each module or operational circuit performs specific control functions such as data processing or input/output/servo control.

The main processor module containing the system microprocessor executes the control program and provides supervisory control over system operations. It additionally performs CNC functions, such as part program data decoding and distribution, arithmetic and logic, and interpolation.

A programmable interface provides the necessary circuitry to interface the CNC with machines and allows the user to define and store in program form his own sequential machine tool logic. A microprocessor on the module executes the programmable interface program, and coordinates functions with the main processor module.

One- or several-part memory circuits provide random access storage for part program data. Such data are read from punched paper tape, or from magnetic media such as tape or disks. In some controllers, part program data are permanently stored in recently developed magnetic bubble memories eliminating the need for interfacing punched tape or magnetic readers.

Lastly the input/output/servo circuitry facilitates the electronic interface between the CNC and the external machine tool. The input/output modules actuate such devices as relays, limit, and proximity switches; they vary in number depending on user needs. One or more servo modules provide the electronic interface with position feedback devices and servo drives.

The concept of feedback is a characteristic common to most computer controlled machines. Feedback involves the interaction of machine servomechanisms and the controller. As an example, figure 18-288 shows a schematic diagram of a feedback loop that determines the position of a movable work table. The part program data storage device first instructs the controller what position is desired. The drive motor and lead screw then move the table until the position transducer reports to the controller through a comparison unit that the correct position has been reached. Various types of feedback transducers are used such as encoders, tachometers, Selsyn motors, variable resistors or in highly accurate machines, optical interferometers. Depending on the sophistication of the feedback system, computer controlled machines are routinely accurate to within a 0.001 inch or less.

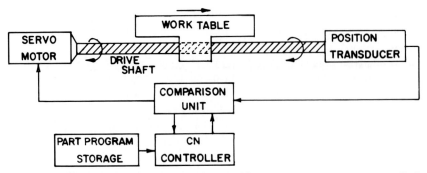

Figure 18-288.—Feedback loop to control worktable position on a computer controlled machine. (Drawing after McMillin[14].)

PROGRAMMING

Computer numerical control programs may be written manually or prepared by ancillary computer systems. The choice of method depends mainly on the type of machining operation and complexity of the part.

The first step in manual programming is determination of operations to be performed and their order. Working from blueprints, the programmer then establishes position coordinates that define the shape of the part or tool path. The coordinates may be in absolute or incremental units. In absolute units all points are in reference to the origin of the coordinate axes (machine zero) with quadrant defined by sign. In the incremental mode, movement is in a step or increment from the present position to the new position. The sign associated with the coordinate indicates if the position is to move in the forward or reverse direction.

The program is then written using a series of statement blocks preceded by a line number. Each statement block contains instructions for the machine to perform a movement and/or function. Typically these include preparatory functions, words to place the control in various modes of operation (i.e., linear interpolation mode, circular interpolation mode), axis movement instructions, feed rate and spindle speed data, and tool offset number.

A typical two-axis program to move the tool in a path as in figure 18-289 is

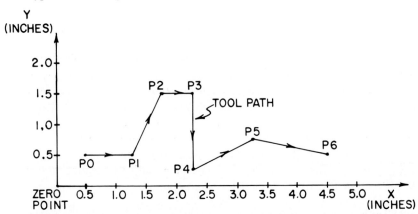

Figure 18-289.—Tool path movement by linear interpolation (Drawing courtesy of Allen Bradley Corp.)

given below with an explanation of each line[15]. The program assumes the controller is in the inch, absolute coordinate, and inch per minute feedrate modes.

n100	g01	×12500		f3000	EOB[16]	(P0 to P1)
n105		×17500	y15000		EOB	(P1 to P2)
n110		×22500			EOB	(P2 to P3)
n115			y2500		EOB	(P3 to P4)
n120		×32500	y7500		EOB	(P4 to P5)
n125		×45000	y5000		EOB	(P5 to P6)

In block n100, a single axis movement of 0.75 inches in the +× direction from P0 to P1 occurs at 30 IPM.

In block n105, a 2-axis linearly interpolated movement of 1.5 inches in the +y direction and 0.5 inches in the +× direction (from P1 to P2) occurs at 30 IPM.

In block n110, a single axis movement of 0.5 inches in the +× direction from P2 to P3 occurs at 30 IPM.

In block n115, a single axis movement of 1.25 inches in the −y direction from P3 to P4 occurs at 30 IPM.

In block n120, a 2-axis linearly interpolated movement of 0.5 inches in the +y direction and 1.0 inches in the +× direction (from P4 to P5) occurs at 30 IPM.

In block n125, a 2-axis linearly interpolated movement of 0.25 inches in the −y direction and 1.25 inches in the +× direction (from P5 to P6) occurs at 30 IPM.

After debugging, the program is typed, reproduced on punched tape or other storage media, and verified by a dry run.

Programs written with the aid of a computer greatly simplify programming since each coordinate point need not be calculated. However, computer-aided programming requires knowledge of a programming language. While about 100 different languages are available, the best known and most widely used is APT (Automatically Programmed Tools).

The APT programmer uses English-like words to define operations involving geometry and motion. Geometry statements include POINT, LINE, PLANE, CIRCLE. Motion statements include GOTO, GOON, GOPAST, and others. Such input commands produce the calculations necessary to define the tool path.

While the APT processor is machine independent, it operates with a machine post-processor specific to individual needs. The post-processor formats numerical data in a manner understood by the controller and processes such commands as DELAY, CYCLE AUXFUN, and REWIND. These commands are oriented to the numerical controller system and always result in execution of the same process. Post-processor machine segment commands differ between machines. Typical commands include SPINDL (spindle on/off), COOLNT (coolant on/off), and GOHOME. This section of the post-processor also defines maximum travel and velocity of each axis and its acceleration limits.

The programmer first assigns alphanumeric names to variables or geometric surfaces (fig. 18-290). This simplifies geometry definition, and variables are more easily located when checking the program against a blueprint. The variable

---

[15]Anonymous. 1980. System 7320 Programming Manual. Allen-Bradley, Cleveland, Ohio.

[16]An end-of-block (EOB) flag is required at end of each statement block.

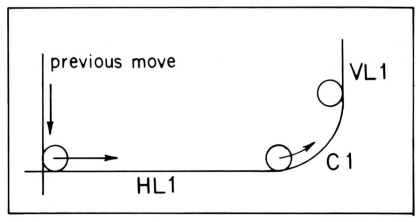

Figure 18-290.—Simple motion with variables labeled for programming by APT. (Drawing courtesy of Allen Bradley Corp.)

names in figure 18-290 are assigned as HL1 (horizontal line 1), C1 (circle 1), and VL1 (vertical line 1). Statements frequently contain a major word preceding a slash(/) followed by minor information. The program statements to move the tool along the path shown are:

    1 GOLFT/HL1, TANTO, C1

    2 GOFWD/C1, TANTO, VL1

Line 1 means based on the previous move, go left along HL1 and stop where HL1 is tangent to C1. Line 2 instructs the controller to continue in a forward direction along C1 and stop where C1 is tangent to VL1. As with manual programming, the completed program is stored on punched tape or magnetic media.

Part program data can also be input to the controller through use of a digitizing graphics table. The shape of the desired part is first drawn on paper and attached to the table. The outline of the part is traced with a hand-held sensor equipped with a cross-hair cursor. Micro-processors in the interface circuitry automatically digitize the x-y coordinates of the workpiece for subsequent processing to machine ready tape. Operation command insertion, editing, program review, curve smoothing, and cutter diameter compensation are activated with appropriate codes at a CRT/keyboard module.

Most manufacturers of CNC woodworking machines provide in-house programming facilities for users preferring to contract some or all of their programming needs.

COMPUTER CONTROLLED MACHINES

A broad variety of woodworking machines can be computer controlled, including routers (fig. 18-203), borers (fig. 18-206), shapers (fig. 18-156), carvers, and panel sizers.

It seems likely that future developments will include machining systems for cutting wood parts with a laser (fig. 18-234) under digital control of an optical image analyzer capable of detecting and plotting defects to be removed (McMillin 1982).

18-31 LITERATURE CITED

Aario, M., P. Haikkala, and A. Lindahl. 1980. Pressure grinding is proceeding. Tappi 63(2):139-142.

Adams, E. L., and D. E. Dunmire. 1978. Solve II users manual: A procedural guide for a sawmill analysis. U.S. Dep. Agric. For. Serv. Gen. Tech. Rep. NE-44. 18 p.

Adkins, W. K., D. B. Richards, D. W. Lewis, and E. H. Bulgrin. 1980. Programs for computer simulation of hardwood log sawing. U.S. Dep. Agric. For. Serv. Res. Pap. FPL 357. 56 p.

Alexander, J. D., Jr. 1978. Make a chair from a tree: An introduction to working green wood. The Taunton Press, Newtown, CT. 128 p.

Allen, F. E. 1975. Some considerations in machining southern hardwoods with multiple bandsaws. FPRS Sep. No. MS-75-S55, 9 p. For. Prod. Res. Soc., Madison, Wis.

Altman, J. A. 1975. Precision whole tree chipping operation. Amer. Pulpwood Assoc. Tech. Release 75-R-8, 4p.

American Society for Testing Materials. 1959. ASTM Standards on wood, wood preservatives, and related materials. ASTM Committee D-7, Amer. Soc. Test. Mater., Philadelphia, Pa. 446 p.

Anderson, D. E. 1979. Value of saw specifications. In Proc., Sixth Wood Machining Seminar, Univ. Calif., For. Prod. Lab., Richmond, Calif. (J. F. Walters, ed.) p. 26-34.

Anderson, W. D. 1981. Economic feasibility of processing low-value hardwood using shaping-lathe headrigs. In Utilization of low-grade southern hardwoods—feasibility studies of 36 enterprises. p. 214-221. (D. A. Stumbo, ed.) For. Prod. Res. Soc., Madison, Wis.

Anderson, G. W., and J. L. Ailport. 1973. Double-taper chipper canters. In Modern Sawmill Techniques, Proc. of the First Sawmill Clinic, Portland, Ore., Feb. (V. S. White, ed.) p. 130-144.

Anonymous. 1975. Rigging for routing. Woodworking and Furniture Digest 77(9):26-29.

Anonymous. 1976a. Planer-sander combination achieves big energy savings. Furn. Methods and Materials 22(8):26.

Anonymous. 1976b. Multifunction N/C router produces more parts. Furniture Des. and Manuf. 48(8):46-47.

Anonymous. 1976c. The continued development of the laser. Woodwork. Tech., May, p. 19-21, 23.

Anonymous. 1977a. 'Electronic sawyer' helps you cut for maximum value. Wood and Wood Prod. 82(7):61-62.

Anonymous. 1977b. Low cost N/C routing of complex parts replaces pin router operation, multiplies output capacity nearly eleven times. Woodwork. and Furniture Dig. 70(10):36-39.

Applefield, M. and P. J. Bois. 1966. Thickness variation of hardwood lumber produced in 1963 by circular sawmills in North Carolina. 5 p. Southeast. For. Exp. Stn., Athens, Ga. (In cooperation with the Hardwood Res. Counc. and Duke Power Co.)

Araman, P. A. 1977. Use of computer simulation in designing and evaluating a proposed rough mill for furniture interior parts. U.S. Dep. Agric. For. Serv. Res. Pap. NE-361. 9 p.

Araman, P. A. and E. L. Lucas. 1975. An automated rough mill for the production of interior furniture parts. U.S. Dep. Agric. For. Serv., Res. Pap. NE-335. 5 p.

Arola, R. A. 1971. Crosscut shearing of roundwood bolts. U.S. Dep. Agric. For. Serv. Res. Pap. NC-68, 21 p.

Arola, R. A. 1972. Estimating force and power requirements for crosscut shearing of roundwood. U.S. Dep. Agric. For. Serv. Res. Pap. NC-73, 8 p.

Arola, R. A., and T. R. Grimm. 1974. Design of thin shear blades for crosscut shearing of wood. U.S. Dep. Agric. For. Serv. Res. Pap. NC-105. 23 p.

Atack, D. 1976. Mechanical pulp-conserver of our forests. Appita 30(2):155-160.

Atack, D., M.I. Stationwala, and A. Karnis. 1982. What happens in refining? PPR No. 407. 12 p. Pulp and Paper Research Institute of Canada.

Aune, J. E. 1977. Computerized sawmill design—model versus reality. In Proc. of the 5th wood machining sem., Univ. of Ca., Berkeley, For. Prod. Lab., Richmond, Ca.

Barwise, R. D., R. A. Arola, J. R. Erickson. 1977. Helical head comminuting shear. U.S. Pat. No. 4,053,004. U.S. Pat. Off., Washington, D.C.

Basile, F. C., and J. B. Matthew. 1978. New soft plates for disk refiners show great promise in mill trials. Pulp Pap. 52(3):120-125.

Behm, R. D. 1974. Hardwood selecting guide for designers & manufacturers. Furniture Des. & Manuf. 46(12):45-54.

Behr-Manning. 1967. Wood sanding with coated abrasives. Behr-Manning Division Norton, Troy, N.Y. 96 p.

Belyea, H. C. 1931. Form and taper. Ch. XI in Forest Measurement. John Wiley and Sons, Inc., New York. p. 191-210.

Bennett, W. 1973. A total system concept—the problem defined. In Wood machining sem., proc., pp. 1-12. Richmond, Ca.: For. Prod. Lab., Univ. of Ca.

Betts, H. 1969. Extra-thin saws increase lumber yield. Wood and Wood Prod. 74(8):28-30.

Bingham, S. A., and J. G. Schroeder. 1976. Short lumber in furniture manufacture. Part I. Short lumber in manufacture. Part II. Bolt and lumber grading. Part III. Drying and handling short lumber. Natl. Hardwood Mag. 50(11):34-35, 48-50; 50(12):90-91, 112-113; 50(13):38-39, 49-50.

Bingham, S. A., and J. G. Schroeder. 1977. Short lumber in furniture manufacture. Integrated plants for production and use of short lumber. Part IV. Natl. Hardwood Mag. 51(1):28-29, 32-33, 35-37.

Birkeland, R. 1968. [Results of measurements of the sawing accuracy of circular saw mills and bandsaw mills.] Norsk Skogindustri 22(6):192-206.

Birkeland, R. 1971. Grit orientation and cutting efficiency. In Proc., Wood Machining Seminar, For. Prod. Lab., Univ. Calif., Richmond, March 24-25, p. 83-91.

Boles, C. L., and R. J. Flanigan. 1976. Noise reducing planer cutter head. U.S. Pat. 3,933,189. U.S. Pat. Off., Washington, D.C. 6 p.

Bolton, W. E. 1975. Veneer value loss due to spin axis misalignment. FPRS Sep. No. PNW-74-S53, 5 p. For. Prod. Res. Soc., Madison, Wis.

Bonac, T. 1975. Measuring of wood surface texture by the pneumatic method. Paperi Ja Puu 57(4):309-312, 315-316, 321-322, 325-326.

Bonac, T. 1979. Wood roughness volume and depth estimated from pneumatic surface measurements. Wood Sci. 11(4):227-232.

Borlew, P. B., and R. L. Miller. 1970. Chip thickness: A critical dimension in kraft pulping. Tappi 53:2107-2111.

Bousquet, D. W. 1972. Sawing pattern and bolt quality effects on yield and productivity from yellow birch. For. Prod. J. 22(11):39-48.

Bousquet, D. W., W. W. Calvert, and R. M. Nacker. 1972. Conversion surplus: Measure of profit from sawing hardwood. Can. For. Ind. 92(2):53-59.

Bousquet, D. W. and I. B. Flann. 1975. Hardwood sawmill productivity for live and around sawing. For. Prod. J. 25(7):32-37.

Bramhall, G., and T. A. McLauchlan. 1970. The preparation of microsections by sawing. Wood and Fiber 2:67-69.

Brecht, W. 1979. Fundamentals of the manufacturing processes of groundwood and refiner pulp. Wochenblatt for Papierfabrikation 107(11/12):394-395.

Bryan, R. W. 1975. "Instant" chip mill chews up 15-20 cords/hour of both pine, hardwoods. For. Ind. 102(6):66-68.

Bryan, R. W. 1978. Portable sawmill proves effective in Lake States hardwood operation. For. Ind. 105(10):48-49.

Bulgrin, E. H. 1974. Ultrasonics to detect wood's defects. In Modern sawmill techniques, vol. 3, pp. 308-318. Proc. of the third sawmill clinic, San Francisco, Ca.: Miller Freeman Publ., Inc.

Burk, B. 1972. Game bird carving. Winchester Press, Tulsa, Okla. 242 p.

Burry, H. W. 1969. Methods of sawing small hardwood logs. The North. Logger and Timber Proc. 18(3):12-13, 30.

Burry, H. W. 1976. Hardwood log lengths. North. Logger 25(1):21.

Butler, R. L. 1974. Wood for wood-carvers and craftsmen. A. S. Barnes and Co., Inc., Cranbury, N. J. 122 p.

Byrd, V. L., R. A. Horn, and D. J. Fahey. 1967. How refining of vessel elements affects offset printing papers. Pap. Trade J. 151(46):55-60.

Calvert, W. W. 1955. Increasing hardwood lumber yields with the short-log bolter. Timber of Can. 15(6):25-43.

Cammack, C. F., and B. Y. Richardson. 1980. Observations on operations of the Pettibone Hydro-slasher PM 800. ED&T 2272 TSI Slash treatment. U.S. Dep. Agric. For. Serv., Equip. Develop. Cent., San Dimas, Calif. 9 p.

Campbell, R. D. 1973. High-speed thin-kerf sawing. Chapter XV in Modern Sawmill Techniques, vol. 2: Proc. of the Second Sawmill Clinic, New Orleans, La., Nov. Miller Freeman Publications, Inc., San Francisco.

Cantin, M. 1965. The machining properties of 16 eastern Canadian woods. Dep. of For. Publ. No. 1111, 27 p. Can. For. Dep., Ottawa, Canada.

Carino, H. F., and J. L. Bowyer. 1979. New tool for solving materials flow problems: A computer-based model for maximizing output at minimum cost. For. Prod. J.29(10):84-90.

Carlson, T. C. 1968. Estimating hourly production of sliced wood. For. Prod. J. 18(11):45-49.

Carlson, A. R. 1977. Building a log house in Alaska. Alaska Coop. Serv. Pub. P-50A, Coop. Ext. Serv., Univ. Alaska. 80 p.

Carroll, D. 1977. Sanding and finishing. Sanding machine technology catches up. Furniture Methods and Materials 23(1):15-18.

Carroll, D. 1978. How thin face veneers can be successfully sanded. Furniture Methods and Materials 24(3):18-19.

Chamberlin, F. B., Jr. 1973. Paperboard application for the high-energy fluid jet cutter. Tappi 56(8):78-80.

Chardin, A. 1973. Field evaluations of sawing accuracy of band headrigs. In Proc., wood machining sem., pp. 126-134. Berkeley, Ca.: Univ. of Ca.

Chardin, A. 1979. Displacement of band saw blades on wheels: An experimental approach. In Proc., Sixth Wood Machining Seminar, Univ. Calif., For. Prod. Lab., Richmond, Calif. (J. F. Walters, ed.) p. 209-221.

Christensen, E. 1976. Advancing the state-of-the-art in screening bark-free and non-bark-free chips. Tappi 59(5):93-95.

Church, T. W., Jr. 1973. The significance of sweep in Appalachian hardwood sawlogs. U.S. Dep. Agric. For. Serv., Res. Note NE-172. 4 p.

Church, T. W., Jr. and L. D. Garrett. 1970. Should a hardwood lumber producer saw ties and timber too? The North. Logger and Timber Proc. 18(10):16, 18, 20-22, 24.

Church, T. W., Jr. and G. R. Niskala. 1966. An elixir for sawmills: cut cants. The North. Logger 14(10):18-19, 44.

Clark, F. 1969. High strain band mills. In Proc. North. Calif. Sect. FPRS, p. 2-5, Calpella, Calif.

Clark, A. Ill. 1976. Sawmill residue yields from yellow-poplar saw logs. For. Prod. J. 26(1):23-27.

Clark, A. C., III, M. A. Taras, and J. G. Schroeder. 1974. Predicted green lumber and residue yields from the merchantable stem of yellow-poplar. U.S. Dep. Agric. For. Serv., Res. Pap. SE-119. 15 p.

Coleman, R. E. and H. W. Reynolds. 1973. Sawing SHOLO logs: three methods. U.S. Dep. Agric. For. Serv., Res. Pap. NE-279. 5 p.

Coleman, R. E. 1977. SEMTAP (Serpentine end match tape program). U.S. Dep. Agric. For. Serv., Res. Pap. NE-384. 5 p.

Collett, B. M., A. Brackley, and J. D. Cumming. 1971. Simplified, highly accurate method of producing high-quality veneer. For. Ind. 98(1):62-65.

Connelly, H. H. 1975a. Solving color problems in the sanding room. Furniture Design and Manuf. 47(4):

Connelly, H. H. 1975b. Choosing and maintaining abrasive planing belts. Furniture Design and Manuf. 47(6):38-39.

Connelly, H. H. 1975c. Abrasive planing for particleboard. Furniture Design and Manuf. 47(5):48.

Connelly, H. H. 1976. A new concept developed for managing abrasives in the furniture factory through scientific testing. Paper presented at Sanding Seminar, Conestoga College, Kitchner, Ontario, May 15. 4 p.

Cooper, G. A. and R. C. Schlesinger. 1974. Yield of furniture dimension from yellow-poplar thinnings. South. Lumberman 229(2848):83-85.

Corbin, J. P. and L. J. Slentz. 1956. Grinding of native New York hardwoods. Tappi 39:169A-174A.

Craft, E. P., and D. M. Emanuel. 1981. Yield of pallet cants and lumber from hardwood poletimber thinnings. Res. Pap. NE 482. U.S. Dep. Agric., For. Serv. 6 p.

Crossley, F. R. E., and J. Sullivan. 1980. Mobile wood chippers, 1980. DOE/ET/20077-1. U.S. Dep. Energy, Solar Energy, Washington, D.C. 129 p.

Cumming, J. D. 1969. Stresses in band saws. In Proc. For. Prod. Res. Soc. North. Calif. Sec., p. 21-23. For. Prod. Res. Soc., Madison, Wis.

Dargan, E. E. 1969. By product production at Roundwood Corporation. In Wood residue utilization, Proc. of Ann. Meet., Mid-South Sect., For. Prod. Res. Soc. (Weldon, D., and C. P. Isaacs, comp.), p. 27-31. Tex. For. Prod. Lab., Tex. For. Serv., Lufkin.

Das, A. K. 1979. Some circular saw design considerations: A systems approach. In Proc., Sixth Wood Machining Seminar, Univ. Calif., For. Prod. Lab., Richmond, Calif. (J. F. Walters, ed.) p. 50-57.

Davis, E. M. 1958. Development of methods for evaluating the machining qualities of wood and wood-base materials. U.S. Dep. Agric. For. Serv., For. Prod. Lab. Rep. No. 2108. 46 p.

Davis, E. M. 1962. Machining and related characteristics of United States hardwoods. U.S. Dep. Agric. For. Serv., Tech. Bull. No. 1267. 68 p. Washington, D. C.

Demsky, S. 1967. Merits of the Klear-Kut Saw. In 22nd Ann. Northwest Wood Prod. Clin. Proc. 1967: 33-35. Spokane, Wash.

Detjen, R. K. 1975. Linebar resawing of small southern hardwoods. Southern Lumberman 230(2855):11-13.

Detjen, R. K. 1978. Small log transport systems for use with multiple band saws and chipping heads. In Complete tree utilization of southern pine: symp. proc., New Orleans, La., April 17-19. C. W. McMillin, ed. For. Prod. Res. Soc., Madison, Wis., p. 325-328.

DeVries, J. 1973. The splitting of pulpwood logs into billets. Leafl. 117, For. and Timber Bureau, Canberra, Australia. 32 p.

DeVries, W. R., D. A. Dornfeld, and S. M. Wu. 1978. Bivariate time series analysis of the effective force variation and friction coefficient distribution in wood grinding. J. Eng. Ind. 100:181-185.

Dilly, C. 1975. Fuel cleaning and grinding. In Wood Residue as an Energy Source. Proc. No. P-75-13. For. Prod. Res. Soc., Madison, Wis. p. 46-49.

Dobie, J. 1976. Economic analysis of finger jointing "green" lumber by the WFPL method. West. For. Prod. Lab. Info. Rep. VP-X-160, 10 p. Can. For. Serv., Vancouver, B.C., Can.

Dokken, M. and V. Godin. 1975. Instrument for measuring knife pitch angle on veneer lathes. For. Prod. J. 25(6):44-45.

Dornfeld, D. A., W. R. DeVries, and S. M. Wu. 1978. An orthomorphic rheological model of the grinding of wood. J. Eng. Ind. 100:153-158.

Dornfeld, D., and S. M. Wu. 1978. Development of an experimental setup for the investigation of grinding of wood and a proposal for a pulsed loading technique. J. Eng. Ind. 100:147-152.

Dowdell, S. A. and H. S. Plotkin. 1975. A pilot operation to determine the feasibility of operating a short log system in western North Carolina hardwoods. N.C. For. Note No. 15, 6 p. Raleigh, N.C.: N.C. For. Serv. Off. of For. Resourc.

DuClos, A. 1967. Thin kerf sawing. In Proc., 22nd Ann. Northwest Wood Prod. Clinic, Spokane, Wash., April 18-19. p. 37-41. Technical Ext. Serv., Washington State University, Pullman.

DuSault, A. 1985. Evaluation of crushing rolls configuration to produce woody biomass. In, proceedings of the symposium "Comminution of wood and bark" held October 1-3, 1984 in Chicago, Illinois. Forest Products Research Society, Madison, Wisconsin.

Dunbar, M. 1976. Windsor chairmaking. Hastings House, New York. 160 p.

Dunfield, J. D. 1974. Log cabin construction. J. D. Dunfield, Ottawa, Can. 92 p.

Dunmire, D. E. and G. H. Englerth. 1967. Development of a computer method for predicting lumber cutting yields. U.S. Dep. Agric. For. Serv., Res. Pap. NC-15. 7 p.

Dutrow, G. F., and J. E. Granskog. 1973. A sawmill manager adapts to change with linear programming. U.S. Dep. Agric. For. Serv. Res. Pap. SO-88. 11 p.

Dziobek, K. 1971a. Sanding with aerostatically supported belts. In Proc., Wood machining seminar, Univ. Calif. For. Prod. Lab., Richmond, March 24-25. p. 106-114.

Dziobek, K. 1971b. A new model for measuring grit distribution in plane and height on sanding belts. In Proc., Wood Machining Seminar, For. Prod. Lab., Univ. Calif., Richmond, March 24-25, p. 92-105.

EUCEPA. 1977. New competitive furnishes for papermaking. Proc., International Mechanical Pulping Conf., Helsinki, Finland, June 6-10. Vols. I-V. EUCEPA, Tech. Assoc. Pulp Pap. Ind., and Can. Pulp and Pap. Assoc.

Edlund, G. 1975. Finger-joint curing developments and marketing. Ch. 12 in Nordic and North American Sawmill Techniques. Proc., 2nd Int. Sawmill Sem. Miller-Freeman Publ., Inc., San Francisco. p. 151-166.

Ekwall, J. 1976. The pursuit of machining quality. Furniture Des. and Manuf. 48(4):23-24.

Ekwall, J. 1977. Getting the most from your tenoner. (In four parts.) Furn. Des. and Manuf. 49(9):14, 16, 18; 49(10):14, 16, 20; 49(11):38-40; 49(12):12-14.

Ekwall. J. 1978. Router developments: Inverted router and intri-shaper. Furniture Des. and Manuf. 50(6):42-45.

Ekwall, J. A. 1979. Anatomy of a shaper jig. Furn. Manuf. 25(1):18, 20-22.

Elmendorf, A., and T. W. Vaughan. 1958. A survey of methods for measuring smoothness of wood. For. Prod. J. 8:275-282.

Emanuel, D. M. 1978. Processing hardwood bark residues by screening. U.S. Dep. Agric. For. Serv. Res. Note NE-260. 3 p.

Endersby, H. J. 1953. The performance of circular plate ripsaws. Dep. Sci. and Indus. Res., For. Prod. Res. Bull. 27, London. 20 p.

Endersby, H. J. 1965. The cutting action of woodworking tools. IUFRO Congr. Proc. 1965, Vol. 3, Sec. 41, 7 p. Melbourne.

Engelgau, W. G. 1977. What is new and old in chipping? Tappi 61(8):77-80.

Erickson, J. E. 1964. An investigation of power requirements for chipping hardwoods under various cutting conditions. Woodland Sect. Index, Can. Pulp and Pap. Assoc. 2301 (B-1) Append. B: 4-5.

Erickson, John R. 1966. Portable pulpwood chipping machines: In: Proceedings of the first annual forest engineering symposium. Tech. Bull. 82. Morgantown, WV: West Virginia University, Engineering Experiment Station: 15-24.

Erickson, J. R. 1967. Crosscut shearing of wood. IUFRO Congr. Proc. 1967, Vol. 8, Sect. 31-32, p. 324-337. Munich.

Erickson, J. R. 1976. Exploratory trials with a spiral-head chipper to make hardwood "fingerling" chips for ring flakers. For. Prod. J. 26(6):50-53.

Fahey, T. D. 1978. Veneer recovery at a high-speed core lathe. For. Prod. J. 28(4):19-20.

Fasick, C. A. and J. D. Lawrence. 1971. An operations research application to furniture round production. For. Prod. J. 21(4):46-52.

Feihl, A. O. 1956. White elm veneer and plywood. Timber of Canada 17(9):95-97.

Feihl, A. O. 1959. Improved profiles for veneer knives. Canadian Woodworker 59(8):37-40.

Feihl, A. O. 1972. Heating frozen and nonfrozen veneer logs. For. Prod. J. 22(10):41-50.

Feihl, A. O. and M. N. Carroll. 1969. Rotary cutting veneer with a floating bar. For. Prod. J. 19(10):28-32.

Feihl, A. O., and V. Godin. 1967. Setting veneer lathes with aid of instruments. Dep. Pub. No. 1206, Canada Dep. For. and Rural Develop., For. Branch. 40 p.

Feihl, O. and V. Godin. 1971. Wear, play, and heat distortion in veneer lathes. Can. For. Serv. Publ. No. 1188, 12 p. Dep. of Fish. and For., Ottawa, Can.

Feihl, A. O., V. Godin, and J. G. T. Whitaker. 1976. Canadian pilot study looks at panel manufacture with thin faces. Plywood and Panel 17(5):18-20.

Feirer, J. L. 1970. Cabinetmaking and millwork. Chas. A. Bennett Co., Inc., Peoria, Ill. 925 p.

Fengel, D., M. Stoll, and G. Wegener. 1978. Studies on milled wood lignin from spruce Part II. Electron microscopic observations on the milled wood. Wood Sci. and Technol. 12:141-148.

Filer and Stowell. 1977. Super slant 5' & 6' linebar resaw systems—single, or twin or quad. For. Ind. 104(11):35-45.

Fine Woodworking Magazine. 1978. Fine Woodworking techniques 1. The Taunton Press, Inc., Newtown, Conn. 189 p.

Firth, R. D. 1972. Chain saw testing by New Zealand Forest Service evaluates noise, vibration, power characteristics. For. Ind. Rev. 3(5):2-9. Auckland.

Fischer, K. 1974. Progress in chip flaking. In Proc. of eighth Washington State Univ. symp. on particleboard, pp. 361-373. T. M. Maloney, Ed. Pullman, Wash.: Washington State Univ.

Flann, I. B. 1974a. Converting hardwood logs. Canadian For. Ind. 94(9):33-38.

Flann, I. B. 1974b. Sawing patterns, log grades, and log lengths for: converting hardwood logs. Can. For. Ind. 94(9):33-38.

Flann, I. B. 1976. Live sawing hardwoods—can it mean dollars for you? In Hardwood sawmill techniques, proc. of the first hardwood sawmill clinic program, pp. 1-23. V. S. White, ed. San Francisco, Ca.: Miller Freeman Publ., Inc.

Flann, I. B. 1978. Short-log processing—1977. Rep. OPX207E, Eastern For. Prod. Lab., Fisheries and Environ. Canada, Ottawa. 12 p.

Flann, I. B. and D. W. Bousquet. 1974. Sawing live vs. around for hard maple logs. Can. For. Serv. Inf. Rep. OPX91E, 107 p. Dep. of the Environ. East. For. Prod. Lab., Ottawa, Canada.

Fleischer, H. O. 1949. Experiments in rotary veneer cutting. For. Prod. Res. Soc. Proc. 3:137-155.

Fleischer, H. O. 1956. Instruments of alining the knife and nosebar on the veneer lathe and slicer. For. Prod. J. 6(1):1-5.

Fleischer, H. O. 1959. Heating rates for logs, bolts, and flitches to be cut into veneer. U.S. Dep. Agric. For. Serv., For. Prod. Lab. Rep. No. 2149. 18 p.

Fobes, E. W. 1959. Wood-chipping equipment and materials handling. U.S. Dep. Agric. For. Serv., For. Prod. Lab. Rep. 2160. 25 p.

Folkema, M. P., and P. Giguere. 1979. Delimbing with a chain flail and a knuckleboom loader. Tech. Rep. TR-35, For. Eng. Res. Inst. Can. (FERIC), Vancouver, B.C., Canada. 28 p.

Fondronnier, J., and J. Guillerm. 1967. Guide pratique de la derouleuse. Centre Technique du Bois, 10 Avenue de Saint-Mande, Paris 12e, France.

Forest Industries. 1981. Knife-like wood slitter said to eliminate sawdust. For. Ind. 108(3):126-127.

Forest Products Research Society. 1959. Wood machining abstracts, 1957-1958. For. Prod. Res. Soc. 20 p.

Forest Products Research Society. 1960. Wood machining abstracts, 1958-1959. For. Prod. Res. Soc. 19 p.

Forest Products Research Society. 1961. Wood machining abstracts, 1959-1961. For. Prod. Res. Soc. 18 p.

Forintek Canada Corportion. 1979. Horizontal gap HG and vertical gap VG computation charts to maintain HG/EG of 1.00. Forintek Canada Corp., Vancouver, B.C., Canada. 4 p.

Foschi, R. O. 1976. Log centering errors and veneer yield. For. Prod. J. 26(2):52-56.

Foschi, R. O. and A. W. Porter. 1970. Lateral and edge stability of high-strain band saws. Can. Dep. Fish. and For. For. Prod. Lab. Inform. Rep. VP-X-68, 13 p. Vancouver, B.C.

Franz, N. C. 1958. An analysis of the woodcutting process. Ph.D. Thesis. Univ. Mich. Press. Ann Arbor. 152 p.

Franz, N. C. 1970a. High velocity liquid jet. U.S. Pat. No. 3,524,367. U.S. Pat. Off., Washington, D.C. 6 p.

Franz, N. C. 1970b. Method for the high velocity liquid jet cutting of soft materials. U.S. Pat. No. 3,532,014. U.S. Pat. Off., Washington, D.C. 7 p.

Franz, N. C., and E. W. Hinken. 1954. Machining wood with coated abrasives. J. For. Prod. Res. Soc. 4:251-254.

Franzen, R.G. 1985. Atmospheric and pressurized disc refining. In, proceedings of symposium "Comminution of wood and bark" held October 1-3, 1984 in Chicago, Illinois. Forest Products Research Society, Madison, Wisconsin.

Freese, F., H. A. Stewart, and R. S. Boone. 1976. Rough thickness requirements for red oak furniture cuttings. U.S. Dep. Agric. For. Serv., Res. Pap. FPL-276. 4 p.

Frid, T. 1979. Tage Frid teaches woodworking. Book 1—Joinery. The Taunton Press, Newtown, CT. 224 p.

Fujii, J. S. 1964. Molecular degradation caused by vibratory ball-milling of wood. Ph.D. Dissertation, State University College of Forestry at Syracuse, Syracuse, N.Y. 120 p.

Fujiwara, K., M. Noguchi, and H. Sugihara. 1976. [Effects of ultrasonic vibration on kinetic friction based on a standpoint of ultrasonic vibratory cutting of wood.] J. Jap. Wood Res. Soc. 22(2):76-81.

Fukui, H., S. Kimura, and T. Yamashita. 1974. Noise reducing properties of helical cutting edge in router bit. J. Jap. Wood Res. Soc. 20(9):411-417.

Gallup, R. M. 1976. Clearing, grubbing, and disposing of road construction slash. U.S. Dep. Agric. For. Serv. Equip. Develop. Cent., San Dimas, Calif., Test Rep. 7700-11. 14 p.

Gambrell, S. C., Jr. and E. F. Byars. 1966. Cutting characteristics of chain saw teeth. For. Prod. J. 16(1):62-71.

Gardner, J. 1977. Building classic small craft. International Marine Publishing Company, Camden, Maine. 300 p.

Gardner, R. B., and D. F. Gibson. 1974. Improved utilization and disposal of logging residues. ASAE 74-1511. Amer. Soc. Agric. Eng., St. Joseph, Mo. 11 p.

Garrett, L. D. 1969. Economic implications of manufacturing sawed ties and timbers. U.S. Dep. Agric. For. Serv., Res. Pap. NE-148. 24 p.

Garrett, L. D. 1970. Physical suitability of Appalachian hardwood sawlogs for sawed timbers. U.S. Dep. Agric. For. Serv., Res. Note NE-121. 6 p.

Gatchell, C. J., R. E. Coleman, and H. W. Reynolds. 1977. Machining the serpentine end-matched joint. Furniture Des. and Manuf. 49(6):30-33.

Gauthier, E. A. 1975. What multi stem chain flail delimbing has done for the pulp and paper industry. Amer. Soc. of Agric. Eng. Pap. No. 75-1590, 5 p.

Gavelin, N. G. 1966. Science and technology of mechanical pulp manufacture. New York: Lockwood Publishing Co., Inc. 245 p.

Gillson, J. L., et al., ed. 1960. Industrial minerals and rocks. 3rd ed. New York: American Inst. Mining, Metallurgical, and Petroleum Eng. 934 p.

Godin, V. 1968. The grinding of veneer knives. Dep. Pub. No. 1236, Canada Dep. For. and Rural Develop., For. Branch, 23 p.

Goho, C. D. and A. J. Martin. 1973. Sawlog sizes: a comparison in two Appalachian areas. U.S. Dep. Agric. For. Serv., Res. Note NE-160. 5 p.

Goho, C. D. and P. S. Wysor. 1970. Characteristics of factory-grade hardwood logs delivered to Appalachian sawmills. U.S. Dep. Agric. For. Serv., Res. Pap. NE-166. 17 p.

Golob, T. B. 1976. Analysis of shear felling of trees. For. Manage. Inst. Inf. Rep. FMR-X-93, 48 p. Can. For. Serv., Ottawa, Ontario, Can.

Goodchild, R. 1955. The machine boring of wood. Bull. No. 35, Dep. Sci. and Ind. Res., For. Prod. Res., London.

Goodman, W. L. 1964. The history of woodworking tools. 208 p. G. Bell and Sons, Ltd., London.

Grant, S. E. 1967. New developments in field manufacture and transportation of wood chips. Tappi 50(5):96A-98A.

Greenlee Bros. & Co. 1962. Round bits . . . square holes. The story of Greenlee Bros. & Co. Greenlee Bros. & Co., Rockford, Ill. 108 p.

Greenwood, J. H. F. 1968. Noise reducing enclosure for a planer and moulder. Woodwork. Ind. 25(11):19-20.

Griffin, G. 1977. Chip-n-saw hardwood run aids land program. Timber Proc. Ind. 2(12):17-20.

Hafley, W. L. and K. W. Hanson. 1973. Optimum sequence of cutting bills. For. Prod. J. 23(8):26-29.

Hailey, J. R. T., W. V. Hancock, and W. G. Warren. 1980. Effect of 4-foot lathe parameters on veneer yield and quality using response-surface analysis. Wood Sci. 12:141-148.

Hakkila, P., and H. Kalaja. 1981. KOPO block chip system. Folia For. 467:1-24.

Hall, S. P., R. A. Wysk, E. M. Wengert, and M. A. Agee. 1980. Yield distributions and cost comparisons of a crosscut-first and a gang-rip-first rough mill producing hardwood dimension stair parts. For. Prod. J. 30(5):34-39.

Hallett, R. M. 1972. A method to analyze rough mill productivity. For. Prod. J. 22(11):22-27.

Hallock, H. 1968. "Taper-tension" saw—a new reduced kerf saw. U.S. Dep. Agric. For. Serv. Res. Note FPL-0185, 7 p.

Hallock, H. 1973. Best opening face for second growth timber. In Modern Sawmill Techniques, Proc. of the First Sawmill Clinic, Portland, Ore., Feb. (V. S. White, ed.), Miller-Freeman Publ., Inc. p. 93-116.

Hallock, H. and L. Galiger. 1971. Grading hardwood lumber by computer. U.S. Dep. Agric. For. Serv., Res. Pap. FPL 157. 16 p.

Hallock, H., and P. Giese. 1980. Does gang ripping hold the potential for higher clear cutting yields. U.S. Dep. Agric. For. Serv., Res. Pap. FPL 369. 6 p.

Hallock, H., and D. W. Lewis. 1971. Increasing softwood dimension yield from small logs—Best Opening Face. U.S. Dep. Agric. For. Serv. Res. Pap. FPL 166. 12 p.

Hallock, H., and D. W. Lewis. 1979. Individual log yields by four centered sawing systems. U.S. Dep. Agric. For. Serv. Supplement to Res. Pap. FPL 321. 48 p.

Hallock, H., A. R. Stern, and D. W. Lewis. 1976. Is there a "best" sawing method? U.S. Dep. Agric. For. Serv., Res. Pap. FPL 280. 12 p.

Hallock, H., A. R. Stern, and D. W. Lewis. 1979. A look at centered vs. offset sawing. U.S. Dep. Agric. For. Serv. Res. Pap. FPL-321. 16 p.

Hallock, H., A. R. Stern, and D. W. Lewis. (n.d.) Individual log yields by eight sawing systems. U.S. Dep. Agric. For. Serv., Suppl. to FPL 280. 81 p.

Hamamoto, K., and M. Mori. 1971. Fundamental studies on low frequency cutting of wood. I. Mechanics of vibratory cutting parallel to feeding direction. J. Jap. Wood Res. Soc. 17:137-146.

Hamamoto, K., and M. Mori. 1973. Fundamental studies on low frequency cutting of wood. V. Effects of vibratory factors on service life of coated abrasives in vibratory sanding. J. Jap. Wood Res. Soc. 19(7):305-310.

Hanchett, K. S., ed. 1946. The Hanchett saw and knife fitting manual. Ed. 6, 471 p. Big Rapids, Mich.: Hanchett Manufacturing Co.

Hancock, W. V. and O. Feihl. 1976. Lathe operators' manual. Part II. Pressure bars: their operation and maintenance. Can. For. Serv. Inf. Rep. VP-X-158, 55 p. West. For. Prod. Lab., Vancouver, B.C., Can.

Hancock, W. V., and O. Feihl. 1980. Lathe operator's manual. Part II. Pressure bars: Their operation and maintenance. Spec. Rep. No. SP-5R. Forintek Canada Corp., Vancouver, B.C. 24 p.

Hancock, W. V. and J. R. T. Hailey. 1975. Lathe operators' manual. Part I. Operating a veneer lathe. West. For. Prod. Lab. Info. Rep. VP-X-130, 43 p. Can. For. Serv., Vancouver, B.C., Can.

Hanks, L. F. 1977. Predicted cubic-foot yields of lumber, sawdust, and sawmill residue from the sawtimber portions of hardwood trees. U.S. Dep. Agric. For. Serv., Res. Pap. NE-380. 23 p.

Hann, R. A. 1957. A method of quantitative topographic analysis of wood surfaces. For. Prod. J. 7:448-452.

Hann, R. A., R. W. Jokerst, R. S. Kurtenacker, C. C. Peters, and J. L. Tschernitz. 1971. Rapid production of pallet deckboards from low-grade logs. U.S. Dep. Agric. For. Serv., Res. Pap. FPL 154. 16 p.

Hanna, J. S. 1975. Marine carving handbook. International Marine Publishing Co., Camden, Maine. 92 p.

Hanover, S. J., W. L. Hafley, A. G. Mullin, and R. K. Perrin. 1973. Linear programming and sensitivity analysis for hardwood dimension production. For. Prod. J. 23(11):47-50.

Harris, P. 1960. Mechanics of sawing: Band and circular saws. Dep. Sci. and Ind. Res., For. Prod. Res. Bull. 30, 30 p. London.

Harris, J. D. and A. J. Nash. 1960. Girard form class estimation from lower bole measurements on upland white oak trees in Missouri. J. of For. 58:534-536.

Harrison, R. T. 1975. Slash—Equipment and methods for treatment and utilization. U.S. Dep. Agric. For. Serv., Equip. Develop. Cen., San Dimas, Calif., ED&T Rep. 7120-7. 47 p.

Hartler, N. 1962. The effect of spout angle as studied in an experimental chipper. Svensk Papperstidn. 65(9):351-362.

Hartler, N. 1977. Influence of chip moisture in mechanical pulping. EUCEPA Intern. Mech. Pulping Conf. (Helsinki) Proc. Vol. 1, Pap. No. 6:19 p.

Hartler, N., and Y. Stade. 1977. Chipper operation for improved chip quality. Svensk Papperstidning 80(14):447-453.

Hartman, H. A. 1975. North American and Scandinavian sawmilling practices and equipment—main characteristics and differences. Ch. 1 in Nordic and North American Sawmill Techniques. Proc., 2nd Int. Sawmill Sem. Miller-Freeman Publ., Inc., San Francisco. p. 9-26.

Hartzell, G. W. 1976. A double cut mill with overhead carriage and rotating end dogs. In Hardwood sawmill techniques, proc. of the first hardwood sawmill clinic program, pp. 1-11. San Francisco, Ca.: Miller Freeman Publ., Inc.

Harvey, H. C. Jr. 1972. Green-wood fibrating means and method. (U.S. Pat. No. 3,674,219) U.S. Pat. Office, Wash., D.C.

Hasenwinkle, E. D. 1974. Log cutting and rejoining process. U.S. Pat. No. 3,903,943. U.S. Pat. Off., Washington, D.C.

Hatton, J. V. 1976. Chip quality evaluation: comparison of a combination slot screen-round hole screen with a round-hole screen classifier. Pulp & Pap. Can. 77(6):T99-T103.

Hatton, J. V. 1977a. Thin chips give highest screened yields in mixed softwood/hardwood kraft pulping. Tappi 60(5):116-117.

Hatton, J. V. 1977b. Screening mill chips for sizeable savings. Pulp and Pap. Can. 78(3):T57-T60.

Hatton, J. V. 1978a. Effect of chip thickness on the delignification of hardwoods by the kraft process. Transactions 79(6):55-62. (Can. Pulp and Pap. Asoc.)

Hatton, J. V. 1978b. Effects of chip size on the kraft pulping of Canadian hardwoods. Transactions 79(6):49-55. (Can. Pulp and Pap. Assoc.)

Hatton, J. V. 1978c. Is TAPPI ready for a new chip-quality testing procedure. Tappi 61(11):71-74.

Hatton, J. V., ed. 1979a. Chip quality monograph. Pulp and Paper Technology Series No. 5, Tech. Assoc. Pulp Pap. Ind., Atlanta, Ga.

Hatton, J. V. 1979b. Screening of pulpwood chips for quality and profit. Pap. Trade J. 163(8):25-27.

Hayashi, D., and O. Hara. 1964. Studies on surface sanding of lauan plywood. Wood Ind. 19(9):1-6. (Japan.)

Heebink, B. G. 1975. Producing flakes for structural particleboard on a ringhead planer. U.S. Dep. Agric. For. Serv., Res. Pap. FPL 246. 12 p.

Hefty, F. V., and J. K. Brooks. 1968. Portable apparatus for measuring surface irregularities in panel products. U.S. Dep. Agric. For. Serv. Res. Note FPL-0192. 9 p.

Helgesson, T. 1978. (Flail delimbing—A bundle delimbing method for wood removed in thinning and cleaning proceses.) Svenska Träforskningsinstitutet, No. 472. 21 p. (Sweden)

Helmers, R. A. 1975. Lumber-yield management: prime opportunity for cost reduction. Furniture Design and Manuf. 47(5):28-36.

Hensel, J. S. 1971. Semi-permanent Nicholson chip mill. Am. Pulpwood Assoc. Tech. Rel. 71-R-51, 4 p. Washington, D.C.

Hensel, J. S. 1976. Chain saw data. Am. Pulpwood Assoc. Tech. Rel. 76-R-38, 32 p. Washington, D.C.: Am. Pulpwood Assoc.

Herrick, O. W. and W. W. Christensen. 1967. A cost analysis of chip manufacture at hardwood sawmills. U.S. Dep. Agric. For. Serv., Res. Pap. NE-69. 14 p.

Hewitt, J. 1978. Armstrong carbide filer's handbook. 70 p. Portland, Ore.: Armstrong Manuf. Co.

Heydt, F., and H. J. Schwarz. 1977. (Noise research and noise abatement on multi-side woodworking planers and moulders.) Holz als Roh- und Werkstoff 35:323-326.

Hilt, D. E., and M. E. Dale. 1979. Stem form changes in upland oaks after thinning. U.S. Dep. Agric. For. Serv. Res. Pap. NE-433. 7 p.

Hirst, K. 1971. Perceptibility of machining marks on painted timber surfaces. CSIRO For. Prod. Newsletter No. 380, Australia, p. 2-4.

Hoadley, R. B. 1980. Understanding wood. A craftsman's guide to wood technology. The Taunton Press, Newtown, Conn. 256 p.

Höglund, H. 1973. Energy uptake by wood in the mechanical pulping process. (Abst.) In International Mechanical Pulping Conf. Proc., Sess. I., Swedish Assoc. Pulp and Pap. Eng., and European Liaison Comm. for Pulp and Pap., Stockholm. p. 15-16.

Holemo, F. J., and P. J. Dyson. 1972. Ratio-delay—A method for analyzing downtime in sawmills. For. Prod. J. 22(8): 56-60.

Howe, J. P. and P. Koch. 1976. Dowel-laminated crossties—performance in service, technology of fabrication, and future promise. For. Prod. J. 26(5):23-30.

Hoyle, R. J., Jr. 1956. The effect of boring speed and feed rate on the strength of glued dowel joints in tension. For. Prod. J. 6:387-393.

Hse, C.Y. 1972. Method for computing a roughness factor for veneer surfaces. Wood Sci. 4:230-233.

Huang, Y., and D. Hayashi. 1973. [Basic analysis of mechanism in woodcutting. Stress analysis in orthogonal cutting parallel to grain.] J. Jap. Wood Res. Soc. 19(1):7-12.

Huber, H. A. 1971. Computerized economic comparison of a conventional furniture rough mill with a new system of processing. For. Prod. J. 21(2):34-38.

Huber, H. A. and S. B. Harsh. 1977. Roughmill improvement program. Woodworking & Furniture Dig. 79(2):26-29.

Huber, H. A., C. W. McMillin and A. Rasher. 1982. Economics of cutting wood parts with a laser under optical image analyzer control. For. Prod. J. 32(3):16-21.

Huffman, J. B. 1973. Timber for Florida today and tomorrow. Comparative yields of furniture parts cut from short bolts and conventional logs of Florida red maple. 25 p. Tallahassee, Fla.: Div. of For., Florida Dep. of Agric.

Huyler, N. K. 1974. Live-sawing: a way to increase lumber grade yield and mill profits. U.S. Dep. Agric. For. Serv., Res. Pap. NE-305. 9 p.

Inoue, H., and M. Mori. 1979. Effects of cutting speed on chip formation and cutting resistance in cutting of wood parallel to the grain. J. Jap. Wood Res. Soc. 25:22-29.

Ivanovskij, E. G., and G. M. Goronok. 1978. (Investigation of the sliding friction between knife and chip in cutting wood.) Holztechnologie 19(1):33-38.

Johansson, K. J. 1979. (Production of fuel chips with four tractormounted chippers.) Rapp. 5/79, Norsk Institutt for Skogforskning, Skogteknologisk Avdeling, 1432 AS-NLH. 37 p.

Johnson, W. W. 1977. Power chain saws—their care and use. Spec. Circ. 228. Pa. State Univ., Coop. Ext. Serv., University Park, Pa. 13 p.

Johnston, J. S. 1967. Investigation of some variables in the crosscutting of small logs by shear blades. IUFRO Congr. Proc. 1967, Vol. 8, Sect. 31-32, p. 338-363. Munich.

Johnston, J. S. 1968a. An experiment in shear-blade cutting of small logs. Pulp and Pap. Mag. Can. 69(3):77-82.

Johnston, J. S. 1968b. Crosscutting trees and logs by shear blades. Can. For. Ind. 88(6):34-37.

Johnston, J. S. 1968c. Experimental crosscut shearing of frozen wood. Proc., For. Eng. Conf. Amer. Soc. Agric. Eng., ASAE Pub. PROC-368, p. 47-49, 55. East Lansing, Mich.

Johnston, J. S. 1968d. Experiments in crosscutting wood with shear blades. For. Prod. J. 18(3):85-89.

Johnston, J. S., and A. St. Laurent. 1970. V-shaped tree-shear blades spell end to butt-splitting. Can. For. Ind. 90(8):39-45.

Johnston, J. S., and A. St. Laurent. 1971. Crosscut shearing of frozen trees. Pulp and Pap. Mag. Can. 72(5):97, 99, 102, 104, 106, 108.

Johnston, J. S., and A. St. Laurent. 1974a. Evaluation of damage in shear-felled sawtimber. Environment Canada, Eastern For. Prod. Lab. Rep. OPX127E. 13 p.

Johnston, J. S. and A. St. Laurent. 1974b. Force and energy required for removal of single branches. East. For. Prod. Lab. Rep. OPX76E, 16 p. Can. For. Serv.

Johnston, J. S., and A. St. Laurent. 1975a. Method for kerfless cutting wood. U.S. Pat. No. 3,916,966. U.S. Pat. Off., Washington, D.C.

Johnston, J. S., and A. St. Laurent. 1975b. Tooth side clearance requirements for high precision saws. For. Prod. J. 25(11):44-49.

Johnston, J. S. and A. St. Laurent. 1978. Compression slicing of wood. For. Prod. J. 28(7):48-53.

Johnston, J. S., and A. St. Laurent. 1979. Development of a full-size lumber slicer. In Proc., Sixth Wood Machining Seminar, Univ. Calif., For. Prod. Lab., Richmond, Calif. (J. F. Walters, ed.) p. 70-80.

Johnstone, J. B., and the Sunset Editorial Staff. 1971. Woodcarving techniques and projects. Lane Publishing Co., Menlo Park, Calif. 80 p.

Jokerst, R. W. and H. A. Stewart. 1976. Knife-versus abrasive-planed wood: quality of adhesive bonds. Wood and Fiber 8(2):107-113.

Jones, D. S. 1963. The performance of thin circular saws in several dense species. Australian Timber J. 29(1):17, 19-20, 24-25.

Jones, D. S. 1965. An experimental analysis of saw tooth stress and deflection. Paper presented at IUFRO Sec. 41 meeting, Melbourne, Oct. 21 p.

Jones, C. W. 1976. Processing southern hardwoods with a double-taper chipper canter. FPRS Sep. No. MS-75-S68, 4 p. For. Prod. Res. Soc., Madison, Wis.

Jones, K. C. 1981a. Energy requirements to produce fuel from harvesting residues. *In* Proc., International Conf., "Harvesting and utilization of wood for energy purposes", ELMIA, Jonkoping, Sweden, Sept. 29-30, 1980. (J. E. Mattsson and P. O. Nilsson, eds.) No. 19, Swedish Univ. Agric. Sci., Dept. Operational Efficiency, Garpenberg, Sweden. p. 184-196.

Jones, K. C. 1981b. Fuel preparation from harvesting residues. Rep. No. P-28, FERIC, For. Eng. Res. Inst. of Canada, Pointe Claire, P.Q.

Kato, K., and I. Asano. 1974a. [Studies on vibration cutting of wood. III. Cutting forces in lateral vibration cutting.] J. Jap. Wood Res. Soc. 20(9):418-423.

Kato, K., and I. Asano. 1974b. [Studies on vibration cutting of wood. IV. The effect of fiber direction and annual ring on vibration cutting forces.] J. Jap. Wood Res. Soc. 20(9):424-429.

Kato, K., and I. Asano. 1974c. [Studies on vibration cutting of wood. II. Cutting forces in longitudinal vibration cutting.] J. Jap. Wood Res. Soc. 20(5):191-195.

Kato, K., K. Tsuzuki, and I. Asano. 1971. [Studies on vibration cutting of wood. I.] J. Jap. Wood Res. Soc. 17(2):57-65.

Kato, C., and H. Fukui. 1976a. The cutting force and the stock removal rate of coated abrasives in sanding wood under constant sanding pressure. J. Jap. Wood Res. Soc. 22:349-357.

Kato, C., and H. Fukui. 1976b. The effect of belt oscillation on sanding performance of belt sander. J. Jap. Wood Res. Soc. 22:550-556.

Kato, C., and H. Fukui. 1977. The distribution status of grain tip heights of coated abrasive belt. I. Measurement of the height distribution of outer grain tip and the effect of the distribution status on the roughness of finished surface of wood. J. Jap. Wood Res. Soc. 23:617-623.

Keating, J. L. 1976. Disc screen cuts hog fuel cost and maintenance requirements. Pulp and Pap. 50(6):139.

Keays, J. 1975. International mechanical pulping conference—a detailed analysis. Pulp & Pap. Can. 76(8):43-47.

Keays, J. L., and R. A. Leask. 1973. Refiner mechanical pulp—past, present, and potential. Pap. Trade J. 157(35):20-29.

Keighley, G. D. 1970. Effects of chainsaw vibration. Timber Trade J. 1970 (April 11, Suppl., Forestry and Home Grown Timber): 39, 41.

Keighley, G. D. 1973. Vibration and noise—risk of injury to chain saw operators. For. and Home Grown Timber 2(1):29-31, Glasgow.

Kempe, C. 1967. (Forces and damage involved in the hydraulic shearing of wood.) Stud. Forest. Suecica 55, 38 p.

Kersavage, P. C. 1972. Sawing method efffect on the production of cherry lumber. For. Prod. J. 22(8):33-40.

Kersavage, P. C. 1978. Continuous recording system for monitoring sawmill production. For. Prod. J. 28(2):21-25.

Kimura, S. 1980. (Cutting with helical cutters. I. Changes of the torque of main spindle during a single engagement.) J. Jap. Wood Res. Soc. 26:254-261.

Kimura, S., and H. Fukui. 1976. Circular saw noise. III. Free running noise. J. Jap. Wood Rep. Soc. 22(3):146-151.

King, W. W. 1956. Effect of sawing method on volume and value of 4/4 lumber from low-grade oak logs—results from test run of 360 logs. Tenn. Val. Auth. No. 24, 15 p. Norris, Tenn.

Kinoshita, N. 1980a. (Oblique cutting of woods. I. Effects of knife inclination angle and depth of cut on cutting force in cutting parallel to the grain.) J. Jap. Wood Res. Soc. 26:241-247.

Kinoshita, N. 1980b. (Oblique cutting of woods. II. Effects of knife inclination angle and depth of cut on cutting force in cutting perpendicular to the grain.) J. Jap. Wood Res. Soc. 26:248-253.

Kirbach, E. [n.d.] The carbide-tipped saw, its use in lumber manufacture. Forintek Canada Corp., Vancouver, B.C., Canada.

Kirbach, E., and T. Bonac. 1978a. An experimental study of the lateral natural frequency of bandsaw blades. Wood and Fiber 10(1):19-27.

Kirbach, E., and T. Bonac. 1978b. The effect of tensioning and wheel tilting on the torsional and lateral fundamental frequencies of bandsaw blades. Wood and Fiber 9:245-251.

Kirbach, E., and T. Bonac. 1979a. Clearance angle of rip-saws. A reduction can pay off. British Columbia Lumberman 63(10):22-23.

Kirbach, E. D., and T. Bonac. 1979b. Minimum clearance angle required for ripsawing some softwoods. In Proc., Sixth Wood Machining Seminar, Univ. Calif., Berkeley. p. 139-153.

Kirbach, E. D., and T. Bonac. 1981. Influence of grinding direction and wheel translation on microsharpness of cemented tungsten-carbide tips. Holz als Roh- und Werkstoff 39:265-270.

Kivimaa, E. 1950. Cutting force in woodworking. State Inst. Tech. Res. Pub. 18, 101 p. Helsinki, Finland.

Kivimaa, E., and J. Kovanen. 1953. (Microsharpening of veneer lathe knives.) Tiedoitus 126, Valtion Teknillinen Tutkimuslaitos, Helsinki, Finland. 24 p.

Klamecki, B. E. 1979. Discontinuous chip formation in woodcutting—A catastrophe theory description. Wood Sci. 12-32-37.

Knokey, E. R. 1979. Computerized block centering increases accuracy, recovery. For. Ind. 106(1):38-39.

Koch. C. B., G. W. Galliger, and N. D. Jackson. 1968. Yield from sprout black cherry trees sawed with a short-log bolter saw. West Virginia Univ. Agric. Exp. Stn. Bull. 569. 11 p. Morgantown.

Koch, P. 1948. Plane talk for better lumber. Two parts. I. Wood 3(10):26, 28, 30; II. 3 (11):34, 36, 38, 53.

Koch, P. 1954. An analysis of the lumber planing process. Ph.D. Thesis, Univ. Washington, Seattle. 339 p. (Available on microfilm from Univ. Michigan, Ann Arbor.)

Koch, P. 1955. An analysis of the lumber planing process: Part. I. For. Prod. J. 5:255-264.

Koch, P. 1956. An analysis of the lumber planing process: Part II. For. Prod. J. 6:393-402.

Koch, P. 1964a. Wood machining processes. 530 p. Ronald Press Co., New York.

Koch, P. 1964b. Square cants from round bolts without slabs or sawdust. For. Prod. J. 14:332-336.

Koch, P. 1967. Development of the chipping headrig. Rocky Mountain Forest Ind. Conf. Proc. 1967: 135-155. Fort Collins, Colo.

Koch, P. 1968a. Wood machining abstracts, 1966 and 1967. Res. Pap. SO-34, U.S. Dep. Agric. For. Serv. 38 p.

Koch, P. 1968b. Converting southern pine with chipping headrigs. Southern Lumberman 217(2704):131-138.

Koch, P. 1971. Force and work to shear green southern pine logs at slow speed. For. Prod. J. 21(3):21-26.

Koch, P. 1972. Utilization of the southern pines. U.S. Dep. Agric. For. Serv., Agric. Handb. 420. 1663 p. 2 vol. U.S. Govt. Print. Off., Washington, D.C.

Koch, P. 1973. Wood machining abstracts, 1970 and 1971. Res. Pap. SO-83, U.S. Dep. Agric. For. Serv. 46 p.

Koch, P. 1974. Development of the shaping-lathe headrig. U.S. Dep. Agric. For. Serv., Res. Pap. SO-98. 20 p.

Koch, P. 1975. Shaping-lathe headrig now commercially available. South. Lumberman 231(2872):93-97.

Koch, P. 1976a. Prototype flaking head smooths surfaces left by headrig or edger chipping heads. For. Prod. J. 26(12):22-27.

Koch, P. 1976b. Making dowel-laminated-crossties with the shaping-lathe headrig—it should be profitable. South. Lumberman. 233(2896):82-83.

Koch, P. 1978a. Five new machines and six products can triple commodity recovery from southern forests. J. For. 76: 767-772.

Koch, P. 1978b. Production opportunities in four southern locations. In Structural flakeboard from forest residues: symp. proc., Kansas City, Mo., June 6-8. U.S. Dep. Agric. For. Serv., Gen. Tech. Rep. WO-5, p. 150-166.

Koch, P. 1978c. Two methods of acquiring residual wood for southern flakeboard plants: the shaping-lathe headrig and the mobile chipper. In Structural flakeboard from forest residues: symp. proc., Kansas City, Mo., June 6-8. U.S. Dep. Agric. For. Serv., Tech. Rep. WO-5, p. 39-46.

Koch, P. 1978d. Helical flaking head with multiple cutting circle diameters. U.S. Pat. No. 4,131,146. U.S. Pat. Off., Washington, D.C.

Koch, P. 1982. Non-pulp utilization of above-ground biomass of mixed-species forests of small trees. Wood and Fiber 14:118-143.

Koch, P., and R. A. Caughey. 1978. Shaping-lathe headrig yields solid and molded-flake hardwood products. For. Prod. J. 28(10):53-61.

Koch, P., and J. Dobie. 1979. Production of chips by chipping headrigs. Ch. 3 *in* Chip Quality Monograph No. 5, Pulp and Paper Technology Series (J. V. Hatton, ed.) p. 33-70. Tech. Assoc. Pulp Pap. Ind., Atlanta, Ga.

Koch, P., and W. Gruenhut. 1981. Turning small-log hardwoods into pallet parts and profits. *In* Utilization of low-grade southern hardwoods. (D. A. Stumbo, ed.) Symposium Proc., Nashville, Tenn., Oct. 1980. For. Prod. Res. Soc., Madison, Wis. p. 113-120.

Koch, P. and D. W. McKenzie. 1976. Machine to harvest slash, brush, and thinnings for fuel and fiber—a concept. J. of For. 74:809-812.

Koch, P., and C. W. McMillin. 1966. Wood machining review, 1963 through 1965. Part I. For. Prod. J. 16(9):76-82, 107-115; Part II. 16(10):43-48.

Koch, P., and T. W. Nicholson. 1978. Harvesting residual biomass and swathe-felling with a mobile chipper. *In* Complete tree utilization of southern pine: symp. proc., New Orleans, La., April 17-19, C. W. McMillin, ed. For. Prod. Res. Soc., Madison, Wis. p. 146-154.

Koch, P. and T. E. Savage. 1980. Development of the swathe-felling mobile chipper. J. For. 78(1):17-21.

Kollmann, F. F. P., and W. A. Côté, Jr. 1968. Principles of wood science and technology. I. Solid wood. 592 p. Springer-Verlag, New York.

Komatsu, M. 1975. Machine boring properties of wood. I. The machine boring tests on Japanese hardwood. J. of the Jap. Wood Res. Soc. 21(10):551-557.

Komatsu, M. 1976. (Machine boring properties of wood. II. The effects of boring conditions on the cutting forces and the accuracy of finishing.) J. Jap. Wood Res. Soc. 22:491-497.

Komatsu, M. 1977. (Machine boring properties of wood. III. The comparison of the machinability of tropical woods in machine boring test with that in sawing test by circular saw.) J. Jap. Wood Res. Soc. 23:640-647.

Komatsu, M. 1978a. (Machine boring properties of wood. IV. The effects of wood moisture content on the boring forces and chip formation.) J. Jap. Wood Res. Soc. 24:26-31.

Komatsu, M. 1978b. (Machine boring properties of wood. VI. The effects of the point angle of twist drill on the accuracy of finishing.) J. Jap. Wood Res. Soc. 24:692-697.

Komatsu, M. 1979a. Machine boring properties of wood. IX. The effects of grain angle of wood on the cutting force of twist drill. J. Jap. Wood Res. Soc. 25(9):573-581.

Komatsu, M. 1979b. Machine boring properties of wood. X. The effects of grain angle of wood on the cutting accuracy of twist drill. J. Jap. Wood Res. Soc. 25(9):582-587.

Kondo, T., H. Fukui, and S. Kimura. 1971. Fundamental machining performance of router and chip producing mechanism in routing (II). Weight distribution of chip. Wood Ind. 26(9):26-29.

Kubler, H. 1960. (Cutting timber with vibrating knives.) Holz Zentralbl. 86:1605-1606.

Kuykendall, R. 1975. Mis-centering veneer block in lathe can cost a $bundle per shift. For. Ind. 102(9):58.

Lambert, M. B. 1974a. Evaluation of power requirements and blade design for slash cutting machinery. Paper No. 74-1570, presented at 1974 Winter Meet., Amer. Soc. Agric. Eng., Chicago, Dec. 10-13. 45 p.

Lambert, H. 1974b. Fiber supply, environment, employment, improved by wood scrap conversion. For. Ind. 101(8):93-95.

Landt, E. F. 1974. Soft maple volume tables for furniture-type, flat, 4/4-inch dimension from small low-quality trees. U.S. Dep. Agric. For. Serv., Res. Note NC-169. 2 p.

Lane, P. H. 1954. Overrun from yellow-poplar sawlogs—some factors affecting overrun at circular mills. For. Invest. Tech. Note No. 21, 16 p. Norris, Tenn.: Div. of For. Relations, Tenn. Vall. Auth.

Larson, P. R. 1963. Stem form development of forest trees. For. Sci. Monograph 5-1963. Soc. Amer. For., Washington, D.C. 42 p.

Leask, R. A. 1977. Mechanical pulp—past and present. South. Pulp Pap. Manuf. 40(3):26-27, 30-33.

Leask, R. A. 1981. The theory of chip refining—a status report. Svensk Papperstidning 84(14):28-30, 33.

Leider, P. J. and J. Rihs. 1977. Understanding the disk refiner. 1. The hydraulic behavior. Tappi 60(9):98-102.

Leitch, W. C. 1976. Hand-hewn. The art of building your own cabin. San Francisco: Chronicle Books. 122 p.

Leney, L. 1960. Mechanism of veneer formation at the cellular level. Mo. Agric. Exp. Stn. Res. Bull. 744, 111 p.

Leslie, H. C. 1976. FS scientists increase lumber yield by listening to boards. U.S. Dep. Agric. For. Serv., Res. News. 3 p. For. Prod. Lab., Madison, Wis.

Little, R. L. 1976. Hardwood sawmills: consideration in their planning and design. Ala. For. Prod. 19(7):57-60.

Lubkin, J. L. 1957. A status report on research in the circular sawing of wood. Cent. Res. Lab. Tech. Rep. CRL-T-12, 193 p. Amer. Mach. and Foundry Co., Greenwich, Conn.

Lucas, E. L. 1973. Long length cuttings from No. 2 Common hardwood lumber. U.S. Dep. Agric. For. Serv., Res. Pap. NE-248. 5 p.

Lucas, E. L., and P. A. Araman. 1975. Manufacturing interior furniture parts: A new look at an old problem. U.S. Dep. Agric. For. Serv., Res. Pap. NE-334. 6 p.

Lucas, E. L., and L. R. R. Catron. 1973. A comprehensive defect data bank for No. 2 Common oak lumber. U.S. Dep. Agric. For. Serv., Res. Pap. NE-262. 7 p.

Lucas, E. L., and C. J. Gatchell. 1976. STUB—A manufacturing system for producing rough dimension cuttings from low-grade lumber. U.S. Dep. Agric. For. Serv. Res. Note NE-222. 5 p.

Lunstrum, S. J. 1972. Circular sawmills and their efficient operation. U.S. Dep. Agric. For. Serv., State and Private For., Southeastern Area, Atlanta, Ga. 86 p.

Lunstrum, S. J. 1975. Origin and background of the sawmill improvement program. FPRS Sep. No. SE-74-S48, 5 p. For. Prod. Res. Soc., Madison, Wis.

Lutz, J. F. 1955. Hickory for veneer and plywood. U.S. Dep. Agric. For. Serv., Hickory Task Force Rep. No. 1. 12 p. Southeast. For. Exp. Stn., Asheville, N.C.

Lutz, J. F. 1956. Effect of wood-structure orientation on smoothness of knife-cut veneers. For. Prod. J. 7:464-468.

Lutz, J. F. 1963. Precompressing flitches to improve the quality of slicewood. For. Prod. J. 13:248-249.

Lutz, J. F. 1975. Manufacture of veneer and plywood from United States hardwoods with special reference to the South. U.S. Dep. Agric. For. Serv., Res. Pap. FPL 255. 8 p.

Lutz, J. F. 1977. Wood veneer: Log selection, cutting, and drying. U.S. Dep. Agric., Tech. Bull. 1577. 137 p. Washington, D.C.

Lutz, J. F., H. H. Haskell, and R. McAlister. 1962. Slicewood—a promising new wood product. For. Prod. J. 12:218-227.

Lutz, J. F., and R. H. McAlister. 1963. Processing variables affect chestnut oak veneer quality. Plywood Magazine 4(3):26-31.

Lutz, J. F., B. G. Heebink, H. R. Panzer, F. V. Hefty, and A. F. Mergen. 1969a. Surfacing softwood dimension lumber to produce good surfaces and high value flakes. For. Prod. J. 19(2):45-51.

Lutz, J. F., A. F. Mergen, and H. R. Panzer. 1969b. Control of veneer thickness during rotary cutting. For. Prod. J. 19(12):21-28.

Lutz, J. F. and R. A. Patzer. 1966. Effects of horizontal roller-bar openings on quality of rotary-cut southern pine and yellow-poplar veneer. For. Prod. J. 16(10):15-25.

Lutz, J. F. and R. A. Patzer. 1976. Spin-out of veneer blocks during rotary cutting of veneer. U.S. Dep. Agric. For. Serv., Res. Pap. FPL 278. 23 p.

McCarty, E. F. 1973. Technology developed for high hardwood groundwood newsprint. Pap. Trade J. 157(24):38-42.

McDonald, K. A. 1975. FPL's defectoscope: experimental device to increase lumber yields. South. Lumberman 231(2872):67-68.

McDonald, K. A. 1978. Lumber defect detection by ultrasonics. U.S. Dep. Agric. For. Serv. Res. Pap. FPL-311. 21 p.

McGovern, J. N., and R. J. Auchter. 1976. Mechanical and semichemical pulping of hardwoods growing on southern pine sites. South. Pulp and Pap. Manuf. 39(3):46, 49, 51, 52.

McIntosh, J. A., and E. L. Kerbes. 1969. Lumber losses in tree shear felling. Brit. Columbia Lumberman 53(10):43-46.

McIntosh, J. A., and T. A. McLauchlan. 1974. Damage reduced in single-blade tree shear tests. British Columbia Lumberman 58(6):60-61.

McKenzie, W. M. 1961. Fundamental analysis of the wood-cutting process. Dep. Wood Technol. Sch. Natur. Resources. Univ. Mich., Ann Arbor. 151 p.

McKenzie, W. M. 1962. The relationship between the cutting properties of wood and its physical and mechanical properties. For. Prod. J. 12:287-294.

McKenzie, W. M. 1970. Choosing your chipper—chipping for pulp production. The Aust. Timber J. 36(4):21, 23, 25, 27, 29, 31.

McKenzie, W. M., and R. L. Cowling. 1971. A factorial experiment in transverse-plane (90/90) cutting of wood. Part I. Cutting force and edge wear. Wood Sci. 3:204-213.

McLauchlan, T. A. 1974. What to choose: Over-arbor or under-arbor. Can. For. Ind. 94(7):42, 43, 45.

McLauchlan, T. A. and D. J. Kusec. 1975. Ribbed blades reduce tree shearing damage. Can. For. Ind. 95(1):67-68, 73.

McLean, H. R. 1974. Treatment techniques for forest residue reduction in the Pacific Northwest. ASAE Pap. No. 74-1531. Amer. Soc. Agric. Eng., St. Joseph, Mo. 11 p.

McMillin, C. W. 1958. The relation of mechanical properties of wood and nosebar pressure in the production of veneer. For. Prod. J. 8: 23-32.

McMillin, C. W. 1970. Wood machining abstracts, 1968 and 1969. Res. Pap. SO-58. U.S. Dep. Agric. For. Serv. 35 p.

McMillin, C. W. 1975. Wood machining highlights, 1972 and 1973. For. Prod. J. 25(8):19-25.

McMillin, C. W. 1982. Application of automatic image analysis to wood science. Wood Sci. 14:97-105.

McMillin, C. W., and J. E. Harry. 1971. Laser machining of southern pine. For. Prod. J. 21(10):34-37.

McMillin, C. W., and J. L. Lubkin. 1959. Circular sawing experiments on a radial arm saw. For. Prod. J. 9:361-367.

McMillin, C. W., and G. E. Woodson. 1972. Moisture content of southern pine as related to thrust, torque, and chip formation in boring. For. Prod. J. 22(11):55-59.

McMillin, C. W., and G. E. Woodson. 1974. Machine boring of southern pine. U.S. Dep. Agric. For. Serv. Tech. Bull. No. 1496, 46 p.

Malcolm, F. B. 1965. A simplified procedure for developing grade lumber from hardwood logs. U.S. Dep. Agric. For. Serv., Res. Note FPL-098. 13 p.

Malcolm, F. B. 1967. Coarse feed sawing of red oak—a limited study. U.S. Dep. Agric. For. Serv., Res. Note FPL-0155. 16 p.

Malcolm, F. B., and A. L. Koster. 1970. Locating maximum stresses in tooth assemblies of inserted-tooth saws. For. Prod. J. 20(10):34-38.

Malito, R. C. 1979. Domtar variable slot chip thickness classifier. In 1979 Pulping Conf. Proc., Tech. Assoc. Pulp and Pap. Ind., Atlanta, Ga. p. 191-192.

Mansfield, J. H. 1952. Woodworking machinery—history of development from 1852-1952. Mech. Eng. 74:983-995.

Marra, A. A., W. A. Hausknecht, and R. F. Day. 1975. Low density composites from high density hardwoods. Bull. No. 610, Mass. Agric. Exp. Stn., Coll. Food and Nat. Resources, Univ. Mass., Amherst.

Martens, D. G. 1965. Log grades—a key to predicting sawmill profits. South. Lumberman 210(2612):29-34.

Martellotti, M. E. 1941. An analysis of the milling process. Part I. Trans. of the ASME 63:677-700.

Martellotti, M. E. 1945. An analysis of the milling process. Part II—Down milling. Trans. of the ASME 67:233-251.

Martyr, N. 1976. Considerations in the installation of enclosures for machines. Woodwork. Ind. 33(8):6-8.

Mason, H. C. 1975. Techniques, machines providing accuracy in sawing softwoods applicable to furniture stock. Crow's For. Prod. Dig. 53(2):16-20.

Massengale, R. 1971. Sawdust, slab and edging weights from mixed oak logs from the Missouri Ozarks. The North. Logger and Timber Proc. 19(10):28-29.

Mattingly, C. Q. 1975. Making cabinets with a computerized router. Furniture Prod. Mag. 38(275):22-23.

Mattson, J. A. 1974. Beneficiation of compression debarked wood chips. U.S. Dep. Agric. For. Serv., Res. Note NC-180. 4 p.

Mattson, J. A. 1980. An investigation of the mechanics of severing trees with auger cutters. Dissertation, Michigan State Univ., East Lansing. 110 p.

Meyer, H. R. 1970. Measurement process for the evaluation of abrasive parameters of abrasive belts for the polishing of wood and wood materials. J. of the C.I.R.P. 18:279-287. (Great Britain)

Miller, W. 1978. Crosscut saw manual. U.S. Dep. Agric. For. Serv., Equip. Develop. Cent., Missoula, Mont. 27 p.

Miller, F., and F. Roppel. 1973. Advantages of headrig chippers. *In* Modern Sawmill Techniques, Proc. of the First Sawmill Clinic, Portland, Ore., Feb. (V. S. White, ed.) p. 130-144.

Miller, R. L., and C. W. Rothrock, Jr. 1963. A history of chip shredding. Tappi 46(7):174A-178A.

Mills, J. 1976. Survey of portable sawmills. World Wood 17(4):31-35.

Mills, C. F. 1980. What ever happened to whole tree chipping? *In* Proc., 1980 Ann. Meet., Tech. Assoc. Pulp and Pap. Ind., Atlanta. p. 365-372.

Minckler, L. S. and A. W. Green. 1958. Upland hardwoods form classes. U.S. Dep. Agric. For. Serv., Stn. Note No. 118. 2 p. Cent. States For. Exp. Stn., Columbus, Ohio.

Moen, A. D. and D. E. Seffens. 1975. Computer-operated drives for log/board processing. *In* Modern sawmill techniques, vol. 5: proc. of the fifth sawmill clinic, pp. 109-124. San Francisco, Ca.: Miller Freeman Publ., Inc.

Mori, M. 1971. An analysis of cutting work in peripheral milling of wood. III. Variation of cutting force in inside cutting of wood with router-bit. IV. The power requirement in inside cutting of wood with router-bit. J. Jap. Wood Res. Soc. 17:437-448.

Morgan, J. T. (Coordinator). 1979. Production of furniture lumber with a bolter saw. A Project by North Carolina Div. of For. Resources in cooperation with Marimont Furniture Co. 34 p.

Morner, B. 1976. Survey of chainsaws for logging. World Wood 17(8):28-40.

Mote, C. D., Jr., and R. Szymani. 1978. Circular saw vibration research. The Shock and Vibration Digest 10(6):15-30.

Mullin, A. G. 1975. Color-coded cutting gauge aids lumber-yield improvement. Furniture Design & Manuf. 47(12):30-40.

Murphey, W. K., and R. S. Cochran. 1972. "Three turn" method for sawing soft maple logs. The North. Logger and Timber Proc. 20(10):22, 24.

Murphey, W. K. and L. E. Rishel. 1972. Finger joint feasibility in furniture production. For. Prod. J. 22(2):30-32.

Murphey, W. K. and G. D. Schneider. 1956. Greater utilization of material obtained from new machine. For. Prod. J. 6:319-320.

Nakamura, G. 1966. Studies of wood sanding by belt-sander. II—Industrial test of plywood sanding by wide belt-sander. Tokyo Univ. Agric. and Technol. Exp. Stn. Bull. 5, 17 p.

Nakamura, G., and H. Takachio. 1960. Relation of light and roughness on sanded surface of wood. J. Jap. Wood Res. Soc. 6:237-242.

Namiki, N. 1973. Manufacture of newsprint with intense utilization of hardwood and other wood resources Tappi 56(10):93-95.

Nearn, W. T., N. A. Norton, and W. K. Murphey. 1953. The strength of dowel joints as affected by hole size and type of dowel. J. of the For. Prod. Res. Soc. 3(4):14-17, 72.

Necasany, V. 1965. (Effect of heat and moisture on the properties of the middle lamella.) Drev. Vyskum 3:149-154.

Neubaumer, H. G. 1971. The horizontal wood hog. *In* Proc., 26th Ann. Northwest Wood Prod. Clinic, Missoula, April 19-21: 75-89. Eng. Ext. Serv., Washington State Univ., Pullman.

Nish, D. L. 1932. Creative woodturning. Brigham Young Univ. Press., Provo, Utah. 248 p.

Niskala, G. R. and T. W. Church, Jr. 1966. Cutting hardwood cants can boost sawmill profits. U.S. Dep. Agric. For. Serv., Res. Note NE-46. 8 p.

Nolan, W. J. 1967. Chip shredding—why not investigate further? Pulp and Pap. 41(44):57-58.

Northcross, S. 1979. Structural finger-jointed dimension becoming viable market commodity. Timber Proc. Ind. 4(10):27-30.

Page, R. H. and R. C. Rodenbach. 1962. Comparison of yield in clear and usable cuttings from different grades of maple lumber. Furniture, Plywood, and Veneer Counc. of the N.C. Assoc. Rep. No. 9, 2 p.

Pahlitzsch, G. 1970. (International state of research in the field of wood sanding.) Holz als Roh- und Werkstoff 28:329-343.

Pahlitzsch, G., and G. Argyropoulos. 1974. (Air-supported sanding belts for wide belt sanders.) Holz als Roh- und Werkstoff 32:425-435.

Pahlitzsch, G., and K. Dziobek. 1959. (Investigation concerning belt polishing of wood using a straight-line cutting movement.) Holz als Roh- und Werkstoff 17:121-134.

Pahlitzsch, G., and K. Dziobek. 1961a. (On the determination of the quality of cutting shaped wood surfaces—Part I: Methods of measurement and estimation for belt-sanded woods.) Holz als Roh- und Werkstoff 19:403-417. (Also transl. by Joint Pub. Res. Serv., U.S. Dep. Commer., Washington, D.C., FPL-650, 44 p. (1966).)

Pahlitzsch, G., and K. Dziobek. 1961b. (On the blunting of sanding belts while sanding wood.) Holz als Roh- und Werkstoff 19:136-149.

Pahlitzsch, G., and K. Dziobek. 1965a. (The evaluation of processed wood surfaces. Part I.) Holztechnologie 6(3):153-160. (Also Transl. by Joint Pub. Res. Serv., U.S. Dep. Commer., Washington, D.C., FPL-706, 18 p. (1970).

Pahlitzsch, G., and K. Dziobek. 1965b. (The evaluation of machined wood surfaces. Part II.) Holztechnologie 6(4):219-224. (Also Transl. by Joint Pub. Res. Serv, U.S. Dep. Commer., Washington, D.C., FPL-668, 15 p. (1966).

Pahlitzsch, G., and K. Puttkammer. 1975. (Cutting tests with bandsawing.) Holz als Roh- und Werkstoff 33:181-186.

Pahlitzsch, G., and K. Puttkammer. 1976. (Cutting tests during bandsawing) Holz als Roh- und Werkstoff 34:17-21.

Pahlitzsch, G., and E. Sandvoss. 1969. (Milling (routing and shaping) of wood and wood materials.) Die Holzbearbeitung 5:23-30.

Paper, Film, and Foil Converter. 1977. Laser used for die-cutting. Paper, Film, and Foil Converter 51(3):72-73.

Paperboard Packaging. 1977. Automated fluid-jet slitting debuts on U.S. corrugator. Paperboard Packag. 63(3):35, 38, 40, 44.

Papworth, R. L. and J. R. Erickson. 1966. Power requirements for producing wood chips. For. Prod. J. 16(10):31-36.

Papworth, R. L. and K. R. Johnson. 1968. Power requirements for producing wood chips with a parallel knife chipper. For. Prod. J. 18(10):42-44.

Park, K. 1978. The "stackwall" log-house, rediscovered. Canadian For. Ind. 98(5):51, 53.

Patterson, D. W. (ed.). 1979. Getting the most grade out of hardwood logs. For. Prod. Notes 4(7): 2 p. For. Prod. Lab., Tex. For. Serv., Lufkin.

Paulapuro, H. 1977a. Prediction of quality parameters of groundwood pulp mixtures. Paperi ja Puu 59(4):297-307.

Paulapuro, H. 1977b. Basic principles and method for control of a grinder room. Paperi ja Puu 59(5):357-371.

Pearson, H. C. 1972. Thin kerf sawing machinery. U.S. Pat. No. 3,703,915. U.S. Pat. Off., Washington, D.C. 11 p.

Pease, D. A. 1972. Planer enclosures cut noise to required levels. For. Ind. 99(3):82-83.

Penburthy, R. J. 1968. Engineering aspects of saw chain cutting. Proc. For. Eng. Conf. Amer. Soc. Agric. Eng., ASAE Pub. PROC-368, p. 44-46. East Lansing, Mich.

Pepke, E. K., H. A. Huber, and A. Sliker. 1977. Pallet lumber makes furniture parts. Woodworking and Furniture Dig. 79(10):28-31.

Perry, H. F. J. 1950. Manufacture of mechanical pulp. Chapter 3 in New Series Pulp and Paper Manufacture, Vol. 1: Preparation and treatment of wood pulp, p. 182-250. New York: McGraw-Hill Book Co., Inc.

Peter, R. K. 1967. Influence of sawing methods on lumber grade yield from yellow-poplar. For. Prod. J. 17(11):19-24.

Peters, C. C. 1968. Multiple-flitch method for thick slicing. For. Prod. J. 18(9):82-83.

Peters, C. C. 1976. Basis and specifications for a lumber slicer. For. Prod. J. 26(2):26-30.

Peters, C.C. and C. M. Banas. 1977. Cutting wood and wood-base products with a multi-kilowatt CO_2 laser. For. Prod. J. 27(11):41-45.

Peters, C. C., and J. D. Cumming. 1970. Measuring wood surface smoothness: A review. For. Prod. J. 20(12):40-43.

Peters, C. C. and H. L. Marshall. 1975. Cutting wood materials by laser. U.S. Dep. Agric. For. Serv., Res. Pap. FPL 250. 10 p.

Peters, C., and A. Mergen. 1971. Measuring wood surface smoothness: A proposed method. For. Prod. J. 21(7):28-30.

Peters, C. C., A. F. Mergen, and H. R. Panzer. 1969a. Effect of cutting speed during thick slicing of wood. For. Prod. J. 19(11):37-42.

Peters, C. C., A. F. Mergen, and H. R. Panzer. 1969b. Slicing wood one-inch thick: four types of pressure bars. For. Prod. J. 19(7):47-52.

Peters, C. C., A. F. Mergen and H. R. Panzer. 1972. Thick slicing of wood: effects of wood and knife inclination angle. For. Prod. J. 22(9):84-91.

Peters, C. C., H. R. Panzer, and A. F. Mergen. 1976. Low-speed effects on thick-slicing. For. Prod. J. 26(5):56-57.

Peters, C. C. and R. A. Patzer. 1976. Thick-slicing of wood: effects of bar-type and speed of cut on quality of slice. For. Prod. J. 26(4):19-24.

Peters, C. C. and R. R. Zenk. 1968. Effect of precompression on sliced wood 1/2 and 1 inch in thickness. U.S. Dep. Agric. For. Serv., Res. Note FPL-0194. 6 p.

Peters, C. C., R. R. Zenk, and A. Mergen. 1968. Effects of roller-bar compression and restraint in slicing wood 1-inch thick. For. Prod. J. 18(1):75-80.

Phillips, D. R. 1975. Lumber and residue yields for black oak saw logs in western North Carolina. For. Prod. J. 25(1):29-33.

Phillips, D. R., J. G. Schroeder, and M. A. Taras. 1974. Predicted green lumber and residue yields from the merchantable stem of black oak trees. U.S. Dep. Agric. For. Serv., Res. Pap. SE-120. 10 p.

Plough, I. L. 1974. The effect of running clearance on flake quality and power requirements of conical knife-ring flakers. In Proc. of eighth Washington State Univ. symp. on particleboard, pp. 287-300. Pullman, Wa.

Plough, I. and P. Koch. 1983. Prototype cutterhead for shaping-lathe headrigs yields pulp chips and cants. Southern Journal of Applied Forestry 7(3): 134-140.

Pnevmaticos, S. M., and D. W. Bousquet. 1972. Sawing pattern effect on the yield of lumber and furniture components from medium and low grade hard maple logs. For. Prod. J. 22(3):34-41.

Pnevmaticos, S. M., I. B. Flann, and F. J. Petro. 1971. How log characteristics relate to sawing profit. Can. For. Ind. 91(1):40-43.

Pnevmaticos, S. M. and P. Mouland. 1978. Hardwood sawing simulation techniques. For. Prod. J. 28(4):51-56.

Porter, A. W. 1970. Some engineering considerations of high-strain band saws. In Proc. Northwest Wood Prod. Clinic, p. 13-32. Spokane, Wash.

Porter, A. W. 1971. Some engineering considerations of high-strain bandsaws. For. Prod. J. 21(4):24-32.

Porter, A. W., D. J. Kusec, and J. L. Sanders. 1971. Air-flow method measures lumber surface roughness. Can. For. Ind. 91(7):42-45.

Potrebic, M. 1969. (Relation between the roughness of beech wood and the grade of sandpaper used to sand it.) Tre Og Mobler 1(6):13-19. (Norway)

Price, Eddie W., and W. F. Lehmann. 1978. Flaking alternatives. In Structural flakeboard from forest residues; symp. proc., Kansas City, Mo., June 6-8. U.S. Dep. Agric. For. Serv., Gen. Tech. Rep. WO-5, P. 47-68.

Prokes, S. 1966. History of woodworking tools. Drevo 21(9):318-320.

Quelch, P. S. 1964. Sawmill feeds and speeds—band and circular rip saws. 46 p. Portland, Ore.: Armstrong Manuf. Co.

Quelch, P. S. 1970. Armstrong saw filer's handbook. Second ed. 103 p. Armstrong Manufacturing Co., Portland, Ore.

Radcliffe, C. J., and C. D. Mote, Jr. 1981. Active control of circular saw vibration using spectral analysis. Wood Sci. 13:129-139.

Rajala, J. 1974. Chipping headrigs and hardwoods. South. Lumberman 228(2831):15-18.

Rajala Timber Company. 1978. Aspen and chipping headrigs. Slide presentation. (Oct. 13, Sawmill Clinic and Machinery Show.) Rajala Timber Co., Deer River, Minn.

Rast, E. D. 1974a. Log and tree sawing times for hardwood mills. U.S. Dep. Agric. For. Serv., Res. Pap. NE-304. 17 p.

Rast, E. D. 1974b. Nonproductive time in hardwood mills. South. Lumberman 229(2843):13-15.

Rast, E. D. 1978. Volume relationships in slicing northern red oak and black cherry logs. U.S. Dep. Agric. For. Serv. Res. Pap. NE-420. 11 p.

Rast, E. D., and V. P. Chebetar, Jr. 1980. Technique for estimating yield of furniture squares and flat stock from flitches. Res. Note NE-287, U.S. Dep. Agric., For. Serv. 4 p.

Ratliff, R. 1973. Operation and maintenance of strob-saws. Ch. IV in Modern Sawmill Techniques, Vol. 2: Proc. of the Second Sawmill Clinic, New Orleans, La., Nov. Miller Freeman. Publications, Inc., San Francisco.

Rau, P., and M. Bauer. 1978/1979. Laser beam precision ensures dimensional accuracy of folding carton blanks. Australian Packaging 26(12):25.

Raymond, A. 1975. Rough mill efficiency and cost cutting. Furniture Prod. 38(281);15-16.

Rea, R. D. 1975. Berkline turns fuel shortage into opportunity. Woodwork. and Furniture Dig. 77(5):26-28.

Redman, G. P. 1957. Short log bolter for furniture stock. Eng. Sch. Bull. No. 62, Dept. Eng. Res., North Carolina State College, Raleigh, N.C.

Reitter, F. J. 1974. A new process for utilization of mixed tropical hardwoods for pulp and paper production. In AITIPE Symp. Utiliz. New Forest Resources (Madrid):113-146. May.

Revilla Perez, A. R. 1972. Rod mill refining of hardwood ammonia plasticized chips. MS Thesis, State Univ. Coll. For. at Syracuse Univ., Syracuse, N.Y.62 p.

Reynolds, H. W. 1968. Grading and unitized seasoning of oak cuttings. Wood and Wood Prod. 73(10):34.

Reynolds, H. W. 1969. The SHOLO (SHOrt LOg) mill. South. Lumberman 219 (2728)182-185.

Reynolds, H. W. 1970. Sawmill simulation: data instructions and computer programs. U.S. Dep. Agric. For. Serv., Res. Pap. NE-152. 41 p.

Reynolds, H. W. 1974. Optimum sawing patterns for low-grade hardwoods. In Use of computers in forestry, 23rd annu. for. symp., pp. 119-134. Baton Rouge, La.: La. State Univ.

Reynolds, H. W. and C. J. Gatchell. 1969. Sawmill simulation: concepts and computer use. U.S. Dep. Agric. For. Serv., Res. Note NE-100. 5 p.

Reynolds, H. W. and C. J. Gatchell. 1970. The Sholo mill: make pallet parts and pulp chips from low-grade hardwoods. U.S. Dep. Agric. For. Serv., Res. Pap. NE-180. 16 p.

Reynolds, H. W. and C. J. Gatchell. 1971. The Sholo mill: return on investment versus mill design. U.S. Dep. Agric. For. Serv., Res. Pap. NE-187. 46 p.

Reynolds, H. W., and C. J. Gatchell. [1982.] New technology for low-grade hardwood utilization: System 6. U.S. Dep. Agric. For. Serv., Res. Pap. NE-504.

Reynolds, J. W., and J. Schroeder. 1977. System 6—A way to use small logs to make grade lumber for furniture cuttings. South. Lumberman 234(2897):9-10.

Reynolds, H. W., and J. Schroeder. 1978. Furniture cuttings made from logging residue: The three-sided cant system. U.S. Dep. Agric. For. Serv., NE-417. 4 p.

Reynolds, D. D., and W. Soedel. 1972. Matching of chain saw power requirements with engine characteristics. For. Prod. J. 22(1): 56-60.

Reynolds, D. D., W. Soedel, and C. Eckelman. 1970. Cutting characteristics and power requirements of chain saws. For. Prod. J. 20(10):28-34.

Richards, D. B. 1973. Hardwood lumber yield by various simulated sawing methods. For. Prod. J. 23(10):50-58.

Richards, D. B. 1977. Value yield from simulated hardwood log sawing. For. Prod. J. 27(12):47-50.

Richards, D. B. 1978. Sawing hardwoods for higher value. Southern Lumberman 237(2942):9-12.

Richards, D. B., W. K. Adkins, H. Hallock, and E. H. Bulgrin. 1979. Simulation of hardwood log sawing. U.S. Dep. Agric. For. Serv. Res. Pap. FPL 355. 8 p.

Richards, D. B., W. K. Adkins, H. Hallock, and E. H. Bulgrin. 1980. Lumber values from computerized simulation of hardwood log sawing. U.S. Dep. Agric. For. Serv. Res. Pap. FPL 356. 28 p.

Richards, D. B., and J. A. Newman. 1979. Sawing high-quality red oak logs. For. Prod. J. 29(9):36-39.

River, B. H., and V. P. Miniutti. 1975. Surface damage before gluing—weak joints. Wood and Wood Prod. 80(7):35-36, 38.

Robichaud, Y. 1975. Band resaws for sawing hardwoods: a comparison between horizontal and vertical resaws. Can. For. Serv. Rep. OPX148E, 13 p. East. For. Prod. Lab., Ottawa, Can.

Robichaud, Y., F. J. Petro, and M. C. S. Kingsley. 1974. Aspen lumber and dimension stock recovery in relation to sawing pattern. For. Prod. J. 24(3):26-30.

Rogers, H. W. 1948. The wood chipper. Pap. Ind. and Pap. World 30:883-888, 1042-.1047.

Rood, J. 1950. Sculpture in wood. The University of Minnesota Press, Minneapolis. 179 p.

Rosen, H. N., H. A. Stewart, and D. J. Polak. 1980. Dimension yields from short logs of low-quality hardwood trees. U.S. Dep. Agric. For. Serv. Res. Pap. NC-184. 22 p.

Roubicek, T. T., and P. Koch. 1981. Economic feasibility of converting whole stems of southern hardwoods into composite panels and pallet parts of solid lumber, using shaping lathes. *In* Utilization of low-grade southern hardwoods. (D. A. Stumbo, ed.) Symposium Proc., Nashville, Tenn., Oct. 1980. For. Prod. Res. Soc. Madison, Wis. p. 183-192.

Rowell, R. M., J. M. Black, L. R. Gjovik, and W. C. Feist. 1977. Protecting log cabins from decay. U.S. Dep. Agric. For. Serv., Gen. Tech. Rep. FPL-11. 11 p.

Rudstrom, L., L.-G. Samuelsson, and K. Uhlin. 1973. Mechanical pulp in graphic papers. *In* Int. Mech. Pulping Conf. Proc., Sess. II, Swedish Assoc. Pulp and Pap. Eng., and European Liaison Comm. for Pulp and Pap., Stockholm, p. 25-26.

St. -Laurent, A. 1973. Improving the surface quality of ripsawn dry lumber. For. Prod. J. 23(12):17-24.

Sage, R. L. and J. Olson. 1975. Circular gang rip-saws for short hardwood pallet cants. South. Lumberman 230(2856):13-14.

Salemme, F. J. 1969. Thin carbide saws in the sawmills. *In* Proc. For. Prod. Res. Soc. North. California Sec., Calpella, Calif., p. 10-13.

Saljé, E., and U. Bartsch. 1976. (Investigating and reducing noise at double-end tenoners.) Holz als Roh- und Werkstoff 34:43-48.

Sarna, R. 1976. Recent evolutions in chain saw design. North. Logger and Timber Proc. 24(11):10-11, 26.

Sarna, R. P. 1979. Chain saw manual. Interstate Printers & Publishers, Inc., Danville, Ill. 118 p.

Saul, J. D. and M. Lagerquist. 1975. American's double enders point the way. Wood and Wood Prod. 80(9):48-49.

Saunders, H. W. 1969. Primary utilization of birch. *In* Birch symp. proc., p. 33-37. U.S. Dep. Agric. For. Serv., Northeast. For. Exp. Stn., Upper Darby, Pa.

Saunders, H. W. 1979. Hardwood sawing process for increasing recovery and improving production. For. Prod. J. 29(2):21-29.

Schafer, E. R. and J. C. Pew. 1942. The grinding of hardwoods. U.S. Dep. Agric. For. Serv., FPL Rep. No. R1419. 16 p.

Schliewe, R. 1969. Thin kerf sawing considerations. *In* Proc. For. Prod. Res. Soc. North. California Sec., Calpella, Calif., p. 14-18.

Schliewe, R. 1973. Lumber transports: functions and importance in sawing systems with special emphasis on edgers. *In* Wood machining sem., p. 109-125. William A. Dost, Ed. Richmond, Ca.: For. Prod. Lab., Univ. of Ca.

Schmidt, J. W., Jr., W. D. Torlone, J. Byrd, and M. R. Fedorko. 1970. Feasibility study on the retrieval and use of primary wood residue. W. Va. Univ. Bull. Ser. 70, No. 8-3, Rep. No. 11, 250 p. Morgantown, W. Va.: W. Va. Univ.

Schmied, J. 1964. (The effect of the size, size nonuniformity, and the shape of chips on the cooking uniformity in pulping.) Pap. a Celulosa 19(4):100-106.

Schmutzler, W. 1970. (Noise prevention in woodworking machines by encapsulation.) Holzindustrie 23(6):170-174.

Scollard, N. 1980. Throwaway chipper knife saves up to 30% in operating costs. Pulp and Pap. 54(2):181-183.

Scully, B. 1975. Sizing panels automatically. Furniture Des. and Manuf. 47(5):50-53.

Scully, B. 1977a. Multiple machining at one work station. Furniture Des. and Manuf. 49(4):26-28, 32, 34.

Scully, B. 1977b. Wood embossing machines cut production steps for "carved" parts. Furniture Des. and Manuf. 49(2):30-33.

Seffens, D. E. 1975. Building a small-log mill for pine and hardwood. *In* Modern sawmill techniques, vol. 4: proc. of the fourth sawmill clinic, p. 77-91. San Francisco, Ca.: Miller Freeman Publ., Inc.

Seto, K., and K. Nozaki. 1966. On the machining of plywood surface with coated abrasives; the performance of drum sander and drum type wide belt sander. Hokkaido For. Prod. Res. Inst. Rep. 49, 27 p. Japan.

Shreve, C. 1972. Vibration cutting in dry wood. Holz als Roh- und Werkstoff 30:325-328.

Sickeler, A. A. 1974. Use and application of a shredding and mulching machine. ASAE Pap. No. 74-1568. Amer. Soc. Agric. Eng., St. Joseph, Mo. 2 p.

Sieminski, R. 1966. (Testing and measuring the surface structure of wood and wood-base materials.) Holz als Roh- und Werkstoff 24:396-404.

Simons, E. N. 1966. The evolution of the saw. Part I. Wood 31(9):33-36; Part II. 31(11):37-40.

Skory, L. D. 1979. Rotary cutter at Ontario operation means clean, virtually shatter-free sawlogs. Can. Pulp Pap. Ind. 32(2):20-22.

Smerdov, V. V., and M. Y. Sotonin. 1978. Double-roller crusher for wood waste products. Lesnaya Promyshlennost (8):23.

Smith, T. C. 1978. Can you afford numeric control? The answer may surprise you. Wood and Wood Prod. 83(6):43-46.

Smith, H. C., G. R. Trimble, Jr., and P. S. DeBald. 1979. Raise cutting diameters for increased returns. Res. Pap. NE-445, U.S. Dep. Agric. For. Serv. 7 p.

Sobieski, C. 1980. The effect of chip thickness on pulping: A selected bibliography with annotations. Rader Companies, Inc., Portland, Ore. 18 p.

Sohn, F. 1973. A computerized studmill: background and concepts, pp. 254-261. *In* Wood machining seminar proc. W. A. Dost, Ed. Berkeley, Ca.: Univ. of Ca.

Sorensson, B. 1969. Vibrations occurring in chainsaws. Timber Trades J. 269(4831):59.

Sperber, R. 1977. Chain-saw lumbering. Cut your wood where it falls. Fine Woodwork. 2(2):50-54.

Sperber Tool Works, Inc. 1977. Portable chain saw mill operation and maintenance manual. Sperber Tool Works, Inc., West Caldwell, N. J.

Springate, N. C. 1978a. Economic prospects for southern manufacture of composite structural sheathing with flake core and veneer faces. *In* Complete tree utilization of southern pine: symp. proc., New Orleans, La., April 17-19, C. W. McMillin, ed. For. Prod. Res. Soc., Madison, Wis., p. 416-426.

Springate, N. C. 1978b. Rotary peeling small diameter logs—a review of techniques and yield. *In* Complete tree utilization of southern pine: symp. Proc., New Orleans, La., April 17-19, C. W. McMillin, Ed. For. Prod. Res. Soc., Madison, Wis., p. 394-398.

Springate, N., I. Plough, and P. Koch. 1978. Shaping-lathe roundup machine is key to profitable manufacture of composite sheathing panels in Massachusetts or Maine. For. Prod. J. 28(10):42-47.

Steele, P. H., H. Hallock and S. Lunstrum. 1981. Procedure and computer program to calculate machine contribution to sawmill recovery. U.S. Dep. Agric., For. Serv. Res. Pap. FPL 383. 19 p.

Steffes, F. J. 1978. Commercial screening for overthick chip removal. Tappi 61(11):63-66.

Steiner, K. 1972. (Chippers for long logs in the particleboard industry. Development and state of technique.) Holz als Roh- und Werkstoff 30:201-214.

Steinhagen, H. P. 1977. Heating times for frozen veneer logs—new experimental data. For. Prod. J. 27(6):24-28.

Stephens, L. G. 1975. Constructing a mill to maximize running time. *In* Modern sawmill techniques, vol. 4, proc. of the fourth sawmill clinic, pp. 63-76. H. G. Lambert, ed. San Francisco, Ca.: Miller Freeman Publ., Inc.

Stern, A. R., H. Hallock, and D. W. Lewis. 1979. Improving sawing accuracy does help. U.S. Dep. Agric. For. Serv. Res. Pap. FPL 320. 12 p.

Stern, A. R., and K. A. McDonald. 1978. Computer optimization of cutting yield from multiple-ripped boards. Res. Pap. FPL 318. U.S. Dep. Agric. For. Serv. 13 p.

Stevens, S. F. 1977. The edge sander—stationary but versatile. Furniture Methods and Materials 23(8):10-11.

Stevens, S. F. 1978. Coated abrasives prove useful in prefinishing hardwood plywood. Plywood and Panel Mag. 18(11):24-25.

Stevens, W. V. 1980. Refiner system selections. A specific example. Tappi 63(2):127-129.

Stewart, H. A. 1969. Effect of cutting direction with respect to grain angle on the quality of machined surface, tool force components, and cutting friction coefficient. For. Prod. J. 19(3):43-46.

Stewart, H. A. 1970a. Effect of cutting angle and depth of cut on the occurrence of chipped grain on sycamore. U.S. Dep. Agric. For. Serv. Res. Note NC-92. 4 p.

Stewart, H. A. 1970b. Cross-grain knife planing hard maple produces high-quality surfaces and flakes. For. Prod. J. 20(10):39-42.

Stewart, H. A. 1970c. Abrasive vs. knife planing. For. Prod. J. 20(7):43-47.

Stewart, H. A. 1971a. Chip formation when orthogonally cutting wood against the grain. Wood Sci. 3:193-203.

Stewart, H. A. 1971b. Rake angle for planing hard maple determined best by depth of chipped grain. U.S. Dep. Agric. For. Serv., Res. Note NC-116. 4 p.

Stewart, H. A. 1971c. Cross-grain knife planing improves surface quality and utilization of aspen. U.S. Dep. Agric. For. Serv. Res. Note NC-127. 4 p.

Stewart, H. A. 1971d. Chips produced with a helical cutter. For. Prod. J. 21(5):44-45.

Stewart, J. S. 1972a. A theoretical and experimental study of wood planer noise and its control. Ph.D. Dissertation. North Carolina State University, Raleigh.

Stewart, H. 1972b. Abrasive vs. knife planing. Wood & Wood Prod. 77(8):73-76.

Stewart, H. A. 1974a. Face milling can be improved for surfacing and flaking. For. Prod. J. 24(2):58-59.

Stewart, H. A. 1974b. Cross-grain knife planing. Woodwork. and Furniture Dig. 76(10):40, 42, 43.

Stewart, H. A. 1974c. A comparison of factors affecting power for abrasive and knife planing of hardwoods. For. Prod. J. 24(3):31-34.

Stewart, J. S. 1974d. Noise-reducing tooling for planer noise control. Chapter 10 *in* Modern Sawmill Techniques 4:174-187. Proc., Fourth Sawmill Clinic, New Orleans. Miller Freeman Publications, Inc., San Francisco.

Stewart, H. A. 1975a. Knife planing across the grain can be applied to hardwoods. USDA For. Serv. Res. Note NC-196, 4 p. North Central For. Exp. Stn., St. Paul, Minn.

Stewart, J. S. 1975b. Noise-reducing tooling for planer noise control *In* Modern sawmill techniques, vol. 4, proc. of the fourth sawmill clinic, pp. 174-187. H. G. Lambert, ed. San Francisco, Ca.: Miller Freeman Publ., Inc.

Stewart, H. A. 1976a. Abrasive planing across the grain with higher grit numbers can reduce finish sanding. For. Prod. J. 26(4):49-51.

Stewart, H. A. 1976b. Preliminary belt life data for abrasive planing of ponderosa pine. U.S. Dep. Agric. For. Serv. Res. Note NC-211. 3 p.

Stewart, H. A. 1977. Optimum rake angle related to selected strength properties of wood. For. Prod. J. 27(1):51-53.

Stewart, J. S. 1978a. Wood planer cutterhead design for reduced noise level. U.S. Pat. No. 4,074,737. U.S. Pat. Off., Washington, D.C.

Stewart, H. A. 1978b. Effects of stock removal rates on belt loading for abrasive planing hardwoods. U.S. Dep. Agric. For. Serv. Res. Note NC-240. 2 p.

Stewart, H. A. 1978c. Stock removal for aluminum oxide- and garnet-coated abrasive belts. For. Prod. J. 28(11):29-31.

Stewart, H. A. 1979a. Analysis of orthogonal woodcutting across the grain. Wood Sci. 12(1):38-45.

Stewart, H. A. 1979b. Some surfacing defects and problems related to wood moisture content. *In* Wood moisture content—temperature and humidity relationships. Symposium proceedings, Blacksburg, Va., Oct. 29. Soc. Wood Sci. and Technol.

Stewart, H. A., and J. B. Crist. 1982. SEM examination of subsurface damage of wood after abrasive and knife planing. Wood Sci. 14(3):106-109.

Stewart, J. A., and F. D. Hart. 1976. Control of industrial wood planer noise through improved cutterhead design. Noise Control Eng. 7(1):4-9.

Stewart, H. A. and W. F. Lehmann. 1973. High-quality particleboard from cross-grain, knife-planed hardwood flakes. For. Prod. J. 23(8):52-60.

Stewart, H. A. and W. F. Lehmann. 1974. Cross-grain cutting with segmented helical cutters produces good surfaces and flakes. For. Prod. J. 24(9):104-106.

Stewart, H. A., and P. D. Parks. 1980. Peripheral milling across the grain with rake angles up to 60°. For. Prod. J. 30 (6):54-57.

Stirling, J. 1979. Extending life of chipper knives: Space age idea used in mills. British Columbia Lumberman 63(5):8, 12.

Stock, S. 1976. Continuous pulpwood shear promises high productivity. Pulp Pap. Can. 77(10):14-17.

Stumbo, D. A. 1974. How to increase profit at the headsaw, edger and trimmer. South. Lumberman 229(2841):13-14, 16, 18.

Suchsland, O. 1980. Operating characteristics and performance limitations of circular saws and band saws. Ext. Bull. E-1353, Coop. Ext. Serv., Michigan State Univ., East Lansing. 6 p.

Sugihara, H. 1970. (Forces acting on a band saw.) Can. Dep. Fisheries and For., Ottawa, Transl. OOFF-130, 32 p. (From Wood Ind. 8(5):38-45. 1951. Toyko.)

Suzuki, R. 1958. The measurement of roughness of cut surfaces by a drop of water. Part I. J. Jap. Wood Res. Soc. 4(4):156-161.

Swan, D. A. 1971a. The revolutionary ring saw. Tech. Rel. 71-R-60, 3 p. Amer. Pulpw. Assoc., New York.

Swan, D. A. 1971b. The Asplundh trailer chipper. Amer. Pulpwood Assoc. Tech. Release 71-R-46. 3 p.

Swartz, J. N. 1960. Newsprint from broadleaf woods. *In* Proc. Fifth World For. Cong. 3:1585-1596.

Sybertz, H. 1974. Energy requirements for the production of cut flakes used for flakeboard. *In* Proc. of eighth Washington State Univ. symp. on particleboard, p. 329-340. T. M. Maloney, ed. Pullman, Wash.: Wash. State Univ.

Szymani, R. 1972. A study of corrugated board cutting by high-velocity liquid jet. For. Prod. J. 22(8):17-25.

Szymani, R. 1978. Technological aspects of thin-kerf circular and band sawing—a review of claasen's system of saw selection. *In* Complete tree utilization of southern pine: symp. proc., New Orleans, La., April 17-19. C. W. McMillin, ed. For. Prod. Res. Soc., Madison, Wis., p. 345-352.

Szymani, R. 1979. Defect detection systems for lumber: current status, trends, and potential applications. For. Prod. Dep., Oregon State Univ., Corvallis. 70 p. (In cooperation with U.S. Dep. Agric. For. Serv., For. Prod. Lab., Madison, Wis.)

Szymani, R. and F. E. Dickinson. 1975. Recent developments in wood machining processes: novel cutting techniques. Wood Sci. and Technol. 9:113-128.

Szymani, R. and C. D. Mote, Jr. 1977. Circular saw stiffness as a measure of tension. For. Prod. J. 27(3):28-32.

Szymani, R., and C. D. Mote, Jr. 1979. Theoretical and experimental analysis of circular saw tensioning. Wood Sci. Technol. 13:211-237.

Tanaka, C., Y. Shiota, A. Takahashi, and M. Nakamura. 1981. Experimental studies on band saw blade vibration. Wood Sci. Technol. 15:145-159.

TAPPI. 1969. Selection of chipper drives. Tech. Info. Sheet 002.02, 15 p. TAPPI Elec. Eng. Comm.

Telford, C. J. 1949. Energy requirements of insert point circular headsaws. South. Lumberman 178(2235): 42, 44-46, 48, 50.

Templer, J., and J. R. G. Bryce. 1979. Shredded chips reduce production costs for bleached kraft. Pulp and Pap. Canada 80(4):61-67.

The Logger and Lumberman. 1980. Do-it-yourself sawmill. The Logger and Lumberman 29(8):39C.

Theien, C. M. 1970. Routing and shaping of particleboard. For. Prod. J. 20(6):30-32.

Thompson, S. E. 1973. Chain saw safety, vibration and noise. Publication Number 730702. 12 p. Society of Automotive Engineers. New York.

Thompson, S. 1976. All about vacuum chucks—how they work, what they do, how they cut costs. Woodwork. and Furniture Dig. 78(9):38-41.

Thrasher, E. W. 1972. Method and apparatus for operating a rotary saw. U.S. Pat. 3,645,304, U.S. Pat. Off., Washington, D.C. 9 p.

Thrasher, A. 1977. Saw lumber: not sawdust. *In* Modern Sawmill Techniques 7:40-46. Miller Freeman Publ., Inc., San Francisco.

Thunell, B. 1967. History of wood sawmilling. Wood Sci. and Technol. 1:174-176.

Timber Processing Industry. 1980. Edger review. Timber Proc. Ind. 5(10): 22, 24, 26, 28-33.

Truax, T. R. 1939. The manufacture of veneer. U.S. Dep. Agric. For. Serv. For. Prod. Lab. Rep. R285.

U.S. Department of Agriculture, Forest Service. 1964. 1963 at the Southern Forest Experiment Station. Ann. Rep., U.S. Dep. Agric., For. Serv., South. For. Exp. Stn., New Orleans, La. p. 60-62.

U.S. Department of Agriculture, Forest Service. 1965. Veneer cutting and drying properties of water oak. U.S. Dep. Agric. For. Serv., Res. Note FPL-0105. 5 p. For. Prod. Lab., Madison, Wis.

U.S. Department of Agriculture, Forest Service. 1970. Results of field trials of the Tree Eater, a tree and brush masticator. ED&T Rep. 7120-1, U.S. Dep. Agric. For. Serv., Equip. Develop. Cent., San Dimas, Calif. 26 p.

Ulsoy, A. G., and C. D. Mote, Jr. 1978. Band saw vibration and stability. The Shock and Vibration Digest 10(11):3-15.

Ulsoy, A. G., and C. D. Mote, Jr. 1980. Analysis of bandsaw vibration. Wood Sci. 13(1):1-10.

Umetsu, J., N. Kinoshita, and D. Hayashi. 1976a. Studies on wood grinding with coated abrasive belts. II. Chip formation and interference phenomenon of groove in cutting by a simulated cutting edge with trigonal pyramid-shaped tool. J. Jap. Wood Res. Soc. 22:217-222.

Umetsu, J., N. Kinoshita, and D. Hayashi. 1976b. Studies on wood grinding with coated abrasive belts. III. Cutting force in cutting process using a simulated cutting edge. J. Jap. Wood Res. Soc. 22(6):337-342.

Umetsu, J., N. Kinoshita, and D. Hayashi. 1979. Studies on wood grinding with coated abrasive belts. VIII. Roughness and spectrum analysis of ground surface produced in belt grinding with contact wheel. J. Jap. Wood Res. Soc. 25(3):197-202.

Upton, B. 1976. CIP beats delimbing costs with chains. Pulp & Pap. Can. 76(9):47-50.

Varbanov. I. 1972. [The possibility of replacing the sanding of wood by scraping with a fixed knife.] Drevo 27(3): 68-69, 76. Prague. (Czechoslovakia.)

Venkateswaran, A. 1972. A comparison of the electrical properties of milled wood, milled wood cellulose, and milled wood lignin. Wood Sci. 4:248-253.

Veuger, F. 1976. Points in selecting a wastewood chipper. In Modern sawmill techniques, proc. of the sixth sawmill clinic, Vol. 6, pp. 162-172. V. S. White, ed. San Francisco, Ca.: Miller Freeman Publ., Inc.

Vick H. 1976. Sawing for grade in hardwoods. In Hardwood sawmill techniques, proc. of the first hardwood sawmill clinic programs, pp. 1-11. V S. White, ed. San Francisco, Ca.: Miller Freeman Publ., Inc.

Vick, H. 1979. Maximize potential from logs by sawing hardwood for grade. For. Ind. 106(8): 52-54.

Wahlman, M. 1967. Importance of chip thickness in alkaline pulping, and chip thickness analysis. Pap. ja Puu 49(3):107-110.

Waller, B., and D. N. Halgrimson. 1980. New technique for removing contaminants from wood chips. For. Prod. J. 30(5):40-43.

Walser, D. C. 1974. Steam-injection knife improves veneer quality. For. Prod. J. 24(9):70-79.

Walser, D. C. 1975. Preloading the pressure-bar assembly for improved veneer-lathe setting accuracy. For. Prod. J. 25(7):44-46.

Walser, D. C. 1978. New developments in veneer peeling. In Modern Plywood Techniques, Proc., Sixth Plywood Clinic, 6:6-18. Miller Freeman Publications, Inc., San Francisco.

Walstad, O. M. 1974. Commercial utlization of the McCartney fluid jet cutting concept. Proc., Second International Symp. on Jet Cutting Technology, BHRA Fluid Engineering, Cranfield, Bedford, England, Pap. E2:11-17.

Ward, D. 1963. Abrasive planing challenges your knife cutting techniques. Hitchcock's Woodwork. Dig. 65(11):29-32.

Ward, D. 1972. Oliver's computer-controlled cutoff machine. Woodwork. and Furniture Dig. 74(8):36-40.

Ward, D. 1975a. Labor reduction and yield exceed expectations. Part 1. Woodwork. and Furn. Dig. 77(11):32-35.

Ward, D. 1975b. Labor reduction and yield exceed expectations. Part 2. Woodwork. and Furn. Dig. 77(12):22-25.

Ward, D. 1975c. Abrasive planing and wide belt sanding. Woodwork. and Furniture Dig. 77(10):30-33.

Ward, D. 1975d. Machines and tools—all about what you probably knew and couldn't prove. Woodwork. and Furniture Dig. 77(7):24-27.

Ward, D. 1976. Grinding out moldings at 40 fpm. Woodwork. and Furniture Dig. 78(6):24-28.

Washburn, E. E., Jr. 1976. Reduce centering errors and increase recovery. Ch. 1 in Modern Plywood Techniques 3: 29-36. Miller-Freeman Publications, Inc., San Francisco.

Welch, R. 1971. Vibration in chain saws. Australian For. 35(4):215-225.

White, J. H. 1964. Groundwood. Chapter 9 in Handbook of pulp and paper technology. (K. W. Britt, ed.), p. 231-265. New York: Reinhold Publishing Corp.

White, V. S., Ed. 1976. Hardwood sawmill techniques, proc. of the first hardwood sawmill clinic program. 300 p. San Francisco, Ca.: Miller Freeman Publ., Inc.

Whitehead, J. C. 1978. Procedures for developing a lumber-size control system. Inf. Rep. VP-X-184, Can. Dep. Environ., West. For. Prod. Lab., Vancouver, B.C. 15 p.

Wiklund, M. 1967. Forces and damage in felling and bucking with knife and shear type tools. Forsk. Skogsarbeten Rep. 9, 39 p.

Wiklund, M. 1971. Temperature measurement in the living tree. Paper presented at IUFRO meeting, Madison, Wis., March. Swedish For. Prod. Lab., Wood Tech. Dept., Stockholm. 12 p.

Wilkins, G. R. 1966. History of circular saws. Timber Trades J. Annu. Spec. Issue: S/1-S/3, S/27, S/29.

Williston, E. 1976. State of the art—hardwood lumber manufacture. In Hardwood sawmill techniques, proc. of the first hardwood sawmill clinic program, pp. 1-19. San Francisco, Ca.: Miller Freeman Publ., Inc.

Williston, E. 1979. State-of-the-art report: Lumber manufacture. For. Prod. J. 29(10):45-49.

Woodfin, R. O., Jr. 1973. Wood losses in plywood production—four species. For. Prod. J. 23(9):98-106.

Woodson, G. E. 1979. Tool forces and chip types in orthogonal cutting of southern hardwoods. Res. Pap. SO-146, U.S. Dep. Agric. For. Serv. 77 p.

Woodson, G. E., and C. W. McMillin. 1972. Boring deep holes in southern pine. For. Prod. J. 22(4):49-53.

Woodson, G. E., and Rigby, S. M. 1978. Headrig chippers and surface quality. In Complete tree utilization of southern pine: symp. proc., New Orleans, La., April 17-19. C. W. McMillin, ed. For. Prod. Res. Soc., Madison, Wis., p. 335-344.

Woodworking Industry. 1978. "You've got to cut noise but don't you dare upset production!" Woodwork. Ind. 35(2):12-13.

Worley, R. 1975. Installation of a precision ball screw carriage with computerized setworks and a high-strain thin-kerf bandmill. FPRS Sep. No. MS-75-S60, 9 p. For. Prod. Res. Soc., Madison, Wis.

Yokochi, H., and H. Fukui. 1978a. Face milling of wood. I. Geometry in formation of face milled surface and calculation of depth of knife mark on ideally machined surface. J. Jap. Wood Res. Soc. 24:815-821.

Yokocki, H., H. Fukui. 1978b. Face milling of wood. II. Effects of basic cutting conditions and the corner shape of tool cutting edge on the roughness of milled surface and power requirement. J. Jap. Wood Res. Soc. 25:14-21.

York, F. W. 1975. Helical heads promise planer noise reduction. Ch. 16 in Modern Sawmill Techniques, Vol. 5. (V. S. White, ed.) Miller Freeman Publications, Inc., San Francisco. p. 249-254.

19
Bending

Chapter 19 is condensed from: Jorgensen, R.N., and P. Koch. 1979. Bending of woods of hardwood species growing on southern pine sites. U.S. Dep. Agric. For. Serv., South For. Exp. Stn., Alexandria, La., Fin. Rep. FS-SO-3201-5.26; September.

Major portions of data were drawn from industry and from research by:

M. Bariska	C.A. Jordan	R.D. Rea
W.G. Baumgardt	P.C. Kitchin	R.O. Rosendahl
H.S. Betts	R.H. Krone	R.D. Runkel
P.M. Butts	E.J. Kubinsky	T.C. Scheffer
F.J. Champion	R.F. Luxford	C. Schuerch
E.T. Choong	H.B. McKean	A.J. Stamm
R.W. Davidson	W.K. Murphey	W.C. Stevens
W.F. Faulkes, Jr.	L.R. Morris	R. Teichgraber
W.J. Finnorn	E.C. Peck	N. Turner
A.D. Freas	M.Y. Pillow	F.F. Wangaard
G. Ifju	A. Rapavi	T.R.C. Wilson

Chapter 19
Bending

CONTENTS

		Page
19-1	Steam Bending	2286
	Wood Selection	2287
	Steaming	2290
	Bending	2290
	Setting the Bends	2295
	Changes in Curvature with Time and Exposure Conditions	2297
	Strength of Steam Bent Wood	2298
19-2	Variability of Wood for Bending	2298
19-3	Laminated Bent Wood	2300
19-4	Plasticizing Wood with Urea	2305
19-5	Plasticizing Wood with Ammonia	2306
	Liquid Ammonia	2306
	Gaseous Ammonia	2307
19-6	Other Plasticizers	2308
19-7	Literature Cited	2309

Chapter 19
Bending[1]

Bending plasticized wood to form is an ancient art which continues today to be of economic importance to several segments of the wood industry. Such bending is simpler and quicker than machining, produces little or no wood waste and requires less machinery investment and processing energy. Bent wood is usually stronger and stiffer than machined parts of the same shape, and needs less surface preparation for finishing. For some furniture parts, sporting goods, canes, and umbrellas, it is the only economic method of production.

Wood in its natural state is an elastic material and within a limited range where stress and strain are proportional it returns to its original shape when stress is removed. With greater stress, the strain is no longer proportional and some permanent deformation occurs. Normal unheated wood is weakened when it is bent to a radius of less than about 320 times its thickness; it usually fractures at a radius of 50 to 80 times the thickness. Thin white oak laminations (less than 1 inch thick) from southern locations are exceptions to this restriction (Rosendahl[2]). Because of the limiting strain capability of wood (its ultimate strength), it must be plasticized for commercial bending to produce satisfactory products. Plasticizing increases the compressibility of wood as much as 30 to 40 percent and, with proper restraint to prevent stretching, permits bending to much smaller radii.

New households for our expanding population and growing interest in physical fitness have kept the industries using bentwood economically healthy. Among factory shipments of household furniture in the United States in 1972 worth $5 billion (U.S. Department of Commerce 1972), about 24 percent valued at about $1.2 billion involved the use of bentwood parts. The sporting goods industry for 1977 made factory shipments of $1.1 billion, and some 22 percent of these products with value of about $242 million were of bentwood. Miscellaneous other bentwood products raise the value of products made by wood bending to some $1.45 billion annually.

Representative species of all of the hardwood genera occurring on southern pine sites have been tested in bending. It is likely that most of the published results are from specimens sampled from better sites for hardwoods than those classed as southern pine sites. In the absence of more specifically applicable information, however, these published data are summarized in table 19-1 as the best available.

---

[1]See acknowledgements on the front page of this chapter.
[2]Rosendahl, R.O. 1945. Breaking radii of white oak and Douglas fir laminations of various thickness. U.S. Dep. Agric. For. Serv. For. Prod. Lab. unpublished report. 9 p.

TABLE 19-1.—*Bending evaluations of 12 southern hardwoods[1] in terms of relative ranking, minimum bending radii, and percent of good bends attainable*

Species	Source of data				
	Peck (1957a)[2]	Stevens and Turner (1970)[3]	Betts (1954)[4]	Choong (1971)[5]	Choong (1971)[6]
	Rank	Inches	Percent	----------Inches----------	
Ash sp.	13	4.5	—	>4	2
Elm, American	10	1.7	—	—	—
Hackberry.	1	—	—	2	2
Hickory sp.	8	1.8	—	—	—
Maple, red	15	—	—	>4	3
Oak, chestnut.	4	—	—	—	—
Oak, northern red	3	1.0	—	—	—
Oak, white	2	.5	—	—	—
Sweetbay	5	—	—	—	—
Sweetgum.	14	—	67	>4	4
Tupelo, black..............	21	—	46	—	—
Yellow-poplar	16	—	58	3	3
Others.	—	—	62[7]	—	—

[1] Wood sampled from hardwood sites which may not have been southern pine sites.

[2] Uniform test without end pressure. Steamed, no time or size reported. Descending order of bending quality 1 to 25. Missing numbers apply to woods outside scope of this paper.

[3] Tests on clear stock, 1 inch thick, steamed at atmospheric pressure, no time given. End pressure and tension strap used in bending. Radii given are minimum safe size so that not more than 5 percent of pieces will break.

[4] Percent of test pieces in good condition after steam bending. No description of test method.

[5] Specimens ¼-inch thick, atmospheric steaming 2 hours. Radii are minimum three replications could be bent without breaking. Hand bending, no restraints.

[6] Same as [5] except specimens were exposed to anhydrous ammonia vapor at 77°F for 1 hour, but were not steamed.

[7] Percent of pieces in good condition after steam bending. 25 unnamed species of southern hardwoods.

19-1 STEAM BENDING

Success in steam bending of hardwoods depends on judicious selection of wood, steaming time, and bending and setting procedures. Steam-bent wood is generally somewhat weakened; its curvature may change with time depending on exposure conditions and restraint during use.

WOOD SELECTION

Efforts by Kitchin[3], Jordan and Peck[4,5,6], Peck[7] (1957a), Faulkes[8], Pillow (1951), Finnorn and Rapavi (1956), and Stevens and Turner (1970) to correlate anatomical, physical, and mechanical properties within species, and between species, with steam bending quality have met with little success. Jordan and Peck[4,5] found that steamed red and white oak with high values for compression deformation indicated better bending material. These results, however, were not correlated with specific gravity or growth rate (rings per inch). Table 19-2 shows the lack of correlation between two strength properties and the ranking by Peck (1957a) of these species in a standardized free-bending test.

Toughness of hardwood, however, may be an indicator of bending properties. Scheffer (1979) found a significant relationship between bending tolerance (breaking radius of curvature) and toughness of fungi-attacked hardwood veneers of four species, including sweetgum and yellow-poplar. He found that tough veneer strips, ½-inch wide and equilibrated to 6 percent moisture content, could be bent at room temperature to smaller radii than less tough veneer, as follows:

Veneer thickness	Toughness, inch-pounds	
	1.25	4
Inch	Breaking radius, inch	
0.044	0.9	0.5
.055	1.4	.7
.066	2.2	.9

The relationship was curvilinear; breaking radius increased rapidly as toughness values decreased from 2 to 1 inch-pounds, but diminished little at toughness values greater than 4 inch-pounds.

---

[3]Kitchin, P.C. 1919. The relation of structure to mechanical properties of wood with special reference to steam bending. U.S. Dep. Agric. For. Serv. For. Prod. Lab. unpublished report. 11 p.

[4]Jordan, C.A., and E.C. Peck. 1946. Relation of endwise compressability to the bending radii for Wisconsin grown white oak, etc. U.S. Dep. Agric. For. Serv. For. Prod. Lab. unpublished report. 16 p.

[5]Jordan, C.A., and E.C. Peck. 1946. Relation of endwise compressibility to bending radii. U.S. Dep. Agric. For. Serv. For. Prod. Lab. unpublished report. 25 p.

[6]Jordan, C.A., and E.C. Peck. 1946. Bending chemically treated and seasoned Appalachian white oak for motor launch frames. U.S. Dep. Agric. For. Serv. For. Prod. Lab. unpublished report. 14 p.

[7]Peck, E.C. 1947. A study of Appalachian white oak to determine the characteristics of high-quality bending stock. U.S. Dep. Agric. For. Serv. For. Prod. Lab. unpublished report. 35 p.

[8]Faulkes, W.F., Jr. 1951. Exploratory tests on the relationship of strength properties to suitability of selected white oak for steam bending. U.S. Dep. Agric. For. Serv. For. Prod. Lab. unpublished report. 10 p.

TABLE 19-2.—*Steam bending ranking of 12 southern hardwoods compared to pore structure and to ranking by crushing strength, modulus of rupture, and specific gravity*[1]

Species	Pore structure	Steam bending[2]	Maximum crushing strength parallel to grain[3]	Modulus of rupture[3]	Specific gravity[4]
		---------- Rank ----------			
Hackberry	Ring	1	11	9	6
Oak, white	Ring	2	2	3	2
Oak, northern red	Ring	3	5	4	4
Oak, chestnut	Ring	4	4	6	3
Sweetbay	Diffuse	5	8	10	8
Hickory	Ring	6	1	1	1
Elm, American	Ring	7	10	8	7
Ash, white	Ring	8	3	2	5
Sweetgum	Diffuse	9	7	7	7
Maple, red	Diffuse	10	6	5	6
Yellow-poplar	Diffuse	11	9	11	9
Tueplo, black	Diffuse	12	6	12	7

[1]These ranking data are for specimens sampled from hardwood sites, and not necessarily from southern pine sites where these species also grow.
[2]Ranked according to Peck (1957a); rank no. 1 is most successfully steam bent.
[3]See table 10-6; rank no. 1 has the most strength when air dry.
[4]See table 10-7; rank no. 1 is most dense based on ovendry weight and volume when green.

Peck[7] found increased bending success in white oak both above and below a range of specific gravity of 0.540 to 0.559 with no logical explanation of the results. Stevens and Turner (1970) and Peck (1957a) conclude that selection of material by age, rate of growth, site characteristics and tree form is of secondary importance to bending quality, provided defect-free, straight-grained material is available.

Investigators agree that quality bending stock must be free of knots, decay, cross and twisted grain, and not excessively low in weight; the latter is usually an indication of brash wood. Figure 19-1 illustrates bending breakage in brash wood and at a knot. For hardwood species found on southern pine sites, growth rates of 6 to 15 rings per inch are preferred, but Jordan[9] and Peck[7] successfully met U.S. Navy requirements for steam bending with both slower and faster grown material. Stevens and Turner (1970) advocate a moisture content of 25 percent for all bends while Peck (1957a) indicates 12-15 percent satisfactory for platen bends and 15-20 percent for form bends. Some companies have successfully bent wood at 4-6 percent moisture content but the bends were on very large radii.[10]

[9]Jordan, C.W. 1945. The bending of 2½-inch by 3-inch by 6-foot Appalachian white oak boat frames for strength tests. U.S. Dep. Agric. For. Serv. For. Prod. Lab. unpublished report. 7 p.
[10]Butts, P.M. 1978. Georgia Forestry Commission case report, with personal correspondence. 2 p.

Bending

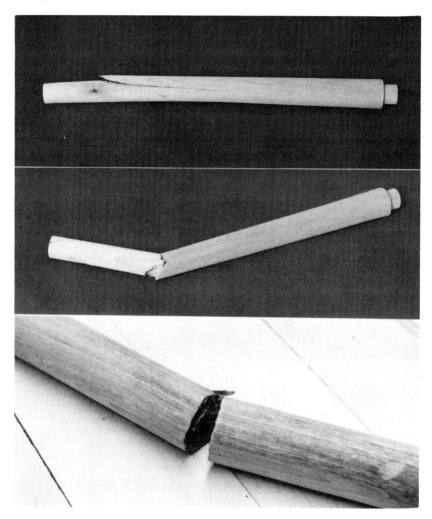

Figure 19-1.—Breakage attributable to defects in bending stock. (Top) Split due to cross grain at pin knot. (Center) Break in somewhat brash wood. (Bottom) Break in extremely brash wood.

Peck (1957a) and Stevens and Turner (1970) agree that stock carefullly machined so that the annual rings are parallel to the bending form (flat sawn) gives slightly better results in bending; in commercial operations the trouble and cost of such careful production would be justified only for unusual or extreme bends. Jordan and Peck[5] found that tangential and radial grain in 1- by 1-inch northern red oak bent with equal success.

STEAMING

Exposure to saturated steam at atmospheric pressure (212°F) or at gauge pressures not in excess of 5 psi (220°F) is the least expensive and easiest method to plasticize wood for bending (fig. 19-2). The principal object of this exposure is to heat the wood thoroughly. Little change in moisture content results. Forty-five minutes of exposure (by either method) per inch of thickness, heats southern hardwoods thoroughly. Both longer and shorter steaming times are in use (Rea 1976), but most commercial bending operations favor somewhat less time, and their bending yields are equally successful.[11]

Kubinsky (1971, 1972) and Kubinsky and Ifju (1973, 1974) have shown that steaming of red oak up to 3 hours reduces drying shrinkage slightly but that longer steaming periods increase shrinkage. No bending tests were performed on their material. Steaming up to 3 hours has negligible effect on the pH of the wood, but long-time steaming reduces pH significantly and changes chemical constituents and physical form. Steaming up to 3 hours lightens the wood color, whereas long-time steaming darkens it. Most of these changes do not affect furniture and bent sports equipment, but if heavy stock is involved, with consequent long steaming periods, they should be considered.

BENDING

Bends in plasticized wood are made without (free), and with, application of end pressure. Free bends are limited to radii between 100 and 300 times the thickness of the piece, the range within which unplasticized wood can be deformed without fracture. Free bending is satisfactory for mild bends because the increased compressibility from plasticization permits strain to be absorbed on the concave side of the bend before tensile strain on the convex side causes failure. Free bends are so near to being within the elastic limits of the wood that they must be slightly overbent as they will tend to revert to original form even when restrained during setting.

Plasticization does not increase the ability of wood to stretch; most bends require end pressure to limit tensile stress and prevent failures on the convex side of the bend (fig. 19-3). Such end pressure is applied by fixed (figs. 19-4 and 19-5) or adjustable (fig. 19-6) end stops. Except in mild platen bends, stops are attached to metal straps running the length of the convex side of the piece. Such straps must be strong enough to retain their original length throughout the bending. Ideally stock should have a 10° end bevel to mid-thickness of the piece toward the concave side so all end pressure is on the convex side of the bend, but this is not done in commercial operations.

---

[11]Jorgensen, R.N. 1978. Present status of the major wood bending industry as determined by visitation. Unpublished report. 5 p.

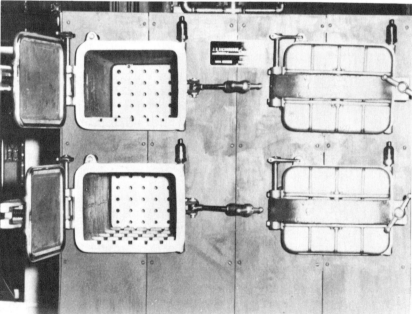

Figure 19-2.—Steam retorts for plasticizing wood. (Top) Blanks for chair backs being removed after 20 minutes of steaming at atmospheric pressure. (Bottom) Door locks for retention of low pressure, and raised floor supports to keep wood above accumulation of hot water on the retort floor.

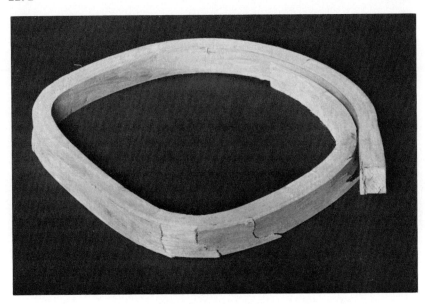

Figure 19-3.—Tensile failures, foreground and right side, in ash seat support as a result of inadequate end pressure. Overlap is for a scarf joint in the finished product.

Figure 19-4.—Heated platen press for mild bends in solid wood. Fixed end-pressure stop indicated by arrow. Pieces being bent are visible between platens.

Bending 2293

Figure 19-5.—Hydraulic press for extreme bends in chair arms and other parts. (Top) Bend partially completed; arrow indicates fixed end stop which provides end restraint. (Bottom) Bend completed; the operator will next install a restraining lock securing the bent wood.

Figure 19-6.—A clasic bentwood chair and workers in the process of producing the major bentwood component, which is attached to all four legs and the seat support. (Top) All pieces, except the front legs, of this Thonet chair are bentwood. (Bottom) A restraining strap, center, follows the convex side of this two-plane bend; an adjustable end-pressure stop is controlled by the operator's hand at lower left.

Fixed end stops are most common in the industry (figs. 19-4 and 19-5), and bending stock must be cut accurately to fit tightly between the stops. Since over-length pieces will not fit within the fixed stops and must be remachined, there is a tendency to undercut the stock length. According to industry representatives, short lengths probably cause more rejections because of tension failures, than any other problem.[11]

Adjustable end stops are most frequently used in hand bending (fig. 19-6). Skilled operators can adjust the end pressure to prevent excessive compression distortions and/or buckling of the parts of the piece not included in the bend while maintaining enough pressure to prevent tension failures. Hand bending remains as much an art as science even today; among five operators with an average of 15 years' experience up to 25 percent rejects are produced on complicated hand bends (fig. 19-6), although much of this is due to poor stock selection, as in figure 19-1.

SETTING THE BENDS

Because of stresses induced by bending, bentwood tends to return to original form unless restrained (figs. 19-7 and 19-8) until set. The setting of a bend is the reverse of plasticization; the wood is cooled, with or without a change in total moisture content. For furniture, setting is usually accompanied by drying to the moisture content expected in service. Setting-room conditions depend on the thickness, species, and changes in moisture content required and can range from simple dry storage to conditions where humidity and temperatures are controlled as closely as in a lumber dry kiln.

Straps on the convex side and end stops should be retained during setting, as earlier removal can cause tension failures. The opening or distance between the ends of a bent piece must also be controlled (figs. 19-7 and 19-8). Compression failures on the concave face of the bend may develop more than ordinary longitudinal shrinkage during setting and, if not controlled, will close the bend, inducing tensile stresses on the convex side, causing distortion and failure. The bend is considered set when it no longer exerts pressure against the restraints. Faulkes and Freas[12] conditioned white oak boat frames for 10 months after bending and found no changes in strength properties from frames tested immediately after setting.

---

[12]Faulkes, W.F., Jr., and A.D. Freas. 1946. Effect of extended conditioning on strength and stiffness of steam-bent white oak frames. U.S. Dep. Agric. For. Serv. For. Prod. Lab. unpublished report. 6 p.

Figure 19-7.—Restraints used for setting of bends. (Top) Pressure locking clamp—in lieu of end-pressure device—holds stainless steel bending strap in place on machine-bent chair-seat support. (Bottom) Cap-like holding clamps secured by diagonal bar maintain degree of bend while also retaining some end pressure on the wood.

Figure 19-8.—Bent boat member prepared for setting and drying. Concave strap retained by the rods maintains end pressure and shape: wood braces reduce twisting. (Photo from U.S. Forest Products Laboratory.)

CHANGES IN CURVATURE WITH TIME AND EXPOSURE CONDITIONS

After setting, a bend will retain the set curvature only if it is restrained by incorporation in some construction or held at a constant moisture content equal to that when set. The compressed fibers on the concave side of all bends respond longitudinally to changes in moisture content, while the stressed fibers on the convex side react only minimally as in normal wood. Thus increases in moisture content after setting will open bends, and decreases will tend to close them. Bends opened by increases in moisture content and subsequently redried to the original moisture content, however, will not close to the original set radius. Peck (1957a) showed that northern red oak steamed and bent at 25 percent moisture content and set at 8 percent opened less and closed more with fluctations in moisture content than specimens bent at 15 percent original moisture content. Kubinsky (1971) reported reduced tangential shrinkage in yellow-poplar and sycamore after 3 hours of steaming at atmospheric pressure, and Kubinsky and Ifju (1974) reported the same true for southern red oak. With certain grain orientations this factor could reduce the post-setting movement of bends but the factor of change is so small that it would be of little value for consideration in commercial operations (Ward 1976).

STRENGTH OF STEAM BENT WOOD

Strength of steam bent wood is affected by both the steaming and the bending. Studies of white ash (Morris and Wilson[13]) and red oak (Kubinsky 1972; Kubinsky and Ifju 1973, 1974) revealed little change in strength properties from atmospheric steaming up to 6 hours. Steaming red oak up to 96 hours impaired physical and mechanical properties significantly, probably from chemical changes in wood constituents. Runkel and Wilke (1951) show similar results with European beech (*Fagus silvatica* L.). Prolonged steaming undoubtedly impairs the properties of southern hardwoods, while the short steaming times involved in plasticization do not. Choong (1971) found that after steaming at atmospheric pressure for 2 hours and reconditioning to the original 20 percent moisture content, specimens of yellow-poplar were slightly stronger and stiffer than unsteamed controls.

Because of induced stresses and compressive failures in bending, bent wood is weaker than straight clear specimens. Wangaard (1952) reported a 32.1-percent loss of strength (modulus of rupture) in steamed 1-inch beech (*Fagus grandiflora* Ehrh.) stock bent to an 8-inch radius and a 26.1-percent loss for white oak similarly bent. The loss in beech was 36.2 percent for a 6-inch radius and 14.8 percent for a 10-inch radius. While modulus of rupture is diminished by bending, ability to absorb energy during deformation (work) is not; Teichgraber (1953) showed increases in impact work for steamed and bent beech in thicknesses up to 4 inches. This is in agreement with McKean et al. (1952) and Luxford and Krone (1962), who found 2-inch white oak boat frames capable of exceptionally larger deflections at maximum load when equivalent laminated frames, while reductions in modulus of rupture and modulus of elasticity occurred.

The literature provides no other data on changes in strength properties of small bent wood pieces (up to 2 inches thick), due perhaps to the variety of shapes produced and the lack of a standard rating for bending success. Furniture and other industries using bends up to 2 inches thick, design with the basic strength values of the species and find that any bend satisfactory for production retains sufficient strength to meet the design requirements.[11]

19-2 VARIABILITY OF WOOD FOR BENDING

As discussed under the heading of WOOD SELECTION, research has found little relationship between success in bending and physical or mechanical properties of wood. Table 19-2 shows the lack of relationship of basic growth structure and specific gravity of hardwoods on southern pine sites to one evaluation of their bending ability. Of these species, the most used in the bending industry are white and red oaks, elm, hickory and ash. The oaks and elm are

---

[13]Morris, L.R., and T.R.C. Wilson. 1919. Some tests of the effect of steaming and bending upon the strength of white ash and sitka spruce. U.S. Dep. Agric. For. Serv. For. Prod. Lab. unpublished report. 7 p.

popular furniture material, and hickory and ash are favored for sporting equipment and tools becuase of their high impact strength. Several evaluations of their bending suitability are summarized in table 19-1. These species, making up about 60 percent of the hardwoods on southern pine sites, constitute an adequate resource to supply the bending industry if needed.

Once the species has been selected for the product, tree selection for good bending stock within that species is even more difficult. Stevens and Turner (1970) and Jorgensen[11] found little agreement within industry on such selection. Finnorn and Rapavi (1956) found specific gravity variations within species to have no effect on bending ability, and Peck (1957a) showed white oak of similar specific gravity and rate of growth to produce both the best and poorest quality bending stock. Luxford and Krone (1962) reported that white oak beams 2 inches wide by 2.6 inches deep from Ohio, which met the Bureau of Ships specifications for bending stock and was successfully bent to a 22.8-inch radius, was considerably weaker than white oak from Arkansas and Louisana. Pillow (1939) found white and green ash wood with large fibril angles to be of poorer bending quality than that with smaller angles. Open grown trees in creek bottoms appear to have more such large angles than those grown on upland sites. Other environmental factors such as dominance, spacing, and site quality also affected size of fibril angles; thus tree selection for bending stock is difficult.

One company has reported[14] 10-percent lower bending yields for ash from southern pine sites than for material from the northern central states. The southern material is bought in relatively small quantities which preclude any tree selection or control over production of the lumber. Results have been poorest with material stored and partially seasoned in log form before manufacture, which may induce stresses adversely affecting bending. No loss of yield is reported when the material is rough sawn immediately after felling and air dried in boards to about 25 percent moisture content. The lower bending yields are not the result of visible seasoning defects formed in the logs, as these are rejected in cutting the bending stock. No research on this facet of the treatment of bending material is reported in the literature; the industry would welcome such work.

Pillow (1939) reported variation in bending material with location in the tree, butt logs being the least satisfactory. Stevens and Turner (1970) confirm this and say that for most bends any part of the tree between the butt swell and the crown is satisfactory if defect free; the second log above the butt swell is usually best, however. They report that wood near the heart of the tree, from overmature trees, and from oval or elliptical trees (which tend to contain tension wood) is not usually satisfactory for bending.

---

[14]In conversations with R.N. Jorgensen, 1979.

19-3 LAMINATED BENT WOOD

As previously noted, unplasticized wood can be bent within certain limits without fracture, particularly if very thin. If a number of concentrically bent pieces are fixed together, usually by adhesive, the bend becomes permanent. **Laminated bending** permits sharper bends with poor steam-bending species, admits limited defects in unstressed areas, permits splicing of laminations, facilitates construction of pieces much larger than available in solid wood, and uses predried material to avoid the seasoning defects common in large timbers.

Bending laminated material requires heavier equipment than solid wood bending; cost of adhesive and the required careful preparation of material also make it more expensive. Glue lines are visible in the completed product, and it is very difficult to produce bends in more than one plane.

Laminated bent wood is used for both structural and non-structural purposes. In furniture and sporting goods (figs. 19-9 and 19-10) veneers up to ⅛-inch thick are the usual lamination material. After adhesive application, the bend is usually formed between male and female platens of the desired shape, often heated to accelerate setting of the glue. For mild bends (fig. 19-9 top) the press may have multiple openings similar to flat plywood production. More acute bends are formed on hydraulic presses of the same type used in solid wood bending (figs. 19-5 and 19-11). Pressures applied by air hoses, water hoses, and the atmosphere on vacuum deflated rubber bags have been used to form laminated material but are rarely in use in the industry today.

Structural laminated material (fig. 19-12) is similarly produced in appropriately larger scale to accommodate larger material. Most large structural members are bent into a fixed female form by mechanical and/or hydraulic jacks placed along the entire length of the form. Laminations are sometimes placed on edge for assembly, to simplify the jig construction. They are usually not thicker than 2 inches to limit degrade, time in pre-assembly drying, and the size of pressure equipment for bending. There is little information on safe bending radii of thick laminations. Most commercial operations use bends less than those causing failure in compression. Finnorn and Rapavi (1956) show a safe bending radius for 1-inch white oak to be half way between a breaking radius of 75 inches and a proportional limit radius of 140 inches. For clear, high quality boards ¼-inch to 1 inch in thickness, they propose a formula for white oak:

$$R_s = 116t - 6 \qquad (19\text{-}1)$$

where:

R_s = safe bending radius, inches
t = board thickness, inches

Figure 19-9.—Laminated furniture parts and completed products. (Top) Chair backs ready for assembly; each back consists of two parts, one thin and one thick, which have been bent together in a hot press but will not be joined until final assembly. (Bottom left) A beautiful Norwegian bentwood chair of laminated beech strips; the seat support is canvas. (Bottom right) An outstanding example of the simplicity possible with laminated bentwood construction.

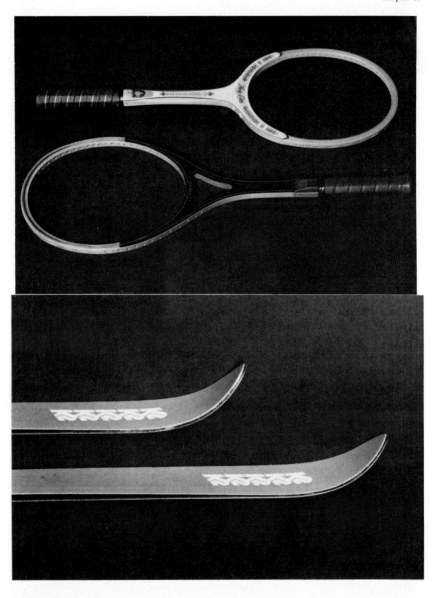

Figure 19-10.—Laminated bentwood sports gear. (Top) Tennis rackets of birch laminations about 1/16-inch thick; lower racket has a graphite insert between laminations at the top bend. (Bottom) Bentwood ski tips with fiberglass overlay.

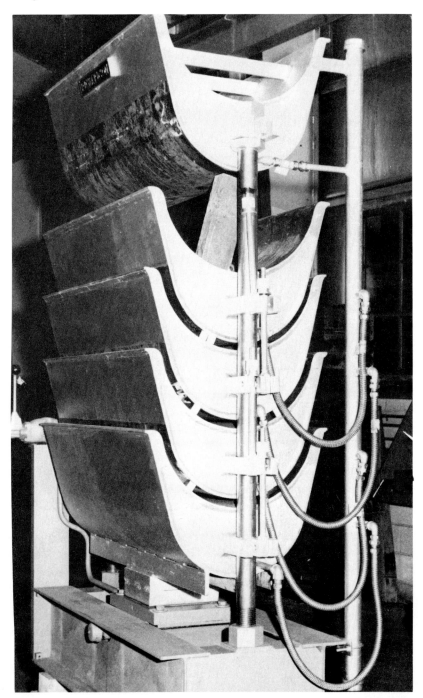

Figure 19-11.—Four-opening hydraulic press used principally for forming veneer laminates for curved plywood. The top opening is propped open for examination. Note absence of end-pressure fixed stops used for solidwood bending as seen in figure 19-4.

Figure 19-12.—Assembly for bending pairs of laminated boat ribs. Ribs are shown clamped in place over a solid wood male form. A flexible top caul distributes the clamping pressure. (Photo from U.S. Forest Products Laboratory.)

Stevens and Turner (1970) have shown that the ratio of radius of curvature to thickness of lamination increases with increasing board thickness and have applied a 9/4 multiplier correction figure to their results from $\frac{1}{8}$-inch veneer bending. For material up to 1 inch thick this results in a radius to thickness ratio of 70 for rock elm (*Ulmus thomasi* Sarg.) and 100 for white oak. The published literature contains no data directly applicable to material 2 inches thick, but Stevens and Turner (1970) indicate further modification of their radius of curvature to thickness ratio may be necessary for poor quality material or material substantially over 1 inch in thickness.

Laminated bent wood is stronger than steam-bent wood of the same size. Luxford and Krone (1962) found that bent laminated material was approximately 50 percent stronger than steam-bent solid wood of the same size (2.0 x 2.6 inches) and 50 to 100 percent stiffer. McKean et al. (1952) found that white oak ship timbers 2 inches thick by $3\frac{1}{2}$ x $40\frac{3}{4}$ inches laminated to a 16-inch radius of curvature were 25 percent stronger than equal size steam bent frames when loaded on the convex side.

When bent to radii that do not cause compression failures, wood is sufficiently elastic that it does not lose strength when incorporated in laminated bent wood. Finnorn and Rapavi (1959) showed no strength differences in white oak beams loaded on the convex sides when the laminations varied between ¼- and ¾-inch on a 72-inch radius. Structural laminates are almost always loaded on the convex side and modern design methods use the unbent working strength values for the species in determining the ultimate strength of the total laminated construction.[15] Laminates loaded on the concave side are not as strong as unbent material (Woodson and Wangaard 1969), but they are rarely loaded in this direction.

Numerous researchers have shown the feasibility of laminating strong species in high stress areas of structural members with lower strength or lower grade material in the low stress areas. Such construction offers exceptional opportunity to utilize a mixed group of the hardwoods occurring on southern pine sites at such time as the seasoning, handling, and availability of cut lumber becomes economically competitive with softwoods.

19-4 PLASTICIZING WOOD WITH UREA

Initial studies (Champion 1941) with blackjack, overcup (*Quercus lyrata* Walt.), and southern red and white oaks showed that urea solutions impregnated into the wood and heated induced plasticity, particularly in thin sections. Jordan and Peck[6] and Peck (1957a) reported that urea-treated 1-inch white oak did not bend as well and was weaker than steam plasticized material. Urea is hygroscopic and in high moisture content situations bends made with urea treated wood have completely returned to their original form (Stevens and Turner 1970). The softening action of the chemical so lowers strength that there is a high incidence of tension failures as bends set. Jordan and Peck[6] and Peck[8] found no benefit in pre-kiln-drying urea treatment of white oak boat frame bending stock over regular air-drying practices and noticably more tension failures occurred.

Plasticization of wood for bending by the use of urea was the first successful attempt to use chemicals instead of steam or hot water and gained considerable publicity. Subsequent testing has shown several disadvantages and it is no longer considered a practical approach.

---

[15]Conversations in 1979 between R.N. Jorgensen and the industry.

19-5 PLASTICIZING WOOD WITH AMMONIA

Wood can be plasticized by immersion in liquid or gaseous ammonia.

LIQUID AMMONIA

In the continuing search for a better way to plasticize wood for bending, Stamm (1955) appears to have been the first to suggest ammonia. Anhydrous ammonia (NH_3) is a commercial chemical sold as a liquid under pressure (about 150 psi) in tanks, as distinguished from the water solution of ammonia (NH_3) used as a household product. At atmospheric pressure, ammonia boils at $-28°F$ ($-33°C$), and freezes at $-108°F$ ($-78°C$). It is used as a fertilizer and in industry. It can be kept as a liquid at atmospheric pressure in loosely covered iron, steel, or glass (but not copper or brass) containers that have cooling coils or are packed in dry ice so the temperature is kept below its boiling point. Caution is advised to protect against the cold and extremely irritating vapor (Schuerch 1964).

Bariska et al. (1969) found some striking similarities between the sorption by wood of ammonia and water. Ammonia appears to penetrate the cellulosic components of wood to a greater extent than water and to also plasticize the lignin fraction.

Schuerch (1964) reported that all species can be plasticized by immersion in liquid ammonia; straight-grained samples of higher density hardwoods were bent with most success. If sufficiently treated and then warmed toward room temperature, thin specimens become pliable and can be readily manipulated with gloved hands. Maximum plasticity is reported to last 8 to 30 minutes; wood will retain extreme bends after being hand-held or clamped in position for a few minutes. Compared to steam-bent wood, bends with ammonia treated wood are much less subject to change in shape with post-set variations in moisture content and return more closely to original bent form when dried.

Required treating time is presently the major problem preventing use of ammonia by the wood bending industry. Schuerch (1964) found that ⅛-inch by 4-inch by 40-inch hardwood slats required in excess of 4 hours to be plasticized. Shorter times could be used and thicker stock treated, by pre-cooling the wood in a carbon dioxide atmosphere until all air is replaced, after which reaction between the carbon dioxide and ammonia moves the liquid into the wood rapidly. Pre-evacuation and impregnation under pressure, as in commercial preservative treating, will also speed plasticization. Since special, expensive equipment is required for these methods, however, it is doubtful that they are presently feasible for industrial use.

GASEOUS AMMONIA

As with water, wood will absorb ammonia in either liquid or vapor form. The amount absorbed is dependent on the vapor pressure of the ammonia atmosphere, an action analogous to wood-water systems where water absorbed varies with relative humidity. Davidson and Baumgardt (1970) reported that treating hard maple at 77°F and about 145 psi after cylinder evacuation produced the same plasticization as liquid ammonia treatment. Absorption rate of the vapor is directly proportional to the initial moisture content, at least up to 20 percent, and maximum softening takes place before maximum possible absorption of ammonia. Thick stock requires longer treating time than thin material. Davidson and Baumgardt (1970) tentatively suggested the following times under pressure:

		Treating time	
Thickness	Ovendry	10-percent moisture content	20-percent moisture content
Inch		------------------------*Hours*------------------------	
1/8	8	1	1/2
1/4	8	2	1

From tests of 12 species, including white ash and northern red oak, Davidson and Baumgardt (1970) concluded that all wood is plasticized by the treatment and can be formed to some extent. Species that are good bending woods by the steaming process (table 19-1) are the most successfully formed with ammonia. They also suggest further investigation since bending radii related to species and thickness have not been determined and little is known of the forces required to shape the ammonia-plasticized wood. Choong (1971) showed some species rated as poor in steam bending to be eminently satisfactory with ammonia treatment (table 19-1).

It is possible to compress and densify wood that has been treated with ammonia vapor. In the United States such modification has been deterred by ample supplies of dense hardwoods. In eastern Europe, however, two *Populus* and two *Alnus* species—densified after ammonia vapor treatment—are being used to supplement supplies of heavier, harder woods (International Academy of Wood Science 1975). Ammonia is favored for such property modification because it is inexpensive and less corrosive than many other chemicals, is absorbed by both wet and dry wood, and has no residual effect after treatment.

19-6 OTHER PLASTICIZERS

Often used in conjunction with steaming, water heated to the boiling point (fig. 19-13) is used to plasticize limited sections of pieces to be bent, e.g., extremities of tool handles, canes and umbrellas (Rea 1976). Immersion times, bending ability, and setting procedures are the same as in steam bending.

Wood can be heated to plasticity by high-frequency electrical energy. Heat generation is directly proportional to the amount of moisture in the wood and localizes in pockets of high moisture content. Turner and Dean (1959) showed that European beech, elm, sap hickory, and red maple could be so heated without developing defects but that ash, black walnut (*Juglans nigra* L.), red-heart beech, oak, and heartwood hickory could not. No data are available on other hardwoods that grow on southern pine sites, but it is likely that only those of lower specific gravity would respond favorably to this method. Peck (1957b) attempted to plasticize northern red oak logs by this method and was unsuccessful because of defect formation.

Figure 19-13.—Hickory tool handles ready to soak in water which will be heated by the introduction of low-pressure steam. Time in the water when hot is about 30 minutes. These are mostly shovel handles about 1⅛ inches in diameter at the end to be bent.

Murphey (1967) reported successful plasticization of ⅛-inch hard maple (*Acer saccharum* Marsh.) and birch (*Betula lenta* L.) veneers by treatment with a 1:1 mix of dimethyl sulfoxide (DMSO) and a molar solution of ammonium chloride and also with a 75-part DMSO and 25-part ammonium hydroxide solution. Samples were soaked 8 hours at 180°F and dried at room conditions for 2 days before bending successfully to thickness:radius ratios of 1:8. Figured black walnut crotch veneers 1/28-inch thick, with the usual extreme cross-grain conditions found in such figure, were successfully bent to a 1-inch radius after only 15 minutes soaking in either solution. DMSO is a molar organic liquid with unusual penetrating qualities. It can be tasted in less than 10 seconds after application to the hands and its toxicological and physiological properties are not fully investigated. Pending further information, it should only be used under the normal precautions for all toxic materials.

Peck[16] concluded that soaking white oak in a 5.1 percent glycerine solution, followed by heating to 212°F and 350°F, was an unsatisfactory method to plasticize wood for bending.

Hardboards, a product of major importance manufactued from pine site hardwoods (see chapter 23) can be successfully bent after heating with platens or formed mandrels (Frost 1952; Johanson and Back 1966).

---

[16]Peck, E.C. 1948. Plasticizing Appalachian white oak with glycerine and high temperature preparatory to bending. U.S. Dep. Agric. For. Serv. For. Prod. Lab. unpublished report. 6 p.

19-7 LITERATURE CITED

Bariska, M., C. Skaar, and R.W. Davidson. 1969. Studies of the wood-anhydrous ammonia system. Wood Sci. 2:65-72.

Betts, H.S. 1954. Sweetgum (*Liquidambar styraciflua*). U.S. Dep. Agric. For. Serv., Amer. Woods Leafl. 8 p. Washington, D.C.

Champion, F.J. 1941. Molding wood to man's will. New plasticizing treatment may open way to use of low-quality timber. Amer. For. 47(4):178-179.

Choong, E. T. 1971. Bending Wood with steam and ammonia. La. Agric. 14(4):4-5. Agric. Exp. Stn., La. State Univ., Baton Rouge.

Davidson, R.W. and W.G. Baumgardt. 1970. Plasticizing wood with ammonia—a progress report. For. Prod. J. 20(3):19-25.

Finnorn, W.J. and A. Rapavi. 1956. Safe bending radii for curved laminates. For. Prod. J. 6:437-442.

Finnorn, W.J. and A. Rapavi. 1959. Effect of lamination thickness on strength of curved laminated beams. For. Prod. J. 9:248-251.

Frost, O.W. 1952. Machinability of hardboard. J. of the For. Prod. Res. Soc. 2(1):42-43.

International Academy of Wood Science. 1975. Structure and utilization of hardwoods, meet. of the European-African Group. Spec. Rep., 22 p. Banská Bystrica.

Johanson, F., and E.L. Back. 1966. Molding dry ligno-cellulosic materials above 325°C. For. Prod. J. 16(9):70.

Kubinsky, E. 1971. Influence of steaming on the properties of *Quercus rubra* L. wood. Holzforschung 25:78-83.

Kubinsky, E.J. 1972. Mechanical behavior of red oak in transverse compression as affected by hydro-thermal treatments and its relations to changes in cell wall structure and composition. Ph.D. Diss., Va. Polytech. Inst. and State Univ., Blacksburg. (Abstr.) Diss. Abstr. Int. B 32(10):5574.

Kubinsky, E. and G. Ifju. 1973. Influence of steaming on the properties of red oak. Part I. Structural and chemical changes. Wood Sci. 6:87-94.

Kubinsky, E.J., and G. Ifju. 1974. Influence of steaming on the properties of red oak. Part II. Changes in shrinkage and related properties. Wood Sci. 7:103-110.

Luxford, R.F. and R.H. Krone. 1962. Laminated oak frames for a 50-foot Navy motor launch compared to steam-bent frames. U.S. Dep. Agric. For. Serv., FPL Rep. No. 1611. 52 p.

McKean, H.B., R.R. Blumenstein, and W.F. Finnorn. 1952. Laminating and steam-bending of treated and untreated oak for ship timbers. South. Lumberman 185(2321):217-222.

Murphey, W.K. 1967. Pretreatment of bending stock. For. Prod. J. 17(9):75-76.

Peck, E.C. 1957a. Bending solid wood to form. U.S. Dep. Agric. For. Serv., Agric. Handb. No. 125. 37 p. U.S. Govt. Print. Off., Washington, D.C.

Peck, E.C. 1957b. Can sweep in saw logs be straightened? The Lumberman 84(2):96.

Pillow, M.Y. 1939. White ash and green ash: their wood structure and properties as influenced by growth. Ph.D. thesis. Univ. Wis., Madison. 161 p.

Pillow, M.Y. 1951 Selecting white oak trees for bending lumber. *In* Proc., For. Prod. Res. Soc. 5:87-92.

Rea, R.D. 1976. Classic bentwood chair. Woodwork. and Furn. Dig. 78(3):44-48.

Runkel, R.O., and K.D. Wilke. 1951. (The thermoplastic behavior of wood. II.) Holz als Roh- und Werkstoff 9:260-270.

Scheffer, T.C. 1979. Bending tolerance of veneer is related to its toughness. For. Prod. J. 29(2):53-54.

Schuerch, C. 1964. Principles and potential of wood plasticization. For. Prod. J. 14:377-381.

Stamm, A.J. 1955. Swelling of wood and fiberboards in liquid ammonia. For. Prod. J. 5:413-416.

Stevens, W.C. and N. Turner. 1970. Wood bending handbook. 110 p. London: Her Majesty's Stationery Off.

Teichgraber, R. 1953. Uber die spannungszustande bei der verformung von holz und die dadurch geanderten. Holzeigenschaften. Diss., Univ. Hamburg, Germany.

Turner, N., and A.R. Dean. 1959. The reaction of some well-known bending timbers when heated in a radio frequency field. Wood 24(3):97-100. (London)

U.S. Department of Commerce. 1972. Census of manufacturers. U.S. Census Bureau. Washington, D.C.: U.S. Gov't. Printing Office. 5 vol.

Wangaard, F.F. 1952. The steam-bending of beech. Beech Util. Ser. No. 3, Northeast. Tech. Comm. on the Util. of Beech and the Northeast. For. Exp. Stn., U.S. Dep. Agric. For. Serv. 26 p.

Ward, D. 1976. Fewer than 2% rejects in bent chair parts. Woodwork. and Furn Dig. 78(1):36-40.

Woodson, G.E., and F.F. Wangaard. 1969. Effect of forming stresses on the strength of curved laminated beams of loblolly pine. For. Prod. J. 19(3):47-58.

20

Drying

This chapter is largely derived from J. M. McMillen and E. M. Wengert's "Drying eastern hardwood lumber" (1978) with additions of sections on drying timbers, poles, posts, firewood, chips, flakes, and fibers. The subsection on veneer drying is condensed from Lutz (1977), that on firewood from Wartluft (1982ab). Readers desiring information on eastern hardwoods other then those important on southern pine sites should consult the original works of these authors.

Major portions of data were also drawn from research by:

E. L. Adams	J. Harrison	E. W. Price
D. G. Arganbright	C. A. Hart	E. F. Rasmussen
R. C. Baltes	J. G. Haygreen	H. Resch
J. S. Bethel	B. G. Heebink	D. B. Richards
P. Bois	R. A. Helmers	R. C. Rietz
R. S. Boone	N.C. Higgins	K. E. Rogers
M. Y. Cech	J. L. Hill	H. N. Rosen
P. Y. S. Chen	M. E. Hittmeier	J. G. Schroeder
W. P. Clark	J. P. Howe	F. S. Shinn
G. L. Comstock	J. B. Huffman	J. E. Shottafer
G. A. Cooper	K. E. Kimball	B. H. Shunk
E. P. Craft	P. Koch	W. T. Simpson
D. G. Cuppett	H. Kubler	C. Skaar
J. Denig	D.P. Lowery	W.R. Smith
P. Deverick	J. W. McMinn	E. L. Springer
R. Finighan	J. F. G. Mackay	M. A. Taras
H. O. Fleischer	R. R. Maeglin	W. C. Thomas
F. Freese	J. S. Mathewson	H. D. Tiemann
L. H. Furman	H. L. Mitchell	K. Townsend
L. D. Garrett	N. Nara	J. L. Tschernitz
H. B. Gatslick	R. H. Page	C. B. Vick
N. B. Goebel	E. C. Peck	J -H. Wang
R. P. Hale	M. K. Peirsol	J. Wartluft
H. Hallock	E. Perem	W. L. Wellford, Jr.
L. F. Hanks	D. M. Post	J. F. White
R. A. Hann	R. G. Potter	C. L. Wolfe

Chapter 20
Drying

CONTENTS

		Page
20-1	INTRODUCTION TO DRYING METHODS	2317
	HOW DRY IS DRY ENOUGH?	2320
20-2	AIR-DRYING	2320
	LUMBER	2320
	Degrade losses in air-drying	2322
	Control of degrade	2324
	Design of air-drying yards	2329
	Time required to air-dry lumber	2332
	Costs in air-drying lumber	2335
	SHORT LUMBER AND SQUARES	2335
	Hickory handle stock	2336
	THICK OAK PLANKING	2339
	From green to about 50-percent moisture content	2339
	From 50- to 25-percent moisture content	2340
	From 25-percent to final moisture content	2340
	Reducing checking in white oak shipbuilding flitches or timbers	2340
	CROSSTIES	2340
	Stack design	2340
	Stack covers	2342
	Occurrence of splits and checks	2343
	Prevention and control of splits and checks	2344
	Time required to air-dry crossties	2346
	Dowel-laminated crossties	2347
	POSTS AND POLES	2350
	Posts 2 to 8 inches in diameter	2351
	Posts 9 to 21 inches in diameter	2354
	Poles and piling	2356
	Transpirational drying	2356

	FIREWOOD..	2356
	Moisture loss in first month........................	2357
	Spring-cut vs. fall-cut wood........................	2358
	Species and stick length............................	2358
	Split wood vs. round wood..........................	2358
	Stacking method	2358
	Cover vs. no cover	2358
	FUEL CHIPS..	2359
	Air-drying of fuel chips on roofed drying grounds; effects of pile depth.............................	2359
	Air-drying of fuel chips on roofed and unroofed drying grounds; effect of turning the chips	2361
20-3	FORCED-AIR FAN PREDRYING.....................	2362
	LUMBER	2363
	CROSSTIES..	2364
20-4	HEATED LOW-TEMPERATURE DRYING............	2365
	SCHEDULES FOR HEATED LOW-TEMPERATURE DRYERS	2368
	Schedules for heated forced-air dryers..............	2368
	Schedules for low-temperature kilns................	2370
	DRYING TIMES	2374
	Yellow-poplar.......................................	2374
	Sweetgum sapwood	2375
	Unstickered hardwood furniture rounds.............	2376
	One-inch lumber of Appalachian red oak sp.	2378
	Some eastern hardwoods that occur on pine sites.....	2378
	ECONOMICS OF HEATED LOW-TEMPERATURE DRYERS......................	2378
20-5	LOW-TEMPERATURE DRYING WITH DEHUMIDIFIERS.............................	2379
	TIME AND ENERGY REQUIRED FOR DEHUMIDIFICATION DRYING.................	2381
20-6	DRYING IN CONVENTIONAL HEATED KILNS	2382
	DRYING PROCEDURES—GENERAL..............	2384
	Basic kiln schedules	2384
	Procedure for 4/4, 5/4, and most 6/4 air-dried stock ..	2391
	Procedure for 8/4 (plus 6/4 oak) air-dried stock	2391
	Procedure for partly air-dried 4/4, 5/4, and most 6/4 stock	2392
	Procedure for partly air-dried 8/4 (plus 6/4 oak) stock ...	2392
	Procedure for including small amounts of one species in large kiln loads of other species	2392

	Kiln schedule acceleration	2394
	Warp and shrinkage reduction	2396
	Adjusting moisture content of kiln-dried wood	2396
	Predicting drying time	2397
	Equalizing and conditioning	2400
	PROCEDURES FOR SPECIAL CLASSES OF PRODUCTS	2402
	Bacterially infected oak lumber	2402
	Presurfaced 1-inch upland red and white oaks	2402
	Hickory handle stock	2403
	Squares of other pine-site hardwoods	2405
	Rounds	2406
	Hardwood dimension parts	2406
	Thin northern red oak lumber	2408
	Thick lumber from pine-site hardwoods	2408
	Crossties	2408
	OPERATIONAL CONSIDERATIONS	2408
	Heating, venting, and air circulation	2408
	Humidity control with high-pressure steam	2409
	Part-time operation	2410
	DRYING TIME	2410
	SPECIAL PREDRYING TREATMENTS	2411
	Steaming to accelerate drying	2411
	Precompression	2412
	Prefreezing	2413
	HEATED-ROOM DRYING	2413
	ECONOMICS OF KILN-DRYING	2415
20-7	HIGH-TEMPERATURE DRYING	2416
	LUMBER	2416
	Basic concepts	2417
	Research and practice to date	2418
	Drying times and species potentialities	2420
	Apparent process requirements	2421
	Jet drying	2422
	Press drying	2422
	CROSSTIES	2428
	VENEER	2431
	Veneer properties that affect drying	2432
	Dryer conditions that affect drying	2434
	Types of veneer dryers	2436
	Drying time	2439
	Control of drying time	2440
	Control of final moisture content	2441
	Control of buckle	2442

	Control of splits.............................	2442
	Control of surface gluability	2442
	Control of collapse, honeycomb, and casehardening ..	2442
	Control of shrinkage...........................	2443
	Control of color..............................	2443
	Control of scorched veneer and dryer fires	2443
	FLAKES, PARTICLES, AND FIBERS.............	2443
	FIBER MATS................................	2443
20-8	ENERGY TO DRY	2444
	LUMBER	2444
	VENEER....................................	2447
20-9	SOLAR DRYING	2448
20-10	HIGH FREQUENCY AND MICROWAVE HEATING ..	2450
20-11	MINOR SPECIAL DRYING METHODS	2451
	IMMERSION IN HOT ORGANIC LIQUIDS.........	2451
	VAPOR DRYING	2452
	SOLVENT SEASONING	2452
	OTHER	2452
20-12	CHEMICAL TREATMENTS TO PREVENT CHECKING	2452
	SODIUM ALGINATE..........................	2453
	SALT PASTE	2453
	POLYETHYLENE GLYCOL	2454
20-13	STORAGE OF LUMBER, PLYWOOD, AND PARTICLEBOARD.........................	2454
	LUMBER	2454
	At the sawmill	2454
	In transit	2455
	At lumber distributing yards	2455
	At woodworking factories........................	2455
	At building sites	2455
	PLYWOOD, FIBERBOARD, PARTICLEBOARD, AND COMPOSITE PANELS	2455
20-14	LITERATURE CITED	2456

Chapter 20
Drying[1]

For most uses, wood must first be dried. Important reasons for this include:

- Because wood shrinks as it loses moisture, it should be dried to the moisture content it will have during use.
- Drying substantially reduces shipping weight of wood.
- Drying reduces the likelihood of stain or decay developing during transit, storage, or use.
- Dry wood is less susceptible to damage by insects than wet wood.
- Most strength properties of wood increase with drying below a moisture content of about 30 percent.
- Nailed and screwed joints are stronger in seasoned wood.
- Glued wood products perform better when assembled from dry wood.
- Prior drying usually makes treatment of wood with preservatives more successful.
- Dry wood takes finishes better than green wood.
- The electrical resistance of dry wood is much greater than that of wet wood.
- Dry wood is a better thermal insulating material than wet wood.

While it is possible to dry many hardwood products in "cook-book" fashion according to published schedules, some knowledge of wood-water relationships is helpful. A discussion of these relationships is contained in chapter 8 under section headings as follows: (8-1) MOISTURE CONTENT IN LIVING TREES; (8-2) FIBER SATURATION POINT; (8-3) EQUILIBRIUM MOISTURE CONTENT; (8-4) SHRINKING AND SWELLING; (8-5) HEAT OF SORPTION; (8-6) PERMEABILITY. For a summary of the mechanism of drying, readers are referred to Agriculture Handbook 420 (Koch 1972, p. 317-328) and to the references listed in section 8-7 of this text. References describing techniques for measuring and computing the moisture content of wood are listed in the opening pages of chapter 8. Width of sapwood in large logs of 23 southern hardwoods is given in table 5-2.

In addition to McMillen and Wengert's (1978) Agriculture Handbook, "Drying eastern hardwood lumber", which is the basis for most of this chapter, much general information needed for air-drying and kiln-drying hardwoods is contained in two other Agriculture Handbooks: "Air drying of lumber" (Rietz and Page 1971), and the "Dry kiln operator's manual" (Rasmussen 1961).

---

[1] See acknowledgements on chapter front page.

This chapter discusses methods, schedules, and equipment required to dry various classes of pine-site hardwood products. Stemwood moisture content of trees 6 inches in dbh, and percentage of pine-site hardwood volume for the 22 important pine-site hardwoods, grouped by ease of drying, are as follows:

Category and species	Stemwood moisture content	Proportion of pine-site hardwood volume
	Percent	Percent
Difficult to dry		
White oak, chestnut, and post oak	61.9-65.6	23.5
All other oaks listed in table 3-1	66.6-74.4	24.3
Easy to dry		
Ash, green	47.4	.9
Ash, white	47.5	
Elm, American	75.5	1.4
Elm, winged	65.6	
Hackberry	72.6	.1
Hickory	51.5	8.5
Maple, red	69.9	3.6
Sweetbay	100.8	.6
Sweetgum	120.4	13.2
Tupelo, black	90.0	5.5
Yellow-poplar	111.7	7.0

The species listed as "easy to dry" lose water readily, in most cases with minimal degrade. Hickory, however, may develop a pink color and tends to end-check and warp on drying. Open-grown elm and sweetgum also tend to warp, yellow-poplar sapwood will develop blue stain, and hackberry will develop a gray-brown stain unless precautionary steps are taken.

The tabulated species account for 88.6 percent of the total volume of hardwoods growing on southern pine sites. The oaks, with 47.8 percent, predominate; sweetgum, hickory, yellow-poplar, black tupelo, and red maple are also significant components. Numerous other hardwoods (including blackjack oak) account for the remaining 11.4 percent.

When cut into 4/4 and 8/4 boards, less than half the lumber yield from most of these pine-site hardwoods is in grades No. 1 Common and better. Crossties pallet and container material, bolter products such as turning squares or rounds, flooring strips and dimension lumber are all major solid wood uses. Trends toward reconstituted products will increase use of hardwoods as veneer for plywood, parallel-laminated veneer and composite panels, and as flakes for structural panels and fibers for board or paper.

20-1 INTRODUCTION TO DRYING METHODS

To dry wood, energy must be supplied to evaporate its water content. At 100-percent efficiency, about 1,000 Btu's are required to evaporate each pound of water from freshly cut pine-site hardwoods (see table 2-4 for moisture content and weight per cubic foot of green stemwood); removal of moisture below 30 percent requires additional energy to release it from the hygroscopic forces that bind it to the wood (fig. 20-1). Also, energy is required to heat the wood and accelerate the rate of moisture movement from board interior to surface; the higher the wood temperature, the faster the drying. The atmosphere adjacent to board surfaces must be able to receive the moisture, i.e., relative humidity must be less than 100 percent. Generally, the lower the relative humidity, the faster the drying. Air movement through a stack of drying lumber must be adequate to remove evaporated moisture, to bring in heat energy, and to maintain desired relative humidity in the atmosphere adjacent to wood surfaces. Because the relative humidity within a lumber pile tends to rise above that of the surrounding atmosphere, the lumber in the interior of the stack dries more slowly than that on the exterior.

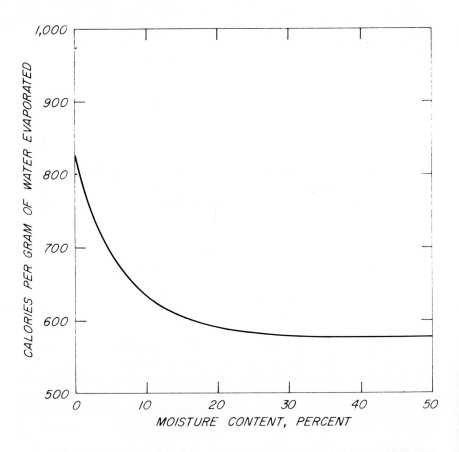

Figure 20-1.—Energy required in drying wood as wood moisture content changes. To convert calories per gram to Btu per pound, multiply by 1.80. (Drawing after Skaar and Simpson 1968.)

Five basic methods were identified by Wengert and White (1979) as being in widespread use to dry hardwoods, as follows:

- In **air-drying,** green wood is exposed to outside environment, preferably protected from direct rainfall and sun by portable roofs or a shed. Temperature is ambient and control of drying rate is minimal. Final moisture content is determined by ambient air temperature, relative humidity (fig. 8-9), and drying time; 20 to 25 percent final moisture content is usual.
- In **fan predryers,** wood protected from rain by a shed roof is exposed to ambient temperatures as in air-drying, but fans force air circulation. Humidistats are arranged to switch fans on and off to achieve some control over relative humidity. No heat is added, so final moisture content is determined by ambient temperatures and relative humidity. As in air-drying, final moisture content is usually 20 to 25 percent.
- **Heated low-temperature dryers** resemble fan predryers but control temperature (70° to 110°F) and humidity through addition of heat or moisture spray. Air velocities are commonly 300 to 600 fpm. By this process lumber is normally dried to 20 to 25 percent moisture content, but some operators dry to 6 percent. In **dehumidifiers,** a type of low-temperature dryer, compressors condense moisture and recycled heats of condensation provide a maximum dry-bulb temperature of about 160°F. With auxiliary heating coils, they can dry wood to 5 or 6 percent moisture content, but most dehumidifiers are used to predry wood from green to 20 or 30 percent moisture content. Kiln walls, which are preferably aluminum or wood, should be well insulated (R = 30 or more). Dehumidifiers are more sensitive to outside relative humidity and temperature than standard kilns, but are considered easy to operate and are especially attractive to the smaller producer who cannot justify a bigger boiler. The **solar-heated kiln** is a third type of low-temperature dryer, but due to the expense of solar energy collection, is largely restricted to outputs of less than 100,000 board feet annually (see sec. 20-9).
- **Standard heated kilns** are permanent insulated structures into which wood is loaded by fork lift or on tracks and exposed to temperatures of 100 to 180°F; humidity is controlled by admission of steam spray and venting to atmosphere. Usual air velocity is 250 to 400 fpm.
- **High-temperature dry kilns** resemble standard heated kilns but operate at temperatures near or above the boiling point of water—usually 200 to 240°F. Initially designed to dry softwoods, they are appropriate for only a few hardwoods that are easy to dry; control of relative humidity is not essential; vents are usually kept closed. Air velocities usually exceed 800 fpm. Preferably, load widths should not exceed 5 feet and kiln sticks should be thicker than those used in standard heated kilns.

Drying by fan predryers, heated low-temperature dryers, and dehumidifiers is classed as **accelerated air-drying.**

Data on these methods of drying lumber, and some additional processes, are provided in sections that follow in this chapter.

HOW DRY IS DRY ENOUGH?

To avoid shrinkage, warping, checking, and splitting in the final product, wood should be dried to a final moisture content about mid-range of expected in-use moisture contents. These moisture contents will vary considerably by product and the location in which the product serves.

As indicated by figure 20-2, wood in exterior exposure, but protected from rain wetting, equilibrates at about 12-percent moisture content in the humid South; in the arid Southwest, wood so exposed equilibrates to a much lower moisture content—perhaps 6 to 9 percent.

Hardwood furniture, all panelling, and other products serving in air-conditioned and heated interiors commonly equilibrate at about 8 percent moisture content. Oak flooring installed over radiantly-heated floors should be installed at even lower moisture content—i.e., about 6 percent (Mathewson 1952).

To attain these low moisture contents economically, lumber destined for remanufacturer into products for interior use is usually first reduced to about 25 percent moisture content by air drying, in a fan predryer, or in a heated low-temperature dryer (with or without dehumidifier). In a second stage, it is kiln-dried to the equilibrium content anticipated for the product in use.

Lumber to be bent to form, as for outdoor furniture, or in constructing unheated barns and garages should be air dried to 20 percent or slightly below. Wood to be bent in a hot press or machined for trim and flooring in boats and buildings that are heated only occasionally should be dried to 12 to 18 percent (McMillen and Wengert 1978).

20-2 AIR-DRYING

Within the southern pine region, it is possible to air-dry wood 10 to 12 months of the year, although drying is usually most rapid in spring because temperatures are warm and relative humidities low (fig. 20-2). Hardwood producers in the South take advantage of the mild climate to air-dry lumber, squares and dimension stock, crossties and timbers, posts and poles, and firewood; some even air-dry fuel chips.

LUMBER

One real advantage of air-drying lumber over drying by other processes is its low initial cost. Although land, installation, and operating costs for air-drying are substantial, the cost of kiln-drying dense hardwoods to levels achieved in air-drying can be much higher.

A second major advantage is substantial energy saving. Each percent moisture removed by air drying saves energy in subsequent kiln drying; the saving is 50 to 85 Btu per board foot. For a conventional kiln with a capacity of 50,000 board feet, this means 2.5 to 4.25 million Btu can be saved for each 1 percent moisture lost in air-drying. E. M. Wengert has computed that at 1980 prices, this amounted to $5.62 to 9.56 per charge per percent.

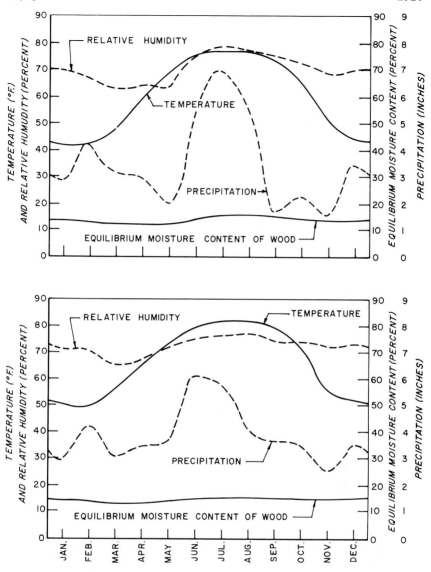

Figure 20-2.—Mean monthly temperatures, relative humidities, precipitation, and equilibrium moisture content of wood. (Top) In North Carolina, South Carolina, Virginia, and West Virginia. (Bottom) In eastern Texas, Louisiana, Mississippi, Georgia, northern Florida, and Arkansas. (Drawing after Peck 1961.)

Air-drying has advantages for large-scale users or producers who must carry inventories to balance periods of low production. Lumber held in drying stacks long enough to meet air-drying standards is ready for kiln-drying at minimum further cost. Even short periods on sticks reduce weight (and shipping costs) and, in white-sapwood species, the probability of oxidative stain during shipment. At most destinations, drying processes are available to accommodate partly air-dried stock. The interest cost of holding a large air-drying inventory can be substantial, however.

In combination with properly applied anti-stain dip treatments (see sec. 11-6 under heading LUMBER), air-drying decreases the chance of mold, stain, and decay degrade during bulk-piled storage and shipment. Rapid air-drying at low relative humidities produces a large amount of set that assists in reducing warp if lumber is stacked properly. If the stock is properly protected in a well-laid-out yard, degrade from checking will be minimal, even for thick stock of dense wood and 1-inch heartwood of especially refractory woods. (Special precautions for thick oak are discussed later.)

Limitations of air-drying are generally associated with the weather—uncontrollable temperature, relative humidity, rainfall, sunshine, and winds.

Lack of absolute control of drying conditions poses some hazards of excessive degrade. Too rapid surface drying, caused by sun or wind and low humidity, can cause surface checking of oak within a day; these surface checks can then be aggravated under even moderate conditions. Brief periods of hot, dry winds may therefore further increase degrade and volume loss due to severe surface, and even interior (honeycomb) checking and end splitting. Blue stain and chemical (oxidation) stain in warm, rainy or sultry periods with little air movement can cause excessive losses unless proper dipping, pile spacing, and piling methods are used.

Excessively large inventories are costly, especially when interest rates are high. Lumber can also deteriorate if air-drying is prolonged beyond the time needed to bring the moisture content down to 25 percent, as repeated wetting and drying cycles deepen checks and splits. If poor markets necessitate holding beyond one summer season, extra measures to avoid wetting may be advisable.

Degrade losses in air-drying.—Hanks and Peirsol (1975) evaluated degrade in air-drying hardwood lumber 1 to 2 inches thick at sawmills in the East. Each board was graded when sawn and again when air-dried to approximately 20 percent moisture content. The difference between the potential and actual dry value established the percent value loss due to degrade (table 20-1). Percentage losses were greatest in No. 1 Common and better grades. Some experienced kiln operators believe that air-drying degrade is frequently more severe than indicated by table 20-1.

TABLE 20-1.—*Percent value loss[1] during air-drying of lumber 1 to 2 inches thick of seven eastern hardwood species that commonly grow on southern pine sites* (Hanks and Peirsol 1975)

Species and lumber thickness (multiples of 1/4-inch)	Loss in value due to degrade
	Percent
Maple, red	
4/4	1.0
5/4	2.2
6/4	1.2
8/4	.0
Oak, black	
4/4	2.9
5/4	2.0
Oak, chestnut	
4/4	1.7
5/4	1.7
6/4	1.6
Oak, northern red	
4/4	2.2
5/4	2.9
6/4	1.6
Oak, scarlet	
2.5/4 (i.e., 5/8-inch)	.9
4/4	2.1
5/4	4.8
Oak, white	
4/4	2.1
5/4	3.1
6/4	4.4
8/4	+5.9[2]
Yellow-poplar	
4/4	2.6
6/4	+.1[2]

[1]Based on lumber price structure in 1973.
[2]Dry value greater than green value due to degrade.

Wengert (1979a) noted that among the commercially important hardwoods of North America, the oaks are the most difficult to dry—particularly southern lowland oaks and especially 2-inch and thicker stock. They must be dried at extremely low rates to avoid degrade, primarily from deep surface checking, but also from splitting, warp, and stain. Because the price of upper grades of oak lumber rose very sharply between 1975 and 1979, the degrade noted in table 20-1 is considerably understated.

Wengert (1979a) applied the information from Hanks and Peirsol (1975) to 1979 lumber costs and found that value losses during air-drying oak were as follows:

Lumber grades	Red oaks	White oaks
	Percent value loss	
FAS and Select	5.9	4.2
No. 1 Common	6.4	5.0

In terms of 1979 prices, these data indicate a degrade during air-drying oak of about $40 per thousand board feet of FAS and Selects, and $15 per thousand board feet of No. 1 Common lumber.

Other studies support these findings. Simpson and Bois[2] estimated losses in air-drying No. 2 Common and Better oak at 6.6 percent of green lumber value. Cuppett (1966) found that air-drying degrade losses averaged 9.6 percent of green lumber value in 18 Appalachian mills cutting hardwoods; his study indicated great potential for improvement since nine of the mills had losses of only 1.2 percent (about $4/MBF), while the remaining nine had losses averaging 18.8 percent (over $60/MBF).

Control of degrade.—Protection against stain, decay, and bacteria is described in section 11-6 (see heading LUMBER). Protection against insect attack is discussed in section 11-9 (see heading PROTECTION OF GREEN LUMBER and CONTROL OF LYCTUS POWDER POST BEETLES). Control of chemical grey stain in hackberry is discussed in this section under the heading SHORT LUMBER AND SQUARES.

Wengert (1979a) described the occurrence of surface checking, splitting, and warp in oak—and summarized procedures to minimize such degrade, as follows:

- **Surface checking**—For 4/4 oak, surface checking occurs primarily during the loss of the first 20 percent of moisture. Once wood has lost this first 20 percent, the risk of formation of new surface checks is minimal. (Surface checks can worsen if the wood is exposed to rain wetting, however.) The risk of internal checking (honeycomb) on the other hand, is high until the wood reaches 30 percent moisture content. However, if the drying rate is controlled (that is, if the rate were slow enough so that surface checking did not occur) and temperatures do not exceed 130°F, internal checking can be controlled. In fact, once 4/4 oak has been dried below 22 percent moisture, no further degrade will result (except under the most extreme or unusual conditions). The key to control of checking is to keep the drying rate below 3 percent loss per day (or even less for southern oaks or 8/4 stock). Commercial control procedures for checking include stacking lumber piles close together in hot, dry weather, covering the piles with burlap (fig. 20-3 bottom), and stacking the lumber in a shed where velocities are low.

---

[2]Simpson, W., and P. Bois. 1978. Sources and causes of drying losses in hardwood lumber. U.S. Dep. Agric. For. Serv. Unnumbered report, For. Prod. Lab., Madison, Wis. 4 p.

- **Splitting**—Splitting, primarily end splitting, develops during the early stages of drying when the ends of lumber are drying faster than the remainder of the board. It is most pronounced on thicker lumber. Control must be initiated before drying begins. The key for control is to slow the drying of the ends. Commercial procedures to control splitting have included end coatings, burlap coverings, and placing lumber piles so that they are tightly stacked end-to-end.
- **Warp**—Warp is usually not much of a problem as the majority of the furniture cuttings are not wide and/or long. Proper stacking—stickers aligned vertically with 4 by 4-inch bolsters under the stickers, level foundations, and uniform lumber thickness—will control warp.

Presurfacing lumber to remove all fine saw marks also reduces surface checking in oak (Rietz and Jenson 1966; McMillen 1969; Wengert and Baltes 1971; Simpson and Baltes 1972). Field trials have been successful (Rice 1971; Cuppett and Craft 1972), but industrial adoption will depend on economic considerations.

To limit end splits almost all 6/4 and thicker hardwood lumber benefits from **end coating,** as do squares 2 inches and larger, and product blanks such as those for gunstocks. End coatings should be applied as soon as possible to freshly cut end surfaces; they may be applied cold or hot. **Cold coatings**—available from kiln manufacturers—are liquid at ordinary temperatures and are applied by spatula, brush, or spray to the ends of logs or lumber; for small-scale use, heavy pastes such as roofing cement can be used. Cold coatings should be allowed to dry a few hours before being subjected to kiln temperatures.

Hot coatings of pitch, asphalt, or paraffin are low in cost and single coats are high in water resistance. Generally applied by dipping, they can also be applied by holding the lumber end against a revolving roller partially submerged in the coating. Paraffin, which softens at elevated temperatures, is suitable only for air-drying or for temporary protection of the ends of squares that will be dried on a kiln schedule that uses a high relative humidity during the first stages. (See Koch 1972, p. 972 for additional discussion of end coatings.)

Sorting is a key element in controlling warp and moisture content during drying. It is essential that lumber be sorted by thickness before stacking. Also, hickory, the red oaks, and the white oaks should be handled individually; and southern lowland oaks should be separated from upland oaks. Oak lumber having streaks of wood infected with bacteria (Ward et al. 1972), if present in large quantity, should be segregated for special drying schedules. Hackberry, which requires special treatment to avoid grey stain should be separately handled; see this section under heading SHORT LUMBER AND SQUARES.

If the lumber is first end racked, end piled, or cribbed for rapid surface drying, it should be repiled in flat piles with stickers within 3 or 4 days. Degrade from warp can be very costly if the stock is end racked or cribbed too long. Proper **pile design** and **stacking** for air-drying is essential to uniform air circulation and minimal warp. Hardwood lumber is most typically piled for drying in packages with horizontal layers using stickers as spacers between each layer (fig. 20-3 top). Space between stickers should be between 12 and 24 inches for 4/4 stock, and not over 48 inches for 8/4 or thicker. Four-inch by 4-inch bolsters and pile foundations should be spaced at 4 feet or less. End and intermediate foundation bolsters are at a prefixed spacing; the middle one is inserted when the pile is built, and removed when it is razed.

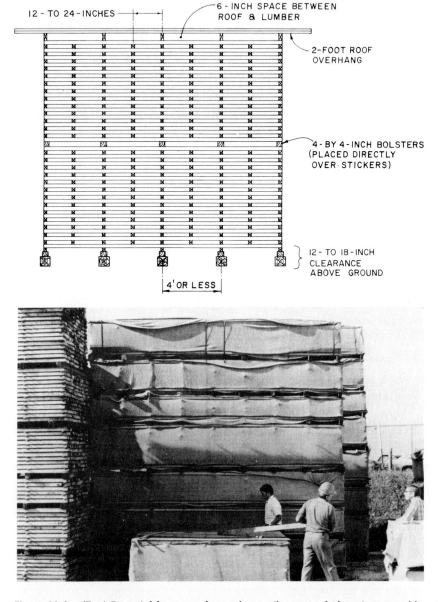

Figure 20-3.—(Top) Essential features of a package pile, center bolster is removable after the pile is razed to permit forklift passage. (Bottom) Burlap-wrapped, end-coated 8/4 oak with plywood roof cover air dries with little end or surface checking. (Photo from Bois 1978.)

Stickers provide columns to support the pile, separate the courses of lumber, and restrain warping by holding the boards in a flat position. They should be perfectly aligned with the bolsters and in good vertical alignment. Allowing end stickers to project slightly beyond the ends of the lumber reduces end checks and splits by retarding drying and sheltering the ends from sunshine and rain. Stickers for packages are commonly a uniform 3/4-inch thick, approximately 1¼ inches wide, and of uniform length (a couple of inches longer than the width of the package); they should be straight. If lumber is thin or prone to warp, it needs more crossbeams and more stickers per course than lumber that is thick and less prone to warp.

The lower levels of a pile tend to dry more slowly than the upper levels. For this reason—and because extreme pile height places excessive weight on foundations and on stickers and boards in the lower part of the pile—package piles are generally limited to 16 or 20 feet in height.

A good pile roof shields the upper courses of lumber, and to a lesser extent the lower part of the pile, from precipitation and direct sunshine. Figure 20-3 illustrates roof construction. Boards may be nailed to a framework of 2x6's and securely covered with roofing paper or corrugated metal. The panel roofs are placed on the top package while it is still on the ground. While the roof for a package pile overhangs the pile at both ends, it can project only on one side—that furthest from the mast of the forklift. A few operators use precast concrete pile roofs to provide shelter and reduce warpage; they are put in place with a forklift.

The foundations for yard piles (pile bottoms) should be well designed and well placed. They must support the pile at least a foot (preferably 18 inches) clear of the yard surface without undue deflection, permit access by handling equipment, and be durable.

The length of a lumber package is usually the same as the lumber being cut; mills cutting several lengths may use corresponding package lengths. Alternatively, lumber is **boxed piled** (fig. 20-4) using full-length boards on each side of the package to keep square corners, short lengths end-to-end where feasible, and keeping voids due to short boards well distributed at both ends. The narrower the pile, the faster the wood will dry. Hand-stacked piles should not exceed 6 feet in width.

In addition to square, level, straight-sided piles on firm level foundations, with perfect vertical alignment of closely spaced stickers, good bolster placement, and good pile roofs, several other practices can help reduce degrade—especially warp. Warp-prone woods, such as open-grown elm, should be cut into the wide boards to reduce crook in dry lumber. Presurfacing lumber on both faces, or blanking it to thickness by planing one face, before stacking, increases the board foot content of a unit package and reduces warp, as does precision sawing to uniform thickness.

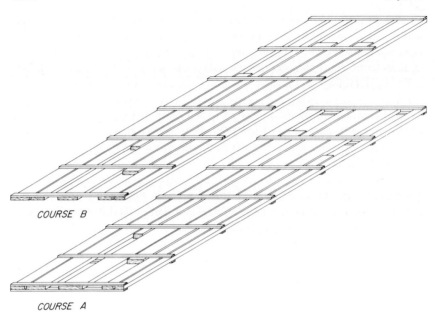

Figure 20-4.—Method of box piling ramdom-length lumber. (Drawing from McMillen and Wengert 1978.)

Top courses of loads (or packages) may be restrained against warp by spring clamps anchored lower in the pile or by added weight. Low density woods require about 50 pounds per square foot; denser woods take more pressure. For example, 150 pounds per square foot top load is more effective than 90 pounds per square foot in restraining black tupelo boards against cupping. Rapid air-drying to develop tension set in the outer shell, which helps hold the lumber flat in latter stages of drying, may be a necessary first step in drying warp-prone woods such as American elm (McMillen 1958).

Control of degrade losses during drying requires measurement of such degrade. There are two approaches to measurement of defect. In the first, lumber is graded dry using National Hardwood Lumber Association rules to yield the present grade and also the grade if drying degrade—splits, warp, stain, and check—were not present (see Cuppett 1965 for details). In the second approach, the dry lumber is diagrammed to locate on an X-Y grid, natural defects such as knots, wane, and shake. These data, together with a schedule of sizes of clear cuttings required in the cut-up plant (rough mill) are processed by computer to indicate yield of clear cutting; this yield is then compared with a computer-obtained yield after including plotted data on drying defects such as checks and splits (see Wengert 1980a for details).

A first step toward reducing such defects is careful measurement of lumber thickness, and if necessary, adjustment of sawing equipment and procedures to minimize variation. Uniform thickness reduces drying time. Assistance in implementing these approaches to degrade measurement and control is available from many of the state forestry staffs in the South (Wengert 1979a).

Design of air-drying yards.—The site for air-drying yards should be high and well drained. Air near low ground, swamps, or bodies of water is likely to be damp. Nearby trees, buildings, or hills are detrimental if they restrict air movement across the site.

Air circulation is the principal means of supplying the heat needed to evaporate water and remove moisture-laden air from the piles. A yard layout for forklift operation includes main alleys, cross alleys, and rows of piles having lateral spaces between rows. Figure 20-5 illustrates this terminology and shows, in the central area, a **row-type** layout with nine piles per row. In a **line-type** layout there are wide alleys between rows only two piles wide.

By custom in the United States, the main alleys for the transport of lumber to the piles are perpendicular to the prevailing wind direction. Recent research, however, shows that for some types of yards, better circulation is obtained by having the main alleys parallel the prevailing wind direction. With either alignment, the air spaces must be adequately large, straight, and continuous across the yard.

Except for those piles exposed directly and perpendicularly to the wind, no wind blows through any pile on an air-drying yard. Rather, all air flows across the boards due to small air pressure and density differences from eddy and aspiration effects. As water evaporates, the air in and round all piles becomes denser, tends to flow downward, and is removed from beneath the piles, being replaced by fresh air from the pile tops. This effect and the consequent drying go on continuously, regardless of wind, unless the air around all the boards is saturated with water vapor. This is why high and open pile foundations are very important.

Thick or especially check-susceptible material should not be located on the windward side of the yard. In central or leeward areas lower air velocities and increased relative humidities temper severity of drying conditions and reduce likelihood of checking. Similarly, in windy, dry weather it would be bad practice to start refilling an empty yard from the leeward side toward the windward side, because each pile in turn would be exposed directly to the wind.

Size and orientation of **alleys,** row and **pile spacing,** and **pile size** are discussed extensively in Rietz and Page (1971). The general function of alleys and pile spaces in removing moisture-laden air from the yard has long been well known; wind-tunnel miniature-model pile studies have clarified their effect on circulation within piles.

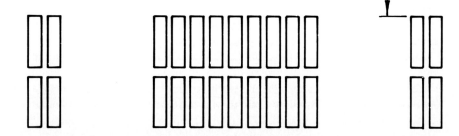

Figure 20-5.—Diagram of a section of a yard with package piles for forklift handling. Alley spacings are minimums.

White (1963) used two orientations of a single block of row-type model piles, with artificial smoke to indicate air movement. With piles oriented perpendicular to the wind, much of the air was deflected over the top of the block. No wind blew directly through any pile except those at the windward edge. Some air flowing between the other piles, however, entered these piles from the rear and left from their fronts, sides, and tops. Double eddies at the rear of each pile moved air up into the turbulent area above the piles. When the piles were oriented parallel with the wind, more air moved through the spaces between the piles and into and out of the piles.

Finighan and Liversidge (1972), using the evaporation rate of solid plugs of moth crystals (paradichlorbenzene) in model piles to stimulate drying rates, also found best combination of average "drying" rate and uniformity when piles and the main alleys were oriented parallel with the wind. Drying rates in line-type layouts were 20 to 40 percent greater than in row-types.

For those species and sizes suitable for most rapid drying without excessive degrade, a line-type layout parallel with the wind seems best. For species such as oak that are subject to checking, a row-type layout with limited length of row and number of rows is best. In row-type layouts, problems of uniform drying among piles in the shortest possible time may be solvable by variable or nonuniform (using moveable pile foundations) pile spacing (McMillen 1964), the spacing between piles in the middle of the row being about double that near the alleys.

Whatever yard layout style is used, the main and cross alleys and the spaces between piles should be well aligned all the way across the yard, and free of all obstructions.

The drying efficiency of a yard depends to some extent on how well the **surface** is graded, paved, and drained. Water standing in a yard after a rain decreases the drying rate. A successful yard described by Minter (1961) was surfaced to a 6-inch depth with crushed rock. Blacktop paving also is used extensively. The yard should also be kept clean. Vegetation and debris, including broken stickers, boards, or pieces of timber from pile foundations, interfere with the movement of air over the ground surface.

The drying rate of lumber is affected by **pile design,** i.e., the way the boards are stacked. Air spaces between edges of boards of unit packages permit greater downflow of air; lumber so spaced dries faster than when laid edge to edge. This is important in calm weather. Boards in unit packages wider than 4 feet should be spaced. In unit packages 4 feet or less in width boards are usually stacked edge to edge, and the downward flow occurs in the spaces between the piles. Minimum pile spacing is 2 feet. When random-length, random-width lumber is box-piled, enough space develops within the pile to make additional board spacing unnecessary. Wide unit packages of even-length lumber and wide hand-stacked piles are often built with flues or central chimneys.

High, openly-designed **pile foundations** are necessary to allow the moist air to pass out readily from beneath the pile, and are essential in obtaining uniform drying from top to bottom. Finighan and Liversidge (1972) found that the evaporative loss from top packages was greater than that from bottom packages due to wind effects. Foundation height for hand-built piles should be at least 18 inches above the yard surface. In well-laid-out and well-drained forklift yards, a 12-inch minimum is satisfactory in many regions, but where rainfall is high, the minimum should be 18 inches. In unpaved yards weeds or debris should not be allowed to block air passage.

Clark and Headlee (1958) showed that **pile roofs** saved enough in degrade and drying time for 4/4 No. 1 Common and Better red oak to pay for the roofs in five uses. Roofs also increased drying rates and lowered final moisture contents in rainy spring weather. Australian air-drying research, reviewed by McMillen (1964), showed similar differences between roofed and nonroofed piles during rainy autumn and winter seasons. In nonroofed piles hardwoods dried to 26 percent moisture content in 70 days, then did not dry any further for the next 150 days.

In regions of high rainfall, **shed roofs,** covering blocks of piles, better protect lumber from rewetting. They can cut customary air-drying times in half or greatly reduce the amount of water to be evaporated in a kiln after a "standard" length of air-drying.

Time required to air-dry lumber.—Drying of stacked green lumber is most rapid during the first 3 to 4 days after it is cut; therefore, when green lumber arrives or is first cut, it should be immediately placed on stickers where air will circulate through the pile. This should be done even if the lumber is soon to be kiln-dried.

Lightweight hardwoods such as yellow-poplar dry rapidly under favorable conditions, but heavier woods such as oak and hickory require longer drying time. Although **specific gravity** (table 7-7) is a rough guide to drying time, the **permeability** of wood (tables 8-14 through 8-17) and the **diffusibility** of the water through it affect that relationship. In general, **heartwood** is less permeable than **sapwood** and takes longer to dry, e.g., oaks, which are predominantly heartwood (table 5-2), dry slowly. Southern **lowland oaks,** both red and white, have drying characteristics similar to each other but different from those of the upland oaks; they are, therefore, separated in drying times and kiln-schedule tables.

Drying time is affected by **sawing patterns;** lumber dries more slowly when quartersawn than when flatsawn. In quartersawn lumber, few wood rays—which aid the movement of moisture—are intersected on the broad surfaces of boards. Flatsawn lumber, with more rays exposed, is more likely to surface check under severe drying conditions.

Thick stock takes longer to dry than thin. One theoretical approach suggests that drying time, under identical or similar drying conditions, is a function of the square of the **thickness.** Since thick stock may take longer than one air-drying season to reach 25 percent moisture content in the South, actual air-drying times for 2-inch stock may be three to four times as long as those for 1-inch. In hardwoods, surface and end checking also tends to increase with thickness.

Green 4/4 red oak lumber should be cut 1/8-inch thicker than required rough dry size to allow for shrinkage. Freese et al. (1976) suggest that rough dry 4/4 red oak lumber can be slightly less than 1 inch thick to produce moderately long, wide cuttings planed on both sides to 13/16-inch. For panels, however, rough dry thickness should be slightly over 1 inch to plane to 13/16-inch.

Air-drying times for 1-inch-thick, pine-site hardwoods vary from 40-75 days for the easy-drying species to 100-280 days for lowland oaks; drying time for 2-inch-thick lumber varies from 170-220 days for easy-drying species to 240-360 days for the upland white oaks and even longer for lowland oaks (table 20-2).

The minimum periods given in table 20-2 apply to lumber piled during the best drying weather, generally spring and early summer. One-inch lumber piled too late in the period of best weather to reach 20 percent that fall, or that piled during the fall or early winter, will not reach 20 percent for a very long time. In fact, during poor drying weather, the times listed may result in moisture contents as high as 23 percent. For 2-inch lumber the times shown may be necessary to bring the wood to somewhere between 23 and 27 percent moisture content.

Because they are based on sketchy information, time estimates in table 20-2 are somewhat speculative. They are, however, the best bases available for calculating the feasibility of revising an old yard or building a new one with a good site, a good yard layout, the best piling practice, high, open pile foundations, and a roof on every pile. Denig and Wengert (1982) described a method of developing regional air-drying calendars based on their observations in the Roanoke, Virginia region; figure 20-5A represents their monthly predictive curves for 1-inch red oak lumber.

The practice of holding 4/4 lumber 90 days on the air-drying yard and then shipping it with the implication that it will safely withstand drastic kiln-drying is faulty on two counts. Ninety days is unnecessarily long and expensive for the South during much of the year. On the other hand, in cold, wet winters, 90 days is not long enough to bring the average moisture content down to 25 percent.

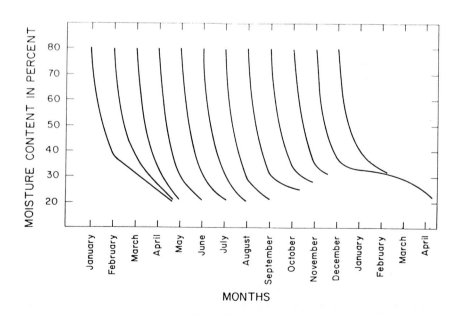

Figure 20-5A.—Monthly prediction drying curves for 1-inch red oak sp. lumber dried in the Roanoke, Virginia region. (Drawing after Denig and Wengert 1982.)

TABLE 20-2.—*Estimated time to air-dry green 1-inch and 2-inch hardwoods of 14 species and species groups found on southern pine sites to approximately 20 percent moisture content* (McMillin and Wengert 1978)[1]

Species and lumber thickness (inches)	South of the 33rd parallel[2]	Southern pine range north of the 33 parallel[3]
	----------Days----------	
Elm, American		
1	40-65	40-75
2	170-200	170-220
Hackberry and sugarberry		
1	40-65	40-75
2	170-200	170-220
Maple, red		
1	40-65	40-75
2	170-200	170-220
Sweetgum sapwood		
1	40-65	40-75
2	170-200	170-220
Yellow-poplar		
1	40-65	40-75
2	170-200	170-220
Sweetbay		
1	40-75	—
2	170-220	—
Ash, sp.		
1	45-70	45-80
2	180-210	180-230
Elm, winged		
1	50-80	50-90
2	190-230	190-250
Hickory, true		
1	50-80	50-95
2	190-230	190-250
Sweetgum heartwood		
1	50-80	50-95
2	190-230	190-250
Tupelo, black		
1	60-110	45-90
2	210-300	180-230
Oak, upland red sp.		
1	60-120	50-100
2	240-360	190-300
Oak, upland white sp.		
1	60-120	50-100
2	240-360	190-300
Oak, lowland sp.		
1	100-280[4]	—
2	—	—

[1] Arranged by time required to dry. Times given are for flat-sawn lumber in unit packages 4 feet or narrower in width, or having central flues.

[2] South of a line running generally east and west from Dallas, Texas on the west through north-central Louisiana, central Mississippi, Alabama, and Georgia, to about Charleston, S.C.; this area generally has 12 months of good drying weather.

[3] Most of the southern pine region north of the 33rd parallel has 10 months of good drying weather; the northernmost portion has 8 months.

[4] To 25 percent moisture content.

There are advantages to keeping track of the moisture of the lumber as it dries. A graph of past runs can provide a good estimate of how long air-drying will take for a given species, thickness, and time of year. The method of using samples described in the Dry Kiln Operator's Manual (Rasmussen 1961), with certain modifications, is suitable for use in air-drying. The sample pocket can be made two boards wide, with the sample placed in the inner space and a dummy board on the outer edge of the package. In preparing the sample, two 1-inch moisture sections are cut at least 6 inches back from each end of the piece. The two sections are weighed, and also the sample blank. The moisture sections should then be ovendried at 214 to 221°F until they come to constant weight and their green moisture content computed. From these data, the ovendry weight of the sample can be computed. From time-to-time during drying, the sample can be weighed and its moisture content computed as a percentage of ovendry weight.

Costs in air-drying lumber.—Changing rates for interest and labor, and rising values of land and lumber make computation of costs useful for short-term periods only. Readers interested in the methodology of computing air-drying costs—as well as related management decision techniques—are referred to McMillen and Wengert (1978, p. 76-80), Wengert (1978), and Wengert and White (1979).

SHORT LUMBER AND SQUARES

Researchers studying utilization of pine-site hardwoods usually conclude that many tree stesm should be cross-cut into short logs or bolts before primary conversion into lumber (e.g., see figs 18-114 and 18-115). Short lengths expose more log ends to rapid drying that may cause splits unless precautions are taken (see secs. 11-6 ROUNDWOOD, and 16-18); preferably such bolts and short logs should be promptly sawn. Of the pine-site hardwood species, hickory is highly prone to end splits; Goebel et al. (1960) concluded split-prone hickory logs contained large numbers of gelatinous fibers (figs. 5-71 and 5-72) across entire cross sections at all levels in the tree. End coatings for slowing the rate of drying to avoid end splits in short thick hickory are also effective on other, less split-prone species. (See Koch 1972, p. 962, for a discusison of end coatings.)

The risk of surface checking is greatest in southern oaks, particularly those from bottomlands. In upland oak lumber these checks can be prevented by controlling early stages of drying to keep drying rates below the following values (Wengert 1979a):

Species	4/4 lumber	8/4 lumber
	----*Percent loss/day*----	
Red oaks	3.5	1.5
White oaks	2.5	1.0

Surface checking in oak occurs primarily during loss of the first 20 percent of moisture.

In oak, the risk of **internal checking** is high until 4/4 lumber reaches 30 percent moisture content (20 percent for 8/4). If the drying rate is controlled to avoid surface checking, and temperatures do not exceed 130°F (to maintain wood strength), internal checking in oak should not occur. Once 4/4 oak is dried below 22 percent moisture content, no further external or internal checking should occur if the wood is not rewetted.

Short lumber of most of the pine-site hardwoods can be stacked to prevent undue **warp** during air-drying. Four-foot-long hickory squares, however, may develop as much as 1 inch of crook as drying proceeds—even if meticulously stacked. Practices that prevent end-splits and limit warping in hickory and prevent surface and internal checks in oak during air-drying of short lumber and squares should yield split and check-free wood if applied to the other pine-site hardwoods.

Gray-brown stain will occur in hackberry if air-drying procedures appropriate for hickory are followed. Price (1982) found that the most effective methods of preventing gray-brown chemical stain in 2.5-in squares sawn from freshly felled hackberry were atmospheric steaming for 60 min before air-drying. A 30-min initial steaming cycle was more effective in preventing stain than air-seasoning. A kiln schedule with high initial temperature (155°F) was about as effective as a 30-min atmospheric steam treatment. A 30-min initial heating cycle in air at 212°F prior to air-seasoning had no benefit. To be effective, steaming or kilning must be initiated promptly after trees are felled and sawn.

Hickory handle stock[3].—Hickory develops a pinkish color if subjected to temperatures above 105°F. (Some experienced operators think the safe temperture range extends to 115°F.) Such **pinking,** diminishes its market value for handles and other specialty items. Air-drying is the simplest way to produce hickory handle stock with no pinking.

Hickory sapwood dried in summer should be dipped to prevent blue stain (see sec. 11-6, under the heading LUMBER); chemicals to protect against insect attack (see sec. 11-9, under the headings DRY LUMBER, WOOD IN USE, and CONTROL OF LYCTUS POWDER POST BEETLES) should be included in the dip when heavy infestations of powder post beetles are known to be present and length of the air-drying period is uncertain. Most handle-blank producers are able to avoid beetle infestations by keeping debris cleaned up and burned.

The blanks should be protected against sun, rain, and wind during production. Temperatures reached under direct exposure to the sun may be high enough to cause pinking; therefore, a mill for hickory handle blanks should be under roof. Similarly, blanks split from bolts in the woods should be stacked under a roof, or the piles themselves should be roofed.

Most blanks have a square or rectangular cross section, which makes them easier to stack and handle. This form, however, rquires great care in drying to prevent pinking of the interior, which will be exposed by final turning. Some handles are rough turned to an oversize dimension before kiln-drying. Turning to final dimension and polishing before final kiln-drying is poor practice because the crook or twist that develops from kiln-drying cannot be removed.

---

[3]Text under this heading is condensed from McMillen (1956a).

Allowance should be made for shrinkage when the blanks are cut. McMillen (1956a) found that normal average radial shrinkage in hickory blanks, from green to kiln-dried, is 5.8 percent, and tangential shrinkage is 9.1 percent. He also found least shrinkage from drying at low temperatures. Maximum shrinkage of small handle blanks was 7.0 percent in thickness and 9.9 percent in width. Supplying blanks in specified air-dried sizes is common industry practice. If a producer shifts to kiln-drying, there will be greater shrinkage. This will necessitate either cutting the blanks slightly larger in the green condition or coming to some agreement with the purchaser as to the sizes he will take.

Handle blanks 2 by 2 inches or less in cross-section do not need to be end coated to control end checking. Larger blanks, however, should be end coated. Dipping ends in molten paraffin is customary and very effective for air-drying. For kiln-drying at temperatures above 105°F, other coatings (U.S. Forest Products Laboratory 1961) should be used. Since end drying is of considerable importance in the drying of short-length wood, the kiln schedules given for kiln-drying squares (sec. 20-6) are designed to hold end checking to a reasonable amount while allowing drying through the ends.

Although bundling of blanks may simplify handling, it is not practical to dry rectangular blanks in bundles.

In the production of hickory handle blanks, all air-drying at permanent installations should be done in a shed (fig. 20-6 top) that will provide adequate air circulation to carry away evaporated moisture, while protecting the stock from sun, rain, and wind. The shed floor should be made of preservative-treated hickory or oak with ½- to ¾-inch spaces between boards. The gound under the shed should be well drained. The stringers and joists should be heavy enough to support the weight of a small forklift truck loaded to capacity with green wood. Alleys and streets between the shed and other buildings must provide for cross ventilation as in a lumber seasoning yard.

If dimension stock must be piled in the open, piles should be roofed as soon as completed, and their sides protected from wind, sun, and driving rain. Pile foundations should hold the material at least 12 inches off the ground.

Handle blanks and other dimension stock should be piled by sizes and lengths; all stock in each layer of the pile should have the same thickness. Handle stock piled for air-drying can be self stickered (fig. 20-6 bottom). Spaces between rows in the drying shed should be lined up as much as possible to promote horizontal air movement within the shed. Piles should be dated when built, and a pocket for moisture-content samples should be incorporated in the middle of the lower third of each pile.

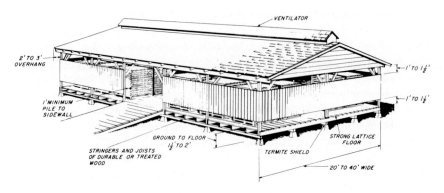
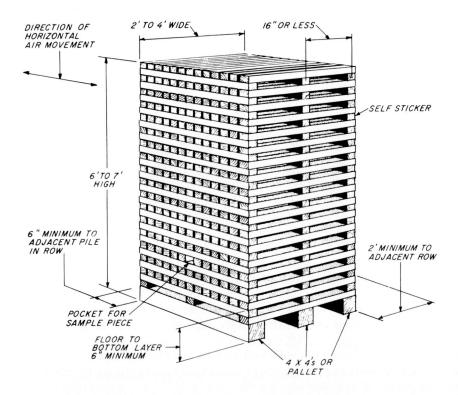

Figure 20-6.—(Top) Air-drying shed for hickory handle stock. (Bottom) Method of piling hickory handle stock for air-drying. (Drawing after McMillen 1956a.)

Blanks to be used for moisture-content samples are selected at the time when the blanks are sawn, and should be cut about 14 inches longer than the blanks will be. This extra length is necessary to provide material for cutting two 1-inch moisture sections taken at least 6 inches back from each end of the piece, leaving the sample the same length as the blanks. The two sections are promptly weighed and also the sample blank. The sample blank should have the same end treatment, either coated or uncoated, as the blanks being dried. The moisture sections then should be ovendried at 214° to 221°F until they come to ovendry weight, and their green moisture content computed. From these data, the ovendry weight of the sample blank can be computed. From time to time during the course of air-drying, the sample blank can be weighed and its moisture content computed as a percentage of its ovendry weight.

Air-drying time for hickory handle stock varies with blank size and the season of the year, as indicated by the following examples. One group of small blanks (1.5 x 2 x 24 inches) air-dried to 20 percent moisture content in 15 days of summer and early fall; drying to 17 percent took an additional 15 days. During winter and early spring, a similar group of small blanks took 110 days to dry from 48 percent to 18 percent moisture content. Another group of small blanks was dried in a kiln at temperatures and relative humidities approximating average summer conditions in the South; drying time to 20 percent moisture content was 14 days. Air-drying time for these small handle blanks thus varied from 14 to 110 days.

One group of large handle blanks (2.5 x 3.5 x 24 inches) without end coatings air-dried from 48 percent to 20 percent in 72 days of late winter and early spring weather. End-check penetration was slightly excessive. Similar blanks, end coated with paraffin, air-dried in colder than average spring weather from 54 to 20 percent moisture content in 105 days. End coating thus appears advisable in air-drying any hickory stock over 2x2 inches in cross-section.

In spite of fast drying above 20 percent moisture content in summertime, McMillen (1965a) concluded that it is not practical to dry handle stock down to conditions suitable for ultimate use by air-drying alone. In a pilot test in north-central Mississippi, total air-drying time was about 290 days; moisture content still averaged 17 percent, which is higher than desirable for handle stock.

THICK OAK PLANKING

Bois (1977a) summarized procedures for air-drying oak planking, 8/4 and thicker, to 30 percent moisture content without causing deep surface checks or internal checks; the control needed (see next two paragraphs) is very difficult to achieve in air-drying.

From green to about 50-percent moisture content.—High relative humidity and low temperatures are needed to prevent surface or interior checking. Air-drying piles should be located in the interior of the air-drying yard and covered (fig. 20-3 bottom). During this period (perhaps 30 to 60 days for 8/4 stock) danger of surface checking is greatest because tension in outer board layers builds up rapidly due to shrinkage. Compression in the core builds, with possible weakening of the wood cells unless temperature is kept low. Temperatures should not exceed 110°F and a wet bulb depression exceeding 4°F is hazardous. In summertime air-drying it may not be possible to hold within these limits, therefore wintertime sawing and drying may be necessary.

From 50- to 25-percent moisture content.—During this period (60 to 120 days for 8/4 stock) shell tension subsides, core compression is maximum. A low temperature should be maintained but relative humidity can be slowly lowered. At the end of the period core compression lessens and some surface compression occurs, closing any small surface checks. Deeper checks caused by initial alternate wetting and drying may not close.

From 25-percent to final moisture content.—This stage in drying should be accomplished in a heated dry kiln as described in section 20-6.

Reducing checking in white oak shipbuilding flitches or timbers.—Since most oak for shipbuilding is received green or nearly green, it is advantageous to accomplish some drying before it is used in construction of a vessel. Unless care is exercised, checking and splitting will occur during this partial drying. Fabricated oak timbers in a vessel are often exposed to alternate sunshine and wetting which may cause excessive checking. Peck (1953) found that two coats of ships hull paint applied to 3- by 12-inch white oak timber prevented all surface checks if the painted timbers were piled on stickers in a roofed pile; after 171 days in the pile, moisture content of the painted timbers had dropped from 72 percent to 59.4 percent. After 309 days moisture content had decreased to 47.4 percent and surface checking was minimal.

CROSSTIES

To be treated effectively, crossties must be seasoned to a moisture content of 20 to 40 percent; the lower the moisture content within this range, the better the treating results (see chapter 21). Air-drying is the simplest and most common method of drying crossties.

In his review of industry practices, Shunk (1976) observed that green crossties, switch ties, and bridge timbers are usually acquired directly from small sawmills and bulk stacked at concentration yards to await transport to treating plants. Such bulk stacking without air circulation is conducive to decay which will continue until the wood seasons well below fiber saturation point.

Shunk (1976) noted that rows of crosstie stacks in air-drying yards varied from 32 to 100 feet long and 13 to 28 feet high; spaces between rows varied from no space up to 5 feet. Good yard layout is an essential in air-drying ties as in air-drying lumber; the same principles recommended in the subsection on LUMBER apply generally to crossties. Stack designs for crossties differ, however, from those used for lumber or squares.

Stack design.—Stringer ties, usually creosoted, are used to keep stacks of air-drying crossties 18 inches off the ground. Parts of switch ties or bridge timbers to be dried that will be in contact with each other in the stack are often brushed with creosote to prevent **stack burn** (initiation of decay).

In most yards studied by Shunk (1976) lift trucks were used for stacking, and average space between rows of stacks was about 3 feet. Crosstie stacks are normally four lifts high (fig. 20-7). Individual piles of crossties vary in number of pieces according to their grade. For example, grades 1, 2, and 3 are usually stacked together in a one-and-nine configuration (fig. 20-7); the larger grades 3A, 4 and 5, are stacked together one-and-eight. Most switch ties measure 7 by 9 inches and are dried one length to a stack. Some operators or air drying yards separate red from white oaks, and sweetgum from black tupelo (Shunk 1976).

An alternative method of stacking crossties is the 2 by 9 method (fig. 20-8).

Figure 20-7.—Stack four lifts high; crossties stacked with one-and-nine construction. (Photo from Shunk 1976.) See also figure 21-10.

Stack covers.—Stack covers, roofs, or sheds shorten drying time, reduce checks, splits, and decay by preventing dry or partially dry crossties from re-absorbing rainwater. Huffman and Post (1962a, 1964) found that uncovered oak crossties in north Florida required 28 to 112 days longer to reach the same moisture content as similar ties under stack covers. Splits over ¼-inch wide developed in 11.3 percent of the ties dried without covers; only 7.1 percent of the covered developed such splits. Light to moderate decay developed in 58.3 percent of the uncovered ties, but in only 13.0 percent of those with covers. Construction of a low-cost disposable cover for crossties during air-drying is illustrated and described in figure 20-9; for such covers Huffman and Post (1962b) recommended polyethylene sheeting of 4-mil thickness meeting Commercial Standard CS 238-61.

Some operators prefer portable reusable covers of more durable sheeting material; storage and handling as well as breakage from windstorms or falls increase maintenance costs of such covers. Permanent sheds have high initial cost but should have low maintenance cost.

Figure 20-8.—Crossties stacked for air-drying using the two-and-nine method.

Figure 20-9.—Disposable stack cover for crossties to be air-dried. (Top) Completed frame ready for 4-mil polyethylene sheeting. (Bottom) Cover stapled to supporting frame; the sheeting material should be quite loose to allow for contraction due to temperature changes or shifting during fork-lift placement of the package on top of the stack. (Photos from Huffman and Post 1962b.)

Occurrence of splits and checks.—Perem (1971) found that in sugar maple (*Acer saccharum* Marsh.), the major seasoning checks usually form on the broad tie face nearest the pith and extend toward the pith in the planes of the rays; the closer a tie face is to the pith, the greater are the lengths and widths of the principal checks. Perem found the most severe checking in ties with the pith about halfway between the broad faces; in such ties, an initial narrow seasoning check, present when installed, creates a plane of weakness along which repeated loading in service develops subsequent splitting. He concluded that the risk of splitting in service is least in boxed-heart ties with the pith less than 1 inch from a broad face, whether placed in track with pith side up or down. No similar data for crossties of southern hardwoods are published.

Prevention and control of splits and checks.—In Shunk's (1976) review of industry practices, he found that some customers require placement of S-irons, C-irons, beegle irons, or dowels (fig. 20-10) in the ends of crossties to retard checking. Others squeeze seasoning splits together with a press, and then bolt or band the end to prevent further splitting. Some firms do not believe that any method has value. Shunk commented that "each railroad has stacks of statistical data supporting its own particular method of preventing checks or splits." Perem (1971) concluded that frequency and dimensions of major checks were not reduced either in seasoning or in service by imbedding S-irons in the ends of ties before seasoning.

Figure 20-10.—Devices to control checking in crossties. (Top) S-irons and C-irons. (Bottom left) Beegle irons. (Bottom right) Steel dowels. (Photos from Shunk 1976.)

Coating ends of crossties or timbers with heavy mastic or paint will reduce end checks, but Shunk (1976) concluded that costs of material and labor exceeded the value of degrade prevented.

Henry (1970) found that **incising** of hardwood crossties and timbers tended to initiate many small checks rather than a few large ones (fig. 20-11). Incising not only reduced check severity, but increased preservative penetration and hastened drying of many refractory hardwoods.

Figure 20-11.—Incisor marks on a large timber showing the beginnings of small checks at the incisions. (Photo from Shunk 1976.)

Shinn (1955) air-dried 626 winter-cut grade 4 and 5 hickory ties to determine the extent that incising would retard checking and splitting. Five-inch C-irons were applied to each end of the ties immediately upon their arrival at an Ohio seasoning yard. Half of the ties (313) were incised; all ties were stacked 15 ties high for air-drying in one-and-nine stacking arrangement. These ties were air-dried from January 1954 to April 1955. Following this 15-month drying period, the incised ties had an average moisture content of 26.4 percent, the unincised ties 26.9 percent. There was less checking and splitting in the incised ties than in the unincised ties, but even the unincised hickory ties displayed less checking and splitting than is normally observed in air-dried oak ties.

Time required to air-dry crossties.—No extensive data are published on the time to dry pine-site hardwood crossties to a moisture content suitable for pressure treatment with creosote. Table 20-2a represents the opinion of eight experienced managers of crosstie air-drying yards in Louisiana, Texas and Oklahoma. Time scheduled for drying is longer if crossties are cut and stacked in November than if stacked in April. Oaks require more drying time than the other species.

Variation of moisture content within an air-dried crosstie is usually substantial. Mathewson et al. (1949) found after 18 months of air-drying 7- by 9-inch northern red oak crossties that moisture content varied from 15 percent at the surface to 57 percent in the core at midlength; in the outer 2 inches of the cross section (72 percent of the cross sectional area), the end-drying effect extended inward only about 6 inches, while in the core it extended inward about 24 inches.

TABLE 20-2a.—*Time required to air-dry heart-center pine-site hardwood crossties to a moisture content suitable for pressure treatment with creosote*[1,2]

Species tie size (inches)	Cut and stacked in April	Cut and stacked in November
	----------------------Months----------------------	
Mixed southern hardwood species, not including oaks[3]		
6 by 8.	4-6	6-8
7 by 9.	4-6	6-8
Oaks[4]		
6 by 8.	7-10	11-14
7 by 9.	7-10	11-14

[1] Based on experience at Texas, Louisiana, and Oklahoma air-drying yards.

[2] Moisture content at treatment averages about 30 percent in the outer 2 inches of the ties; moisture content of the entire crosstie averages about 45 percent.

[3] Includes ashes, black tupelo, elms, hackberry, red maple, sweetbay, sweetgum, and yellow-poplar. May also include hickories.

[4] Includes both upland and lowland species of red and white oaks. Hickory crossties may be grouped for air drying with oak crossties.

Drying

Dowel-laminated crossties.—It is practical to construct—without use of glue—a serviceable dowel-laminated 7- by 9-inch crosstie from two lengths of hardwood measuring 7 inches wide and only 4.5 inches thick (fig. 20-12). Employment of such smaller timbers promotes utilization of small logs and may hasten air-drying (Howe and Koch 1976). The pairs of timbers comprising a single tie can be air-dried separately and then dowelled, or they can be dowelled when green. (See sect. 28-18 for an economic feasibility study of their manufacture.)

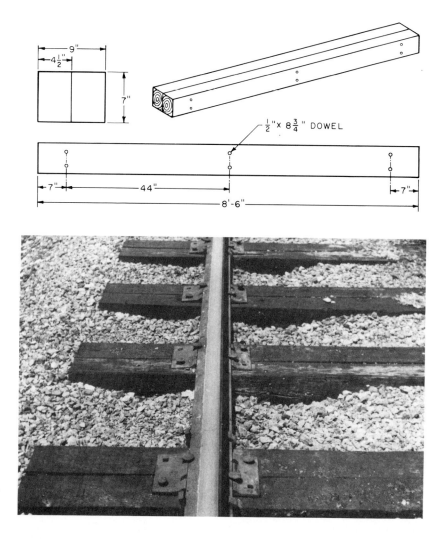

Figure 20-12.—(Top) Mainline 7- by 9-inch crosstie laminated from 4.5- by 7-inch cants with six 0.5-inch spiral steel dowels; no adhesive required. (Bottom) Dowel-laminated ties in mainline track of the Chicago and Northwestern Railway between Deval and Shermer, Ill.; at time of photo, the ties had been in service 7 years. (Photo from Howe and Koch 1976.)

Hale and Howe[4], in experiments with northern red oak, red maple, beech (*Fagus grandifolia* Ehrh.), yellow birch (*Betula alleghaniensis* Britton), and sugar maple (*Acer saccharum* Marsh.), dowel-laminated 7- by 9-inch crossties when green, and air-dried them 5 months to assess width of crack at the interface between tie halves; the green tie halves were pressed tightly together and ½-inch spirally fluted steel dowels were driven into six ⅜-inch holes bored two at each end and two in the center of each tie (fig. 20-12). They found that increase in crack width during air-drying (fig. 20-13) was significantly less for ties with pith to the outside (0.031 inch), as compared with pith centered (0.037 inch), or pith to inside (0.094 inch).

---

[4]Hale, R. P., and J. P. Howe. Technical practicality of dowel-laminating crossties before drying. Fin. Rep. FS-SO-3201-2.85, dated Jan. 28, 1982. U.S. Dep. Agric. For. Serv., South. For. Exp. Stn., Pineville, La. (See also article with same authors and title submitted to For. Prod. J.)

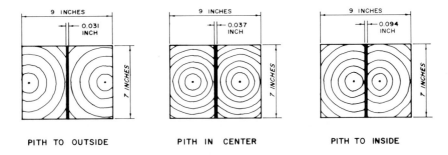

Figure 20-13.—Average increase, after 5 months of air-drying, in crack width at interface of crossties dowel-laminated from green 4.5- by 7-inch timbers sawn from northern red oak, red maple, beech, yellow birch, and sugar maple. (Data from Hale and Howe, text footnote [4].)

After establishing that half-ties with off-center piths should be turned so that piths would be to the outsides, Hale and Howe[4] conducted another experiment using only northern red oak; all wood was incised when green. Factors in the experiment included:

Dowelling dry vs. dowelling green
Treatment of interface of ties dowelled green (no treatment, pentachlorophenol applied to one face, pentachlorophenol applied to both faces of the interface)
All above compared to one-piece ties

Response variables were moisture content (before and after 15 months of air-drying in Nashua, New Hampshire), crack width at interface (when dowelled, and after air-drying), warp when air-dry (crook and bow), distribution of moisture content in the ties after air-drying, and occurrence of **stack burn** (incipient decay) in the interface.

After air-drying, all crossties were pressure-treated with a 50/50 creosote-petroleum solution by the Rueping process; treating pressure was 175 psi. Preservative penetrations in outer faces and inner faces were measured.

From this experiment, Hale and Howe[4] concluded that the quality of crossties dowel-laminated before air-drying is at least as good as for those dowelled after drying if preservative is applied before dowelling to one or both crack faces. The width of the crack at the interface of ties dowelled green, and then air-dried, is only a little over 1/32-inch if pith is centered or toward the outside. After air-drying, the moisture content at the interface of green-dowelled ties is about the same as that 1 inch in from outer faces. Tie-halves air-dried 15 months before dowelling attained somewhat lower average moisture content (31 percent) than one-piece ties or those dowel-laminated when green (about 40 percent).

After pressure treatment preservative penetration from the interface of ties dowelled green was nearly as deep as that from the outside surface. Average retention of preservative in ties dowelled green was comparable to that in one-piece ties, but less than that of ties dowelled dry.

Service tests of these dowel-laminated ties were initiated in single-track mainline routes of the Boston and Maine Railroad in September and October of 1979 on one curve west of Baldwinville, Mass. and on another at Wells Beach Station, Maine. At the first site every fourth tie was replaced with a test tie. At the second site test ties were used as random replacements. All ties were inspected by R. P. Hale and B. & M. Engineer of Track in May of 1982. The Engineer found that the ties were all functionally completely satisfactory. With one exception, there was no indication of spike pulling, and no sign of plate cutting. The general condition of the wood compared favorably with solid ties placed in the track in 1978; crook and bow were negligible.

Crack widths were measured at the dowel points on each tie. Measurements outside the rail on the inner side of the curve averaged 0.148 inch with minimum of 0.0625 inch and maximum of 0.219 inch. Average crack width between rails was 0.170 inch with range from 0.094 to 0.375 inch. Measurements outside the rail on the outer side of the curve reflected the coverage of some of the ties with up to 5 inches of crushed stone ballast, and averaged 0.100 inch with range from 0 to 0.250 inch. The tapered gage used to measure the crack width tended to bottom out at about half depth on many ties indicating that crack width at the bottom of the tie was much less than at the top. The warm, dry month preceding inspection probably dried the exposed upper portions of ties more than portions covered by ballast.

The original average crack width after drying of the ties dowelled green was 0.043 inch. Comparing this with the average for the center measurement (0.170 inch), the increase in crack width was approximately ⅛-inch. The crack width at the time of treating the ties dowelled dry was 0.021 inch, and on the same ties in track 0.162 inch, again approximately ⅛-inch greater.

As long as the crack widths remain in the current state, probably increasing and decreasing with the seasons, the interface crack should pose no problems. With dirty or fine ballast, however, solids in the crack could conceivably cause additional widening.

POSTS AND POLES

Roundwood posts of hardwood are used for fencing, for highway guard rail supports, and also in small farm structures. Numerous methods of accomplishing preservative treatment are practiced (see sec. 21-3). For effective pressure treatment, posts should be air-dried to 20 to 40 percent moisture content. Figure 20-14 illustrates essential features of a stack desgined for air-drying small and round posts. The foundation of treated timbers should support the posts at least 12 inches clear of the ground (preferably 18 inches). Treated 10-foot-long uprights placed in the ground (not exceeding 8 feet center-to-center) contain the stack. Posts are then bulk piled with horizontal spaces to separate the posts into fork-lift units. Temporary cross members can be rigged to support stack covers to protect the posts from rain and sun. Cross-stacking of alternate layers of posts can accelerate drying and diminish growth of mold, but precludes forklift handling.

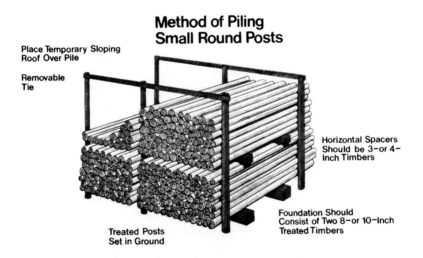

Figure 20-14.—Method of stacking small round posts for air seasoning; such stacks can be easily roofed after they are in place. Foundation timbers should rest on blocks to achieve ground clearance of 12 to 18 inches. (Drawing after Shunk 1976.)

Posts 2 to 8 inches in diameter.—Time required to air-dry round posts 8 feet in length varies with degree of bark removal, species, post diameter, final moisture content, and time of year cut and stacked. To ascertain the relationships Koch[5], in early November 1980, cut fifteen 8-foot posts of each of three species from trees growing among southern pines in central Louisiana; these were machine debarked with a rosser head. Another 15 were cut and debarked in early April 1981. At the same times, an equal number of posts were cut and bark left in place. At least six trees of each species were felled to supply the posts, which had top diameters randomly in the range from 2 to 8 inches. In November 1980, and again in April 1981 the freshly cut bark-in-place and bark-free posts were randomly piled without sticks in three mobile roofed racks—one species to a rack (fig. 20-15A). The racks were positioned in an open field adjacent to Highway 71 on the grounds of the Alexandria Forestry Center, Pineville, Louisiana. At this location, air current were relatively unimpeded. Posts were weighed at approximately 2-week intervals for 6 months and moisture contents recorded as a percent of ovendry weight. Post position within each rack changed randomly at each weighing.

---

[5]Koch, P. 1981. Time to air-dry 8-foot posts of upland pine-site sweetgum, true hickory sp., and southern red oak during winter and summer in central Louisiana. Fin. Rep. FS-SO-3201-9, dated Dec. 14, 1981. U.S. Dep. Agric. For. Serv., South For. Exp. Stn., Pineville, La. See also: Koch, P. 1982. Time to dry 8-foot posts of sweetgum, hickory, and oak in Louisiana. So. Lumberman. Vol.234 (December): 32-33.

Figure 20-15A.—Method of stacking 8-foot-long posts for air-drying in covered mobile racks to facilitate transport to weighing scale. Each rack contained posts of a single species and of varying diameters—half with bark in place and half bark free. (Photo from Koch; see text footnote [5].)

Green moisture contents of the posts averaged highest for sweetgum (117 percent) and lowest for hickory (53 percent); southern red oak was intermediate (72 percent). Posts with bark in place dried faster in summer than in winter. Hickory and southern red oak bark-free posts dried faster in summer than in winter, but sweetgum did not, as follows (figs. 20-15BCD):

Species and condition	115 days of winter drying		115 days of summer drying	
	Average dib at mid-length	Moisture content (ending	Average dib at mid-length	Moisture content (ending)
	Inches	Percent (OD basis)	Inches	Percent (OD basis)
True hickory				
Bark-free	4.4	25.1	4.1	19.1
Bark in place	5.1	36.6	3.6	23.9
Southern red oak				
Bark-free	4.6	30.8	4.0	22.6
Bark in place	5.1	49.1	4.4	30.1
Sweetgum				
Bark-free	4.4	12.2	4.8	17.9
Bark in place	4.8	69.7	4.4	24.4

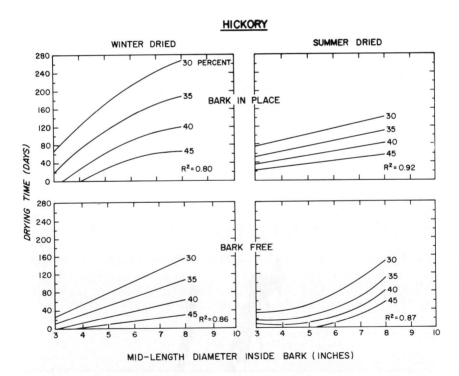

Figure 20-15B.—Time required in central Louisiana to dry 8-foot long freshly cut true hickory posts to four moisture contents (ovendry-weight-basis) related to diameter, summer and winter, with bark in place and bark free. Each curve was produced by regression analyses of data from 15 posts; R^2 values indicated. (Drawing from Koch; see text footnote [5].)

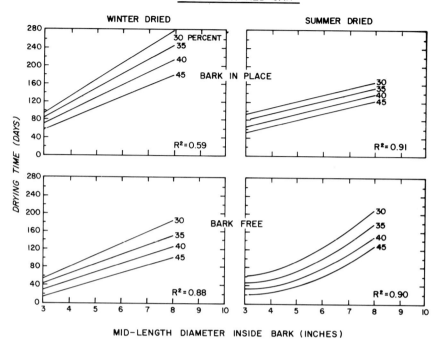

Figure 20-15C.—Time required in central Louisiana to dry 8-foot long freshly cut southern red oak posts to four moisture contents (ovendry-weight basis) related to diameter, summer and winter, with bark in place and bark free. Each curve was produced by regression analyses of data from 15 posts; R^2 values indicated. (Drawing from Koch; see text footnote [5].)

Time to dry to a particular moisture content (e.g., 30, 35, 40 or 45 percent of ovendry weight) was proportional to post diameter except for bark-free winter-dried sweetgum. Slowest drying was sweetgum with bark in place, winter-dried; such a 5-inch post might require about 180 winter days to reach 30 percent moisture content. Fastest drying was bark-free hickory and sweetgum, summer-dried; such 5-inch posts might require only about 60 summer days to reach 30 percent moisture content. For all three species, removal of bark significantly accelerated winter drying in posts of all diameters. In summer, sweetgum dried much more rapidly with bark removed; however, large hickory and southern red oak posts with bark removed did not dry in summer more rapidly with bark removed than with bark in place.

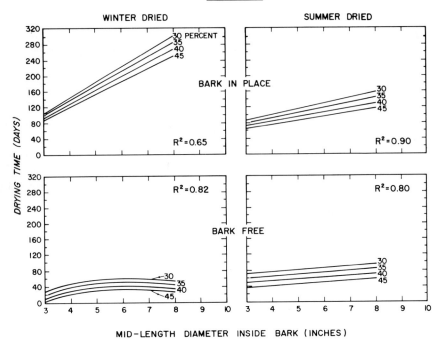

Figure 20-15D.—Time required to dry 8-foot long freshly cut sweetgum posts to four moisture contents (ovendry-weight basis) related to mid-length diameter inside bark, summer and winter with bark in place and bark free. Each curve was produced by regression analyses of data from 15 posts; R^2 values indicated. (Drawing from Koch; see text footnote [5].)

Posts 9 to 21 inches in diameter.—Larger posts and piles air-dry more slowly. Adams (1971) air-dried freshly cut red oak logs with bark in place measuring 9 to 21 inchs in scaling diameter and 8 to 14 feet long by supporting them in a single layer on 6- by 6-inch stringers on an asphalt slab. The 12-week drying period extended from May 30 to August 22, 1969 in Princeton, West Virginia. Weight losses during the 12-week period varied from 5.3 to 14.5 percent of initial green weight; as log diameter increased, percent weight loss decreased. Logs 9 to 14 inches in diameter generally lost 10 to 12 percent of their weight; those 14 to 21 inches in diameter lost 6 to 8 percent of their weight. Since red oak stemwood has a moisture content when green of about 70 percent, the smaller logs dried to a moisture content of about 51 percent, and the larger ones to about 58 percent moisture content after 12 weeks. Figure 20-16 shows the pattern of weight loss for the period, and the corresponding relative humidity of ambient atmosphere.

If it is essential to minimize drying checks in posts, a saw kerf can be cut longitudinally from post surface to pith, thereby concentrating action caused by tangential shrinkage (Ruddick and Ross 1979).

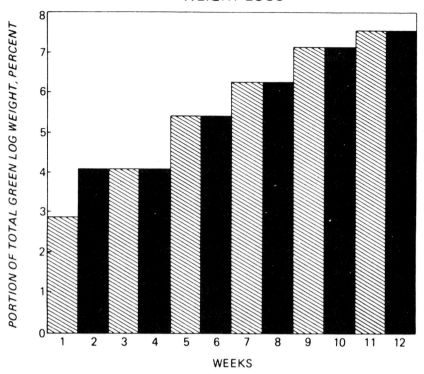

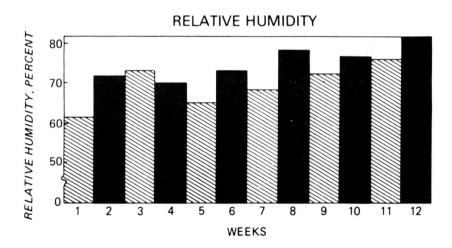

Figure 20-16.—Average cumulative weekly percentage weight loss (due to moisture loss), and relative humidity (recorded 24 hours a day) observed during summer air-drying in West Virginia of 21 freshly cut red oak saw logs measuring 9 to 14 inches in diameter and 8 to 14 feet long. (Drawing after Adams 1971.)

Poles and piling.—Air-drying procedures for long poles and piling are described and illustrated on pages 959-961 of Koch (1972).

Transpirational drying.—**Transpirational drying** is accomplished by delaying bucking and limbing of trees felled while in full foliage. Data specific to pine-site hardwoods are scarce, but some pertinent information has been published.

Smith and Goebel (1952) found that hickory trees felled in August near Pickens, S.C., and left with crown and foliage intact, did not lose stemwood moisture content significantly faster than those felled and trimmed free of branches but not bucked, or those bucked as well as trimmed.

McMinn and Taras (1983) felled 6-inch, 7-year-old *Eucalyptus grandis* Hill ex Maid. growing in full foliage near LaBelle, Fla., and observed resulting transpirational drying. They found that trees felled in November had initial moisture content of 120 percent of dry weight, but less than 80 percent 2 weeks after felling. After 4 weeks moisture content was reduced more than 50 percentage points, i.e., the initial moisture content of 120-135 percent was reduced to 60-85 percent, depending on stem position.

Garrett (1983) felled soft hardwoods during summer in Vermont and found that they lost 5 to 10 percent of their water content in 7 days.

In East Texas, Rogers (1981) felled white oak and sweetgum trees November 12, 1979 and monitored moisture content of their heartwood and sapwood with results as follows:

Weeks after felling	White oak		Sweetgum	
	Heartwood	Sapwood	Heartwood	Sapwood
	------------- Moisture content, percent of OD weight -------------			
Initial	71.6	72.7	118.7	114.7
1	71.2	70.3	116.2	108.3
2	70.9	68.4	114.0	102.6
3	70.6	66.5	112.0	97.5
6	69.5	61.9	107.6	86.7
9	68.6	58.5	105.3	82.0
12	66.6	55.4	105.4	83.8

FIREWOOD[6]

For efficient burning in fireplaces, stoves, and furnaces, firewood should be dried to about 17 percent moisture content, wet-weight basis. Most freshly cut southern hardwoods contain 30 to 55 percent moisture, wet-weight basis; drying to 17 percent removes about 1,500 pounds of water per cord.

Wartluft (1982; 1983) seasoned Appalachian hardwoods to determine the effects of stack location, cover type, splitting, stack design, species, and stick length on the time to dry firewood. His experiment was conducted in West Virginia at an elevation of 2,500 feet.

---

[6] Text under this heading is slightly condensed from Wartluft (1982; 1983).

Drying 2357

Moisture loss in first month.—Wartluft found that the average moisture content of the firewood decreased from 38 percent to 20 percent, wet-weight basis, over the period of the study (fig. 20-17). It dropped sharply in the first month of seasoning; only a few species and longer lengths retained 30 percent or more moisture after the first month. Thereafter rate of drying slowed.

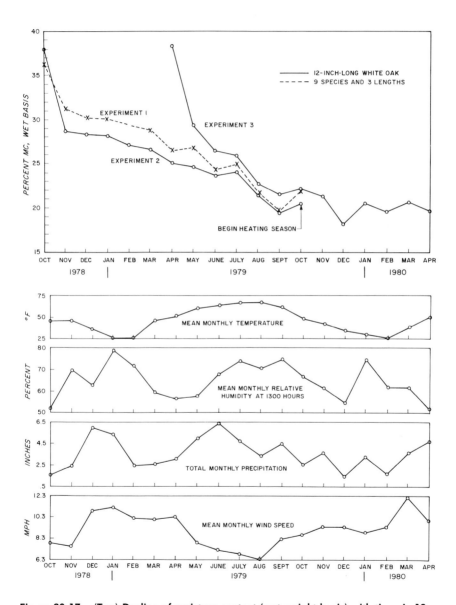

Figure 20-17.—(Top) Decline of moisture content (wet-weight basis) with time, in 12-inch-long white oak firewood and in wood 12, 18, and 24 inches long of nine hardwood species. Wood in experiments 1 and 2 was winter-cut; that in experiment 3 was cut in the spring. (Bottom) Meteorological data during the period. (Drawings after Wartluft 1982; 1983.)

Spring-cut vs. fall-cut wood.—Wartluft found that spring-cut wood averaged 22 percent moisture content after 6 months of seasoning, just 1 percent wetter than autumn-cut firewood after 12 months' seasoning.

Species and stick length.—Final moisture content after 12 months' seasoning (to October 1979) was affected by both species and piece length, as follows; red maple ended driest and hickory wettest:

Species	Fresh-cut	After 12 months' drying		
		12-inch	18-inch	24-inch
	---Percent moisture content, wet weight basis---			
Red maple	35	13	14	15
Sugar maple	36	20	19	19
Ash sp	31	20	21	24
Red oak sp	39	20	22	24
White oak	38	19	23	26
Birch	40	26	22	23
Beech	42	25	23	23
Black cherry	35	26	22	24
Hickory	32	25	26	26

Within the range of lengths tested, Wartluft found that length made little practical difference in seasoning time.

Split wood vs. round wood.—Wartluft's experiment included round and split white oak from 3 to 12 inches in diameter. Only pieces 7 inches or larger in diameter were split. After both 6 and 12 months of drying (to October 1979) split wood was 2 to 3 percent drier than unsplit wood:

Description	Unsplit	Split
	Percent moisture content	
Autumn-cut wood dried 12 months	22	19
Spring-cut wood dried 6 months	23	21

Stacking method.—The type of pile, parallel-stacked or ricked in criss-cross pattern, did not affect final moisture content of 12-inch white oak.

Cover vs. no cover.—Wartluft (1982; 1983) found that covering stacks with tar paper hastened drying of spring-cut white oak; the covered wood had a final moisture content after 6 months of 22 percent, compared to a final moisture content of 25 percent for uncovered wood. Placement of stacks under a clear fiberglass roof and over a black base, or use of a more expensive solar dryer further accelerated drying.

Wartluft also compared time to dry oak stacked in the open with time to dry oak stacked in the woods; he found that once uncovered wood gets wet from snow or rain, it takes longer to redry in the woods than in the open.

FUEL CHIPS

Springer (1979) reviewed the advantages and disadvantages of drying whole-tree chips for fuel before placing them in storage. Such chips (fig. 17-10) are about half water, by weight, when harvested. Because of the danger of spontaneous ignition, fresh, moist, whole-tree chips can be stored in outdoor piles for only short periods. Maintaining an inventory of such chips necessitates frequent expensive pile rotation. The ignition hazard can be eliminated by drying the chips and maintaining them in a dry condition. Springer concluded that in many instances drying costs can be entirely recovered when whole-tree chips are burned for fuel. The cost of maintaining an inventory for fuel purposes in these cases is simply the cost of providing a cover for the dry chips and moving them in and out of storage.

The economic rationale for drying fuel chips is depicted in figure 26-6.

Fuel chips can be air-dried, pre-dried (figs. 26-3 and 26-4), dried en route to the combustion zone (fig. 26-21), or dried in the combustion chamber as they are burned (fig. 26-9).

Sawyer (1983) suggests that fuel chips should be dried at the point of harvest by gasifying a portion of them to provide heat energy. Brace Research Institute (1975) reviewed solar agricultural dryers, some of which should be effective on fuel chips. Haygreen (1981) has demonstrated on a laboratory-scale that moisture content of green chips can be reduced to 35 percent or less by application of very high mechanical pressure.

Air-drying of fuel chips on roofed drying grounds; effect of pile depth.— Koch[7] studied the effect of pile depth on time to dry whole-tree chips under roof. In November 1980 and in April 1981, whole-tree chips of southern red oak and hickory trees 5 to 7 inches in diameter freshly cut in central Louisiana were placed in open-top boxes with solid bottoms (drain holes provided) and expanded-metal screen sides. Trays measured 6 feet square with sides 4, 8, and 12 inches high, and were replicated. Each tray had a rigid cover (with 1-foot overhang all around) supported about 12 inches above the top of the tray, and was elevated about 24 inches clear of the ground in a single row with about 8 feet between trays.

---

[7]Koch, P. 1982. Time to air-dry hickory and southern red oak whole-tree chips spread to various depths in roofed drying trays in November 1980 and April 1981 in Central Louisiana. Fin. Rep. FSSO-3201-10, dated May 19, 1982. U.S. Dept. Agric. For. Serv., South. For. Exp. Stn., Pineville, La. See also: Koch, P. 1983. Moisture changes in oak and hickory fuel chips on roofed and unroofed Louisiana air-drying grounds as affected by pile depth and turning of chips. Forest Products Journal 3(6):59-61.

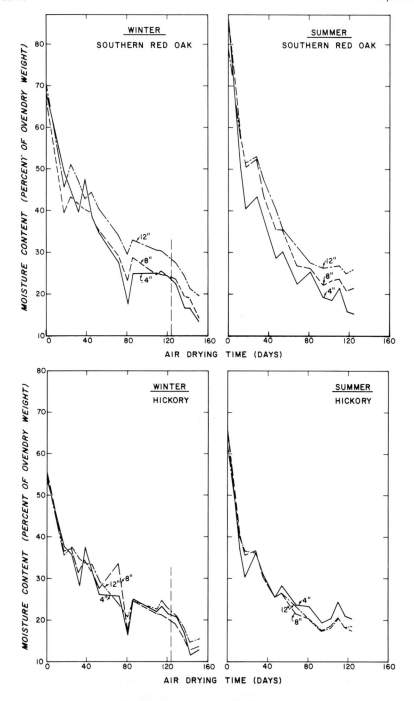

Figure 20-18A.—Moisture content of southern red oak (top) and true hickory sp. (bottom) whole-tree chips vs. time in roofed 6-foot-square drying trays related to depth of chips in trays (4, 8, and 12 inches) and season. Winter-dried chips were placed in the trays in November 1980; those summer-dried were placed in the trays in April 1981. (Drawing after Koch; text footnote [7].)

Moisture contents of the green whole-tree chips when loaded in the trays were as follows: southern red oak cut in November 1980 (69 percent) and April 1981 (85 percent); hickory cut November 1980 (55 percent) and April 1981 (65 percent). Times to reach 40, 30, and 20 percent moisture content (ovendry-weight basis) were as follows: see figure 20-18A:

Moisture content and pile depth	Southern red oak		Hickory	
	Winter	Summer	Winter	Summer
Inches	----------------*Days required to dry*----------------			
40 percent				
12	54	48	14	12
8	18	40	15	14
4	32	33	15	11
30 percent				
12	81	73	52	38
8	70	62	49	35
4	65	47	31	38
20 percent				
12	150	—	81	124
8	136	124	79	82
4	79	94	78	92

In most instances, chips spread 4 inches deep dried more rapidly than if spread 8 or 12 inches deep. Hickory chips had lower green moisture content initially and dried more rapidly than southern red oak. Chips spread in April 1981 did not dry to 20 percent moisture content faster than those spread in October 1980.

Air-drying of fuel chips on roofed and unroofed drying grounds; effect of turning the chips.—Paved drying grounds might be roofed or unroofed, and chips could be turned with plow or bulldozer or left unturned. To compare roofed to unroofed exposure in Louisiana and to evaluate the effect of turning the chips, Koch[7] performed an experiment in which southern red oak whole-tree chips were piled 12 inches deep in trays similar to those described under the previous paragraph heading (except that tray sides and bottoms were of solid sheet metal drilled for drainage). Factors in the experiment were as follows:

- chips turned weekly with a shovel vs. chips not turned
- chips under roof vs. chips not roofed
- replications: 2

The whole-tree southern red oak chips were harvested in central Louisiana on October 15, 1981 and immediately placed in trays, weighed weekly, and after 151 days ovendried and their weight recorded. Initial moisture content of the chips in roofed trays was 75 percent of ovendry weight; that of chips in the unroofed trays was 77 to 81 percent.

Chips in unroofed trays increased in moisture content with time, but less so if turned weekly (fig. 20-18B).

Chips in roofed trays dried considerably faster if turned. Even when in roofed trays and turned weekly, however, 87 days of drying were required for these southern red oak whole-tree chips in 12-inch-deep piles to reach 31 percent moisture content. At the end of the 151-day experiment, the turned chips had 29 percent moisture content; those not turned had 41 percent moisture content, ovendry-weight basis (fig. 20-18B)

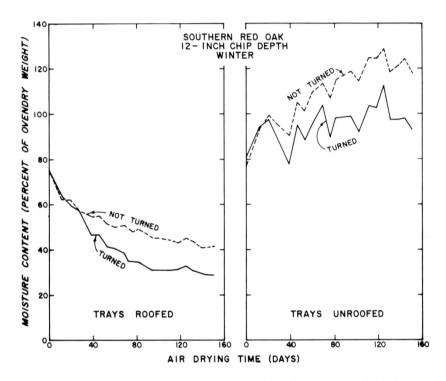

Figure 20-18B.—Moisture content of southern red oak whole-tree chips piled 12 inches deep related to time in roofed and unroofed 6-foot-square trays with solid (but drained) sides and bottoms—half turned weekly with a shovel, and half unturned. The freshly harvested chips were loaded in the trays October 15, 1981. (Drawing after Koch; see text footnote [7].)

20-3 FORCED-AIR FAN PREDRYING

In fan predryers, green wood is protected from rain by a shed roof; fans are used to force air circulation, but no heat is added. Humidistats are arranged to switch fans on when relative humidity is low (e.g., below 75 percent) and off when humidity is high. As no heat is added, final moisture content is determined by ambient air temperatures and relative humidities. As in air-drying, final moisture content for lumber is usually 20 to 25 percent; for crossties, as high as 45 percent.

The **shed-fan dryer** has a permanent roof and a canvas baffle that can be let down at each end of the dryer (fig. 20-19). One side wall is permanent and furnished with a large number of fans. The entering-air side of the dryer is open to the yard. The fans always turn in one direction, to pull the air through the lumber. As heat in the ambient air is an important factor in drying the lumber, the shed-fan dryer works very well in the South.

The **yard-fan dryer** is similar to the shed-fan dryer in construction, operation, and applicability except that different fan wall configurations can be used and the fans may be portable. The roof also is temporary and portable and may be made of canvas, plywood, or sheet metal panels. Effectiveness of a yard-fan dryer is greatly reduced if additional stacks of lumber are placed in front of the fans in an effort to push air through them.

LUMBER

Wengert and White (1979) cautioned that 4/4 and 8/4 oak must be dried with great care in a forced-air predryer; drying rates are usually too fast and surface checks develop. Huber and Klimaszewski (1963) found that 1-inch southern red oak boards dried in a forced-air fan predryer during August and September in Memphis, Tenn., reached 28.7 percent moisture content after 30 days; some surface checking occurred. Pile width usually does not exceed 9 feet, but may be as much as 24 feet if fork lift units are stacked side by side.

M 143 544

Figure 20-19.—Shed-fan forced-air predryer; lumber stacks are carried on tracks. Fans are located in the rear wall to pull air through the lumber stack. (Photo from McMillen and Wengert 1978.)

One-inch elm, sap sweetgum, hackberry, and yellow-poplar have been dried from green to 29 percent moisture content in 6 to 7 days in a shed-fan dryer (Helmers 1959). To attain 20 to 25 percent moisture content in red maple and yellow-poplar during spring in the South, Wengert and White (1979) estimated drying times in a shed-fan dryer as follows:

Lumber thickness	Drying time
	Days
1 inch	20 to 30 days for both species
2 inches	25 to 45 days for yellow-poplar
	35 to 45 days for red maple

Hart (1979) found that during forced-air drying of 6/4 oak, checking of surfaces could be controlled by inserting thin sheets of permeable plywood between stickers and the lumber. He evaluated yield as the percent of board surface occupied by check-free strips 1 inch or wider along the middle 2 feet of board length. In his trials, Hart used two-ply, 1/8-inch yellow-poplar, three-ply, 3/16-inch yellow-poplar, and three-ply, 1/4-inch Douglas-fir plywood. Yields were respectively, 71, 88, and 96 percent, as compared with 42, 61, and 49 for conventional stick placement. Air-drying times were increased about 30, 50, and 100 percent by insertion of the three types of plywood sheets.

CROSSTIES

Huffman and Post (1961, 1962a) found that use of a forced-air fan predryer (fig. 20-20) in Gainesville, Fla. reduced time to season southern red oak 7- by 9-inch crossties to 121 days from 170 days when conventionally air-dried in covered stacks or 240 days in uncovered stacks; in these experiments all ties were seasoned to a core moisture content of 59 percent. Sweetgum and black tupelo crossties required only 54 days to attain a core moisture content of 62 percent in the forced-air fan predryer. These core moisture contents corresponded to an average moisture content for entire crossties of about 42 percent. The fans, arranged to move air through stacked ties at 350 fpm, operated only when relative humidity fell below 80 percent, 8 or 9 hours per day.

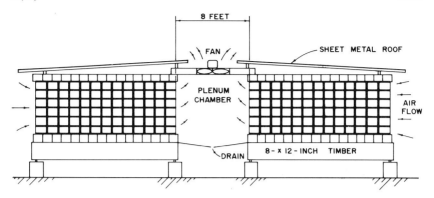

Figure 20-20.—Cross-section through a forced-air fan predryer for crossties. In the experiments of Huffman and Post (1962a), 8½-foot-long ties were stacked 15 courses high with 16 7- by 9-inch ties per course separated by 1- by 4-inch rough-sawn stickers. Ties were spaced 3 inches apart. Sides of the dryer (i.e., at crosstie ends) were enclosed with plastic sheeting. The plenum measured 8 feet wide; 350 fpm air movement was provided with a ½-hp, 42-inch fan operated when relative humidity fell below 80 percent. (Drawing after Huffman and Post 1960.)

20-4 HEATED LOW-TEMPERATURE DRYING

Addition of heat to forced-air fan predryers further accelerates drying. To be economical, the heated air must be recirculated. Two types of recirculating dryers are used: the **forced-air dryer** that only partially controls relative humidity by control of venting (fig. 20-21 top), and the **low-temperature kiln** that fully controls relative humidity (fig. 20-21 bottom). Both types usually consist of the necessary heating and controlling equipment installed in an inexpensive structure. Temperatures in both range from 70° to 120°F. Both may have vents to exhaust excessive moisture, but the vents generally are kept closed, opening only when lower humidity is prescribed. In the forced-air dryer, temperature is controlled by a thermostat. Both the dry-bulb temperature and the wet-bulb temperature are observed with a hygrometer. In the low-temperature kiln, a steam spray line supplies humidity when needed, and a kiln-type recorder-controller regulates the heating coils and the steam spray and records the dry- and wet-bulb temperatures.

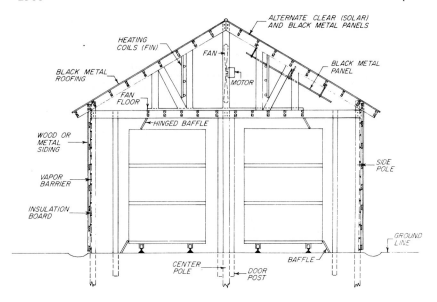

M 144 692, M 140 966-7

Figure 20-21.—Recirculating, heated, low-temperature dryers. (Top) Cross section of a typical forced-air dryer with fans located overhead. (Drawing after Cuppett and Craft 1975.) (Bottom) Low-temperature kiln. (Photo from McMillen and Wengert 1978.)

A third type of heated low-temperature dryer is the controlled-air dryer (fig. 20-22), a large warehouse-type building of tight design capable of drying large quantities of green lumber to 20 percent moisture content. Common sizes hold 200,000 to 1,000,000 board feet. A convenient constant temperature, such as 80° or 85°F, is maintained throughout the building. Green lumber is trucked in through a single large door at one end and transferred to kiln trucks on which it is removed, after drying, through a similar door in the other end. Most of the year in the South these doors can be left open during operations without upsetting interior conditions. Relative humidity in each section, or 50,000-board-foot bay of the building, can be controlled at the desired level (for example, 50 percent) by power intake and exhaust vents, which open when the humidity gets too high. This procedure is, essentially, a one-step drying schedule.

Unit heaters or heating coils located at or near each intake vent are turned on when the vents open to heat incoming air to the dryer temperature. Air is fan-circulated through the lumber at a minimum velocity of 100 feet per minute. User-furnished data indicate that drying times for common items are about 40 percent longer than those of heated low-temperature kilns. Controlled air-dryers have found application in the northernmost States, in Canada, and most recently in the South.

Care must be used in piling lumber in a package-loaded dryer so that packages are not tight against each other or air will not flow freely through the sticker spaces.

M 140 966-23

Figure 20-22.—Interior of a controlled air-dryer. (Photo from McMillen and Wengert 1978.)

SCHEDULES FOR HEATED LOW-TEMPERATURE DRYERS

Two major types of schedules for accelerated-air drying have been tested. In forced-air dryers, with only partial control of relative humidity, two-step temperature schedules are used. For low-temperature kilns with steam spray and automatic recorder-controllers, multiple-step schedules are more appropriate.

Schedules for heated forced-air dryers.—From research at the Forest Products Marketing Laboratory, Princeton, West Va., Cuppett and Craft (1975) developed three forced-air drying schedules (table 20-3). Table 20-4 lists species groups for which they are appropriate. If the dryer has vents, they are kept closed as the dryer is warmed to the first-step temperature. Both dry-bulb and wet-bulb temperatures are observed on the entering-air side of the lumber stack.

When the specified wet-bulb depression is reached, the thermostat is set at the dry-bulb temperature then prevailing. This dry-bulb temperature is maintained throughout the first schedule step. The wet-bulb depression tends to remain about the same for the first day or two because of moisture coming from the wood, but thereafter increases gradually. When the moisture content of the wettest half of the drying samples (chapter 6, Dry Kiln Operator's Manual, Rasmussen 1961), falls to the moisture content for starting the second step of the schedule, the thermostat is raised to the second temperature step. Generally, the vents can be opened for a few days at this time to lower the equilibrium moisture content and accelerate drying. When the wet-bulb temperature settles to a nearly constant level, the vents can be closed again to conserve energy. Drying is continued through successive steps until the moisture content of the wettest sample is down to the desired average moisture content for the whole dryer charge. Vents might be used earlier in the schedule when drying yellow-poplar. Cuppett and Craft (1975) suggested hygrostat-controlled power venting for yellow-poplar and other easy-drying species.

TABLE 20-3.—*Heated-forced-air dryer schedules*[1] *for selected hardwoods*[2]

Step number and schedule designation	Moisture content at start of step	Dry-bulb temperature	Wet-bulb depression
	Percent	°F	
Step number 1			
FA-1	Above	70-80	3
FA-2	moisture content	75-85	4
FA-3	of step 2	80-90	7
Step number 2			
FA-1	40	90	[3]
FA-2	40	95	[3]
FA-3	45	100	[3]

[1] Developed by Forest Products Marketing Laboratory, Princeton, W.Va. (Cuppett and Craft 1975). All operate at an air velocity of 450-600 feet per minute except for 6/4 and 8/4 oak and 8/4 beech and hickory, which use 250-350 feet per minute.
[2] See table 20-4 for pine-site hardwood species appropriate for each schedule.
[3] No control of wet-bulb depression.

TABLE 20-4.—*Index of heated-forced-air dryer schedules appropriate for species (or species groups) of pine-site hardwoods*[1,2] (McMillen and Wengert 1978)

Species	Lumber thickness			
	4/4	5/4	6/4	8/4
Ash sp	FA-3	FA-3	FA-2	FA-1
Elm, sp[4]	—	—	—	—
Hackberry[4]	—	—	—	—
Hickory, sp	FA-2	FA-2	FA-1	FA-1[3]
Maple, red	FA-3	FA-3	FA-2	FA-1
Oak, red sp	FA-2	FA-1	FA-1[3]	FA-1[3]
Oak, white sp	FA-2	FA-1	FA-1[3]	FA-1[3]
Sweetbay[4]	—	—	—	—
Sweetgum (sapwood)[4]	—	—	—	—
Sweetgum (heartwood)[4]	—	—	—	—
Tupelo, black[4]	—	—	—	—
Yellow-poplar	FA-3	FA-3	FA-2	FA-2

[1]See table 20-3 for definition of the three schedules.
[2]Air velocity through stock 450-600 feet per minute except where noted by footnote 3.
[3]Air velocity through stock 250-300 feet per minute.
[4]Schedules for these species are not included in the source publication, but E. M. Wengert suggests that the yellow-poplar schedule is appropriate for them.

A final schedule consideration for this type of dryer is air velocity. Two-inch oak and hickory require a relatively low air velocity—250 to 350 feet per minute. The same low air velocity probably should be used for 2-inch stock of other species for which no schedule has been established, until gradual increases in velocity show the higher rate can be used without causing drying defects. All 4/4/ and 5/4 stock, plus the 6/4 of all species except the oaks, can use a higher air velocity—450 to 600 feet per minute.

Items and species having the same schedule designation (table 20-4) can be dried together in the same dryer load, but samples should be selected for the thickest, slowest drying item to determine when to remove the stock from the dryer.

During the first 6 months of development of a schedule for forced-air drying of a particular item of a particular species, a moisture meter test (James 1975) should be made of 5 percent of the material in packages of lumber from various parts of the dryer. If all the boards do not have core moisture content values at or below 30 percent, a longer drying time should be used the next time this item is dried.

The schedules are not ironclad rules. A user should gradually approach a more severe schedule for items that develop no defects when drying by the indicated schedules. For warmer climates, slightly higher dry-bulb temperatures could be used if the desired wet-bulb depression can be maintained during the first step. Solar radiation may raise the dry-bulb temperature above the set point for 6 to 8 hours. This should be satisfactory unless the dry-bulb temperature exceeds the thermostat setting by more than 5° during step 1. If that should occur, lower the thermostat setting 5°F throughout step 1.

Schedules for low-temperature kilns.—In low-temperature kiln schedules (table 20-5), the desired wet-bulb temperature is controlled during all steps in the schedule. Partial control is maintained over the dry-bulb temperature. As with the forced-air dryer, the vents are kept closed as the kiln is heated. When both the dry-bulb and the wet-bulb temperatures reach the scheduled set points, the recorder-controller maintains the wet-bulb temperature at its prescribed level and holds dry-bulb temperature at or above the set point. If solar radiation raises the dry-bulb temperature slightly above the set point during the day, no change in settings is made.

These schedules were developed with air velocity through the load of 350 feet per minute and presumably would work equally well with air velocities up to 450 feet per minute. The vents would be controlled automatically by the recorder-controller and presumably would stay closed until 10°F or larger wet-bulb depressions are scheduled. Then the vents might be opened a short time at the start of each drying step but would automatically close most of the time and conserve energy. Drying condition changes would be made on the basis of the wettest half of the drying samples (chapter 6, Dry Kiln Operator's Manual, Rasmussen 1961). Determination of when to terminate a run would be on the same basis as for the forced-air dryer.

TABLE 20-5.—*Low-temperature kiln schedules*[1] *for selected eastern hardwoods*[2]

Step number and schedule designation	Moisture content at start of step	Dry-bulb temperature	Wet-bulb depression
	Percent	°F	°F
Step number 1			
LT-1	Above	90	2
LT-2	moisture content	90	2
LT-3	of step 2	90	5
Step number 2			
LT-1	50	95	2
LT-2	60	95	2
LT-3	60	100	5
Step number 2A			
LT-2	50	100	4
LT-3	50	100	10
Step number 3			
LT-1	40	100	5
LT-2	45	100	5
LT-3	40	100	15
Step number 3A			
LT-2 only	40	100	7
Step number 4			
LT-1	35	100	10
LT-2	35	100	10
Step number 5			
LT-1	30	100	15
LT-2	30	100	15

[1] Developed by Robert G. Potter, Potter Lumber Co., Allegany, N.Y.; reproduced from McMillen and Wengert (1978).

[2] See table 20-6 for pine-site species appropriate for each schedule.

Schedule selections appropriate for 4/4 red maple and upland red oaks of four thickenesses are indicated in table 20-6. The table is incomplete for other major southern species because test data are lacking. Items and species having the same schedule designation can be dried together, but samples should be selected for the thickest, slowest-drying item to determine when to remove the stock from the dryer.

Kiln-drying of 8/4 lowland southern oaks from the green condition is difficult. Bois (1978) conclued that no kiln-drying schedule—even at low temperature—is available for drying 8/4 lowland red and white oaks green from the saw without the possibility of heavy surface checking. He recommended gentle air-drying to 25 percent moisture content before kiln-drying (see sec. 20-6 for schedules).

Upland oaks from the northern portion of the southern pinery are easier to dry than lowland southern oaks. In studying the benefits of presurfacing oak before drying, McMillen (1969) and McMillen and Baltes (1972) used low-temperature kiln-drying schedules that increased temperature stepwise from 85°F to 105°F. Satisfactory results were obtained with white oak and red oak using an air velocity of 575 to 600 feet per minute. Cherrybark oak, however, surface checked excessively even at air velocity of only 375 to 400 feet per minute. For cautious trial on central and northern-region rough-sawn 4/4 red and white oak, McMillen and Wengert (1978) suggest the schedule in table 20-7.

TABLE 20-6.—*Index of low-temperature kiln schedules appropriate for species (or species groups) of pine-site hardwoods*[1,2]

Species	Lumber thickness			
	4/4	5/4	6/4	8/4
Ash sp				
Elm sp				
Hackberry				
Hickory sp				
Maple, red	LT-3[3]			
Oak, red sp. (upland)	LT-2	LT-2	LT-1[4]	LT-1[5]
Oak, white sp. (upland)				
Sweetbay				
Sweetgum (sapwood)				
Sweetgum (heartwood)				
Tupelo, black				
Yellow-poplar				

[1]See table 20-5 for definition of schedules; air velocity through the load, 350-450 feet per minute.
[2]Based on personal communication with E. M. Wengert, Virginia Polytechnic Institute and State University, Blacksburg, Va.
[3]Minimum time in each step, 24 hours.
[4]Minimum time on first step, 36 to 48 hours.
[5]Part-time drying (fans on and kiln under control only 50 percent of time) during first step.

TABLE 20-7.—*Suggested low-temperature kiln schedule[1] for rough-sawed 4/4 red and white oak species from the central and northern regions of the United States* (McMillen and Wengert 1978)

Species and step number	Moisture content at start of step	Dry-bulb temperature	Wet-bulb depression
	Percent	°F	°F
Red oak sp.			
1.	Above 50	85	5
2.	50	90	8
3.	35	95	13
4.	28	105	25
White oak sp.			
1.	Above 40	85	5
2.	40	90	8
3.	33	95	13
4.	28	105	25

[1] Postulated from results with presurfaced oak (McMillen 1969; McMillen and Baltes 1972) and rough-sawed northern red oak (Gatslick 1962).

Bacterially infected heartwood of oak—upland as well as southern lowland—presents special drying problems. Infected oak is more likely to develop honeycomb and ring failure than healthy oak, and infected heartwood has a greater tendency to surface check. Such infected oak is characterized by a vinegary smell in early stages and a rancid smell in later stages of infection.

McMillen et al. (1979), after extensive experiments on 4/4 (but not 8/4) infected northern red oak winter cut in Wisconsin and Illinois, concluded that bacterially infected 4/4 oak should be dried green from the saw by a low-temperature, forced-air schedule to at least 25 percent moisture content, and preferably to 20 percent. The forced-air-dried lumber can then be kiln-dried to 6 to 8 percent moisture content with good results by using a milder kiln schedule (table 20-8 and fig. 20-23) than normally recommended for air-dried oak.

McMillen and his associates found that this schedule eliminated honeycomb except in boards with advanced, rancid stages of bacterial infection, but even these boards had minimal honeycomb. Severe surface checking was not reduced as effectively in bacterially infected oak as was honeycombing. Ring failure was not greatly minimized because it is an incipient form of ring shake that begins in the living tree. They found that shrinkage was greater in bacterially infected heartwood than in normal non-infected heartwood.

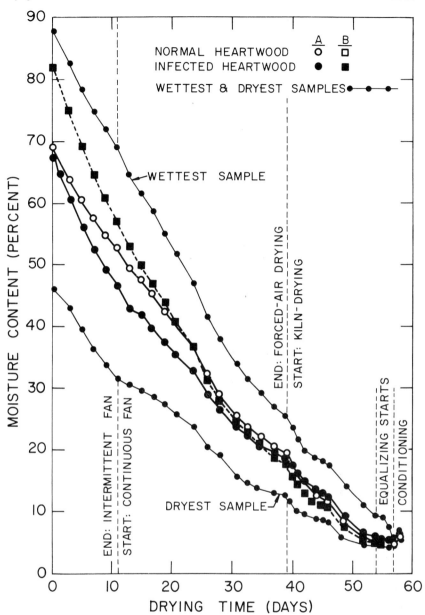

Figure 20-23.—Drying curves for winter-cut normal and bacterially infected 4/4 rough green heartwood of northern red oak. Source A was in Wisconsin; source B was in Illinois. (Drawing after McMillen et al. 1979.)

TABLE 20-8.—*Drying conditions and times for a mixed charge of rough green 4/4 red oak lumber with normal and bacterially infected heartwood dried by a combination of low-temperature forced-air, and conventional kiln-drying procedures* (McMillen et al. 1979)

Moisture content range for each temperature step[1] (percent)	Kiln temperatures Dry-bulb	Wet-bulb	Equilibrium moisture content	Cumulative drying time by temperature steps
	----------------°F----------------		Percent	Days
LOW-TEMPERATURE, FORCED-AIR[2]				
Green to 60	80	77	18.7	[2]0-11
60 to 53	82	78	17.1	11-15
53 to 48	83	78	15.6	15-19
48 to 40	83	77	14.4	19-24
40 to 23	[3]89-90	[3]70-75	6.8-9.3	24-39
KILN DRYING				
23 to 20	120	88	5.0	39-41
20 to 15	128	100	6.1	41-46
15 to 11	148	103	3.6	46-54
EQUALIZING[4]				
11 to 5	169	127	4.1	54-57
CONDITIONING				
5 to 7	178	166	10.2	[5]57-58

[1]Average moisture of five wettest kiln samples used to control step change.
[2]From start of drying to 11 days, fans were used only between 8 a.m. and 8 p.m. with 3-hour reversal. Continuous 24-hour fan operation after 11 days with 3-hour reversal.
[3]No control of wet-bulb unless equilibrium moisture content conditions drop below 6.8 percent.
[4]Equalizing started when driest sample reached 4 percent moisture content.
[5]Total conditioning time was 15 hours.

DRYING TIMES

Research data on drying times for low-temperature drying of several pine-site species are available.

Yellow-poplar.—Data from an experiment by Vick (1965a) were used by McMillen and Wengert (1978) to show the influence of temperature on drying time at a constant equilibrium moisture content condition (fig. 20-24). Air velocity was 550 feet per minute, and load width 8 feet. Yellow-poplar has fast-drying heartwood but the sapwood dried somewhat faster. At the 10 percent equilibrium moisture content used, the effect of temperature is significant. A 100°F temperature dries 4/4 sapwood to 20 percent moisture content in three-fifths of the time required at 80°F. The original paper also shows the effects on drying time and degrade of different equilibrium moisture contents. Degrade was very limited (maximum of $1.87 per thousand board feet) except when a 6 percent equilibrium moisture content was used at 120°F. The paper also shows the mathematical basis by which the actual results were found to agree with theoretical expectations.

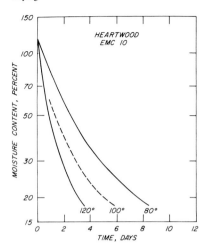 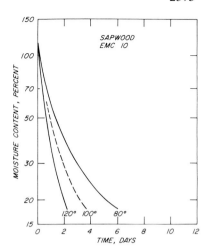

Figure 20-24.—Effect of temperature on drying 1-inch yellow-poplar lumber in a low-temperature forced-air dryer under conditions of 10-percent equilibrium moisture content. (Drawing from McMillen and Wengert 1978, derived from Vick 1965a.)

Sweetgum sapwood.—Vick (1968a) also showed the influence of temperature on drying time of 4/4 sweetgum sapwood, with equilibrium moisture content conditions held constant. As in the yellow-poplar study, air velocity was 550 feet per minute and load width was 8 feet. With conditions set for 10 percent equilibrium moisture content, about 6 days were required at 80°F to reach 20 percent moisture content; at 120°F, only about 2.4 days were needed (fig. 20-25). The original paper also shows the effects of different equilibrium moisture contents on drying time, summarized as follows:

Dry-bulb temperature and equilibrium moisture content	Time to 20 percent moisture content
Percent	*Days*
80°F	
6.........................	5.0
10.........................	6.0
14.........................	7.3
18.........................	9 (approx.)
100°F	
6.........................	3.2
10.........................	4.0
14.........................	4.7
18.........................	6 (approx.)
120°F	
6.........................	2.0
10.........................	2.4
14.........................	2.8
18.........................	4 (approx.)

With conditions set for 10-percent equilibrium moisture content, degrade did not exceed $1.49 per thousand board feet, and uniformity of moisture content was good.

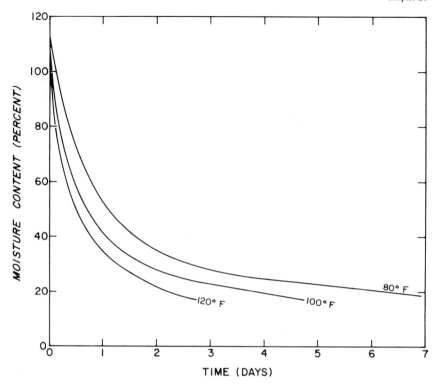

Figure 20-25.—Effect of temperature on drying 1-inch sweetgum sapwood lumber in a low-temperature, forced-air dryer under conditions of 10 percent equilibrium moisture content. (Drawing after Vick 1968a.)

Unstickered hardwood furniture rounds.—The idea of piling furniture rounds in solid stacks without conventional stickering (fig. 18-26 top) was investigated by Vick (1965b) who found that 2-inch-diameter, 30-inch-long, sapwood rounds of tupelo sp. could be dried under laboratory control from green condition to 20 percent moisture content within 6 days at 120°F and 8 percent equilibrium moisture content. At 100°F and 10 percent equilibrium moisture content green sap tupelo dried to 20 percent moisture content in 7.5 days and green red maple in 8 days. In these tests (Vick 1965b), leaving-air velocity at the small voids was about 1,000 feet per minute; the only degrade was hair-line end-checking.

Vick (1968b) observed drying curves of solid-stacked 2-inch by 30- to 40-inch furniture rounds of five species dried for 6 days in a commercial forced-air dryer operated at 100°F with wet-bulb depression set for 10-percent equilibrium moisture content (fig. 20-26). Leaving air velocity was 350 feet per minute. Red oak rounds reached 46 percent moisture content, sweetgum heartwood 40, sweetgum sapwood 27, tupelo sapwood 27, red maple 24, and ash sp. 22 percent.

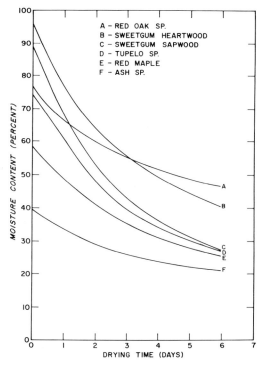

Figure 20-26.—(Top) Furniture rounds solid-stacked for drying in a heated forced-air dryer. (Photo from Vick 1965b.) (Bottom) Drying curves observed in a commercial operation drying solid-stacked 2-inch-diameter, 30- to 44-inch-long furniture rounds of several hardwood species at 100°F dry-bulb temperature and conditions set for 10 percent equilibrium moisture content with air circulation velocity of 350 fpm. (Drawing after Vick 1968b.)

Sapwood of sweetgum, sapwood of tupelo sp., red maple, and ash sp. dried satisfactorily in terms of drying rate, degrade, and moisture content uniformity. Red oak sp. and heartwood of sweetgum developed considerable degrade. Hourly fan reversal and spray application of a wax emulsion end coating were suggested to reduce end checking.

One-inch lumber of Appalachian red oak sp.—In Princeton, West Va., Cuppett and Craft (1971) dried matched lots of rough green red oak sp. 4/4 lumber to a 20 percent moisture content in a heated forced-air dryer and in an adjacent covered air-drying yard. Forced-air drying required 31 days on a 70°F dry bulb schedule and 19 days on an 85°F schedule; during the schedules, initial equilibrium moisture conditions of about 15 percent were decreased to near 10 percent. Air flow was 450 to 550 feet per minute through loads 8 feet wide; fans were reversed every 6 hours. Stickers were 25/32-inch thick.

Air drying required 61 days for lumber piled during June and July, and 138 days for lumber piled in November and December. Degrade losses for low temperature drying (about 0.7 percent of lumber value) averaged less than half those for air-drying under cover.

Some eastern hardwoods that occur on pine sites.—Drying times for rough green lumber in heated forced-air dryers and low-temperature kilns depend somewhat on ambient weather conditions and the dryer temperature selected by the operator. Equally important are the dryer design and the quality of stacking and loading. No two dryers will give the same performance. Gatslick (1962) and Vick (1965a, 1968a) give extensive drying data for low-temperature kilns that completely control equilibrium moisture conditions during the early part of the drying cycle. Cuppett and Craft (1975) give results of drying several eastern hardwoods in a dryer where the only control of equilibrium moisture content is by the dry-bulb temperature.

McMillen and Wengert (1978) tabulated drying times (table 20-9) based on observation of forced-air dryers and low-temperature kilns in operation; they caution that the times should be considered only as rough estimates. They estimate that drying times in controlled-air dryers (fig. 20-22) are about 40 percent longer than those shown in table 20-9.

ECONOMICS OF HEATED LOW-TEMPERATURE DRYERS

Because of changing lumber prices, and the costs of labor, energy, and interest, no economic analysis of drying costs is presented. Readers needing such analyses will find the following references useful: Gatslick (1962), Davenport and Wilson (1969), Catterick (1970), Cuppett and Craft (1971), Wengert and White (1979), and Wengert (1980b).

TABLE 20-9.—*Drying times for 10 species, or species groups, of hardwood lumber in slightly heated forced-air dryers or low-temperature kilns* (Data from McMillen and Wengert 1978).

Species and thickness	Final moisture content	Drying time
	Percent	*Days*
Elm, American		
4/4	20	7
8/4	35	7
Hickory sp.		
4/4	20	16-18
5/4	20	22-24
Maple, red		
4/4	20	6-7
5/4	20	7-8
5/4	12	11
Oak, Appalachian red sp.		
4/4 (at 70°F)	20	31
4/4 (at 85°F)	20	19
Oak, northern red		
4/4	20	18-23
5/4	25	14-35
6/4	25	15-40
8/4	25	40-62
8/4	20	55
Oak, white		
4/4 through 8/4		1 or 2 days less than for red oak sp.
Sweetbay (*magnolia* sp.)		
4/4	20	8-9
8/4	20	15-18
Sweetgum sapwood		
4/4 (at 80°F)	20	7
4/4/ (at 100°F)	20	4½
Yellow-poplar sapwood		
4/4 (at 80°F)	20	6
4/4 (at 100°F)	20	4
Yellow-poplar heartwood		
4/4 (at 80°F)	20	9
4/4 (at 100°F)	20	6½

20-5 LOW-TEMPERATURE DRYING WITH DEHUMIDIFIERS

Dehumidification drying is one way of achieving the required controlled drying rate needed to dry thick hardwood with minimum degrade. Dryers operating on the dehumidification principle are not greatly dependent on outside climate, and they consume less energy than low-temperature dry kilns (Wengert 1980b). The energy they do consume is electrical energy, however, so they may not be more energy-cost efficient; i.e., electrical energy for dehumidification dryers frequently costs more than the heat energy derived from waste wood used in dry kilns.

Dehumidification dryers are potentially energy efficient because the latent heat of vaporization in water evaporated from wood during the drying process—instead of being vented to the atmosphere as in a conventional dry kiln—is recovered by condensing the water vapor in a dehumidifier, thus releasing this heat of vaporization to evaporate more moisture from the lumber. In warm climates it is sometimes necessary to dump (waste) a portion of this heat to avoid excessive temperature buildup in the kiln, so efficiency is lowered in warm weather.

Arganbright (1979) described a dehumidification dryer (fig. 20-27) as consisting of a well insulated drying chamber, a refrigerator whose evaporator works at a temperature less than the dew point of the circulating air, the refrigerator's condenser, and often an auxiliary condenser, and resistance-type heaters. At the beginning of drying, the resistance heaters are actuated until a high enough temperature is reached (70°F) for the compressor to operate. Heated dry air passing through the lumber stack becomes laden with moisture evaporated from the wood. This moisture-laden air passes through the evaporator coil and is dehumidified by cooling condensation, reheated by passing it through the condenser coil, and finally sent back to the drying chamber. Thermal energy equivalent to the compressor's work is removed with the auxiliary condenser coil in warm weather. Outside air passing the exterior condenser can be admitted directly into the negative side of the interior fan together with air from the leaving side of the charge. Since the energy required for moisture condensation is recovered by the condenser, energy consumption of dehumidification dryers is about 50 percent less than that of conventional dry kilns (Wengert 1980c).

Wengert and Wellford (1978) outlined essential operating characteristics of dehumidification dryers as follows:

- The units work best at dryer temperatures between 70° and 160°F.
- Species and thicknesses can be mixed because there are only two or three "schedules".
- Drying times are substantially less than in air-drying, but slightly longer than normal kiln-drying for all but the most difficult species.
- Energy requirements can be reduced perhaps 50 to 70 percent, especially when it is cooler than 70°F outside; but the dehumidifier uses electricity, the high cost of which may offset the energy saving when compared with wood-residue energy systems.
- Degrade is potentially less than in air-drying or normal kiln-drying for some degrade-prone species.
- Conditioning (stress relief) must still be done at high humidities and temperatures (160°F or higher) in another dry kiln, or else the dehumidifier must be removed. Because thick lumber is usually machined or resawn after drying, such stress relief is essential.
- Operation is so simple that labor costs are low.
- The by-product of drying is warm water mixed with extractives and having an acid pH of 4 to 4.5; for each thousand board feet of oak lumber dried to 20 percent moisture content, nearly 200 gallons of water are removed. Disposal of this water may cause pollution problems.

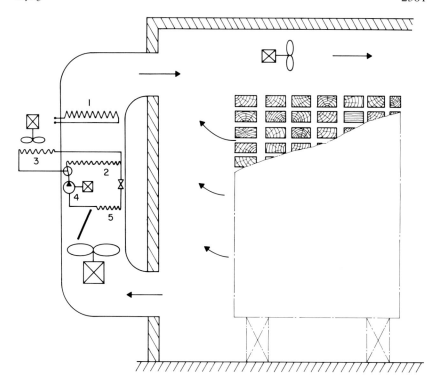

Figure 20-27.—Schematic drawing of a dehumidification dryer. (1) Electric air heater. (2) Condenser. (3) Exterior condenser. (4) Compressor. (5) Evaporator. (Drawing after Arganbright 1979.)

TIME AND ENERGY REQUIRED FOR DEHUMIDIFICATION DRYING

It is possible to dry to as low as 6 percent moisture content with a dehumidification dryer (but not to condition the lumber). Townsend (1979a) provided some data on drying times and energy consumption (including fan energy) when drying oak to 8 percent moisture content, as follows:

Statistic	White oak	Northern red oak
Lumber thickness	6/4	4/4
Initial moisture content (percent)	45	70
Time to dry (days)	40	42
Energy consumption (kWhr/MBF)	501	595

Wengert and White (1979) found that 4/4 oak could be dried in a dehumidifier from 30 percent moisture content to 6 percent in 13 days, 5/4 oak from 50 percent to 6 percent in 34 days, and 4/4 yellow-poplar from green to 6 percent in 15 days.

Townsend (1979b) observed that 4/4 northern red oak dried by dehumidification during August in Vermont required 21 days and 449 kWhr/MBF to dry from 70 percent moisture content to 15 percent. For additional data on energy consumption see Wengert (1980c).

20-6 DRYING IN CONVENTIONAL HEATED KILNS

Kiln-drying is carried out in a closed chamber or building in which air is rapidly circulated over the surface of the wood being dried. Conventional dry kilns use initial drying temperatures from 100° to 180°F and final temperatures from 150° to 200°F. These higher air temperatures and faster circulations accelerate drying greatly beyond the rates of air-drying and accelerated air-drying. Control of relative humidity (and hence, equilibrium moisture content) is necessary to avoid shrinkage-associated defects and to equalize and condition the wood for optimum stability. Air velocities through the load in drying hardwoods generally are between 200 and 450 feet per minute. Temperature and relative humidity are controlled by semi-automatic dry- and wet-bulb temperature recorder-controllers.

Two types of conventional kilns are in general use for hardwoods—package-loaded compartment and track-loaded compartment kilns (fig. 20-28). There are two basic heating systems, steam and hot air (or direct-fired). Well over three-fourths of the kilns designed to dry hardwoods are steam-heated, internal-fan, forced-circulation, track- and package-loaded kilns. In recent years, direct-fired kilns, with supplemental steam or water spray for humidification, have been used occasionally for hardwoods; direct-fired kilns without provision for supplemental humidification are suitable for only the most readily dried hardwoods—and then only with considerable modification of published schedules.

Several conditions must be satisfied to consistently achieve uniform drying in a conventional heated kiln. Good stacking practices, the principles of which are explained in the section on air-drying, are essential to good kiln-drying; flow of heated air of controlled humidity must move at uniform velocity through all parts of the entire kiln-load. As this air moves through the load, it drops in temperature and its humidity rises. Load width (generally 8 feet or less) must not be excessive or the cool humid air on the leaving side of the load will be ineffective in drying. To minimize temperature drop across two loads in a double-track kiln, "booster" reheat steam coils are placed between the loads (fig. 20-28 bottom). To further diminish the effect of air temperature drop across kiln loads, fans are reversed periodically to reverse the direction of air flow through the kiln loads.

Knowledge of progress of drying is essential to good kiln practice. This knowledge is obtained through the use of kiln samples prepared and placed somewhat differently than those previously described for air-drying or low-temperature drying.

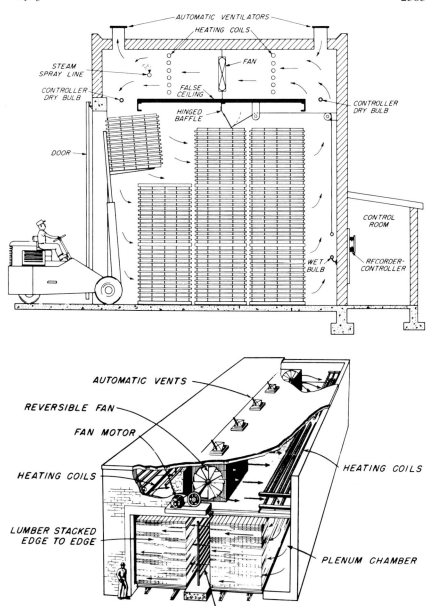

Figure 20-28.—Conventional heated dry kilns. (Top) Package-loaded compartment kiln for charging by fork lift. (Bottom) Track-loaded compartment kiln with alternately opposed fans mounted on a long shaft. Steam "booster" coils are located between the two tracks to raise temperature and lower humidity of air before it enters the second pile. Many fan arrangements, besides the one shown, are in use. (Drawings after Rasmussen 1961.)

Boards from which knot- and bark-free single kiln samples 30 inches or longer are to be cut should be 7 to 9 inches wide and selected during stacking as representative heavier, wetter, thicker stock containing a comparatively high proportion of heartwood; these slow-drying samples are the **controlling kiln samples.** Also, some samples should be cut from flatsawn, narrow, thin, sapwood boards expected to dry rapidly. Two 1-inch moisture sections are sawn adjacent to sample ends (a foot or more from original board end), weighed immediately and ovendried at 214°F to 221°F to constant weight and their green moisture contents computed. From these data the ovendry weight of the kiln sample can be computed. After samples are cut, end coated, and weighed, they are placed in sample pockets built into the piles or packages of lumber during the stacking operation. These pockets are located at mid-length of the pile at perhaps three levels, and on the edge of the pile, but not projecting. The sample should be supported on sticks to duplicate the spacing between boards in the rest of the pile. From time to time during the course of the schedule the kiln samples can be weighed and their moisture contents computed as a percentage of ovendry weight. For further details on use of kiln samples see Rasmussen (1961, p. 99-111).

Kiln-drying defects and methods of minimizing them are illustrated and described extensively in the Dry Kiln Operator's Manual (Rasmussen 1961).

DRYING PROCEDURES—GENERAL

Basic kiln schedules.—Kiln schedules for hardwoods are, for the most part, moisture content schedules, i.e., sequential levels of dry- and wet-bulb temperatures are determined by observed moisture contents of sample boards in the kiln charge. Changes in wet-bulb depression are made gradually—and large depressions are avoided until the average moisture content of the wettest half of the kiln samples (**controlling kiln samples**) reaches 25 percent. The initial temperature is maintained until the controlling kiln samples have an average moisture content of 30 percent.

The **recommended kiln schedules** for hardwoods in the Dry Kiln Operator's Manual are carefully worked out from research and experience, to dry any hardwood item at a generally satisfactory rate without causing objectionable drying defects. They were presented as guides from whch the well-trained, experienced kiln operator was expected to derive optimum schedules best suited for his material, equipment, and operations. In all, the Manual indexed 77 hardwood schedules, but a total of 672 schedules were available for changing schedules as circumstances and his experience dictated. (Code numbers in tables 20-26 and 20-27 identify schedules in the Manual and in this text.)

Because much future kiln-drying of hardwoods will involve air-dried or partly air-dried stock, it is feasible to select 10 pairs of **basic kiln schedules** for drying pine-site hardwoods or species groups. Identical or very similar dry-bulb temperatures are specified for species in each schedule group and drying times for the basic schedules should be very close to those for the recommended schedules. In a few cases, longer times will be required for drying green stock, and for a few of the species the basic schedule also may be slightly too severe for green stock (see footnotes in the basic schedule tables). When drying large quantities of green stock, the kiln operator is encouraged to revert to the originally recommended schedule from the Manual.

For the basic kiln schedules, pine-site species are grouped and listed alphabetically in table 20-10; other eastern hardwoods are grouped in the tabulation under the paragraph describing *Procedure for including small amounts of one species in large kiln loads of other species*. The basic schedules are defined in tables 20-11 through 20-20, along with the schedule code numbers for green stock recommended in the Dry Kiln Operator's Manual; tables 20-26 and 20-27 show temperatures for these schedule code numbers. Each basic table includes temperatures for equalization and conditioning.

TABLE 20-10.—*Alphabetical list of major hardwood species that are found on southern pine sites, appropriate basic kiln-schedule table number, and indication of schedule severity*[1] (Adapted from McMillen and Wengert)

Species	Severity of schedule	Table number
Ash, green and white[2]	Moderate	20-17[3]
Elm, American	Moderate	20-15
Elm, winged	Mild	20-14
Hackberry[2]	Moderate	20-17
Hickory	Mild	20-16
Maple, red[2]	Moderate	20-17
Oak, red (northern or upland)	Mild	20-12[3]
Oak, white (northern or upland)	Mild	20-13[3]
Oak, red or white (southern lowland)	Irregular	20-11[3]
Sweetbay	Moderate	20-18
Sweetgum (heartwood)[2]	Moderate	20-17
Sweetgum (sapwood)[4]	Moderate	20-20
Tupelo, black[4]	Moderate	20-20
Yellow-poplar	Moderate	20-19

[1]For schedules of all the major species of eastern hardwoods, the source publication should be consulted.

[2]Green and white ash, hackberry, red maple, and sweetgum heartwood can be grouped for kiln-drying.

[3]See special note in table indicated.

[4]Sweetgum sapwood and black tupelo can be grouped for kiln-drying.

TABLE 20-11.—*Basic kiln schedules for red and white oak, southern lowland*[1] (McMillen and Wengert 1978)

Moisture content at start of tep (percent)	4/4, 5/4 (T2-C1)[2]			6/4, 8/4 (Irregular)		
	Dry-bulb temper- ature	Wet-bulb depres- sion	Wet-bulb temper- ature	Dry-bulb temper- ature	Wet-bulb depres- sion	Wet-bulb temper- ature
	---------------------------------°F----------------------------------					
Above 40	100	3	97			
40	100	4	96	(Air dry to 25 pct MC)		
35	100	6	94			
30	110	10	100	105	8	97
25	120	25	95	110	11	99
20	130	40	90	120	15	105
15	150	45	105	130	30	100
11	160	50	110	160	50	110
Equalize	173	43	130	173	43	130
Condition	180	10	170	180	10	170

[1] For all oak species, 6/4 stock usually is dried by the 8/4 schedule.
[2] The recommended green-stock code number is the same.

TABLE 20-12.—*Basic kiln schedules for red oak, northern or upland*[1] (McMillen and Wengert 1978)

Moisture content at start of step (percent)	4/4, 5/4 (T4-D2)[2]			6/4, 8/4 (T3-D1)[2]		
	Dry-bulb temper- ature	Wet-bulb depres- sion	Wet-bulb temper- ature	Dry-bulb temper- ature	Wet-bulb depres- sion	Wet-bulb temper- ature
	---------------------------------°F----------------------------------					
Above 50	110	4	106	110	3	107
50	110	5	105	110	4	106
40	110	8	102	110	6	104
35	110	14	96	110	10	100
30	120	30	90	120	25	95
25	130	40	90	130	40	90
20	140	45	95	140	45	95
15	180	50	130	160	50	110
Equalize	173	43	130	173	43	130
Condition	180	10	170	180	10	170

[1] For all oak species, 6/4 stock usually is dried by the 8/4 schedule.
[2] The recommended green-stock code numbers are the same.

TABLE 20-13.—*Basic kiln schedules for white oak, northern or upland*[1] (McMillen and Wengert 1978)

Moisture content at start of step (percent)	4/4, 5/4 (T4-C2)[2]			6/4, 8/4 (T3-C1)[2]		
	Dry-bulb temperature	Wet-bulb depression	Wet-bulb temperature	Dry-bulb temperature	Wet-bulb depression	Wet-bulb temperature
	°F					
Above 40	110	4	106	110	4	107
40	110	5	105	110	4	106
35	110	8	102	110	6	104
30	120	14	106	120	10	110
25	130	30	100	130	25	105
20	140	45	95	140	40	100
15	180	50	130	160	50	110
Equalize	173	43	130	173	43	130
Condition	180	10	170	180	10	170

[1] For all oak species, 6/4 stock usually is dried by the 8/4 schedule.
[2] The recommended green-stock code numbers are the same.

TABLE 20-14.—*Basic kiln schedule for winged elm*[1] (McMillen and Wengert 1978)

Moisture content at start of step (percent)	4/4, 5/4, 6/4 (T6-B3)			8/4 (T3-B2)		
	Dry-bulb temperature	Wet-bulb depression	Wet-bulb temperature	Dry-bulb temperature	Wet-bulb depression	Wet-bulb temperature
	°F					
Above 35	120	5	115	110	4	106
35	120	7	113	110	5	105
30	130	11	119	120	8	112
25	140	19	121	130	14	116
20	150	35	115	140	30	110
15	180	50	130	160	50	110
Equalize	173	43	130	173	43	130
Condition	180	10	170	180	10	170

[1] The recommended green-stock code numbers are the same.

TABLE 20-15.—*Basic kiln schedules for American elm*[1,2] (McMillen and Wengert 1978)

Moisture content at start of step (percent)	4/4, 5/4, 6/4 (T6-C4)			8/4 (T3-C3)		
	Dry-bulb temperature	Wet-bulb depression	Wet-bulb temperature	Dry-bulb temperature	Wet-bulb depression	Wet-bulb temperature
	----------------------------------°F----------------------------------					
Above 40	120	7	113	110	5	105
40	120	10	110	110	7	103
35	120	15	105	110	10	100
30	130	25	105	120	19	101
25	140	35	105	130	35	95
20	150	40	110	140	40	100
15	180	50	130	160	50	110
Equalize	173	43	130	173	43	130
Condition	180	10	170	180	10	170

[1]The recommended green-stock code numbers are: T6-D4 (for 4/4, 5/4, and 6/4), and T5-D3 (for 8/4).

[2]McMillen and Wengert (1978) comment that if 4/4, 5/4, or 6/4 American elm is of the soft type (forest-grown with narrow rings, specific gravity—dry-weight, green-volume basis—of 0.40 to 0.48) schedule T8-D4 should be used; if of the hard type (open-grown, wide rings, specific gravity of 0.44 to 0.52), the lumber should be air-dried first and then schedule T2-D5 used.

TABLE 20-16.—*Basic kiln schedules for hickory*[1] (McMillen and Wengert 1978)

Moisture content at start of step (percent)	4/4, 5/4, 6/4 (T8-C2)			8/4 (T5-C1)		
	Dry-bulb temperature	Wet-bulb depression	Wet-bulb temperature	Dry-bulb temperature	Wet-bulb depression	Wet-bulb temperature
	----------------------------------°F----------------------------------					
Above 40	130	4	126	120	3	117
40	130	5	125	120	4	116
35	130	8	122	120	6	114
30	140	14	126	130	10	120
25	150	30	120	140	25	115
20	160	40	120	150	35	115
15	180	50	130	160	50	110
Equalize	173	43	130	173	43	130
Condition	180	10	170	180	10	170

[1]The recommended green-stock code numbers are: T8-D3 (for 4/4, 5/4 and 6/4), and T6-D1 (for 8/4). See also table 17 in the Dry Kiln Operator's Manual, and kiln schedules for hickory lumber of McMillen (1956a).

TABLE 20-17.—*Basic kiln schedules for green and white ash,[1] hackberry, red maple, and sweetgum heartwood[2]* (McMillen and Wengert 1978)

Moisture content at start of step (percent)	4/4, 5/4, 6/4 (T8-C4)			8/4 (T5-C3)		
	Dry-bulb temperature	Wet-bulb depression	Wet-bulb temperature	Dry-bulb temperature	Wet-bulb depression	Wet-bulb temperature
	---------- °F ----------					
Above 40	130	7	123	120	5	115
40	130	10	120	120	7	113
35	130	15	115	120	11	109
30	140	25	115	130	19	111
25	150	35	115	140	30	110
20	160	45	115	150	40	110
15	180	50	130	160	50	110
Equalize	173	43	130	173	43	130
Condition	180	10	170	180	10	170

[1] For green or white ash over 35 percent moisture content, use the green-stock code numbers recommended below.

[2] Recommended green-stock code numbers are:

Species	4/4, 5/4, 6/4	8/4
	---------- Code numbers ----------	
Ash, green and white	T8-B4	T5-B3
Hackberry	T8-C4	T6-C3
Maple, red	T8-D4	T6-C3
Sweetgum (heartwood)	T8-C4	T5-C3

TABLE 20-18.—*Basic kiln schedule for sweetbay.* (McMillen and Wengert 1978)

Moisture content at start of step (percent)	4/4, 5/4, 6/4 (T10-D4)[1]			8/4 (T8-D3)[1]		
	Dry-bulb temperature	Wet-bulb depression	Wet-bulb temperature	Dry-bulb temperature	Wet-bulb depression	Wet-bulb temperature
	---------- °F ----------					
Above 50	140	7	133	130	5	125
50	140	10	130	130	7	123
40	140	15	125	130	11	119
35	140	25	115	130	19	111
30	150	35	115	140	35	105
25	160	40	120	150	40	110
20	170	45	125	160	45	115
15	180	50	130	180	50	130
Equalize	173	43	130	173	43	130
Condition	180	10	170	180	10	170

[1] The recommended green-stock code number is the same.

TABLE 20-19.—*Basic kiln schedule for yellow-poplar* (McMillen and Wengert 1978)

Moisture content at start of step (percent)	4/4, 5/4, 6/4 (T11-D4)[1]			8/4 (T10-D3)[1]		
	Dry-bulb temperature	Wet-bulb depression	Wet-bulb temperature	Dry-bulb temperature	Wet-bulb depression	Wet-bulb temperature
	------------------------------------°F------------------------------------					
Above 50	150	7	143	140	5	135
50	150	10	140	140	7	133
40	150	15	135	140	11	129
35	150	25	125	140	19	121
30	160	35	125	150	35	115
25	160	40	120	160	40	120
20	170	45	125	170	45	125
15	180	50	130	180	50	130
Equalize	173	43	130	173	43	130
Condition	180	10	170	180	10	170

[1]The recommended green-stock code number is the same.

TABLE 20-20.—*Basic kiln schedule for sweetgum sapwood and black tupelo*[1,2] McMillen and Wengert 1978)

Moisture content at start of step (percent)	4/4, 5/4, 6/4 (T12-E5)			8/4 (T11-D3)		
	Dry-bulb temperature	Wet-bulb depression	Wet-bulb temperature	Dry-bulb temperature	Wet-bulb depression	Wet-bulb temperature
	------------------------------------°F------------------------------------					
Above 60	160	10	150	150	5	145
60	160	14	146	150	5	145
50	160	20	140	150	7	143
40	160	30	130	150	11	139
35	160	40	120	150	19	131
30	170	45	125	160	35	125
25	170	50	120	160	40	120
20	180	50	130	170	45	125
15	180	50	130	180	50	130
Equalize	173	43	130	173	43	130
Condition	180	10	170	180	10	170

[1]Lower grades of sweetgum heartwood often are included with sweetgum sapwood; if stock is more than 15 percent heartwood use schedule in table 20-17.

[2]Recommended green-stock code numbers are:

Species	4/4, 5/4, 6/4	8/4
Sweetgum (sapwood)	T12-F5	T11-D4
Black tupelo	T12-E5	T11-D3

Also, special schedules for tupelo species are given in table 17 of the Dry Kiln Operator's Manual.

Temperatures and depressions in the basic schedules differ slightly from the original Manual recommendations; temperature and wet-bulb depression values have also been added for equalizing and conditioning. These specifics are designed to conserve energy and are essentially what a prudent kiln operator would already be doing to accomplish this end.

The intermediate and final-stage wet-bulb temperatures are close to those that are naturally attained when the vents are kept closed and the steam spray is turned off. Close control of wet-bulb temperature is not necessary during the final drying stage when the schedule shows depression values of 45°F or larger; depressions resulting from closed vents without steam spray are acceptable. During equalizing and conditioning, however, a clean wet-bulb wick and close control are essential.

When starting up the kiln the average moisture content of air-dried stock should be 25 percent or lower with no material over 30 percent. For partly air-dried stock, no material should be over 50 percent moisture content. For green lumber follow specific procedures for starting up in chapter 10 of the Dry Kiln Operator's Manual. Although the same schedule generally is prescribed for 4/4, 5/4, and 6/4 of each species, the 6/4 will take considerably longer to dry than the 4/4, and therefore the best practice is to dry each size separately. If 4/4 and 5/4 must be dried together, most kiln samples should be taken from the 5/4 stock.

Procedure for 4/4, 5/4, and most 6/4 air-dried stock.—First, bring dry-bulb temperature up to the value prescribed by the schedule for the average moisture content of the controlling kiln samples (wettest 50 percent), keeping the vents closed and the steam spray turned off.

Second, after prescribed dry-bulb temperature has been reached:

(a) If the air-dried stock has not recently undergone surface wetting or long exposure to high humidity, set the wet-bulb controller at the prescribed wet-bulb temperature. Turn on the steam spray only if necessary to start equalizing.

(b) If there has been surface moisture regain, set the wet-bulb controller for a 10°F wet-bulb depression and turn on the steam spray. Let the kiln run 12-18 hours at this wet-bulb setting, then change to the dry- and wet-bulb settings prescribed by the schedule.

Procedure for 8/4 (plus 6/4 oak) air-dried stock.—First, bring dry-bulb temperature up to the value prescribed by the schedule for the average moisture content of the controlling kiln samples, keeping the vents closed. Use steam spray only as needed to keep wet-bulb depression from exceeding 12°F.

Second, after the prescribed dry-bulb temperature has been reached:

(a) If there has been no surface moisture regain, set the wet-bulb controller at the prescribed wet-bulb temperature. Turn on the steam spray only if necessary.

(b) If there has been surface moisture regain, set the wet-bulb controller for an 8°F wet-bulb depression and turn on the steam spray. Let the kiln run 18 to 24 hours at this setting. Then set for a 12°F depression and run for 18 to 24 hours more before changing to the conditions prescribed by the schedule.

Procedure for partly air-dried 4/4, 5/4, and most 6/4 stock.—First, bring dry-bulb temperature up to the value prescribed by the schedule for the average moisture content of the controlling kiln samples, with vents closed. Use steam spray only as needed to keep wet-bulb depression from exceeding 10°F. Do not, however, allow the depression to become less than 5°F or moisture will condense on the lumber.

Second, after the prescribed dry-bulb temperature has been reached, run a minimum of 12 hours on each of the first three wet-bulb depression steps of the whole schedule, but still observe the 5°F minimum wet-bulb depression. Then change to the conditions prescribed for the moisture content of the controlling samples.

Procedure for partly air-dried 8/4 (plus 6/4 oak) stock.—First, bring dry-bulb temperature up to the value prescribed by the schedule for the average moisture content of the controlling kiln samples, with vents closed. Use steam spray only as needed to keep wet-bulb depression from exceeding 8°F. Do not, however, allow depression to become less than 5°F.

Second, after the prescribed dry-bulb temperature has been reached, run a minimum of 18 hours on each of the first three wet-bulb depression steps of the schedule, but still observing the 5°F minimum wet-bulb depression. When the kiln conditions coincide with those prescribed by the schedule for the average moisture content of the controlling samples, change to the moisture content basis of operation.

Procedure for including small amounts of one species in large kiln loads of other species.—Individuals who have lumber from a log or two of their own often wish to use this lumber for paneling, cabinetry, or fine furniture. Such wood should be carefully air-dried first under a good pile roof or in an unheated shed. Then, if possible, it should be kiln-dried so as to bring the moisture content to about 7 percent and relieve drying stress. Such air-dried lumber can be kiln-dried with the same thickness of air-dried stock of another species in the same basic kiln schedule group. See table 20-10 and the following tabulation:

Table number and species	Botannical name
20-14	
Apple	*Malus* sp.
Dogwood	*Cornus florida* L.
Rock elm	*Ulmus thomasii* Sarg.
Hophornbean	*Ostrya virginiana* (Mill.) K. Koch
Black locust	*Robinia pseudoacacia* L.
Osage orange	*Maclura pomifera* (Raf.) Schneid.
Persimmon	*Diospyros virginiana* L.
Sycamore	*Platanus occidentalis* L.
20-15	
Slippery elm	*Ulmus rubra* Muhl.
Holly	*Ilex opaca* Ait.
Mahogany	*Swietenia macrophylla* King
Black walnut	*Juglans nigra* L.

Table number and species	Botannical name
20-16	
Beech	*Fagus grandifolia* Ehrh.
Sugar maple	*Acer saccharum* Marsh.
Pecan	*Carya* sp. (see Chapter 3)
20-17	
Black ash	*Fraxinus nigra* Marsh.
Yellow birch	*Betula alleghaniensis* Britton
Cherry	*Prunus serotina* Ehrh.
Cottonwood (wet-streak)	*Populus deltoides* Bartr. ex Marsh.
Silver maple	*Acer saccharinum* L.
Sassafras	*Sassafras albidum* (Nutt.) Nees
20-18	
Basswood (light color)	*Tilia americana* L.
Paper birch	*Betula papyrifera* Marsh.
Buckeye	*Aesculus* sp.
Butternut	*Juglans cinerea* L.
Chestnut	*Castanea dentata* (Marsh.) Borkh.
Cottonwood (normal)	*Populus deltoides* Bartr. ex Marsh.
Magnolia	*Magnolia grandiflora* L.
Swamp tupelo	*Nyssa sylvatica* var. *biflora* (Walt.) Sarg.
Black willow	*Salix nigra* Marsh.
20-19	
Cucumber tree	*Magnolia acuminata* L.
20-20	
Aspen	*Populus tremuloides* Michx.
Basswood	*Tilia americana* L.

Both basswood and aspen (sapwood or box lumber) can be dried faster on a more severe schedule given in McMillen and Wengert (1978, table 22).

The kiln schedule will have to be operated on the basis of kiln samples representing the major kind of wood in the kiln, but final moisture content and stress condition should be satisfactory with proper equalizing and conditioning treatment.

Small amounts of fully air-dried lumber (20 percent moisture content or less) can be kiln dried with air-dried stock of species that take other basic kiln schedules, but final moisture content and stress relief may not be as satisfactory as when dried with a species of its own group. If the only hardwood kiln available is drying green or partly air-dried stock, arrangements should be made to put the small amount in the kiln sometime after it is started, when the major stock and the air-dried small amount of material have about the same moisture content.

Small amounts of air-dried stock also can have their moisture content reduced to the proper level by heated room drying, which is described later. Such drying may leave the wood with unrelieved drying stresses, which can cause warping problems unless special care is exercised in the woodworking operations.

Kiln schedule acceleration.—Shortening drying time can save significant energy and cost—a possibility that should be explored by every commercial kiln operator. A step-by-step approach to kiln schedule acceleration for green stock is given on pages 124-126, Dry Kiln Operator's Manual. Under certain circumstances, several steps can be combined. For instance, the recommended schedule for red oak (4/4 thickness) is T4-D2. This is essentially the same as the basic schedule for 4/4 in table 20-12.

If the green moisture content is very high, the first acceleration is to change from T4-D2 to schedule T4-E2, applying the same wet-bulb depressions at higher moisture contents. The dry bulb value is unchanged.

If T4-E2 works well for several charges, with no surface checking, the next change should be to schedule T4-E3. In this change from E2 to E3, the initial and subsequent wet-bulb depressions are increased. In a kiln with well-calibrated instruments and good construction, T4-E3 should work well, but a slight amount of surface checking could occur on the entering-air edges of the load.

The 50°F depression prescribed in the Manual for the latter stages of drying is only a guide. Generally the kiln operator should turn off the steam spray by a hand shut-off valve and set the wet-bulb controller so that the vents stay closed during the latter half of the drying schedule. If the wet-bulb temperature does not come down to the value shown for each step in the basic schedules, the kiln operator may want to open the vents for short periods only. When dry-bulb temperatures of 160°F or higher are reached, however, the vents should be kept closed.

Finally, one temperature acceleration is suggested for 4/4 thickness only—when all the samples are below 20 percent, go to the final 180°F temperature. The results of all the modifications outlined above are summarized in table 20-21. If the oak cited above is from stock that normally is very wet when green, but has had some slight air-drying, it may come down to 50 percent moisture content very rapidly. In this situation, extend the time on the first schedule step to a total of 2 days.

Not all the generally recommended kiln schedules are so conservative that they can be modified as much as the oak schedule. A general modification that applies to all species when drying 8/4 stock, however, is to use a final dry-bulb temperature of 180°F when the average moisture content of the controlling samples reaches 11 percent.

TABLE 20-21.—*Accelerated kiln schedule for 4/4 northern or upland red oak, and 5/4 with high initial moisture content with some risk of slight surface checking* (McMillen and Wengert 1978)

Moisture content (percent)	Dry-bulb temperature	Wet-bulb temperature	Depression
	°F		
Above 60	110	105	5
60-50	110	107	7
50-40	110	99	11
40-35	110	91	19
35-30	110	(^{1})	^{1}30
30-25	120	(^{1})	^{1}40
25-20	130	(^{1})	^{1}50
20-18	140	(^{1})	^{1}50
18-(F-3)^2	180	(^{1})	^{1}50

[1] Vents closed, steam spray shut off; accept whatever depression occurs.
[2] Equalize and condition as in table 20-12; F is desired final moisture content.

The question is sometimes asked: "What can be done when drying seems to stop in the middle of the kiln run?" Drying will never actually stop if the equilibrium moisture content of the kiln atmosphere is lower than the core moisture content of the lumber. Slight increases of dry-bulb temperature during the intermediate stages of drying can be made, but generally they are not recommended. An increase of 5°F can be tolerated, however, by some woods. In table 20-21, it might be satisfactory to use 115°F dry-bulb temperature at 35 percent moisture content. A temperature of 125°F dry-bulb has been used at 35 percent moisture content for 8/4 hickory (table 17, Dry Kiln Operator's Manual). Before trying such increases of temperature at moisture contents above 30 percent, the kiln operator should make sure that his kiln recorder-controller is properly calibrated.

Experimentally, drying of oak has been somewhat accelerated by automation in connection with presurfacing and an accelerated kiln schedule (Wengert and Baltes 1971). Automation has not been adopted commercially. Pilot tests have been made, however, on presurfacing and the accelerated kiln schedule it permits (Rice 1971; Cuppett and Craft 1972). Drying time savings were estimated to be 24 percent or higher and kiln capacity was increased 8 to 12 percent. Presurfacing does not fit in with current hardwood processing practice, and there would, of course, be some added costs for the presurfacing.

Simpson (1980a) kiln-dried 4/4 northern red oak and white oak from green to 7-percent moisture content in 9 to 12 days—an approximate 50-percent reduction in kiln time from conventional kiln schedules. His procedure for obtaining such accelerated drying started with presurfacing the green lumber to 1-1/32-inch thickness, then presteaming the lumber at 190°F for 3 to 4 hours to attain 190°F and then holding for 2 hours at a constant temperature of 190°F, then using a smoothed accelerated schedule with small temperature changes made often and with air circulation velocity of 600 fpm (the schedule began with dry-bulb temperature of 115°F with wet-bulb tempertaure of 111°F), and finally drying at 230°F with no wet-bulb temperature control from 18-percent to 7-percent lumber moisture contents. He found that in most casees lumber quality was acceptable, but enough honeycomb and surface checking were present to cause concern, i.e., some 4/4 oak would not tolerate the rapid drying.

Warp and shrinkage reduction.—Precision sawing and stacking are important factors in reducing warp. Another effective measure is to use rapid natural air-drying or accelerated air-drying of properly piled lumber. Research on red oak species has shown that shrinkage is reduced by using lower temperatures and rapid reduction of relative humidity (McMillen 1958). The effect is uniform from 140°F down to 95°F. From 95°F down to 80°F the effect is greater because more of the wood is affected by **tension set** in outer board portions. Tension set tends to resist shrinkage of the board and thus reduces warping. This is a major part of the reason why air-drying gives the least shrinkage and warping. **Compression** set in the interior of the wood is also important, as it tends to increase board shrinkage and warping. Thus, the drying of green oak with an initial temperature above 110°F or 115°F would be expected to produce more than normal warping.

The effects of uniform thickness, good stacking, and restraint may overcome the effects of temperature differences with 4/4 and 5/4 stock. For thicker stock, in which compression set probably has more influence, use of lower than customary initial temperature and relative humidity probably is helpful.

Simpson (1982) designed a restraining device consisting of hydraulic cylinders and steel cables to encircle a load of short lumber at three places, in his tests on 5/4 mixed oak and hickory at different levels of restraining force he found that cup and bow were not reduced, but twist was reduced by 25 to 50 percent, and crook was reduced up to 35 percent.

No series of low-shrinkage, low-warp schedules has been developed. The kiln operator who has the kiln time available to take slightly longer in drying can experiment with lower relative humidity schedules by using slightly larger initial wet-bulb depressions. The lower the wet-bulb temperature, the easier it is to achieve a lower dry-bulb temperature in the kiln. Changes from recommended or basic schedules should be slight at first, and the kiln operator will need to observe the stock in the kiln frequently to see that surface checking is not developing.

Adjusting moisture content of kiln-dried wood.—Once wood has been kiln-dried to a moisture content suitable for interior purposes, it should be stored in a heated or dehumidified shed or room (see sec. 20-14). Situations occur, however, when the moisture content of the lumber must be changed: (a) the moisture content is too low for use in steam bending, boat construction, or the like, or (b) the wood has not been properly stored and must be redried.

If the moisture content is too low, a two-step procedure is advised. The lumber must be stickered in properly built piles and void spaces of the kiln baffled. The exact schedule will depend on the moisture content desired and the time available. Allow 2 days for each step for 4/4 stock, longer for thicker material or large moisture differences. For the first half of the time required, use a kiln equilibrium moisture content (EMC) equal to the desired moisture content. For the second half use an EMC 3 percent higher. If enough time is available, use a kiln temperature of 130°F or 140°F. For quicker results, 160°F or 180°F can be used. In either case, the vents should be closed during kiln warm-up. Steam spray should be used intermittently to avoid EMC's lower than the present moisture content or higher than the desired moisture content during the warm-up period. Do not condense moisture on the lumber. Use the kiln sample procedure to monitor the moisture pick-up. If the rate is too slow, use a higher temperature with the same EMC's. The moisture level to which hardwoods can be easily raised is limited to about 13 percent, for it is difficult to hold an EMC above 16 percent in most kilns.

To redry kiln-dried lumber that has been kept in uncontrolled storage, great care is needed to avoid permanently opening tight checks or causing internal hairline checks. If the storage period has been short, tightly bundled lumber can be redried in the bundles because most of the moisture pick-up will have been on the board ends and the surfaces of exposed boards. After longer storage, or for lumber that has been kept outdoors on stickers, the lumber must be stickered for redrying.

Two temperature steps are suggested for the redrying operation; the first about 1 day long at 130° or 140°F. The second should be at the final temperature of the drying phase of the basic schedule (tables 20-11 to 20-20) for the species and size involved. Do not use steam spray during kiln warm-up. Surface checks already present may open up, but will close again as the wood dries. When the first kiln temperature is reached, set the wet-bulb controller to achieve an EMC somewhere between the current moisture content of the stock and the moisture content desired. For the final step, set the controller to give an EMC 2 percent below the desired moisture content. When the wettest kiln sample reaches the desired moisture content, stop the drying. No conditioning should be needed.

Predicting drying time.—Frequently management is faced with determining how long it will take to dry a load or when the next change in kiln conditions can be made.

"How long before the next change?" can be easily answered if the operator will graph the moisture content of the samples as they dry (fig. 20-29 top). Special graph paper with horizontal markings in tenths of an inch and vertical markings in twelfths makes it easy to show drying time to days and hours. Graphing the moisture content of the samples as they are weighed shows a smooth drying curve, any portion of which can be extended 24 hours ahead with a straight line to predict the next day's moisture content. When the same species and thickness is dried repeatedly, the total drying time can be reasonably well estimated.

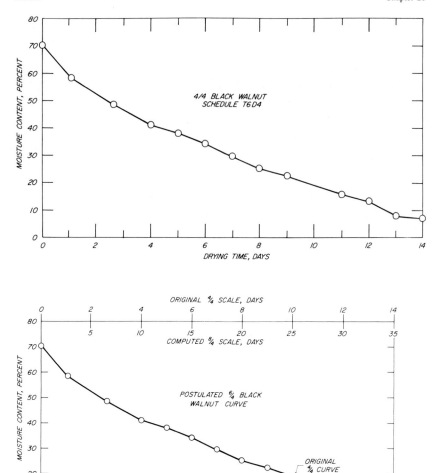

Figure 20-29.—(Top) Drying curve for 4/4 black walnut. (Bottom) Same curve with computed new time scales for 6/4 and 8/4 thicknesses. (Drawings after McMillen and Wengert 1978.)

Early in a kiln operator's career at a given location, he may find it helpful to develop basic drying curves. He should use his own sample data for green or nearly green stock of various species and thicknesses. For any subsequent charge of lumber dried from a lower initial moisture content, the drying curve of the first day or two would generally parallel the line for the first day or two of the green curve. After that, the drying of air-dried or partly air-dried stock will follow fairly closely the curve for the green stock.

Since it will not be practical to establish green stock drying curves for all thicknesses of lumber right away, approximate curves can be estimated. These can be set up using drying time factors for the most common thicknesses. One such set of factors was developed by N. C. Higgins (Michigan State University, East Lansing) to predict total drying time. Such factors have a theoretical base related to diffusion and other aspects of wood drying, but these were empirically developed after several years of experience in commercial drying of foreign and domestic woods. They are roughly corroborated by other commercial drying time data. The drying time factor for 8/4 stock was set at 1.00 because the best approximations for other thicknesses were obtained when the 8/4 drying time data were used as a base.

Until more precise factors become available, the ones developed by Higgins can be used in setting up the preliminary curves to estimate drying time from various moisture levels. These factors are as follows:

Thickness	Drying time factor
3/4	0.25
4/4	.40
5/4	.55
6/4	.70
8/4	1.00
10/4	1.35
12/4	1.75
14/4	2.25
16/4	2.85

The factors for 4/4 and 8/4 indicate that 2½ times as long will be required to dry 8/4 material as 4/4.

To employ these factors in developing a set of drying curves for a species, the kiln operator should use either his own 8/4 curve or a 4/4 curve obtained with a conservative schedule. To develop curves for other thicknesses from the available curve, list times and moisture contents from the available curve and compute times for these moisture contents for 8/4 lumber by applying the appropriate drying-time factor. From figure 20-29 (top), some of the times would be:

Thickness	Factor	Drying time			
		----------------------Days----------------------			
4/4	(given)	2	6	10	14
8/4	(*)	5	15	25	35
5/4	0.55	2.75	8.75	13.75	19.25
6/4	.70	3.50	10.50	17.50	24.50
10.4	1.35	6.75	20.25	33.75	47.25

*Time for 8/4 = 1.00 ÷ (factor for given thickness) x given time; in this example, 8/4 time = $\frac{1.00}{0.40}$ x given time = 2.5 x given time.

Times for other thicknesses can be computed in this way, or more simply, by multiplying times for 8/4 lumber by appropriate factors. Values for each thickness can then be plotted opposite corresponding moisture contents, and points for each thickness connected (fig. 20-29 bottom). As successive charges are dried and schedules improved, new lines can be drawn. A crucial point, however, is the fact that hardwoods of varying quality cannot be dried by time schedules. The kiln operator must revert to the moisture content schedule when he is in doubt about the drying characteristics of a particular charge.

A side benefit of such a finalized drying graph is that any subsequent charge of normal stock having a slower drying rate will indicate the kiln may not be operating properly. It may have a malfunctioning instrument, a water- or air-bound steam line, an inoperative fan, or some similar fault, and may need maintenance. A comparison of graphs over a period of time should also show any slow loss of efficiency in the kiln.

Readers interested in a more theoretical approach to predicting the effect of thickness on drying rate of 4/4 red oak sp., will find useful work by Tschernitz and Simpson (1979).

Equalizing and conditioning.—Frequently in kiln-drying, and especially with mixed species, sizes, or initial moisture levels, some lumber in a charge is as dry as required while other lumber is still too wet. An **equalizing treatment** reduces the spread between the wettest and driest boards, and prepares lumber for conditioning. Drying stresses and sets (often called **case-hardening**) are always present at the end of kiln-drying and equalizing. Any lumber that will be resawed, ripped, or machined nonuniformly should be **conditioned** to relieve stresses. Failure to do so will result in warping (cupping, crooking, or bowing) *during* or *immediately after* machining and will cause difficulty in boring.

The normal procedures for equalizing and drying stress relief are described in the last parts of chapters 8 and 10 of the Dry Kiln Operator's Manual. Figure 20-30 shows a method of cutting specimens for final moisture content and longitudinal as well as transverse stress tests. The kiln operator's attention is especially drawn to the method of evaluation of case-hardening in chapter 10 of the Manual. A final analysis for freedom from stress cannot be made until the test prongs have air-dried for 16 to 24 hours, but a significant turning out of the transverse test prongs immediately after they are cut often indicates that the transverse stresses have been relieved.

It was formerly considered that dense hardwood 4/4 lumber required 16 to 24 hours to condition. Conditioning has high energy demands, so the time should be no longer than needed to get stress relief. Several low-density imported woods have been succesfully conditioned by equalizing at 1 percent lower moisture content than the recommended procedure in the Manual, then conditioned 6 hours at the regular equilibrium moisture content (McMillen and Boone 1974). This short-cut has been successfully used on sugar maple sapwood in kiln-drying demonstrations, taking about 8 hours. Red oak 4/4 reportedly has been conditioned in 13 hours.

Drying

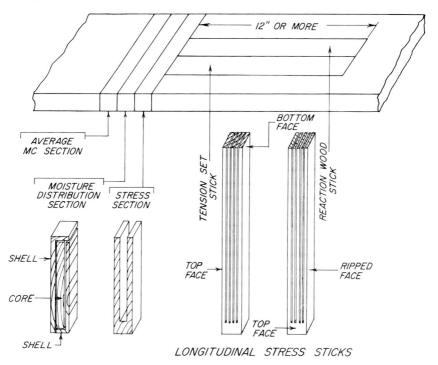

FINAL MOISTURE CONTENT AND STRESS TESTS

Figure 20-30.—Method of cutting specimens for final moisture content and drying stress tests. (Drawing after McMillen and Wengert 1978.)

If average moisture content determinations are made immediately after the conditioning treatment, the moisture content obtained will be about 1 to 1.5 percent above the desired value because of the surface moisture regain. After cooling, the average moisture content of the lumber should be close to that desired. The general rules of the conditioning procedure do not apply to moisture content values above 11 percent. Conditioning is hard to accomplish at the higher values.

The important factor in conditioning is to obtain the specified equilibrium moisture content. The simplest way is to use an equalizing dry-bulb temperature that is lower than the dry-bulb temperature for conditioning. At the end of equalizing, the heating coils are shut off. The wet-bulb controller is then set up for the conditioning wet-bulb temperature. As the steam spray raises the wet-bulb temperature, the dry-bulb temperature also will rise some. When the wet-bulb temperature reaches the desired level, the dry-bulb controller is set to its prescribed value. Only one heating coil should be turned on. Both the dry-bulb and the wet-bulb temperatures should reach and remain at the desired levels for conditioning.

The equalizing and conditioning temperatures shown in the basic kiln schedules (tables 20-11 to 20-20) are selected to utilize the above procedure and give stress-free wood at a final 7 percent moisture content. For other desired final moisture content levels, use temperatures to equalize at an equilibrium moisture content 3 percent below the desired moisture content and to condition at 4 percent above the desired moisture content. If equalizing is expected to be prolonged, use the reduced dry-bulb temperature only the last 12 to 18 hours of equalizing.

The conditioning time could be unduly lengthened if high-pressure steam is used for conditioning and no special measures were taken to reach the correct wet-bulb depression. A desuperheater should be used in the steam line to remove the excess heat. (See OPERATIONAL CONSIDERATIONS.)

Occasionally, the transverse prong test will show no stress, but the lumber will bow when resawed. The cause of the bowing is longitudinal stress resulting from either longitudinal tension set in the surface zones or longitudinal shrinkage differentials due to reaction wood (tension wood in hardwoods). These stresses are most likely to be unrelieved when conditioning temperature or equilibrium moisture content is too low, or when conditioning time is too short. The longitudinal stress sticks in figure 20-30 will show whether such stresses are present. If stresses are a problem, conditioning should be 180°F or higher. The lumber must have been equalized, and the recording instrument must be in calibration. If longitudinal stresses are still a problem, the wet-bulb setting can be raised 1°F over the recommended value. Also, the conditioning period can be extended about 4 hours per inch of thickness. If tension wood stresses are very severe, they may not yield to any conditioning treatment.

PROCEDURES FOR SPECIAL CLASSES OF PRODUCTS

Bacterially infected oak lumber.—Lumber from northern red oak and black oak infected with anaerobic heartwood bacteria is very susceptible to honeycombing and ring separation when dried by normal or accelerated oak schedules. Lumber in advanced stages of infection is especially subject to surface checking and honeycombing; such wood should be sawed into 4/4 rather than thicker lumber and dried by a special schedule (table 20-8).

Presurfaced 1-inch upland red and white oak.—The suitability of an accelerated kiln schedule for fully presurfaced upland red and white oak was confirmed by research at the U.S. Forest Products Laboratory (McMillen 1969; Wengert and Baltes 1971; McMillen and Baltes 1972). Drying time can be reduced 25 to 50 percent by the fully accelerated schedule and kiln capacity is increased 8 to 12 percent by the presurfacing. Slight additional benefits are obtainable by automation. The schedule is not applicable to presurfaced green cherrybark oak, a common southern oak, but can be applied to such stock carefully dried by other means to 40 percent moisture content. Successful field tests of the presurfaced oak schedule have been carried out with northern red and mixed Appalachian red oak in Massachusetts and West Virginia (Rice 1971; Cuppett and Craft 1972).

The accelerated schedule, modified for kiln sample operation (table 20-22), is for oak fully surfaced while green to remove all saw marks, not to material that is merely blanked or skip dressed to obtain uniform size. The thickness after complete surfacing probably should be 33/32-inch to insure kiln-dry dimension cuttings 13/16-inch thick. Rough sawing would have to be to a 1-5/32-inch thickness (see McMillen (1969) for further suggestions).

TABLE 20-22.—*Dry-kiln schedule for presurfaced 4/4 upland oak*[1] (McMillen and Wengert 1978)

Percent of moisture content for:		Dry-bulb temperature	Wet-bulb depression	Wet-bulb temperature
White oak[2]	Red oak[2]			
		---------°F---------		
BASIS—AVERAGE MOISTURE CONTENT, ALL SAMPLES				
Above 42	Above 53	115	4	111
42	53	115	5	110
37	43	115	8	107
33	37	115	14	101
BASIS—AVERAGE MOISTURE CONTENT, WETTEST HALF OF SAMPLES				
35	35	120	35	85
30	30	125	40	85
27	27	130	45	85
21	21	140	50	90
17	17	180	50	130

[1] Accelerations achieved by presurfacing depend partly on 400 ft/min air velocity through the load.

[2] When there is a mixture of red and white oak in the kiln, operate on the basis of the moisture content of the samples of the predominating species. If either the red oak or the white oak is not strictly green, operate on the basis of the samples of the species closest to the green moisture content.

Hickory handle stock.[8]**.**—The principal defects to be avoided by proper kiln schedules for hickory handle stock are pinking and end checking. **Pinking** is a pinkish to reddish-brown coloration of the sapwood caused by chemical changes in the sap normally present in the wood. It can be prevented, except for a slight discoloration in the cores of a small percentage of boards, by keeping the dry-bulb temperature down to 105°F or below until the stock has dried to an average moisture content of 20 percent. From 20 to 12 percent moisture content, any kiln temperature can be used.

In short stock, like handle blanks, much moisture leaves the blanks from end-grain surfaces. Mild initial and intermediate kiln conditions can largely prevent formation of **end checks** exceeding 0.5 inch and hold maximum end-check penetration to 1 inch. Checks of this depth are removed in normal end-trimming operations. If less end checking is desired, still milder kiln conditions should be used during the first and second steps, or else a good end coating should be used.

---

[8] Text under this heading is condensed from McMillen (1956[a]); readers interested in more detail should consult the source text.

Surface checking is not severe in narrow items if drying conditions are controlled to limit end checking. Warping occurs in long handle stock, but kiln conditions cannot be adjusted much to influence it. Stock that is prone to warp should be very carefully stickered at 12-inch intervals, and loaded with top weights.

Kiln schedules shown in table 20-23 were developed with 1½- by 2- by 24-inch handle blanks with uncoated ends, and apply to green blanks 2 by 2 inches in cross section and smaller. Under the nonpinking schedule pinking was either absent or restricted to a slight amount in the inner one-third of the piece. In applying this schedule, pieces of various cross sections and length should be segregated as much as possible. Stock that has been air-dried only a short time should be started on the second step of the schedule. Stock air-dried to 20 percent moisture content or less may be started on the next-to-last temperature step of either schedule using a 20°F wet-bulb depression.

When the driest sample comes to a moisture content equal to 1 percent below the desired average, the kiln conditions should be changed to an equilibrium moisture content the same as the moisture content of the driest sample. For instance, if the desired minimum and maximum values permitted are 7 and 10 percent, the kiln should be changed to 7 percent equilibrium moisture content when the driest sample reaches that moisture content; this condition should be held until the wettest sample is down to 10 percent moisture content.

In pilot tests of these schedules (table 20-23), small blanks 32 to 38 inches long, required 24 days on the nonpinking schedule and 9 days on the pinking-no-defect schedule. A charge previously air-dried to 17 percent moisture content required 7 days to complete drying to 8 percent on the nonpinking schedule.

TABLE 20-23.—*Kiln schedules for small hickory blanks*[1] (McMillen 1956a)

Moisture content (percent)	Dry-bulb temperature	Wet-bulb temperature	Wet-bulb depression
	---------- °F ----------		
	NON-PINKING SCHEDULE		
100 to 60	105	101	4
60 to 40	105	100	5
40 to 30	105	92	13
30 to 20	105	80	25
20 to 8	115	79	36
	PINKING-NO-DEFECT SCHEDULE		
100 to 60	130	127	3
60 to 40	130	126	4
40 to 30	130	120	10
30 to 20	150	121	29
20 to 8	180	135	45

[1] 2 by 2 inches in cross section and smaller.

Kiln schedules shown in table 20-24 were developed with 2½- by 3½- by 24-inch blanks with uncoated ends for use on large handle blanks measuring more than 2 by 2 inches in cross section. End-check penetration was limited to an average of 0.6 inch for fast-drying stock and 0.5 inch for slow-drying stock; maximum end-check penetration was 2.0 inches. The fast-drying stock (initial moisture content 52 percent) required 23 days and the slow-drying stock (initial moisture content 67 percent) 36 days to dry to 8 percent moisture content. Somewhat longer times would be required in commercial drying. Large handle blanks that have been partially or fully air-dried may be kiln-dried following recommendations given under small blanks, but a smaller wet-bulb depression should be used until the moisture content of the stock is known.

For additional discussion of drying squares and hickory handle stock, see figure 20-6 and section 20-2 AIR-DRYING, under the subsection SHORT LUMBER AND SQUARES.

Squares of other pine-site hardwoods.—For kiln-drying of 10/4, 12/4, and 16/4 squares of pine-site hardwoods other than hickory, see schedules in table 20-25 (as modified by table footnote 3).

Also, Rasmussen (1961, table 12) has suggested the following schedules for small squares of sweetgum sapwood:

Size	Schedule
1-inch square	T12-F6
2-inch square	T11-D5

Explanation of these schedule numbers and those of table 20-25 is given in table 20-26 and 20-27.

TABLE 20-24.—*Kiln schedules for large hickory blanks*[1] (McMillen 1956a)

Moisture content (percent)	Dry-bulb temperature	Wet-bulb temperature	Wet-bulb depression
	------°F------		
	NON-PINKING SCHEDULE		
Green to 40	105	101	4
40 to 35	105	100	5
35 to 30	105	95	10
30 to 25	105	86	19
25 to 20	105	80	25
20 to 8	115	79	36
	PINKING-NO-DEFECT SCHEDULE		
Green to 40	120	117	3
40 to 35	120	116	4
35 to 30	130	124	6
30 to 25	140	125	15
25 to 20	150	112	38
20 to 8	180	135	45

[1] More than 2 by 2 inches in cross section.

TABLE 20-25.—*Code number[1] index of kiln schedules suggested for drying thick lumber and squares cut from pine-site hardwoods* (Rasmussen 1961, table 13)[2,3]

Species	10/4	12/4	16/4
	---------------Schedule code number---------------		
Ash, white	T5-B3	T3-B2	T3-A1
Elm, American	T5-D2	T3-C2	—
Hackberry	T6-C3	T5-C2	T3-B1
Maple, red	T5-C2	T3-B2	—
Oak, red sp	T3-C1	T3-C1	—
Oak, white sp	T3-B1	T3-B1	—
Sweetgum (sapwood)	T11-D3	T9-C3	—
Sweetgum (heartwood)	T5-C2	T5-B2	—
Tupelo, black	T11-D3	T9-C2	T7-C2
Yellow-poplar	T9-C3	T7-C2	T5-C2

[1] See tables 20-26 and 20-27 for explanation of schedule code numbers.
[2] For most of these species, an effective end coating should be applied to the green lumber.
[3] For squares use a wet-bulb depression code number one unit higher than the one suggested for lumber; thus for 3- by 3-inch red maple use T3-B3.

Rounds.—Hickory blanks measuring less than 2 by 2 inches in cross section, if rough-turned and air-dried until core moisture contents are 15 percent or less, can be further kiln-dried at temperatures up to 130°F, without pinking (McMillen 1956a).

For additional discussion of drying furniture rounds, see section 20-4 HEATED LOW-TEMPERATURE DRYING, under the paragraph heading *Unstickered hardwood furniture rounds*.

Hardwood dimension parts.—Simpson and Schroeder (1980) examined the feasibility of drying only usable cuttings for furniture rather than long lumber cut from small low-grade bolts. In their experiment the level of mill rejects of furniture cuttings was too high to be tolerable. Crook in the long cuttings and cup in the wide cuttings were the major reasons for rejection. Other shorter and narrower cuttings had a much lower rejection rate—one that could probably be tolerated. Although some end checking occurred (there was some evidence that end coating would reduce this occurrence), end checking caused no rejections. Other drying defects such as honeycomb or surface checking were practically non-existent regardless of kiln schedule, suggesting that accelerated drying schedules may be feasible for interior frame parts of furniture. Regardless of kiln schedule used, successful kiln-drying of long or wide cuttings, or of both, from lumber cut from small low-grade bolts will require special measures to control warp.

Simpson (1982) described a restraining device consisting of hydraulic cylinders and steel cables to encircle a load of short lumber at three places; in his tests on 5/4 mixed oak and hickory at different levels of restraining force he found that cup and bow were not reduced, but twist was reduced by 25 to 50 percent, and crook was reduced up to 35 percent.

Drying

TABLE 20-26.—*General temperature schedules for hardwoods related to codes T1 through T14 in the Dry Kiln Operator's Manual* (Rasmussen 1961, p. 119)

Temperature step No.	Moisture content at start of step	Dry-bulb temperatures for temperature schedule No.													
		T1	T2	T3	T4	T5	T6	T7	T8	T9	T10	T11	T12	T13	T14
	Percent	°F
1	Above 30	100	100	110	110	120	120	130	130	140	140	150	160	170	180
2	30	105	110	120	120	130	130	140	140	150	150	160	170	180	190
3	25	105	120	130	130	140	140	150	150	160	160	160	170	180	190
4	20	115	130	140	140	150	150	160	160	160	170	170	180	190	200
5	15	120	150	160	180	160	180	160	180	160	180	180	180	190	200

TABLE 20-27.—*General wet-bulb depression schedules for hardwoods related to wet-bulb depression codes A1 through F8 in the Dry Kiln Operator's Manual* (Rasmussen 1961, p. 119)

Wet-bulb depression step No.	Moisture content at start of step for moisture content class						Wet-bulb depressions for wet-bulb depression schedule No.							
	A	B	C	D	E	F	1	2	3	4	5	6	7	8
	Percent						°F							
1	Above 30	Above 35	Above 40	Above 50	Above 60	Above 70	3	4	5	7	10	15	20	25
2	30	35	40	50	60	70	4	5	7	10	14	20	30	35
3	25	30	35	40	50	60	6	8	11	15	20	30	40	50
4	20	25	30	35	40	50	10	14	19	25	35	50	50	50
5	15	20	25	30	35	40	25	30	35	40	50	50	50	50
6	10	15	20	25	30	35	50	50	50	50	50	50	50	50

Thin northern red oak lumber.—Considerable oak is used in ½-inch thickness. For such thin stock, drying time is substantially less than for 4/4 lumber; moreover, if kiln-dried 4/4 lumber is resawn to yield ½-inch stock, the freshly sawn surfaces may check and pieces may warp. Kimball (1955) experimentally determined a schedule (table 20-28) that successfully dried ½-inch-thick northern red oak from a Wisconsin source in 36 hours.

Thick lumber from pine-site hardwoods.—The Dry Kiln Operator's Manual lists kiln schedules appropriate for thick lumber of 22 hardwood species or species groups; those for species found on southern pine sites are indicated in table 20-25.

Crossties.—Shunk (1976), in his survey of industry drying practices, found that crossties are seldom kiln-dried due to the long schedules required. Huffman (1958) concluded that if ties of southern hardwood are to be kiln-dried, the kiln should be heated to about 230°F; results of Huffman's experiments are summarized in section 20-7.

Data on air-drying of crossties are summarized in connection with figures 20-7 through 20-13; those on forced-air predrying are discussed in connection with figure 20-20.

TABLE 20-28.—*Dry-kiln schedule for ½-inch-thick northern red oak*[1] (Kimball 1955)

Moisture content (percent)	Dry-bulb temperature	Wet-bulb depression	Drying time
	------°F------		Hours
Green to E = 0.7	170	165	7
E = 0.7 to 20	170	128	17
20 to 7	200	150	8
Equalization	170	141	4

[1]Data based on Wisconsin source of oak.
[2]E = (Moisture content, percent − EMC, percent)/(moisture content, original − EMC, percent). For example, if the original moisture content was 80 percent and the kiln maintained at 18 percent EMC, E would equal 0.7 when the moisture content dropped to 61 percent.

OPERATIONAL CONSIDERATIONS

Considerations for operating a dry kiln and minimizing energy use are discussed in chapter 10 of the Dry Kiln Operator's Manual (Rasmussen 1961).

Following are additional suggestions on heating, venting, and air circulation equipment, on humidity control with high-pressure steam, and part-time drying.

Heating, venting, and air circulation.—Many older kilns designed for drying air-dried hardwoods have inadequate heating, humidification, and circulation systems and insufficient venting for the rapid drying of green stock of faster drying species.

Heating capacity is short if a kiln takes more than 4 hours to reach the desired elevated temperature. If circulation is adequate and other equipment is functioning properly, the kiln manufacturer can add more heating pipe or ducts.

A kiln with inadequate humidification will be unable to condition lumber adequately—the steam or water spray will be running all the time but a 10°F depression cannot be obtained. If there is little or no superheat, if the vents are closed tightly, if there are no leaks in the structure, and if the spray line and holes or nozzles are not plugged, a larger steam supply is needed. Increasing only the pressure will increase the superheat and may not solve the problem.

Adequate venting capacity is important. Failure to exhaust moist air rapidly causes higher wet-bulb temperature than the schedule requires. The result is slow drying. For some species, like red maple, the risk of stain and discoloration is increased. Restricted use of steam spray during kiln warm-up is very helpful, but vents are inadequate if desired depressions cannot be easily obtained a few hours after they are set on the instrument. Venting can be increased by increasing vent size or by power venting.

Power venting is the discharging of humid air from or the injection of air into a kiln by fans or blowers and suitable duct work installed by the kiln manufacturer. One aim is to increase the venting rate of a kiln when the regular vents cannot conveniently be enlarged. Another aim is to conserve energy. To conserve energy, place the power vents so that the humid air is discharged before it passes over the heating coils.

A poor air circulating system is evidenced by a wide range of final moisture content values in the charge and uneven drying. If air velocity measuring instruments are available, poor circulation will be indicated by velocities below 200 feet per minute (fpm) or by differences between the highest and lowest velocities greater than 150 fpm (measured on the leaving-air side of the load). The recommended kiln schedules are based on a velocity through the load of 200 to 400 fpm and probably are as satisfactory with velocities up to 450 fpm. Any improvement in circulation will improve heating and venting, however. Increasing the velocities in the early stages of drying to 500 fpm or so will accelerate drying, but will increase the risk of surface checking with check-prone species like oak or beech. This can be compensated for by decreasing wet-bulb depressions 1°F. Thorough baffling to prevent "short circuiting" of air will greatly improve circulation. Changes in fan speed and design usually require advice of a kiln engineer.

Humidity control with high-pressure steam.—Some hardwood dry kilns are heated with high-pressure steam (80 to 100 pounds per square inch (psi) or higher). High-pressure steam is excellent for economy in heating coils because of lower initial cost, less fin radiation area required, and smaller steam feed lines and heat control valves than a low-pressure kiln operating at 10 psi. However, high-pressure steam is too hot for use in the humidity spray system. It raises dry-bulb temperature excessively when the wet-bulb temperature is raised, making it impossible to carry a 4° or 5°F wet-bulb depression during the initial stages of some schedules, or to get a 10°F depression for conditioning and stress relief at the end of a charge. If the steam pressure is not too high, modifying the conditioning procedure, as described in a preceding section, is helpful. For steam pressures of 100 psi and over, however, a desuperheater is needed.

A desuperheater unit supplies low-pressure wet steam, cooled to the saturation point, for the humidity spray system. A pressure-reducing valve provides hot steam at approximately 5 psi, which is then cooled with a water spray.

Part-time operation.—In past years some mills and factories which generate their own electricity with steam developed by burning wood waste used the exhaust steam to operate the dry kilns. Since steam was available only during working hours, the kilns were on a part-time basis, 8 or 9 hours per day, 5 days per week. Other firms generating kiln steam from wood waste used coal, oil, or gas during nonworking hours. Recent costs of fossil fuel have caused such firms to consider part-time operation for periods of slack product demand. Such operation could be considered more widely if a greater proportion of the kiln input were air-dried instead of green.

Part-time drying is technically feasible on air-dried stock and can be successful with green or partly air-dried stock with proper schedules and operating procedures. Rasmussen (1961) recommended full-time drying during the first stages of drying refractory hardwoods when excessive checking occurs during part-time operation.

Part-time drying takes almost twice as long to dry an item as does full-time operation. Under the experimental conditions used, Rasmussen and Avanzado (1961) found that full-time operation required slightly less total energy (4,510 British thermal units per pound of water (Btu/lb H_2O) evaporated) than part-time operation, either with fans running all the time (5,490 Btu/lb H_2O) or only during heat and spray control periods (5,560 Btu/lb H_2O). The material dried was rough 4/4 red oak lumber. Drying quality was the same by all methods. Drying times, from 70 to 7 percent moisture content, were 18 days for full-time, 35½ for part-time with fans continually running, and 46½ days for part-time with restricted fan use.

In their report, the authors concluded that a more severe kiln schedule for part-time operation would not be expected to reduce drying time greatly.

Wolfe (1962, 1963) used temperatures during the "operating" hours as much as 15°F above established kiln schedule temperature. He also used the fans and some venting during "off" hours. He concluded part-time operation was economical under regular production circumstances.

The technology of part-time drying has not been well enough established to include schedules and drying time in this publication.

DRYING TIME

The time required to kiln-dry a given species and thickness depends upon the character of the wood, the type of kiln, and the kiln schedule used. The time estimates given in table 20-29 are generally minimum times that can be obtained in well maintained commercial kilns with relatively short air travel. They also are based on the assumption that the operator will take some steps to increase drying rate. The times will vary a half to a full day from kiln to kiln. Kilns with longer air travel, with less than 200 fpm air velocity, or in a poor state of maintenance will take longer in drying green stock. They may not take much longer than the table values for air-dried stock.

By drying to 6 percent average moisture content or slightly lower, moisture content will finalize at about 7 percent after equalizing and conditioning. If a greatly different moisture content is desired, proper adjustments in conditions must be made, and total times will differ from the table values. The time estimates in table 20-29 are for the drying of stock for high-quality uses. If considerably more severe conditions than the basic schedules are used, time will be shortened, but quality may be decreased.

TABLE 20-29.—*Approximate kiln-drying times[1] for 4/4 hardwood lumber in conventional internal-fan kilns* (McMillen and Wengert 1978)

Species	Time required to kiln-dry from	
	Air-dried[2] condition	Green condition
	----------------*Days*----------------	
Ash, green and white	4	10
Elm, American	4	9
Elm, winged	5	13
Hackberry	4	7
Hickory	4	10
Maple, red	4	7
Oak, red sp. (northern or upland)	5	21
Oak, white sp. (northern or upland)	5	23
Oak, red or white sp. (southern lowland)	6	[3]
Sweetbay	4	8
Sweetgum (heartwood)	6	15
Sweetgum (sapwood)	4	10
Tupelo, black	4	8
Yellow-poplar	3	6

[1] Approximate times to dry to 6 percent moisture content, prior to equalizing and conditioning, in kilns having air velocities through the load of 200 to 450 feet per minute.
[2] 20 percent moisture content for most woods; 25 percent for slow-drying woods like oak and hickory.
[3] These items should be air-dried before kiln-drying.

The drying times in the table are for precisely sawed rough green material 1 1/8 inches thick, plus or minus 1/8-inch. Miscut lumber with some of greater thickness will, of course, take longer to dry. For estimates of the effect of other thicknesses on drying time, see the paragraph *Predicting drying time* under the subsection DRYING PROCEDURES.

SPECIAL PREDRYING TREATMENTS

Treatments used before or early in kiln-drying to accelerate drying rate include steaming, precompression, and prefreezing.

Steaming to accelerate drying.—Early in the study of wood drying at the U.S. Forest Products Laboratory, observations were made that steaming hardwoods before drying sometimes reduced drying time. On air-dried stock, however, surface checks deepened and widened and sometimes became bottleneck or honeycomb checks. The steaming used at that time was relatively long. The conclusion was drawn that such steaming represented a delay during which no drying occurred and general use of the practice was stopped.

More recent studies have shown benefits to both softwoods and hardwoods from short-period steaming. Moisture migration rates are increased significantly, and drying times are reduced. Prefabricated aluminum kilns have made possible the use of steaming in commercial drying operations.

Simpson (1975) accelerated the drying of small specimens of several species of wood by steaming them at 212°F. The drying rates at 50-percent moisture content for these small specimens increased 34 to 75 percent for northern red oak and cherrybark oak, and 11 to 36 percent for sweetgum heartwood. Steaming time ranged from 0.5 to 5 hours; for sweetgum heartwood, the 5-hour period was best. The drying rate of sweetgum sapwood was slightly reduced by steaming.

In another study with 1-inch-thick oak, Simpson (1976a) found that the moisture gradients, during drying after steaming 4 hours at 212°F, were smooth curves, contrasting with inflected gradients of unsteamed controls. He concluded that free water migration from the center toward the surface was enhanced by steaming.

The above results were achieved with saturated steam at 212°F, a condition difficult to obtain in comercial kilns. In a larger scale study Simpson (1976b) used both green and partly air-dried rough 4/4 northern red oak pretreated with nearly saturated steam at 185°F for 4 hours. Drying time was reduced about 17 percent for both classes of lumber. No defects occurred in the green lumber nor in one batch of the partly air-dried material. The other partly air-dried batch, however, had been severely surface checked during air-drying. The steaming appeared to deepen unobserved surface checks and change them into bottle neck or honeycomb checks. This confirms the admonition not to use steam spray during warmup of a kiln charge of fully air-dried oak.

Simpson's exploratory studies give some prospect for practically accelerating the drying of eastern hardwoods by presteaming treatments. Additional studies are needed to confirm benefits and determine limitations, and to establish commercial operating procedures, comparative energy demands, and economics.

Precompression.—Cech (1971) established that dynamic transverse compression of 2-inch yellow birch lumber, before drying by a severe schedule, significantly improved drying behavior. The drying was carried out at 215°F. Momentary thickness compression of 7 to 8.5 percent in a roller device greatly reduced collapse and honeycombing compared with uncompressed material dried by conventional schedules. Drying time was only 8⅓ days compared with a customary 18 days for noncompressed material dried by conventional schedules. Accompanying increases in permeability suggest that precompression may develop small failures in cell walls or pits that permit faster drying.

Cech and Pfaff (1975) also found that a 7.5-percent compression of 4/4 red oak reduced drying time 5 percent with both conventional and mild accelerated kiln schedules.

Precompression has not been accepted commercially on any large scale, but experimental results to date appear to merit more study.

Prefreezing.—Freezing of green wood, followed by thawing before drying, also increases drying rate and decreases shrinkage and seasoning defects in some species. Most research on hardwoods has been with black walnut (Cooper et al. 1970). Favorable results have been attained with black cherry, American elm, and white oak (Cooper and Barham 1972). It also decreased shrinkage in black tupelo.

The best prefreezing temperature for black walnut is $-100°F$, but significant improvement is found at $-10°F$, a temperature readily attained in commercial freezing equipment. A freezing time of 24 hours is adequate for thickness up to 3 inches. A similar length of time has been used for thawing.

Although prefreezing would require substantial additional investment, it could shorten drying time for walnut gunstock blanks to half that required by some of the industry (Cooper et al. 1976). Of 600 prefrozen gunstock blanks dried from green in 103 days by a slightly accelerated kiln schedule, only 2.66 percent were defective from collapse, checking, or warp, while 6.50 percent of the 400 unfrozen blanks in the pilot test had such defects.

All requirements for prefreezing treatments are not yet known, nor is the reason for observed improvement in drying. Commercial application of the technique requires fuller knowledge.

HEATED-ROOM DRYING

Conventional kiln-drying circulates air rapidly over the wood to be dried and controls both dry-bulb and wet-bulb temperatures. In **heated-room drying,** which in a strict sense cannot be termed kiln-drying, air is not force-circulated, and the wet-bulb temperature is not controlled. Instead, a small amount of heat is used to lower the relative humidity, thereby lowering the equilibrium moisture content. This method is suitable only for wood that has been air-dried first; green lumber may check and split if so dried. The method does not dry lumber rapidly, but it is suitable for small amounts of lumber.

Before air-drying is started, the lumber should be cut as close as possible to the size it will have in the product, with allowance for some shrinkage and warping and for a small amount to be removed during planing and machining. If long air-dried pieces are shortened before heated-room drying is started, the fresh cuts should be end coated to prevent checks, splits, and honeycomb.

For reasonably fast heated-room drying, equilibrium moisture content should be about 2 percent below the moisture content of use. The wood is left in the room just long enough to come to the desired average moisture content. It is then removed and stored in a solid pile, preferably in an area with the same equilibrium moisture content as the area in which the wood will be used.

The amount that the temperature must be raised above the average outdoor temperature depends upon the average outdoor relative humidity. Typical values are given in table 20-30. Do not attempt to use more heating with this method.

Any ordinary room or shed can be used and any ordinary means of heating the room should be satisfactory. A slight amount of air circulation is desirable to achieve temperature uniformity. If the material is relatively small-sized, it can be piled in small, stickered piles on a strong floor. It also could be sticker-piled on carts that can be pushed in and out of the room. Long lumber should be box piled on strong, raised supports (fig. 20-31).

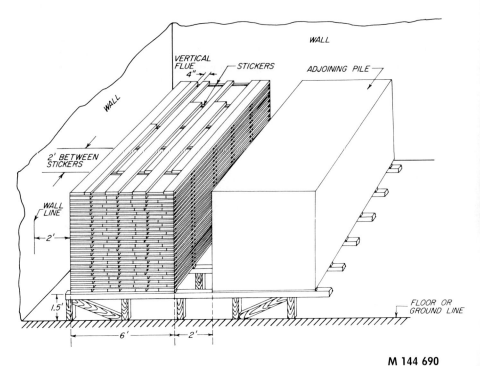

Figure 20-31.—Lumber piled for drying in a heated room. (Drawing after McMillen and Wengert 1978.)

TABLE 20-30.—*Amount temperature must be raised (at three relative humidities) above average outdoor temperature for various equilibrium moisture content values in heated-room drying* (McMillen and Wengert 1978)

EMC value desired (percent)	Degrees above average[1] outdoor temperature at:			Degrees above the morning's outside low temperature[2]
	70 percent relative humidity	75 percent relative humidity	80 percent relative humidity	
	----------------------------------°F----------------------------------			
4	38	40	42	42
5	31	33	35	33
6	23	25	27	27
7	18	20	23	22
8	13	15	17	18
9	10	12	14	15
10	6	8	10	—

[1]Daytime and nighttime.
[2]Assumes outside relative humidity of near 100 percent.

The lumber should be marked or records kept as to when it entered the room and when to expect to remove it. Any amount of air-dry lumber can be put in or dried lumber removed without upsetting conditions. The doors should be kept closed as much as possible to help maintain desired temperature. Variations of the average outdoor temperature from night to day and small variations from day to day do not affect the process.

ECONOMICS OF KILN-DRYING

The cost of drying hardwoods includes the cost of stacking, handling, air-drying, and kiln-drying. Degrade cost should be assigned to each of the drying phases, but often these costs are overlooked.

As wood drying involves the evaporation of large quantities of water, energy can be a very important element of total cost. Traditionally, the lowest cost and lowest energy usage in hardwood lumber drying has been achieved with air-drying followed by kiln-drying. Accelerated air-drying is an attractive alternative to air-drying when time to dry is short, the good air-drying season is short, good land for an air-drying site is not available, or when minimal inventories are necessary.

Estimates of the cost of drying hardwood lumber are meaningful only if the year of computation is specified, because energy costs and rates for interest, and inflation are fluctuating constantly. McMillen and Wengert (1978) provided, from a review of the literature, the following cost data estimated for the year 1976:

Procedure	Cost per thousand board feet
	Dollars
Handling	18.45
Air drying	
For 3 months	6.40
For 6 months	10.85
Low-temperature drying	
For 10 days	16.30
For 20 days	19.80
Kiln-drying	
For 5 to 7 days	23.80
For 10 to 15 days	35.00

Wengert and White (1979) estimated the cost of dehumidification drying of 4/4 oak (for 13 days) from air-dry to 5 or 6 percent moisture content at $17.64 per thousand board feet, plus $18.45 handling cost. These costs, and those tabulated above, do not include general company selling and administrative expenses, nor do they include costs of degrade.

For 4/4 and 5/4 Appalachian hardwood lumber, Cuppett and Craft (1975) estimated minimum drying degrade—from air-drying followed by kiln-drying—at 0.5 to 2.0 percent of the value of the lumber when air-dry; for 6/4 and 8/4 he estimated minimum degrade at 1 to 3 percent. If not controlled meticulously, by the procedures described by Rietz and Page (1971, ch. 6) for air-drying, and Rasmussen (1961, ch. 9) for kiln-drying, degrade costs can be much higher.

Readers interested in computational methods and computer programs for establishing drying costs will find the following references useful:

Cuppett (1965, 1966)
Goulet and Ouimet (1970)
Catterick (1970)
Engalichev and Eddy (1970)
Cuppett and Craft (1971)
Hanks and Peirsol (1975)
McMillen and Wengert (1978, ch. 10)
Wengert and White (1979)

20-7 HIGH-TEMPERATURE DRYING

LUMBER

High-temperature kiln-drying of wood has been explored for many years. It is similar to conventional kiln-drying, but operates at temperatures of 212°F or higher. Technically, high-temperature drying includes two processes. In **superheated steam drying,** the wet-bulb temperature is maintained at 212°F or above and air is excluded[9]. In the other, a mixture of air and steam is used, and the wet-bulb temperature is below 212°F, often with no precise control. Current use of high-temperature drying in the United States involves only the latter process, hereafter simply termed **high-temperature drying.**

In the last 20 years, high-temperature drying has become acceptable to accelerate the drying of many softwoods, and most new installations for southern pine lumber are high-temperature kilns.

High-temperature drying technology for hardwoods has not advanced sufficiently to permit recommendations, but the prospects appear good enough to justify considering equipment suitable for ultimate use of the process in new kilns. Drying rates two to five times those of conventional methods make high-temperature drying very attractive. With increased drying speed, smaller kilns could handle the same annual volume of lumber. Smaller kilns would permit more flexibility in drying (less mixing of species, faster loading and unloading) and require less space. Due to faster turnover, inventory could be reduced, saving interest, insurance, and taxes. Last, but not least, high-temperature drying would require 25 to 60 percent less energy than conventional kiln-drying. Added capital investment for extra insulation, kiln tightness, larger capacity heating system, and fans would decrease cost reductions, but it has been estimated cost would decrease by 20 percent or more (Wengert 1972).

High-temperature drying has been commercially used in Europe for air-dried hardwood stock. Research in Eastern Canada and the United States has indicated that the process would be technically feasible for air-dried hardwoods of many U.S. species. It is technically applicable to green hardwoods of very permeable species.

---

[9]For an experiment in which 27-mm-thick yellow-poplar lumber was dried from 110 to 5 percent moisture content in 30 hours using temperatures to 127°C and pressures to 1.27 atmospheres, see: Rosen, H.N. Drying of lumber in superheated steam. Submitted to American Institute of Chemical Engineers. 1980.

Most of the 20 species of hardwoods listed in a review of high-temperature drying (Wengert 1972) developed more drying defects under high-temperature drying than at conventional temperatures. The principal defects were collapse, end checking, and honeycombing. If dried from the air-dried condition, such defects were absent or almost so. Another defect that is common to high-temperature drying is discoloration, usually a toast-brown color, very thin in some species, thicker in others. While these defects may largely preclude high-temperature drying for highest quality hardwood uses, the process should be applicable for hardwoods suitable for less exacting uses.

Basic concepts.—Six concepts seem central to understanding the process of high-temperature drying of hardwood lumber.

- In theory the high-temperature kiln-drying process has three distinct stages when used on green wood: (1) the constant drying rate; (2) the first falling rate; and (3) the second falling rate (Lowery et al. 1968). In the first stage, moisture moves toward the wood surface predominately by mass flow. Rapid evaporation of moisture at or near the surface keeps the wood temperature at the wet-bulb temperature. Only the sapwoods of low- to medium-density hardwoods are likely to be permeable enough to sustain the first stage very long. When the wood can no longer supply free water to the entire surface area (usually quite soon after drying has started), the first falling rate starts. It continues until the free water is gone. The temperature of the wood is between the wet-bulb and the dry-bulb temperatures of the kiln. The length of both the constant and the first falling rate periods are governed by the permeability of the wood, while the final stage is more influenced by the diffusivity (as it is in normal drying). When the free water is gone, the second falling rate starts. The drying slows considerably more, and the wood temperature rises approximately to the dry-bulb temperature.
- Both the rate of moisture movement and the rate of evaporation are greatly increased by the high temperatures. During the first two stages (and perhaps longer), the rate of drying is controlled almost solely by the rate of heat transferred to the wood. Very high air velocities are needed to maintain this high rate of heat transfer as well as to rapidly remove the evaporated water from the surface.
- No matter what the relative humidity is in the high-temperature kiln, the kiln conditions are severe (see table 8-7; equilibrium moisture content cannot exceed 7 percent at 230°F). As a result, a high-temperature kiln, after being heated to 212°F at the start, can be operated without undue concern for the wet-bulb temperature until the conditioning. Unless the wood being dried is very wet and very permeable, venting is unnecessary. This is one of the major energy-saving aspects of the process.
- After the first stage of drying, steep moisture gradients can develop; these gradients can, in some cases, cause severe drying stresses and checking. Stresses in unchecked high-temperature-dried wood can be relieved easily, but stress relief must be done below 212°F and with properly equalized material, as is normally required.

- The equilibrium moisture content of high-temperature-dried wood is somewhat lower than that of conventionally dried wood.
- The exposure of wet wood to high temperature causes both temporary and permanent reductions in strength of the wood. The shortness of the drying time achieved by the use of high-air velocities in present-day kilns is an alleviating factor and makes obsolete results of some older studies on strength. This area needs further evaluation; few data pertaining to hardwood lumber have been published.

Research and practice to date.—The first research on high-temperature drying of hardwoods in the United States, by Tiemann (1918), was done with superheated steam. Basswood and sweetgum sapwood were dried successfully, but other hardwoods were not. Commercial superheated steam kilns were used for softwoods on the West Coast about 1918, but they deteriorated so badly under the severe drying conditions that their use was discontinued. Subsequent research by Richards (1958) and John L. Hill[10] indicated good results when drying a variety of air-dried southern hardwoods but poor results when drying the same hardwoods from the green condition. Their drying equipment had no humidity supply. These results and the likelihood of discoloration, strength losses, and other degrading factors discouraged further research on high-temperature drying of hardwoods in the United States at that time.

Research continued elsewhere, confirming the general suitability of the method for air-dried hardwoods and bringing out other favorable aspects of the process. Noteworthy reviews of this research were prepared by Kollmann (1961); Lowery et al. (1968); Wengert (1972); Cech (1973); and Cassens (1979). Kollmann (1961) reported use of high-temperature dryers in many German woodworking plants. These generally used air-steam mixtures rather than superheated steam. Rosen[11], using a pressure steam dryer operated at temperatures of 260°F and pressures to 1.27 atmospheres, dried 4/4 yellow-poplar from 110 to 5 percent moisture content in 30 hours; the lumber had minimal defects and no drop in grade, but was darker than conventionally dried lumber.

The commercial use of high-temperature drying for softwoods began slowly in the United States (Lowery and Kimball 1966; Kimball and Lowery 1967). Research by Koch (1964; 1969; 1972, p. 988-1002; 1973) provided significant impetus for rapid industrial acceptance by southern pine producers. Availability of tight prefabricated kilns, and the mass manufacture of 2- by 4-inch studs for houses, further accelerated adoption of high-temperature kilns for permeable softwoods such as southern pine. During this development, very rapid heating of the kiln to operating temperature was found necessary to save time and minimize degrade. Meanwhile, research by various investigators as reviewed by Lowery et al. (1968), indicated some hardwoods could be high-temperature-dried from the green condition if heating up was rapid under saturated steam conditions.

---

[10]Presented at Forest Products Research Society Annual Meeting, Madison, Wis., June 1958.
[11]Rosen, H.N. 1980. Drying lumber above atmospheric pressure—development of a prototype dryer. Paper presented at IUFRO Div. 5 Conf., Oxford, England. April. 13 p.

A commercial trial of high-temperature drying of hardwoods confirmed the applicability of the process to air-dried sapwood of sweetgum in which the redgum (heartwood) portions are below 25 percent moisture content[12]. The Eastern Forest Products Laboratory of Canada has done considerable research on high-temperature drying of birch and maple over the years; Cech (1973) reported a most favorable procedure was to force-air-dry the stock to 20 percent moisture content then kiln-dry at 212°F. Similar results were obtained with red oak when the high-temperature portion was kept at 200°F. Some of the research on birch, however, showed possibilities for high-temperature drying after very brief air-drying or pre-drying periods.

To determine safe starting moisture contents for high-temperature drying a wide variety of U.S. hardwoods, Wengert (1974b) heated the kiln very rapidly to 200°F. To do this, the steam spray was used continuously and the heating coils intermittently. Then the temperature was increased almost immediately to 230°F. Somewhat surprisingly, most of the wood tested tolerated high-temperature drying satisfactorily from the green condition. Red and white oak and sweetgum heartwood required air-drying to 25 percent moisture content, or conventional kiln-drying to 20 percent moisture content, before high temperature could be used without causing honeycombing and collapse.

Mackay (1974) designed a high-temperature-drying procedure for mixed aspen and balsam poplar studs involving a mid-process conditioning period. After 4 days total time the studs were dry enough in the outer zones to have 19 percent moisture content or less by ordinary moisture meter techniques. Follow-up research, however, showed that wet spots still present continued to dry, producing delayed shrinkage and collapse (MacKay 1976).

To solve this problem of delayed shrinkage, several industries drying aspen in the Lake States have high-temperature-dried for 2 days and then gone to an equalization setting at lower temperatures (160°F dry-bulb, 140°F wet-bulb temperature) until the required moisture contents are obtained. This appears to be a possible procedure for hardwoods that dry readily at first but have persistent wet spots.

Its success in experimental tests suggests that high-temperature drying of hardwoods has great potential for commercial use in the United States. Data under semi-commercial conditions are needed, both to determine maximum safe initial moisture content conditions to avoid degrade and to determine procedures acceptable for commercial operation. Also, some method of determining the moisture content during drying must be developed; without such a method, some overdrying or underdrying is likely. There also is need for research on color, strength, and other aspects of quality for the various uses to which hardwoods can be put.

---

[12]Presentation by P. Deverick, Fairchild Chair Co., Lenoir, N.C., to Southern Dry Kiln Club, Clyde, N.C.; November 1972.

Drying times and species potentialities.—Table 20-31 lists 10 hardwoods species, or species groups, found on southern pine sites for which information on high-temperature drying is available. Included in the tabulation are some very rough estimates of drying time for 4/4 lumber to 8-percent moisture content with minimum equalizing and conditioning.

Because of its ease of drying and its ready availability in diameters appropriate for small-log mills, yellow-poplar has received particular study for conversion into 2- by 4-inch studs for house walls. In the method proposed by Maeglin (1978) and Hallock and Bulgrin (1977), yellow-poplar logs are live-sawn through and through into 7/4 flitches. The flitches are rough edged to make a compact kiln load, dried to about 11 percent moisture content, and then ripped into studs. In a study to determine the most economic kilning procedure, Boone and Maeglin (1980) successfully dried such 7/4 flitches in 28 hours at a dry-bulb temperature of 235°F (wet-bulb temperature of 190°F) followed by an equalizing period of 48 hours at 200°F with wet-bulb temperature of 188°F. Load width in these experiments was limited to 5 to 6 feet; stickers were ¾-inch thick and air was circulated at about 1,000 feet per minute. At conclusion of the 76-hour schedule, average flitch moisture content was 11 percent with range from 6.4 to 19.4 percent. Boone and Maeglin estimated a 20- to 25-percent increase in kiln time when using an industrial kiln.

TABLE 20-31.–*Principal hardwood species found on southern pine sites for which high-temperature kiln-drying potential has been indicated, and estimated drying time for 1-inch-thick lumber*[1] (Adapted from McMillen and Wengert 1978)

Species or species group	Drying time from:		Reference
	Green condition	Air-dried condition[2]	
	----------*Hours*----------		
Elm, American	72	30	Wengert (1972)
Hackberry	72	30	Wengert (1974b)
Hickory, sp	—	36	Wengert (1972)
Maple, red	66	24	Wengert (1974b)
Oak, northern red	—	42	Cech (1973), Wengert (1972, 1974b)
Oak, white	—	42	Wengert (1972, 1974b)
Sweetgum (sapwood)	66	24	Wengert (1972, 1974b)
Sweetgum (heartwood)	—	36	Wengert (1972, 1974b)
Tupelo, black	66	24	Wengert (1972)
Yellow-poplar	66	24	Wengert (1974b)

[1] Estimated time for drying to 8-percent moisture content, including 6 hours for cooling, equalizing, and conditioning.
[2] 22 percent moisture content for oaks; 20 percent moisture content or lower for other woods.

Apparent process requirements.—If high-temperature drying of hardwoods is to be applied commercially, the process requirements are somewhat different from those of softwoods. When planning to install a high-temperature kiln for softwoods now and for hardwoods later on or when installing a hardwood kiln for conventional drying now which eventually may be converted to high-temperature drying, these requirements should be kept in mind, as follows:

- Target dry-bulb temperature should be very rapidly attained. The usual target is 230°F although higher temperatures have been tried.
- The initial heating medium would depend on the material being dried:
    1. Green—steam spray and heated air.
    2. Partly air-dried—undetermined at present.
    3. Air-dried—heated air.

    On green stock dry-bulb and wet-bulb temperatures should be brought up as close together as possible, with the wet-bulb temperatures raised to 200°F. To attain the high wet-bulb temperatures needed, vents, wall panels, and doors must seal very well.
- Air temperature should be uniform when at high temperatures, with no area below boiling, especially on the leaving-air side of the load.
- Air velocity should be high through the load—a minimum of approximately 600 feet per minute with 4-foot-wide loads and of 900 feet per mintues for wider loads. Research still in progress is using velocities up to 2,000 feet per minute. Narrow loads will yield more uniform moisture contents than wide loads.
- Equalizing and conditioning, when necessary, must be done below 212°F. In the high-temperature drying of 7/4 yellow-poplar flitches, 28 hours of drying was followed by 48 hours of equalizing at 200°F with wet-bulb temperature of 188°F. (See sec. 20-6 for general equalizing and conditioning procedures.)

In view of these process requirements, the kiln must be operable at both normal and high temperatures. The high heat and humidity shorten average life expectancy of masonary structures. Such structures would have to be well insulated, and the interior walls and ceiling sealed. Cracks would have to be repaired immediately. Prefabricated kilns available with insulated aluminum-paneled walls and roofs are expected to have long service lives. Corners, joints, floor stills, and door frames must be well designed and properly installed to prevent leakage and undue heat loss. Tight, well-insulated roofs are essential. Some thought should be given to a lightweight aggregate to provide more thermally efficient concrete floors.

Heating capacity must be sufficient to attain target temperatures rapidly. Since the kiln will require steam for humidification, heat for major operations should be provided by fin-type steam coils. In experimental high-temperature drying of softwoods, extra heat for the heating up period has been provided by direct-firing equipment. Steam pressure should be 150 pounds per square inch.

Air travel length should be 8 feet or less[13], or 16 feet or less in a double-track kiln with booster coils midway between the loads. In package-loaded kilns, 10 to 12 feet would be the maximum air travel unless the kiln can be loaded from both sides and has booster coils along the middle. To obtain the high air velocities necessary, a direct-connected fan system (rather than a life shaft) is required. Two-speed fan motor operation is desirable when both normal and high-temperature drying will be done, to save electrical energy. Commonly available in-kiln motors are designed to operate safely up to 250°F, but in commercial practice motor failures are frequent. New ceramic windings may lengthen motor life. Special attention must be paid to baffling.

Unlike some softwood kilns, a humidification system is essential. A proper system should be able to deliver large quantities of saturated steam at low pressure. High-pressure steam suitable for the heating system must go through a desuperheater before being used for humidification.

If a high-temperature kiln is to be designed for both hardwoods and softwoods, provisions can be made for opening the vents on just the high-pressure side of the kiln. This helps to maintain uniform drying conditions. Otherwise, the venting that is necessary when drying a very wet permeable wood will take place through leaks, tending to deteriorate the kiln.

Jet drying.—When air at a temperature above 212°F is impinged vertically at high velocity onto the surface of lumber, drying can be very rapid. For example, Koch (1964, 1972, p. 1010) found that at 300°F sawn 7/16-inch-thick southern pine dried to 10 percent moisture content in about 75 mintues when air flow was parallel, but only 50 minutes were required for impingement flow.

Two hardwood species have been evaluated as candidates for impingement (jet) drying. Rosen and Bodkin (1978) found that 4/4 silver maple (*Acer saccharinum* L.) lumber could be jet dried at 300°F to 10 percent moisture content in a little over 2 hours with an air velocity of 3,000 fpm; the surface of the lumber was darkened to a depth of about 0.01 inch by so drying, and some internal checking occurred.

In experiments with yellow-poplar, Rosen and Bodkin (1981)[14] found that acceptable results were obtained with a two-step schedule at temperatures to 360°F with impingement velocity of 3,000 fpm; under these conditions, drying time from green to 8 percent moisture content was about 24 hours, surface darkening was not severe, and no internal or external checking occurred.

Press-drying.—Press-drying of lumber or veneer is accomplished by applying a pair of heated platens (250 to 450°F) to the board or veneer—one to each face. A platen pressure of 25 to 85 psi assures good thermal contact between the heated platens and the board.

---

[13]Experience in drying 8/4 southern pine has shown that attainment of uniform moisture content in lumber stacked in 8-foot-wide loads is more difficult than with narrower loads; in the opinion of many researchers knowledgeable in high-temperature drying, load width should not exeed 5 feet when uniformity of moisture content is required.

[14]See also Rosen (1977).

During drying, heat transferred from the platens to the wood causes air in the wood to expand and water to vaporize. A mixture of vapor and liquid then moves to the surface of the board where it escapes. Ventilated cauls and wire screens have been interposed between the platens and the board to help vapor escape from the board faces (fig. 20-32).

Heated platens hold the board flat during drying and reduce width shrinkage; but shrinkage in thickness may be greater than that caused by conventional hot air drying. In cyclic shrinking and swelling tests (Hittmeier et al. 1968), press-dried hardwood lumber was significantly more stable in width but (except for the oaks) less stable in thickness, than conventionally kiln-dried material.

The heartwood of some species tends to darken and develop checks and honeycombing during press-drying. These defects, however, may not adversely affect the board for many applications, and the darker color is often more appealing than the original color. Some experimental data on press drying hardwood species found on southern pine sites is summarized in the following paragraphs.

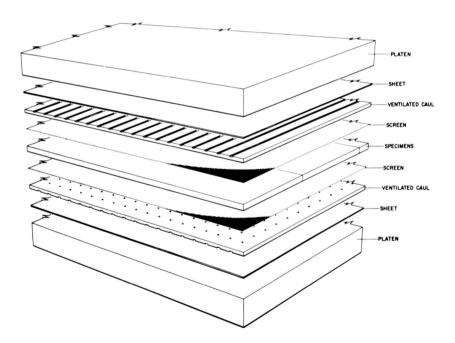

Figure 20-32.—System of ventilated cauls for press drying lumber or veneer. The aluminum protector sheets are 0.064 inch thick. Top and bottom cauls are of aluminum and measure ¼-inch by 26 by 104 inches. Rectangular grooves 1/16-inch deep by 3/16-inch wide were milled on 1-inch centers on the back of each caul. One-eighth-inch holes were drilled at 1-inch intervals along each groove. A 75-mesh Fourdrinier wire screen was interposed between the veneer and each ventilated caul. (Drawing after Koch 1964.)

Heebink and Compton (1966) press-dried 8-foot-long, 0.6-inch-thick, live-sawn boards of **red oak species** (probably northern red oak) with platen temperatures of 300°F and 350°F and established approximate drying curves (fig. 20-33); they found that when moisture content averaged 8 pecent, board surfaces were almost ovendry and board centers were near 16 percent moisture content. After moisture in the boards had diffused and become more or less uniform throughout the thickness, there was a marked tendency for the boards to cup, check, and in some places collapse. A green rough thickness of 0.6 inch was found adequate for a dry planed thickness of 7/16-inch (0.438 inch).

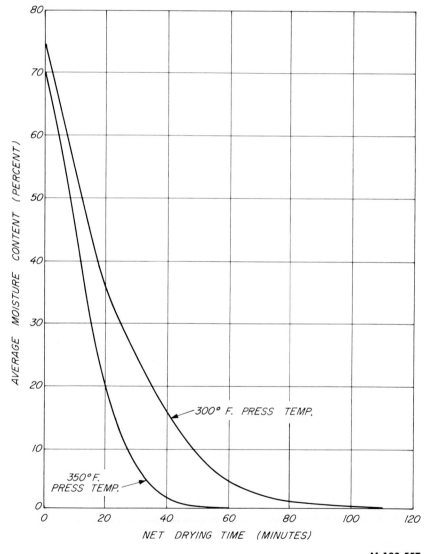

Figure 20-33.—Drying rate curves for 0.6-inch-thick red oak boards dried in a hot press with platen temperatures of 300° and 350°F and specific pressure of 50 psi. (Drawing after Heebink and Compton 1966.)

Wang and Beall (1975) press-dried short lengths of **northern red oak** at 200°C with specific platen pressure of 35 psi. For 0.59-inch-thick boards 5.5 inches long they found that the ratio of average moisture content at any time during drying to the initial board moisture content was a function of the square root of time in the press, i.e., ratio = $a - b\sqrt{\text{time in press}}$; for these conditions constant (a) had a value of 1.097 and constant (b) was 0.223, when time in press was expressed in minutes. They also found that drying time is positively correlated with board thickness and inversely correlated with platen temperature and pressure (fig. 20-34).

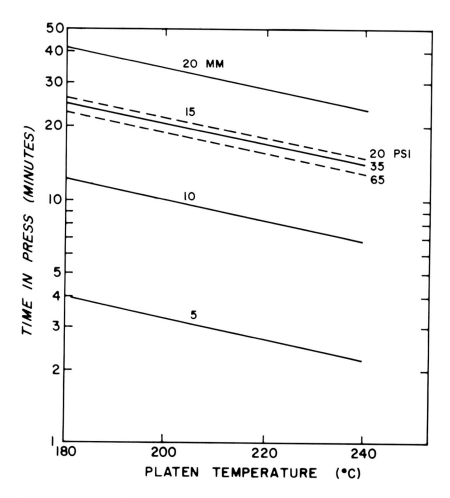

Figure 20-34.—Press drying times to 5 percent average moisture content for 5.5-inch long northern red oak boards of four thicknesses as influenced by platen temperature and specific pressure. (Drawing after Wang and Beall 1975.)

Wang and Beall (1975) observed that thickness shrinkage of northern red oak was linearly related to pressure, initial moisture content, and to final moisture content; i.e., shrinkage was greatest if platen pressure and initial board moisture content were high and final moisture content low. Thickness shrinkage of edge-grain boards was about twice that of flat-grain. Thickness swelling of press-dried boards placed in a humid atmosphere was least in boards dried at the highest temperatures.

Hittmeier et al. (1968) provided press-drying data on nine species, eight of which are hardwoods found on southern pine sites. They dried 8-foot-long boards that had been live-sawn and surfaced when green to 0.5-inch and 1.0-inch thickness. These boards were dried at a platen temperature of 345°F and specific platen pressure of 50 psi until a thermocouple at mid-thickness of the board indicated a temperature 20°F below the platen temperature; at this temperature the average moisture content of the boards was below 6 percent.

Hittmeier et al. found that internal temperatures during drying varied with species (fig. 20-35), but all passed through a temperature plateau when the rate of heat transfer to the wood was exactly in balance with rate of heat consumption by evaporation of water and heating of the wood. While free water is evaporating, the plateau temperature is about 212°F, unless pressure builds up inside the wood. The sweetgum plateau temperatures were very close to 212°F; in impermeable species such as white oak and post oak, temperatures rose higher, indicating internal pressures above atmospheric pressure (fig. 20-35 and table 20-32).

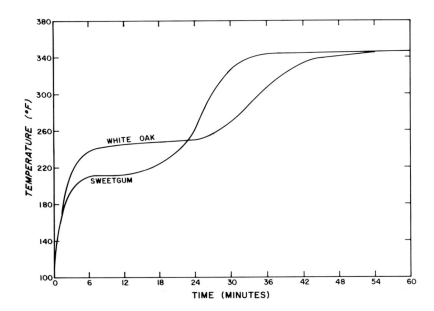

Figure 20-35.—Temperature during press drying at mid-thickness of ½-inch-thick sweetgum and white oak boards showing two levels of plateau temperatures; platen temperature was 345°F and specific pressure was 50 psi. (Drawing after Hittmeier et al. 1968.)

For ½-inch boards, drying times to a mid-thickness temperature of 20°F less than the 345°F platen temperature varied from 20 to 76 minutes; those for 1-inch-thick blocks varied from 108 to 200 minutes. Thickness shrinkage was from 8.2 to 23.7 percent in ½-inch boards and 10.3 to 19.7 in 1-inch boards. Width shrinkage was generally less than 5 percent (table 20-32).

TABLE 20-32.—*Drying time, shrinkage, plateau temperature, and degrade in press-dried, surfaced hardwood boards of two thicknesses and eight species[1]* (Hittmeier et al. 1968)

Species	Initial moisture content	Drying time[2]	Shrinkage		Plateau temperature	Drying degrade[3]
			Thickness	Width		
	Percent	*Minutes*	---------*Percent*---------		°*F*	
		½-INCH-THICK BOARDS[4]				
Ash sp............	41-45	26-43	10.6-14.7	1.5-3.1	234-275	Minor
Elm, rock[5].........	52-87	30-70	11.5-18.4	1.3-4.7	219-300	Moderate
Hickory sp	34-52	31-63	14.9-23.7	1.2-4.9	264-300	Moderate
Oak, post	40-68	33-66	10.0-15.1	.3-3.5	250-280	Major
Oak, red sp........	64-83	36-60	10.1-19.6	1.4-4.3	219-243	Moderate
Oak, white	41-71	21-76	9.7-16.9	1.0-5.6	244-275	Major
Sweetgum	103-130	28-32	11.0-16.2	1.9-4.8	210-215	Minor
Tupelo, black	58-123	20-41	10.2-17.6	.9-3.9	212-219	Minor
		1-INCH-THICK BOARDS[6]				
Ash, sp	43-107	130-134	10.9-13.0	2.0-2.1	259-270	Moderate
Elm, rock[5].........	47-57	108-126	13.8-14.5	1.9-2.5	214-271	Major
Hickory sp	46-54	122-179	16.7-18.3	1.6-2.8	252-277	Major
Oak, post	66-67	122-200	10.4-11.2	1.3-2.3	246-279	Major
Oak, red sp........	67-80	152-203	15.3-19.7	1.8-3.0	232-290	Major
Oak, white	56-75	132-184	12.1-16.9	.5-3.4	243-304	Major
Sweetgum	111-126	154-181	13.2-14.5	2.8-3.2	212-215	Moderate
Tupelo, black	90-122	136-168	10.3-12.8	1.8-3.1	210-220	Moderate

[1]Press temperature 345°F; 50 psi specific platen pressure.
[2]Time for mid-thickness of board to attain 325°F, corresponding to an average board moisture content of less than 6 percent.
[3]Based on visual inspection for checking, honeycombing, and collapse.
[4]9 to 40 boards of each species were dried.
[5]*Ulmus thomasii* Sarg.
[6]Two to five boards of each species were dried.

Width stability with humidity changes of the boards press-dried by Hittmeier et al. (1968) was 30 to 60 percent greater than matching kiln-dried specimens. Thickness stability of the press-dried boards, however, was generally 10 to 80 percent lower than boards conventionally kiln-dried; the press-dried oaks were an exception to this generalization and were more stable in thickness than kiln-dried oak.

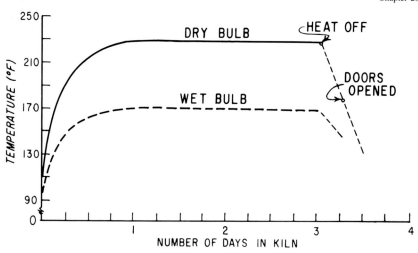

Figure 20-36.—Approximate kiln conditions used to dry incised 7- by 9-inch hardwood crossties in an industrial-size dry kiln. All steam spray lines were turned off and vents were kept closed. (Drawing after Huffman 1958.)

In the experiments of Hittmeier et al. (1968), 1-inch-thick ash, hickory, sweetgum and black tupelo dried with only moderate degrade; the other species were, in general, severely riddled with surface checks and honeycombing. The thermal effects of press-drying did not seem to significantly affect quality of glue bonds obtainable, but checks and honeycombing appeared to lower block shear strength in oak boards. Wood machinability and finish-holding capability seemed unaffected by the press-drying. Species that press-dried without severe checking and honeycombing had strength values essentially equal to those of matched kiln-dried specimens; this included all tested species except rock elm, white oak, and post oak.

Chen (1980) press-dried **yellow-poplar** boards ½-inch and 1 inch thick at two platen temperatures (250° and 350°F), two platen pressures (50 and 150 psi), with two types of caul perforation (144 and 245 holes per square foot). He found that increasing platen temperature decreased drying time, but increasing platen pressure did not. Increasing thickness of boards increased drying time more than the proportional increase in thickness. Both higher platen temperature and pressure caused greater thickness shrinkage. No difference was found between the two cauls in drying rate and time.

CROSSTIES

Shunk (1976) reported that crossties are seldom kiln-dried because of the long time required. One of the first efforts to shorten this time by high-temperature drying was made by Huffman (1957) at the University of Florida. He showed, in laboratory-scale tests, that it is technically feasible to kiln-dry incised sweetgum and tupelo (*Nyssa* sp.) crossties to a treatable moisture content in 3 days without objectionable seasoning defects.

In a second study in an industrial-size kiln Huffman (1958) dried a single charge of 2,016, 9-foot-long, 7- by 9-inch crossties, i.e., 1,875 of sweetgum and black gum (*Nyssa* sp.), 54 of red oak (*Quercus* sp.), 54 of hickory (*Carya* sp.), 11 of elm (*Ulmus* sp.), and 22 of beech (*Fagus grandifolia* Ehrh.). The incised ties were dried in a double-track, cross-circulation, 135-foot-long kiln at a dry-bulb temperature of 230°F and a wetpbulb temperature of 170°F (fig. 20-36). Tiers of ties were separated by 2- by 4-inch stickers. Load width was 8 feet; air velocity was unstated but was probably less than 400 feet per minutes.

The ties were segregated into heartwood volume classes. Moisture content before and after kiln-drying was determined from increment cores taken at the midpoint of 11 ties of each group; after drying, core moisture contents ranged from 31.7 percent for sapwood hickory to 77.3 percent for sweetgum with 30 percent heartwood (table 20-33), and were probably higher than the true average moisture contents of the ties. In general, moisture contents in the outer 1 inch layer were below fiber saturation. Seasoning checks, for the most part, were smaller in size (table 20-33) than those generally observed in air-seasoned ties.

TABLE 20-33.—*Moisture content and check dimensions in 7- by 9-inch hardwood crossties kiln-dried for 3 days at 230°F according to species group and heartwood content*[1,2]
(Data from Huffman 1958)

Species or species group and heartwood volume (percent)	Moisture content when kiln-dry[3]	Average dimension of most objectionable checks[4]		Dimension of widest check observed[4]	
		Width	Length	Width	Length
	Percent	------Inch------			
Beech, American					
20..........................	39.2	0.25	7.9	0.50	12
50..........................	42.6	.20	8.7	.38	12
Elm sp.					
100.........................	48.3	.22	9.4	.50	34
Hickory sp.					
0...........................	31.7	.26	8.4	.50	7
30..........................	46.8	.24	12.1	.38	20
Oak, water and laurel					
100.........................	61.0	.27	7.1	.63	9
Oak, red sp. excluding water and laurel					
100.........................	60.7	.13	4.8	.19	6
Sweetgum					
10..........................	69.7	.15	6.6	.38	7
30..........................	77.3	.16	7.5	.25	6
60..........................	70.6	.27	9.4	.75	18
90..........................	58.6	.20	6.3	.38	4
Tupelo sp.					
0...........................	42.5	.08	6.5	.13	10
10..........................	49.7	.10	6.2	.19	10
30..........................	58.1	.13	8.2	.19	18

[1]See figure 20-36 for schedule.
[2]All observations except last two columns based on average of 10 ties.
[3]Determined by ovendrying borings taken at midpoint on broad face of ties.
[4]Observations made on top face of ties.

None of the 2,016 kiln-dried ties were rejected because of seasoning degrade. It was necessary to dowel some of the ties before they met acceptance qualifications. The percent of ties requiring dowelling was greatest for American beech and water and laurel oak, as follows:

Species	Ties requiring dowelling
	Percent
Sweetgum	0
Tupelo sp	0
Red oaks (except water and laurel)	0
Elm sp	9.1
Hickory sp	9.3
Beech, American	31.8
Oak, water and laurel	32.6

Drying each crosstie required consumption of 105 pounds of steam for heating, and 0.65 kilowatt hours of electricity to drive the fans. The average water loss per tie during kiln-drying was 55.4 pounds; green weight of the crossties averaged 255.5 pounds.

To determine the treatability of these crossties that were kiln-dried at high temperature, a representative sample (all with boxed hearts) were treated with a 80/20 mixture of creosote and coal tar. Conditions during the pressure treatment were as follows:

Statistic	Value
Initial air pressure	75 psi
Treating pressure	195 psi
Average treating temperature	202°F
Pressure period	4.5 hours

This schedule was appropriate for the sweetgum and tupelo ties but did not yield adequate retention in hickory, breech, and oak ties (4.3 to 5.9 pounds per cubic foot); penetration of preservative was excellent in hickory, ranging from 1.9 to 2.6 inches. Higher retentions would likely have resulted if the treating schedule had been adjusted appropriately for the more difficult species. For all species, gross absorption and net retentions of preservatives were 11.94 and 7.04 pounds per cubic foot (table 20-34).

TABLE 20-34.—*Retention and penetration of creosote and coal tar in pressure-treated, kiln-dried, 7- by 9-inch, heart-center hardwood crossties of six species groups according to heartwood percentage*[1] (Data from Huffman 1958)

Species or species group and heartwood volume (percent)	Preservative retention	Preservative penetration
	Pounds per cubic foot	Inches
Beech, American		
20	5.3	1.8
50	5.1	1.6
Elm sp.		
100	7.2	2.4
Hickory sp.		
0	5.9	2.6
30	5.0	1.9
Oak, water and laurel		
100	4.3	.8
Oak, red sp. excluding water and laurel		
100	4.3	1.5
Sweetgum		
10	6.7	2.5
30	6.7 1.3	
60	6.3	1.4
90	6.6	1.8
Tueplo sp.		
0	7.6	1.9
10	7.0	1.9
30	6.8	1.5

[1] Values based on averages of ten crossties.

VENEER[15]

Hardwood veneers are dried to various moisture contents according to their use. Veeners for bushel baskets and fruit containers may be dried only enough to prevent mold—to about 20 percent moisture content. Decorative hardwood face veeners for the furniture trade may be dried to 8 or 10 percent. Commercial hardwood veneers that are to be bonded with a urea glue may be dried to 6 or 8. Veneers for faces and backs to be applied over flakeboard cores to form phenolic-bonded structural composite hardwood panels (figs. 24-53 and sec. 28-6) are dried to about 3 percent.

---

[15] With some additions, text under this heading is condensed from Lutz (1977).

Most veneer is rotary peeled or sliced on a production-line basis so the drying process must be rapid and is usually continuous. Dried veneer should display the following:

- Uniform moisture content
- Freedom from buckle or end waviness
- Freedom from splits
- Surfaces in good condition for gluing
- Desirable color
- Minimum shrinkage
- Freedom from collapse and honeycomb
- Minimum casehardening where outer layers are in compression and core layers are in tension.

Veneer properties that affect drying.—Thick veneer dries more slowly than thin veneer (figs. 20-37 and 20-38) and veneer varying in thickness will not dry to a uniform moisture content. Because moisture moves more readily parallel to the grain than perpendicular to it, ends of veneer sheets tend to dry more rapidly than mid-portions. In curly-grained or other figured veneer, end grain is exposed on the broad surface and dries faster than the rest of the sheet, causing stresses and buckling. Quarter-sliced veneer (fig. 18-236G) will take slightly longer to dry than rotary-cut veneer of the same thickness, and flat-sliced veneer (fig. 18-273F) may dry slower on the near-quarter-sliced edges than in the flat-grain central area.

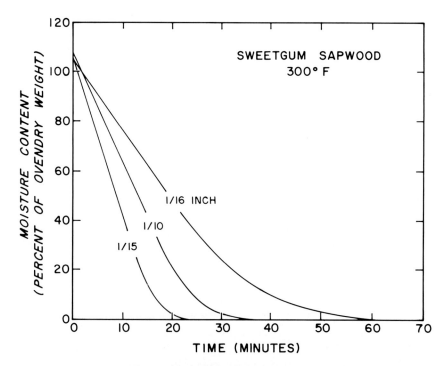

Figure 20-37.—Time to dry rotary-peeled sweetgum sapwood veneer of three thicknesses in a laboratory kiln with air flow at 300°F parallel to the surface of the veneer. (Drawings after Bethel and Hader 1952.)

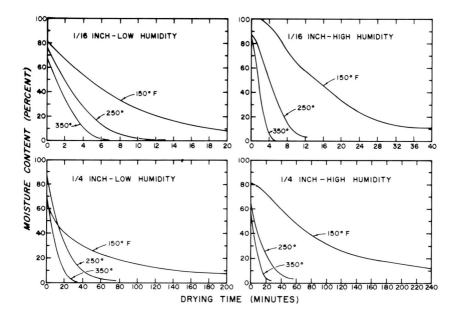

Figure 20-38.—Drying curves relating moisture content to time for yellow-poplar heartwood 1/16- and 1/4-inch thick dried at three temperatures and at high and low humidity with air circulated parallel to veneer surfaces at 600 fpm. (Drawing after Fleischer 1953.)

Veneer of high moisture content usually dries slower than veneer of low moisture content (fig. 20-39 top). Variation of moisture content within and among trees can be substantial (see fig. 8-2 and tables 8-1 and 8-2).

In veneers ⅛-inch and thicker, permeability across the grain (see table 8-6) may affect time to dry, so that thick post oak heartwood veneer, for example, may require more time to dry than equally thick yellow-poplar sapwood. Within-species variation may also be significant; Bethel and Hader (1952) found that the drying rate (in a parallel-flow dryer) of sapwood was much faster than that of heartwood in sweetgum, black tupelo, and yellow-poplar. These differences are probably more observable in older parallel-flow dryers with low rates of heat transfer than in more modern impingement jet dryers or press dryers that have high rates of heat transfer.

In veneers thiner than ⅛-inch thick, time to dry appears to be controlled by rate of heat transfer to the veneer. In such thin veneer, the presence or absence of knife checks (fig. 18-14) appears not to affect time to dry; loosely cut veneer is easier to flatten after drying, however.

Tension wood in hardwoods (figs. 5-71 and 5-72) shrinks more longitudinally than typical wood. As a result, sheets of veneer containing tension wood tend to buckle during drying.

Drying characteristics of pine-site species are summarized in table 20-35.

Dryer conditions that affect drying.—In general, dryers are operated to hold the veneer flat and transfer as much heat as possible to it. Buckle will be greates in veneer hung from the ends and allowed to dry at ambient room conditions. Next will be veneer restrained by stickers and dried in a batch-type kiln. Veneer dried in a mechanical dryer with a roller or wire-mesh conveyor will buckle less than matched material dried in the kiln. The least buckled will be veneer dried between flat hotplates. Drying time is inversely proportional to dryer temperature (fig. 20-39 bottom), and air circulation velocity. Hot air impinged vertically on veneer surfaces dries the veneer more rapidly than air of the same temperature circulated parallel to the surface (fig. 20-40).

TABLE 20-35.—*Drying times and tendencies to suffer drying defects of veneers from 20 hardwoods found on southern pine sites* (Lutz 1977)

Species	Drying time[1]		Defects in drying[2]		
	Sapwood	Heartwood	Buckle	Splits	Collapse
Ash, white	B	B	B	B	A
Elm, American	C	C	C	B	A
Elm, winged	C	C	—	—	—
Hackberry	B	B	A	A	A
Hickory, true	B	C	B[3]	B	A
Maple, red	C	C	A	A	A
Oak, black	C	C	A	B	A
Oak, cherrybark	C	C	A	B	B
Oak, chestnut	C	C	A	B	A
Oak, laurel[4]	C	C	B	C	C
Oak, northern red	C	C	B	B	B
Oak, post	C	C	—	—	—
Oak, scarlet	C	C	A	B	—
Oak, southern red	C	X	A	B	B
Oak, water	C	C	A	C	C
Oak, white	C	C	A	B	B
Sweetbay	—	—	A	A	A
Sweetgum	C	C	A	B	B
Tueplo, black	C	C	B	A	B
Yellow-poplar	B	B	A	A	A

[1] A, dries faster than average; B, dries in average time; C, dries more slowly than average.
[2] A, species property very suitable for veneer; B, intermediate; C, less suitable for veneer.
[3] Pignut hickory has a greater tendency to buckle than the other true hickories.
[4] Laurel oak has a pronounced tendency to check during drying (Lutz 1972).

The roller conveyor or wire-mesh conveyor in conventional mechanical mechanical vener dryers transfer heat by conduction to veneer surfaces; heat transfer from the rolls may be as much as 20 percent of the total heat transferred to the veneer. This heat transfer from the rolls is very obvious when comparing the drying rates of veneer through essentiall empty dryer and one full of veneer. In the full dryer, the rolls are cooled by the wet veneer and the required drying time for a given final moisture content increases. It follows that the first veneer through an empty dryer will emerge much drier than veneer coming from a full dryer.

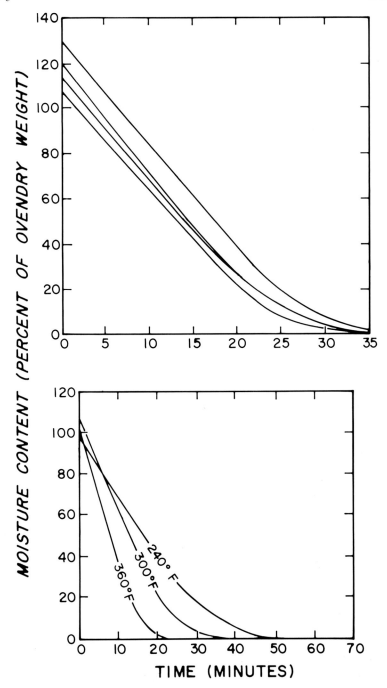

Figure 20-39.—Time to dry 1/10-inch-thick, rotary-peeled, sweetgum sapwood veneer in a laboratory kiln with air flow parallel to the surface of the veneer. (Top) Effect of initial veneer moisture content; kiln temperature 300°F. (Bottom) Effect of kiln temperature. (Drawings after Bethel and Hader 1952.)

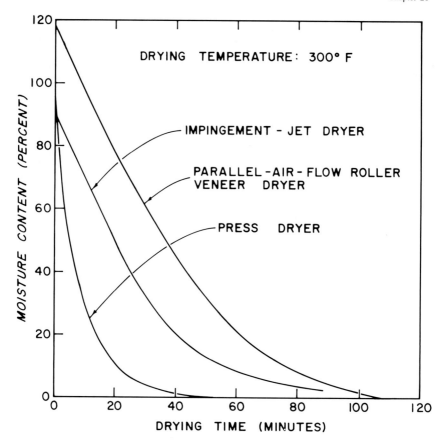

Figure 20-40.—Drying curves for S4S, 7/16-inch-thick, sawn southern pine veneers dried at 300°F with air counter-flowing parallel to the veneer at about 600 fpm, with air impinging vertically on the veneer at about 3,500 fpm, and in a press dryer (fig. 20-32) with specific platen pressure of 82.6 psi. (Drawing after Koch 1964.)

When veneer is dried at temperatures below 200°F in a dry kiln, the relative humidity can be controlled to control the final moisture content (table 8-6). Most veneer is dried, however, in mechanical dryers at temperatures above 250°F where wet-bulb temperature has relatively small effect on equilibrium moisture content (fig. 8-14).

Types of veneer dryers.—By far the most common veneer dryer is the direct-fired or steam or hot-water-heated progressive conveyor type. These dryers are made in two styles. The **roller conveyor** is used most commonly with rotary-cut veneer. The rollers are generally hollow tubes that rest directly on the veneer; sheets of veneer are fed endwise. A **wire-mesh conveyor** is used for drying continuous ribbons of rotary-cut veneer and for sliced half-round veneer; it permits feeding the veneer sidewise so that the sheets can be kept in sequence for matching. The wire-mesh conveyor is reported to work most satisfactorily with a restraint weight of about 5 pounds per square foot when drying thin face veneer.

Longitudinal cross-circulation, amd impingement air movement are used in these progressive dryers. Most common in new veneer plants is the jet dryer (fig. 20-41) with air impinging on the veneer surfce at velocities of 2,000 to 10,000 fpm.

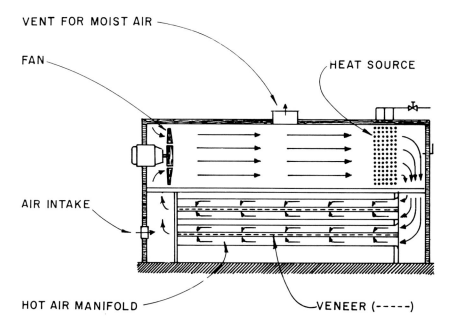

Figure 20-41.—Schematic cross section (transverse to veneer flow) through a two-deck, impingement-jet dryer. Heat can be applied by steam coils, but direct-gas-fired dryers are more common. Long slits in the hot-air manifolds cause curtains of hot air to impinge vertically at high velocity on both top and bottom surfaces of the moving veneer. (Drawing after Fessel 1964.)

Lahtinen (1975) described a continuous veneer dryer used on birch (*Betula* sp.) up to 0.06 inch (1.5 mm) thick. The veneer moves directly from the lathe through the dryer to dry clippers. In Finland, this continuous dryer has six levels with veneer enter at the top and feeding continuously in zigzag fashion between wire screen conveyor belts through the heated chamber.

Some veneer is dried in kilns operated at temperatures below 212°F so that relative humidity in the kiln and equilibrium moisture content of the veneer can be closely controlled. Control of the final moisture content and improved gluability are two of the main advantages of such kilns.

Heated tunnels equipped with conveyors may be used to dry baskets assembled from green veneer to a moisture content of about 20 percent.

Press-drying (figs. 20-32 through 20-35, 20-40, 20-42, and 20-43) of veneer has been studied for many years in numerous laboratories. Heebink (1952) made flooring from press-dried, 1/8-inch yellow-birch veneer, and Hann et al. (1971) produced press-dried pallet deckboards from thick-peeled rotary peeled hardwoods (fig. 20-42). Lutz (1974) successfully controlled the final moisture content of press-dried 1/24-inch, rotary-peeled oak by steaming the veneer as it was being press-dried. Final moisture contents of 4, 5½, 7 and 11 percent ($\pm$ 1 percent) were obtained by control of the steaming tempeature; the curve for control to 7 percent is shown in figure 20-43. Lutz cautioned that specific platen pressure for oak should be more than 50 psi and that too much superheat in the steam resulted in lower equilibrium moisture contents than those indicated in the literature (fig. 8-14); the successful system used a 15-pound steam supply, reduced to 1 psi pressure in a 6-foot-long uncovered line ahead of the pressure, followed by steam separators coupled as closely as possible to the heated aluminum cauls through which the steam was admitted to the veneer.

Progress toward commercialization of press dryers for veneer has been reported by Mustakallio and Paaki (1977) in Finland, and by Weyerhaeuser Company in this country (Timber Processing Industry 1980; Pease 1980).

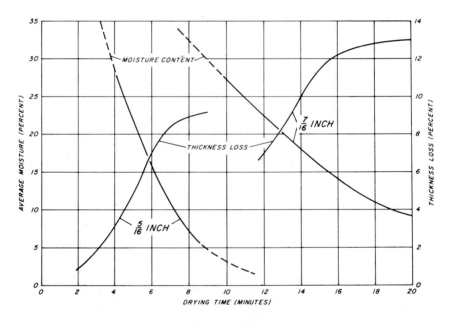

Figure 20-42.—The relationship between average final moisture content or thickness loss and drying time for 5/16- and 7/16-inch-thick, rotary-peeled red oak, (Wisconsin source) veneer dried between platens at 375°F platen temperature and 50 psi platen pressure. Initial moisture content was about 85 percent. (Drawing after Hann et al. 1971.)

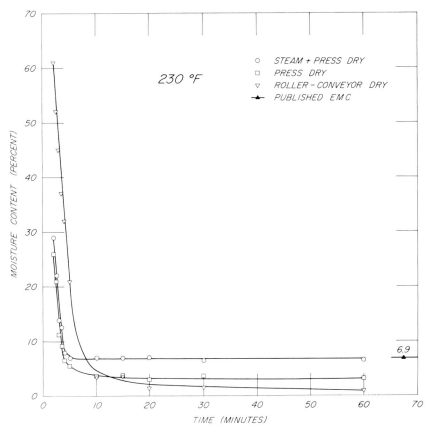

Figure 20-43.—Comparison of drying curves for 1/24-inch-thick, rotary-peeled red oak sp. heartwood veneer dried at 230°F in a conventional roller veneer dryer with parallel air flow, a platen press without steaming, and a platen press with steaming. At 230°F, with steam, equilibrium moisture content is about 6.9 percent. (Drawing after Lutz 1974.)

Gas-fired **infrared heaters** have been used ahead of green veneer dryers to boost veneer temperature and shorten drying time, but are not in wide use. To equalize moisture at the end of the drying cycle, **high-frequency** and **microwave energy** have been used as part of the veneer drying system; high equipment and energy costs have deterred general use, however.

Drying time.—Comstock (1971), on the basis of tests of softwood veneer dried in an impingement-jet laboratory dryer, concluded that drying times for different species appear to be predictable purely on the basis of the density and green moisture content, other things being equal. Such a relationship implies that heat transfer rate is of overriding importance, and that, in veneer thicknesses, the permeability of wood is of lesser importance. Functional relationships were developed by Bethel and Hader (1952), Fleischer (1953), and

Keylwerth (153), but they appear to have limited applicability to the impingement-jet dryers now widely used by industry. Studies of impingement-jet dryers have been limited to softwood veneer (Milligan and Davies 1963; Comstock 1971). In the absence of more specific information, some data from the literature are presented in graphic form to allow the reader to estimate likely drying times, as follows:

Relationship described by drying curves	Figure number
Comparison of parallel-flow, impingement-jet, and press-drying of 7/16-inch-thick-southern pine at 300°F	20-40
Comparison of parallel-flow and press drying of 1/24-inch-thick red oak at 230°F	20-43
Parallel-flow drying of 1/10-inch sweetgum sapwood related to initial moisture content	20-39 (top)
Parallel-flow drying of 1/10-inch sweetgum sapwood related to kiln temperature	20-39 (bottom)
Parallel-flow drying of sweetgum sapwood at 300°F related to veneer thickness	20-37
Parallel-flow drying of yellow-poplar heartwood related to veneer thickness and kiln temperature and humidity	20-38
Press-drying of 0.6-inch red oak at 300° and 350°F	20-33
Press-drying of red oak at 375°F relatd to veneer thickness	20-42
Press-drying of northern red oak related to wood thickness and platen temperature and pressure	20-34

Control of drying time.—Veneer, especially light-colored wood, should be dried as soon as practical after cutting to minimize end splits, oxidation stain, mold, and blue stain.

Maximum temperatures consistent with good glue bonds and wood color will minimize drying time. In general, this will be about 400°F at the green end and 360°F at the dry end of the dryer. If gluing or veneer color are problems, lower the dryer temperature. Decreasing the dryer temperature by 100°F (for example, from 350° to 250°F) will approximately double the drying time.

To reduce the energy consumed and veneer dryer emissions, dryer vents should be as nearly closed as practical. To reduce condensation and haze in the building, open vents the minimum amount needed to correct the problem.

The dryer should be operated with the maximum air circulation possible. Lowe air velocity is sometimes necessary to prevent overdrying and splitting of very thin veneer.

Dryers should operate as full of veneer as possible. Dryer schedules should be based on a full dryer operating at a steady temperature and air movement.

Green veneer should be segregated by required drying time; sorts should be by veneer thickness and species, and—possibly—by sapwood and heartwood. Doubling the veneer thickness will more than double the drying time.

The veneer drying time should be regulated by the kind of veneer being ed in the green end. It is tempting for the dryer operator to change the drying time from the dry end, depending on whether the emergying veneer seems too wet or too dry, thereby constantly shifting drying times and average moisture content of the output veneer. A bette method is to carefully determine the proper time to dry veneer of a given thickness, species, and sapwood or heartwood and use this schedule when similar veneer is dried again.

Control of final moisture content.—Probably the most universal problem in drying veneer in a progressive mechanical veneer-type dryer operating above 220°F is the nonuniform moisture content in the veneer as it comes from the dryer. This is true of a dryer having longitudinal circulation, cross circulation, or jet impingement circulation. It is similarly true for a progressive platen-type dryer.

For example, veneer dried to an average moisture content of 8 percent will generally have a range of moisture content from 2 to 20 percent. This is because the equilibrium moisture conditions in the dryer are for all pratical purposes 2 percent or less. When drying to an average moisture content of 8 percent, the faster drying veneer may come to 2 percent and the slower drying to 20 percent moisture content. Different drying rates in different areas of the same sheet of veneer produce a wide range in final moisture content in the veneer as it comes from the dryer.

To keep this problem to a minimum, the green veneer should be sorted for thickness, moisture content, and ensity. Control will probably be best if the green veneer is also sorted for sapwood and heartwood and by species. If drying is still uneven when veneer of one type is being dried, uniformity of drying conditions in different parts of the dryer must be checked.

One method is to run matched sasmples of veneer through different portions of hte dryer, one through the left side of the upper conveyor, another through the left side of the lower conveyor, and so on. If moisture content immediately out of the dryer shows that one portion is consistently drying veneer faster than the other, drying rates can sometimes be equalized by adding steam coils, baffles, or fans where needed in the dryer.

Another way of controlling the final moisture content is to dry all of the veneer to 5 percent moisture content or less. This may result in overdrying of some of the veneer, but it will result in a narrow range of veneer moisture.

A very common method of reducing the spread of moisture in the veneer is to mark and pull separately for further drying all pieces that electronically measure too wett. Leaving this wet veneer in a solid stack overnight will help to equalize the moisture content, and resorting through the moisture detector the next day will reduce the number of pieces that need to be redried.

Some moisture meters are sensitive to wood temperature as well as moisture content. They should be calibrated under the conditions in which they will be used.

Another method that is sometimes used when nonuniform moisture content is a serious problem is to dry in two stages. In the first pass, the veneer is brought to an average moisture content of about 20 percent. It is then stacked overnight to allow some equaliztion and rerun the next day to the average moisture content desired.

High-frequency or microwave units have been used experimentally at the dry end of the dryer to equalize the moisture content of the veneer. Both these methods work on the principle that the more moist areas in the veneer absorb more energy. Heating and drying are proportional to this absorption of energy. Both of these methods do equalize moisture content in the veneer, but they have not been generally adopted because of cost (Resch et al. 1970).

It is possible to dry veneer to controlled moisture contents in superheated steam at atmospheric pressure. To date this method has not been used commercially.

Control of buckle.—Buckle in veneer may be caused by stresses in the wood, by reaction wood, by irregular grain with resulting irregular drying rates and shrinkage, and possibly also by improper setting of the lathe or slicer. Maximum restraint to hold veneer flat without causing shrinkage splits, and anything that improves uniformity of dry-veneer moisture content will reduce buckling. In most cases, buckling can be minimized by redrying in a **plate dryer** (a type of press dryer). The redrying temperature and time will depend on the moisture content of the veneer (Lutz 1970).

The dried veneer should be neatly stacked on flat skids and the top of the pile weighted. Flitches of sliced veneer should be promptly strapped in flat crates.

Control of splits.—Splits in veneer that has been dried in a progressive mechanical dryer are generally related to splits htat were in the green veneer or result from rough handling. If stacks of green veneer must be held before drying, the ends should be protected from end drying by covering them with a plastic sheet (such as polyethylene) or if necessary by spraying them with water.

A recent development to control splits from rough handling is green veneer taping. Tape is applied at the lathe primarily to veneer thinner than 1/26-inch (1 mm). Taping reportedly improves veneer grade, reduces need to splice and repair veneer, and reduces end waviness.

Another method of reducing handling splits is to dry rotary-cut veneer in a continuous ribbon using a wire-mesh conveyor in a mechanical dryer. The method was used as early as 1950 with birch veneer which was reeled as it came from the lathe and then unreeled into the dryer. The dry veneer was then clipped for grade.

More recently, in a system widely used for softwood, veneer is stored on long trays and then fed in line to the dryer. In addition to reducing splits, recover is reportedly improved because the veneer is clipped dry and need not be oversize to compensate for variability in shrinkage.

Control of surface gluability.—Poor glue bonds have been reported with veneer dried in direct oil-fired dryers operating at temperatures as high as 550°F. This is les sof a problem with direct gas-fired dryers and les syet with steam-heated dryers. Weakening of the surfces and extractives brought to the wood surface during high-temperature drying appear to be the causes. At any rate, lower drying temperatures and prevention of overdrying the veneer are the common means of overcoming veneer gluing problems.

Control of collapse, honeycomb, and casehardening.—Collapse and honeycomb may occur in species that are relatively impermeable. Typical examples would be ⅛-inch and thicker heartwood of sweetgum and overcup oak (*Quercus lyrata* Walt.). Collapse in sweetgum heartwood is likely to occur in early stages of the drying. Sweetgum heartwood dried at 350°F had much moree honeycomb than that dried at 150°F. Experiments at the U.S. Forest Products Laboratory showed that ⅛-inch overcup oak dried at 320°F might shrink as much as 20 percent in thickness. The solution to these drying problems in all cases appears to be the use of a lower drying temperature.

Casehardening was at a maximum in ⅛-inch heartwood of sweetgum when dried at temperatures of 120° to 160°F. Casehardening can be removed by use of high temperature, particularly if the veneer has a high moisture content.

Control of shrinkage.—Width shrinkage of flat-grain veneer generally decreases with increasing drying temperature. For example, ⅛-inch yellow-poplar dried at 150°F shrank 6 percent; when dried at 250°F it shrank 5½ percent; and when dried at 350°F it shrank 4½ percent. Shrinkage in thickness, however, tends to increase with an increase in drying temperature.

Control of color.—Color in face veneer can often be controlled to some degree by varying the time the green veneer is held in a stack prior to drying. In general, the wet veneer tends to oxidize and darken in storage. Consequently, if a light color is desired, the veneer should be dried as soon as possible after cutting.

Control of scorched veneer and dryer fires.—High drying temperatures may cause scorched veneer and possibly fires in the dryer. At temperatures from 200° to 300°F, extraneous materials volatize from wood. From 300° to 400°F, there is scorching and slow evolution of flammable gases from the wood. This progressively becomes more rapid until at about 600° to 650°F the wood can ignite spontaneously.

Even if wood does not ignite spontaneously until the temperature at its surface reaches about 650°F, if the surface becomes charred, charcoal gases may ignite at a temperature as low as 450°F.

Veneer being dried in dryers operating at 400°F or less sometimes ignites in the dryer, probably by a static spark that ignites flammable gases of volatile extraneous materials.

Avoidance of overdrying and use of lower drying temperatures are the primary means of preventing dryer fires and scorched veneer.

FLAKES, PARTICLES, AND FIBERS

Dryers for particulate wood and bark may be classified according to the form and end use of the particles; these dryers are discussed in product-oriented chapters, as follows:

Classification	Chapter and section
Fibers for fiberboard.	23-8 and 23-9
Flakes and particles of flakeboard and particleboard. .	24-4
Bark and hogged wood for energy	26-3 (fig. 26-3)

FIBER MATS

Drying of thick fiber mats is discussed in section 23-8. For a brief discussion fo drying thin, as well as thick, fiber mats, see Koch (1972, p. 1016-1020). Optimizing dryer performance on thin fiber webs is discussed by Lee and Hinds (1979). Design and application of open-mesh fabrics for dryers is discussed by Lachmann (1979, 1980).

The developing technology of press-drying thin sheets under a compressive force to induce greater conformability and bonding to stiff fibers from high-yield pulp is described by Setterholm and Benson (1977), Back and Anderson (1979), Horn (1979), and Setterholm (1979).

Readers needing further information on the technology of drying paper are referred to Dreshfield et al. (1956), Sundberg and Osterberg (1966), and Technical Association of the Pulp and Paper Industry (1974).

20-8 ENERGY TO DRY

Drying lumber, veneer, particles, and fiber mats uses more than half the energy consumed in most wood conversion plants (Boyd et al. 1976). Discussion in this section is limited to lumber and veneer. For data on energy to dry particles, see section 24-4. Roller dryers for fiber mats are typically 50 to 75 percent efficient. Energy to dry paper is not discussed in this text; for treatment of this subject the reader is referred to Evans (1979) and Wallace (1979).

LUMBER

Shottafer and Shuler (1974) presented a method of computing the components of energy usage in kiln-dryingl in one of their sample computations; energy consumption was estimated for a 25,000-board-foot charge of 4/4 sugar maple (*Acer saccharum* Marsh.) dried from 70 to 10 percent moisture content in 14 days in a prefabricated, aluminum, insulated kiln, as follows (not including fan energy):

Component of energy use	Energy/25,000 board feet	Portion of total
	Million Btu	*Percent*
Heating wood and residual water	4.3	4.0
Water evaporation	49.8	46.2
Venting	18.0	16.7
Building heat losses	35.7	33.1
Total	107.8	100

In the foregoing tabulation, building losses were broken down by Shottafer and Shuler into four componetns: through walls, 29 percent; through the roof, 7 percent; through the door, 12 percent; and through the floor, 52 percent. Some heat transfer engineers believe that heat loss through the floor accounts for a lesser percentage of total loss.

The energy used for heating the wood and evaporating the water cannot be reduced in conventional kilns, except through air-drying. In a properly operating and well adjusted kiln, vent losses can be reduced somewhat, but not greatly. However, savings can be made in building losses—for example, the floor can be insulated or the kiln can be enclosed in another building (but vented to the outside). With the enclosure, the heat loss through the kiln walls can be utilized to heat the building that encloses the kiln.

Kiln wall insulation can also be increased to reduce wall losses. If the insulation and wall thickness is increased from 2 to 4 inches in an aluminum prefab-type wall, the savings in the above example would be about 5 million Btu's per 25,000 board feet. However, thicker walls and added insulation would add several thousand dollars to the building's cost.

Total energy use is greatly dependent on initial moisture content of the lumber; for example, 4/4 hardwood dried from green condition to 7 percent moisture content will require about 6,900,000 Btu per thousand board feet dried, wheeas air-dried lumber dried to 7 percent moisture content requires only about 2,587,000 Btu, or 37 percent as much (table 20-36). Fan energy to circulate the air, which is not included in these totals, is widely variable, depending on kiln schedule and design, but may amount to 40 kwHr/thousand board feet of 4/4 hardwood dried to 7 percent moisture content from air-dry condition (Wengert 1974a).

TABLE 20-36.—*Theoretical and actual energy consumption in kiln-drying air-dried, partially air-dried, and green 4/4 rough hardwood lumber to 7 percent moisture content (McMillen and Wengert 1978)[1,2]*

Lumber condition and percentage points dried	Theoretical energy per thousand board feet[3]	Actual total energy used per thousand board feet[4]	
		Efficient	Moderately efficient
	------------------------------Million Btu------------------------		
Air-dried			
22 − 7 = 15 percent...........	0.518	1.294	2.587
Partly air-dried			
37 − 7 = 30 percent...........	1.035	2.587	5.175
Green			
47 − 7 = 40 percent...........	1.380	3.450	6.900

[1]Assumptions: (a) Average specific gravity of wood = 0.55; green volume, ovendry weight basis.
(b) Water/1 percent moisture content/thousand board feet = 33.95 pounts.
(c) Energy to evaporate 1 percent moisture content/thousand board feet = 34,561 Btu.
[2]Does not include electrical energy for fans.
[3]Assumes 100-percent kiln efficiency.
[4]Kiln efficiencies of 40 percent for "good" and 20 percent for "intermediate" are assumed; if 50-percent efficiency is maximum possible in well-insulated, steam-heated kilns in virtually perfect conditions, these estimates seem practical and are borne out generally by industry experience.

The following suggestions (Wengert 1974a) should be helpful in increasing the efficiency with which energy is used in kiln-drying.

- Use as much air-drying or forced-air-drying as possible—preferably drying to 25 percent moisture content or less.
- Do not use steam spray or water spray in the kiln except during the conditioning. Let moisture coming from the wood build humidity to desired levels. Steam may have to be used, however, when very small wet-bulb depressions are required. (See Rasumssen 1961, p. 160-162.)

- Repair and calk all leaks, cracks, and holes in the kiln structure and doors to prevent unnecessary venting and loss of heat. Make sure the door close tightly, especially at the top. Temporarily plug any leaks around the doors with rags, and order new gaskets, shimming strips, or hangers if necessary. In a track kiln, use sawdust-filled burlap bags to plug leaks around tracks. Adjust and repair the vents so that, when they are closed, they close tightly.
- For brick or cinder block kilns, maintain the moisture-vapor-resistant kiln coating in the best possible condition. This will prevent the walls and the roofs from absorbing water. Dry walls conduct less heat to the outside.
- For outdoor aluminum kilns only, paint the exterior walls and roof a dark color to increase the wall temperature by solar heat and reduce the the loss from the kiln. Check to insure that weep holes are open, not plugged. (Painting would be disastrous on permeable walls like brick or cinder block.)
- In many kilns, more heat is lost through the roof than through the walls, due mainly to wet insulation. To reduce heat losses, consider installing a new roof or repairing an old one. Add additional insulation if necessary. Make sure the interior vapor barrier or coating is intact.
- Install or repair baffling to obtain a high, uniform air velocity through the lumber and prevent short circuiting the air travel. This pays off in saving energy. Reverse air circulation only every 6 hours.
- Research has shown that in the early stages of drying, high air velocities (more than 600 fpm) can accelerate drying. In the late stages, low velocities (250 fpm) are as effective as high velocities and use less energy. Therefore, arrange to adjust fan speeds if possible during a run.
- Have the recorder-controller calibrated and checked for efficient operation. The kiln should not oscillate between periods of venting and steam spraying and should not vent and steam at the same time (Rasmussen 1961, p. 67-73).
- Check the remainder of the equipment. Are traps working? Do traps eject mostly hot water with little, if any, steam? Do valves close tightly? Are heating coils free of debris? Is valve packing tight? Is there adequate water for the wet bulb?
- Accurately determine the moisture content of the wood you are drying. Do not waste energy by overdrying or by taking too long because your samples do not represent the load. Try to plan your loads so that, when they are sufficiently dry, someone will be available to shut off the kiln (and, if possible, to unload it, reload it, and start it again). Do not allow a kiln load of dry lumber to continue to run overnight or through a weekend.
- Unload and reload the kiln as fast as possible. Avoid doing this until the air temperature has warmed up from the morning low—do not coll the kiln unnecessarily.
- In a battery of adjacent kilns, avoid unloading or loading one killn while the adjacent kiln is at 180°F or other high temperature.

- During nonuse periods, close all valves tightly and keep kiln doors closed. Use a small amount of heat, if necessary, to prevent freezing of steamlines and waterlines.
- Use accelerated schedules where possible. (See section 20-6 for methods of accelerating schedules with minimum risk.) The higher the temperature for drying, the more efficiently energy is used.
- If possible, reduce the length of time used for conditioning; some low-density hardwoods can be conditioned in 6 hours.
- Finally, check with the manufacturer of your equipment and find out if you can lower steam pressures or reudce gas or oil flow rates during perios of constant dry-bulb temperature. Also have the manufactuer adjust the burner to attain its top efficiency.

Because water evaporation and venting account for most of the energy expended in kiln-drying lumber, researchers have studied the potential for energy recovery from humid air streams (Rosen 1979), conservation through vapor recompression (Miller 1977), and desiccant heat recovery.[16] Research is continuing, but practical industrial application of these techniques is not yet accomplished.

VENEER

Veneer drying differs from lumber drying in that veneer is usually not thicker than 1/6-inch and veneer dryer temperatures generally exceed 300°F. In veneer drying, heat transfer rate governs drying rate, rather than the diffusion process. At constant temperature, veneer drying is slowest in parallel-flow dryers, somewhat faster in impingement-jet dryers, and most rapid in press dryers.

Lumber kilns may be 20 to 50 percent efficient in applying heat energy to dry wood. While firm experimental data are not published, it is probable that continuous-flow impingement-jet veneer dryers are somewhat more efficent than batch-loaded lumber dry kilns that operate at lower temperatures. Press dryers for veneer should be more energy efficient than impingement-jet dryers.

High-temperature flash dryers for particles expend 1,500 to 1,900 Btu per pound of water evaporated (Stillinger 1967); the most efficient veneer dryers probably require somewhat greater energy expenditure to evaporate a pound of water from hardwood veneer. Comstock (1975) indicated a range from 1,600 to 3,000 Btu per pound of water evaporated from softwood veneer.

---

[16]Moore International-Memphis and Lockheed-Huntsville. 1980. Proposal MMW 5457-1 to U.S. Department of Energy to use a liquid dessicant to absorb water from moist air normally vented from a lumber kiln.

Furman and Desmon,[17] in a case study of the saings possible through conversion of a propane-fired veneer dryer to direct-wood-fired, indicated an energy consumption of 2.1 million Btu to dry veneer for 1,000 square feet of softwood plywood—3/8-inch basis. Comstock (1975) estimated that 1 million to 2.8 million Btu are required to dry softwood veneer sufficient for 1,000 square feet of 3/8-inch plywood.

Veneer dryer operators interested in determining the energy efficiency of their dryers should find Cory's (1975) analysis useful.

20-9 SOLAR DRYING

A thousand board feet of green hardwood lumber requires 3 to 7 million Btu for kiln-drying from green to 7 percent moisture content (table 20-36). In southern latitudes a solar panel tipped toward the sun at an angle with the ground equal to the latitude (e.g., 37.2° at Blacksburg, Va.) receives average insolation of about 1,500 to 1,800 Btu daily per square foot of panel (Thomas 1977). To solar-dry a thousand board feet of 4/4 hardwood in 30 days should therefore require 87 to 202 square feet of solar panel capturing 1155 Btu/square foot/day (i.e., 70 percent of insolation of 1,650 Btu).

Most of the experimental solar kilns built have employed air systems because they are cheaper to construct and maintain than liquid systems, and are easily built locally. Active air systems, however, require larger and more expensive insulated ducts and fans than comparable components of liquid systems. If thermal storage is considered, pebble-bin storage bins typically used with air systems are about three times the volume of comparable hot-water thermal storage tanks, but installed costs of the two thermal storage systems are about equal (Thomas 1977).

Air-system solar-heated kilns typically have no provision for control of wet-bulb depression at the end of the drying cycle, and therefore lumber dried in them is neither equalized or conditioned. More sophisticated (and expensive) designs might incorporate humidity control.

Experience suggests that simple air systems may be suitable for craft operations needing only small quantities of kiln-dried hardwood lumber. Readers interested in construction of such simple kilns will find useful the designs and operating data provided by Bois (1977b), Simpson and Tschernitz (1977), Troxell (1977), Chen et al. (1978), Wengert (1980d) whose design is illustrated in figure 20-44, and Oliveira et al. (1982).

For readers interested in drying cordwood for use in home heating, Wengert (1979b) has described a low-cost solar dryer that should dry a cord of green oak in about 45 days.

---

[17]Furman, L.H., and L.G. Desmon. 1976. Conserving energy by burning wood residue to dry veneer. Paper prepared for presentation at the First Ann. Tech. Session, Energy Conservation Tech. Comm., For. Prod. Res. Soc., July 13, Toronto, Can.

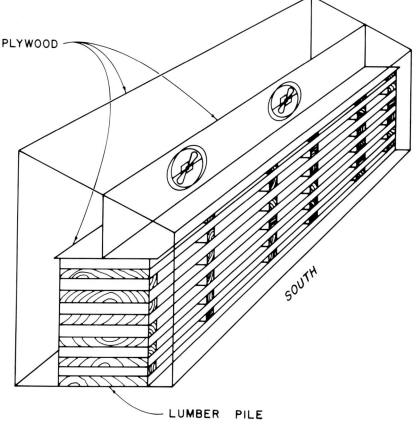

Figure 20-44.—Solar-heated lumber dryer for 2,000 board feet of 4/4 lumber. (Top) Dryer framed in 2 by 4's, roofed with clear platic or fiberglass, floored with 2 by 6's (plywood covered and insulated), and with interior plywood walls, baffles, and fan housings painted black to absorb solar energy. (Bottom) Kiln in use. (Drawing after Wengert 1980b.)

It seems likely that the major contribution of solar energy to drying lumber will be through conventional air-drying of green lumber to about 20 percent moisture content before it is kiln-dried. Also, low temperature forced-air dryers for hardwoods can be modified to use solar heat to supplement a steam heating system, as described by Cuppett and Craft (1975).

To take a broader view of utilization of solar energy in the United States, it is evident that forest residual chips harvested from cull trees and logging slash will be a major component of the fuel consumed in the future by the forest products industry. These forest residual chips—and other wood residues—will be burned in wood-fired furnaces to generate process heat and electrical energy. Other plants will produce charcoal and liquid and gaseous fuels from forest and mill woody residues. (See chapter 26.)

20-10 HIGH FREQUENCY AND MICROWAVE HEATING

While there are technical and cost problems to be surmounted, it is possible to season wood by placing it in an electrical field that oscillates at high frequency (e.g., more than 1 million cycles per second) between condenser plates or electrodes. Such a field heats the free water more quickly than the wood because of the polarity of the water molecules; the water is supplied sufficient energy to vaporize it and also heat the wood.

If there were no resistance to the movement of free water or vapor through a wood, it could be dried rapidly. In general, however, wood—including most southern hardwood—is not very permeable. Moisture movement may be so impeded that high internal pressures and temperatures well above the boiling point of water result. Local explosions or ruptures may occur. In permeable woods, such as yellow-poplar sapwood, temperature levels off slightly above the boiling point as long as free water is present. When only bound water remains, the temperature rises. Prolonged high temperatures weaken wood and lower its resistance to internal pressures.

High-frequency (radio frequency) and microwave-frequency drying differ by the wavelength and frequencies of the electromagnetic energy used, and by the method of generating the energy. **High-frequency** energy is generated by vacuum tubes, has wavelength of several to 10 meters, and frequency of 1 to 5 megahertz. **Microwave energy** is generated by a magnatron, has wave length of a few centimenters, and frequency of 3,000 to 10,000 megahertz.

Apart from technical difficulties, high-frequency heating is generally expensive. The main costs are for high-frequency generators, power tube replacement, and electricity.

For example, drying 10,000 board feet of green hardwood sapwood from 70 to 8 percent moisture content in 24 hours calls for expenditure of about 263 kW to heat the lumber and remove 764 pounds of water per hour. Because a high-frequency generator is only about 50 percent efficient and a certain amount of heat would be lost, the power line would have to deliver at least 600 kW to the generator. At 3¢/kWh, the cost would be $43.20/1,000 bd ft of lumber.

High-frequency heating has been used for drying persimmon (*Diospyros virginiana* L.) sapwood golf club heads. A company in New England tried the method for drying turning squares of birch (*Betula* sp.) and maple (*Acer* sp.) but then abandoned it.

Microwave heating has been used to season tanoak (*Lithocarpus densiflorus* (Hook & Arn.) Rehd.) baseball bats in Oregon. The principles and constraints are similar to those of high-frequency drying. Although the tanoak was believed to be principally sapwood, degrade was high (20 to 30 percent) and there was considerable collapse; thus the method was abandoned.

Simpson (1980b) studied the feasibility of drying clear hardwood cuttings by radio-frequency drying. Northern red oak boards 1 inch thick by 4 inches wide by 24 inches long were dried from an initial moisture content of about 80 percent. Drying occurred rapidly above the fiber saturation point, but slowed considerably below this moisture content. With the drying system used in Simpson's investigation, the average moisture content could be reduced in 15 minutes from 80 to 24 percent. He found, however, that the level of degrade in boards dried to 25 percent or below was probably too high for many oak products in which honeycomb is an important defect. He concluded that radio-frequency drying has some potential as a predrying process to be followed by lower temperature drying.

Industrial trials combining radio-frequency drying with vacuum drying of southern hardwoods were in process in 1982; feasibility analyses of these trials, when published, should be of interest.

20-11 MINOR SPECIAL DRYING METHODS

IMMERSION IN HOT ORGANIC LIQUIDS

Wood can be rapidly dried in an oily liquid maintained at a temperature high enough to boil off the water. A complicated variation of the method is called azeotropic drying. Recent research (Eckelman and Galezewski 1970; Huffman et al. 1972) has demonstrated that the process does not have good prospects for drying hardwoods unless a considerably reduced pressure (partial vacuum) is used. Although simple boiling-in-oil is not suitable for hardwoods because of checking, a review of the details (McMillen 1956b) would be valuable to anyone interested in the process.

In Shunk's (1976) review of industry practices, he noted that the double Rueping process (Bescher 1958) is used by the wood treating industry. As part of the treating cycle, intermittent pressure and vacuum are used. The process depends on heating green, or partially air-dried, hardwood crossties or timbers by impregnating them with hot creosote and then utilizing this heat to evaporate and redistribute moisture during a subsequent vacuum period. This permits deeper preservative penetration during a second pressure cycle. The entire process requires 15 to 17 hours in a treating retort.

VAPOR DRYING

Vapor drying is accomplished by condensing the vapors of high-boiling (280° to 320°F) organic solvents on the surface of the wood. As condensation occurs, the latent heat of the solvent is given up and moisture in the wood vaporizes (McMillen 1961). This method has had considerable use in drying oak, hickory, and other hardwood railroad crossties before preservative treatment. It would also be suitable for crossing plank and car decking, but is not suitable for hardwoods to be used at low moisture content values for fine purposes.

SOLVENT SEASONING

In the solvent seasoning process, wood is continuously sprayed, or immersed, in hot acetone or other solvent miscible with water. Treatment continues for a number of hours until most of the water is extracted from the wood. Then the solvent and some additional water are removed by steaming or by application of vacuum.

While this method has not been applied to eastern hardwoods, extensive research has been done in California on drying tanoak sapwood in this way. One-inch lumber has been dried in 30 hours. A few boards suffered streaks of collapse, probably because of the presence of heartwood.

OTHER

Infrared radiation has been proposed in the past for drying wood, but the depth of ray penetration is slight and the method is not suitable for drying hardwood lumber.

Vacuum drying and freeze-vacuum drying alone are of no practical value for hardwoods because any heat available is quickly used up in initial evaporation and no further evaporation can occur without additional heat; air must be present to conduct the heat. Patents have been granted on combining heated-platen or high-frequency heating with vacuum drying; such systems have received only limited application in the United States, however, because of high cost.

20-12 CHEMICAL TREATMENTS TO PREVENT CHECKING

Presurfacing is one method of reducing surface checking. Check reduction can also be accomplished through application of certain chemicals. Some that have been used successfully are wax, sodium alginate, salt paste, and polyethylene glycol. These materials either retard moisture movement or alter the vapor pressure at the surface. Few details are available on use of waxes, but a thick emulsion of microcrystalline wax has been applied to the sides of highly figured gunstocks in California to prevent checking, and it is frequently used in commercial practice for end coatings of thick, valuable hardwoods.

SODIUM ALGINATE

In Australia, Harrison (1968) investigated the use of very viscous sodium alginate solutions or emulsions as dip treatments on a variety of hardwoods up to 2 inches thick. When the lumber is air-dried, the alginate dries to form a porous skin over the surface. In all of the woods tried under severe air-drying conditions, it is effective in preventing checking. The method has not been tried in the United States but might have some benefit for thick oak.

Sodium alginate is a dry powder obtained from seaweed and is used in a variety of products in the United States. Considerable care must be used in mixing it to form the 1½-percent solution found most effective in Australia. To effectively prevent checking, the wood must be quite green; air-drying piles must be carefully roofed to keep the alginate from being washed off by rain. For 2-inch stock, a drying shed is necessary to protect the treating chemical and provide mild conditions for air-drying.

SALT PASTE

The U.S. Bicentennial celebration inspired widespread interest in seasoning disks or thick sections of large trees. The disks were desired for exhibits or usable items on which the chronology of important events could be shown.

To cut tree disks without formation of checks at the time they are cut, Kubler and Chen (1974) suggest that logs be first cut into short sections and then divided into disks; in check-prone stems with considerable internal growth stresses, they found that checks could be prevented entirely by cutting many disks simultaneously from a log.

It is very difficult to season disks successfully with even the least-difficult-to-dry species. The most damaging defect is the large V-shaped check that is likely to develop because tangential shrinkage is usually much greater than radial shrinkage. In addition, many small end checks tend to appear over the entire surface.

A simple and inexpensive method developed by W. T. Simpson and J. L. Tschernitz of the U.S. Forest Products Laboratory is a thick-paste modification of the old "salt seasoning" method (Loughborough 1939; Haygreen 1962), combining some bulking of the surface zone with retention of moisture at the disk surface by hygroscopic action. The bark should be left on the disk. Before applying the paste, the surfaces of the green disk are alternatively brushed with a concentrated table salt solution—3 pounds salt per gallon of water, with excess salt crystals visible in the solution after thorough mixing. Several hours are allowed for the salt to penetrate by diffusion and gravity.

Then the wood is treated with the paste. To make the paste, the concentrated solution above is mixed with enough cornstarch to get the right consistency to build up a thick layer on the disk surface. The addition of several egg whites acts as a convenient binder to reduce the flaking of the paste after it dries. The treated disks can be air-dried in a room with plenty of ventilation or kiln-dried using a moderate schedule.

POLYETHYLENE GLYCOL

Saturating wood in polyethylene glycol (PEG) is also an effective way of treating green wood so that it will not check or split when dried (Mitchell 1972); the technique is slow, however, requires a moderately to fully permeable wood, is costly, and the temptation always exists to terminate treatment before all wood cells are fully saturated with PEG

Nara et al. (1978) soaked boxed-heart specimens of ash, elm, and oak in 60-percent solutions of PEG 1,000, restrained them against warping, and dried them to 8 percent moisture content. The PEG treatment prevented the development of checks that formed in untreated wood. With ash and oak, drying times were reduced compared to untreated wood; time required to dry elm was extended, however.

20-13 STORAGE OF LUMBER, PLYWOOD, AND PARTICLEBOARD

Protection against stain, decay, and bacteria is discussed in section 11-6 (see headings LUMBER and COMPOSITE PRODUCTS). Protection against insect attack is discussed in section 11-9 (see headings DRY LUMBER, WOOD IN USE, and CONTROL OF LYCTUS POWDER POST BEETLES).

Aside from the economics and materials handling aspects, the primary objectives of storage are to keep the lumber, millwork, or panels, clean, undamaged, and to maintain them at a moisture content approximating that which they will reach in use. Solid-piled wood changes moisture content slowly if protected from the elements. Protection afforded commonly ranges from a simple roof to an enclosed and heated warehouse. Fluctuations of moisture content in southern pine under various storage situations were measured a number of years ago; these data, abstracted in Koch (1972, p. 746-752), are still highly applicable to southern pine, and in the absence of other information, to pine-site hardwoods. Data on equilibrium moisture contents of hardwood lumber and reconstituted products are given in section 8-3 of this text.

LUMBER

In his handbook on the storage of lumber Rietz (1978) makes recommendations for handling hardwood lumber during the major stages of its manufacture and use; his recommendations are abstracted in the following paragraphs.

At the sawmill.—Store air-dried lumber under cover to prevent rewetting of surfaces or moisture regain. Covers on outside bulk-piled stock should effectively exclude rainwater, and the piles should be clear of the ground to allow ventilation. Air-dried lumber can be stored in closed sheds, either bulked or on sticks; shed doors should be kept closed when possible. Kiln-dried lumber should be stored solid-piled in closed heated sheds.

In transit.—Green lumber shipments can be made without protection when the transit periods are short. Lumber should be air-dried prior to long periods of rail, truck, and ship transport. Air-dried lumber can generally be shipped by truck, rail, or ocean-going vessel without protection; if lumber is rewetted, however, it may take on a compression set that will cause any checks to stay open when the wood is finally kiln-dried. Kiln-dried lumber requires protection during transit. Use tight boxcars or package wrappers for rail shipments. Use watertight tarpaulins on open truck shipments. Stow below decks on vessels.

At lumber distributing yards.—Green or partially air-dried rough lumber should be stickered and stacked for air-drying. High-value lumber should be air-dried in an open shed. Air-dried lumber, when bulked, requires protection from weathering and moisture regain. Outdoor storage requires suitable raintight cover and some elevation off the ground. Storage in open sheds is ideal. Kiln-dried lumber should be bulk piled and stored in a closed heated shed.

At woodworking factories.—Green lumber must be stickered for drying. If the stickered units are air-dried outdoors, pile covers should be used. Storage of stickered lumber in an open shed gives better protection than pile covers. Air-dried lumber can be bulked and stored in an open shed. If the lumber exceeds 20 percent moisture content, stickering and further drying in a conventional air-drying yard or an open shed is beneficial. Kiln-dried lumber should be stored in a heated shed if the stock is left on stickers pending transport to the rough mill or cutup plant. If the stock is bulked for storage outdoors, protection from the weather is essential and the storage period must be limited. Bulked indoor storage should be in a shed with temperature and relative humidity controlled to yield the desired moisture content.

At building sites.—Do not unload lumber or millwork in the rain, or directly onto the ground. Protect dry wood from the weather and keep exposed storage time as short as possible.

PLYWOOD, FIBERBOARD, PARTICLEBOARD, AND COMPOSITE PANELS

Equilibrium moisture contents of plywood, fiberboard, and flakeboard panels are given in table 8-8 and figures 8-15 through 8-19.

Plywood and other reconstituted panels for interior use in heated and cooled buildings should be stored in closed heated sheds with conditions controlled to yield an equilibrium moisture content approximating that they will attain in use. Generally in the United States these conditions will be near 50 percent relative humidity at 72°F.

Composite panels, comprised of hardwood veneers over flake cores, and structural panels comprised entirely of flakes, will be used increasingly in the future. Produced with phenol-formaldehyde adhesives, these panels generally are at less than 10 percent moisture content when manufactured. Because they are used in concealed structural, rather than decorative, applications, storage conditions are less critical than for appearance-grade fine hardwood products. To prevent sag, the panels should be stored on a rigid platform, and a platform should also be used between fork-lift units in a pile. Closed unheated sheds should be satisfactory for structural panels, except during the rainy seasons in the humid South when heated closed sheds would be better.

20-14 LITERATURE CITED

Adams, E. L. 1971. Effect of moisture loss on red oak sawlog weight. U.S. Dep. Agric. For. Serv., Res. Note NE-133. 4 p.

Arganbright, D. G. 1979. Developments in applied drying technology, 1971-1977. For. Prod. J. 29(12):14-20.

Back, E., and R. Anderson. 1979. Multistage press drying of paper. Svensk Papperstidning 82(2):35-39.

Bescher, R. H., Chairman. 1958. Report of sub-committee 3 on checking and splitting of crossties. Railway Tie Association, St. Louis, Mo.

Bethel, J. S. and R. J. Hader. 1952. Hardwood veneer drying. J. For. Prod. Res. Soc. 2(5):205-215.

Bois, P. J. 1977a. The four stages of drying in thick oak. FPRS News-Dig. File: J-1.3, 4 p. Madison, Wis.: For. Prod. Res. Soc.

Bois, P. J. 1977b. Constructing and operating a small solar-heated lumber dryer. U.S. Dep. Agric. For. Serv., For. Prod. Util. Tech. Rep. No. 7. 12 p. State and Private For., Madison, Wis.

Bois, P. J. 1978. Handling, drying, and storing heavy oak lumber. U.S. Dep. Agric. For. Serv., For. Prod. Util. Tech. Rep. No. 8. 28 p. State and Private For., Madison, Wis.

Boone, R. S., and R. R. Maeglin. 1980. High-temperature drying of 7/4 yellow-poplar flitches for S-D-R studs. U.S. Dep. Agric. For. Serv. Res. Pap. FPL 365. 9 p.

Boyd, C. W., P. Koch, H. B. McKean, C. R. Morschauser, S. B. Preston, F. F. Wangaard. 1976. Wood for structural and architectural purposes. Wood and Fiber 8:1-72.

Brace Research Institute. 1975. A survey of solar agricultural dryers. Tech. Rep. T99. Brace Res. Inst., Macdonald College of McGill University, St. Anne de Bellevue, Quebec, Can.

Cassens, D. L. (compiler). 1979. High-temperature drying of hardwoods. Symp. Proc. (New Albany, Ind., March 22.) Coop. Ext. Serv., Purdue Univ., West Lafayette, Ind., 79 p.

Catterick, J. 1970. Costs of low temperature drying. North. Logger and Timber Proc. 18(9):18, 50, 60

Cech, M. Y. 1971. Dynamic transverse compression treatment to improve drying behavior of yellow birch. For. Prod. J. 21(2):41-50.

Cech, M. Y. 1973. Status of high-temperature kiln drying in eastern Canada. Can. For. Ind. 93(8): 63-65, 67, 69, 71.

Cech, M. Y. and F. Pfaff. 1975. Kiln-drying of 1-inch red oak. For. Prod. J. 25(8):30-37.

Chen, P. Y. S. 1980. Press conditions affect drying rate and shrinkage of hardwood boards. For. Prod. J. 30(7):43-47.

Chen, P. Y. S., E. C. Workman, Jr. and C. E. Helton. 1978. How to build a collector for a solar kiln with aluminum cans. FPRS News-Dig. File G-1.14, 4 p.

Clark, W. P., and T. M. Headlee. 1958. Roof your lumber and increase your profits. For. Prod. J. 8(12):19A-21A.

Comstock, G. L. 1971. The kinetics of veneer jet drying. For. Prod. J. 21(9):104-111.

Comstock, G. L. 1975. Energy requirements for drying of wood products. *In* Wood residue as an energy source. Proc. No. P-75-13, For. Prod. Res. Soc., Madison, Wis., p. 8-12.

Cooper, G. A., and S. H. Barham. 1972. Prefreezing effects on three hardwoods. For. Prod. J. 22(2):24-25.

Cooper, G. L., P. J. Bois, and R. W. Erickson. 1976. Progress report on kiln-drying prefrozen walnut gunstocks—techniques and results. FPRS Sep. No. MW-75-S70, 6 p. For. Prod. Res. Soc., Madison, Wis.

Cooper, G. A., R. W. Erickson, and J. G. Haygreen. 1970. Drying behavior of prefrozen black walnut. For. Prod. J. 20(1):30-35.

Cory, W. N. 1975. How to determine how effectively your dryer is using energy input. For. Ind. 102(8):54-55.

Cuppett, D. G. 1965. How to determine seasoning degade losses in sawmill lumberyards. U.S. Dep. Agric. For. Serv., Res. Note NE-32. 7 p.

Cuppett, D. G. 1966. Air-drying practices in the central Appalachians. U.S. Dep. Agric. For. Serv., Res. Pap. NE-56. 19 p.

Cuppett, D. G. and E. P. Craft, 1971. Low-temperature drying of 4/4 Appalachian red oak. For. Prod. J. 21(1):34-38.

Cuppett, D. G., and E. P. Craft. 1972. Kiln-drying of presurfaced 4/4 Appalachian oak. For. Prod. J. 22(6):36-41.

Cuppett, D. G. and E. P. Craft. 1975. Low-temperature forced-air drying of Appalachian hardwoods. U.S. Dept. Agric. For. Serv., Res. Pap. NE-328. 10 p.

Davenport, K. L. and L. E. Wilson. 1969. A cost analysis of yard drying and low-temperature forced-air drying of lumber. Circ. 170, 19 p. Agric. Exp. Stn., Auburn Univ., Auburn, Ala.

Denig, J., and E. M. Wengert. 1982. Estimating air-drying moisture content losses for red oak and yellow-poplar lumber. For. Prod. J. 32(2):26-31.

Dreshfield. A. C., Jr., and S. T. Han 1956. The drying of paper. Tappi 39:449-455.

Eckelman, C. A. and J. A. Galezewski. 1970. Azeotropic drying of hardwoods under vacuum. For. Prod. J. 20(6):33-36.

Engalichev, N., and W. E. Eddy. 1970. Economic analysis of low temperature kilns in processing softwood lumber for market. Ext. Bull. No. 178, Coop. Ext. Serv., Univ. New Hampshire, Durham. 107 p.

Evans, J. C. W. 1979. Major savings are possible with improved linerboard drying. Pulp and Pap. 53(12):97-104.

Fessel, F. 1964. Continuous veneer drying with jet ventilation. Holz als Roh- und Werkstoff 22:129-139.

Finighan, R., and R. M. Liversidge. 1972. Improving the performance of air seasoning yards. Some factors affecting air-flow. The Australian Timber J. 38(3):12, 13, 15, 16, 18, 19, 23, 25, 27.

Fleischer, H. O. 1953. Veneer drying rates and factors affecting them. J. For. Prod. Res. Soc. 3(3):27-32, 91.

Freese, F., H. A. Stewart, and R. S. Boone. 1976. Rough thickness requirements for red oak furniture cuttings. U.S. Dep. Agric. For. Serv., Res. Pap. FPL 276. 4 p.

Garrett, L. D. 1983. Moisture loss from felled eastern hardwood and softwood trees. *In* Proc., Wood Industrial Energy Forum '82, Vol. I. Washington, D.C., March 8-10, 1982. For Prod. Res. Soc., Madison, Wis., p. 210-214.

Gatslick, H. B. 1962. The potential of the forced-air drying of northern hardwoods. For. Prod. J. 12:385-388.

Goebel, N. B., M. A. Taras, and W. R. Smith. 1960. Tension wood and its relation to splitting in hickory. S. C. Agric. Exp. Stn. Bull 480, 21 p. Clemson Agric. Coll., Clemson, S.C.

Goulet, M., and M. Ouimet. 1970. (Determination of costs in wood seasoning.) Univ. Laval (Quebec) Tech. Note No. 5, 52 p.

Hallock, H. and E. H. Bulgrin. 1977. A look at yellow-poplar for studs. U.S. Dep. Agric. For. Serv., Res. Note FPL-0238. 7 p.

Hanks, L. F., and M. K. Peirsol. 1975. Value loss of hardwood lumber during air-drying. U.S. Dep. Agric. For. Serv. Res. Pap. NE-309. 10 p.

Hann, R. A., R. W. Jokerst, R. S. Kurtenacker, C. C. Peters, and J. L. Tschernitz. 1971. Rapid production of pallet deckboards from low-grade logs. U. S. Dep. Agric. For. Serv. Res. Pap. FPL 154. 16 p.

Harrison, J. 1968. Reducing checking in timber by use of alginates. Austral. Timber J. 34(7):24-25.

Hart, C. A. 1979. Moisture pallets—A new approach to reduce degrade in air drying. *In* Management and utilization of oak. Proc., Seventh Ann. Hardwood Symp., Hardwood Res. Counc. p. 91-99.

Haygreen, J. G. 1962. A study of the kiln-drying of chemically seasoned lumber. For. Prod. J. 12:11-16.

Haygreen, J. G. 1981. Potential for compression drying of green wood chip fuel. For. Prod. J. 31(8):43-54.

Haygreen, J. G. 1981. Potential for compression drying of green wood chi fuel. For. Prod. J. 31(8):43-54.

Heebink, B. G. 1952. Veneer flooring. J. of the For. Prod. Res. Soc. 2(3):138-140.

Heebink, B. G. and K. C. Compton. 1966. Paneling and flooring from low-grade hardwood logs. U.S. Dep. Agric. For. Serv., Res. Note FPL-0122. 24 p.

Helmers, R. A. 1959. New way to speed air drying of hardwoods. Furn. Des. and Manuf. 31(11):30-31, 35.

Henry, W. T. 1970. Effect of incising on the reduction of serious checking in incised vs. unincised hardwood crossties. In Proc., Amer. Wood Preservers' Assoc. 66:171-178.

Hittmeier, M. E., G. L. Comstock, and R. A. Hann. 1968. Press drying nine species of wood. For. Prod. J. 18(9):91-96.

Horn, R. A. 1979. Bonding of press-dried high-yield pulps—role of lignin and hemicellulose. Tappi 62(7):77-80.

Howe, J. P., and P. Koch. 1976. Dowel-laminated crossties. Performance in service, technology of fabrication, and future promise. For. Prod. J. 26(5):23-30.

Huber, H. A. and A. W. Klimaszewski. 1963. Studies on predrying and polyethylene glycol treatment of green oak. For. Prod. J. 13:439-442.

Huffman, D. R., F. Pfaff, and S. M. Shah. 1972. Azeotropic drying of yellow birch and hard maple lumber. For. Prod. J. 22(8):53-56.

Huffman, J. B. 1957. The kiln drying of gum crossties. Proc., Amer. Wood-Preserv. Assoc. 53:163-178.

Huffman, J. B. 1958. Kiln drying of southern hardwood crossties. For. Prod. J. 8:165-172.

Huffman, J. B., and D. M. Post. 1960. Forced-air drying of gum and oak crossties. South Lumberman 200(2505): 33-34, 36-37.

Huffman, J. B. and D. M. Post. 1961. An exploratory study—the forced air drying of gum and oak crossties. FPRS News-Digest E-1.3, 2 p. For. Prod. Res. Soc., Madison, Wis.

Huffman, J. B. and D. M. Post. 1962a. The use of covers and fans to improve the seasoning of oak cross ties. Cross Tie Bull. 43(12):28-37.

Huffman, J. B. and D. M. Post. 1962b. Practical covers for protecting crossties during air seasoning. Res. Rep. No. 8, 8 p. Univ. of Fla., Sch. of For., Gainesville, Fla.

Huffman, J. B. and D. M. Post. 1964. Covers for protecting crossties during air seasoning. FPRS News-Digest No. D 4.2, 2 p. For. Prod. Res. Soc., Madison, Wis.

James W. L. 1975. Electric moisture meters for wood. FPRS News-Digest B-4.7, 4 p. For. Prod. Res. Soc., Madison, Wis.

Keylwerth, R. 1953. (Experiments on veneer drying.) Holz als Roh- und Werkstoff 11(1):11-17.

Kimball, K. E. 1955. Drying schedule for thin red oak. South. Lumberman 191(2393):252, 254.

Kimball, K. E., and D. P. Lowery. 1967. High-temperature and conventional temperature methods for drying lodgepole pine and western larch studs. For. Prod. J. 17(4):32-40.

Koch, P. 1964. Techniques for drying thick southern pine veneer. For. Prod. J. 14(9):382-386.

Koch, P. 1969. At 240°F southern pine studs can be dried and steam-straightened in 24 hours. South. Lumberman 219(2723):26, 28-29.

Koch, P. 1972. Utilization of the southern pines. U.S. Dep. Agric. For. Serv., Agric. Handb. 420. 1663 p. 2 vol. U.S. Govt. Print. Off., Washington, D.C.

Koch, P. 1973. High-temperature kilning of southern pine poles, timbers, lumber, and thick veneer. Proc. Amer. Wood Preserv. Assoc. 69:123-140.

Kollman, F. F. P. 1961. High-temperature drying—research, application, and experience in Germany. For. Prod. J. 11(11):508-515.

Kubler, H. and T-H. Chen. 1974. How to cut tree disks without formation of checks. For. Prod. J. 24(7):57-59.

Lachmann, T. E. 1979. Open-mesh clothing for dryers. General design parameters. Tappi 62(12):85-87.

Lachmann, T. E. 1980. Open-mesh clothing for dryers. Application guidelines. Tappi 63(1):64-66.

Lahtinen, P. 1975. Continuous veneer drying reduces production costs. Crow's 53(6):20-21.

Lee, P. F., and J. A. Hinds. 1979. Optimizing dryer performance: A technique for measuring the drying characteristics of fiber webs. Tappi 62(4):45-48.

Loughborough, W. K. 1939. Chemical seasoning of overcup oak. South. Lumberman 159(2009):137-140.

Lowery, D. P., and K. E. Kimball. 1966. Evaluation of high-temperature and conventional methods of stud drying. In Proc., 21st Ann. Northwest Wood Prod. Clinic, Spokane, Wash., April 19-20. p. 55-65. Washington State Univ., Tech. Ext. Serv., Pullman.

Lowery, D. P., J.P. Krier, and R. A. Hann. 1968. High-temperature drying of lumber—a review. U.S. Dep. Agric. For Serv. Res. Pap. INT-48. 10 p.

Lutz, J. F. 1970. Buckle in veneer. U.S. Dep. Agric. For. Serv., Res. Note FPL-0207. 11 p.

Lutz, J. F. 1972. Veneer species that grow in the United States. U.S. Dep. Agric. For. Serv., Res. Pap. FPL 167. 127 p.

Lutz, J. F. 1974. Drying veneer to a controlled final moisture content by hot pressing and steaming. U.S. Dep. Agric. For. Serv., Res. Pap. FPL 227. 8 p.

Lutz, J. F. 1977. Wood veneer: Log selection, cutting, and drying U.S. Dep. Agric. Tech. Bull. 1577, 137 p.

McMillen, J. M. 1956a. Seasoning hickory lumber and handle blanks. U.S. Dep. Agric. For. Serv., Hickory Task Force Rep. No. 4. 36 p. Southeast. For. Exp. Stn., Asheville, N.C.

McMillen, J. M. 1956b. Special methods of seasoning wood. Boiling in oily liquids. Rep. No. 1665 (rev.) U.S. Dep. Agric., For. Serv., For. Prod. Lab., Madison, Wis. 4 p.

McMillen, J. M. 1958. Stresses in wood during drying. U.S. Dep. Agric. For. Serv., For. Prod. Lab. Rep. No. 1652. 52 p.

McMillen, J. M. 1961. Special methods of seasoning wood. Vapor drying. U.S. Dep. Agric., For. Serv., For. Prod. Lab. Rep. 1665-3 (rev.). 3 p.

McMillen, J. M. 1964. Wood drying—techniques and economics. South. Lumberman 208 (2589):25, 26, 28, 32, 34.

McMillen, J. M. 1969. Accelerated kiln drying of presurfaced 1-inch northern red oak. U.S. Dep. Agric. For. Serv. Res. Pap. FPL 122. 29 p.

McMillen, J. M. and R. C. Baltes. 1972. New kiln schedule for presurfaced oak lumber. For. Prod. J. 22(5):19-26.

McMillen, J. M., and R. S. Boone. 1974. Kiln-drying selected Colombian woods. For. Prod. J. 24(4):31-36.

McMillen, J. M., J. C. Ward, and J. Chern. 1979. Drying procedures for bacterially infected northern red oak lumber. U.S. Dep. Agric. For. Serv. Res. Pap. FPL 345. 15 p.

McMillen, J. M., and E. M. Wengert. 1978. Drying eastern hardwood lumber. U.S. Dep. Agric. Handb. 528, 104 p.

McMinn, J. W., and M. A. Taras. 1983. Transpirational drying. In Proc., Wood Industrial Energy Forum '82 Vol. I. Washington, D.C., March 8-10, 1982. For Prod. Res. Soc., Madison, Wis. p. 206-207.

Mackay, J. F. G. 1974. High-temperature kiln-drying of northern aspen 2- by 4-inch light-framing lumber For. Prod. J. 24(10):32-35.

Mackay, J. F. G. 1976. Delayed shrinkage after surfacing of high-temperature kiln-dried northern aspen dimension lumber. For. Prod. J. 26(2)33-36.

Maeglin, R. R. 1978. Yellow-poplar studs by S-D-R. South. Lumberman 237(2944):58-60.

Mathewson, J. S. 1952. Radiant heated hardwood floors with coil between subfloor and finish floor. Heating and Ventilating 49(1):72-73.

Mathewson, J. S., D. S. Morton, and R. H. Bescher. 1949. Air seasoning of red oak crossties. In Proc., Amer. Wood Preservers' Assoc. 45: 216-231.

Miller, W. 1977. Energy conservation in timber-drying kilns by vapor recompression. For. Prod. J. 27(9):54-58.

Milligan, F. H., and R. D. Davies. 1963. High speed drying of western softwoods for exterior plywood. For. Prod. J. 13(1):23-29.

Minter, J. L. 1961. 1 yard replaces 3 in model lumber storage, seasoning and grading setup. Wood and Wood Prod. 66(8):23, 24, 66, 70.

Mitchell, H. L. 1972. How PEG helps the hobbyist who works with wood. U.S. Dep. Agric. For. Serv., For. Prod. Lab. Madison, Wis. 20 p.

Mustakallio, K. and M. Paaki. 1977. Finns leading way in eliminating hot-air methods of veneer drying. For. Ind. 104(12): F-18-F-19.

Nara, N., M. Yonede, M. Chiba, S. Kanno, and Y. Oyama. 1978. (Drying of hardwood from small log—PEG treatment of boxed heart.) Rep. Hokkaido For. Prod. Res. Inst. 67:129-170. (Japan)

Oliveira, L. C. de S., C. Skaar, and E. M. Wengert. 1982. Solar and air lumber drying during winter in Virginia. For. Prod. J. 32(1):37-44.

Pease, D. A. 1980. Platen veneer drying system cuts shrinkage, energy costs. For. Ind. 107(7):29-31.

Peck, E. C. 1953. Reducing checking in heavy white oak shipbuilding material during storage and construction. J. of the For. Prod. Res. Soc. 3(4):22-23.

Peck, E. C. 1961. Air-drying of lumber. U.S. Dep. Agric. For. Serv., For. Prod. Lab. Rep. 1657, 21 p.

Perem, E. 1971. Checking and splitting of hardwood rail ties in seasoning and service. Can. For. Serv. Dep. of Fish. and For. Publ. No. 1293, 32 p.

Price, E. W. 1982. Chemical stains in hackberry can be prevented. South. Lumberman 243(3019):13-15.

Rasmussen, E. F. 1961. Dry kiln operator's manual. Agric. Handb. 188. U.S. Dep. Agric., For. Serv., For. Prod. Lab., Madison, Wis. 197 p.

Rasmussen, E. F., and M. B. Avanzado. 1961. Full-time or part-time kiln-drying? Comparison of the energy, drying time, and costs. South. Lumberman 203(2537):99-104.

Resch, H., C. A. Lofdahl, F. J. Smith, and C. Erb. 1970. Moisture leveling in veneer by microwaves and hot air. For. Prod. J. 20(10):50-58.

Rice, W. W. 1971. Field test of a schedule for accelerated kiln-drying of presurfaced 1-inch northern red oak. Res. Bull. 595, Univ. Mass., Coll. Agric., Agric. Exp. Stn. 19 p.

Richards, D. B. 1958. High temperature drying of southern hardwoods. Agric. Exp. Stn. Circ. 123, 11 p. Ala. Polytech. Inst., Auburn, Ala.

Rietz, R. C. 1978. Storage of lumber. U. S. Dep. Agric. For. Serv., Agric. Handb. 531. 63 p.

Rietz, R. C., and J. A. Jenson. 1966. Presurfacing green oak lumber to reduce surface checking. U.S. Dep. Agric. For. Serv. Res. Note FPL-0146. 2 p.

Rietz, R. C. and R. H. Page. 1971. Air drying of lumber: a guide to industrial practices. U.S. Dep. Agric. For. Serv., Agric. Handb. 402. 110 p. U.S. Govt. Print. Off., Washington, D.C.

Rogers, K. E. 1981. Preharvest drying of logging residues. For. Prod. J. 31(12):32-36.

Rosen, H. N. 1977. Impinging air jets and high temperatures to dry lumber. In Proc., 5th Ann. Hardwood Symp. of the Hardwood Res. Counc., Asheville, N.C., p. 18-27.

Rosen, H. N. 1979. Potential for energy recovery from humid air streams. U.S. Dep. Agric. For. Serv. Res. Pap. NC-170. 10 p.

Rosen, H. N., and R. E. Bodkin. 1978. Drying curves and wood quality of silver maple jet-dried at high temperature. For. Prod. J. 28(9):37-43.

Rosen, H. N., and R. E. Bodkin. 1981. Development of a schedule for jet-drying yellow-poplar. For. Prod. J. 31(3):39-44.

Ruddick, J. N. R., and N. A. Ross. 1979. Effect of kerfing on checking of untreated Douglas-fir pole sections. For. Prod. J. 29(9):27-30.

Sawyer, W. C. 1983. In-forest drying of forest residue (IFD). In Proc., Wood Industrial Energy Forum '82, Vol. I. Washington, D.C., March 8-10, 1982. For. Prod. Res. Soc., Madison, Wis. p. 192-196.

Setterholm, V. C. 1979. An overview of drying. Tappi 62(3):45-46.

Setterholm, V. C. and R. E. Benson. 1977. Variables in press drying pulps from sweetgum and red oak. U.S. Dep. Agric. For. Serv., Res. Pap. FPL 295. 16 p.

Shinn, F. S. 1955. A progress report on hickory cross ties. Cross Tie Bull. 36(11):70-72.

Shottafer, J. E., and C. E. Shuler. 1974. Estimating heat consumption in kiln-drying lumber. Tech. Bull. 73, Life Sci. and Agric. Exp. Stn., Univ. Maine, Orono. 25 p.

Shunk, B. H. 1976. Drying southern hardwood timbers, crossties, and posts. For. Prod. J. 26(4):51-57.

Simpson, W. T. 1975. Effect of steaming on the drying rate of several species of wood. Wood Sci. 7:247-255.

Simpson, W. T. 1976a. Effect of presteaming on moisture gradient of northern red oak during drying. Wood Sci. 8:272-276.

Simpson, W. T. 1976b. Steaming northern red oak to reduce kiln-drying time. For. Prod. J. 26(10):35-36.

Simpson, W. T. 1980a. Accelerating the kiln-drying of oak. Res. Pap. FPL-378, U.S. Dep. Agric., For. Serv., 9 p.

Simpson, W. T. 1980b. Radio-frequency dielectric drying of short lengths of northern red oak. U.S. Dep. Agric., For. Serv., Res. Pap. FPL 377. 8 p.

Simpson, W. T. 1982. Warp reduction in kiln-drying hardwood dimension. For Prod. J. 32(5):29-32.

Simpson, W. T., and R. C. Baltes. 1972. Accelerating oak air drying by presurfacing. U.S. Dep. Agric. For. Serv. Res. Note FPL-0223. 12 p.

Simpson, W. T., and J. G. Schroeder. 1980. Kiln-drying hardwood dimenstion parts. U.S. Dep. Agric., For. Serv., Res. Pap. FPL 388. 8 p.

Simpson, W. T., and J. L. Tschernitz. 1977. Solar lumber dryer designs for developing countries. In Practical application for wood processing. Proc. of a Workshop at Virginia Polytech. Inst. and State Univ., Blacksburg, Va. Jan. 6-7, p. 56-61.

Skaar, C., and W. Simpson. 1968. Thermodynamics of water sorption by wood. For. Prod. J. 18(7):49-58.

Smith, W. R. and N. B. Goebel. 1952. The moisture content of green hickory. J. of For. 50(8):616-618.

Springer, E. L. 1979. Should whole-tree chips for fuel be dried before storage. U.S. Dep. Agric. For. Serv. Res. Note FPL-0241. 5 p.

Stillinger, J. R. 1967. The Heil dryer. In Proc., First Particleboard Symposium 1:205-215. Washington State Univ., Pullman, Wash.

Sundberg, T., and L. Osterberg. 1966. Thermal resistance between drying cylinders and paper. Svensk Papperstidning 69:854-856.

Taras, M. A. and M. Hudson. 1959. Seasoning and preservative treatment of hickory crossties. U.S. Dep. Agric. For. Serv., Hickory Task Force Rep. No. 8. 24 p. Southeast. For. Exp. Stn., Asheville, N.C.

Technical Association of the Pulp and Paper Industry. 1974. Paper machine drying rate. TIS 014-39, TIS 014-40, TIS 014-41, TIS 014-42, TIS 014-43, and TIS 014-44. TAPPI, Atlanta, Ga.

Thomas, W. C. 1977. Introduction to solar energy application for wood processing. In Practical application of solar energy to wood processing. Proc. of a Workshop at Virginia Polytech. Inst. and State Univ., Blacksburg, Va. Jan. 6-7, p. 5-13.

Tiemann, H. D. 1918. Dry kiln. U.S. Pat No. 1,268,180. U.S. Pat. Off., Washington, D.C. 5 p.

Timber Processing Industry. 1980. Green veneer drying process developed by Weyerhaeuser. Timber Proc. Ind. 5(5):34, 36.

Townsend, K. 1979a. Dehumidification drying. In Management and utilization of oak. Proc., Seventh Ann. Hardwood Symp., Hardwood Res. Counc., p. 100-116.

Townsend, K. 1979b. Dehumidification drying—Part II. Views of the equipment manufacturer. National Hardwood Magazine 52(13):38-39, 61-64.

Troxell, H. E. 1977. An application of solar energy for drying lumber in the central Rocky Mountain region. In Practical application for wood processing. Proc. of a Workshop at Virginia Polytech. Inst. and State Univ., Blacksburg, Va. Jan. 6-7, p. 49-55.

Tschernitz, J. L., and W. T. Simpson. 1979. Drying rate of northern red oak lumber as an analytical function of temperature, relative humidity, and thickness. Wood Sci. 11:202-208.

U.S. Forest Products Laboratory. 1961. List of manufacturers and dealers for log and lumber end coatings. U.S. Dep. Agric. For. Serv. For. Prod. Lab. Rep. 1954. 1 p.

Vick, C. B. 1965a. Drying-rate curves for one-inch yellow-poplar lumber in low-temperature forced-air dryers. For. Prod. J. 15:500-504.

Vick, C. B. 1965b. Forced air drying sweet gum furniture billets in solid stacks. For. Prod. J. 15:121.

Vick, C. B. 1968a. Low-temperature drying of 1-inch sweetgum. For. Prod. J. 18(4):29-32.

Vick, C. B. 1968b. Commercial forced-air drying of unstickered hardwood furniture rounds. For. Prod. J. 18(7):46-48.

Wallace, B. W. 1979. Controlling moisture on paper machines results in significant energy savings. Pap. Trad J. 163(21):36, 38-40.

Wang, J.-H. and F. C. Beall. 1975. Laboratory press-drying of red oak. Wood Sci. 8:131-140.

Ward, J. C., R. A. Hann, R. C. Baltes, and E. H. Bulgrin. 1972. Honeycomb and ring failure in bacterially infected red oak lumber after kiln drying. U.S. Dep. Agric. For. Serv. Res. Pap. FPL 165. 36 p.

Wartluft, J. 1982. Air-drying Appalachian hardwoods. Wood 'N Energy 2(1):37-38.

Wartluft, J. [1983.] Seasoning Appalachian hardwood firewood. In Proc., Wood Industrial Energy Forum '82, Vol. I. Washington, D.C., March 8-10, 1982. For. Prod. Res. Soc., Madison, Wis. p. 175-186.

Wengert, E. M. 1972. Review of high-temperature kiln-drying of hardwoods. South. Lumberman 225(2794):17-18, 20.

Wengert, E. M. 1974a. How to reduce energy consumption in kiln-drying lumber. U.S. Dep. Agric. For. Serv. Res. Note FPL-0228. 4 p.

Wengert, E. M. 1974b. Maximum initial moisture contents for kiln-drying 4/4 hardwoods at high temperatures. For. Prod. J. 24(8):54-56.

Wengert, E. M. 1978. Making management decisions in lumber drying. Lumber Manufacturers' Assoc. of Virginia, Sandston, Va. 41 p.

Wengert, E. M. 1979a. Quality control in drying oak lumber. *In* Management and utilization of oak. Proc., Seventh Ann. Hardwood Symp., Hardwood Res. Counc. p. 91-99.

Wengert, E. M. 1979b. Solar heated firewood dryer. Renewable Natural Resources MT #13 C Utilization and Marketing, Coop. Ext. Serv., Virginia Polytech. Inst. and State Univ., Blacksburg, 6 p.

Wengert, E. 1980a. Rx for drying quality: DQA once a year. Furn. Des. and Manuf. 52(4):31-33.

Wengert, E. 1980b. The best way to dry hardwood lumber from green to 25% MC. Furn. Des. and Manuf. 52(6):60, 63, 64.

Wengert, G. 1980c. Drying at the sawmill with dehumidifiers. North. Logger and Timb. Proc. 28(9):20-21.

Wengert, E. M. 1980d. Solar heated lumber dryer for the small business. MT #20C Utilization and Marketing, Virginia Coop. Ext. Serv., Virginia Polytech. Inst. and State Univ., Blacksburg, 16 p.

Wengert, E. M. and R. C. Baltes. 1971. Accelerating oak drying by presurfacing, accelerated schedules, and kiln automation. U.S. Dep. Agric. For. Serv. Res. Note FPL-0214. 10 p.

Wengert, G., and W. L. Wellford, Jr. 1978. The use of dehumidifiers for hardwood lumber drying. South Lumberman 236(2923):13-14.

Wengert, G., and M. White. 1979. Lumber drying cost comparisons. Timber Processing Ind. 4(1):12-14, 32.

White, J. F. 1963. Air circulation within a lumber yard—An analysis by model techniques. M.S. Thesis, Univ. Georgia, Athens. 38 p.

Wolfe, C. L. 1962. Part-time kiln operation. Now let's try using a little different method. South. Lumberman 205(2555):41-42.

Wolfe, C. L. 1963. Another method of part-time kiln operation. South. Lumberman 206(2569):33, 36.

21

Treating

This chapter is taken, with minor editorial changes, from W. S. Thompson and P. Koch's (1981) General Technical Report SO-35, "Preservative treatment of hardwoods: A review", available from the Southern Forest Experiment Station, U.S. Department of Agriculture, Forest Service, New Orleans, La.

Major portions of data were drawn from research by:

M. A. Akhtar	H. Greaves	C. H. Nethercote
American Wood Preservers' Association	H. J. Hall	D. D. Nicholas
	H. Hallock	G. G. Olson
R. D. Arsenault	D. Hatcher	P. W. Perrin
D. Aston	M. E. Hedley	D. F. Purslow
R. H. Baechler	W. T. Henry	J. A. Putnam
W. B. Banks	C. A. Holmes	R. H. Rawson
R. H. Beal	H. A. Huber	E. T. Reese
E. A. Behr	M. S. Hudson	H. N. Rosen
D. S. Belford	D. R. Huffman	R. M. Rowell
E. G. Bergin	M. A. Hulme	T. C. Scheffer
R. H. Bescher	R. E. James	O. H. Schrader, Jr.
J. O. Blew, Jr.	B. R. Johnson	M. L. Selbo
S. Boone	P. E. Jurazs	C. E. Shuler
K. Borkin	A. J. Kass	J. F. Siau
H. H. Bosshard	W. C. Kelso	H. H. Smith
J. J. Brenden	J. Krzyzewski	N. A. Sorkhoh
S. J. Buckman	E. Kubinsky	A. J. Stamm
B. E. Carpenter, Jr.	J. W. Kulp	N. Tamblyn
F. C. Cech	S.-T. Lan	M. A. Taras
E. T. Choong	J. F. Laundrie	C. H. Teesdale
M. Chudnoff	M. P. Levi	F. O. Tesoro
D. B. Cook	C. R. Levy	R. J. Thomas
W. A. Côté, Jr.	J. F. Levy	W. S. Thompson
H. L. Davidson	W. Liese	E. R. Toole
W. H. Davis	W. E. Loos	J. L. Tschernitz
D. J. Dickinson	A. E. Lund	U.S. Department of Agriculture, Forest Service
A. A. Eddy	R. F. Luxford	
K. E. B. Ernst	H. B. MacFarland	A. F. Verrall
G. F. Franciosi	J. D. MacLean	C. B. Vick
R. O. Gertjejansen	R. W. Merz	C. S. Walters
L. R. Gjovik	J. A. Meyer	L. W. Wood
R. D. Graham	G. C. Meyers	R. L. Youngs

Chapter 21
Treating

CONTENTS

	Page
21-1 FACTORS AFFECTING TREATMENT QUALITY	2468
WOOD RELATED FACTORS	2468
Anatomical structure	2468
Species	2470
Permeability	2471
PROCESSING-RELATED FACTORS	2474
Treating conditions	2474
Incising	2476
Conditioning	2476
Boulton drying	2478
Vapor drying	2479
Steaming	2480
21-1 WOOD PRESERVATIVES	2480
PRESERVATIVES OF THE PAST AND PRESENT	2480
Retentions needed to protect hardwoods	2483
NEW PRESERVATIVES AND PRESERVATIVE SYSTEMS	2484
21-3 METHODS OF PRESERVATIVE TREATMENT	2485
PRESSURE PROCESSES	2485
Solvent-recovery processes	2488
Modified empty-cell process	2489
Oscillating-pressure techniques	2490
Use of shock waves and ultrasonics	2491
Pressure treatment of crossties	2491
Pressure treatment of posts	2492
NON-PRESSURE PROCESSES	2492
Thermal process	2494
Cold-soaking	2495
Diffusion treatments	2498
Dip, brush, and soak treatments	2500

21-4	**EFFICACY OF PRESERVATIVE TREATMENTS**	2501
	NATURAL DECAY AND TERMITE RESISTANCE	2501
	INTERNATIONAL OVERVIEW	2502
	SERVICE DATA IN NORTH AMERICA	2504
	Crossties	2506
	Fence posts	2508
21-5	**EFFECTS OF TREATMENTS ON WOOD PROPERTIES**	2516
	INCISING	2516
	CONDITIONING	2517
	Steaming	2517
	Boulton drying	2519
	Vapor drying	2520
	EFFECT OF TREATING CYCLE	2522
	STABILIZATION TREATMENTS	2522
	Recommendations for future research	2525
	EFFECT OF PRESERVATIVES	2525
	Solid-wood products	2525
	Reconstituted and composite wood products	2528
21-6	**LITERATURE CITED**	2536

Chapter 21
Treating[1,2]

Wood is given preservative treatment to make it durable. Resulting extended useful life reduces the need for replacements, a substantial part of the annual cost of wood in service. Preservative treatment also reduces need for oversize design of structural members to compensate for anticipated deterioration.

Increasing alternative demands for softwoods, now supplying two-thirds of the 330 million cu ft of wood annually treated with preservative in the United States, presage major increases in hardwoods treated by this industry. Currently hardwoods, mostly crossties, plus some mine timbers, highway posts, and oak piling, account for about one-third of the total. The wood industries of Europe, Asia, and Africa, however, are treating hardwoods for a much wider range of products.

A substantial body of literature has evolved from long-term and continued interest in the problem of getting uniform adequate penetration and retention in the many species of hardwoods that grow in the world, and there is much need for continued research on this problem (Hatfield 1959).

This chapter reviews information from the literature pertinent to treatment of the 23 major hardwood species (see section 2-5) found on southern pine sites, and studies of 31 additional American hardwood species (table 21-1), by variations of five pressure and four non-pressure treatment methods, together with data from softwood tests of materials and methods for which hardwood test results are not available. It also reviews data on impregnation of hardwoods for stabilization and to improve resistance to fire.

Pine-site hardwoods available for treatment are primarily in sizes and qualities unsuitable for high grade sawn products, but usable for products such as cross ties, landscape timbers, posts, mine timbers, piling, pallets, laminated wood, fiberboards, and flakeboards. Some hardwood species with low specific gravities are easily treated at pressures and for durations similar to those used for softwoods of like density; heavy hardwoods are more difficult to treat and require longer impregnation times and higher pressures. Hardwood species differ widely, however, in their individual requirements.

---

[1]See acknowledgements on chapter front page.

[2]This text reports research involving fungicides and pesticides. It does not contain recommendations for their use, nor does it imply that the uses discussed here have been registered. All uses of fungicides and pesticides must be registered by appropriate State and/or Federal agencies before they can be recommended.
CAUTION: Fungicides and pesticides can be injurious to humans, domestic animals, desirable plants, and fish or other wildlife—if they are not handled or applied properly. Use all fungicides and pesticides selectively and carefully. Follow recommended practices for the disposal of surplus fungicides, pesticides, and their containers.

TABLE 21-1.—*Hardwood species common on non-pine sites, for which preservative treatment data are available*

Common name	Botanical name
Aspen, bigtooth	*Populus grandidentata* Michx.
Aspen, trembling	*P. tremuloides* Michx.
Basswood	*Tilia* sp.
Beech, American	*Fagus grandifolia* Ehrh.
Birch, river	*Betula nigra* L.
Birch, sweet (or black)	*B, lenta* L.
Birch, white	*B. papyrifera* Marsh.
Birch, yellow	*B. alleghaniensis* Britton
Box elder	*Acer negundo* L.
Butternut	*Juglans cinerea* L.
Cherry, black	*Prunus serotina* Ehrh.
Cottonwood	*Populus deltoides* Bartr.
Elm, slippery	*Ulmus rubra* Muhl.
Elm, rock	*U. thomasii* Sarg.
Elm, cedar	*U. crassifolia* Nutt.
Honey locust	*Gleditsia triacanthos* L.
Maple, silver[1]	*Acer saccharinum* L.
Maple, sugar	*A. saccharum* Marsh.
Oak, bur	*Quercus macrocarpa* Michx.
Oak, California blue	*Q. douglasii* Hook. & Arn.
Oak, overcup	*Q. lyrata* Walt.
Oak, pin	*Q. palustris* Muenchh.
Osage-orange	*Maclura pomifera* (Raf.) Schneid.
Pecan	*Carya illinoensis* (Wangenh.) K. Koch.
Pecan, bitter	*C. aquatica* (Michx. f.) Nutt.
Persimmon	*Diospyros virginiana* L.
Poplar, aspen	*Populus* sp.
Sycamore	*Platanus occidentalis* L.
Tupelo, water	*Nyssa aquatica* L.
Walnut, black	*Juglans nigra* L.
Willow, black	*Salix nigra* Marsh.

[1]With red maple, termed soft maples.

Available chemicals and technology protect hardwood crossties from decay for service lives of 25 to 50 years; much current replacement in rail lines results from physical wear rather than decay. Somewhat shorter service lives may be expected from posts, which are normally more severely exposed, but for most species treatment can extend service life by 10 to 25 years. A number of nonpressure treatments and a variety of preservatives are effectively used in post treatment.

For products in less severe exposure, the same chemicals can be used in lighter applications, or other chemicals selected to minimize undesirable effects such as altered color, stickiness, or toxicity to mammals. Thus life of boxes and pallets used for storage in damp situations can be greatly extended, and exposed woodwork can be impregnated for protection against insects and decay, and to improve fire resistance.

For applications in plywoods, particleboards, flakeboards, and fireboards, chemicals must be compatible with adhesives which give strength to the panels. This requirement limits the choices of usable chemicals, but for most applications treatments with selected chemicals appear feasible, though some loss of strength may be involved.

21-1 FACTORS AFFECTING TREATMENT QUALITY

Amenability of wood to preservative treatment depends upon numerous factors related to pre-treatment processing, such gross characteristics as sapwood thickness, and anatomical characteristics related to type, number, distribution, and size of its various structural elements. Variation in permeability to preservatives among species is almost always associated in part with differences in anatomical structure and the presence or absence of infiltrated materials. Variation within species may be due to many factors, among which are ratio of sapwood to heartwood, moisture content differences, and tree-to-tree differences in the amount and distribution of tyloses.

The fact that heartwood of all species is much less permeable than sapwood has a profound influence on the quality of preservative treatments of sawn products. Properly seasoned wood of all species is more easily penetrated by pressure processes than unseasoned wood; the drier the wood the deeper the penetration. Radial penetration of water-borne preservatives virtually stops at the point where free water is present. Freshly cut wood of most species can be properly treated only with difficulty, if at all.

WOOD-RELATED FACTORS

Anatomical structure.—The size, distribution, and condition of vessels are the most important basic factors affecting the preservative treatment of hardwoods. (See chapter 5 for an illustrated description of anatomical features of stemwood from the pine-site hardwoods; and sec. 8-6 for a discussion of permeability.) Open pores (vessels) are the principal avenues for initial penetration in such species as red oak. White oak and other species whose vessels are wholly or partly occluded by *tyloses* (figs. 5-22 and 8-34, and sect. 5-6), or infiltrated substances are difficult to treat. The degree to which tyloses block penetration of preservatives varies with species. Tyloses in a few species, including American beech and some species of oak and hickory, are thin-walled and can be ruptured during treatment of the wood, thus permitting penetration. In other species tyloses are scattered and their effect on preservative treatment is proportional to their frequency (Teesdale and MacLean 1918).

When vessels are blocked by tyloses or extraneous deposits, penetration may occur through such prosenchymatous tissues as fibers. Teesdale and MacLean (1918) concluded that hardwood rays, unlike those of softwoods, are not important in transverse distribution of preservatives. Behr et al. (1969), however, reported that, while longitudinal parenchyma is unimportant as a reservoir for preservative in most species, ray parenchyma retains significant amounts and is an important avenue for radial penetration in some species. Similarly, Greaves and Levy (1978) found CCA-type preservative in relatively high concentrations in ray tissue of several hardwoods. Bosshard (1961) likewise found creosote in ray tissue of beech—both as droplets and as a film on lumen walls. In areas of low retention the rays held most of the creosote. Narrow rays seem more important in preservative movement than wide rays such as occur in oak and beech.

Passage of preservatives through fibers and other prosenchymatous tissue is much slower and far less extensive than through vessels because hardwood **pits** (figs. 5-62 and 8-31 through 8-35, and sect. 5-6) are poorly developed. Teesdale and MacLean (1918), however, found it relatively important in certain species, such as hickory. They found the type and arrangement of the various structural elements more important to treatability than earlywood-latewood differences.

All recent work confirms Teesdale and MacLean's (1918) finding that vessels are important in the preservative treatment of hardwoods. Behr et al. (1969) reported latewood vessels more important as reservoirs of preservative than earlywood vessels, presumably because of the smaller diameter of the former (fig. 5-22); the ability of a capillary, such as the lumen of a vessel, to hold liquid varies inversely with diameter of the capillary. **Perforation plates** between vessels (figs. 5-52, 8-33D, 8-34AB, and 8-35CD) were often coated with or contained preservative when the vessel was not filled. These investigators stated that the magnitude of preservative adhesion to wood may be of importance in determining its effectiveness and contribute to differences in efficacy among species.

Teesdale and MacLean (1918) found fibers to be unimportant as avenues for preservative penetration unless the vessels were occluded. Reports by Liese (1957) that hardwood pit membranes are incapable of transmitting preservative because they contain no pores tend to support this conclusion. Côté (1963), however, observed that, while no openings were visible in basswood pits, the pits were nonetheless permeable to treating solution. Ernst (1964) found that fibers are important in determining the permeability of certain hardwoods, and Behr et al. (1969) found evidence of fiber-to-fiber movement of liquids via pits and concluded that fibers, as well as vasicentric and vascular tracheids in species containing these elements, are important reservoirs of preservative in treated wood. It appears, therefore, that the permeability of most hardwoods is attributable largely to their vessels, the fibers contributing little because of small lumen size and poor network of interconnecting pits—but this generalization is not true of all hardwoods.

Lack of agreement among investigators on the importance of different cell and tissue types to wood permeability may be due in part to natural variations among species. Thus, rays apparently are important avenues for preservative penetration in some species and not in others (Behr et al. 1969). Work of Levy and Greaves (1978) and Dickinson et al. (1976), who studied the distribution of preservative among tissue types in several species, revealed that each species has its own characteristic distribution pattern. Some of the ten hardwoods studied treated primarily through the rays, for others the principal avenues for penetration were the vessels, and for still others both rays and vessels were important.

Species.—Teesdale and MacLean (1918) studied creosote penetration and retention in heartwood and a few sapwood samples of 25 hardwood species. All samples were similarly seasoned to identical moisture contents, and were similarly treated. On the basis of their results they grouped the species into three treatability classes (table 21-2).

Sribahiono et al. (1974) reported that species could be placed less laboriously in the three classes defined by Teesdale and MacLean (1918) by treating specimens with 0.5 percent keystone oil red dye in mineral spirits by a vacuum impregnation technique and determining longitudinal penetration of the dissected specimens.

TABLE 21-2—*Tested species grouped according to difficulty of impregnation* (Teesdale and MacLean 1918)

Basis for grouping	Group I	Group II	Group III
Depth of penetration at 100 psi[1]			
Lateral	Complete	Variable	—
Longitudinal, inches	>8	4 - 8	2.5-
Retention of creosote, pcf [2]	12+	7 - 10	6 −
Tyloses	Absent	Present, closure incomplete	Vessels occluded by tyloses
	Ash, green	Aspen, bigtooth	Beech
	Ash, white	Chestnut	Oak, bur
	Basswood	Elm, rock	Oak, white
	Beech	Hackberry	Sweetgum
	Birch, river	Hickory[3]	
	Birch, sweet	Maple, silver	
	Birch, yellow	Maple, sugar	
	Cherry	Sycamore	
	Elm, American	Willow, black	
	Elm, slippery		
	Hackberry[4]		
	Maple, silver[4]		
	Oak, chestnut		
	Oak, red		
	Sweetgum[4]		
	Tupelo gum [4,5]		

[1]Psi—pounds per square inch.
[2]Pcf—pounds per cubic foot.
[3]Vessels closed by tyloses, but creosote penetrated through fibers and tracheids.
[4]Samples were sapwood; all others were heartwood.
[5]Probably *Nyssa sylvatica* Marsh.

Siau et al. (1978) classified southern hardwoods into similar groups based on the ease of treatment with methyl methacrylate monomer using a vacuum process. Ease of treatment was evaluated in terms of the fractional void volume filled by monomer (table 21-3).

They reported that treatability was generally well correlated with permeability and varied inversely with specific gravity. Thus, the average specific gravities of the "easy", "moderate", and "difficult" groups were 0.51, 0.69, and 0.74, respectively.

Cook (1978) studied directional penetration by pressure impregnation of short bolts in a commercial retort with a 5.3 percent solution of pentachlorophenol in a petroleum solvent, using resin coatings to restrict preservative movement to the transverse or longitudinal directions. Although his treating schedule was too mild to provide for penetration into refractory species, the method has promise for use in future studies.

Permeability.—Permeability of wood to liquids (see also sec. 8-6) is a function of anatomical structure. The presence of cell deposits and incrustations on the membranes of pits connecting wood cells is a major factor in resistance to penetration of liquids. Various methods of improving permeability have been tried, including chemical treatments to extract material from the pit membrane or degrade the pit membrane itself. Sodium chlorite, pulping liquors, acids and bases have been tested (Nicholas 1977). All degrade the wood and reduce its strength. None are effective in heartwood, where permeability is most limited.

TABLE 21-3.—*Southern hardwoods grouped by difficulty of impregnation with methyl methacrylate monomer by vacuum process* (Siau et al. 1978)

Easy[1]	Moderate[2]	Difficult[3]
Maple, red	Ash, green	Hackberry
Sweetbay	Elm, American	Oak, black
Sweetgum	Elm, winged	Oak, blackjack
Tupelo, black	Hickory	Oak, post
Yellow-poplar	Oak, cherrybark	Oak, white
	Oak, laurel	
	Oak, northern red	
	Oak, scarlet	
	Oak, Shumard	
	Oak, southern red	
	Oak, water	

[1] 0.8+ of voids filled.
[2] 0.4-0.8 of voids filled.
[3] Less than 0.4 of voids filled.

Biological treatments using fungi and bacteria, while used successfully on southern pine, are also effective only in sapwood. Enzymes (cellulase, pectinase, etc.) have been successfully used to improve permeability, without strength loss, of sapwood but not of heartwood. Greaves and Barnacle (1970), for example, reported that infection by microorganisms increased the permeability of slash pine and *Eucaluptus diversicolor* to creosote. This effect was attributed in both instances to depletion of cell contents in ray tissue and alterations of bordered pits by enzymes of invading organisms.

Cech (1971) and Cech and Huffman (1972) improved the seasoning and treating properties of refractory woods such as spruce (*Picea* sp.) by compressing boards from 5 to 15 percent of their thickness between pressure rolls (table 21-4). Nicholas and Siau (1973) applied the process to Douglas-Fir (*Pseudotsuga menziesii* (Mirb.) Franco) with equal success. Retention and penetration of creosote in spruce increased more than 50 percent when Cech and Huffman (1972) compressed boards 15 to 20 percent preparatory to treatment.

Cech (1971) hypothesized that the compression treatment increased permeability by damaging pit membranes. This theory is supported by electron micrographs showing splits in some pit membranes, and by Nicholas and Siau's (1973) observation that radial compression is more effective than tangential.

Tesoro and Choong (1976) measured the permeability of nine hardwoods and three softwoods to both liquids and gas. For every species except cottonwood, heartwood was less permeable than sapwood. Longitudinal water permeability of the hardwood specimens that had been dried and resaturated was nine times greater than that of specimens that had never been dried. For soft maple the increase was 59-fold. Similar results were not obtained with softwoods because of extensive pit aspiration during drying.

TABLE 21-4.—*Effect of transverse compression on retention and penetration of creosote into eastern white spruce (Picea glauca (Moench) Voss) (Cech and Huffman 1972)*

Compression (percent)	Retention		Penetration	
		Fraction of controls	Fraction of cross section	Fraction of controls
	pcf^1	---------Percent---------		
0	3.4	—	20.8	—
10	4.6	135.3	24.6	118.3
15	5.5	161.8	32.8	157.7
20	5.2	152.9	31.5	151.4

[1]Pounds/cu ft.

In the Tesoro and Choong (1976) study, correlation was high between various measures of permeability and treatability of all hardwoods except semi-ring porous species, notably black willow and cottonwood. This was not true, however, for softwoods. Interestingly, permeability to water was the best indicator of treatability with copper sulfate, while permeability to gas was the best index for predicting treatability with creosote.

In other studies involving nine hardwoods and six softwoods Tesoro et al. (1966) found positive linear relationships between the logarithm of lateral air permeability and both creosote retention and depth of penetration. The correlation coefficient for permeability and retention level for the pooled results of hardwoods and softwoods was 0.847; that for permeability and depth of penetration was 0.687. Both values were significant at the 1 percent level. There was no significant difference between the regression curves for hardwoods and softwoods at the 1 percent level. Their data are presented in table 21-5.

Thomas (1976) conducted anatomical studies in an effort to determine the differences among sweetgum, hickory, and blackgum heartwood in degree of penetrability to fluids. He reported that blocking of intervessel pits, isolation of vessels from fibers, parenchyma, tyloses, and low vessel volume induced flow along heartwood fibers. He attributed the fact that hickory is moderately difficult to treat to the minor role played by vessels in preservative distribution, due to tyloses which obstruct liquid flow.

Tyloses and encrusted pit membranes in both vessels and fibers of sweetgum heartwood were found to halt liquid flow effectively, thus accounting for the difficulty of treating this species. By contrast, black tupelo heartwood lacks tyloses, contains little incrustation, and is relatively easy to treat.

TABLE 21-5.—*Mean transverse air permeabilities and retentions and depth of penetration of creosote for selected wood species[1]* (Adapted from Tesoro et al. 1966)

Species	Type of wood	Permeability	Retention	Depth of penetration
		$cm^3/sec\text{-}cm\text{-}atm$	g/cc	mm
Balsam fir (*Abies balsamea* L. Mill.)............	Sapwood	0.0094	0.19	2.86
Maple...............	Heartwood	.0110	.31	6.94
Hickory	Sapwood	.0036	.15	4.34
Hickory	Heartwood	.0032	.19	5.15
Ash.................	Sapwood	.0130	.23	4.70
Larch (*Larix* sp.)......	Sapwood	.0037	.13	2.05
Larch	Heartwood	.0031	.05	0.66
Spruce	Sapwood	.0072	.23	3.88
Pine (*Pinus* sp.).......	Sapwood	.0350	.36	—
Pine	Heartwood	.0068	.14	—
Aspen..............	Heartwood	.0590	.44	4.95
Cherry	Heartwood	.0062	.06	1.08
Douglas-fir...........	Sapwood	.0440	.17	5.78
Douglas-fir...........	Heartwood	.0022	.14	2.17
Oak.................	Heartwood	.0062	.05	.73
Mahogany (*Swietenia mahogani* Jacq.)	Heartwood	.0018	.02	.19
Baldcypress (*Taxodium distichum* (L.) Rich.)	Heartwood	.0390	.44	6.03
Teak (*Tectona grandis* L. F.)..............	Sapwood	.0008	.04	.17
Teak	Heartwood	.0000	.03	.20
Basswood...........	Heartwood	.1900	.51	6.36

[1]Each measurement is the average for four replicated samples.

Longitudinal and transverse permeabilities of cottonwood were determined with nitrogen gas by Isaacs et al. (1971) to fall within the range of 0.109-13.81 darcys and 0.0014-0.0715 darcys, respectively.[3] Surprisingly, sapwood of cottonwood was less permeable than its heartwood. Permeability was correlated with specific gravity in heartwood but not in sapwood. Permeability was not correlated with extractive content nor with treatability with creosote. Retentions of creosote varied widely among samples.

Measurements of the relative and absolute permeability[3] of willow, sweetgum, and tupelo were made by Tesoro et al. (1972). They reported that the pattern of relative permeability of these woods resembled that of oil-bearing rock. When the relative permeability to water was zero, that to oil was not equal to one. For any given level of saturation, the sum of the relative permeabilities of the two fluids was always less than unity except where the flowing phase—either oil or water—completely saturated the specimen. A small increase in the level of oil saturation resulted in a major increase in the relative permeability to oil; likewise, a small decrease in water saturation of a specimen occasioned a sharp decrease in the relative permeability of wood to that fluid. Oil flow through a specimen displaced only a small portion of the water.

Movement of liquids into hardwoods and the anatomical and physical factors affecting it have been reported on by Rosen (1974a, 1974b, and 1975a) and Rosen and vanEtten (1974). They found that bound moisture uptake is proportional to the square root of the time required for bound water to increase from the initial moisture content to two-thirds of the fiber saturation point. Bound and free uptake of organic solvents by red oak and black walnut were linear with the square root of time up to two-thirds of maximum adsorption.

PROCESSING-RELATED FACTORS

Quality of treatment, as judged by preservative distribution and retention, varies widely among species because of the factors discussed above. Variation in treatment quality within a species, while perhaps due in part to some of these same factors, is most influenced by pretreatment processing and by treating conditions themselves. Wood may be incised and must be dried preparatory to all except diffusion treatments. The results achieved in pressure treatments are functions of the pressure and temperature employed. Treatment duration is important in both pressure and non-pressure processes.

Treating conditions.—Within limits, preservative retentions achieved by pressure treatments vary inversely with moisture content of the wood and directly with temperature and pressure. Thus, for example, a direct relationship has been found between treating pressure within the range of 0 to 800 psi and preservative retention (Walters 1967). Retentions also varied directly with preservative temperature, so that increasing temperature from 100 to 200°F effected

---

[3]The Darcy unit is a measure of permeability; high values indicate greater permeability than low numbers. Relative permeability values are not corrected for viscosity of the flowing medium; absolute permeability is relative permeability multiplied by the viscosity of the flowing medium.

a 28 percent increase in adsorption. Siau (1970) likewise has shown that adsorption of preservative by wood species having a range of permeabilities increases directly with treating pressure and inversely with solution viscosity. Retentions obtained with the various species were in the same order as their permeabilities. Retentions of 91 and 78 percent of void volume were achieved by the full-cell process with Douglas-fir and white oak heartwood, respectively, when a pressure of 1,000 psi was used.

The relationship between treating pressure and preservtive adsorption is anchored to wood structure. The size openings available for preservative movement in wood varies widely among species and between sapwood and heartwood. Size of openings determines the pressure required to insure an adequate treatment. Siau (1971) calculated the pressure needed to force various liquids through different pore sizes. He found that the smallest pore that could effectively be filled with water using the highest gauge pressure available at most commercial treating plants (200 psi) has a radius of 0.10 μm (Nicholas and Siau 1973). Stamm (1967, 1970) showed that many species have pore radii that are considerably smaller than that value. The effect of pore size on adsorption is shown dramatically in figure 21-1. The breaks in the curves (e.g., Douglas-fir) were shown by Nicholas and Siau (1973) to correspond to radii of pores through which the liquid could be forced by the pressure involved.

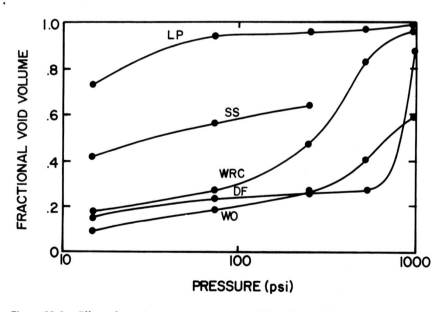

Figure 21-1.—Effect of treating pressure on permeability of several species to a hydrocarbon oil with a viscosity of 1.0 poise. LP denotes longleaf pine (*Pinus palustris* Mill.); SS, sitka spruce (*Picea sitchensis (Bong.)* Carr.); WRC, western redcedar (*Thuja plicata* Donn.); DF, Douglas-fir; WO, white oak. (Drawing after Siau 1970.)

The effect of pressure is not necessarily the same for all species. Akhtar and Walters (1974) reported changes in retention in white oak from 4.59 pcf to 8.67 pcf, a difference of 89 percent, accompanied pressure increases in the range of 150 to 600 psi. For red oak, however, the corresponding increase over the same pressure range was 17 percent, from 16.66 pcf to 19.48 pcf. Differences between these species were confirmed by Rosen (1975b) at pressures from 200 to 2,015 psi; increased retention with higher pressure was more pronounced for white oak than for red oak. Degree of impregnation, expressed as a percentage of the theoretical maximum retention, was linearly related to the logarithm of treating pressure.

Incising.—Relatively few wood species are sufficiently permeable for consistently uniform penetration and desired retention of preservatives. To expedite penetration and retention of preservative solution, wood surfaces of moderately to difficultly permeable species may be incised by roller-mounted knives, leaving variously patterned longitudinal cuts, usually less than 1 inch long and 1 inch deep (fig. 21-2). Many sawn products, hardwood crossties among them, are routinely incised, regardless of species, because sapwood of some species and heartwood of most hardwoods are difficult to treat adequately. Round stock may or may not be incised depending upon species.

Perrin (1978) has reviewed the history and practice of incising. That it has accomplished its original purpose to increase penetration and retention into refractory species of wood (Bellman 1968) is attested to by numerous reports. Increases in treatment quality effected by incising of eucalyptus (Johanson 1975), Japanese oak (Anonymous 1955), and other hardwood species (Kanehira and Taniguchi 1960; Chudnoff and Goytia 1967) clearly demonstrate the importance of the process. Typical data from Graham (1956) on the effects of incising on preservative treatment of six softwoods are given in table 21-6. Incising improved both penetration and retention of creosote in several of the species. Similar results have been reported earlier by MacLean (1930).

Beneficial effects of incising on checking were reported by Harkom (1932) who found incising at the rate of 56 incisions per square foot before air-drying reduced serious checking by 21, 23, and 61 percent in birch, maple, and beech ties. Henry (1970) and Franciosi (1956, 1967) found that the beneficial effects of incising were still evident after 10 to 18 years in service. Graham and Estep (1966), however, reported that checking after treatment in all incised Douglas-fir spar crossarms in their study exposed untreated heartwood.

Conditioning.—Maximum preservative retention is a negative function of moisture content for woods of the same specific gravity. Walters (1967) found that sweetgum specimens with a moisture content of 8 percent adsorbed 2.3 times as much preservative during high-pressure treatments as specimens with a moisture content of 25 percent.

Treating

Figure 21-2.—Incising of refractory species to increase retention and depth of penetration of preservative. (Top) Greenlee drums designed for 56 incisions per square foot. (Bottom) Incisions measure about ¾-inch long and ¾-inch deep.

TABLE 21-6.—*The effect of incising on the creosote treatment of several Oregon conifers* (Graham 1956)[1]

Species	Average preservative penetration		Average preservative retention	
	Incised	Nonincised	Incised	Nonincised
	----------Inches----------		---------Pounds/cu ft---------	
Libocedrus decurrens Torr.	1.0	0.7	17.1	12.1
Abies magnifica A. Murr.	.9	.6	14.7	10.7
Abies concolor (Gord. & Glend.) Lindl. ex Hildebr.	1.5	1.5	17.8	19.5
Tsuga heterophylla (Raf.) Sarg.	1.2	1.1	16.1	16.0
Tsuga mertensiana (Bong.) Carr.	.7	.6	14.3	14.0
Picea sitchensis (Bong.) Carr. ...	1.4	1.3	13.5	11.0

[1]These conifers, like most pine-site hardwoods, are refractory and must be incised to get acceptable retentions of preservatives. The retention values reflect heavy preservative adsorption in a thin band of sapwood.

Conditioning treatments are applied to reduce the moisture content of stock to a level at which it will accept the requisite amount of preservative and permit radial penetration at least to the deepest limit of the zone within which samples are collected for retention determinations. Air-drying is the most common method of conditioning large items, such as crossties. Because air-drying requires large inventories of stock, Boulton drying and vapor drying are being used by an increasing number of plants.

Boulton drying is applicable to any stock destined for treatment with an oily preservative. Basically, the process involves the use of the preservative itself as a heat-transfer medium and a vacuum to reduce the boiling point of water. Temperatures up to 220°F and vacuum of 14 to 24 inches of mercury are employed (Graham 1980). Vacuums in this range correspond to boiling points of water of 152°F to 182°F. Duration of the Boulton process varies with product size and initial moisture content, but typically ranges from 20 to 60 hours. Wood can be dried to a moisture content below the fiber saturation point by this method. Representative moisture content data for Boulton drying of Douglas-fir pole sections for 43 hours are shown below (Graham and Womack 1972; Graham 1980). Kiln-drying for 43 hours provided an almost identical moisture gradient.

Inches from surface	Initial moisture content	Moisture content after Boulton drying
	----------------------Percent----------------------	
1	103	31
2	71	31
3	44	35
4	34	33
5	34	32
6	33	30

Similar results can be obtained with hardwood crossties, although the target moisture content for this product is generally 40 to 45 percent, somewhat higher than that shown for Douglas-fir.

Vapor drying has been used almost exclusively to condition hardwood ties since its development in the 1940's. Carried out in the preservative treatment retort, the process exposes the wood to hot vapors of organic solvents such as xylene. Drying temperature depends upon the composition of the heat transfer medium used, but generally is within the range of 280° to 320°F. Drying times are short, generally 12 hours or less. Typical data showing the reduction in moisture content with time effected by vapor drying of a charge of green oak crossties are shown in table 21-7. According to Hudson (1949), moisture content bears a linear relationship to vapor drying time that is expressed by the equation:

$$\text{Log } C = -0.0304t + \text{Log } I, \tag{21-1}$$

where C is final moisture content, t is drying time, and I is initial moisture content.

TABLE 21-7.—*Data for typical charge of oak crossties seasoned by the vapor-drying process* (Hudson 1949)[1]

Time (hours)	Cumulative water removed	Moisture content
	Pounds/cu ft	*Percent*
	WOOD EXPOSED TO HOT VAPOR	
0	—	77.6
1	.26	76.9
2	1.12	74.5
3	2.13	71.7
4	3.12	69.0
5	3.86	66.7
6	4.86	64.2
7	5.61	62.1
8	6.42	59.9
9	7.19	57.8
10	7.93	55.8
11	8.51	54.2
12	9.14	52.5
	VACUUM AFTER HEATING IN VAPOR	
2	11.50	46.0
	FINAL VACUUM	
1	12.38	43.6

[1] Charge was composed of 2,804 cu ft (814 ties) and retained an average of 0.48 pcf of drying agent.

Steaming, though less efficient than Boulton or vapor drying, may also be used to condition certain species of wood. The unseasoned wood is exposed to live steam at 245°F, followed by vacuum. Duration varies with size and moisture content, but is usually 8 to 20 hours. A volume of water equivalent to 4 to 8 pounds per cubic foot is typically removed. Assuming an initial moisture content of 80 percent, an 8-pound removal per cubic foot would leave wood having a specific gravity of 0.52 with a residual moisture content of 56 percent. Steaming is not permitted with oaks, the species group that comprises at least 50 percent of the volume of hardwoods that receive commercial preservative treatments annually.

Stock intended for non-pressure treatment is normally air-dried to 30 percent moisture content or less.

21-2 WOOD PRESERVATIVES

PRESERVATIVES OF THE PAST AND PRESENT

The first successful wood preserving process was patented by Kyan in 1832. It involved immersing wood in a solution of mercuric chloride. Similar use of copper sulfate and zinc chloride followed in 1837 and 1838, respectively. Creosote was patented as a preservative in 1836; but it was not used commercially until 1839, when Bethell developed a pressure process for injecting preservatives into wood.

Creosote was first used in the United States in 1865 to treat railroad ties, and it and its solutions with coal tar and petroleum continue to be used for that purpose almost to the exclusion of other preservatives. Pentachlorophenol was patented as a preservative in 1931, but was little used until World War II, when creosote was in short supply. Other oil-borne wood preservatives include copper naphthenate and copper-8-quinolinolate, patented in 1948 and 1963, respectively, but little used commercially at present. Copper naphthenate has the potential for "heavy-duty" use in ground-contact service, but its undesirable physical properties and high cost have limited its use. Copper-8-quinolinolate is also too expensive for wide-scale commercial use, but is used to treat container stock used by the food industry, since it is the only preservative that is permitted in contact with food products.

Fluor-chrome-arsenate-phenol is the oldest of the commercially successful water-borne salt formulations, having been patented in 1918. It was followed, in order, by acid copper chromate (ACC) in 1928, chromated copper arsenate (CCA) in 1938, and chromated zinc chloride (CZC) and ammoniacal copper arsenate (ACA) in 1939. The commercial use of fluor-chrome-arsenate-phenol has decreased from about 7 million pounds in 1966 to less than 250 thousand pounds in 1977. Chromated zinc chloride and acid copper chromate are used mainly on lumber for above-ground service; together they accounted for less than 1 percent of wood treated in 1977 (American Wood Preservers' Association 1978). CCA and ACA are widely used by the wood preserving industry today.

TABLE 21-8—*Volume of wood treated in 1978 by product and preservative*[1]

Product	Preservative			Total
	Creosote	Penta	CCA and ACA	
	----------------Thousand cu ft----------------			
Railroad ties	103,138	449	3,498	106,085
Poles	18,237	41,905	4,038	64,180
Piling	9,993	1,154	943	12,090
Lumber, timbers	10,780	21,209	73,317	105,306
Fence posts	4,584	10,983	4,461	20,028
Other	7,855	4,296	7,646	19,797
Total	154,587	79,996	92,903	327,486

[1] U.S. Department of Agriculture. 1980. Biological and economic assessment of pentachlorophenol, inorganic arsenicals, and creosote. Report of the USDA-EPA Preservative Chemicals Assessment Team. Final draft report.

Creosote, pentachlorophenol, and two inorganic salt formulations, CCA and ACA, the only preservatives widely used commercially for ground contact service in the United States, accounted for 47, 24, and 28 percent of the 330 million cubic feet of wood treated in the United States in 1978.[4] Creosote is used principally to treat railroad ties, piling, and poles, accounting for 97, 83, and 28 percent of these products in 1978 (table 21-8). The principal products treated with pentachlorophenol are poles, posts, and sawn products, other than ties, 64, 55, and 20 percent of which are treated with that preservative. CCA and ACA jointly account for 70 percent of all lumber treated in the United States and the use of these preservatives for poles, posts, and piling is increasing rapidly.

Creosote is used straight or blended with coal-tar or petroleum. Its use with petroleum is confined almost entirely to tie production. Pentachlorophenol may be applied as a solution in any of a wide range of petroleum solvents, depending upon the end use of the product. It is also applied, mainly on utility poles, in such volatile solvents as methylene chloride and liquified petroleum gas.

ACA is an ammoniacal solution of arsenic and copper which, upon evaporation of ammonia, forms an insoluble precipitate of copper arsenate in the wood. CCA is an acid solution of copper oxide, chromic acid, and arsenic acid which react *in situ* to form a complex of relatively insoluble precipitates. Various salts of chromium and copper may be used instead of chromic acid and copper oxide

---

[4] U.S. Department of Agriculture. 1980. Biological and economic assessment of pentachlorophenol, inorganic arsenicals, and creosote. Report of the USDA-States-EPA Preservative Chemicals Assessment Team. Final draft report.

in formulation of CCA preservatives. Three formulations, designated as Type A, Type B, and Type C, are currently recognized by the American Wood-Preservers' Association (1977). Their compositions are shown below.

Ingredient	Percent composition		
	Type A	Type B	Type C
CrO_3	65.5	35.3	47.5
CuO	18.1	19.6	18.5
As_2O_5	16.4	45.1	34.0

CCA Type C is most widely used and it probably will, in time, entirely displace the other two formulations.

While creosote and waterborne salts are equally efficacious in preventing biological deterioration, tests show that salt-treated wood is much more prone to split, check, and deteriorate with weathering. Either preservative prevents decay in crossties, but most replacements are due to mechanical failure; creosote treatment delays the need for such replacements. Creosote appears to impart water repellency that lowers moisture content and reduces moisture-content cycling between wet and dry periods, especially while ties are exuding preservative (Hedley 1973). Over a period of years, however, this effect declines and water adsorption of creosote-treated ties approaches that of salt-treated ties.

Salt-treated ties are also reported to cause more corrosion of iron fasteners, and to be more conductive of electricity than creosoted ties, a matter of concern in railroads where electrical-signal systems are used.

Efforts to improve the weathering properties of salt-treated ties include a CCA emulsion to reduce rapid uptake of water (Belford and Nicholson 1968, 1969). The product reduced splitting of ties, performing in a manner comparable to a heavy oil treatment. The CCA addition is described as a solution of long-chain, petroleum-hydrocarbon fractions containing non-ionic surface-active agents and a light-petroleum fraction solvent that is self-dispersing in the CCA solution. After injection into wood, the emulsion is reported to break and the hydrophobic materials to be deposited on the inner surfaces of the cell cavities. Levi et al. (1970) reported that autoradiography revealed these hydrophobic materials in the earlywood rays of pine sapwood. The water repellents were effective to a depth of at least 1 inch.

Pretreatments with petroleum solvents have not improved the weathering characteristics of salt-treated ties (Hedley 1973). However, a light treatment of hardwood ties with a heavy oil or with creosote did (data from Hickson's Timber Products, Ltd. *in* Cooper 1976). Salt-treated ties subsequently treated to a 2.5 pcf retention with 3:1 diesel oil in creosote performed better in long-term service tests than ties conventionally treated with a 50:50 creosote-petroleum blend. Whether this double-treatment process is currently used commercially is unknown.

Laboratory and field tests indicate that 2.5 to 3.0 percent concentrations of pentachlorophenol in a heavy petroleum solvent protect hardwoods against decay as well as 50-50 blends of creosote and petroleum or coal tar (Baechler 1947a; Blew 1950; Olson et al. 1958). Long-term weathering characteristics of ties treated with this preservative need further study.

Treating 2483

Retentions needed to protect hardwoods.—The amount of chemical required to adequately protect wood from decay and insect damage varies with wood species, exposure conditions, and type of preservative. (See discussion under heading 21-4 EFFICACY OF PRESERVATIVE TREATMENTS.) Table 21-9 is intended to give the reader some indication of equivalency—with important reservations, as follows. There are only three heavy-duty preservatives (four if the relatively expensive copper naphthenate is included) suitable for soil-contact service: pentachlorophenol, CCA-ACA, and creosote. The other two listed in table 21-9 are recommended only for above-ground service. Acid copper chromate (ACC), however, is permitted for ground-contact service of non-critical items such as fence posts.

Of the preservatives listed in table 21-9, only creosote and CCA-ACA are suitable for softwoods and only creosote for hardwoods in marine (salt water) use. Retentions and assay zones for this type of service are (American Wood-Preservers' Association 1977):

Preservative and species or species group	Retention in assay zone	Assay zone
	Pcf	Inches
Creosote		
Southern pine	20	0 to 3.0
Douglas-fir and oak	10	0 to 2.0
CCA-ACA		
Southern pine	2.5	0 to 0.5
	1.5	0.5 to 2.0
Douglas-fir	2.5	0 to 1.0

CCA and ACA are not recommended for hardwood marine piling.

TABLE 21-9.—*Retentions of six classes of preservatives in assay zones for approximately equivalent protection of hardwoods*[1]

Preservative	Service above ground	Service in soil contact
	------Retention, pounds/cu ft------	
Creosote[2]	6.00	8.00
Pentachlorophenol[3]	.25	.40
Chromated copper arsenate (CCA)[4] or ammoniacal copper arsenate (ACA)[4]	.25	.40
Copper naphthenate[5]	.40	.75
Acid copper chromate (ACC)[4]	.25	.50[6]
Chromated zinc chloride (CZC)[3]	.45	not recommended

[1]Protection of piling against marine organisms requires higher preservative retentions of creosote than those shown. None of the other preservatives is recommended for use in salt-water exposure of hardwoods.
[2]Retention of whole creosote.
[3]Retention of anhydrous chemical.
[4]Retention of anhydrous chemical, oxide basis.
[5]Retention of copper, as metal.
[6]Recommended only for non-critical items in soil content, e.g., for fence posts.

Retention by hardwood lumber, timbers, and crossties is determined by gauge readings or weight gain rather than by assay. Only two species groups are covered by current standards of the American Wood-Preservers' Association: oak (white and red) and gum (sweetgum and black tupelo).

While assay zones are defined in standards for southern pine poles, these definitions have no application to hardwood poles; hardwood poles are not used in the United States and no standards have been written for them.

NEW PRESERVATIVES AND PRESERVATIVE SYSTEMS

Other preservative systems, or modifications of existing ones, are being developed in response to the increasing cost of oil-borne preservatives and their petroleum solvents. Concern about environmental restrictions that may be imposed on the four major preservatives has further expedited the search for new ones with low mammalian toxicity which meet requirements of economy and effectiveness.

An oil-water-emulsion solvent system for pentachlorophenol that significantly reduces petroleum requirements has been developed. Initial test results of this new solvent system are promising.[5] Alkylammonium compounds developed in New Zealand are being studied for possible use in the United States. They have proven acceptable for above-ground use and are being tested for ground-contact service. A water-solubilized form of copper-8-quinolinolate appears to have promise for use in above-ground applications.

Preservative systems whose toxic mechanisms are operative outside the body of the invading organism have been investigated. The use of chelating agents to inactivate the micronutrients required by fungi for growth was first suggested by Zentmyer (1944) and tested against wood-decay fungi by Baechler (1947b) and Thompson (1964a). Success of this method depends upon inactivation of the trace elements in wood, as well as those from other sources, such as the soil, to which the fungus has access. Hence, it does not work. This is likewise true—for the same reason—of attempts to dethiaminize wood and and hence deny to fungi this essential growth factor (Gjovik and Baechler 1968). Efforts to treat wood with materials that chemically inhibit the cellulase enzymes secreted by fungal hyphae, thus preventing decay from occurring, have also failed (Reese and Mandels 1957; Thompson 1965).

---

[5]Hatcher, D. 1979. Private communication.

21-3 METHODS OF PRESERVATIVE TREATMENT

Wood may be preservative treated by either pressure or non-pressure processes. Pressure impregnations involve the application of 50 psi pressure or more in large cylinders to force preservative liquid into the wood (figs. 21-3 and 21-4). Pressure treatments account for over 95 percent of the wood treated in the United States by all processes except brush, dip, and spray. The remainder is treated by various non-pressure processes that include thermal, diffusion, cold soak, and vacuum methods. Brush, dip, and spray methods are also used, but not for material intended for ground-contact service. Products treated by these latter methods include millwork, which alone accounts for 60 million cubic feet of wood.[4]

PRESSURE PROCESSES

Pressure processes are differentiated by the air pressure applied to the wood before impregnation with a preservative. The full-cell process employs an initial vacuum to obtain maximum retention of preservative (fig. 21-5). The empty-cell processes employ either atmospheric pressure (Lowry Process) or higher air pressures of 15-75 psi (Rueping Process) (fig. 21-6). Hydrostatic or pneumatic pressures of 50 to 200 psi are imposed on the system and maintained until the desired gross injection of preservative has been achieved. The pressure is then gradually returned to atmospheric level and the excess preservative returned to storage. The treated wood may be subjected to a final vacuum or steaming and vacuum of short duration to remove surplus preservative from the surface of the wood.

The treating process and conditions employed and the preservative retention are determined by the species of wood, the preservative used, and the service conditions to which the wood will be exposed. Refractory species, such as white oak, require higher treating pressures than permeable species like southern pine. Treating temperatures of 210°F to 239°F are used to reduce the viscosity of creosote, whole those used with pentachlorophenol-hydrocarbon solutions range from ambient to 220°F, depending upon the hydrocarbon (heavy petroleum, mineral spirits, or liquified petroleum gas). Target adsorption determines which of the two pressure processes is used. Products such as marine piling must be treated by the full-cell process to obtain the high preservative retention of 20 pcf required. Water-borne preservatives also are applied by the full-cell process. The empty-cell process is used for products, such as utility and construction poles, for which retentions of 6 to 12 pcf are needed.

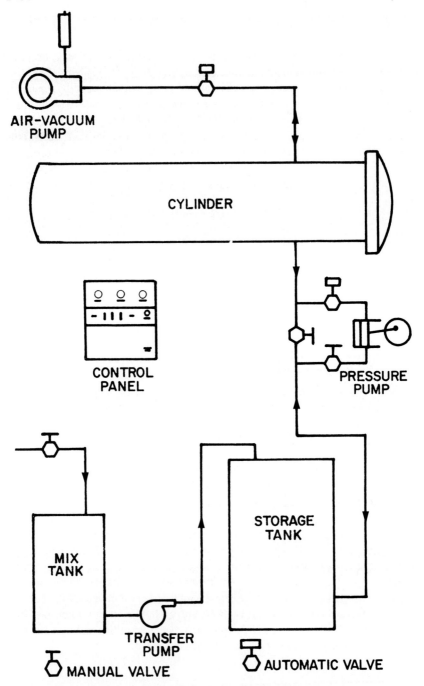

Figure 21-3.—Schematic diagram of piping, storage tanks, and treating cylinder (retort) for pressure-impregnating wood with preservative. The air vacuum pump permits the cylinder to be evacuated before admission of the preservative, during a drying cycle (boiling under vacuum), or following the pressure cycle and drainage of the preservative.

Figure 21-4.—Commercial operation for pressure-impregnating wood with preservative. (Top) Frontal view of 7- by 130-foot retort ready for discharge. (Bottom) Freshly pulled charge of creosote-treated hardwood crossties.

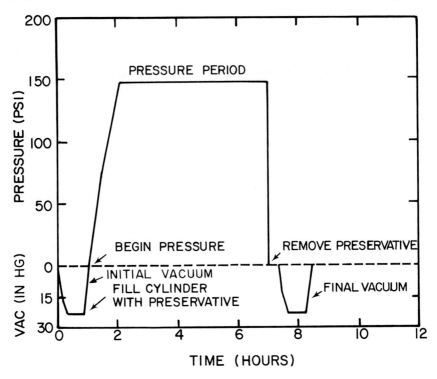

Figure 21-5.—Schematic showing the treating schedule for full-cell treatment.

Modifications of the basic pressure treating process that involve different pressure and temperature regimes from those in common usage have been explored at pilot-scale levels. Pressures in the range of 600 to over 2,000 psi have been used in these studies in an effort to improve the treatment quality achieved with refractory hardwood species (Walters and Guiher 1970; Akhtar and Walters 1974) and to obtain maximum loading of monomers in stabilization treatments (Rosen 1975a). In addition, oscillating pressure and vacuum treatments and other variations have been studied for possible beneficial effect on treatment quality. Some innovations and special applications of pressure methods that have been tried—some successfully and some not—are discussed in following sections.

Solvent recovery processes.—Two relatively new pressure treating processes receiving commercial application are unique in that the preservative—pentachlorophenol—is injected into wood in volatile solvents which are subsequently recovered and reused. The more commercially successful of these is the "Cellon Process" developed by Koppers Company, Inc., that employs liquified petroleum gas as the solvent. The other process was developed by Dow Chemical Company and uses methylene chloride as the solvent. The treating cycles employed with both processes are similar to those used with pentachlorophenol in oil, except that each has a solvent-recovery step following the pressure period.

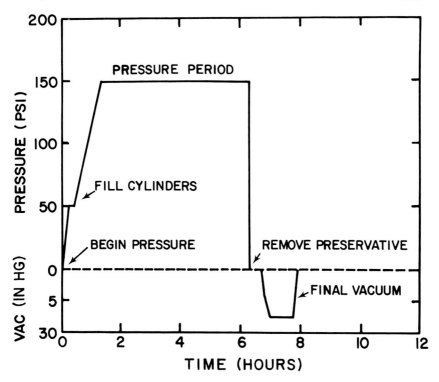

Figure 21-6.—Schematic showing the treating schedule for empty cell treatment.

Wood treated by these processes retains its natural appearance and can be used where cleanliness and paintability are requirements. Neither method is used to any significant extent in the treatment of hardwoods.

Modified empty-cell process.—A modified empty-cell treatment for use with CCA-type preservatives was developed by Kelso[6] as an alternative to the conventional full-cell treatment currently used by industry (fig. 21-7). Impregnation is accomplished by either the Rueping or Lowry processes. After the desired gross adsorption is achieved, the preservative is removed from the retort while simultaneously maintaining a pressure such that kickback of the preservative solution in the wood is prevented. Fixation of the preservative is accelerated *in situ* by admitting to the retort either water or steam and maintaining a temperature of about 220°F for a period of 1 to 4 hours, depending upon the size of the items in the charge. Upon completion of the fixation period, the pressure is reduced and kickout of the spent preservative solution allowed to occur. Moisture contents of stock treated by this process are typically only one-third that of wood treated by the conventional full-cell process. In addition, the process permits rapid sequential treatment of wood with dyes, water repellents, stabilizing chemicals, or other preservatives.

---

[6]W. C. Kelso, Jr. Unpublished data. Mississippi For. Prod. Lab., Mississippi State Univ., Mississippi State.

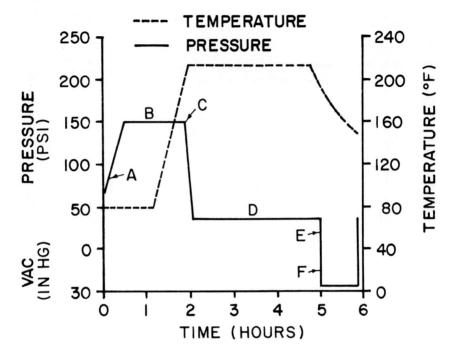

Figure 21-7.—Treating cycle used to treat southern pine pole sections with CCA by a modified empty cell process.
A. Initial air pressure and full cylinder.
B. Pressure period.
C. Maintain pressure sufficient to prevent loss of preservative from wood during fixation period.
D. Release pressure and recover kickback from wood.
E. Final vacuum.

Oscillating-pressure techniques.—Efforts to improve treatment quality of refractory species by sequential changes in pressure have generally been unsuccessful. An oscillating pressure technique, in which pressure (120 psi) and vacuum (28 inches of mercury) were alternated repeatedly throughout a 22-hour impregnation cycle, did not improve retention and penetration in the heartwood of either hardwoods or softwoods (Blew et al. 1961). However, the process did improve sapwood treatment in sweetgum over that achieved by conventional treating processes.

Work by McQuire (1964) indicates that the oscillating pressure method may have potential in the treatment of green wood, but not for dry wood. Its efficacy with green wood is attributed to exchange of wood water and preservative during the alternate pressure and vacuum cycles and to an increasing flow rate which accompanies the removal of air bubbles from the pores during the vacuum phase (Nicholas and Siau 1973).

Use of shock waves and ultrasonics.—Low frequency shock waves and ultrasonics were investigated to improve the treatment of wood. The use of shock waves, induced by striking the retort with a hammer, produced some improvement in the treatment of oak crossties (Burdell and Barnett 1969). This effect was attributed to deaspiration of pits and dislodging of air bubbles from the pores (Nicholas and Siau 1973). Penetration of refractory woods was not improved.

The use of ultrasonic waves while wood is immersed in a liquid at atmospheric pressure can result in a significant increase in adsorption of certain liquids (Borkin and Corbett 1970). The improvement mechanism is thought to be the formation and removal of minute air bubbles, resulting in the removal of air-liquid interfaces which block penetration of liquids.

Pressure treatment of crossties.—Although hardwoods, in the form of crossties, constitute a large fraction of the total volume of wood pressure treated annually in the United States, treatment-response data for individual species are limited. Current standards (American Wood-Preservers' Association 1980a) for the treatment of hardwood crossties with creosote and its solutions specify the following minimum retention and penetration values for the species groups shown:

Species group	Retention	Penetration
	Pounds/cu ft	
Red oak	8	65 percent of annual rings
White oak	Refusal	95 percent of sapwood
Sweetgum and black tupelo	10	1.5 inches or 75 percent of sapwood
Beech, birch, maple	7	85 percent of sapwood
Ash, hickory	Refusal	95 percent of sapwood

Determination of actual retention values by assay is not part of this standard. Hence, data are usually available only for cylinder charges, based on calculated retentions and, where required, checks of penetration on sample pieces.

Federal specification TTW-571 is more explicit and provides that, for species not covered either in that document or in AWPA specifications, penetration of heartwood faces will be not less than 0.4 inch in lumber and 0.5 inch in timbers.

The relative abundance and favorable mechanical properties of hickory have prompted extensive research on its suitability for crossties. Its reputation for excessive checking and splitting and for being difficult to treat has concentrated much research on seasoning and treating properties (Taras and Hudson 1959).

Investigators generally agree that moisture content of hickory should be between 20 and 40 percent for adequate preservative retention and penetration (table 21-10). Seasoning and treating properties of hickory appear to be similar to those of the red oaks. Huffman (1958) found that degrade of hickory ties during seasoning in some studies was no worse—and in some instances was less—than that for other dense species such as the oaks.

In contrast, uneven distribution of creosote in pressure-treated, 4-inch-square hickory timbers was reported by Lund (1966). Despite an average retention of 9.6 pcf, only 46 percent of the treated stock had commercially acceptable penetration. Eighty-five percent had penetration skips, mostly associated with mineral streaks at worm galleries and other injuries. Skips present following treatment were not reduced by preservative migration over a 5-year period.

TABLE 21-10.—*Summary of creosote penetration and retention in hickory crossties by method of seasoning* (Taras and Hudson 1959)

Investigator	Seasoning method and time	Moisture content at treatment	Penetration	Retention
		Percent	*Inches*	*Pounds/cu ft*
Hudson (1954).......	Air drying (6 months)	34	1.0	4.8
Shinn (1955)	Air drying (15 months)	26	1.3	6.0
Vaughan[1]	Controlled air seasoning[2] (11 days)	43	1.3	6.0
Hudson (1951).......	Vapor drying (12 hours)	35	1.7	6.4
Collister (1955)	Vapor drying (15 hours)	20	1.8	11.7
Diechman[3]	Vapor drying (15 hours)	32	2.2	9.0
Huffman (1958)......	Vapor drying (3 days)	32-47	2.2	5.5

[1]Vaughan, J. A. 1954. The use of hickory for railroad crossties. Southern Wood Preserving Company Rep, File 11-A-5 HTS.
[2]Process for slow drying of poles and crossties. See Vaughan (1954, 1955).
[3]Diechman, M. W. 1954. Vapor drying and treatment of hickory crossties. Unpublished report, Taylor-Colquitt Company.

Pressure treatment of posts.—While the literature on treatment of hardwood posts deals principally with non-pressure treatments, some retention values for pressure treatments of post-size material with creosote by the empty-cell process are available (table 21-11). No penetration data accompanied these values. The retention values for all species met or exceeded the minimum recommended retention for softwood posts (American Wood Preservers' Association 1980b); there are no standards for hardwood posts.

Cook (1978) pressure treated 22 species of hardwood (table 21-12) with pentachlorophenol in a petroleum solvent conforming to P-9, Type A (American Wood-Preservers' Association 1977) by an empty-cell process with a maximum pressure of 100 psi for 50 minutes. This treatment regime is very mild for hardwoods, in terms of both pressure and duration. Retentions in the outer 0.25 inch varied greatly, the soft hardwoods generally having greatest retentions in this zone. In only six species was penetration deep enough to estimate retention in the 0.25- to 1.0-inch zone.

NON-PRESSURE PROCESSES

Non-pressure processes are employed both commercially and by individuals for farm and ranch uses. Conditioning of stock is usually by air-drying and for most purposes the wood must be seasoned to a moisture content of 30 percent or less prior to treatment. Diffusion-type treatments are applied to unseasoned wood. Thermal and cold-soak treatments are the non-pressure processes most frequently employed commercially. In both, the wood is exposed to the preservative in an open vessel.

TABLE 21-11.—*Creosote retentions in hardwood posts treated by a pressure process* (Kulp 1961)

Species	Retention
	Pounds/cu ft
Elm, American	9.1
Hickory	9.7
Maple	6.5
Oak, black	6.6
Oak, blackjack	7.9
Oak, northern red	5.9
Oak, white	6.0
Sweetgum	9.7

TABLE 21-12.—*Retentions of pentachlorophenol by hardwood post sections treated by a pressure process*[1] (Cook 1978)

Species	Retention of pentachlorophenol	
	0.25 inch	0.25-1.0 inch
	---------*Pounds/cu ft*--------	
1. Sweetbay	0.56	—
2. Black tupelo	.52	0.17
3. Yellow-poplar	.50	—
4. Sweetgum	.48	.23
5. Red maple	.37	—
6. American elm	.36	—
7. Post oak	.34	.06
8. Scarlet oak	.26	—
9. Shumard oak	.26	—
10. Black oak	.26	—
11. Green ash	.26	—
12. Hackberry	.26	—
13. Blackjack oak	.26	—
14. Cherrybark oak	.25	—
15. Water oak	.23	—
16. Hickory, true	.22	.02
17. White ash	.22	—
18. Winged elm	.21	—
19. Southern red oak	.20	.05
20. Northern red oak	.20	—
21. White oak	.18	.05
22. Laurel oak	.17	—

[1]Because treatments were very mild, these values reflect relative difficulty of treatment rather than typical effective treatments for these species.

Thermal process.—In this process wood is immersed in hot perservative for a fixed period of time—usually 4 to 12 hours—followed by exposure to preservative at ambient conditions (fig. 21-8). Variations of this treatment include use of separate vessels for the hot and cold baths and using a single vessel to which hot or cold preservative can be pumped as appropriate. A few small operators in the South leave the stock in the hot tank so that wood and preservative cool together.

Pentachlorophenol in petroleum is the preservative normally used with the thermal process, but other preservatives, notably creosote, may also be used. In the past this process was used extensively because creosote, the principal preservative, was too viscous at ambient temperature. Availability of pentachlorophenol and copper naphthenate which can be applied by cold-soaking has reduced use of the thermal process; thus the data base for thermal treatment is limited. Retentions of creosote and 5-percent pentachlorophenol solution achieved with thermal and cold-soak treatments, respectively, are shown in table 21-13. The numerous sources of these data do not permit more than a general interpretation of results. Retentions obtained with the thermal process for creosote are generally higher than those for pentachlorophenol obtained by cold soaking. By both processes the soft hardwoods appear to be more easily treated than the dense hardwoods.

Figure 21-8.—Typical thermal treating plant. A boiler (not shown in photo) supplies steam to heat coils in the bottom of the tank at left. The tank at right is for cool preservative. Wood is transferred between tanks manually or by hoist mounted on adjacent pole.

TABLE 21-13.—*Retention of creosote applied by the thermal process and 5 percent pentachlorophenol by cold-soak treatments in hardwood fence posts* (Data from Kulp 1966; Gjovik and Davidson 1975)

Species	Retention	
	Thermal	Cold soak
	---------------*Pounds/cu ft*---------------	
Ash, green	8.9-11.9	6.1 (2)[1]
Ash, white (butts only)	5.1[2]	—
Basswood	6.1[2]	2.4[2]
Beech	8.5	—
Elm, American	8.0	—
Hackberry	8.0-12.7	3.5 (5)
Maple, red	8.2	7.0 (2.6)
Oak, blackjack	4.1	4.5 (8.1)
Oak, overcup	6.1-9.5	5.2 (5)
Oak, northern red	2.4	4.2 (2.8)
Oak, water	8.4-9.6	—
Oak, white	2.4	—
Pecan, bitter	6.0-8.9	5.2 (5)
Persimmon	7.9	3.9 (2)
Sweetgum	9.9-11.0	5.6 (2.7)
Tupelo, water	8.6	5.8 (4.5)
Tupelo, black	7.2	—
Willow, black	8.4	3.8 (2)
Cottonwood	10.0	6.3 (4)

[1] Numbers in parentheses are treating times in days.
[2] Pounds per post.

Putnam (1947) investigated the treatability of bottomland hardwoods in Mississippi, using residence times of 1½ to 2 hours in the hot bath and 15 to 60 minutes in the cold bath. Satisfactory treatments were obtained in posts of ash, cypress (*Taxodium distichum* (L.) Rich.), elm, hackberry, honeylocust, maple, overcup oak, water oak, pin oak, pecan, persimmon, sweetgum, and willow. Results with cottonwood were unsatisfactory.

Cold-soaking.—This process involves simply placing wood in a vessel, covering it with preservative solution, and allowing it to soak until it adsorbs the appropriate amount of preservative, as determined by weight gains (fig. 21-9). It has been used extensively to treat fence posts and lumber for farm and ranch uses, and in some research tests. The process is normally used with pentachlorophenol solutions, but other preservatives, including water-borne salts, may also be used.

M89992F

Figure 21-9.—Preservative treating plant equipped to employ the cold-soaking process. Tank tops are equipped with latch mechanism to hold wood submerged. Piping arrangement maintains preservative level desired in tanks. Wood is charged manually or with the aid of a hoist.

Preservative retentions obtained by cold soaking vary widely among species and with moisture content. Some control over retention by permeable species can be exercised by adjusting the period of immersion. Retentions achieved with refractory species usually represent a compromise between the adsorption desired and the maximum feasible soaking period. Retention of 6 pcf of 5 percent pentachlorophenol solutions, the minimum allowed for ground-contact service by the American Wood Preservers' Association (1980b), is seldom achieved in oak and certain other species by non-pressure treatments. Retentions after cold soaking obtained in selected studies are summarized in table 21-14 for 16 of the hardwood species commonly found growing on southern pine sites.

The retention data presented are for pentachlorophenol in petroleum solvents and represent many different studies. Pretreatment moisture content and treating solution temperature are unknown. Differences among species are apparent. Retentions for the oaks, elm, and hickory were generally less than 5.0 pcf while those for the soft hardwoods—red maple, sweetgum, black tupelo, and yellow-poplar—were generally greater than 5.0 pcf. These grouping agree reasonably well with the categories in tables 21-2 and 21-3 developed by Teesdale and MacLean (1918) and Siau et al. (1978), respectively. There is a tendency for the retention values to vary inversely with specific gravity. Green ash is an exception in all three tables, its retention of preservative being greater than predicted by its specific gravity.

TABLE 21-14.—*Preservative retention in hardwood posts treated by cold soaking in pentachlorophenol-petroleum solutions* (Data from Kulp 1966; Gjovik and Davidson 1975)

Species and treating period (days)	Retention
	Pounds/cu ft
Ash, green	
2	6.1
Elm, American	
2	3.2
4	3.2
5	2.5
Hackberry	
2	4.4
5	3.5
Hickory	
2	3.8
3.7	4.3
5	3.6
Maple, red	
0.5	2.5
1	5.4
2	5.8
2.2	5.7
3.5	9.5
Oak, black	
2	2.7
Oak, blackjack	
6.3	4.1
8.1	4.5
Oak, chestnut	
2	8.7
2.8	5.6
Oak, northern red	
2.8	2.6
4	4.4
Oak, post	
2	3.2
3	3.1
—	5.7
Oak, scarlet	
1.8	3.9
Oak, southern red	
2	5.9
—	4.8
Oak, white	
2	2.4
3.9	5.3
—	6.1

Table continued next page.

TABLE 21-14.—*Preservative retention in hardwood posts treated by cold soaking in pentachlorophenol-petroleum solutions* (Data from Kulp 1966; Gjovik and Davidson 1975)—Continued

Species and treating period (days)	Retention
	Pounds/cu ft
Sweetgum	
0.25	5.0
1	5.7
2	5.1
3	6.0
4	4.5
6	7.2
Tupelo, black	
0.2	7.4
0.5	5.5
4	4.4
Yellow-poplar	
1	5.4
1.4	5.2
2	6.6
3	6.1
6	8.8

Diffusion treatments.—These processes apply water-borne salt formulations to unseasoned wood by diffusion. In the traditional application, the treating solution contains a single salt of a metal, such as copper sulfate or zinc chloride. Such treatments have generally been unsatisfactory for exterior exposure because of rapid loss of the preservative by leaching.

In a variation called **double diffusion** (Baechler et al. 1959), the wood is soaked successively in two different aqueous inorganic salt solutions. The two components react within the wood to form relatively insoluble, toxic precipitates. A retention of 0.4 pcf by double diffusion is adequate for soil-contact service and is not subject to leaching. One pcf of leachable salts (e.g., zinc chloride) is more than adequate for protection, but an excess is needed to compensate for leaching.

Another diffusion process studied extensively in the South (Anonymous 1960) employs a solution of a single salt, such as zinc chloride, and green, unpeeled posts. The posts are placed vertically in a shallow pail containing the quantity of solution calculated to provide a net retention of about 1 pcf of the salt in the posts (fig. 21-9). The ends of the posts are reversed when about one-half of the solution has been adsorbed and the process is continued until the solution is adsorbed. This process provides somewhat better protection for wood in exterior service than conventional diffusion treatments with zinc chloride because the bark inhibits leaching of the preservative as long as it remains intact.

Yet another variation of the diffusion process employs as the preservative a water-soluble salt or salt formulation in dry or paste form. The perservative is spread over the surface of unseasoned, peeled wood, which is then dead-piled and covered to reduce loss of moisture. The preservative diffuses into the wood during a treating period that ranges from a few days to more than 2 weeks, depending upon the product involved.

Diffusion treatments have potential in the treatment of hardwoods because retentions comparable to those obtained by pressure processes are attainable. The double-diffusion process, in particular, is promising because of the long-term protection afforded by the heavy metal complexes formed in the wood.

In studies conducted by Baechler et al. (1959), fence posts cut from several hardwood species were soaked first in a solution of zinc sulfate and arsenic acid and then in a solution of sodium chromate. Exposure periods were 1, 2, and 3 days in each solution. Total retentions of anhydrous salt by species for each exposure period are shown below (Vick et al. 1967):

Species	Retention		
	1 day	2 days	3 days
	---------*Pounds/cu ft*--------		
Yellow-poplar	1.99	2.97	3.15
Sweetgum	1.93	2.41	2.88
White oak	1.27	1.50	2.26
Red oak	1.04	1.57	2.07
Hickory	.39	.70	1.03

Retentions for cypress and southern pine, which were also included in the study, were of the same order of magnitude as those for the hardwoods. Heating the first solution accelerated chemical uptake and, except for white oak (1.69 pcf), red oak (1.60 pcf), and hickory (0.78 pcf), the adsorption was higher than would be desired in practice.

In other work reported by Baechler and Roth (1964), the double-diffusion treatment with copper sulfate and sodium arsenate provided good retention and adequate to excellent penetration in birch, maple, basswood, California blue oak, and bluegum (*Eucalyptus globulus* Labill.) when these species were soaked one day in each of the two solutions. They concluded that this process seems to offer promise as a method of treating hardwoods. However, the authors emphasized that several investigators have shown that certain decay fungi frequently associated with decay of hardwoods are relatively resistant to inorganic preservatives. They surmised that chemical costs probably will be higher in treating hardwoods than in treating softwoods because of that fact.

Boone et al. (1976) found that penetration and retentions by aspen, soft maple, and red oak lumber were generally not satisfactory when the conventional double-diffusion process was used. However, a modified process, using a heated first bath followed by partial drying, gave very good results. Retentions achieved in aspen and maple exceeded those obtained by pressure treatment.

Modified double-diffusion processes were also used successfully by Johnson and Gonzales (1976) to treat several tropical hardwood species. The first solution was 10 percent copper sulfate and the second was 2.25 percent sodium arsenate and 3.25 percent sodium chromate for green material and 4.5 percent sodium arsenate and 6.5 percent sodium chromate for partially seasoned material. Retentions in the outer ½-inch of sawn material, principally sapwood, ranged from 0.14 to over 1.0 pcf, oxide basis, depending upon treatment schedule, species, and whether or not the specimens were incised. Retentions for all species were considered adequate to provide a high degree of protection for ground-contact service in the tropics. Similar results were obtained by Smith and

Baechler (1961) in exploratory treatment studies of Hawaiian-grown fence posts. Promising results were obtained with silk oak (*Grevillea robusta* A. Cunn.), bluegum (*Eucalyptus globulus* Labill.), *Eucalyptus saligna* Sm., and *Eucalyptus robusta* Sm.). Skolmen (1962, 1971), however, treated these same species by double diffusion and reported only a marginal improvement in service life after 10 years of field testing in Hawaii.

Retention values for diffusion treatments of posts of some of the hardwood species that grow on southern pine sites are given in table 21-15. End diffusion values, being easily controlled to target retentions, tend to be lower than those for double diffusion. Posts treated by double diffusion tend to be overtreated. For both diffusion treatments, differences between refractory and permeable species are much less pronounced than with cold-soak and thermal treatments.

Behr (1964) treated American elm and aspen fence posts with anhydrous salts of copper sulfate, followed by disodium arsenate and/or sodium dichromate. The chemicals were applied under a plastic bandage to the debarked butt portion of green posts. Copper sulfate was applied first and left in contact for 6 to 13 days. Disodium arsenate and sodium dichromate were then applied to a wet paper towel inside the plastic bandage and the bandage rewrapped. The posts were then stored for several months. Distribution data for the preservative are given below; values have been converted to an oxide basis.

Species and assay zone	CuO	As_2O_5	CrO_3
	------------Pounds/cu ft------------		
American elm			
First ¼-inch	0.273	0.285	0.065
Second ¼-inch	.023	0	.004
Third ¼-inch	.004	0	0
Aspen			
First ¼-inch	.233	.396	.160
Second ¼-inch	.021	.003	.010
Third ¼-inch	.004	.001	0

Total retention was comparable to that attained by double-diffusion treatments, but most of the salts were concentrated in the outer ¼-inch of wood. Better results were achieved with aspen than with American elm. Behr recommended the procedure for groundline treatments of poles in service.

Dip, brush, and soak treatments.—Extensive, long-term testing has shown that surface treatments applied by dipping, brushing, and spraying are efficacious in preventing deterioration of wood in above-ground exposure (Verrall 1961). Benefits vary with the preservative, treating method, solvent, presence or absence of a water repellent, wood finish, and exposure conditions. Two- to threefold increases in service life in above-ground, exterior exposure can be achieved by 3-minute dips with such preservatives as pentachlorophenol and copper naphthenate (2 percent copper) (Verrall 1953, 1959, 1961). The efficacy of such treatments for suchitems as pallets has also been verified (Nethercote et al. 1977).

Superficial treatments are also extensively used on unseasoned lumber to prevent stain, mold, and decay during seasoning (Lindgren 1930; Lindgren et al. 1932; Verrall and Scheffer 1949). Such treatments fall outside those activities

Treating   2501

normally considered to be wood preservation, and hence are not covered here. Likewise, protection of logs (Burkhardt and Wagner 1978) and chips (Bois et al. 1962; Silberman 1970; Eslyn 1973) in storage fall outside the subject matter considered. For further discussion of wood deterioration in storage, and protective measures, see sections 11-6 (under the heading ROUNDWOOD) and 16-18.

TABLE 21–15—*Preservative retentions of inorganic salts by hardwood posts treated by diffusion processes* (Data from Kulp 1966; Gjovik and Davidson 1975; Baechler et al. 1959)

Species	Retention
	Pounds/cu ft of anhydrous chemical
Elm	0.90
Hickory	.30[1]
Hickory	.70[1]
Maple, red	1.10
Oak, black	.65
Oak, blackjack	.70
Oak, post	.90
Oak, red	.89
Oak, red	1.56[1]
Oak, white	.87
Oak, white	1.70[1]
Sweetbay	.94
Sweetgum	1.14
Tupelo, black	.89
Tupelo, black	.52[1]
Yellow-poplar	1.43
Yellow-poplar	2.95

[1] Denotes double-diffusion treatments. All other values are for end-diffusion.

21-4 EFFICACY OF PRESERVATIVE TREATMENTS

NATURAL DECAY AND TERMITE RESISTANCE

Data on the natural resistance of pine-site hardwoods to decay and termites can be located in this text as follows:

Subject	Table, section, or page
Discussion of natural decay resistance	NATURAL DECAY RESISTANCE under section 11-1
Decay weight-loss ranking, 22 species	Table 11-1
Termite resistance ranking, 22 species	Table 11-6
Service life of untreated crossties	Page 852, and figure 22-55 top
Service life of untreated posts	Table 21-18, and page 2687

INTERNATIONAL OVERVIEW

Hardwood crossties (fig. 21-10) constitute a significant fraction of the total volume of wood treated annually in the United States. Practically all crossties are treated with creosote (fig. 22-53). Almost a century of use has shown that a broad range of hardwood species, when properly treated with creosote solutions, perform satisfactoily in rail lines. Average service life of crossties ranges from 25 to 50 years, depending upon location; increasingly, tie failure is caused by factors unrelated to biological deterioration.

Performance data for hardwood products treated with preservatives other than creosote and for hardwoods in other ground-contact service is less extensive. Available data show an uneven pattern of performance ranging from good to bad, perhaps reflecting in part differences among species and among treating methods employed.

Definitive information is generally lacking on the specific organism responsible for failure. In general, the white-rot fungi are the most important causes of decay in hardwoods. Some of these organisms (e.g., *Coriolus versicolor* L. per Fr.) are much more tolerant of certain preservatives than are brown-rot organisms (Thompson[7]; Gjovik[8]). Soft-rot fungi, which destroy wood despite preservative loadings that prevent decay by *Basidiomycetes*, are increasingly recognized as important in decay of hardwoods in ground-contact service.

Figure 21-10.—Hardwood crossties, shown here air-drying prior to treatment, account for about one-third of the 330 million cu ft of wood treated annually in the United States.

---

[7]Thompson, W. S. 1975. Unpublished data. Miss. For. Prod. Lab., Miss. State Univ., Mississippi State, Miss.
[8]Gjovik, L. R. 1980. Personal communication with W. S. Thompson.

In Europe and Australia hardwoods with high adsorption and deep penetration of water-borne salts are rapidly attacked by soft rot. Some authorities recommend that salt-treated hardwoods not be used in ground contact (Bergman 1977).

Noting that in tropical climates soft rot is the primary cause of failure of treated wood in ground contact, especially hardwoods, Levy (1978) recommended use of softwoods for poles wherever a choice is available. Soft-rot damage to improperly treated softwood poles, however, has also been reported (Henningsson et al. 1976).

Greaves and Levy (1978) found evidence that low levels of CCA preservative in the fibers of hardwoods may be responsible for poor performance. Eight of ten hardwoods studied showed skewed CCA distribution patterns among vessels, fibers, and ray tissues. The pattern evident for CCA as a whole was also found individually for copper, chromium, and arsenic. Fibers contained much less CCA than either vessels or rays at all depths studied and ranged as low as 9 percent. The two species which showed uniform distribution performed satisfactorily in field tests.

On a different level, Dickinson et al. (1976) investigated the distribution of copper, chromium, and arsenic within individual wood cells where a normal CCA treatment protected softwood from both soft rot and *Basidiomycetes* and hardwood from *Basidiomycetes* only. Even a double treatment failed to protect the hardwood from soft rot. Elemental distribution of copper within tracheid and fiber walls, determined by electron-probe microanalysis, was as follows (data for chromium were similar):

Wall layer	Relative amount of copper	
	Softwood tracheids	Hardwood fibers
	X-rays emitted/second	
S_3	239	6.8
S_2	222	4.9
S_1	224	6.1
Middle lamella	183	5.4

Unlike the uniformly well treated tracheids, the fibers were poorly treated, with lowest quantities of preservative in the S_2 layer, where soft-rot fungi attack wood by forming distinctive cavities. Concentrations of copper, chromium, and arsenic were much higher in ray cells than in fibers.

Levy (1978) cited three reasons for the apparent preferential attack of hardwoods by soft rot.

- High content of pentosan and other hemicelluloses easily utilized by soft rot.
- Poor penetration of the fiber cell walls by the preservative.
- High wall-to-lumen volume ratios in fiber cells resulting in low retention of preservative in fiber walls even at normally adequate retentions in pounds/cu ft of wood.

Hulme and Butcher (1977), however, found no major differences in the gross or cellular distribution of CCA preservatives between hardwood and softwood specimens.

Tamblyn (1973) reported that in field trials at eight sites in Australia and New Guinea involving both eucalyptus and pine poles, and several preservatives, soft rot had seriously deteriorated the hardwood after only 6 to 8 years. He concluded that retentions of the best preservatives providing satisfactory protection in temperate climates are inadequate for some hardwoods in ground-contact service in the tropics.

By contrast Aston and Watson (1976) reported that in numerous parts of Africa creosote-treated eucalyptus has performed extremely well. They cited a power line near Pretoria, in which no poles failed during a 40-year period. They also reported similar results in Malaysis with *Shoria* sp treated with CCA. Purslow (1975) reported no failures in 44-year-old beech stakes treated with creosote or CCA and installed at Princes Risborough in England. In field trials in Scandinavia, however, beech and other hardwood species performed poorly when treated with either organic or inorganic preservatives (Henningsson 1974).

Investigations by Sorkhoh and Dickinson (1976) on the cause of early failure of treated hardwoods confirmed the findings of others (Tamblyn 1973; Henningsson et al. 1976; Levy and Greaves 1978; Greaves and Levy 1978), that soft-rot fungi are the causal organisms. Cavities were forming in the S_2 layer of the cell walls of six species within 3 days to 2 weeks following innoculation of samples. Treatment with CCA reduced neither the incidence of cavity formation nor the time required for their development. In similar studies Liese and Peters (1977) found that hardwood specimens treated with either copper or chromium were more heavily attacked by soft-rot fungi than those treated with both metals.

Those interested in performance records of salt-treated railroad ties outside the United States will find the following publications useful:

    Cooper 1976   Borup (1961)
    Ellwood (1956)   Hedley (1973)
    Krzyzcwski (1973)

SERVICE DATA IN NORTH AMERICA

North America service data for treated hardwoods are concerned mainly with crossties and fence posts. These two products represent distinctly different levels of exposure. Crossties placed on well ballasted roadbeds (fig. 22-58) are subject to less biological hazard than fence posts in direct contact with soil (fig. 21-11). On the other hand, they are subjected to much higher levels of mechanical stress (fig. 22-56) and may require replacement for causes unrelated to biological factors (fig. 22-58). Service data for the two products are thus not interchangeable.

Figure 21-11.—(Top) Peeled hardwood posts air drying prior to treatment. (Bottom) Treated posts installed during the years 1950-1954 near Starkville, Miss., to determine service life; some treated stakes are visible in the foreground.

Current estimates of service life of crossties in line (fig. 22-55) are based on tests begun 30 to 50 years ago. Many preservatives then in actual or potential use have since been discontinued. For all practical purposes, only creosote and its coal-tar or petroleum solutions are employed in crosstie treatment today. A somewhat similar situation exists in the case of fence posts, although the time involved is usually shorter.

Crossties.—Average service lives of 50 years or more were reported by Blew (1963) for red oak and hard maple crossties treated with 5-12 pcf of creosote in tracks in Wisconsin compared to 6 to 10 years for untreated ties. Surprisingly, semi-refined paraffin oil provided a service life for both species nearly as long as that for coal-tar creosote, as did a 20:80 emulsion of zinc chloride solution and creosote.

Blew attributed differences in average service life in two Wisconsin test tracks to the use of tie plates on some ties, better drainage, and less traffic at one site. In related studies, 50:50 blends of coal-tar creosote and wood-tar creosote at retentions of 7 to 10 pcf provided service lives of 23 to 44 years for slippery elm, red oak, white oak, cherry, and butternut crossties. Red oak ties treated with zinc chloride failed after an average service life of 17 years.

In other work, Blew (1966) found that service conditions as well as treatments affect the life of oak crossties. Average service life in Maryland for red oak and white oak crossties representing a variety of preservatives was 20 to 34 years and 21 to 31 years, respectively. Red oak ties treated to retentions of 8 to 10 pcf with creosote remained serviceable for about 34, 33, 21, and 19 years in Ohio, Pennsylvania, Texas, and Arkansas, respectively.

Reports of tie service life in Northern Pacific Railroad show that red oak ties treated either with straight creosote or a 50:50 creosote-petroleum blend have a service life throughout the system of 30 to 50 years (Radkey 1960). Mechanical failure was a contributing factor in the replacement of over 50 percent of these ties (fig. 22-59).

A comprehensive test of crossties in track has been underway in Canada for many years. Forty-year tests data reported by Krzyzewski (1969) are summarized in table 21-16 by species and type of preservative treatment. It is of interest in this table that brush treatments with Osmolit (sodium fluoride and dinitrophenol) and Osmotite pastes (sodium fluoride, dinitrophenol and chromium salt) performed about as well as pressure treatments with Boliden salt or pentachlorophenol in petroleum. Only about 25 percent of all failures resulted from biological deterioration. Blew (1963, 1966) found that rail plates greatly reduced mechanical failure (fig. 22-54).

The service life of properly treated hickory ties has been shown to be equal to that of other more widely used tie species. Taras and Hudson (1959) summarized data from several studies which indicate a service life for hickory ties from 25 to 35 years. This range is very close to the estimated service life of 31 years for red oak ties included in the same track.

Bescher (1977) summarized available data on the life of creosoted crossties in mainline tracks. Tests beginning in 1909 and 1910 by the Chicago, Burlington, and Quincy Railroad showed that service life varied from 25 to 35 years, except for ash, which had only a 20-year life (fig. 22-55 top). Beech, sycamore, elm, red oaks, sweetgum, and black tupelo all had service lives averaging 30 to 35 years. Birch, cottonwood, hickory, pine, hard and soft maple, and white oak had a 25- to 30-year service life. Crosstie life also varied significantly according to the quality of track and ballast maintenance, and with severity of traffic; crossties in tracks of a midwestern railroad that traditionally made a profit and maintained its tracks had service lives varying from 25 to 60 years depending on the amount of traffic (fig. 22-55 bottom).

TABLE 21–16—*Service life of hardwood crossties in Canadian railways by species, preservative, and retention* (Krzyzewski 1969)

Species	Retention	Preservative	Average service life
	Pounds/cu ft		Years
Aspen	7.2	Creosote	32.9
Red oak	.26	Zinc-meta-arsenite	23.9
Beech	.20	Zinc-meta-arsenite	21.9
Hard maple	.31	Zinc-meta-arsenite	23.5
Yellow birch	.25	Zinc-meta-arsenite	22.5
White birch	.26	Zinc-meta-arsenite	21.7
Beech, hard maple, red oak	.43	Osmotite (brush treated)	21.2
Red oak, yellow birch, hard maple	.43	Osmolit (brush treated)	21.2
Yellow birch, hard maple, beech	5.3–8.8	70/30 creosote-coal tar	27.0–31.7+
Yellow birch	8.1	50/50 creosote-petroleum	21.0+
Aspen	—	70/30 creosote-coal tar	18.0+
Beech, birch, maple	—	Copper naphthenate	20.0+
Birch	.40	Pentachlorophenol	17–18+
Birch	.60	Boliden salt	16.7+
Maple	.60	Boliden salt	15.7+
Red oak	.60	Boliden salt	17.0+
Aspen	7	50/50 creosote-petroleum	15–16+
Yellow birch	.30	Pentachlorophenol	15+
Hard maple	.42	Pentachlorophenol	15+

Tie life is longest in side and yard tracks (60-year average), intermediate in branch lines (about 30 to 35 years), and shortest on heavily used mainline tracks (about 25 years—possibly 35 years); on curved portions of mainline tracks, crosstie life is shortest.

In the future crosstie life may be further extended by fabricating them from two or more pieces of wood smaller than 7 by 9 inches, thus permitting more complete penetration by preservatives. One such fabrication procedure now in limited commercial use calls for dowel-lamination (no glue required) of two pieces measuring 4.5 by 7 inches into a 7- by 9-inch assembly (figs. 20-12, 20-13, and 18-104D bottom). Such two-piece crossties are more readily treated than one-piece ties. Preservative penetration from the two-piece interface, nearly equal to that from the outside surfaces, should augment protection against decay. (See discussion related to figures 20-12 and 20-13).

Tschernitz et al. (1979) described the **Press-Lam** process for making crossties (fig. 22-57 bottom) which embodies stored-heat glue-laminating of thick-sliced veneer (0.25 inch or more thick) using residual heat of the wood as removed from the veneer dryer. They found that creosote preservative treatment was adequate for all species and treating time was only about half that for solid wood crossties.

Fence posts.—Many service-life data for treated hardwood posts are from field installations (fig. 21-11) made between 1940 and 1960 at Land Grant universities and by the United States Forest Service. They involved a great number of preservative materials, methods of application, species of woods, and exposure sites.

The variables of site and method of application have complicated interpretation of these data, making it difficult to reconcile inconsistencies in results. This problem is aggravated by the large number of researchers involved and resultant lack of uniformity in treating solutions, applications, and in interpretation of inspection results (Gjovik and Davidson 1975). It often is impossible after the fact to distinguish between procedural errors and differences that are site related.

The reports available cover a range of non-pressure treating methods. Among these, the cold-soak, hot-and-cold bath, and diffusion treatments were most frequently used, but superficial brush, spray, and dip applications of preservative are also well represented.

Treatments of selected hardwood species to retentions averaging between 6 and 12 pcf by cold-soak and hot-and-cold bath process have proved satisfactory. In a study reported on by Carpenter and Bouler (1962), 85 percent of ash, American elm, honeylocust, mixed oak, bitter pecan, and sweetgum posts treated by this process with coal-tar creosote were still serviceable after 19 years' exposure in Mississippi. Sixty-three to 68 percent of cottonwood, red maple, and hackberry, but only 15 percent of the willow posts were serviceable after the same period.

Retentions of 5 percent pentachlorophenol in fuel oil comparable to those obtained with coal-tar creosote were also obtained by the same authors using the hot-and-cold-bath process (Carpenter and Bouler 1962). After 13 years' exposure, 83 to 100 percent (average 94) of posts of nine hardwood species, representing both heartwood and sapwood were serviceable. Results for sweetgum were poorest, with 59 percent serviceable after 13 years. After being pressure-treated with creosote to a retention of 12 pcf, cottonwood posts were 100 percent and sweetgum 89 percent serviceable after 23 years; with retentions of 6 pcf, however, only 42 percent of sweetgum and 8 percent of cottonwood posts were serviceable. Carpenter and Bouler (1962) also reported that soaking of black willow and cottonwood in chromated zinc chloride provided a service life of 7 to 10 years.

Among 398 hardwood posts included in a study reported on by Walters and Peterson (1965), at least 87 percent of the green ash, black cherry, shagbark hickory, black oak, and red oak were still serviceable 21 years following a cold-soak treatment in a 5 percent solution of pentachlorophenol in fuel oil.

Green, unpeeled posts treated by end-diffusion with an aqueous solution of zinc chloride protected pine and some hardwood species for over 20 years (Toole and Thompson 1973). Over 80 percent of red oak, post oak, sweetgum, and pine posts in certain size classes were still serviceable after 20 years of exposure in a test plot in Mississippi following end-diffusion treatments that yielded retention of about 1.0 pcf (Thompson 1954). Time of harvest had some effect on serviceability, with posts cut when dormant out-performing those cut during the growing season, a result attributed to retention of bark by fall- and winter-cut posts and consequent reduced leaching of preservative.

Verrall (1959) studied the protection afforded ammunition boxes by dipping in a 5 percent pentachlorophenol solution containing a water repellent. After 4.4 years' exposure near Gulfport, Miss., the decay ratings were 2.3, 3.0, and 2.3 for sapwood of yellow-poplar, sweetgum, and tupelo gum, respectively, on a scale where 0 = no decay and 5 = destroyed. The comparable rating for southern pine sapwood was 1.8. Untreated wood of all species had ratings of 4.8 to 5.0.

Scheffer (1953) tested several fungicides against common wood-decay fungi on oak and several other species, for possible use in bilgewater in boats. The order of effectiveness was: phenylmercuric acetate, sodium pentachlorophenate, 2,4,5-trichlorophenol, orthophenylphenol, pentachlorophenol, boric acid- + sodium dichromate, and boric acid. All were effective, but orthophenylphenol and pentachlorophenol were recommended because of their low solubility, safety in use, and low cost.

Superficial treatments of pentachlorophenol in light solvent containing a water repellent were found by Scheffer and Browne (1954) to provide a high degree of protection to sweetgum blocks exposed to *Coriolus versicolor*. This work was later verified by other studies (Scheffer and Clark 1967).

Compilations of service-life data for posts are available from the following sources:

Gjovik and Davidson (1975)	Reports on over 800 combinations of preservatives, retention levels, treatment methods, species, and exposure sites.
American Wood Preservers' Association Proceedings	Reports of Committee U-5, Post Service Records: Annually through 1969.
Anonymous (1960)	Resuts of studies by members of Wood Preservation Council
Krzyzewski and Spicer (1974)	Results of Canadian post studies, 34 years.
Krzyzewski and Sedziak (1975)	Performance of treated posts.

Service-life and treating data gleaned from the afore-mentioned sources are shown in table 21-17 for selected species. Service-life for untreated wood of the same species is included in table 21-18. Extension of service life by preservative treatment varied principally with the preservative used and exposure site. Zinc chloride, chromated zinc chloride, and other water-soluble, non-reactive formulations generally gave poorer results than creosote or the oil-borne preservatives.

Exposure site greatly affected service-life with these light-duty preservatives. Red oak posts treated by end-diffusion had a service life of 21 years in Wisconsin, but only 9 years in Mississippi. Retention by the Mississippi posts at 0.95 pcf was 27 percent higher than that for the Wisconsin posts.

The double-diffusion process generally protected posts better than one-step diffusion. Service life of red oak, sweetgum, and hickory posts treated with zinc sulfate followed by arsenic acid + sodium chromate and exposed in Georgia ranged from 16 to 22 years and averaged almost 20 years (table 21-17). This is near the average service life of 23 years for pentachlorophenol-treated oak posts exposed in Mississippi and Alabama. Results are even better when copper sulfate replaces zinc sulfate in double-diffusion treatment (Gjovik and Davidson 1979).

TABLE 21-17.—*Service life of hardwood fence posts by species, treating method, preservative, and exposure site*
(Data from Anonymous 1960; Kulp 1966; Gjovik and Davidson 1975)

Species and treatment method (and time, hours)	Preservative Material	Retention	Exposure state	Year installed	Service life
		Pounds/cu ft			*Years*[1]
American elm					
Hot-and-cold bath...	Creosote	11.1	Mississippi	1941	—
Cold-soaking (48)...	Pentachlorophenol (10 percent)	3.2	Illinois	1942	20
Cold-soaking (120) .	Pentachlorophenol	2.5	Mississippi	1953	14
Beech					
Hot-and-cold bath...	Creosote	8.5	Maryland	1908	48
River birch					
Hot-and-cold bath...	Creosote	—	Maryland	1908	30
Bitter pecan					
Hot-and-cold bath...	Creosote	6.0 to 8.9[2]	Wisconsin	1941	18+
Hot-and-cold bath...	Pentachlorophenol	4.3 to 6.7	Mississippi	1949	13
Cold-soaking.......	Pentachlorophenol	5.2	Mississippi	1953	21
Blackjack oak					
Pressure	Creosote	6 to 11	Missouri	1938	50+
Pressure	Zinc chloride	.84	Wisconsin	1938	34
Hot-and-cold bath..	Creosote	4.1	Wisconsin	1938	40
Cold-soaking (105) .	Pentachlorophenol	4.1	Tennessee	1949	—
Cold-soaking (195) .	Pentachlorophenol	4.5	Tennessee	1949	—
Cold-soaking (47)...	Copper naphthenate	4.4	Tennessee	1953	8
Cold-soaking (48) ..	Copper naphthenate	3.4	Tennessee	1953	12
Black willow					
Steeping..........	Chromated zinc chloride	—	Mississippi	1948	5
Steeping..........	Chromated zinc chloride	—	Mississippi	1947	10
Hot-and-cold bath ..	Creosote	8.4[2]	Mississippi	1941	17
Box elder					
Cold-soaking (120) .	Pentachlorophenol	9.8	Mississippi	1953	18

TABLE 21-17.—*Service life of hardwood fence posts by species, treating method, preservative, and exposure site*
(Data from Anonymous 1960; Kulp 1966; Gjovik and Davidson 1975)—Continued

Species and treatment method (and time, hours)	Preservative Material	Retention	Exposure state	Year installed	Service life
		Pounds/cu ft			Years[1]
Cedar elm					
Cold-soaking (120) .	Pentachlorophenol	5.1	Mississippi	1953	26
Cottonwood					
Pressure	Creosote	6.0	Mississippi	1937	18
Pressure	Creosote	12.0	Mississippi	1937	—
Hot-and-cold bath ..	Creosote	12.2[3]	Minnesota	1909	30
Steeping	Sodium fluoride	.28	Montana	1926	9
Steeping	Sodium fluoride	.67	Montana	1926	13
Hackberry					
Hot-and-cold bath ..	Creosote	8–12	Mississippi	1941	23
Hot-and-cold bath ..	Pentachlorophenol	10–13	Mississippi	1948	17
Cold-soaking (120) .	Pentachlorophenol	3.5	Mississippi	1953	11
Hickory					
Double diffusion ...	[4]	.20	Mississippi	1953	14
Cold-soaking (120) .	Pentachlorophenol	3.6	Mississippi	1953	21
Double diffusion ...	[5]	.39	Georgia	1955	19
Double diffusion ...	[5]	1.0	Georgia	1955	19
Northern red oak					
Pressure	Creosote	5.8	Wisconsin	1960	—
Hot-and-cold bath ..	Creosote	—	Maryland	1908	36
End diffusion	Chromated zinc chloride	.75	Wisconsin	1946	21
Cold-soaking (8) ...	Pentachlorophenol	3.2	Wisconsin	1943	14
Cold-soaking (24) ..	Pentachlorophenol	3.5	Wisconsin	1943	20
Cold-soaking (48) ..	Pentachlorophenol	4.5	Wisconsin	1943	28
Cold-soaking (96) ..	Pentachlorophenol	5.8	Wisconsin	1943	36
Cold-soaking (168) .	**Pentachlorophenol**	5.7	Wisconsin	1943	36
Cold-soaking	Pentachlorophenol	1.8[2]	Minnesota	—	33
Cold-soaking (66) ..	Pentachlorophenol	2.6	Illinois	1942	—
Overcup oak					
Hot-and-cold bath ..	Creosote	6.1 to 9.5[2]	Mississippi	1941	30+
Hot-and-cold bath ..	Pentachlorophenol	2.7 to 7.9[2]	Mississippi	1947	18
Post oak					
Cold-soaking (72) ..	Pentachlorophenol	4.2	Alabama	1948	23
Cold-soaking (46) ..	Pentachlorophenol	3.2	Alabama	1948	17
Cold-soaking	Pentachlorophenol	5.7	Mississippi	1956	—
Cold-soaking (48) ..	Pentachlorophenol	2.0[3]	Texas	1963	—
Cold-soaking (123) .	Copper naphthenate	7.1	Tennessee	1954	11
Cold-soaking (72) ..	Copper naphthenate	7.1	Tennessee	1954	—
Cold-soaking	Copper naphthenate	9.6	Mississippi	1956	—
Red maple					
Hot-and-cold bath ..	Creosote	—	Maryland	1908	32
Cold-soaking (23) ..	Pentachlorophenol	6.1	Tennessee	1951	15
Cold-soaking (53) ..	Pentachlorophenol	5.6	Tennessee	1949	22
Cold-soaking (46) ..	Copper naphthenate	4.3	Tennessee	1953	7

See footnotes page 2513.

TABLE 21-17.—*Service life of hardwood fence posts by species, treating method, preservative, and exposure site* (Data from Anonymous 1960; Kulp 1966; Gjovik and Davidson 1975)—Continued

Species and treatment method (and time, hours)	Preservative Material	Retention	Exposure state	Year installed	Service life
		Pounds/cu ft			Years[1]
Scarlet oak					
Cold-soaking (44) ..	Pentachlorophenol	3.9	Tennessee	1949	—
Cold-soaking (48) ..	Copper naphthenate	4.7	Tennessee	1953	12
Cold-soaking (55) ..	Copper naphthenate	5.7	Tennessee	1953	14
Cold-soaking (47) ..	Copper naphthenate	5.2	Tennessee	1954	—
Southern red oak					
Cold-soaking	Pentachlorophenol	5.6	Mississippi	1953	—
Cold-soaking	Pentachlorophenol	3.9	Mississippi	1947	23
End diffusion	Zinc chloride	.95	Mississippi	1947	9
Double diffusion ...	[5]	1.0	Georgia	1955	21
Double diffusion ...	[5]	1.6	Georgia	1955	22
Double diffusion ...	[5]	2.1	Georgia	1955	16
Cold-soaking (48) ..	Pentachlorophenol	6.2	Mississippi	1948	—
Cold-soaking (48) ..	Pentachlorophenol	5.0	Mississippi	1956	—
Cold-soaking (48) ..	Pentachlorophenol	6.4	Mississippi	1951	—
Cold-soaking (84) ..	Copper naphthenate	7.6	Mississippi	1951	11
Cold-soaking	Copper naphthenate	9.2	Mississippi	1956	—
Sweetgum					
Pressure	Boliden salt	.78	Mississippi	1957	—
Pressure	Creosote	9.7	Mississippi	1957	
Hot-and-cold bath ..	Creosote	7.7	Louisiana	1909	23
Hot-and-cold bath ..	Creosote	—	Maryland	1908	14
Pressure	Creosote	6.0	Mississippi	1937	23
Pressure	Creosote	12.0	Mississippi	1937	34
Hot-and-cold bath ..	Pentachlorophenol	2.3–10.0	Mississippi	1947	15
Cold-soaking (48) ..	Pentachlorophenol	4.9	Mississippi	1947	21
End diffusion	Zinc chloride	.95	Mississippi	1947	7
Double diffusion ...	[5]	1.9	Georgia	1955	19
Water oak					
Hot-and-cold bath ..	Creosote	8.4 to 9.6[2]	Mississippi	1941	27
Hot-and-cold bath ..	Pentachlorophenol	2.7 to 5.6[2]	Mississippi	1947	—
Water tupelo					
Hot-and-cold bath ..	Creosote	8.6	Louisiana	1908–10	23
Cold-soaking (120) .	Pentachlorophenol	6.5	Mississippi	1953	22
End diffusion	Zinc chloride	.94	Mississippi	1947	5
White oak					
Pressure	Chromated copper arsenate	.28	Mississippi	1957	—
End diffusion	Chromated zinc chloride	.75	Wisconsin	1946	29
Hot-and-cold bath ..	Creosote	—	Maryland	1908	37
Pressure	Creosote	8.1	Mississippi	1957	—
Steeping	Zinc chloride	1.1 to 1.8	Wisconsin	1940	16
Double-diffusion ...	[5]	1.3 to 2.3	Georgia	1955	16–21

TABLE 21-17.—*Service life of hardwood fence posts by species, treating method, preservative, and exposure site*
(Data from Anonymous 1960; Kulp 1966; Gjovik and Davidson 1975)—Continued

Species and treatment method (and time, hours)	Preservative		Exposure state	Year installed	Service life
	Material	Retention			
		Pounds/cu ft			*Years*[1]
Yellow-poplar					
Double-diffusion ...	5	2.0	Georgia	1955	19

[1]Estimated average life of posts in tests. The most common method of determination of service life is use of mortality curves (such as those developed for crossties). An estimate of service life can be made when as few as 11 percent of the posts have failed. Other authors whose data are cited in this table made no estimate of service life until 60 percent of the posts had failed. A dash entry (—) in this column indicates than an insufficient number of posts had failed at the last inspection to permit an estimate of service life.
[2]Mixed round and split posts.
[3]Split posts.
[4]Copper sulfate-sodium chromate.
[5]Zinc sulfate and arsenic acid-sodium chromate.

TABLE 21-18.—*Service life of untreated hardwood posts of 16 species*
(Gjovik and Davidson 1975)

Species	Form	Location of test	Service life
			Years
Ash, green	Round	Halsey, Nebraska	18.7
Ash, green	Round	Miles City, Minnesota	8.6
Ash, white	Round	Madison, Wisconsin	4.3
Elm, slippery	Round	Oregon, Wisconsin	3.8
Hickory, mockernut	Round	Saucier, Mississippi	3.5
Hickory, shagbark	Round	Oregon, Wisconsin	4.0
Hickory, sp.	Round	Norris, Tennessee	4.7
Hickory, sp.	Round	Saucier, Mississippi	2.8
Maple, red	Round	College Park, Maryland	3.8
Oak, black	Round	Norris, Tennessee	2.8
Oak, blackjack	Square	Madison, Wisconsin	8.4
Oak, blackjack	Round	Ava, Missouri	6.0
Oak, blackjack	Square	Madison, Wisconsin	7.9
Oak, southern red	Round	Saucier, Mississippi	2.8
Oak, red sp.	Round	Athens, Georgia	4.3
Oak, overcup	Round	Stoneville, Mississippi	4.3
Oak, overcup	Split	Stoneville, Mississippi	5.0
Osage orange	Round and split	Childress, Texas	43.0
Sweetbay.................	Round	Saucier, Mississippi	1.6
Sweetgum	Round	Athens, Georgia	2.2
Sweetgum	Round	College Park, Maryland	4.2
Sweetgum	Round	Wilson Dam, Alabama	2.3
Sweetgum	Round	Saucier, Mississippi	1.8
Tupelo, black	Round	Norris, Tennessee	3.4
Tupelo, black	Round	College Park, Maryland	4.2
Tupelo, water	Round	Saucier, Mississippi	2.1

The performance of copper naphthenate-treated posts was generally inferior to that for posts treated with creosote and pentachlorophenol. For example, oak posts treated with this preservative and exposed in Tennessee and Mississippi failed after only 8 to 14 years. These results indicate that solution concentration—in terms of percent copper metal—was inadequate even though gross adsorption of solution was comparable to that for other preservatives. A retention of copper naphthenate equivalent to 0.40 pcf of copper metal is needed for long-term protection of wood.[4] The copper content of treating solutions used in the studies was usually about 0.5 percent, too low by a factor of about 10 to provide the desired copper retention when gross adsorption was typically 4 to 6 pcf.

Service life for oaks treated with creosote, mainly by the thermal process, ranged from 27 years in Mississippi (pcf = 8.4 to 9.2), 50 years in Missouri (pcf = 6 to 11), and 40 years in Wisconsin (pcf = 4.1). These high performances appear to reflect above average treatment quality rather than the superiority of the preservative. For any given retention, pressure and thermal treatments should provide better penetration of preservative than cold-soak treatments. This trend was also evident with species other than the oaks.

Species prone to develop deep checks or splits may decay faster than split-resistant species. There is some evidence that cottonwood and willow perform more poorly than could be anticipated based on preservative retention (Hopkins 1951; Furnival 1954), probably attributable to uneven distribution of preservative.

Sweetgum is the hardwood species that has received the most attention from researchers in establishing field tests of wood preservatives. Results of tests conducted by eight state agricultural experiment stations located within the natural range of this speices are listed in table 21-19. These data show that the two-year natural service life of this species is increased by a factor of seven or more by preservative treatment with pentachlorophenol and copper naphthenate and by a factor of about three by treatment with zinc chloride.

TABLE 21-19.—*Treating and service-life data for sweetgum fence posts* (Data from Kulp 1966; Anonymous 1960)

Treating method (and time, hours)	Perservative		Age at last inspection	Exposure state	Number in test	Failure	Service life[1]
	Material	Retention					
		Pounds/cu ft	*Years*			*Percent*	*Years*
Cold-soaking (26)........	Pentachlorophenol	6.8	11	Mississippi	23	44	13
Cold-soaking (—)........	Pentachlorophenol	6.3	8	Mississippi	292	41	9
Cold-soaking (48)........	Pentachlorophenol	5.1	14	Georgia	23	39	16
Cold-soaking (24)........	Pentachlorophenol	4.7	14	Georgia	18	33	17
Cold-soaking (72)........	Pentachlorophenol	5.8	14	Georgia	25	28	18
Cold-soaking (72)........	Pentachlorophenol	5.7	14	Georgia	24	4	—
Cold-soaking (72)........	Pentachlorophenol	6.4	13	Georgia	26	8	—
Cold-soaking (144).......	Pentachlorophenol	7.2	14	Georgia	25	1	—
Cold-soaking (24)........	Pentachlorophenol	5.5	15	Alabama	28	14	22
Cold-soaking (6).........	Pentachlorophenol	5.0	15	Alabama	27	85	11
Cold-soaking (72)........	Pentachlorophenol	5.6	12	Texas	65	0	—
Cold-soaking (72)........	Pentachlorophenol	5.6	13	Texas	98	0	—
Cold-soaking (72)........	Pentachlorophenol	5.6	13	Texas	97	0	—
Cold-soaking (72)........	Pentachlorophenol	5.6	12	Texas	30	0	—
Cold-soaking (12)........	Pentachlorophenol	5.1	12	Virginia	5	80	10
Cold-soaking (48)........	Pentachlorophenol	4.1	12	Virginia	11	48	13
Cold-soaking (105).......	Copper naphthenate	7.0	8	Mississippi	50	18	11
Cold-soaking (24)........	Copper naphthenate	4.4	14	Georgia	47	56	15
Cold-soaking (48)........	Copper naphthenate	5.4	14	Georgia	24	38	16
Cold-soaking (72)........	Copper naphthenate	4.2	14	Georgia	22	64	14
Cold-soaking (144).......	Copper naphthenate	5.7	15	Georgia	23	52	14

See footnote page 2516.

TABLE 21-19.—*Treating and service-life data for sweetgum fence posts* (Data from Kulp 1966; Anonymous 1960)—Continued

Treatment method (and time, hours)	Preservative		Age at last inspection	Exposure state	Number in test	Failure	Service life[1]
	Material	Retention					
		Pounds/cu ft	*Years*			*Percent*	*Years*
Cold-soaking (0.3)	Copper naphthenate	—	6	Tennessee	17	35	7
End diffusion ..	Zinc chloride	.9	6	Mississippi	200	34	7
End diffusion ..	Zinc chloride	.4	12	Georgia	50	82	6.5
End diffusion ..	Zinc chloride	.6	10	Georgia	24	96	6
	None	—		Georgia	24	100	3.0
	None	—		South Carolina	21	100	1.0
	None	—		Mississippi	24	100	2.2
	None	—		Louisiana	25	100	2.0

[1] Estimated average life of posts in test. Blank entries (—) in this column indicate that an insufficient number of posts had failed at the last inspection to permit an estimate of service life.

21-5 EFFECTS OF TREATMENTS ON WOOD PROPERTIES

Some of the treatments to which wood is subjected during sizing, preserving, and stabilization treatments adversely affect its mechanical properties. Certain preservatives are themselves reported to reduce wood strength as do squaring from round (table 22-26), incising, and conditioning preparatory to preservative treatment. The beneficial effects of these operations—improved quality of preservative treatments and extended service life—usually outweigh the negative effects.

Most studies of the relationship between wood strength and processing operations deal with softwood species. There are some data for hardwoods, however, and much can be inferred about hardwoods from softwood studies.

INCISING

Perrin (1978) concluded that incising probably has a negligible effect on the strength properties of large items such as poles and crossties. For timbers smaller than crossties, however, the results were mixed, as indicated for Douglas-fir in table 21-20.

Schrader (1945) determined the effect of incising on the strength of Douglas-fir laminated beams. Beams 8 x 18 inches in cross section fabricated from 1-inch lumber and beams 8 x 16 inches prepared from 2-inch lumber sustained reductions in bonding strength of from 10 to 20 percent as a result of ⅝-inch-deep incisions applied at a rate of 65 per square foot. Banks (1973) found that 2- by 2-inch samples of Norway spruce (*Picea abies* L. Karst.) lost 16 and 13 percent of their MOR and MOE, respectively, following incising to a depth of ¼-inch with

860 incisions per square foot. Similar depths and frequencies of incisions on redwood (*Sequoia sempervirens* (D. Don) Endl.) dimension sized for use in cooling towers reduced bending strength by 7 to 28 percent (Kass 1975).

From the limited data available on the effect of incising on the strength of hardwoods, it can be surmised that effects are similar to those for softwoods. A study of strength losses sustained by small hardwood posts prepared for non-pressure preservative treatment (Chudnoff and Goytia 1967) tends to confirm this. Incising at a rate of 160 incisions per square foot caused a reduction of MOR of about 14 percent.

TABLE 21–20—*The effect of incising on the strength of Douglas-fir timbers and ties*
(Adapted from Perrin 1978)

Dimensions (inches)	Incisions per square foot	Change in strength						
		Bending[1]				Compression[3]		
		PL	MOR	MOE	SH[2]	CS	MOE	FS
	Number	----Percent----						
4 by 8	64	−8	−15	−4	—	−4	—	−7
4 by 8	64	−18	−15	−14	—	−17	—	−33
4 by 8	64	−4	−9	−4	—	−3	—	+3
4 by 8	64	+2	+11	−6	—	−5	—	−5
6 by 12	56	−2	−2	—	−2	—	—	—
6 by 12	56	+1	−2	—	−2	—	—	—
6 by 12	75	−2	0	−7	0	—	—	—
6 by 12	75	−4	−7	−4	−7	—	—	—
7 by 8	75	—	—	—	—	—	—	−10
7 by 9	76	—	—	—	—	—	—	−8
7 by 9	76	+6	+4	0	0	+3	0	+6
7 by 10	90	—	—	—	—	—	—	+17

[1] PL—proportional limit stress, MOR—modulus of rupture, MOE—modulus of elasticity.
[2] SH—shear parallel to grain.
[3] CS—crushing strength parallel to grain; MOE in compression parallel to grain; FS—compression strength perpendicular to grain.

CONDITIONING

Steaming.—Steaming of wood reduces water content in green stock and renders it more permeable to preservatives. It is routinely applied to southern pine and certain other conifers but less frequently applied to hardwoods. Current industry standards (American Wood Preservers' Association 1977) do not permit steam conditioning of oak because of its susceptibility to damage from this process. While steaming is permitted with other hardwoods, those that receive preservative treatment are mainly for crossties, conditioned usually by air drying, Boultonizing, or vapor drying.

Steam conditioning at the pressures and durations permitted under existing standards may cause significant strength losses in treated products. Buckman and Reese (1938), Davis and Thompson (1964), MacLean (1951), and Wood et al. (1960) show conclusively that above certain temperatures wood undergoes chemical degradation and sustains losses in strength to a degree dependent upon the duration and severity of exposure.

Wood heated in an atmosphere of steam loses weight, becomes discolored, sustains reductions in strength, and undergoes chemical degradation (Baechler 1954). At atmospheric pressure, effects are those of a mild hydrolysis, catalyzed by natural acids in the wood, becoming progressively more severe with increasing temperature and time of exposure. The carbohydrate fraction of wood, especially the hemicellulose component, is particularly susceptible to hydrolysis (MacLean 1951).

Much of the original work on the effect of steaming on the mechanical properties of wood was conducted by MacLean (1951, 1952, 1953). Following a study in which small specimens were steamed at temperatures of 250°F to 350°F for 8 to 32 hours, he concluded that shock resistance was the property most seriously affected, followed in order by MOR, fiber stress at the proportional limit, and MOE (MacLean 1953). An identical order of effect was reported by Thompson (1969a) based on bending tests of Class 6, 30-foot southern pine poles. In the latter study, MOR was reduced about 37 percent, from 7,902 psi to 5,707 psi, as a result of steaming 14 hours at 245°F.

The limited data available on hardwoods indicate they are more susceptible to damage from steaming than are softwoods. Davis and Thompson (1964) found that the residual toughness of small red oak specimens following steaming at 138°C for 120 minutes was only 60 percent of the control values, a greater reduction than those for southern pine and Douglas-fir specimens similarly treated. Chemical analyses of these specimens revealed that the reduction in strength was well correlated with changes in chemical composition. The carbohydrate fraction was more seriously degraded in oak than in either of the two coniferous species.

Kubinsky (1971) reported that compressive strength of small red oak cubes was reduced by 20 and 23 percent in the tangential and radial directions after steaming for 6 hours at atmospheric pressure. Reductions after exposure for 96 hours were 55 and 49 percent, respectively. Thompson (1969b) found a 14 percent reduction in compressive strength for southern pine piling sections steamed 16 hours at 245°F. While specimen size differed, results suggest that hardwoods are more sensitive to steaming than softwoods.

Further substantiation of this statement is provided by extensive research conducted by MacLean (1951, 1953). Weight losses resulting from heating small specimens for 17.4 days in water at 250°F were as follows for the species indicated:

Species	Weight loss
	Percent
Yellow birch	38
Yellow-poplar	37
Basswood	36
White oak	36
Sweetgum	34
Hard maple	33
Southern pine	28
Douglas-fir	25
White pine	24
Sitka spruce	24

Although this order of effect of thermal treatments was not maintained for all combinations of temperatures and heating mediums employed by MacLean (1951), the hardwoods averaged far more sensitive than the softwoods in all tests.

Boulton drying.—Conditioning by the Boulton process has less deleterious effects on wood than steaming because of the lower temperatures (180–210°F) used. As in the case of steaming, strength reductions caused by the process at a given temperature are determined by the duration of the conditioning process, by species, and by the size of items involved.

Data on the effects of Boulton-drying are lacking for hardwood species. Based on the experience with other forms of thermal treatment, it is probable that they would be at least equal to those for softwoods. Data compiled by Graham (1980) on the effect of three conditioning processes on the strength of Douglas-fir sawn products are of interest (table 21-21). Reductions in MOR for timbers in the size range from 6 x 12 to 8 x 16 inches that were Boultonized at temperatures of 190° to 215°F ranged between 5 and 18 percent and averaged about 10 percent in tests conducted by Rawson (1927), MacFarland (1916), Luxford and MacLean (1951), and Harkom and Rochester (1930). Reductions in MOR for 1-inch and 2-inch stock exposed to the same temperature averaged almost 12 percent; items in this size class that were kiln-dried or vapor-dried in the temperature range of 220° to 250°F sustained reductions in MOR of 18 to 21 percent. Rawson's study of the effect of Boulton drying on Douglas-fir timbers was superimposed on a study of the effect of incising. The modest reduction in MOR of 6 percent attributed by the authors to the Boulton process was higher than the apparent reduction caused by incising—about 2 percent.

TABLE 21-21—*Effect of conditioning temperature on the modulus of rupture of Douglas-fir* (Graham 1980)

Reference	Specimen size	\multicolumn{4}{c}{Reduction of MOR as fraction of control values[1]}				Heat source
		140 to 170°F	190 to 215°F	220 to 230°F	250°F	
	Inches	----------Percent----------				
Eddy and Graham (1955)	2 x 2		9	18	21	Organic vapors
	2 x 2	4				Kiln drying
Graham (1980)........	1 x 1			16		Organic vapors
	1 x 1	7	16			Kiln drying
Harkom and Rochester (1930)...............	6 x 12		13			Boulton drying
Kozlik (1968).........	2 x 6	1	10	21		Kiln drying
Luxford and MacLean (1951)...............	4 x 8		4			Boulton drying
	8 x 16		9			Boulton drying
	8 x 16		12			Boulton drying
	8 x 16		7			Boulton drying
	8 x 16		8			Boulton drying
MacFarland (1916)	7 x 16		18			Boulton drying
Rawson (1927)........	6 x 12		5			Boulton drying
	6 x 12		7			Boulton drying

[1] Adjusted for differences in moisture content between treated and control specimens.

Reductions in MOR caused by Boultonizing for 30-foot Douglas-fir and western larch poles were reported by Wood et al. (1960) to be of the same order of magnitude as those for sawn products (table 21-22). Larch was the most seriously affected, showing a decrease in MOR and MOE of 17 and 20 percent, respectively. However, these values were smaller than the reductions sustained by steam conditioned southern pine, which ranged from 23 to 34 percent for MOR and 12 to 16 percent for MOE.

Vapor drying.—The effect of this conditioning process on wood strength is somewhat greater than that of Boulton-drying because of the higher temperatures employed. Eddy and Graham (1955) found reductions in MOR of 9, 18, and 21 percent following vapor drying of 2- by 2-inch Douglas-fir at 190, 225, and 250°F, respectively. Reductions in work to maximum load were, in order, 17, 36, and 49 percent.

Strength reductions in oak and gum crossties attributed to vapor drying are of the same order of magnitude (table 21-23) notwithstanding the large difference in cross-sectional dimension of the test specimens: 2 by 2 inches compared to 7 by 9 inches. The MOR of untreated vapor-dried ties at 8,025 psi was about 8.3 percent less than that for matched air-dried ties. The comparable reduction for fiber stress at the proportional limit in compression perpendicular to the grain was 11.9 percent. For reasons that are unclear, the reduction in both strength properties was increased if the ties were water soaked prior to testing. Thus, the MOR of air-dried, water-soaked oak ties was 8,320 psi compared to 6,760 psi for vapor-dried, water-soaked ties, an apparent reduction due to vapor drying of almost 19 percent. The same effect is evident in the data for gum crossties.

TABLE 21-22.—*Effect of conditioning method on the strength of 30-foot poles* (Wood et al. 1960)

Species	Conditioning method	Modulus of rupture (Fraction of controls)	Modulus of elasticity unseasoned
		-----------Percent-----------	
Longleaf and slash pines (*Pinus palustris* Mill. and *P. elliottii* Engelm.)	Steam and vacuum[1]	−23	−12
Shortleaf and loblolly pines (*Pinus echinata* Mill. and *P. taeda* L.)	Steam and vacuum[1]	−34	−16
Douglas-fir (*Pseudotsuga menziesii* (Mirb.) Franco)	Boulton drying[2]	−7	−4
Western larch (*Larix occidentalis* Nutt.)	Boulton drying[2]	−17	−20
Lodgepole pine (*Pinus contorta* Dougl.)	Air drying	+4	+3
Western redcedar (*Thuja plicata* Donn)	Air drying	−4	−1

[1]Steaming conditioning was conducted at 259°F for 8.5 to 13.5 hours.
[2]Boulton drying was conducted at 195 to 210°F for 16 to 30 hours with a vacuum of 17 to 25 inches of mercury.

TABLE 21-23. *Strength of green, air-seasoned, and vapor-dried 7- by 9-inch oak and gum crossties*[1]

Species group and condition at time of test	Modulus of rupture	Fiber stress at proportional limit in compression perpendicular to the grain
	----------------Psi----------------	
Oak (*Quercus* sp.)[2]		
Green	9,560	804
Air-dried, water-soaked	8,320	706
Air-dried	8,754	687
Air-dried, creosoted, water-soaked	8,360	800
Air-dried, creosoted	8,920	811
Vapor-dried, water-soaked	6,760	474
Vapor-dried	8,025	605
Vapor-dried, creosoted, water-soaked	6,360	459
Vapor-dried, creosoted	8,250	611
Gum (sweetgum and black tupelo)[3]		
Green	8,604	685
Vapor-dried, water-soaked	7,500	446
Vapor-dried	7,828	490
Vapor-dried, creosoted, water-soaked	6,010	490
Vapor-dried, creosoted	7,390	792

[1]Data provided by M. S. Hudson, Spartanburg, S.C.
[2]Each value is the average of 32 tests.
[3]Each value is the average of 12 tests.

EFFECT OF TREATING CYCLE

Woods with low permeability sometimes sustain cell collapse under high temperatures and pressures in wood preserving. The incidence of collapse varies both within and among species, as well as with the treating conditions imposed. Thus, Rosen (1975b) found that both white oak and red oak collapsed at a pressure of 1,000 psi when the temperature was 75°F and at 500 psi when the temperature was 200°F. James (1961), however, found no collapse in red oak following pressure treatments conducted at 1,000 psi and 200°F. Likewise, Walters and Guiher (1970) successfully treated redgum (sweetgum) at 800 psi and 200°F without inducing collapse. Walters (1967) found collapse of sweetgum only in specimens subjected to pressures greater than 400 psi. It varied from slight to severe depending upon treating temperature and wood moisture content.

In red oak specimens, following treatments covering pressure and temperature ranges of 200 to 800 psi and 100 to 200°F, respectively, James (1961) detected no collapse in any of the specimens. Toughness reductions increased with the severity of treating conditions and ranged from about 6 percent (200 psi, 100°F) to about 11 percent (800 psi, 200°F).

While temperature and pressure are important, the refractory nature of collapse-prone species is a basic cause of collapse during preservative treatment. Unlike easily penetrated species, in which pressures are rapidly equalized by the flow of preservative into cell lumens, refractory species permit large pressure differentials to develop, inducing collapse of cells.

STABILIZATION TREATMENTS

Rowell and Youngs (1981) noted that: "There are two basic types of wood treatments for dimensional stability: (1) those which reduce the rate of water vapor or liquid absorption but do not reduce the extent of swelling to any great degree, and (2) those which reduce the extent of swelling and may or may not reduce the rate of water absorption (fig. 21-12). Terms most often used to describe the effectiveness of the first type of treatment are moisture-excluding effectiveness (MEE), which can be determined in either water or water vapor form, and water repellency (WR), which is a specific liquid test. The term used to describe the effectiveness of the second type of treatment is reduction in swelling (R) or antishrink efficiency (ASE). Most of the type (1) treatments have very low R or ASE values. The R or ASE values can be determined in water vapor tests or single-soak liquid test for water-leachable treatments, or in double-soak liquid tests for nonleachable treatments."

Rowell and Youngs further conclude that: "In selecting a treatment to achieve product stability to moisture, at least three factors must be considered. The environment of the end product is the most important factor. If the product will come into contact with water, nonleachable—and perhaps even bonded—treatments will be needed. If, however, the product will be subjected to changes in

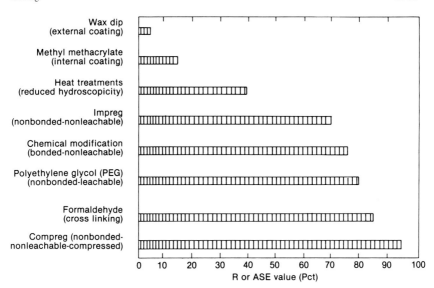

Figure 21-12.—Comparison of wood treatments and the degree of dimensional stability achieved. Impreg is laminated from veneers impregnated with phenol-formaldehyde resin. Compreg is made by densifying layers of resin-impregnated veneer by application of heat and pressure up to 1,000 psi. (Drawing after Rowell and Youngs 1981.)

relative humidity in the indoor environment, a leachable or water-repellent treatment might be satisfactory. The degree of dimensional stability must also be considered. If very rigid tolerances are required in a product—as in pattern wood dies—a treatment with very high R or ASE values is needed. If, on the other hand, only a moderate degree of dimensional stability is satisfactory, a less rigorous treatment will suffice. A final consideration is the cost effectiveness of a treatment. For example, the millwork industry uses a simple wax dip treatment to achieve a moderate degree of water-repellency. They would, no doubt, like a higher degree of water repellency or dimensional stability, but the cost to achieve this may not be recoverable in the marketplace. On the other hand, musical instrument makers require a very high degree of dimensional stability and the value of the final instrument can absorb the high cost to accomplish the desired results."

Readers interested in a general review of stabilization treatments such as cross lamination, water resistance coatings, hygroscopicity reduction, crosslinking, and bulking should read Rowell and Youngs (1981). Some research results specific to eastern hardwoods follow. See also section 22-13.

Typically, small specimens have been treated with polyethylene glycol or monomers, such as styrene and methyl methacrylate, which are subsequently polymerized *in situ* by heat (Siau and Meyer 1966) or gamma radiation (Siau et al. 1965). Treatment efficiency, as assessed by reduction in dimensional changes or volumetric swelling associated with changing moisture conditions, is a function of the fractional void volume of the wood filled by the chemical—a function of wood permeability and moisture content at time of treatment. Impregnation by pressure and vacuum processes have been used for treatments with monomers, while polyethylene glycol is applied by soaking unseasoned wood in a 30 to 50 percent aqueous solution of the chemical for several days or weeks, depending upon specimen size (Hallock and Bulgrin 1972).

Siau and Meyer (1966) reported that mechanical properties of yellow birch specimens impregnated (average loading 97 percent) with methyl methacrylate differed by curing method. Curing was by heat (68°C for 19.5 hours) or gamma radiation (0.64 megarads/hour for 4.7 to 15.6 hours). Mean compression strength was 11,700 psi for the 3 to 10 megarads treatment and 12,180 psi for heat. Values for control specimens averaged 2,905 psi. Thus, impregnations with methyl methacrylate improved compressive strength by an average factor of 4.11. Shear strength was not affected by treatment. Surface hardness was 25 percent greater in irradiated than in heat-cured specimens, ascribed by the authors to a preferential loss of monomer near the surface during heat curing. Antishrinkage efficiency was only 7.3 percent, thus suggesting that only small quantities of the monomer penetrated cell walls.

Mechanical properties of small static-bending specimens of white ash and red oak impregnated with methyl methacrylate and cured, compared with untreated controls, are given in table 22-29.

Much higher antishrinkage efficiencies—up to 80 percent—were reported by Siau et al. (1965) following irradiation curing of styrene-impregnated yellow-poplar. Retentions of 80 to 200 percent (based on ovendry wood weight) were obtained using solvent exchange and vacuum methods. Best results were achieved with dioxane, methanol, or ethanol as solvents. In later work, Siau (1969) found that volumetric swelling of up to 9 percent may occur in basswood following prolonged immersion in methyl methacrylate and styrene and after wood-polymer composites have been made using these materials. The amount of swelling was reported to be a function of temperature and moisture content of wood. Siau et al. (1975) found that both smoke evolution and flamespread are significantly increased by the presence in wood of polymers whose structure includes benzene rings.

Unlike Siau and Meyer (1966), Loos and Kent (1968) found significant increases in shear strength of yellow-poplar following loadings of 50 to 100 percent with methyl methacrylate.

The addition of waxes to styrene and methyl methacrylate monomers prior to impregnation of wood was reported by Lan and Rosen (1978) to affect adversely the polymerized properties of the wood-polymer composite. Application of waxes to the surface of composites provided water and water vapor resistance superior to all wax-monomer combinations.

Antishrink efficiencies of 60 to 70 percent have been reported for softwood maples treated with methyl isocyanate to weight gains of 16 to 28 percent (Rowell and Ellis 1979). Volumetric changes resulting from treatment were approximately equal to the volume of chemical absorbed, thus indicating that a chemical addition took place within the cell wall. Decay resistance was imparted by the treatment. At weight gains greater than about 26 percent, part of the chemical could be leached from the wood, and electron micrographs revealed cell-wall splitting.

Hallock and Bulgrin (1972) and Merz and Cooper (1968) have shown the efficiency of polyethylene glycol (molecular weight 1000) in reducing shrinkage and warpage of wood during drying and in service. The latter authors, who worked with black oak specimens 3 x 1½ x 1½ inches long, reported that treatment durations of 96 hours at 130°F to 150°F provide antishrink efficiencies of 50 to 65 percent and that further gains from longer treatments are of little practical significance. Hallock and Bulgrin subjected maple flooring to a treatment regime in polyethylene glycol that included solution temperatures as high as 200°F. Exposure at this temperature for 10 days produced a 22 percent retention and essentially no shrinkage upon drying.

Irradiation of wood has been employed commercially in the production of wood-polymer composites. The effect of irradiation on the mechanical properties of the wood varies with dosage and also apparently with species. Loos (1962) reported an increase in toughness of specimens of yellow-poplar for gamma radiation dosages up to about 0.85×10^5 rads. Toughness was reduced by as much as 28 percent, however, by levels in the range of 1.0×10^7 rads.

A significant increase in MOR was reported by Shuler et al. (1975) for specimens of American elm cut from saplings that had been exposed for 5 years to a gamma radiation level of 22,000 roentgens. By contrast, maximum work was decreased by all radiation levels studied by these authors.

Recommendations for future research.—Rowell and Youngs (1981) identified some specific research avenues that need further investigation. They noted that "the properties of coatings can be tailored to perform more duties than just water repellency. A water-repellent coating could also serve as an ultraviolet screen or flameproofing shield. It could also contain bound functional insecticides or fungicides that could protect the wood from attack by insects and decay organisms.

"In bonded bulking treatments, the bound chemical could be a fire retardant if the bonded chemical has an adequate distribution in the wood structure; thus, the treated wood would be both dimensionally stabilized and resistant to attack by termites, decay organisms, and marine organisms. Chemical modification of wood could provide a variety of nonleachable treated wood products that are both dimensionally stabilized and fire retardant or nonbiodegradable or acid and base resistant.

"Complete dimensional stability of wood (R or ASE = 100) has never been achieved, and perhaps never will be; there is much yet to be learned about wood-moisture relationships" (Rowell and Youngs 1981).

EFFECT OF PRESERVATIVES

Solid wood products.—The pH of treating solutions of some water-borne preservative formulations must be maintained within certain limits to prevent precipitation of the heavy-metal salts of which they are composed. Some of the solutions may be quite acid. Thus, for example, chromated copper arsenate (CCA) solutions may have a pH as low as 1.9, and the acidity of acid copper chromate (ACC) solutions may range from pH 2.0 to 3.9. By contrast, ammoniacal copper arsenate (ACA) is quite alkaline since solutions of this preservative must contain a weight of ammonia equal to 1.5 to 2.0 times the weight of the copper oxide; copper oxide comprises 47.7 percent of the dry weight of the formulation.

The effect of the pH of the treating solution and, indeed, the effect of the preservative salts themselves on wood properties have not been clearly defined. Thompson (1964b) investigated the effect of CCA, ACA, and ACC on the toughness of sweetgum, yellow-poplar, and black tupelo veneer for retention levels of 1 to 4 pounds per cubic foot. Toughness of sweetgum and yellow-poplar was not significantly affected by the treatments. Black tupelo, however, sustained important reductions in toughness, the values varying with retention. At retentions greater than about 1.0 pound per cubic foot, embrittlement was observed in specimens of all species; yellow-poplar was least affected. Subsequent analyses of the specimens revealed that the chemical composition of the wood, particularly carbohydrate content, was altered by high retention of all three preservatives.

Additional evidence that high retentions of salt-type preservatives may reduce the shock-resistant properties of timbers was supplied by Wood et al. (1980). Specimens cut from southern pine pole sections treated with CCA to a retention of 2.5 pcf had significantly lower values of toughness and work to maximum load than untreated control specimens. Although not significant statistically, there was a trend toward lower bending strength among specimens treated to this retention. Wood's data (table 21-24) showed no effect on strength properties at retention values lower than 2.5 pcf. This is consistent with Kelso's[6] data showing no deleterious effects of CCA retentions less than 1 pcf on strength of wood.

How reduced shock resistance revealed by impact studies of small specimens translates to full-size structural members is unknown. Marine piling is the only item for which retentions of CCA-type preservatives in excess of 2.0 pcf are employed. Definitive data on the effect of such treatments on piling are not available, but it is generally conceded within the industry that piling treated with CCA tend to break during driving more frequently than those treated with creosote. Likewise, data are much too limited to permit more than speculation on how the strength properties of load-bearing hardwood members, such as railroad ties, are affected by preservative salts. Even if it is assumed that these chemicals would have a more serious effect on hardwoods than softwoods, it is unlikely that the effect would have practical significance because of the relatively low retentions used.

TABLE 21-24.—*Effect of high retentions of a CCA-type preservative on the mechanical properties of southern pine* (Wood et al. 1980)

Strength value	Retention (pounds/cu ft)[1]			
	0	1.0	1.2	2.5
Modulus of rupture (psi)	16,350	16,450	16,350	15,500
Modulus of elasticity (million psi)	2.21	2.18	2.06	2.08
Fiber stress at proportional limit (psi)	8,775	8,215	8,580	8,450
Work to proportional limit (in-lb/cu in)	1.98	1.75	2.04	1.96
Work to maximum load (in-lb/cu in)	14.4	15.2	14.6	10.7
Toughness (in-lb)	236	223	219	165
Compressive strength parallel-to-grain (psi)	3,355	3,485	3,530	4,500

[1]Each value is the average of 32 measurements.

Strength reductions associated with treatments with oil-type or oil-borne preservatives are attributed to conditioning and not to the preservatives themselves. For example, it has been shown that the crushing strength of 3-foot piling sections cut from kiln-dried southern pine stock and treated with creosote was essentially the same as that for untreated, matched controls (Thompson 1969b). Reductions in this strength property occurred only among specimens that were steam conditioned preparatory to treatment. Similarly, the data shown in table 21-23 for oak and gum crossties show essentially no effect of the creosote per se on strength properties.

Published data also indicate that preservatives interfere with bonding of treated wood (Selbo 1959a; Blew and Olson 1950). Reductions in both shear strength and wood failure (i.e., increased glueline failure) have been reported for treated compared to untreated laminated wood. These reductions are attributed to changes in the surface properties of wood brought about by the preservative chemicals.

The effect of preservative salts on gluing properties and bond quality of sweetgum was evaluated by Thompson (1962). One-eighth-inch veneer pieces treated to a gradient series of retentions with each of four water-borne preservatives were bonded to form three-ply samples using three different adhesives (table 21-25). Retentions of 3.0 pcf greatly reduced both shear strength and wood failure for all combinations of adhesives and preservatives. Retentions of sodium pentachlorophenate between 0.37 and 1.50 pcf had little deleterious effect on bond quality for any of the adhesives. The effect of lower retentions of the three inorganic salt formulations varied among adhesives, with phenol formaldehyde resin showing the poorest results. Among preservatives, acid copper chromate most seriously reduced bond quality. With the exception of specimens treated with sodium pentachlorophenate, most of the average wood failure values (table 21-25) were much too low to meet applicable standards.

TABLE 21-25.—*Effect of preservative retention of 0.37 pcf on wood failure in sweetgum veneer specimens bonded with three adhesives and tested both wet and dry* (Adapted from Thompson 1962)

	Adhesives					
	Resorcinol-phenol		Phenol		Melamine	
Preservative	Wet	Dry	Wet	Dry	Wet	Dry
	----------Percent wood failure----------					
Sodium pentachlorophenate (PCP)...	90	89	58	92	100	100
Acid copper chromate (ACC).......	15	3	72	28	4	5
Ammoniacal copper arsenate (ACA).	85	22	48	56	31	12
Chromated copper arsenate (CCA)...	48	40	3	35	42	48
Controls........................	94	90	88	89	100	100

Similar results were obtained by Bergin (1962) who studied the gluability of birch veneer treated with fire-retardant chemicals to retentions of 6.2 to 9.5 pcf of anyhydrous chemical. None of the 11 adhesives employed in the study produced exerior-quality bonds when the veneer was treated with a fire retardant composed of zinc chloride, ammonium sulfate, and boric acid. Serious interference with gluing was also recorded for fire retardants composed of ammonium phosphate and ammonium sulfate, but a special resorcinol adhesive met specifications for exterior-quality bonds in plywood treated with that formulation.

Block-shear values of red oak treated with a range of preservatives before bonding were reported by Selbo (1959b) to be lower after each exposure period (3 to 36 months) than matched controls. Reductions in shear values attributed to the preservatives ranged from about 4 to 16 percent.

Reconstituted and composite wood products.—The increased use of reconstituted board products, with or without veneer overlays, as paneling products in home, hospital, and school construction and in applications where insect or decay hazards exist has prompted several studies of fire-retardant and preservative treatments of these products (Anonymous 1978; Surdyk 1975). Treatments are usually applied to the wood particles prior to application of adhesive and forming, because concurrent application of the preservative or fire-retardant with the resin adversely affects adhesive properties (Johnson 1964). Huber (1958), however, successfully treated both hardboard and particleboard by adding sodium pentachlorophenate to the adhesive in amounts equivalent to 0.65 percent of dry wood weight, without adversely affecting MOR or dimensional properties of the boards produced.

Brenden (1974) reported reduced peak heat release from building materials commercially treated with fire retardant salts: three-fourths-inch Douglas-fir plywood from 611 to 132 Btu/min/ft,^{2} and a gypsum wallboard-Douglas-fir stud assembly from 206 to 105 Btu/min/ft.^{2}

Flame-spread ratings ranging from good to excellent have been achieved with particleboard by spraying a volume of solution equivalent to 3 to 5 pcf on the furnish and redrying subsequent to boardmaking (Gilbert 1962). Arsenault (1964), however, found that two commercial fire-retardants added to aspen furnish prior to pressing into flakeboard seriously interfered with bonding. Sharp reductions in internal bond and bending strength were induced by all retentions between 2 to 8 pcf. Reductions in these strength properties occurred only in the case of urea-bonded boards treated with zinc borate; a formulation composed of dicyandiamide and ortho-phosphoric acid appeared to increase MOR.

Efforts to impart fire resistivity to hardboard by a bromination process have been unsuccessful because of the adverse effect of the treatment on board strength and water adsorption properties (Jurazs and Paszner 1978). Use of bromination to impart decay resistance to hardboard has been reported by Hong et al. (1978).

Reductions in hardboard flame spread of up to 60 percent were achieved by Mayers and Holmes (1975) by treating the fiber furnish with a series of fire retardants in amounts equal to 10 percent of dry fiber weight (table 21-26). Ten of the formulations tested reduced MOR by 20 percent or less, while three reduced this property by 40 to 50 percent.

In later work Myers and Holmes (1977) tested 4- and 8-foot panels of a hardboard following treatment of the fiber furnish with disodium octaborate tetrahydrate-boric acid (DOT-BA) or dicyandiamide-phosphoric acid-formaldehyde (DPF). An application rate of 20 percent based on dry fiber weight was used. Both treatments gave average flame-spread values that met criteria for Class B material—75 or under. Smoke development with the DOT-BA treatment was quite low—17 in the 25-foot tunnel test, compared to 399 for untreated controls. This treatment reduced internal bond strength (IB) by 30 percent but had no effect on or increased other strength properties.

TABLE 21-26.—Strength properties and fire performance of fire-retardant treated hardboards (Myers and Holmes 1975)

Board treatment[3]	Bending properties[1]		Internal bond[1]	Fire performance				
				8-foot tunnel furnace			2-foot tunnel furnace[2]	
	MOE	MOR		Flame-spread index	Fuel-contributed index	Smoke density index	Flame-spread index	
	Thousand psi	Psi	Psi					
Untreated (control)								
Test 1	694	6,120	439	119	128	302	111	
Test 2	702	6,110	336	122	124	289	112	
Test 3	752	6,490	299	119	151	375	122	
Average	716	6,240	358	120	134	322	115	
Untreated (¼-inch thick) (control)[2]	676	6,000	289	96	140	474	113	
10-percent fire retardant treatment								
Water-soluble salts								
Disodium octaborate tetrahydrate	748	5,640	398	103	63	143	42	
Disodium octaborate tetrahydrate-boric acid 94:1	772	5,910	369	97	66	82	42	
Monoammonium phosphate[4]	680	4,600	307	92	50	708	75	
Monoammonium phosphate (¼-inch thick)	—	—	—	83	46	594	86	
Ammonium sulfate	725	4,760	349	90	86	254	62	
Monoammonium phosphate-ammonium sulfate (1:1)	677	4,260	293	90	52	428	69	
Diammonium phosphate	691	4,970	277	62	37	732	69	
Diammonium phosphate-ammonium sulfate (1:1)	682	4,590	273	95	56	276	70	
Borax	840	5,950	432	79	53	159	44	
Borax-monoammonium phosphate (2:1) ..	634	5,000	385	83	72	141	62	
Borax-boric acid (1:1)	739	6,040	257	83	55	221	50	
AWPA Type C	659	4,190	248	109	88	89	71	
AWPA Type D	624	3,040	233	93	76	176	74	

TABLE 21-26.—Strength properties and fire performance of fire-retardant treated hardboards (Myers and Holmes 1975)—Continued

Board treatment[3]	Bending properties[1]		Internal bond[1]	Fire performance				2-foot tunnel furnace[2]
				8-foot tunnel furnace				
	MOE	MOR		Flame-spread index	Fuel-contributed index	Smoke density index		Flame-spread index
	Thousand psi	Psi	Psi					
Liquid ammonium polyphosphates[5]								
11-37-0	620	4,360	242	59	43	958		64
11-37-0 and ammonium sulfate (1:1)	684	4,220	264	93	60	568		61
12-44-0	586	3,730	178	48	23	954		61
Curing-type-organic phosphates								
THPC[6]	652	4,830	295	121	140	395		108
THPOH[7]	758	4,250	316	127	134	397		109
Dicyandiamide-phosphoric acid	822	5,430	281	79	45	641		64
Dicyandiamide-phosphoric acid-formaldehyde (pre-reacted)	797	5,630	265	83	53	485		67
MDP[8]	791	6,050	356	109	108	164		85
MDP[9]	627	3,700	232	79	69	694		81
Guanylurea phosphate	727	5,240	276	93	59	571		73

See footnotes page 2532.

TABLE 21-26.—*Strength properties and fire performance of fire-retardant treated hardboards* (Myers and Holmes 1975)—Continued

Board treatment[3]	Bending properties[1]		Internal bond[1]	Fire performance				
				8-foot tunnel furnace				2-foot tunnel furnace[2]
	MOE	MOR		Flame-spread index	Fuel-contributed index	Smoke density index		Flame-spread index
	Thousand p/s	*Psi*	*Psi*					
20-percent fire-retardant treatment								
Disodium octaborate tetrahydrate-boric acid (4:1)	796	6,610	349	69	22	281		44
12-44-0 liquid ammonium polyphosphate	634	4,090	274	7	1	986		—
Dicyandiamide-phosphoric acid-formaldehyde (pre-reacted)	725	5,100	260	21	4	909		46

[1]Adjusted to 60 lb/cu ft density.
[2]Values are averages of two tests.
[3]Specimens 1/8-inch thick unless indicated otherwise.
[4]Values for 8-foot tunnel furnace are averages of two tests; values for 2-foot tunnel furnace are averages of four tests.
[5]Products bearing code numbers listed were formulated as commercial fertilizers; the numbers refer to percent assay of nitrogen, phosphate, and potash, respectively.
[6]THPC means tetrakis (hydroxymethyl) phosphonium chloride.
[7]THPOH means tetrakis (hydroxymethyl) phosphonium hydroxide.
[8]MDP means melamine dicyanidiamide phosphoric acid.
[9]Contained 15 percent fire-retardant chemical and no phenolic resin.

While it is the usual procedure either to add fire-retardant chemicals to particleboard or fiberboard furnish prior to consolidation or to employ post-treatment in a pressure retort, Shen and Fung (1972) used a hot-pressing technique to accomplish this goal. Fire retardant chemicals (ammonium dihydrogen orthophosphate or liquid ammonium polyphosphate) were added to the surface of panels and forced into the surface by pressing the treated panels for a short period of time using temperatures and pressures of about 500°F and 250 psi, respectively. Surface loadings of up to 50 g/ft^2 were achieved in this manner. Flame spread was reduced from over 100 for untreated panels to less than 30 at loadings of 40 to 50 g/ft^2.

Strength and dimensional properties of phenolic-bonded particleboard prepared from ACA-treated flakes of 22 Ghanaian hardwood species were evaluated by Hall and Gertjejansen (1978). The effect of preservative retentions of 0, 0.2, 0.4, and 0.6 pcf on MOR, MOE, and IB (internal bond) for two resin levels and both the vacuum-pressure-soak-dry and accelerated-aging tests are summarized in figure 21-13. Percent thickness swelling is shown in figure 21-14 as a function of preservative retention, resin content, and type of test. This figure shows mechanical properties expressed as percentages of respective test values for control specimens—specimens which were not subjected to the vacuum-pressure-soak-dry or accelerated aging test, but which were in other respects treated like the remaining specimens.

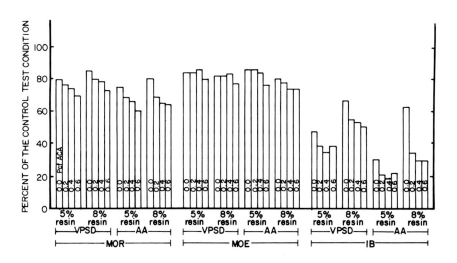

Figure 21-13.—Average modulus of rupture, modulus of elasticity and internal bond values following exposure of vacuum-pressure-soak-dry (VPSD) and accelerated aging (AA) conditions, as percents of their control condition values. Samples were from ACA-treated phenolic-bonded particleboard manufactured from flakes of Ghanaian hardwood species. (Drawing after Hall and Gertjejansen 1978.)

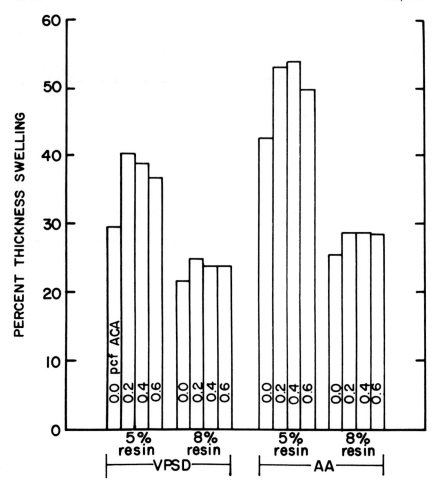

Figure 21-14.—Average percent thickness swelling values for test samples similar to those in figure 21-13. (Drawing after Hall and Gertjejanson 1978.)

All panels manufactured from flakes treated with preservative had lower MOR's and IB's than corresponding controls. MOR was reduced by preservative treatment by an amount equivalent to 5 percent for each 0.2-pcf increase in preservative retention. The effect of preservative treatment on IB was magnified by the weathering tests to an extent greater than that for the other strength properties. MOE was the strength property least affected by the preservatives.

Thickness swelling was influenced more by resin content (either 5 or 8 percent) and method of test than by preservative retention. Treated boards had substantially higher thickness swelling than untreated boards; but within the former, there was a trend toward a direct relationship between swelling and retention. A similar trend was evident for irreversible thickness swelling and irreversible linear expansion.

The condition of 3-¼- by 18-inch stakes cut from these phenolic-bonded particleboard panels and exposed for one year in a test plot in Puerto Rico is shown in table 21-27. All treated stakes, including those containing the lowest retention of preservative, sustained little or no decay and termite damage following exposure. Controls suffered substantial decay damage, but those bonded with 8 percent resin had much less damage than those bonded with 5 percent resin. Results with fiberboard specimens treated and exposed in the same manner as that for the particleboard specimens were very similar.

TABLE 21-27.—*Condition of particleboard stakes from mixed Ghanaian hardwoods after 1 year's exposure in the Caribbean National Forest of Puerto Rico*[1]

Retention of ammoniacal copper arsenate (Pounds/cu ft)	5 percent phenolic resin		8 percent phenolic resin	
	Decay	Termite	Decay	Termite
	----------Ratings[2]----------			
0	49	86	87	91
0	46	89	77	92
.2	86	96	92	100
.2	91	100	90	98
.4	97	100	100	100
.4	100	99	100	99
.6	100	100	100	100
.6	100	100	100	100

[1]Laundrie, J. F., G. C. Myers, and L. R. Gjovik. 1980. Evaluation of particleboards and hardboards from mixed Ghanaian hardwoods after a one-year exposure in the Caribbean National Forest of Puerto Rico. U.S. For. Prod. Lab., U.S. Dep. Agric. For. Serv., Madison, Wis. Interoffice Rep. prepared by the Univ. Minnesota. 20 p.
[2]Ratings based on a perfect score of 100.

Internal bond losses of specimens after field exposure were reported by the authors to parallel those induced by the vacuum-pressure-soak-dry test. Reductions in internal bond attributable to exposure averaged 62 and 45 percent for 5 and 8 percent resin levels, respectively. The magnitude of the reductions appeared to be related to preservative retention only in the case of specimens containing 8 percent resin.

Beal[9] studied the efficacy against termites of insecticides applied at several concentrations in the glue of plywood, particleboard, and hardboard by exposing them in southern Mississippi and in the Panama Canal Zone. Chlordane at 0.05, 0.10, and 0.20 percent, and heptachlor at 0.02, 0.05, and 0.10 percent protected all three materials from subterranean termites over 4 years. The higher levels also prevented damage by drywood termites in laboratory tests.

[9]Beal, R. H. 1980. Final office report summary FS-SO-7.303, Southern Forest Experiment Station, U.S. Dep. Agric. For. Serv., Gulfport, Mississippi.

21-6 LITERATURE CITED

Akhtar, M. A. and C. S. Walters. 1974. High-pressure treatment of white oak and red oak. In Proc., Amer. Wood Preserv. Assoc. 70:209-234.

American Wood Preservers' Association. 1977. Book of standards. Amer. Wood Preserv. Assoc., Washington, D.C.

American Wood Preservers' Association. 1978. Wood preservation statistics, 1977. Proc., American Wood Preserv. Assoc. 74:285-323.

American Wood Preservers' Association. 1980a. Crossties—Pressure treated. Standard C6-80. In Book of Standards. Amer. Wood Preserv. Assoc., Washington, D.C.

American Wood Preservers' Association. 1980b. Posts—Pressure treated. Standard C5-80. In Book of Standards. Amer. Wood Preserv. Assoc., Washington, D.C.

Anonymous. 1955. The value of incising for creosote treatment of wooden products. Wood Ind. 10(4):175-176. (Tokyo)

Anonymous. 1960. Fence posts service life tests. Coordinated Wood Perservation Council. Group Rep. No. 3, Tennessee Valley Authority, Norris, Tenn.

Anonymous. 1978. Directions for making fire-retardant particleboard bonded with UF resins. Bulletin SB-96, Borden Chemical Co., New York. 6 p.

Arsenault, R. D. 1964. Fire-retardant particleboard from treated flakes. For. Prod. J. 14(1):33-39.

Aston, D., and R. W. Watson. 1976. The performance of preservative treated hardwoods in ground contact. In Proc., British Wood Preserv. Assoc. 1976:41-57.

Baechler, R. H., Chairman. 1947a. Report of Committee 4—Preservatives. In Proc., Amer. Wood Preserv. Assoc. 43:58-93.

Baechler, R. H. 1947b. Relations between the chemical constitution and toxicity of aliphatic compounds. In Proc., Amer. Wood Preserv. Assoc. 43:94-111.

Baechler, R. H. 1954. Wood in chemical engineering construction. For. Prod. J. 4:332-336.

Baechler, R. H., E. Conway, and H. G. Roth. 1959. Treating hardwood posts by the double-diffusion method. For. Prod. J. 9:216-220.

Baechler, R. H. and H. G. Roth. 1964. The double-diffusion method of treating wood: a review of studies. For. Prod. J. 14:171-178.

Banks, W. B. 1973. Preservative penetration of spruce—close spaced incising an improvement. Timber Trades J. 285:51-53.

Behr, E. A. 1964. Preservative treatment of posts with dry chemicals. For. Prod. J. 14:511-515.

Behr, E. A., I. B. Sachs, B. F. Kukachka, and J. O. Blew. 1969. Microscopic examination of pressure-treated wood. For. Prod. J. 19(8):31-40.

Belford, D. S., and J. Nicholson. 1968. Emulsion additives: A new concept in copper-chrome-arsenate treatment. In Proc., British Wood Preserv. Assoc. 1968:69-101.

Belford, D. S., and J. Nicholson. 1969. Emulson additives for CCA preservatives to control weathering. In Proc., Amer. Wood-Preserv. Assoc. 65:38-51.

Bellman, H. 1968. Pretreatment of wood for pressure impregnation. J. Inst. Wood Sci. 21:54-62.

Bergin, E. G. 1962. The gluability of fire-retardant-treated birch veneer. Can. Dep. For., For. Prod. Res. Branch, Rep. 191. 23 p.

Bergman, O. 1977. (Factors affecting the permeability of hardwoods. A literature study.) Meddelande, Svenska Träskyddsinstitutet No. 126, 57 p.

Bescher, R. H. 1977. Creosote crossties. In Proc., Amer. Wood Preserv. Assoc. 73:117-125.

Blew, J. O., Jr. 1950. Comparison of wood preservatives in Mississippi post study (1950 progress report). U.S. For. Prod. Lab., Madison, Wis. Rep. No. R1757. 13 p.

Blew, J. O., Jr. 1963. A half-century of service testing crossties. In Proc., Amer. Wood Preserv. Assoc. 59:138-146.

Blew, J. O. 1966. Performance of treated oak crossties. Crosstie Bull. 47(3):7-10, 12-18, 20.

Blew, J. O., Jr., M. S. Hudson, and S. T. Henriksson. 1961. Oscillating pressure treatment of 10 U.S. woods. For. Prod. J. 11:275-282.

Blew, J. O., Jr., and W. Z. Olson. 1950. The durability of birch plywood treated with wood preservatives and fire retarding chemicals. *In* Proc., Amer. Wood Preserv. Assoc. 46:323-338.

Bois, P. J., R. A. Flick, and W. D. Gilmer. 1962. A study of outside storage of hardwood pulp chips in the Southeast. Tappi 45: 609-618.

Boone, S., L. R. Gjovik, and H. L. Davidson. 1976. Sawn hardwood stock treated by double-diffusion and modified double-diffusion. USDA For. Serv. Res. Pap. FPL 265, 11 p. For. Prod. Lab., Madison, Wis.

Borkin, K., and K. Corbett. 1970. Improvement of capillary penetration of liquids into wood by use of supersonic waves. Wood Sci. and Technol. 4:189-194.

Borup, L. 1961. A comparison between creosote oil and Swedish salt preservatives K33 and KP. Timb. Technol. 69(2260):53-55; 69(2261):95-98.

Bosshard, H. H. 1961. (On the Taroil-impregnation of railway sleepers from beech- and oak wood with temperatures of 100°C and 130°C—Part I: Microscopical observation of changes in the structure and moisture content of the impregnated woods.) Holz als Roh- und Werkstoff 19:357-370.

Brenden, J. J. 1974. Rate of heat release from wood-base building materials exposed to fire. U.S. Dep. Agric. For. Serv., Res. Pap. FPL 230. 16 p.

Buckman, S. J., and L. W. Reese. 1938. Effect of steaming on the strength of southern yellow pine. *In* Proc., Amer. Wood Preserv. Assoc. 34:264-300.

Burdell, C. A., and J. H. Barnett, Jr. 1969. Pilot plant evaluation of shock-wave pressure treatments. *In* Proc., Amer. Wood Preserv. Assoc. 65:174-190.

Burkhardt, E. C. and F. C. Wagner, Jr. 1978. End treatment of hackberry logs to prevent blue stain. For. Prod. J. 28(1):36-38.

Carpenter, B. E., Jr. and T. P. Bouler. 1962. Hardwood fence posts give good service. Agric. Exp. Stn. Inf. Sheet 782, 2 p. Miss. State Univ., State Coll., Miss.

Cech, F. C. 1971. Tree improvement research in oak species *In* Oak symp. proc., pp. 55-59. USDA Northeast. For. Exp. Stn. Upper Darby, Pa.

Cech, M. Y. and D. R. Huffman. 1972. Dynamic compression results in greatly increased creosote retention in spruce heartwood. For. Prod. J. 22(4):21-25.

Chudnoff, M., and E. Goytia. 1967. The effect of incising on drying treatability, and bending strength of posts. U.S. Dep. Agric. For. Serv. Res. Pap. ITF-5. 20 p.

Collister, L. C. 1955. Experience with hickory ties on the Santa Fe. Cross Tie Bull. 36(11):74-76.

Cook, D. B. 1978. Retention and penetration of pentachlorophenol in four-inch bolts of twenty-two hardwood species cut from southern pine sites. M.S. thesis Miss. State Univ., Mississippi State. 79 p.

Cooper, P. A. 1976. Waterborne preservative-treated railway ties—an annotated bibliography. Can. For. Serv. Inf. Rep. VP-X-143, 12 p. West For. Prod. Lab., Vancouver, B.C., Can.

Côté, W. A., Jr. 1963. Structural factors affecting the permeability of wood. J. Poly. Sci. Part C(2):231-242.

Davis, W. H., and W. S. Thompson. 1964. Influence of thermal treatments of short duration on the toughness and chemical composition of wood. For. Prod. J. 14:350-356.

Dickinson, D. J., N. A. A. Sorkhoh, and J. F. Levy. 1976. The effect of the microdistribution of wood preservatives on the performance of treated wood. *In* Proc., British Wood Preserv. Assoc., Cambridge, June 29-July 2.

Eddy, A. A., and R. D. Graham. 1955. The effect of drying conditions on strength of coast-type Douglas-fir. For. Prod. J. 5(4):226-229.

Ellwood, E. L. 1956. Preservative treatment for Australian rail sleepers. Commonwealth Engineer, Melbourne 44(1):6-10.

Ernst, K. E. B. 1964. Uber die Impragnierbarkeit einheimische Nadel- und Laubhölzer mit Steinhohlen- Teeröl. Schweiz. Anstalt für das Forstliche Versuchswesn Mitt. 40 (2):187-244.

Eslyn, W.E., 1973. Propionic acid—a potential control for deterioration of wood chips stored in outside piles. Tappi 56(4):152-153.

Franciosi, G. F. 1956. Effect of incising beech cross ties. For. Prod. J. 6:264-270.

Franciosi, G. F. 1967. Behavior of incised beech crossties in track. For. Prod. J. 17(2):48-50.

Furnival, G. M. 1954. Study cold-soaking treatment of posts of Delta hardwoods. Miss. Farm. Res. 17(8):5.

Gilbert, W. E. 1962. Fire-retardant testing of particleboard. Tech. Serv. Rep., Reichhold Chem., Inc.

Gjovik, L. R. and R. H. Baechler. 1968. Field tests on wood dethiaminized for protection against decay. For. Prod. J. 18(1):25-27.

Gjovik, L. R., and H. L. Davidson. 1975. Service records on treated and untreated fenceposts. U.S. Dep. Agric. For. Serv. FPL-068. 45 p.

Gjovik, L. R., and H. L. Davidson. 1979. Comparisons of wood preservatives in stake tests (1979 Progress Report). U.S. Dep. Agric. For Serv. Res. Note FPL-02. 81 p.

Graham, R. D. 1956. The preservative treatment of eight Oregon conifers by pressure processes. *In* Proc., Amer. Wood Preserv. Assn. 51:118-138.

Graham, R. D. 1980. Boulton drying: A review of its effects on wood. *In* Proc., Amer. Wood Preservers' Assoc., Rep. of Committee T-4: Poles, Appendix C. Vol. 76:85-88.

Graham, R. D., and R. J. Womack. 1972. Kiln- and Boulton-drying Douglas-fir pole sections at 220° to 290°F. For. Prod. J. 22(10):50-55.

Graham, R. D., and E. M. Estep. 1966. Effect of incising and saw kerfs on checking of pressure-treated Douglas-fir spar crossarms. *In* Proc., Amer. Wood Preserv. Assoc. 62:155-160.

Greaves, H. and J. E. Barnacle. 1970. A note on the effect of microorganisms on creosote penetration in *Pinus elliottii* sapwood and *Eucalyptus diversicolor* heartwood. For. Prod. J. 20(8):47-51.

Greaves, H., and J. F. Levy. 1978. Penetration and distribution of cooper-chrome-arsenic preservative in selected wood species. 1. Influence of gross anatomy on penetration, as determined by X-ray microanalysis. Holzforschung 32(6):200-208.

Hall, H. J., and R. O. Gertjejansen. 1978. An evaluation of the weatherability of phenolic bonded flake-type particleboard from ammoniacal copper arsenite (ACA) treated Ghanaian hardwood flakes. AID Project No. TA(AG)03-75, Dep. For. Prod., Univ. Minnesota St. Paul. 13 p., illus.

Hallock, H. and E. Bulgrin. 1972. Stabilization of hard maple flooring with polyethylene glycol 1000. U.S. Dep. Agric. For. Serv., Res. Pap FPL 187. 8 p.

Harkom, J. F. 1932. Experimental treatment of hardwood ties. *In* Proc., Amer. Wood Preserv. Assoc. 28:268-282.

Harkom, J. F., and G. H. Rochester. 1930. Strength tests of creosoted Douglas-fir beams. Circ. 28, Dep. Interior, Canada, For. Serv., 14 p.

Hatfield, I. 1959. Research needed on major problems in wood preservation. For. Prod. J. 9(5):26A-29A.

Hedley, M. E. 1973. Service tests of softwood railway sleepers in New Zealand. Reprint No. 675, New Zealand For. Serv. 2 p.

Henningsson, B. 1974. NWCP field test No. 1 with preservatives applied by pressure treatment. Results after 5 years. Information, Nordiska Traskyddradet No. 6, 18 p.

Henningsson, G., T. Nilsson, P. Hoffmeyer, H. Friis-Hansen, L. Schmidt, and S. Jakobsson. 1976. Soft rot in utility poles salt-treated in the years 1940-1954. Meddelande, Svenska Traskyddsinstitutet, No. 117E. 137 p. (Sweden)

Henry, W. T. 1970. Effect of incising on the reduction of serious checking in incised vs. unincised hardwood crossties. *In* Proc., Amer. Wood Preserv. Assoc. 66:171-178.

Hong, H.-M., R. S. Smith, L. Paszner, and P. E. Jurazs. 1978. Decay resistance of some brominated and non-brominated hardboards. Material und Organismen 13(3):233-239.

Hopkins, W. C. 1951. Cottonwood posts difficult to treat. Inf. Sheet 465, Mississippi State College, Agric. Exp. Stn., State College, Miss. 1 p.

Huber, H. A. 1958. Preservation of particleboard and hardboard with pentachlorophenol. For. Prod. J. 8:357-360.

Hudson, M. S. 1949. Vapor-drying of oak cross ties. *In* Proc. Amer. Railway Eng. Assoc. 50:403-418.

Hudson, M. S. 1951. Hickory for cross ties: Vapor drying of hickory ties. Cross Tie Bull. 32(11):5-7.

Hudson, M. S. 1954. Cross ties can help in solving the hickory problem. Railway Purchases and Stores 47(12):76-78.

Huffman, J. B. 1958. Kiln-drying of southern hardwood crossties. For. Prod. J. 8(6):165-172.

Hulme, M. A., and J. A. Butcher. 1977. Soft-rot control in hardwoods treated with chromated copper arsenate preservatives. **Material und Organismen** 12(2):81-95.

Isaacs, C. P., E. T. Choong, and P. J. Fogg. 1971. Permeability variation within a cottonwood tree. Wood Sci. 3:231-237.

James, R. E. 1961. A preliminary experiment with high pressure preservative treatments. *In* Proc., Amer. Wood Preserv. Assoc. 57:108-114.

Johanson, R. 1975. Arsenical diffusion treatment of *Eucalyptus diversicolor* rail sleepers. Holzforschung 29:187-191.

Johnson, W. P. 1964. Flame-retardant particleboard. For. Prod. J. 14:273-276.

Johnson, B. R. and G. E. Gonzalez. 1976. Experimental preservative treatment of three tropical hardwoods by double-diffusion processes. For. Prod. J. 26(1):39-46.

Jurazs, P. E. and L. Paszner. 1978. Fire-retardant treatment of hardwoods—mechanical and physical properties of slurry brominated hardboards. Wood Sci. 10:128-138.

Kanehira, Y., and T. Taniguchi. 1960. (Studies on preservative treatment of Itajii (*Pasania Sieboldii* Makino) ties.) Wood Ind. 15(8):16-21.

Kass, A. J. 1975. Effect of incising on bending properties of redwood dimension lumber. U.S. Dep. Agric. For. Serv., Res. Pap. FPL 259. 8 p.

Kozlik, C. J. 1968. Effect of kiln temperatures on strength of Douglas fir and western hemlock dimension lumber. Report D-11, For. Res. Lab., Oregon State Univ., Corvallis. 20 p.

Krzyzewski, J. 1969. Durability data on treated and untreated railway ties. Pub. No. 1113, Dep. For., Canada. 53 p.

Krzyzewski, J. 1973. Performance of diffusion treated railway crossties. Cross Ties 54(9):80-86.

Krzyzewski, J. and H. P. Sedziak. 1975. Preservation and performance of fence posts. Can. For. Serv. Rep. OPX82E, 30 p. East. For. Prod. Lab., Ottawa, Can.

Krzyzewski, J. and B. G. Spicer. 1974. 34 years of field testing of fence posts—results of final inspection. Can. For. Serv. Rep. OPX95E, 17 p. East. For. Prod. Lab., Ottawa, Can.

Kubinsky, E. 1971. Influence of steaming on the properties of *Quercus rubra* L. wood. Holzforschung 25:78-83.

Kulp, J. W. (Chairman). 1961. Report of Committee U-5, Post Service Records. *In* Proc., Amer. Wood Preserv. Assoc. 62:60-130.

Kulp, J. W. (Chairman). 1966. Report of Committee U-5, Post Service Records. *In* Proc., Amer. Wood-Preserv. Assoc. 62:60-130.

Lan, S.-T., and H. N. Rosen. 1978. Effect of the addition of waxes to wood-polymer composite. For. Prod., J. 28(2):36-39.

Levi, M. P., C. Coupe, and J. Nicholson. 1970. Distribution and effectiveness in *Pinus* sp. of water-repellent additive for water-borne wood perservatives. For. Prod. J. 20(11):32-37.

Levy, C. R. 1978. Soft rot. *In* Proc., Amer. Wood Preserv. Assoc. 74:145-164.

Levy, J. F., and H. Greaves. 1978. Penetration and distribution of copper-chrome-arsenic preservative in selected wood species. 2. Detailed microanalysis of vessels, rays, and fibers. Holzforschung 32(6):209-213.

Liese, W. 1957. (The fine structure of hardwood pits.) Holz als Roh- und Werkstoff 15:449-453.

Liese, W. and G.-A. Peters. 1977. (On probable causes of soft-rot attack of CCA-impregnated hardwoods.) Material und Organismen 12(4):263-270.

Lindgren, R. M. 1930. Preliminary experiments on control of sap-stain and mold in southern pine and sap gum by chemical treatment. Lumber Trade J. 97(9):25-26.

Lindgren, R. M., T. C. Scheffer, and A. D. Chapman. 1932. Recent chemical treatments for the control of sap stain and mold in southern pine hardwood lumber. South. Lumberman 8(87):43-46.

Loos, W. E. 1962. Effect of gamma radiation on the toughness of wood. For. Prod. J. 12:261-264.

Loos, W. E. and J. A. Kent. 1968. Shear strength of radiation produced wood-plastic combinations. Wood Sci. 1:23-28.

Lund, A. E. 1966. Preservative penetration variations in hickory. For. Prod. J. 16(2):28-32.

Luxford, R. F., and J. D. MacLean. 1951. Effect of pressure treatment with coal-tar creosote on the strength of Douglas-fir structural timbers. Report D179g, U.S. Dep. Agric. For. Serv., For. Prod. Lab.

MacFarland, H. B. 1916. Tests of Douglas-fir bridge stringers. *In* Proc., Amer. Railway Eng. Assoc. 17 (Part II):281-467.

MacLean, J. D. 1930. Preservative treatment of Engelmann spruce ties. *In* Proc., Amer. Wood Preserv. Assoc., 26:164-183.

MacLean, J. D. 1951. Rate of disintegration of wood under different heating conditions. *In* Proc., Amer. Wood Preserv. Assoc. 47:155-169.

MacLean, J. D. 1952. Preservative treatment of wood by pressure methods. U. S. Dep. Agric. For. Serv., Agric. Handb. No. 40. 160 p. U.S. Govt. Print. Off., Washington, D.C.

MacLean, J. D. 1953. Effect of steaming on the strength of wood. *In* Proc., Amer. Wood-Preserv. Assoc. 49:88-112.

McQuire, Z. J. 1964. The oscillating pressure method for the impregnation of New Zealand-grown timber. Tech. Pap. No. 44, New Zealand For. Serv., Wellington.

Merz, R. W., and G. A. Cooper. 1968. Effect of polyethylene glycol on stabilization of black oak blocks. For. Prod. J. 18(3):55-59.

Myers, G. C. and C. A. Holmes. 1975. Fire-retardant treatments for dry-formed hardboard. For. Prod. J. 25(1):20-28.

Myers, G. C. and C. A. Holmes. 1977. A commercial application of fire retardants to dry-formed hardboards. U.S. Dep. Agric. For. Serv., Res. Pap. FPL 298. 8 p.

Nethercote, C. H., J. K. Shields, and A. Manseau. 1977. Performance of pallets treated with a zinc salt polymer system. Rep. No. OPX187E, East. For. Prod. Lab., Canada, Ottawa. 19 p.

Nicholas, D. D. 1977. Chemical methods of improving the permeability of wood. *In* Wood Technology: Chemical Aspects. I. S. Goldstein, ed. ACS Ser. No. 43:33-46. Amer. Chem. Soc., Washington, D.C.

Nicholas, D. D., and J. F. Siau. 1973. Factors influencing the treatability of wood. *In* Wood deterioration and its prevention by preservative treatments. Vol. II. Preservatives and preservative systems. D. D. Nicholas, Ed. Syracuse University Press, Syracuse, N.Y. p. 299-343.

Olson, E. G., F. J. Meyer, and R. M. Gooch. 1958. Pentachlorophenol and heavy petroleum for the preservative treatment of railway crossties. For. Prod. J. 8(3):87-90.

Perrin, P. W. 1978. Review of incising and its effects on strength and preservative treatment of wood. For. Prod. J. 28(9):27-33.

Purslow, D. F. 1975. Results of stake tests on wood preservatives. Progress Report to 1974. Current Paper No. CP 86/75. Building Res. Establishment, Princes Risborough, England. 30 p.

Putnam, J. A. 1947. Fence posts from bottomland hardwoods. *In* U.S. Dep. Agric. For. Serv., South. For. Notes No. 52, p. 2-3. South. For. Exp. Stn., New Orleans, La.

Radkey, R. B. (Chairman). 1960. Report of Committee U-3, Tie Service Records. *In* Proc., Amer. Wood Preserv. Assoc. 56:236-237.

Rawson, R. H. 1927. A study of the treatment of six inch by twelve inch Douglas fir beams, covering boiling-under-vacuum-pressure process and influence of incising. *In* Proc., Amer. Wood Preserv. Assn. 23:203-215.

Reese, E. T., and M. Mandels. 1957. Chemical inhibition of cellulases and β-glucosidases. Res. Rep., Microbiol. Ser. No. 17. QM Res. and Eng. Ctr., Natick, Mass. 60 p.

Rosen, H. N. 1974a. Distribution of water in hardwoods: a mathematical model. Wood Sci. and Technol. 8:283-299.

Rosen, H. N. 1974b. Penetration of water into hardwoods. Wood and Fiber 5:275-287.

Rosen, H. N. 1975a. Penetration of two organic liquids into hardwoods. Wood and Fiber 6:290-297.

Rosen, H. N. 1975b. High pressure penetration of dry hardwoods. Wood Sci. 8:355-363.

Rosen, H. N. and C. H. vanEtten. 1974. Fiber saturation point determination of once-dried wood in solvents. Wood Sci. 7:149-152.

Rowell, R. M., and W. D. Ellis. 1979. Chemical modification of wood: Reaction of methyl isocyanate with southern pine. Wood. Sci. 12(1):52-57.

Rowell, R. M., and R. L. Youngs. 1981. Dimensional stabilization of wood in use. Res. Note FPL-0243. U.S. Dep. Agric., For. Serv. 8 p.

Scheffer, T. C. 1953. Treatment of bilgewater to control decay in the bilge area of wooden boats. J. of the For. Prod. Res. Soc. 3(3):72-95.

Scheffer, T. C. and F. L. Browne. 1954. Tests of some superficial treatments of exposed wood surfaces for their protection against fungus attack. J. of the For. Prod. Res. Soc. 4:131-132.

Scheffer, T. C. and J. W. Clark. 1967. On-site preservative treatments for exterior wood of buildings. For. Prod. J. 17(12):21-29.

Schrader, O. H., Jr. 1945. Tests on creosoted laminated stringers. Eng. News-Record 135(20):80-83.

Selbo, M. L. 1959a. Summary of information on gluing of treated wood. Rep. No. 1789 (rev.), U.S. Dep. Agric., For. Serv., For. Prod. Lab. 32 p.

Selbo, M. L. 1959b. Effect of perservatives on block-shear values of laminated red oak over a 3-year period. *In* Proc., Amer. Wood Preservers' Assoc. 55:155-164.

Shen, K. C., and D. P. Fung. 1972. A new method for making particle board fire-retardant. For. Prod. J. 22(7):46-52.

Shinn, F. S. 1955. A progress report on hickory cross ties. Cross Tie Bull. 36(11):70-72.

Shuler, C. E., J. E. Shottafer, and R. J. Campana. 1975. Effect of gamma irradiation *in vivo* on the flexural properties of American elm. Wood Sci. 7:209-212.

Siau, J. F. 1969. The swelling of basswood by vinyl monomers. Wood Sci. 1:250-253.

Siau, J. F. 1970. Pressure impregnation of refractory woods. Wood. Sci. 3:1-7.

Siau, J. F. 1971. Flow in wood. 131 p. New York, N.Y.: Syracuse Univ. Press.

Siau, J. F., G. S. Campos, and J. A. Meyer. 1975. Fire behavior of treated wood and wood-polymer composites. Wood Sci. 8:375-383.

Siau, J. F. and J. A. Meyer. 1966. Comparison of the properties of heat and radiation cured wood-polymer combinations. For. Prod. J. 16:47-56.

Siau, J. F., J. A. Meyer, and C. Skaar. 1965. Wood-polymer combinations using radiation techniques. For. Prod. J. 15:426-434.

Siau, J. F., W. B. Smith, and J. A. Meyer. 1978. Wood-polymer composites from southern hardwoods. Wood Sci. 10:158-164.

Silberman, R. A. 1970. Chip brightness and chip brightness preservation: how they affect pulp. Pulp & Pap. 44(6):67-69.

Skolmen, R. G. 1962. Treating costs and durability tests of Hawaii-grown wood posts treated by double-diffusion. Res. Note 198, U.S. Dep. Agric. For. Serv., Pacific Southwest For. Exp. Stn. 5 p.

Skolmen, R. G. 1971. A durability test of wood posts in Hawaii—third progress report. U.S. Dep. Agric., For. Serv., Res. Note PSW-260. 4p.

Smith, H. H., and R. H. Baechler. 1961. Treatment of Hawaiian grown woods posts by the double-diffusion wood preservation process. U.S. Dep. Agric. For. Serv. Res. Note No. 187. 8 p.

Sorkhoh, N. A., and D. J. Dickinson. 1976. An effect of wood preservation on the colonization and decay of wood by microorganisms. Mat. Und. Organism. Supplement (Feb. 1976), p. 287-293.

Sribahiono, U. I., J. N. McGovern and B. R. Johnson. 1974. A procedure for estimating the pressure treatability of hardwoods. Univ. of Wis. For. Res. Notes Note 188, 4 p.

Stamm, A. J. 1967. Movement of fluids in wood—Part I: flow of fluids in wood. Wood Sci. and Technol. 1:122-141.

Stamm, A. J. 1970. Maximum effective pit pore radii of the heartwood and sapwood of six softwoods as affected by drying and resoaking. Wood and Fiber 1:263-269.

Surdyk, L. V. 1975. Flame-retardant particleboard and process for making same. U.S. Pat. No. 3,874,990. U.S. Pat. Off., Washington, D.C.

Tamblyn, N. 1973. Soft rot in treated wood. Sixteenth For. Prod. Res. Conf., Melbourne, C.S.I.R.O., Div. Building Res., Australia.

Taras, M. A. and M. Hudson. 1959. Seasoning and preservative treatment of hickory crossties. U.S. Dep. Agric. For. Serv., Hickory Task Force Rep. No. 8. 24 p. Southeast. For. Exp. Stn., Asheville, N.C.

Teesdale, C. H. and J. D. MacLean. 1918. Relative resistance of various hardwoods to injection with creosote. U.S. Dep. Agric. For. Serv., Bull. No. 606. 36 p. Washington, D.C.

Tesoro, F. O. and E. T. Choong. 1976. Relationship of longitudinal permeability to treatability of wood. Holzforschung 30(3):86-90.

Tesoro, F. O., E. T. Choong, and C. Skaar. 1966. Transverse air permeability of wood: as in indicator of treatability with creosote. For. Prod. J. 16(3):57-59.

Tesoro, F. O., O. K. Kimbler, and E. T. Choong. 1972. Determination of the relative permeability of wood to oil and water. Wood Sci. 5:21-26.

Thomas, R. J. 1976. Anatomical features affecting liquid penetrability in three hardwood species. Wood and Fiber 7:256-263.

Thompson, W. S. 1954. End diffusion method of treating fence posts. Circ. 193, Miss. State Coll., Agric. Exp. Stn., State College, Miss. 7 p.

Thompson, W. S. 1962. Gluing characteristics of treated sweetgum veneer. For. Prod. J. 12:431-436.

Thompson, W. S. 1964a. Response of two wood decay fungi to metal-binding compounds. LSU Wood Util. Notes No. 3, 4 p. La. State Univ., Baton Rouge, La.

Thompson, W. S. 1964b. Effect of preservative salts on properties of hardwood veneer. For. Prod. J. 14:124-128.

Thompson, W. S. 1965. Response of *Poria monticola* and *Polyporus versicolor* to aliphatic amines. For. Prod. J. 15:282-284.

Thompson, W. S. 1969a. Effect of steaming and kiln drying on the properties of southern pine poles. Part I. Mechanical properties. For. Prod. J. 19(1):21-28.

Thompson, W. S. 1969b. Factors affecting the variation in compressive strength of southern pine piling. *In* Proc., Amer. Wood Preserv. Assoc. 65:133-144.

Thompson, W. S., and P. Koch. 1981. Preservative treatment of hardwoods: A review. U.S. Dep. Agric. For. Serv., Gen. Tech. Rep. SO-35. 47 p.

Toole, E. R. and W. S. Thompson. 1973. End-diffusion treated fence posts—after 20 years. Miss. For. Prod. Util. Lab. Inf. Ser. No. 17, 3 p. Miss. State, Miss.

Tschernitz, J. L., E. L. Schaffer, R. C. Moody, R. W. Jokerst, D. S. Gromala, C. C. Peters, and W. T. Henry. 1979. Hardwood Press-Lam crossties: Processing and performance. U.S. Dep. Agric. For. Serv. Res. Pap. FPL 313. 22 p.

Vaughan, J. A. 1954. Controlled-air-seasoning. *In* Proc., Amer. Wood Preserv. Assoc. 50:282-287.

Vaughn, J. A. 1955. Controlled air-seasoning of wood for preservative treatment. For. Prod. J. 5(5):45A-46A.

Verrall, A. F. 1953. Decay prevention in wooden steps and porches through proper design and protective treatment. J. For. Prod. Res. Soc. 3(4):54-60.

Verrall, A. F. 1959. Preservative moisture-repellent treatments for wooden packing boxes. For. Prod. J. 9(1):1-22.

Verrall, A. F. 1961. Brush, dip, and soak treatments with water-repellent preservatives. For. Prod. J. 11:23-26.

Verrall, A. F. and T. C. Scheffer. 1949. Control of stain, mold, and decay in green lumber and other wood products. For. Prod. Res. Soc. 1949 Prepr. 58, 9 p.

Vick, C. B., L. I. Gaby, and R. H. Baechler, 1967. Treatment of hardwood fence posts by the double-diffusion process. For. Prod. J. 17(12):33-35.

Walters, C.S. 1967. The effect of treating pressure on the mechanical properties of wood: I. Red gum. *In* Proc., Amer. Wood Preserv. Assoc. 63:166-186.

Walters, C. S., and J. K. Guiher. 1970. Treating gum crossties at high Lowry pressure. *In* Proc., Amer. Wood Preserv. Assoc. 66:258-259.

Walters, C. S. and K. R. Peterson. 1965. Report on Project 331-A. Preservative treatment of fence posts by cold-soaking in pentachlorophenol—fuel-oil solutions. Univ. of Ill. Agric. Exp. Stn. For. Note No. 117, 4 p.

Wood, L. W., E. C. O. Erickson, and A. W. Dohr. 1960. Strength and related properties of wood poles. ASTM Wood Pole Res. Prog. Final Rep., Amer. Soc. Testing Materials, Philadelphia, Pa.

Wood, M. W., K. C. Kelso, Jr., H. M. Barnes, and S. Parkikh. 1980. Effects of the MSU process and high preservative retentions on southern pine treated with CCA-type C. *In* Proc., Amer. Wood Preserv. Assoc. 76:22-37.

Zentmyer, G. A. 1944. Inhibition of metal catalysis as a fungistatic mechanism. Sci. 100(2596):294-295.